Cancer survival trends in England and Wales, 1971-1995: deprivation and NHS Region

Michel P Coleman

Penny Babb

Philippe Damiecki

Pascale Grosclaude

Satoshi Honjo

Jennifer Jones

Gerhardt Knerer

Alexandre Pitard

Mike Quinn

Andy Sloggett

Bianca De Stavola

Studies in Medical and Population Subjects no. 61

London: The Stationery Office

Office for National Statistics

The Office for National Statistics works in partnership with other agencies in the Government Statistical Service to provide Parliament, government and the wider community with the statistical information, analysis and advice needed to improve decision-making, stimulate research and inform debate. It aims to provide an authoritative and impartial picture of society and a window on the work and performance of government, allowing the impact of government policies and actions to be assessed.

London School of Hygiene and Tropical Medicine

The London School of Hygiene and Tropical Medicine aims to contribute to the improvement of health world-wide through the pursuit of excellence in research, postgraduate teaching, advanced training and consultancy in international public health and tropical medicine. To achieve this mission, the School will enhance its role as Britain's national school of public health, as a leading institution in Europe for research and postgraduate education in public health and tropical medicine, and as an international centre of excellence in public health and medicine in developing countries.

Cancer Research Campaign

The Cancer Research Campaign is a research organisation funded solely from public donation and supports a comprehensive programme of research in institutes, hospitals, universities and medical schools throughout Britain and Northern Ireland. We are European leaders in anti-cancer drug development. Our mission is to attack and defeat the disease of cancer in all its forms, to investigate its causes, distribution, symptoms, pathology and treatment, and to promote its cure.

How to obtain more information

Enquiries about this publication, the accompanying CDROM and related statistics should be addressed to Dr P Babb Office for National Statistics, 1 Drummond Gate, London SW1V 2QQ (tel 0171 533 5266)

General enquiries about official statistics should be directed to the National Statistics Public Enquiry Service, room DG/19 at the above address. Telephone contacts:

Business statistics:	01633 812973	(Textphone (Minicom): 01633 812 399)
Economic statistics:	0171 533 6363/6364	
Social statistics:	0171 533 6262	(Textphone (Minicom): 0171 533 6260)

You can also visit us on the Internet at **http://www.ons.gov.uk**

Subscription orders and enquiries

For products published through The Stationery Office, please contact:
The Stationery Office Publications Centre, PO Box 276, London SW8 5DT (tel 0171 873 9090, fax 0171 873 8200)

ISBN 0 11 621031 1

For cancer patients

LIST OF TABLES

Note on Figures in Chapters 5-62

For some cancers, Figures 2 and 3 include only data for England and Wales

Figure 5 (age-specific survival rates) is not presented for childhood cancers (Chapters 50-61)

LIST OF AUTHORS

Michel P Coleman
Professor of Epidemiology and Vital Statistics
Head of Cancer and Public Health Unit
Department of Epidemiology and Population Health, London School of Hygiene and Tropical Medicine; and
Deputy Chief Medical Statistician, Demography and Health Division, Office for National Statistics

Penny Babb
Epidemiologist
Demography and Health Division
Office for National Statistics

Philippe Damiecki
Ingénieur en informatique
International Agency for Research on Cancer, Lyon, France

Pascale Grosclaude
Visiting Research Fellow in epidemiology
Cancer and Public Health Unit, Department of Epidemiology and Population Health
London School of Hygiene and Tropical Medicine; and
Registre des cancers du Tarn, Albi, France

Satoshi Honjo
Visiting Research Fellow in epidemiology
Cancer and Public Health Unit, Department of Epidemiology and Population Health
London School of Hygiene and Tropical Medicine; and
National Defence Medical College, Tokorozawa, Japan

Jennifer Jones
Epidemiologist
Demography and Health Division
Office for National Statistics

Gerhardt Knerer
Research Assistant
Cancer and Public Health Unit, Department of Epidemiology and Population Health
London School of Hygiene and Tropical Medicine

Alexandre Pitard
Visiting Research Fellow in biostatistics
Cancer and Public Health Unit, Department of Epidemiology and Population Health
London School of Hygiene and Tropical Medicine; and
Registre des Tumeurs du Doubs, Besançon, France

Mike Quinn
Director, National Cancer Registration Bureau
Demography and Health Division
Office for National Statistics

Andy Sloggett
Lecturer in Medical Demography
Centre for Population Studies, Department of Epidemiology and Population Health
London School of Hygiene and Tropical Medicine

Bianca De Stavola
Senior Lecturer in Biostatistics
Cancer and Public Health Unit, Department of Epidemiology and Population Health
London School of Hygiene and Tropical Medicine

Office for National Statistics

In February 1998, the Rt Hon Margaret Beckett MP, then Secretary of State for Industry and the cabinet minister responsible for science, stressed the government's desire to strengthen the role of scientific information in the formulation of policy. The Rt Hon Frank Dobson MP, Secretary of State for Health, announced in April 1998 his intention "to even out huge inequalities in care between the richest and poorest parts of the capital". In the same month, the Prime Minister wrote of his government's "commitment to improve the health of the nation by attacking inequalities in health, [and its] determination to achieve the highest standards of care in all hospitals".

This book provides some of the information required to meet the government's objectives. It describes the survival patterns of almost three million cancer patients who were diagnosed in England and Wales between 1971 and 1990 and were followed up until 31 December 1995. This is the first comprehensive national set of information on trends in the outcome of cancer treatment by age, sex, geographic region and degree of material deprivation. It should prove a valuable source of information for government, public health authorities, researchers and the wider public. The results in this book will provide a baseline against which the reorganisation of NHS cancer treatment services can be evaluated.

The Office for National Statistics is pleased to make this authoritative resource available as part of its mission to stimulate research and inform debate for Parliament, the government and the wider public.

Tim Holt, Director

London School of Hygiene and Tropical Medicine

Cancer is a worldwide public health problem, and reliable information on the outcome of its management is invaluable. This book contains the results of one of the largest and most detailed analyses of cancer survival ever carried out. It includes survival rates for 58 different cancers in adults and children, thus covering more than 90% of all cancers diagnosed in England and Wales.

National cancer survival statistics provide a key yardstick by which cancer control can be assessed. The trends in cancer survival show just how much has been achieved for many cancers in the last 25 years, yet how far we still have to go before some of the most common cancers can be considered anything other than a death sentence. The innovative approach to estimating the survival gradient between cancer patients in different socioeconomic groups adds considerable weight to one of the main conclusions from the analyses: that cancer survival is worse for the poor than the rich.

This major new publication fits the London School's research objectives: to inform health policy in order to improve the health of the population. The book and the accompanying CDROM should provide a rich seam of information for public health researchers, and they will become an essential tool for public health planners in government and the National Health Service.

Professor Harrison Spencer, Dean

Cancer Research Campaign

Better cancer treatments require new insights into cancer biology, and large clinical trials to decide if a new drug or therapeutic regime is better than the best so far. A key test of public access to better treatment is whether the survival of cancer patients who were diagnosed in the most recent period for which information is available is higher than for patients who were diagnosed in earlier periods.

This report provides a comprehensive picture of trends in the survival of cancer patients over the last 25 years. It confirms that there have been substantial gains in survival throughout England and Wales for children with acute lymphoid leukaemia and other malignancies, and for adults with Hodgkin's disease and testicular cancer, among many others. But survival for cancers such as those of lung, oesophagus and pancreas has not improved as much. For many cancers there are unexplained differences in survival between socioeconomic groups. This requires more research. Improving the survival of *all* patients with a given cancer will require that advances in diagnosis and treatment become rapidly accessible to the whole population.

The Campaign is the major funder of cancer research in universities and medical schools throughout Britain and Northern Ireland. Income for the Campaign comes entirely from voluntary donations. For over 20 years, the Campaign has provided health professionals with up-to-date statistics and information on cancer; its support for this important project continues the tradition. The Campaign warmly welcomes the much-needed information on survival which this joint publication provides.

Professor Gordon McVie, Director General

This book provides extremely valuable summary data relating to the survival of cancer patients in England and Wales. The information is particularly valuable in that it is based on Cancer Registry data rather than on selected series of patients, and thus reflects a true measure of the success, or lack of success, of current medical endeavours to combat cancer. The detailed collation of geographical and time trends for the different types of cancer will enable strategic planning of resources and will be a guide to priorities in addressing the burden of disease in society.

The volume contains information on survival patterns for just under three million individuals with a diagnosis of cancer registered during the twenty-year period between January 1971 and December 1990. Survival rates up to ten years are provided for 47 types of adult cancer and 11 types of paediatric cancer.

The analyses display provocative trends in survival when presented by sociological patterns such as the level of material deprivation or in the comparisons of survival in England and Wales with survival reported from elsewhere in Europe or the United States. In particular, the patterns of survival highlighted in the EUROCARE study in 1995 are brought up to date. They continue to suggest that cancer survival may be relatively low in the UK compared to a number of other European countries.

A particular value of this book is the details provided of methodology for population-based surveys. The long period of study enables survival figures to be based on substantial numbers even for cancers which are relatively uncommon in England and Wales. Examples of this include more than 16,000 patients with testicular tumours, more than 22,000 with Hodgkin's disease and more than 20,000 with cancer of the gallbladder.

Clear differences in the pattern of survival in different patient groups raise important questions for a health care system. It would clearly be unacceptable for these differences to be due to an unfair distribution of health care resources. The figures provided in this book challenge those concerned with public health to investigate and provide an explanation for apparent differences in outcome. Thus, this volume succeeds in a key parameter of any scientific enterprise, namely the urgent stimulation of further research.

Alan Horwich
Professor of Radiotherapy & Director of Clinical Research and Development
The Royal Marsden Hospital & Institute of Cancer Research

This national study stems directly from three decades of meticulous data collection, checking, coding and classification by the staff of the regional cancer registries in England and Wales and by the staff of the Office for National Statistics in the National Health Service Central Register, Southport, and the Cancer Section in Titchfield.

We gratefully acknowledge here the dedication of all these staff. They play a crucial but largely unsung role in creating the information that our society must have if it is to improve the control of cancer.

Many other people have given generously of their time and expertise during the preparation of this book. We hope they will see the book as some repayment of the enormous debt we owe them. We would particularly like to mention the following:

Professor Jacques Estève
(Université Claude Bernard, Lyon, France), who gave us the benefit of his immense experience in the methodology of cancer survival analysis, and who wrote the annotated GAUSS algorithm which formed the basis of the STATA algorithm used to carry out the survival analyses reported here

Dr Michael Hills
(formerly at the London School of Hygiene and Tropical Medicine), who wrote the initial STATA algorithm to a precise specification and to a very tight schedule

Mr Adrian Mander
(Department of Epidemiology and Population Health, London School of Hygiene and Tropical Medicine), who provided valuable help in the development of the survival algorithm

Professor Timo Hakulinen
(Helsinki, Finland), who kindly made available the latest version of his widely used cancer survival analysis package for testing and comparison

Dr Guy Hédelin
(Strasbourg, France), who also kindly made available the latest version of his survival analysis program, Relsurv, for testing and comparison

Dr Gerald Draper
(Childhood Cancer Research Group, Oxford), and

Dr Charles Stiller
(Childhood Cancer Research Group, Oxford), who generously shared their expertise and insights into trends in survival from cancer in children

Dr Kate Ryan
(Central Middlesex Hospital, London), who kindly commented on the trends in survival for adult leukaemias and lymphomas

Mrs Jane Toms and Mrs Julia Maidment
(Cancer Research Campaign), who gave us the benefit of their experience in presenting complex information about cancer to a wide audience

Mr David Mayer
(Department of Epidemiology and Population Health, London School of Hygiene and Tropical Medicine), and

Mr Chris Ault
(Office for National Statistics), who both gave invaluable help with editorial tables and in checking drafts of the chapters

Dr Basia Zaba
(Department of Epidemiology and Population Health, London School of Hygiene and Tropical Medicine), who gave valuable advice on the preparation of life tables and in setting the precision limits for relative survival

Dr Sam Pattenden
(Department of Public Health and Policy, London School of Hygiene and Tropical Medicine), who provided raw data for the deprivation-specific life tables

Dr Chris Grundy
(Department of Public Health and Policy, London School of Hygiene and Tropical Medicine), who kindly prepared the maps

Mr Andy Reid, Ms Kate Outhwaite, Mr James Sanderson and Ms Sheena Wakefield (London School of Hygiene and Tropical Medicine), for indefatigable technical support

Ms Pat Broad
(Office for National Statistics), and

Ms Lorraine Piper
(Colourscript), for their patience, expertise and artistic skill

Funding

Dr Pascale Grosclaude was a Visiting Research Fellow from the Registre des cancers du Tarn, Albi, France, funded by the Ligue contre le cancer du département du Tarn, by Fondation de France and by the London School of Hygiene and Tropical Medicine.

Dr Satoshi Honjo was a Visiting Research Fellow funded by the National Defense Medical College, Tokorozawa, Japan.

Dr Alexandre Pitard was a Visiting Research Fellow from the Registre du cancer du Doubs, Besançon, France, funded by the Association pour la Recherche contre le Cancer (ARC, France).

This work was supported by the Cancer Research Campaign (project grant no. SP2381/0101).

Cancer and public health

Cancer is a major public health problem. At current rates of incidence in the UK it will afflict one person in three before their seventy-fifth birthday. Almost a quarter of a million new cancers are diagnosed every year, and there are 140,000 cancer deaths each year, about one in four of all deaths. The financial cost of diagnosis, treatment and long-term care and support for cancer patients is immense. The emotional cost is incalculable.

Cancer control measures are aimed at reducing the burden of cancer – reducing the number of new cases and deaths in the population as a whole, and improving survival and the quality of life of cancer patients.

In order to cut the number of deaths from cancer in the long term, a society must either reduce the number of new cases that occur each year (incidence), or improve cancer survival, or both. It would clearly be preferable to prevent a cancer rather than having to treat it once diagnosed, and there is scope for improved cancer prevention in the UK. Most cancers develop over years or decades, however, and any effect of cancer control measures taken today will only be seen in reduced incidence or mortality five, ten or more years from now. Many cancers are not preventable in the current state of knowledge. For the foreseeable future, therefore, hundreds of thousands of patients in the UK will need treatment for new or recurrent cancer each year. In turn, it will remain necessary to deploy substantial resources for the investigation, treatment and care of these patients.

Cancer is a chronic, progressive disease, and long-term surveillance is needed to assess the effectiveness of efforts to control it. This is equally true for clinical surveillance – in the individual cancer patient – and for public health surveillance in the population as a whole. Clinical treatment and care of the individual cancer patient are designed to eradicate malignant disease, or to achieve permanent remission with minimal loss of function, or to provide palliation of symptoms.

The long-term goal in public health is a continuing decline in cancer incidence and mortality, but more short-term measures of the effectiveness of society's efforts in cancer control are also required.

This book is about cancer survival.

Contrary to popular belief[1], cancer survival is not a lottery. Lotteries are fair. A lottery ticket buys the same chance of winning for rich and poor, and in Leeds or London. The chances of surviving cancer are not the same for all patients with a given cancer, however, or in all regions of the country. For most major cancers, there is evidence that patients from affluent neighbourhoods have better survival than patients from deprived neighbourhoods, and that this is not simply due to chance[2-7] or the extent of disease at the time of diagnosis[8-10]. The underlying mechanisms are complex[11] and are likely to be difficult to address[12].

Survival from many cancers in the UK is not as high as in comparable European countries[13]. Regional differences in survival within the UK have been reported[14,15]. Geographic and socio-economic differences in investigation and treatment have also been reported for patients with cancers of the breast[16-20] and cervix[21], and variations in investigation or departure from treatment guidelines have caused inequity in survival for cancers of the breast[22-24], uterus[25], ovary[26] and large bowel[27]. Inequitable access to high-quality cancer treatment and care in the NHS has been acknowledged by government[28], and government commitment has been made to overcoming it[29].

The proportion of cancer patients who survive five years has become the most widely reported figure for cancer survival, and it is sometimes referred to (incorrectly) as the "five-year cure rate". Survivors at five years are not necessarily cured. They may still relapse and need further treatment, and they may still die of their cancer. For many cancers, however, a substantial proportion of patients who do survive to the fifth anniversary of diagnosis will be able to lead a normal life, and many of them will prove to have a life expectancy similar to that of men or women of the same age who have never had cancer.

Survival rates at one year and five years after diagnosis of each of the main types of cancer comprise the main content of this book.

Cancer and malignant disease

"Cancer" is widely understood to mean one of many different types of malignant disease (tumour, neoplasm, or new growth). Strictly speaking, the term "cancer" applies only to tumours of organs with an epithelial surface, such as the breast, lung, bowel, prostate or skin, and it excludes other types of malignant neoplasm such as melanoma, leukaemia, lymphoma, sarcoma of soft tissue or bone, and brain tumours. For brevity, however, and where the context allows, we will often use the familiar term "cancer" to include all these diseases.

The need for cancer survival rates

Reliable information on cancer survival is needed for several reasons. Cancer patients and their families have an obvious and acute interest in their own chances of survival[30]. They may overestimate their chances, and this may in turn influence their choice of treatment[31]. Doctors need cancer survival information in order to advise patients when planning their treatment. Cancer survival

rates provide a baseline for oncologists carrying out clinical trials to test the efficacy of new cancer treatment regimes. For health care planners and administrators, cancer survival is one overall measure of the efficiency with which cancer treatment services are organised, as well as of the extent to which patients can be diagnosed and treated at an early stage of disease. In turn, early diagnosis depends on public education about the warning symptoms of cancer, on the attitude of members of the public towards seeking professional help for symptoms suggestive of possible cancer, and on ready access to efficient diagnostic services. Trends over time in cancer survival also reflect improvements in diagnosis and treatment and the extent to which these have become available to all patients.

The risk of developing cancer is known to vary between sub-groups of a population, between different regions of a country, and between countries of broadly similar wealth and development. Similar differences in the chance of surviving a cancer once it is diagnosed may be less readily accepted, and they should be an incentive to search for remediable causes.

Measuring progress against cancer

Cancer patient survival is a key indicator of the effectiveness of cancer control in the population. Others include the annual number of new cases (incidence) of cancer and the annual number of deaths from cancer (mortality). Incidence and mortality can be expressed as the number of new cases or deaths per 100,000 population per year (annual incidence or mortality rates), or as a risk over a given age range, e.g. a 12% risk up to age 75 years.

Incidence, mortality and survival each have distinct characteristics as indicators of progress against cancer.

Incidence

Cancer prevention is designed to reduce the number of people who develop cancer, and a fall in incidence is the key measure of its effectiveness. The lifetime risk of lung cancer up to the age of 75 in England and Wales, for example, has fallen by more than half between two successive generations in men, from 9% (one in 11) for men born in 1915 to 4% for those born in 1940[32]. The corresponding figures for women, about 2.2% and 2.0%, are much lower, but show much less progress.

Incidence is the best measure of trends in cancer risk. The raw data for cancer incidence in the population are collected by cancer registries directly from the medical records: diagnostic precision is generally good, and quality control at the registry often enables errors to be corrected at source. Incidence trends can be affected by changes in the completeness of registration, however, as well as by changes in diagnostic methods, pathological definitions or the coding of malignancy, and by the introduction of screening programmes.

Since complete prevention is unlikely to be achievable for most cancers, measures of remission, survival and mortality are also needed.

Mortality

Long-term reduction in cancer mortality is a key measure of progress against cancer, but mortality rates have several limitations for this purpose. Cancer mortality rates in a given year depend mainly on the numbers of patients diagnosed recently, and on their survival, but also on the numbers of patients diagnosed in the previous ten years or more, and the extent to which their treatment led to cure or to postponement of death from cancer. About a quarter of cancer patients die of other causes, and cancer mortality rates do not reflect their experience of the disease at all.

Time trends and geographic differences in cancer mortality rates are difficult to interpret without information about any underlying trends or differences in incidence. Mortality from testicular cancer is falling in England and Wales and many other countries, for instance, but incidence continues to increase[33].

Cancer mortality is therefore an incomplete public health indicator of recent progress against cancer, and the interpretation of mortality trends is blurred by their complex dependence on trends in incidence and survival over the previous ten or more years.

Survival

Information on cancer survival is essential for monitoring the efficacy and equity of a national cancer treatment programme. The two sources of reliable information about cancer survival are randomised controlled clinical trials and population-based studies. Although the statistical methods may be identical, survival rates from these two sources require completely different interpretation[34]. It is crucial to understand why this is so.

Randomised trials measure the best survival that is achievable, while population studies measure the average survival actually achieved.

Improved survival in a well-conducted clinical trial can confidently be ascribed to the treatment. Improvement in survival for all cancer patients is more difficult to achieve and more complex to interpret[35]. Depending on the cancer, improvements in population survival may require one or more of several components. These include public

education about the availability of screening programmes or the need to obtain early medical advice for suspicious symptoms; higher diagnostic suspicion by the general practitioner; more prompt referral; more effective investigation and staging of disease, and the widespread availability of more effective treatment.

Randomised controlled clinical trials

Clinical trials in which patients are randomly allocated to receive either a new and supposedly improved treatment, or the best treatment known thus far, are the most reliable and ethical way to decide if the new treatment is indeed a significant improvement on the old one. Clinical trials are carefully designed experimental studies, and they are an essential component of cancer control. Trials have proved that many modern cancer treatment protocols are indeed more effective than earlier protocols or, just as importantly, less effective – for example, that some postoperative radiotherapy protocols for non-small-cell lung cancer have an adverse effect on survival[36].

With the exception of cancers in children, however, clinical trials of cancer treatment in the UK rarely include more than a small fraction of all patients with a given cancer. This is not a disadvantage in deciding if a new treatment is better than previous regimes. Nor would it matter, for the purpose of estimating survival rates among all cancer patients, if the small proportion of patients included in clinical trials were broadly representative of all cancer patients in all clinical settings: but they are not, and cannot be expected to be.

Clinical trials for cancer are often carried out in hospitals that specialise in cancer treatment, although multi-centre trials using a common protocol have become more routine. Such trials show what improvements in survival are possible under ideal conditions, with the latest diagnostic techniques and treatment regimes, usually for selected cancer patients in the care of experienced specialist oncologists.

The results of clinical trials may not provide ideal guidance for the treatment of all cancer patients. Elderly patients, for example, are often excluded from trials by an upper age limit for recruitment such as 70 or 75 years, either because they more frequently have concurrent illnesses which may complicate evaluation of the new treatment, or in order to minimise the risk of side-effects[37,38]. There is a wide variety of opinion about how to treat elderly women with breast cancer, for example[39], and older patients are less likely to be referred for specialist advice[40]. Among adults diagnosed with cancer in England and Wales in 1990, however, some 52% of men and 47% of women were aged 70 or more at diagnosis, and this proportion is likely to increase as the population continues to age. Survival is not the same at all ages, and clinical trials with upper age limits for recruitment cannot reliably estimate any gain in survival that the new treatment may confer on elderly patients.

Information is therefore needed on survival at different ages among all cancer patients, not just those recruited in clinical trials.

Hospital survival studies

Randomised controlled clinical trials apart, it is not generally possible to provide reliable comparisons of survival rates between individual hospitals. One detailed comparison of ovarian cancer survival between women treated in university and non-teaching hospitals showed differences in treatment patterns and survival for early disease, for example, but interpretation of the results for individual hospitals was not straightforward[41]. For the purposes of evaluating overall progress against cancer, hospital-based cancer survival estimates also have several disadvantages.

An increasing number of cancer patients are treated without admission to hospital, and hospital-based survival estimates do not include them. Few hospitals in the UK have systematically followed up all their cancer patients until death, perhaps many years later, and losses to follow-up can lead to bias in survival estimates. Such follow-up has been a routine component of hospital accreditation in the USA for many years[42]. The patients treated at any given hospital are not a representative sample of all cancer patients, and case-mix varies between hospitals. Valid comparison of hospital-specific survival rates is complicated by the need to adjust for differences in the severity of disease at diagnosis (clinical stage) among the patients seen in each hospital. Finally, cancer patients in England and Wales are often treated in more than one hospital and, in contrast with Scotland, hospital information systems have not yet been adequate to enable the data for individual patients to be satisfactorily linked. Survival estimates from two or more hospitals may therefore include the same patients. A more logical unit of analysis might be the group of hospitals in a particular referral network, but these also overlap in urban areas.

Since the clinical stage of disease at diagnosis is such a key factor in the success of treatment, it would be preferable to adjust hospital-specific survival estimates for case-mix when making comparisons. This is difficult to do satisfactorily. An explicit statement of the clinical stage is frequently missing from the medical records[6], and the stage can be difficult and time-consuming to reconstruct from other information in the record, even for expert medical staff. Further, since the stage of disease that is recorded at diagnosis may depend on investigations such

as magnetic resonance imaging, or staging laparotomy, which may not be available (or may not be used) for all cancer patients, the diagnostic basis for the clinical stage should also be taken into account in the adjustment[43,44]. Unfortunately, the information required to enable adjustment of hospital-specific survival rates for differences between hospitals in the stage of disease at diagnosis – as well as for the investigative techniques used to determine the stage at each hospital – is not routinely available.

In short, while survival estimates from clinical trials are invaluable for assessing new treatments, and hospital-specific survival estimates can be equally useful for audit, neither can tell us how effective society has become at improving the survival of all cancer patients. For this, the data from population-based cancer registries are indispensable[45,46]

Population-based studies

Population studies of cancer survival are the public health counterpart of clinical trials. They are observational studies of the survival rates actually being achieved in the population as a whole.

Population studies require cancer registries that record suitable information for all, or virtually all, cancer patients living in the region or country they cover, and can arrange for the survival of those patients to be monitored accurately over long periods of time. The cancer registration system in England and Wales does this. It provides *"essential basic information needed by regional and district health authorities to ensure that the health needs of the population for which they are responsible are met. This information should provide a basis for plans and evaluation of the high quality, effective and efficient services needed for the prevention and treatment of cancer"*[47]

The survival of cancer patients can be compared between geographic regions or social groups with data from population-based cancer registries. Trends in cancer survival over time can also be measured. If a substantially more effective new treatment has been introduced, the speed with which cancer survival improves in the population as a whole can indicate how quickly the benefits of the new treatment become available to all patients, and whether that improvement is equally available in all geographic regions and to all groups of the population. Similarly, survival gains would be expected from a mass screening programme or a major reorganisation of cancer treatment services, and trends in population survival rates should be an essential component of their overall evaluation. Randomised clinical trials and non-randomised studies of cancer survival in individual hospitals or groups of hospitals do not address these questions.

It would be unwise to expect that the best survival results obtained in clinical trials can be achieved among all cancer patients, at least in the short term. Even after convincing survival benefits have been shown in randomised clinical trials, the adoption of new treatment protocols into routine clinical practice can be slow, either because of delayed acceptance of the results or for lack of equipment or resources[48]. For this reason, and because elderly cancer patients often have lower survival than younger patients, the survival achieved by cancer patients who are recruited to clinical trials will almost inevitably be higher than that experienced by the generality of cancer patients diagnosed during the same calendar period.

Crude and relative survival

The measures used to report cancer survival in clinical trials and population studies also require some comment. Cancer survival rates in clinical trials are usually expressed as the percentage (or cumulative probability) of survival at a given time since diagnosis. This observed (or "crude") survival rate is sufficient to compare two or more groups of patients under medical observation, often over a relatively short period, and for whom precise details on other prognostic factors will be available. By contrast, population survival rates are usually expressed as a ratio of the crude survival and the corresponding expected survival in the general population, since cancer patients do not all die of cancer. "Relative" survival rates obtained in this way are numerically *higher* than crude survival rates precisely because of this adjustment for background mortality[34,42]. These terms are explained in Chapter 3.

This intrinsic difference between crude and relative survival rates must be borne in mind when interpreting the population-based cancer survival figures presented here.

The other differences between cancer survival rates obtained from randomised trials and population studies, in particular the selection by age and other factors of patients included in trials, should also be noted when comparing the results in this book with those from clinical studies.

Both crude and relative survival rates are provided in the main tables.

National policies on cancer

The "Health of the Nation" strategy

The *Health of the Nation* strategy for England in 1992 identified cancer as one of five key areas in which progress was required[49]. Cancers of the lung, breast, cervix and skin

were singled out for special attention, and the importance of high quality and timely information on cancer was stressed. Numerical targets were set for reductions in the incidence or mortality of these cancers within a period of 10-15 years. Professional[50], regional[51] and local[52] guidelines on cancer management were published to promote these objectives.

Cancer survival was not used as a measure of progress.

Reorganisation of treatment services

In 1991, the Royal College of Radiologists reported that there were too few oncologists in the UK and that their workload was too high[53]. In 1994, the Association of Cancer Physicians argued for a wide-ranging reorganisation of cancer treatment services[54]. In 1995, the EUROCARE report[13] showed that survival from many cancers in England and Scotland was lower than in comparable western European countries. Inequalities in access to radiotherapy services have been reported in the North & West Region[55]. A national audit recently attributed the fact that 28% of cancer patients waited longer to receive radiotherapy than the maximum acceptable delay set out in professional guidelines to lack of equipment and staff[56].

By 1995 there was clear evidence of socio-economic inequalities in cancer survival between geographic areas and social groups in England and Scotland, and that these could be at least partly due to inequity of access to treatment. In the West of Scotland, for example, melanoma survival rates were lower for patients from deprived areas than for those from affluent areas[57]. Inequalities in survival have also been found between patients living in affluent and deprived areas who were diagnosed with cancers of the lung, breast, large bowel, bladder, prostate, uterus or cervix in south-east England during the decade 1980-89[9,10].

In 1995, an Expert Advisory Group on Cancer to the Chief Medical Officers of England and Wales recommended[28] that the commissioning of cancer treatment services should in future be based on the principle that *"All patients should have access to a uniformly high quality of care in the community or hospital wherever they may live to ensure the maximum possible cure rates and best quality of life"*. That report appears to be the first official acknowledgment of concern about the organisation of cancer treatment. It explicitly sets out the principle of equity of access to the highest standard of cancer treatment for all cancer patients, regardless of where they live.

The Chief Medical Officers (Drs Calman and Hine) subsequently proposed reorganisation of cancer treatment services along "hub and spoke" lines in order to achieve this objective. Cancer units in a number of district general hospitals would each liaise closely with a cancer centre (the hub), a large hospital with a complete range of specialist oncology services and serving a population of a million or so. For acute lymphoblastic leukaemia and other malignancies in children, there is evidence that survival is better in such centres[58,59]. The Calman-Hine proposals were intuitively reasonable, and there is evidence, for example, that access to a cancer specialist or a multi-disciplinary team is associated with better outcome for cancers of the breast[60] and ovary[61,62]. The evidence that the reorganisation of cancer treatment services will improve equity of access to optimal treatment for adults and thus, in due course, cancer survival in the population, remains somewhat limited[46,63,64].

One objective of the analyses reported in this book was to provide a baseline for monitoring the future impact on cancer survival of the current reorganisation of cancer treatment services. The book includes survival rates for most types of cancer among patients diagnosed during 1986-90, in each of five socio-economic groups, and in each of the NHS Regions in England and in Wales.

Strategy for "Our Healthier Nation"

Cancer has also been chosen as a priority in the most recent government health strategy, *Our Healthier Nation*. In contrast to the 1992 strategy, only a single overall target has been set: *"to reduce the death rate from cancer amongst people aged under 65 years by at least a further fifth (20%) by 2010 from a baseline at 1996"*. The document noted that such a reduction in 1996 would have represented 6,000 fewer deaths from cancer in that year[29].

One government commitment towards this target is that, by April 1999 for breast cancer and a year later for other cancers, *"everyone with suspected cancer will be able to see a specialist within two weeks of their GP deciding they need to be seen urgently, and requesting an appointment"*[65]. This important guarantee carries major implications for the organisation of referral systems and hospital treatment services, and it may prove difficult to achieve[66,67].

In Devon (South & West Region), the median delay between seeing the general practitioner and the first appointment with the hospital consultant was 11 days for breast cancer during 1986-90, but the longest delay exceeded a year. For five other major cancers, even the *median* delay ranged from five to 12 weeks[68]. One university hospital recently reported a median delay of 17 weeks between first symptoms and histological diagnosis for cancers of the stomach and oesophagus, of which 15 weeks followed first attendance for medical advice[69]. Another reported a median delay of 20 weeks, although

only four weeks elapsed between first attendance and diagnosis[70]. The medical component of this delay appeared longer than in Germany[71] or Japan[72]. The delay between referral by the GP and attendance in the specialist clinic may also be longer for more deprived patients[19]. Median delays for cancer of the large bowel have ranged from one month to four months or more, although the evidence that longer delays translate into poorer survival has been judged as equivocal[73].

Guidelines and performance management

Evidence and guidance designed to improve the management and commissioning of cancer treatment services has been published, both nationally[73,74] and regionally[75]. Implementation of the guidelines for breast cancer has been slow[20].

Guidelines for cancer treatment have also been developed by various professional bodies, although few rigorous evaluations of the effect of such guidelines on cancer treatment appear to have been published[76].

Health outcome measures have recently been used as clinical performance indicators in the NHS[77]. Fifteen indicators of hospital performance aimed at the public, the NHS and government have been published at Health Authority level in England[78], and the use of population-based cancer survival rates has been piloted. Cervical cancer mortality rates at Health Board level were published in this context in Scotland in 1994[79], and one-year and five-year survival rates for cancers of the lung, large bowel, breast and ovary were used in 1996[80].

Cancer survival in England and Wales

In this book we report on the survival of almost three million cancer patients up to 31 December 1995. These patients were resident in England and Wales when diagnosed with cancer during the 20 years from 1 January 1971 to 31 December 1990. All the patients were followed up for at least five years, and for those diagnosed in the early 1970s, over 20 years. The final data set became available in July 1997.

Age and sex

Cancer survival rates up to ten years after diagnosis are provided for 47 different types of cancer in adults (40 cancers that occur in both sexes, three only in men and four only in women) and 11 different types of cancer in boys and girls. We have followed convention in defining children for this purpose as aged under 15 at diagnosis. From the age of 15, most cancers become more common with advancing age, but only a tiny proportion of all cancers arises after the age of 100. Survival analyses for adults have therefore been carried out for the age range 15-99 years.

Cancer survival in adults in England and Wales has been examined with respect to age at diagnosis. Information on cancer survival for patients aged 80-99 years at diagnosis is included, in view of the need to plan for increasing numbers of elderly patients[81]. Overall survival (all ages combined) is also reported.

Affluence and deprivation

Inequalities in cancer survival have been examined between groups of cancer patients defined as belonging to one of five categories of material deprivation, from affluent to deprived, on the basis of census-derived characteristics of the small area (census enumeration district) in which they were living when diagnosed.

The extent to which any inequalities in survival between affluent and deprived patients diagnosed during the late 1980s were more – or less – pronounced than for patients diagnosed in the early 1980s has also been explored.

Groups of cancer patients will sometimes be described as affluent or deprived for ease of reference. These are terms that apply, strictly, to the population of the census enumeration district in which the patient was living at diagnosis, and there will be variation in the degree of material deprivation of individuals living within such districts.

NHS Region

Geographic patterns and temporal trends in cancer survival up to ten years after diagnosis have also been examined for the eight current NHS Regions of England, and for Wales.

Trends over time

Changes in population cancer survival with the passage of time are an important measure of progress in cancer control, both at regional and national level, and between sub-groups of the population defined by age, sex or socio-economic status.

Effective treatments have become available for some cancers during the last 20 years. Chemotherapy for leukaemia and other malignancies in children, and for testicular cancer and Hodgkin's disease in adults, has led to dramatically improved survival. The most well-known example is that of acute lymphoid leukaemia in children, for which five-year survival increased from less than 10% in the 1960s to around 35% for children diagnosed in the early 1970s, and to nearly 75% for those diagnosed in the

late 1980s[58,82-84]. This dramatic improvement in population-based survival from the commonest childhood malignancy is welcome evidence both of the efficacy of modern chemotherapy and of the fact that it is available to almost all children who are diagnosed with leukaemia.

Acute lymphocytic leukaemia in children represents about one in 500 of all malignancies, however, and the adult cancers for which there have been large increases in survival within the last 25 years, such as testicular cancer and Hodgkin's disease, also comprise only a small proportion of all cancers.

Temporal trends in cancer survival may also be influenced by change over time in factors other than the efficacy of the best available treatments. This becomes clear when trends in survival are not the same in all sub-groups of the population, or improvements in survival are seen earlier in some countries than others. Such factors include the speed with which patients seek medical help when experiencing symptoms that could suggest cancer; the quality of preliminary diagnosis by the primary care physician; the speed of referral for specialist attention; the thoroughness of investigation and diagnosis; the availability of staff and equipment; the quality of treatment, and the general health of patients who develop cancer.

National information on cancer survival trends in adults has not so far been available. In this report, cancer survival trends in England and Wales have been examined for each sex and for most cancers. For 18 of the most frequent cancers, trends in survival have also been examined separately for each NHS Region and for Wales.

Large numbers of patients are required to obtain stable estimates of long-term trends in cancer survival. In analyses for the less common cancers, patients diagnosed in each of the four quinquennia 1971-75, 1976-80, 1981-85 and 1986-90 were grouped together. Survival has not changed dramatically in the last 25 years for most of the common cancers, so this grouping was retained in all analyses, for consistency. Cancer survival trends are thus presented for four consecutive five-year periods of diagnosis up to 1990, with follow-up to the end of 1995.

Key sources

Several sources are used repeatedly throughout this book, without formal citation on each occasion. The main sources for oncology and epidemiology were the *Oxford textbook of oncology*[85], *Oncology: a multidisciplinary textbook*[86], *Clinical oncology*[87], *Cancer epidemiology and prevention*[88], and *Cancer: causes, occurrence and control*[89].

International convention in cancer registration practice is

set out in the 1995 monograph *Cancer registration: principles and methods*[90]. In the national cancer registry of England and Wales, the anatomic location of tumours diagnosed during 1971-90 was coded to the eighth[91] or ninth[92] revisions of the World Health Organisation's International Classification of Diseases (ICD-8 and ICD-9). The morphology of tumours was coded according to the Manual of Tumor Nomenclature and Coding[93] (MOTNAC), or to the International Classification of Diseases for Oncology[94] (ICD-O). For childhood malignancies, we adapted the widely used 1987 classification by Birch and Marsden[95] to cover the period 1971-90.

Cancer incidence patterns in England and Wales in 1990 were taken from the corresponding annual reference volume *Cancer statistics: registrations of cancer diagnosed in 1990, England and Wales*[96]. Mortality data for England and Wales have been published[97]. Information has also been used on recent patterns of cancer incidence in Scotland[98], on international patterns of incidence[99] and on long-term trends in cancer incidence and mortality[32,100].

Cancer survival rates for patients diagnosed between 1960 and 1974 in several regions of England and Wales were reported in 1982[15]. National survival rates for patients diagnosed between 1971 and 1981 were published up to 1988[101-107]. In 1998, national survival rates for the 15 most common cancers were published for patients diagnosed in 1981 and 1989[108]. For children with cancer, national survival rates are available for patients diagnosed between 1971 and 1991[82,109]. Most cancer registries have also published regional survival rates for the most common cancers, either regularly or at intervals during this period.

The survival rates reported here for England and Wales are variously compared with those for Scotland, Italy and other European countries, and the USA. Trends in survival in Scotland have been reported for 22 cancers in adults (35-84 years) and seven cancers in children and young adults (0-34 years) who were diagnosed during the period 1968-1987 and followed up to the end of 1990[110]. The ITACARE group has published detailed estimates of cancer survival in Italy for 44 cancers in adults diagnosed during the period 1978-89[111]. Survival for cancers diagnosed during 1986-90 in the US states participating in the SEER programme (Surveillance Epidemiology and End Results) has also been published on disk[112].

Finally, the EUROCARE study is the most comprehensive international study of cancer survival patterns to date, and it is the main source for comparisons of survival with other European countries in this book. These data have been published by Berrino *et al.* in the monograph *Survival of cancer patients in Europe: the EUROCARE study*[13]. Further survival rates for cancer patients diagnosed up to 1989 in 17 European countries are about to be published from the EUROCARE study[113]. We are most grateful to the authors of this work for permission to cite some of their results.

1. Sikora K. Breast cancer: why Britain's women deserve better. Reader's Digest November 1994, 145: 55-61

2. Kogevinas M. *Longitudinal study. Socio-economic differences in cancer survival. Series LS No.5.* London: OPCS, 1990

3. Kogevinas M, Marmot MG, Fox AJ, Goldblatt PO. Socioeconomic differences in cancer survival. J Epidemiol Comm Hlth 1991; 45: 216-219

4. Sharp L, Finlayson AR, Black RJ. Cancer survival and deprivation in Scotland. J Epidemiol Comm Hlth 1995; 49: s79 (Abstract)

5. Schrijvers CTM. Socioeconomic inequalities in cancer survival in the Netherlands and Great Britain: small-area based studies using cancer registry data. PhD dissertation. Erasmus University, Rotterdam, 1996

6. Pollock AM, Vickers N. Breast, lung and colorectal cancer incidence and survival in South Thames Region, 1987-1992: the effect of social deprivation. J Publ Hlth Med 1997; 19: 288-294

7. Kidd J. Socioeconomic variations in breast cancer incidence, survival and the uptake of screening: a case study in Merseyside. PhD dissertation. University of Liverpool, 1997

8. Carnon AG, Ssemwogerere A, Lamont DW, Hole DJ, Mallon E, George WD, Gillis CR. Relation between socioeconomic deprivation and pathological prognostic factors in women with breast cancer. Br Med J 1994; 309: 1054-1057

9. Schrijvers CTM, Mackenbach J, Lutz J-M, Quinn MJ, Coleman MP. Deprivation and survival from breast cancer. Br J Cancer 1995; 72: 738-743

10. Schrijvers CTM, Mackenbach J, Lutz J-M, Quinn MJ, Coleman MP. Deprivation, stage at diagnosis and cancer survival. Int J Cancer 1995; 63: 324-329

11. Greenwald HP, Borgatta EF, McCorkle R, Polissar N. Explaining reduced cancer survival among the disadvantaged. Milbank Q 1996; 74: 215-238

12. Tomatis L. Socioeconomic factors and human cancer. Int J Cancer 1995; 62: 121-125

13. Berrino F, Sant M, Verdecchia A, Capocaccia R, Hakulinen T, Estève J, (eds.) *Survival of cancer patients in Europe: the EUROCARE study. (IARC Scientific Publications No. 132).* Lyon: International Agency for Research on Cancer, 1995

14. Silman AJ, Evans SJ. Regional differences in survival from cancer. Comm Med 1981; 3: 291-297

15. Cancer Research Campaign. *Trends in cancer survival in Great Britain: cases registered between 1960 and 1974.* London: CRC, 1982

16. Chouillet AM, Bell CMJ, Hiscox JG. Management of breast cancer in southeast England. Br Med J 1994; 308: 168-171

17. Macleod U, Twelves CJ, Ross S, Gillis C, Watt G. A comparison of the care received by women with breast cancer living in affluent and deprived areas. Br J Cancer 1998; 78: s15 (Abstract)

18. Landon M, Wilkinson P, Grundy C, Elliott P. Deprivation related differentials in mortality and hospital admission ratios. J Epidemiol Comm Hlth 1995; 49: s79 (Abstract)

19. Macleod U, Ross S, Twelves CJ, Gillis C, Watt GCM. A comparison of the care received from primary and secondary care by women with breast cancer living in affluent and deprived areas. J Epidemiol Comm Hlth 1998; 52: 687 (Abstract)

20. All-Party Parliamentary Group on Breast Cancer. *Improving outcomes in breast cancer.* London: House of Commons, 1998

21. Wolfe C, Tilling K, Bourne HM, Raju KS. Variations in the screening history and appropriateness of management of cervical cancer in South East England. Eur J Cancer 1996; 32A: 1198-1204

22. Twelves CJ, Thomson CS, Gould A, Dewar JA. Variation in the survival of women with breast cancer in Scotland. Br J Cancer 1998; 78: 566-71

23. Gillis CR, Hole DJ. Survival outcome of care by specialist surgeons in breast cancer: a study of 3786 patients in the west of Scotland. Br Med J 1996; 312: 145-148

24. Richards MA, Wolfe C, Tilling K, Barton J, Bourne HM, Gregory WM. Variations in the management and survival of women under 50 years with breast cancer in the South East Thames Region. Br J Cancer 1996; 73: 751-757

25. Tilling K, Wolfe C, Raju KS. Variations in the management and survival of women with endometrial cancer in south east England. Eur J Gynaecol Oncol 1998; 19: 64-68

26. Wolfe C, Tilling K, Raju KS. Management and survival of ovarian cancer patients in south east England. Eur J Cancer 1997; 33: 1835-1840

27. McArdle CS, Hole D. Impact of variability among surgeons on postoperative morbidity and mortality and ultimate survival. Br Med J 1991; 302: 1501-1505

28. Expert Advisory Group on Cancer. *A policy framework for commissioning cancer services.* London: Department of Health, 1995

29. Department of Health. *Our healthier nation. A contract for health.* London: The Stationery Office, 1998

30. Anon. "Patient-centred cancer services"? What patients say. National Cancer Alliance News 1997; 1: 4-5

31. Weeks JC, Cook EF, O'Day SJ, Peterson LM, Wenger N, Reding D, Harrell FE, Kussin P, Dawson NV, Connors AF, Lynn J, Phillips RS. Relationship between cancer patients' predictions of prognosis and their treatment preferences. J Am Med Assoc 1998; 279: 1709-1714

32. Coleman MP, Estève J, Damiecki P, Arslan A, Renard H. *Trends in cancer incidence and mortality (IARC Scientific Publications No. 121).* Lyon: International Agency for Research on Cancer, 1993

33. Coleman MP, Estève J. Trends in cancer incidence and mortality in the United Kingdom. In: *Cancer statistics: registrations of cancer diagnosed in 1989, England and Wales. Series MB1 no. 22.* London: HMSO, 1994, pp8-13

34. Estève J, Benhamou E, Raymond L. *Statistical methods in cancer research, volume IV. Descriptive epidemiology. (IARC Scientific Publications No. 128).* Lyon: International Agency for Research on Cancer, 1994

35. Berrino F, Micheli A, Sant M, Capocaccia R. Interpreting survival differences and trends. Tumori 1997; 83: 9-16

36. PORT Meta-analysis Trialists Group. Postoperative radiotherapy in non-small-cell lung cancer: systematic review and meta-analysis of individual patient data from nine randomised controlled trials. Lancet 1998; 352: 257-263

37. Fentiman IS, Tirelli U, Monfardini S, Schneider M, Festen J, Cognetti F, Aapro MS. Cancer in the elderly: why so badly treated? Lancet 1990; 335: 1020-1022

38. Fentiman IS. Are the elderly receiving appropriate treatment for cancer? Ann Oncol 1996; 7: 657-658

39. Harries SA, Lawrence RN, Scrivener R, Fieldman NR, Kissin MW. A survey of the management of breast cancer in England and Wales. Ann R Coll Surg Engl 1996; 78: 197-202

40. Newcomb PA, Carbone PP. Cancer treatment and age: patient perspectives. J Natl Cancer Inst 1993; 85: 1580-1584

41. Hole DJ, Gillis CR. Use of cancer registry data to evaluate the treatment of ovarian cancer on a hospital basis. Health Rep 1993; 5: 117-119

42. Ederer F, Axtell LM, Cutler SJ. The relative survival: a statistical methodology. Natl Cancer Inst Monogr 1961; 6: 101-121

43. Berrino F, Estève J, Coleman MP. Basic issues in the estimation and comparison of cancer patient survival. In: Berrino F, Sant M, Verdecchia A, Capocaccia R, Hakulinen T, Estève J, (eds.) *Survival of cancer patients in Europe: the EUROCARE study. (IARC Scientific Publications No. 132).* Lyon: International Agency for Research on Cancer, 1995, pp1-14

44. Feinstein AR, Sosin DM, Wells CK. The Will Rogers phenomenon: stage migration and new diagnostic techniques as a source of misleading statistics for survival in cancer. N Engl J Med 1985; 312: 1604-1608

45. Pollock AM. Off the record: cancer data. Health Serv J 1994; 3 November: 28-29

46. Selby P, Gillis C, Haward R. Benefits from specialised cancer care. Lancet 1996; 348: 313-318

47. Office of Population Censuses and Surveys. *A review of the national cancer registration system in England and Wales. Series MB1 no. 17.* London: HMSO, 1990

48. Anon. Clinical trials and clinical practice. Lancet 1993; 342: 877-878

49. Department of Health. *The Health of the Nation: a strategy for health in England.* London: HMSO, 1992

50. Joint Council for Clinical Oncology. *Cancer care and treatment services: advice for purchasers and providers.* London: Royal Colleges of Physicians & Radiologists, 1991

51. SW Thames RHA. *Towards a strategic framework for reducing ill-health caused by cancers.* London: South West Thames Regional Health Authority, 1992

52. Anon. *Purchasing and providing cancer services: a guide to good practice.* London: Royal Marsden Hospital, 1994

53. Royal College of Radiologists. *Medical manpower and workload in clinical oncology in the United Kingdom.* London: Royal College of Radiologists, 1991

54. Association of Cancer Physicians. *Review of the pattern of cancer services in England and Wales.* Southampton: Association of Cancer Physicians, 1994

55. NHS Executive North West. *Inequalities in health in the North West.* Warrington: NHS Executive, 1998

56. Royal College of Radiologists. *A national audit of waiting times for radiotherapy.* London: Royal College of Radiologists, 1998

57. MacKie R, Hole D. Incidence and thickness of primary tumours and survival of patients with cutaneous malignant melanoma in relation to socioeconomic status. Br Med J 1997; 312: 1125-1128

58. Stiller CA, Draper GJ. Treatment centre size, entry to trials, and survival in acute lymphoblastic leukaemia. Arch Dis Child 1989; 64: 657-661

59. Stiller CA. Centralisation of treatment and survival rates for cancer. Arch Dis Child 1988; 63: 23-30

60. Sainsbury R, Haward R, Rider L, Johnston C, Round C. Influence of clinician workload and patterns of treatment on survival from breast cancer. Lancet 1995; 345: 1265-1270

61. Junor EJ, Hole DJ, Gillis CR. Management of ovarian cancer: referral to a multi-disciplinary team matters. Br J Cancer 1994; 70: 363-370

62. Woodman C, Baghdady A, Collins S, Clyma J. What changes in the organisation of cancer services will improve the outcome for women with ovarian cancer? Br J Obstet Gynaecol 1998; 105: 135-139

63. Stiller CA. Centralised treatment, entry to trials and survival. Br J Cancer 1994; 70: 352-362

64. Kingston RD, Walsh S, Jeacock J. Colorectal surgeons in district general hospitals produce similar outcomes to their teaching hospital colleagues: review of 5-year survivals in Manchester. J Roy Coll Surg Edinb 1992; 37: 235-237

65. NHS Executive. *The new NHS. Modern. Dependable.* London: Department of Health, 1997

66. Joint Council for Clinical Oncology. *Reducing delays in cancer treatment: some targets.* London: Royal Colleges of Physicians & Radiologists, 1993

67.　Anon. Waiting for Dobbo – The government's confidence that it can cut hospital waiting lists may be misplaced. It was a silly promise to make in the first place. Economist 23 May 1998: 33-34

68.　Jones RVH, Dudgeon TA. Time between presentation and treatment of six common cancers: a study in Devon. Br J Gen Pract 1992; 42: 419-422

69.　Martin IG, Young S, Sue-Ling H, Johnston D. Delays in the diagnosis of oesophagogastric cancer: a consecutive case series. Br Med J 1997; 314: 467-471

70.　Wayman J, Raimes S, Griffin SM, Hayes N. Small study found that four fifths of delay was due to patients. Br Med J 1997; 315: 428

71.　Siewert JR, Fink U. Delays in the diagnosis of oesophagogastric cancer: a consecutive case series. Commentary: Britain does better than Germany before patients reach hospital. Br Med J 1997; 314: 471

72.　Sano T, Maruyama K. Delays in the diagnosis of oesophagogastric cancer: a consecutive case series. Commentary: Japanese point of view. Br Med J 1997; 314: 470-471

73.　NHS Executive. *Improving outcomes in colorectal cancer: the research evidence.* London: NHS Executive, 1997

74.　NHS Executive. *Improving outcomes in colorectal cancer: the manual.* London: NHS Executive, 1997

75.　Anglia & Oxford Cancer Steering Group. *Standards of cancer care in Anglia and Oxford: a report describing target standards of care agreed by health authorities in Anglia and Oxford region and key Calman implementation issues.* Milton Keynes: Anglia & Oxford Cancer Project, 1996

76.　Grimshaw JM, Russell IT. Effect of clinical guidelines on medical practice: a systematic review of rigorous evaluations. Lancet 1993; 342: 1317-1322

77.　NHS Executive. *Clinical indicators for the NHS (1994-95). A consultation document.* Leeds: NHS Executive, 1997

78.　NHS Executive. *The new NHS. Modern and dependable: a national framework for assessing performance. Consultation document.* London: Department of Health, 1998

79.　Clinical Outcomes Working Group. *Clinical outcome indicators, December 1994.* Edinburgh: The Scottish Office, 1996

80.　Clinical Outcomes Working Group. *Clinical outcome indicators, July 1996.* Edinburgh: The Scottish Office, 1996

81.　Coleman MP, Lutz J-M. Tendances évolutives du cancer et prévision des besoins: vers une meilleure utilisation des registres du cancer. Rev épidémiol santé publ 1996; 44: s2-s6

82.　Stiller CA. Population-based survival rates for childhood cancer in Britain, 1980-91. Br Med J 1994; 309: 1612-1616

83.　Hammond GD. The cure of childhood cancers. Cancer 1986; 58: 407-413

84.　Schillinger JA, Grosclaude PC, Honjo S, Quinn MJ, Sloggett A, Coleman MP. Survival from acute lymphocytic leukaemia: socioeconomic status and geographic region. Arch Dis Child 1999; (in press)

85.　Peckham M, Pinedo HM, Veronesi U, (eds.) *Oxford textbook of oncology.* Oxford: Oxford Medical Publications, 1995

86.　Horwich A (ed.) *Oncology: a multidisciplinary textbook.* London: Chapman & Hall Medical, 1995

87.　Neal AJ, Hoskin PJ. *Clinical oncology.* London: Edward Arnold, 1994

88.　Schottenfeld D, Fraumeni JF, (eds.) *Cancer epidemiology and prevention.* 2nd edn. Oxford: Oxford University Press, 1996

89.　Tomatis L, Aitio A, Day NE, Heseltine E, Kaldor JM, Miller AB, Parkin DM, Riboli E, (eds.) *Cancer: causes, occurrence and control (IARC Scientific Publications No. 100).* Lyon: International Agency for Research on Cancer, 1990

90.　Jensen OM, Parkin DM, MacLennan R, Muir CS, Skeet RG, (eds.) *Cancer registration: principles and methods. (IARC Scientific Publications No. 95).* Lyon: International Agency for Research on Cancer, 1991

91.　World Health Organisation. *Manual of the International Statistical Classification of Diseases, Injuries and Causes of Death, 8th Revision.* Geneva: WHO, 1965

92.　World Health Organisation. *International Classification of Diseases, 1975, 9th revision.* Geneva: WHO, 1977

93.　American Cancer Society. *Manual of tumor nomenclature and coding.* Washington, DC: American Cancer Society, 1951

94.　World Health Organisation. *International Classification of Diseases for Oncology (ICD-O).* Geneva: WHO, 1976

95.　Birch JM, Marsden HB. A classification scheme for childhood cancer. Int J Cancer 1987; 40: 620-624

96.　Office for National Statistics. *Cancer statistics, registrations: England and Wales, 1990. Series MB1 no. 23.* London: The Stationery Office, 1997

97.　Office for National Statistics. *Deaths registered in 1995 by cause, and by area of residence. Series DH2 96/2.* London: HMSO, 1996

98.　Harris V, Sandridge AL, Black RJ, Brewster DH, Gould A. *Cancer registration statistics, Scotland, 1986-1995.* Edinburgh: Information and Statistics Division, 1998

99.　Parkin DM, Whelan SL, Ferlay J, Raymond L, Young J, (eds.) *Cancer incidence in five continents, volume VII (IARC Scientific Publications No. 143).* Lyon: International Agency for Research on Cancer, 1997

100.　Doll R, Fraumeni JF, Muir CS, (eds.) Trends in cancer incidence and mortality. Cancer Surv 1994; 19/20

101.　Office of Population Censuses and Surveys. *Cancer statistics, survival. Survival up to 1978 of cases of diagnosed cancer registered in England and Wales during 1971-3, with additional survival statistics for some previous years. Series MB1 no. 3.* London: HMSO, 1980

102.　Office of Population Censuses and Surveys. *Cancer statistics, survival. Survival up to 1980 of cases of diagnosed cancer registered in England and Wales during 1971-5. Series MB1 no. 9.* London: HMSO, 1982

103.　Office of Population Censuses and Surveys. *Cancer survival 1976-78. OPCS Monitor MB1 84/1.* London: OPCS, 1984

104.　Office of Population Censuses and Surveys. *Cancer survival 1977-79. OPCS Monitor MB1 85/1.* London: OPCS, 1985

105.　Office of Population Censuses and Surveys. *Cancer survival 1978-80. OPCS Monitor MB1 86/1.* London: OPCS, 1986

106.　Office of Population Censuses and Surveys. *Cancer survival 1979-81. OPCS Monitor MB1 86/2.* London: OPCS, 1986

107.　Office of Population Censuses and Surveys. *Cancer survival 1981. OPCS Monitor MB1 88/1.* London: OPCS, 1988

108.　Reeves GK, Appleby P, Beral V, Quinn MJ, Babb P, Jones J. *Cancer survival in England and Wales: 1981 and 1989 registrations. Monitor MB1 98/1.* London: Office for National Statistics, 1998

109.　Stiller CA, Bunch KJ. Trends in survival for childhood cancer in Britain diagnosed 1971-85. Br J Cancer 1990; 62: 806-815

110.　Black RJ, Sharp L, Kendrick SW. *Trends in cancer survival in Scotland, 1968-1990.* Edinburgh: Information and Statistics Division, 1993

111.　Sant M, Gatta G, Valente F, Barchielli A, Ramazzotti V, Serventi L, Rosso S. The ITACARE study. Tumori 1997; 83: 17-24

112.　National Cancer Institute. SEER Stat – cancer incidence public use database 1973-95. Release 1.1. Bethesda, MD: National Cancer Institute, 1998

113.　Berrino F, Capocaccia R, Estève J, Gatta G, Hakulinen T, Micheli M, Sant M, Verdecchia A, (eds.) *Survival of cancer patients in Europe: the EUROCARE study, II. (IARC Scientific Publications No. 151).* Lyon: International Agency for Research on Cancer, 1999

National information about cancer

Cancer registries collect data for all cancer patients in a defined population. The data for each patient are less detailed than in clinical studies, but they provide a detailed and evolving picture of the public health burden of cancer over a long period of time.

A regional cancer registration system has covered the entire population of England and Wales since 1962, and the national cancer registry has collated the regional data to produce national information on cancer patterns since then. The national cancer registry has been linked to the National Health Service Central Register (NHSCR) since 1971. It receives notification of the eventual death of all cancer patients registered since 1 January 1971 whose record was successfully "flagged" at NHSCR with details of the initial cancer registration. It also receives information about the death of all people for whom cancer was mentioned on the death certificate. In addition to cancer incidence, therefore, the national cancer registry can also provide a broad overview of geographic patterns and long-term trends in cancer survival.

This chapter provides an outline of cancer registration in England and Wales, an explanation of how the cancers, geographic regions and deprivation groups were defined for the analyses, and a description of how the data were prepared for analysis.

The national cancer registry

The national cancer registry has been substantially improved since 1990. Computer systems were revised, separate annual files were converted to a person-based database, and probability matching was used to identify duplicate registrations and link the registrations of true multiple primary cancers in the same person. Information on linked registrations was sent to the cancer registries for deletion or amendment of their records as appropriate. Large numbers of records were added, amended or cancelled. All the records in the new database were then revalidated with more stringent logical checks than had been applied previously.

Before creation of the data set for survival analysis, special efforts were made to ensure that the national cancer registry was as accurate and up-to-date as possible. Data sets that had been recently submitted by regional cancer registries were checked for late registrations of cancer patients diagnosed up to 31 December 1990. The National Health Service Central Register was checked to ensure that any deaths occurring in cancer patients eligible for inclusion in the analysis had been linked to the corresponding cancer registration. Linkage of deaths up to 31 December 1995 was considered satisfactory, and this was taken as the closing date for all analyses.

Of particular relevance for this survival analysis, the records of some 50,000 patients were updated with the date of death or emigration; or marked as being untraceable for their vital status; or corrected for the date of birth or death; or, in a few cases, corrected by removal of the date of death where the patient had been recorded as dead but was found to be alive. This extensive exercise should largely have removed the so-called "immortals" – patients registered alive who later died, but whose death had not previously been linked to the cancer registration. Such patients can bias survival estimates upwards[1].

Completeness of registration

The regional cancer registries differ considerably in their methods of data collection. Some employ peripatetic clerks, others use hospital record staff to extract data for the registry, and several rely heavily on other organisations' computer systems, including those in hospitals and pathology laboratories. The principal sources for cancer notification are hospital in-patient records and pathology records. Registries also obtain data on out-patients and from radiotherapy and cytology departments, chest clinics, and GPs.

The registries probably differ in the completeness of their data, but the best are known to have very high levels of completeness. Direct measures are only available from occasional special studies. One study[2] estimated the incompleteness of registration of childhood cancers in the regional registries at just under five per cent. Incompleteness may be greater for adults, for whom cancer registration and record linkage at NHSCR may be more difficult than for children. An indication of the completeness of registration for a given cancer can be obtained from the ratio of the number of registrations and the number of deaths certified as due to that cancer in residents of the same geographical area in the same period. Such incidence-to-mortality ratios are presented by sex and site in annual reference volumes on cancer registrations[3]. These ratios have limitations as an indicator of completeness, but the ratios vary between cancer registries and over time in a way that would be difficult to explain without some corresponding variations in completeness.

The basis of cancer registration

The reporting of cancer is not a legal requirement in the UK, although reporting of many common infectious diseases has been a statutory requirement for more than 100 years. In 1990, a review of NHS information systems[4] concluded that *"National registers, such as the Cancer Registry, [are] a very important data source for public*

Table 2.1 Data items submitted by regional cancer registries to the national cancer registry

For tumours diagnosed 1971-90

Cancer registry number
Tumour identification number
Names
Address
Postcode
Sex
NHS number
Date of diagnosis
Anatomic location (site)
Morphologic type
Behaviour of tumour

Multiple tumour indicator
Date of death (if dead)

For tumours diagnosed in 1993 or later

Basis of diagnosis
Death-certificate-only indicator
Side (laterality)
Treatment indicators[1]
Stage of disease at diagnosis[2]
Grade[2]

[1] Surgery, radiotherapy, chemotherapy, hormonal therapy, other (yes/no)
[2] Only for cancers of breast and cervix initially

Data items typically recorded in regional cancer registries

Person	names, sex, date of birth, address and postcode, marital status, occupation, NHS number
Tumour	site, morphology, date of diagnosis, behaviour (benign, *in situ*, malignant, etc.), stage (TNM, other), laterality, basis of diagnosis (histology, cytology, etc.), mode of presentation (symptoms, screening, autopsy, etc.)
Treatment	treatment initiated within six months of diagnosis (chemotherapy, surgery, radiotherapy, other, none), surgical procedure, consultant(s) and hospital(s) with hospital case number, GP and GP's address
Outcome	cause, place and date of death, post-mortem (yes/no)

Source: adapted from ONS and cancer registry publications

health and outcome measures ... It was proposed therefore that hospitals ... should be obliged to provide data for those registries seen as having a national significance".

The NHS Management Executive later instructed all NHS health care providers to supply the national minimum data set for cancer registration to their respective regional cancer registry from July 1993[5]. Provision of the data for cancer registration has in effect become mandatory within the NHS since then.

The basic items of information needed to initiate a cancer registration are shown in Table 2.1. They have become part of the minimum data set required by the Department of Health for every cancer patient treated in hospital.

Geographic regions

In 1990, twelve regional cancer registries were in operation (Figure 2.1). Ten cancer registries each covered the territory of one Regional Health Authority (RHA), the eleventh covered the four Thames RHAs, and Wales was covered by one registry.

The organisation of the NHS and of the cancer registration system has changed substantially since 1990. These changes influenced our approach to the geographic analyses and the presentation of cancer survival rates.

The 14 RHAs in England were responsible for cancer registration in their populations for most of the period 1971-90, but they were merged into eight RHAs in 1994. In 1996, these eight new RHAs were themselves replaced, largely on the same territories, by new Regional Offices of the NHS National Executive. Responsibility for cancer registration was devolved to one or two district health authorities within each NHS Region. Cancer registries are now managed by these health authorities, which in theory remain accountable to their NHS Regional Office for cancer registration.

The territories covered by cancer registries for data collection have been realigned with the boundaries of the new NHS Regions, but the relationship between cancer registry and NHS Region is not as simple as before.

In both Northern & Yorkshire and South & West Regions, the two cancer registries that previously covered their territory have merged since 1996, and these regions are now covered by a single cancer registry. In Anglia & Oxford and North & West Regions, the two cancer registries in each region continue to operate on largely the same territory as before: Anglia & Oxford gained Bedfordshire from North Thames, and North & West gained part of Cumbria from Northern & Yorkshire. In the Trent and West Midlands Regions, there has been little or

no boundary change, and each region retains the original cancer registry. Finally, the North Thames and South Thames NHS Regions are covered by a single registry, as were their four predecessor RHAs. The position in Wales is also unchanged.

NHS Region (1998)	Cancer registry (up to 1990)
Northern & Yorkshire	Northern (except W Cumbria)
	Yorkshire
Trent	Trent
Anglia & Oxford	East Anglian
	Oxford
North Thames	Thames (except Bedfordshire)
South Thames	Thames
North & West	North Western
	Merseyside and Cheshire
West Midlands	West Midlands
South & West	South Western
	Wessex
Wales	Wales

Nine cancer registries now cover the eight NHS Regions in England – four regions each have one cancer registry, two regions each have two registries, and two regions share a single cancer registry (Figure 2.2).

NHS Regions

From this brief account, it can be seen that neither the regional cancer registries, nor the territories they covered, nor the NHS health authorities themselves are the same as they were during the period 1971-90. We judged it impractical to present geographic patterns of cancer survival aligned to the regional boundaries for which the data were originally collected. The Regional Offices of the NHS Executive are currently responsible for strategic supervision of health care delivery in the NHS Regions and of the regional cancer registration system(s) in their territory, however organised. Geographic patterns of cancer survival aligned to these territories seemed more likely to contribute to the development of health policy.

We have therefore chosen to present long-term trends in cancer survival for the eight current NHS Regions of England, and for Wales, even though these regions did not exist during the period 1971-95.

As a consequence, the survival rates presented here for four of the eight NHS Regions are derived from data originally supplied to the national cancer registry since 1971 by two or more regional cancer registries. For those regions, the data represent an amalgam of the various differences between the regional cancer registries over 25 years in operating practices and efficiency.

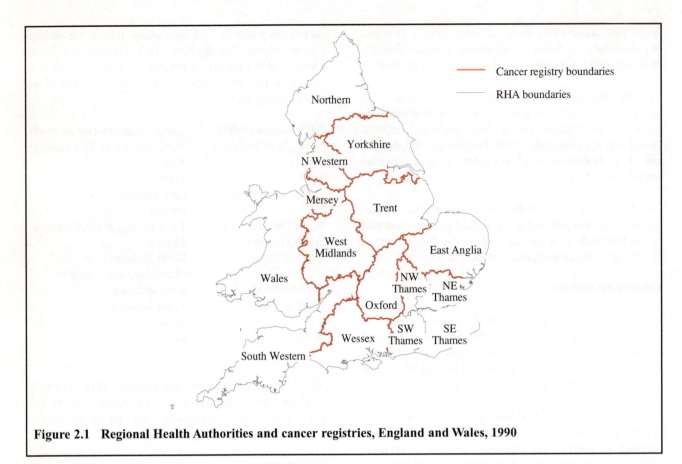

Figure 2.1   Regional Health Authorities and cancer registries, England and Wales, 1990

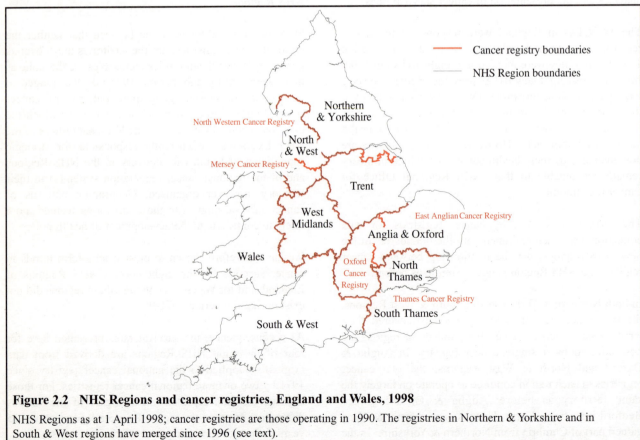

Figure 2.2   NHS Regions and cancer registries, England and Wales, 1998

NHS Regions as at 1 April 1998; cancer registries are those operating in 1990. The registries in Northern & Yorkshire and in South & West regions have merged since 1996 (see text).

Boundary data copyright Automobile Association

For the Trent, North Thames, South Thames and West Midlands NHS Regions, the survival rates should be more directly comparable with those previously published by the corresponding cancer registries, since these regions cover the same (or virtually the same) territories as before. The way in which the data for these analyses have been extracted, prepared and analysed, however (see below and Chapter 3), makes it unlikely that the results presented here would be identical to those published previously by the regional cancer registries, although the overall patterns will be broadly similar.

The analyses reported here represent the first time that cancer survival has been examined for the whole of England and Wales at regional level and according to a standard protocol. Differences between the results published here and those previously published at regional level must therefore be expected.

Allocation of patients to NHS Regions

For each cancer patient resident in England, the current NHS Region, as defined at 1 April 1997, was determined by the postcode of the patient's usual address at the time of diagnosis (1971-90). Patients who could not be assigned to any region were excluded. Patients who could be assigned to an NHS Region of England, or to Wales, but not to a census enumeration district, were included in overall regional and national analyses but excluded from the analyses by deprivation category, since the enumeration district is required to determine the patient's deprivation category (see below).

In geographic analyses of survival, NHS Region of residence is implicitly a proxy for regional aspects of the health service such as access to and the quality of primary care, the speed of referral and the quality of treatment services. It is also a proxy, however, for other factors relevant to survival that may also vary between regions, such as the degree of education and public understanding of cancer. None of these factors is explicitly taken into account in the geographic analyses.

Regional differences in cancer survival may be due to deficiencies in the data (artefact), variations in the extent of disease at diagnosis, or regional deficiencies in aspects of the health service and cancer treatment.

The possible influence on survival patterns of regional differences in data quality is mentioned where relevant in the chapters for each cancer. There is no national information on the extent of disease at diagnosis (clinical stage) for patients diagnosed during 1971-90; information on stage has been collected nationally for breast and cervical cancer since 1993. The distribution of clinical stage may have varied between regions during 1971-90. If any regional differences in survival were shown to be due to regional differences in the distribution of stage of disease, however, this would need to be seen as part of the public health problem. Finally, regional differences in cancer survival may also *"turn out to be due to the persistence of outmoded practices, or other remediable deficiencies in service provision or therapeutic regimes, and correcting these should lead to an improvement in overall standards of care"*[6].

Some cancer patients will move to another town or region in the five or ten years after diagnosis, but there is no practical method in the UK of collecting the information to incorporate such changes of residence in the analysis, nor is it obvious how such an adjustment should be done. Most patients would be likely to remain at their usual residence for at least six months after diagnosis, during which period most primary treatment for the malignancy would be given, and health authorities are responsible for the provision of NHS treatment to their resident population.

For all these reasons, it would be unwise to draw firm conclusions about differences in the efficacy of cancer treatment services between NHS Regions solely on the basis of differences in the survival rates reported here. The intent is to provide a coherent set of regional cancer survival rates as the basis for comparisons that would otherwise be difficult or impossible to make.

Deprivation and affluence

We have explored the relationship between material deprivation and cancer survival in England and Wales, both within NHS Regions and nationally, and over time.

The Carstairs index was used as a measure of material deprivation[7,8]. Other similar measures exist, but they are strongly correlated with one another[9] and they produce similar patterns of health in relation to social disadvantage, however defined[10].

The Carstairs index is a score of material deprivation derived from census data. It is used at the level of the census enumeration district (ED). There were 109,578 census enumeration districts for the 1991 census in Great Britain, with an average of 200 households and 500 persons in each. For each census ED, percentage values are obtained for four variables measured in the national census: car ownership, household overcrowding, head of household in social class IV or V, and male unemployment. Overcrowding is defined as a density of more than one person per room. The proportion of households headed by a person of social class IV or V is calculated at the more aggregated level of the census ward, and that value is assigned to each of the 10 or so enumeration districts within the ward. The log-transformed value of percentage unemployment is used, because the distribution of this variable is highly skewed. Percentage values for each ED are then standardised by subtraction of the mean value for all EDs in Great Britain and division of the result by the population standard deviation. Standardised scores are derived in this way for each of the four variables. The Carstairs score for an ED is then the sum of the four standardised scores[11,12].

Five deprivation categories were defined. In the strict sense, the range is from the least deprived to the most deprived, but the terms affluent and deprived, respectively, are often used for convenience[13]. The Carstairs scores for all the EDs in Great Britain were put in rank order and divided into five quintiles, from the most affluent (least deprived) 20% of the enumeration districts to the most deprived 20% of districts. A deprivation category was then assigned to each patient on the basis of the census ED in which he or she was resident at diagnosis, using contemporary dictionaries of correspondence between postcodes of residence and EDs. The analysis was limited to cancer patients resident in England and Wales, but we used the distribution of Carstairs scores for all the EDs of Great Britain (England, Wales, and Scotland) to define the deprivation categories, so that results produced in similar fashion for Scotland will then be directly comparable.

The numbers of cancer patients included in the survival analyses are not the same in each deprivation category. There are four reasons for this. First, the average deprivation score for EDs in Scotland is higher (more deprived) than in England, so slightly more than 20% of the EDs in England and Wales fall in the most affluent category, and slightly fewer than 20% in the most deprived category, than for Great Britain as a whole. Second, the resident population varies between EDs. Third, for many of the common cancers, the risk of developing the cancer also varies with deprivation. Finally, there were difficulties in assigning some patients to an enumeration district from their postcode.

Data from the 1971 census were not available to calculate Carstairs indices, so data from the 1981 census were used to assign the deprivation category to patients diagnosed between 1971 and 1985. Postcodes for a given address can change over time. The correspondence between postcodes and census enumeration districts can also change. For these reasons, both the proportion of cancer patients who could not be reliably assigned to a census enumeration district and the proportion of districts to which no Carstairs score could be assigned were higher for patients diagnosed in the early 1970s than for patients diagnosed in later periods. This proportion also varied between regions.

The degree of deprivation of a given census enumeration district may change between decennial censuses. Its Carstairs score and the quintile to which it is assigned may therefore also change between censuses. The appropriateness of assigning a deprivation category to a patient resident in a given district becomes less secure the longer the interval between the date of diagnosis and the date of the corresponding census. Although the shifts in correspondence between postcodes and EDs and the movement of EDs between deprivation categories at successive censuses are not large, these influences would also tend to blur the precision with which patients are assigned to a suitable deprivation category. In turn, they would tend to blur any underlying association between deprivation and survival for patients diagnosed in the early 1970s.

For patients diagnosed in the period 1986-90, Carstairs scores were derived from the 1991 census.

Patients who could not be assigned to a deprivation category through their census enumeration district were excluded from analyses by deprivation category, but retained in overall analyses at regional and national level. For this reason, the number of patients included in regional and national analyses usually exceeds the total number of patients included in analyses for the five deprivation categories. The differences were more marked for some regions than others, but they were always greatest for the period 1971-75, and the differences are generally very small in all regions for the decade 1980-89.

With these caveats, the great majority of cancer patients diagnosed in each NHS Region and in each five-year period were assigned to a deprivation category for survival analysis.

The uncertainty inherent in attributing the characteristics of a group (in this case, residents of a census ED) to individual members of the group will lead to some patients being assigned to a deprivation category that does not reflect their own socio-economic status accurately. This would be expected to blur any underlying association between deprivation and cancer survival. No reliable measure of individual deprivation or socio-economic status was available for cancer patients, however: less than 20% of cancer patient records contained an informative social class. There is also evidence from studies combining personal and areal measures of socio-economic status that small-area indicators of deprivation do reflect individual disadvantage in relation to health[14].

Definition of cancers

When examining cancer survival trends for patients diagnosed over a 20-year period, it is important that the tumours being grouped for analysis are as similar as possible. We have tried to define groups of cancers that are clinically relevant, coherent across the various classification systems in use since 1971, and comparable with cancers included in other major reports on cancer survival, particularly those in Scotland, Italy and Europe (Table 2.2). Cancers were variously defined on the basis of their anatomic location (site), their microscopic appearance (morphology) and whether they were classified as benign, *in situ* or malignant (behaviour): see Tables 2.3-2.5.

Anatomic location

During the period 1971-90, two coding systems for the anatomic location of tumours were used by the national cancer registry. Between 1971 and 1978, the anatomic site was coded to the eighth revision of the International Classification of Diseases (ICD-8). For cases diagnosed in 1979 or later, this was replaced by the ninth revision (ICD-9). In general, ICD-9 codes enable more detailed recording of the anatomic location of tumours, but include the same tumours as the corresponding code in ICD-8. Thus for example, ICD-9 enables the upper, middle and lower parts of the oesophagus to be separately coded (150.3, 150.4 and 150.5), whereas in ICD-8 all oesophageal cancers were coded to the same rubric (150), without further detail.

The change from ICD-8 to ICD-9 has implications for the analysis of time trends in cancer survival. Most of the analyses are for cancers defined at the broader (three-digit) level of the classifications. Some transfers between three-digit rubrics were necessary, because tumours defined at the level of the fourth digit are not always assigned to the same three-digit rubric in both revisions of the ICD (see below). To ensure that the tumours grouped for analysis were comparable, despite the changes in classification, an ICD-9 equivalent of the ICD-8 site code was assigned to all tumours diagnosed up to 1978 (Table 2.3).

Code conventions

We have used the following shorthand in describing codes and groups of codes in this chapter:

"Valid"   codes are those published in the eighth and ninth revisions of the International Classification of Diseases (ICD-8 and ICD-9), or in the International Classification of Diseases for Oncology (ICD-O), or in the Manual of Tumor Nomenclature and Coding (MOTNAC)

NCR   special codes used for site, morphology or behaviour by the national cancer registry and the regional cancer registries, in addition to valid codes

999x   includes all valid four-digit codes within this three-digit rubric

999-   means there is no valid four-digit code within this rubric (although the dash character "-" is actually held as a "9" in the data to simplify numerical manipulation)

990-999   includes all valid three-digit and four-digit codes within this range

NOS   not otherwise specified

Special site codes

During the period 1971-78, covered by the eighth revision of the ICD, the national cancer registry stored information about the anatomic location of some cancers of the tongue, oesophagus, colon, breast, cervix and eye in more detail than was available in ICD-8, by using some additional codes (Table 2.6). The implications for grouping of tumours for analysis are described below.

Morphologic type

Between 1971 and 1978, the morphological classification used for cancer registration by the national cancer registry was based on the Manual of Tumor Nomenclature and Coding. From 1979 onwards, this was replaced by the morphological component of the International Classification of Diseases for Oncology. In order to ensure that the groups of cancers defined for survival analysis were as homogeneous as possible throughout the period 1971-90, MOTNAC morphology codes were re-coded to ICD-O. As with the classifications for anatomic site, the MOTNAC classification is less detailed than ICD-O, and this occasionally constrained the choice of cancers to be studied.

Solid tumours

In the MOTNAC classification, the first three digits encode the tumour morphology for solid tumours, and the fourth designates the degree of malignancy (behaviour):

0   Benign
1   Uncertain whether benign or malignant
2   Carcinoma *in situ*, intra-epithelial, non-infiltrating
3   Malignant, primary site
6   Malignant, metastatic site
9   Malignant, uncertain whether primary or metastatic site

MATERIAL

Table 2.2 Correspondence with recent reports on cancer survival up to five years (period of diagnosis)

Chapter	England and Wales (1971-90) Adults	Europe (1978-89) Adults	Italy (1978-89) Adults	Scotland (1968-87) Adults
5	Lip	Lip*	Lip	-
6	Tongue	Tongue	Tongue	-
7	Salivary glands	Salivary glands*	Salivary glands (pooled)	-
8	Oral cavity	Oral cavity	Oral cavity -	-
9	Oropharynx	Oropharynx	Oropharynx	-
10	Nasopharynx	Nasopharynx	Nasopharynx (pooled)	-
11	Hypopharynx	Hypopharynx	Hypopharynx	-
	-	Head and neck	Head and neck	-
12	Oesophagus	Oesophagus	Oesophagus	Oesophagus
13	Stomach	Stomach	Stomach	Stomach
14	Small intestine	Small intestine*	Small intestine (pooled)	-
15	Colon	Colon	Colon	Large bowel
16	Rectum	Rectum	Rectum	Large bowel
17	Liver	Liver*	Liver, primary	-
18	Gallbladder	Gallbladder*	Gallbladder	-
19	Pancreas	Pancreas	Pancreas	Pancreas
20	Nasal cavities and paranasal sinuses	Nasal cavities*	Nasal cavities (pooled)	-
21	Larynx	Larynx	Larynx	Larynx
22	Lung	Lung	Lung	Lung
23	Pleura	Pleura*	Pleura	-
24	Thymus	-	-	-
25	Bone	Bone	Bone (pooled)	Bone and connective tissue
26	Connective tissue	Soft tissue*	Soft tissue	Bone and connective tissue
27	Melanoma of skin	Melanoma of skin*	Melanoma of skin	Melanoma
28	Breast	Breast (female)	Breast	Breast (female)
29	Cervix	Cervix	Cervix	Cervix
30	Uterus	Uterus	Uterus	Uterus
31	Ovary	Ovary	Ovary	Ovary
32	Vagina and vulva	Vagina and vulva	Vagina and vulva	-
33	Prostate	Prostate*	Prostate	Prostate
34	Testis	Testis	Testis	Testis
35	Penis	Penis, male genital organs	Penis (pooled)	-
36	Bladder	Bladder*	Bladder	Bladder
37	Kidney	Kidney	Kidney	Kidney
38	Eye	Choroid (melanoma)*	Choroid (melanoma) (pooled)	-
39	Brain	Brain	Brain	Brain and CNS
40	Thyroid	Thyroid*	Thyroid	Thyroid
41	Non-Hodgkin lymphoma	Non-Hodgkin lymphoma*	Non-Hodgkin lymphoma	Non-Hodgkin lymphoma
42	Hodgkin's disease	Hodgkin's disease	Hodgkin's disease	Hodgkin's disease
43	Multiple myeloma	Multiple myeloma*	Multiple myeloma	Multiple myeloma
44	Acute lymphoid leukaemia	Acute lymphoid leukaemia	Acute lymphoid leukaemia	-
45	Chronic lymphoid leukaemia	Chronic lymphoid leukaemia	Chronic lymphoid leukaemia	-
46	Acute myeloid leukaemia	Acute myeloid leukaemia	Acute myeloid leukaemia	-
47	Chronic myeloid leukaemia	Chronic myeloid leukaemia	Chronic myeloid leukaemia	-
48	Monocytic leukaemia	-	-	-
49	All leukaemias	Leukaemia	Leukaemia	Leukaemia
62	Spinal cord	-	-	-
62	Adrenal	-	-	-
62	Pituitary	-	-	-
4	All malignancies	All malignancies*	All malignancies	All malignancies
	Children	**Children**		**Children**
50	All leukaemias	Leukaemia		Leukaemia
51	Acute lymphoid leukaemia	-		-
52	Hodgkin's disease	Hodgkin's disease		Hodgkin's disease
53	Non-Hodgkin lymphoma	-		Non-Hodgkin lymphoma
54	Brain and spinal cord	Brain		Brain and other CNS
55	Retinoblastoma	-		-
56	Wilms' tumour	Kidney		Kidney
57	Hepatoblastoma	-		-
58	Osteosarcoma	Bone		Bone and connective tissue
59	Ewing's sarcoma	Bone		Bone and connective tissue
60	Soft tissue sarcoma	-		Bone and connective tissue
61	Germ-cell and gonadal	Ovary, testis		Testis
-	-	Nasopharynx		
-	-	-		All malignancies

*Only analysed in EUROCARE II study (1985-89 diagnoses). See Chapter 1 for references.

Table 2.3 Definition of solid tumours

	Short title	Full description	Site code (ICD-9)	Site code (ICD-8)
5	Lip	Lip (excluding skin of lip)	140x	140x
6	Tongue	Tongue .	141x, *excluding 141.6*	141x (including NCR 141.8)
7	Salivary glands	Major salivary glands	142x	142x
8	Oral cavity	Gum, floor of mouth, other and unspecified parts of mouth	143x, 144x, 145x	143x, 144-, 145x
9	Oropharynx	Oropharynx	146x, 141.6	146x
10	Nasopharynx	Nasopharynx	147x	147-
11	Hypopharynx	Hypopharynx	148x	148x
12	Oesophagus	Oesophagus	150x	150- (plus NCR 150.0-150.2, 150.9)
13	Stomach	Stomach	151x	151x
14	Small intestine	Small intestine, including duodenum	152x	152x
15	Colon	Colon, *excluding intestinal tract, part unspecified*	153x, *not 159.0*	153.0-153.3, (plus NCR 153.7),153.8, exclude 153.9
16	Rectum	Rectum, rectosigmoid junction and anus	154x	154x
17	Liver	Liver, primary *(excluding tumours not specified as primary or secondary)*, and intrahepatic bile ducts	155.0, 155.1, *exclude155.2*	155.0, 155.1, *not 197.8*
18	Gallbladder	Gallbladder and extrahepatic bile ducts	156x	156x
19	Pancreas	Pancreas	157x	157x
20	Nasal cavities and paranasal sinuses	Nasal cavities, middle ear and accessory sinuses	160x	160x
21	Larynx	Larynx	161x	161x
22	Lung	Trachea, bronchus and lung	162x	162x
23	Pleura	Pleura	163x	163.0
24	Thymus	Thymus	164.0	194.2
25	Bone	Bone and articular cartilage	170x	170x
26	Connective tisssue	Connective and other soft tissue	171x	171x
27	Melanoma of skin	Melanoma of skin, *includes skin of breast, scrotum*	172x, plus 187.7 if morphology 8720-8780	172x, plus NCR 174.2 if morphology 8723-8783
28	Breast	Breast, *excluding skin of breast and carcinoma in situ*	Women 174x, men 175-, *not 233.0*	174- (plus NCR 174.0, 174.1, 174.3, *not 174.2), exclude behaviour code 2 (in situ)*
29	Cervix	Cervix uteri, *excluding carcinoma in situ*	180x, *not 233.1*	180- (plus NCR 180.1, 180.2) *not 234.0*
30	Uterus	Corpus uteri, *excluding tumours not specified as cervix or corpus uteri*	182x, *not 179-*	182.0, *exclude 182.9*
31	Ovary	Ovary and other uterine adnexa	183x	183x
32	Vagina and vulva	Vagina, vulva, other and unspecified female genital organs	184x	184x
33	Prostate	Prostate	185-	185-
34	Testis	Testis	186x	186-
35	Penis	Penis, *excluding other male genital organs and scrotum*	187.1-187.4	187.0
36	Bladder	Bladder	188x	188-
37	Kidney	Kidney, other and unspecified urinary organs	189x	189x
38	Eye	Eye	190x	190- (plus NCR 190.0-190.2, 190.7-190.9)
39	Brain	Brain, *excluding cranial nerves, spinal cord and meninges*	*191x*	*191-*
40	Thyroid	Thyroid	193-	193-
62	Spinal cord	Spinal cord	192.2	192.2
62	Adrenal	Adrenal gland	194.0	194.0
62	Pituitary	Pituitary gland and craniopharyngeal duct	194.3	194.3

Neoplasms with morphology codes for lymphoma, myeloma or leukaemia (ICD-O M9590-9940) were excluded from all solid tumour categories (see text). NCR - special codes used by the National Cancer Registry (see Table 2.6)

MATERIAL

Table 2.4 Definition of lymphomas, multiple myeloma and leukaemias (adults)

	Description	Site code (ICD-9)	Morphology code (ICD-O)	Site code (ICD-8)	Morphology code (MOTNAC)
41	Non-Hodgkin lymphoma	200x, 202x, 140x-199x	9590-9642, 9690-9701, 9710-9722, 9740-9750, 9940	200x, 202x, 140x-199x	9593-9643, 9693-9753
42	Hodgkin's disease	201x	9650-9662	201-	9653-9683
43	Multiple myeloma	203x, 170x	9730, 9731, 9830	203-, 170x	9733, 9833
44	Acute lymphoid leukaemia	204.0	9821	204.0	9825
45	Chronic lymphoid leukaemia	204.1	9823	204.1	9827
46	Acute myeloid leukaemia	205.0	9861, 9866, 9870, 9880	205.0	9865
47	Chronic myeloid leukaemia	205.1	9863	205.1	9867
48	Monocytic leukaemia	206x	9890-9894	206x	9893, 9895, 9897-9899
49	All leukaemias	204x-208x	9800-9804, 9810, 9820-9825, 9840-9842, 9850, 9860-9866, 9870, 9880, 9890-9894, 9900, 9910, 9920, 9930	204x-207x	9803, 9805, 9807-9809, 9813, 9823, 9825, 9827-9829, 9853, 9863, 9865, 9867-9869, 9873, 9883, 9893, 9895, 9897-9899, 9903, 9913

Solid tumours with morphology codes ICD-O M9590-9642, M9660-9722 or M9940 were included with non-Hodgkin lymphoma.
Bone neoplasms with morphology codes ICD-O M9730-9731 were included with multiple myeloma. Plasma cell leukaemias (ICD-O M9830) were included with myeloma.

Table 2.5 Definition of childhood malignancies

	Group[1]	Description	Morphology (ICD-O)	Site (ICD-9)	Morphology (MOTNAC)	Site (ICD-8)
50	I	All leukaemias	9800-9804, 9810, 9820-9825, 9830, 9840-9842, 9850, 9860-9866, 9870, 9880, 9890-9894, 9900-9940	204x-208x	9803-9809, 9813, 9823-9829, 9833, 9853, 9863-9869, 9873, 9883, 9893-9899, 9903-9933	204x-207x
51	Ia	Acute lymphoid leukaemia	9821, 9824	204.0	9825, 9829	204.0
52	IIa	Hodgkin's disease	9650-9662	201x	9653-9683	201-
53	IIb & IId	Non-Hodgkin lymphoma and unspecified lymphoma	9590-9642, 9690-9701	140x- 200x, 202x	9593-9643, 9693-9703	140x-200x, 202x
54	III	Central nervous system & miscellaneous intracranial or spinal neoplasms	8000-8004	191x, 192x, 194.3, 194.4	800x (behaviour 3-9)	191x-192x, 194.3, 194.4
			9990	191x, 192x, 194.3, 194.4, 225x, 227.3, 227.4, 237.0, 237.1, 237.5, 237.6, 237.9, 239.6	999x (behaviour 0, 1, 3-9)	191x-192x, 194.3, 194.4, 225x, 226.2, 226.3, 238.x
			8270-8281, 8300, 9350-9362, 9380-9384, 9390-9394, 9505, 9530-9539	140x-239x	827x-828x, 830x, 935x-936x, 938x, 939x, 950x, 953x (behaviour 0, 1, 3-9)	140x-239x
			9400-9441, 9470-9480	140x-199x	940x-943x, 947x (behaviour 3-9)	140x-199x
			9442-9460, 9481	140x-199x, 230x-239x	944x-946x, 948x (behaviour 1, 3-9)	140x-199x, 230x-239x
			9060-9102	225x, 227.3, 227.4, 237.0, 237.1, 237.5, 237.6, 237.9, 239.6	906x-910x (behaviour 0, 1)	225x, 226.2, 226.3, 238x
55	V	Retinoblastoma	9510-9512	140x-199x	951x (behaviour 3-9)	140x-199x
56	VIa	Wilms' tumour	8960	140x-199x	896x (behaviour 3-9)	140x-199x
57	VIIa	Hepatoblastoma	8970	140x-199x	897x (behaviour 3-9)	140x-199x
58	VIIIa	Osteosarcoma	9180-9190	140x-199x	918x, 919x (behaviour 3-9)	140x-199x
59	VIIIb	Ewing's sarcoma	9260	170x, 196x-199x	926x (behaviour 3-9)	170x, 196x, 197x, 198x, 199x
60	IX	Soft tissue sarcoma	8810, 8811, 8813-8832, 8840-8895, 8900-8920, 8990, 8991, 9040-9044, 9120-9170, 9251, 9540, 9560, 9581	140x-199x	881x, 884x-889x, 890x-892x, 899x, 904x, 912x-917x, 925x, 954x, 956x, 958x (behaviour 3-9)	140x-199x
			8800-8804	140x-169x, 171x-199x	880x (behaviour 3-9)	140x-169x, 171x-199x
			9240, 9260	140x-169x, 171x-195x	924x, 926x (behaviour 3-9)	140x-169x, 171x-195x
61	X	Germ-cell, trophoblastic & other gonadal neoplasms	8381, 8441-8471, 8600-8650, 9000, 9060-9102	140x- 199x	838x, 844x-847x, 860x-865x, 900x, 906x-910x (behaviour 3-9)	140x- 199x
			8000-8041, 8043, 8140, 8230, 8231, 8260, 8310, 8440, 8480, 8481, 9990	183.0, 186x	800x-804x, 814x, 823x, 826x, 831x, 844x, 848x, 999x (behaviour 3-9)	183.0, 186x,

[1] Birch and Marsden group (see text). Not all groups were included.

Table 2.6 Non-standard codes used in the national cancer registry

Anatomic location (codes used in addition to ICD-8, 1971-1978)

ICD-8	NCR code	Description
141.0		Base of tongue
141.1		Dorsal surface of tongue
141.2		Borders and tip of tongue
141.3		Ventral surface of tongue
	141.8	Tongue, other specified parts
150.-		Oesophagus
	150.0	Upper third of oesophagus
	150.1	Middle third of oesophagus
	150.2	Lower third of oesophagus
	150.9	Oesophagus, unspecified
153.0		Caecum, appendix and ascending colon
153.1		Transverse colon, including hepatic and splenic flexures
153.2		Descending colon
153.3		Sigmoid colon
	153.7	Multiple parts
153.8		Large intestine (including colon), part unspecified
174.-		Breast
	174.0	Breast NOS, areola and axilla
	174.1	Breast, bilateral
	174.2	Breast, skin of breast
	174.3	Breast, nipple
180.-		Cervix
	180.1	Cervix uteri
	180.2	Endocervix
190.-		Eye
	190.1	Eye - retina
	190.2	Eye - choroid
	190.3	Eye - ciliary body
	190.7	Eye - multiple parts
	190.8	Eye - other specified parts
	190.9	Eye - part unspecified

Morphology (codes used in addition to ICD-O, 1979-90)

9092	Germ cell tumours: mixed teratoma and seminoma
9093	Germ cell tumours: mixed type tumours
9826	B-cell leukaemias
9827	T-cell leukaemias

Behaviour (codes used for degree of differentiation, not behaviour, in addition to MOTNAC, 1971-78)

4	Malignant, primary site, well (or moderately) differentiated
8	Malignant, primary site, undifferentiated (or poorly differentiated, or anaplastic)

Thus a malignant transitional cell (urothelial) carcinoma of the bladder would have the MOTNAC code 8123, in which 812 codes the morphologic type (transitional cell carcinoma), while the fourth digit, encoding behaviour, indicates that it is a primary malignant tumour (code 3), originating in urothelial tissue.

As for tumour site, however, the national cancer registry also used additional codes on the behaviour axis during the period 1971-78 (Table 2.6), this time to encode the degree of morphologic differentiation of malignant primary tumours, rather than their behaviour:

4   Malignant, primary site, well (or moderately) differentiated
8   Malignant, primary site, undifferentiated (or poorly differentiated, or anaplastic)

The International Classification of Diseases for Oncology (ICD-O), in use from 1979, represents a major expansion of MOTNAC, with many new and more precise terms. A fourth digit was added to the morphology code to incorporate this detail, and the single-digit code for behaviour again comes last. In ICD-O, the morphology of a malignant primary urothelial carcinoma would be coded to 8120/3, where 8120 represents the morphologic type and the fifth digit (/3) again indicates that it is a primary malignant tumour. The fourth digit allows the various subtypes of transitional cell carcinoma to be coded separately (M8120-8124), a level of detail not provided by MOTNAC.

In order to convert morphology codes from MOTNAC to ICD-O, we first added a fourth digit of zero. The last digit of the MOTNAC code was then used as the behaviour code (fifth digit), since the structure of this code was the same in both classifications for solid tumours. The NCR special behaviour codes 4 and 8 in MOTNAC (Table 2.6) were converted to 3 in ICD-O. The information on the degree of tumour differentiation was therefore lost, but these codes were not widely used, and no attempt has been made to separate well-differentiated and poorly-differentiated tumours in the analysis.

Leukaemias

Leukaemias are malignant by definition, and the fourth digit of the morphology code in MOTNAC designated the type of leukaemia, not the behaviour. ICD-O retains the terminology of leukaemia types, but the codes are different:

	Last digit of morphology code in	
Type of leukaemia	*MOTNAC*	*ICD-O*
	(1971-78)	(1979-90)
Not otherwise specified	3	0
Acute	5	1
Subacute	8	2
Chronic	7	3
Subleukaemic or aleukaemic	9	4

This correspondence was used to convert the last digit of morphology codes for leukaemias from MOTNAC to ICD-O.

Hodgkin's disease

The codes for the different sub-types of Hodgkin's disease recognised within ICD-O and ICD-9 do not follow the simple rules above, and they were re-coded as follows:

	Morphology code in	
Type of Hodgkin's disease	*MOTNAC*	*ICD-O*
	(1971-78)	(1979-90)
Not otherwise specified	9653	9650/3
Lymphocytic predominance	9654	9651/3
Mixed cellularity	9655	9652/3
Lymphocytic depletion, NOS	9657	9653/3
Nodular sclerosis	9658	9656/3
Hodgkin's paragranuloma	9663	9660/3
Hodgkin's granuloma	9673	9661/3
Hodgkin's sarcoma	9683	9662/3

Selection of cancers

The large numbers of cases available for analysis derive from the incidence of cancer in a national population of some 50 million, under observation for 20 years. This has made it possible to examine trends in survival for relatively uncommon cancers, for some of which no population-based survival estimates appear to have been published previously. Skin cancers other than melanoma were not analysed, and they are not included in any figures given in this book.

We selected 58 cancers for analysis: 39 solid tumours and eight types of leukaemia and lymphoma in adults, and 11 malignancies in childhood. Survival analyses are also presented for all types of leukaemia combined, both for adults and for children (Tables 2.3-2.5).

These groups account for more than 90% of the malignant neoplasms that occur in England and Wales.

Table 2.7 Re-classification of solid tumours with morphology of lymphoma, myeloma or leukaemia

Malignancy	Morphology codes (ICD-O)	Anatomic site codes (ICD-9) Original	Re-classified
Non-Hodgkin lymphoma		140-199	
Lymphosarcoma and reticulosarcoma	9640-9642		200.0
Other malignant neoplasms of lymphoid			
and histiocytic tissue	9590-9633		202.8
	9690-9722		202
	9740-9750		202
	9940		202.4
Hodgkin's disease	9650-9662	140-199	Excluded
Multiple myeloma	9730-9731	140-199	203
Plasma cell leukaemia	9830	169	203.1
Leukaemia	9800-9825	140-199	Excluded
	9840-9930		Excluded

Solid tumours in adults

Only malignant tumours were analysed: tumours with benign, *in situ* or uncertain behaviour were excluded. Solid tumours were generally grouped for analysis on the basis of their anatomic location (site code). Malignancies that were coded to a solid tumour site but with the morphology of lymphoma, myeloma or leukaemia were excluded from any solid tumour group. Those with the morphology of non-Hodgkin lymphoma and some of those with the morphology of myeloma were re-classified for analysis (Table 2.7). Solid tumours with a leukaemia morphology were excluded. Melanoma of the skin was defined for analysis both by anatomic site and by morphology. Tumour behaviour was also used in defining cancers of the breast and cervix (see below).

Tongue and oropharynx

The lingual tonsil is included with the oropharynx in ICD-8 (146.0, lingual and faucial tonsils), but it is coded with the tongue in ICD-9 (141.6). To ensure comparability of the data for both tongue and oropharynx across the 20-year period 1971-90, tumours of the lingual tonsil diagnosed during 1979-90 were excluded from the rubric for tumours of the tongue (ICD-9 141) and analysed with those coded to the oropharynx (ICD-9 146).

Colon and rectum

Tumours of the colon and rectum were analysed separately. Colon and rectum were assigned to the three-digit rubrics 153 and 154, respectively, in both ICD-8 and ICD-9. Large bowel cancers not specified as either colon or rectum fell within the three-digit rubric for colon cancer

in ICD-8 (153.9), however. These tumours were excluded, since they are coded to a different three-digit rubric in ICD-9 (159.0).

Liver

The liver is a common location for secondary spread from cancers arising in other organs. Our intention was to analyse the survival for primary tumours of the liver (ICD-9 155.0) and intra-hepatic bile ducts (155.1). Unspecified tumours of the liver were assigned to a separate rubric in ICD-8 (197.8), and these were not analysed. Liver tumours that are not specified as primary or secondary fell within the same three-digit rubric as primary tumours of the liver in ICD-9 (155.2), however, and these were excluded.

Melanoma of the skin

Melanomas of the skin have been analysed as a single group. Melanomas of internal organs have been analysed together with other histologic types of malignancy in those organs.

Breast; melanoma of the skin of the breast

Tumours of the skin of the breast diagnosed during 1971-78 were assigned a special site code in ICD-8 (NCR 174.2; Table 2.6). There were 381 such cancers, between 0.01% and 0.9% of the regional totals: in all, some 0.2% of the 168,000 breast cancers registered in England and Wales in that period. Skin of the breast is specifically excluded from the ICD-9 breast cancer rubrics, so these tumours were excluded from the data set for breast cancer, for consistency of definition with breast cancers diagnosed during 1979-90 (Table 2.3). Melanomas of the skin of the breast diagnosed during 1971-78 were then re-assigned for analysis with all other melanomas of the skin.

Breast and cervix

Only malignant tumours of the breast and uterine cervix were included. *In situ* carcinomas of the breast were excluded: for the period 1971-78, these were defined on the basis of a behaviour code 2, and for 1979-90 on the basis of the site code (ICD-9 233.0). *In situ* carcinomas of the cervix (ICD-8 234.0, ICD-9 233.1) were also excluded.

Uterus

Cancers of the cervix and body of the uterus are coded to separate three-digit rubrics (180 and 182, respectively) in both the eighth and ninth ICD revisions, and all uterine cancers should be registered to one or other of these sites[15]. Uterine cancers not specified as having arisen either in the cervix or the body of the uterus fell within the three-digit rubric for the body of the uterus in ICD-8 (182.9), however, and these were excluded. Such tumours are coded to a separate rubric in ICD-9 (179-): they were not analysed.

Penis; melanoma of the scrotum

Cancers of the penis are coded to 187.0 in ICD-8 and, in more detail, to 187.1-187.4 in ICD-9. In most reports of cancer incidence and survival, cancers of the penis and other male genital organs (but not testis or prostate, which have separate three-digit rubrics) are analysed together. Here, we have chosen to analyse the survival for cancers of the penis, excluding tumours of other male genital organs such as the epididymis (ICD-9 187.5), spermatic cord (187.6) and seminal vesicle (187.8).

Cancers of the skin of the scrotum (187.7) were also excluded from this group, but melanomas of the scrotal skin were re-assigned and analysed with all other melanomas of the skin (ICD-9 172), for consistency with ICD-8: this is similar to the reassignment for melanomas of the skin of the breast. Tumours of the skin of the penis were included with other tumours of the penis in both ICD-8 and ICD-9, and these tumours were not re-coded: there were only 36 melanomas, 0.7% of all the cancers at this site.

Brain

Benign tumours of the brain and central nervous system are usually combined with malignant tumours in survival analyses because they may also cause death through intracranial expansion. We chose to analyse only malignant tumours of the brain (ICD-8 and ICD-9 191).

Most benign brain tumours were correctly allocated to the separate rubric 225 in ICD-8 or ICD-9, but 84 (0.4%) of the 22,000 tumours coded as malignant tumours of the brain during the 1970s (ICD-8 191), had been assigned a benign behaviour code, and these were included in the analyses.

Tumours of other parts of the central nervous system (cranial nerves, spinal cord and meninges, ICD-9 192x) were not included with brain tumours. Tumours of the spinal cord (192.2) were analysed separately. A few patients with brain tumours assigned to the morphology code for malignant meningeal sarcoma (ICD-O M9530/3) were also included, but tumours explicitly coded to the meninges (ICD-9 192.1) were excluded.

Haematopoietic malignancies

Lymphomas, multiple myeloma and the leukaemias are largely defined on the basis of their microscopic appearance. They usually arise in lymphatic tissue or bone marrow, when their anatomic location is coded accordingly (Table 2.4).

Lymphomas

Lymphomas sometimes arise in (or are diagnosed from) the lymphatic tissue contained in organs such as bowel or stomach, and they may be assigned to the anatomic site code for that organ at cancer registration. Lymphomas coded to an anatomic site other than lymphatic tissue (extra-nodal lymphomas) were not included with other tumours at the extra-nodal site, but transferred for analysis with non-Hodgkin lymphoma. In all, 1,192 extra-nodal lymphomas of non-Hodgkin type (1.8% of the total) were re-classified in this way.

Multiple myeloma

Multiple myeloma, from which bone deposits are so common that they have been considered radiologically diagnostic, is sometimes actually coded to bone (ICD-9 170), although the malignant cell line arises in the marrow. Neoplasms coded to bone with the morphology of multiple myeloma (ICD-O M9730-9731) were included in the analyses with multiple myeloma (Table 2.7), but those at other anatomic sites were not accepted. Seven cases of myeloma that had been coded to bone were re-assigned to multiple myeloma in this way. Plasma cell leukaemias (ICD-O M9830/3) are classified among the immunoproliferative neoplasms in ICD-9, and they were re-assigned for analysis with multiple myeloma rather than with the leukaemias.

Leukaemias

Hairy cell leukaemia in adults was classified in ICD-9 as a malignant neoplasm of lymphoid and histiocytic tissue, among the non-Hodgkin lymphomas (ICD-9 202.4 ICD-O M9940/3). A total of 698 hairy-cell leukaemias were transferred for analysis with the non-Hodgkin lymphomas, comprising 1.0% of the total.

Malignancies with a morphology code for leukaemia that had been assigned to an anatomic site code not specified for leukaemia (204-208 in ICD-9) were excluded.

Childhood malignancies

Cancer in childhood (under 15 years of age) accounts for about 0.6% of all cancers in England and Wales. Childhood malignancies are conventionally grouped by their morphology, rather than by the anatomic site codes used for adult cancers.

We adapted the widely used Birch and Marsden classification[16] to define 12 groups of childhood malignancies for analysis. The groups and their code definitions are shown in Table 2.5. Despite the relatively large numbers of tumours available for analysis, it was not possible to define groups that are consistent with the Birch and Marsden classification for the entire period 1971-90, because the MOTNAC morphology codes used from 1971 to 1978 are not specific enough. We used combinations of ICD-8 anatomic site codes and MOTNAC morphology codes to match the ICD-O site and morphology codes of the classification as closely as possible.

We chose to report the survival for some specific childhood malignancies of particular interest, such as hepatoblastoma, Wilms' tumour, Ewing's sarcoma and osteosarcoma, without including all other tumours of the liver, kidney and bone, respectively.

Data extraction on 6 July 1997

The data files analysed here were created from the national cancer registry database on 6 July 1997. Selected data items (Table 2.8) were extracted from all registrations held in the registry on that date for cancers diagnosed during the 20-year period 1971-90. The data set was treated as a frozen file. Errors in individual records identified during preparation of the data for analysis were corrected if possible. No cancer registrations received by the national cancer registry after 6 July 1997 were added to the data, however, since they had not been included in the same systematic preparations for analysis.

Table 2.8 Variables extracted from national cancer registry

To define the tumour (names were not extracted)
- Location of tumour (site: ICD-8 or ICD-9)
- Morphology of tumour (type: MOTNAC or ICD-O)
- Behaviour of tumour

To calculate survival
- Date of diagnosis
- Date of emigration (where applicable)
- Date of death (where applicable)
- Source of the date of death (cancer registry or NHSCR)
- Vital status (alive; emigrated or lost to follow-up; or dead)

Covariates for comparative analyses
- Date of birth
- Sex
- Deprivation category: Carstairs score based on standardised distribution for Great Britain
- Social class (Registrar General's classification)
- Region of residence: three separate codes for the various territorial configurations in force during 1971-73, 1974-80 and 1981-90; and an expanded area code to enable transfer to a new region of patients living in areas subsequently transferred to a different health authority
- County and county district of residence
- Country of birth

To enable correction of errors
- Tumour numbers (cancer registry, year of registration, identification number)
- Person number (some persons have more than one tumour)
- Data quality indicators

Criteria for eligibility and exclusion

All tumours diagnosed in the period 1971-90 and meeting the basic definitions given above for each cancer were provisionally included in the data set for that cancer.

Standard criteria were defined for all cancers to determine the eligibility and selection of tumour records for analysis. For some cancers, extra criteria were defined. For childhood malignancies, some exclusion criteria (age, morphology) were not applicable, since these tumours were defined on the basis of age and morphology in the first place. The criteria were applied to the provisional data set for each cancer in order to produce the final data set for survival analysis. This was done in two phases.

First, all the tumour records in each data file were checked against each of the exclusion criteria and assigned one or more error codes, if appropriate. A set of error files identifying the records in each category was produced at the same time, to enable amendment of the national cancer registry database.

Next, all records meeting one or more exclusion criteria were excluded from the data set by applying the criteria in a defined sequence. Since a number of records had more than one defect, the order in which the criteria were applied had an impact on the number of cancers actually excluded in each category. We chose a standard sequence of descending severity, so that the most basic reasons for excluding a record from analysis were applied first.

For each error category, a count was made of the number of cancer *records affected* by that error. A separate count was made of the number of *patients excluded* from analysis because of that error in the record. The numbers might differ for two reasons: a patient may be registered more than once, with the same or different cancers, and the exclusion criteria were applied in a hierarchical sequence. A record might contain several errors but could only be excluded from analysis once. The number of records in a given error category is therefore always greater than (or equal to) the number of patients actually excluded for that reason. For the first category in the sequence of exclusions, the two numbers must always be the same.

Ineligible records

Tumour records in the provisional data set for each cancer that would not have been considered eligible for survival analysis (even if the record itself had been internally logical) were excluded first. The main criteria were data quality, residence and tumour morphology or behaviour. Many of the tumours considered ineligible for these analyses because of their morphology or behaviour would be perfectly eligible for some analyses of cancer incidence.

Incomplete data

The entire national data set had previously been subjected to stringent quality controls. Many cancer records had been returned to regional cancer registries for checking and correction. Records that, at the time of data extraction on 6 July 1997, still contained invalid or incompatible values for key variables – name, sex, date of birth, date of diagnosis, date of emigration (if present), date of death (if present), postcode, cancer site, morphology or behaviour –

had been assigned a status code of 3 by the national cancer registry. This implies that they should be excluded from both incidence and survival analyses unless and until appropriate corrections could be made at a later date. Such records were excluded.

Resident outside England and Wales

A number of cancer patients treated in England and Wales are resident in Scotland or Northern Ireland, or outside the United Kingdom. These patients are recorded by regional cancer registries for inclusion in hospital workload statistics, and, in the case of UK residents, for transmission to the cancer registry that covers the patient's place of residence. They are excluded from the calculation of incidence rates, however, because as non-residents of England and Wales, they are also excluded from the census populations used as the denominators for those incidence rates.

Cancers in non-resident patients were also excluded from survival analyses, even if the tumour record passed all other tests. Such patients often return to their country of origin after treatment. Since most do not have a primary care physician (GP) in the National Health Service, they would not be flagged for follow-up in the National Health Service Central Register, and there is little prospect of systematically ascertaining their vital status over long periods of time. Patients known to be resident in England and Wales but who could not be reliably assigned to a particular region were also excluded.

Tumour behaviour

Most analyses were confined to primary malignant tumours. For most of the adult cancers, we excluded those with behaviour codes for benign, *in situ*, or uncertain behaviour, and those registered as metastases in a particular organ rather than as a primary malignant tumour of that organ.

Other criteria

For some cancers, other criteria of anatomic location, morphology or behaviour were applied. These are described in the corresponding chapter. An example is the exclusion of uterine cancers that were not specified as originating either in the cervix or the body of the uterus.

Morphology

Solid tumours with a morphology code implying lymphoma, myeloma or leukaemia were excluded from analysis with the solid tumours. Some of these were then included with non-Hodgkin lymphoma or multiple myeloma (see above). Solid tumours with leukaemia morphology were excluded as errors.

Exclusions

The number of records passing the first set of checks for each cancer, considered potentially eligible for analysis, was taken as a baseline of 100%. A second group of exclusion criteria, described below, was then applied to obtain the final data set for analysis.

Age

Age at diagnosis was calculated from the dates of birth and diagnosis. If the day of a date was missing, the central day of the month was imputed, i.e. the 16th (15th for February). If both day and month were missing, the central day of the year was imputed, i.e. 2 July (183rd day of non-leap year).

If the patient was born in the same calendar year as diagnosis, but the month of birth was unknown, the date of birth was imputed as mid-way between 1 January of that year and the date of diagnosis.

If the patient died in the same calendar year as diagnosis, but the month of death was unknown, the date of death was imputed as mid-way between the date of diagnosis and 31 December of that year.

Patients were excluded from adult tumour groups if the date of diagnosis preceded their 15th birthday, or occurred on or after their 100th birthday. For childhood tumours, the initial data sets were created on the basis of diagnosis before the 15th birthday, as well as on the morphology and behaviour of the tumour, and the age criterion was not re-applied during preparation of these data.

Vital status unknown

Records for which the patient's vital status was unknown were excluded from analysis. These were patients who were not known to be dead, but whose record at the National Health Service Central Register had not been traced to enable "flagging" with details of the cancer registration by 6 July 1997. An untraced registration cannot be linked to the eventual death of the patient through NHSCR. These patients were of two types: those for whom previous attempts to trace the NHSCR record had failed, and those for whom no information was available on 6 July 1997 as to whether the record had been traced for flagging or not. The cancer registrations for the second group, mostly diagnosed in the late 1980s, had been sent to NHSCR only recently. This group is unlikely to be biased with respect to survival.

Duplicate registration

Tumour records from the same cancer registry with the same four-digit site code and cancer registry number were treated as duplicate registrations, even if the dates of diagnosis were not identical. Occasionally there were three or, very rarely, more duplicate records in such a set, but the vast majority of duplicate sets were pairs. If two or more records in a duplicate set had a date of death, the records were sorted in ascending order of date of death, and the earlier date of death was used. If the record in the duplicate set with the earliest date of death apparently had zero survival, however, this record was discarded, and the next record with a later date of death was retained for analysis. Records in duplicate sets that implied death on the day of diagnosis were judged likely to have been made originally from a death certificate only, and submitted to ONS, then not cancelled when a full registration had been established later, after successful tracing of the clinical record by the regional registry. In all cases, we retained only one record for analysis from each duplicate set identified.

Multiple primary

Tumour records with the same site code and person number as an earlier registration were treated as multiple primary tumours. Many of these had been confirmed as such by the regional cancer registries after further enquiries. Those that had not been directly confirmed as multiple primaries were nevertheless considered as such. The record with the earlier date of diagnosis (if three or more, the earliest) was retained for analysis.

Synchronous tumours

Tumours arising in the same organ (same three-digit site code) and with the same sex, date of birth and date of diagnosis as another registration were identified. These were mostly pairs (occasionally triplets) of separate, synchronous cancers in the same individual, previously unlinked, in organs such as the colon or stomach. Bilateral cancers in paired organs such as breast, salivary gland or ovary also occurred. A few were previously undetected duplicate registrations. In a data set this large, it is likely that a few pairs of records linked in this way, particularly for the most common tumours, were actually for two different persons of the same sex with the same date of birth and the same date of diagnosis of a tumour at the same site. We did not extract names or NHS numbers, however, and logical distinction between these possibilities was impossible without asking for further checks by the regional cancer registries concerned.

Theoretically, it would have been possible to ask each of the cancer registries to examine listings of these records, in order to distinguish synchronous or bilateral cancers from previously undetected duplicate registrations or cancers in two different persons, and thus to enable the files to be amended. Synchronous tumours comprised less than 0.3% of adult tumours eligible for analysis, however. The scale of the clerical task and the likely timescale for its completion were judged to be disproportionate.

In each case, only one of the records was retained for analysis, on the same basis as for duplicate registrations (see above).

Sex not known

Records with a missing or invalid code for sex were excluded.

Sex incompatible with site

In addition to the obvious potential errors of gender in cancers of sex-specific organs, we checked for incompatibility between sex and site where ICD-9 made it possible to do so. Tumour records for tumours of the female (ICD-9 174) and male (175) breast were checked. Tumours of the skin of the breast (174.2) and the scrotum (187.7) were only included in the analyses if they were melanomas, but those tumours were also checked for sex-site compatibility. Since we did not extract the names of tumour patients, it was not possible to cross-check the name with the sex of the patient where there was a conflict between the codes for sex and tumour site. All records with sex-site incompatibility were therefore excluded.

Invalid dates

A series of checks was made on the completeness, validity and logical coherence of dates in each tumour record. In the event, some of these checks failed to trap even one tumour record, but in a data set of this size we judged it necessary to check for all possible logical errors. Some records were excluded if an imputed date clashed with a known date. Records of the following types were excluded:

• Year of birth unknown
• Emigrated or dead but year of emigration or death unknown
• Date of diagnosis, emigration or death earlier than date of birth
• Date of emigration or death earlier than date of diagnosis
• Date of death invalid or later than when the data files were extracted for analysis (6 July 1997)

Records with both a date of emigration and a date of death can arise when a patient first emigrates, then later returns to the UK and registers once more with the NHS, before eventually dying. Such patients were censored alive from the analysis at the date of emigration in order to avoid bias, since there is reason to believe that not all such re-entries to the UK and the NHS are successfully recorded.

Patients whose survival is zero or unknown

Where the date of diagnosis and the date of death are the same, this may be because the patient actually died on the same day as the diagnosis was confirmed. Some 20% of large bowel cancers present as clinical emergencies, for example with bowel obstruction for previously undiagnosed cancer, when operative mortality is high[17]. More often, however, the information on which the cancer registration is based was obtained solely from the death certificate, and no information about the date of diagnosis was available. For the purposes of measuring cancer incidence, and since no other date is available, such "death certificate only" (DCO) registrations are assigned the date of death as the date of diagnosis. This is in line with international cancer registration practice[15], and it enables DCO cases to be included in the estimation of cancer incidence.

Patients registered solely from a death certificate comprise an appreciable proportion of cases for some cancers and in some regions. They cannot be included in survival calculations, because the date of diagnosis and thus the duration of follow-up is unknown. Survival estimates may be too high as a result of their exclusion[18], but unless special estimates of their survival time can be obtained, it is generally impossible to include such cases in survival analyses, and the bias is generally minor[1].

Regional cancer registries have always noted which patients are registered solely from a death certificate, and they can therefore include patients with true zero survival in their survival analyses. This is one reason why the regional survival rates derived from the national cancer registry may differ from those published by the corresponding regional cancer registries. Since we have presented results on the basis of modern NHS Regions, however, such differences are likely to be apparent only for those regions for which all or nearly all of their current territory is still covered by a single cancer registry – namely Trent, North Thames, South Thames and West Midlands.

The national cancer registry for England and Wales does not include an indicator to show which cases were recorded solely from a death certificate for patients diagnosed before 1993. We therefore excluded from analysis all patients diagnosed during the period 1971-90 for whom the dates of diagnosis and death were the same. Most of these were in fact registered solely from a death certificate, but a few will have died on the day of diagnosis.

The effect of this exclusion on survival estimates needs to be considered. Exclusion of patients whose true recorded survival was zero would tend to inflate overall estimates of survival, but such patients comprise a small proportion of cases for most cancers, and the effect is likely to be minor.

A check was carried out to evaluate the extent to which patients whose recorded survival is zero (date of death equal to date of diagnosis) could be divided into those registered solely from a death certificate (date of diagnosis unknown) and those who apparently died on the day of diagnosis. For this purpose we used data in the national cancer registry for patients diagnosed in 1993, since DCO indicators were not submitted by regional registries for patients diagnosed in earlier years. Patients whose records do not contain a DCO indicator (Table 2.9, last column) represent up to a third of all patients with zero survival. This proportion is broadly similar for four selected cancers, both highly lethal and less lethal. Patients with zero recorded survival were excluded from all analyses because – for most of them – the date of diagnosis was unknown, even to the regional cancer registry.

If the proportion of patients who did die on the day of diagnosis during the period up to 1990 was similar to these preliminary estimates for 1993, our results will slightly overestimate true survival, because a very small proportion of patients who died almost immediately will have been excluded.

Interpretation

Stage at diagnosis

The clinical stage of disease at diagnosis is not held in the national cancer registry for patients diagnosed before 1993. The percentage of registrations held at regional cancer registries for which the clinical stage is recorded directly from the medical record was in any case often less than 20% in the period up to 1990[19]. Regional cancer registries used simplified staging systems for solid tumours, such as the WHO standard system[20], including local disease, extension beyond the organ of origin, regional lymph node involvement and the presence of metastases. The clinical or pathological stage of cancer at diagnosis is important for evaluating the prognosis and treatment of individual patients. Unfortunately, incomplete recording of stage in the medical records and lack of standardisation of the coding of stage between cancer registries would make national comparisons of survival by stage of disease unreliable for patients diagnosed up to 1990. This position should improve for patients diagnosed more recently.

It is thus possible that differences in patient survival between calendar periods, regions or deprivation

Table 2.9 Cancer patients with zero or unknown survival: England and Wales, adults diagnosed 1993, selected cancers

Cancer	No. of patients [1]		Patients with survival zero or unknown		
	Total	Deaths	Total [2]	Zero [3]	% [4]
Lung	26,424	25,942	4,043	1,758	43
Colon	10,059	8,372	1,417	507	36
Breast (F)	15,418	8,686	1,412	376	27
Testis	376	156	9	3	33

[1] Registered at ONS by 1 April 1998
[2] All those with zero or unknown survival (i.e. including those with DCO indicator)
[3] "True" zero survival - date of diagnosis same as date of death, but no DCO indicator
[4] As a percentage of total with zero or unknown survival

categories could be due at least in part to differences in the distribution of disease stage between such groups. Substantial differences in survival may remain even after adjustment for stage at diagnosis, however, for example with cancers of the breast and other common malignancies in south-east England[21,22].

Follow-up

Analysis of cancer survival requires linkage of information about the death of cancer patients to the original cancer registration. Up to 1971, each regional cancer registry actively sought information on the vital status of every registered patient on successive anniversaries of the date of diagnosis (e.g. at 1, 2, 5, 7, 10 and 15 years after diagnosis), for up to 15 years or more. This was mainly done by writing to the GP or hospital physician responsible for the patient.

In 1971, the national cancer registry introduced a system of passive follow-up, using the National Health Service Central Register (NHSCR). The NHSCR is a national person index, created in 1939, and used since 1948 to administer the National Health Service. It contains a record for each person ever registered with an NHS doctor, and the person's record is in due course closed with details of death. Details of each new cancer registration received by ONS are matched against the database at NHSCR, using the NHS number, name(s), address and date of birth. The individual's record is then flagged with details of the cancer registration. The procedures for tracing and flagging cancer patients at NHSCR enable linkage, in due course, of the information about cancer diagnosis with the date and cause of death. The relevant regional cancer registry is then notified of the death of registered cancer patients, whatever the cause of death.

A second, parallel system has also operated since 1971.

The Cancer Section at ONS sends to cancer registries in England and Wales an extract of the death certificate for all persons dying in the registry's territory whose death certificate mentions cancer or malignant disease, usually within three weeks of death, and regardless of whether the patient was known to have been previously registered with cancer.

Information about death may thus reach the cancer registry in several ways. The patient may die during a hospital admission, either the one from which clinical details are abstracted for the initial cancer registration, or a later admission to the same or another hospital. Information about such deaths will often be actively obtained by the regional cancer registry as part of its routine data collection procedures. The two passive systems for notifying the registry of the death of a registered cancer patient (via NHSCR) and of any death due to malignant disease (via death certificates) complete the process.

Introduction of the passive follow-up system for cancer patients removed the need for cancer registries to perform active follow-up. Only the East Anglian Cancer Registry still follows up all registered patients actively, to determine their vital status at three and five years after diagnosis, and after that at five-yearly intervals[23,24.]

Direct linkage of the national cancer registry files with the national mortality data files is not currently done. The role of the NHSCR in enabling both regional and national cancer survival analyses will therefore remain vital for the foreseeable future.

Where tracing at NHSCR is unsuccessful, the relevant regional cancer registry is contacted. If the cancer record can be corrected or completed with additional information, the record is again forwarded to NHSCR to repeat the tracing process. If the NHS record of a registered cancer patient remains untraced, the patient is considered as lost to follow-up, since details of the person's eventual death may well not be linked to the cancer registration. These patients are excluded from survival analyses in order to avoid bias, since they may appear to the national cancer registry to survive indefinitely ("immortals"). Those known to have emigrated from the UK are censored from survival analyses on the day of departure, because their death would not be routinely notified after emigration.

Up to four per cent of cancer registrations remain untraced at NHSCR[25]. About a quarter of cancer patients are certified as dying from a cause other than cancer, and for some of these patients, cancer may well not appear on their death certificate at all. If such a death occurs in a cancer patient whose NHSCR record could not be traced for flagging with the cancer registration, then the death is unlikely to be notified to the cancer registry. The loss of information is likely to be small, but it may affect estimates of long-term survival, particularly for rare or highly lethal cancers, for which only a few long-term survivors would be expected. For such cancers, even a few individuals who have in fact died but who appear to the national cancer registry to "survive" indefinitely could substantially distort the true picture.

Taken together, these two approaches to the follow-up of cancer patients from diagnosis to registration and death are very effective, and the small losses to follow-up that do occur have been shown to have only an extremely small effect on the estimates of cancer survival rates[26].

Data quality

Six standard editorial tables were produced for each cancer, to enable scrutiny of data quality by NHS Region, deprivation category and calendar period of diagnosis:

• Number of eligible cases (see above), by "original" cancer registry, single year of diagnosis, and sex
• Number of otherwise eligible cases excluded from analysis as death-certificate-only (DCO) cases, by calendar period of diagnosis and NHS Region
• Number of otherwise eligible cases excluded from analysis as DCOs, by calendar period of diagnosis and deprivation category
• Number of patients analysed with unknown deprivation category, by original cancer registry and calendar period of diagnosis
• Number of patients analysed, by detailed anatomic localisation (four-digit site code) ICD revision and calendar period of diagnosis
• Number of patients analysed, by morphologic type (four-digit ICD-O code) and single year of diagnosis

The "original" cancer registry is the one that registered the patient. The territories covered by these registries in 1971-90 are in most cases no longer coterminous with NHS Regions in 1998, for which survival results are presented. The purpose of these tables was to help identify data anomalies in relation to the organisation that collected them, where such anomalies might underpin unexpected survival patterns.

Patients included in the analyses

More than 3.6 million adults and 21,000 children were diagnosed in the 20-year period 1971-90 and registered in the national cancer registry with one of the 58 different cancers selected for survival analysis. Over 770,000 adults were diagnosed during 1971-75. The number increased by more than a third to 1,062,000 during the five-year period

1986-90. The number of children diagnosed in the same five-year periods fell from 5,600 to 5,100.

The number and percentage of records excluded from analysis is shown for each five-year period of diagnosis (1971-75, ... 1986-90) in Table 2.10 for adults and Table 2.11 for children. The corresponding figures for each cancer are given in Table 1 of Chapters 5-62.

In all, some 250,000 of the adult patients were ineligible for the survival analyses carried out here. Of these, about 180,000 (70%) were registered with *in situ* neoplasms, mainly of the cervix. About 40,000 were either resident outside England and Wales, or could not be assigned to an NHS Region, or else the tumour record contained a serious data defect. A further 21,000 patients were excluded because the tumour was benign, or because the tumour was metastatic but the organ of origin was unknown, or because it was uncertain whether the tumour was benign or malignant. Some 13,000 patients were excluded for reasons that are specific to a particular cancer. These are described in more detail in the corresponding chapter.

About 1,000 of the records for children with cancer were excluded because of a serious defect in the data or because the child did not live in England or Wales.

The records for 3.3 million adults and 20,000 children were considered eligible for analysis.

Nine per cent of the eligible adult patients were excluded with zero survival. The proportion of patients in this category increased from 6.6% in 1971-75 to 11.2% in 1986-90.

A further 4.2% of adults were excluded because their vital status was unknown at 6 July 1997: this proportion also increased for patients diagnosed more recently, from 3.1% to 5.3% of eligible patients. Among those tumours selected because they arise principally in adults, 0.6% were excluded as having arisen in children, or at age 100 or more, but the majority were in children, and most of these were included in analyses with the other childhood cancers. Only 0.03% of all tumours were diagnosed in centenarians.

Almost 9,000 (0.27%) tumours that would otherwise have been eligible for analysis were excluded because they were the second or subsequent primary cancer in an adult with two or more cancers. The number excluded for this reason rose sharply in the four successive quinquennia, as would be expected, since the pool of surviving cancer patients on the national cancer registry and at risk of a second or subsequent cancer rises with the passage of time. Similarly, 11,000 (0.34%) records were for the second

(rarely, third etc.) of two or more synchronous cancers diagnosed on the same day in the same individual, either in the same organ (e.g. colon), or in bilateral organs such as the breast, lung, ovary, kidney or testis. The number of such records did not change markedly with time.

Strictly speaking, the exclusion of 20,000 duplicate registrations, multiple primary cancers and synchronous cancers refers to tumour records rather than patients, since in each case a single record representing the patient's first tumour was retained for analysis. For all three categories, one record was retained, so that – insofar as was possible in a data set without names or unique identifiers – each cancer patient was represented once and only once in the survival analyses.

Most of the 1,000 children excluded as ineligible were not resident in England or Wales, or else their cancer record in the national cancer registry was defective – but in any case more than half of these children could not have been included in analyses because their vital status was unknown at 6 July 1997. Table 2.11 shows that vital status was unknown in 1,922 childhood cancer records, but only 1,385 (6.9%) children were actually excluded for this reason. Thus the records for 537 children (the difference between the two numbers) were excluded under higher criteria in the sequence, mainly as non-resident or with data defects. As for adults, the proportion of children excluded from analysis because their vital status was unknown on 6 July 1997 increased from 5% for those diagnosed 1971-75 to slightly over 10% for those diagnosed in 1986-90. About 3% of children were excluded because the date of death was the same as the date of diagnosis. This proportion was substantially lower than for adults, and did not change much over the 20 years 1971-90.

In total, therefore, it was possible to analyse survival for almost 2.9 million (86.5%) of the eligible adult patients and for about 18,000 (89.3%) of the children.

MATERIAL

Table 2.10 Ineligible and excluded records, and adults (15-99 years) included in analyses, by calendar period of diagnosis

	Calendar period of diagnosis														
	1971-75			1976-80			1981-85			1986-90			Overall		
Total registered	**772,666**			**839,652**			**940,450**			**1,063,555**			**3,616,323**		
Ineligible	*Records*	*Patients*	*%*	*Records*	*Patients*	*%*	*Records*	*Patients*	*%*	*Records*	*Patients*	*%*	*Records*	*Patients*	*%*
Incomplete data[1]	2,181	2,181		3,470	3,470		4,919	4,919		8,170	8,170		18,740	**18,740**	
Resident outside England and Wales	7,993	7,917		5,908	5,793		4,500	3,983		4,386	3,344		22,787	**21,037**	
In situ neoplasm[2]	19,032	18,789		24,496	24,318		44,210	43,870		95,825	92,078		183,563	**179,055**	
Benign or uncertain behaviour[2]	8,464	5,799		5,077	3,435		45	42		29	25		13,615	**9,301**	
Metastatic[2]	8,000	7,924		3,598	3,547		555	546		669	665		12,822	**12,682**	
Otherwise ineligible[3]	3,161	2,906		3,805	3,398		3,894	3,380		3,969	3,479		14,829	**13,163**	
Lymphoma[4]	750	715		480	425		204	57		153	19		1,587	**1,216**	
Leukaemia or myeloma[4]	16	13		16	5		32	2		25	2		89	**22**	
		46,244			44,391			56,799			107,782			255,216	
Total eligible		**726,422**	**100**		**795,261**	**100**		**883,651**	**100**		**955,773**	**100**		**3,361,107**	**100**
Aged under 15 or 100 years or over at diagnosis	5,817	5,402	0.7	5,784	5,428	0.7	5,786	5,402	0.6	6,149	5,900	0.6	23,536	**22,132**	0.7
Vital status unknown[5]	28,548	22,490	3.1	38,743	32,849	4.1	40,500	35,407	4.0	60,793	51,070	5.3	168,584	**141,816**	4.2
Duplicate registration[6]	5	5	<0.1	28	17	<0.1	8	3	<0.1	29	25	<0.1	70	**50**	<0.1
Multiple primary[7]	155	70	<0.1	977	644	<0.1	4,089	3,143	0.4	6,145	5,079	0.5	11,366	**8,936**	0.3
Synchronous tumours[8]	3,804	2,629	0.4	3,420	2,127	0.3	4,460	3,089	0.3	6,609	3,631	0.4	18,293	**11,476**	0.3
Sex not known	0	0	0.0	0	0	0.0	1	0	0.0	4	2	<0.1	5	**2**	<0.1
Sex incompatible with site code[9]	2	1	<0.1	13	7	<0.1	15	10	<0.1	65	42	<0.1	95	**60**	<0.1
Invalid dates[10]	37	29	<0.1	95	58	<0.1	270	176	<0.1	939	331	<0.1	1,341	**594**	<0.1
Zero survival[11]	53,547	48,267	6.6	65,683	60,468	7.6	98,535	90,504	10.2	115,306	107,091	11.2	333,071	**306,330**	9.1
Total exclusions		**78,893**	10.9		**101,598**	12.8		**137,734**	15.6		**173,171**	18.1		**491,396**	14.6
Patients included in analysis		**647,529**	**89.1**		**693,663**	**87.2**		**745,917**	**84.4**		**782,602**	**81.9**		**2,869,711**	**85.4**
Percentage of eligible adults[12]			**90.1**			**88.1**			**85.5**			**83.2**			**86.5**

[1] Main data item(s) invalid or incompatible with one another: name, sex, date of birth, date of diagnosis and (if present) date of death, postcode, site, morphology and behaviour

[2] Tumour either *in situ* (behaviour code 2), benign (0), uncertain if benign or malignant (1) or metastatic (6 or 9)

[3] Various: see corresponding table in the chapter for each cancer

[4] Morphology that of lymphoma (included with non-Hodgkin lymphoma, Ch. 41) or myeloma (included with multiple myeloma, Ch. 43), or leukaemia (excluded)

[5] Tracing of vital status at National Health Service Central Register incomplete at 6 July 1997

[6] Same site code, sex, cancer registry and cancer registry number as an earlier registration

[7] Same site code and person number as an earlier registration(s): mostly confirmed multiple primary tumours at this site, some unresolved duplicate registrations

[8] Same site code, sex, date of birth and date of diagnosis as another registration(s): mostly synchronous or (in paired organs) bilateral tumours in same anatomic site in one individual, not previously linked: also some duplicate registrations

[9] Site code specific to one sex: tumour registered to opposite sex

[10] Impossible sequence of dates (birth, diagnosis and, if present, emigration or death); or date of death after 6 July 1997

[11] Date of diagnosis same as date of death: some are patients who did die on the day of diagnosis, but most were registered solely from a death certificate (death certificate only, or DCO), and their survival time is thus unknown

[12] Total eligible patients, less children and centenarians, and less records for duplicate, multiple and synchronous registrations (see text)

Table 2.11 Ineligible and excluded records, and children (0-14 years) included in analyses, by calendar period of diagnosis

	Calendar period of diagnosis														
	1971-75			1976-80			1981-85			1986-90			Overall		
Total registered	5,593			5,371			5,104			5,102			21,170		
Ineligible	Records	Patients	%	Records	Patients	%	Records	Patients	%	Records	Patients	%	Records	Patients	%
Incomplete data[1]	69	69		68	68		159	159		131	131		427	**427**	
Resident outside England and Wales	197	193		171	169		135	129		41	35		544	**526**	
In situ neoplasm[2]	1	0		0	0		0	0		0	0		1	**0**	
Benign or uncertain behaviour[2]	98	63		41	27		0	0		1	1		140	**91**	
Metastatic[2]	0	0		0	0		0	0		0	0		0	**0**	
Otherwise ineligible[3]	-	-		-	-		-	-		-	-		-	**-**	
Lymphoma[4]	-	-		-	-		-	-		-	-		-	**-**	
Leukaemia or myeloma[4]	-	-		-	-		-	-		-	-		-	**-**	
		325			264			288			167			1,044	
Total eligible		5,268	100		5,107	100		4,816	100		4,935	100		20,126	100
Aged under 15 or 100 years or over at diagnosis[4]	-	-	-	-	-	-	-	-	-	-	-	-	-	**-**	-
Vital status unknown[5]	445	255	4.8	466	297	5.8	446	317	6.6	565	516	10.5	1,922	**1,385**	6.9
Duplicate registration[6]	0	0	0.0	0	0	0.0	2	1	<0.1	0	0	0.0	2	**1**	<0.1
Multiple primary[7]	10	10	0.2	14	14	0.3	29	26	0.5	16	13	0.3	69	**63**	0.3
Synchronous tumours[8]	16	11	0.2	29	21	0.4	30	26	0.5	40	20	0.4	115	**78**	0.4
Sex not known	0	0	0.0	0	0	0.0	0	0	0.0	1	0	0.0	1	**0**	0.0
Sex incompatible with site code[9]	-	-	-	-	-	-	-	-	-	-	-	-	-	**-**	-
Invalid dates[10]	1	0	0.0	1	0	0.0	1	0	0.0	2	1	<0.1	5	**1**	<0.1
Zero survival[11]	172	148	2.8	195	149	2.9	217	167	3.5	178	155	3.1	762	**619**	3.1
Total exclusions		424	8.0		481	9.4		537	11.2		705	14.3		**2,147**	10.7
Patients included in analysis		4,844	92.0		4,626	90.6		4,279	88.8		4,230	85.7		**17,979**	89.3

[1] Main data item(s) invalid or incompatible with one another: name, sex, date of birth, date of diagnosis and (if present) date of death, postcode, site, morphology and behaviour

[2] Tumour either in situ (behaviour code 2), benign (0), uncertain if benign or malignant (1) or metastatic (6 or 9)

[3] Not applicable for these malignancies

[4] Not applicable for these malignancies

[5] Tracing of vital status at National Health Service Central Register incomplete at 6 July 1997

[6] Same site code, sex, cancer registry and cancer registry number as an earlier registration

[7] Same site code and person number as an earlier registration(s): mostly confirmed multiple primary tumours at this site, some unresolved duplicate registrations

[8] Same site code, sex, date of birth and date of diagnosis as another registration(s): mostly synchronous or (in paired organs) bilateral tumours in same anatomic site in one individual, not previously linked: also some duplicate registrations

[9] Not applicable for these malignancies

[10] Impossible sequence of dates (birth, diagnosis and, if present, emigration or death); or date of death after 6 July 1997

[11] Date of diagnosis same as date of death: some are patients who did die on the day of diagnosis, but most were registered solely from a death certificate (death certificate only, or DCO), and their survival time is thus unknown

MATERIAL

1. Berrino F, Estève J, Coleman MP. Basic issues in the estimation and comparison of cancer patient survival. In: Berrino F, Sant M, Verdecchia A, Capocaccia R, Hakulinen T, Estève J, (eds.) *Survival of cancer patients in Europe: the EUROCARE study. (IARC Scientific Publications No. 132).* Lyon: International Agency for Research on Cancer, 1995, pp1-14

2. Hawkins MM, Swerdlow AJ. Completeness of cancer and death follow-up obtained through the National Health Service Central Register for England and Wales. Br J Cancer 1992; 66: 408-413

3. Office for National Statistics. *Cancer statistics, registrations: England and Wales, 1990. Series MB1 no. 23.* London: The Stationery Office, 1997

4. Office of Population Censuses and Surveys. *A review of the national cancer registration system in England and Wales. Series MB1 no. 17.* London: HMSO, 1990

5. NHS Executive. *Cancer registration in the NHS. NHS EL(92)95.* London: Department of Health, 1992

6. Clinical Outcomes Working Group. *Clinical outcome indicators, July 1996.* Edinburgh: The Scottish Office, 1996

7. Carstairs V, Morris R. Deprivation and mortality: an alternative to social class? Comm Med 1989; 11: 210-219

8. Carstairs V, Morris R. *Deprivation and health in Scotland.* Aberdeen: Aberdeen University Press, 1991

9. Morris R, Carstairs V. Which deprivation? A comparison of selected deprivation indexes. J Publ Hlth Med 1991; 13: 318-326

10. Jarman B, Townsend P, Carstairs V. Deprivation indices. Br Med J 1991; 303: 523

11. Carstairs V, Morris R. Deprivation: explaining differences in mortality between Scotland and England and Wales. Br Med J 1989; 299: 886-889

12. Dolk H, Mertens B, Kleinschmidt I, Walls P, Shaddick G, Elliott P. A standardisation approach to the control of socioeconomic confounding in small area studies of environment and health. J Epidemiol Comm Hlth 1995; 49: s9-s14

13. Carstairs V. Deprivation indices: their interpretation and use in relation to health. J Epidemiol Comm Hlth 1995; 49: s3-s8

14. Sloggett A, Joshi H. Deprivation indicators as predictors of life events 1981-1992 based on the UK ONS longitudinal study. J Epidemiol Comm Hlth 1998; 52: 228-233

15. Jensen OM, Parkin DM, MacLennan R, Muir CS, Skeet RG, (eds.) *Cancer registration: principles and methods. (IARC Scientific Publications No. 95).* Lyon: International Agency for Research on Cancer, 1991

16. Birch JM, Marsden HB. A classification scheme for childhood cancer. Int J Cancer 1987; 40: 620-624

17. NHS Executive. *Improving outcomes in colorectal cancer: the research evidence.* London: NHS Executive, 1997

18. Pollock AM, Vickers N. The impact on colorectal cancer survival of cases registered by 'death certificate only': implications for national survival rates. Br J Cancer 1994; 70: 1229-1231

19. Chouillet AM, Bell CMJ, Hiscox JG. Management of breast cancer in southeast England. Br Med J 1994; 308: 168-171

20. World Health Organisation. *WHO Handbook for Standardised Cancer Registries.* Geneva: WHO, 1976

21. Schrijvers CTM, Mackenbach J, Lutz J-M, Quinn MJ, Coleman MP. Deprivation and survival from breast cancer. Br J Cancer 1995; 72: 738-743

22. Schrijvers CTM, Mackenbach J, Lutz J-M, Quinn MJ, Coleman MP. Deprivation, stage at diagnosis and cancer survival. Int J Cancer 1995; 63: 324-329

23. Wilson S, Bell CMJ, Black RJ, Coleman MP, Cummins C, Lawrence G, Page M, Rider L, Smith J, Youngson J. Health care system, cancer registration and follow-up of cancer patients in the United Kingdom. In: Berrino F, Sant M, Verdecchia A, Capocaccia R, Hakulinen T, Estève J, (eds.) *Survival of cancer patients in Europe: the EUROCARE study. (IARC Scientific Publications No. 132).* Lyon: International Agency for Research on Cancer, 1995, pp71-74

24. East Anglian Cancer Registry. *Cancer incidence and survival in East Anglia 1995.* Cambridge: East Anglian Cancer Registry, 1997

25. Swerdlow AJ. Cancer registration in England and Wales: some aspects relevant to interpretation of the data. J R Statist Soc Series A 1986; 149: 146-160

26. Sant M, Gatta G. The EUROCARE database. In: Berrino F, Sant M, Verdecchia A, Capocaccia R, Hakulinen T, Estève J, (eds.) *Survival of cancer patients in Europe: the EUROCARE study. (IARC Scientific Publications No. 132).* Lyon: International Agency for Research on Cancer, 1995, pp15-31

3 METHOD

Basic principles

Time from diagnosis to death

The period of time between diagnosis and death of each individual cancer patient is the basic element of information in survival analysis. In analysing the survival times of a large group of cancer patients, the object is to obtain a satisfactory estimate of the chances of survival at given times since diagnosis, such as one or five years, and to take due account of available information on factors known or likely to influence the pattern of survival, such as age, sex, year of diagnosis, socio-economic status and geographic region.

There are three main approaches to estimating cancer survival in population studies[1,2].

The observed, or crude, survival is simply the estimated probability of survival at the end of some specified period of time. It takes no account of the cause of death, or of the background risk of death in the general population, to which cancer patients are also subject. It can be interpreted as the probability of survival from cancer and all other causes of death combined.

The concept of net (or corrected) survival allows the overall mortality among cancer patients to be separated into two components, a background risk of death applicable to everyone, and an extra risk of death due to cancer, which is to be estimated. The risks are assumed to act independently of one another. Those certified as dying of cancer provide the endpoints for analysis, while those certified as dying of other causes are treated as censored observations, in effect as lost to follow-up at the time of death. This approach requires agreement on which deaths should be considered attributable to the cancer, and suitably accurate information on the cause of death of all cancer patients. With these caveats, net survival can be interpreted as the probability of survival from cancer in the absence of other causes of death.

Relative survival, the third approach, also assumes additivity between the risk of death due to the cancer and the background (or competing) risk of death from other causes, but it does not require information about the cause of death in the cancer patients. On the assumption that the two risks of death may be considered to act independently, the impact of other causes of death can be estimated from routine vital statistics, i.e. the mortality rates in the general population from which the cancer patients are drawn[3]. Relative survival is then defined as the ratio of the observed survival in the group of cancer patients under study and the survival that would have been expected had they been subject only to the mortality rates of the general population[1,4].

Relative survival has become the most widely used method for analysing the survival of cancer patients in population studies. Because it does not require information on the cause of death, it can be used in countries where the death certificate (or its medical component) is not publicly available, or certification of the cause of death is not deemed sufficiently reliable. It avoids the need for attribution of death to the cancer or another cause[3]. It is also important in analysing survival over long periods, because the extra risk of death related to the cancer tends to decay with time, while the background risk from other causes of death rises inexorably as the surviving cancer patients become older[5].

Both crude and relative survival rates are presented in this book.

Rate, probability and proportion

A "rate" in epidemiology is usually taken to mean the number of events (diagnoses, deaths) that occur per unit of time, and is expressed per 100,000 persons per year or per 100,000 person-years at risk. By contrast, the chances of survival are estimated from the observed distribution of survival times, in terms of the statistical probability of survival at specified times since diagnosis. These can be interpreted as estimates of the proportion of patients who survive, after correction for attenuation of the patient group by death or loss to follow-up with the passage of time. Relative survival, the ratio of observed and expected survival, is thus a ratio of two proportions: it is usually expressed as a percentage (e.g. 0.4/0.8=50%), and it can also be interpreted as the proportion of survivors (see below).

The term "survival *rate*" has become common currency, however, and it will be used here.

Crude survival

The time since diagnosis of cancer for each patient is arbitrarily divided into a number of successive intervals, within each of which an estimate of survival for the group of patients will be made. Crude or observed survival is estimated as the proportion of subjects alive at the beginning of the jth time interval since diagnosis, at time t_{j-1}, who are still alive at the end of the interval, at time t_j. The probability of survival at the end of the jth interval, conditional on the subject being alive at the start of the interval, can be expressed as:

$$s(t_j) = (N_j - d_j)/N_j \tag{1}$$

where N_j is the number of individuals at risk at the start of the interval (corrected for losses to follow-up during the

interval) and d_j is the number of deaths that occur during the interval. This is simply the number of survivors as a proportion of the initial number of subjects.

By multiplying the conditional probabilities of surviving to the end of each successive interval, we can estimate the overall probability of surviving from the beginning of observation (i.e. at diagnosis, time t_0) to the end of the ith interval[1]:

$$S(t_i) = \prod_{j=1}^{i} s(t_j) \tag{2}$$

Within a very short time interval starting at time t, and of duration Δt, the probability of death, or the instantaneous mortality rate, is given by:

$$\lambda(t) = \lim_{\Delta t \Rightarrow 0} \frac{1}{\Delta t} \cdot prob\,(death\,during\,[t, t+\Delta t], \atop given\,alive\,at\,time\,t) \tag{3}$$

In the actuarial method, the maximum duration of survival is divided into a series of time intervals of specified (but not necessarily equal) length. The instantaneous death rate in the jth interval is taken to be constant throughout the interval, and is denoted λ_j.

A subject may become lost to follow up during the interval, in which case the end of observation for that subject is considered to be the last date for which information on vital status is known. Patients diagnosed more recently may also be withdrawn from follow-up because their duration of observation at the end of the study period is still less than $t + \Delta t$. If the date of death or loss to follow-up is known for all individuals, then the total person-time of observation m_j in that interval is exactly known. The maximum likelihood estimate of the instantaneous mortality rate for each interval is given by the number of deaths occurring in the interval divided by the person-years at risk:

$$\lambda_j = d_j/m_j \tag{4}$$

The equivalent probability of survival to time t within the jth interval, conditional on being alive at the start of the interval, is:

$$s(t_j) = \exp(-\lambda_j \cdot t) \tag{5}$$

The cumulative rate of death from diagnosis up to time t is the sum of the death rates (λ_j) for each of the preceding intervals and for the last (*i*th) interval. The cumulative probability of survival to time t is the exponent of this cumulative death rate:

$$S(t) = \exp-\left\{\sum_{j=1}^{i-1}((d_j/m_j)(t_j - t_{j-1})) + (d_i/m_i)(t - t_{i-1})\right\} \quad (6)$$

with t occurring in the last interval:

$$t_{i-1} < t \leq t_i$$

Background mortality

In addition to their risk of death due to the cancer, cancer patients remain subject to a background risk of death from causes unrelated to the cancer, as do members of the general population. This competing risk of death varies not only with age, sex, and calendar time, but also with other characteristics of the patient population, such as the region of the country in which they live and whether they live in an area that is affluent or deprived[6,7].

Differences between cancer patients in the background risk of death cannot be addressed with crude survival. Net survival can be used in clinical studies where there is agreement on which deaths should be attributed to the cancer, and precise information is available on the cause of death for all patients. Patients may die of the cancer, of complications of its treatment, or of other causes[8]. There may be inadequate information to attribute each death to cancer or to other causes, however, and the distinction is arguable in any case, even in clinical studies[3,4,8].

For estimating cancer survival in the population, net survival would require that certification of the underlying cause of death be sufficiently reliable for such a distinction to be made for cancer patients in all the regions and groups of the population for which survival comparisons were required, and over the relevant time period. There is evidence that this condition is not met, either in the UK[9-13] or in other countries[14-20].

Relative survival

Berkson and Gage originally proposed comparing the survival of cancer patients with that of the general population in order to separate the effect of cancer from the competing risk of death from other causes[4]. The main purpose was to compare the survival of groups of hospital patients with differing age distributions, and for whom the background risk differed substantially. The concept has wider application, however.

The overall force of mortality for cancer patients can be considered as the sum of two components, a death rate due to the cancer under investigation, λ_c, and an expected death rate from all causes, λ_e:

$$\lambda_o(t) = \lambda_c(t) + \lambda_e(t) \quad (7)$$

The observable probability of survival, S_o, is then the exponent of the sum of the two cumulative rates, integrated across the period t:

$$S_o(t) = \exp\left\{-\int_0^t (\lambda_c(u) + \lambda_e(u))\,du\right\} \quad (8)$$

The overall survival can be expressed as the product of the probabilities of survival from cancer and from other causes, respectively, on the assumption that they can be considered as independent:

$$S_o(t) = S_c(t) . S_e(t) \quad (9)$$

The probability of survival from the cancer is thus[1,4]:

$$S_c(t) = S_o(t) / S_e(t) \quad (10)$$

Overall mortality has a component due to cancer, but it is unnecessary to subtract the component due to cancer mortality. The basis of relative survival is a comparison of mortality in cancer patients with that of the general population, regardless of cause. It avoids reliance on the precision of death certification for individuals and on the accuracy of death rates for any particular cancer. In the estimation of lung cancer survival, even adjustment of the expected survival for the excess mortality related to smoking only increased the estimate of relative survival by one per cent or less[3].

Relative survival has the advantage of allowing an estimate of the proportion of patients who may be considered as "cured". When a plot of relative survival has reached a plateau and is no longer falling with time, the group of cancer patients is no longer subject to higher overall mortality than the general population. The survival rate at that time may be taken as an estimate of the proportion of patients who are "cured". This is a public health concept of "cure" applied to the population of cancer patients, not to the individual cancer patient. It should not be mis-interpreted: some surviving patients will still die from their cancer, just as some patients who died earlier will have died from other causes.

Life tables

The estimation of relative survival requires a comparison of observed and expected death rates. Our intention was to compare relative survival rates between groups of patients defined by their region of residence and their degree of material deprivation, and for whom general mortality is known to differ.

For such analyses, the use of the same life table for all deprivation groups can lead to bias in estimating the survival in each deprivation group[21]. If affluent groups have better relative survival, the true gradient in survival between affluent and deprived groups will be exaggerated.

This is because the life table represents background mortality averaged across all socioeconomic groups, and it ignores the known gradient in mortality between them,

Figure 3.1 Deprivation life tables reduce bias in estimating the relative survival gradient

(a) General life table

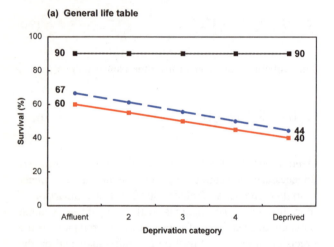

(b) Deprivation life tables

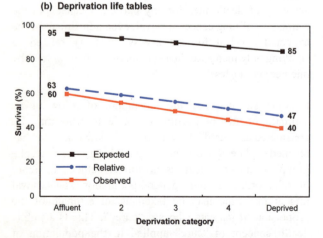

Hypothetical example with observed survival of 60% in affluent and 40% in deprived group: (a) 90% expected survival from general life table, relative survival gradient 67% to 44%; (b) 95% to 85% expected survival from deprivation-specific life tables, relative survival gradient 63% to 47%

which favours the more affluent. The denominator of the relative survival rate, the expected survival, will therefore be systematically *under*-estimated for the more affluent groups, which have better than average survival, and over-estimated for the more deprived groups. This leads in turn to *over*-estimation of relative survival among more affluent groups, and under-estimation for more deprived groups. A deprivation gradient in survival favouring the more affluent groups would therefore tend to be exaggerated. Similarly, a gradient favouring more deprived groups would tend to be flattened.

A hypothetical example is shown in Figure 3.1. There is a 20% difference in crude survival between the most affluent and the most deprived groups. When a single life table is used, giving rise to an expected survival of 90% in all deprivation categories, the difference in relative survival is 23% (i.e. 60/90=67% and 40/90=44%). When deprivation-specific life tables are used, with expected survival ranging from 95% to 85%, the gap in relative survival between the two categories falls to 16%.

The use of separate life tables for each deprivation category would therefore improve the estimation of expected mortality in each category, and any deprivation gradient in survival would be estimated with less inherent bias. The utility of this approach has recently been demonstrated for cancer survival in Finland[22].

There are substantial differences in mortality by geographic region and socio-economic status in Great Britain[23,24], but life tables were not available for the deprivation categories or the geographic regions for which we wished to examine cancer survival. We investigated some life tables for different socio-economic groups that had been prepared for other studies, but these were mostly tables of death rates for five-year age groups up to age 85. They would have required the application of smoothing procedures, particularly at older ages, or were based on relatively small numbers of deaths. None had the required degree of detail by age, sex, deprivation group and calendar period. The NHS Regions were created in 1996 and no regional life tables were available.

Regional life tables by deprivation

We therefore constructed new tables of mortality rates for England and Wales by single year of age at death up to age 99, for every combination of NHS Region, deprivation category and sex, and for each of two time periods. The annual numbers of deaths were averaged across 1980-82 and 1990-92, and the population denominators were derived from the 1981 and 1991 censuses.

The numbers of deaths and the populations in each census enumeration district were available in tabular form for five-year age groups up to age 84, and for 85 years and over, in electronic databases of small area and local base statistics from the 1981 and 1991 censuses. These data are published by the Office for National Statistics and are available from MIDAS (Manchester Information Datasets and Associated Services). The Environmental Epidemiology Unit of the London School of Hygiene and Tropical Medicine kindly re-classified these data to provide the numbers of deaths and population denominators by the same five-year age groups at death, by NHS Region as defined at 1 April 1998, by sex and calendar period, and by the deprivation category of the census enumeration district of residence at death. Five deprivation categories were defined on the basis of the quintiles of the distribution of the Carstairs score in Great Britain (see Chapter 2).

Abridged life tables (in five-year age groups up to age 84, and age 85 and over) were first prepared from these tabular raw data. These life tables were then smoothed, converted to single-year tables, and extended to age 99 with the four-parameter relational model life table system. This technique is a refinement of the two-parameter system devised by Brass[25].

Relational model life table system

The approach depends on the empirical finding that the relationship of mortality with age (the survivorship (ℓ_x) function of life tables) is inherently similar in shape for all human populations. As a result, a linear relation can theoretically be expected between any two survivorship functions. If a logit transformation is applied, a good linear relation is indeed usually found, especially if the two life tables are not drastically dissimilar in overall level of mortality, and do not show marked idiosyncrasies[26].

If the life table for a given sex, calendar period and region is taken as a standard, any other table can be described as a variant from the standard using two parameters, an intercept and slope, which define the linear relationship with the standard life table. This is usually expressed as:

$$Y_x = \alpha + \beta Y_{sx} \qquad (11)$$

where:
Y is the logit-transformed ℓ_x value that we wish to relate to the standard life table;
the subscript x signifies age;
the subscript s signifies a value from the standard life table, and
α and β are the intercept and slope parameters, respectively, of the linear relationship between the two life tables.

The basic relational life table system enables two prime

sources of variation to be modelled in the reconstruction of mortality tables. The intercept (alpha) relates to the overall level of mortality, and the slope (beta) to the balance between mortality at younger and older ages: these two parameters largely define the variation in human mortality patterns. They can be estimated by maximum likelihood methods.

The procedure has two steps. A linear relationship is first established between a suitably detailed standard life table, preferably a national life table, and another observed life table (the index table). The statistical stability and detail (e.g. single year of age) of the standard table, and its linear relationship with the observed table, can then be used to smooth and, if necessary, extend the index table, e.g. from 85 to 99 years. The official decennial life tables for England and Wales for 1981[27] and 1991[28] were used as the standard life tables. These life tables are smoothed, and are published by sex and single year of age at death up to age 114 years.

The refined version of the relational model life table system was used[26]. This allows variation between the index and standard mortality tables at younger and older ages, respectively, to be independently modelled, in addition to the *balance* in mortality between younger and older ages, represented by the parameter beta. This is done with two extra parameters, kappa and lambda.

Application to England and Wales

Application of the four-parameter system to the data for England and Wales gave rise to some complications. The parameter lambda, representing mortality at older ages, applies only to that part of the mortality table where survival is less than half the original population (radix), i.e., after the median age at death. In the 1991 data for females in England and Wales, this point occurred at an advanced age (81 years), in the penultimate age group (80-84 years) for which raw data were available. Consequently there was only one data point (85 years and over) on which to base the maximisation procedure for estimating lambda. The values derived for lambda to extend the abridged life tables from age 85 to age 99 were therefore unstable. This produced anomalous age-mortality curves that were fairly parallel for the different deprivation categories up to age 85 but crossed over irregularly at more advanced ages.

The procedure was therefore reduced to a three-parameter system by constraining the lambda values to be a simple factor of alpha, as described by Ewbank *et al.*[26]. This factor was approximated from the relation between alpha and lambda as derived from the decennial life tables for 1981 and 1991, when compared with the 1971 life table. In effect, therefore, the abridged five-year life tables were fitted to the official standard life tables by single year of age up to 85 years, in order to estimate the parameters, and

METHOD

Table 3.1 General mortality by deprivation and NHS Region, England and Wales, 1990-92

	Deprivation category						Deprivation category					
	Affluent	*2*	*3*	*4*	*Deprived*	*All*	*Affluent*	*2*	*3*	*4*	*Deprived*	*All*

Risk (%) of dying between 15th and 60th birthdays ($_{45}q_{15}$)

	Men						Women					
England & Wales	8	10	12	14	18	**12**	5	6	7	9	11	**7**
England	8	10	12	14	18	**12**	5	6	7	9	11	**7**
Northern & Yorkshire	9	10	12	15	18	**13**	6	6	7	9	11	**8**
Trent	8	10	11	14	17	**12**	5	6	7	8	11	**7**
Anglia & Oxford	8	9	11	14	16	**10**	5	6	7	8	10	**6**
North Thames	8	10	11	14	17	**12**	5	6	7	8	9	**7**
South Thames	8	10	12	14	18	**11**	5	6	7	8	10	**6**
South & West	8	10	11	14	18	**10**	5	6	7	8	10	**6**
West Midlands	8	10	11	14	17	**12**	5	6	7	9	11	**7**
North & West	9	11	13	16	20	**14**	5	7	8	9	12	**8**
Wales	9	10	12	14	17	**12**	5	6	7	8	10	**7**

Risk (%) of dying between 60th and 85th birthdays ($_{25}q_{60}$)

	Men						Women					
England & Wales	71	74	77	80	82	**77**	55	55	57	60	63	**58**
England	71	74	76	80	82	**77**	55	55	57	60	63	**58**
Northern & Yorkshire	73	76	79	82	84	**81**	57	58	61	63	65	**63**
Trent	72	75	78	80	82	**79**	55	58	59	60	62	**61**
Anglia & Oxford	71	73	75	77	79	**75**	55	55	57	58	59	**58**
North Thames	70	73	75	78	80	**76**	53	54	55	57	60	**59**
South Thames	69	73	75	78	81	**75**	52	54	56	57	60	**58**
South & West	70	71	74	78	80	**74**	54	53	55	56	59	**57**
West Midlands	73	76	78	81	83	**79**	56	57	58	60	64	**61**
North & West	74	77	79	82	84	**81**	58	59	61	63	66	**64**
Wales	72	75	78	80	82	**79**	55	56	58	60	63	**61**

Expected years of life at birth (e_0)

	Males						Females					
England & Wales	76.3	75.2	73.9	72.0	69.9	**73.4**	80.6	80.1	79.3	78.2	76.6	**78.9**
England	76.3	75.2	73.9	72.0	69.9	**73.4**	80.6	80.1	79.3	78.2	76.6	**78.9**
Northern & Yorkshire	75.8	74.8	73.5	71.6	69.5	**72.5**	80.2	79.6	78.7	77.4	76.0	**77.7**
Trent	76.0	75.0	73.8	72.4	70.3	**73.3**	80.3	79.8	79.2	78.1	76.8	**78.4**
Anglia & Oxford	76.4	75.5	74.6	72.6	71.4	**74.7**	80.6	80.3	79.6	78.8	77.7	**79.4**
North Thames	76.7	75.3	74.2	72.3	70.8	**73.7**	81.1	80.4	80.0	79.0	77.8	**79.2**
South Thames	76.7	75.3	73.9	72.1	70.0	**74.2**	81.2	80.4	79.5	78.7	77.4	**79.4**
South & West	76.5	75.5	74.5	72.6	70.4	**74.6**	80.7	80.8	80.0	79.1	77.9	**79.7**
West Midlands	75.9	74.9	73.7	71.8	70.1	**73.1**	80.2	79.7	79.2	77.9	76.3	**78.2**
North & West	75.7	74.4	73.1	71.2	68.9	**72.2**	80.0	79.3	78.3	77.0	75.6	**77.6**
Wales	76.1	74.8	73.4	72.4	70.2	**73.2**	80.7	80.1	79.1	78.3	77.0	**78.6**

Figure 3.2 Death rates per 100,000 per year by single year of age and deprivation category, England and Wales, males, 1990-92

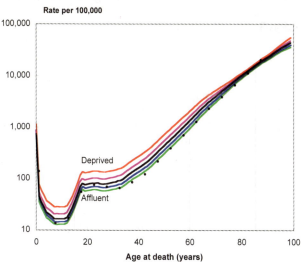

Smoothed life tables; the raw (un-smoothed) five-year age-specific mortality rates for the affluent group are shown as data points (see text)

these were then used to extrapolate the observed life tables from age 85 to age 99.

This modification amounts to assuming that the relative mortality between deprivation categories observed up to age 85 is not likely to change dramatically at older ages. In every case, the three-parameter fit was notably superior to a two-parameter fit. The residual error (root mean squared difference between the standard life table rate and the fitted death rate) was usually reduced by more than half with the additional parameter.

The fitting procedure was automated with macros using general function minimisation routines in EXCEL to optimise the values of alpha, beta and kappa for each life table against the corresponding standard life table. Resulting survivorship (l_x) values were converted to death rates. These are expressed as the number of deaths at each single year of age at death divided by the corresponding number of person-years at risk.

Variation in background mortality

Figure 3.2 shows an example of the resultant age-mortality curves by deprivation category for men in England and Wales during the period 1990-92. The graph shows the clear differences in mortality that have been reported in many studies. The precise shape of the curves above age 85 is speculative – it has been imputed by the modelling process – but the mortality curves in old age remain broadly parallel. The original age-specific mortality rates for five-year age groups are also shown for the affluent category, to show the overall fit of the smoothed age-mortality curve to the raw data.

Summary measures of the differences in background mortality between the deprivation categories across were calculated for several age ranges, in males and females and in each NHS Region. These were the probabilities of death between the 15th and 60th birthdays ($_{45}q_{15}$), and between the 60th and 85th birthdays ($_{25}q_{60}$), as well as the expectation of life at birth (e_o).

The differences in overall mortality between deprivation categories and NHS Regions are quite marked (Table 3.1). For men in England and Wales in 1990-92, for example, the risk of death between 15 and 59 years was 18% for the most deprived group and 8% for the most affluent group, a two-fold difference. Mortality in women was lower, but the discrepancy was similar (11% vs. 5%). The average risk of death between 60 and 84 years was much higher (77% for men and 58% for women), but the absolute difference in this risk between deprivation categories remained similar, about 11% in men and 8% in women. Within a given deprivation category, mortality varied between regions over an absolute range of 3% in both sexes between 15 and 59 years, and over a 5% range between 60 and 84 years.

There were equivalent differences in the expectation of life at birth between the deprivation categories and regions. In England and Wales, the expectation of life for both sexes was four years longer in affluent groups than in deprived groups. Within a given deprivation category, there were regional differences in life expectancy of up to a year. Similar differences in mortality between deprivation categories and regions were observed for the period around the 1981 census (data not shown).

Age, sex, period, region and deprivation

These substantial differences in background mortality were taken into account in the estimation of relative survival by the use of life tables that are specific for each deprivation category and NHS Region.

Separate sets of mortality rates by single year of age were derived for each of five deprivation categories in each of eleven geographic areas (England and Wales, England, Wales, and eight NHS Regions). Nine additional life tables were constructed for the eight NHS Regions and for Wales, for all deprivation categories combined.

The absence of regional life tables to estimate expected mortality over the age of 85 may help to explain the paucity of information on cancer survival in this age group. Extension of the life tables used here to the age of 99 enabled relative survival to be examined for very elderly cancer patients. Life tables were not extended beyond age 99, however, and death rates for 99-year-olds were used to estimate expected mortality for patients surviving to higher ages. Competing risks of death from

Table 3.2 Typical output of algorithm

Count of cases
267

NUMBER OF NON-CONVERGENCES DETECTED = 1 :REGROUPING

Table of crude and relative survival probabilities
(expressed as percentages with 95% confidence intervals)

left	right	deaths	NetR	Crude	Cr_lo	Cr_up	RelR	Re_lo	Re_up
0.00	0.50	83	0.7166	67.40	61.31	72.75	69.89	63.53	75.36
0.50	1.00	43	0.4724	51.86	45.71	57.66	55.18	48.61	61.27
1.00	1.50	29	0.4001	41.28	35.37	47.07	45.18	38.69	51.43
1.50	2.00	16	0.2630	35.30	29.64	41.00	39.61	33.24	45.90
2.00	2.50	15	0.3088	29.76	24.41	35.27	33.94	27.83	40.14
2.50	3.00	7	0.1390	27.16	22.00	32.57	31.67	25.63	37.85
3.00	4.00	12	0.1041	22.70	17.91	27.85	28.53	22.51	34.83
4.00	5.00	4	0.0077	21.25	16.59	26.30	28.31	22.14	34.78
5.00	10.00	14	0.0112	15.92	11.84	20.55	26.77	20.45	33.49

Non-convergence led to regrouping of the last two intervals (up to 7 years and up to 10 years)

Figure 3.3 Schema of life tables by deprivation category and NHS Region

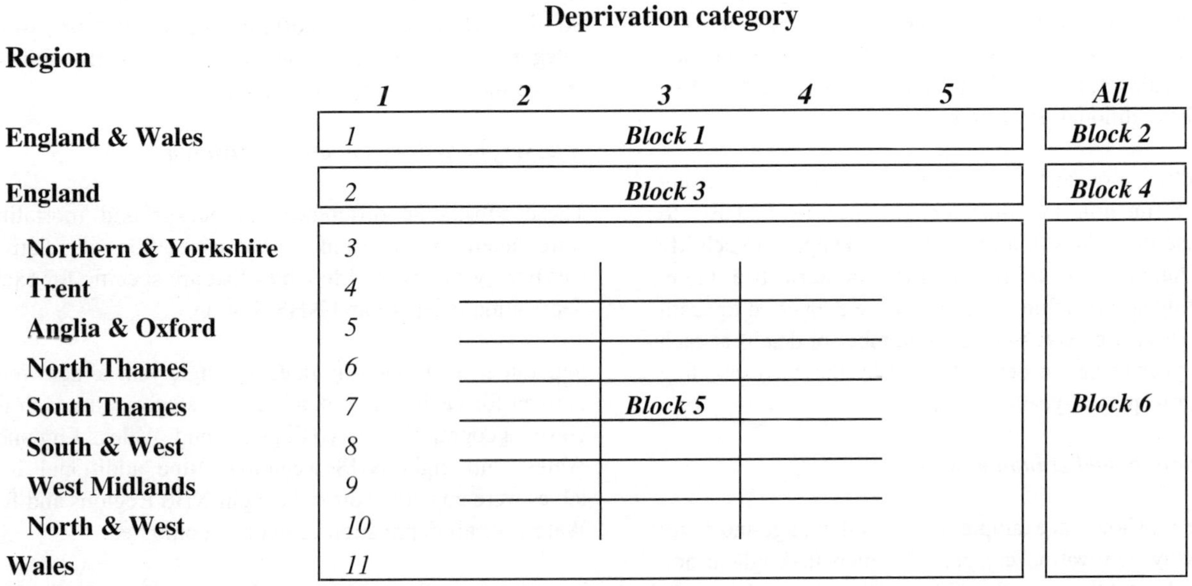

The official England and Wales life tables were used for overall analyses (blocks 2 and 4). Four matrices of 64 life tables (blocks 1, 3, 5 and 6) were constructed for each sex and each peri-censal period (see text)

other causes are high at this age, but the effect on overall expected mortality among patients aged 80-99 at diagnosis would be very small. Of the total person-time lived at age 80 and over by the general population, the proportion lived by those aged 100 or more was 0.2% for men and 0.7% for women, on the basis of the 1991 life table for England and Wales. The corresponding proportion in cancer patients would be much lower. Any bias in estimation of the relative survival for patients aged 80-99 would therefore be extremely small.

In all, 64 sets of mortality rates were derived for each sex and for the three-year period around each of the 1981 and 1991 censuses, a total of 256 life tables (Figure 3.3: blocks 1, 3, 5 and 6). The standard life tables used to smooth and extend the abridged life tables for each NHS Region and deprivation category were the corresponding national life tables for England and Wales for males and females for 1981 and 1991 (the published $_nq_x$ values). The 1981 life tables were used as the basis for death rates for the period 1971-1985; the 1991 tables were used for the period 1986-1995. All the sets of mortality rates were plotted graphically for visual checking before use.

The official decennial life tables (cited above) for 1981 and 1991 by single year of age and sex were used without modification to provide expected mortality rates in the overall analyses for the corresponding periods, both for England and Wales and for England (Figure 3.3: blocks 2 and 4). The population of Wales is less than 6% of the population of England and Wales combined, and death rates were so similar that separate rates were not derived for England alone.

Trends in mortality by deprivation

Trends over time in the mortality gradient between affluent and deprived groups of the population can also lead to bias in estimating the survival gradient between these groups. We used deprivation-specific life tables derived from the period 1990-92 to estimate expected mortality during the period 1986-95, but any change in relative mortality between the deprivation categories during this period would not be reflected in the life tables. This could affect estimates of trend in deprivation-related survival.

Mortality differentials between deprivation categories did in fact increase between 1981 and 1994[6,7]. For example, the difference in life expectancy at birth between the most affluent and the most deprived areas of England during 1984-86 was 2.8 years for men and 1.6 years for women, while the corresponding figures for 1992-94 were 4.0 and 2.4 years[7].

We estimated the slope of the survival gradient across deprivation categories at one year and five years after

diagnosis. This was done both for patients diagnosed during 1981-85 and followed up until the end of 1990, and for patients diagnosed during 1986-90, who were followed up to the end of 1995. Patients diagnosed in both periods accrued substantial time at risk in the late 1980s, but in these particular analyses, only patients diagnosed during 1986-90 accrued time at risk after 1990.

The expected mortality of both sets of patients is estimated on the basis of this time at risk. However, the life tables used to estimate the expected deaths for each deprivation category *throughout* the ten-year period 1986-95 were constructed from mortality data from the period around the 1991 census, because more recent life tables were unavailable. Since the gradient in background mortality between affluent and deprived groups was increasing up to 1994, expected survival for the more affluent groups could be somewhat under-estimated towards the end of this period. In turn, the gradient in relative survival between affluent and deprived patients diagnosed during 1986-90 may be slightly over-estimated, by comparison with the corresponding gradient for patients diagnosed during the period 1981-85.

The position is analogous to the example shown in Figure 3.1(b), as if a further slight increase in the gradient in background mortality had not been taken into account. The magnitude of any resultant bias in estimating the deprivation gradient in survival will be difficult to quantify until further life tables become available. It is clearly very small compared to the bias from using a single life table for all deprivation categories, which has been the only option until now.

Differences in background mortality related to age, sex, calendar time, geographic region and category of deprivation are thus largely taken into account in this approach to the estimation of relative survival.

Estimation of relative survival

Estève *et al.*[29] proposed an approach to estimating the net mortality from cancer when the cause of death of the cancer patients is unknown or unreliable. They showed that it led to relative survival estimates that are closely similar both to conventional estimates of relative survival and to estimates of net survival obtained by Cox regression analysis of cancer registry data in which the cause of death of each patient was precisely known, and in which non-cancer deaths could therefore be treated as censored observations.

The death rate from cancer in each of a series of time intervals since diagnosis is estimated from the survival times of individual patients with a proportional hazards model. The intervals are defined in advance, as with the actuarial method for crude survival. The approach allows

METHOD

information on prognostic factors (covariates) in individuals to be taken into account. The background mortality rate within a given interval of time since diagnosis is assumed to be constant.

The observable mortality of the cancer patients is treated as the sum of two components. The background mortality is related to age, sex, calendar period and other variables for which suitable life tables are available, while the cancer-related mortality λ_c is treated as a proportional hazard function dependent on these and any other covariates for which information is available. This allows multivariate modelling of the relative hazard of death between various patient groups, using covariates for which there are no corresponding life tables to estimate expected mortality. The model is written:

$$\lambda_o(t,z,x) = \lambda_c(t,z) + \lambda_e(x+t,z_1) \qquad (12)$$

where:

λ_o is the overall mortality of the patient group at time t after diagnosis, x is the age at diagnosis, and z is a set of covariates;

λ_e is the expected mortality at age $(x + t)$, which is known from vital statistics for a sub-set z_1 of the z covariates – for the analyses reported here, these z_1 covariates were sex, calendar period, geographic region and deprivation category, and no additional covariates for the cancer patients were used (i.e. $z_1 = z$), and

λ_c is the net hazard due to cancer at time t after diagnosis, dependent on the same vector of covariates z.

The expected mortality is taken from the life tables, leaving the net hazard rate λ_c of the patient cohort at time t after diagnosis to be estimated. It is treated as a multiplicative function of the baseline hazard in each of the pre-specified intervals of time since diagnosis, given the set of z covariates, with their corresponding parameters, β:

$$\lambda_c(t,z) = \exp(\beta z)\sum_{k=1}^{m}\tau_k I_k(t) \qquad (13)$$

where:

τ_k is the net mortality rate in the kth interval, assumed to be constant within that interval, for patients with the reference values for each of the covariates ($z = 0$), and

$I_k(t)$ is an indicator function which equals one in the kth interval and is otherwise zero

The method then consists of estimating the net mortality rate τ_k in each interval, and the parameters β, given a group of patients for each of whom data are available on time since diagnosis, vital status, age at diagnosis and any other covariates. This is done with an iterative procedure to maximise the likelihood of the parameters and the net mortality rates, given the observed data[29].

If t_0, t_1, ... t_{i-1} are the starting points of a series of i time intervals, the cumulative relative survival at time t, in the ith interval, can then be expressed as:

$$S_t = \exp -\left\{\sum_{k=1}^{i-1}(\tau_k(t_k - t_{k-1})) + \tau_i(t - t_{i-1})\right\} \qquad (14)$$

with t occurring within the last interval:

$$t_{i-1} < t \leq t_i$$

and where:

the summation term is the sum of the net mortality rates in each of the $(i - 1)$ preceding time intervals;

the second term is the component of net mortality within the ith time interval up to time t, and

the cumulative relative survival rate S_t is the exponent of this cumulative mortality rate: note that the term for covariates z is not required here because no additional variables were used for individual patients.

This is similar to the expression for cumulative survival obtained from the actuarial approach (equation 6).

The standard error and confidence interval of the estimate are obtained from the second derivative of the log-likelihood.

Under this approach, person-time contributed within each of the defined intervals is aggregated for all those who were alive at the start of the interval, and the death rate in the interval is estimated. The death rate for single year of age at death, sex and other covariates is attached to the record of each individual who dies in an interval. Individuals are grouped into the interval in which they die, and the net mortality of the cancer patients is estimated from the patients at risk during that interval, adjusted for background mortality. This amounts to subtracting the number of expected deaths in the interval from the number of observed deaths, and estimating the net mortality rate from the difference[1].

Relative survival is thus estimated from net mortality. This has theoretical and empirical advantages, since the estimate of relative survival is closer in theory to the net survival that one would ideally want to know, and empirical evidence shows that it approximates the net survival more closely than other methods[29].

Algorithm in STATA

At the time these analyses were planned, the two best-known software packages for computing relative survival from cancer registry data presented constraints that limited their applicability. The Hédelin package[30] is based on the approach outlined by Estève et al.[29], and enables modelling of relative survival with patient covariates, but the size of the patient group was limited to 65,500, and there was a requirement to generate birth-cohort-specific mortality tables. The most widely used package for analysing relative survival, by Hakulinen[31], offers several methods of estimating expected and relative survival and various statistical tests, but the number of standard death rates in any analysis was limited to 5,000 (e.g. 50 life tables with 100 single-year death rates). For most purposes this would not pose a problem, but we needed to use more than 250 life tables in each series of analyses. The Hakulinen package also required an operator to supply information at the screen for each analysis, and it could not be easily adapted for unattended (batch) operation on very large numbers of consecutive analyses.

We therefore developed a new algorithm for the analysis of relative survival as a command within the STATA language[32]. The original algorithm to apply the method was written in GAUSS[29]. It was kindly provided by the author, but it proved difficult to manipulate for large data sets. The core of the algorithm was re-written in STATA, using vectorial arithmetic for speed of computation. The algorithm is very fast and can handle large data sets easily. There is no limit to the number of standard mortality rates that can be used, or to the number of patients that may be included in the analysis, subject only to adequate computer memory being available. The use of prognostic covariates for individuals, other than those for which life tables are available, was not incorporated at this stage.

The algorithm was then extended to provide a range of additional functions and options. The main enhancements were:

• to enable a series of analyses to be concatenated for unattended (batch) handling of data sets;
• to set the number and length of the time intervals to be used in each type of analysis, depending on the number of deaths
• to set the minimum number of patients and the minimum number of deaths before allowing an analysis to begin;
• to provide a standard table of results for each time interval;
• to indicate when the survival estimate for a given interval fell outside pre-set limits of precision, or was extrapolated, so that the results could be presented appropriately (see below);
• to provide a complete log of the options and initial values used in each analysis;
• to set the initial value for τ_k in the iteration;
• to set the level of precision (i.e. the size of successive increments in τ_k used in the iteration), and the maximum number of iterations, and
• to enable automated handling of the re-grouping of time intervals when non-convergence occurs

The aim was to create a statistical tool for general use by integrating the required functions into a flexible and widely used statistical programming language, with fast calculation. This also enables the user to take advantage of other inbuilt STATA functions.

Time intervals

The number of intervals used in the survival analyses was set on the basis of three criteria – the frequency of the cancer, the level of detail (region, deprivation, age, sex etc.) and the maximum time of follow-up (five or ten years). For a given analysis to be carried out, there had to be at least as many eligible patients as the number of time intervals for that analysis. If fewer patients were eligible, no analysis was done and the results are represented by dashes ("-") in the tables.

Common cancers

Large numbers of cases were usually available for regional and national analyses (Figure 3: blocks 1-4 and 6). For these analyses, 17 time intervals were used for survival up to ten years (patients diagnosed 1971-85): three-month intervals for the first year, six-month intervals from one to five years, and annual intervals from five to ten years.

For analyses by deprivation and region (Figure 3.2: block 5), numbers of patients were generally smaller. Ten time intervals were used for survival up to ten years: six-month intervals up to three years, one-year intervals up to four and five years, then intervals of two and three years up to seven and ten years, respectively.

For patients diagnosed during 1986-90, survival was only analysed for up to five years, to 31 December 1995. For regional and national analyses, the first 12 of the 17 intervals defined above were used. For analyses by deprivation and region, the first eight of the ten intervals defined above were used.

Less common cancers

The same set of ten intervals (up to ten years since diagnosis) were used in all analyses – national, regional and by deprivation – for patients diagnosed up to 1985. The same set of eight intervals (up to five years since diagnosis) was used in all analyses for patients diagnosed during 1986-90.

METHOD

Non-convergence

The estimations of survival for all the intervals in a given analysis usually converged in about 12 iterations of the maximisation procedure. When data are sparse, the algorithm may fail to converge on a maximum likelihood for one or more time intervals. A routine was incorporated in the STATA algorithm to check for non-convergence: the data for intervals with non-convergent estimates were regrouped with data for the preceding time interval and the analysis repeated. Constraints were imposed to ensure that the one-year, five-year and ten-year boundaries were not lost in this process, so that survival estimates would always be available for those time-points. No more than two attempts to avoid non-convergence were allowed in any analysis. If non-convergence still occurred under these constraints, no estimate of relative survival was made for that time interval.

Extrapolation

When no death occurred in a given time interval, the crude and relative survival estimates were extrapolated from the previous interval for which there was a valid estimate. Since there is always a background mortality, this amounts to accepting that the negative estimate of net mortality for the given interval should be ignored[29]. Survival curves (Figures 1 and 2 in Chapters 5-62) were smoothed by exclusion of such extrapolated results, since it was considered that these apparent plateaux would not represent true inflexion in the plots of survival with time.

The exception here was if the last result in an analysis was extrapolated, when there was in effect no later information. The "last" result for this purpose was defined as either the ten-year survival rate (1971-85 diagnoses) or the five-year survival rate (1986-90 diagnoses), or, if earlier, the result for the last interval in which there were still surviving patients at risk. No estimates of survival beyond ten years were made, and the algorithm does not produce survival estimates for time intervals in which there are no patients at risk. The last result was used in the survival curves, even if it was extrapolated; this allows a plateau to occur at the end of a curve.

Example results

A typical set of results from one analysis is shown in Table 3.2. This was a regional analysis of survival up to ten years for a relatively uncommon cancer in men, diagnosed during 1971-75. Ten time intervals were therefore specified in advance (see above). Non-convergence in the last interval forced a regrouping of the last two intervals (up to seven years and up to ten years), and the second analysis was successful for the nine remaining time intervals.

The variables "left" and "right" define the start and end of successive time intervals after diagnosis, in this case in years. "NetR", "Crude" and "RelR" are, respectively, the net mortality rate, the cumulative crude survival, and the cumulative relative survival, while "Cr_lo", "Cr_up", "Re_lo" and "Re_up" are the lower and upper bounds of the 95% confidence intervals for the crude and relative survival rates, respectively. For a small data set such as this, the analysis would take about four seconds on a modern desk-top computer.

Results obtained from this application of the method proposed by Estève et al. are similar to those obtained from the same data with the Hakulinen package (Table 3.3). Crude and relative survival rates obtained from the two methods are usually within 1% up to five years after diagnosis. At longer intervals since diagnosis, the relative survival estimate may be up to 5% lower with the Estève et al. method[29]. The survival curves are also similar (Figure 3.4), and the Hakulinen estimates fall within the confidence intervals of the Estève estimates at every time point.

Table 3.3 Comparison with results from Hakulinen package

Patient group: 220 men with cancer of the oral cavity

| Years since | | Cumulative survival rates (%) | | | | | | | | | | | |
| | | STATA algorithm (Esteve method) | | | | | | Hakulinen package | | | | | |
diagnosis	Deaths	Crude	Low	High	RSR	Low	High	Crude	Low	High	RSR	Low	High
0.5	47	78.6	72.6	83.5	82.2	75.8	87.1	78.6	73.1	84.2	81.2	75.5	87.0
1.0	32	63.9	57.2	69.9	69.3	61.9	75.5	64.5	58.1	71.0	68.8	61.9	75.7
1.5	22	53.6	46.8	60.0	59.6	51.9	66.4	54.1	47.4	60.8	59.6	52.2	67.0
2.0	19	44.9	38.2	51.4	51.5	43.8	58.7	46.4	39.6	53.1	52.8	45.1	60.4
2.5	8	41.3	34.8	47.8	48.3	40.5	55.5	42.3	35.6	48.9	49.8	41.9	57.6
3.0	5	39.1	32.6	45.5	47.3	39.4	54.7	39.5	33.0	46.1	48.1	40.1	56.1
4.0	10	34.6	28.4	40.9	45.3	37.2	53.1	35.5	29.0	41.9	46.2	37.8	54.6
5.0	11	29.7	23.8	35.8	41.1	33.0	49.1	30.0	23.8	36.2	41.9	33.2	50.5
7.0	10	25.1	19.6	31.0	38.6	30.2	47.0	25.5	19.6	31.3	41.0	31.5	50.5
10.0	14	18.9	14.0	24.3	34.9	26.1	43.9	19.5	14.2	24.9	39.6	28.8	50.4

RSR - relative survival rate; low, high - 95% confidence bounds

Figure 3.4 Survival curves obtained with alternative methods

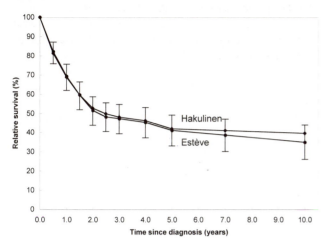

Precision

Given the volume of results, it was not possible to present the confidence interval for every survival estimate in the tables. Precision is indicated by the use of italics, using rules set out below. The precision of individual survival estimates and the overall precision of survival estimates for a given cancer were considered separately.

Precision of each survival estimate

The precision of a survival estimate depends on the number of deaths in the corresponding time interval, which in turn depends on both the lethality and the frequency of the cancer. For a given number of deaths, precision is maximal at 50% survival, tailing off for both higher and lower survival rates.

The degree of "acceptable" precision is essentially subjective. We considered that the most direct measure of precision for this purpose was the width of the 95% confidence interval (i.e. the difference between the upper and lower bounds); that the maximum acceptable width of the interval would be progressively narrower as survival values approached the extremes of 0% and 100%; and that this change in maximum acceptable width of the interval should be symmetrical around 50% survival, with the same width at, say, 30% survival as at 70%.

Acceptable levels of precision for individual survival estimates were set after examination of a large sample of results for both common and rare cancers of high, moderate and low survival. The longer arm of the 95% confidence interval was used for this, because the intervals are asymmetric for survival values other than 50%.

The criteria chosen for acceptable precision of a single survival estimate were that the longer arm of the confidence interval should be 12.5% or less at 50%

survival, and 7.5% or less near the extremes. Binomial, quadratic and quartic functions were explored as a means of allocating acceptable bounds to survival estimates between these limits. A quartic function was eventually chosen as offering the best properties. This provided a symmetrical banana-shaped envelope of acceptable precision (Figure 3.5). Survival estimates were flagged as imprecise if the confidence interval exceeded this envelope.

Imprecise survival estimates are printed in italics. A survival rate of 50% with a 95% confidence interval no wider than 37.5% to 62.5% would thus be presented in normal type, while a wider interval would entail italicisation. For a survival rate of 10%, the corresponding interval would be 2.5% to 17.5% – a relatively wide interval, but narrower in absolute terms than the 25% range accepted for survival estimates near 50%. Intervals were truncated at zero and 100% survival.

Thus, although the degree of acceptable precision was arbitrarily defined, this approach enabled us to determine objectively whether each of a very large number of survival estimates was acceptably precise, and to flag imprecise results in each analysis for presentation accordingly. It also ensured that the degree of acceptable precision varied naturally with the survival rate.

Precision and detail of results

A key objective of the analyses was to enable differences in survival between deprivation categories to be examined within each NHS Region, but not all the results were sufficiently precise to make this useful. The level of detail at which results are presented was decided on the basis of the precision of the most detailed analyses, the five-year survival rates for deprivation categories within NHS Regions.

Figure 3.5 Envelope of acceptable precision for estimates of relative survival

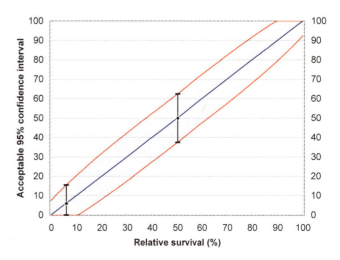

METHOD

We chose to evaluate the precision of five-year survival estimates for men and women diagnosed in 1986-90 in each deprivation category in West Midlands. This region had a population of just over five million in mid-1998. Wales and two of the English NHS Regions have smaller populations, but West Midlands is the smallest of the six large English regions. The ranked populations of the NHS Regions (in millions) in mid-1998 were:

	Men	Women
North Thames	3.530	3.717
Northern & Yorkshire	3.259	3.433
South Thames	3.193	3.421
North & West	3.106	3.287
South & West	2.989	3.175
West Midlands	2.570	2.637
Trent	2.294	2.369
Anglia & Oxford	2.249	2.310
Wales	1.387	1.471

Ignoring any difference in regional incidence and survival rates, it could be assumed that if the results for West Midlands were considered sufficiently precise to enable examination of regional survival patterns by deprivation category, then the results for at least the six largest English regions would all be sufficiently precise.

If five-year survival rates for patients of either sex diagnosed in West Midlands during 1986-90 were sufficiently precise (i.e. they would not be italicised: see above) for *at least three* of the five deprivation categories, then survival results are presented for that sex by deprivation category, for all NHS Regions and for all calendar periods. If not, deprivation-specific results at the regional level (Figure 3.3: block 5) are suppressed in the tables.

National estimates of survival for each deprivation category are nevertheless presented for each cancer.

The global precision of survival estimates for each cancer was evaluated in this way for each sex and for both sexes combined. Given the disparities in incidence between men and women for cancers such as those of the breast, larynx and gallbladder, this led in some cases to presentation of results at different levels of detail for men and for women.

An exception to the rule was made for acute lymphoid leukaemia in children, because of the rapid improvement in survival for this malignancy since the 1970s, and the interest in equity of that improvement between deprivation categories[33]. Acute lymphoid leukaemia is the commonest malignancy in children, but it did not pass the test of global precision for the most detailed survival rates: results are nevertheless presented in full detail. Results for all leukaemias in childhood are therefore also presented in full.

Age-standardisation of relative survival

Mortality increases with age. Observed (crude) cancer survival declines with age, largely because of this overall force of mortality, but also from the additional risk of death due to the cancer.

A relative survival rate is not age-adjusted *per se*. It adjusts for age-specific mortality from other causes, but as with most biological and clinical risks, the excess hazard of death from cancer is *itself* often age-dependent. For many cancers, relative survival declines with age. The difference in relative survival between young and old patient groups may be large.

Simple summary estimates of relative survival may therefore provide a misleading comparison of cancer survival between groups of patients if their age distribution differs markedly. Age-specific relative survival rates should be examined, but if one summary measure of overall survival is needed, age-standardisation is desirable (the effect of age can also be modelled). The rationale is the same as for standardisation in the comparison of incidence or mortality rates, only here it is the age distribution of cancer patients that is important, not the age distribution of the general population.

Age-adjustment is also important for the analysis of time trends in relative survival, since if survival varies markedly with age, a change in the age distribution of cancer patients over time can produce spurious survival trends or obscure real trends.

The age-standardised relative survival is interpretable as the overall survival rate that would have occurred, given the same survival rate at each age, if the age distribution of the patient group under study had been the same as that of the standard population. The survival rate does not change greatly with standardisation when the age distributions are similar, but the conceptual advantage of standardisation even when the numerical effect is small is that the effect of age can be largely discounted in the interpretation of any remaining differences in survival between the groups being compared.

Method

The directly age-standardised relative survival rate is a weighted average of the relative survival rates in each age group in the index population. The weights for each age are provided by the proportion of cancer patients in the corresponding age group in the standard population. The age-standardised survival rate (ASR) for a given cancer in a given region or time period is given by:

$$ASR = \sum_{j=1}^{n} w_j . S_j \qquad (15)$$

where

w_j is the proportion of all individuals in the standard population who are in the jth age group;

S_j is the cumulative relative survival rate for that age group in the index population, and

the summation is across all n age groups.

The sum of all the n age-specific weights is unity:

$$\sum_{j=1}^{n} w_j = 1 \qquad (16)$$

The standard error of the age-standardised survival rate is a similarly weighted average of the standard errors of the age-specific relative survival rates:

$$SE(ASR) = \sqrt{\sum_{j=1}^{n} w_j (SE(S_j))^2} \qquad (17)$$

Standard population

There is no ideal standard population for cancer survival. Incidence or mortality rates for events that occur in the general population can be standardised to the structure of a national population census, or to a well-known hypothetical standard such as the European or world standard population[34]. The event of concern in cancer survival, however, occurs only in cancer patients. Their age distribution is very different from that of the general population, and for cancer patients there has been no widely accepted standard age distribution to enable the calculation of internationally comparable survival rates. Such a standard has been proposed very recently[35], but it was not available when these analyses were carried out.

All cancer patients

Ideally, age-standardised relative survival rates would enable direct comparison of survival rates for every cancer, after adjusting for differences in their age distribution at diagnosis. This would suggest the age distribution for *all* cancer patients as the standard weights, an approach that has occasionally been used for international comparisons of survival[36]. We explored its use here, but the age distribution of patients varies widely between cancers, and a single age distribution for all cancers leads to age-adjusted survival rates that are widely different from the unadjusted rates, particularly for cancers with a relatively young age distribution.

EUROCARE standard

It would have been possible to adopt the standard weights for each cancer published from the EUROCARE study[37]. This would have enabled direct comparison with the EUROCARE results (and with those of the ITACARE study, which used the same standards). There were several disadvantages, however. The EUROCARE standard is inevitably an arbitrary population, comprising the patients from 30 contributing cancer registries in 12 countries. Not all the cancers studied here were included in the EUROCARE study, and no standard population would have been available for those cancers. Some cancers in the EUROCARE study are different from those defined here for the analysis of survival trends over a 20-year period (see Chapter 2). The time periods covered by the two studies were also different. Finally, the age groups chosen for analysis were different. The same problems remain with the EUROCARE II study[38], although it now includes 17 countries and more cancers than the original study (see Table 2.2).

Uniform standard

It would also have been possible to use equal weights for each age group in the standardisation, as in the cumulative rate for incidence or mortality. This has the advantage that the age-standardised rate becomes a simple unweighted average of the age-specific survival rates. The disadvantage is that while a cumulative rate can be interpreted as the overall incidence (say) that would occur if the group under study were to experience the same incidence at each age as is currently observed in the general population[39], there is no such intuitive interpretation of an unweighted average of age-specific survival rates.

England and Wales, 1986-90

The main object of the age-standardised analyses was to enable age-adjusted comparisons of cancer survival to be made between NHS Regions in England and Wales, between men and women, and over time. This aim largely dictated the choice of standard population: it remains arbitrary, but no ideal standard was available.

Survival rates for each cancer were therefore directly standardised to the age distribution of all patients who were diagnosed with that cancer in England and Wales during the period 1986-90 and who were included in the analyses. The age groups used for standardisation were the same as those used to present the age-specific survival rates: 15-39 years, 40-49, 50-59, 60-69, 70-79 and 80-99 years. The percentages of patients in each group were used as weights in the standardisation: these are shown, rounded to the nearest integer, in Table 3.4.

METHOD

Table 3.4   Weights for age-standardisation: numbers (%) of adults included in analyses, England and Wales, patients diagnosed 1986-90

| | \multicolumn{14}{c}{Age group (years)} | | | | | | | | | | | | |
| | 15-39 | | 40-49 | | 50-59 | | 60-69 | | 70-79 | | 80-99 | | All ages | |
	No.	%	No.	%	No.	%	No.	%	No.	%	No.	%	No.	%
Lip	23	2	61	5	145	13	307	27	381	33	228	20	1,145	100
Tongue	132	5	281	10	514	19	772	28	711	26	368	13	2,778	100
Salivary glands	142	9	146	9	235	15	358	23	384	25	279	18	1,544	100
Oral cavity	124	3	336	9	721	20	1,128	31	912	25	473	13	3,694	100
Oropharynx	46	3	185	11	375	23	545	34	326	20	143	9	1,620	100
Nasopharynx	108	14	100	13	147	19	220	28	153	20	46	6	774	100
Hypopharynx	15	1	110	7	312	19	522	32	455	28	198	12	1,612	100
Oesophagus	167	1	701	3	2,442	12	5,885	28	7,184	35	4,351	21	20,730	100
Stomach	450	1	1,281	3	4,293	10	11,492	26	15,905	36	10,164	23	43,585	100
Small intestine	46	3	111	7	256	17	428	29	423	28	224	15	1,488	100
Colon	894	1	2,670	4	7,435	11	17,461	25	23,626	35	16,395	24	68,481	100
Rectum	541	1	1,871	4	5,671	13	12,574	28	14,901	34	8,830	20	44,388	100
Liver	145	4	180	5	544	15	1,083	30	1,141	31	573	16	3,666	100
Gallbladder	40	1	184	4	541	11	1,270	25	1,752	35	1,198	24	4,985	100
Pancreas	202	1	774	3	2,446	11	6,154	27	8,047	35	5,194	23	22,817	100
Nasal cavities, sinuses	81	5	122	7	271	16	479	28	493	29	265	15	1,711	100
Larynx	96	1	538	6	1,676	19	3,282	37	2,393	27	802	9	8,787	100
Lung	876	1	4,585	3	17,796	12	51,364	35	51,775	35	19,679	13	146,075	100
Pleura	38	1	256	8	651	20	1,125	35	862	27	272	8	3,204	100
Thymus	32	19	30	18	36	22	39	23	23	14	7	4	167	100
Bone	607	37	162	10	196	12	272	17	232	14	161	10	1,630	100
Connective tissue	796	19	459	11	613	15	901	22	859	21	498	12	4,126	100
Melanoma of skin	3,245	20	2,784	17	2,759	17	3,244	20	2,578	16	1,330	8	15,940	100
Breast	7,092	6	18,255	16	23,667	20	29,809	25	24,459	21	14,457	12	117,739	100
Cervix	5,370	28	3,650	19	2,866	15	3,671	19	2,473	13	1,078	6	19,108	100
Uterus	275	2	1,114	7	3,926	24	5,149	31	4,106	25	1,938	12	16,508	100
Ovary	1,270	6	2,424	11	4,466	21	5,949	28	4,784	23	2,348	11	21,241	100
Vagina and vulva	162	3	198	4	408	9	902	19	1,533	33	1,457	31	4,660	100
Prostate	33	0	171	0	2,214	4	12,598	24	23,384	45	13,510	26	51,910	100
Testis	3,848	69	1,027	18	390	7	170	3	104	2	42	1	5,581	100
Penis	58	4	109	8	203	15	339	26	398	30	218	16	1,325	100
Bladder	650	1	1,653	3	5,756	12	14,269	29	17,460	35	9,530	19	49,318	100
Kidney	455	3	1,109	7	2,766	18	4,712	31	4,344	29	1,784	12	15,170	100
Eye	148	9	190	11	319	18	493	28	409	24	178	10	1,737	100
Brain	1,951	16	1,600	13	2,510	21	3,541	30	1,996	17	403	3	12,001	100
Thyroid	913	25	478	13	517	14	698	19	635	18	342	10	3,583	100
Non-Hodgkin lymphoma	2,015	8	2,305	10	3,706	16	6,097	26	6,374	27	3,222	14	23,719	100
Hodgkin's disease	2,756	55	612	12	560	11	547	11	393	8	153	3	5,021	100
Multiple myeloma	83	1	390	4	1,362	13	2,899	28	3,601	34	2,195	21	10,530	100
Acute lymphoid leukaemia	495	44	120	11	108	10	144	13	158	14	95	8	1,120	100
Chronic lymphoid leukaemia	36	1	154	2	655	10	1,607	25	2,241	35	1,633	26	6,326	100
Acute myeloid leukaemia	649	12	441	8	712	13	1,362	24	1,582	28	880	16	5,626	100
Chronic myeloid leukaemia	284	10	215	7	331	11	551	19	825	28	690	24	2,896	100
Monocytic leukaemia	21	7	19	6	31	10	51	17	106	34	81	26	309	100
All leukaemias	1,571	9	1,020	6	1,942	11	4,020	23	5,413	30	3,791	21	17,757	100
Spinal cord	73	38	38	20	35	18	31	16	13	7	3	2	193	100
Adrenal	68	21	38	12	71	22	83	25	54	17	13	4	327	100
Pituitary	47	21	34	15	50	22	61	27	25	11	10	4	227	100
All cancers analysed	37,684	5	54,342	7	107,809	14	216,943	28	237,474	30	128,350	16	782,602	100

Men and women were combined in the standard population for each cancer. This enables direct comparison of standardised survival rates for a given cancer between men and women. Survival rates for children were not age-standardised.

Trends in survival

Crude and relative survival rates were estimated for each of four successive five-year periods of diagnosis (1971-75, 1976-80, 1981-85, 1986-90), by NHS Region, sex and, for the more common cancers, by deprivation category. Detailed results are given in the chapter for each cancer.

The improvement in cancer survival since the early 1970s was quantified as the average increase every five years between the four successive five-year periods.

Trends were quantified by fitting a weighted least squares regression to the age-standardised relative survival rates for the four time periods. The variance of each estimate was used as the weight. The regression takes the form of an ordinary least squares regression:

$$S_i = a + bP_i + e_i \qquad (18)$$

where:

$i \ (= 1, \ldots 4)$ indexes the calendar period of diagnosis 1971-75, ... 1986-90;

a is the intercept, a constant;

b is the slope of the regression;

S_i is the estimate of age-standardised relative survival for the ith period;

P_i is an index for the period which takes the values 1, ... 4, and

e_i is the error term, but in which each term is divided by the standard error of the relative survival rate for the ith period of diagnosis.

This leads to an estimate of the slope of the regression, b:

$$\hat{b} = \frac{\sum_{i=1}^{n} \frac{1}{\sigma_i^2} \cdot (P_i - \overline{P}) \cdot (S_i - \overline{S})}{\sum_{i=1}^{n} \frac{1}{\sigma_i^2} \cdot (P_i - \overline{P})^2} \qquad (19)$$

and the standard error of the slope:

$$SE(\hat{b}) = \sqrt{\frac{\frac{1}{n-2}\left\{\sum_{i=1}^{n} (\frac{S_i}{\sigma_i} - \overline{S})^2 - b^2 \sum_{i=1}^{n} (\frac{P_i}{\sigma_i} - \overline{P})^2\right\}}{\sum_{i=1}^{n} (\frac{P_i}{\sigma_i} - \overline{P})^2}} \qquad (20)$$

where:

n is the number of data points in the regression – four for calendar periods in these analyses, and five for deprivation categories (see below);

σ_i is the standard error of the age-standardised relative survival estimate for the ith period or deprivation category;

S bar is the average of the four (or five) relative survival estimates, weighted by their variances, and

P bar is the average value for the index P_i, again weighted by the variances of the relative survival estimates in each of the four calendar periods (or five deprivation categories).

The slope b provides an estimate of the average change in age-standardised relative survival between successive five-year periods of diagnosis, with a 95% confidence interval derived from its standard error.

This is expressed as an absolute percentage change every five years. For example, an increase in five-year survival from 50% for patients diagnosed in 1971-75 to 65% for those diagnosed in 1986-90 would be expressed as a gain of 5% every five years.

Cancer survival trends were estimated in this way for survival at one and five years since diagnosis, separately for males and females, and for England and Wales. For 17 of the most common cancers in adults, and for acute lymphocytic leukaemia in children, similar estimates of trend were also made for each NHS Region.

This approach implies a linear trend in survival with time. For several childhood malignancies[40,41], testicular cancer and some adult lymphomas, there were fairly rapid gains in survival over a relatively short period, after the introduction of combined chemotherapy and radiotherapy regimes in the 1970s. The impact of these more rapid improvements is not well reflected by a single average covering a 20-year period of diagnosis. For most cancers, however, any gains in survival have been fairly regular. A consistent approach to estimating these gains for each cancer was considered more appropriate than more complex approaches for selected cancers. The estimates of trend in survival for each cancer have all been estimated as an average linear change over five years, and they are thus directly comparable.

METHOD

Deprivation gradient in survival

The deprivation gradient in survival at one year and five years after diagnosis was estimated for patients diagnosed during 1981-85 and 1986-90 in England and Wales. For five-year survival, the gradient was also plotted for each NHS Region.

Variance-weighted least squares regression was used to estimate the gradient in relative survival across the five deprivation categories, in the same way as for the estimation of trends over time. The regression was only carried out if there were at least four out of a possible five valid survival estimates for that region and time period.

This approach implies a linear relationship between survival and the degree of deprivation, as categorised in five groups of the Carstairs index. It is doubtless a simplification, but there appeared to be no strong argument to fit more complex models. For most of the common cancers, the actual and fitted estimates of relative survival in the affluent and deprived groups were in any case very similar.

The slope b provides an estimate of the change in relative survival between adjacent deprivation categories.

The "gap" in survival between patients in the most affluent (least deprived) and the most deprived groups (the first and fifth groups) was therefore estimated as a simple multiple of the slope:

$$Gap = 4b \qquad (21)$$

This represents the absolute difference between the relative survival values obtained from the regression for the most affluent and most deprived categories. Since it depends on the survival rates in all five deprivation categories, the "gap" is more readily interpretable as the deprivation gradient in survival than just the difference in survival rates between the most affluent and most deprived groups.

The confidence interval for the gap in survival between affluent and deprived groups was derived from the standard error of the slope of the regression, as before.

Negative values for the gap in survival imply higher survival in the more affluent groups.

The deprivation gradient is shown graphically in Figure 3 in the chapter for each cancer. The regression line is shown as a solid line if the survival gradient across deprivation categories is statistically significant at the 5% level, and a dotted line otherwise.

Quantitative estimates of the gradient were made only at national level.

Presentation of results

Results for each cancer – Chapters 5 to 62

These chapters each contain a brief epidemiological and clinical outline, with a résumé of the most important results. The format of the results is standard, but there are several variations for single-sex and less common cancers:

Tables

1   Ineligible and excluded records, and patients included in analyses, by calendar period of diagnosis
2   Crude and relative survival at one, five and ten years since diagnosis, by NHS Region, deprivation category and calendar period of diagnosis

Figures

1   Relative survival up to ten years: trends by calendar period of diagnosis (1971-90) and NHS Region
2   Relative survival up to five years, by deprivation category and NHS Region: patients diagnosed 1986-90
3   Relative survival at five years by deprivation category, period of diagnosis (1981-90) and NHS Region
4   Relative survival by NHS Region and sex: one-year and five-year survival for patients diagnosed 1986-90, with 95% confidence intervals
5   Relative survival at five years, with 95% confidence interval, by age at diagnosis and sex: England and Wales, patients diagnosed 1986-90

Variant presentations

Table 2 varies from one to six pages, depending on the overall precision of the results and whether the cancer affects one or both sexes. Survival rates are only presented for each deprivation category and calendar period in each NHS Region if the results are sufficiently precise (see above): this decision is made separately for each sex and for both sexes combined. Individual estimates of survival are italicised in Table 2 if they are considered imprecise. A dash ("-") is placed in the tables if no estimate of survival was made, either because there were too few patients to carry out an analysis at all, or because non-convergence for that particular estimate could not be resolved within the pre-set constraints.

Figure 1 (trends in survival) is always presented for England and Wales and for the eight NHS Regions and Wales (ten graphics).

Figure 2 (survival curves by deprivation) and Figure 3 (deprivation gradient) are always presented for England and Wales, but the nine regional graphics are only presented in these Figures if the results are sufficiently precise (see above). Extrapolated survival estimates were skipped in survival curves (Figures 1 and 2).

In Figure 3 (deprivation gradient in survival), a five-year survival estimate may be omitted for one or more deprivation categories in some regions (too few patients, or non-convergence). This arises frequently for the less common cancers.

In Figure 4 (regional patterns of survival) an asterisk is used if no estimate was made (too few patients, or unresolved non-convergence).

Similarly, in Figure 5 (age-specific relative survival rates), a double backslash ("\\") indicates that no estimate is available for that age group. If the age group(s) concerned are the youngest or oldest groups, the "\\" symbol is placed at the same level as the nearest valid survival estimate in the graphic, otherwise it is placed on the line joining the valid estimates for the younger and older adjacent age groups. Figure 5 is not presented for children.

Summary – Chapter 4

Chapter 4 contains a summary of the numbers of patients included in the analyses for each cancer, by calendar period (Table 4.1), NHS Region (Table 4.2) and deprivation category (Table 4.3). Table 4.3 also shows the percentage of patients who could not be assigned to a deprivation category in each NHS Region and calendar period.

National trends in survival at one year and five years are presented for all cancers in Table 4.4. These are presented as relative survival rates, with their 95% confidence intervals, standardised to the age distribution of all patients included in the analysis for that cancer in England and Wales for the period 1986-90. The trend is shown as the average change (gain) in the survival rate in successive five-year periods over the period 1971-90. The 95% confidence interval for the gain in survival is also shown.

For 17 of the most common cancers in adults, and for acute lymphocytic leukaemia in children, time trends in survival have been estimated separately for each NHS Region (Table and Figure 4.5).

Cancers have been ranked in order of the five-year survival rate in each sex, and grouped into three categories of survival – "good" (five-year survival 50% or more), moderate (10% to 49%) and poor (less than 10%). This is done separately for men and women and for boys and girls (Table 4.6)

Age-specific relative survival rates are shown in Table 4.7, with their 95% confidence intervals.

The gradient or "gap" in survival between deprivation categories at one and five years after diagnosis is shown in Table 4.9, with its 95% confidence interval, for all patients

diagnosed in England and Wales during 1981-85 and 1986-90.

The more common cancers (1,000 or more patients diagnosed in England and Wales during 1986-90) have also been ranked in order of the "gap" in national survival between affluent and deprived categories (Table 4.10).

International comparisons with Scotland, Europe and the USA are summarised in Table 4.12.

Construction of data pages

The methodological principles of the analyses carried out here were well known, but the method had not been used on a data set this large or complex. The algorithm and the life tables were both developed as part of the project. Over 100,000 analyses were done. Many had to be repeated while the algorithm was being developed, or after data errors were corrected, or as the various analyses were being refined.

In order to avoid transcription errors for this very large volume of results, the pages of tables and graphics were each prepared as Postscript files directly from the results of the STATA algorithm. A set of FORTRAN programs was developed to read the STATA output, to select the required information and then to transform it into Postscript format. As well as avoiding transcription errors, this approach also:

- enabled the seven different chapter formats to be constructed automatically from standard files of results produce by the algorithm;
- enabled all the variable aspects of data pages (titles, headers, etc.) to be parameterised;
- allowed automated italicisation of imprecise results (these were flagged during execution of the algorithm);
- reduced proof-reading to a minimum

This procedure substantially reduced the time required to complete the project.

CDROM

A compact disk containing the raw data, the results and various metadata files is available from the Office for National Statistics (see page ii).

The data are provided in several formats, both as files corresponding to each chapter of this book and as combined data files for adults and children. This should enable users to conduct additional analyses.

The results shown in the book are provided as tables or graphics in various common software formats.

Two additional types of material are provided. The results

METHOD

for the less common cancers, presented selectively in the book, are presented in full on the CDROM. The complete results for each analysis are also provided, in the form shown in Table 3.2.

1.   Estève J, Benhamou E, Raymond L. *Statistical methods in cancer research, volume IV. Descriptive epidemiology. (IARC Scientific Publications No. 128).* Lyon: International Agency for Research on Cancer, 1994

2.   Berrino F, Estève J, Coleman MP. Basic issues in the estimation and comparison of cancer patient survival. In: Berrino F, Sant M, Verdecchia A, Capocaccia R, Hakulinen T, Estève J, (eds.) *Survival of cancer patients in Europe: the EUROCARE study. (IARC Scientific Publications No. 132).* Lyon: International Agency for Research on Cancer, 1995, pp1-14

3.   Ederer F, Axtell LM, Cutler SJ. The relative survival: a statistical methodology. Natl Cancer Inst Monogr 1961; 6: 101-121

4.   Berkson J, Gage RP. Calculation of survival rates for cancer. Proc Staff Meet Mayo Clinic 1950; 25: 270-286

5.   Hakulinen T. On long-term relative survival rates. J Chron Dis 1977; 30: 431-443

6.   Phillimore P, Beattie A, Townsend P. Widening inequality of health in northern England, 1981-91. Br Med J 1994; 308: 1125-1128

7.   Raleigh VS, Kiri VA. Life expectancy in England: variation and trends by gender, health authority and level of deprivation. J Epidemiol Comm Hlth 1997; 51: 649-658

8.   Astrow AB. Rethinking cancer. Lancet 1994; 343: 494-495

9.   Heasman MA, Lipworth L. *Accuracy of certification of cause of death. Studies on Medical and Population Subjects No. 20.* London: HMSO, 1966

10.   Alderson MR, Meade TW. Accuracy of diagnosis on death certificates compared with that in hospital records. Br J Prev Soc Med 1967; 21: 22-29

11.   Grulich AE, Swerdlow AJ, dos Santos Silva I, Beral V. Is the apparent rise in cancer mortality in the elderly real? Analysis of changes in certification and coding of cause of death in England and Wales, 1970-1990. Int J Cancer 1995; 63: 164-168

12.   Percy CL, Muir CS. The international comparability of cancer mortality data: results of an international death certificate study. Am J Epidemiol 1989; 129: 934-946

13.   Ashworth TG. Inadequacy of death certification: proposal for change. J Clin Pathol 1991; 44: 265-268

14.   Percy CL, Dolman AB. Comparison of the coding of death certificates related to cancer in seven countries. Public Health Reports 1978; 93: 335-350

15.   Percy CL, Stanek E, Gloeckler Ries LA. Accuracy of cancer death certificates and its effect on cancer mortality statistics. Am J Public Health 1981; 71: 242-250

16.   Percy CL, Miller BA, Gloeckler Ries LA. Effect of changes in cancer classification and the accuracy of cancer death certificates on trends in cancer mortality. Ann N Y Acad Sci 1990; 609: 87-97

17.   Hoel DG, Ron E, Carter R, Mabuchi K. Influence of death certificate errors on cancer mortality trends. J Natl Cancer Inst 1993; 85: 1063-1068

18.   Lindahl BI, Johansson LA. Multiple cause-of-death data as a tool for detecting artificial trends in the underlying cause statistics: a methodological study. Scand J Soc Med 1994; 22: 145-158

19.   Garne JP, Aspegren K, Balldin G. Breast cancer as cause of death—a study over the validity of the officially registered cause of death in 2631 breast cancer patients dying in Malmö, Sweden 1964-1992. Acta Oncol 1996; 35: 671-675

20.   Jablon S, Thompson D, McConney M, Mabuchi K. Accuracy of cause-of-death certification in Hiroshima and Nagasaki, Japan. Ann N Y Acad Sci 1990; 609: 100-109

21.   Schrijvers CTM, Mackenbach J, Lutz J-M, Quinn MJ, Coleman MP. Deprivation and survival from breast cancer. Br J Cancer 1995; 72: 738-743

22.   Dickman PW, Auvinen A, Voutilainen ET, Hakulinen T. Measuring social class differences in cancer patient survival: development of a gold standard and comparison with other methods. J Epidemiol Comm Hlth 1999; (in press)

23.   Law MR, Morris JK. Why is mortality higher in poorer areas and in more northern areas of England and Wales? J Epidemiol Comm Hlth 1998; 52: 344-352

24.   Marang-van de Mheen PJ, Davey Smith G, Hart CL, Gunning-Schepers LJ. Socioeconomic differentials in mortality among men in Great Britain: time trends and contributory causes. J Epidemiol Comm Hlth 1998; 52: 214-218

25.   Brass W. On the scale of mortality. In: Brass W (ed.) *Biological aspects of demography.* London: Taylor and Francis, 1971, pp69-110

26.   Ewbank DC, Gomez de Leon JC, Stoto MA. A reducible four-parameter system of model life tables. Pop Studies 1983; 37: 105-129

27.   Office of Population Censuses and Surveys. *English Life Tables No. 14 1980-1982. Series DS no. 7.* London: HMSO, 1987

28.   Office of Population Censuses and Surveys. *English Life Tables No. 15 1990-1992. Series DS no. 14.* London: HMSO, 1997

29.   Estève J, Benhamou E, Croasdale M, Raymond L. Relative survival and the estimation of net survival: elements for further discussion. Stat Med 1990; 9: 529-538

30.   Hédelin G. *RELSURV 1.0, a program for relative survival. Technical report of the Department of Epidemiology and Public Health, Faculty of Medicine.* Strasbourg: Louis Pasteur University, 1995

31.   Hakulinen T, Abeywickrama KH. A computer program package for relative survival analysis. Comp Prog in Biomed 1985; 19: 197-207

32.   StataCorp. STATA statistical software. Release 5.0. College Station,Texas: Stata Corporation, 1997

33.   Schillinger JA, Grosclaude PC, Honjo S, Quinn MJ, Sloggett A, Coleman MP. Survival from acute lymphocytic leukaemia: socioeconomic status and geographic region. Arch Dis Child 1999; (in press)

34.   Smith PG. Comparison between registries: age-standardized rates. In: Parkin DM, Muir CS, Whelan SL, Gao Y-T, Ferlay J, Powell J, (eds.) *Cancer Incidence in Five Continents. (IARC Scientific Publications No. 120).* Lyon: International Agency for Research on Cancer, 1992, pp865-870

35.   Black RJ, Bashir SA. World standard cancer patient populations: a resource for comparative analysis of survival data. In: *(IARC Scientific Publications, in press).* Lyon: International Agency for Research on Cancer, 1999

36.   Hakulinen T. A comparison of nationwide cancer survival statistics in Finland and Norway. World Health Stat Q 1983; 36: 35-46

37.   Verdecchia A, Capocaccia R, Hakulinen T. Methods of data analysis. In: Berrino F, Sant M, Verdecchia A, Capocaccia R, Hakulinen T, Estève J, (eds.) *The survival of cancer patients in Europe: the Eurocare study. (IARC Scientific Publications No. 132).* Lyon: IARC, 1995, pp32-37

38.   Berrino F, Capocaccia R, Estève J, Gatta G, Hakulinen T, Micheli M, Sant M, Verdecchia A, (eds.) *Survival of cancer patients in Europe: the EUROCARE study, II. (IARC Scientific Publications No. 151).* Lyon: International Agency for Research on Cancer, 1999

39.   Day NE. Cumulative rate and cumulative risk. In: Parkin DM, Muir CS, Whelan SL, Gao Y-T, Ferlay J, Powell J, (eds.) *Cancer incidence in five continents, volume VI (IARC Scientific Publications No. 120).* Lyon: International Agency for Research on Cancer, 1992, pp862-864

40.   Stiller CA, Bunch KJ. Trends in survival for childhood cancer in Britain diagnosed 1971-85. Br J Cancer 1990; 62: 806-815

41.   Hammond GD. The cure of childhood cancers. Cancer 1986; 58: 407-413

Cancer survival trends

The survival of cancer patients is a key indicator of the effectiveness of cancer control.

Randomised clinical trials measure the best survival that is achievable for a given cancer at a given time. Average survival rates actually achieved in the population as a whole are inevitably lower than survival rates obtained in clinical trials, but they are arguably just as important.

Population-based cancer registries enable us to estimate average survival rates for all patients diagnosed with a given cancer, and to examine trends in those survival rates with the passage of time for each of the many different types of cancer. It is also possible to explore differences in survival between men and women, young and old, north and south, affluent and deprived.

Surveillance of the overall efficacy of cancer treatment programmes and the equity of access to care requires information on cancer survival in the whole population.

This book is the result of a detailed examination of cancer survival trends and patterns among adults and children in England and Wales. The data were collected by population-based cancer registries that have covered the entire territory of England and Wales since the early 1960s. Cancer registries regularly provide data extracts to the Office for National Statistics, where the data are collated for national publication. The data for these analyses became available in July 1997 (see Chapter 2).

In this chapter, selected results for each cancer are brought together for comparison and an overview. More detailed results on the patterns and trends in survival for each type of cancer, together with a short commentary, can be found in Chapters 5 to 62.

Cancers in adults and children

For adults, the analyses cover 47 different types of cancer: 39 solid tumours and eight types of leukaemia and lymphoma. Of these, 40 occur in both men and women, four only in women and three only in men. Collectively, these cancers account for about 93% of all diagnoses of malignant disease in adults in England and Wales. The remaining 7% of adult tumours largely comprise malignant neoplasms of unspecified or ill-defined origin.

The 11 types of malignancy examined for children represent more than 80% of all childhood malignancies[1].

Caution

It should be borne in mind in the interpretation of the

SUMMARY

results that information on the stage of disease at diagnosis and on the treatment given to each patient was not available (see Chapter 2). The survival estimates reflect the overall average survival of all patients diagnosed with a given cancer in a given time period. Where the number of eligible patients enabled more detailed analyses, survival rates are also provided for each deprivation category and NHS Region, and for each calendar period. More detailed studies will be required to help interpret the geographic and other differences in cancer survival identified here, and several such studies are already in progress[2]. If it were to be shown that differences in the stage of disease at diagnosis do contribute to regional variations in survival, this should help to focus the attention of clinical and public health specialists on suitable measures to redress the balance.

The most recent results included in this book relate to patients who were diagnosed during 1986-90 and followed up to 31 December 1995 – thus allowing the possibility of survival up to at least five years after diagnosis for all patients included in the analyses. National data with complete follow-up were not available for more recently diagnosed patients when these analyses began in July 1997. Many of the increasing trends in survival reported below are likely to have continued in the intervening period. It is therefore probable that for many cancers the average survival rates of all patients diagnosed in the late 1990s will be higher than those reported here: only future analyses will show this.

Survival estimates for rare cancers are not as precise as those for the common cancers. The precision of survival rates or survival trends presented in this chapter is generally indicated by the 95% confidence interval around the estimate. In the chapters for each cancer (Chapters 5-62), the survival rates reported in the main tables are italicised if precision falls outside specified limits, and the results should be interpreted accordingly. The tables also show the number of patients on which each survival estimate is based. Survival rates were only estimated if a specified minimum number of patients was available for the analysis (see Chapter 3).

Patients included in survival analyses

The analyses are based on about 2,870,000 adults aged 15-99 at diagnosis and about 18,000 children (0-14 years) who were diagnosed with a malignant tumour in England and Wales between 1971 and 1990 and followed up to the end of 1995.

For ease of presentation, patients who were diagnosed in each of four successive time periods have been grouped together: 1971-75, 1976-80, 1981-85 and 1986-90.

For adults, the number of patients included in the survival analyses for each five-year period increased by 21%, from 647,529 who were diagnosed during 1971-75 to 782,602 diagnosed during 1986-90 (Table 4.1). The number of eligible adults increased by somewhat more than this, but higher proportions of adults diagnosed during the 1980s were excluded from the analyses (see Table 2.10).

The number of children included in the analyses for each five-year calendar period fell by 13% from 4,844 for the period 1971-75 to 4,230 diagnosed during 1986-90. The number of children diagnosed in each period who were actually eligible for the analyses fell by rather less, but again higher proportions of children diagnosed in the 1980s were excluded (see Table 2.11).

Cancer type

Survival has been examined for all the most common cancers, but also for some of the least common. For lung cancer, more than 590,000 adults diagnosed between 1971 and 1990 were included in the analyses, about a thousand times as many as for the thymus (583 adults). There was almost a 50-fold range in frequency for the childhood tumours, from over 5,200 with acute lymphoid leukaemia to about 120 with hepatoblastoma (Table 4.1).

For each of the seven most common cancers (those of lung, breast, colon, stomach, rectum, bladder and prostate), the analyses include from 160,000 to almost 600,000 adults diagnosed during the period 1971-90. For each of these cancers, 43,000 to 146,000 adults were diagnosed during 1986-90 alone.

For each of the next ten most common malignancies (non-Hodgkin lymphoma, pancreas, ovary, oesophagus, cervix, uterus, melanoma of the skin, kidney, brain and myeloma), from 10,000 to 24,000 adults diagnosed during 1986-90 were included in the analyses.

Fewer than 1,000 adults were diagnosed during 1986-90 for each of the six least common malignancies examined (nasopharynx, spinal cord, adrenal, pituitary, monocytic leukaemia and thymus).

NHS Region

The numbers of patients included in the analyses for each cancer in each NHS Region of England and in Wales are given in Table 4.2. They reflect differences between the regions both in population size and in cancer incidence, as well as smaller regional differences in the proportion of patients excluded from analysis.

Deprivation category

The distribution of patients by sex and deprivation category within each NHS Region and five-year period is shown separately for adults and children (Table 4.3). For 1986-90, the analyses include 395,550 women and 387,052 men in England and Wales. Of these, 19% were in the most affluent category, 21% in each of the intermediate deprivation categories (2, 3 and 4), and 17% in the most deprived category. The categories were defined by the quintiles of distribution of the Carstairs deprivation score for all the census enumeration districts in Great Britain (England and Wales, and Scotland). The reasons why this distribution differs slightly from what might have been expected (20% of patients in each category) are given in Chapter 2.

For almost a fifth (19%) of adult cancer patients who were diagnosed in England and Wales during 1971-75, it was impossible to assign a deprivation category to the patient's residence at cancer diagnosis. This proportion reached 60% for West Midlands and 29% for Northern & Yorkshire, and was about 20% for Wales and North & West. The patterns were very similar for children (Table 4.3). These difficulties arose because postcoding of addresses at the time of cancer registration during the period 1971-73 was incomplete, and retrospective postcoding had only been done as far back as 1974: as a result, neither a census enumeration district nor a deprivation category could be assigned for many patients diagnosed during 1971-73. The proportion of patients for whom no postcode was available fell to 4% in England for the period 1976-80, though it was still high (13%) in Wales. For the period 1986-90, a deprivation category was available for almost 99% of patients in England and Wales.

Patients who could not be assigned to a deprivation category were excluded from survival analyses by deprivation, but were retained in overall analyses by age, sex and NHS Region, and in national analyses. These exclusions reduced the number of patients available for the analysis (and the precision of survival estimates) in each deprivation category for the early 1970s.

Survival rates and trends

To assist with interpretation of this summary of results, a brief résumé of the key measures used for cancer survival and trends in survival is provided below: full details are given in Chapter 3.

Relative survival

Cancer patients have higher death rates than the general population, but they do not all die of cancer. The extra risk of death conferred by the cancer can be estimated by comparing the overall mortality of cancer patients with mortality in the general population. Relative survival rates reflect this excess risk of death. If a group of cancer patients were to have a relative survival rate of 100%, it would imply that they had the same overall mortality as the general population, not that they all survived. Relative survival is thus a measure of the overall survival of cancer patients *relative* to that of the general population.

When relative survival is no longer falling with time since diagnosis, the implication is that, as a group, the remaining survivors no longer have higher mortality than the general population. This background mortality itself varies widely, however. Mortality rates are generally higher for the old than the young, for men than for women, and in the north than the south. Mortality also differs between groups of the population defined by their socioeconomic status or their degree of material deprivation. Overall mortality in the general population has also declined with time since the 1970s.

In estimating trends in relative survival of cancer patients and the differences in relative survival by age, sex, NHS Region and degree of deprivation, both the trends over time in general population mortality and the differences in mortality between subgroups of the population have been taken into account.

Measures of trend in survival

To obtain quantitative measures of survival trends, the relative survival rates for each cancer in adults are first standardised to the combined age distribution of men and women included in the analyses who were diagnosed with that cancer during 1986-90. This enables direct comparison of age-standardised relative survival rates for a given cancer between men and women and between successive five-year periods. Survival rates for men and women are shown, with their 95% confidence intervals.

The overall rate of change (usually improvement) in survival in England and Wales is then estimated as the average change in the age-standardised relative survival rate over the four successive five-year periods of diagnosis. The rate of change in survival is expressed as the absolute percentage change every five years – if relative survival increased by, say, 12% from 50% for patients diagnosed during 1971-75 to 62% for those diagnosed in 1986-90, three five-year periods later, this is expressed as an average gain of 4.0% every five years. This measure does not reflect well the sharp improvements in survival that occurred in the space of a few years for some cancers, but these are described in more detail in the relevant chapters, and improvements over time were fairly regular for many cancers. For 17 of the most common cancers in adults, the national trends in one-year and five-

SUMMARY

Table 4.1 Numbers of patients included in analyses, by calendar period of diagnosis

Chapter	Adults	1971-75	1976-80	1981-85	1986-90	Total
5	Lip	1,989	1,463	1,339	1,145	5,936
6	Tongue	2,507	2,496	2,700	2,778	10,481
7	Salivary glands	1,803	1,655	1,544	1,544	6,546
8	Oral cavity	3,072	3,132	3,537	3,694	13,435
9	Oropharynx	1,358	1,393	1,630	1,620	6,001
10	Nasopharynx	747	742	863	774	3,126
11	Hypopharynx	1,950	1,770	1,855	1,612	7,187
12	Oesophagus	14,910	16,455	18,025	20,730	70,120
13	Stomach	51,549	50,633	48,292	43,585	194,059
14	Small intestine	1,294	1,365	1,465	1,488	5,612
15	Colon	55,723	60,288	65,190	68,481	249,682
16	Rectum	38,412	41,120	44,657	44,388	168,577
17	Liver	2,597	2,788	3,242	3,666	12,293
18	Gallbladder	4,859	5,156	5,100	4,985	20,100
19	Pancreas	21,161	23,261	23,108	22,817	90,347
20	Nasal cavities, sinuses	2,138	1,788	1,655	1,711	7,292
21	Larynx	7,245	7,817	8,308	8,787	32,157
22	Lung	142,399	151,288	152,552	146,075	592,314
23	Pleura	872	1,377	1,992	3,204	7,445
24	Thymus	110	141	165	167	583
25	Bone	1,578	1,539	1,686	1,630	6,433
26	Connective tissue	3,349	3,434	3,611	4,126	14,520
27	Melanoma of skin	6,554	8,396	11,589	15,940	42,479
28	Breast	91,250	96,656	105,356	117,739	411,001
29	Cervix	18,152	17,895	18,826	19,108	73,981
30	Uterus	15,390	16,317	16,666	16,508	64,881
31	Ovary	17,782	19,009	20,480	21,241	78,512
32	Vagina and vulva	4,584	4,740	4,744	4,660	18,728
33	Prostate	30,428	35,785	43,987	51,910	162,110
34	Testis	2,901	3,434	4,421	5,581	16,337
35	Penis	1,165	1,272	1,314	1,325	5,076
36	Bladder	34,267	38,939	45,682	49,318	168,206
37	Kidney	9,904	11,102	13,232	15,170	49,408
38	Eye	1,431	1,382	1,556	1,737	6,106
39	Brain	8,928	9,380	10,803	12,001	41,112
40	Thyroid	2,969	3,335	3,568	3,583	13,455
41	Non-Hodgkin lymphoma	11,964	14,214	18,021	23,719	67,918
42	Hodgkin's disease	6,278	5,905	5,421	5,021	22,625
43	Multiple myeloma	6,727	8,280	9,768	10,530	35,305
44	Acute lymphoid leukaemia	851	990	1,117	1,120	4,078
45	Chronic lymphoid leukaemia	4,493	4,976	6,038	6,326	21,833
46	Acute myeloid leukaemia	3,938	4,554	5,029	5,626	19,147
47	Chronic myeloid leukaemia	2,210	2,390	2,517	2,896	10,013
48	Monocytic leukaemia	464	491	415	309	1,679
49	All leukaemias	14,662	15,908	17,321	17,757	65,648
62	Spinal cord, adrenal, pituitary	571	613	646	747	2,577
		647,529	693,663	745,917	782,602	2,869,711
	Children					
50	All leukaemias	1,931	1,744	1,649	1,640	6,964
51	Acute lymphoid leukaemia	1,316	1,367	1,312	1,294	5,289
52	Hodgkin's disease	229	260	232	191	912
53	Non-Hodgkin lymphoma	315	337	266	277	1,195
54	Brain and spinal cord	1,195	1,141	1,005	1,063	4,404
55	Retinoblastoma	171	128	102	106	507
56	Wilms' tumour	322	314	286	259	1,181
57	Hepatoblastoma	32	27	25	32	116
58	Osteosarcoma	163	141	138	117	559
59	Ewing's sarcoma	78	111	127	97	413
60	Soft tissue sarcoma	300	304	331	324	1,259
61	Germ cell and gonadal neoplasms	108	119	118	124	469
		4,844	4,626	4,279	4,230	17,979
	Total	652,373	698,289	750,196	786,832	2,887,690

Table 4.2 Numbers of patients included in analyses, by NHS Region, 1971-90

	England & Wales	Northern & Yorkshire	Trent	Anglia & Oxford	North Thames	South Thames	South & West	West Midlands	North & West	Wales
Adults										
Lip	5,936	663	609	1,196	585	654	911	407	470	441
Tongue	10,481	1,485	976	897	1,291	1,341	1,307	971	1,571	642
Salivary glands	6,546	936	618	720	619	958	823	550	862	460
Oral cavity	13,435	2,251	1,338	998	1,274	1,618	1,322	1,421	2,398	815
Oropharynx	6,001	821	510	474	602	726	626	681	1,143	418
Nasopharynx	3,126	385	298	286	444	425	338	309	451	190
Hypopharynx	7,187	927	615	617	688	946	748	891	1,252	503
Oesophagus	70,120	9,279	6,864	5,698	6,649	8,822	9,651	7,006	11,398	4,753
Stomach	194,059	28,787	19,754	16,871	17,316	23,206	21,397	21,646	31,756	13,326
Small intestine	5,612	669	527	582	467	759	742	688	807	371
Colon	249,682	33,043	24,533	24,890	23,227	33,375	35,007	25,871	36,196	13,540
Rectum	168,577	23,066	17,880	15,446	15,537	21,415	21,047	19,043	23,973	11,170
Liver	12,293	1,512	1,330	986	1,336	1,447	1,561	1,076	2,016	1,029
Gallbladder	20,100	2,901	2,306	1,976	1,801	2,407	2,337	2,276	2,977	1,119
Pancreas	90,347	12,561	9,019	9,017	8,289	12,654	11,261	8,556	13,647	5,343
Nasal cavities, sinuses	7,292	871	733	738	900	945	906	707	1,013	479
Larynx	32,157	4,699	3,266	2,561	3,710	3,905	3,544	3,152	5,202	2,118
Lung	592,314	88,378	58,065	51,844	63,287	80,432	63,889	58,503	96,245	31,671
Pleura	7,445	1,322	473	619	877	846	1,221	434	1,384	269
Thymus	583	88	51	79	75	75	66	58	62	29
Bone	6,433	720	656	561	729	825	805	603	868	666
Connective tissue	14,520	2,118	1,453	1,514	1,567	1,519	2,101	1,496	1,818	934
Melanoma of skin	42,479	4,520	3,782	4,843	4,315	6,338	7,929	3,937	4,587	2,228
Breast	411,001	49,357	40,511	39,944	46,030	56,988	55,218	43,753	55,211	23,989
Cervix	73,981	10,858	7,856	5,814	6,853	7,958	8,794	7,991	12,398	5,459
Uterus	64,881	7,055	6,664	7,129	7,221	9,611	9,636	7,220	6,954	3,391
Ovary	78,512	9,796	7,793	7,756	7,930	11,549	10,548	7,942	10,708	4,490
Vagina and vulva	18,728	2,554	1,985	1,771	1,673	2,250	2,543	1,931	2,756	1,265
Prostate	162,110	19,747	15,994	17,664	16,013	22,328	25,542	15,159	20,527	9,136
Testis	16,337	1,951	1,596	1,821	1,768	2,258	2,207	1,645	2,234	857
Penis	5,076	762	552	471	480	557	626	548	763	317
Bladder	168,206	22,906	16,708	14,936	19,030	23,810	22,847	14,965	23,790	9,214
Kidney	49,408	6,935	5,063	4,823	5,075	6,258	6,532	4,871	7,078	2,773
Eye	6,106	860	612	561	795	863	886	488	642	399
Brain	41,112	5,252	3,880	4,545	4,480	5,913	5,283	3,717	5,583	2,459
Thyroid	13,455	1,614	1,303	1,328	1,520	1,785	1,958	1,335	1,812	800
Non-Hodgkin lymphoma	67,918	7,715	6,844	6,536	8,442	9,893	10,668	6,021	8,416	3,383
Hodgkin's disease	22,625	2,628	2,430	2,165	2,775	2,873	2,877	2,268	3,180	1,429
Multiple myeloma	35,305	4,675	3,718	3,508	3,441	4,622	5,534	3,219	4,348	2,240
Acute lymphoid leukaemia	4,078	535	379	374	440	521	600	415	596	218
Chronic lymphoid leukaemia	21,833	3,132	2,430	2,119	1,866	2,825	3,422	2,004	2,866	1,169
Acute myeloid leukaemia	19,147	2,354	1,989	2,009	1,993	2,537	2,668	1,868	2,687	1,042
Chronic myeloid leukaemia	10,013	1,372	1,002	1,012	929	1,139	1,581	988	1,420	570
Monocytic leukaemia	1,679	196	218	165	130	231	277	156	228	78
All leukaemias	65,648	8,725	6,987	6,352	6,310	8,572	9,773	5,975	9,063	3,891
Other sites	2,577	245	199	265	265	507	296	294	304	202
	2,869,711	*385,637*	*286,351*	*270,802*	*295,686*	*384,233*	*371,307*	*289,624*	*417,863*	*168,208*
Children										
All leukaemias	6,964	854	758	732	860	855	881	697	955	372
Acute lymphoid leukaemia	5,289	657	588	581	654	630	663	537	743	236
Hodgkin's disease	912	116	91	99	127	102	95	103	124	55
Non-Hodgkin lymphoma	1,195	159	129	125	154	145	164	103	144	72
Brain and spinal cord	4,404	582	484	468	437	463	563	505	663	239
Retinoblastoma	507	72	54	56	37	53	84	56	52	43
Wilms' tumour	1,181	148	114	127	160	136	163	102	166	65
Hepatoblastoma	116	15	11	9	14	12	17	14	22	2
Osteosarcoma	559	84	60	58	85	71	57	53	69	22
Ewing's sarcoma	413	49	44	51	58	54	50	31	53	23
Soft tissue sarcoma	1,259	151	146	147	166	166	159	117	140	67
Germ cell and gonadal	469	64	41	50	41	61	59	55	82	16
	17,979	*2,294*	*1,932*	*1,922*	*2,139*	*2,118*	*2,292*	*1,836*	*2,470*	*976*
Total	**2,887,690**	387,931	288,283	272,724	297,825	386,351	373,599	291,460	420,333	169,184

Table 4.3 Numbers of patients included in analyses, by period of diagnosis, deprivation category, NHS Region and sex

ADULTS

Period of diagnosis

	1971-75 Deprivation category								1976-80 Deprivation category							
	Affluent	*2*	*3*	*4*	*Deprived*	*Not assigned No.*	*%*	*All*	*Affluent*	*2*	*3*	*4*	*Deprived*	*Not assigned No.*	*%*	*All*
ENGLAND & WALES																
Men	45,410	51,939	57,199	58,483	53,460	64,378	19.5	330,869	58,654	66,211	71,776	73,302	68,232	15,018	4.3	353,193
Women	47,766	52,553	55,401	53,842	45,713	61,385	19.4	316,660	61,392	66,561	70,238	68,155	59,157	14,967	4.4	340,470
Adults	93,176	104,492	112,600	112,325	99,173	125,763	19.4	647,529	120,046	132,772	142,014	141,457	127,389	29,985	4.3	693,663
ENGLAND																
Men	43,671	49,618	53,246	53,650	50,539	60,387	19.4	311,111	56,588	63,277	67,141	67,975	65,032	12,271	3.7	332,284
Women	45,980	49,997	51,602	49,336	43,232	57,394	19.3	297,541	59,210	63,460	65,619	62,953	56,245	12,334	3.9	319,821
Adults	89,651	99,615	104,848	102,986	93,771	117,781	19.4	608,652	115,798	126,737	132,760	130,928	121,277	24,605	3.8	652,105
Northern & Yorkshire																
Men	3,157	4,442	5,791	7,360	10,396	12,775	29.1	43,921	5,023	6,703	8,718	11,359	15,807	1,236	2.5	48,846
Women	3,257	4,354	5,566	6,688	8,628	11,271	28.3	39,764	5,106	6,724	8,484	10,353	13,319	1,142	2.5	45,128
Adults	6,414	8,796	11,357	14,048	19,024	24,046	28.7	83,685	10,129	13,427	17,202	21,712	29,126	2,378	2.5	93,974
Trent																
Men	3,198	4,500	5,994	7,765	6,725	2,875	9.3	31,057	4,179	5,919	7,384	9,004	7,577	1,042	3.0	35,105
Women	3,403	4,646	5,790	7,024	5,743	2,662	9.1	29,268	4,372	5,660	7,057	8,164	6,565	1,076	3.3	32,894
Adults	6,601	9,146	11,784	14,789	12,468	5,537	9.2	60,325	8,551	11,579	14,441	17,168	14,142	2,118	3.1	67,999
Anglia & Oxford																
Men	6,015	6,353	6,298	4,679	2,298	3,755	12.8	29,398	8,139	8,084	7,443	5,284	2,567	1,318	4.0	32,835
Women	6,043	6,050	5,887	4,095	1,856	3,476	12.7	27,407	8,268	7,921	7,054	4,886	2,163	1,369	4.3	31,661
Adults	12,058	12,403	12,185	8,774	4,154	7,231	12.7	56,805	16,407	16,005	14,497	10,170	4,730	2,687	4.2	64,496
North Thames																
Men	6,923	7,246	7,177	7,872	6,656	4,697	11.6	40,571	6,338	6,675	6,882	7,446	6,310	1,366	3.9	35,017
Women	7,490	7,121	6,746	7,101	5,620	4,512	11.7	38,590	6,693	6,498	6,699	6,934	5,549	1,365	4.0	33,738
Adults	14,413	14,367	13,923	14,973	12,276	9,209	11.6	79,161	13,031	13,173	13,581	14,380	11,859	2,731	4.0	68,755
South Thames																
Men	10,807	9,380	9,495	8,781	4,891	4,019	8.5	47,373	12,503	10,815	9,867	8,574	4,104	2,597	5.4	48,460
Women	11,322	9,572	9,235	8,209	4,397	4,450	9.4	47,185	13,225	11,065	9,846	8,177	3,644	2,864	5.9	48,821
Adults	22,129	18,952	18,730	16,990	9,288	8,469	9.0	94,558	25,728	21,880	19,713	16,751	7,748	5,461	5.6	97,281
South & West																
Men	7,663	10,633	9,926	5,357	2,116	3,635	9.2	39,330	9,557	13,120	12,047	6,406	2,537	1,769	3.9	45,436
Women	7,739	10,795	9,832	5,235	1,845	3,675	9.4	39,121	9,935	13,375	11,856	6,186	2,154	1,774	3.9	45,280
Adults	15,402	21,428	19,758	10,592	3,961	7,310	9.3	78,451	19,492	26,495	23,903	12,592	4,691	3,543	3.9	90,716
West Midlands																
Men	1,507	1,846	2,442	3,002	3,818	19,171	60.3	31,786	4,479	5,053	6,560	8,323	9,700	475	1.4	34,590
Women	1,787	1,939	2,373	2,812	3,327	18,330	60.0	30,568	4,750	5,138	6,485	7,659	8,276	547	1.7	32,855
Adults	3,294	3,785	4,815	5,814	7,145	37,501	60.1	62,354	9,229	10,191	13,045	15,982	17,976	1,022	1.5	67,445
North & West																
Men	4,401	5,218	6,123	8,834	13,639	9,460	19.8	47,675	6,370	6,908	8,240	11,579	16,430	2,468	4.7	51,995
Women	4,939	5,520	6,173	8,172	11,816	9,018	19.8	45,638	6,861	7,079	8,138	10,594	14,575	2,197	4.4	49,444
Adults	9,340	10,738	12,296	17,006	25,455	18,478	19.8	93,313	13,231	13,987	16,378	22,173	31,005	4,665	4.6	101,439
WALES																
Men	1,739	2,321	3,953	4,833	2,921	3,991	20.2	19,758	2,066	2,934	4,635	5,327	3,200	2,747	13.1	20,909
Women	1,786	2,556	3,799	4,506	2,481	3,991	20.9	19,119	2,182	3,101	4,619	5,202	2,912	2,633	12.8	20,649
Adults	3,525	4,877	7,752	9,339	5,402	7,982	20.5	38,877	4,248	6,035	9,254	10,529	6,112	5,380	12.9	41,558

Table 4.3 (cont.) Numbers of patients included in analyses, by period of diagnosis, deprivation category, NHS Region and sex

						Period of diagnosis										**All periods**	
		1981-85								**1986-90**						**1971-90**	
		Deprivation category								**Deprivation category**							
Affluent	*2*	*3*	*4*	*Deprived*	*Not assigned*		*All*	*Affluent*	*2*	*3*	*4*	*Deprived*	*Not assigned*		*All*	*Total*	
					No.	*%*							*No.*	*%*			**ENGLAND & WALES**
67,171	72,972	77,477	78,546	70,515	8,548	2.3	**375,229**	70,473	79,568	82,315	81,343	69,127	4,226	1.1	**387,052**	1,446,343	*Men*
71,523	74,618	77,130	75,283	63,391	8,743	2.4	**370,688**	76,651	84,279	85,228	80,254	64,256	4,882	1.2	**395,550**	1,423,368	*Women*
138,694	147,590	154,607	153,829	133,906	17,291	2.3	**745,917**	147,124	163,847	167,543	161,597	133,383	9,108	1.2	**782,602**	2,869,711	*Adults*
								19	*21*	*21*	*21*	*17*	*1*				*%*
																	ENGLAND
64,813	69,888	72,340	72,750	66,938	6,540	1.9	**353,269**	67,490	76,020	77,283	75,529	65,828	3,431	0.9	**365,581**	1,362,245	*Men*
69,030	71,293	72,098	69,512	60,091	6,712	1.9	**348,736**	73,487	80,382	80,002	74,466	61,002	3,821	1.0	**373,160**	1,339,258	*Women*
133,843	141,181	144,438	142,262	127,029	13,252	1.9	**702,005**	140,977	156,402	157,285	149,995	126,830	7,252	1.0	**738,741**	2,701,503	*Adults*
								19	*21*	*21*	*20*	*17*	*1*				*%*
																	Northern & Yorkshire
6,027	7,782	9,688	12,290	16,363	766	1.4	**52,916**	6,677	8,679	10,088	12,126	15,093	352	0.7	**53,015**	198,698	*Men*
6,170	7,706	9,288	11,535	14,381	711	1.4	**49,791**	7,262	9,126	10,379	11,597	13,554	338	0.6	**52,256**	186,939	*Women*
12,197	15,488	18,976	23,825	30,744	1,477	1.4	**102,707**	13,939	17,805	20,467	23,723	28,647	690	0.7	**105,271**	385,637	*Adults*
								13	*17*	*19*	*23*	*27*	*1*				*%*
																	Trent
5,184	6,888	8,634	10,612	8,029	268	0.7	**39,615**	5,448	7,521	9,160	10,098	9,062	118	0.3	**41,407**	147,184	*Men*
5,441	6,777	8,117	9,602	7,001	258	0.7	**37,196**	5,622	7,669	8,937	9,359	8,079	143	0.4	**39,809**	139,167	*Women*
10,625	13,665	16,751	20,214	15,030	526	0.7	**76,811**	11,070	15,190	18,097	19,457	17,141	261	0.3	**81,216**	286,351	*Adults*
								14	*19*	*22*	*24*	*21*	*0*				*%*
																	Anglia & Oxford
9,300	9,075	8,275	5,576	2,517	765	2.2	**35,508**	10,247	10,683	9,007	6,180	2,547	306	0.8	**38,970**	136,711	*Men*
9,787	9,066	8,027	5,329	2,278	774	2.2	**35,261**	10,907	10,893	8,967	6,170	2,468	357	0.9	**39,762**	134,091	*Women*
19,087	18,141	16,302	10,905	4,795	1,539	2.2	**70,769**	21,154	21,576	17,974	12,350	5,015	663	0.8	**78,732**	270,802	*Adults*
								27	*27*	*23*	*16*	*6*	*1*				*%*
																	North Thames
6,479	6,493	6,448	6,708	5,555	588	1.8	**32,271**	6,871	7,742	8,451	8,517	7,316	994	2.5	**39,891**	147,750	*Men*
7,286	6,862	6,504	6,691	5,082	626	1.9	**33,051**	7,837	8,624	8,843	8,849	7,222	1,182	2.8	**42,557**	147,936	*Women*
13,765	13,355	12,952	13,399	10,637	1,214	1.9	**65,322**	14,708	16,366	17,294	17,366	14,538	2,176	2.6	**82,448**	295,686	*Adults*
								18	*20*	*21*	*21*	*18*	*3*				*%*
																	South Thames
13,548	10,950	9,654	8,146	3,645	1,844	3.9	**47,787**	12,040	11,097	9,386	7,909	3,989	751	1.7	**45,172**	188,792	*Men*
14,754	11,554	10,240	8,225	3,549	1,946	3.9	**50,268**	13,535	12,144	10,378	8,239	4,000	871	1.8	**49,167**	195,441	*Women*
28,302	22,504	19,894	16,371	7,194	3,790	3.9	**98,055**	25,575	23,241	19,764	16,148	7,989	1,622	1.7	**94,339**	384,233	*Adults*
								27	*25*	*21*	*17*	*8*	*2*				*%*
																	South & West
10,975	14,380	12,557	6,687	2,453	1,400	2.9	**48,452**	10,963	14,758	13,143	8,665	3,165	377	0.7	**51,071**	184,289	*Men*
11,311	14,471	12,941	6,656	2,285	1,471	3.0	**49,135**	11,611	15,389	14,049	8,807	3,274	352	0.7	**53,482**	187,018	*Women*
22,286	28,851	25,498	13,343	4,738	2,871	2.9	**97,587**	22,574	30,147	27,192	17,472	6,439	729	0.7	**104,553**	371,307	*Adults*
								22	*29*	*26*	*17*	*6*	*1*				*%*
																	West Midlands
5,595	6,112	7,959	9,829	10,377	329	0.8	**40,201**	6,430	6,710	8,650	9,949	8,430	224	0.6	**40,393**	146,970	*Men*
6,117	6,205	7,729	9,042	9,269	361	0.9	**38,723**	7,213	7,147	8,646	9,522	7,718	262	0.6	**40,508**	142,654	*Women*
11,712	12,317	15,688	18,871	19,646	690	0.9	**78,924**	13,643	13,857	17,296	19,471	16,148	486	0.6	**80,901**	289,624	*Adults*
								17	*17*	*21*	*24*	*20*	*1*				*%*
																	North & West
7,705	8,208	9,125	12,902	17,999	580	1.0	**56,519**	8,814	8,830	9,398	12,085	16,226	309	0.6	**55,662**	211,851	*Men*
8,164	8,652	9,252	12,432	16,246	565	1.0	**55,311**	9,500	9,390	9,803	11,923	14,687	316	0.6	**55,619**	206,012	*Women*
15,869	16,860	18,377	25,334	34,245	1,145	1.0	**111,830**	18,314	18,220	19,201	24,008	30,913	625	0.6	**111,281**	417,863	*Adults*
								16	*16*	*17*	*22*	*28*	*1*				*%*
																	WALES
2,358	3,084	5,137	5,796	3,577	2,008	9.1	**21,960**	2,983	3,548	5,032	5,814	3,299	795	3.7	**21,471**	84,098	*Men*
2,493	3,325	5,032	5,771	3,300	2,031	9.3	**21,952**	3,164	3,897	5,226	5,788	3,254	1,061	4.7	**22,390**	84,110	*Women*
4,851	6,409	10,169	11,567	6,877	4,039	9.2	**43,912**	6,147	7,445	10,258	11,602	6,553	1,856	4.2	**43,861**	168,208	*Adults*
								14	*17*	*23*	*26*	*15*	*4*				*%*

SUMMARY

Table 4.3 (cont.) Numbers of patients included in analyses, by period of diagnosis, deprivation category, NHS Region and sex

CHILDREN

Period of diagnosis

	Affluent	2	3	4	Deprived	Not assigned No.	%	All	Affluent	2	3	4	Deprived	Not assigned No.	%	All
	1971-75 — Deprivation category								**1976-80** — Deprivation category							
ENGLAND & WALES																
Boys	510	448	426	462	419	554	19.7	**2,819**	612	475	487	467	492	105	4.0	**2,638**
Girls	356	297	325	314	306	427	21.1	**2,025**	480	400	327	346	363	72	3.6	**1,988**
Children	866	745	751	776	725	981	20.3	**4,844**	1,092	875	814	813	855	177	3.8	**4,626**
ENGLAND																
Boys	484	427	401	422	390	523	19.8	**2,647**	591	457	456	438	471	82	3.3	**2,495**
Girls	347	283	295	292	281	414	21.7	**1,912**	459	380	308	332	342	57	3.0	**1,878**
Children	831	710	696	714	671	937	20.6	**4,559**	1,050	837	764	770	813	139	3.2	**4,373**
Northern & Yorkshire																
Boys	29	32	43	37	77	117	34.9	**335**	69	44	52	64	107	13	3.7	**349**
Girls	19	24	31	38	54	84	33.6	**250**	37	41	37	50	81	5	2.0	**251**
Children	48	56	74	75	131	201	34.4	**585**	106	85	89	114	188	18	3.0	**600**
Trent																
Boys	46	40	58	80	53	31	10.1	**308**	47	49	52	56	68	8	2.9	**280**
Girls	35	35	47	51	34	29	12.6	**231**	44	46	52	63	41	8	3.1	**254**
Children	81	75	105	131	87	60	11.1	**539**	91	95	104	119	109	16	3.0	**534**
Anglia & Oxford																
Boys	84	52	54	31	21	46	16.0	**288**	93	72	65	35	20	9	3.1	**294**
Girls	62	44	42	33	16	33	14.3	**230**	65	65	33	31	13	7	3.3	**214**
Children	146	96	96	64	37	79	15.3	**518**	158	137	98	66	33	16	3.1	**508**
North Thames																
Boys	70	73	72	59	55	37	10.1	**366**	60	53	54	57	44	13	4.6	**281**
Girls	59	41	43	48	27	25	10.3	**243**	50	38	33	38	32	8	4.0	**199**
Children	129	114	115	107	82	62	10.2	**609**	110	91	87	95	76	21	4.4	**480**
South Thames																
Boys	92	57	48	66	28	30	9.3	**321**	104	54	43	49	26	13	4.5	**289**
Girls	59	44	31	41	33	27	11.5	**235**	91	51	39	35	20	8	3.3	**244**
Children	151	101	79	107	61	57	10.3	**556**	195	105	82	84	46	21	3.9	**533**
South & West																
Boys	86	79	62	50	17	32	9.8	**326**	104	94	88	51	28	12	3.2	**377**
Girls	61	52	50	28	17	24	10.3	**232**	79	73	56	31	15	12	4.5	**266**
Children	147	131	112	78	34	56	10.0	**558**	183	167	144	82	43	24	3.7	**643**
West Midlands																
Boys	19	31	24	28	31	157	54.1	**290**	52	39	51	56	79	1	0.4	**278**
Girls	18	15	17	15	16	124	60.5	**205**	37	24	25	30	52	2	1.2	**170**
Children	37	46	41	43	47	281	56.8	**495**	89	63	76	86	131	3	0.7	**448**
North & West																
Boys	58	63	40	71	108	73	17.7	**413**	62	52	51	70	99	13	3.7	**347**
Girls	34	28	34	38	84	68	23.8	**286**	56	42	33	54	88	7	2.5	**280**
Children	92	91	74	109	192	141	20.2	**699**	118	94	84	124	187	20	3.2	**627**
WALES																
Boys	26	21	25	40	29	31	18.0	**172**	21	18	31	29	21	23	16.1	**143**
Girls	9	14	30	22	25	13	11.5	**113**	21	20	19	14	21	15	13.6	**110**
Children	35	35	55	62	54	44	15.4	**285**	42	38	50	43	42	38	15.0	**253**

Table 4.3 (cont.) Numbers of patients included in analyses, by period of diagnosis, deprivation category, NHS Region and sex

Period of diagnosis

1981-86 Deprivation category								1986-90 Deprivation category								All periods 1971-90	
Affluent	2	3	4	Deprived	Not assigned No.	%	All	Affluent	2	3	4	Deprived	Not assigned No.	%	All	Total	
																	ENGLAND & WALES
572	472	461	402	455	53	2.2	**2,415**	504	535	418	463	409	27	1.1	**2,356**	10,228	*Boys*
435	351	353	351	341	33	1.8	**1,864**	411	419	324	355	340	25	1.3	**1,874**	7,751	*Girls*
1,007	823	814	753	796	86	2.0	**4,279**	915	954	742	818	749	52	1.2	**4,230**	17,979	*Children*
								22	*23*	*18*	*19*	*18*	*1*				*%*
																	ENGLAND
554	454	434	368	426	44	1.9	**2,280**	493	514	402	440	392	21	0.9	**2,262**	9,684	*Boys*
416	336	334	330	328	20	1.1	**1,764**	398	401	309	319	318	20	1.1	**1,765**	7,319	*Girls*
970	790	768	698	754	64	1.6	**4,044**	891	915	711	759	710	41	1.0	**4,027**	17,003	*Children*
								22	*23*	*18*	*19*	*18*	*1*				*%*
																	Northern & Yorkshire
48	53	60	61	95	6	1.9	**323**	56	48	59	66	73	2	0.7	**304**	1,311	*Boys*
41	38	43	48	62	0	0.0	**232**	39	49	45	48	65	4	1.6	**250**	983	*Girls*
89	91	103	109	157	6	1.1	**555**	95	97	104	114	138	6	1.1	**554**	2,294	*Children*
								17	*18*	*19*	*21*	*25*	*1*				*%*
																	Trent
41	52	53	40	45	2	0.9	**233**	35	69	46	53	57	0	0.0	**260**	1,081	*Boys*
33	29	39	48	36	1	0.5	**186**	31	34	28	46	41	0	0.0	**180**	851	*Girls*
74	81	92	88	81	3	0.7	**419**	66	103	74	99	98	0	0.0	**440**	1,932	*Children*
								15	*23*	*17*	*23*	*22*	*0*				*%*
																	Anglia & Oxford
94	72	42	34	19	7	2.6	**268**	68	63	39	44	19	0	0.0	**233**	1,083	*Boys*
60	43	46	32	17	0	0.0	**198**	61	55	32	31	15	3	1.5	**197**	839	*Girls*
154	115	88	66	36	7	1.5	**466**	129	118	71	75	34	3	0.7	**430**	1,922	*Children*
								30	*27*	*17*	*17*	*8*	*1*				*%*
																	North Thames
60	55	58	46	40	5	1.9	**264**	52	61	71	77	71	8	2.4	**340**	1,251	*Boys*
44	49	39	36	25	2	1.0	**195**	57	49	40	56	45	4	1.6	**251**	888	*Girls*
104	104	97	82	65	7	1.5	**459**	109	110	111	133	116	12	2.0	**591**	2,139	*Children*
								18	*19*	*19*	*23*	*20*	*2*				*%*
																	South Thames
97	56	51	47	27	9	3.1	**287**	99	76	40	51	14	5	1.8	**285**	1,182	*Boys*
83	46	44	43	18	10	4.1	**244**	71	48	47	29	13	5	2.3	**213**	936	*Girls*
180	102	95	90	45	19	3.6	**531**	170	124	87	80	27	10	2.0	**498**	2,118	*Children*
								34	*25*	*17*	*16*	*5*	*2*				*%*
																	South & West
101	66	55	35	23	12	4.1	**292**	69	99	66	44	16	3	1.0	**297**	1,292	*Boys*
75	67	56	42	19	3	1.1	**262**	56	68	53	39	22	2	0.8	**240**	1,000	*Girls*
176	133	111	77	42	15	2.7	**554**	125	167	119	83	38	5	0.9	**537**	2,292	*Children*
								23	*31*	*22*	*15*	*7*	*1*				*%*
																	West Midlands
53	55	47	47	75	1	0.4	**278**	45	35	37	42	50	2	0.9	**211**	1,057	*Boys*
40	39	27	43	65	2	0.9	**216**	29	39	32	30	56	2	1.1	**188**	779	*Girls*
93	94	74	90	140	3	0.6	**494**	74	74	69	72	106	4	1.0	**399**	1,836	*Children*
								19	*19*	*17*	*18*	*27*	*1*				*%*
																	North & West
60	45	68	58	102	2	0.6	**335**	69	63	44	63	92	1	0.3	**332**	1,427	*Boys*
40	25	40	38	86	2	0.9	**231**	54	59	32	40	61	0	0.0	**246**	1,043	*Girls*
100	70	108	96	188	4	0.7	**566**	123	122	76	103	153	1	0.2	**578**	2,470	*Children*
								21	*21*	*13*	*18*	*26*	*0*				*%*
																	WALES
18	18	27	34	29	9	6.7	**135**	11	21	16	23	17	6	6.4	**94**	544	*Boys*
19	15	19	21	13	13	13.0	**100**	13	18	15	36	22	5	4.6	**109**	432	*Girls*
37	33	46	55	42	22	9.4	**235**	24	39	31	59	39	11	5.4	**203**	976	*Children*
								12	*19*	*15*	*29*	*19*	*5*				*%*

year survival between successive periods are also shown graphically, together with trend estimates for each region (see below). The 95% confidence interval for the average gain in survival between successive five-year periods is also shown in the tables.

Age-standardised relative survival rates for most cancers did not differ greatly from the unstandardised rates. They are nevertheless more appropriate for quantitative estimation of survival trends, because the survival rates for the four time periods and for men and women were standardised to a common age distribution for a given cancer.

Estimates of survival trends over time are directly comparable between different cancers, because they were estimated as the average improvement in the age-standardised relative survival rate between successive five-year periods. There is no particular advantage in using age-standardised rates to compare survival rates for different cancers within a given period, however, because the survival rates for each adult cancer were standardised with a different set of weights (see Table 3.4). Relative survival rates for children were not age-standardised.

Survival trends

Adults

Trends in one-year and five-year survival for men and women diagnosed in England and Wales during the four consecutive five-year periods from 1971-75 to 1986-90 are summarised for each type of cancer in Table 4.4. The overall increase in five-year survival between adults diagnosed during 1971-75 and those diagnosed during 1986-90 is ranked for men and women separately in Figure 4.4.

Cancers of the large bowel have shown some of the most consistent improvements in survival among the common solid tumours. For colon cancer, average gains in both one-year and five-year survival between successive five-year periods have been 5% or more for both men and women, and survival from cancer of the rectum has improved almost as much, with average gains of 5% every five years in one-year survival, and about 4% for five-year survival, again in both sexes.

Survival from both acute and chronic forms of lymphoid leukaemia has also improved substantially for both men and women, with an average increase every five years in the range 4-7% for both one-year and five-year survival. One-year survival rates for acute myeloid leukaemia in adults have risen by an average of about 5% every five years since the early 1970s, although five-year survival rates have risen less quickly (2-3% every five years).

One-year and five-year survival rates from malignant bone tumours rose in both sexes by an average of 4-7% between successive five-year periods.

Longer-term survival has risen more quickly than short-term survival for several cancers, strongly suggesting an increase in the proportion of patients effectively cured of their disease. For melanoma of the skin, five-year survival rose by an average of 7.6% every five years for men, and by 5.8% for women. One-year survival also increased in both sexes, but the average gains were lower: 3.9% every five years for men and 2.1% for women.

For breast cancer, five-year survival in England and Wales rose from 52% for women diagnosed during 1971-75 to 66% for those diagnosed during 1986-90, an average increase of 4.4% every five years, while one-year survival rose by 2.2% every five years. The decline in breast cancer mortality since the early 1990s is too early to be attributable to the national screening programme[3], but improved survival in East Anglia has been attributed to earlier diagnosis[4].

Five-year survival also increased substantially more than one-year survival for testicular cancer.

Improvements in average survival for several other major cancers in England and Wales have been very small. In men, five-year survival rates from cancers of lung, oesophagus, pancreas, liver and the pleura all increased by an average of less than 1% every five years between 1971-75 and 1986-90, although in each case one-year survival has improved more than five-year survival. These five cancers account for about a third (32%) of all cancers in men: for all five cancers, relative survival rates for men diagnosed during 1986-90 were well below 10%.

Survival trends for the same five cancers in women are also uniformly poor, but they represent a smaller proportion of all cancers in women (17%).

All cancers combined in adults

Overall trends in survival for all adult cancers that were included in the analyses have also been estimated. The results in Table 4.4A are provided as a rough summary of combined survival trends from most cancers diagnosed in adults in England and Wales during the period 1971-90.

Special care is required in interpreting combined survival rates and trends for such a broad group of diseases. Several major cancers became more (or less) common during the 1970s and 1980s, altering the composition of "all cancers combined", and this would affect overall survival from all cancers combined even if survival rates for individual cancers did not change. The effect will differ between men and women because cancer trends were not the same for both sexes. Thus lung cancer became less common in men and more common in women; breast cancer became more common in women; stomach cancer fell in both sexes, and lymphoma and melanoma increased in both sexes. Survival rates for these common cancers differ widely.

Trends in survival from all cancers combined therefore reflect changes over time in the relative frequency of each cancer, as well as a weighted average of widely differing trends in survival for both common and rare tumours. Further, the survival trends shown here for all cancers cannot be directly compared with survival trends for all malignancies in other countries, because the cancers included in the analyses are not the same. In these analyses, "all cancers combined" includes all 47 malignancies examined in adults (Table 4.1), representing about 93% of all cancers in adults in England and Wales.

It is nevertheless of some interest to have survival rates available as an overall public health measure of trends in average survival for all malignant disease, providing these points are taken into account in their interpretation.

With these caveats, one-year survival for all cancers combined in England and Wales has improved from 52% for women diagnosed in the early 1970s to 62% for those diagnosed during 1986-90, while the corresponding gain has been from 37% to 51% for men (Table 4.4A).

Five-year survival from all cancers combined rose from 32% to 43% for women and from 19% to 31% for men over the same period.

Survival from all cancers combined has increased slightly but significantly more quickly for men than for women. Averaged over the four quinquennia of diagnosis 1971-75 to 1986-90 for England and Wales, one-year survival rose by 4.9% every five years for men and by 3.4% for women. The corresponding national average rates of improvement in five-year survival were 4.2% every five years for men and 4.0% for women.

Among men, one-year survival rose slightly but significantly more quickly than five-year survival (4.9% every five years vs. 4.2%). The converse was observed for women: five-year survival rose slightly but significantly more quickly than one-year survival (4.0% every five years vs. 3.4%).

The largest single component of the overall increase in survival occurred between patients diagnosed during 1976-80 and those diagnosed during 1981-85. Between those two quinquennia, the national average one-year survival rate for all cancers combined rose by 6.6% for

men and 4.7% for women. Five-year survival rates increased by 5.1% in men and 4.5% in women over the same period.

Survival curves for each five-year period of diagnosis from 1971-75 to 1986-90 are shown for all cancers combined, for ten years after diagnosis for adults diagnosed up to 1985, and up to five years after diagnosis for those diagnosed during 1986-90 (Figure 4.4A).

The differences in survival trends between men and women are likely to be attributable to at least two components. Changes over time in the composition of "all cancers combined" between men and women have occurred – lung cancer has declined in men and increased in women, for example – and this alone would tend to produce more favourable survival trends for men, since lung cancer is the most common cancer, and the low survival rates from lung cancer have improved very little. Breast cancer is the most common cancer among women, and incidence has increased, but survival at five years improved significantly faster than survival at one year, and this has contributed to the more favourable trend in longer-term survival for all cancers combined among women.

Finally, overall survival from all cancers combined has been 10-15% higher for women than for men, both one year and five years after diagnosis, and in all time periods. These substantial differences are again largely due to the composition of "all cancers combined": the most common cancers in men have much lower survival than the most common cancers in women. For many cancers, however, women do also have a small to moderate survival advantage over men. Both these points are discussed further below.

Children

The survival trends for children confirm the well-known and substantial improvements for most of the common malignancies of childhood[5-9]. For five childhood malignancies (non-Hodgkin lymphoma, acute lymphoid leukaemia, osteosarcoma, hepatoblastoma and germ-cell and gonadal tumours combined), five-year survival in England and Wales increased by an average of 10% or more every five years for both boys and girls between 1971-75 and 1986-90 (Table 4.4). The most rapid increase in five-year survival for any adult cancer over the same four quinquennia was for melanoma of the skin in men (7.6% every five years).

For acute lymphoid leukaemia, five-year survival rates improved by an average of 12-13% every five years, to 73% in boys and 79% in girls diagnosed during 1986-90. In both sexes, five-year survival improved much more rapidly than one-year survival (4-6% every five years).

SUMMARY

Table 4.4 Cancer survival trends in England and Wales, by period of diagnosis and sex: age-standardised relative survival at one and five years, with 95% confidence intervals (CI), and average increase in relative survival between successive five-year periods of diagnosis

ADULTS		One-year survival (%), patients diagnosed in				Average increase every five years	Five-year survival (%), patients diagnosed in				Average increase every five years
		1971-75 % (CI)	1976-80 % (CI)	1981-85 % (CI)	1986-90 % (CI)	% (CI)	1971-75 % (CI)	1976-80 % (CI)	1981-85 % (CI)	1986-90 % (CI)	% (CI)
Lip	Men	96 (94-98)	95 (93-97)	98 (96-100)	96 (94-98)	0.3 (-0.6, 1.1)	85 (80-89)	87 (82-92)	86 (82-91)	86 (82-91)	0.6 (-1.4, 2.5)
	Women	94 (90-98)	92 (88-97)	97 (94-100)	94 (90-98)	0.7 (-1.1, 2.4)	73 (66-80)	86 (77-94)	80 (73-87)	85 (78-92)	3.0 (-0.2, 6.2)
Tongue	Men	60 (58-63)	60 (57-62)	60 (57-62)	61 (58-63)	0.1 (-1.1, 1.2)	32 (29-34)	34 (31-37)	36 (33-38)	36 (33-38)	1.3 (0.1, 2.6)
	Women	67 (64-70)	65 (62-68)	68 (65-71)	75 (72-77)	2.5 (1.2, 3.8)	44 (40-47)	41 (38-44)	45 (42-49)	50 (46-53)	2.2 (0.7, 3.8)
Salivary glands	Men	69 (66-73)	71 (68-75)	73 (69-76)	74 (71-78)	1.7 (0.2, 3.2)	47 (43-52)	48 (45-52)	49 (45-54)	47 (43-51)	-0.1 (-2.0, 1.8)
	Women	79 (76-82)	79 (76-82)	80 (77-84)	81 (78-84)	0.7 (-0.7, 2.0)	62 (58-66)	64 (60-68)	62 (58-67)	62 (58-66)	0.0 (-1.8, 1.9)
Oral cavity	Men	68 (66-70)	67 (65-69)	70 (68-72)	72 (70-74)	1.4 (0.5, 2.3)	38 (35-40)	41 (39-44)	42 (40-45)	43 (40-45)	1.6 (0.4, 2.7)
	Women	72 (69-74)	70 (67-73)	76 (74-78)	74 (72-77)	1.4 (0.2, 2.5)	49 (46-52)	48 (45-51)	54 (51-58)	52 (49-55)	1.5 (0.1, 2.9)
Oropharynx	Men	57 (54-60)	57 (53-60)	58 (55-61)	64 (61-67)	2.3 (0.9, 3.7)	26 (22-29)	28 (25-32)	28 (25-32)	33 (30-36)	2.2 (0.8, 3.7)
	Women	65 (60-69)	63 (59-68)	67 (63-71)	65 (61-69)	0.6 (-1.4, 2.6)	40 (34-45)	37 (32-42)	40 (35-45)	37 (32-42)	-0.5 (-2.7, 1.7)
Nasopharynx	Men	60 (55-64)	65 (61-69)	59 (54-63)	62 (58-66)	0.2 (-1.7, 2.0)	27 (22-31)	32 (28-37)	30 (26-35)	29 (25-33)	0.5 (-1.3, 2.4)
	Women	64 (58-70)	63 (58-68)	64 (59-69)	64 (59-70)	0.2 (-2.4, 2.7)	29 (23-34)	38 (32-44)	33 (27-38)	38 (32-43)	2.1 (-0.5, 4.7)
Hypopharynx	Men	37 (34-40)	42 (39-46)	47 (44-50)	48 (45-51)	3.9 (2.5, 5.3)	13 (10-15)	15 (12-17)	17 (15-20)	20 (17-22)	2.4 (1.2, 3.6)
	Women	34 (31-37)	36 (33-39)	43 (40-46)	49 (45-53)	4.9 (3.3, 6.4)	14 (11-16)	14 (12-17)	19 (16-22)	20 (17-24)	2.4 (1.1, 3.7)
Oesophagus	Men	12 (12-13)	13 (12-13)	17 (17-18)	21 (20-22)	3.1 (2.7, 3.4)	3 (3-3)	3 (3-4)	4 (4-5)	5 (5-6)	0.8 (0.7, 1.0)
	Women	16 (15-17)	17 (16-18)	23 (22-24)	25 (24-26)	3.2 (2.8, 3.6)	5 (5-6)	6 (5-6)	8 (7-8)	8 (7-9)	1.1 (0.8, 1.3)
Stomach	Men	13 (12-13)	14 (13-14)	20 (19-20)	23 (23-24)	3.6 (3.5, 3.8)	4 (4-4)	5 (5-5)	7 (7-8)	9 (8-9)	1.6 (1.5, 1.8)
	Women	15 (15-16)	16 (16-17)	22 (21-23)	26 (25-26)	3.5 (3.3, 3.8)	5 (5-5)	6 (6-6)	9 (8-9)	11 (10-12)	2.0 (1.8, 2.2)
Small intestine	Men	35 (32-39)	40 (36-44)	43 (39-47)	50 (46-53)	4.6 (2.9, 6.2)	20 (16-23)	21 (17-24)	21 (18-24)	23 (20-27)	1.1 (-0.3, 2.6)
	Women	43 (39-47)	41 (38-45)	51 (47-54)	50 (46-53)	3.1 (1.4, 4.8)	21 (17-24)	23 (20-26)	27 (23-30)	24 (20-27)	1.2 (-0.3, 2.7)
Colon	Men	39 (39-40)	45 (44-45)	54 (54-55)	59 (58-59)	6.7 (6.4, 7.0)	22 (22-23)	28 (27-29)	35 (34-36)	38 (37-39)	5.3 (5.0, 5.6)
	Women	40 (40-41)	45 (45-46)	54 (54-55)	59 (59-60)	6.5 (6.3, 6.8)	23 (23-24)	28 (28-29)	35 (34-36)	39 (39-40)	5.4 (5.1, 5.7)
Rectum	Men	50 (50-51)	54 (53-55)	61 (61-62)	65 (64-65)	5.0 (4.6, 5.3)	25 (24-26)	29 (29-30)	34 (33-35)	36 (36-37)	3.9 (3.5, 4.2)
	Women	51 (51-52)	55 (54-56)	62 (61-62)	66 (65-66)	5.0 (4.6, 5.3)	27 (27-28)	31 (30-32)	36 (35-37)	39 (38-40)	4.0 (3.6, 4.3)
Liver	Men	3 (2-4)	3 (3-4)	5 (4-6)	9 (8-10)	1.6 (1.3, 2.0)	1 (1-1)	1 (1-1)	1 (1-2)	2 (2-3)	0.4 (0.2, 0.7)
	Women	4 (3-5)	6 (5-7)	9 (8-11)	10 (9-11)	2.1 (1.6, 2.7)	1 (1-2)	1 (1-2)	3 (2-4)	2 (2-3)	0.5 (0.2, 0.8)
Gallbladder	Men	12 (11-13)	13 (11-14)	18 (17-20)	23 (21-25)	3.8 (3.1, 4.4)	3 (2-4)	4 (4-5)	7 (6-8)	9 (8-10)	1.9 (1.4, 2.3)
	Women	11 (10-12)	12 (11-13)	17 (16-18)	21 (19-22)	3.3 (2.8, 3.8)	3 (3-4)	4 (4-5)	6 (5-7)	8 (7-9)	1.6 (1.3, 2.0)
Pancreas	Men	6 (6-6)	7 (6-7)	8 (8-8)	9 (9-10)	1.1 (0.9, 1.3)	2 (1-2)	2 (1-2)	2 (2-2)	2 (2-3)	0.3 (0.2, 0.4)
	Women	7 (7-8)	7 (7-8)	9 (9-10)	10 (9-10)	0.9 (0.6, 1.1)	2 (2-2)	2 (1-2)	2 (2-2)	2 (2-3)	0.2 (0.1, 0.3)
Nasal cavities, sinuses	Men	66 (63-70)	66 (63-69)	63 (60-67)	66 (62-69)	-0.5 (-1.9, 0.9)	38 (34-42)	37 (33-42)	36 (32-40)	40 (36-44)	0.5 (-1.2, 2.3)
	Women	67 (64-70)	62 (58-65)	66 (62-69)	66 (63-70)	-0.1 (-1.5, 1.4)	39 (35-43)	36 (32-40)	39 (35-43)	40 (36-44)	0.5 (-1.2, 2.3)
Larynx	Men	78 (77-79)	78 (77-79)	82 (81-83)	83 (82-84)	2.0 (1.5, 2.4)	57 (56-59)	58 (57-60)	63 (61-64)	63 (62-65)	2.1 (1.4, 2.9)
	Women	73 (70-75)	75 (72-77)	78 (76-80)	79 (77-81)	2.1 (1.1, 3.2)	48 (44-51)	52 (48-55)	56 (53-59)	57 (54-59)	3.1 (1.7, 4.4)
Lung	Men	15 (15-15)	15 (15-16)	18 (18-18)	19 (19-19)	1.4 (1.3, 1.5)	4 (4-4)	5 (4-5)	5 (5-5)	5 (5-5)	0.4 (0.3, 0.4)
	Women	13 (13-14)	14 (14-15)	17 (17-18)	19 (18-19)	1.9 (1.8, 2.1)	4 (4-4)	4 (4-4)	5 (5-5)	5 (5-6)	0.5 (0.4, 0.6)
Pleura	Men	21 (18-24)	20 (18-22)	25 (23-27)	26 (24-27)	1.9 (0.9, 2.9)	3 (1-4)	2 (1-3)	3 (3-4)	4 (3-5)	0.6 (0.2, 1.1)
	Women	26 (20-31)	22 (18-27)	29 (24-33)	29 (25-32)	1.8 (-0.2, 3.8)	6 (3-10)	4 (2-7)	3 (1-4)	5 (3-7)	0.0 (-1.2, 1.2)
Thymus	Men	60 (47-73)	71 (60-81)	72 (62-83)	66 (57-75)	1.0 (-3.8, 5.8)	38 (23-54)	43 (32-55)	45 (33-57)	44 (34-54)	1.4 (-4.2, 6.9)
	Women	61 (47-76)	67 (54-80)	74 (64-84)	76 (66-85)	4.9 (-0.4, 10.2)	37 (23-51)	43 (29-57)	48 (35-60)	48 (36-59)	3.4 (-2.3, 9.1)
Bone	Men	53 (50-56)	56 (53-59)	62 (59-65)	69 (66-72)	5.4 (4.1, 6.8)	27 (24-30)	35 (32-39)	39 (36-43)	47 (44-50)	6.5 (5.1, 7.9)
	Women	57 (53-60)	63 (60-67)	65 (62-68)	71 (68-74)	4.5 (3.0, 6.0)	36 (31-40)	40 (36-44)	39 (36-43)	53 (49-57)	5.3 (3.5, 7.0)
Connective tissue	Men	65 (63-68)	68 (65-70)	69 (67-71)	72 (70-74)	2.1 (1.1, 3.1)	40 (37-43)	46 (43-49)	47 (45-50)	50 (47-52)	2.9 (1.7, 4.2)
	Women	66 (63-68)	67 (65-69)	71 (69-74)	72 (70-74)	2.3 (1.3, 3.2)	44 (41-46)	44 (41-47)	49 (47-52)	50 (47-52)	2.3 (1.2, 3.4)
Melanoma of skin	Men	78 (76-80)	82 (80-84)	85 (83-86)	90 (89-90)	3.9 (3.3, 4.5)	46 (43-49)	51 (48-53)	58 (56-60)	68 (66-69)	7.6 (6.6, 8.5)
	Women	89 (88-90)	89 (88-90)	93 (92-93)	94 (94-95)	2.1 (1.7, 2.4)	65 (63-66)	69 (68-71)	76 (74-77)	82 (81-83)	5.8 (5.2, 6.4)

Table 4.4 (cont.)   Cancer survival trends in England and Wales, by period of diagnosis and sex: age-standardised relative survival at one and five years, with 95% confidence intervals (CI), and average increase in relative survival between successive five-year periods of diagnosis

ADULTS		One-year survival (%), patients diagnosed in				Average increase every five years		Five-year survival (%), patients diagnosed in				Average increase every five years	
		1971-75	1976-80	1981-85	1986-90			1971-75	1976-80	1981-85	1986-90		
		% CI	% CI	% CI	% CI	%	CI	% CI	% CI	% CI	% CI	%	CI
Breast	Men	82 (80-85)	84 (82-87)	85 (82-88)	90 (88-93)	2.5	(1.3, 3.7)	57 (52-62)	61 (57-66)	60 (55-64)	70 (66-74)	3.9	(2.0, 5.9)
	Women	82 (82-82)	85 (84-85)	87 (87-87)	89 (88-89)	2.2	(2.1, 2.2)	52 (52-53)	59 (58-59)	62 (62-62)	66 (66-66)	4.4	(4.3, 4.6)
Cervix	Women	75 (75-76)	76 (76-77)	80 (79-80)	82 (81-82)	2.3	(2.0, 2.6)	52 (51-53)	54 (54-55)	58 (57-58)	61 (61-62)	3.2	(2.8, 3.5)
Uterus	Women	78 (77-79)	80 (80-81)	82 (82-83)	84 (83-85)	2.0	(1.7, 2.3)	61 (60-62)	65 (64-66)	68 (67-69)	70 (69-70)	2.9	(2.5, 3.3)
Ovary	Women	42 (41-43)	44 (44-45)	51 (50-52)	54 (53-55)	4.3	(4.0, 4.6)	21 (20-21)	22 (22-23)	27 (26-27)	28 (28-29)	2.7	(2.4, 3.0)
Vagina and vulva	Women	62 (60-63)	64 (62-65)	67 (66-69)	71 (69-72)	3.0	(2.3, 3.6)	40 (38-42)	45 (44-47)	48 (47-50)	51 (49-52)	3.4	(2.6, 4.3)
Prostate	Men	65 (64-65)	69 (68-69)	76 (75-76)	76 (76-77)	4.1	(3.8, 4.3)	31 (30-32)	37 (36-37)	41 (40-41)	41 (41-42)	3.5	(3.2, 3.7)
Testis	Men	82 (81-84)	87 (86-88)	94 (94-95)	95 (94-95)	3.8	(3.4, 4.3)	69 (67-71)	78 (77-79)	88 (87-89)	90 (89-91)	6.6	(6.1, 7.2)
Penis	Men	76 (73-79)	79 (77-82)	80 (77-82)	83 (81-85)	2.2	(1.0, 3.3)	60 (56-64)	61 (57-65)	63 (59-67)	67 (64-71)	2.4	(0.6, 4.1)
Bladder	Men	66 (65-67)	71 (71-72)	77 (77-78)	80 (79-80)	4.6	(4.3, 4.8)	44 (43-45)	53 (52-54)	59 (58-60)	62 (61-63)	6.0	(5.6, 6.3)
	Women	59 (58-60)	63 (62-64)	70 (69-71)	72 (71-73)	4.5	(4.1, 4.9)	42 (40-43)	48 (47-50)	56 (55-57)	57 (56-58)	5.4	(4.9, 5.8)
Kidney	Men	45 (44-46)	47 (45-48)	52 (51-54)	56 (55-57)	3.8	(3.3, 4.4)	28 (27-29)	31 (29-32)	35 (34-36)	38 (36-39)	3.3	(2.8, 3.9)
	Women	45 (44-47)	45 (44-47)	51 (49-52)	53 (52-54)	2.9	(2.3, 3.6)	28 (27-30)	29 (28-31)	33 (32-34)	35 (34-37)	2.6	(1.9, 3.2)
Eye	Men	87 (84-91)	88 (85-91)	90 (87-92)	92 (90-95)	1.9	(0.6, 3.1)	60 (55-65)	64 (59-69)	65 (60-70)	67 (63-71)	2.2	(0.2, 4.2)
	Women	89 (86-91)	92 (89-94)	92 (89-94)	93 (91-95)	1.2	(0.2, 2.2)	60 (56-64)	63 (59-67)	66 (62-70)	70 (66-73)	3.2	(1.4, 5.0)
Brain	Men	19 (18-20)	22 (21-23)	25 (24-26)	28 (28-29)	3.2	(2.8, 3.6)	7 (6-8)	9 (8-10)	11 (11-12)	13 (12-14)	1.9	(1.6, 2.3)
	Women	20 (19-21)	23 (22-24)	26 (25-27)	30 (29-31)	3.3	(2.8, 3.8)	9 (8-10)	10 (9-11)	12 (11-13)	15 (15-16)	2.0	(1.6, 2.4)
Thyroid	Men	62 (59-65)	62 (60-65)	72 (69-75)	74 (72-77)	4.7	(3.5, 6.0)	50 (46-53)	52 (49-55)	59 (56-62)	64 (61-67)	5.0	(3.5, 6.4)
	Women	65 (64-67)	69 (67-70)	74 (73-76)	77 (75-78)	4.0	(3.3, 4.7)	56 (55-58)	60 (59-62)	66 (65-68)	70 (68-72)	4.7	(4.0, 5.4)
Spinal cord	Men	63 (54-72)	72 (63-82)	72 (64-79)	85 (78-92)	6.9	(3.3, 10.5)	48 (37-59)	57 (46-67)	56 (47-65)	67 (58-76)	5.7	(1.3, 10.1)
	Women	65 (52-78)	77 (67-87)	81 (73-90)	79 (70-88)	3.8	(-0.9, 8.5)	42 (29-55)	56 (44-68)	66 (57-76)	62 (52-72)	6.4	(1.3, 11.6)
Adrenal	Men	25 (17-34)	29 (21-38)	41 (33-48)	52 (44-60)	9.3	(5.6, 12.9)	13 (6-20)	24 (16-33)	28 (20-35)	33 (26-41)	6.6	(3.4, 9.8)
	Women	38 (29-47)	35 (27-43)	38 (30-47)	55 (47-62)	6.0	(2.4, 9.6)	22 (14-31)	26 (17-34)	25 (17-32)	37 (29-45)	4.4	(0.8, 8.1)
Pituitary	Men	83 (75-90)	83 (76-91)	91 (83-98)	85 (79-92)	1.4	(-1.9, 4.6)	67 (57-77)	73 (65-81)	74 (64-84)	73 (65-81)	1.8	(-2.1, 5.7)
	Women	73 (64-81)	77 (68-85)	83 (74-93)	87 (81-93)	4.9	(1.6, 8.3)	58 (50-67)	68 (60-77)	75 (65-85)	72 (63-81)	4.7	(0.9, 8.5)
Non-Hodgkin lymphoma	Men	49 (47-50)	52 (51-53)	59 (58-60)	62 (61-63)	4.7	(4.2, 5.2)	27 (26-29)	32 (31-33)	37 (36-38)	41 (40-42)	4.6	(4.1, 5.2)
	Women	51 (49-52)	55 (54-56)	62 (61-63)	65 (64-66)	4.9	(4.4, 5.4)	31 (30-32)	37 (35-38)	41 (40-42)	45 (44-46)	4.5	(4.0, 5.0)
Hodgkin's disease	Men	75 (74-76)	78 (77-79)	84 (83-85)	87 (86-88)	4.1	(3.5, 4.6)	55 (54-57)	61 (59-62)	67 (65-68)	71 (70-73)	5.4	(4.7, 6.1)
	Women	79 (77-80)	81 (80-82)	85 (83-86)	86 (85-88)	2.6	(2.0, 3.2)	60 (58-62)	65 (63-67)	72 (70-74)	73 (71-75)	4.5	(3.6, 5.3)
Multiple myeloma	Men	32 (31-34)	38 (36-39)	46 (45-48)	51 (50-53)	6.5	(5.8, 7.1)	10 (9-11)	12 (11-14)	15 (14-16)	17 (16-18)	2.2	(1.7, 2.7)
	Women	38 (36-40)	43 (42-45)	51 (50-53)	55 (54-56)	5.9	(5.2, 6.5)	11 (10-13)	14 (13-16)	17 (16-19)	19 (18-20)	2.7	(2.1, 3.2)
Acute lymphoid leukaemia	Men	30 (26-34)	41 (37-44)	44 (41-48)	50 (47-53)	6.4	(4.8, 7.9)	10 (8-13)	13 (11-16)	23 (19-26)	22 (19-25)	4.4	(3.1, 5.6)
	Women	31 (26-35)	45 (40-49)	48 (44-52)	54 (50-58)	7.1	(5.2, 9.1)	11 (8-15)	20 (16-23)	21 (17-25)	29 (25-33)	5.5	(3.8, 7.2)
Chronic lymphoid leukaemia	Men	56 (54-58)	57 (55-59)	68 (66-69)	73 (72-75)	6.3	(5.5, 7.1)	26 (24-28)	31 (29-34)	41 (39-43)	44 (42-46)	6.2	(5.2, 7.2)
	Women	61 (59-63)	65 (63-67)	73 (71-75)	77 (75-79)	5.5	(4.6, 6.4)	33 (30-35)	39 (36-41)	49 (46-51)	52 (50-54)	6.7	(5.7, 7.8)
Acute myeloid leukaemia	Men	10 (9-11)	15 (13-16)	20 (19-21)	25 (23-26)	5.0	(4.4, 5.5)	2 (1-3)	3 (3-4)	7 (6-7)	8 (7-9)	2.1	(1.7, 2.4)
	Women	10 (9-11)	15 (14-16)	21 (20-23)	25 (23-26)	5.2	(4.6, 5.8)	2 (1-2)	4 (3-5)	8 (7-9)	9 (8-10)	2.6	(2.3, 3.0)
Chronic myeloid leukaemia	Men	43 (40-46)	44 (42-47)	54 (51-57)	59 (56-61)	5.8	(4.5, 7.0)	10 (8-12)	12 (9-14)	15 (13-17)	20 (17-22)	3.3	(2.3, 4.3)
	Women	47 (44-50)	47 (45-50)	55 (52-58)	60 (58-63)	4.8	(3.5, 6.0)	11 (9-12)	12 (10-14)	16 (14-18)	23 (20-25)	3.9	(2.9, 4.8)
Monocytic leukaemia	Men	8 (6-11)	17 (13-22)	15 (11-18)	18 (13-23)	3.1	(1.4, 4.7)	4 (0-8)	5 (2-8)	6 (3-9)	11 (6-17)	1.9	(0.1, 3.8)
	Women	10 (7-13)	13 (9-17)	13 (9-18)	17 (12-23)	2.1	(0.2, 4.1)	4 (1-7)	5 (3-8)	4 (1-6)	7 (3-11)	0.5	(-1.0, 1.9)
All leukaemias	Men	33 (32-34)	37 (36-38)	45 (44-46)	50 (49-51)	5.9	(5.5, 6.4)	12 (12-13)	16 (15-17)	22 (21-23)	24 (23-25)	4.2	(3.8, 4.6)
	Women	32 (31-33)	37 (36-38)	44 (43-45)	49 (47-50)	5.6	(5.1, 6.1)	13 (12-13)	17 (16-18)	23 (22-24)	26 (25-27)	4.6	(4.2, 5.0)

Table 4.4 (cont.)   Cancer survival trends in England and Wales, by period of diagnosis and sex:
relative survival at one and five years, with 95% confidence intervals (CI),
and average increase in relative survival between successive five-year periods of diagnosis

CHILDREN[1]		One-year survival (%), patients diagnosed in								Average increase every five years		Five-year survival (%), patients diagnosed in								Average increase every five years	
		1971-75		1976-80		1981-85		1986-90				1971-75		1976-80		1981-85		1986-90			
		%	CI	%	CI	%	CI	%	CI	%	CI	%	CI	%	CI	%	CI	%	CI	%	CI
All leukaemias	Boys	65	(62-68)	74	(71-76)	81	(78-83)	88	(86-90)	7.5	(6.4, 8.6)	30	(28-33)	45	(42-48)	61	(58-64)	66	(63-69)	12.6	(11.3, 13.9)
	Girls	67	(64-71)	76	(73-79)	79	(76-82)	85	(82-87)	5.4	(4.1, 6.7)	36	(33-39)	48	(44-51)	62	(59-66)	72	(69-75)	12.1	(10.6, 13.6)
Acute lymphoid leukaemia	Boys	75	(71-78)	81	(78-84)	88	(85-90)	94	(92-96)	6.4	(5.4, 7.5)	36	(33-40)	51	(47-54)	69	(66-72)	73	(70-76)	12.8	(11.3, 14.2)
	Girls	78	(74-81)	84	(81-87)	87	(85-90)	92	(89-94)	4.4	(3.1, 5.7)	46	(42-50)	54	(50-58)	71	(67-75)	79	(76-82)	11.5	(9.8, 13.2)
Hodgkin's disease	Boys	94	(90-98)	96	(94-99)	98	(96-100)	100	(99-100)	2.0	(0.9, 3.0)	78	(72-84)	91	(86-95)	89	(84-93)	96	(92-99)	4.4	(2.3, 6.5)
	Girls	93	(86-99)	89	(82-96)	97	(93-100)	97	(92-100)	1.9	(-0.5, 4.4)	81	(71-90)	75	(66-85)	91	(85-98)	85	(76-94)	3.2	(-1.0, 7.3)
Non-Hodgkin lymphoma	Boys	41	(35-48)	67	(61-73)	78	(72-84)	84	(79-89)	13.2	(10.6, 15.8)	25	(19-31)	45	(39-52)	67	(61-74)	76	(70-82)	17.5	(14.8, 20.1)
	Girls	53	(44-63)	60	(50-70)	82	(74-91)	86	(79-94)	11.9	(8.1, 15.7)	28	(20-37)	37	(27-47)	71	(60-81)	77	(67-86)	17.6	(13.6, 21.6)
Brain and spinal cord	Boys	61	(57-64)	64	(60-67)	75	(71-78)	77	(74-81)	6.0	(4.4, 7.6)	43	(39-47)	45	(41-49)	59	(55-63)	57	(53-62)	5.7	(4.0, 7.5)
	Girls	59	(55-63)	64	(60-68)	74	(70-78)	74	(71-78)	5.5	(3.7, 7.3)	43	(38-47)	51	(47-55)	57	(53-62)	62	(57-66)	6.4	(4.5, 8.4)
Retinoblastoma	Boys	96	(92-100)	97	(92-100)	100	(95-100)	100	(96-100)	1.5	(-0.3, 3.2)	89	(83-96)	94	87-100)	91	(83-99)	98	95-100	2.8	(0.6, 5.0)
	Girls	93	(87-98)	94	(89-100)	96	(90-100)	96	(91-100)	1.2	(-1.3, 3.6)	86	(79-94)	85	(77-94)	90	(81-98)	88	(79-97)	1.0	(-2.7, 4.6)
Wilms' tumour	Boys	79	(73-85)	85	(79-90)	89	(84-95)	92	(87-97)	4.2	(1.7, 6.7)	63	(55-70)	76	(69-82)	81	(75-88)	84	(77-91)	6.8	(3.7, 9.9)
	Girls	73	(66-81)	87	(82-92)	88	(83-94)	96	(92-99)	6.3	(4.1, 8.5)	59	(51-67)	74	(67-81)	74	(67-81)	84	(78-90)	7.2	(4.1, 10.3)
Hepatoblastoma	Boys	42	(18-66)	14	(0-33)	48	(20-76)	55	(33-77)	8.5	(-1.7, 18.7)	13	(0-30)	7	(0-21)	48	(20-76)	39	(17-61)	11.3	(2.4, 20.1)
	Girls	15	(0-31)	24	(4-43)	44	(16-71)	66	(38-93)	16.2	(6.5, 25.8)	15	(0-31)	19	(2-37)	37	(11-63)	49	(21-78)	11.2	(1.4, 21.0)
Osteosarcoma	Boys	54	(45-64)	67	(55-78)	81	(71-91)	85	(77-94)	10.6	(6.4, 14.7)	18	(11-25)	22	(12-32)	55	(42-67)	47	(34-59)	11.9	(7.4, 16.3)
	Girls	57	(44-69)	69	(59-79)	82	(74-91)	91	(83-99)	11.4	(7.0, 15.8)	17	(7-26)	33	(23-43)	42	(30-53)	56	(43-69)	12.8	(7.8, 17.8)
Ewing's sarcoma	Boys	83	(71-94)	77	(67-87)	85	(76-94)	93	(86-100)	4.8	(0.8, 8.8)	37	(22-51)	32	(20-43)	35	(23-48)	69	(57-81)	11.3	(5.4, 17.1)
	Girls	74	(60-88)	87	(78-97)	91	(85-98)	83	(71-94)	2.9	(-2.5, 8.2)	30	(15-45)	43	(29-57)	40	(28-51)	50	(34-65)	5.3	(-1.4, 11.9)
Soft tissue sarcoma	Boys	61	(54-68)	75	(69-81)	81	(75-87)	87	(82-92)	7.8	(5.1, 10.5)	41	(34-49)	50	(43-57)	61	(53-68)	68	(61-75)	9.1	(5.9, 12.3)
	Girls	58	(50-67)	72	(64-80)	81	(74-87)	84	(78-90)	8.2	(5.0, 11.5)	40	(31-48)	52	(43-61)	58	(50-66)	64	(56-71)	7.8	(4.1, 11.4)
Germ-cell and gonadal neoplasms	Boys	68	(55-80)	84	(74-93)	85	(76-94)	94	(88-100)	7.6	(3.6, 11.5)	53	(40-67)	73	(61-85)	74	(64-85)	88	(80-96)	10.3	(5.6, 15.0)
	Girls	51	(38-65)	68	(57-80)	81	(70-91)	95	(90-100)	14.2	(10.1, 18.4)	40	(26-53)	56	(44-68)	79	(67-90)	78	(68-89)	13.4	(8.1, 18.7)

[1] Survival rates for children are not age-standardised (see text)

Relative survival from non-Hodgkin lymphoma has trebled, from 25-28% for children diagnosed in the early 1970s to 76-77% for those diagnosed during 1986-90, an average improvement of about 18% every five years. The data are also compatible with more rapid improvement in five-year survival, even though one-year survival also rose swiftly, by about 12-13% every five years for boys and girls. This provides supporting evidence for increases in cure[9].

Regional trends

Regional trends in survival are summarised for 17 of the most common malignancies in adults, and for acute lymphoid leukaemia in children, in Table-Figure 4.5. Differences in survival for patients diagnosed during 1986-90 are not generally large, and no region shows systematically lower or higher survival for all the common cancers.

Regional trends in survival since the early 1970s for some of the most common cancers do differ significantly from the overall trend in England and Wales, or from the trends in other regions. For some malignancies, the calendar period in which survival began to increase also differs between regions. These patterns are discussed in the chapters for each cancer.

Cancer survival rates

Type of cancer

For adults diagnosed during 1986-90 and followed up to the end of 1995, the highest five-year survival rates were for testis (90%), lip (85-86%) and for women, melanoma of the skin (82%).

The lowest five-year survival rates in both men and women were those for cancers of the oesophagus (5-8%), lung (5%), pleura (essentially mesothelioma) (4-5%), liver (2%) and pancreas (2%).

A broad summary of survival patterns in men and women may be obtained by grouping five-year relative survival rates into three bands: 50% or higher, 10% to 49%, and below 10% (Table 4.6).

Figure 4.4   Overall increase in relative survival[1] at five years between patients diagnosed in 1971-75 and those diagnosed in 1986-90[2]

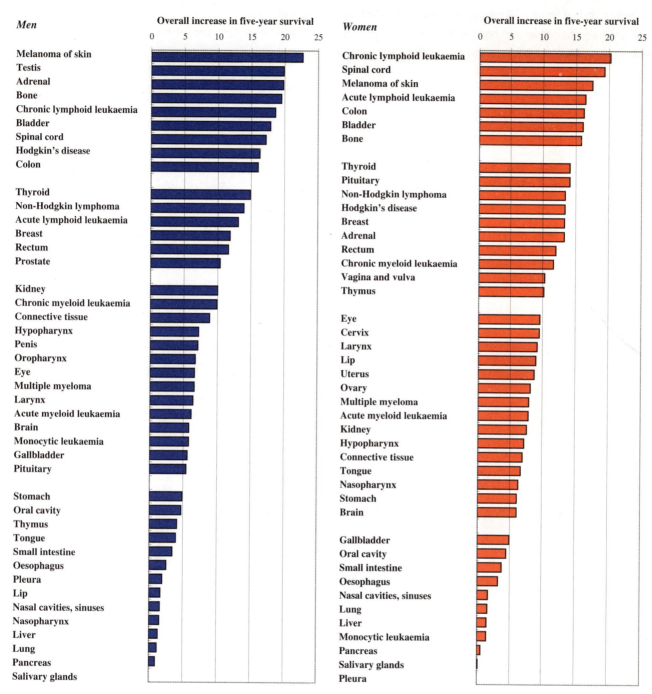

[1] Estimated from the linear regression of age-standardised relative survival rates between successive five year periods 1971-75 to 1986-90:
the overall increase is three times the average increase every five years (see Table 4.4).

[2] The overall change in five-year survival for cancers of the oropharynx in women was -1.4% (not shown above).

Table 4.4A

Cancer survival trends, all cancers combined in adults[1], England and Wales:
one-year and five-year survival rates[2] by sex and period of diagnosis, with 95% confidence intervals (CI),
and average increase between successive five-year periods

		Relative survival (%), adults diagnosed during								Average increase every five years	
		1971-75		1976-80		1981-85		1986-90			
		%	CI	%	CI	%	CI	%	CI	%	CI
Time since diagnosis											
One year	*Men*	37.0	(36.8-37.2)	39.8	(39.7-40.0)	46.4	(46.2-46.6)	50.9	(50.8-51.1)	4.9	(4.8, 4.9)
	Women	52.4	(52.2-52.6)	54.4	(54.2-54.6)	59.1	(58.9-59.2)	62.1	(62.0-62.3)	3.4	(3.3, 3.5)
Five years	*Men*	19.1	(18.9-19.3)	22.5	(22.4-22.7)	27.6	(27.4-27.8)	31.4	(31.2-31.6)	4.2	(4.1, 4.3)
	Women	31.7	(31.5-31.8)	35.2	(35.0-35.4)	39.7	(39.5-39.8)	43.4	(43.2-43.6)	4.0	(3.9, 4.0)

[1] All malignant neoplasms included in analyses: these represent about 93% of all cancers in adults (see text).

[2] Relative survival rate (%), standardised to the age distribution of adult cancer patients diagnosed during 1986-90 (see text).

Figure 4.4A Relative survival up to ten years after diagnosis, all cancers combined:
trends by calendar period of diagnosis (1971-90), England and Wales

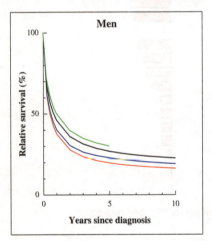
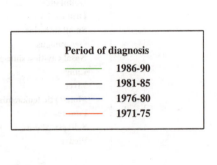
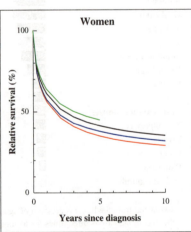

Relative survival at five years was 50% or higher for 13 different types of cancer among men diagnosed in England and Wales during 1986-90: together, these account for about 15% of all cancers in men. Among women, five-year survival was 50% or higher for 19 different types of cancer, but these accounted for 45% of all cancers occurring in women.

For 22 types of cancer in men (39% of all cancers) and for 17 cancers in women (29% of all cancers), five-year survival rates were in the intermediate range, between 10% and 49%.

For eight types of cancer in each sex, five-year survival rates for those diagnosed during 1986-90 were below 10%. The cancers with the lowest survival rates are largely the same cancers in both sexes (Table 4.6). These cancers account for about 18% of all malignant tumours in women, but 39% of all cancers occurring in men fall into this category:

Five-year survival bands, adults diagnosed 1986-90

	Men		Women	
Survival rate	Types of cancer	% of all cancers[1]	Types of cancer	% of all cancers[1]
50% or more	**13**	15	**19**	45
10% – 49%	**22**	39	**17**	29
Less than 10%	**8**	39	**8**	18
	43	93	**44**	93

[1] *Percentage of all cancers arising in men (or in women) in England and Wales: cancers not included in the analyses represent the other 7% of all malignancies.*

The corresponding summary of survival for cancers in children is much more favourable. For all but two of the 11 cancers examined, which in total represent over 80% of all childhood malignancies, the five-year survival rate for children diagnosed in England and Wales during 1986-90 was 50% or higher (Table 4.6).

For six of these cancers, representing about half of all malignancies in childhood (retinoblastoma, Hodgkin's disease, Wilms' tumour, acute lymphocytic leukaemia, non-Hodgkin lymphoma, and germ-cell and gonadal tumours combined), five-year survival rates were 73% or higher:

Five-year survival bands, children diagnosed 1986-90

	Boys		Girls	
Survival rate	Types of cancer	% of all cancers[1]	Types of cancer	% of all cancers[1]
50% or more	**9**	80	**10**	80
10% – 49%	**2**	3	**1**	1
Less than 10%	-	-	-	-
	11	83	**11**	81

[1] *Percentage of all cancers arising in boys (or in girls) in England and Wales: cancers not included in the analyses represent the other 17-19% of all childhood malignancies.*

Age at diagnosis

The survival rates for adults diagnosed in England and Wales during 1986-90 are shown for men and women for each of six age groups at diagnosis (15-39, 40-49, ... 70-79, 80-99) in Table 4.7, together with the 95% confidence intervals and the number of patients included in the analysis for each age group. Age-specific survival rates and confidence intervals are also presented graphically in the chapter for each cancer (Chapters 5-62).

Age-specific relative survival rates for almost all cancers are lower (often much lower) among elderly patients than among younger patients, even though they take account of general population mortality for the corresponding age group.

Breast cancer in women and prostate cancer are exceptions to the general pattern. Relative survival for the youngest group of patients is lower than for older patients, although the most elderly patients also have low survival. For breast cancer, women under 40 at diagnosis during 1986-90 had five-year survival of 66% (95% confidence interval 65% to 67%), some 7% lower than for women aged 40-49 (73%, 95%CI 72-74%) and similar to that for women aged 70-79 at diagnosis (see also Figure 28.5). Similar results have been published for Sweden[10] and for various other countries in Europe[11].

For prostate cancer, a similar relationship of survival with age was observed, but at a higher age: men aged 40-49 at diagnosis in 1986-90 had five-year relative survival of 31% (95%CI 24-38%), some 12% lower than for men aged 50-59 (see also Figure 33.5). Men aged 60-69 at diagnosis had the highest five-year survival, at 46%.

Thyroid cancer shows a particularly steep dependency of survival on age: relative survival at five years is 95-99% in patients aged 15-39 at diagnosis, but only 16% in women and 34% in men aged 80-99. This pattern is largely due to differences in the morphological types of thyroid cancer at different ages. Papillary carcinoma of the thyroid, for which survival is very high, is the most common type at

SUMMARY

Table 4.5 Cancer survival trends by NHS Region, selected cancers, patients diagnosed 1971-90:
age-standardised relative survival rates (with 95% confidence intervals) at one and five years after diagnosis,
and average increase in relative survival between successive five-year periods of diagnosis

Table 4.5 A

| | | One-year survival (%), patients diagnosed in | | | | Average increase every five years | Five-year survival (%), patients diagnosed in | | | | Average increase every five years |
| | | 1971-75 | 1976-80 | 1981-85 | 1986-90 | | 1971-75 | 1976-80 | 1981-85 | 1986-90 | |
		% CI	% CI	% CI	% CI	% CI	% CI	% CI	% CI	% CI	% CI

OESOPHAGUS

Region	Sex	1971-75	1976-80	1981-85	1986-90	Avg inc 5yr	1971-75	1976-80	1981-85	1986-90	Avg inc 5yr
ENGLAND & WALES	Men	12 (12-13)	13 (12-13)	17 (17-18)	21 (20-22)	3.1 (2.7, 3.4)	3 (3-3)	3 (3-4)	4 (4-5)	5 (5-6)	0.8 (0.7, 1.0)
	Women	16 (15-17)	17 (16-18)	23 (22-24)	25 (24-26)	3.2 (2.8, 3.6)	5 (5-6)	6 (5-6)	8 (7-8)	8 (7-9)	1.1 (0.8, 1.3)
ENGLAND	Men	12 (12-13)	13 (12-13)	17 (17-18)	21 (20-22)	3.1 (2.8, 3.4)	3 (2-3)	3 (3-4)	4 (4-5)	5 (5-6)	0.8 (0.6, 1.0)
	Women	16 (15-17)	17 (16-18)	23 (22-24)	25 (24-26)	3.2 (2.8, 3.6)	5 (5-6)	6 (5-6)	7 (7-8)	8 (7-9)	1.0 (0.7, 1.3)
Northern & Yorkshire	Men	10 (9-12)	11 (10-13)	14 (12-16)	18 (16-20)	2.6 (1.8, 3.3)	2 (1-2)	3 (2-3)	3 (2-5)	4 (3-5)	0.9 (0.5, 1.3)
	Women	14 (12-16)	16 (14-19)	19 (17-21)	21 (19-24)	2.5 (1.5, 3.5)	4 (3-6)	5 (4-6)	5 (3-6)	6 (4-8)	0.5 (-0.2, 1.1)
Trent	Men	14 (11-16)	11 (9-13)	16 (14-18)	19 (17-21)	2.3 (1.3, 3.2)	2 (1-3)	3 (2-4)	4 (2-5)	5 (4-7)	1.0 (0.5, 1.5)
	Women	18 (15-21)	19 (16-22)	23 (20-26)	24 (21-26)	1.9 (0.6, 3.3)	5 (3-8)	6 (4-8)	6 (4-8)	8 (6-10)	0.7 (-0.3, 1.6)
Anglia & Oxford	Men	14 (12-16)	15 (12-17)	18 (15-20)	20 (18-23)	2.2 (1.2, 3.3)	4 (3-6)	3 (1-4)	4 (2-5)	5 (3-6)	0.4 (-0.2, 1.0)
	Women	17 (14-21)	20 (16-23)	23 (20-27)	27 (24-30)	3.2 (1.7, 4.7)	7 (4-9)	6 (4-8)	10 (7-13)	10 (7-12)	1.3 (0.3, 2.4)
North Thames	Men	14 (12-16)	15 (13-17)	23 (20-25)	27 (24-30)	4.6 (3.5, 5.6)	4 (3-5)	5 (3-6)	6 (5-8)	7 (5-9)	1.1 (0.5, 1.7)
	Women	16 (14-19)	19 (16-22)	27 (23-30)	29 (26-33)	4.7 (3.3, 6.0)	6 (4-8)	6 (4-8)	7 (5-9)	9 (7-11)	1.0 (0.0, 1.9)
South Thames	Men	13 (12-15)	14 (12-16)	21 (18-23)	23 (21-26)	3.6 (2.6, 4.5)	3 (2-4)	4 (3-5)	5 (3-6)	4 (3-6)	0.6 (0.1, 1.1)
	Women	15 (12-17)	16 (14-18)	28 (25-31)	29 (26-32)	5.6 (4.4, 6.7)	3 (2-5)	4 (3-6)	10 (8-12)	8 (6-10)	2.1 (1.4, 2.8)
South & West	Men	12 (10-14)	14 (12-16)	19 (17-21)	25 (23-27)	4.4 (3.6, 5.3)	3 (2-4)	4 (3-5)	5 (4-6)	7 (5-8)	1.1 (0.5, 1.6)
	Women	19 (16-21)	18 (16-20)	26 (24-29)	28 (25-30)	3.6 (2.4, 4.7)	7 (5-9)	7 (5-8)	10 (8-12)	9 (7-11)	0.9 (0.1, 1.7)
West Midlands	Men	10 (8-12)	11 (9-12)	14 (12-16)	19 (17-21)	3.1 (2.2, 4.0)	2 (1-4)	3 (2-4)	5 (4-7)	6 (5-7)	1.3 (0.7, 1.9)
	Women	15 (12-18)	16 (14-19)	21 (18-23)	23 (21-26)	2.9 (1.7, 4.2)	5 (3-7)	6 (4-7)	7 (5-9)	7 (5-9)	0.8 (0.0, 1.5)
North & West	Men	11 (10-13)	11 (9-12)	15 (13-16)	18 (16-20)	2.4 (1.6, 3.1)	4 (3-5)	3 (2-4)	3 (2-4)	5 (4-6)	0.4 (-0.1, 0.8)
	Women	16 (14-18)	14 (12-16)	18 (16-21)	21 (19-23)	1.9 (1.0, 2.9)	5 (4-7)	5 (4-6)	5 (4-7)	7 (6-9)	0.5 (-0.1, 1.1)
WALES	Men	13 (10-15)	16 (13-18)	20 (17-23)	20 (17-23)	2.6 (1.4, 3.8)	3 (1-4)	4 (2-5)	7 (5-10)	6 (4-8)	1.3 (0.6, 2.0)
	Women	15 (12-17)	17 (14-20)	23 (19-27)	23 (20-27)	3.2 (1.8, 4.7)	5 (3-7)	8 (6-10)	10 (7-13)	7 (5-10)	1.0 (0.1, 2.0)

Table 4.5 B

STOMACH

Region	Sex	1971-75	1976-80	1981-85	1986-90	Avg inc 5yr	1971-75	1976-80	1981-85	1986-90	Avg inc 5yr
ENGLAND & WALES	Men	13 (12-13)	14 (13-14)	20 (19-20)	23 (23-24)	3.6 (3.5, 3.8)	4 (4-4)	5 (5-5)	7 (7-8)	9 (8-9)	1.6 (1.5, 1.8)
	Women	15 (15-16)	16 (16-17)	22 (21-23)	26 (25-26)	3.5 (3.3, 3.8)	5 (5-5)	6 (6-6)	9 (8-9)	11 (10-12)	2.0 (1.8, 2.2)
ENGLAND	Men	13 (12-13)	14 (13-14)	20 (19-20)	23 (23-24)	3.7 (3.5, 3.9)	4 (4-4)	5 (5-5)	7 (7-8)	9 (8-9)	1.6 (1.5, 1.8)
	Women	15 (15-16)	16 (16-17)	22 (21-22)	26 (25-27)	3.6 (3.3, 3.8)	5 (4-5)	6 (5-6)	9 (8-9)	11 (10-11)	2.0 (1.8, 2.2)
Northern & Yorkshire	Men	11 (10-12)	13 (13-14)	18 (17-19)	21 (20-22)	3.3 (2.9, 3.8)	3 (3-4)	5 (4-5)	7 (6-8)	8 (7-9)	1.7 (1.3, 2.0)
	Women	15 (14-16)	15 (14-16)	22 (21-24)	24 (22-26)	3.4 (2.8, 4.1)	4 (4-5)	5 (4-6)	9 (8-10)	10 (9-11)	2.1 (1.6, 2.5)
Trent	Men	12 (11-13)	12 (11-13)	19 (18-21)	22 (21-24)	3.7 (3.2, 4.3)	4 (3-5)	5 (4-5)	7 (6-8)	8 (7-9)	1.4 (1.0, 1.8)
	Women	15 (14-17)	17 (15-18)	21 (19-23)	25 (23-28)	3.4 (2.6, 4.2)	5 (4-6)	7 (5-8)	8 (7-10)	12 (10-14)	2.2 (1.6, 2.8)
Anglia & Oxford	Men	15 (14-16)	16 (14-17)	21 (19-22)	23 (21-24)	2.9 (2.3, 3.5)	5 (4-6)	5 (4-6)	7 (6-8)	8 (7-9)	0.9 (0.5, 1.3)
	Women	18 (16-20)	18 (16-20)	22 (20-24)	25 (22-27)	2.4 (1.5, 3.3)	7 (5-8)	7 (6-8)	8 (7-10)	9 (7-11)	0.9 (0.2, 1.5)
North Thames	Men	15 (14-16)	17 (15-18)	24 (22-26)	31 (29-33)	5.2 (4.5, 5.8)	5 (5-6)	7 (6-8)	10 (9-12)	11 (10-13)	2.1 (1.6, 2.6)
	Women	18 (16-19)	22 (20-23)	27 (25-30)	31 (28-33)	4.5 (3.6, 5.4)	7 (6-8)	8 (7-9)	12 (10-14)	13 (11-15)	2.2 (1.5, 2.9)
South Thames	Men	13 (12-14)	14 (13-15)	22 (21-24)	27 (25-28)	4.7 (4.1, 5.3)	3 (3-4)	5 (4-5)	8 (7-10)	9 (8-10)	2.2 (1.8, 2.6)
	Women	15 (14-17)	16 (15-17)	25 (23-27)	29 (27-32)	4.8 (3.9, 5.6)	5 (4-5)	6 (5-7)	10 (9-12)	12 (10-14)	2.6 (2.1, 3.2)
South & West	Men	14 (13-15)	15 (13-16)	22 (21-24)	25 (24-27)	4.1 (3.5, 4.7)	5 (4-6)	5 (5-6)	9 (8-11)	10 (9-11)	1.9 (1.4, 2.3)
	Women	14 (13-16)	18 (17-20)	24 (22-26)	29 (27-31)	4.8 (4.0, 5.6)	5 (4-6)	7 (6-9)	10 (9-12)	14 (12-16)	2.9 (2.3, 3.5)
West Midlands	Men	12 (11-13)	14 (13-15)	20 (18-21)	21 (19-22)	3.1 (2.6, 3.6)	4 (3-4)	4 (4-5)	7 (6-8)	9 (8-10)	1.7 (1.3, 2.1)
	Women	17 (15-18)	16 (14-17)	20 (18-22)	25 (23-27)	2.7 (1.9, 3.5)	5 (4-6)	5 (4-6)	8 (7-9)	10 (9-12)	1.6 (1.1, 2.2)
North & West	Men	10 (9-11)	12 (11-13)	16 (15-17)	20 (18-21)	3.1 (2.7, 3.6)	3 (2-3)	4 (3-4)	5 (5-6)	7 (6-8)	1.5 (1.2, 1.8)
	Women	13 (12-14)	14 (12-15)	18 (17-19)	23 (21-25)	3.2 (2.6, 3.7)	4 (3-4)	4 (4-5)	7 (6-8)	10 (9-11)	2.0 (1.6, 2.4)
WALES	Men	12 (11-14)	15 (14-16)	20 (19-22)	21 (19-22)	3.1 (2.4, 3.8)	4 (3-5)	7 (6-9)	8 (7-10)	8 (6-9)	1.3 (0.8, 1.8)
	Women	15 (13-16)	18 (16-19)	23 (21-25)	22 (20-25)	3.0 (2.1, 4.0)	6 (5-8)	8 (7-10)	10 (9-12)	12 (10-15)	2.0 (1.2, 2.8)

OESOPHAGUS   Figure 4.5A

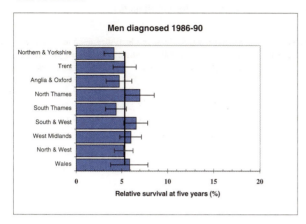

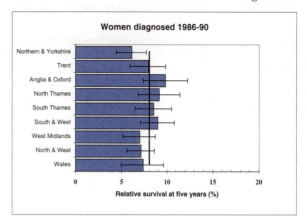

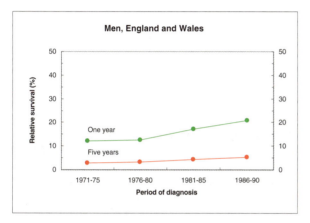

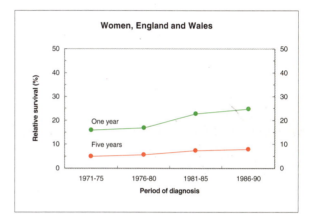

STOMACH   Figure 4.5B

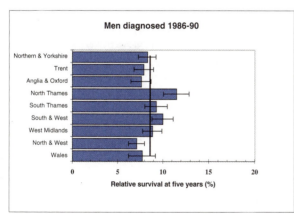

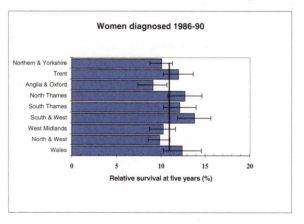

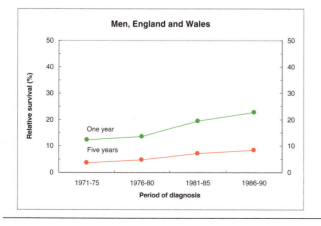

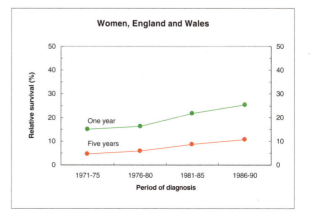

Table 4.5 (cont.) Cancer survival trends by NHS Region, selected cancers, patients diagnosed 1971-90: age-standardised relative survival rates (with 95% confidence intervals) at one and five years after diagnosis, and average increase in relative survival between successive five-year periods of diagnosis

Table 4.5 C

| | | One-year survival (%), patients diagnosed in | | | | | | | | Average increase every five years | | Five-year survival (%), patients diagnosed in | | | | | | | | Average increase every five years | |
| | | 1971-75 | | 1976-80 | | 1981-85 | | 1986-90 | | | | 1971-75 | | 1976-80 | | 1981-85 | | 1986-90 | | | |
		%	CI	%	CI	%	CI	%	CI	%	CI	%	CI	%	CI	%	CI	%	CI	%	CI
COLON																					
ENGLAND & WALES	*Men*	**39**	**(39-40)**	**45**	**(44-45)**	**54**	**(54-55)**	**59**	**(58-59)**	**6.7**	**(6.4, 7.0)**	**22**	**(22-23)**	**28**	**(27-29)**	**35**	**(34-36)**	**38**	**(37-39)**	**5.3**	**(5.0, 5.6)**
	Women	**40**	**(40-41)**	**45**	**(45-46)**	**54**	**(54-55)**	**59**	**(59-60)**	**6.5**	**(6.3, 6.8)**	**23**	**(23-24)**	**28**	**(28-29)**	**35**	**(34-36)**	**39**	**(39-40)**	**5.4**	**(5.1, 5.7)**
ENGLAND	*Men*	39	(39-40)	45	(44-45)	54	(54-55)	59	(58-59)	6.8	(6.5, 7.1)	22	(22-23)	28	(27-29)	35	(34-36)	38	(37-39)	5.4	(5.1, 5.7)
	Women	40	(40-41)	45	(45-46)	54	(54-55)	59	(59-60)	6.6	(6.4, 6.9)	23	(23-24)	28	(28-29)	35	(34-36)	39	(39-40)	5.5	(5.2, 5.7)
Northern & Yorkshire	*Men*	36	(34-38)	39	(37-41)	53	(52-55)	56	(54-58)	7.4	(6.6, 8.1)	20	(18-22)	25	(23-27)	35	(33-37)	36	(34-38)	5.9	(5.1, 6.8)
	Women	37	(36-39)	44	(42-45)	52	(51-54)	56	(55-58)	6.7	(6.0, 7.3)	20	(19-21)	27	(25-28)	34	(32-35)	37	(36-39)	6.0	(5.3, 6.6)
Trent	*Men*	37	(35-39)	43	(41-45)	54	(52-56)	57	(55-59)	7.0	(6.2, 7.9)	21	(19-23)	26	(24-28)	33	(30-35)	37	(35-39)	5.5	(4.5, 6.5)
	Women	38	(37-40)	45	(43-47)	52	(50-54)	59	(57-60)	6.8	(6.0, 7.5)	22	(20-24)	29	(28-31)	33	(31-34)	39	(37-40)	5.3	(4.5, 6.1)
Anglia & Oxford	*Men*	43	(41-45)	48	(46-50)	55	(53-57)	56	(54-58)	4.5	(3.6, 5.3)	26	(23-28)	30	(28-32)	36	(34-38)	36	(34-38)	3.5	(2.5, 4.5)
	Women	43	(41-45)	47	(45-49)	58	(56-59)	57	(55-58)	5.0	(4.3, 5.8)	25	(24-27)	29	(27-31)	39	(37-41)	38	(37-40)	4.9	(4.1, 5.7)
North Thames	*Men*	45	(42-47)	51	(49-53)	61	(58-63)	64	(62-66)	6.8	(5.8, 7.7)	27	(25-30)	33	(30-36)	42	(39-45)	41	(39-44)	5.0	(3.9, 6.1)
	Women	46	(44-48)	50	(48-52)	62	(60-64)	65	(64-67)	6.9	(6.2, 7.6)	29	(27-30)	33	(31-34)	42	(39-44)	43	(41-45)	5.1	(4.3, 5.9)
South Thames	*Men*	39	(37-41)	45	(43-47)	59	(57-60)	64	(62-65)	8.7	(7.9, 9.4)	21	(19-23)	29	(27-31)	39	(36-41)	41	(39-43)	7.1	(6.2, 7.9)
	Women	41	(40-42)	45	(43-46)	56	(55-58)	65	(63-66)	8.3	(7.6, 8.9)	23	(21-24)	27	(26-29)	38	(36-39)	43	(41-45)	7.0	(6.4, 7.7)
South & West	*Men*	44	(42-46)	51	(49-53)	59	(57-60)	64	(62-65)	6.7	(5.9, 7.4)	27	(25-29)	35	(32-37)	39	(37-41)	43	(41-45)	5.2	(4.3, 6.0)
	Women	45	(43-46)	51	(49-52)	59	(58-61)	64	(63-65)	6.6	(5.9, 7.2)	29	(27-30)	34	(32-35)	40	(38-42)	45	(43-46)	5.5	(4.8, 6.3)
West Midlands	*Men*	41	(39-43)	46	(44-48)	52	(50-53)	57	(56-59)	5.5	(4.6, 6.3)	24	(22-27)	30	(27-32)	31	(29-33)	38	(35-40)	4.2	(3.2, 5.2)
	Women	41	(39-43)	45	(43-46)	52	(51-54)	58	(56-59)	5.8	(5.0, 6.5)	25	(23-26)	28	(26-30)	33	(32-35)	40	(38-41)	5.0	(4.2, 5.8)
North & West	*Men*	32	(31-34)	36	(34-37)	47	(45-48)	53	(52-55)	7.5	(6.8, 8.2)	18	(16-19)	21	(19-23)	28	(27-30)	34	(32-36)	5.5	(4.7, 6.3)
	Women	34	(33-35)	40	(39-41)	48	(47-50)	54	(52-55)	6.7	(6.1, 7.3)	20	(18-21)	25	(23-26)	31	(29-32)	35	(33-37)	5.3	(4.6, 5.9)
WALES	*Men*	40	(37-42)	45	(43-48)	53	(51-56)	51	(48-54)	4.3	(3.0, 5.5)	23	(20-27)	30	(27-33)	34	(31-37)	35	(32-39)	4.0	(2.6, 5.5)
	Women	41	(39-43)	46	(43-48)	54	(51-56)	53	(51-56)	4.6	(3.6, 5.6)	25	(23-27)	30	(28-33)	35	(32-38)	37	(35-40)	4.3	(3.1, 5.4)

Table 4.5 D

| | | One-year survival (%), patients diagnosed in | | | | | | | | Average increase every five years | | Five-year survival (%), patients diagnosed in | | | | | | | | Average increase every five years | |
| | | 1971-75 | | 1976-80 | | 1981-85 | | 1986-90 | | | | 1971-75 | | 1976-80 | | 1981-85 | | 1986-90 | | | |
		%	CI	%	CI	%	CI	%	CI	%	CI	%	CI	%	CI	%	CI	%	CI	%	CI
RECTUM																					
ENGLAND & WALES	*Men*	**50**	**(50-51)**	**54**	**(53-55)**	**61**	**(61-62)**	**65**	**(64-65)**	**5.0**	**(4.6, 5.3)**	**25**	**(24-26)**	**29**	**(29-30)**	**34**	**(33-35)**	**36**	**(36-37)**	**3.9**	**(3.5, 4.2)**
	Women	**51**	**(51-52)**	**55**	**(54-56)**	**62**	**(61-62)**	**66**	**(65-66)**	**5.0**	**(4.6, 5.3)**	**27**	**(27-28)**	**31**	**(30-32)**	**36**	**(35-37)**	**39**	**(38-40)**	**4.0**	**(3.6, 4.3)**
ENGLAND	*Men*	51	(50-51)	54	(53-55)	62	(61-62)	65	(64-66)	5.0	(4.7, 5.3)	25	(25-26)	29	(29-30)	34	(34-35)	37	(36-37)	3.8	(3.5, 4.2)
	Women	52	(51-52)	55	(54-56)	62	(61-63)	66	(65-67)	5.0	(4.7, 5.4)	28	(27-28)	31	(30-32)	36	(35-37)	39	(38-40)	4.0	(3.7, 4.4)
Northern & Yorkshire	*Men*	48	(46-50)	53	(51-55)	60	(58-61)	63	(61-65)	5.0	(4.2, 5.9)	25	(22-27)	29	(27-31)	32	(30-34)	35	(33-37)	3.4	(2.4, 4.3)
	Women	50	(47-52)	54	(52-56)	61	(59-63)	64	(62-66)	5.0	(4.1, 5.9)	26	(24-29)	29	(27-31)	36	(34-38)	38	(36-41)	4.4	(3.4, 5.4)
Trent	*Men*	51	(49-53)	53	(51-55)	61	(59-63)	64	(62-66)	4.7	(3.7, 5.6)	25	(23-28)	27	(24-29)	34	(32-37)	36	(33-38)	3.8	(2.8, 4.8)
	Women	49	(47-52)	53	(51-56)	59	(57-61)	63	(61-65)	4.7	(3.6, 5.7)	27	(24-29)	30	(27-32)	35	(33-37)	38	(36-40)	4.0	(2.9, 5.1)
Anglia & Oxford	*Men*	56	(54-58)	55	(53-58)	63	(61-65)	66	(64-68)	3.7	(2.7, 4.7)	29	(26-32)	31	(28-34)	38	(35-40)	38	(35-40)	3.3	(2.2, 4.4)
	Women	55	(53-58)	57	(54-59)	64	(62-67)	66	(64-68)	4.0	(2.9, 5.1)	30	(27-32)	33	(31-36)	40	(37-42)	43	(40-45)	4.5	(3.3, 5.7)
North Thames	*Men*	56	(53-58)	58	(55-60)	67	(64-69)	70	(68-72)	5.3	(4.3, 6.3)	30	(28-33)	34	(31-37)	40	(36-43)	38	(36-41)	2.8	(1.6, 4.0)
	Women	58	(55-60)	60	(58-62)	69	(67-72)	72	(70-74)	5.3	(4.3, 6.2)	32	(29-34)	36	(33-38)	43	(40-46)	43	(40-45)	4.0	(2.8, 5.1)
South Thames	*Men*	51	(49-53)	55	(53-57)	65	(63-67)	69	(67-71)	6.5	(5.6, 7.3)	25	(23-27)	31	(29-33)	37	(34-39)	39	(36-41)	4.8	(3.8, 5.8)
	Women	53	(51-55)	57	(55-59)	64	(62-66)	70	(68-71)	5.7	(4.9, 6.6)	26	(25-28)	32	(30-34)	36	(34-38)	40	(38-42)	4.5	(3.5, 5.4)
South & West	*Men*	54	(51-56)	59	(58-61)	65	(63-67)	67	(65-68)	4.4	(3.6, 5.3)	29	(27-31)	34	(32-36)	39	(36-41)	39	(37-41)	3.5	(2.5, 4.5)
	Women	55	(52-57)	59	(57-61)	68	(66-69)	69	(67-71)	5.1	(4.2, 6.0)	31	(29-34)	34	(32-36)	42	(39-44)	43	(41-45)	4.3	(3.3, 5.3)
West Midlands	*Men*	50	(48-53)	54	(52-56)	61	(59-63)	64	(63-66)	4.9	(4.0, 5.8)	25	(23-27)	29	(26-31)	35	(32-37)	37	(34-39)	4.1	(3.0, 5.1)
	Women	51	(49-54)	54	(52-57)	61	(59-63)	65	(63-67)	4.9	(3.9, 5.8)	28	(26-31)	31	(28-33)	34	(32-36)	39	(36-41)	3.4	(2.3, 4.5)
North & West	*Men*	43	(42-45)	48	(46-50)	56	(54-58)	59	(58-61)	5.6	(4.7, 6.4)	20	(18-22)	26	(24-28)	29	(27-31)	33	(31-35)	4.2	(3.3, 5.1)
	Women	46	(44-48)	50	(48-52)	56	(55-58)	61	(59-63)	5.2	(4.4, 6.1)	24	(22-26)	28	(26-30)	31	(29-33)	35	(33-37)	3.5	(2.7, 4.4)
WALES	*Men*	48	(45-50)	52	(50-55)	56	(54-59)	59	(57-62)	3.9	(2.7, 5.1)	23	(20-26)	30	(26-33)	33	(30-36)	36	(33-39)	4.3	(3.0, 5.7)
	Women	48	(45-51)	54	(51-57)	57	(54-59)	63	(60-65)	4.6	(3.3, 5.9)	26	(23-29)	32	(29-35)	36	(33-39)	38	(35-41)	4.0	(2.6, 5.4)

COLON

Figure 4.5C

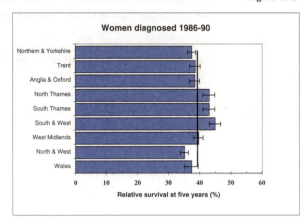

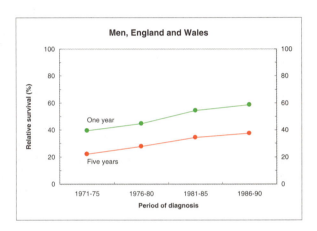

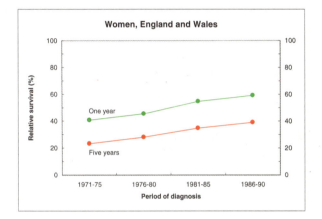

RECTUM

Figure 4.5D

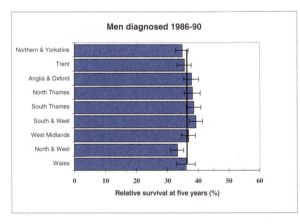

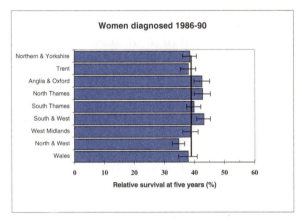

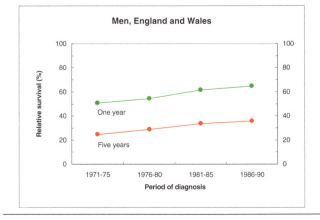

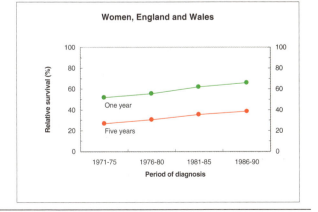

SUMMARY

Table 4.5 (cont.) Cancer survival trends by NHS Region, selected cancers, patients diagnosed 1971-90: age-standardised relative survival rates (with 95% confidence intervals) at one and five years after diagnosis, and average increase in relative survival between successive five-year periods of diagnosis

Table 4.5 E

	One-year survival (%), patients diagnosed in								Average increase every five years		Five-year survival (%), patients diagnosed in								Average increase every five years	
	1971-75		1976-80		1981-85		1986-90				1971-75		1976-80		1981-85		1986-90			
	%	CI	%	CI	%	CI	%	CI	%	CI	%	CI	%	CI	%	CI	%	CI	%	CI
PANCREAS																				
ENGLAND & WALES Men	6	(6-6)	7	(6-7)	8	(8-8)	9	(9-10)	1.1	(0.9, 1.3)	2	(1-2)	2	(1-2)	2	(2-2)	2	(2-3)	0.3	(0.2, 0.4)
Women	7	(7-8)	7	(7-8)	9	(9-10)	10	(9-10)	0.9	(0.6, 1.1)	2	(2-2)	2	(1-2)	2	(2-2)	2	(2-3)	0.2	(0.1, 0.3)
ENGLAND Men	6	(6-6)	7	(6-7)	8	(7-8)	9	(9-10)	1.1	(0.9, 1.3)	1	(1-2)	2	(1-2)	2	(2-2)	2	(2-3)	0.3	(0.2, 0.4)
Women	7	(7-8)	7	(7-8)	9	(9-10)	10	(9-10)	0.9	(0.6, 1.1)	2	(1-2)	2	(1-2)	2	(2-2)	2	(2-3)	0.2	(0.1, 0.3)
Northern & Yorkshire Men	6	(5-6)	6	(5-7)	8	(7-9)	9	(8-11)	1.4	(0.9, 1.9)	1	(1-2)	2	(1-2)	2	(1-3)	2	(1-3)	0.2	(0.0, 0.5)
Women	7	(6-8)	6	(5-7)	10	(8-11)	9	(7-10)	0.9	(0.3, 1.5)	2	(1-2)	1	(1-2)	2	(1-2)	2	(1-2)	0.1	(-0.2, 0.4)
Trent Men	7	(6-8)	8	(6-9)	8	(7-9)	8	(7-10)	0.4	(-0.1, 1.0)	1	(1-2)	2	(2-3)	2	(1-3)	2	(2-3)	0.2	(-0.1, 0.6)
Women	8	(6-9)	7	(6-9)	7	(5-8)	8	(7-10)	0.1	(-0.6, 0.7)	2	(1-3)	2	(1-3)	1	(1-2)	2	(1-3)	-0.1	(-0.5, 0.2)
Anglia & Oxford Men	5	(4-6)	6	(5-7)	7	(6-9)	9	(8-10)	1.2	(0.7, 1.8)	2	(1-2)	1	(1-2)	2	(1-3)	2	(1-3)	0.3	(-0.1, 0.6)
Women	8	(6-9)	8	(6-9)	9	(7-10)	8	(7-10)	0.2	(-0.5, 0.9)	2	(1-3)	1	(1-2)	2	(1-3)	2	(1-2)	-0.1	(-0.4, 0.3)
North Thames Men	7	(6-9)	8	(6-9)	11	(9-13)	14	(12-16)	2.2	(1.5, 2.9)	2	(1-3)	2	(1-3)	4	(2-5)	4	(3-6)	0.8	(0.3, 1.2)
Women	8	(7-9)	8	(7-10)	12	(10-14)	14	(12-16)	1.9	(1.1, 2.6)	2	(1-3)	2	(1-2)	3	(2-4)	3	(2-4)	0.5	(0.0, 0.9)
South Thames Men	6	(5-7)	7	(6-7)	10	(9-11)	12	(10-13)	2.0	(1.5, 2.6)	1	(1-2)	1	(1-2)	2	(2-3)	3	(2-3)	0.5	(0.2, 0.8)
Women	8	(7-9)	7	(6-9)	12	(11-14)	13	(11-15)	2.0	(1.4, 2.7)	2	(1-2)	2	(1-2)	2	(2-3)	3	(2-4)	0.5	(0.2, 0.8)
South & West Men	6	(5-7)	7	(6-9)	9	(7-10)	10	(9-12)	1.3	(0.7, 1.8)	2	(1-3)	2	(2-3)	2	(2-3)	3	(2-3)	0.2	(-0.1, 0.5)
Women	8	(6-9)	9	(7-10)	11	(9-12)	10	(8-11)	0.8	(0.2, 1.5)	2	(1-3)	2	(2-3)	3	(2-4)	2	(1-3)	0.1	(-0.3, 0.4)
West Midlands Men	6	(5-7)	6	(5-7)	6	(5-7)	7	(5-8)	0.2	(-0.3, 0.8)	1	(1-2)	1	(0-1)	1	(0-1)	1	(1-2)	0.1	(-0.2, 0.4)
Women	6	(5-8)	7	(5-8)	8	(6-9)	8	(7-9)	0.6	(-0.1, 1.2)	1	(0-2)	1	(0-1)	1	(0-1)	2	(1-3)	0.2	(-0.1, 0.5)
North & West Men	5	(4-6)	6	(5-7)	6	(5-7)	7	(6-8)	0.4	(-0.1, 0.8)	1	(1-1)	1	(1-2)	1	(1-2)	1	(1-1)	0.0	(-0.2, 0.2)
Women	7	(6-8)	6	(5-7)	8	(6-9)	8	(7-9)	0.5	(0.0, 1.0)	1	(1-2)	1	(1-2)	1	(1-2)	2	(1-3)	0.3	(0.0, 0.5)
WALES Men	7	(6-9)	8	(6-9)	9	(7-11)	10	(8-12)	1.0	(0.1, 1.8)	3	(2-4)	2	(1-3)	4	(3-6)	4	(3-6)	0.7	(0.1, 1.2)
Women	8	(6-11)	7	(6-9)	11	(8-13)	10	(8-12)	0.9	(-0.1, 1.9)	3	(1-4)	3	(1-4)	3	(1-4)	3	(2-5)	0.2	(-0.4, 0.8)

Table 4.5 F

	One-year survival (%), patients diagnosed in								Average increase every five years		Five-year survival (%), patients diagnosed in								Average increase every five years	
	1971-75		1976-80		1981-85		1986-90				1971-75		1976-80		1981-85		1986-90			
	%	CI	%	CI	%	CI	%	CI	%	CI	%	CI	%	CI	%	CI	%	CI	%	CI
LUNG																				
ENGLAND & WALES Men	15	(15-15)	15	(15-16)	18	(18-18)	19	(19-19)	1.4	(1.3, 1.5)	4	(4-4)	5	(4-5)	5	(5-5)	5	(5-5)	0.4	(0.3, 0.4)
Women	13	(13-14)	14	(14-15)	17	(17-18)	19	(18-19)	1.9	(1.8, 2.1)	4	(4-4)	4	(4-4)	5	(5-5)	5	(5-6)	0.5	(0.4, 0.6)
ENGLAND Men	15	(15-16)	16	(15-16)	18	(18-18)	19	(19-19)	1.4	(1.3, 1.5)	4	(4-4)	4	(4-5)	5	(5-5)	5	(5-5)	0.3	(0.3, 0.4)
Women	13	(13-14)	14	(14-14)	17	(17-18)	19	(18-19)	1.9	(1.8, 2.1)	4	(4-4)	4	(4-4)	5	(5-5)	5	(5-5)	0.5	(0.4, 0.6)
Northern & Yorkshire Men	13	(13-14)	13	(12-13)	16	(15-16)	17	(16-17)	1.3	(1.0, 1.5)	3	(3-4)	3	(3-4)	4	(4-5)	5	(4-5)	0.5	(0.3, 0.6)
Women	11	(10-12)	12	(11-13)	15	(14-16)	16	(15-17)	1.7	(1.3, 2.1)	3	(2-3)	3	(3-4)	4	(4-5)	4	(4-5)	0.4	(0.2, 0.7)
Trent Men	14	(13-15)	14	(13-14)	16	(15-16)	17	(16-17)	1.0	(0.7, 1.3)	4	(3-4)	4	(3-4)	4	(4-4)	5	(4-5)	0.3	(0.1, 0.5)
Women	13	(12-14)	13	(12-14)	16	(15-17)	16	(15-17)	1.1	(0.6, 1.6)	4	(3-5)	5	(4-6)	4	(4-5)	4	(3-4)	-0.1	(-0.4, 0.2)
Anglia & Oxford Men	18	(17-18)	18	(17-18)	18	(17-19)	19	(18-20)	0.4	(0.1, 0.8)	5	(4-5)	4	(4-5)	5	(4-5)	5	(5-6)	0.2	(0.0, 0.4)
Women	15	(13-16)	15	(13-16)	16	(15-17)	18	(17-19)	1.4	(0.8, 1.9)	4	(3-4)	4	(3-5)	4	(3-4)	5	(5-6)	0.5	(0.1, 0.8)
North Thames Men	19	(18-19)	21	(20-21)	25	(24-26)	27	(26-28)	2.9	(2.6, 3.3)	6	(5-6)	6	(6-7)	8	(8-9)	7	(6-7)	0.6	(0.4, 0.8)
Women	16	(15-17)	17	(16-18)	24	(23-25)	27	(26-28)	4.0	(3.5, 4.5)	5	(4-6)	5	(4-6)	7	(7-8)	7	(6-8)	0.9	(0.6, 1.2)
South Thames Men	18	(17-18)	18	(17-18)	22	(21-22)	24	(23-25)	2.2	(1.9, 2.5)	5	(4-5)	5	(5-6)	6	(6-6)	6	(5-6)	0.4	(0.2, 0.6)
Women	15	(14-16)	16	(15-17)	21	(20-22)	23	(22-24)	2.8	(2.4, 3.3)	4	(3-4)	5	(4-5)	6	(6-7)	6	(5-7)	0.8	(0.6, 1.1)
South & West Men	16	(15-16)	17	(16-17)	19	(18-19)	19	(18-20)	1.2	(0.9, 1.5)	5	(5-5)	6	(5-6)	6	(5-6)	6	(5-6)	0.2	(0.0, 0.4)
Women	14	(13-15)	16	(15-17)	18	(17-19)	18	(17-19)	1.5	(1.0, 2.0)	5	(4-6)	5	(4-6)	5	(5-6)	6	(5-6)	0.3	(0.0, 0.6)
West Midlands Men	15	(14-16)	15	(15-16)	16	(16-17)	18	(17-18)	0.9	(0.6, 1.2)	4	(3-4)	4	(3-4)	4	(4-4)	5	(4-5)	0.3	(0.1, 0.5)
Women	12	(11-14)	14	(13-15)	16	(14-17)	17	(16-18)	1.7	(1.2, 2.2)	4	(3-4)	3	(3-4)	4	(3-4)	5	(4-6)	0.5	(0.2, 0.8)
North & West Men	12	(11-12)	13	(12-13)	15	(15-16)	17	(16-17)	1.8	(1.6, 2.1)	4	(3-4)	4	(4-4)	5	(4-5)	5	(5-5)	0.5	(0.4, 0.7)
Women	11	(10-12)	11	(11-12)	15	(14-15)	17	(16-18)	2.3	(1.9, 2.6)	4	(3-4)	3	(3-4)	4	(4-5)	5	(5-6)	0.6	(0.4, 0.9)
WALES Men	14	(13-15)	15	(14-16)	18	(17-19)	18	(17-19)	1.7	(1.3, 2.1)	4	(3-4)	6	(5-6)	7	(6-8)	7	(6-8)	1.1	(0.8, 1.4)
Women	14	(12-16)	16	(14-18)	18	(17-20)	20	(19-22)	2.0	(1.3, 2.8)	4	(3-5)	6	(5-7)	7	(5-8)	7	(6-8)	1.0	(0.5, 1.5)

PANCREAS

Figure 4.5E

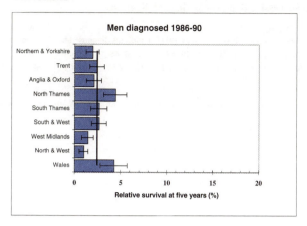

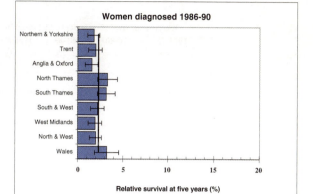

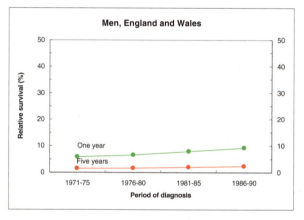

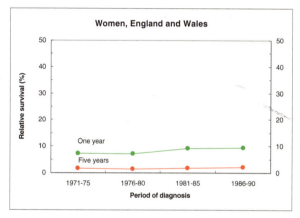

LUNG

Figure 4.5F

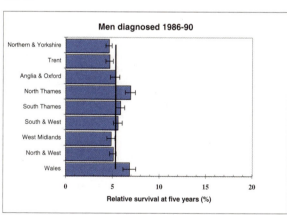

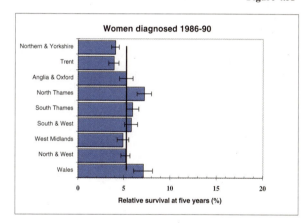

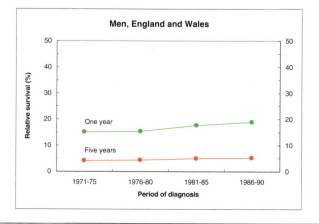

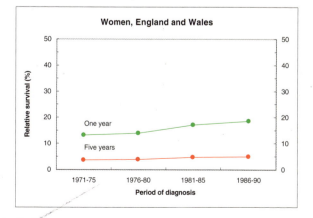

SUMMARY

Table 4.5 (cont.)   Cancer survival trends by NHS Region, selected cancers, patients diagnosed 1971-90: age-standardised relative survival rates (with 95% confidence intervals) at one and five years after diagnosis, and average increase in relative survival between successive five-year periods of diagnosis

Table 4.5 G		One-year survival (%), patients diagnosed in								Average increase every five years		Five-year survival (%), patients diagnosed in								Average increase every five years	
		1971-75		1976-80		1981-85		1986-90				1971-75		1976-80		1981-85		1986-90			
		%	CI	%	CI	%	CI	%	CI	%	CI	%	CI	%	CI	%	CI	%	CI	%	CI
MELANOMA OF SKIN																					
ENGLAND & WALES	Men	78	(76-80)	82	(80-84)	85	(83-86)	90	(89-90)	3.9	(3.3, 4.5)	46	(43-49)	51	(48-53)	58	(56-60)	68	(66-69)	7.6	(6.6, 8.5)
	Women	89	(88-90)	89	(88-90)	93	(92-93)	94	(94-95)	2.1	(1.7, 2.4)	65	(63-66)	69	(68-71)	76	(74-77)	82	(81-83)	5.8	(5.2, 6.4)
ENGLAND	Men	78	(76-80)	82	(81-84)	85	(84-86)	90	(89-91)	3.9	(3.3, 4.6)	47	(44-49)	51	(48-53)	58	(56-60)	68	(67-70)	7.7	(6.7, 8.6)
	Women	89	(88-90)	90	(89-90)	93	(92-94)	95	(94-95)	2.1	(1.7, 2.4)	65	(63-66)	70	(68-71)	76	(74-77)	82	(81-83)	5.9	(5.3, 6.5)
Northern & Yorkshire	Men	79	(73-85)	76	(71-82)	87	(83-91)	90	(87-93)	4.6	(2.7, 6.6)	47	(38-56)	49	(42-56)	57	(51-64)	69	(64-74)	8.3	(5.3, 11.2)
	Women	87	(84-90)	87	(84-90)	93	(91-95)	94	(93-96)	2.7	(1.6, 3.8)	59	(54-64)	71	(66-76)	76	(72-79)	83	(80-86)	7.3	(5.5, 9.1)
Trent	Men	75	(68-81)	74	(69-80)	81	(77-86)	88	(85-91)	5.1	(3.0, 7.2)	45	(37-52)	40	(34-47)	50	(44-57)	66	(61-71)	8.4	(5.6, 11.2)
	Women	84	(80-88)	88	(84-91)	90	(87-92)	92	(90-94)	2.7	(1.4, 4.0)	60	(54-66)	67	(62-73)	71	(67-76)	82	(79-85)	7.4	(5.4, 9.5)
Anglia & Oxford	Men	75	(69-81)	84	(79-88)	88	(84-91)	92	(89-94)	4.9	(3.1, 6.6)	43	(36-49)	53	(46-59)	62	(56-67)	72	(68-76)	9.6	(7.2, 12.1)
	Women	92	(89-95)	92	(89-94)	95	(93-97)	95	(94-97)	1.4	(0.4, 2.3)	65	(60-70)	72	(68-77)	78	(75-82)	82	(79-85)	5.4	(3.7, 7.1)
North Thames	Men	74	(69-80)	85	(80-90)	81	(76-85)	92	(90-95)	5.3	(3.5, 7.0)	40	(34-47)	46	(39-53)	47	(41-53)	66	(61-70)	8.7	(6.1, 11.2)
	Women	88	(85-91)	88	(85-91)	93	(90-95)	95	(94-97)	2.7	(1.7, 3.7)	64	(59-69)	67	(62-72)	77	(73-81)	83	(80-86)	6.8	(5.0, 8.6)
South Thames	Men	83	(78-87)	89	(85-93)	88	(85-91)	88	(85-90)	0.9	(-0.6, 2.4)	49	(43-56)	57	(51-63)	59	(54-63)	66	(62-70)	5.2	(2.8, 7.6)
	Women	91	(88-93)	92	(90-94)	95	(93-96)	96	(94-97)	1.6	(0.8, 2.4)	68	(64-72)	73	(70-77)	75	(72-78)	82	(80-85)	4.6	(3.1, 6.1)
South & West	Men	83	(78-87)	84	(81-88)	86	(83-88)	90	(88-92)	2.8	(1.5, 4.2)	50	(43-56)	53	(48-58)	64	(59-68)	70	(67-73)	7.3	(5.2, 9.4)
	Women	91	(89-93)	89	(87-91)	94	(92-95)	95	(94-96)	1.8	(1.0, 2.5)	71	(67-75)	72	(69-75)	80	(77-82)	81	(79-83)	3.9	(2.6, 5.2)
West Midlands	Men	78	(71-84)	84	(79-89)	87	(83-91)	92	(89-95)	4.6	(2.7, 6.5)	45	(37-54)	43	(36-51)	55	(49-62)	71	(66-76)	10.1	(7.2, 13.0)
	Women	89	(85-92)	91	(88-94)	94	(92-96)	96	(94-98)	2.4	(1.3, 3.5)	64	(58-69)	68	(63-74)	76	(72-80)	84	(81-88)	7.3	(5.3, 9.2)
North & West	Men	77	(71-83)	79	(74-85)	82	(78-86)	89	(86-91)	4.2	(2.3, 6.1)	41	(33-48)	50	(43-56)	52	(46-58)	63	(59-68)	7.4	(4.6, 10.1)
	Women	90	(87-93)	89	(87-92)	90	(88-93)	94	(93-96)	1.8	(0.7, 2.8)	60	(55-66)	64	(60-68)	69	(65-73)	81	(78-84)	7.4	(5.6, 9.3)
WALES	Men	72	(63-81)	76	(69-82)	77	(70-83)	79	(73-84)	2.0	(-1.0, 5.1)	37	(27-47)	48	(40-55)	51	(43-59)	55	(48-62)	5.1	(1.4, 8.8)
	Women	85	(80-89)	82	(77-86)	87	(83-90)	87	(84-91)	1.4	(-0.4, 3.2)	63	(56-70)	64	(58-70)	72	(66-77)	74	(70-79)	4.3	(1.8, 6.8)

Table 4.5 H

BREAST (WOMEN)	1971-75		1976-80		1981-85		1986-90		Avg inc		1971-75		1976-80		1981-85		1986-90		Avg inc	
ENGLAND & WALES	82	(82-82)	85	(84-85)	87	(87-87)	89	(88-89)	2.2	(2.1, 2.2)	52	(52-53)	59	(58-59)	62	(62-62)	66	(66-66)	4.4	(4.3, 4.6)
ENGLAND	82	(82-83)	85	(85-85)	87	(87-88)	89	(89-89)	2.2	(2.1, 2.3)	52	(52-53)	59	(58-59)	62	(62-62)	66	(66-67)	4.4	(4.3, 4.6)
Northern & Yorkshire	81	(80-82)	84	(83-84)	87	(86-88)	88	(88-89)	2.4	(2.1, 2.7)	52	(51-53)	57	(56-59)	62	(61-63)	66	(65-67)	4.6	(4.1, 5.0)
Trent	82	(81-83)	84	(83-85)	86	(85-87)	87	(86-88)	1.5	(1.2, 1.9)	52	(51-54)	59	(58-60)	60	(59-61)	64	(63-65)	3.7	(3.1, 4.2)
Anglia & Oxford	84	(83-85)	87	(86-88)	88	(87-89)	89	(88-90)	1.5	(1.1, 1.8)	54	(53-56)	60	(59-61)	65	(63-66)	67	(66-68)	4.1	(3.6, 4.6)
North Thames	83	(82-84)	86	(86-87)	90	(89-91)	91	(91-92)	2.8	(2.5, 3.1)	54	(53-55)	61	(59-62)	65	(64-66)	69	(68-70)	4.9	(4.4, 5.4)
South Thames	82	(82-83)	86	(85-86)	89	(89-90)	91	(91-92)	3.0	(2.7, 3.2)	52	(51-53)	60	(59-61)	65	(64-66)	69	(68-70)	5.5	(5.1, 5.9)
South & West	83	(82-84)	85	(84-86)	87	(87-88)	88	(88-89)	1.8	(1.5, 2.1)	55	(54-56)	61	(60-62)	64	(63-65)	67	(66-68)	3.7	(3.3, 4.1)
West Midlands	84	(83-85)	85	(84-86)	88	(87-88)	89	(89-90)	1.9	(1.5, 2.2)	53	(52-54)	60	(59-61)	62	(60-63)	67	(66-68)	4.3	(3.8, 4.8)
North & West	80	(79-81)	83	(82-84)	85	(85-86)	87	(86-88)	2.2	(1.9, 2.5)	51	(50-52)	58	(57-59)	59	(58-60)	64	(63-65)	4.1	(3.6, 4.5)
WALES	79	(78-80)	82	(81-83)	82	(81-83)	83	(82-84)	1.2	(0.7, 1.7)	51	(50-53)	56	(54-58)	60	(59-62)	64	(63-66)	4.3	(3.6, 4.9)

Table 4.5 J

CERVIX	1971-75		1976-80		1981-85		1986-90		Avg inc		1971-75		1976-80		1981-85		1986-90		Avg inc	
ENGLAND & WALES	75	(75-76)	76	(76-77)	80	(79-80)	82	(81-82)	2.3	(2.0, 2.6)	52	(51-53)	54	(54-55)	58	(57-58)	61	(61-62)	3.2	(2.8, 3.5)
ENGLAND	75	(75-76)	76	(76-77)	80	(79-80)	82	(81-82)	2.3	(2.0, 2.6)	52	(51-53)	54	(53-55)	58	(57-58)	61	(60-62)	3.2	(2.9, 3.6)
Northern & Yorkshire	71	(69-73)	73	(72-75)	78	(76-79)	82	(81-83)	3.8	(3.1, 4.6)	48	(45-50)	51	(49-53)	56	(54-58)	64	(62-66)	5.5	(4.6, 6.4)
Trent	75	(73-78)	76	(74-78)	79	(77-81)	81	(80-83)	2.0	(1.2, 2.9)	52	(50-55)	54	(52-56)	57	(55-60)	60	(58-62)	2.8	(1.7, 3.8)
Anglia & Oxford	77	(75-79)	79	(77-81)	82	(81-84)	82	(81-84)	1.9	(1.0, 2.8)	56	(53-59)	56	(54-59)	63	(60-65)	63	(61-66)	2.9	(1.7, 4.0)
North Thames	78	(76-80)	77	(75-79)	83	(81-84)	83	(81-85)	2.0	(1.1, 2.8)	55	(53-57)	53	(50-56)	61	(59-64)	59	(57-62)	2.1	(1.0, 3.2)
South Thames	76	(74-78)	78	(76-80)	81	(80-83)	83	(82-85)	2.6	(1.8, 3.4)	50	(48-53)	55	(52-57)	58	(56-60)	59	(57-61)	2.9	(1.9, 3.9)
South & West	76	(74-78)	77	(75-79)	81	(79-82)	83	(81-84)	2.4	(1.6, 3.2)	52	(50-55)	56	(54-59)	59	(57-61)	62	(60-64)	3.1	(2.2, 4.1)
West Midlands	76	(74-78)	79	(77-80)	80	(78-82)	81	(79-82)	1.5	(0.6, 2.3)	53	(50-55)	58	(55-60)	59	(57-61)	63	(61-65)	3.1	(2.1, 4.2)
North & West	75	(74-77)	75	(73-76)	78	(76-79)	80	(78-81)	1.7	(1.0, 2.4)	52	(50-54)	51	(50-53)	53	(51-55)	59	(57-61)	2.5	(1.7, 3.3)
WALES	75	(73-78)	77	(75-79)	77	(75-79)	81	(79-83)	1.9	(0.9, 2.9)	56	(53-59)	59	(56-62)	59	(56-61)	62	(60-65)	1.7	(0.5, 3.0)

MELANOMA OF SKIN

Figure 4.5G

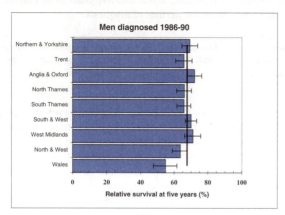

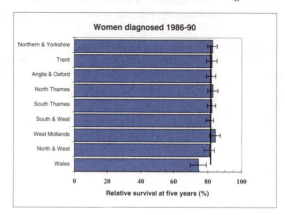

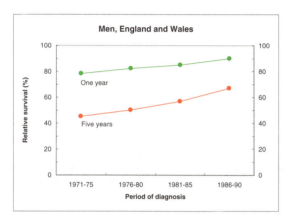

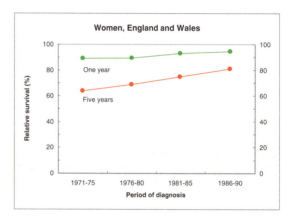

BREAST

Figure 4.5H

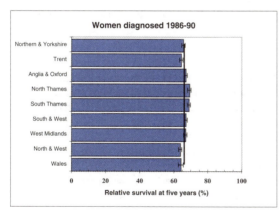

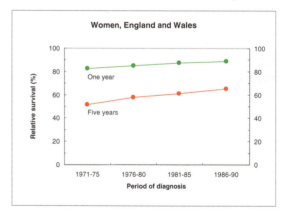

CERVIX

Figure 4.5J

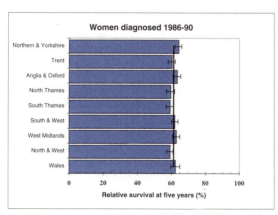

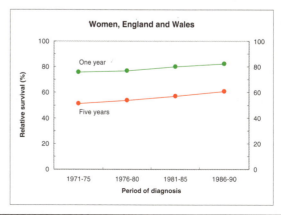

SUMMARY

Table 4.5 (cont.) Cancer survival trends by NHS Region, selected cancers, patients diagnosed 1971-90: age-standardised relative survival rates (with 95% confidence intervals) at one and five years after diagnosis, and average increase in relative survival between successive five-year periods of diagnosis

Table 4.5 K	One-year survival (%), patients diagnosed in				Average increase every five years	Five-year survival (%), patients diagnosed in				Average increase every five years
	1971-75	1976-80	1981-85	1986-90		1971-75	1976-80	1981-85	1986-90	
	% CI	% CI	% CI	% CI	% CI	% CI	% CI	% CI	% CI	% CI
UTERUS										
ENGLAND & WALES	**78 (77-79)**	**80 (80-81)**	**82 (82-83)**	**84 (83-85)**	**2.0 (1.7, 2.3)**	**61 (60-62)**	**65 (64-66)**	**68 (67-69)**	**70 (69-70)**	**2.9 (2.5, 3.3)**
ENGLAND	78 (78-79)	80 (80-81)	82 (82-83)	84 (84-85)	2.0 (1.7, 2.3)	61 (60-62)	65 (64-66)	68 (68-69)	70 (69-71)	3.0 (2.6, 3.4)
Northern & Yorkshire	77 (75-80)	79 (77-81)	83 (81-85)	84 (82-86)	2.3 (1.5, 3.2)	62 (59-65)	63 (60-66)	70 (67-72)	71 (69-74)	3.4 (2.1, 4.6)
Trent	76 (74-78)	81 (79-83)	82 (80-84)	82 (80-84)	1.8 (0.9, 2.8)	59 (56-62)	68 (65-71)	66 (64-69)	68 (65-70)	2.5 (1.2, 3.7)
Anglia & Oxford	81 (79-84)	82 (81-84)	82 (81-84)	82 (80-84)	0.2 (-0.7, 1.1)	63 (60-66)	70 (67-72)	69 (67-72)	68 (65-70)	1.0 (-0.2, 2.2)
North Thames	80 (78-82)	82 (80-85)	84 (82-86)	87 (85-88)	2.0 (1.2, 2.8)	63 (60-66)	66 (63-69)	68 (66-71)	70 (68-73)	2.5 (1.3, 3.6)
South Thames	76 (74-77)	80 (78-82)	83 (82-85)	86 (84-87)	3.4 (2.6, 4.1)	57 (55-60)	63 (61-65)	70 (67-72)	73 (71-75)	5.3 (4.3, 6.3)
South & West	80 (79-82)	81 (79-82)	85 (83-86)	87 (85-88)	2.3 (1.6, 3.0)	66 (64-68)	67 (65-69)	71 (68-73)	74 (72-76)	2.8 (1.7, 3.8)
West Midlands	81 (79-83)	82 (80-83)	82 (80-84)	83 (82-85)	0.8 (-0.1, 1.7)	64 (61-67)	66 (64-69)	69 (66-71)	70 (67-72)	2.1 (0.9, 3.4)
North & West	75 (73-77)	76 (74-79)	80 (78-82)	81 (79-83)	2.0 (1.1, 3.0)	55 (53-58)	59 (56-62)	66 (63-69)	65 (63-68)	3.6 (2.4, 4.7)
WALES	74 (71-77)	84 (81-87)	82 (79-84)	82 (79-85)	1.9 (0.6, 3.2)	58 (54-62)	67 (62-71)	64 (60-68)	65 (61-69)	1.9 (0.1, 3.7)

Table 4.5 L

OVARY										
ENGLAND & WALES	**42 (41-43)**	**44 (44-45)**	**51 (50-52)**	**54 (53-55)**	**4.3 (4.0, 4.6)**	**21 (20-21)**	**22 (22-23)**	**27 (26-27)**	**28 (28-29)**	**2.7 (2.4, 3.0)**
ENGLAND	42 (41-43)	44 (44-45)	51 (50-52)	54 (54-55)	4.4 (4.1, 4.7)	20 (20-21)	22 (22-23)	27 (26-27)	28 (28-29)	2.8 (2.5, 3.0)
Northern & Yorkshire	38 (36-40)	40 (38-42)	49 (47-51)	51 (50-53)	4.9 (4.0, 5.7)	17 (15-19)	19 (17-20)	25 (24-27)	26 (25-28)	3.4 (2.6, 4.2)
Trent	43 (41-46)	45 (43-47)	51 (49-53)	55 (53-57)	4.1 (3.1, 5.1)	22 (20-24)	24 (22-26)	27 (25-29)	30 (28-32)	2.6 (1.7, 3.5)
Anglia & Oxford	45 (42-47)	44 (42-46)	52 (50-54)	54 (52-56)	3.6 (2.7, 4.6)	23 (21-25)	23 (21-25)	29 (27-31)	30 (28-32)	2.8 (1.9, 3.7)
North Thames	45 (43-47)	48 (46-51)	56 (53-58)	60 (58-62)	5.2 (4.3, 6.1)	22 (20-24)	23 (21-25)	29 (27-31)	31 (29-33)	3.3 (2.5, 4.1)
South Thames	41 (39-43)	44 (42-46)	55 (53-56)	59 (57-61)	6.5 (5.7, 7.2)	19 (17-20)	22 (20-23)	28 (26-30)	30 (28-32)	4.1 (3.4, 4.8)
South & West	41 (39-43)	48 (46-50)	51 (49-53)	51 (49-53)	3.1 (2.3, 4.0)	22 (21-24)	25 (23-27)	27 (25-29)	25 (23-26)	0.9 (0.2, 1.7)
West Midlands	48 (46-51)	47 (45-49)	52 (50-54)	54 (52-56)	2.3 (1.3, 3.2)	25 (22-27)	26 (24-29)	28 (26-29)	29 (27-31)	1.4 (0.4, 2.3)
North & West	38 (36-40)	40 (38-41)	45 (43-46)	50 (49-52)	4.3 (3.5, 5.1)	18 (16-19)	18 (17-20)	22 (21-24)	26 (24-28)	2.8 (2.1, 3.5)
WALES	45 (41-48)	45 (42-48)	51 (48-54)	51 (48-54)	2.5 (1.2, 3.8)	23 (20-26)	26 (23-29)	28 (26-31)	29 (26-31)	1.9 (0.7, 3.1)

Table 4.5 M

PROSTATE										
ENGLAND & WALES	**65 (64-65)**	**69 (68-69)**	**76 (75-76)**	**76 (76-77)**	**4.1 (3.8, 4.3)**	**31 (30-32)**	**37 (36-37)**	**41 (40-41)**	**41 (41-42)**	**3.5 (3.2, 3.7)**
ENGLAND	65 (64-66)	69 (68-69)	76 (76-76)	77 (76-77)	4.1 (3.8, 4.3)	31 (30-32)	37 (36-38)	41 (40-42)	41 (41-42)	3.4 (3.1, 3.7)
Northern & Yorkshire	64 (63-66)	69 (68-71)	75 (74-77)	77 (76-78)	4.2 (3.5, 4.8)	28 (26-30)	38 (36-40)	40 (38-42)	42 (41-44)	4.3 (3.5, 5.2)
Trent	60 (58-62)	63 (61-65)	75 (73-76)	75 (73-76)	5.3 (4.6, 6.1)	30 (27-32)	36 (33-38)	42 (39-44)	41 (40-43)	3.9 (2.9, 4.8)
Anglia & Oxford	72 (70-73)	72 (70-74)	79 (77-80)	77 (76-78)	2.2 (1.5, 2.8)	36 (34-39)	38 (36-41)	43 (41-45)	42 (40-44)	2.0 (1.0, 2.9)
North Thames	68 (66-70)	75 (73-76)	80 (78-81)	82 (81-83)	4.5 (3.9, 5.2)	36 (33-38)	42 (40-45)	47 (44-49)	45 (43-47)	3.1 (2.1, 4.0)
South Thames	65 (64-67)	69 (68-71)	80 (79-81)	82 (81-83)	5.9 (5.3, 6.5)	29 (27-30)	39 (37-41)	44 (42-46)	43 (41-44)	4.6 (3.8, 5.4)
South & West	67 (66-69)	70 (69-72)	76 (75-77)	74 (73-75)	2.3 (1.8, 2.9)	36 (34-38)	39 (37-41)	40 (38-41)	40 (38-41)	1.3 (0.5, 2.0)
West Midlands	65 (63-67)	71 (69-73)	76 (74-77)	76 (74-77)	3.4 (2.7, 4.2)	31 (29-34)	37 (35-40)	40 (38-43)	41 (39-43)	3.0 (2.0, 4.0)
North & West	59 (57-61)	64 (63-66)	71 (70-73)	73 (72-75)	4.9 (4.2, 5.5)	27 (25-29)	32 (30-34)	39 (37-41)	40 (38-42)	4.6 (3.8, 5.4)
WALES	59 (57-62)	64 (62-66)	69 (67-71)	70 (68-72)	3.5 (2.5, 4.5)	30 (26-33)	37 (34-40)	39 (36-42)	41 (39-44)	3.7 (2.4, 4.9)

Table 4.5 N

TESTIS										
ENGLAND & WALES	**82 (81-84)**	**87 (86-88)**	**94 (94-95)**	**95 (94-95)**	**3.8 (3.4, 4.3)**	**69 (67-71)**	**78 (77-79)**	**88 (87-89)**	**90 (89-91)**	**6.6 (6.1, 7.2)**
ENGLAND	82 (81-84)	87 (86-88)	94 (94-95)	95 (94-96)	3.9 (3.5, 4.4)	69 (67-71)	78 (76-79)	88 (87-89)	90 (90-91)	6.7 (6.1, 7.3)
Northern & Yorkshire	81 (77-86)	90 (87-93)	93 (91-95)	94 (92-96)	3.1 (1.8, 4.3)	67 (61-72)	77 (73-81)	85 (82-88)	88 (86-91)	6.6 (4.9, 8.3)
Trent	83 (78-87)	84 (79-88)	93 (91-95)	94 (92-96)	4.1 (2.7, 5.5)	71 (65-76)	73 (68-78)	84 (81-88)	88 (85-91)	6.2 (4.4, 8.0)
Anglia & Oxford	89 (85-93)	89 (85-92)	97 (96-99)	94 (93-96)	1.8 (0.5, 3.0)	75 (69-80)	79 (75-83)	93 (91-95)	89 (86-91)	4.2 (2.5, 5.9)
North Thames	82 (77-86)	89 (86-92)	95 (93-97)	98 (96-99)	4.5 (3.4, 5.7)	70 (65-75)	81 (77-85)	92 (89-95)	93 (91-95)	6.6 (5.0, 8.1)
South Thames	82 (78-86)	90 (87-92)	97 (96-99)	96 (95-98)	3.9 (2.8, 4.9)	69 (64-73)	82 (78-85)	89 (87-92)	92 (90-94)	6.8 (5.4, 8.1)
South & West	84 (81-88)	87 (84-90)	94 (93-96)	95 (94-97)	3.6 (2.4, 4.7)	70 (65-75)	77 (72-81)	88 (86-91)	90 (88-92)	6.5 (5.0, 8.0)
West Midlands	81 (77-85)	86 (82-90)	94 (92-96)	94 (92-96)	4.2 (2.8, 5.5)	67 (62-72)	77 (73-82)	87 (83-90)	90 (87-93)	7.2 (5.5, 8.9)
North & West	82 (78-85)	87 (84-90)	93 (91-95)	96 (94-97)	4.5 (3.4, 5.6)	68 (64-73)	77 (73-81)	87 (84-90)	91 (89-94)	7.5 (6.0, 8.9)
WALES	83 (77-89)	88 (84-93)	95 (91-98)	93 (90-95)	2.7 (0.7, 4.6)	70 (63-78)	80 (74-86)	88 (83-92)	87 (84-91)	4.9 (2.5, 7.4)

UTERUS

Figure 4.5K

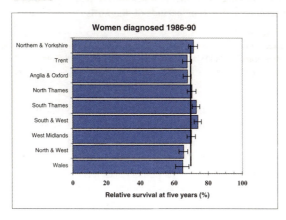

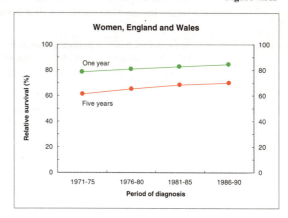

OVARY

Figure 4.5L

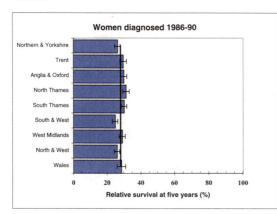

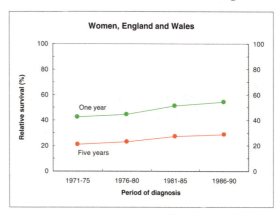

PROSTATE

Figure 4.5M

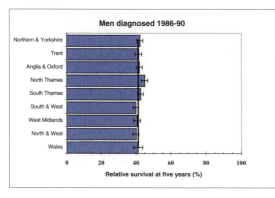

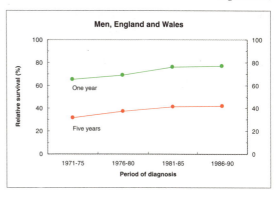

TESTIS

Figure 4.5N

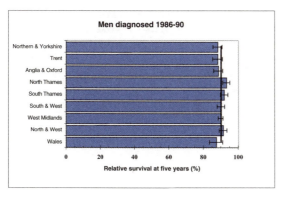

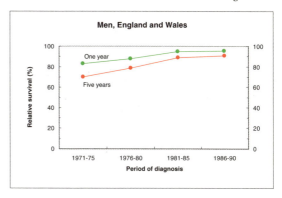

Table 4.5 (cont.)   Cancer survival trends by NHS Region, selected cancers, patients diagnosed 1971-90:
age-standardised relative survival rates (with 95% confidence intervals) at one and five years after diagnosis,
and average increase in relative survival between successive five-year periods of diagnosis

Table 4.5 P

| | | One-year survival (%), patients diagnosed in | | | | | | | | | Average increase every | | Five-year survival (%), patients diagnosed in | | | | | | | | | Average increase every | |
| | | 1971-75 | | 1976-80 | | 1981-85 | | 1986-90 | | five years | | 1971-75 | | 1976-80 | | 1981-85 | | 1986-90 | | five years | |
		%	CI	%	CI	%	CI	%	CI	%	CI	%	CI	%	CI	%	CI	%	CI	%	CI
BLADDER																					
ENGLAND & WALES	Men	66	(65-67)	71	(71-72)	77	(77-78)	80	(79-80)	**4.6**	**(4.3, 4.8)**	44	(43-45)	53	(52-54)	59	(58-60)	62	(61-63)	**6.0**	**(5.6, 6.3)**
	Women	59	(58-60)	63	(62-64)	70	(69-71)	72	(71-73)	**4.5**	**(4.1, 4.9)**	42	(40-43)	48	(47-50)	56	(55-57)	57	(56-58)	**5.4**	**(4.9, 5.8)**
ENGLAND	Men	66	(66-67)	72	(71-72)	77	(77-78)	80	(79-80)	**4.4**	**(4.2, 4.7)**	44	(43-45)	54	(53-54)	60	(59-60)	62	(61-62)	**5.8**	**(5.4, 6.2)**
	Women	59	(58-60)	64	(63-65)	70	(69-71)	72	(71-73)	**4.4**	**(3.9, 4.8)**	42	(40-43)	49	(48-50)	56	(55-57)	57	(56-58)	**5.2**	**(4.7, 5.7)**
Northern & Yorkshire	Men	64	(62-66)	71	(69-72)	77	(75-78)	76	(74-77)	**3.9**	**(3.1, 4.6)**	43	(41-46)	54	(51-56)	59	(57-61)	59	(57-62)	**5.1**	**(4.1, 6.2)**
	Women	55	(52-58)	62	(59-64)	69	(67-72)	70	(68-72)	**4.9**	**(3.8, 6.0)**	39	(36-42)	46	(43-49)	57	(54-60)	53	(50-56)	**5.1**	**(3.7, 6.4)**
Trent	Men	67	(64-69)	74	(72-76)	74	(72-76)	79	(77-80)	**3.6**	**(2.7, 4.4)**	43	(40-46)	56	(53-59)	57	(55-60)	61	(59-63)	**5.5**	**(4.3, 6.6)**
	Women	58	(55-61)	64	(61-67)	69	(66-72)	72	(69-74)	**4.6**	**(3.3, 5.9)**	39	(35-42)	51	(47-54)	56	(52-59)	57	(53-60)	**5.7**	**(4.1, 7.3)**
Anglia & Oxford	Men	70	(68-72)	73	(71-75)	78	(76-80)	79	(77-80)	**3.0**	**(2.2, 3.8)**	51	(48-54)	55	(53-58)	62	(59-65)	61	(58-63)	**3.4**	**(2.2, 4.6)**
	Women	61	(58-65)	65	(62-68)	71	(68-74)	70	(67-73)	**3.0**	**(1.6, 4.4)**	46	(41-50)	49	(45-53)	56	(52-60)	57	(54-61)	**4.2**	**(2.4, 5.9)**
North Thames	Men	69	(67-71)	74	(72-76)	80	(78-81)	83	(82-85)	**4.9**	**(4.2, 5.6)**	47	(45-50)	57	(54-59)	63	(61-66)	64	(62-66)	**5.6**	**(4.6, 6.7)**
	Women	61	(58-64)	67	(65-70)	75	(72-77)	77	(75-79)	**5.4**	**(4.3, 6.5)**	45	(42-48)	54	(50-57)	58	(54-61)	64	(61-67)	**6.0**	**(4.6, 7.5)**
South Thames	Men	68	(66-70)	74	(72-75)	81	(80-82)	83	(82-84)	**5.0**	**(4.4, 5.7)**	45	(43-48)	56	(54-58)	62	(60-65)	64	(62-66)	**6.0**	**(5.1, 7.0)**
	Women	62	(59-64)	65	(63-68)	74	(72-76)	77	(75-79)	**5.4**	**(4.3, 6.4)**	43	(40-47)	50	(47-53)	59	(57-62)	60	(57-63)	**5.9**	**(4.6, 7.2)**
South & West	Men	70	(68-72)	75	(73-76)	80	(79-82)	81	(80-83)	**3.8**	**(3.1, 4.4)**	49	(47-52)	58	(56-61)	63	(61-65)	65	(63-66)	**4.8**	**(3.9, 5.8)**
	Women	64	(61-67)	70	(67-72)	73	(70-75)	72	(70-74)	**2.4**	**(1.3, 3.5)**	51	(47-55)	57	(54-61)	60	(57-63)	60	(57-63)	**2.7**	**(1.3, 4.1)**
West Midlands	Men	64	(62-67)	69	(66-71)	75	(73-77)	78	(77-80)	**4.8**	**(3.9, 5.7)**	40	(37-43)	50	(47-54)	54	(52-57)	61	(58-63)	**6.6**	**(5.3, 7.9)**
	Women	58	(54-62)	62	(58-65)	66	(63-69)	69	(67-72)	**3.9**	**(2.4, 5.3)**	41	(37-45)	43	(39-47)	53	(50-57)	56	(53-60)	**5.8**	**(4.1, 7.4)**
North & West	Men	60	(58-61)	66	(64-67)	74	(73-76)	77	(76-79)	**6.0**	**(5.3, 6.7)**	36	(34-38)	45	(43-48)	57	(55-60)	60	(58-62)	**8.2**	**(7.2, 9.1)**
	Women	54	(51-56)	59	(56-61)	66	(64-68)	69	(67-71)	**5.2**	**(4.1, 6.2)**	34	(31-37)	44	(41-47)	53	(50-55)	56	(53-58)	**7.2**	**(6.0, 8.5)**
WALES	Men	58	(55-61)	67	(64-70)	73	(71-75)	80	(78-82)	**7.1**	**(6.0, 8.2)**	41	(37-45)	48	(44-52)	55	(51-58)	68	(65-71)	**8.8**	**(7.3, 10.4)**
	Women	57	(52-62)	53	(48-57)	65	(62-69)	75	(72-78)	**7.5**	**(5.7, 9.3)**	39	(33-44)	39	(34-44)	52	(48-57)	61	(57-65)	**8.3**	**(6.2, 10.4)**

Table 4.5 Q

| | | 1971-75 | | 1976-80 | | 1981-85 | | 1986-90 | | five years | | 1971-75 | | 1976-80 | | 1981-85 | | 1986-90 | | five years | |
		%	CI	%	CI	%	CI	%	CI	%	CI	%	CI	%	CI	%	CI	%	CI	%	CI
KIDNEY																					
ENGLAND & WALES	Men	45	(44-46)	47	(45-48)	52	(51-54)	56	(55-57)	**3.8**	**(3.3, 4.4)**	28	(27-29)	31	(29-32)	35	(34-36)	38	(36-39)	**3.3**	**(2.8, 3.9)**
	Women	45	(44-47)	45	(44-47)	51	(49-52)	53	(52-54)	**2.9**	**(2.3, 3.6)**	28	(27-30)	29	(28-31)	33	(32-34)	35	(34-37)	**2.6**	**(1.9, 3.2)**
ENGLAND	Men	45	(44-46)	47	(46-48)	53	(51-54)	56	(55-57)	**3.9**	**(3.4, 4.4)**	28	(26-29)	31	(30-32)	35	(34-36)	38	(37-39)	**3.4**	**(2.8, 3.9)**
	Women	45	(44-47)	45	(44-47)	51	(49-52)	54	(52-55)	**3.0**	**(2.4, 3.7)**	28	(27-30)	29	(28-31)	33	(31-34)	36	(34-37)	**2.6**	**(1.9, 3.2)**
Northern & Yorkshire	Men	44	(41-48)	46	(43-49)	50	(47-53)	52	(49-55)	**2.8**	**(1.3, 4.2)**	24	(20-28)	29	(26-33)	33	(30-37)	36	(33-39)	**3.9**	**(2.3, 5.4)**
	Women	43	(39-47)	42	(39-46)	48	(44-51)	52	(49-55)	**3.5**	**(1.8, 5.1)**	23	(19-26)	29	(25-33)	30	(27-34)	36	(33-40)	**4.2**	**(2.5, 5.8)**
Trent	Men	41	(37-45)	44	(40-48)	49	(46-52)	55	(52-58)	**5.0**	**(3.4, 6.6)**	27	(23-31)	29	(25-33)	32	(28-35)	38	(35-42)	**3.8**	**(2.1, 5.6)**
	Women	43	(37-48)	49	(44-54)	48	(44-52)	55	(51-59)	**3.7**	**(1.5, 5.8)**	29	(24-35)	30	(25-35)	30	(26-34)	36	(32-40)	**2.3**	**(0.2, 4.4)**
Anglia & Oxford	Men	50	(46-54)	51	(47-55)	56	(53-60)	52	(49-55)	**1.0**	**(-0.7, 2.6)**	34	(30-39)	35	(31-39)	40	(36-44)	35	(31-38)	**0.4**	**(-1.4, 2.1)**
	Women	46	(41-52)	48	(43-53)	53	(49-58)	52	(48-56)	**2.1**	**(-0.1, 4.3)**	31	(26-37)	28	(23-33)	35	(30-40)	32	(28-36)	**0.7**	**(-1.4, 2.9)**
North Thames	Men	48	(44-52)	48	(44-52)	59	(55-63)	64	(61-67)	**5.9**	**(4.3, 7.4)**	30	(26-34)	31	(27-34)	37	(33-41)	44	(40-47)	**4.7**	**(3.0, 6.5)**
	Women	53	(48-57)	46	(41-51)	56	(51-62)	60	(56-64)	**3.2**	**(1.2, 5.2)**	35	(30-40)	30	(25-35)	39	(33-44)	39	(34-44)	**2.1**	**(-0.1, 4.2)**
South Thames	Men	45	(41-48)	47	(44-50)	58	(55-61)	61	(58-64)	**6.1**	**(4.6, 7.5)**	26	(23-29)	31	(28-35)	39	(36-43)	40	(36-43)	**5.0**	**(3.4, 6.5)**
	Women	48	(43-52)	43	(39-47)	54	(50-58)	57	(53-61)	**4.2**	**(2.3, 6.0)**	27	(23-31)	27	(23-31)	37	(33-41)	36	(32-41)	**3.9**	**(2.1, 5.7)**
South & West	Men	47	(44-51)	50	(47-53)	55	(52-58)	57	(54-60)	**3.3**	**(1.9, 4.7)**	31	(28-35)	34	(30-37)	37	(34-40)	39	(36-43)	**2.8**	**(1.3, 4.3)**
	Women	45	(41-49)	49	(45-54)	55	(51-59)	54	(50-58)	**3.1**	**(1.3, 4.9)**	32	(28-37)	32	(28-37)	36	(32-40)	38	(34-42)	**2.1**	**(0.3, 4.0)**
West Midlands	Men	45	(41-50)	48	(44-52)	51	(47-55)	52	(48-55)	**2.1**	**(0.4, 3.8)**	27	(23-30)	31	(27-35)	32	(28-35)	34	(31-37)	**2.3**	**(0.6, 3.9)**
	Women	47	(42-52)	49	(44-54)	50	(45-54)	52	(48-56)	**1.6**	**(-0.5, 3.7)**	29	(24-34)	32	(27-38)	31	(26-35)	37	(33-42)	**2.3**	**(0.1, 4.4)**
North & West	Men	40	(37-43)	42	(39-45)	47	(44-49)	54	(51-57)	**4.9**	**(3.5, 6.3)**	22	(19-26)	27	(24-31)	30	(27-33)	36	(33-39)	**4.2**	**(2.8, 5.7)**
	Women	41	(37-45)	41	(37-45)	47	(43-50)	49	(46-53)	**3.2**	**(1.5, 4.9)**	23	(19-27)	27	(23-30)	29	(25-32)	32	(29-36)	**2.9**	**(1.3, 4.6)**
WALES	Men	45	(40-51)	44	(38-49)	49	(44-54)	52	(47-56)	**2.6**	**(0.3, 4.8)**	28	(23-33)	27	(22-33)	31	(26-36)	35	(30-39)	**2.4**	**(0.2, 4.6)**
	Women	38	(31-45)	43	(37-50)	50	(43-56)	45	(39-50)	**2.4**	**(-0.4, 5.1)**	24	(17-31)	30	(24-36)	38	(32-44)	28	(23-34)	**1.7**	**(-1.0, 4.4)**

BLADDER

Figure 4.5P

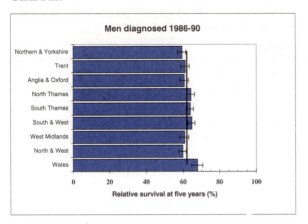

Men diagnosed 1986-90

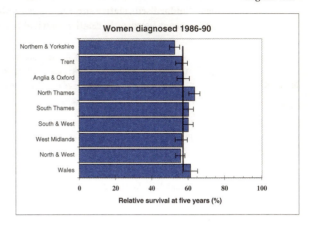

Women diagnosed 1986-90

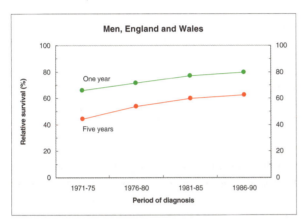

Men, England and Wales

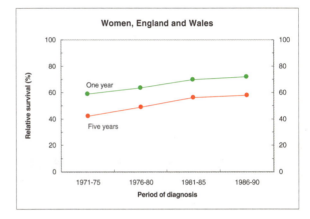

Women, England and Wales

KIDNEY

Figure 4.5Q

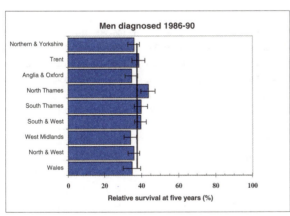

Men diagnosed 1986-90

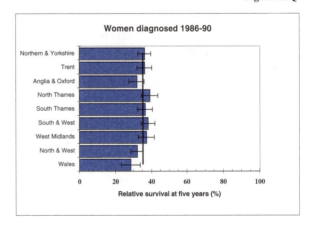

Women diagnosed 1986-90

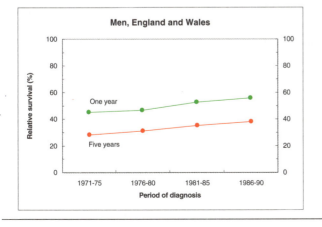

Men, England and Wales

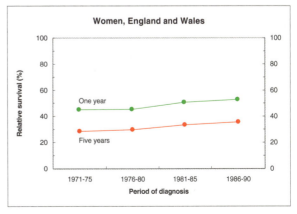

Women, England and Wales

Table 4.5 (cont.) Cancer survival trends by NHS Region, selected cancers, patients diagnosed 1971-90: age-standardised relative survival rates (with 95% confidence intervals) at one and five years after diagnosis, and average increase in relative survival between successive five-year periods of diagnosis

Table 4.5 R		One-year survival (%), patients diagnosed in								Average increase every five years		Five-year survival (%), patients diagnosed in								Average increase every five years	
		1971-75		1976-80		1981-85		1986-90				1971-75		1976-80		1981-85		1986-90			
		%	CI	%	CI	%	CI	%	CI	%	CI	%	CI	%	CI	%	CI	%	CI	%	CI
BRAIN																					
ENGLAND & WALES	Men	19	(18-20)	22	(21-23)	25	(24-26)	28	(28-29)	3.2	(2.8, 3.6)	7	(6-8)	9	(8-10)	11	(11-12)	13	(12-14)	1.9	(1.6, 2.3)
	Women	20	(19-21)	23	(22-24)	26	(25-27)	30	(29-31)	3.3	(2.8, 3.8)	9	(8-10)	10	(9-11)	12	(11-13)	15	(15-16)	2.0	(1.6, 2.4)
ENGLAND	Men	19	(18-20)	22	(21-23)	25	(24-26)	29	(28-30)	3.3	(2.9, 3.8)	7	(6-8)	9	(8-10)	11	(10-12)	13	(12-14)	1.9	(1.6, 2.3)
	Women	20	(19-21)	23	(22-24)	26	(25-27)	30	(29-31)	3.4	(2.9, 3.9)	9	(8-10)	10	(9-11)	12	(11-13)	15	(14-16)	2.1	(1.7, 2.5)
Northern & Yorkshire	Men	18	(15-21)	20	(17-22)	23	(20-25)	27	(24-29)	3.0	(1.8, 4.2)	6	(4-9)	9	(7-10)	10	(9-12)	14	(12-16)	2.3	(1.4, 3.2)
	Women	19	(15-22)	21	(19-24)	23	(20-25)	28	(24-31)	2.8	(1.4, 4.1)	8	(5-10)	9	(6-11)	9	(7-11)	12	(9-14)	1.2	(0.1, 2.3)
Trent	Men	16	(13-19)	22	(19-25)	22	(19-25)	24	(22-27)	2.4	(1.2, 3.7)	8	(6-11)	10	(7-13)	10	(8-12)	11	(9-13)	0.8	(-0.2, 1.8)
	Women	17	(14-20)	26	(22-29)	27	(24-31)	29	(26-33)	4.1	(2.6, 5.5)	11	(8-14)	15	(11-19)	14	(11-17)	15	(12-18)	1.0	(-0.4, 2.3)
Anglia & Oxford	Men	23	(20-25)	25	(22-28)	24	(22-27)	31	(28-33)	2.4	(1.2, 3.7)	9	(7-11)	9	(7-11)	9	(7-11)	15	(12-17)	1.7	(0.8, 2.7)
	Women	27	(23-30)	24	(21-27)	28	(24-31)	26	(23-29)	0.4	(-1.1, 1.8)	14	(11-17)	10	(7-14)	14	(11-18)	14	(12-17)	0.5	(-0.8, 1.9)
North Thames	Men	25	(22-28)	26	(23-29)	33	(30-37)	36	(33-39)	4.0	(2.8, 5.3)	11	(9-14)	11	(8-13)	16	(13-19)	16	(14-19)	2.1	(1.0, 3.1)
	Women	25	(22-29)	30	(27-34)	30	(27-34)	34	(30-37)	2.5	(1.0, 3.9)	12	(9-15)	15	(11-18)	16	(13-19)	18	(15-21)	1.9	(0.6, 3.1)
South Thames	Men	22	(19-24)	25	(22-27)	28	(25-30)	32	(29-35)	3.4	(2.2, 4.6)	9	(7-12)	10	(8-12)	11	(9-13)	14	(12-16)	1.6	(0.7, 2.6)
	Women	20	(17-23)	25	(22-28)	30	(27-33)	32	(29-34)	4.0	(2.7, 5.3)	8	(5-10)	12	(10-15)	15	(12-17)	18	(15-20)	3.4	(2.3, 4.4)
South & West	Men	15	(12-17)	21	(19-24)	26	(23-28)	27	(25-29)	4.0	(2.9, 5.1)	6	(4-8)	10	(8-12)	11	(9-13)	12	(10-14)	1.7	(0.9, 2.6)
	Women	19	(16-23)	20	(17-23)	26	(23-29)	31	(28-34)	4.1	(2.7, 5.6)	12	(8-15)	9	(7-12)	13	(10-15)	15	(13-18)	1.7	(0.5, 3.0)
West Midlands	Men	17	(15-20)	21	(18-25)	25	(22-27)	28	(25-31)	3.6	(2.3, 4.9)	6	(4-8)	17	(12-21)	10	(8-12)	14	(12-17)	2.3	(1.3, 3.3)
	Women	17	(14-20)	22	(18-25)	24	(21-28)	32	(28-35)	4.7	(3.2, 6.2)	10	(7-13)	9	(6-12)	12	(9-15)	15	(12-18)	1.8	(0.5, 3.1)
North & West	Men	17	(14-19)	18	(15-20)	22	(19-24)	24	(22-27)	2.7	(1.6, 3.7)	8	(6-10)	8	(6-9)	11	(8-13)	10	(8-12)	1.0	(0.2, 1.8)
	Women	19	(16-22)	20	(17-23)	20	(18-23)	32	(29-34)	3.7	(2.5, 5.0)	9	(7-12)	12	(10-15)	11	(9-13)	19	(16-22)	2.5	(1.4, 3.6)
WALES	Men	23	(19-27)	25	(21-29)	30	(26-34)	25	(21-29)	0.8	(-0.9, 2.6)	9	(6-13)	14	(10-18)	16	(13-19)	15	(12-19)	2.2	(0.7, 3.7)
	Women	25	(20-30)	28	(23-34)	31	(26-36)	29	(25-34)	1.4	(-0.6, 3.5)	16	(12-21)	15	(10-20)	15	(11-19)	16	(12-19)	-0.1	(-1.9, 1.7)

Table 4.5 S

		%	CI	%	CI	%	CI	%	CI	%	CI	%	CI	%	CI	%	CI	%	CI	%	CI
NON-HODGKIN LYMPHOMA																					
ENGLAND & WALES	Men	49	(47-50)	52	(51-53)	59	(58-60)	62	(61-63)	4.7	(4.2, 5.2)	27	(26-29)	32	(31-33)	37	(36-38)	41	(40-42)	4.6	(4.1, 5.2)
	Women	51	(49-52)	55	(54-56)	62	(61-63)	65	(64-66)	4.9	(4.4, 5.4)	31	(30-32)	37	(35-38)	41	(40-42)	45	(44-46)	4.5	(4.0, 5.0)
ENGLAND	Men	49	(47-50)	52	(51-54)	59	(58-60)	62	(61-63)	4.7	(4.2, 5.2)	27	(26-29)	32	(31-34)	37	(36-38)	41	(40-42)	4.6	(4.1, 5.2)
	Women	51	(49-52)	55	(54-57)	62	(61-63)	66	(65-67)	5.1	(4.6, 5.6)	31	(30-32)	37	(35-38)	41	(40-42)	45	(44-46)	4.6	(4.1, 5.1)
Northern & Yorkshire	Men	47	(43-51)	47	(43-50)	57	(54-60)	56	(54-59)	3.7	(2.3, 5.2)	27	(23-31)	27	(24-30)	37	(33-40)	37	(34-39)	3.9	(2.4, 5.3)
	Women	48	(44-52)	54	(51-57)	60	(57-63)	61	(58-63)	4.2	(2.7, 5.6)	30	(26-33)	34	(31-38)	42	(38-45)	43	(40-46)	4.5	(3.0, 6.0)
Trent	Men	45	(41-49)	51	(48-55)	54	(51-57)	58	(55-61)	4.1	(2.5, 5.6)	25	(21-29)	31	(27-35)	33	(29-36)	38	(35-41)	4.0	(2.5, 5.6)
	Women	52	(48-57)	52	(48-56)	60	(57-63)	60	(57-63)	3.1	(1.4, 4.7)	30	(26-35)	36	(32-40)	38	(35-41)	41	(38-44)	3.2	(1.6, 4.8)
Anglia & Oxford	Men	52	(48-56)	54	(51-58)	57	(54-61)	60	(57-63)	2.8	(1.2, 4.3)	31	(27-35)	32	(29-36)	35	(32-39)	39	(36-42)	2.7	(1.1, 4.3)
	Women	57	(52-61)	56	(53-60)	58	(54-61)	64	(61-67)	2.9	(1.2, 4.5)	38	(34-43)	38	(34-42)	39	(36-43)	43	(39-46)	1.7	(-0.1, 3.4)
North Thames	Men	54	(50-58)	57	(53-61)	63	(60-67)	67	(65-70)	4.7	(3.2, 6.1)	31	(27-35)	36	(32-41)	43	(40-47)	47	(44-50)	5.4	(3.7, 7.0)
	Women	53	(50-57)	63	(59-66)	68	(65-71)	71	(68-73)	5.5	(4.2, 6.8)	34	(30-37)	44	(40-47)	47	(44-51)	51	(48-54)	5.4	(4.0, 6.9)
South Thames	Men	50	(46-53)	53	(50-57)	65	(62-67)	68	(66-70)	6.5	(5.2, 7.8)	27	(24-30)	33	(30-37)	40	(37-43)	44	(41-47)	5.7	(4.3, 7.1)
	Women	51	(48-54)	54	(51-57)	67	(64-70)	72	(70-74)	7.7	(6.5, 8.9)	30	(27-34)	34	(31-37)	45	(42-48)	49	(46-52)	6.6	(5.3, 8.0)
South & West	Men	49	(46-53)	53	(50-56)	61	(59-64)	67	(65-69)	6.1	(4.9, 7.3)	27	(24-31)	34	(31-37)	37	(35-40)	45	(42-47)	5.5	(4.2, 6.8)
	Women	49	(46-53)	57	(54-61)	61	(58-63)	70	(68-72)	6.6	(5.4, 7.9)	29	(26-32)	38	(35-41)	38	(36-41)	48	(45-50)	5.8	(4.5, 7.1)
West Midlands	Men	46	(42-50)	50	(46-55)	57	(53-61)	60	(57-63)	4.8	(3.1, 6.4)	24	(20-28)	34	(29-39)	35	(31-39)	39	(35-42)	4.6	(2.8, 6.3)
	Women	46	(42-50)	52	(48-56)	58	(54-62)	60	(57-63)	4.7	(3.1, 6.3)	26	(22-30)	33	(29-37)	38	(34-42)	40	(37-43)	4.5	(2.9, 6.1)
North & West	Men	46	(42-50)	50	(46-53)	55	(52-58)	58	(56-61)	4.1	(2.6, 5.5)	24	(21-28)	30	(26-33)	35	(32-38)	38	(36-41)	4.7	(3.3, 6.0)
	Women	50	(46-54)	54	(51-57)	61	(58-64)	63	(61-66)	4.5	(3.0, 5.9)	30	(26-33)	36	(32-39)	41	(38-45)	43	(40-45)	4.3	(2.9, 5.8)
WALES	Men	49	(43-55)	49	(44-55)	57	(52-63)	57	(53-61)	3.2	(1.0, 5.4)	30	(24-35)	28	(24-33)	35	(29-40)	38	(34-43)	3.5	(1.2, 5.7)
	Women	51	(46-57)	54	(49-60)	62	(57-66)	56	(52-60)	1.8	(-0.4, 3.9)	32	(26-38)	36	(31-42)	38	(33-43)	39	(35-43)	2.3	(0.1, 4.5)

BRAIN

Figure 4.5R

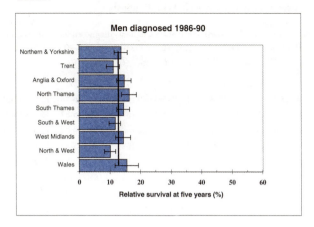

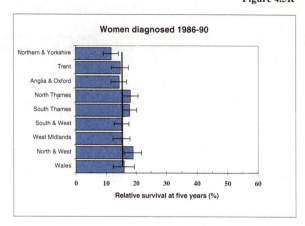

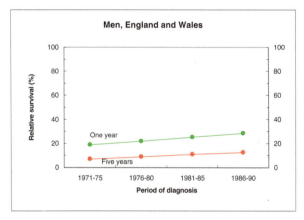

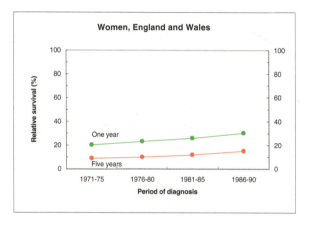

NON-HODGKIN LYMPHOMA

Figure 4.5S

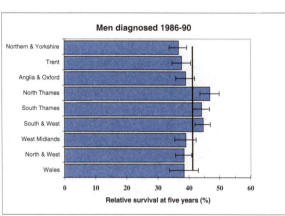

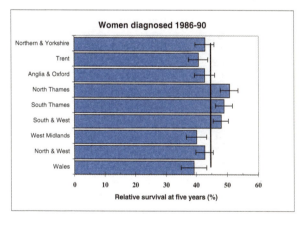

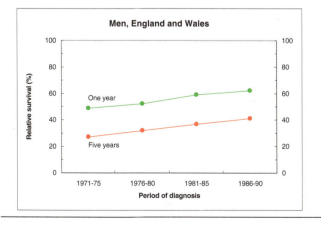

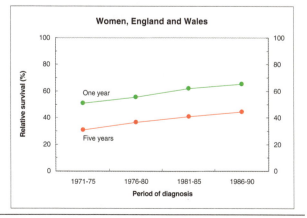

Table 4.5 (cont.) Cancer survival trends by NHS Region, selected cancers, children diagnosed 1971-90: relative survival rates (with 95% confidence intervals) at one and five years after diagnosis, and average increase in relative survival between successive five-year periods of diagnosis

Table 4.5 T		One-year survival (%), patients diagnosed in						Average increase every five years		Five-year survival (%), patients diagnosed in						Average increase every five years	
		1971-75		1976-80		1981-85		1986-90				1971-75		1976-80		1981-85		1986-90			
		%	CI	%	CI	%	CI	%	CI	%	CI	%	CI	%	CI	%	CI	%	CI	%	CI

Table 4.5 T		1971-75 %	CI	1976-80 %	CI	1981-85 %	CI	1986-90 %	CI	Avg incr %	CI	1971-75 %	CI	1976-80 %	CI	1981-85 %	CI	1986-90 %	CI	Avg incr %	CI
ACUTE LYMPHOID LEUKAEMIA (Children) [1]																					
ENGLAND & WALES	Boys	75	(71-78)	81	(78-84)	88	(85-90)	94	(92-96)	6.4	(5.4, 7.5)	36	(33-40)	51	(47-54)	69	(66-72)	73	(70-76)	12.8	(11.3, 14.2)
	Girls	78	(74-81)	84	(81-87)	87	(85-90)	92	(89-94)	4.4	(3.1, 5.7)	46	(42-50)	54	(50-58)	71	(67-75)	79	(76-82)	11.5	(9.8, 13.2)
ENGLAND	Boys	75	(71-78)	81	(78-83)	88	(85-90)	94	(92-96)	6.5	(5.4, 7.6)	36	(33-40)	50	(47-54)	69	(66-73)	73	(70-77)	12.9	(11.4, 14.4)
	Girls	79	(75-82)	84	(81-87)	87	(84-90)	92	(90-94)	4.3	(3.0, 5.6)	46	(42-50)	54	(50-58)	72	(68-75)	79	(75-82)	11.5	(9.7, 13.2)
Northern & Yorkshire	Boys	72	(63-81)	78	(69-86)	84	(76-92)	99	(97-100)	9.8	(7.3, 12.2)	35	(26-44)	49	(39-59)	65	(54-75)	76	(67-85)	13.9	(9.7, 18.0)
	Girls	65	(53-77)	89	(82-97)	87	(80-94)	90	(83-97)	4.8	(0.9, 8.7)	38	(26-51)	60	(48-72)	75	(65-84)	81	(73-90)	13.7	(9.0, 18.4)
Trent	Boys	69	(60-79)	73	(64-82)	92	(86-98)	92	(86-98)	8.1	(4.6, 11.6)	37	(27-47)	44	(33-54)	71	(61-82)	71	(61-81)	12.8	(8.3, 17.3)
	Girls	80	(70-91)	79	(70-88)	85	(76-94)	93	(87-100)	5.0	(1.3, 8.8)	44	(31-57)	49	(38-60)	69	(57-80)	81	(71-91)	13.4	(8.3, 18.4)
Anglia & Oxford	Boys	69	(59-79)	85	(77-92)	84	(76-92)	91	(85-98)	6.0	(2.4, 9.6)	35	(25-45)	58	(48-68)	71	(61-81)	73	(62-83)	12.7	(8.1, 17.4)
	Girls	82	(73-92)	87	(79-94)	87	(78-96)	90	(83-98)	2.4	(-1.4, 6.1)	50	(37-62)	55	(44-66)	68	(55-81)	81	(71-91)	10.9	(6.0, 15.9)
North Thames	Boys	78	(70-87)	86	(78-94)	93	(88-98)	94	(90-98)	4.8	(2.0, 7.7)	44	(34-54)	58	(47-69)	76	(67-85)	75	(67-83)	10.5	(6.4, 14.6)
	Girls	86	(78-95)	81	(71-91)	92	(85-98)	87	(80-95)	1.2	(-2.4, 4.8)	55	(42-68)	56	(44-69)	79	(69-89)	76	(67-85)	8.1	(3.2, 13.0)
South Thames	Boys	78	(70-87)	79	(70-87)	93	(88-99)	95	(91-100)	6.3	(3.4, 9.2)	32	(23-41)	41	(31-51)	67	(57-77)	74	(64-83)	15.0	(10.7, 19.2)
	Girls	75	(64-86)	82	(74-91)	92	(85-98)	97	(92-100)	7.1	(3.8, 10.4)	36	(24-48)	51	(40-63)	77	(67-87)	85	(75-94)	16.9	(12.2, 21.6)
South & West	Boys	80	(71-88)	82	(74-89)	90	(84-96)	95	(90-99)	5.4	(2.5, 8.3)	41	(31-51)	54	(44-63)	77	(68-85)	71	(62-81)	11.4	(7.0, 15.7)
	Girls	84	(74-93)	83	(75-92)	89	(82-96)	94	(89-100)	4.2	(0.9, 7.6)	58	(44-71)	50	(39-62)	77	(68-86)	75	(65-85)	8.4	(3.3, 13.5)
West Midlands	Boys	69	(59-79)	77	(67-86)	78	(70-87)	98	(95-100)	10.8	(7.9, 13.7)	32	(22-42)	46	(35-57)	59	(49-69)	80	(70-90)	15.7	(11.2, 20.2)
	Girls	71	(58-83)	79	(68-90)	83	(73-92)	96	(91-100)	8.7	(4.8, 12.6)	35	(22-48)	67	(55-79)	62	(49-75)	86	(77-96)	14.8	(9.8, 19.8)
North & West	Boys	78	(71-86)	85	(78-91)	87	(81-94)	88	(82-95)	3.1	(0.0, 6.2)	35	(26-44)	53	(43-62)	68	(59-78)	67	(58-77)	11.6	(7.5, 15.6)
	Girls	82	(74-89)	89	(82-96)	83	(74-92)	90	(83-97)	2.0	(-1.2, 5.2)	50	(40-59)	50	(39-61)	61	(49-73)	71	(61-81)	7.4	(2.9, 11.8)
WALES	Boys	74	(61-88)	93	(85-100)	90	(80-100)	92	(82-103)	3.6	(-1.5, 8.7)	35	(21-50)	65	(48-82)	64	(47-81)	65	(47-84)	9.8	(2.4, 17.1)
	Girls	66	(48-84)	86	(72-100)	89	(78-100)	88	(76-99)	5.4	(-0.9, 11.8)	46	(27-66)	54	(33-75)	63	(45-82)	82	(68-95)	11.9	(4.5, 19.3)

[1] Survival rates for children are not age-standardised (see text)

ACUTE LYMPHOID LEUKAEMIA   **Figure 4.5T**

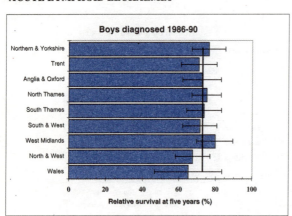

Boys diagnosed 1986-90

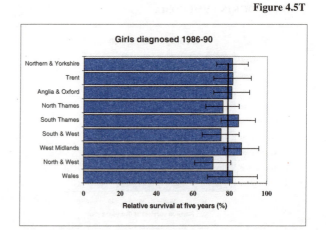

Girls diagnosed 1986-90

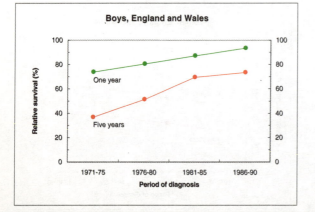

Boys, England and Wales

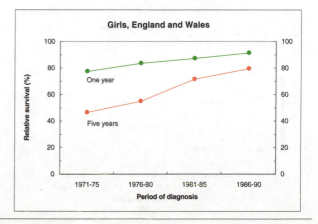

Girls, England and Wales

younger ages, while anaplastic carcinomas that are common at older ages have very poor survival. Part of the increase over time in survival from thyroid cancer among all women (aged 15-99 at diagnosis) is almost certainly due to the increase in papillary carcinoma in younger women.

Survival from the acute leukaemias is much worse in patients over 60 than for younger patients. Five-year relative survival for patients diagnosed during 1986-90 was generally less than 10% for acute myeloid leukaemia, monocytic leukaemia and acute lymphoid leukaemia. Large numbers of adults with acute myeloid leukaemia have been recruited into successive MRC trials since the early 1970s, and these have shown incremental improvements in survival for younger patients. The increases have been attributed to more intensive chemotherapy that can be better tolerated with improved supportive care. The poorer outcome in patients over 60 may be due to a lesser ability to withstand intensive chemotherapy than younger patients, but older patients are less frequently recruited into clinical trials of chemotherapy[12].

Sex

Women have higher survival than men for many cancers (Table 4.8).

For melanoma of the skin, five-year survival rates for women diagnosed during 1986-90 were more than 10% higher than for men in every age group (Table 4.7). For cancers of the tongue and salivary glands, survival was also at least 6% higher for women at every age, although not all of the age-specific differences are individually significant.

Survival for women was 5-10% higher than among men for cancers of the thyroid and the oral cavity, and for chronic lymphoid leukaemia, as well as for several less common cancers.

Although the differences between the sexes were less than 5%, five-year survival was significantly higher among women than men for several common cancers, including those of the oesophagus, stomach and rectum, as well as for non-Hodgkin lymphoma and brain cancer.

Men had significantly higher five-year survival than women for cancers of the larynx (7% higher in men) and bladder (5%). For laryngeal cancer, the sex difference in survival has been consistent since the early 1970s. It is due to the distribution of cancers within the larynx: glottal cancers are more common in men, and survival is higher than for other parts of the larynx (see Chapter 21). Breast cancer is very rare in men, but five-year survival for men

diagnosed during 1986-90 was 70%, slightly higher than for women (66%): the difference is of borderline significance.

All but one of these sex differences in survival were also observed in the EUROCARE II study for patients diagnosed during 1985-89, including patients from the UK[2]. The only exception was for multiple myeloma, for which average survival in Europe was lower in women than in men. In England and Wales, one-year survival from multiple myeloma was significantly higher for women diagnosed during 1986-90 than for men (55% vs. 51%), although five-year survival was similar.

Deprivation

Differences in survival between patients in the most affluent and most deprived groups who were diagnosed during 1981-90 in England and Wales are summarised in Table 4.9. One- and five-year survival rates for patients in the most affluent category and the difference in survival between the most affluent and most deprived groups (the deprivation gap) are estimated from linear regression of the relative survival rates for all the deprivation groups, weighted for their precision; in most cases, the regression estimate of survival for the most affluent group is close to the original value (see Chapter 3). The numbers of patients of known deprivation status included in the analyses for each five-year period are also shown.

The deprivation gap is shown as negative if survival among patients in the most deprived group is lower than survival for patients in the most affluent group. The 95% confidence interval around the deprivation gap is also given. If this range includes zero, the deprivation gap is not statistically significant at the 5% level. In interpreting deprivation gradients in survival it is also helpful to consider the wider picture, including the consistency of the survival gradient across the five deprivation groups between the sexes, at one and five years after diagnosis, and in different calendar periods. Results are provided in more detail in the chapters for each cancer.

Adults

None of the 47 different types of cancer for which survival has been examined revealed a significant survival advantage for patients in the most deprived group. This applies to one-year and five-year survival, and for patients diagnosed either during 1981-85 or during 1986-90.

For 44 of the 47 cancers, there was either limited or strong evidence of a negative survival gradient (lower survival in the most deprived group) among patients diagnosed during either 1981-85 or 1986-90.

SUMMARY

Table 4.6   Ranked five-year survival rates, England and Wales, patients diagnosed 1986-90

ADULTS

MEN		Survival rate (%)	% of all cancers[1]		WOMEN		Survival rate (%)	% of all cancers[1]
	Five-year survival 50% or more					*Five-year survival 50% or more*		
1	Testis	90	1.2		1	Lip	85	0.1
2	Lip	86	0.2		2	Melanoma of skin	82	2.1
3	Pituitary	73	0.0		3	Hodgkin's disease	73	0.4
4	Hodgkin's disease	71	0.7		4	Pituitary	72	0.0
5	Breast	70	0.2		5	Thyroid	70	0.6
6	Melanoma of skin	68	1.5		6	Eye	70	0.2
7	Penis	67	0.4		7	Uterus	70	3.6
8	Eye	67	0.2		8	Breast	66	27.7
9	Spinal cord	67	0.0		9	Salivary glands	62	0.2
10	Thyroid	64	0.2		10	Spinal cord	62	0.0
11	Larynx	63	1.6		11	Cervix	61	4.1
12	Bladder	62	8.1		12	Bladder	57	3.0
13	Connective tissue	50	0.5		13	Larynx	57	0.4
	Subtotal		14.7		14	Bone	53	0.2
					15	Chronic lymphoid leukaemia	52	0.6
					16	Oral cavity	52	0.3
					17	Vagina and vulva	51	1.0
					18	Connective tissue	50	0.5
					19	Tongue	50	0.2
						Subtotal		45.3
	Five-year survival 10% to 49%					*Five-year survival 10% to 49%*		
14	Bone	47	0.2		20	Thymus	48	0.0
15	Salivary glands	47	0.2		21	Non-Hodgkin lymphoma	45	2.6
16	Chronic lymphoid leukaemia	44	0.9		22	Nasal cavities, sinuses	40	0.2
17	Thymus	44	0.0		23	Colon	39	8.9
18	Oral cavity	43	0.5		24	Rectum	39	4.3
19	Prostate	41	12.9		25	Nasopharynx	38	0.1
20	Non-Hodgkin lymphoma	41	3.3		26	Oropharynx	37	0.1
21	Nasal cavities, sinuses	40	0.2		27	Adrenal	37	0.0
22	Colon	38	7.8		28	Kidney	35	1.5
23	Kidney	38	2.5		29	Acute lymphoid leukaemia	29	0.1
24	Rectum	36	5.8		30	Ovary	28	4.8
25	Tongue	36	0.4		31	Small intestine	24	0.2
26	Adrenal	33	0.0		32	Chronic myeloid leukaemia	23	0.4
27	Oropharynx	33	0.3		33	Hypopharynx	20	0.1
28	Nasopharynx	29	0.1		34	Multiple myeloma	19	1.2
29	Small intestine	23	0.2		35	Brain	15	1.2
30	Acute lymphoid leukaemia	22	0.2		36	Stomach	11	3.8
31	Chronic myeloid leukaemia	20	0.4			*Subtotal*		29.4
32	Hypopharynx	20	0.2					
33	Multiple myeloma	17	1.3					
34	Brain	13	1.6					
35	Monocytic leukaemia	11	0.0					
	Subtotal		39.2					
	Five-year survival less than 10%					*Five-year survival less than 10%*		
36	Gallbladder	9	0.5		37	Acute myeloid leukaemia	9	0.7
37	Stomach	9	6.3		38	Gallbladder	8	0.7
38	Acute myeloid leukaemia	8	0.8		39	Oesophagus	8	2.1
39	Oesophagus	5	3.0		40	Monocytic leukaemia	7	0.0
40	Lung	5	24.3		41	Pleura	5	0.2
41	Pleura	4	0.8		42	Lung	5	11.0
42	Liver	2	0.7		43	Liver	2	0.4
43	Pancreas	2	2.9		44	Pancreas	2	3.0
	Subtotal		39.3			*Subtotal*		18.1
	Total		**93.2**			**Total**		**92.8**

[1] Percentage of all malignant neoplasms excluding non-melanoma skin cancer registered in adults (15 and over)
in England and Wales in 1990. If less than 0.05%, the table shows 0.0

Table 4.6 (cont.) Ranked five-year survival rates, England and Wales, patients diagnosed 1986-90

CHILDREN

BOYS		Survival rate (%)	% of all cancers[2]	GIRLS		Survival rate (%)	% of all cancers[2]
	Five-year survival 50% or more				*Five-year survival 50% or more*		
1	Retinoblastoma	98	2	1	Retinoblastoma	88	3
2	Hodgkin's disease	96	6	2	Hodgkin's disease	85	3
3	Germ-cell and gonadal	88	2	3	Wilms' tumour	84	7
4	Wilms' tumour	84	5	4	Acute lymphoid leukaemia	79	25
5	Non-Hodgkin lymphoma	76	7	5	Germ-cell and gonadal	78	3
6	Acute lymphoid leukaemia	73	27	6	Non-Hodgkin lymphoma	77	4
7	Ewing's sarcoma	69	2	7	Soft tissue sarcoma	64	6
8	Soft tissue sarcoma	68	6	8	Brain and spinal cord	62	24
9	Brain and spinal cord	57	23	9	Osteosarcoma	56	3
	Subtotal		80	10	Ewing's sarcoma	50	2
					Subtotal		80
	Five-year survival 10% to 49%				*Five-year survival 10% to 49%*		
10	Osteosarcoma	47	3				
11	Hepatoblastoma	39	1	11	Hepatoblastoma	49	1
	Subtotal		3				
	Total		83		**Total**		81

[2] Percentage of malignant neoplasms in children (0-14 years) in England and Wales in 1978-87

For 36 types of cancer, there is some evidence of lower five-year survival among patients in the most deprived group, both for those diagnosed during 1981-85 and for those diagnosed during 1986-90. Some of the deprivation gaps are small, but for 22 cancers, the gap in five-year survival for patients diagnosed during 1986-90 was statistically significant. For 20 of these cancers, both one-year and five-year survival were significantly lower among patients in the most deprived group, both those diagnosed during 1981-85 and those diagnosed during 1986-90 (Table 4.9).

Among eight of the other 11 cancers examined for adults, there is some evidence for lower one-year survival among patients in the most deprived group, both for those diagnosed during 1981-85 and for those diagnosed during 1986-90. Some of the deprivation gaps in one-year survival are also small, but for cancers of the stomach, pancreas and liver, the deprivation gap was statistically significant both for patients diagnosed in 1981-85 and for those diagnosed during 1986-90. No significant difference between deprivation categories was found in five-year survival for these three cancers, but overall five-year survival is only 9-11% for cancer of the stomach, and only 2% for cancers of the pancreas and liver.

For three cancers, those of the lip, hypopharynx and spinal cord, there is no consistent evidence of a deprivation gap in survival.

The relative survival rates for each deprivation category take account of the underlying mortality in the same deprivation category of the general population at each single year of age and for each sex, region and calendar period. This approach has been described as the gold standard for evaluating socioeconomic differences in cancer survival[13].

Taken together, the results from analysis of this very large data set provide overwhelming evidence that cancer survival in adults in England and Wales is generally lower among patients in more deprived groups than those in more affluent groups, even after allowance is made for the higher mortality from all causes of death in the more deprived groups.

All cancers combined in adults

The relative survival curves for all cancers combined up to five years after diagnosis are shown in Figure 4.9, for men and for women in each deprivation category who were diagnosed in England and Wales during 1986-90 and followed up to the end of 1995. The survival curves are clearly separate before the end of the first year after diagnosis. Each curve is based on 65,000 to 75,000 men or women.

Relative survival rates for all cancers combined in each deprivation category reflect this general pattern (Table 4.9A).

SUMMARY

Special caution is required in interpreting the patterns of survival between deprivation groups for all cancers combined, as for trends in survival over time (see above). The composition of "all cancers combined" is not the same for all deprivation groups, and this would affect overall survival from all cancers combined even if survival rates for each cancer were the same in all deprivation groups. Breast cancer is more common in affluent groups than in the deprived, for example, and cancers of the stomach and lung are more common in deprived groups: survival rates for these cancers differ widely.

The survival gradient for all cancers combined between patients in the various deprivation categories clearly represents a weighted average of the widely differing gradients for both common and rare cancers, as well as the difference in composition of "all cancers combined" in each deprivation category. Since the deprivation gradients in survival occur for so many individual cancers, however, it seems reasonable to assess the overall influence of deprivation on survival from all cancers combined, as a public health measure of overall differences in cancer survival in each deprivation group, given the current profile of cancer in each group.

One-year and five-year survival rates for all cancers combined are 2-4% lower among patients in each successive deprivation category, from the most affluent to the most deprived, both for those diagnosed in 1981-85 and those diagnosed during 1986-90.

Overall, the estimated difference in survival for all cancers combined between patients in the most affluent and most deprived groups was in the range 11-13%. This difference is partly attributable to differences in the composition of "all cancers combined" between deprivation categories, as explained above. Again, however, there is evidence of a deprivation gradient in survival for many adult cancers when they are examined separately.

There is no evidence that the differences in survival between deprivation categories for all cancers combined became any smaller for patients diagnosed during 1986-90 compared to those diagnosed during 1981-85.

The deprivation gradients in one-year and five-year survival for all cancers in adults combined are shown for adults diagnosed during the decade 1981-90 (Figure 4.10). The almost linear relationship between relative survival and the five deprivation categories is apparent.

Children

Among children, the deprivation gaps in one-year or five-year survival are not statistically significant for any of the malignancies examined here among patients diagnosed

either during 1981-85 or during 1986-90. For all leukaemias combined, the deprivation gap in one-year survival was just significant at 4.7% for children diagnosed during 1986-90, but acute lymphoid leukaemia is the most common leukaemia in children, and the evidence for any difference in survival between deprivation categories for this leukaemia is not strong. Both one-year and five-year survival were 2-4% lower among the most deprived group during the 1980s, but these differences are not significant. More detailed analyses confirm the absence of a significant deprivation gradient after adjustment for age, sex and region of residence[8].

Ranked differences in survival

For 41 adult cancers and two childhood cancers for which more than 1,000 patients of known deprivation status were included in the analyses for the period 1986-90, the deprivation gaps in five-year survival are ranked in Table 4.10.

The difference in survival between patients in the most affluent and most deprived groups exceeds 5% for 16 cancers in adults, and is significant for all but two of these: cancers of the salivary glands and of the nasal cavities and paranasal sinuses.

Among 20 adult malignancies for which the deprivation gap in five-year survival is less than 5%, it is significant for seven of the most common cancers; those of lung, colon, prostate, ovary, cervix, uterus and kidney. Apart from lung cancer, the relative survival rates at five years for patients in the most affluent group are 30% or higher for each of these cancers. The relative survival rate for lung cancer at five years in the most affluent group is only about 6%, and the estimated difference in five-year survival between the most affluent and most deprived groups is only 1%. The statistical significance of this small difference is partly due to the very large numbers of patients.

For five malignancies (acute lymphoid leukaemia and cancers of the stomach, lip, bone and the pleura), there was either no difference in five-year survival between deprivation groups, or survival was slightly higher among patients in the most deprived group, but none of these differences was significant. For stomach cancer, the most common of these five malignancies, survival at *one* year was 2-3% lower in the most deprived group than the most affluent group for patients diagnosed during both 1981-85 and 1986-90 (Table 4.9).

Avoidable deaths

The overall consistency of cancer survival differences

between adults in the various deprivation categories raises questions for public health. The reasons for these differences are difficult to identify with confidence[14]. They are part of a wide spectrum of socioeconomic inequalities in health and disease[15, 16].

Wide cancer survival differences between socioeconomic groups defined by housing tenure have been observed in England and Wales for 11 of 13 neoplasms in men and 12 of 15 in women diagnosed during 1971-81[17]. Similar differences in survival have been identified for many cancers in the UK[18-26] and in other countries[27-34] with various measures of socioeconomic status, and artefact is an unlikely explanation.

Possible explanations for lower survival in more deprived groups include longer delay in diagnosis or more advanced disease. at diagnosis, worse general health or resistance to malignancy, different histological type or more aggressive disease, poorer access to optimal care, and lower compliance with treatment[35,36]. Differences in stage of disease did not account for the survival differences observed between deprivation categories for breast cancer in Glasgow[18], however, or for several common cancers including breast cancer in south-east England[20]: in both studies, the deprivation categories were defined in the same way as for the analyses reported here. Geographic or socioeconomic differences in investigation, treatment or survival have also been reported for a number of common cancers[37-48]. One population-based study reported that five-year survival from breast cancer might be expected to increase by 4-5% if best surgical practice were used by all clinicians[49].

The public health impact of the differences in survival between deprivation groups can be addressed by considering the number of deaths attributable to cancer that might be avoidable if patients in all groups of society were to have the same relative survival as that actually observed for patients in the most affluent category.

This is equivalent to asking how many deaths would occur among patients with a given cancer if excess mortality in each deprivation category were the same as that in the most affluent category. Excess mortality is defined for these purposes as the number of deaths observed among patients of a given age at diagnosis in excess of the number that would have occurred, within a given time from diagnosis, if the cancer patients had only had the same mortality as the corresponding group of the general population. The corresponding group is defined by age, sex, deprivation category and calendar period of death. Similar approaches have been adopted in estimating the potential reduction in cancer deaths from eliminating regional and social class variation in survival in the Nordic countries[50].

The working assumption is that cancer patients in deprivation categories 2, 3, 4, and 5 (most deprived) have the same relative survival rates as those observed for category 1 (most affluent), at each age. The number of deaths in excess of general population mortality that would be expected under this assumption can then be compared with the total number of excess deaths actually observed. The difference between the number of excess deaths actually observed within (say) five years and the number expected, under the assumption of equality of five-year relative survival in all deprivation categories, is an estimate of the number of avoidable deaths for that cancer within five years of diagnosis.

Estimates of the number of avoidable deaths within five years of diagnosis were made for 41 of the 47 cancers examined for adults. The six least common malignancies (nasopharynx, thymus, spinal cord, adrenal, pituitary, monocytic leukaemia) were excluded: collectively, they account for only 0.25% of adult malignancies in England and Wales. The estimates were based on the age-specific relative survival rates of patients in each deprivation category who were diagnosed in England and Wales during 1986-90. The numbers of avoidable deaths were summed across all 41 cancers and within each deprivation category.

Almost 770,000 adults diagnosed in England and Wales during 1986-90 and of known deprivation status were included in the analyses. Among these patients, some 493,000 deaths were observed in excess of the number that would have occurred had they been subject only to the same mortality rates as the general population in each category of age, sex, deprivation and calendar period. When the calculation was repeated after patients in each deprivation category were assumed to have the same relative survival rate at each age as patients in the most affluent category, a total of 480,000 excess deaths would have been expected by the end of 1995.

The total number of avoidable deaths, calculated as the difference between the two estimates, is about 12,700, representing some 2.6% of all deaths attributable to the 41 adult cancers within five years of diagnosis (Table 4.11A).

For breast cancer, the 2,800 avoidable deaths represent some 7% of the approximately 40,000 deaths attributable to breast cancer within five years of diagnosis. One previous study suggested 7.4% of all deaths among 30,000 women with breast cancer diagnosed during 1980-89 might be avoidable on a similar basis[19].

Even though the deprivation gap among lung cancer patients was only 1%, the public health significance is large in terms of the number of avoidable deaths, because lung cancer is so common and survival is so low. Among

SUMMARY

Table 4.7   Five-year survival by age at diagnosis and sex, England and Wales, adults diagnosed 1986-90:
number of patients, relative survival (%) and 95% confidence interval (CI)

SITE		No.	15-39 %	CI	No.	40-49 %	CI	No.	50-59 %	CI	No.	60-69 %	CI	No.	70-79 %	CI	No.	80-99 %	CI
Lip	Men	19	95	(67-99)	52	91	(79-97)	118	90	(81-95)	258	91	(83-95)	301	84	(74-90)	145	80	(57-92)
	Women	4	(a)	-	9	100	(a)	27	93	(73-98)	49	91	(73-97)	80	83	(66-92)	83	74	(49-88)
Tongue	Men	84	61	(50-71)	200	42	(35-49)	367	41	(35-46)	524	34	(29-38)	389	33	(27-39)	138	23	(13-34)
	Women	48	67	(52-79)	81	57	(45-67)	147	58	(49-66)	248	49	(42-55)	322	49	(42-55)	230	30	(22-39)
Salivary glands	Men	73	80	(69-87)	71	75	(62-83)	128	46	(37-55)	204	46	(38-53)	209	40	(31-49)	125	27	(15-41)
	Women	69	90	(80-95)	75	87	(77-93)	107	72	(62-80)	154	58	(49-66)	175	58	(48-66)	154	39	(27-50)
Oral cavity	Men	69	54	(41-65)	250	50	(43-56)	538	50	(46-55)	717	44	(40-48)	518	34	(29-40)	209	37	(26-48)
	Women	55	75	(61-84)	86	70	(59-79)	183	57	(49-64)	411	53	(47-58)	394	47	(41-53)	264	30	(22-38)
Oropharynx	Men	30	70	(50-83)	139	44	(35-52)	286	31	(25-36)	384	32	(27-37)	212	36	(28-44)	79	14	(6-26)
	Women	16	70	(41-86)	46	46	(31-60)	89	51	(40-61)	161	31	(24-39)	114	29	(20-39)	64	19	(8-33)
Nasopharynx	Men	63	42	(30-54)	68	40	(28-51)	101	31	(22-40)	145	32	(24-40)	100	12	(6-21)	20	7	(1-23)
	Women	45	65	(49-77)	32	47	(29-63)	46	37	(23-51)	75	30	(20-41)	53	35	(21-49)	26	0	(0-8)
Hypopharynx	Men	10	30	(7-58)	72	27	(18-38)	214	17	(12-23)	339	25	(20-30)	262	16	(11-22)	98	14	(6-25)
	Women	5	(a)	-	38	24	(12-38)	98	27	(18-36)	183	25	(18-32)	193	14	(9-20)	100	8	(3-17)
Oesophagus	Men	114	14	(8-21)	528	9	(7-12)	1,710	8	(7-10)	3,980	7	(6-8)	4,134	5	(4-6)	1,696	2	(1-2)
	Women	53	26	(15-38)	173	18	(13-24)	732	12	(10-15)	1,905	11	(10-12)	3,050	6	(5-7)	2,655	3	(2-4)
Stomach	Men	261	20	(16-25)	883	18	(15-20)	3,179	14	(13-15)	8,283	11	(10-11)	10,036	8	(7-8)	4,652	4	(3-5)
	Women	189	23	(17-29)	398	19	(15-23)	1,114	18	(16-21)	3,209	15	(13-16)	5,869	9	(8-10)	5,512	5	(4-6)
Small intestine	Men	27	44	(25-61)	68	37	(26-49)	145	29	(21-36)	211	30	(24-37)	200	15	(10-22)	81	8	(2-16)
	Women	19	42	(20-62)	43	27	(15-41)	111	29	(20-37)	217	22	(16-28)	223	21	(15-27)	143	21	(13-30)
Colon	Men	446	48	(43-52)	1,344	45	(42-47)	3,809	42	(40-44)	9,043	40	(38-41)	11,289	39	(38-40)	5,720	32	(30-34)
	Women	448	49	(44-54)	1,326	45	(43-48)	3,626	44	(42-45)	8,418	43	(42-44)	12,337	41	(40-42)	10,675	30	(29-31)
Rectum	Men	279	44	(38-49)	1,085	43	(40-46)	3,514	41	(39-42)	7,835	39	(37-40)	8,518	36	(35-37)	3,833	30	(27-32)
	Women	262	48	(42-54)	786	46	(43-50)	2,157	45	(43-48)	4,739	43	(41-44)	6,383	40	(38-41)	4,997	26	(24-28)
Liver	Men	77	15	(8-24)	112	3	(1-7)	368	4	(3-6)	760	2	(1-3)	687	2	(1-3)	249	0	(0-2)
	Women	68	11	(5-19)	68	4	(1-11)	176	4	(2-8)	323	3	(1-5)	454	1	(0-2)	324	0	(0-1)
Gallbladder	Men	18	29	(11-50)	76	25	(16-36)	246	16	(12-21)	598	11	(9-14)	726	6	(4-8)	352	5	(2-8)
	Women	22	15	(4-31)	108	18	(11-26)	295	11	(8-15)	672	11	(9-14)	1,026	7	(6-9)	846	4	(3-6)
Pancreas	Men	105	12	(7-18)	474	5	(3-7)	1,462	4	(3-5)	3,410	3	(2-4)	3,938	2	(1-2)	1,825	2	(1-3)
	Women	97	12	(7-20)	300	6	(4-9)	984	4	(3-6)	2,744	2	(2-3)	4,109	2	(2-2)	3,369	1	(1-2)
Nasal cavities, sinuses	Men	43	49	(33-63)	77	52	(40-62)	186	50	(43-58)	317	37	(31-43)	296	36	(29-43)	99	35	(20-50)
	Women	38	50	(33-65)	45	56	(40-69)	85	50	(39-60)	162	42	(34-50)	197	37	(29-45)	166	21	(13-30)
Larynx	Men	69	83	(72-90)	431	69	(65-74)	1,391	66	(63-68)	2,725	63	(61-65)	1,963	62	(59-66)	616	53	(45-60)
	Women	27	82	(61-92)	107	71	(61-79)	285	69	(63-74)	557	57	(52-61)	430	49	(44-55)	186	37	(27-47)
Lung	Men	467	18	(15-22)	2,941	12	(11-13)	12,227	9	(8-10)	35,687	7	(6-7)	36,984	3	(3-4)	13,382	1	(1-2)
	Women	409	24	(20-28)	1,644	11	(9-12)	5,569	9	(9-10)	15,677	6	(6-7)	14,791	3	(3-4)	6,297	2	(2-2)
Pleura	Men	29	5	(1-16)	211	6	(3-9)	561	4	(3-6)	919	3	(2-4)	686	4	(3-6)	198	4	(1-9)
	Women	9	11	(1-38)	45	3	(0-11)	90	8	(4-15)	206	7	(4-11)	176	3	(1-7)	74	2	(0-9)
Thymus	Men	19	58	(33-76)	17	53	(27-73)	23	46	(25-65)	22	37	(17-57)	9	20	(3-49)	2	(a)	-
	Women	13	77	(44-92)	13	61	(30-82)	13	47	(19-71)	17	36	(14-59)	14	8	(0-31)	5	(a)	-
Bone	Men	375	61	(55-65)	94	58	(48-68)	122	42	(33-51)	156	36	(28-45)	115	41	(29-52)	56	17	(7-32)
	Women	232	66	(60-72)	68	70	(57-79)	74	58	(45-68)	116	41	(32-51)	117	38	(28-48)	105	24	(14-36)
Connective tissue	Men	439	57	(52-61)	244	60	(53-66)	330	50	(44-55)	508	46	(41-51)	466	44	(38-50)	191	46	(33-58)
	Women	357	65	(60-70)	215	62	(55-68)	283	55	(49-61)	393	45	(40-50)	393	40	(34-46)	307	33	(25-41)
Melanoma of skin	Men	1,112	76	(73-78)	1,050	74	(71-77)	1,104	70	(67-73)	1,331	65	(62-68)	974	63	(58-68)	393	44	(35-53)
	Women	2,133	90	(89-91)	1,734	87	(85-89)	1,655	85	(83-86)	1,913	80	(78-82)	1,604	74	(71-77)	937	61	(55-67)

(a)   Too few patients for reliable estimation of survival rate, or confidence interval not computed

Table 4.7 (cont.) Five-year survival by age at diagnosis and sex, England and Wales, adults diagnosed 1986-90: number of patients, relative survival (%) and 95% confidence interval (CI)

SITE		15-39 No.	%	CI	40-49 No.	%	CI	50-59 No.	%	CI	60-69 No.	%	CI	70-79 No.	%	CI	80-99 No.	%	CI
Breast	Men	16	88	(59-97)	60	68	(54-78)	134	72	(62-79)	216	74	(65-80)	304	78	(69-85)	126	41	(26-56)
	Women	7,076	66	(65-67)	18,195	73	(72-74)	23,533	70	(69-71)	29,593	70	(70-71)	24,155	64	(63-65)	14,331	46	(44-47)
Cervix	Women	5,370	78	(77-79)	3,650	72	(70-73)	2,866	61	(59-62)	3,671	54	(53-56)	2,473	39	(36-41)	1,078	20	(17-24)
Uterus	Women	275	86	(81-90)	1,114	84	(82-86)	3,926	83	(81-84)	5,149	74	(73-76)	4,106	59	(57-61)	1,938	42	(39-45)
Ovary	Women	1,270	69	(66-72)	2,424	43	(41-45)	4,466	34	(33-36)	5,949	25	(23-26)	4,784	17	(16-18)	2,348	12	(10-13)
Vagina and vulva	Women	162	81	(74-86)	198	79	(72-84)	408	66	(61-70)	902	56	(52-59)	1,533	48	(45-51)	1,457	38	(34-42)
Prostate	Men	33	39	(23-55)	171	31	(24-38)	2,214	43	(40-45)	12,598	46	(45-47)	23,384	43	(42-44)	13,510	34	(33-36)
Testis	Men	3,848	92	(91-93)	1,027	92	(90-94)	390	87	(82-90)	170	74	(65-81)	104	69	(53-81)	42	36	(15-57)
Penis	Men	58	78	(65-87)	109	79	(70-86)	203	70	(62-76)	339	66	(60-72)	398	65	(57-71)	218	63	(48-75)
Bladder	Men	448	88	(85-91)	1,273	83	(80-85)	4,370	74	(72-75)	10,820	67	(66-68)	12,715	59	(58-60)	5,913	48	(45-50)
	Women	202	77	(71-83)	380	77	(72-81)	1,386	75	(72-77)	3,449	64	(62-66)	4,745	53	(51-54)	3,617	40	(38-43)
Kidney	Men	272	57	(51-63)	717	50	(47-54)	1,916	42	(40-45)	3,098	39	(37-41)	2,637	33	(31-36)	887	25	(20-29)
	Women	183	51	(44-58)	392	45	(40-49)	850	42	(38-45)	1,614	39	(36-41)	1,707	31	(28-33)	897	19	(16-22)
Eye	Men	75	82	(71-89)	92	73	(62-81)	174	72	(64-79)	257	68	(61-75)	209	67	(56-76)	65	38	(20-56)
	Women	73	84	(73-91)	98	84	(74-90)	145	72	(63-79)	236	64	(57-71)	200	67	(57-75)	113	59	(42-73)
Brain	Men	1,115	43	(40-46)	977	22	(19-24)	1,526	8	(6-9)	2,082	3	(3-4)	1,066	2	(1-3)	176	1	(0-4)
	Women	836	49	(45-52)	623	27	(24-31)	984	10	(8-12)	1,459	3	(3-4)	930	3	(2-5)	227	5	(3-9)
Thyroid	Men	192	95	(90-97)	109	78	(68-85)	168	61	(52-68)	260	58	(51-65)	179	33	(25-42)	84	34	(19-49)
	Women	721	99	(98-100)	369	96	(92-97)	349	80	(75-84)	438	56	(51-61)	456	46	(40-51)	258	16	(11-22)
Spinal cord	Men	50	70	(55-81)	21	81	(56-93)	17	60	(33-79)	19	50	(24-71)	7	(a)	-	0	(a)	-
	Women	23	78	(55-90)	17	77	(49-91)	18	52	(28-72)	12	16	(2-42)	6	(a)	-	3	(a)	-
Adrenal	Men	28	53	(33-69)	20	22	(8-42)	38	31	(17-46)	48	32	(19-46)	28	20	(7-38)	2	(a)	-
	Women	40	52	(36-66)	18	56	(30-75)	33	32	(17-48)	35	29	(15-44)	26	24	(9-42)	11	0	(a)
Pituitary	Men	24	92	(70-98)	18	95	(63-99)	28	73	(51-86)	31	69	(45-84)	11	21	(3-50)	6	(a)	-
	Women	23	87	(65-96)	16	87	(58-97)	22	72	(48-87)	30	64	(43-80)	14	40	(14-66)	4	(a)	-
Non-Hodgkin lymphoma	Men	1,289	63	(60-65)	1,369	59	(56-61)	2,105	53	(50-55)	3,444	43	(41-44)	3,134	30	(28-32)	1,298	22	(19-26)
	Women	726	67	(64-71)	936	66	(62-69)	1,601	57	(54-59)	2,653	46	(43-48)	3,240	33	(31-35)	1,924	23	(20-26)
Hodgkin's disease	Men	1,534	86	(84-87)	432	76	(72-80)	387	66	(60-70)	327	49	(42-55)	198	29	(21-37)	53	13	(4-27)
	Women	1,222	86	(84-88)	180	80	(73-85)	173	69	(61-75)	220	45	(38-52)	195	31	(24-39)	100	20	(10-32)
Multiple myeloma	Men	43	51	(35-65)	218	32	(26-38)	796	28	(25-31)	1,602	21	(19-23)	1,820	13	(11-14)	913	7	(5-10)
	Women	40	55	(39-69)	172	40	(33-48)	566	32	(28-36)	1,297	20	(18-23)	1,781	16	(15-18)	1,282	9	(7-12)
Acute lymphoid leukaemia	Men	336	38	(32-43)	67	24	(15-35)	59	15	(7-25)	85	7	(3-13)	82	4	(1-10)	45	1	(0-4)
	Women	159	44	(36-52)	53	36	(23-49)	49	24	(13-37)	59	16	(8-26)	76	8	(3-16)	50	3	(1-10)
Chronic lymphoid leukaemia	Men	23	79	(56-91)	101	78	(68-85)	437	62	(57-66)	1,072	54	(51-58)	1,335	42	(38-45)	724	25	(20-30)
	Women	13	84	(51-96)	53	82	(68-90)	218	71	(64-77)	535	65	(61-70)	906	49	(45-53)	909	32	(27-36)
Acute myeloid leukaemia	Men	326	25	(20-29)	218	18	(14-24)	374	12	(9-15)	785	6	(5-8)	854	2	(1-3)	358	0	(0-2)
	Women	323	37	(32-42)	223	22	(17-28)	338	11	(8-15)	577	6	(4-8)	728	1	(1-2)	522	0	(0-1)
Chronic myeloid leukaemia	Men	164	41	(33-48)	128	33	(25-42)	187	22	(16-29)	322	20	(15-24)	466	15	(11-19)	300	12	(7-18)
	Women	120	54	(45-63)	87	35	(25-45)	144	24	(17-31)	229	26	(20-32)	359	17	(13-21)	390	10	(7-15)
Monocytic leukaemia	Men	14	32	(11-56)	8	49	(15-77)	12	8	(1-27)	29	8	(2-21)	56	0	(a)	36	0	(0-8)
	Women	7	(a)	-	11	17	(3-43)	19	12	(3-28)	22	4	(0-15)	50	9	(3-19)	45	2	(0-9)
All leukaemias	Men	916	35	(32-38)	560	34	(30-38)	1,127	34	(31-37)	2,473	29	(27-31)	3,077	21	(19-23)	1,626	13	(10-15)
	Women	655	43	(39-46)	460	34	(30-38)	815	30	(27-34)	1,547	30	(28-33)	2,336	22	(20-24)	2,165	15	(13-17)

(a) Too few patients for reliable estimation of survival rate, or confidence interval not computed

Table 4.8   Differences in survival between women and men, England and Wales, patients diagnosed during 1986-90: five-year relative survival[1] (%) and number of patients included in analyses

	Women		Men		Difference in five-year survival (%)
	No. of patients	Five-year survival	No. of patients	Five-year survival	
Salivary glands	734	62	810	47	
Tongue	1,076	50	1,702	36	
Melanoma of skin	9,976	82	5,964	68	
Oral cavity	1,393	52	2,301	43	
Nasopharynx	277	38	497	29	
Chronic lymphoid leukaemia	2,634	52	3,692	44	
Acute lymphoid leukaemia	446	29	674	22	
Thyroid	2,591	70	992	64	
Bone	712	53	918	47	
Oropharynx	490	37	1,130	33	
Thymus	75	48	92	44	
Adrenal	163	37	164	33	
Non-Hodgkin lymphoma	11,080	45	12,639	41	
Chronic myeloid leukaemia	1,329	23	1,567	20	
Oesophagus	8,568	8	12,162	5	
Brain	5,059	15	6,942	13	
Rectum	19,324	39	25,064	36	
Multiple myeloma	5,138	19	5,392	17	
Eye	865	70	872	67	
Stomach	16,291	11	27,294	9	
Pleura	600	5	2,604	4	
Hodgkin's disease	2,090	73	2,931	71	
Colon	36,830	39	31,651	38	
Acute myeloid leukaemia	2,711	9	2,915	8	
Hypopharynx	617	20	995	20	
Small intestine	756	24	732	23	
Connective tissue	1,948	50	2,178	50	
Lung	44,387	5	101,688	5	
Liver	1,413	2	2,253	2	
Pancreas	11,603	2	11,214	2	
Nasal cavities, sinuses	693	40	1,018	40	
Gallbladder	2,969	8	2,016	9	
Pituitary	109	72	118	73	
Lip	252	85	893	86	
Kidney	5,643	35	9,527	38	
Breast	116,883	66	856	70	
Monocytic leukaemia	154	7	155	11	
Bladder	13,779	57	35,539	62	
Spinal cord	79	62	114	67	
Larynx	1,592	57	7,195	63	

Chart: horizontal bar chart of "Difference in five-year survival (%)" with axis from 10 on left (labelled 10, 5, 0, 5, 10, 15, 20). Bars to the right labelled "Survival higher in women"; bars to the left labelled "Survival higher in men".

[1] Age-standardised relative survival rates, rounded to nearest whole number: where values are equal for men and women, actual difference may be from -1 to +1.
See Table 4.4 for confidence intervals around survival rate for each sex.

more than 144,000 patients diagnosed during 1986-90, there were almost 137,000 deaths in excess of general population mortality by the end of 1995, some 1,300 more than would have been expected if all patients had experienced the same relative survival as patients in the most affluent category.

The deprivation gradient in survival for many adult cancers is reflected in the total numbers of avoidable deaths from these 41 cancers among adults in the various deprivation categories. Defined as above, avoidable deaths are by definition zero for the most affluent group. The proportion of all deaths attributable to cancer that would be avoidable if all patients had the same relative survival as those in the most affluent group was progressively higher among the other deprivation groups (Table 4.11B). More than 5,000 avoidable deaths occurred in the most deprived category, or 5.5% of all deaths attributable to cancer in this group of patients within five years of diagnosis.

International comparison

Five-year relative survival rates for men and women diagnosed in England and Wales during 1986-90 were compared with the corresponding survival rates for Europe (1985-89) and the USA (1986-90). These comparisons are also referred to in the chapters for many individual cancers (Chapters 5-62).

For England and Wales, the national average five-year relative survival rate is given for patients diagnosed during 1986-90, together with the survival rate for patients in the most affluent category in England and Wales, and the highest survival rate observed for the same period for any of the NHS Regions of England, or for Wales (Table 4.12). The national survival rate for Scotland (1985-89) and the average survival rate for European countries participating in the second EUROCARE study were taken, with permission, from the forthcoming monograph[2]. Relative survival rates for the states and areas of the USA covered by the Surveillance Epidemiology and End Results (SEER) programme were computed from the SEER data set for public use[51].

It was not possible to use a common set of standard age weights for each cancer for England and Wales, Europe and the USA, so relative survival rates are not age-standardised for this comparison. The survival rates are summarised in Table 4.12 for the purposes of simple comparison. Confidence intervals are not shown in the table, but these are very large data sets, and precision is indicated by the 95% confidence intervals for selected cancers for England and Wales and the USA in Figure 4.12.

Europe

For most of the common cancers, five-year survival for patients diagnosed in England Wales during 1978-85 was lower than in several comparable European countries[52]. The results in Table 4.12 suggest that this pattern has not changed greatly for patients diagnosed during 1986-90.

The average five-year survival rate for men in Europe was at least 5% higher than for men in England and Wales for ten of 38 cancers for which suitable data were available, including those of prostate (14% higher, i.e. 56% vs. 42%), kidney (9%) and colon (8%). Five-year survival in England and Wales was no more than 5% below the European average for 19 cancers, including those of the lung (4%) and bladder (3%). For the remaining nine cancers, estimated five-year survival for men in England and Wales was the same as or higher than the average for Europe, including cancer of the oropharynx (5% higher than the European average) and Hodgkin's disease.

For women, the average five-year survival rate for Europe was at least 5% higher than in England and Wales for 17 of 39 cancers for which suitable data were available, including those of kidney (14%), stomach (13%), colon (8%), ovary (6%), breast (5%) and lung (5%). The deficit compared to the European average was less than 5% for 15 cancers, including those of rectum (4%), brain (4%), gallbladder (4%) and non-Hodgkin lymphoma (3%). For the last seven of the 39 malignancies, survival in England and Wales was the same as or slightly higher than the European average, including cervical cancer (2%) and melanoma of the skin (2%). Survival from melanoma of the skin is 4% higher in Scotland than in England and Wales for men and women: similar differences have been noted since the early 1970s[53].

USA

Relative survival at five years for men in the USA SEER programme area was at least 5% higher than for men in England and Wales for 25 of 39 cancers for which suitable data were available, including those of prostate (44% higher, i.e. 86% vs. 42%), colon (25% higher), rectum (23%), kidney (19%), bladder (19%) and melanoma of the skin (15%) (Table 4.12). Five-year survival estimates for men in the USA were up to 4% higher than in England and Wales for eight cancers, including those of oesophagus (4%), larynx (4%) and testis (4%) and non-Hodgkin lymphoma (4%). For the remaining six cancers, including cancers of the oropharynx and pancreas, estimated five-year survival for men in England and Wales was the same as or higher than in the USA.

For women, five-year survival in the USA was 5% or more higher than in England and Wales for 33 of 40 cancers for

SUMMARY

Table 4.9

Difference (gap) in survival between affluent and deprived groups in England and Wales, by period of diagnosis: total number of patients, survival in affluent group, gap and 95% confidence interval around the gap (low, high)

ADULTS	Period of diagnosis	No. of patients[1]	One-year survival (%)				Five-year survival (%)			
			Affluent	Gap[2]	Low	High	Affluent	Gap[2]	Low	High
Lip	1981-85	1,287	98.3	0.7	-2.8 to	4.2	88.9	-0.2	-9.2 to	8.9
	1986-90	1,131	95.5	2.7	-1.9 to	7.3	89.0	0.6	-8.6 to	9.9
Tongue	1981-85	2,629	71.0	-12.4	-17.9 to	-6.8	46.9	-11.3	-17.4 to	-5.2
	1986-90	2,749	75.5	-16.1	-21.2 to	-10.9	50.4	-16.3	-22.2 to	-10.4
Salivary glands	1981-85	1,497	84.3	-9.4	-15.9 to	-3.0	64.0	-7.7	-16.2 to	0.8
	1986-90	1,521	81.8	-5.6	-12.2 to	1.0	61.2	-6.2	-15.0 to	2.6
Oral cavity	1981-85	3,448	77.1	-7.2	-11.7 to	-2.6	55.2	-12.5	-18.1 to	-6.9
	1986-90	3,663	76.6	-5.0	-9.3 to	-0.6	53.9	-11.6	-17.0 to	-6.2
Oropharynx	1981-85	1,578	67.7	-11.7	-18.8 to	-4.6	38.5	-10.7	-18.1 to	-3.4
	1986-90	1,603	73.5	-15.1	-22.0 to	-8.3	41.9	-12.0	-19.4 to	-4.5
Nasopharynx	1981-85	847	63.7	-5.8	-15.6 to	4.1	34.4	-4.7	-15.1 to	5.8
	1986-90	762	72.2	-13.8	-24.0 to	-3.7	42.0	-16.9	-27.6 to	-6.3
Hypopharynx	1981-85	1,811	44.5	2.2	-4.8 to	9.2	18.6	0.3	-5.6 to	6.3
	1986-90	1,588	47.5	1.9	-5.5 to	9.2	21.0	-2.6	-9.2 to	3.9
Oesophagus	1981-85	17,575	22.2	-4.7	-6.4 to	-3.0	6.7	-1.8	-2.9 to	-0.8
	1986-90	20,475	24.3	-3.7	-5.4 to	-2.0	6.4	-0.2	-1.2 to	0.9
Stomach	1981-85	47,152	22.8	-3.1	-4.2 to	-2.1	8.7	-1.0	-1.8 to	-0.2
	1986-90	43,100	25.1	-2.4	-3.6 to	-1.2	9.3	0.4	-0.5 to	1.3
Small intestine	1981-85	1,421	52.7	-9.9	-17.8 to	-2.0	27.2	-5.9	-13.1 to	1.4
	1986-90	1,475	60.0	-20.8	-28.2 to	-13.3	27.9	-7.3	-14.1 to	-0.4
Colon	1981-85	63,599	59.4	-7.3	-8.5 to	-6.1	37.6	-4.8	-6.1 to	-3.5
	1986-90	67,741	62.1	-5.6	-6.8 to	-4.5	40.3	-4.3	-5.5 to	-3.0
Rectum	1981-85	43,544	66.8	-7.8	-9.3 to	-6.4	39.0	-6.7	-8.3 to	-5.2
	1986-90	43,870	69.1	-6.8	-8.2 to	-5.4	40.2	-5.4	-6.9 to	-3.8
Liver	1981-85	3,195	7.4	-2.2	-4.3 to	-0.1	1.7	0.0	-1.1 to	1.0
	1986-90	3,624	10.6	-4.1	-6.4 to	-1.7	2.6	-0.8	-2.1 to	0.5
Gallbladder	1981-85	5,030	19.1	-2.7	-5.7 to	0.3	6.8	-1.1	-3.2 to	0.9
	1986-90	4,946	24.4	-5.6	-8.9 to	-2.3	9.3	-2.0	-4.3 to	0.3
Pancreas	1981-85	22,566	9.7	-2.0	-3.0 to	-1.0	2.0	0.0	-0.5 to	0.5
	1986-90	22,566	10.3	-2.1	-3.1 to	-1.1	2.6	-0.6	-1.2 to	0.0
Nasal cavities, sinuses	1981-85	1,616	70.5	-10.8	-17.9 to	-3.7	43.6	-10.6	-18.5 to	-2.7
	1986-90	1,681	69.5	-6.2	-13.1 to	0.8	43.9	-6.6	-14.4 to	1.1
Larynx	1981-85	8,113	85.8	-4.9	-7.5 to	-2.3	68.8	-8.8	-12.5 to	-5.1
	1986-90	8,671	85.9	-4.8	-7.3 to	-2.3	68.4	-9.3	-12.9 to	-5.7
Lung	1981-85	149,376	20.0	-3.1	-3.7 to	-2.5	5.9	-0.8	-1.1 to	-0.4
	1986-90	144,604	20.7	-3.0	-3.6 to	-2.5	6.0	-1.0	-1.3 to	-0.6
Pleura	1981-85	1,948	27.3	-2.2	-7.9 to	3.4	3.7	-0.1	-2.6 to	2.4
	1986-90	3,177	26.8	-1.3	-5.7 to	3.1	3.1	1.4	-0.5 to	3.4
Thymus	1981-85	163	73.0	-4.9	-26.3 to	16.5	45.5	4.0	-20.9 to	28.9
	1986-90	162	75.8	-10.2	-31.1 to	10.7	52.4	-14.6	-39.1 to	9.9
Bone	1981-85	1,626	68.2	-9.4	-16.3 to	-2.5	41.2	-3.0	-10.3 to	4.2
	1986-90	1,603	72.0	-0.7	-7.4 to	6.1	49.1	4.3	-3.4 to	12.1
Connective tissue	1981-85	3,530	75.8	-8.8	-13.3 to	-4.3	54.2	-8.7	-13.9 to	-3.4
	1986-90	4,061	76.5	-8.1	-12.3 to	-3.9	52.4	-4.8	-9.8 to	0.2
Melanoma of skin	1981-85	11,309	93.5	-6.2	-8.0 to	-4.4	76.4	-12.5	-15.4 to	-9.6
	1986-90	15,703	94.9	-3.7	-5.1 to	-2.3	81.8	-8.1	-10.5 to	-5.6

[1] Patients included in analyses and with a known deprivation category (see text)

[2] Difference in relative survival between affluent and deprived groups - a negative value means survival is lower in the deprived group

96

Table 4.9 (cont.)

Difference (gap) in survival between affluent and deprived groups in England and Wales, by period of diagnosis: total number of patients, survival in affluent group, gap and 95% confidence interval around the gap (low, high)

ADULTS	Period of diagnosis	No. of patients[1]	One-year survival (%)				Five-year survival (%)			
			Affluent	Gap[2]	Low	High	Affluent	Gap[2]	Low	High
Breast	1981-85	102,800	90.7	-5.5	-6.1 to	-4.8	67.1	-9.0	-10.0 to	-8.0
	1986-90	116,169	91.8	-4.7	-5.3 to	-4.1	70.9	-7.6	-8.5 to	-6.6
Cervix	1981-85	18,421	83.3	-4.7	-6.4 to	-2.9	63.6	-7.8	-10.1 to	-5.6
	1986-90	18,868	85.2	-3.8	-5.5 to	-2.2	66.4	-4.4	-6.6 to	-2.2
Uterus	1981-85	16,289	86.4	-4.7	-6.5 to	-2.8	74.3	-5.8	-8.2 to	-3.3
	1986-90	16,261	86.9	-3.5	-5.3 to	-1.7	74.1	-3.6	-6.1 to	-1.1
Ovary	1981-85	20,002	56.3	-8.6	-10.6 to	-6.5	28.9	-3.0	-4.9 to	-1.0
	1986-90	20,999	57.9	-6.9	-9.0 to	-4.9	29.8	-2.2	-4.1 to	-0.3
Vagina and vulva	1981-85	4,621	73.0	-7.9	-12.2 to	-3.6	53.4	-4.6	-9.7 to	0.5
	1986-90	4,600	74.2	-5.0	-9.3 to	-0.7	53.8	-2.0	-7.3 to	3.3
Prostate	1981-85	42,864	80.3	-7.2	-8.6 to	-5.8	44.2	-5.6	-7.4 to	-3.7
	1986-90	51,354	78.9	-3.8	-5.1 to	-2.6	43.0	-2.9	-4.6 to	-1.2
Testis	1981-85	4,308	96.3	-3.7	-5.8 to	-1.6	92.4	-6.9	-9.8 to	-3.9
	1986-90	5,460	96.9	-3.3	-5.1 to	-1.5	93.9	-6.2	-8.7 to	-3.7
Penis	1981-85	1,282	81.8	-1.3	-8.7 to	6.2	65.4	-1.2	-11.1 to	8.8
	1986-90	1,305	86.6	-4.9	-11.6 to	1.8	70.8	-3.0	-12.8 to	6.9
Bladder	1981-85	44,710	80.6	-7.1	-8.4 to	-5.8	64.5	-5.7	-7.4 to	-4.0
	1986-90	48,722	81.4	-6.0	-7.2 to	-4.8	66.3	-7.0	-8.6 to	-5.3
Kidney	1981-85	12,921	57.7	-10.0	-12.6 to	-7.4	37.8	-4.9	-7.6 to	-2.2
	1986-90	15,020	58.1	-6.9	-9.3 to	-4.4	39.3	-4.3	-6.8 to	-1.8
Eye	1981-85	1,514	93.8	-3.8	-8.4 to	0.7	71.7	-8.5	-16.7 to	-0.2
	1986-90	1,712	93.3	0.4	-3.8 to	4.6	71.7	-2.3	-10.0 to	5.5
Brain	1981-85	10,546	26.9	-3.6	-5.9 to	-1.2	11.4	-0.1	-1.8 to	1.7
	1986-90	11,853	30.4	-4.3	-6.6 to	-1.9	14.4	-1.7	-3.6 to	0.1
Thyroid	1981-85	3,481	78.8	-8.9	-13.4 to	-4.3	73.2	-11.2	-16.4 to	-6.0
	1986-90	3,532	81.0	-5.9	-10.2 to	-1.6	75.8	-5.5	-10.5 to	-0.6
Spinal cord	1981-85	206	73.1	3.3	-13.4 to	20.0	58.2	4.2	-15.1 to	23.4
	1986-90	192	84.5	-3.9	-19.9 to	12.2	69.0	-9.5	-29.4 to	10.4
Adrenal	1981-85	266	46.5	-26.0	-41.7 to	-10.2	23.9	-10.6	-23.6 to	2.3
	1986-90	323	55.9	-7.1	-23.4 to	9.2	33.1	2.3	-13.6 to	18.2
Pituitary	1981-85	152	85.9	1.8	-14.0 to	17.6	82.4	-12.9	-34.9 to	9.1
	1986-90	222	92.8	-13.2	-26.1 to	-0.2	82.8	-17.0	-35.7 to	1.7
Non-Hodgkin lymphoma	1981-85	17,640	66.1	-9.1	-11.3 to	-6.9	44.5	-6.7	-9.1 to	-4.3
	1986-90	23,477	67.7	-6.7	-8.6 to	-4.8	47.8	-6.9	-9.0 to	-4.8
Hodgkin's disease	1981-85	5,292	87.5	-5.9	-8.9 to	-2.8	72.7	-5.9	-9.8 to	-2.0
	1986-90	4,950	90.8	-5.7	-8.5 to	-2.9	78.9	-8.4	-12.3 to	-4.5
Multiple myeloma	1981-85	9,537	53.6	-6.6	-9.7 to	-3.4	18.1	-2.3	-4.8 to	0.3
	1986-90	10,426	57.4	-7.4	-10.4 to	-4.4	19.0	-0.7	-3.2 to	1.8
Acute lymphoid leukaemia	1981-85	1,092	47.1	-7.0	-15.7 to	1.8	19.5	0.3	-6.8 to	7.4
	1986-90	1,109	53.9	-1.9	-10.9 to	7.2	25.5	0.0	-7.8 to	7.9
Chronic lymphoid leukaemia	1981-85	5,924	75.9	-6.7	-10.5 to	-2.9	51.8	-9.7	-14.3 to	-5.0
	1986-90	6,267	81.1	-10.0	-13.6 to	-6.4	55.2	-10.7	-15.5 to	-6.0
Acute myeloid leukaemia	1981-85	4,908	24.2	-6.8	-10.0 to	-3.7	7.4	-0.4	-2.4 to	1.7
	1986-90	5,569	27.4	-6.9	-10.2 to	-3.5	9.1	-1.8	-4.1 to	0.4
Chronic myeloid leukaemia	1981-85	2,456	60.5	-4.4	-10.4 to	1.7	18.7	-3.0	-7.7 to	1.7
	1986-90	2,863	63.6	-6.5	-12.2 to	-0.8	23.6	-3.2	-8.3 to	1.8
Monocytic leukaemia	1981-85	405	17.8	-10.7	-19.5 to	-1.9	8.0	-8.0	-11.6 to	-4.4
	1986-90	304	21.2	-6.2	-18.6 to	6.3	9.8	-3.7	-12.5 to	5.2
All leukaemias	1981-85	16,914	48.9	-6.6	-8.9 to	-4.3	25.1	-4.3	-6.4 to	-2.3
	1986-90	17,573	53.8	-8.1	-10.4 to	-5.8	28.3	-4.9	-7.1 to	-2.8

[1] Patients included in analyses and with a known deprivation category (see text)

[2] Difference in relative survival between affluent and deprived groups - a negative value means survival is lower in the deprived group

SUMMARY

Table 4.9 (cont.)

Difference (gap) in survival between affluent and deprived groups in England and Wales, by period of diagnosis: total number of patients, survival in affluent group, gap and 95% confidence interval around the gap (low, high)

CHILDREN	Period of diagnosis	No. of patients[1]	One-year survival (%)				Five-year survival (%)			
			Affluent	Gap[2]	Low	High	Affluent	Gap[2]	Low	High
All leukaemias	1981-85	1,614	81.7	-3.7	-9.3 to	1.8	62.2	-1.8	-8.5 to	4.9
	1986-90	1,619	88.8	-4.7	-9.3 to	-0.1	70.9	-4.1	-10.4 to	2.3
Acute lymphoid leukaemia	1981-85	1,283	89.1	-3.4	-8.5 to	1.7	71.8	-4.4	-11.5 to	2.7
	1986-90	1,278	94.1	-2.3	-6.3 to	1.7	77.2	-3.2	-9.8 to	3.4
Hodgkin's disease	1981-85	228	97.5	2.8	-2.8 to	8.3	90.2	0.1	-9.9 to	10.0
	1986-90	189	99.1	-0.5	-7.0 to	6.0	95.0	-3.9	-13.4 to	5.6
Non-Hodgkin lymphoma	1981-85	262	75.9	6.8	-7.1 to	20.6	64.9	6.3	-9.2 to	21.8
	1986-90	273	89.7	-8.6	-20.7 to	3.6	80.9	-9.0	-23.4 to	5.5
Brain and spinal cord	1981-85	984	75.7	-3.6	-11.0 to	3.9	59.8	-2.9	-11.3 to	5.5
	1986-90	1,050	77.5	-3.5	-10.9 to	3.9	61.9	-6.0	-14.4 to	2.3
Retinoblastoma	1981-85	100	-	-	-	-	-	-	-	-
	1986-90	104	-	-	-	-	-	-	-	-
Wilms' tumour	1981-85	278	92.8	-4.9	-13.6 to	3.9	81.1	-5.5	-18.6 to	7.6
	1986-90	257	95.4	-0.9	-9.1 to	7.4	83.0	4.2	-9.7 to	18.1
Hepatoblastoma	1981-85	25	-	-	-	-	-	-	-	-
	1986-90	32	-	-	-	-	-	-	-	-
Osteosarcoma	1981-85	134	75.7	16.5	-0.6 to	33.5	39.8	16.0	-6.9 to	38.8
	1986-90	117	92.6	-8.7	-26.5 to	9.2	57.3	-15.6	-40.0 to	8.8
Ewing's sarcoma	1981-85	125	93.2	-10.9	-26.4 to	4.7	36.4	3.8	-20.4 to	27.9
	1986-90	97	93.6	-8.8	-25.8 to	8.2	60.9	-10.1	-38.3 to	18.2
Soft tissue sarcoma	1981-85	325	79.5	7.0	-4.3 to	18.4	59.4	1.8	-13.4 to	17.0
	1986-90	319	82.9	8.1	-2.9 to	19.1	62.3	8.0	-7.3 to	23.3
Germ-cell and gonadal neoplasms	1981-85	118	80.0	-1.9	-21.5 to	17.7	71.6	-1.4	-23.0 to	20.2
	1986-90	121	93.7	4.8	-6.2 to	15.9	78.5	12.4	-3.2 to	28.0

[1] Patients included in analyses and with a known deprivation category (see text)

[2] Difference in relative survival between affluent and deprived groups - a negative value means survival is lower in the deprived group

---

Table 4.9A

Relative survival (%) by deprivation category, all cancers combined[1], adults diagnosed in England and Wales, 1981-85 and 1986-90, and difference (gap) between most affluent and most deprived categories, with 95% confidence interval

ADULTS DIAGNOSED 1981-85 — Deprivation category

Time since diagnosis	Affluent %	CI	2 %	CI	3 %	CI	4 %	CI	Deprived %	CI	Deprivation gap[3] %	CI
(No. of patients)[2]	(138,694)		(147,590)		(154,607)		(153,829)		(133,906)		(728,626)	
One year	59.1	(58.8-59.4)	55.6	(55.3-55.8)	52.8	(52.6-53.1)	49.8	(49.5-50.0)	46.1	(45.8-46.3)	-12.8	(-13.1 to -12.4)
Five years	39.6	(39.3-39.9)	36.3	(36.0-36.6)	34.2	(33.9-34.4)	31.6	(31.4-31.9)	28.6	(28.3-28.9)	-10.6	(-11.0 to -10.3)

ADULTS DIAGNOSED 1986-90 — Deprivation category

Time since diagnosis	Affluent %	CI	2 %	CI	3 %	CI	4 %	CI	Deprived %	CI	Deprivation gap[3] %	CI
(No. of patients)[2]	(147,124)		(163,847)		(167,543)		(161,597)		(133,383)		(773,494)	
One year	62.6	(62.4-62.9)	59.8	(59.5-60.0)	57.1	(56.8-57.3)	53.3	(53.1-53.6)	49.9	(49.6-50.2)	-12.7	(-13.1 to -12.4)
Five years	43.4	(43.1-43.7)	40.8	(40.6-41.1)	38.6	(38.3-38.9)	35.3	(35.0-35.6)	32.3	(32.0-32.6)	-11.1	(-11.5 to -10.8)

[1] All malignant neoplasms included in analyses, representing 93% of all cancers in adults (see text).

[2] Patients included in analyses and with a known deprivation status.

[3] Regression estimate of difference (gap) in relative survival between most affluent and most deprived groups.
A negative value for the gap means survival is lower in the most deprived group.

Figure 4.9 Relative survival up to five years after diagnosis, by deprivation category, all cancers combined: England and Wales, adults diagnosed 1986-90

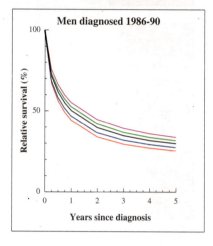

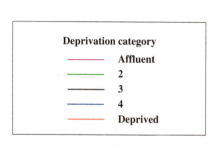

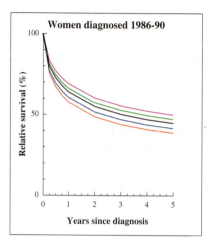

Figure 4.10 Relative survival at one year and five years by deprivation category, all cancers combined: England and Wales, adults diagnosed 1981-85 and 1986-90

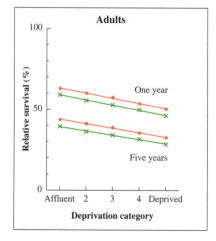

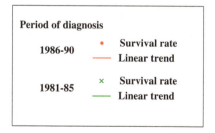

SUMMARY

Table 4.10 Five-year survival rate in most affluent group, and difference (gap) between most affluent and most deprived groups, for cancers with more than 1,000 patients diagnosed in England and Wales during 1986-90

	No. of patients[1]	Five-year survival (%) Affluent	Gap[2]
Tongue	2,749	50.4	**-16.3**
Oropharynx	1,603	41.9	**-12.0**
Oral cavity	3,663	53.9	**-11.6**
Chronic lymphoid leukaemia	6,267	55.2	**-10.7**
Larynx	8,671	68.4	**-9.3**
Hodgkin's disease	4,950	78.9	**-8.4**
Melanoma of skin	15,703	81.8	**-8.1**
Breast	116,169	70.9	**-7.6**
Small intestine	1,475	27.9	**-7.3**
Bladder	48,722	66.3	**-7.0**
Non-Hodgkin lymphoma	23,477	47.8	**-6.9**
Nasal cavities and sinuses	1,681	43.9	-6.6
Testis	5,460	93.9	**-6.2**
Salivary glands	1,521	61.2	-6.2
Brain and spinal cord (C)	1,050	61.9	-6.0
Thyroid	3,532	75.8	**-5.5**
Rectum	43,870	40.2	**-5.4**
Connective tissue	4,061	52.4	-4.8
Cervix	18,868	66.4	**-4.4**
Kidney	15,020	39.3	**-4.3**
Colon	67,741	40.3	**-4.3**
Uterus	16,261	74.1	**-3.6**
Chronic myeloid leukaemia	2,863	23.6	-3.2
Acute lymphoid leukaemia (C)	1,278	77.2	-3.2
Penis	1,305	70.8	-3.0
Prostate	51,354	43.0	**-2.9**
Hypopharynx	1,588	21.0	-2.6
Eye	1,712	71.7	-2.3
Ovary	20,999	29.8	**-2.2**
Gallbladder	4,946	9.3	-2.0
Vagina and vulva	4,600	53.8	-2.0
Acute myeloid leukaemia	5,569	9.1	-1.8
Brain	11,853	14.4	-1.7
Lung	144,604	6.0	**-1.0**
Liver	3,624	2.6	-0.8
Multiple myeloma	10,426	19.0	-0.7
Pancreas	22,566	2.6	-0.6
Oesophagus	20,475	6.4	-0.2
Acute lymphoid leukaemia	1,109	25.5	0.0
Stomach	43,100	9.3	0.4
Lip	1,131	89.0	0.6
Pleura	3,177	3.1	1.4
Bone	1,603	49.1	4.3

[1] Patients included in analyses and with a known deprivation category (see text). Adults (15-99 years) unless marked (C) - children (0-14 years)

[2] Difference in five-year relative survival between affluent and deprived groups - a negative value means survival is lower in the deprived group.

Values in italics are not statistically significant at the 5% level (see Table 4.9 for confidence intervals)

Table 4.11   Avoidable deaths within five years of diagnosis: selected cancers, England and Wales, adults diagnosed 1986-90

(A) Avoidable deaths within five years of diagnosis, by type of cancer

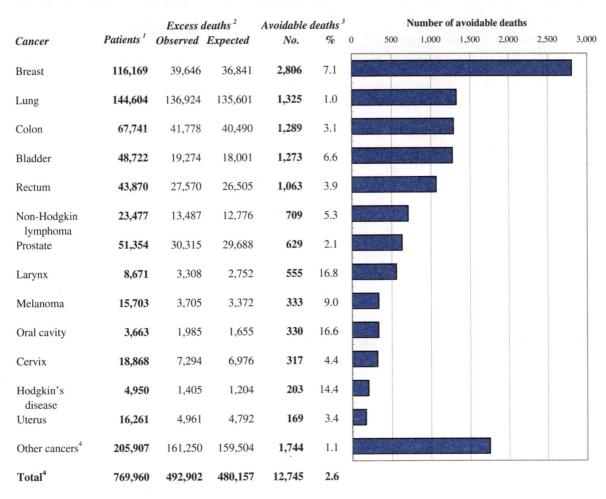

Cancer	Patients [1]	Excess deaths [2] Observed	Excess deaths [2] Expected	Avoidable deaths [3] No.	Avoidable deaths [3] %
Breast	116,169	39,646	36,841	**2,806**	7.1
Lung	144,604	136,924	135,601	**1,325**	1.0
Colon	67,741	41,778	40,490	**1,289**	3.1
Bladder	48,722	19,274	18,001	**1,273**	6.6
Rectum	43,870	27,570	26,505	**1,063**	3.9
Non-Hodgkin lymphoma	23,477	13,487	12,776	**709**	5.3
Prostate	51,354	30,315	29,688	**629**	2.1
Larynx	8,671	3,308	2,752	**555**	16.8
Melanoma	15,703	3,705	3,372	**333**	9.0
Oral cavity	3,663	1,985	1,655	**330**	16.6
Cervix	18,868	7,294	6,976	**317**	4.4
Hodgkin's disease	4,950	1,405	1,204	**203**	14.4
Uterus	16,261	4,961	4,792	**169**	3.4
Other cancers[4]	205,907	161,250	159,504	**1,744**	1.1
Total[4]	**769,960**	**492,902**	**480,157**	**12,745**	**2.6**

(B) Avoidable deaths within five years of diagnosis, by deprivation category

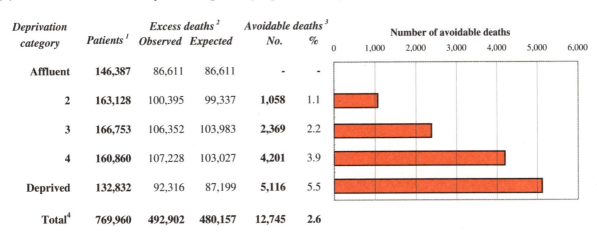

Deprivation category	Patients [1]	Excess deaths [2] Observed	Excess deaths [2] Expected	Avoidable deaths [3] No.	Avoidable deaths [3] %
Affluent	146,387	86,611	86,611	-	-
2	163,128	100,395	99,337	**1,058**	1.1
3	166,753	106,352	103,983	**2,369**	2.2
4	160,860	107,228	103,027	**4,201**	3.9
Deprived	132,832	92,316	87,199	**5,116**	5.5
Total[4]	**769,960**	**492,902**	**480,157**	**12,745**	**2.6**

[1]   Patients included in analyses and with a known deprivation category (see text).

[2]   Number of deaths in excess of general population mortality within five years of cancer diagnosis. Observed - actual number. Expected - estimated on the basis that patients in each deprivation category were to experience the same relative survival rate as patients in the most affluent category.

[3]   Avoidable deaths - difference between observed and expected. Percentage of observed excess deaths for a given cancer (Table 4.11A) or for a given deprivation category (Table 4.11B). Estimated numbers of deaths are all rounded to the nearest integer: small discrepancies in totals or subtractions are due to rounding.

[4]   Excluding six rare malignancies (nasopharynx, thymus, spinal cord, adrenal, pituitary, monocytic leukaemia). Total still represents 93.0% of all adult cancers.

Table 4.12

International comparison of cancer survival in adults (aged 15-99 at diagnosis):

five-year relative survival (%) in England and Wales[1] (adults diagnosed 1986-90), Europe[2] (1985-89) and the USA[3] (1986-90)

SITE		National average	Affluent group	Highest region	Scotland	Average	USA	SITE		National average	Affluent group	Highest region	Scotland	Average	USA
		England & Wales			Europe					England & Wales			Europe		
Lip	Men	89	91	95	94	91	95	Breast	Men	72	75	78	65	72	85
	Women	89	84	92	89	91	99		Women	68	70	70	66	73	84
Tongue	Men	37	46	43	36	37	44	Cervix	Women	64	66	68	62	62	70
	Women	49	57	59	41	52	55	Uterus	Women	73	74	76	72	75	85
Salivary glands	Men	51	52	69	59	54	68	Ovary	Women	29	31	34	30	35	45
	Women	66	68	75	86	72	81	Vagina and vulva	Women	53	52	59	55	52	66
Oral cavity	Men	44	52	50	45	44	45	Prostate	Men	42	43	46	48	56	86
	Women	52	60	60	52	52	68	Testis	Men	91	93	95	92	90	95
Oropharynx	Men	34	40	45	34	29	23	Penis	Men	69	71	86	66	72	66
	Women	37	45	48	27	44	30								
Nasopharynx	Men	30	52	46	30	34	48	Bladder	Men	65	67	70	65	68	84
	Women	38	44	58	21	36	51		Women	58	62	65	59	60	75
Hypopharynx	Men	20	29	32	13	24	25	Kidney	Men	40	41	45	38	49	59
	Women	20	17	32	20	27	32		Women	36	37	39	35	50	56
Oesophagus	Men	6	5	8	6	8	10	Eye	Men	70	75	78	89	74	77
	Women	7	8	9	9	12	10		Women	71	68	77	82	75	76
Stomach	Men	10	10	13	10	21	18	Brain	Men	13	13	18	13	18	22
	Women	11	11	14	12	24	25		Women	16	17	20	13	20	25
Small intestine	Men	26	31	32	33	40	50	Thyroid	Men	65	66	83	72	69	94
	Women	23	22	30	20	32	55		Women	76	81	83	74	80	96
Colon	Men	39	40	43	41	47	64	Spinal cord	Men	68	85	(a)	(a)	(a)	(a)
	Women	39	41	45	41	47	63		Women	58	67	(a)	(a)	(a)	(a)
Rectum	Men	37	40	40	38	43	60	Adrenal	Men	34	28	(a)	(a)	(a)	(a)
	Women	39	41	42	39	43	60		Women	36	48	(a)	(a)	(a)	(a)
Liver	Men	2	3	8	1	5	3	Pituitary	Men	77	91	(a)	(a)	(a)	(a)
	Women	2	2	8	4	5	8		Women	74	75	(a)	(a)	(a)	(a)
Gallbladder	Men	10	10	14	8	12	18								
	Women	8	7	14	8	12	14								
Pancreas	Men	3	3	6	4	4	3	Non-Hodgkin	Men	45	48	52	45	48	49
	Women	2	3	4	3	4	4	lymphoma	Women	45	50	52	42	48	56
Nasal cavities,	Men	42	45	56	46	45	53	Hodgkin's	Men	75	79	81	68	72	78
sinuses	Women	39	41	51	44	44	55	disease	Women	75	79	82	67	74	83
Larynx	Men	64	71	70	64	63	68	Multiple	Men	19	20	24	17	30	30
	Women	58	57	66	58	63	62	myeloma	Women	20	19	28	23	28	28
Lung	Men	6	6	8	6	10	13	Acute lymphoid	Men	25	23	35	34	28	24
	Women	6	7	9	7	11	16	leukaemia	Women	27	27	36	24	29	21
Pleura	Men	4	3	7	3	6	3	Chronic lymphoid	Men	49	54	59	54	62	73
	Women	5	3	17	2	8	12	leukaemia	Women	53	58	64	52	67	74
Thymus	Men	46	55	65	(a)	(a)	(a)	Acute myeloid	Men	8	8	11	6	11	10
	Women	47	57	63	(a)	(a)	(a)	leukaemia	Women	9	13	13	10	12	13
Bone	Men	50	50	61	42	48	60	Chronic myeloid	Men	22	23	34	26	29	30
	Women	53	48	71	44	51	71	leukaemia	Women	23	25	34	22	32	31
Connective tissue	Men	50	54	57	52	54	62	Monocytic	Men	9	4	16	(a)	(a)	9
	Women	51	55	65	51	61	65	leukaemia	Women	9	12	13	(a)	(a)	10
Melanoma of skin	Men	70	71	74	74	70	85	All leukaemias	Men	27	28	33	30	34	41
	Women	84	87	87	88	82	91		Women	27	29	34	25	35	38

[1] Average survival rate for England and Wales; survival rate for affluent group (England & Wales); and highest survival rate among NHS Regions (all deprivation groups combined)

[2] Data from EUROCARE II study. Average survival rate for European countries (incl. Scotland) and regions (incl. parts of England and Wales) covered by EUROCARE II study

[3] Average survival rate for US states covered by SEER programme

(a) Too few cases for reliable estimation, or cancer not included in analyses

Figure 4.12
International comparison of five-year relative survival (%), selected cancers:
England and Wales[1] (adults diagnosed 1986-90), Europe[2] (1985-89) and the USA[3] (1986-90)

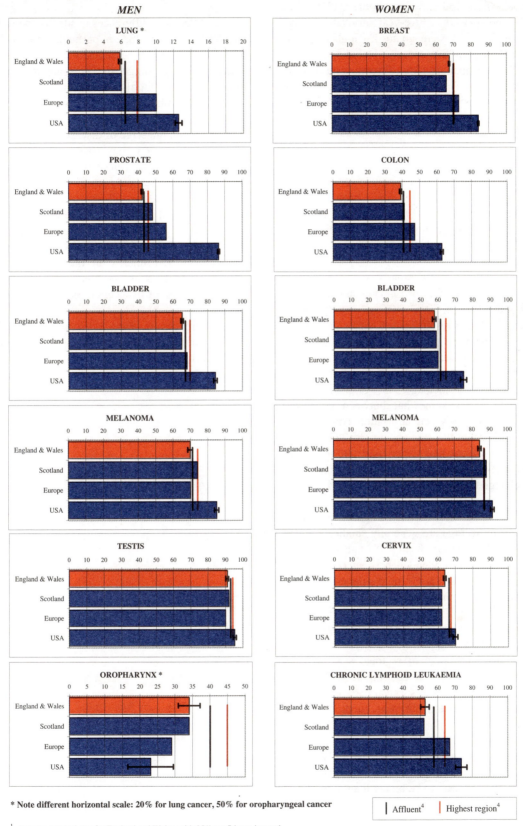

* Note different horizontal scale: 20% for lung cancer, 50% for oropharyngeal cancer

Affluent[4]   Highest region[4]

[1] Average survival rate for England and Wales, with 95% confidence interval

[2] Average survival rate for European countries (incl. Scotland) and regions (incl. parts of England and Wales) covered by EUROCARE II study

[3] Average survival rate for US states covered by SEER programme, with 95% confidence interval

[4] Survival rate for affluent group (England & Wales); and highest survival rate among NHS Regions (all deprivation groups combined)

which data were available, including those of colon (24%), rectum (21%), kidney (20%), bladder (17%), ovary (16%), breast (16%), non-Hodgkin lymphoma (11%) and melanoma of the skin (7%). Five-year survival estimates for women in the USA were up to 4% higher than in England and Wales for five cancers, including those of larynx (4%) and oesophagus (3%). Five-year survival for women in England and Wales was higher than in the USA for two malignancies: oropharynx and acute lymphoid leukaemia.

Comment

Relative survival estimates are generally lower in England and Wales than the average for Europe or the corresponding survival rates for the area of the USA covered by the SEER programme, which represents about 10% of the US population.

Several caveats apply to this comparison. The method of survival estimation used here for England and Wales was different from that used in the EUROCARE study[54], but it produces closely similar results[55] (see Chapter 3). Relative survival rates for men and women incorporate the general population mortality at each age in each country, and they correctly reflect the excess mortality that is attributable to cancer in that country at each age. The age distribution of cancer patients varies somewhat between countries, however, and suitable age-standardisation could not be carried out to adjust for this (see above): this will become possible shortly[56]. Meanwhile, experience with these data suggests that age-standardisation of relative survival does not usually change the estimate by more than 5%.

The differences in survival between England and Wales and the European average are generally smaller than the corresponding differences from US survival rates. Survival rates in Europe vary widely between countries for many cancers[2], and these should be examined for more detailed comparison than is possible here.

When the national survival rate for patients diagnosed in England and Wales during 1986-90 was lower than the European average, the range of regional survival rates often included the European average (Figure 4.12), which suggests that the differences in survival between England and Wales and other western European countries are not ineluctable.

Survival from acute lymphoid leukaemia is similar to that in the USA. Survival from cancer of the oropharynx in England and Wales appears to be higher than in the USA, and close to the European average. This may be partly due to a more favourable sub-site distribution: cancers of the tonsil comprise about 70% of cancers of the oropharynx in England and Wales (see Table 9A), as opposed to about

40% in France and Italy, and they have higher survival than for other cancers of the oropharynx[57].

Some of the large differences in survival between England and Wales and the USA, in particular, are likely to be due to the definition of disease and the approaches commonly used for cancer diagnosis. This applies particularly to prostate cancer, but also to chronic lymphoid leukaemia, myeloma and bladder cancer[58].

Differences in the stage of disease at diagnosis and in the treatment provided may also underlie some of the observed differences. Survival rates for testicular cancer, Hodgkin's disease and acute lymphoid leukaemia in England and Wales are all very close to the average levels in Europe and the USA. The differences in survival from cancer of the breast in women and cancer of the colon between England and Wales and other western European countries arise primarily in the first six months after diagnosis, which suggests international differences in the stage of disease or in access to optimal treatment[11,59].

The differences in cancer survival between England and Wales, comparable countries in western Europe and the USA warrant further investigation.

1. Stiller CA, Allen MB, Eatock EM. Childhood cancer in Britain: the national registry of childhood tumours and incidence rates 1978-1987. Eur J Cancer 1995; 31A: 2028-2034

2. Berrino F, Capocaccia R, Estève J, Gatta G, Hakulinen T, Micheli M, Sant M, Verdecchia A, (eds.) *Survival of cancer patients in Europe: the EUROCARE study, II. (IARC Scientific Publications No. 151)*. Lyon: International Agency for Research on Cancer, 1999

3. Quinn MJ, Allen E, UK Association of Cancer Registries. Changes in incidence of and mortality from breast cancer in England and Wales since introduction of screening. Br Med J 1995; 311: 1391-1393

4. Stockton D, Davies TW, Day NE, McCann J. Retrospective study of reasons for improved survival in patients with breast cancer in East Anglia: earlier diagnosis or better treatment? Br Med J 1997; 314: 472-475

5. Stiller CA, Bunch KJ. Trends in survival for childhood cancer in Britain diagnosed 1971-85. Br J Cancer 1990; 62: 806-815

6. Stiller CA. Population-based survival rates for childhood cancer in Britain, 1980-91. Br Med J 1994; 309: 1612-1616

7. Stiller CA, Draper GJ. Treatment centre size, entry to trials, and survival in acute lymphoblastic leukaemia. Arch Dis Child 1989; 64: 657-661

8. Schillinger JA, Grosclaude PC, Honjo S, Quinn MJ, Sloggett A, Coleman MP. Survival from acute lymphocytic leukaemia: socioeconomic status and geographic region. Arch Dis Child 1999; (in press)

9. Hawkins MM. Long term survival and cure after childhood cancer. Arch Dis Child 1989; 64: 798-807

10. Adami H-O, Malker B, Holmberg L, Persson I, Stone BJ. The relation between survival and age at diagnosis in breast cancer. N Engl J Med 1986; 315: 559-563

11. Sant M, Capocaccia R, Verdecchia A, Estève J, Gatta G, Micheli A, Coleman MP, Berrino F, EUROCARE Working Group. Survival of women with breast cancer in Europe: variation with age, year of diagnosis and country. Int J Cancer 1998; 77: 679-683

12. Löwenberg B, Burnett AK. Acute myeloid leukaemia in adults. In: Degos L, Linch DC, Löwenberg B, (eds.) *Textbook of malignant haematology*. London: Martin Dunitz, 1999, pp743-769

13. Dickman PW, Auvinen A, Voutilainen ET, Hakulinen T. Measuring social class differences in cancer patient survival: development of a gold standard and comparison with other methods. J Epidemiol Comm Hlth 1999; (in press)

14. Tomatis L. Socioeconomic factors and human cancer. Int J Cancer 1995; 62: 121-125

15. Drever F, Whitehead M, (eds.) *Health inequalities. Series DS no.15*. London: Office for National Statistics, 1997

16. Department of Health. *Independent inquiry into inequalities in health (Acheson report)*. London: The Stationery Office, 1998

17. Kogevinas M, Marmot MG, Fox AJ, Goldblatt PO. Socioeconomic differences in cancer survival. J Epidemiol Comm Hlth 1991; 45: 216-219

18. Carnon AG, Ssemwogerere A, Lamont DW, Hole DJ, Mallon E, George WD, Gillis CR. Relation between socioeconomic deprivation and pathological prognostic factors in women with breast cancer. Br Med J 1994; 309: 1054-1057

19. Schrijvers CTM, Mackenbach J, Lutz J-M, Quinn MJ, Coleman MP. Deprivation and survival from breast cancer. Br J Cancer 1995; 72: 738-743

20. Schrijvers CTM, Mackenbach J, Lutz J-M, Quinn MJ, Coleman MP. Deprivation, stage at diagnosis and cancer survival. Int J Cancer 1995; 63: 324-329

21. Kogevinas M. *Longitudinal study. Socio-economic differences in cancer survival. Series LS No.5*. London: OPCS, 1990

22. Lamont DW, Symonds RP, Brodie MM, Nwabineli NJ, Gillis CR. Age, socio-economic status and survival from cancer of cervix in the West of Scotland 1980-87. Br J Cancer 1993; 67: 351-357

23. Pollock AM, Vickers N. Breast, lung and colorectal cancer incidence and survival in South Thames Region, 1987-1992: the effect of social deprivation. J Publ Hlth Med 1997; 19: 288-294

24. Kidd J. Socioeconomic variations in breast cancer incidence, survival and the uptake of screening: a case study in Merseyside. PhD dissertation. University of Liverpool, 1997

25. Schrijvers CTM. Socioeconomic inequalities in cancer survival in the Netherlands and Great Britain: small-area based studies using cancer registry data. PhD dissertation. Erasmus University, Rotterdam, 1996

26. Sharp L, Finlayson AR, Black RJ. Cancer survival and deprivation in Scotland. J Epidemiol Comm Hlth 1995; 49: s79 (Abstract)

27. Bassett MT, Kreiger N. Social class and black-white differences in breast cancer survival. Am J Public Health 1986; 76: 1400-1403

28. Dayal HH, Power RN, Chiu C. Race and socio-economic status in survival from breast cancer. J Chron Dis 1982; 35: 675-683

29. Cella DF, Orav EJ, Kornblith AB, Holland JC, Silberfarb PM, Lee WK, Comis RL, Perry M, Cooper R, Maurer LH, Hoth DF, Perloff M, Bloomfield CD, McIntyre OR, Leone L, Lesnick G, Nissen N, Glicksman A, Henderson E, Barcos M, Crichlow R, Faulkner CS, Eaton W, North W, Schein PS, Chu F, King G, Chahinian AP. Socioeconomic status and cancer survival. J Clin Oncol 1991; 9: 1500-1509

30. Auvinen A. Social class and colon cancer survival in Finland. Cancer 1992; 70: 402-409

31. Gordon NH, Crowe JP, Brumberg DJ, Berger NA. Socioeconomic factors and race in breast cancer recurrence and survival. Am J Epidemiol 1992; 135: 609-618

32. Auvinen A, Karjalainen S, Pukkala E. Social class and cancer patient survival in Finland. Am J Epidemiol 1995; 142: 1089-1102

33. Karjalainen S, Pukkala E. Social class as a prognostic factor in breast cancer survival. Cancer 1990; 66: 819-826

34. Kogevinas M, Porta M. Socioeconomic differences in cancer survival: a review of the evidence. In: Kogevinas M, Pearce N, Susser M, Boffetta P, (eds.) *Social inequalities and cancer. (IARC Scientific Publications No. 138)*. Lyon: IARC, 1997, pp177-206

35. Auvinen A, Karjalainen S. Possible explanations for social class differences in cancer patient survival. In: Kogevinas M, Pearce N, Susser M, Boffetta P, (eds.) *Social inequalities and cancer. (IARC Scientific Publications No. 138)*. Lyon: IARC, 1997, pp377-397

36. Leon DA, Wilkinson RG. Inequalities in prognosis: socioeconomic differences in cancer and heart disease survival. In: Fox J (ed.) *Health inequalities in European countries*. Aldershot: Gower, 1989, pp280-300

37. Chouillet AM, Bell CMJ, Hiscox JG. Management of breast cancer in southeast England. Br Med J 1994; 308: 168-171

38. McArdle CS, Hole D. Impact of variability among surgeons on postoperative morbidity and mortality and ultimate survival. Br Med J 1991; 302: 1501-1505

39. Gillis CR, Hole DJ. Survival outcome of care by specialist surgeons in breast cancer: a study of 3786 patients in the west of Scotland. Br Med J 1996; 312: 145-148

40. Macleod U, Twelves CJ, Ross S, Gillis C, Watt G. A comparison of the care received by women with breast cancer living in affluent and deprived areas. Br J Cancer 1998; 78: s15 (Abstract)

41. Twelves CJ, Thomson CS, Gould A, Dewar JA. Variation in the survival of women with breast cancer in Scotland. Br J Cancer 1998; 78: 566-571

42. Wolfe C, Tilling K, Bourne HM, Raju KS. Variations in the screening history and appropriateness of management of cervical cancer in South East England. Eur J Cancer 1996; 32A: 1198-1204

43. Richards MA, Wolfe C, Tilling K, Barton J, Bourne HM, Gregory WM. Variations in the management and survival of women under 50 years with breast cancer in the South East Thames Region. Br J Cancer 1996; 73: 751-757

44. Tilling K, Wolfe C, Raju KS. Variations in the management and

survival of women with endometrial cancer in south east England. Eur J Gynaecol Oncol 1998; 19: 64-68

45. Wolfe C, Tilling K, Raju KS. Management and survival of ovarian cancer patients in south east England. Eur J Cancer 1997; 33: 1835-1840

46. Landon M, Wilkinson P, Grundy C, Elliott P. Deprivation related differentials in mortality and hospital admission ratios. J Epidemiol Comm Hlth 1995; 49: s79 (Abstract)

47. Macleod U, Ross S, Twelves CJ, Gillis C, Watt GCM. A comparison of the care received from primary and secondary care by women with breast cancer living in affluent and deprived areas. J Epidemiol Comm Hlth 1998; 52: 687 (Abstract)

48. All-Party Parliamentary Group on Breast Cancer. Improving outcomes in breast cancer. London: House of Commons, 1998

49. Sainsbury R, Haward R, Rider L, Johnston C, Round C. Influence of clinician workload and patterns of treatment on survival from breast cancer. Lancet 1995; 345: 1265-1270

50. Dickman PW, Gibberd RW, Hakulinen T. Estimating potential savings in cancer deaths by eliminating regional and social class variation in cancer survival in the Nordic countries. J Epidemiol Comm Hlth 1997; 51: 289-298

51. National Cancer Institute. SEER Stat – cancer incidence public use database 1973-95. Release 1.1. Bethesda, MD: National Cancer Institute, 1998

52. Berrino F, Sant M, Verdecchia A, Capocaccia R, Hakulinen T, Estève J, (eds.) *Survival of cancer patients in Europe: the EUROCARE study. (IARC Scientific Publications No. 132)*. Lyon: International Agency for Research on Cancer, 1995

53. Black RJ, Clarke JA, Warner JM. Malignant melanoma in Scotland: trends in incidence and survival, 1968-87. Health Bull (Edinb) 1991; 49: 97-105

54. Verdecchia A, Capocaccia R, Santaquilani M, Hakulinen T. Methods of survival data analysis and presentation issues: EUROCARE 1985-89. In: Berrino F, Capocaccia R, Estève J, Gatta G, Hakulinen T, Micheli M, Sant M, Verdecchia A, (eds.) *Survival of cancer patients in Europe: the EUROCARE study, II. (IARC Scientific Publications No. 151)*. Lyon: International Agency for Research on Cancer, 1999

55. Estève J, Benhamou E, Croasdale M, Raymond L. Relative survival and the estimation of net survival: elements for further discussion. Stat Med 1990; 9: 529-538

56. Black RJ, Bashir SA. World standard cancer patient populations: a resource for comparative analysis of survival data. In: *(IARC Scientific Publications, in press)*. Lyon: International Agency for Research on Cancer, 1999

57. Berrino F, Micheli A, Sant M, Capocaccia R. Interpreting survival differences and trends. Tumori 1997; 83: 9-16

58. Berrino F, Estève J, Coleman MP. Basic issues in the estimation and comparison of cancer patient survival. In: Berrino F, Sant M, Verdecchia A, Capocaccia R, Hakulinen T, Estève J, (eds.) *Survival of cancer patients in Europe: the EUROCARE study. (IARC Scientific Publications No. 132)*. Lyon: International Agency for Research on Cancer, 1995, pp1-14

59. Sant M, Capocaccia R, Verdecchia A, Gatta G, Micheli A, Mariotto A, Hakulinen T, Berrino F, EUROCARE Working Group. Comparisons of colon cancer survival among European countries: the EUROCARE study. Int J Cancer 1995; 63: 43-48

Chapter 5

LIP

Cancer of the lip is an uncommon tumour, with about 220 cases diagnosed each year in England and Wales. Over the last 20 years, there has been a decline in incidence in many parts of the world, particularly in men. The highest incidence rates occur in Australia, parts of Canada, and Spain; the lowest rates are seen in Asia and amongst blacks in the USA. The lip is defined for these purposes as the vermilion border (lipstick area) and the internal (mucosal) surface of the lip, and it excludes the skin of the lip. Cancers of the skin of the lip may nevertheless be misclassified to the lip or the anterior part of the oral cavity. Improvement in histological classification and coding may explain some of the decline in incidence, but the trends are widespread. Most lip cancers occur on the lower lip. The majority are well-differentiated squamous cell carcinomas. Sunlight is the main risk factor. High-risk occupations are predominantly those associated with the outdoors, such as fishermen, farmers and sailors. Tobacco and alcohol are two other important risk factors. Lip cancer has been specifically associated with pipe smoking. Infection with herpes viruses may also be a risk factor.

Lip cancer is less common in England and Wales than in other countries in Europe. Between 1971 and 1990, incidence in men fell by two-thirds to 0.7 per 100,000, a more rapid decline than in Scotland or other European countries. Incidence remained virtually unchanged among women at 0.2 per 100,000. The male-to-female sex ratio is about 4 to 1. Mortality from lip cancer has fallen in parallel with incidence for both sexes, and in England and Wales in 1995, there were only 25 deaths in men and three deaths in women.

Of some 6,300 eligible patients, about 5% were excluded because their vital status was unknown on 6 July 1997, and 1% because of zero or unknown survival (Table 5.1). Exclusions in both categories were twice as common in 1986-90 as in 1971-75. Survival analyses are based on almost 6,000 patients, some 93% of those eligible. Two-thirds of the cancers were coded to the vermilion border of the lower lip (Table 5A). Papillary and squamous cell carcinomas accounted for around 85% of cases.

Survival trends

Survival from lip cancer is very good: relative survival among adults diagnosed in England and Wales during 1986-90 was 97% at one year and 89% at five years, with little difference between men and women (Table 5.2). Age-standardised survival at one year has been 90% or higher since 1971: for patients diagnosed during 1986-90, one-year survival was 96% for men and 94% for women (see Table 4.4). Survival in men has been consistently high. For patients diagnosed during 1986-90, five-year survival was 86% and 85% in men and women, respectively, compared to 85% and 73% for those diagnosed during 1971-75. The difference in survival between men and women has effectively disappeared: from 12% for patients diagnosed in 1971-75 to 1% for 1986-90.

Regional variation in survival is not marked, but survival estimates for individual regions are generally imprecise, particularly for patients diagnosed during the 1980s, because there were fewer cases than during the 1970s. Trends in survival have varied between regions, but the range in five-year survival between regions has changed very little (Table 5.2, Figure 5.1).

Under the age of 70 at diagnosis, the average five-year relative survival was similar at 90% or higher in men and

Table 5A Sub-site distribution of lip cancer[1]

ICD-9 code	Site description	1971-75 No.	%	1976-80 No.	%	1981-85 No.	%	1986-90 No.	%	Total No.	%
140.0	Upper lip, vermilion border	172	8.6	127	8.7	97	7.2	98	8.6	**494**	**8.3**
140.1	Lower lip, vermilion border	1,398	70.3	1,047	71.6	807	60.3	639	55.8	**3,891**	**65.5**
140.2	Both lips (1971-78 only)	5	0.3	6	0.4	-	-	-	-	**11**	**0.2**
140.3	Upper lip, inner aspect	-	-	7	0.5	24	1.8	29	2.5	**60**	**1.0**
140.4	Lower lip, inner aspect	-	-	25	1.7	75	5.6	60	5.2	**160**	**2.7**
140.5	Lip, unspecified, inner aspect	-	-	2	0.1	10	0.7	6	0.5	**18**	**0.3**
140.6	Commissure of lip	-	-	4	0.3	10	0.7	8	0.7	**22**	**0.4**
140.8	Other	-	-	1	0.1	3	0.2	4	0.3	**8**	**0.1**
140.9	Unspecified	414	20.8	244	16.7	313	23.4	301	26.3	**1,272**	**21.4**
	Total	**1,989**		**1,463**		**1,339**		**1,145**		**5,936**	

[1] Upper lip was coded 140.0 in ICD-8 (1971-78), including the inner aspect (ICD-9 140.3); similarly, lower lip was coded 140.1 in ICD-8, including the inner aspect (ICD-9 140.4).

women diagnosed in England and Wales during 1986-90. For older patients, relative survival was lower (Figure 5.5, see Table 4.7).

Deprivation

There was no significant gradient in five-year survival with deprivation during the 1980s in England and Wales (Figure 5.3, see Table 4.9).

International comparison

The average five-year survival in England and Wales, 89% in men and women, is similar to the average for Europe. The highest regional survival rates, 95% and 92%, are similar to those in Scotland and the USA (see Table 4.12).

Table 5.1 Ineligible and excluded records, and patients included in analyses, by calendar period of diagnosis

	Calendar period of diagnosis														
	1971-75			**1976-80**			**1981-85**			**1986-90**			**Overall**		
Total registered	**2,241**			**1,619**			**1,438**			**1,288**			**6,586**		
Ineligible	*Records*	*Patients*	*%*	*Records*	*Patients*	*%*	*Records*	*Patients*	*%*	*Records*	*Patients*	*%*	*Records*	*Patients*	*%*
Incomplete data[1]	1	1		2	2		2	2		0	0		5	**5**	
Resident outside England and Wales	21	21		12	11		7	7		10	10		50	**49**	
In situ neoplasm[2]	20	20		10	10		0	0		0	0		30	**30**	
Benign or uncertain behaviour[2]	110	108		40	37		0	0		0	0		150	**145**	
Metastatic[2]	5	5		1	1		0	0		0	0		6	**6**	
Otherwise ineligible[3]	-	-		-	-		-	-		-	-		-	**-**	
Lymphoma[4]	0	0		0	0		0	0		0	0		0	**0**	
Leukaemia or myeloma[4]	0	0		0	0		0	0		0	0		0	**0**	
		155			**61**			**9**			**10**			**235**	
Total eligible	**2,086**	**100**		**1,558**	**100**		**1,429**	**100**		**1,278**	**100**		**6,351**	**100**	
Aged under 15 or 100 years or over at diagnosis	4	4	0.2	1	1	<0.1	2	2	0.1	0	0	0.0	7	**7**	0.1
Vital status unknown[5]	107	81	3.9	93	83	5.3	70	67	4.7	108	106	8.3	378	**337**	5.3
Duplicate registration[6]	0	0	0.0	0	0	0.0	0	0	0.0	0	0	0.0	0	**0**	0.0
Multiple primary[7]	0	0	0.0	4	2	0.1	2	2	0.1	4	4	0.3	10	**8**	0.1
Synchronous tumours[8]	2	1	<0.1	5	2	0.1	4	4	0.3	5	3	0.2	16	**10**	0.2
Sex not known	0	0	0.0	0	0	0.0	0	0	0.0	0	0	0.0	0	**0**	0.0
Sex incompatible with site code[9]	-	-	-	-	-	-	-	-	-	-	-	-	-	**-**	-
Invalid dates[10]	1	1	<0.1	1	0	0.0	0	0	0.0	0	0	0.0	2	**1**	<0.1
Zero survival[11]	13	10	0.5	8	7	0.4	18	15	1.0	20	20	1.6	59	**52**	0.8
Total exclusions		**97**	**4.7**		**95**	**6.1**		**90**	**6.3**		**133**	**10.4**		**415**	**6.5**
Patients included in analysis		**1,989**	**95.3**		**1,463**	**93.9**		**1,339**	**93.7**		**1,145**	**89.6**		**5,936**	**93.5**

[1] Main data item(s) invalid or incompatible with one another: name, sex, date of birth, date of diagnosis and (if present) date of death, postcode, site, morphology and behaviour

[2] Tumour either *in situ* (behaviour code 2), benign (0), uncertain if benign or malignant (1) or metastatic (6 or 9)

[3] Not applicable for this malignancy

[4] Morphology that of lymphoma (included with non-Hodgkin lymphoma, Ch. 41) or myeloma (included with multiple myeloma, Ch. 43), or leukaemia (excluded)

[5] Tracing of vital status at National Health Service Central Register incomplete at 6 July 1997

[6] Same site code, sex, cancer registry and cancer registry number as an earlier registration

[7] Same site code and person number as an earlier registration(s): mostly confirmed multiple primary tumours at this site, some unresolved duplicate registrations

[8] Same site code, sex, date of birth and date of diagnosis as another registration(s): mostly synchronous or (in paired organs) bilateral tumours in same anatomic site in one individual, not previously linked: also some duplicate registrations

[9] Not applicable for this malignancy

[10] Impossible sequence of dates (birth, diagnosis and, if present, emigration or death); or date of death after 6 July 1997

[11] Date of diagnosis same as date of death: some are patients who did die on the day of diagnosis, but most were registered solely from a death certificate (death certificate only, or DCO), and their survival time is thus unknown

Table 5.2   Crude and relative survival (%) at one, five and ten years since diagnosis: NHS Region, deprivation category and calendar period of diagnosis

		1971-75						1976-80					
		Affluent	2	3	4	Deprived	All[1]	Affluent	2	3	4	Deprived	All[1]
		C R	C R	C R	C R	C R	C R	C R	C R	C R	C R	C R	C R
ENGLAND & WALES													
Men		(222)	(288)	(326)	(293)	(208)	(1,707)	(181)	(278)	(296)	(220)	(193)	(1,241)
	One year	91 97	89 94	93 99	90 96	92 98	91 97	90 96	90 97	92 98	95 100	88 95	91 98
	Five years	62 85	60 84	65 92	68 91	66 89	65 89	65 87	62 88	70 94	74 95	69 91	68 93
	Ten years	40 78	43 80	49 88	49 87	44 83	45 85	44 83	47 88	50 91	49 90	52 89	48 91
Women		(39)	(58)	(55)	(47)	(29)	(282)	(40)	(47)	(56)	(38)	(30)	(222)
	One year	82 91	88 95	91 95	91 95	93 98	89 95	92 95	87 96	91 96	84 88	90 93	88 93
	Five years	53 69	62 90	64 90	62 79	48 65	59 83	70 84	66 87	73 91	58 79	67 84	67 88
	Ten years	38 66	43 75	56 90	42 71	27 52	43 79	48 73	49 87	53 88	44 79	44 84	48 88
Adults		(261)	(346)	(381)	(340)	(237)	(1,989)	(221)	(325)	(352)	(258)	(223)	(1,463)
	One year	90 97	89 94	92 98	91 96	92 98	91 97	90 96	90 97	91 98	93 98	88 95	91 97
	Five years	60 84	60 85	65 92	67 91	64 87	64 89	66 87	63 88	71 94	72 93	69 91	67 92
	Ten years	40 77	43 80	50 89	48 87	42 80	45 84	45 83	47 88	50 91	48 90	51 89	48 91

		Men	Women	Adults			Men	Women	Adults
Northern & Yorkshire		(212)	(34)	(246)			(117)	(27)	(144)
	One year	90 97	88 100	89 97			91 100	89 95	91 99
	Five years	59 83	44 70	57 82			67 97	63 86	66 96
	Ten years	44 82	29 70	42 81			42 90	51 86	43 91
Trent		(126)	(25)	(151)			(112)	(20)	(132)
	One year	90 96	88 90	89 95			90 99	95 96	91 99
	Five years	68 91	59 74	67 88			70 97	70 89	70 97
	Ten years	55 90	43 68	53 87			56 97	40 76	54 97
Anglia & Oxford		(314)	(37)	(351)			(238)	(31)	(269)
	One year	92 98	89 95	91 98			89 96	90 95	90 96
	Five years	65 90	62 88	65 90			68 90	64 81	67 91
	Ten years	47 90	47 88	47 89			47 89	49 81	47 88
North Thames		(190)	(35)	(225)			(129)	(23)	(152)
	One year	89 97	83 89	88 96			93 98	96 97	93 98
	Five years	61 83	62 78	61 83			66 87	74 92	67 89
	Ten years	42 81	51 76	43 82			58 87	56 87	58 89
South Thames		(142)	(45)	(187)			(171)	(34)	(205)
	One year	97 100	96 97	97 100			95 100	86 90	93 98
	Five years	71 91	55 76	67 88			74 96	59 73	71 93
	Ten years	47 79	33 63	44 76			54 93	38 73	51 91
South & West		(237)	(50)	(287)			(190)	(35)	(225)
	One year	92 96	90 99	92 97			89 95	89 96	89 96
	Five years	69 92	64 92	68 92			64 90	69 85	65 89
	Ten years	46 86	52 86	47 85			41 81	51 85	43 83
West Midlands		(164)	(18)	(182)			(101)	(16)	(117)
	One year	93 99	94 95	93 98			86 95	81 85	85 93
	Five years	69 96	56 77	68 95			62 87	62 81	62 86
	Ten years	42 83	35 77	41 82			45 84	36 79	44 83
North & West		(188)	(19)	(207)			(115)	(23)	(138)
	One year	89 95	84 88	89 95			94 100	87 94	93 99
	Five years	61 89	63 71	61 87			68 94	74 86	69 93
	Ten years	43 83	47 71	43 83			42 84	53 83	43 85
WALES		(134)	(19)	(153)			(68)	(13)	(81)
	One year	89 95	85 91	88 94			93 99	78 83	90 97
	Five years	60 84	74 89	61 86			70 89	70 81	70 90
	Ten years	39 76	53 78	41 78			50 85	55 72	51 85

[1] Including those for whom no deprivation category was assigned. C - crude survival, R - relative survival (see Chapter 3). Number of patients contributing to each analysis in parentheses.

ENGLAND & WALES

	1981-85						1986-90						
	Affluent C R	*2* C R	*3* C R	*4* C R	*Deprived* C R	*All*[1] C R	*Affluent* C R	*2* C R	*3* C R	*4* C R	*Deprived* C R	*All*[1] C R	
	(183)	*(247)*	*(231)*	*(204)*	*(152)*	*(1,057)*	*(169)*	*(187)*	*(178)*	*(206)*	*(144)*	*(893)*	**Men**
One year	93 99	92 97	94 99	91 97	92 98	93 98	92 98	90 96	89 96	94 99	92 97	91 97	
Five years	65 87	64 88	66 91	67 92	68 91	66 91	72 91	63 89	63 85	67 93	66 87	66 89	
	46 83	46 88	45 90	50 91	43 78	46 87							
	(58)	*(70)*	*(57)*	*(52)*	*(33)*	*(282)*	*(41)*	*(61)*	*(65)*	*(52)*	*(28)*	*(252)*	**Women**
One year	95 97	91 100	90 99	87 96	100 100	91 100	85 91	85 90	86 92	94 97	93 98	88 94	
Five years	72 91	59 84	61 87	54 69	59 73	61 85	61 84	59 86	65 87	77 92	64 80	65 89	
	55 89	41 76	49 84	39 66	43 69	46 83							
	(241)	*(317)*	*(288)*	*(256)*	*(185)*	*(1,339)*	*(210)*	*(248)*	*(243)*	*(258)*	*(172)*	*(1,145)*	**Adults**
One year	94 98	91 98	93 100	90 97	93 99	92 99	91 97	89 95	88 95	94 99	92 97	91 97	
Five years	67 89	63 87	65 91	64 87	66 89	65 89	69 90	62 89	64 85	69 94	66 87	66 89	
	48 86	45 85	46 91	48 87	43 78	46 86							

	1981-85			1986-90			
	Men C R	*Women* C R	*Adults* C R	*Men* C R	*Women* C R	*Adults* C R	
	(108)	*(24)*	*(132)*	*(112)*	*(29)*	*(141)*	**Northern & Yorkshire**
One year	91 97	100 100	92 98	91 97	97 100	92 98	
Five years	58 85	61 72	58 84	62 81	62 82	62 83	
	35 77	52 67	37 78				
	(129)	*(27)*	*(156)*	*(142)*	*(28)*	*(170)*	**Trent**
One year	86 92	96 99	88 94	93 99	78 84	91 97	
Five years	60 84	55 74	59 82	66 88	53 67	64 85	
	43 81	33 56	41 79				
	(245)	*(62)*	*(307)*	*(219)*	*(50)*	*(269)*	**Anglia & Oxford**
One year	94 100	92 100	93 100	91 99	92 95	91 98	
Five years	70 93	74 98	71 94	70 94	72 92	71 95	
	48 89	56 93	50 92				
	(108)	*(27)*	*(135)*	*(55)*	*(18)*	*(73)*	**North Thames**
One year	94 99	89 97	93 98	93 97	95 96	93 97	
Five years	62 87	52 69	60 82	69 90	72 91	70 92	
	45 84	37 66	44 79				
	(118)	*(37)*	*(155)*	*(75)*	*(32)*	*(107)*	**South Thames**
One year	96 99	95 100	95 99	92 95	90 96	92 96	
Five years	70 90	70 95	70 91	73 90	66 90	71 90	
	47 85	40 75	46 84				
	(169)	*(53)*	*(222)*	*(132)*	*(45)*	*(177)*	**South & West**
One year	95 99	89 100	93 99	88 94	83 91	86 93	
Five years	73 93	49 75	67 90	61 86	63 89	62 87	
	54 93	37 73	50 88				
	(46)	*(13)*	*(59)*	*(34)*	*(15)*	*(49)*	**West Midlands**
One year	89 95	100 100	92 97	94 99	100 100	96 100	
Five years	59 74	69 73	61 78	71 95	74 84	72 96	
	42 65	69 73	49 71				
	(53)	*(11)*	*(64)*	*(45)*	*(16)*	*(61)*	**North & West**
One year	96 100	82 100	94 100	96 100	75 83	90 98	
Five years	70 95	55 80	67 93	64 95	44 55	59 86	
	47 83	55 80	48 85				
	(81)	*(28)*	*(109)*	*(79)*	*(19)*	*(98)*	**WALES**
One year	91 97	82 92	89 96	90 95	89 90	90 94	
Five years	67 90	61 80	65 91	59 80	84 90	64 83	
	48 87	46 80	48 88				

[1] Including those for whom no deprivation category was assigned. C - crude survival, R - relative survival (see Chapter 3). Number of patients contributing to each analysis in parentheses.

Fig. 5.1 Relative survival up to ten years: trends by calendar period of diagnosis (1971-90) and NHS Region

ADULTS

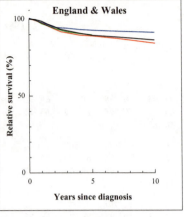

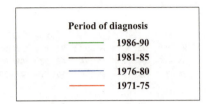

Period of diagnosis
- 1986-90
- 1981-85
- 1976-80
- 1971-75

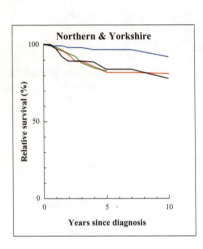

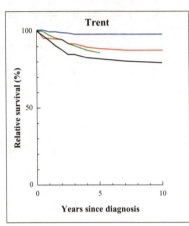

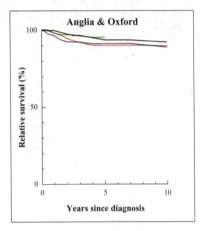

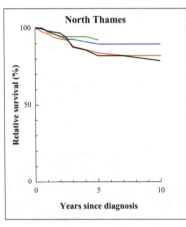

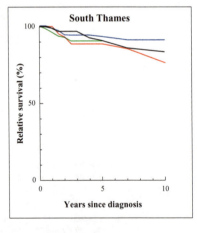

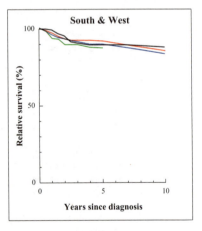

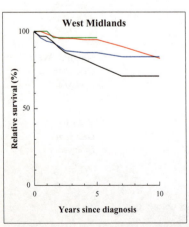

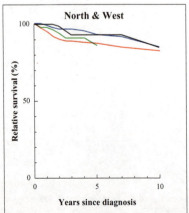

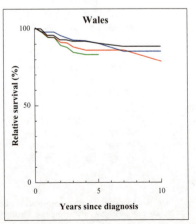

Fig. 5.2 Relative survival up to five years, by deprivation category: England and Wales, patients diagnosed 1986-90

ADULTS

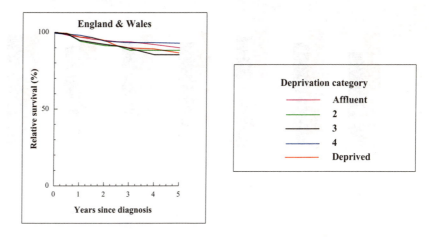

Fig. 5.3 Relative survival at five years, by deprivation category and period of diagnosis (1981-90): England and Wales

ADULTS

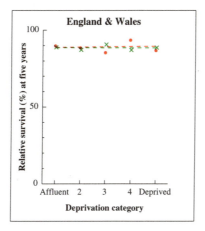

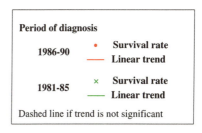

Fig. 5.4   Relative survival (%) by NHS Region and sex:
one-year and five-year survival for patients diagnosed 1986-90, with 95% confidence intervals

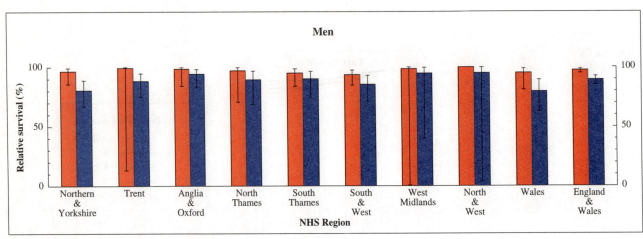

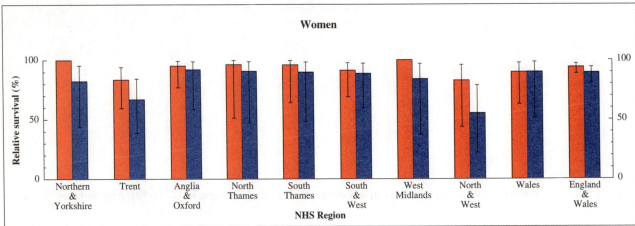

Fig. 5.5   Relative survival (%) at five years, with 95% confidence interval, by age at diagnosis and sex:
England and Wales, patients diagnosed 1986-90

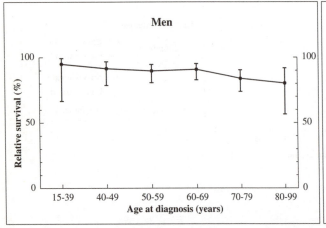

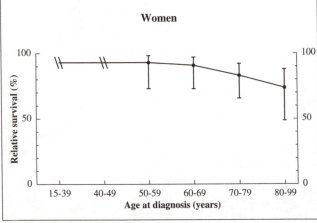

Chapter 6

TONGUE

Cancers of the tongue account for one in five cancers in the oral cavity. About 650 new cases are diagnosed each year in England and Wales. They are usually well-differentiated squamous cell carcinomas. Spread within the oral cavity occurs to adjacent structures such as the floor of the mouth and tonsillar pillars. Early diagnosis is essential for the success of radical treatment, but the regional lymph nodes are already involved in 70% of cases at diagnosis.

Between 1971 and 1990, the age-standardised incidence of tongue cancer in England and Wales increased slightly from 1.4 to 1.7 per 100,000 in men, and from 0.6 to 0.8 in women. These increases have been less marked than in Scotland, Germany and Denmark. Cancer of the tongue is very rare before age 35. As elsewhere, however, the increase in England and Wales since 1971 has occurred at ages below 75; incidence at older ages has fallen. The male-to-female sex ratio has remained at around 2 to 1. Since 1971, mortality has been around 0.9 per 100,000 per year in men and around 0.4 in women. As with other cancers of the oral cavity, the main risk factors are tobacco smoking (especially pipe smoking), tobacco and betel chewing, and alcohol.

The survival analyses are based on 10,500 patients, some 90% of those eligible (Table 6.1). About 5% were excluded because their vital status was unknown on 6 July 1997, and 4% because of zero or unknown survival. Almost two-thirds of the cases were coded to the tongue, without specification of the part of the tongue involved (Table 6A). Cancers of the lingual tonsil are not included here: they were analysed with cancers of the oropharynx, for consistency with the classification used during the 1970s (see Chapter 2). Papillary and squamous cell carcinomas accounted for more than three-quarters of tongue cancers, and most of the remainder were also classified as epithelial cancers but without further detail.

Survival trends

Survival from tongue cancer is only moderate. For patients diagnosed in 1986-90 in England and Wales, relative survival at one year was 63% in men and 73% in women. Five-year survival was 37% and 49%, respectively (Table 6.2).

Survival in England and Wales has improved slightly since the early 1970s (Figure 6.1). Age-standardised survival for women one year after diagnosis increased from 67% to 75%, and survival at five years rose from 44% to 50%. This steady but significant improvement in both short-

term and longer-term survival among women with tongue cancer represents a gain of about 2.5% every five years (see Table 4.4). It stands in sharp contrast to the almost negligible change in survival for men. Age-adjusted survival one year after diagnosis for men has remained at 60-61% since the early 1970s, although five-year survival has increased by about 1% between successive quinquennia, reaching 36% for men diagnosed during 1986-90.

There are regional differences in recent survival rates and trends over time (Figure 6.1). North Thames showed an improvement of about 10% in five-year and ten-year survival during the 1970s. For patients diagnosed during 1986-90, there was a 15% difference in one-year survival between the regions with the highest survival (Anglia & Oxford, 76%) and the lowest (North & West, 61%) (Table 6.2). Differences in five-year survival were smaller, but there was still an absolute range of about 10% between the three regions with the highest survival (48%) and the four regions where five-year survival stands at 38% – Northern & Yorkshire, West Midlands, North & West and Wales.

Survival from cancer of the tongue falls sharply with age at diagnosis in both sexes (Figure 6.5). For men diagnosed during 1986-90, five-year survival was 61% for those aged 15-39 at diagnosis, compared with 23% in men aged 80-99 years. For women, the corresponding survival rates were 67% and 30%. Women had five-year survival rates at least 15% better than those for men across the age range 40-79 years, with smaller differences among the very young and the very old (see Table 4.7).

As a result of these differences in time trends and age-specific survival, the age-adjusted one-year survival advantage for women has increased to some 14%, the same as at five years after diagnosis.

Deprivation

There was a steep survival gradient across the deprivation groups in England and Wales for patients diagnosed during the 1980s (Figures 6.2 and 6.3). The affluent group showed an advantage of around 12% in both one- and five-year survival for patients diagnosed during 1981-85. The gap between affluent and deprived groups was 16% for patients diagnosed during 1986-90 (see Table 4.9). Similarly large differences were seen in most regions (data not shown). Five-year survival in England and Wales was similar for the two most affluent groups (deprivation categories 1 and 2).

Table 6A Sub-site distribution of cancer of the tongue

ICD-9 code	Site description	1971-75 No.	%	1976-80 No.	%	1981-85 No.	%	1986-90 No.	%	Total No.	%
141.0	Base of tongue	366	14.6	337	13.5	430	15.9	434	15.6	1,567	15.0
141.1	Dorsal surface of tongue	75	3.0	64	2.6	71	2.6	41	1.5	251	2.4
141.2	Tip and lateral border of tongue	268	10.7	310	12.4	343	12.7	304	10.9	1,225	11.7
141.3	Ventral surface of tongue	60	2.4	52	2.1	50	1.9	59	2.1	221	2.1
141.4	Anterior two-thirds of tongue, part unspecified	-	-	70	2.8	163	6.0	103	3.7	336	3.2
141.5	Junctional zone	-	-	14	0.6	15	0.6	6	0.2	35	0.3
141.8	Other[1]	94	3.7	143	5.7	95	3.5	45	1.6	377	3.6
141.9	Unspecified	1,644	65.6	1,506	60.3	1,533	56.8	1,786	64.3	6,469	61.7
	Total	**2,507**		**2,496**		**2,700**		**2,278**		**10,481**	

[1] Special NCR code 141.8 used for "other specified parts" during 1971-78 (see Chapter 2).

The survival gradient between affluent and deprived patient groups was amongst the steepest seen for any cancer. Only cancers of the small bowel showed larger differentials in survival at one year, and only cancers of the nasopharynx and pituitary for survival at five years.

Among the more common cancers – those for which more than 1,000 patients were included in the analyses for 1986-90 – tongue cancer showed the largest gap in survival between affluent and deprived groups (see Table 4.10).

International comparison

Average five-year survival rates for tongue cancer diagnosed in England and Wales during 1986-90 were 37% for men and 49% for women, similar to those in Scotland and to the average survival rates in Europe (see Table 4.12). Survival rates in the affluent group, and in the regions with the highest survival rates (Anglia & Oxford and North Thames), were about 45% for men and almost 60% for women; these survival rates were as high as in the USA.

Table 6.1 Ineligible and excluded records, and patients included in analyses, by calendar period of diagnosis

	Calendar period of diagnosis														
	1971-75			1976-80			1981-85			1986-90			Overall		
Total registered	2,772			2,870			3,123			3,316			12,081		
Ineligible	Records	Patients	%	Records	Patients	%	Records	Patients	%	Records	Patients	%	Records	Patients	%
Incomplete data[1]	5	5		10	10		13	13		14	14		42	**42**	
Resident outside England and Wales	59	59		47	47		25	22		6	6		137	**134**	
In situ neoplasm[2]	14	12		3	3		0	0		0	0		17	**15**	
Benign or uncertain behaviour[2]	32	24		34	27		0	0		0	0		66	**51**	
Metastatic[2]	11	11		7	7		0	0		0	0		18	**18**	
Otherwise ineligible[3]	0	0		21	21		84	82		68	66		173	**169**	
Lymphoma[4]	15	14		9	7		1	0		2	0		27	**21**	
Leukaemia or myeloma[4]	0	0		0	0		0	0		0	0		0	**0**	
		125			122			117			86			450	
Total eligible		2,647	100		2,748	100		3,006	100		3,230	100		11,631	100
Aged under 15 or 100 years or over at diagnosis	1	1	<0.1	6	4	0.1	6	6	0.2	4	4	0.1	17	**15**	0.1
Vital status unknown[5]	145	91	3.4	206	162	5.9	170	146	4.9	244	236	7.3	765	**635**	5.5
Duplicate registration[6]	0	0	0.0	0	0	0.0	0	0	0.0	0	0	0.0	0	**0**	0.0
Multiple primary[7]	0	0	0.0	5	4	0.1	16	15	0.5	26	23	0.7	47	**42**	0.4
Synchronous tumours[8]	10	8	0.3	16	6	0.2	6	5	0.2	19	10	0.3	51	**29**	0.2
Sex not known	0	0	0.0	0	0	0.0	0	0	0.0	0	0	0.0	0	**0**	0.0
Sex incompatible with site code[9]	-	-	-	-	-	-	-	-	-	-	-	-	-	**-**	-
Invalid dates[10]	0	0	0.0	0	0	0.0	1	0	0.0	5	4	0.1	6	**4**	<0.1
Zero survival[11]	48	40	1.5	90	76	2.8	155	134	4.5	196	175	5.4	489	**425**	3.7
Total exclusions		140	5.3		252	9.2		306	10.2		452	14.0		**1,150**	9.9
Patients included in analysis		2,507	94.7		2,496	90.8		2,700	89.8		2,778	86.0		**10,481**	90.1

[1] Main data item(s) invalid or incompatible with one another: name, sex, date of birth, date of diagnosis and (if present) date of death, postcode, site, morphology and behaviour

[2] Tumour either *in situ* (behaviour code 2), benign (0), uncertain if benign or malignant (1) or metastatic (6 or 9)

[3] Lingual tonsil (ICD-9 141.6), included with oropharynx (Ch. 9)

[4] Morphology that of lymphoma (included with non-Hodgkin lymphoma, Ch. 41) or myeloma (included with multiple myeloma, Ch. 43), or leukaemia (excluded)

[5] Tracing of vital status at National Health Service Central Register incomplete at 6 July 1997

[6] Same site code, sex, cancer registry and cancer registry number as an earlier registration

[7] Same site code and person number as an earlier registration(s): mostly confirmed multiple primary tumours at this site, some unresolved duplicate registrations

[8] Same site code, sex, date of birth and date of diagnosis as another registration(s): mostly synchronous or (in paired organs) bilateral tumours in same anatomic site in one individual, not previously linked: also some duplicate registrations

[9] Not applicable for this malignancy

[10] Impossible sequence of dates (birth, diagnosis and, if present, emigration or death); or date of death after 6 July 1997

[11] Date of diagnosis same as date of death: some are patients who did die on the day of diagnosis, but most were registered solely from a death certificate (death certificate only, or DCO), and their survival time is thus unknown

Table 6.2 Crude and relative survival (%) at one, five and ten years since diagnosis: NHS Region, deprivation category and calendar period of diagnosis

	1971-75 Affluent C	R	2 C	R	3 C	R	4 C	R	Deprived C	R	All[1] C	R	1976-80 Affluent C	R	2 C	R	3 C	R	4 C	R	Deprived C	R	All[1] C	R
ENGLAND & WALES																								
Men	(210)		(199)		(242)		(235)		(236)		(1,435)		(207)		(251)		(280)		(287)		(342)		(1,442)	
One year	64	67	63	66	59	63	53	56	54	58	57	61	69	73	58	61	56	59	54	58	54	57	58	61
Five years	31	39	29	36	21	29	25	34	19	28	25	33	36	45	28	35	28	38	26	34	23	29	28	36
Ten years	22	34	21	35	14	26	19	32	11	22	17	29	28	44	17	27	20	34	16	28	14	23	19	30
Women	(164)		(169)		(198)		(172)		(159)		(1,072)		(184)		(235)		(199)		(215)		(176)		(1,054)	
One year	69	72	67	71	61	64	61	64	58	61	63	66	66	70	62	65	63	66	57	60	54	57	61	64
Five years	40	48	38	49	32	40	32	39	35	46	35	43	35	41	34	43	33	40	34	42	25	31	33	40
Ten years	30	44	26	43	23	37	20	34	24	37	24	38	25	36	25	40	23	35	22	35	19	30	23	36
Adults	(374)		(368)		(440)		(407)		(395)		(2,507)		(391)		(486)		(479)		(502)		(518)		(2,496)	
One year	66	69	65	69	60	64	56	60	55	60	59	63	68	72	60	63	59	62	55	59	54	57	59	62
Five years	35	43	33	41	26	34	28	36	26	36	29	37	36	44	31	39	30	39	30	38	24	30	30	38
Ten years	25	38	23	38	18	32	19	33	16	29	20	33	26	40	21	33	22	35	19	31	16	25	21	33

	1971-75 Men C	R	Women C	R	Adults C	R	1976-80 Men C	R	Women C	R	Adults C	R
Northern & Yorkshire	(223)		(150)		(373)		(216)		(141)		(357)	
One year	52	57	56	58	54	58	54	57	56	59	54	58
Five years	19	26	31	40	24	32	24	31	35	44	29	36
Ten years	12	23	19	35	15	29	16	27	24	37	19	32
Trent	(131)		(84)		(215)		(144)		(102)		(246)	
One year	55	59	65	68	59	63	61	65	58	62	60	64
Five years	26	35	40	50	31	41	27	34	25	32	26	33
Ten years	16	29	28	45	20	35	18	28	17	31	18	29
Anglia & Oxford	(116)		(104)		(220)		(97)		(88)		(185)	
One year	68	74	62	66	65	70	59	63	62	67	61	65
Five years	39	50	36	44	38	47	32	40	38	49	35	45
Ten years	29	46	26	40	28	43	24	37	30	47	27	43
North Thames	(186)		(146)		(332)		(185)		(130)		(315)	
One year	63	67	61	65	62	66	58	61	69	73	62	66
Five years	24	33	29	37	26	35	36	47	38	45	37	46
Ten years	15	25	23	34	19	29	24	39	27	39	25	39
South Thames	(184)		(136)		(320)		(158)		(187)		(345)	
One year	59	63	66	70	62	66	56	59	56	60	56	59
Five years	23	30	34	44	28	36	27	34	28	35	27	35
Ten years	16	27	24	39	19	32	19	31	17	28	18	30
South & West	(160)		(144)		(304)		(209)		(129)		(338)	
One year	62	66	69	74	65	69	57	60	71	75	62	66
Five years	30	40	37	48	33	43	26	33	37	46	30	38
Ten years	23	37	23	40	23	39	15	25	28	43	20	33
West Midlands	(129)		(95)		(224)		(131)		(75)		(206)	
One year	48	51	57	60	52	55	60	64	65	69	62	66
Five years	22	30	30	36	25	33	28	36	32	38	29	37
Ten years	13	23	23	35	17	29	18	30	25	37	21	33
North & West	(213)		(160)		(373)		(203)		(140)		(343)	
One year	51	55	61	65	55	59	56	60	56	60	56	60
Five years	22	28	38	46	29	36	26	34	31	43	28	37
Ten years	14	25	29	41	20	32	19	30	27	43	22	35
WALES	(93)		(53)		(146)		(99)		(62)		(161)	
One year	60	65	70	75	64	68	64	68	58	61	62	65
Five years	31	39	42	51	35	44	30	39	29	35	30	37
Ten years	21	32	22	40	22	36	16	29	14	22	15	26

[1] Including those for whom no deprivation category was assigned. C - crude survival, R - relative survival (see Chapter 3). Number of patients contributing to each analysis in parentheses.

1981-85

	Affluent		2		3		4		Deprived		All[1]		
	C	R	C	R	C	R	C	R	C	R	C	R	
	(228)		(269)		(313)		(349)		(387)		(1,597)		**ENGLAND & WALES** *Men*
One year	66	69	61	65	58	61	57	60	54	57	58	62	
Five years	35	43	34	40	31	40	28	35	26	32	30	37	
	28	40	21	33	21	32	20	32	18	27	21	32	
	(226)		(200)		(228)		(244)		(185)		(1,103)		*Women*
One year	69	72	67	71	69	73	60	62	60	62	65	68	
Five years	37	46	42	51	41	51	34	41	34	40	37	45	
	29	45	26	39	28	43	24	36	21	30	25	38	
	(454)		(469)		(541)		(593)		(572)		(2,700)		*Adults*
One year	68	71	64	67	62	66	58	61	56	58	61	64	
Five years	36	44	37	45	35	44	31	38	29	35	33	41	
	28	43	23	36	24	37	22	33	19	28	23	35	

1986-90

	Affluent		2		3		4		Deprived		All[1]		
	C	R	C	R	C	R	C	R	C	R	C	R	
	(277)		(297)		(311)		(369)		(432)		(1,702)		**ENGLAND & WALES** *Men*
One year	71	75	64	67	58	61	57	59	56	58	60	63	
Five years	39	46	37	45	29	35	27	32	28	34	31	37	
	(216)		(231)		(249)		(182)		(185)		(1,076)		*Women*
One year	76	79	71	75	72	76	65	67	64	68	70	73	
Five years	46	57	44	52	40	49	36	43	33	41	40	49	
	(493)		(528)		(560)		(551)		(617)		(2,778)		*Adults*
One year	73	77	67	70	64	67	59	62	58	61	64	67	
Five years	42	50	40	48	34	41	30	36	29	36	35	42	

Regional (1981-85 and 1986-90)

	1981-85 Men		Women		Adults		1986-90 Men		Women		Adults		
	C	R	C	R	C	R	C	R	C	R	C	R	
	(232)		(142)		(374)		(230)		(151)		(381)		**Northern & Yorkshire**
One year	59	62	59	63	59	63	56	58	68	71	61	63	
Five years	31	39	34	41	32	40	28	34	36	44	31	38	
	23	36	22	31	23	34							
	(168)		(93)		(261)		(151)		(103)		(254)		**Trent**
One year	54	57	64	66	57	60	62	64	74	78	67	70	
Five years	24	30	35	42	28	34	30	37	39	49	34	42	
	11	18	21	31	15	22							
	(133)		(123)		(256)		(132)		(104)		(236)		**Anglia & Oxford**
One year	61	65	70	75	66	70	66	70	79	82	72	76	
Five years	36	43	44	56	40	49	35	43	43	55	38	48	
	27	39	34	54	31	46							
	(147)		(145)		(292)		(220)		(132)		(352)		**North Thames**
One year	62	66	67	71	65	68	62	65	74	77	67	69	
Five years	35	42	40	48	38	45	36	42	51	59	42	48	
	27	38	27	41	27	40							
	(208)		(140)		(348)		(186)		(142)		(328)		**South Thames**
One year	62	65	68	71	64	68	64	67	78	83	70	74	
Five years	32	39	39	50	35	44	33	41	43	55	38	47	
	24	35	26	40	25	37							
	(187)		(149)		(336)		(180)		(149)		(329)		**South & West**
One year	57	60	70	74	63	66	57	60	69	73	62	66	
Five years	33	39	40	49	36	44	29	35	41	48	34	41	
	20	31	26	41	23	35							
	(146)		(99)		(245)		(185)		(111)		(296)		**West Midlands**
One year	57	60	57	60	57	60	60	62	65	69	62	65	
Five years	28	36	27	34	28	35	29	35	36	45	32	38	
	22	31	18	27	20	30							
	(259)		(149)		(408)		(315)		(132)		(447)		**North & West**
One year	58	61	63	66	60	63	58	60	59	62	58	61	
Five years	28	34	34	42	30	37	30	35	35	45	32	38	
	18	30	23	39	20	33							
	(117)		(63)		(180)		(103)		(52)		(155)		**WALES**
One year	57	61	53	55	56	59	63	65	60	61	62	64	
Five years	27	35	37	45	31	40	31	37	30	39	30	38	
	18	28	32	45	23	36							

[1] Including those for whom no deprivation category was assigned. C - crude survival, R - relative survival (see Chapter 3). Number of patients contributing to each analysis in parentheses.

Fig. 6.1   Relative survival up to ten years: trends by calendar period of diagnosis (1971-90) and NHS Region

ADULTS

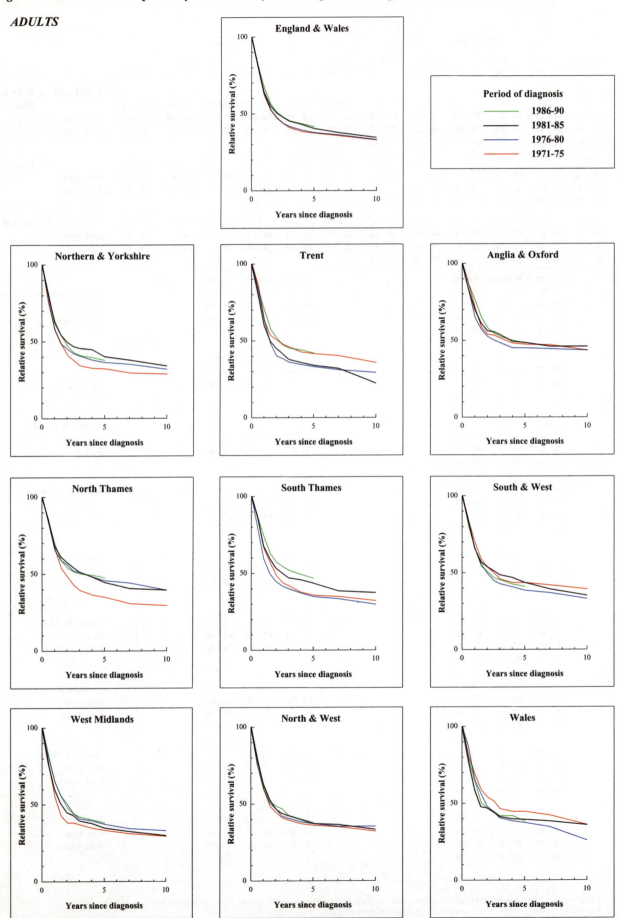

Fig. 6.2 Relative survival up to five years, by deprivation category: England and Wales, patients diagnosed 1986-90

ADULTS

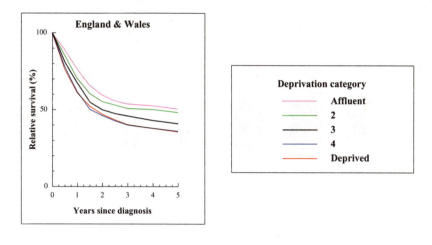

Fig. 6.3 Relative survival at five years, by deprivation category and period of diagnosis (1981-90): England and Wales

ADULTS

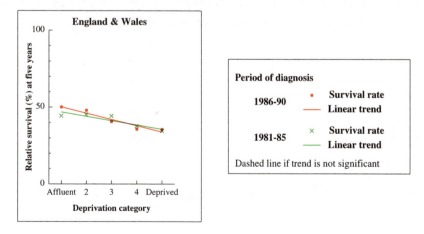

Fig. 6.4   Relative survival (%) by NHS Region and sex:
one-year and five-year survival for patients diagnosed 1986-90, with 95% confidence intervals

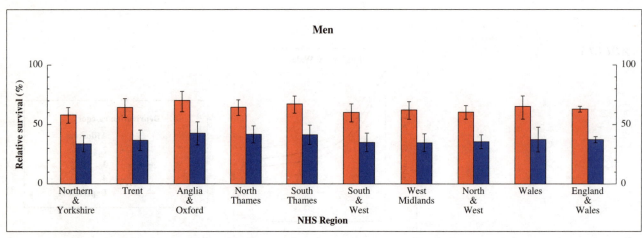

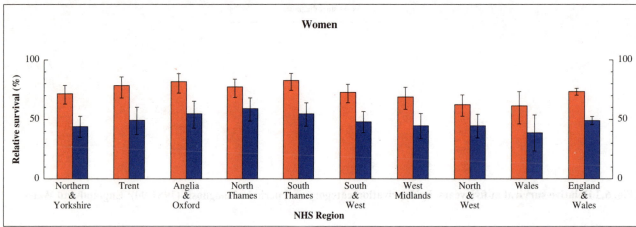

Fig. 6.5   Relative survival (%) at five years, with 95% confidence interval, by age at diagnosis and sex:
England and Wales, patients diagnosed 1986-90

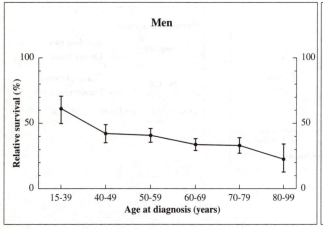

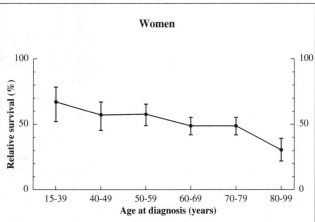

Chapter 7

SALIVARY GLANDS

Malignant neoplasms of the major salivary glands are uncommon: about 400 new cases are diagnosed each year in England and Wales. The tumours are mostly adenocarcinomas of the parotid, the largest salivary gland. These tumours are rare under the age of 40, and incidence at older ages is higher in men than women. Five-year survival in clinical series is moderate, and varies widely – from 20% to 80% – with the stage of disease at diagnosis, the histological type and the degree of differentiation (grade). Radical treatment involves resection of the primary followed by radiotherapy.

Salivary gland cancers are less common in England and Wales than in countries such as Australia, the USA and Brazil. Age-standardised (Europe) incidence in England and Wales in 1990 was 0.7 per 100,000 per year in men and 0.5 in women. Mortality has not changed over the past 25 years, with about 200 deaths each year. In 1995, age-standardised mortality rates were 0.4 per 100,000 in men and 0.2 in women.

A striking shift occurred in the age distribution of salivary gland cancers in both men and women over the 20 years 1971-90. Overall incidence has not changed markedly, but the rates have fallen in younger age groups and increased in older groups. The proportion of patients aged 70 or over was 30% in 1971-75, but for those diagnosed during 1986-90, this proportion had risen to 41% in men and 45% in women. This shift had an impact on the observed trends in survival.

The causes of cancer of the salivary glands are largely unknown. Unlike most other cancers in the head and neck region, tobacco and alcohol have not been implicated. Higher risks have been found in several occupational groups, including rubber workers, woodworkers in the car manufacturing industry, and farmers. Ionising radiation from exposure to repeated medical or dental X-rays may also increase risk. The decline in incidence under the age of 70 in England and Wales is consistent with the reduction in the use of such medical irradiation.

Benign mixed tumours of the parotid are relatively common: they were not included in the analyses. About 600 cases diagnosed during 1971-78 were excluded because of benign or uncertain behaviour. During that period, ICD-8 and MOTNAC were used to classify the tumour site and morphology, respectively.

Survival analyses were based on 6,500 patients, 90% of those eligible (Table 7.1). Some 5% of eligible patients were excluded because their vital status was unknown at 6 July 1997, and a further 3% because of zero or unknown survival. Three-quarters of all salivary gland cancers arise in the parotid gland (Table 7A). From 1979, with the introduction of ICD-9, some tumours of the minor salivary glands were classified to other three-digit rubrics in the oral cavity, while submandibular and sublingual glands were separately identified for the first time. It seems likely that about 10-15% of salivary gland cancers have occurred in the submandibular gland throughout the study period. Several morphologic types are frequent: almost one-third of cases were adenocarcinomas, with a further 40% of squamous, mucoepidermoid and other epithelial carcinomas.

Survival trends

Survival from cancer of the salivary glands in England and Wales for adults diagnosed in 1986-90 was 75-80% at one year and 50-60% at five years (Table 7.2). Relative survival rates one year after diagnosis have remained around 80% since the early 1970s in both sexes, but by the late 1980s there had been a remarkable apparent decline in survival at five years, from 59% to 51% in men and from 74% to 66% in women. This decline represents the national average, but similar patterns are seen in most regions, with the apparent decline in five-year survival exceeding 20% in some regions (Table 7.2, Figure 7.1). The trends in West Midlands and Wales do not fit this

Table 7A Sub-site distribution of cancers of the salivary glands

ICD-9 code	Site description	1971-75 No.	1971-75 %	1976-80 No.	1976-80 %	1981-85 No.	1981-85 %	1986-90 No.	1986-90 %	Total No.	Total %
142.0	Parotid gland	1,373	76.2	1,257	76.0	1,156	74.9	1,164	75.4	**4,950**	**75.6**
142.1	Submandibular gland	-	-	71	4.3	238	15.4	237	15.3	**546**	**8.3**
142.2	Sublingual gland	-	-	2	0.1	9	0.6	10	0.6	**21**	**0.3**
142.8	Other	220	12.2	153	9.2	4	0.3	2	0.1	**379**	**5.8**
142.9	Site unspecified	210	11.6	172	10.4	137	8.9	131	8.5	**650**	**9.9**
	Total	**1,803**		**1,655**		**1,544**		**1,544**		**6,546**	

pattern: survival in these regions was 20% lower than elsewhere in the 1970s, and increased to around the national average for patients diagnosed in the late 1980s.

The declining national trend in overall survival is not adjusted for the age distribution of the patients, however, and there is a marked effect of age on survival (Figure 7.5). Age-specific relative survival rates were therefore examined more closely. For patients diagnosed during 1986-90, five-year relative survival was 80% in men aged 15-39 at diagnosis and 90% in women, compared with 27% and 39%, respectively, at ages 80-99 (see Table 4.7). These survival rates and the corresponding rates for earlier periods are shown in Table 7B. There is in fact little change in survival over time within each age group, and the trends in age-standardised five-year survival rates since 1971 are negligible (see Table 4.4). The apparent decline in (unstandardised) relative survival with time is thus an artefact, resulting from the steep age-dependence of survival and the marked shift in age-incidence toward the elderly since 1971. There has been little or no change in five-year survival.

There was an average increase every five years of around 2% in relative survival at one year for men, but no significant increase for women.

Women have a survival advantage of 10% or more compared with men, at all ages over 40 and for both one-year and five-year survival.

Deprivation

There was a deprivation gradient in survival during the 1980s, with lower survival in the most deprived group. Figure 7.2 illustrates survival up to five years by deprivation category for patients diagnosed during 1986-90. For patients diagnosed in 1981-85, the gap between the most affluent and the most deprived groups was 9% for one-year survival and 8% for five-year survival. For those diagnosed during 1986-90, the gap was 6% at both one year and five years since diagnosis (see Table 4.9). Although there has been an increase in the deprivation gradient in five-year survival, since the late 1970s it was not statistically significant in either 1981-85 or 1986-90 (Figure 7.3).

International comparison

Average five-year survival for men and women diagnosed during 1986-90 in England and Wales was 3% to 6% below the average for Europe. The highest regional survival rates exceeded the European average and the survival rates in Scotland, however, and are comparable with those in the USA (see Table 4.12).

Table 7B Survival trends for cancers of the major salivary glands: numbers of patients and five-year relative survival rate (%) by period of diagnosis, age at diagnosis and sex

Age at diagnosis	1971-75		1976-80		1981-85		1986-90	
	No.	Rate	No.	Rate	No.	Rate	No.	Rate
MEN 15-39	134	91	98	94	70	83	73	80
40-49	107	77	91	74	74	64	71	75
50-59	148	50	137	56	139	51	128	46
60-69	229	50	240	48	216	52	204	46
70-79	166	30	220	38	241	45	209	40
80-99	93	27	91	20	86	26	125	27
All ages	*877*	*59*	*877*	*57*	*826*	*54*	*810*	*51*
Age-standardised		47		48		49		47
WOMEN 15-39	146	93	93	84	90	89	69	90
40-49	125	87	85	90	73	79	75	87
50-59	179	78	128	72	118	76	107	72
60-69	196	66	167	71	168	63	154	58
70-79	170	51	179	58	152	52	175	58
80-99	110	28	126	34	117	42	154	39
All ages	*926*	*74*	*778*	*70*	*718*	*68*	*734*	*66*
Age-standardised		62		64		62		62

Table 7.1 Ineligible and excluded records, and patients included in analyses, by calendar period of diagnosis

	Calendar period of diagnosis														
	1971-75			**1976-80**			**1981-85**			**1986-90**			**Overall**		
Total registered	**2,429**			**2,110**			**1,769**			**1,839**			**8,147**		
Ineligible	*Records*	*Patients*	*%*	*Records*	*Patients*	*%*	*Records*	*Patients*	*%*	*Records*	*Patients*	*%*	*Records*	*Patients*	*%*
Incomplete data[1]	3	3		15	15		27	27		30	30		75	**75**	
Resident outside England and Wales	47	47		28	26		13	12		8	5		96	**90**	
In situ neoplasm[2]	16	16		0	0		0	0		0	0		16	**16**	
Benign or uncertain behaviour[2]	382	364		271	263		0	0		0	0		653	**627**	
Metastatic[2]	14	14		1	1		1	1		1	1		17	**17**	
Otherwise ineligible[3]	-	-		-	-		-	-		-	-		-	**-**	
Lymphoma[4]	37	35		18	17		5	0		7	0		67	**52**	
Leukaemia or myeloma[4]	0	0		0	0		0	0		1	0		1	**0**	
		479			**322**			**40**			**36**			**877**	
Total eligible		**1,950**	**100**		**1,788**	**100**		**1,729**	**100**		**1,803**	**100**		**7,270**	**100**
Aged under 15 or 100 years or over at diagnosis	27	23	1.2	21	15	0.8	19	18	1.0	15	15	0.8	82	**71**	1.0
Vital status unknown[5]	125	78	4.0	108	80	4.5	77	68	3.9	142	136	7.5	452	**362**	5.0
Duplicate registration[6]	1	1	<0.1	0	0	0.0	0	0	0.0	1	1	<0.1	2	**2**	<0.1
Multiple primary[7]	0	0	0.0	2	2	0.1	6	5	0.3	16	14	0.8	24	**21**	0.3
Synchronous tumours[8]	12	2	0.1	9	3	0.2	16	12	0.7	12	4	0.2	49	**21**	0.3
Sex not known	0	0	0.0	0	0	0.0	0	0	0.0	0	0	0.0	0	**0**	0.0
Sex incompatible with site code[9]	-	-	-	-	-	-	-	-	-	-	-	-	-	**-**	-
Invalid dates[10]	0	0	0.0	0	0	0.0	1	1	<0.1	4	1	<0.1	5	**2**	<0.1
Zero survival[11]	53	43	2.2	44	33	1.8	91	81	4.7	96	88	4.9	284	**245**	3.4
Total exclusions		**147**	**7.5**		**133**	**7.4**		**185**	**10.7**		**259**	**14.4**		**724**	**10.0**
Patients included in analysis		**1,803**	**92.5**		**1,655**	**92.6**		**1,544**	**89.3**		**1,544**	**85.6**		**6,546**	**90.0**

[1] Main data item(s) invalid or incompatible with one another: name, sex, date of birth, date of diagnosis and (if present) date of death, postcode, site, morphology and behaviour

[2] Tumour either *in situ* (behaviour code 2), benign (0), uncertain if benign or malignant (1) or metastatic (6 or 9)

[3] Not applicable for this malignancy

[4] Morphology that of lymphoma (included with non-Hodgkin lymphoma, Ch. 41) or myeloma (included with multiple myeloma, Ch. 43), or leukaemia (excluded)

[5] Tracing of vital status at National Health Service Central Register incomplete at 6 July 1997

[6] Same site code, sex, cancer registry and cancer registry number as an earlier registration

[7] Same site code and person number as an earlier registration(s): mostly confirmed multiple primary tumours at this site, some unresolved duplicate registrations

[8] Same site code, sex, date of birth and date of diagnosis as another registration(s): mostly synchronous or (in paired organs) bilateral tumours in same anatomic site in one individual, not previously linked: also some duplicate registrations

[9] Not applicable for this malignancy

[10] Impossible sequence of dates (birth, diagnosis and, if present, emigration or death); or date of death after 6 July 1997

[11] Date of diagnosis same as date of death: some are patients who did die on the day of diagnosis, but most were registered solely from a death certificate (death certificate only, or DCO), and their survival time is thus unknown

Table 7.2   Crude and relative survival (%) at one, five and ten years since diagnosis: NHS Region, deprivation category and calendar period of diagnosis

ENGLAND & WALES

	1971-75						1976-80					
	Affluent	*2*	*3*	*4*	*Deprived*	*All*[1]	*Affluent*	*2*	*3*	*4*	*Deprived*	*All*[1]
	C R	C R	C R	C R	C R	C R	C R	C R	C R	C R	C R	C R
Men	*(133)*	*(153)*	*(152)*	*(141)*	*(112)*	*(877)*	*(177)*	*(196)*	*(164)*	*(155)*	*(139)*	*(877)*
One year	73 76	72 77	67 73	67 71	71 76	71 76	73 78	76 80	65 69	69 74	70 75	71 75
Five years	42 51	47 60	48 61	40 50	45 59	47 59	48 59	48 59	46 55	43 57	37 51	45 57
Ten years	37 50	36 55	36 58	31 46	35 58	37 57	35 49	33 50	36 53	31 53	31 49	33 51
Women	*(136)*	*(166)*	*(172)*	*(126)*	*(140)*	*(926)*	*(160)*	*(154)*	*(150)*	*(167)*	*(112)*	*(778)*
One year	84 88	81 84	80 84	79 84	78 82	81 85	84 88	78 82	81 85	71 76	78 84	78 83
Five years	68 78	60 72	60 72	64 76	60 72	62 74	60 70	62 73	61 75	53 65	60 71	58 70
Ten years	58 74	49 69	51 70	57 76	51 68	53 72	48 66	48 71	48 71	43 62	47 67	46 66
Adults	*(269)*	*(319)*	*(324)*	*(267)*	*(252)*	*(1,803)*	*(337)*	*(350)*	*(314)*	*(322)*	*(251)*	*(1,655)*
One year	79 83	77 81	74 79	72 77	75 80	76 80	78 83	77 81	73 77	70 75	74 79	74 79
Five years	55 65	54 67	54 68	51 63	53 67	55 67	53 64	54 66	53 65	48 62	47 61	51 63
Ten years	48 63	42 63	44 64	43 61	44 64	45 65	41 58	40 59	42 62	37 58	38 59	39 59

	1971-75			1976-80		
	Men	*Women*	*Adults*	*Men*	*Women*	*Adults*
Northern & Yorkshire	*(151)*	*(149)*	*(300)*	*(133)*	*(123)*	*(256)*
One year	77 83	82 85	80 84	68 73	80 85	74 79
Five years	51 66	67 78	59 73	44 58	65 75	54 67
Ten years	45 66	60 77	53 73	36 55	53 75	44 66
Trent	*(80)*	*(73)*	*(153)*	*(101)*	*(69)*	*(170)*
One year	68 73	89 92	78 82	76 81	74 80	75 81
Five years	47 55	73 82	59 70	35 45	59 73	45 57
Ten years	37 49	62 80	49 66	21 34	40 62	29 46
Anglia & Oxford	*(116)*	*(131)*	*(247)*	*(98)*	*(86)*	*(184)*
One year	82 88	82 87	82 87	83 86	77 80	80 84
Five years	59 75	66 80	63 77	58 70	57 66	58 68
Ten years	49 73	59 79	54 76	48 70	41 59	45 65
North Thames	*(109)*	*(111)*	*(220)*	*(68)*	*(56)*	*(124)*
One year	69 73	76 79	72 76	71 76	74 79	72 77
Five years	48 59	59 68	53 64	46 58	60 71	52 65
Ten years	40 57	52 67	46 62	40 54	52 69	45 63
South Thames	*(121)*	*(130)*	*(251)*	*(129)*	*(135)*	*(264)*
One year	70 75	82 87	77 81	72 76	81 89	77 82
Five years	45 57	58 72	52 65	50 64	61 75	56 69
Ten years	39 55	49 70	44 63	37 59	50 74	44 66
South & West	*(103)*	*(114)*	*(217)*	*(93)*	*(93)*	*(186)*
One year	66 69	84 88	75 79	67 71	79 83	73 77
Five years	40 54	67 79	54 67	44 55	47 61	46 59
Ten years	29 53	51 74	41 63	27 41	34 53	31 48
West Midlands	*(58)*	*(56)*	*(114)*	*(87)*	*(67)*	*(154)*
One year	55 59	75 80	65 69	66 71	83 90	74 79
Five years	33 40	46 62	39 50	41 54	64 77	51 64
Ten years	24 40	39 57	31 47	29 47	52 71	39 58
North & West	*(91)*	*(105)*	*(196)*	*(114)*	*(104)*	*(218)*
One year	73 79	82 87	78 83	64 69	75 81	69 75
Five years	46 63	64 76	56 70	35 46	52 63	43 55
Ten years	32 57	55 75	44 67	23 37	41 60	32 49
WALES	*(48)*	*(57)*	*(105)*	*(54)*	*(45)*	*(99)*
One year	64 68	68 71	66 70	76 80	69 72	72 76
Five years	36 45	47 55	42 51	51 69	58 64	54 66
Ten years	19 33	31 44	26 40	42 69	46 62	44 64

[1] Including those for whom no deprivation category was assigned. C - crude survival, R - relative survival (see Chapter 3). Number of patients contributing to each analysis in parentheses.

	1981-85					1986-90						
	Affluent	*2*	*3*	*4*	*Deprived*	*All*[1]	*Affluent*	*2*	*3*	*4*	*Deprived*	*All*[1]	
	C R	C R	C R	C R	C R	C R	C R	C R	C R	C R	C R	C R	
													ENGLAND & WALES
	(175)	*(150)*	*(180)*	*(157)*	*(140)*	*(826)*	*(165)*	*(188)*	*(157)*	*(153)*	*(138)*	*(810)*	**Men**
	79 83	68 74	73 78	69 74	65 70	71 76	73 79	71 75	71 76	75 79	67 71	72 76	*One year*
	46 59	40 50	42 55	41 52	37 51	42 54	38 52	38 51	40 51	39 51	37 48	39 51	*Five years*
	33 50	27 42	29 48	29 45	24 48	29 47							
	(164)	*(138)*	*(132)*	*(137)*	*(124)*	*(718)*	*(148)*	*(153)*	*(166)*	*(133)*	*(120)*	*(734)*	**Women**
	81 86	83 88	79 83	80 85	73 76	79 84	82 86	79 85	77 81	75 79	76 81	78 82	*One year*
	60 74	56 68	52 65	57 71	51 60	56 68	57 68	56 69	54 68	54 65	47 59	54 66	*Five years*
	50 66	45 65	39 58	44 64	40 55	44 62							
	(339)	*(288)*	*(312)*	*(294)*	*(264)*	*(1,544)*	*(313)*	*(341)*	*(323)*	*(286)*	*(258)*	*(1,544)*	**Adults**
	80 85	75 81	76 80	74 79	69 73	75 80	77 82	75 79	74 79	75 79	71 75	75 79	*One year*
	53 66	48 59	46 59	49 61	44 55	49 61	47 60	46 59	47 60	46 58	42 53	46 59	*Five years*
	41 58	35 54	33 53	36 54	31 52	36 54							

	Men	Women	Adults		Men	Women	Adults	
	(109)	*(103)*	*(212)*		*(94)*	*(74)*	*(168)*	**Northern & Yorkshire**
	67 73	76 81	71 77		72 76	79 85	75 80	*One year*
	41 54	53 66	47 60		38 51	51 67	43 57	*Five years*
	25 44	37 54	31 49					
	(70)	*(68)*	*(138)*		*(87)*	*(70)*	*(157)*	**Trent**
	67 72	82 88	74 80		75 78	67 71	71 75	*One year*
	44 57	63 77	54 68		36 45	46 58	40 50	*Five years*
	30 51	50 69	40 60					
	(75)	*(75)*	*(150)*		*(82)*	*(57)*	*(139)*	**Anglia & Oxford**
	77 83	76 80	77 82		62 67	81 84	70 75	*One year*
	48 65	55 64	51 64		38 49	56 72	45 59	*Five years*
	35 57	44 62	39 60					
	(74)	*(46)*	*(120)*		*(82)*	*(73)*	*(155)*	**North Thames**
	69 73	76 78	71 75		73 76	84 89	78 82	*One year*
	38 48	56 63	45 55		32 43	59 68	45 56	*Five years*
	19 33	46 62	29 47					
	(121)	*(116)*	*(237)*		*(101)*	*(105)*	*(206)*	**South Thames**
	78 82	81 86	79 84		75 80	80 85	78 82	*One year*
	45 58	58 74	51 66		44 58	60 71	52 66	*Five years*
	35 55	50 74	42 65					
	(114)	*(89)*	*(203)*		*(111)*	*(106)*	*(217)*	**South & West**
	77 83	81 86	79 85		73 78	83 88	78 83	*One year*
	41 53	54 67	47 59		38 50	54 69	45 60	*Five years*
	26 41	36 51	31 45					
	(69)	*(67)*	*(136)*		*(69)*	*(77)*	*(146)*	**West Midlands**
	80 83	83 88	82 86		81 86	78 83	79 84	*One year*
	45 57	61 73	53 66		52 69	51 65	51 66	*Five years*
	32 51	51 70	41 62					
	(117)	*(110)*	*(227)*		*(113)*	*(108)*	*(221)*	**North & West**
	65 69	77 83	71 75		65 69	65 70	65 69	*One year*
	38 50	52 65	45 58		34 46	47 58	41 52	*Five years*
	25 48	41 59	33 53					
	(77)	*(44)*	*(121)*		*(71)*	*(64)*	*(135)*	**WALES**
	64 68	82 86	70 75		72 76	88 95	79 85	*One year*
	38 48	59 69	46 57		45 57	63 75	53 66	*Five years*
	30 46	52 67	38 55					

[1] Including those for whom no deprivation category was assigned. C - crude survival, R - relative survival (see Chapter 3).
Number of patients contributing to each analysis in parentheses.

Fig. 7.1 Relative survival up to ten years: trends by calendar period of diagnosis (1971-90) and NHS Region

ADULTS

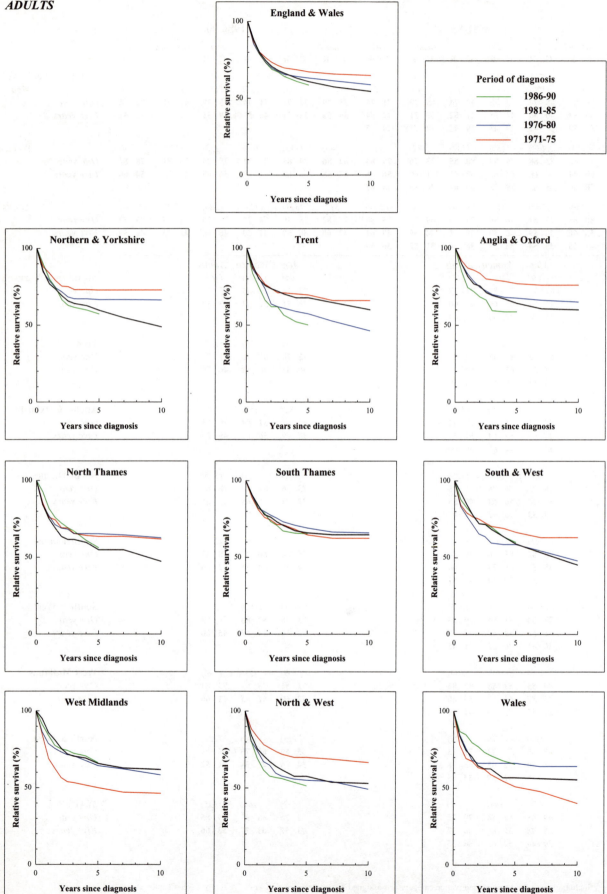

Fig. 7.2 Relative survival up to five years, by deprivation category: England and Wales, patients diagnosed 1986-90

ADULTS

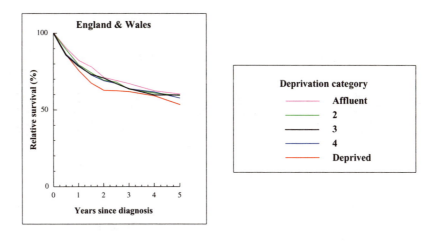

Fig. 7.3 Relative survival at five years, by deprivation category and period of diagnosis (1981-90): England and Wales

ADULTS

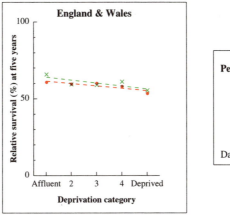
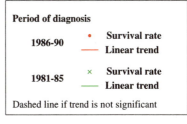

Fig. 7.4   Relative survival (%) by NHS Region and sex:
one-year and five-year survival for patients diagnosed 1986-90, with 95% confidence intervals

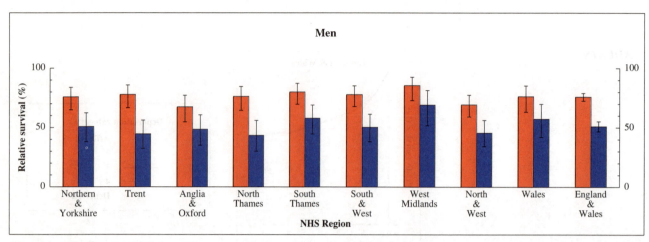

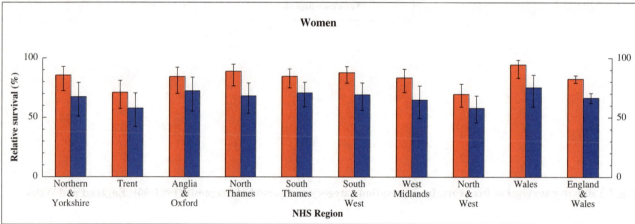

Fig. 7.5   Relative survival (%) at five years, with 95% confidence interval, by age at diagnosis and sex:
England and Wales, patients diagnosed 1986-90

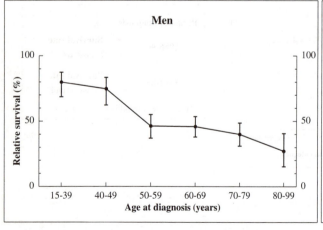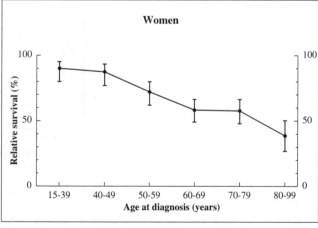

Chapter 8

ORAL CAVITY

Tumours of the oral cavity are defined here as those of the gum, the floor of the mouth, the cheek, the hard and soft palate, and other and unspecified parts of the mouth (ICD-9 143-145). Tumours of the tongue, the salivary glands and the pharynx are described in Chapters 6, 7 and 9-11, respectively.

Cancers of the mouth are relatively uncommon in England and Wales, with some 500 new cases a year. They represent about 1 in 200 of all cancers in adults. Among men, these cancers comprise 25% of all cancers of the lip, oral cavity and pharynx combined (ICD-9 140-149), and the proportion is 30% among women. The male-to-female sex ratio in incidence is about 1.5 to 1. Incidence has been rising steadily since 1971, and the average number of new cases has risen from 650 to 900 a year. The main risk factors for cancers of the mouth are tobacco smoking, the chewing of tobacco and betel quid, and alcohol consumption. These cancers are largely preventable, and regular inspection of the mouth offers opportunities for early diagnosis.

Cancers of the oral cavity have a wide range of clinical presentations. Many are only diagnosed late: spread to the regional lymph nodes of the neck and invasion of the lower jaw are common features at the time of diagnosis. The range of treatment options includes surgery, radiotherapy, chemotherapy, immunotherapy and other modalities, but there appears to be no consensus as to the best strategy for treating many of these tumours[1].

The survival analyses here are based on over 13,000 patients diagnosed with a primary malignant cancer of the oral cavity in England and Wales during the period 1971-90, about 91% of those eligible (Table 8.1). Of the 14,800 eligible patients, 9% were excluded, mostly because their vital status was unknown at 6 July 1997 (5%) or because of zero or unknown survival (3%). The proportion of patients excluded in these two categories rose in successive calendar periods to reach 7% and 4%, respectively, in 1986-90. For about 1% of patients, a subsequent primary malignancy had also been registered: only survival from the date of diagnosis of the first tumour is analysed.

About a third of the tumours were cancers of the floor of the mouth, 20% were cancers of the gum and alveolar margin, another 13% were cancers of the cheek, and the rest (30%) were cancers of other or unspecified parts of the mouth (Table 8A). Cancers of the floor of the mouth have became more common since the early 1970s, while those of the gum and cheek have become less common,

falling from 40% of all oral cavity tumours in 1971-75 to 25% in 1986-90. More than 70% of the tumours were squamous carcinomas, a further 15% were classified as unspecified carcinoma, and less than 2% were adenocarcinomas.

Survival trends

Survival has improved slowly for both men and women. In England and Wales, relative survival at one year rose from 69% for adults diagnosed in the early 1970s to 74% for those diagnosed in the late 1980s. Five-year survival improved from 43% to 47% during these two decades (Table 8.2). Most of the increase occurred between the late 1970s and the early 1980s (Figure 8.1).

The age-standardised trends show small but statistically significant average gains of 1.5% in one-year and five-year survival between successive five-year periods since 1971-75, both for men and for women (see Table 4.4). Throughout this period, one-year survival has been similar in men and women, in sharp contrast to the 10% advantage in five-year survival for women.

The age-dependence of survival is more marked for women than for men (Figure 8.5). Survival is 70% or more in women aged less than 50 at diagnosis, but it declines steeply at older ages to 30% for those aged 80-99. Survival in men is about 50% below the age of 60, falling to around 35% over age 70 (see Table 4.7).

Deprivation

Almost all the overall improvement in survival in England and Wales can be ascribed to the increase for patients in the more affluent groups. There was little difference in survival between the deprivation categories for patients diagnosed in the early 1970s, but substantial differences were evident by the early 1980s (Table 8.2). Whilst average five-year survival in England and Wales has improved from 43% to 55% for the most affluent group, it has barely changed for the most deprived group (42% to 44%).

The difference in survival between the five deprivation categories for adults diagnosed in England and Wales is marked (Figure 8.2). The gradient is statistically significant, and it has changed very little for patients diagnosed during 1986-90 compared with those diagnosed five years earlier (Figure 8.3).

Table 8A Sub-site distribution of cancers of the oral cavity

ICD-9 code	Site description	1971-75 No.	%	1976-80 No.	%	1981-85 No.	%	1986-90 No.	%	Total No.	%
143	Gum	768	25.0	660	21.1	645	18.2	557	15.1	**2,630**	**19.6**
144	Floor of mouth	1,024	33.3	1,113	35.5	1,325	37.5	1,359	36.8	**4,821**	**35.9**
145	Mouth, other and unspecified[1]	1,280	41.7	1,359	43.4	1,567	44.3	1,778	48.1	**5,984**	**44.5**
	Total	**3,072**		**3,132**		**3,537**		**3,694**		**13,435**	

[1] Includes minor salivary gland of unspecified site, and oral cavity, not otherwise specified

Compared to the most affluent group of patients, the estimated deficit in one-year survival for the most deprived group was 7% for those diagnosed in 1981-85, and 5% for those diagnosed during 1986-90 (see Table 4.9). The corresponding deficit in five-year survival was about 12% throughout the 1980s.

The differences in survival between men and women occur among both affluent and deprived patients. For example, among patients diagnosed in 1986-90, relative survival at five years in the two most affluent categories was 60% in women, compared to about 50% in men, while five-year survival in the deprived group was 51% in women and 42% in men.

International comparison

Average five-year survival rates for adults diagnosed in England and Wales during 1986-90 were similar to those in Scotland and to the average survival for Europe, at 44% and 52% for men and women, respectively (see Table 4.12). For men, the national survival rate in the affluent group (52%) and the highest regional survival rate (50%, Anglia & Oxford), were higher than the survival rates in the USA. For women, the highest regional survival rate at five years was 60% – higher than the European average, but still lower than the survival rate in the USA (68%).

1.   Peckham M, Pinedo HM, Veronesi U, (eds.) *Oxford textbook of oncology*. Oxford: Oxford Medical Publications, 1995

Table 8.1 Ineligible and excluded records, and patients included in analyses, by calendar period of diagnosis

	Calendar period of diagnosis														
	1971-75			1976-80			1981-85			1986-90			Overall		
Total registered	**3,473**			**3,508**			**3,962**			**4,277**			**15,220**		
Ineligible	*Records*	*Patients*	*%*	*Records*	*Patients*	*%*	*Records*	*Patients*	*%*	*Records*	*Patients*	*%*	*Records*	*Patients*	*%*
Incomplete data[1]	8	8		24	24		28	28		27	27		87	**87**	
Resident outside England and Wales	77	76		36	35		31	28		16	14		160	**153**	
In situ neoplasm[2]	29	29		6	5		0	0		2	2		37	**36**	
Benign or uncertain behaviour[2]	67	51		52	42		0	0		1	1		120	**94**	
Metastatic[2]	7	6		3	3		0	0		0	0		10	**9**	
Otherwise ineligible[3]	-	-		-	-		-	-		-	-		-	**-**	
Lymphoma[4]	9	9		5	4		2	0		4	0		20	**13**	
Leukaemia or myeloma[4]	0	0		0	0		0	0		0	0		0	**0**	
		179			**113**			**56**			**44**			**392**	
Total eligible		**3,294**	**100**		**3,395**	**100**		**3,906**	**100**		**4,233**	**100**		**14,828**	**100**
Aged under 15 or 100 years or over at diagnosis	19	18	0.5	14	13	0.4	13	12	0.3	8	8	0.2	54	**51**	0.3
Vital status unknown[5]	186	132	4.0	212	181	5.3	177	156	4.0	327	313	7.4	902	**782**	5.3
Duplicate registration[6]	0	0	0.0	0	0	0.0	0	0	0.0	0	0	0.0	0	**0**	0.0
Multiple primary[7]	14	13	0.4	17	16	0.5	51	50	1.3	44	39	0.9	126	**118**	0.8
Synchronous tumours[8]	13	4	0.1	16	9	0.3	29	19	0.5	17	8	0.2	75	**40**	0.3
Sex not known	0	0	0.0	0	0	0.0	0	0	0.0	0	0	0.0	0	**0**	0.0
Sex incompatible with site code[9]	-	-	-	-	-	-	-	-	-	-	-	-	-	**-**	-
Invalid dates[10]	1	0	0.0	0	0	0.0	0	0	0.0	5	2	<0.1	6	**2**	<0.1
Zero survival[11]	64	55	1.7	54	44	1.3	151	132	3.4	183	169	4.0	452	**400**	2.7
Total exclusions		**222**	**6.7**		**263**	**7.7**		**369**	**9.4**		**539**	**12.7**		**1,393**	**9.4**
Patients included in analysis		**3,072**	**93.3**		**3,132**	**92.3**		**3,537**	**90.6**		**3,694**	**87.3**		**13,435**	**90.6**

[1] Main data item(s) invalid or incompatible with one another: name, sex, date of birth, date of diagnosis and (if present) date of death, postcode, site, morphology and behaviour

[2] Tumour either *in situ* (behaviour code 2), benign (0), uncertain if benign or malignant (1) or metastatic (6 or 9)

[3] Not applicable for this malignancy

[4] Morphology that of lymphoma (included with non-Hodgkin lymphoma, Ch. 41) or myeloma (included with multiple myeloma, Ch. 43), or leukaemia (excluded)

[5] Tracing of vital status at National Health Service Central Register incomplete at 6 July 1997

[6] Same site code, sex, cancer registry and cancer registry number as an earlier registration

[7] Same site code and person number as an earlier registration(s): mostly confirmed multiple primary tumours at this site, some unresolved duplicate registrations

[8] Same site code, sex, date of birth and date of diagnosis as another registration(s): mostly synchronous or (in paired organs) bilateral tumours in same anatomic site in one individual, not previously linked: also some duplicate registrations

[9] Not applicable for this malignancy

[10] Impossible sequence of dates (birth, diagnosis and, if present, emigration or death); or date of death after 6 July 1997

[11] Date of diagnosis same as date of death: some are patients who did die on the day of diagnosis, but most were registered solely from a death certificate (death certificate only, or DCO), and their survival time is thus unknown

Table 8.2 Crude and relative survival (%) at one, five and ten years since diagnosis: NHS Region, deprivation category and calendar period of diagnosis

		1971-75						1976-80					
		Affluent	*2*	*3*	*4*	*Deprived*	*All[1]*	*Affluent*	*2*	*3*	*4*	*Deprived*	*All[1]*
		C R	C R	C R	C R	C R	C R	C R	C R	C R	C R	C R	C R
ENGLAND & WALES													
Men		*(206)*	*(257)*	*(315)*	*(360)*	*(369)*	*(1,968)*	*(293)*	*(285)*	*(372)*	*(435)*	*(493)*	*(1,975)*
	One year	62 67	66 72	65 71	63 68	62 66	64 69	70 74	67 70	63 67	64 68	61 65	64 68
	Five years	28 36	27 37	29 42	28 39	27 38	29 39	38 48	36 45	36 47	29 39	30 39	33 42
	Ten years	16 27	16 30	18 35	16 30	15 28	17 31	26 42	22 35	22 38	15 27	20 33	20 33
Women		*(153)*	*(165)*	*(158)*	*(200)*	*(187)*	*(1,104)*	*(159)*	*(202)*	*(241)*	*(257)*	*(227)*	*(1,157)*
	One year	69 74	70 75	66 71	67 71	71 74	67 71	65 68	64 68	65 69	64 68	68 72	65 69
	Five years	40 51	40 50	43 51	41 51	40 51	39 49	35 42	39 50	40 50	39 48	40 51	38 47
	Ten years	28 45	28 43	26 39	27 44	25 39	26 40	25 36	25 43	29 42	27 43	28 43	27 41
Adults		*(359)*	*(422)*	*(473)*	*(560)*	*(556)*	*(3,072)*	*(452)*	*(487)*	*(613)*	*(692)*	*(720)*	*(3,132)*
	One year	65 70	67 73	66 71	64 69	65 69	65 69	68 72	66 69	64 68	64 68	63 67	65 68
	Five years	33 43	32 42	34 46	33 44	31 42	32 43	37 46	37 47	38 48	33 42	33 43	35 44
	Ten years	21 35	21 36	21 37	20 36	18 32	20 35	26 40	23 38	25 40	20 33	23 36	23 36

		Men	Women	Adults		Men	Women	Adults
Northern & Yorkshire		*(321)*	*(171)*	*(492)*		*(372)*	*(170)*	*(542)*
	One year	65 71	69 73	67 72		59 64	57 61	59 63
	Five years	25 36	45 55	32 43		29 38	31 39	30 38
	Ten years	13 27	32 47	19 35		17 29	24 38	19 32
Trent		*(199)*	*(101)*	*(300)*		*(186)*	*(106)*	*(292)*
	One year	64 68	71 75	66 71		59 63	73 76	64 68
	Five years	24 34	39 49	29 40		33 42	39 48	35 44
	Ten years	13 24	22 38	16 30		17 30	24 38	20 33
Anglia & Oxford		*(112)*	*(83)*	*(195)*		*(130)*	*(100)*	*(230)*
	One year	74 80	76 82	75 80		63 67	65 68	64 68
	Five years	38 51	52 63	44 57		39 50	41 48	40 49
	Ten years	23 40	30 51	26 45		29 45	32 46	30 46
North Thames		*(159)*	*(122)*	*(281)*		*(172)*	*(125)*	*(297)*
	One year	68 72	60 64	65 69		69 73	69 73	69 73
	Five years	28 37	32 40	30 39		37 47	35 45	36 46
	Ten years	20 33	21 34	20 34		27 42	21 34	24 38
South Thames		*(220)*	*(163)*	*(383)*		*(225)*	*(153)*	*(378)*
	One year	64 69	61 65	63 67		65 69	66 70	65 69
	Five years	30 41	36 45	32 43		33 45	37 47	35 46
	Ten years	19 35	23 36	21 35		21 39	24 37	22 38
South & West		*(207)*	*(105)*	*(312)*		*(212)*	*(128)*	*(340)*
	One year	61 66	71 77	64 70		64 68	60 64	62 66
	Five years	28 37	37 46	31 40		36 47	40 50	38 48
	Ten years	19 34	30 42	23 37		20 33	29 43	24 37
West Midlands		*(227)*	*(108)*	*(335)*		*(213)*	*(123)*	*(336)*
	One year	62 67	71 76	65 70		65 69	70 75	67 71
	Five years	28 38	32 41	29 39		28 36	48 59	35 44
	Ten years	20 34	*19 34*	19 33		18 31	34 52	24 39
North & West		*(409)*	*(180)*	*(589)*		*(349)*	*(183)*	*(532)*
	One year	62 67	67 72	64 68		65 69	62 66	64 68
	Five years	31 43	43 56	35 47		30 39	38 48	33 42
	Ten years	16 30	29 47	20 36		18 31	24 41	20 34
WALES		*(114)*	*(71)*	*(185)*		*(116)*	*(69)*	*(185)*
	One year	60 65	*57 60*	59 63		78 82	*75 81*	77 81
	Five years	27 40	*39 48*	32 42		45 57	*43 52*	44 55
	Ten years	18 37	25 42	20 39		24 37	*36 52*	29 44

[1] Including those for whom no deprivation category was assigned. C - crude survival, R - relative survival (see Chapter 3).
Number of patients contributing to each analysis in parentheses.

ENGLAND & WALES

	1981-85						1986-90						
	Affluent	2	3	4	Deprived	All[1]	Affluent	2	3	4	Deprived	All[1]	
	C R	C R	C R	C R	C R	C R	C R	C R	C R	C R	C R	C R	
Men	(289)	(350)	(388)	(515)	(625)	(2,220)	(298)	(364)	(422)	(530)	(661)	(2,301)	
One year	77 81	68 72	67 71	65 69	65 69	67 71	72 75	72 76	66 70	72 76	71 74	71 74	
Five years	48 58	36 48	37 46	32 41	30 37	35 44	43 52	38 48	35 42	36 44	34 42	36 44	
	32 49	24 39	22 34	19 32	18 29	22 34							
Women	(223)	(223)	(261)	(318)	(256)	(1,317)	(235)	(289)	(303)	(276)	(285)	(1,393)	
One year	73 78	64 67	77 82	71 75	70 74	71 75	77 81	73 77	66 69	68 71	66 69	70 73	
Five years	41 52	39 48	51 64	46 56	41 51	44 54	49 60	49 60	37 46	36 43	42 51	42 52	
	28 44	26 40	34 53	29 46	33 49	30 47							
Adults	(512)	(573)	(649)	(833)	(881)	(3,537)	(533)	(653)	(725)	(806)	(946)	(3,694)	
One year	75 80	66 70	71 75	67 71	66 70	69 73	74 78	73 76	66 70	71 74	69 72	70 74	
Five years	45 56	37 48	43 53	37 47	33 41	38 48	46 55	43 53	36 44	36 44	36 44	39 47	
	30 47	25 39	27 42	23 37	22 35	25 39							

	1981-85			1986-90			
	Men	Women	Adults	Men	Women	Adults	
	C R	C R	C R	C R	C R	C R	
Northern & Yorkshire	(404)	(210)	(614)	(401)	(202)	(603)	
One year	66 70	70 75	67 72	69 73	69 73	69 73	
Five years	33 42	43 52	36 46	33 41	44 54	37 46	
	18 31	34 51	24 39				
Trent	(242)	(117)	(359)	(221)	(166)	(387)	
One year	70 74	65 70	69 73	72 76	69 73	71 75	
Five years	36 44	41 52	38 46	40 47	43 55	41 50	
	20 30	29 45	23 35				
Anglia & Oxford	(147)	(124)	(271)	(182)	(120)	(302)	
One year	71 75	74 80	72 77	74 78	73 78	74 78	
Five years	46 59	42 52	44 56	38 50	42 54	40 51	
	40 59	29 47	35 56				
North Thames	(201)	(128)	(329)	(222)	(145)	(367)	
One year	71 75	71 74	71 75	73 77	76 80	75 78	
Five years	36 45	43 53	39 48	40 48	45 54	42 50	
	23 36	32 48	27 41				
South Thames	(254)	(171)	(425)	(234)	(198)	(432)	
One year	69 74	72 77	71 75	73 76	69 73	71 75	
Five years	38 46	44 55	40 50	39 49	40 49	39 49	
	24 35	30 50	26 41				
South & West	(204)	(131)	(335)	(190)	(145)	(335)	
One year	69 73	70 74	69 73	70 73	70 74	70 74	
Five years	37 47	41 53	38 50	37 46	48 60	42 52	
	29 44	26 42	28 44				
West Midlands	(243)	(129)	(372)	(240)	(138)	(378)	
One year	58 61	67 72	61 65	67 70	69 73	68 71	
Five years	28 36	42 51	33 41	34 42	39 47	36 44	
	16 28	30 45	21 34				
North & West	(387)	(215)	(602)	(476)	(199)	(675)	
One year	68 72	72 77	69 74	71 75	68 71	70 74	
Five years	34 43	46 58	38 49	36 43	42 51	38 46	
	20 32	29 47	23 38				
WALES	(138)	(92)	(230)	(135)	(80)	(215)	
One year	67 71	76 80	70 75	64 67	59 61	62 65	
Five years	33 42	52 68	41 52	33 40	34 41	33 40	
	16 25	32 49	22 35				

[1] Including those for whom no deprivation category was assigned. C - crude survival, R - relative survival (see Chapter 3).
Number of patients contributing to each analysis in parentheses.

Fig. 8.1   Relative survival up to ten years: trends by calendar period of diagnosis (1971-90) and NHS Region

ADULTS

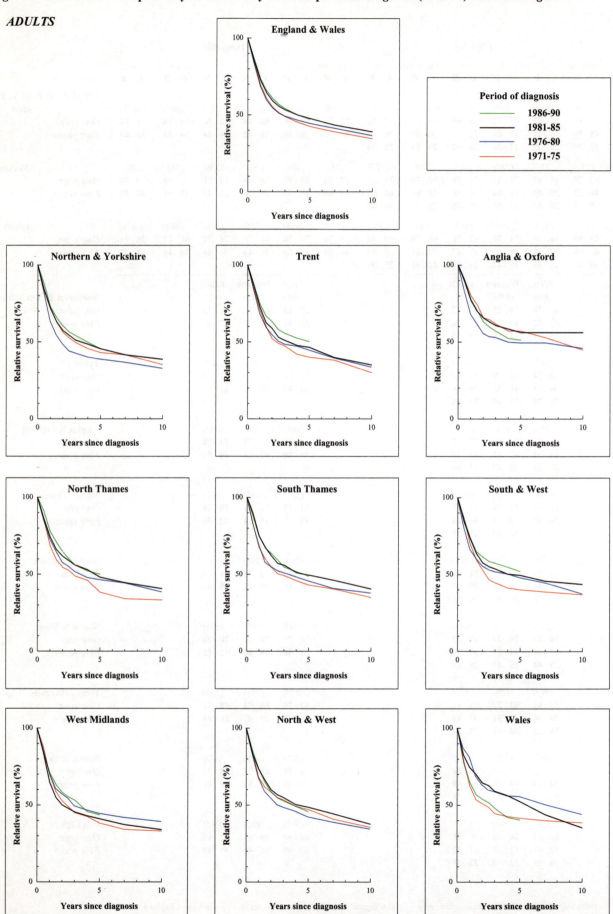

Fig. 8.2 Relative survival up to five years, by deprivation category: England and Wales, patients diagnosed 1986-90

ADULTS

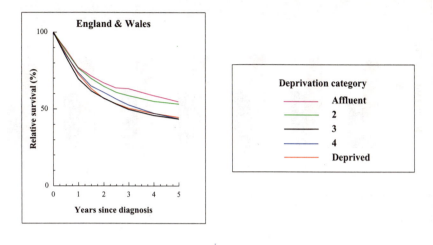

Fig. 8.3 Relative survival at five years, by deprivation category and period of diagnosis (1981-90): England and Wales

ADULTS

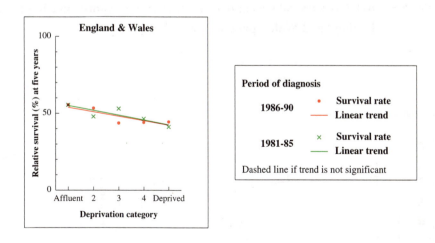

Fig. 8.4   Relative survival (%) by NHS Region and sex:
one-year and five-year survival for patients diagnosed 1986-90, with 95% confidence intervals

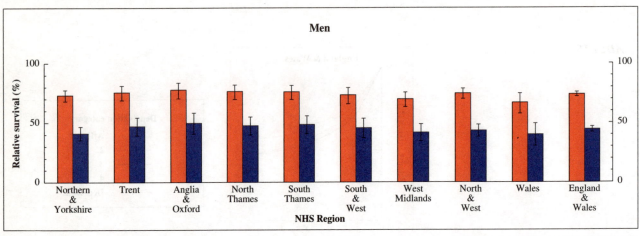

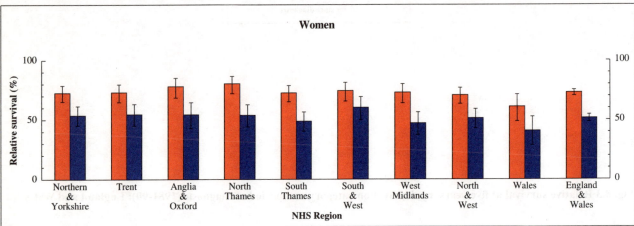

Fig. 8.5   Relative survival (%) at five years, with 95% confidence interval, by age at diagnosis and sex:
England and Wales, patients diagnosed 1986-90

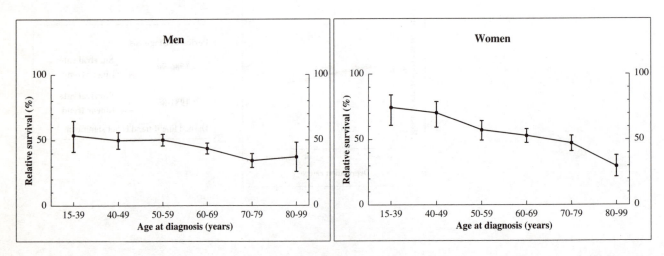

Chapter 9

OROPHARYNX

Cancers of the pharynx are often described together, because of the difficulty in determining the precise anatomical origin of the tumour. Misclassification occurs at death certification. Cancers of the oropharynx and hypopharynx are generally more common. In western Europe, cancers of the oropharynx and hypopharynx account for about 80% of pharyngeal cancers.

The oropharynx includes the tonsils and tonsillar pillars and the lateral and posterior walls of the pharynx. Early symptoms of discomfort and difficulty in swallowing can be vague, and diagnosis often occurs at an advanced stage of disease, with spread to lymph nodes in the neck. Most tumours are poorly differentiated squamous cell carcinomas. Radical surgery for early disease is often disfiguring, and radiotherapy is the usual method of treatment for most cancers of the oropharynx.

The main risk factors are alcohol and tobacco. Changes in alcohol and tobacco consumption are likely to underpin the trends in incidence. Trends in mortality have been more closely related to trends in alcohol consumption than to those for the use of tobacco.

The incidence of oropharyngeal cancer in England and Wales has remained stable since 1971 for women: there were 114 new cases in 1990 (age-standardised rate 0.4 per 100,000). For men the overall trend has been a slight increase: about 300 cases were registered in 1990 (1.2 per 100,000). The pattern suggests a cohort effect, with increasing incidence at younger ages, and a fall at older ages. The male-to-female sex ratio in incidence is about 3 to 1. The age-standardised mortality in 1995 was 0.6 per 100,000 in men and 0.2 in women.

The survival analyses reported here are based on 6,000 patients, 91% of those eligible (Table 9.1). Patients were excluded from analysis mainly because their vital status was unknown at 6 July 1997 (5%) or because their survival was zero or unknown (3%). Two-thirds of the cancers were coded to the tonsil (Table 9A). Papillary, squamous and other epithelial carcinomas accounted for almost all the cases.

Survival trends

For patients diagnosed with oropharyngeal cancer in England and Wales during 1986-90, one-year relative survival was about 64% in both sexes. Five-year survival was 34% for men and 37% for women (Table 9.2).

There has been a small improvement in survival since the early 1970s in England and Wales, but the pattern is rather variable between the regions (Figure 9.1).

Women have slightly better survival than men, but these differences were much less marked for adults diagnosed during the late 1980s than for those diagnosed during the early 1970s, when the difference between the sexes in five-year survival was often 10% or more (Table 9.2). This pattern arises because survival for men was much lower but has improved. The age-standardised trends show an average increase of 2% every five years in one-year and five-year survival for men since the early 1970s, while survival for women has shown no improvement (see Table 4.4).

For patients diagnosed during 1986-90, there was wide variation in survival between regions. Anglia & Oxford

Table 9A Sub-site distribution of cancers of the oropharynx

ICD-9 code	Site description	1971-75 No.	%	1976-80 No.	%	1981-85 No.	%	1986-90 No.	%	Total No.	%
146.0	Tonsil[1]	882	64.9	993	71.3	1,092	70.0	1,091	67.3	**4,058**	**67.6**
146.1	Tonsillar fossa	-	-	19	1.4	91	5.6	63	3.9	**173**	**2.9**
146.2	Tonsillar pillars	-	-	22	1.6	39	2.4	34	2.1	**95**	**1.6**
146.3	Vallecula	-	-	32	2.3	91	5.6	76	4.7	**199**	**3.3**
146.4	Anterior aspect of epiglottis	-	-	13	0.9	41	2.5	20	1.2	**74**	**1.2**
146.5	Junctional region	-	-	0	0.0	0	0.0	1	0.1	**1**	**0.0**
146.6	Lateral wall of oropharynx	-	-	3	0.2	10	0.6	3	0.2	**16**	**0.3**
146.7	Posterior wall of oropharynx	-	-	9	0.6	21	1.3	12	0.7	**42**	**0.7**
146.8	Other	251	18.5	120	8.6	23	1.4	34	2.1	**428**	**7.1**
146.9	Oropharynx, unspecified	225	16.6	182	13.1	222	13.6	286	17.7	**915**	**15.2**
	Total	**1,358**		**1,393**		**1,630**		**1,620**		**6,001**	

[1] Includes 153 cancers of the lingual tonsil (ICD-9 141.6) registered during 1979-90 (see Chapter 2)

had the highest survival, 71% at one year and 46% at five years after diagnosis, while Wales had the lowest (58% and 28%, respectively) (Table 9.2, Figure 9.4).

Survival was higher at younger ages than in the oldest age groups for both men and women (Figure 9.5). In 1986-90, relative survival at five years was 70% for men and women aged 15-39 at diagnosis, but less than 40% in those aged 60 and over (see Table 4.7).

Deprivation

The national differences in survival between deprivation categories were larger for patients diagnosed during the 1980s than for those diagnosed during the 1970s (Table 9.2). Survival up to five years differed widely for adults diagnosed in during 1986-90 (Figure 9.2). There was a significant deprivation gradient in five-year survival in both the early and late 1980s (Figure 9.3). The gap in one-year survival between the most affluent and the most deprived groups was 12% for those diagnosed during

1981-85 and 15% for patients diagnosed in 1986-90 (see Table 4.9). There were similarly wide differences in five-year survival: 11% in 1981-85 and 12% in 1986-90.

International comparison

The five-year survival rate of 34% for men diagnosed with cancer of the oropharynx in England and Wales during 1986-90 is similar to that in Scotland, and higher than the average survival rate in Europe (see Table 4.12). The corresponding figure for women, 37%, is lower than the European average.

Survival in England and Wales is 7% to 9% higher than in the USA, however, where survival is unusually low by comparison with Europe. Five-year survival rates for the affluent group in England and Wales, at 40-45%, are similar to the highest regional survival rates for all deprivation categories combined (Anglia & Oxford). For men, these are substantially higher than the European average or the survival rates in the USA.

Table 9.1 Ineligible and excluded records, and patients included in analyses, by calendar period of diagnosis

	Calendar period of diagnosis														
	1971-75			1976-80			1981-85			1986-90			Overall		
Total registered	**1,583**			**1,589**			**1,843**			**1,896**			**6,911**		
Ineligible	*Records*	*Patients*	*%*	*Records*	*Patients*	*%*	*Records*	*Patients*	*%*	*Records*	*Patients*	*%*	*Records*	*Patients*	*%*
Incomplete data[1]	0	0		8	8		30	30		33	33		71	**71**	
Resident outside England and Wales	49	49		24	24		12	11		9	6		94	**90**	
In situ neoplasm[2]	10	10		2	2		0	0		0	0		12	**12**	
Benign or uncertain behaviour[2]	10	9		7	6		0	0		0	0		17	**15**	
Metastatic[2]	9	9		9	7		0	0		0	0		18	**16**	
Otherwise ineligible[3]	-	-		-	-		-	-		-	-		-	**-**	
Lymphoma[4]	78	73		60	52		15	0		15	0		168	**125**	
Leukaemia or myeloma[4]	1	0		0	0		1	0		1	0		3	**0**	
		150			99			41			39			329	
Total eligible		**1,433**	**100**		**1,490**	**100**		**1,802**	**100**		**1,857**	**100**		**6,582**	**100**
Aged under 15 or 100 years or over at diagnosis	7	3	0.2	9	6	0.4	4	4	0.2	5	5	0.3	25	**18**	0.3
Vital status unknown[5]	98	57	4.0	89	70	4.7	99	87	4.8	143	130	7.0	429	**344**	5.2
Duplicate registration[6]	0	0	0.0	0	0	0.0	0	0	0.0	0	0	0.0	0	**0**	0.0
Multiple primary[7]	0	0	0.0	0	0	0.0	15	15	0.8	4	4	0.2	19	**19**	0.3
Synchronous tumours[8]	3	1	<0.1	5	3	0.2	9	6	0.3	10	6	0.3	27	**16**	0.2
Sex not known	0	0	0.0	0	0	0.0	0	0	0.0	0	0	0.0	0	**0**	0.0
Sex incompatible with site code[9]	-	-	-	-	-	-	-	-	-	-	-	-	-	**-**	-
Invalid dates[10]	0	0	0.0	1	0	0.0	0	0	0.0	4	0	0.0	5	**0**	0.0
Zero survival[11]	15	14	1.0	26	18	1.2	70	60	3.3	100	92	5.0	211	**184**	2.8
Total exclusions		**75**	**5.2**		**97**	**6.5**		**172**	**9.5**		**237**	**12.8**		**581**	**8.8**
Patients included in analysis		**1,358**	**94.8**		**1,393**	**93.5**		**1,630**	**90.5**		**1,620**	**87.2**		**6,001**	**91.2**

---

[1]   Main data item(s) invalid or incompatible with one another: name, sex, date of birth, date of diagnosis and (if present) date of death, postcode, site, morphology and behaviour

[2]   Tumour either *in situ* (behaviour code 2), benign (0), uncertain if benign or malignant (1) or metastatic (6 or 9)

[3]   Not applicable for this malignancy

[4]   Morphology that of lymphoma (included with non-Hodgkin lymphoma, Ch. 41) or myeloma (included with multiple myeloma, Ch. 43), or leukaemia (excluded)

[5]   Tracing of vital status at National Health Service Central Register incomplete at 6 July 1997

[6]   Same site code, sex, cancer registry and cancer registry number as an earlier registration

[7]   Same site code and person number as an earlier registration(s): mostly confirmed multiple primary tumours at this site, some unresolved duplicate registrations

[8]   Same site code, sex, date of birth and date of diagnosis as another registration(s): mostly synchronous or (in paired organs) bilateral tumours in same anatomic site in one individual, not previously linked: also some duplicate registrations

[9]   Not applicable for this malignancy

[10]   Impossible sequence of dates (birth, diagnosis and, if present, emigration or death); or date of death after 6 July 1997

[11]   Date of diagnosis same as date of death: some are patients who did die on the day of diagnosis, but most were registered solely from a death certificate (death certificate only, or DCO), and their survival time is thus unknown

Table 9.2 Crude and relative survival (%) at one, five and ten years since diagnosis: NHS Region, deprivation category and calendar period of diagnosis

	1971-75											1976-80										
	Affluent		2		3		4		Deprived		All[1]		Affluent		2		3		4		Deprived		All[1]
	C R		C R		C R		C R		C R		C R		C R		C R		C R		C R		C R		C R
ENGLAND & WALES																						
Men	*(105)*		*(130)*		*(143)*		*(152)*		*(186)*		*(938)*		*(137)*		*(158)*		*(156)*		*(203)*		*(245)*		*(955)*
One year	52 55		52 56		51 54		53 56		53 56		54 57		59 61		51 54		52 55		60 63		52 55		54 57
Five years	19 27		24 30		12 16		20 26		17 23		19 25		30 36		23 28		21 27		27 35		23 29		24 30
Ten years	13 22		12 23		8 15		14 26		11 18		12 22		20 28		12 19		14 23		18 27		16 25		15 24
Women	*(56)*		*(66)*		*(70)*		*(87)*		*(53)*		*(420)*		*(64)*		*(65)*		*(100)*		*(82)*		*(110)*		*(438)*
One year	53 55		57 60		63 66		57 60		61 64		59 62		51 53		70 74		60 62		64 66		59 60		60 62
Five years	28 32		31 37		36 44		39 46		25 28		32 38		34 40		36 42		34 40		25 31		28 32		31 36
Ten years	25 32		20 27		26 42		25 36		16 24		21 31		23 31		22 34		24 33		18 25		16 22		20 28
Adults	*(161)*		*(196)*		*(213)*		*(239)*		*(239)*		*(1,358)*		*(201)*		*(223)*		*(256)*		*(285)*		*(355)*		*(1,393)*
One year	52 55		54 57		55 58		55 58		55 58		55 58		56 59		57 60		55 58		61 64		54 57		56 59
Five years	23 29		26 32		20 26		27 34		19 24		23 29		32 38		27 33		26 32		26 34		25 30		26 32
Ten years	17 27		15 24		14 24		18 29		12 19		15 25		21 29		15 24		18 27		18 27		16 24		17 25

	Men	Women	Adults					Men	Women	Adults
Northern & Yorkshire	*(124)*	*(44)*	*(168)*					*(114)*	*(67)*	*(181)*
One year	50 53	63 66	53 57					57 61	60 63	58 61
Five years	17 23	30 37	20 26					24 30	34 43	28 35
Ten years	11 23	20 31	13 26					13 20	18 28	15 23
Trent	*(80)*	*(38)*	*(118)*					*(76)*	*(41)*	*(117)*
One year	62 67	63 66	62 66					55 59	66 69	59 63
Five years	23 33	38 41	28 36					21 28	32 36	25 32
Ten years	17 32	24 32	20 32					16 26	27 34	20 31
Anglia & Oxford	*(69)*	*(39)*	*(108)*					*(81)*	*(37)*	*(118)*
One year	58 60	62 65	59 62					58 61	49 52	55 58
Five years	24 33	29 33	26 33					26 33	31 35	28 35
Ten years	13 25	18 24	15 25					15 23	15 23	15 23
North Thames	*(100)*	*(55)*	*(155)*					*(96)*	*(35)*	*(131)*
One year	68 72	54 57	63 67					53 56	75 80	59 62
Five years	31 38	29 32	30 36					25 33	37 43	28 35
Ten years	26 38	21 32	25 36					20 32	29 37	22 33
South Thames	*(99)*	*(54)*	*(153)*					*(123)*	*(61)*	*(184)*
One year	53 56	50 52	52 55					58 61	54 56	56 59
Five years	23 30	36 43	28 35					24 30	24 28	24 30
Ten years	13 23	26 38	18 29					18 27	17 24	18 27
South & West	*(103)*	*(43)*	*(146)*					*(106)*	*(52)*	*(158)*
One year	54 58	60 62	56 59					44 46	68 70	52 54
Five years	17 23	30 43	21 28					18 22	41 48	25 31
Ten years	8 14	23 41	13 23					6 10	27 36	13 19
West Midlands	*(120)*	*(45)*	*(165)*					*(122)*	*(38)*	*(160)*
One year	57 60	63 66	59 62					50 53	63 65	53 56
Five years	17 24	34 39	22 28					24 29	28 35	25 31
Ten years	10 19	15 25	12 21					16 26	16 28	16 27
North & West	*(187)*	*(69)*	*(256)*					*(161)*	*(76)*	*(237)*
One year	45 48	64 67	50 53					55 58	57 59	56 58
Five years	12 17	35 41	18 24					28 37	24 27	27 34
Ten years	8 14	20 27	11 18					17 28	14 19	16 25
WALES	*(56)*	*(33)*	*(89)*					*(76)*	*(31)*	*(107)*
One year	45 47	50 53	47 50					59 63	51 53	56 60
Five years	17 20	24 28	20 24					23 28	32 36	25 31
Ten years	7 11	24 28	13 19					16 23	22 31	18 26

[1] Including those for whom no deprivation category was assigned. C - crude survival, R - relative survival (see Chapter 3). Number of patients contributing to each analysis in parentheses.

ENGLAND & WALES

1981-85

	Affluent C	R	2 C	R	3 C	R	4 C	R	Deprived C	R	All[1] C	R
Men (n)	(155)		(162)		(204)		(241)		(335)		(1,129)	
One year	64	68	51	54	65	68	54	56	52	55	56	59
Five years	30	38	23	29	30	37	21	26	19	24	23	29
	21	33	13	20	18	28	13	20	11	19	14	23
Women (n)	(91)		(92)		(97)		(112)		(89)		(501)	
One year	67	70	69	72	60	63	57	59	58	61	62	65
Five years	30	36	37	45	31	37	32	38	34	38	33	39
	24	32	23	35	21	27	21	31	23	31	22	31
Adults (n)	(246)		(254)		(301)		(353)		(424)		(1,630)	
One year	65	69	57	60	63	66	55	57	53	56	58	61
Five years	30	37	28	35	30	37	24	30	22	27	26	32
	22	33	16	26	19	28	15	24	14	21	17	25

1986-90

	Affluent C	R	2 C	R	3 C	R	4 C	R	Deprived C	R	All[1] C	R
Men (n)	(146)		(167)		(205)		(284)		(316)		(1,130)	
One year	69	72	66	68	65	67	60	62	58	61	62	65
Five years	34	40	29	34	33	40	26	32	26	30	29	34
Women (n)	(82)		(91)		(100)		(98)		(114)		(490)	
One year	82	84	56	59	65	68	56	58	54	55	62	64
Five years	40	45	33	38	37	43	29	34	23	26	32	37
Adults (n)	(228)		(258)		(305)		(382)		(430)		(1,620)	
One year	74	76	63	65	65	67	59	61	57	59	62	64
Five years	36	42	31	36	34	41	27	32	25	29	30	35

Regions — 1981-85

Region	Men C	R	Women C	R	Adults C	R	
Northern & Yorkshire (n)	(176)		(70)		(246)		
One year	55	58	62	65	57	60	
Five years	20	26	32	39	24	29	
	10	17	23	33	14	22	
Trent (n)	(92)		(46)		(138)		
One year	53	56	61	63	56	58	
Five years	23	29	37	43	28	35	
	17	28	24	33	20	30	
Anglia & Oxford (n)	(70)		(44)		(114)		
One year	66	70	64	67	65	69	
Five years	34	45	39	50	36	47	
	21	32	22	36	22	33	
North Thames (n)	(106)		(44)		(150)		
One year	50	52	68	72	55	58	
Five years	25	32	27	32	26	31	
	20	31	18	26	20	30	
South Thames (n)	(130)		(60)		(190)		
One year	61	65	63	67	62	65	
Five years	26	33	38	46	30	37	
	14	25	23	34	17	28	
South & West (n)	(124)		(51)		(175)		
One year	48	50	66	69	53	56	
Five years	19	24	33	41	23	29	
	13	20	28	39	18	27	
West Midlands (n)	(137)		(57)		(194)		
One year	58	61	65	68	60	63	
Five years	21	25	31	36	24	29	
	9	14	20	27	12	19	
North & West (n)	(224)		(93)		(317)		
One year	59	61	54	57	57	60	
Five years	24	30	30	35	26	32	
	15	24	18	26	16	24	
WALES (n)	(70)		(36)		(106)		
One year	52	54	64	68	56	59	
Five years	23	29	30	40	26	32	
	13	21	25	35	17	27	

Regions — 1986-90

Region	Men C	R	Women C	R	Adults C	R
Northern & Yorkshire (n)	(166)		(60)		(226)	
One year	57	59	55	57	56	58
Five years	30	35	27	31	29	34
Trent (n)	(98)		(39)		(137)	
One year	64	67	64	67	64	67
Five years	28	31	25	31	27	31
Anglia & Oxford (n)	(91)		(43)		(134)	
One year	67	70	69	73	68	71
Five years	37	45	44	48	39	46
North Thames (n)	(110)		(56)		(166)	
One year	65	68	67	69	66	68
Five years	35	40	32	38	34	40
South Thames (n)	(133)		(66)		(199)	
One year	65	68	64	66	65	68
Five years	30	36	35	38	32	37
South & West (n)	(93)		(54)		(147)	
One year	70	73	60	63	67	70
Five years	33	40	31	37	32	39
West Midlands (n)	(121)		(41)		(162)	
One year	59	62	67	69	61	64
Five years	26	30	32	36	27	31
North & West (n)	(233)		(100)		(333)	
One year	61	63	58	60	60	62
Five years	25	29	31	36	27	32
WALES (n)	(85)		(31)		(116)	
One year	57	59	52	54	56	58
Five years	20	26	27	32	22	28

[1] Including those for whom no deprivation category was assigned. C - crude survival, R - relative survival (see Chapter 3). Number of patients contributing to each analysis in parentheses.

Fig. 9.1   Relative survival up to ten years: trends by calendar period of diagnosis (1971-90) and NHS Region

ADULTS

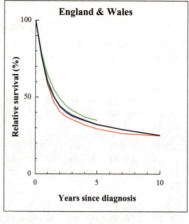

England & Wales

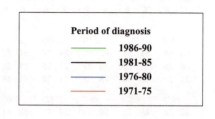

Period of diagnosis	
1986-90	
1981-85	
1976-80	
1971-75	

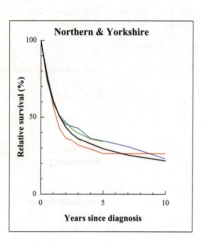

Northern & Yorkshire

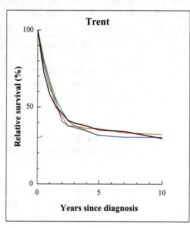

Trent

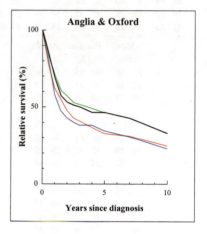

Anglia & Oxford

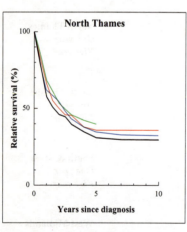

North Thames

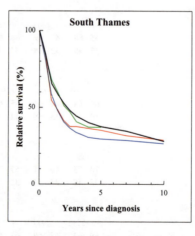

South Thames

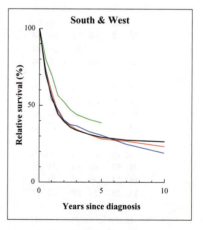

South & West

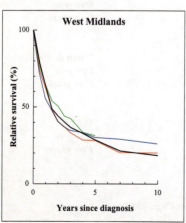

West Midlands

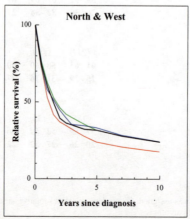

North & West

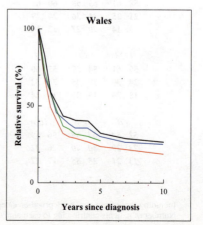

Wales

Fig. 9.2 Relative survival up to five years, by deprivation category: England and Wales, patients diagnosed 1986-90

ADULTS

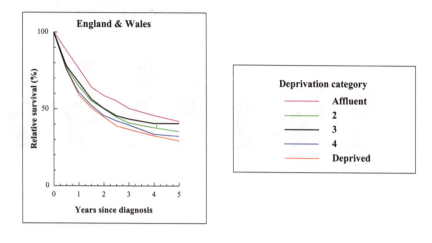

Fig. 9.3 Relative survival at five years, by deprivation category and period of diagnosis (1981-90): England and Wales

ADULTS

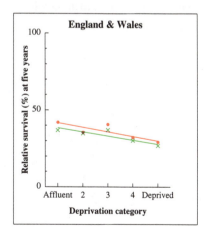
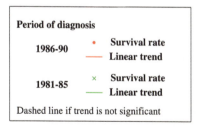

Fig. 9.4   Relative survival (%) by NHS Region and sex:
one-year and five-year survival for patients diagnosed 1986-90, with 95% confidence intervals

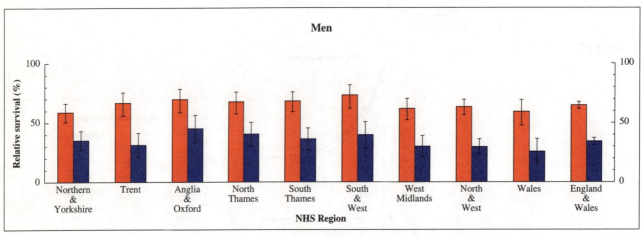

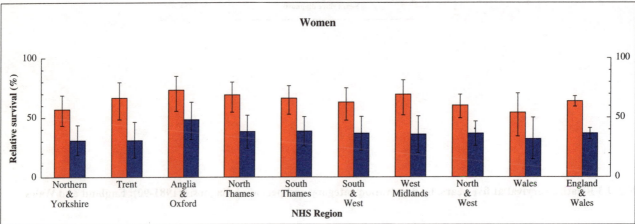

Fig. 9.5   Relative survival (%) at five years, with 95% confidence interval, by age at diagnosis and sex:
England and Wales, patients diagnosed 1986-90

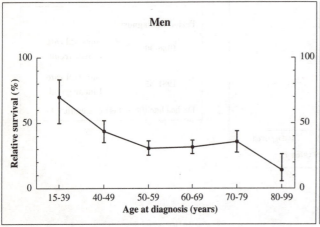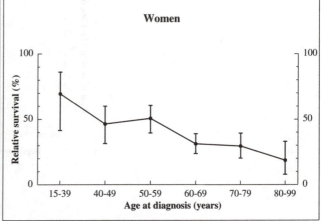

Chapter 10

NASOPHARYNX

Cancer of the nasopharynx is rare in England and Wales, with about 200 new cases a year in adults. Annual incidence rates are about 0.4 per 100,000 in men and 0.2 in women. The male-to-female sex ratio in incidence is 1.8 to 1.

Symptoms of nasopharyngeal cancer are usually non-specific, and diagnosis often occurs at a late stage. Spread to adjacent structures, involvement of regional lymph nodes and metastasis occur in about 70% of cases by the time of diagnosis. Surgery of curative intent is rarely possible, and radiotherapy is the mainstay of treatment.

The survival analyses here are based on 3,100 patients. After the exclusion of 128 cases arising in children (0-14 years), these patients represent some 87% of all adults eligible for analysis (Table 10.1). Of the 3,700 eligible patients, 8% overall were excluded because their vital status was unknown at 6 July 1997, and 4% because of zero or unknown survival. The proportion of patients in these two categories rose to 10% and 6%, respectively, for patients diagnosed during 1986-90. Synchronous and multiple primary cancers were rare.

During the period 1971-78, covered by the eighth revision of the ICD, no anatomic detail was available within the single rubric 147-, but since 1979 the great majority of tumours (86%) have also been assigned to the nasopharynx, without further specification (ICD-9 147.9; Table 10A). Throughout the period 1971-90, about 40% of tumours have been classified as squamous carcinomas, and another 35-40% simply as carcinomas, without further specification.

Survival trends

Relative survival at five years was 33% for adults diagnosed in England and Wales during 1986-90. Survival was higher for women (38%) than for men (30%) (Table 10.2).

There has been no substantial improvement in survival since the early 1970s. In England and Wales, relative survival for patients diagnosed during 1976-80 was 65% at one year and 36% at five years, some 1-3% *higher* than for those diagnosed 10 years later (Table 10.2, Figure 10.1). This pattern is consistent in most regions: only in South Thames and West Midlands was survival any higher in the late 1980s than in the late 1970s. The same pattern is also seen in crude survival, so the apparent lack of improvement in survival is not due to the choice of life tables used to adjust for intercurrent mortality.

The trend in age-standardised relative survival rates tells the same story. There has been no change at all in one-year survival rates for men or women, and although there was a gain of 5-10% in five-year survival between patients diagnosed in 1971-75 and those diagnosed during 1976-80, there has been no subsequent improvement (see Table 4.4). These results are age-standardised to the same reference population, that of all adults with nasopharyngeal cancer included in the analyses for England and Wales during 1986-90; they are thus directly comparable over time and between the sexes. The average gain in five-year survival between successive quinquennia has been small, and it is not statistically significant: 0.5% for men and 2.0% for women (95% CI -0.6% to 4.7%).

There was a difference of 8-9% in five-year survival between men and women diagnosed during 1986-90, but little difference in survival at one year. In earlier periods, the differences in survival between men and women were minor.

The effect of age on survival is marked (Figure 10.5). Survival at five years was 40% or higher for patients aged under 50 at diagnosis during 1986-90, but less than 10% for those aged 80-99 (see Table 4.7).

Deprivation

The survival curves for adults diagnosed in 1986-90 show a striking difference between the most affluent group and all other deprivation categories (Figure 10.2).

The difference in five-year survival between the most affluent and most deprived groups was 19%, but this is mainly due to the disparity in men (20-30%), who contribute two-thirds of the cases (Table 10.2). The range in survival between deprivation groups for women was about 15%.

Survival in the most deprived group and the intermediate groups (categories 2-4) is fairly similar. Linear regression is thus less suitable for estimating the survival gradient between deprivation categories for this cancer than for many other cancers, for which any differences between the groups tend to be more regular. Nonetheless, the fitted linear gradient in survival between deprivation categories for patients diagnosed during 1986-90 was significant, with an estimated gap of 14-17% between the most affluent and most deprived groups in survival at one and five years (see Table 4.9). This estimate is based on 762 patients for whom the deprivation category was known, but it depends mostly on the much higher survival rate in

Table 10A Sub-site distribution of nasopharyngeal cancer

ICD-9 code	Site description	1971-75 No.	%	1976-80 No.	%	1981-85 No.	%	1986-90 No.	%	Total No.	%
147.0	Superior wall	-	-	1	0.1	6	0.7	7	0.9	**14**	**0.4**
147.1	Posterior wall	-	-	16	2.2	44	5.1	39	5.0	**99**	**3.2**
147.2	Lateral wall	-	-	4	0.5	28	3.2	18	2.3	**50**	**1.6**
147.3	Anterior wall	-	-	18	2.4	44	5.1	33	4.3	**95**	**3.0**
147.8	Other	-	-	4	0.5	11	1.3	4	0.5	**19**	**0.6**
147.9	Unspecified	747	100.0	699	94.2	730	84.6	673	87.0	**2,849**	**91.1**
	Total	**747**		**742**		**863**		**774**		**3,126**	

the most affluent category. The survival gradient for patients diagnosed in 1981-85 was less steep, and not statistically significant (Figure 10.3).

There was a wide range in survival between NHS Regions for patients diagnosed during 1986-90, from 53% to 76% at one year and from 16% to 44% at five years (Figure 10.4). The regional estimates for each five-year period are each based on about 100 patients. The differences in survival were less marked for the period 1981-85 (Table 10.2).

International comparison

Five-year survival for patients diagnosed in England and Wales during 1986-90 was 30% for men and 38% for women, fairly close to the average survival rates for Europe, and higher than those in Scotland (see Table 4.12). Survival at five years in the most affluent group in England and Wales, and the highest regional survival rates (Thames regions), at 45-50%, were similar to those in the USA.

Calendar period of diagnosis

	1971-75			1976-80			1981-85			1986-90			Overall		
Total registered	**985**			**959**			**1,053**			**967**			**3,964**		
Ineligible	Records	Patients	%	Records	Patients	%	Records	Patients	%	Records	Patients	%	Records	Patients	%
Incomplete data[1]	2	2		6	6		6	6		10	10		24	**24**	
Resident outside England and Wales	87	86		45	43		27	24		4	1		163	**154**	
In situ neoplasm[2]	1	1		0	0		0	0		0	0		1	**1**	
Benign or uncertain behaviour[2]	15	13		11	6		0	0		0	0		26	**19**	
Metastatic[2]	16	14		4	4		1	1		0	0		21	**19**	
Otherwise ineligible[3]	-	-		-	-		-	-		-	-		-	**-**	
Lymphoma[4]	35	30		14	14		2	0		4	0		55	**44**	
Leukaemia or myeloma[4]	0	0		0	0		0	0		0	0		0	**0**	
		146			73			31			11			261	
Total eligible		**839**	**100**		**886**	**100**		**1,022**	**100**		**956**	**100**		**3,703**	**100**
Aged under 15 or 100 years or over at diagnosis	29	25	3.0	47	42	4.7	34	33	3.2	28	28	2.9	138	**128**	3.5
Vital status unknown[5]	136	54	6.4	131	77	8.7	90	68	6.7	99	93	9.7	456	**292**	7.9
Duplicate registration[6]	0	0	0.0	0	0	0.0	0	0	0.0	0	0	0.0	0	**0**	0.0
Multiple primary[7]	0	0	0.0	0	0	0.0	3	3	0.3	7	6	0.6	10	**9**	0.2
Synchronous tumours[8]	0	0	0.0	0	0	0.0	3	3	0.3	4	1	0.1	7	**4**	0.1
Sex not known	0	0	0.0	0	0	0.0	0	0	0.0	0	0	0.0	0	**0**	0.0
Sex incompatible with site code[9]	-	-	-	-	-	-	-	-	-	-	-	-	-	**-**	-
Invalid dates[10]	0	0	0.0	0	0	0.0	0	0	0.0	2	0	0.0	2	**0**	0.0
Zero survival[11]	14	13	1.5	35	25	2.8	56	52	5.1	57	54	5.6	162	**144**	3.9
Total exclusions		**92**	**11.0**		**144**	**16.3**		**159**	**15.6**		**182**	**19.0**		**577**	**15.6**
Patients included in analysis		**747**	**89.0**		**742**	**83.7**		**863**	**84.4**		**774**	**81.0**		**3,126**	**84.4**

[1]   Main data item(s) invalid or incompatible with one another: name, sex, date of birth, date of diagnosis and (if present) date of death, postcode, site, morphology and behaviour

[2]   Tumour either *in situ* (behaviour code 2), benign (0), uncertain if benign or malignant (1) or metastatic (6 or 9)

[3]   Not applicable for this malignancy

[4]   Morphology that of lymphoma (included with non-Hodgkin lymphoma, Ch. 41) or myeloma (included with multiple myeloma, Ch. 43), or leukaemia (excluded)

[5]   Tracing of vital status at National Health Service Central Register incomplete at 6 July 1997

[6]   Same site code, sex, cancer registry and cancer registry number as an earlier registration

[7]   Same site code and person number as an earlier registration(s): mostly confirmed multiple primary tumours at this site, some unresolved duplicate registrations

[8]   Same site code, sex, date of birth and date of diagnosis as another registration(s): mostly synchronous or (in paired organs) bilateral tumours in same anatomic site in one individual, not previously linked: also some duplicate registrations

[9]   Not applicable for this malignancy

[10]   Impossible sequence of dates (birth, diagnosis and, if present, emigration or death); or date of death after 6 July 1997

[11]   Date of diagnosis same as date of death: some are patients who did die on the day of diagnosis, but most were registered solely from a death certificate (death certificate only, or DCO), and their survival time is thus unknown

Table 10.2   Crude and relative survival (%) at one, five and ten years since diagnosis: NHS Region, deprivation category and calendar period of diagnosis

		1971-75												1976-80											
		Affluent		*2*		*3*		*4*		*Deprived*		*All[1]*		*Affluent*		*2*		*3*		*4*		*Deprived*		*All[1]*	
		C	R	C	R	C	R	C	R	C	R	C	R	C	R	C	R	C	R	C	R	C	R	C	R
ENGLAND & WALES																									
Men		*(62)*		*(66)*		*(90)*		*(92)*		*(87)*		*(494)*		*(71)*		*(72)*		*(93)*		*(110)*		*(88)*		*(459)*	
	One year	59	61	66	69	60	63	61	63	52	54	58	61	66	70	63	64	63	66	66	69	60	63	64	67
	Five years	16	19	29	33	21	26	28	33	19	22	23	27	31	37	20	24	28	32	31	39	36	42	30	35
	Ten years	13	17	21	27	16	22	22	29	15	22	17	24	23	30	16	20	17	23	22	31	23	33	21	28
Women		*(29)*		*(36)*		*(46)*		*(42)*		*(46)*		*(253)*		*(47)*		*(52)*		*(53)*		*(54)*		*(64)*		*(283)*	
	One year	63	64	66	69	63	65	59	61	50	52	62	64	64	65	69	71	67	68	53	55	50	51	60	62
	Five years	31	33	31	35	28	34	20	22	20	22	25	29	34	38	41	47	46	50	22	24	30	34	34	38
	Ten years	13	20	12	17	17	25	15	17	13	18	15	20	21	25	31	36	30	37	13	15	22	29	23	29
Adults		*(91)*		*(102)*		*(136)*		*(134)*		*(133)*		*(747)*		*(118)*		*(124)*		*(146)*		*(164)*		*(152)*		*(742)*	
	One year	60	62	66	69	61	63	60	62	51	53	60	62	65	68	65	67	65	67	62	64	56	58	62	65
	Five years	21	24	29	34	24	28	26	30	19	22	24	28	32	38	28	33	35	39	28	33	33	39	31	36
	Ten years	13	18	18	23	16	22	20	26	14	20	17	22	22	28	22	27	22	29	19	25	23	32	22	28

		Men		*Women*		*Adults*								*Men*		*Women*		*Adults*							
Northern & Yorkshire		*(66)*		*(36)*		*(102)*								*(56)*		*(42)*		*(98)*							
	One year	58	61	54	56	57	59							64	67	48	50	57	60						
	Five years	21	26	25	28	22	27							30	36	21	26	26	32						
	Ten years	15	21	20	23	17	23							15	22	14	19	15	20						
Trent		*(38)*		*(16)*		*(54)*								*(46)*		*(20)*		*(66)*							
	One year	49	51	49	51	49	51							59	62	75	76	64	67						
	Five years	21	23	31	32	24	26							30	35	35	39	32	38						
	Ten years	18	20	4	5	14	18							21	28	14	17	19	25						
Anglia & Oxford		*(37)*		*(17)*		*(54)*								*(43)*		*(17)*		*(60)*							
	One year	65	69	76	79	68	72							75	77	70	72	73	75						
	Five years	30	37	58	68	39	48							40	45	58	62	45	51						
	Ten years	25	34	35	43	28	37							28	37	30	37	28	37						
North Thames		*(89)*		*(27)*		*(116)*								*(64)*		*(27)*		*(91)*							
	One year	66	68	63	64	65	67							66	69	70	72	67	70						
	Five years	30	35	26	30	29	34							38	41	40	43	39	42						
	Ten years	23	31	19	24	22	29							31	41	40	43	34	42						
South Thames		*(78)*		*(39)*		*(117)*								*(71)*		*(45)*		*(116)*							
	One year	57	60	72	73	62	65							66	69	55	57	62	64						
	Five years	16	19	28	30	20	23							29	37	22	24	26	32						
	Ten years	13	16	15	21	13	18							18	25	17	20	18	23						
South & West		*(41)*		*(25)*		*(66)*								*(45)*		*(44)*		*(89)*							
	One year	62	65	72	72	66	68							64	67	61	65	63	66						
	Five years	27	32	12	14	21	25							24	29	41	45	32	38						
	Ten years	17	25	5	10	12	19							15	23	22	25	19	24						
West Midlands		*(45)*		*(31)*		*(76)*								*(41)*		*(28)*		*(69)*							
	One year	47	49	61	62	53	55							64	66	60	62	62	65						
	Five years	21	26	27	30	24	27							19	21	26	28	22	24						
	Ten years	15	21	16	18	16	20							17	20	14	22	16	21						
North & West		*(72)*		*(42)*		*(114)*								*(76)*		*(46)*		*(122)*							
	One year	51	54	57	59	53	56							58	62	54	55	57	59						
	Five years	18	24	19	21	18	23							27	32	41	45	32	37						
	Ten years	14	22	9	14	12	19							17	22	34	42	23	31						
WALES		*(28)*		*(20)*		*(48)*								*(17)*		*(14)*		*(31)*							
	One year	70	72	52	54	62	64							64	67	63	69	63	67						
	Five years	24	27	16	18	21	23							35	38	36	43	36	41						
	Ten years	24	27	12	13	19	22							29	38	29	36	29	41						

[1]   Including those for whom no deprivation category was assigned. C - crude survival, R - relative survival (see Chapter 3). Number of patients contributing to each analysis in parentheses.

ENGLAND & WALES

	1981-85						**1986-90**					
	Affluent	*2*	*3*	*4*	*Deprived*	*All*[1]	*Affluent*	*2*	*3*	*4*	*Deprived*	*All*[1]
	C R	C R	C R	C R	C R	C R	C R	C R	C R	C R	C R	C R

Men

	(84)	(95)	(121)	(125)	(121)	(554)	(66)	(84)	(114)	(122)	(103)	(497)
One year	70 73	59 61	47 49	53 55	63 65	57 59	76 78	61 63	61 64	58 60	60 63	62 64
Five years	38 42	30 34	18 21	21 26	34 40	27 32	47 52	24 27	26 29	19 21	26 30	26 30
	29 35	15 21	13 17	12 17	20 28	17 23						

Women

	(47)	(53)	(70)	(69)	(62)	(309)	(49)	(50)	(61)	(55)	(58)	(277)
One year	66 68	55 57	61 63	56 59	60 63	60 62	75 77	57 59	68 70	54 56	59 61	63 65
Five years	34 37	34 41	27 30	29 32	27 30	29 33	40 44	41 45	35 38	31 35	27 29	35 38
	32 37	25 38	26 30	19 23	18 23	23 29						

Adults

	(131)	(148)	(191)	(194)	(183)	(863)	(115)	(134)	(175)	(177)	(161)	(774)
One year	68 71	58 60	52 54	54 56	62 64	58 60	75 78	59 61	64 66	57 59	60 62	62 64
Five years	37 40	32 37	22 24	24 29	31 37	28 32	44 48	30 34	29 33	22 26	26 29	29 33
	30 36	18 27	17 22	14 19	19 26	19 25						

	1981-85			**1986-90**			
	Men	Women	Adults	Men	Women	Adults	

Northern & Yorkshire

	Men	Women	Adults	Men	Women	Adults
	(62)	(34)	(96)	(60)	(29)	(89)
One year	41 43	59 63	48 50	49 51	51 52	50 52
Five years	20 25	24 27	21 26	20 22	24 25	21 23
	12 18	21 27	15 21			

Trent

	Men	Women	Adults	Men	Women	Adults
	(59)	(35)	(94)	(51)	(33)	(84)
One year	64 67	57 60	61 64	66 69	52 54	61 63
Five years	25 29	32 38	28 32	25 28	37 45	31 36
	12 20	23 32	16 25			

Anglia & Oxford

	Men	Women	Adults	Men	Women	Adults
	(57)	(37)	(94)	(48)	(30)	(78)
One year	63 65	55 57	60 62	73 75	62 63	69 71
Five years	35 41	26 29	31 36	38 44	31 35	35 41
	19 24	21 26	20 26			

North Thames

	Men	Women	Adults	Men	Women	Adults
	(71)	(42)	(113)	(77)	(47)	(124)
One year	70 72	68 71	69 72	62 64	85 87	70 73
Five years	45 49	47 49	46 49	31 34	55 58	40 44
	31 38	38 45	33 41			

South Thames

	Men	Women	Adults	Men	Women	Adults
	(63)	(31)	(94)	(67)	(31)	(98)
One year	61 62	47 49	56 58	74 78	71 72	73 76
Five years	25 30	22 25	24 29	39 46	28 31	36 40
	14 18	19 22	16 20			

South & West

	Men	Women	Adults	Men	Women	Adults
	(68)	(35)	(103)	(50)	(30)	(80)
One year	49 51	51 54	50 52	58 60	67 68	61 63
Five years	20 23	26 28	22 25	18 20	43 47	27 31
	14 19	14 21	14 20			

West Midlands

	Men	Women	Adults	Men	Women	Adults
	(57)	(25)	(82)	(58)	(24)	(82)
One year	59 61	63 65	60 63	62 65	62 64	62 64
Five years	21 26	24 27	22 26	26 30	29 32	27 30
	10 18	20 23	13 20			

North & West

	Men	Women	Adults	Men	Women	Adults
	(74)	(46)	(120)	(58)	(37)	(95)
One year	56 59	59 61	57 60	55 57	45 47	51 53
Five years	25 29	23 25	24 27	12 13	19 20	15 16
	18 23	17 22	18 22			

WALES

	Men	Women	Adults	Men	Women	Adults
	(43)	(24)	(67)	(28)	(16)	(44)
One year	51 53	83 89	63 65	51 53	60 62	54 57
Five years	30 34	37 42	33 37	28 32	37 40	31 38
	17 23	33 38	22 29			

[1] Including those for whom no deprivation category was assigned. C - crude survival, R - relative survival (see Chapter 3). Number of patients contributing to each analysis in parentheses.

Fig. 10.1   Relative survival up to ten years: trends by calendar period of diagnosis (1971-90) and NHS Region

ADULTS

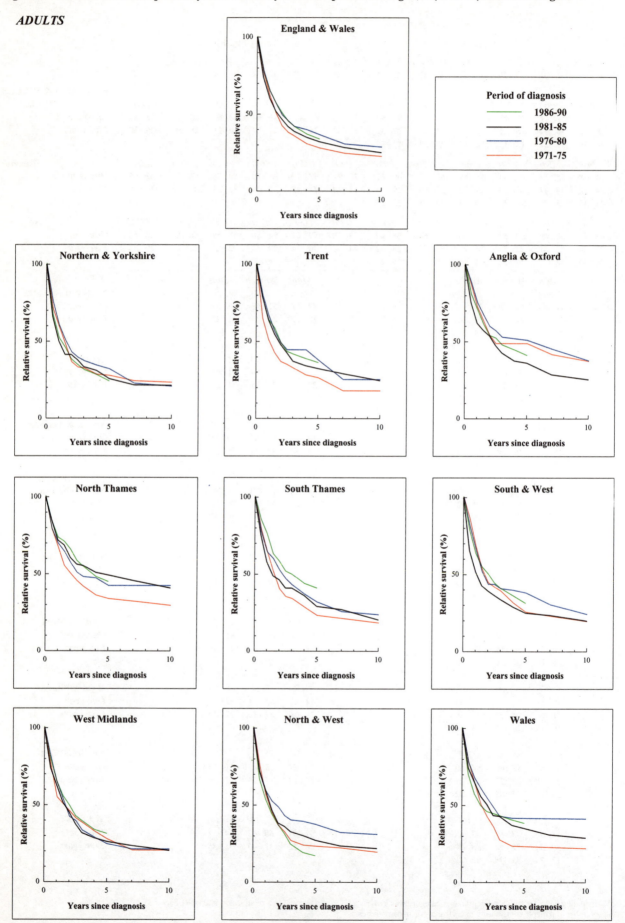

Fig. 10.2 Relative survival up to five years, by deprivation category: England and Wales, patients diagnosed 1986-90

ADULTS

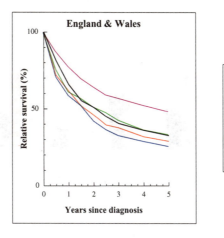

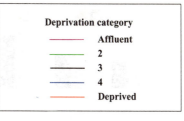

Fig. 10.3 Relative survival at five years, by deprivation category and period of diagnosis (1981-90): England and Wales

ADULTS

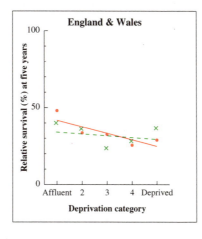

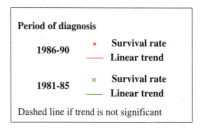

Fig. 10.4 Relative survival (%) by NHS Region and sex:
one-year and five-year survival for patients diagnosed 1986-90, with 95% confidence intervals

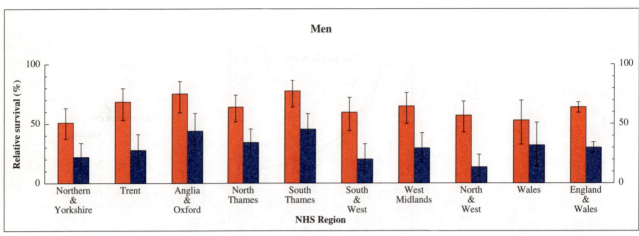

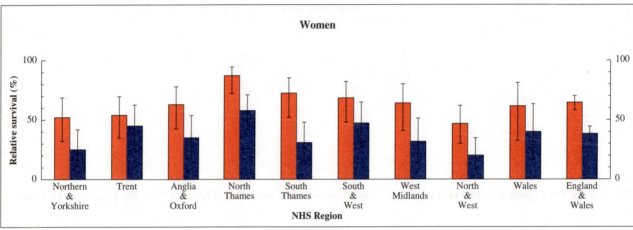

Fig. 10.5 Relative survival (%) at five years, with 95% confidence interval, by age at diagnosis and sex:
England and Wales, patients diagnosed 1986-90

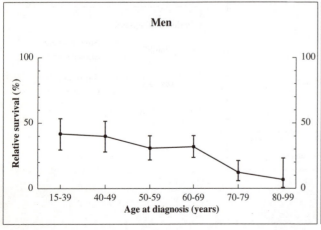

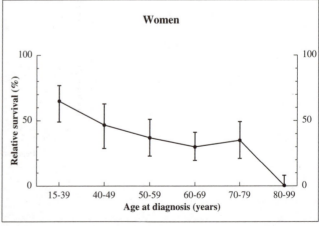

Chapter 11

HYPOPHARYNX

Cancers of the hypopharynx comprise a disparate group of malignancies affecting the pyriform fossa, the postcricoid region and the posterior wall of the pharynx. Most are squamous carcinomas. Tobacco and alcohol have a multiplicative effect on the risk of hypopharyngeal cancer[1], although iron deficiency is also associated with postcricoid tumours. Mild symptoms of difficulty or pain on swallowing may be ignored until deep infiltration or spread to regional lymph nodes or adjacent structures has occurred, and diagnosis often occurs late. Radical surgery and reconstruction of the hypopharynx is extensive, and it carries significant mortality. Radiotherapy is used as an adjunct to surgery, but may be the only treatment for advanced tumours.

There are about 350 new cases of hypopharyngeal carcinoma diagnosed in England and Wales each year. Incidence rates are about 0.9 per 100,000 in men and 0.6 in women, and the rates in men exceed those in women by about 1.5-fold at each age. Incidence has been broadly stable for men, but fell slightly for women in the 20 years up to 1990. Hypopharyngeal cancer is much less common in the UK than in the Latin countries of southern Europe.

The survival analyses here include 7,000 patients diagnosed with cancer of the hypopharynx in England and Wales during 1986-90 and followed up to the end of 1995, about 92% of those eligible (Table 11.1). Five per cent of patients were excluded because their vital status was unknown at 6 July 1997, and 3% because their survival was zero or unknown.

Most tumours were assigned to the pyriform sinus (47%), but this proportion exceeded 50% during the 1980s (Table 11A). Tumours of the postcricoid region accounted for 35%, and only 1% of tumours were assigned to the posterior pharyngeal wall. It can be difficult to locate the anatomic origin of pharyngeal tumours precisely, and for 12% no detail was available. At least three-quarters (76%) were squamous carcinomas, but more than 20% were poorly specified epithelial tumours; less than 1% were adenocarcinomas.

Survival trends

For patients diagnosed in England and Wales during 1986-90, relative survival from carcinoma of the hypopharynx was 49% at one year and 20% at five years (Table 11.2). There has been a national improvement in survival since the early 1970s, particularly in the early 1980s, but the regional pattern of improvement is very variable (Figure 11.1).

Substantial gains have been recorded in Trent, South Thames, West Midlands and Wales, but very little in North Thames or in Anglia & Oxford. The improvements in survival also appear to occur earlier in some regions than others. The regional survival curves each represent 150-200 patients diagnosed in a given five-year period. Regional survival rates for patients diagnosed during 1986-90 are within a 10% range around the national average (Figure 11.4).

After adjustment for differences in the age distribution of patients, there was a national average gain of 4% to 5% in one-year survival between successive five-year periods, and the corresponding gain in five-year survival was over 2% (see Table 4.4). Gains in survival were broadly similar in men and women.

Survival declines with age, more for men than women, but the differences between age groups are much less marked than with many epithelial tumours (Figure 11.5). Up to age 70 at diagnosis, five-year relative survival rates are in the range 20% to 30% in both men and women, but above that age they are generally in the range 10% to 20% (see Table 4.7). There is no systematic difference in survival between men and women.

Deprivation

There was little difference in survival between the various deprivation categories for patients who were diagnosed during the 1970s (Table 11.2). Differences became apparent during the 1980s, but the pattern is unusual, in that patients in the most deprived group have survival rates at one and five years that are close to those for patients in the most affluent group, while patients in deprivation category 4 have the lowest survival (Figure 11.2). Regional patterns of survival by deprivation group are variable because of smaller numbers of patients (data not shown), but the national survival estimates for each deprivation group each depend on 300-400 cases, and they are precise.

This pattern of survival is seen for patients diagnosed during 1981-85 and those diagnosed during 1986-90 (Figure 11.3). A simple linear gradient does not provide an adequate description of the survival rates in the various deprivation groups, and it is not significant in either period.

Table 11A Sub-site distribution of cancer of the hypopharynx

ICD-9 code	Site description	1971-75 No.	%	1976-80 No.	%	1981-85 No.	%	1986-90 No.	%	Total No.	%
148.0	Postcricoid region	769	39.4	679	38.4	609	32.8	471	29.2	2,528	35.2
148.1	Pyriform sinus	840	43.1	758	42.8	922	49.7	881	54.7	3,401	47.3
148.2	Aryepiglottic fold, hypopharyngeal aspect	-	-	14	0.8	45	2.4	39	2.4	98	1.4
148.3	Posterior hypopharyngeal wall	-	-	12	0.7	47	2.5	23	1.4	82	1.1
148.8	Other	97	5.0	60	3.4	11	0.6	9	0.6	177	2.5
148.9	Unspecified	244	12.5	247	14.0	221	11.9	189	11.7	901	12.5
	Total	**1,950**		**1,770**		**1,855**		**1,612**		**7,187**	

International comparison

Five-year survival for patients diagnosed in England and Wales during 1986-90 was 20%, some 4% to 7% below the average for Europe (see Table 4.12). In Scotland, five-year survival for men diagnosed during 1985-89 was lower than in England and Wales at 13% (based on 134 cases), but survival in women was also 20%. The highest regional survival rates of 32% in England and Wales were both based on fewer than 50 cases, and are imprecise (see Chapter 3), but they are similar to the survival rates reported from the USA.

1.   Tuyns A, Estève J, Raymond L, Berrino F, Benhamou E, Blanchet F, Boffetta P, Crosignani P, del Moral Aldaz A, Lehmann W, Merletti F, Péquignot G, Riboli E, Sancho-Garnier H, Terracini B, Zubiri A, Zubiri L. Cancer of the larynx/hypopharynx, tobacco and alcohol: IARC international case-control study in Turin and Varese (Italy), Zaragoza and Navarra (Spain), Geneva (Switzerland) and Calvados (France). Int J Cancer 1988; 41: 483-491

Table 11.1 Ineligible and excluded records, and patients included in analyses, by calendar period of diagnosis

	Calendar period of diagnosis														
	1971-75			**1976-80**			**1981-85**			**1986-90**			**Overall**		
Total registered	**2,136**			**1,948**			**2,063**			**1,839**			**7,986**		
Ineligible	*Records*	*Patients*	*%*	*Records*	*Patients*	*%*	*Records*	*Patients*	*%*	*Records*	*Patients*	*%*	*Records*	*Patients*	*%*
Incomplete data[1]	0	0		1	1		5	5		4	4		10	**10**	
Resident outside England and Wales	50	50		24	24		15	14		9	8		98	**96**	
In situ neoplasm[2]	8	8		0	0		0	0		0	0		8	**8**	
Benign or uncertain behaviour[2]	12	11		16	15		0	0		0	0		28	**26**	
Metastatic[2]	10	9		5	5		1	1		0	0		16	**15**	
Otherwise ineligible[3]	-	-		-	-		-	-		-	-		-	**-**	
Lymphoma[4]	3	3		0	0		0	0		0	0		3	**3**	
Leukaemia or myeloma[4]	0	0		0	0		0	0		0	0		0	**0**	
		81			45			20			12			158	
Total eligible	.	**2,055**	**100**		**1,903**	**100**		**2,043**	**100**		**1,827**	**100**		**7,828**	**100**
Aged under 15 or 100 years or over at diagnosis	3	3	0.1	3	3	0.2	0	0	0.0	0	0	0.0	6	**6**	<0.1
Vital status unknown[5]	100	63	3.1	104	86	4.5	107	101	4.9	137	132	7.2	448	**382**	4.9
Duplicate registration[6]	0	0	0.0	0	0	0.0	0	0	0.0	0	0	0.0	0	**0**	0.0
Multiple primary[7]	1	0	0.0	1	1	<0.1	10	6	0.3	4	2	0.1	16	**9**	0.1
Synchronous tumours[8]	5	3	0.1	5	2	0.1	6	5	0.2	4	2	0.1	20	**12**	0.2
Sex not known	0	0	0.0	0	0	0.0	0	0	0.0	0	0	0.0	0	**0**	0.0
Sex incompatible with site code[9]	-	-	-	-	-	-	-	-	-	-	-	-	-	**-**	-
Invalid dates[10]	0	0	0.0	0	0	0.0	0	0	0.0	1	1	<0.1	1	**1**	<0.1
Zero survival[11]	44	36	1.8	53	41	2.2	89	76	3.7	86	78	4.3	272	**231**	3.0
Total exclusions		**105**	5.1		**133**	7.0		**188**	9.2		**215**	11.8		**641**	8.2
Patients included in analysis		**1,950**	94.9		**1,770**	93.0		**1,855**	90.8		**1,612**	88.2		**7,187**	91.8

[1] Main data item(s) invalid or incompatible with one another: name, sex, date of birth, date of diagnosis and (if present) date of death, postcode, site, morphology and behaviour

[2] Tumour either *in situ* (behaviour code 2), benign (0), uncertain if benign or malignant (1) or metastatic (6 or 9)

[3] Not applicable for this malignancy

[4] Morphology that of lymphoma (included with non-Hodgkin lymphoma, Ch. 41) or myeloma (included with multiple myeloma, Ch. 43), or leukaemia (excluded)

[5] Tracing of vital status at National Health Service Central Register incomplete at 6 July 1997

[6] Same site code, sex, cancer registry and cancer registry number as an earlier registration

[7] Same site code and person number as an earlier registration(s): mostly confirmed multiple primary tumours at this site, some unresolved duplicate registrations

[8] Same site code, sex, date of birth and date of diagnosis as another registration(s): mostly synchronous or (in paired organs) bilateral tumours in same anatomic site in one individual, not previously linked: also some duplicate registrations

[9] Not applicable for this malignancy

[10] Impossible sequence of dates (birth, diagnosis and, if present, emigration or death); or date of death after 6 July 1997

[11] Date of diagnosis same as date of death: some are patients who did die on the day of diagnosis, but most were registered solely from a death certificate (death certificate only, or DCO), and their survival time is thus unknown

Table 11.2   Crude and relative survival (%) at one, five and ten years since diagnosis: NHS Region, deprivation category and calendar period of diagnosis

1971-75 / 1976-80

		1971-75																	1976-80																	
		Affluent		*2*		*3*		*4*		*Deprived*		*All[1]*		*Affluent*		*2*		*3*		*4*		*Deprived*		*All[1]*	
		C	R	C	R	C	R	C	R	C	R	C	R	C	R	C	R	C	R	C	R	C	R	C	R
ENGLAND & WALES																									
Men		*(105)*		*(136)*		*(153)*		*(165)*		*(169)*		*(965)*		*(154)*		*(144)*		*(151)*		*(192)*		*(181)*		*(869)*	
	One year	40	42	35	37	36	39	32	34	37	40	35	37	43	46	51	55	39	41	38	40	37	39	40	43
	Five years	11	15	11	14	8	12	11	15	11	16	10	13	10	12	12	15	10	14	10	13	11	14	10	13
	Ten years	7	11	6	10	5	9	8	13	6	13	6	10	4	8	7	12	6	10	6	10	8	12	6	10
Women		*(120)*		*(151)*		*(161)*		*(204)*		*(146)*		*(985)*		*(141)*		*(151)*		*(187)*		*(207)*		*(167)*		*(901)*	
	One year	31	32	39	40	36	37	36	37	28	29	34	35	42	43	37	38	38	39	32	33	35	37	35	37
	Five years	10	11	17	20	11	13	16	18	8	9	12	14	13	15	10	12	15	17	13	14	13	15	12	15
	Ten years	7	11	13	18	8	12	10	15	7	8	9	13	8	12	6	8	11	15	9	11	9	12	8	12
Adults		*(225)*		*(287)*		*(314)*		*(369)*		*(315)*		*(1,950)*		*(295)*		*(295)*		*(338)*		*(399)*		*(348)*		*(1,770)*	
	One year	35	36	37	39	36	38	34	36	33	34	35	36	43	45	44	46	38	40	35	36	36	38	38	40
	Five years	10	13	14	18	10	12	14	17	10	13	11	14	11	14	11	13	13	16	11	14	12	15	11	14
	Ten years	7	11	10	15	7	11	9	14	6	11	7	12	6	10	7	10	9	13	8	11	8	12	7	11

		Men		*Women*		*Adults*						*Men*		*Women*		*Adults*	
Northern & Yorkshire		*(123)*		*(117)*		*(240)*						*(100)*		*(106)*		*(206)*	
	One year	26	28	33	35	30	31					30	32	34	35	32	33
	Five years	11	13	9	11	10	12					9	11	10	12	9	12
	Ten years	5	10	5	9	5	9					6	10	4	6	5	8
Trent		*(65)*		*(74)*		*(139)*						*(74)*		*(83)*		*(157)*	
	One year	28	29	35	37	32	33					45	48	33	34	39	40
	Five years	9	13	15	17	13	16					4	5	7	9	6	7
	Ten years	5	9	13	17	9	15					3	4	6	7	4	6
Anglia & Oxford		*(63)*		*(79)*		*(142)*						*(90)*		*(78)*		*(168)*	
	One year	32	34	48	50	41	43					42	44	45	47	43	45
	Five years	11	15	19	22	15	20					12	16	21	24	16	20
	Ten years	8	12	15	19	12	17					4	7	12	16	8	12
North Thames		*(106)*		*(80)*		*(186)*						*(88)*		*(78)*		*(166)*	
	One year	44	46	44	45	44	46					46	49	39	40	43	45
	Five years	18	24	18	21	18	22					16	21	19	21	17	22
	Ten years	13	20	13	21	13	20					11	17	14	17	12	18
South Thames		*(150)*		*(128)*		*(278)*						*(119)*		*(130)*		*(249)*	
	One year	42	44	31	33	37	39					31	33	33	35	32	34
	Five years	10	13	15	17	12	15					7	9	7	9	7	9
	Ten years	4	7	11	17	8	13					4	7	5	9	4	8
South & West		*(112)*		*(110)*		*(222)*						*(85)*		*(87)*		*(172)*	
	One year	38	40	27	28	32	34					47	49	35	37	41	43
	Five years	9	12	9	11	9	12					10	13	12	13	11	14
	Ten years	6	11	7	11	7	11					7	10	9	12	8	11
West Midlands		*(121)*		*(125)*		*(246)*						*(117)*		*(106)*		*(223)*	
	One year	30	32	36	37	33	35					41	43	35	36	38	40
	Five years	5	7	9	10	7	8					11	14	14	15	12	15
	Ten years	2	5	6	8	4	6					6	10	8	12	7	11
North & West		*(158)*		*(186)*		*(344)*						*(155)*		*(153)*		*(308)*	
	One year	34	37	29	30	31	33					43	46	36	38	40	42
	Five years	8	11	10	12	9	12					13	18	13	16	13	17
	Ten years	4	8	8	11	6	11					8	14	10	15	9	15
WALES		*(67)*		*(86)*		*(153)*						*(41)*		*(80)*		*(121)*	
	One year	42	45	34	35	38	39					44	46	29	31	34	36
	Five years	12	16	13	15	12	16					8	9	13	15	11	13
	Ten years	6	9	9	13	8	12					8	9	8	10	8	10

[1] Including those for whom no deprivation category was assigned. C - crude survival, R - relative survival (see Chapter 3).
Number of patients contributing to each analysis in parentheses.

1981-85

	Affluent C R	2 C R	3 C R	4 C R	Deprived C R	All[1] C R		
ENGLAND & WALES								**Men**
	(128)	(165)	(188)	(250)	(274)	(1,032)		
One year	43 46	49 51	46 49	43 46	48 50	46 48		
Five years	13 17	17 21	12 16	13 17	17 21	14 18		
	6 10	13 20	7 12	6 10	9 15	8 13		
								Women
	(129)	(141)	(186)	(180)	(170)	(823)		
One year	45 47	37 39	38 39	39 41	46 48	41 42		
Five years	19 22	13 16	17 20	13 16	19 22	16 19		
	13 18	9 14	11 16	9 13	14 18	11 16		
								Adults
	(257)	(306)	(374)	(430)	(444)	(1,855)		
One year	44 46	43 45	42 44	42 44	47 49	44 46		
Five years	16 20	15 19	15 18	13 17	18 21	15 19		
	9 14	11 17	9 14	7 12	11 17	10 15		

1986-90

	Affluent C R	2 C R	3 C R	4 C R	Deprived C R	All[1] C R
Men						
(116)	(164)	(190)	(219)	(292)	(995)	
49 51	48 51	47 49	42 44	51 52	47 49	
25 29	14 18	16 20	12 15	18 22	16 20	
Women						
(97)	(108)	(136)	(127)	(139)	(617)	
44 46	45 46	47 49	38 39	53 54	46 47	
13 17	18 21	19 22	11 13	20 24	17 20	
Adults						
(213)	(272)	(326)	(346)	(431)	(1,612)	
47 49	47 49	47 49	41 42	51 53	47 49	
19 24	16 19	17 21	12 14	19 22	16 20	

Region	Men (1981-85)	Women	Adults	Men (1986-90)	Women	Adults	
Northern & Yorkshire	(127)	(120)	(247)	(151)	(83)	(234)	
One year	39 42	43 45	41 43	44 46	43 45	43 46	
Five years	10 13	16 19	13 16	14 16	10 12	12 15	
	5 8	10 14	7 11				
Trent	(94)	(77)	(171)	(90)	(58)	(148)	
One year	45 48	37 39	42 44	57 60	47 49	53 55	
Five years	14 18	11 12	13 15	20 24	21 24	20 24	
	8 13	6 8	7 10				
Anglia & Oxford	(83)	(75)	(158)	(84)	(65)	(149)	
One year	45 47	40 41	42 44	44 47	40 43	43 45	
Five years	15 20	13 15	14 18	14 16	13 15	14 16	
	9 16	8 10	9 13				
North Thames	(104)	(85)	(189)	(103)	(44)	(147)	
One year	57 59	51 52	54 56	53 55	60 61	55 56	
Five years	16 20	25 31	20 25	17 21	27 32	20 24	
	12 16	15 24	13 20				
South Thames	(142)	(88)	(230)	(113)	(76)	(189)	
One year	48 50	45 46	46 49	52 55	49 52	51 54	
Five years	16 20	19 22	17 21	24 30	17 20	21 26	
	8 14	12 15	10 16				
South & West	(111)	(74)	(185)	(95)	(74)	(169)	
One year	46 50	37 38	42 45	52 54	37 39	45 48	
Five years	11 14	17 20	14 16	15 19	17 22	16 21	
	7 11	11 16	8 13				
West Midlands	(161)	(88)	(249)	(111)	(62)	(173)	
One year	43 45	37 39	41 43	47 49	47 48	47 49	
Five years	15 19	10 12	14 17	15 19	21 24	17 21	
	7 12	8 11	8 13				
North & West	(132)	(154)	(286)	(200)	(114)	(314)	
One year	38 40	40 41	39 41	39 41	43 45	40 42	
Five years	13 17	15 18	14 17	13 15	15 18	13 16	
	9 13	12 18	11 16				
WALES	(78)	(62)	(140)	(48)	(41)	(89)	
One year	59 63	38 40	50 52	48 51	56 58	52 54	
Five years	21 27	22 26	21 28	25 32	15 18	20 25	
	12 22	22 26	16 28				

[1] Including those for whom no deprivation category was assigned. C - crude survival, R - relative survival (see Chapter 3). Number of patients contributing to each analysis in parentheses.

Fig. 11.1   Relative survival up to ten years: trends by calendar period of diagnosis (1971-90) and NHS Region

ADULTS

England & Wales

Relative survival (%)

Years since diagnosis

Period of diagnosis
1986-90
1981-85
1976-80
1971-75

Northern & Yorkshire

Relative survival (%)

Years since diagnosis

Trent

Relative survival (%)

Years since diagnosis

Anglia & Oxford

Relative survival (%)

Years since diagnosis

North Thames

Relative survival (%)

Years since diagnosis

South Thames

Relative survival (%)

Years since diagnosis

South & West

Relative survival (%)

Years since diagnosis

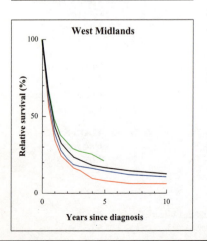

West Midlands

Relative survival (%)

Years since diagnosis

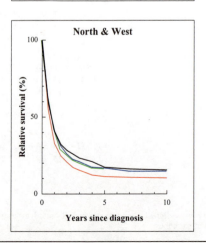

North & West

Relative survival (%)

Years since diagnosis

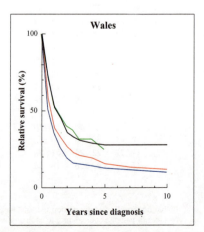

Wales

Relative survival (%)

Years since diagnosis

Fig. 11.2 Relative survival up to five years, by deprivation category: England and Wales, patients diagnosed 1986-90

ADULTS

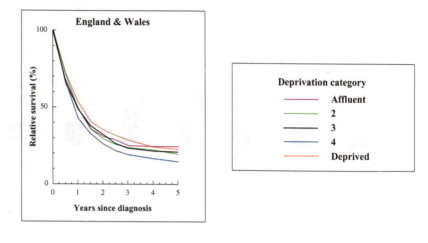

Fig. 11.3 Relative survival at five years, by deprivation category and period of diagnosis (1981-90): England and Wales

ADULTS
(Note vertical scale)

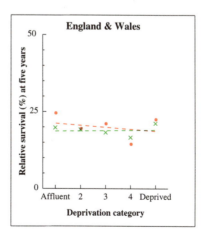
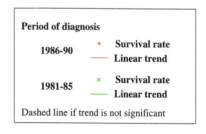

Fig. 11.4 Relative survival (%) by NHS Region and sex:
one-year and five-year survival for patients diagnosed 1986-90, with 95% confidence intervals

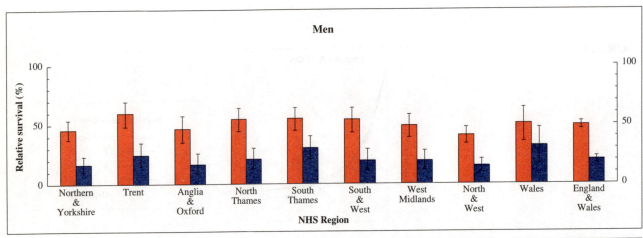

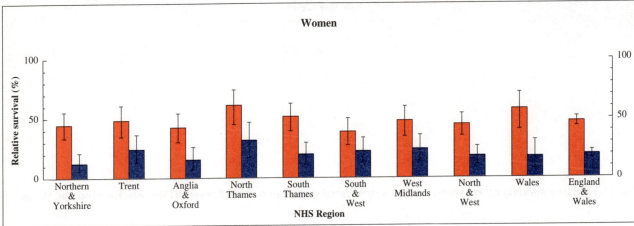

Fig. 11.5 Relative survival (%) at five years, with 95% confidence interval, by age at diagnosis and sex:
England and Wales, patients diagnosed 1986-90

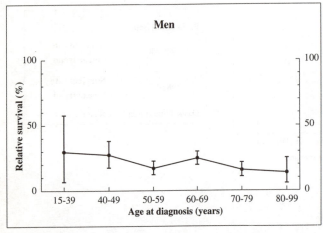
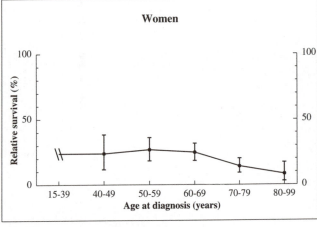

Chapter 12

OESOPHAGUS

Cancer of the oesophagus accounts for about 1 in 40 of all cancers in adults in England and Wales, and it ranks as the eleventh most frequent tumour in both sexes combined. Lifetime risks up to age 74 are of the order of 1% in men and 0.5% in women.

The incidence of oesophageal cancer in England and Wales has been rising steadily since the 1960s in both sexes. There has been an unexplained increase of adenocarcinoma of the middle and lower thirds of the oesophagus, and of the adjacent part of the stomach[1]. The male-to-female sex ratio in incidence is about 1.5 to 1. Women in the UK have 2- to 3-fold higher risks than women in other western European countries. The higher risk for men is still average for men in western Europe; in north-eastern France, the alcohol-related incidence in men is four times higher than in the UK. Alcohol and tobacco consumption are the main risk factors in Europe, but mortality trends may be partially explained by higher consumption of fruit and vegetables, which are protective[2]. Nutritional deficiencies and opium have been implicated in the extremely high risks of oesophageal cancer in parts of China and Iran. Mortality trends resemble those for incidence, since survival has been poor for many years, typically less than 10% at five years.

Clinical presentation with difficulty in swallowing implies an advanced tumour, but pain and vague upper gastrointestinal symptoms of earlier disease may not lead to diagnosis. Delays of 1-12 weeks between onset of symptoms and seeing the GP have been reported[3], and median delay from presentation to treatment may be 3 months or more[3,4], but the impact of such delay on survival has been questioned[5]. Surgery may be directed to cure or palliation, with or without radiotherapy.

The survival analyses here relate to some 70,000 adults registered in England and Wales during 1971-90 and followed up to the end of 1995. About 4% of eligible patients were excluded because their vital status was unknown at 6 July 1997, and 10% because their recorded survival was zero or unknown (Table 12.1). The proportion of patients excluded for this reason rose from 7% to 12% in successive quinquennia. Almost 370 patients (0.5%) had synchronous or subsequent multiple primary tumours: their survival was estimated from the date of diagnosis of the first primary cancer. Overall, survival was analysed for about 85% of eligible patients.

Most (62%) of the cancers had simply been coded to the oesophagus, without further detail. Among the remainder, 24% (17,000 tumours) had been assigned to the lower third of the oesophagus, 10% to the middle third, and 3% to the upper third (Table 12A). Overall, 38% of the tumours were classified as squamous carcinoma: this proportion was fairly constant at 40% between 1971 and 1986, but fell to 32% by 1990. The proportion of adenocarcinomas rose steadily from 15% in the early 1970s to reach 31% by 1990, and they accounted for 22% overall. Another 30% were unspecified carcinomas.

Survival trends

Relative survival for adults diagnosed in England and Wales during 1986-90 was 23% at one year and only 7% at five years. Improvement in survival since the early 1970s has been slow: for those diagnosed during 1971-75, the corresponding estimates were 16% at one year and 5% at five years (Table 12.2).

The only discernible improvement since the early 1970s is this small gain in survival within a year of diagnosis, visible as a slight lifting of the heel of the survival curves in successive time periods, both in England and Wales overall and in most of the NHS Regions (Figure 12.1).

The trend in age-standardised survival suggests a national average gain of 3% in one-year survival between successive quinquennia, increasing from 12% to 21% in men and from 16% to 25% in women (see Table 4.4). The average gain in five-year survival has only been about 1% in successive five-year periods, also in both sexes. These improvements in survival are numerically small, but they are systematic, and because of the very large numbers of tumours involved, they are statistically significant. They may be due to reductions in peri-operative mortality.

There is little variation in survival between the regions for patients diagnosed during 1986-90 (Figure 12.4). Survival trends have varied between the regions, however, with higher increases in one-year survival in the Thames regions (about 5% every five years) than in North & West (about 2%). Significant gains in one-year survival have occurred in all NHS Regions, but regional trends in five-year survival are uniformly small (see Table-Figure 4.5A).

Survival for women has been persistently better than for men, by about 3-4% at one year, and 2-3% at five years. Since five-year survival in men is only 5%, this is a substantial difference.

Survival declines with age at diagnosis (Figure 12.5), but age-specific survival is higher in women at all ages. Under age 50, survival for women is almost twice that of men (see Table 4.7).

Table 12A Sub-site distribution of oesophageal cancer[1]

ICD-9 code	Site description	1971-75 No.	%	1976-80 No.	%	1981-85 No.	%	1986-90 No.	%	Total No.	%
150.3(.0)	Upper third (cervical part)	350	2.3	547	3.3	642	3.6	718	3.5	**2,257**	**3.2**
150.4(.1)	Middle third (thoracic part)	1,131	7.6	1,861	11.3	2,170	12.0	2,074	10.0	**7,236**	**10.3**
150.5(.2)	Lower third (abdominal part)	2,472	16.6	4,149	25.2	4,953	27.5	5,478	26.4	**17,052**	**24.3**
150.8	Other	-	-	71	0.4	135	0.7	94	0.5	**300**	**0.4**
150.9	Unspecified	10,957	73.5	9,827	59.7	10,125	56.2	12,366	59.7	**43,275**	**61.7**
	Total	**14,910**		**16,455**		**18,025**		**20,730**		**70,120**	

[1] The ICD-8 code (1971-78) for oesophagus was 150-, without sub-division. During this period, ONS assigned the special codes 150.0-150.2 to the upper, middle and lower thirds, respectively (see Chapter 2). These tumours have been included with the corresponding ICD-9 codes (150.3-150.5) for the purposes of this table. Only about 60 tumours were actually coded to each of the ICD-9 rubrics 150.0-150.2 (cervical, thoracic and abdominal parts) during 1979-90; they have also been included in the corresponding rubric for upper, middle and lower thirds.

Deprivation

Survival at five years is very low for all deprivation categories (Table 12.2). There is little difference between deprivation categories for patients diagnosed during 1986-90, either in England and Wales as a whole or in any of the NHS Regions or Wales (Figure 12.2). The deprivation gradient in survival declined slightly during the 1980s. For patients diagnosed during 1981-85, the gradient was small but statistically significant (Figure 12.3 - note expanded y-axis scale).

Quantitative estimates of the gradient in survival at one year and five years after diagnosis are given in Table 4.9. For the 17,000 patients with known deprivation status who were diagnosed in 1981-85, the gap in one-year survival between the most affluent and most deprived categories was almost 5%. The gap in five-year survival was smaller at around 2%, but also statistically significant. For the 20,000 patients with known deprivation status diagnosed during 1986-90, the deficit in one-year survival for the most deprived group was only slightly smaller at 4%. There was no deprivation gradient in five-year survival for patients diagnosed in the late 1980s.

International comparison

Five-year survival in England and Wales is similar to that in Scotland (6% and 7% in men and women, respectively) but slightly lower than the corresponding average survival rates of 8% and 12% in Europe (see Table 4.12). Five-year survival in the USA is also poor (10%).

1. Powell J, McConkey CC. Increasing incidence of adenocarcinoma of the gastric cardia and adjacent sites. Br J Cancer 1990; 62: 440-443

2. Cheng KK, Day NE, Davies TW. Oesophageal cancer mortality in Europe: paradoxical time trend in relation to smoking and drinking. Br J Cancer 1992; 65: 613-617

3. Martin IG, Young S, Sue-Ling H, Johnston D. Delays in the diagnosis of oesophagogastric cancer: a consecutive case series. Br Med J 1997; 314: 467-471

4. Jones RVH, Dudgeon TA. Time between presentation and treatment of six common cancers: a study in Devon. Br J Gen Pract 1992; 42: 419-422

5. Renehan A, Tweedle DE. Misconceptions exist over whether delay in diagnosis influences survival. Br Med J 1997; 315: 427-428

Table 12.1 Ineligible and excluded records, and patients included in analyses, by calendar period of diagnosis

	Calendar period of diagnosis														
	1971-75			**1976-80**			**1981-85**			**1986-90**			**Overall**		
Total registered	**17,014**			**19,127**			**21,677**			**25,413**			**83,231**		
Ineligible	Records	Patients	%	Records	Patients	%	Records	Patients	%	Records	Patients	%	Records	Patients	%
Incomplete data[1]	15	15		31	31		107	107		109	109		262	**262**	
Resident outside England and Wales	190	188		157	157		100	89		79	64		526	**498**	
In situ neoplasm[2]	33	30		10	9		0	0		0	0		43	**39**	
Benign or uncertain behaviour[2]	69	20		38	25		0	0		0	0		107	**45**	
Metastatic[2]	71	70		41	40		0	0		4	4		116	**114**	
Otherwise ineligible[3]	-	-		-	-		-	-		-	-		-	**-**	
Lymphoma[4]	0	0		0	0		0	0		1	0		1	**0**	
Leukaemia or myeloma[4]	0	0		0	0		1	0		0	0		1	**0**	
		323			262			196			177			958	
Total eligible	**16,691**	100		**18,865**	100		**21,481**	100		**25,236**	100		**82,273**	100	
Aged under 15 or 100 years or over at diagnosis	1	1	<0.1	10	10	<0.1	8	6	<0.1	17	17	<0.1	36	**34**	<0.1
Vital status unknown[5]	635	539	3.2	896	793	4.2	977	923	4.3	1,211	1,167	4.6	3,719	**3,422**	4.2
Duplicate registration[6]	0	0	0.0	0	0	0.0	0	0	0.0	1	1	<0.1	1	**1**	<0.1
Multiple primary[7]	0	0	0.0	20	12	<0.1	97	79	0.4	136	116	0.5	253	**207**	0.3
Synchronous tumours[8]	43	37	0.2	40	29	0.2	53	42	0.2	85	54	0.2	221	**162**	0.2
Sex not known	0	0	0.0	0	0	0.0	0	0	0.0	0	0	0.0	0	**0**	0.0
Sex incompatible with site code[9]	-	-	-	-	-	-	-	-	-	-	-	-	-	**-**	-
Invalid dates[10]	0	0	0.0	2	1	<0.1	4	3	<0.1	15	10	<0.1	21	**14**	<0.1
Zero survival[11]	1,326	1,204	7.2	1,701	1,565	8.3	2,618	2,403	11.2	3,359	3,141	12.4	9,004	**8,313**	10.1
Total exclusions		**1,781**	10.7		**2,410**	12.8		**3,456**	16.1		**4,506**	17.9		**12,153**	14.8
Patients included in analysis		**14,910**	89.3		**16,455**	87.2		**18,025**	83.9		**20,730**	82.1		**70,120**	85.2

[1] Main data item(s) invalid or incompatible with one another: name, sex, date of birth, date of diagnosis and (if present) date of death, postcode, site, morphology and behaviour

[2] Tumour either *in situ* (behaviour code 2), benign (0), uncertain if benign or malignant (1) or metastatic (6 or 9)

[3] Not applicable for this malignancy

[4] Morphology that of lymphoma (included with non-Hodgkin lymphoma, Ch. 41) or myeloma (included with multiple myeloma, Ch. 43), or leukaemia (excluded)

[5] Tracing of vital status at National Health Service Central Register incomplete at 6 July 1997

[6] Same site code, sex, cancer registry and cancer registry number as an earlier registration

[7] Same site code and person number as an earlier registration(s): mostly confirmed multiple primary tumours at this site, some unresolved duplicate registrations

[8] Same site code, sex, date of birth and date of diagnosis as another registration(s): mostly synchronous or (in paired organs) bilateral tumours in same anatomic site in one individual, not previously linked: also some duplicate registrations

[9] Not applicable for this malignancy

[10] Impossible sequence of dates (birth, diagnosis and, if present, emigration or death); or date of death after 6 July 1997

[11] Date of diagnosis same as date of death: some are patients who did die on the day of diagnosis, but most were registered solely from a death certificate (death certificate only, or DCO), and their survival time is thus unknown

Table 12.2 Crude and relative survival (%) at one, five and ten years since diagnosis: NHS Region, deprivation category and calendar period of diagnosis

MEN	1971-75												1976-80											
	Affluent		2		3		4		Deprived		All[1]		Affluent		2		3		4		Deprived		All[1]	
	C	R	C	R	C	R	C	R	C	R	C	R	C	R	C	R	C	R	C	R	C	R	C	R
ENGLAND & WALES	(1,147)		(1,286)		(1,467)		(1,434)		(1,316)		(8,299)		(1,504)		(1,705)		(1,865)		(1,864)		(1,813)		(9,167)	
One year	18	19	15	16	15	16	15	16	13	14	15	15	18	19	15	16	15	16	14	15	12	13	15	16
Five years	3	4	3	4	2	3	3	4	3	4	3	4	3	4	4	5	4	5	3	4	3	4	3	4
Ten years	2	4	2	3	1	2	2	3	2	3	2	3	2	3	2	4	3	4	2	3	2	3	2	4
ENGLAND	(1,100)		(1,220)		(1,348)		(1,301)		(1,226)		(7,736)		(1,446)		(1,615)		(1,719)		(1,717)		(1,709)		(8,541)	
One year	18	19	15	16	15	16	14	15	13	14	15	15	18	19	15	16	15	16	13	14	12	13	15	15
Five years	3	4	3	4	3	3	3	4	3	4	3	4	3	4	4	5	3	4	3	4	3	4	3	4
Ten years	2	4	2	3	2	3	2	3	2	3	2	3	2	3	2	4	2	4	2	3	2	3	2	4
Northern & Yorkshire	(104)		(114)		(155)		(198)		(259)		(1,136)		(132)		(177)		(266)		(285)		(390)		(1,289)	
One year	21	22	14	14	12	13	9	10	8	9	13	14	13	13	16	17	13	14	10	11	12	12	13	14
Five years	2	2	4	4	1	1	2	2	1	1	2	2	1	1	4	5	3	4	1	1	3	3	3	3
Ten years	0	0	2	4	0	0	1	1	0	0	1	1	0	0	2	3	2	4	1	1	2	3	2	3
Trent	(92)		(113)		(146)		(166)		(143)		(734)		(123)		(155)		(182)		(254)		(188)		(935)	
One year	18	19	5	5	19	20	19	19	16	17	17	18	16	16	13	14	15	16	10	11	11	11	13	14
Five years	2	2	1	1	2	2	3	3	2	3	2	3	3	3	4	5	3	4	3	4	2	2	3	4
Ten years	1	1	1	1	2	2	2	2	1	2	1	2	2	3	2	4	2	3	2	3	1	2	2	3
Anglia & Oxford	(134)		(165)		(144)		(123)		(43)		(702)		(171)		(198)		(161)		(106)		(48)		(720)	
One year	18	19	16	17	11	12	10	11	16	17	16	17	16	17	13	14	13	14	16	17	17	18	16	17
Five years	4	5	5	7	2	3	3	3	4	5	4	5	2	3	3	3	2	3	2	3	2	2	2	3
Ten years	1	3	4	7	1	2	2	3	2	2	3	4	1	2	2	3	1	3	2	2	2	2	2	3
North Thames	(147)		(148)		(159)		(161)		(167)		(892)		(194)		(182)		(162)		(174)		(160)		(911)	
One year	17	18	13	14	14	15	14	15	15	16	17	18	19	20	16	17	11	12	14	15	15	16	17	18
Five years	2	2	2	2	3	4	6	6	6	7	4	5	4	4	3	4	4	4	6	7	6	6	5	6
Ten years	1	2	1	2	2	3	4	5	4	5	3	5	2	4	2	3	4	4	3	5	4	6	3	5
South Thames	(254)		(208)		(216)		(201)		(85)		(1,070)		(303)		(282)		(218)		(191)		(111)		(1,176)	
One year	14	15	14	15	13	14	14	15	12	13	15	16	18	19	13	14	16	16	11	12	15	16	16	17
Five years	4	5	1	1	2	2	2	3	1	1	2	3	5	5	3	4	3	4	2	2	4	4	4	5
Ten years	2	3	0	0	1	1	2	2	1	1	1	2	3	4	2	3	2	3	2	2	0	1	2	4
South & West	(188)		(275)		(270)		(121)		(43)		(999)		(256)		(342)		(343)		(172)		(64)		(1,225)	
One year	16	17	13	14	12	12	8	9	8	9	14	15	17	17	16	17	13	14	15	16	9	10	16	17
Five years	4	4	3	4	2	3	2	2	5	6	3	4	4	4	5	6	4	5	4	5	5	5	4	6
Ten years	3	4	1	3	1	2	1	2	5	6	2	4	2	3	3	5	2	4	2	3	5	5	3	5
West Midlands	(38)		(64)		(71)		(71)		(86)		(810)		(98)		(104)		(151)		(214)		(281)		(870)	
One year	11	11	10	10	15	16	9	10	10	11	12	12	18	19	9	10	11	12	9	9	11	12	12	13
Five years	0	0	2	2	2	2	3	4	2	2	2	2	2	3	2	2	1	2	2	2	3	4	2	3
Ten years	0	0	0	0	1	1	2	2	1	1	1	2	2	3	2	2	1	2	1	2	2	3	2	3
North & West	(143)		(133)		(187)		(260)		(400)		(1,393)		(169)		(175)		(236)		(321)		(467)		(1,415)	
One year	16	16	13	14	16	17	14	15	11	11	14	15	18	18	10	10	18	19	14	15	8	8	14	14
Five years	5	6	4	6	5	7	3	4	2	3	4	5	3	4	3	4	4	5	2	3	2	2	3	4
Ten years	3	6	2	3	3	5	2	3	1	2	2	4	2	3	2	4	3	4	2	3	1	2	2	3
WALES	(47)		(66)		(119)		(133)		(90)		(563)		(58)		(90)		(146)		(147)		(104)		(626)	
One year	19	20	13	13	14	15	16	17	11	11	16	17	23	25	10	10	17	18	17	18	13	13	17	18
Five years	2	2	3	3	1	1	2	3	2	2	3	4	4	5	1	1	6	8	3	3	3	3	4	5
Ten years	2	2	2	3	0	0	1	2	1	1	2	3	3	5	1	1	4	7	2	3	2	2	3	4

[1] Including those for whom no deprivation category was assigned. C - crude survival, R - relative survival (see Chapter 3).
Number of patients contributing to each analysis in parentheses.

	1981-85						1986-90						MEN
	Affluent C R	*2* C R	*3* C R	*4* C R	*Deprived* C R	*All¹* C R	*Affluent* C R	*2* C R	*3* C R	*4* C R	*Deprived* C R	*All¹* C R	
	(1,772)	*(2,037)*	*(2,103)*	*(2,092)*	*(1,928)*	*(10,172)*	*(2,120)*	*(2,501)*	*(2,633)*	*(2,528)*	*(2,246)*	*(12,162)*	**ENGLAND & WALES**
One year	**21 22**	**18 19**	**20 21**	**18 19**	**17 18**	**19 20**	**23 24**	**24 25**	**22 23**	**21 22**	**21 22**	**22 23**	*One year*
Five years	**4 5**	**4 5**	**5 6**	**4 5**	**4 5**	**4 5**	**4 5**	**5 6**	**5 7**	**5 6**	**5 7**	**5 6**	*Five years*
	3 4	2 4	3 5	2 4	3 4	3 4							
	(1,695)	*(1,941)*	*(1,960)*	*(1,929)*	*(1,821)*	*(9,530)*	*(2,034)*	*(2,387)*	*(2,483)*	*(2,323)*	*(2,134)*	*(11,467)*	**ENGLAND**
	21 22	18 19	19 21	18 19	17 18	**18 20**	23 24	24 25	22 23	21 22	21 22	**22 23**	*One year*
	4 5	4 5	4 6	3 4	4 5	**4 5**	5 6	5 6	5 7	4 6	5 7	**5 6**	*Five years*
	3 4	2 4	3 5	2 4	3 4	**3 4**							
	(143)	*(197)*	*(253)*	*(317)*	*(426)*	*(1,351)*	*(186)*	*(255)*	*(296)*	*(338)*	*(473)*	*(1,556)*	**Northern & Yorkshire**
	24 26	13 13	18 19	14 15	10 11	**16 16**	22 23	19 21	18 19	18 19	19 20	**20 21**	*One year*
	7 9	2 3	2 3	1 2	2 3	**3 4**	2 2	4 5	5 6	3 4	3 4	**4 5**	*Five years*
	4 8	1 2	1 2	*1 2*	1 3	**2 3**							
	(138)	*(229)*	*(228)*	*(288)*	*(209)*	*(1,103)*	*(198)*	*(238)*	*(302)*	*(310)*	*(302)*	*(1,353)*	**Trent**
	12 13	15 16	17 18	16 17	20 22	**17 18**	22 23	21 22	20 21	18 19	16 17	**20 21**	*One year*
	2 2	4 5	3 4	4 5	4 5	**4 5**	4 5	5 6	7 8	5 6	4 5	**5 6**	*Five years*
	1 1	2 4	1 3	3 4	2 4	**2 3**							
	(197)	*(216)*	*(185)*	*(124)*	*(58)*	*(800)*	*(239)*	*(286)*	*(237)*	*(155)*	*(81)*	*(1,011)*	**Anglia & Oxford**
	16 17	17 18	21 22	20 22	8 9	**19 20**	22 23	20 22	19 21	16 17	25 27	**20 22**	*One year*
	3 3	3 3	5 6	5 6	*0 0*	**3 4**	4 5	4 5	4 5	2 3	9 11	**4 5**	*Five years*
	2 3	2 3	3 5	4 6	*0 0*	**2 4**							
	(176)	*(179)*	*(181)*	*(180)*	*(169)*	*(903)*	*(192)*	*(201)*	*(232)*	*(225)*	*(210)*	*(1,088)*	**North Thames**
	23 24	22 23	27 28	21 23	23 24	**24 25**	32 34	26 27	27 28	25 27	29 31	**28 30**	*One year*
	5 6	4 5	7 9	7 9	6 8	**6 7**	8 9	6 8	6 8	5 7	7 9	**7 8**	*Five years*
	3 5	*4 5*	4 6	4 6	5 8	**4 6**							
	(399)	*(282)*	*(279)*	*(218)*	*(113)*	*(1,337)*	*(327)*	*(331)*	*(273)*	*(242)*	*(126)*	*(1,326)*	**South Thames**
	20 21	21 23	18 19	21 22	23 25	**21 22**	22 23	27 29	19 20	21 22	26 29	**24 25**	*One year*
	4 5	3 5	4 6	4 5	5 7	**4 6**	4 5	5 6	4 5	3 4	5 5	**4 5**	*Five years*
	3 5	3 5	3 6	3 4	3 6	**3 5**							
	(312)	*(453)*	*(404)*	*(205)*	*(69)*	*(1,498)*	*(382)*	*(506)*	*(457)*	*(309)*	*(109)*	*(1,773)*	**South & West**
	25 26	18 19	18 19	16 17	29 31	**20 22**	28 30	24 26	24 26	22 24	27 29	**25 27**	*One year*
	5 6	4 6	4 5	4 5	6 8	**4 6**	6 7	5 6	5 7	6 8	8 10	**6 7**	*Five years*
	3 5	2 4	3 5	2 4	4 6	**3 5**							
	(112)	*(132)*	*(182)*	*(220)*	*(251)*	*(901)*	*(203)*	*(244)*	*(316)*	*(336)*	*(292)*	*(1,402)*	**West Midlands**
	20 21	13 14	13 14	12 13	15 16	**16 17**	17 19	25 26	22 24	18 19	17 18	**21 22**	*One year*
	3 5	4 4	4 6	2 2	5 6	**4 5**	3 3	4 6	6 7	4 6	4 5	**5 6**	*Five years*
	2 4	2 3	4 5	1 2	3 6	**3 4**							
	(218)	*(253)*	*(248)*	*(377)*	*(526)*	*(1,637)*	*(307)*	*(326)*	*(370)*	*(408)*	*(541)*	*(1,958)*	**North & West**
	17 17	12 13	17 18	16 17	14 15	**16 17**	16 17	21 23	18 19	20 22	18 19	**19 20**	*One year*
	3 4	2 3	3 4	2 3	3 4	**3 4**	5 7	5 7	4 5	4 6	5 7	**5 6**	*Five years*
	2 2	1 2	2 4	2 3	2 3	**2 4**							
	(77)	*(96)*	*(143)*	*(163)*	*(107)*	*(642)*	*(86)*	*(114)*	*(150)*	*(205)*	*(112)*	*(695)*	**WALES**
	18 19	22 23	20 21	19 20	17 18	**21 22**	14 14	20 21	21 22	24 26	16 16	**20 22**	*One year*
	4 5	7 9	- -	4 6	5 7	**6 8**	2 3	2 3	6 8	6 8	7 9	**6 7**	*Five years*
	3 4	*7 9*	*8 12*	*3 5*	*3 5*	**5 8**							

¹ Including those for whom no deprivation category was assigned. C - crude survival, R - relative survival (see Chapter 3).
Number of patients contributing to each analysis in parentheses.

Table 12.2 Crude and relative survival (%) at one, five and ten years since diagnosis:
NHS Region, deprivation category and calendar period of diagnosis

	1971-75						1976-80					
WOMEN	Affluent	2	3	4	Deprived	All[1]	Affluent	2	3	4	Deprived	All[1]
	C R	C R	C R	C R	C R	C R	C R	C R	C R	C R	C R	C R
ENGLAND & WALES	(905)	(1,029)	(1,186)	(1,196)	(992)	(6,611)	(1,215)	(1,404)	(1,550)	(1,444)	(1,302)	(7,288)
One year	19 20	17 18	18 19	15 16	16 17	17 17	17 18	18 19	17 18	17 18	17 17	17 18
Five years	5 6	5 6	4 5	4 5	5 6	5 6	5 6	6 7	4 5	5 6	5 6	5 6
Ten years	3 4	3 5	3 4	3 4	3 5	3 4	4 5	3 6	3 4	3 4	4 6	3 5
ENGLAND	(855)	(954)	(1,065)	(1,056)	(908)	(6,032)	(1,146)	(1,324)	(1,406)	(1,316)	(1,226)	(6,716)
One year	19 20	17 18	18 19	15 16	16 17	17 18	17 18	17 18	17 18	17 18	16 17	17 18
Five years	5 6	5 6	4 5	4 5	5 6	5 6	5 6	5 7	4 5	5 6	5 6	5 6
Ten years	3 4	3 5	3 4	3 5	3 5	3 5	3 5	3 5	3 4	3 4	3 5	3 5
Northern & Yorkshire	(56)	(96)	(114)	(158)	(187)	(843)	(84)	(128)	(189)	(200)	(300)	(925)
One year	19 20	19 20	9 9	16 16	12 13	15 15	16 17	12 13	16 17	16 17	13 14	16 17
Five years	6 8	4 5	1 1	5 6	5 5	4 5	6 6	3 4	3 4	7 8	2 3	4 5
Ten years	3 4	4 4	0 0	4 6	2 3	2 4	4 5	2 3	2 3	2 4	2 3	2 4
Trent	(67)	(83)	(90)	(132)	(108)	(527)	(72)	(115)	(132)	(149)	(133)	(625)
One year	18 19	25 26	24 25	15 16	13 14	20 21	22 23	20 22	17 18	14 15	18 18	19 20
Five years	0 0	5 6	3 3	3 4	5 5	4 5	7 7	5 5	5 6	3 4	7 8	6 7
Ten years	0 0	4 5	2 3	2 3	2 2	2 3	3 5	3 3	2 4	2 3	7 8	3 5
Anglia & Oxford	(98)	(100)	(117)	(78)	(38)	(503)	(151)	(160)	(153)	(95)	(33)	(614)
One year	20 21	18 19	12 13	12 12	13 13	17 18	19 20	16 17	15 16	15 16	13 14	18 20
Five years	6 7	7 8	3 4	4 5	2 2	5 6	4 5	5 6	5 6	3 3	2 3	5 6
Ten years	3 5	5 7	2 3	4 5	0 0	3 5	4 5	4 6	4 5	2 2	2 3	4 6
North Thames	(119)	(143)	(126)	(123)	(114)	(724)	(152)	(119)	(133)	(129)	(104)	(670)
One year	16 17	11 11	20 21	13 14	17 18	16 17	13 14	21 22	12 13	23 24	20 21	19 20
Five years	4 5	5 6	5 5	4 6	7 8	5 7	2 3	4 5	3 4	9 11	6 7	5 6
Ten years	4 5	5 6	3 4	4 5	4 6	4 6	2 3	3 4	2 4	5 7	4 5	3 5
South Thames	(215)	(172)	(185)	(165)	(76)	(904)	(260)	(215)	(199)	(166)	(71)	(976)
One year	13 14	14 14	14 15	9 9	18 19	15 16	13 13	12 13	17 18	17 18	11 12	15 16
Five years	4 5	2 2	1 2	1 2	4 5	3 4	4 5	3 4	2 3	3 4	2 2	4 4
Ten years	1 1	1 1	1 2	1 1	2 3	1 2	3 5	1 2	1 2	1 2	2 2	2 4
South & West	(157)	(194)	(220)	(134)	(34)	(838)	(178)	(313)	(273)	(127)	(62)	(1,003)
One year	21 21	18 19	17 18	15 15	12 12	19 20	15 16	17 18	12 12	19 20	22 23	18 19
Five years	6 7	5 6	6 7	6 8	2 2	6 7	5 6	7 8	3 4	6 7	8 9	6 7
Ten years	5 6	4 6	4 6	6 8	1 1	5 7	2 4	5 8	2 3	4 5	4 6	4 6
West Midlands	(36)	(32)	(47)	(60)	(55)	(577)	(88)	(117)	(141)	(167)	(190)	(718)
One year	15 15	18 19	23 24	11 12	13 14	17 18	16 17	21 22	20 21	11 12	13 14	18 19
Five years	3 3	3 3	9 11	2 3	4 4	5 6	4 5	9 11	4 5	4 4	3 4	5 6
Ten years	3 3	3 3	6 8	2 3	2 2	3 4	2 3	4 6	2 2	3 4	3 4	3 5
North & West	(107)	(134)	(166)	(206)	(296)	(1,116)	(161)	(157)	(186)	(283)	(333)	(1,185)
One year	17 18	13 13	15 16	13 13	15 15	17 17	13 14	13 14	12 13	14 14	13 13	15 16
Five years	5 6	4 5	5 6	3 4	5 6	5 6	4 5	4 5	3 4	2 3	4 6	4 5
Ten years	4 6	1 1	3 5	1 2	4 6	3 5	3 4	3 5	3 4	2 2	3 4	3 5
WALES	(50)	(75)	(121)	(140)	(84)	(579)	(69)	(80)	(144)	(128)	(76)	(572)
One year	21 22	12 13	15 16	17 18	12 12	16 17	18 19	18 19	19 20	10 10	20 21	18 18
Five years	4 5	4 4	2 3	2 3	5 5	4 4	9 10	7 10	6 7	4 5	8 9	7 9
Ten years	3 3	3 3	1 2	1 2	5 5	3 4	8 9	4 7	5 7	4 5	8 9	6 9

[1] Including those for whom no deprivation category was assigned. C - crude survival, R - relative survival (see Chapter 3).
Number of patients contributing to each analysis in parentheses.

	1981-85 Affluent C	R	2 C	R	3 C	R	4 C	R	Deprived C	R	All[1] C	R	1986-90 Affluent C	R	2 C	R	3 C	R	4 C	R	Deprived C	R	All[1] C	R	WOMEN
n	*(1,323)*		*(1,514)*		*(1,674)*		*(1,663)*		*(1,469)*		*(7,853)*		*(1,459)*		*(1,820)*		*(1,892)*		*(1,855)*		*(1,421)*		*(8,568)*		**ENGLAND & WALES**
One year	25	26	23	25	21	22	19	20	20	21	21	23	25	26	24	26	22	23	21	23	21	22	23	24	*One year*
Five years	7	9	7	9	6	7	6	7	4	5	6	7	7	8	7	9	5	6	5	6	6	7	6	7	*Five years*
	5	8	5	8	4	6	3	5	3	4	4	6													
n	*(1,269)*		*(1,446)*		*(1,559)*		*(1,526)*		*(1,383)*		*(7,337)*		*(1,381)*		*(1,711)*		*(1,755)*		*(1,722)*		*(1,339)*		*(8,008)*		**ENGLAND**
One year	25	27	23	24	21	22	19	20	19	20	21	22	25	26	25	26	22	23	21	22	21	22	23	24	*One year*
Five years	7	9	7	8	6	7	5	7	4	5	6	7	7	8	7	9	5	6	5	7	6	7	6	7	*Five years*
	5	8	5	8	4	6	3	5	2	4	4	6													
n	*(123)*		*(151)*		*(190)*		*(238)*		*(349)*		*(1,073)*		*(147)*		*(202)*		*(224)*		*(255)*		*(269)*		*(1,106)*		**Northern & Yorkshire**
One year	22	23	19	20	16	17	18	19	16	17	19	20	26	27	24	25	17	18	15	16	20	21	21	22	*One year*
Five years	6	8	5	7	3	3	4	4	3	3	4	5	7	10	7	8	2	3	3	3	4	5	5	6	*Five years*
	4	6	5	7	1	2	1	2	1	1	2	3													
n	*(89)*		*(141)*		*(159)*		*(208)*		*(160)*		*(760)*		*(87)*		*(164)*		*(193)*		*(203)*		*(169)*		*(827)*		**Trent**
One year	22	23	24	25	23	24	21	22	18	19	22	24	27	28	19	20	18	19	18	19	24	25	22	23	*One year*
Five years	6	7	3	4	6	7	6	7	3	3	5	6	5	6	8	9	4	5	5	6	8	10	6	8	*Five years*
	3	4	3	4	4	5	3	5	2	3	3	5													
n	*(150)*		*(153)*		*(137)*		*(104)*		*(36)*		*(597)*		*(180)*		*(189)*		*(186)*		*(132)*		*(51)*		*(751)*		**Anglia & Oxford**
One year	26	27	18	19	18	19	19	20	4	4	20	22	27	29	22	23	20	21	16	16	24	25	23	24	*One year*
Five years	9	11	7	9	8	9	7	8	0	0	8	9	10	14	4	5	5	6	5	6	9	10	7	9	*Five years*
	4	7	5	8	3	5	5	6	0	0	5	7													
n	*(141)*		*(142)*		*(120)*		*(158)*		*(108)*		*(683)*		*(137)*		*(148)*		*(184)*		*(156)*		*(136)*		*(778)*		**North Thames**
One year	28	30	25	27	20	21	23	24	24	25	25	26	24	26	29	30	30	31	26	27	23	24	27	28	*One year*
Five years	5	6	5	6	5	6	6	7	8	10	6	7	6	8	6	8	9	10	6	8	6	8	7	9	*Five years*
	5	6	4	6	4	6	4	6	5	9	4	7													
n	*(276)*		*(225)*		*(245)*		*(168)*		*(87)*		*(1,041)*		*(231)*		*(260)*		*(201)*		*(201)*		*(77)*		*(992)*		**South Thames**
One year	27	28	29	30	22	23	21	23	28	29	26	27	25	27	31	32	25	26	25	27	18	18	26	28	*One year*
Five years	7	8	10	13	8	10	5	6	7	9	8	10	4	5	7	9	6	8	8	9	6	8	6	8	*Five years*
	4	6	8	12	5	9	4	5	5	7	6	9													
n	*(215)*		*(304)*		*(304)*		*(162)*		*(57)*		*(1,083)*		*(249)*		*(344)*		*(327)*		*(226)*		*(74)*		*(1,232)*		**South & West**
One year	23	25	23	25	23	25	17	19	22	23	23	25	23	25	22	23	20	21	28	30	21	22	24	25	*One year*
Five years	10	11	8	10	5	6	6	8	6	7	7	9	7	8	6	7	5	6	8	11	4	5	6	8	*Five years*
	7	10	5	9	3	5	3	4	5	6	5	7													
n	*(112)*		*(129)*		*(174)*		*(198)*		*(187)*		*(801)*		*(136)*		*(157)*		*(186)*		*(243)*		*(199)*		*(927)*		**West Midlands**
One year	18	19	21	22	19	20	18	19	20	21	20	21	23	24	21	22	21	22	18	19	23	24	22	23	*One year*
Five years	6	8	6	8	6	7	6	8	5	6	6	8	6	6	8	10	6	7	3	3	6	7	6	7	*Five years*
	4	5	4	6	4	5	4	5	4	5	4	6													
n	*(163)*		*(201)*		*(230)*		*(290)*		*(399)*		*(1,299)*		*(214)*		*(247)*		*(254)*		*(306)*		*(364)*		*(1,395)*		**North & West**
One year	24	26	16	17	16	17	13	14	16	17	17	18	18	19	21	22	18	20	18	19	15	16	19	21	*One year*
Five years	7	9	5	6	3	4	4	5	2	3	4	5	5	7	8	9	3	4	5	6	5	6	6	7	*Five years*
	4	8	4	6	3	4	3	5	1	2	3	5													
n	*(54)*		*(68)*		*(115)*		*(137)*		*(86)*		*(516)*		*(78)*		*(109)*		*(137)*		*(133)*		*(82)*		*(560)*		**WALES**
One year	13	14	27	28	20	21	17	19	23	24	22	23	16	17	19	20	15	15	21	22	21	22	21	22	*One year*
Five years	7	7	7	9	6	9	6	7	10	12	8	11	5	5	7	8	3	4	3	4	6	7	6	8	*Five years*
	3	5	7	9	5	8	3	4	8	12	6	9													

[1] Including those for whom no deprivation category was assigned. C - crude survival, R - relative survival (see Chapter 3).
Number of patients contributing to each analysis in parentheses.

Table 12.2 Crude and relative survival (%) at one, five and ten years since diagnosis: NHS Region, deprivation category and calendar period of diagnosis

ADULTS	1971-75 Affluent C	R	2 C	R	3 C	R	4 C	R	Deprived C	R	All[1] C	R	1976-80 Affluent C	R	2 C	R	3 C	R	4 C	R	Deprived C	R	All[1] C	R
ENGLAND & WALES	(2,052)		(2,315)		(2,653)		(2,630)		(2,308)		(14,910)		(2,719)		(3,109)		(3,415)		(3,308)		(3,115)		(16,455)	
One year	18	19	16	17	16	17	15	16	14	15	16	16	18	19	16	17	16	17	15	16	14	15	16	17
Five years	4	5	4	5	3	4	4	5	4	5	4	5	4	5	5	6	4	5	4	5	4	5	4	5
Ten years	2	4	3	4	2	3	2	4	2	4	2	4	3	4	3	5	3	4	2	3	3	4	3	4
ENGLAND	(1,955)		(2,174)		(2,413)		(2,357)		(2,134)		(13,768)		(2,592)		(2,939)		(3,125)		(3,033)		(2,935)		(15,257)	
One year	18	19	16	17	16	17	15	15	14	15	16	16	18	19	16	17	16	17	15	16	14	15	16	16
Five years	4	5	4	5	3	4	4	5	4	5	4	5	4	5	5	6	4	5	4	5	4	5	4	5
Ten years	2	4	3	4	2	3	3	4	2	4	2	4	3	4	3	5	3	4	2	3	3	4	3	4
Northern & Yorkshire	(160)		(210)		(269)		(356)		(446)		(1,979)		(216)		(305)		(455)		(485)		(690)		(2,214)	
One year	20	21	16	17	11	11	12	13	10	11	14	14	14	15	14	15	14	15	13	13	12	13	14	15
Five years	3	4	4	5	1	1	3	4	2	3	3	3	3	3	3	4	3	4	3	4	3	3	3	4
Ten years	1	2	3	5	0	0	2	3	1	2	1	2	1	2	2	3	2	3	2	2	2	3	2	3
Trent	(159)		(196)		(236)		(298)		(251)		(1,261)		(195)		(270)		(314)		(403)		(321)		(1,560)	
One year	18	19	13	13	21	22	17	18	15	16	18	19	18	19	17	18	16	17	12	12	13	14	16	17
Five years	1	1	2	2	2	3	3	3	4	4	3	4	4	5	4	5	4	5	3	4	4	5	4	5
Ten years	0	1	2	2	2	3	2	3	2	3	2	3	2	3	2	5	2	4	2	3	3	5	2	4
Anglia & Oxford	(232)		(265)		(261)		(201)		(81)		(1,205)		(322)		(358)		(314)		(201)		(81)		(1,334)	
One year	19	20	17	18	11	12	11	11	15	15	17	17	17	18	15	16	14	15	15	16	15	16	17	18
Five years	5	6	6	7	3	3	3	4	3	4	5	6	3	4	4	5	4	5	2	3	2	2	4	5
Ten years	2	4	5	7	2	2	3	4	1	1	3	5	2	3	3	4	3	4	2	3	2	2	3	4
North Thames	(266)		(291)		(285)		(284)		(281)		(1,616)		(346)		(301)		(295)		(303)		(264)		(1,581)	
One year	17	17	12	13	16	17	14	14	16	16	17	18	16	17	18	19	12	12	18	19	17	18	18	19
Five years	3	4	4	4	4	5	5	6	6	8	5	6	3	4	3	4	3	4	7	9	6	7	5	6
Ten years	2	4	3	4	2	5	4	6	4	6	3	5	2	4	2	4	3	4	4	6	4	6	3	5
South Thames	(469)		(380)		(401)		(366)		(161)		(1,974)		(563)		(497)		(417)		(357)		(182)		(2,152)	
One year	14	15	14	15	13	14	12	12	15	16	15	16	15	16	13	13	16	17	14	15	14	14	16	17
Five years	4	5	1	1	1	2	2	2	2	3	3	3	4	5	3	4	3	3	2	3	3	3	4	4
Ten years	1	2	1	1	1	2	1	2	1	2	1	2	3	5	2	3	2	2	2	2	1	2	2	4
South & West	(345)		(469)		(490)		(255)		(77)		(1,837)		(434)		(655)		(616)		(299)		(126)		(2,228)	
One year	18	19	15	16	14	15	12	12	10	10	16	17	16	17	16	17	13	13	16	17	15	16	17	18
Five years	5	6	4	5	4	5	4	5	4	4	4	6	4	5	6	7	3	4	5	6	6	7	5	6
Ten years	4	5	3	5	2	4	4	5	3	4	3	5	2	3	4	7	2	4	3	5	4	6	3	5
West Midlands	(74)		(96)		(118)		(131)		(141)		(1,387)		(186)		(221)		(292)		(381)		(471)		(1,588)	
One year	13	13	12	13	18	19	10	10	11	12	14	15	17	18	15	16	16	16	10	10	12	13	15	16
Five years	1	1	2	2	5	6	3	3	3	3	3	4	3	4	5	6	3	3	3	3	3	4	4	5
Ten years	1	1	1	1	3	4	2	2	1	2	2	3	2	3	3	4	1	2	2	3	2	4	2	4
North & West	(250)		(267)		(353)		(466)		(696)		(2,509)		(330)		(332)		(422)		(604)		(800)		(2,600)	
One year	16	17	13	14	15	16	13	14	12	13	15	16	15	16	11	12	15	16	14	15	10	10	14	15
Five years	5	6	4	5	5	6	3	4	3	4	4	5	4	4	4	4	3	4	2	3	3	4	3	4
Ten years	4	6	1	2	3	5	2	3	2	4	3	4	3	4	3	4	3	4	2	3	2	3	3	4
WALES	(97)		(141)		(240)		(273)		(174)		(1,142)		(127)		(170)		(290)		(275)		(180)		(1,198)	
One year	20	21	12	13	15	15	17	18	11	12	16	17	20	21	14	14	18	19	14	14	16	16	17	18
Five years	3	4	3	4	2	2	2	3	3	4	3	4	7	8	4	5	6	8	3	4	5	6	5	7
Ten years	2	3	2	3	1	1	1	2	3	4	2	3	5	7	3	4	4	7	3	4	4	6	4	7

[1] Including those for whom no deprivation category was assigned. C - crude survival, R - relative survival (see Chapter 3). Number of patients contributing to each analysis in parentheses.

	1981-85						1986-90						
	Affluent	*2*	*3*	*4*	*Deprived*	*All*[1]	*Affluent*	*2*	*3*	*4*	*Deprived*	*All*[1]	**ADULTS**
	C R	C R	C R	C R	C R	C R	C R	C R	C R	C R	C R	C R	
	(3,095)	(3,551)	(3,777)	(3,755)	(3,397)	(18,025)	(3,579)	(4,321)	(4,525)	(4,383)	(3,667)	(20,730)	**ENGLAND & WALES**
	23 24	**20 21**	**20 21**	**19 20**	**18 19**	**20 21**	**23 25**	**24 25**	**22 23**	**21 22**	**21 22**	**22 23**	*One year*
	6 7	**5 6**	**5 7**	**4 6**	**4 5**	**5 6**	**5 7**	**6 7**	**5 7**	**5 6**	**6 7**	**5 7**	*Five years*
	4 6	**4 6**	**3 6**	**3 4**	**3 4**	**3 5**							
	(2,964)	(3,387)	(3,519)	(3,455)	(3,204)	(16,867)	(3,415)	(4,098)	(4,238)	(4,045)	(3,473)	(19,475)	**ENGLAND**
	23 24	20 21	20 21	18 20	18 19	**20 21**	24 25	24 25	22 23	21 22	21 22	**22 24**	*One year*
	6 7	5 6	5 6	4 6	4 5	**5 6**	5 7	6 7	5 7	5 6	6 7	**5 7**	*Five years*
	4 6	4 6	3 5	3 4	3 4	**3 5**							
	(266)	(348)	(443)	(555)	(775)	(2,424)	(333)	(457)	(520)	(593)	(742)	(2,662)	**Northern & Yorkshire**
	23 25	15 16	17 18	15 16	13 13	**17 18**	23 25	21 23	18 19	16 17	19 20	**20 21**	*One year*
	6 8	4 5	2 3	2 3	2 3	**3 4**	4 6	5 6	4 5	3 4	4 5	**4 5**	*Five years*
	4 7	3 4	1 2	1 2	1 2	**2 3**							
	(227)	(370)	(387)	(496)	(369)	(1,863)	(285)	(402)	(495)	(513)	(471)	(2,180)	**Trent**
	16 17	19 20	19 20	18 19	19 20	**19 20**	24 25	20 22	19 20	18 19	19 20	**21 22**	*One year*
	3 4	4 5	4 5	5 6	3 4	**4 5**	4 5	6 7	6 7	5 6	5 7	**6 7**	*Five years*
	1 3	2 4	2 4	3 4	2 4	**3 4**							
	(347)	(369)	(322)	(228)	(94)	(1,397)	(419)	(475)	(423)	(287)	(132)	(1,762)	**Anglia & Oxford**
	20 21	18 18	20 21	19 21	6 7	**19 20**	24 25	21 22	20 21	16 17	25 26	**21 23**	*One year*
	5 7	5 6	6 7	6 7	0 0	**5 7**	7 9	4 5	5 6	4 5	9 11	**5 7**	*Five years*
	3 5	3 5	3 5	5 6	0 0	**3 5**							
	(317)	(321)	(301)	(338)	(277)	(1,586)	(329)	(349)	(416)	(381)	(346)	(1,866)	**North Thames**
	25 27	23 25	24 25	22 24	23 25	**24 26**	29 31	27 29	28 30	25 27	27 28	**28 29**	*One year*
	5 6	- -	6 8	6 8	7 9	**6 7**	7 9	6 8	7 9	6 7	7 9	**7 8**	*Five years*
	4 6	4 5	4 6	4 6	5 8	**4 7**							
	(675)	(507)	(524)	(386)	(200)	(2,378)	(558)	(591)	(474)	(443)	(203)	(2,318)	**South Thames**
	23 24	25 26	20 21	21 22	25 27	**23 25**	23 25	29 30	21 23	23 24	23 25	**25 26**	*One year*
	5 7	6 8	6 8	4 6	6 8	**6 7**	4 5	6 8	5 6	5 6	5 7	**5 6**	*Five years*
	4 5	5 8	4 8	3 5	4 6	**4 7**							
	(527)	(757)	(708)	(367)	(126)	(2,581)	(631)	(850)	(784)	(535)	(183)	(3,005)	**South & West**
	24 25	20 21	20 22	17 17	26 27	**21 23**	26 28	23 25	22 24	25 26	24 26	**25 26**	*One year*
	7 8	6 7	4 6	5 6	6 8	**6 7**	6 8	5 6	5 7	7 9	6 8	**6 7**	*Five years*
	4 7	3 6	3 5	3 4	5 7	**4 6**							
	(224)	(261)	(356)	(418)	(438)	(1,702)	(339)	(401)	(502)	(579)	(491)	(2,329)	**West Midlands**
	19 20	17 18	16 17	15 16	17 18	**18 19**	19 21	23 25	22 23	18 19	19 21	**21 23**	*One year*
	5 6	5 6	5 7	4 5	5 6	**5 6**	4 5	6 8	6 7	4 5	5 6	**5 6**	*Five years*
	3 4	3 5	4 5	2 3	4 6	**3 5**							
	(381)	(454)	(478)	(667)	(925)	(2,936)	(521)	(573)	(624)	(714)	(905)	(3,353)	**North & West**
	20 21	14 15	17 18	15 16	15 16	**17 18**	17 18	21 23	18 19	19 20	17 18	**19 21**	*One year*
	5 6	3 4	3 4	3 4	3 3	**4 5**	5 7	6 8	4 5	4 6	5 7	**5 7**	*Five years*
	3 5	3 4	2 4	3 4	2 3	**3 4**							
	(131)	(164)	(258)	(300)	(193)	(1,158)	(164)	(223)	(287)	(338)	(194)	(1,255)	**WALES**
	16 17	24 25	20 21	18 19	19 20	**21 22**	15 16	19 21	18 19	23 25	18 19	**21 22**	*One year*
	5 6	7 9	7 10	5 6	7 10	**7 9**	3 4	5 6	5 6	5 7	7 8	**6 7**	*Five years*
	3 5	7 9	6 10	3 5	5 9	**5 8**							

[1] Including those for whom no deprivation category was assigned. C - crude survival, R - relative survival (see Chapter 3). Number of patients contributing to each analysis in parentheses.

Fig. 12.1   Relative survival up to ten years: trends by calendar period of diagnosis (1971-90) and NHS Region

ADULTS

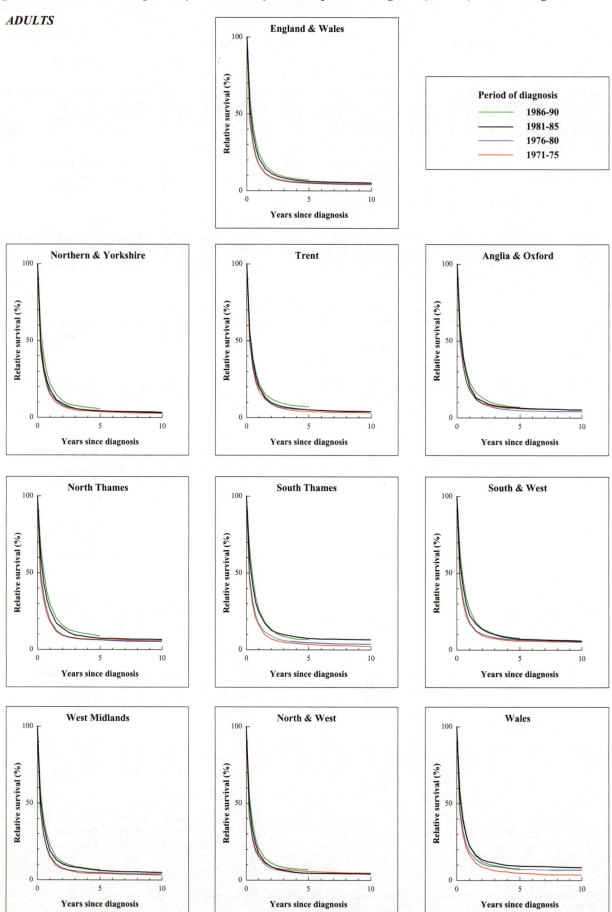

Fig. 12.2   Relative survival up to five years, by deprivation category and NHS Region: patients diagnosed 1986-90

ADULTS

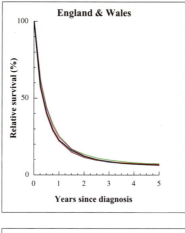

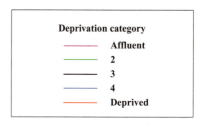

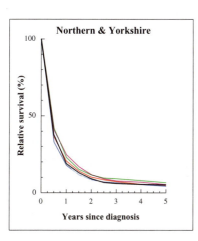

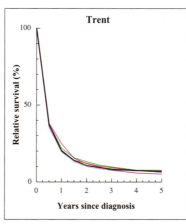

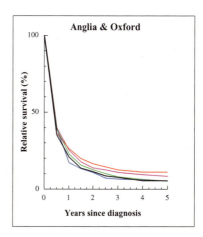

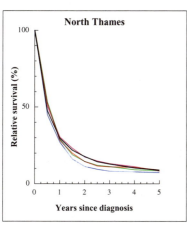

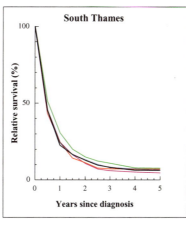

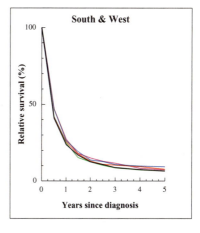

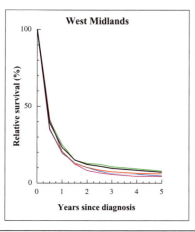

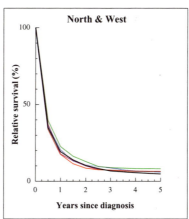

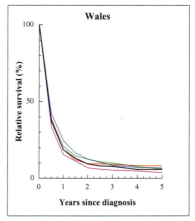

Fig. 12.3 Relative survival at five years by deprivation category, period of diagnosis (1981-90) and NHS Region

ADULTS
(Note vertical scale)

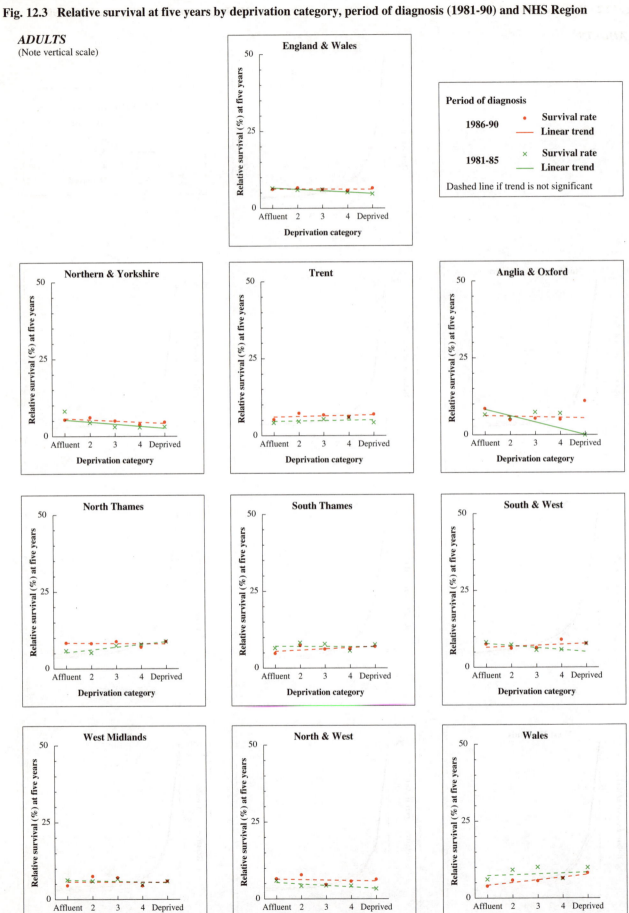

Fig. 12.4 Relative survival (%) by NHS Region and sex:
one-year and five-year survival for patients diagnosed 1986-90, with 95% confidence intervals

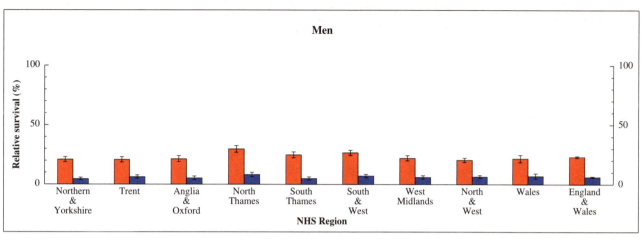

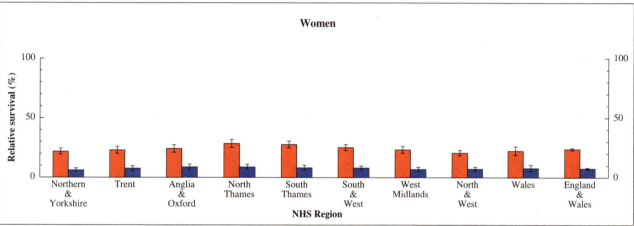

Fig. 12.5 Relative survival (%) at five years, with 95% confidence interval, by age at diagnosis and sex:
England and Wales, patients diagnosed 1986-90

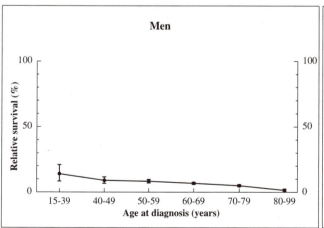

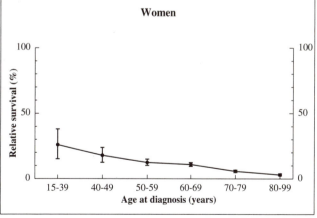

Chapter 13

STOMACH

Stomach cancer is one of the most common cancers in England and Wales, ranking fifth among men and seventh among women. Each year there are some 6,500 new cases in men and 4,000 in women, representing about 5% of all malignant neoplasms. The male-to-female ratio of incidence rates is about 2.5 to 1. Incidence is higher in lower social classes, but has been declining steadily by about 5% every five years for many years; the decline has been more rapid for women than for men. Mortality has declined slightly faster than incidence. Similar trends have been observed in many countries. Incidence is higher in the north (Northern & Yorkshire, North & West) than the south-east (Thames, Anglia & Oxford).

The trends in incidence alone suggest a strong environmental influence on stomach cancer risk. Chronic infection with *Helicobacter pylori* has been identified as a carcinogen. Dietary salt and nitrites are also risk factors, and fruit and vegetables are protective.

Survival is heavily dependent on the stage of disease at diagnosis, but most tumours are not diagnosed until they have spread beyond the stomach, because symptoms are usually non-specific. The increase in cancers of the cardia and gastro-oesophageal junction[1] may influence survival trends, because these proximal tumours are more difficult to remove. About 90% of tumours are adenocarcinomas; the remainder are mostly lymphomas.

The survival analyses here include 194,000 patients, 82% of those eligible (Table 13.1). About 4% of patients were excluded because their vital status was unknown at 6 July 1997. Over 1,000 patients had synchronous or multiple primary cancers. A further 14% of eligible patients were excluded from analysis because their recorded survival was zero; at least two-thirds of these patients are likely to have been registered from a death certificate, and their survival is thus unknown (see Chapter 2). Overall, 10% of cancers arose in the cardia, but the proportion increased from 6% to 15% by 1986-90 (Table 13A). Another 10% arose in the pylorus or pyloric antrum, but most cancers were assigned to the stomach without anatomic detail. About 45% were adenocarcinomas and another 40% were poorly specified carcinomas; for 12% the histology was not specified or not known.

Survival trends

Relative survival from stomach cancer for patients diagnosed in England and Wales in 1986-90 was poor in both sexes, about 26% at one year and 10% at five years (Table 13.2). Survival within the first year after diagnosis has improved, particularly between the late 1970s and the early 1980s (Figure 13.1).

The age-standardised trends show an average increase of 3-4% in one-year survival in both sexes between successive five-year periods of diagnosis in England and Wales. Most (6%) of the overall 10% improvement occurred around 1980 in both sexes (see Table 4.4 and Table-Figure 4.5B). Improvements in five-year survival have been smaller – an average gain of about 2% every five years. These patterns may reflect the decline in peri-operative mortality.

Regional variation is not marked, although survival in North Thames has been slightly higher than in other regions since the 1970s (Table 13.2). The main improvement in all regions occurs for patients diagnosed in the early 1980s, but the average gain in one-year survival in South & West and both Thames regions (about 5% every five years) is larger than in other regions.

Relative survival falls with increasing age at diagnosis in both sexes (Figure 13.5). Five-year survival was around 20% for patients aged under 50 who were diagnosed during 1986-90, but 5% or less for those aged 80-99 (see Table 4.7). The age distribution of patients has also shifted: in the early 1970s, the proportion of adults aged 75 and over was 25% in men and 45% in women, but by the late 1980s the corresponding figures were 40% and 60%. Women have a consistent 2% survival advantage both one year and five years after diagnosis, once these differences in age distribution are taken into account (see Table 4.4).

Deprivation

One-year survival is slightly higher in more affluent groups, but five-year survival differs very little between the various deprivation categories. This pattern is remarkably consistent between the regions (Figure 13.2), and did not change much during the 1980s (Figure 13.3). The gap between the most affluent and most deprived groups was 2-3% for one-year survival. For patients diagnosed during 1981-85 the gap in five-year survival was significant, but extremely small (1%), and for those diagnosed in the period 1986-90 there was virtually no difference (see Table 4.9).

Similar patterns have been reported for patients diagnosed during 1980-89 in South Thames Region[2]. The proportion of patients with advanced disease was very similar in all deprivation groups. The analytic method was different

Table 13A Sub-site distribution of stomach cancer[1]

ICD-9 code	Site description	1971-75 No.	%	1976-80 No.	%	1981-85 No.	%	1986-90 No.	%	Total No.	%
151.0	Cardia	3,285	6.4	4,365	8.6	6,043	12.5	6,563	15.1	**20,256**	**10.4**
151.1	Pylorus	5,623	10.9	3,508	6.9	1,480	3.1	954	2.2	**11,565**	**6.0**
151.2	Pyloric antrum	-	-	1,674	3.3	4,084	8.5	2,892	6.6	**8,650**	**4.5**
151.3	Fundus of stomach	-	-	522	1.0	929	1.9	638	1.5	**2,089**	**1.1**
151.4	Body of stomach	-	-	838	1.7	1,624	3.4	1,198	2.7	**3,660**	**1.9**
151.5	Lesser curvature, unspecified	-	-	1,170	2.3	2,986	6.2	2,380	5.5	**6,536**	**3.4**
151.6	Greater curvature, unspecified	-	-	423	0.8	899	1.9	788	1.8	**2,110**	**1.1**
151.8	Other[1]	9,256	18.0	7,858	15.5	1,069	2.2	725	1.7	**18,908**	**9.7**
151.9	Unspecified	33,385	64.8	30,275	59.8	29,178	60.4	27,447	63.0	**120,285**	**62.0**
	Total	**51,549**		**50,633**		**48,292**		**43,585**		**194,059**	

[1] From 1971 to 1978, all specified parts except cardia and pylorus (ICD-9 151.2-151.6) were included in the rubric "other" (ICD-8 151.8).

(Hakulinen) and national rather than deprivation-specific life tables were used to estimate expected survival (see Chapter 3), but the relative survival rates (10-12%) and the absence of a deprivation gradient are remarkably similar to the results here. The 11,000 patients in that report are included here.

International comparison

The five-year survival rate of 10-11% for patients diagnosed during 1986-90 in England and Wales was only half the average for Europe. Survival is similar to that in Denmark, Scotland, Slovakia, Slovenia, Poland and Estonia. Even the highest regional survival rates in England and Wales (about 14%) are well below the levels seen in Europe and the USA (see Table 4.12).

1.   Powell J, McConkey CC. Increasing incidence of adenocarcinoma of the gastric cardia and adjacent sites. Br J Cancer 1990; 62: 440-443

2.   Schrijvers CTM, Mackenbach J, Lutz J-M, Quinn MJ, Coleman MP. Deprivation, stage at diagnosis and cancer survival. Int J Cancer 1995; 63: 324-329

Table 13.1 Ineligible and excluded records, and patients included in analyses, by calendar period of diagnosis

	Calendar period of diagnosis														
	1971-75			**1976-80**			**1981-85**			**1986-90**			**Overall**		
Total registered	**63,093**			**61,569**			**60,310**			**55,575**			**240,547**		
Ineligible	*Records*	*Patients*	*%*	*Records*	*Patients*	*%*	*Records*	*Patients*	*%*	*Records*	*Patients*	*%*	*Records*	*Patients*	*%*
Incomplete data[1]	20	20		48	48		113	113		97	97		278	**278**	
Resident outside England and Wales	609	605		325	323		203	179		198	163		1,335	**1,270**	
In situ neoplasm[2]	46	46		17	17		0	0		0	0		63	**63**	
Benign or uncertain behaviour[2]	394	99		130	104		2	2		4	4		530	**209**	
Metastatic[2]	1,287	1,279		542	534		2	2		1	1		1,832	**1,816**	
Otherwise ineligible[3]	-	-		-	-		-	-		-	-		-	**-**	
Lymphoma[4]	87	83		47	39		19	0		18	0		171	**122**	
Leukaemia or myeloma[4]	1	1		0	0		0	0		0	0		1	**1**	
		2,133			1,065			296			265			3,759	
Total eligible		**60,960**	**100**		**60,504**	**100**		**60,014**	**100**		**55,310**	**100**		**236,788**	**100**
Aged under 15 or 100 years or over at diagnosis	18	17	<0.1	27	27	<0.1	24	24	<0.1	42	41	<0.1	111	**109**	<0.1
Vital status unknown[5]	1,927	1,722	2.8	2,310	2,127	3.5	2,411	2,321	3.9	2,466	2,385	4.3	9,114	**8,555**	3.6
Duplicate registration[6]	1	1	<0.1	0	0	0.0	1	1	<0.1	3	3	<0.1	5	**5**	<0.1
Multiple primary[7]	0	0	0.0	46	42	<0.1	181	158	0.3	204	177	0.3	431	**377**	0.2
Synchronous tumours[8]	347	290	0.5	172	123	0.2	183	155	0.3	199	144	0.3	901	**712**	0.3
Sex not known	0	0	0.0	0	0	0.0	0	0	0.0	0	0	0.0	0	**0**	0.0
Sex incompatible with site code[9]	-	-	-	-	-	-	-	-	-	-	-	-	-	**-**	-
Invalid dates[10]	1	1	<0.1	5	2	<0.1	15	11	<0.1	20	12	<0.1	41	**26**	<0.1
Zero survival[11]	8,173	7,380	12.1	8,002	7,550	12.5	9,652	9,052	15.1	9,464	8,963	16.2	35,291	**32,945**	13.9
Total exclusions		**9,411**	**15.4**		**9,871**	**16.3**		**11,722**	**19.5**		**11,725**	**21.2**		**42,729**	**18.0**
Patients included in analysis		**51,549**	**84.6**		**50,633**	**83.7**		**48,292**	**80.5**		**43,585**	**78.8**		**194,059**	**82.0**

[1] Main data item(s) invalid or incompatible with one another: name, sex, date of birth, date of diagnosis and (if present) date of death, postcode, site, morphology and behaviour

[2] Tumour either *in situ* (behaviour code 2), benign (0), uncertain if benign or malignant (1) or metastatic (6 or 9)

[3] Not applicable for this malignancy

[4] Morphology that of lymphoma (included with non-Hodgkin lymphoma, Ch. 41) or myeloma (included with multiple myeloma, Ch. 43), or leukaemia (excluded)

[5] Tracing of vital status at National Health Service Central Register incomplete at 6 July 1997

[6] Same site code, sex, cancer registry and cancer registry number as an earlier registration

[7] Same site code and person number as an earlier registration(s): mostly confirmed multiple primary tumours at this site, some unresolved duplicate registrations

[8] Same site code, sex, date of birth and date of diagnosis as another registration(s): mostly synchronous or (in paired organs) bilateral tumours in same anatomic site in one individual, not previously linked: also some duplicate registrations

[9] Not applicable for this malignancy

[10] Impossible sequence of dates (birth, diagnosis and, if present, emigration or death); or date of death after 6 July 1997

[11] Date of diagnosis same as date of death: some are patients who did die on the day of diagnosis, but most were registered solely from a death certificate (death certificate only, or DCO), and their survival time is thus unknown

Table 13.2 Crude and relative survival (%) at one, five and ten years since diagnosis: NHS Region, deprivation category and calendar period of diagnosis

	1971-75												1976-80											
MEN	Affluent		2		3		4		Deprived		All[1]		Affluent		2		3		4		Deprived		All[1]	
	C	R	C	R	C	R	C	R	C	R	C	R	C	R	C	R	C	R	C	R	C	R	C	R
ENGLAND & WALES	*(3,660)*		*(4,503)*		*(5,289)*		*(5,836)*		*(5,385)*		*(30,889)*		*(4,390)*		*(5,328)*		*(6,254)*		*(6,970)*		*(6,424)*		*(30,645)*	
One year	17	18	17	18	16	17	16	17	17	18	16	17	18	19	17	18	17	18	17	18	17	17	17	18
Five years	4	5	4	6	4	5	4	5	5	6	4	5	5	7	5	6	5	6	5	7	5	7	5	7
Ten years	3	4	3	4	3	4	3	4	3	5	3	4	3	5	3	5	3	5	3	6	3	6	3	5
ENGLAND	*(3,475)*		*(4,244)*		*(4,846)*		*(5,272)*		*(5,076)*		*(28,663)*		*(4,206)*		*(5,050)*		*(5,778)*		*(6,382)*		*(6,081)*		*(28,493)*	
One year	17	17	17	18	16	17	16	17	17	18	16	17	18	18	18	19	17	18	17	18	16	17	17	18
Five years	4	5	4	6	4	5	4	5	5	6	4	5	5	6	5	6	5	6	5	7	5	7	5	6
Ten years	3	4	3	4	3	4	3	4	3	5	3	4	3	5	3	5	3	5	3	6	3	6	3	5
Northern & Yorkshire	*(265)*		*(422)*		*(557)*		*(727)*		*(1,065)*		*(4,365)*		*(369)*		*(550)*		*(757)*		*(1,034)*		*(1,505)*		*(4,318)*	
One year	14	14	13	14	13	14	11	12	14	15	15	16	18	19	16	17	14	15	15	16	14	15	17	18
Five years	3	4	2	3	4	5	3	3	3	4	4	5	5	6	3	4	4	6	4	5	4	5	5	6
Ten years	2	3	1	3	2	3	2	2	2	3	2	4	3	5	2	4	3	5	2	4	3	4	3	5
Trent	*(236)*		*(414)*		*(569)*		*(786)*		*(676)*		*(2,947)*		*(346)*		*(541)*		*(688)*		*(864)*		*(692)*		*(3,214)*	
One year	15	16	14	15	14	15	14	14	14	15	16	17	14	15	14	15	15	15	15	16	12	13	16	17
Five years	2	3	5	6	4	6	3	3	5	6	5	6	5	6	4	5	5	7	5	6	4	5	5	7
Ten years	1	1	3	5	2	4	2	3	3	6	3	5	3	5	3	5	4	6	3	5	2	5	4	6
Anglia & Oxford	*(501)*		*(515)*		*(518)*		*(452)*		*(248)*		*(2,520)*		*(618)*		*(633)*		*(635)*		*(536)*		*(266)*		*(2,790)*	
One year	15	16	17	18	16	16	17	18	15	16	18	19	16	17	17	18	16	17	16	17	17	18	18	19
Five years	4	5	4	5	5	7	5	6	5	7	5	7	6	7	4	5	5	6	4	5	5	6	5	7
Ten years	3	4	2	3	4	6	3	5	3	5	3	5	4	6	3	5	3	5	3	4	3	6	3	6
North Thames	*(497)*		*(561)*		*(601)*		*(701)*		*(617)*		*(3,389)*		*(428)*		*(515)*		*(548)*		*(602)*		*(507)*		*(2,712)*	
One year	16	17	16	17	16	17	18	19	18	19	19	20	17	18	18	19	18	19	18	19	21	22	21	22
Five years	5	7	4	6	4	4	6	7	6	8	6	7	5	7	6	7	6	8	6	8	8	11	7	9
Ten years	3	5	3	5	2	4	3	5	4	7	4	6	4	7	4	5	3	6	4	7	6	10	5	8
South Thames	*(826)*		*(777)*		*(887)*		*(864)*		*(484)*		*(4,210)*		*(844)*		*(753)*		*(791)*		*(772)*		*(407)*		*(3,757)*	
One year	14	14	13	14	13	14	14	14	17	18	16	17	15	15	13	14	12	13	14	15	15	16	16	17
Five years	2	3	3	4	2	3	2	3	5	7	3	4	4	5	4	5	2	3	4	5	6	8	4	6
Ten years	1	2	2	3	1	2	1	2	3	5	2	3	2	4	3	5	2	3	2	4	3	6	3	5
South & West	*(610)*		*(880)*		*(857)*		*(512)*		*(212)*		*(3,381)*		*(711)*		*(994)*		*(963)*		*(532)*		*(226)*		*(3,548)*	
One year	14	15	15	16	16	17	16	17	12	13	17	18	14	15	15	16	15	16	17	18	14	15	17	18
Five years	4	5	5	7	4	6	5	7	*1*	*1*	5	6	5	6	5	7	3	4	6	8	3	5	5	7
Ten years	3	5	3	5	3	5	3	6	*1*	*1*	3	5	3	5	3	6	2	4	4	7	1	3	3	6
West Midlands	*(119)*		*(163)*		*(232)*		*(324)*		*(327)*		*(3,026)*		*(382)*		*(464)*		*(647)*		*(869)*		*(941)*		*(3,343)*	
One year	16	17	15	16	14	15	15	16	14	15	17	17	17	18	17	18	15	16	15	15	15	16	17	19
Five years	5	8	4	4	5	6	3	3	4	5	4	6	4	5	4	5	5	6	4	5	4	6	5	6
Ten years	4	8	2	3	4	5	2	3	3	5	3	4	2	3	3	4	3	5	2	4	3	5	3	5
North & West	*(421)*		*(512)*		*(625)*		*(906)*		*(1,447)*		*(4,825)*		*(508)*		*(600)*		*(749)*		*(1,173)*		*(1,537)*		*(4,811)*	
One year	14	15	13	14	10	11	10	11	13	14	14	15	14	15	15	16	13	14	12	13	12	13	15	16
Five years	3	4	3	3	2	2	3	4	3	4	3	4	2	3	4	5	3	4	4	5	3	4	4	5
Ten years	3	4	2	3	1	2	2	3	2	3	2	3	1	2	2	4	2	4	3	5	2	4	3	4
WALES	*(185)*		*(259)*		*(443)*		*(564)*		*(309)*		*(2,226)*		*(184)*		*(278)*		*(476)*		*(588)*		*(343)*		*(2,152)*	
One year	17	18	14	15	12	12	14	15	14	15	16	17	17	18	13	14	17	18	15	16	17	18	18	19
Five years	4	5	4	6	3	4	4	5	5	6	5	6	-	-	4	5	7	9	6	8	6	8	7	10
Ten years	3	4	2	4	1	2	2	5	4	6	3	5	6	9	3	5	4	7	4	7	4	6	5	8

[1] Including those for whom no deprivation category was assigned. C - crude survival, R - relative survival (see Chapter 3).
Number of patients contributing to each analysis in parentheses.

MEN

1981-85

Region	Period	Affluent C	R	2 C	R	3 C	R	4 C	R	Deprived C	R	All¹ C	R
ENGLAND & WALES	(n)	(4,589)	(5,554)		(6,194)		(6,724)		(6,101)		(29,819)		
	One year	25	26	23	24	22	23	21	22	21	22	22	23
	Five years	8	10	7	9	6	8	6	8	7	9	7	9
		5	8	5	8	4	7	4	7	4	8	4	8
ENGLAND	(n)	(4,405)	(5,261)		(5,706)		(6,150)		(5,746)		(27,738)		
	One year	24	26	23	24	22	23	21	22	21	22	22	23
	Five years	8	10	7	9	6	8	6	8	7	9	7	9
		5	8	5	8	4	7	4	7	4	8	4	7
Northern & Yorkshire	(n)	(413)	(602)		(789)		(1,105)		(1,408)		(4,375)		
	One year	25	26	20	21	18	19	18	20	17	19	21	22
	Five years	9	11	6	8	4	6	5	7	6	9	6	8
		6	9	4	7	3	5	3	6	4	8	4	7
Trent	(n)	(388)	(568)		(697)		(930)		(659)		(3,270)		
	One year	24	25	22	23	18	19	20	21	17	19	21	23
	Five years	7	9	6	7	6	8	6	8	5	6	6	8
		5	8	4	7	3	6	3	6	2	5	3	6
Anglia & Oxford	(n)	(664)	(653)		(607)		(480)		(245)		(2,699)		
	One year	21	23	21	22	20	21	20	21	21	23	23	24
	Five years	7	9	7	9	6	8	5	7	3	4	6	8
		4	7	5	8	4	7	3	6	2	3	4	7
North Thames	(n)	(380)	(449)		(446)		(480)		(409)		(2,194)		
	One year	23	24	28	29	27	28	24	26	25	27	27	29
	Five years	8	11	9	12	11	14	9	12	10	12	10	13
		6	10	6	10	8	13	7	12	7	11	7	12
South Thames	(n)	(823)	(748)		(680)		(584)		(280)		(3,263)		
	One year	24	25	21	22	21	22	21	23	27	29	24	25
	Five years	8	9	7	9	6	8	7	10	8	12	8	10
		5	8	4	8	4	7	5	8	6	10	5	9
South & West	(n)	(683)	(995)		(931)		(498)		(208)		(3,393)		
	One year	24	26	22	23	20	22	20	21	21	23	24	25
	Five years	8	11	9	11	5	7	7	9	9	13	8	11
		6	9	6	11	3	6	5	9	7	12	5	9
West Midlands	(n)	(438)	(559)		(764)		(974)		(969)		(3,735)		
	One year	22	24	21	22	21	23	20	21	19	20	22	24
	Five years	5	7	6	8	7	9	5	7	7	9	7	9
		3	5	3	5	5	9	3	7	5	8	4	7
North & West	(n)	(616)	(687)		(792)		(1,099)		(1,568)		(4,809)		
	One year	19	20	16	17	16	17	15	16	17	18	18	19
	Five years	5	6	5	6	4	5	4	5	4	6	5	6
		2	4	3	6	3	5	2	4	3	6	3	5
WALES	(n)	(184)	(293)		(488)		(574)		(355)		(2,081)		
	One year	24	26	21	23	20	21	21	22	19	20	23	24
	Five years	9	12	7	9	6	8	7	9	7	10	8	10
		6	12	4	8	4	7	5	9	5	8	5	9

1986-90

Region	Period	Affluent C	R	2 C	R	3 C	R	4 C	R	Deprived C	R	All¹ C	R
ENGLAND & WALES	(n)	(4,256)	(5,091)		(5,856)		(6,288)		(5,512)		(27,294)		
	One year	26	28	25	26	24	26	23	25	23	25	24	26
	Five years	8	10	8	10	7	10	7	10	8	10	8	10
ENGLAND	(n)	(4,063)	(4,817)		(5,463)		(5,849)		(5,259)		(25,692)		
	One year	27	28	25	26	24	26	23	25	23	25	25	26
	Five years	8	10	8	10	8	10	7	10	8	10	8	10
Northern & Yorkshire	(n)	(428)	(601)		(768)		(948)		(1,239)		(4,020)		
	One year	20	21	22	23	23	25	21	23	20	22	23	24
	Five years	6	8	6	8	7	10	7	9	8	11	8	10
Trent	(n)	(374)	(485)		(649)		(810)		(759)		(3,083)		
	One year	31	33	24	26	22	23	19	21	19	21	24	25
	Five years	11	14	7	10	7	10	5	8	5	7	7	9
Anglia & Oxford	(n)	(632)	(701)		(647)		(463)		(228)		(2,692)		
	One year	23	24	22	24	24	26	24	26	15	16	24	25
	Five years	6	8	5	7	7	9	8	11	5	8	7	8
North Thames	(n)	(345)	(457)		(533)		(590)		(502)		(2,488)		
	One year	36	39	28	30	30	32	28	30	33	36	32	34
	Five years	9	12	10	13	8	11	9	12	11	16	10	13
South Thames	(n)	(635)	(607)		(603)		(534)		(288)		(2,717)		
	One year	30	32	27	28	22	23	25	27	26	29	27	29
	Five years	8	11	8	10	6	8	7	10	10	14	8	10
South & West	(n)	(643)	(821)		(874)		(627)		(241)		(3,227)		
	One year	25	26	25	27	23	25	26	27	24	26	26	28
	Five years	8	11	8	11	7	10	9	13	4	6	8	11
West Midlands	(n)	(417)	(533)		(709)		(889)		(737)		(3,307)		
	One year	21	23	23	24	23	25	19	20	20	22	23	24
	Five years	6	8	8	10	7	10	7	10	8	11	8	10
North & West	(n)	(589)	(612)		(680)		(988)		(1,265)		(4,158)		
	One year	21	22	20	22	17	19	18	20	21	22	21	22
	Five years	7	8	6	9	6	8	5	7	6	9	6	8
WALES	(n)	(193)	(274)		(393)		(439)		(253)		(1,602)		
	One year	19	20	21	22	21	22	17	18	20	21	22	23
	Five years	4	7	7	10	7	10	6	8	4	6	7	9

¹ Including those for whom no deprivation category was assigned. C - crude survival, R - relative survival (see Chapter 3).
Number of patients contributing to each analysis in parentheses.

Table 13.2 Crude and relative survival (%) at one, five and ten years since diagnosis: NHS Region, deprivation category and calendar period of diagnosis

	1971-75											1976-80											
WOMEN	Affluent		2		3		4		Deprived		All[1]		Affluent		2		3		4		Deprived		All[1]	
	C	R	C	R	C	R	C	R	C	R	C	R	C	R	C	R	C	R	C	R	C	R	C	R
ENGLAND & WALES	*(2,354)*		*(2,989)*		*(3,470)*		*(3,876)*		*(3,614)*		*(20,660)*		*(2,865)*		*(3,339)*		*(4,041)*		*(4,478)*		*(4,392)*		*(19,988)*	
One year	17	17	16	17	15	16	17	18	17	17	**16**	**17**	18	19	17	18	16	17	16	17	17	18	**17**	**18**
Five years	4	5	4	5	4	5	5	6	5	6	4	6	5	7	5	7	5	6	5	6	5	7	5	6
Ten years	2	4	3	4	3	4	3	5	3	4	3	5	4	6	4	6	3	5	3	5	3	5	3	5
ENGLAND	*(2,241)*		*(2,825)*		*(3,199)*		*(3,481)*		*(3,419)*		*(19,199)*		*(2,738)*		*(3,140)*		*(3,708)*		*(4,071)*		*(4,173)*		*(18,521)*	
One year	17	17	16	17	15	16	17	18	17	18	**16**	**17**	18	19	17	18	16	17	16	17	17	18	**17**	**18**
Five years	4	5	4	5	4	5	5	6	4	6	**4**	**5**	5	6	5	7	5	6	5	6	5	6	**5**	**6**
Ten years	2	4	2	4	3	4	3	5	3	4	**3**	**4**	3	6	4	6	3	5	3	5	3	5	**3**	**5**
Northern & Yorkshire	*(203)*		*(271)*		*(404)*		*(567)*		*(738)*		*(3,125)*		*(282)*		*(427)*		*(541)*		*(775)*		*(1,068)*		*(3,173)*	
One year	14	15	13	14	13	14	13	14	14	14	**16**	**17**	14	15	12	13	11	12	14	15	14	15	**15**	**16**
Five years	2	3	4	5	3	4	3	4	4	5	**4**	**5**	4	5	3	4	4	4	4	5	4	5	**4**	**6**
Ten years	2	2	2	3	2	4	2	3	2	3	**2**	**4**	*3*	*4*	2	4	2	3	2	3	3	4	**3**	**5**
Trent	*(156)*		*(287)*		*(345)*		*(410)*		*(403)*		*(1,770)*		*(197)*		*(277)*		*(380)*		*(521)*		*(476)*		*(1,899)*	
One year	13	14	15	15	12	12	13	14	19	20	**17**	**18**	17	18	15	15	18	19	13	14	16	16	**18**	**19**
Five years	3	3	2	3	4	4	3	4	6	7	**5**	**6**	5	6	5	6	6	7	5	6	5	6	**6**	**7**
Ten years	2	3	2	2	2	3	2	4	3	4	**3**	**4**	3	5	3	5	4	6	3	5	3	5	**4**	**6**
Anglia & Oxford	*(264)*		*(324)*		*(351)*		*(248)*		*(132)*		*(1,518)*		*(386)*		*(364)*		*(359)*		*(296)*		*(142)*		*(1,614)*	
One year	15	16	21	22	13	13	18	19	15	16	**19**	**20**	13	14	17	18	10	11	19	21	17	18	**17**	**18**
Five years	4	5	8	10	3	4	7	8	4	5	**6**	**7**	3	4	5	6	4	5	6	8	7	9	**5**	**7**
Ten years	2	3	5	8	2	3	5	8	*3*	*5*	**4**	**6**	2	4	3	4	2	4	5	7	5	8	**3**	**6**
North Thames	*(316)*		*(349)*		*(392)*		*(423)*		*(380)*		*(2,135)*		*(283)*		*(282)*		*(328)*		*(354)*		*(318)*		*(1,632)*	
One year	19	20	17	18	16	16	19	20	15	16	**20**	**20**	19	20	22	23	17	18	19	20	23	24	**22**	**23**
Five years	7	7	6	7	5	6	6	7	6	7	**6**	**8**	6	7	7	9	-	-	6	8	7	10	**7**	**8**
Ten years	3	5	3	5	4	6	4	6	4	6	**4**	**7**	4	6	5	8	5	6	3	6	4	7	**4**	**7**
South Thames	*(579)*		*(585)*		*(598)*		*(616)*		*(345)*		*(3,012)*		*(617)*		*(537)*		*(542)*		*(517)*		*(238)*		*(2,607)*	
One year	14	15	10	10	14	14	14	15	13	14	**15**	**16**	17	17	13	14	12	12	11	12	19	20	**16**	**17**
Five years	3	4	2	3	4	5	4	5	2	3	**4**	**5**	5	6	5	6	3	3	3	3	7	10	**5**	**6**
Ten years	2	3	1	2	3	4	3	5	2	2	**2**	**4**	3	5	3	5	2	3	2	3	5	8	**3**	**5**
South & West	*(336)*		*(549)*		*(509)*		*(308)*		*(114)*		*(2,011)*		*(413)*		*(572)*		*(611)*		*(311)*		*(114)*		*(2,108)*	
One year	14	15	12	12	12	12	16	17	20	21	**16**	**16**	18	19	16	16	17	17	16	17	17	18	**19**	**20**
Five years	3	4	3	4	3	4	5	7	9	12	**4**	**6**	5	6	5	6	6	7	6	7	10	13	**6**	**8**
Ten years	2	3	1	2	3	4	4	6	7	10	**3**	**5**	3	5	4	6	3	6	4	6	4	8	**4**	**7**
West Midlands	*(82)*		*(102)*		*(148)*		*(230)*		*(250)*		*(2,114)*		*(224)*		*(271)*		*(378)*		*(479)*		*(615)*		*(1,999)*	
One year	15	16	11	12	13	14	19	20	15	16	**17**	**18**	15	16	15	15	16	17	14	15	14	14	**17**	**17**
Five years	4	5	2	2	2	3	4	4	4	5	**4**	**6**	4	5	5	6	4	5	4	5	3	4	**4**	**6**
Ten years	3	4	2	2	2	2	3	4	3	4	**3**	**5**	3	4	3	6	3	5	2	4	2	3	**3**	**5**
North & West	*(305)*		*(358)*		*(452)*		*(679)*		*(1,057)*		*(3,514)*		*(336)*		*(410)*		*(569)*		*(818)*		*(1,202)*		*(3,489)*	
One year	12	12	13	14	11	12	13	13	14	14	**15**	**15**	15	15	13	14	10	11	10	11	13	13	**14**	**15**
Five years	4	5	2	3	2	2	3	4	3	3	**3**	**4**	4	5	4	5	3	4	3	4	3	4	**4**	**5**
Ten years	2	3	2	2	1	2	2	4	2	2	**2**	**3**	3	4	3	4	2	4	2	3	2	3	**3**	**5**
WALES	*(113)*		*(164)*		*(271)*		*(395)*		*(195)*		*(1,461)*		*(127)*		*(199)*		*(333)*		*(407)*		*(219)*		*(1,467)*	
One year	14	15	11	12	11	11	14	15	13	13	**16**	**17**	17	18	14	14	18	18	15	15	17	18	**18**	**19**
Five years	4	6	5	5	4	5	5	6	5	6	**6**	**7**	8	10	6	8	6	8	6	7	6	7	**7**	**9**
Ten years	4	5	5	5	3	4	2	5	3	5	**4**	**6**	5	8	5	7	5	8	4	7	3	5	**5**	**8**

[1] Including those for whom no deprivation category was assigned. C - crude survival, R - relative survival (see Chapter 3). Number of patients contributing to each analysis in parentheses.

WOMEN

Crude (C) and relative (R) survival, regions. C - crude survival, R - relative survival.

1981-85

	Affluent C	R	2 C	R	3 C	R	4 C	R	Deprived C	R	All[1] C	R
ENGLAND & WALES (n)	(2,697)		(3,243)		(3,860)		(4,256)		(3,934)		(18,473)	
One year	22	23	23	24	21	23	21	22	20	22	**21**	**23**
Five years	7	9	8	10	7	9	7	9	7	9	**7**	**9**
	5	8	5	8	5	8	5	7	4	7	**5**	**8**
ENGLAND (n)	(2,586)		(3,061)		(3,561)		(3,894)		(3,732)		(17,183)	
One year	22	23	23	24	21	22	21	22	20	22	**21**	**23**
Five years	7	9	8	10	7	9	7	9	7	9	**7**	**9**
	5	7	5	8	5	8	4	7	4	7	**5**	**8**
Northern & Yorkshire (n)	(264)		(388)		(525)		(709)		(920)		(2,859)	
One year	19	20	22	24	18	19	19	20	20	22	**22**	**23**
Five years	5	6	9	12	5	7	8	9	7	9	**8**	**10**
	2	4	7	10	4	6	6	9	4	8	**5**	**8**
Trent (n)	(209)		(285)		(376)		(551)		(420)		(1,858)	
One year	18	19	21	22	18	19	19	20	18	19	**21**	**22**
Five years	6	7	5	6	7	8	6	8	6	8	**7**	**9**
	5	7	3	6	5	7	4	6	3	5	**4**	**7**
Anglia & Oxford (n)	(375)		(388)		(384)		(278)		(119)		(1,572)	
One year	21	22	20	21	18	19	21	22	15	15	**21**	**22**
Five years	6	8	7	9	6	8	4	4	4	5	**6**	**8**
	4	7	3	6	5	8	3	4	3	4	**4**	**7**
North Thames (n)	(231)		(258)		(289)		(260)		(244)		(1,309)	
One year	29	30	22	24	26	28	21	22	21	23	**26**	**28**
Five years	9	11	9	12	11	13	8	11	8	11	**10**	**13**
	7	9	6	10	8	13	6	10	6	10	**7**	**12**
South Thames (n)	(497)		(463)		(449)		(389)		(162)		(2,065)	
One year	22	24	20	21	24	25	21	22	20	22	**24**	**25**
Five years	8	10	8	10	7	9	6	8	7	9	**8**	**11**
	5	8	5	7	5	9	4	7	6	8	**5**	**9**
South & West (n)	(375)		(531)		(577)		(306)		(117)		(1,973)	
One year	20	21	22	23	20	21	21	23	20	21	**23**	**24**
Five years	6	8	8	10	7	10	7	9	7	9	**8**	**11**
	4	7	5	8	4	7	5	7	3	5	**5**	**8**
West Midlands (n)	(268)		(295)		(410)		(571)		(630)		(2,194)	
One year	17	18	22	24	17	18	17	18	18	19	**20**	**22**
Five years	6	8	7	8	5	6	5	6	6	8	**7**	**8**
	4	7	5	7	4	6	3	5	3	5	**4**	**6**
North & West (n)	(367)		(453)		(551)		(830)		(1,120)		(3,353)	
One year	14	15	17	18	16	17	16	17	16	17	**18**	**19**
Five years	4	6	6	7	5	6	5	7	5	7	**6**	**8**
	3	4	2	4	2	4	3	6	3	6	**3**	**6**
WALES (n)	(111)		(182)		(299)		(362)		(202)		(1,290)	
One year	24	25	17	18	21	22	21	23	20	21	**23**	**24**
Five years	11	13	8	10	7	10	8	10	5	7	**8**	**11**
	8	10	6	10	5	8	5	9	3	6	**6**	**10**

1986-90

	Affluent C	R	2 C	R	3 C	R	4 C	R	Deprived C	R	All[1] C	R
ENGLAND & WALES (n)	(2,408)		(3,060)		(3,424)		(3,770)		(3,435)		(16,291)	
One year	25	26	24	25	23	25	24	25	24	26	24	25
Five years	9	11	8	10	8	10	8	11	9	12	9	11
ENGLAND (n)	(2,272)		(2,904)		(3,152)		(3,504)		(3,267)		(15,244)	
One year	25	26	24	26	24	25	24	25	24	26	**24**	**26**
Five years	9	11	8	10	8	10	8	11	9	12	**8**	**11**
Northern & Yorkshire (n)	(275)		(349)		(490)		(653)		(771)		(2,552)	
One year	19	20	19	21	22	23	24	25	21	22	**23**	**25**
Five years	7	8	6	7	7	9	8	11	7	9	**8**	**10**
Trent (n)	(168)		(262)		(363)		(469)		(443)		(1,713)	
One year	28	29	19	20	22	23	22	23	23	24	**24**	**26**
Five years	10	12	6	9	7	8	10	12	10	13	**9**	**12**
Anglia & Oxford (n)	(336)		(408)		(331)		(251)		(125)		(1,466)	
One year	21	22	23	24	18	19	20	22	19	21	**22**	**23**
Five years	8	10	5	7	5	7	7	9	7	8	**7**	**8**
North Thames (n)	(196)		(279)		(287)		(352)		(302)		(1,457)	
One year	25	26	30	32	25	26	29	31	28	29	**29**	**31**
Five years	10	11	11	14	7	9	11	13	9	11	**10**	**13**
South Thames (n)	(354)		(400)		(322)		(294)		(183)		(1,575)	
One year	25	27	23	24	26	27	23	24	32	34	**27**	**28**
Five years	9	11	6	8	8	10	7	9	13	17	**9**	**12**
South & West (n)	(324)		(488)		(482)		(327)		(124)		(1,756)	
One year	24	25	23	24	25	26	25	26	26	27	**26**	**27**
Five years	9	12	9	12	8	11	9	12	13	16	**10**	**13**
West Midlands (n)	(264)		(301)		(390)		(516)		(440)		(1,928)	
One year	23	25	25	26	23	24	19	20	21	22	**23**	**25**
Five years	8	9	10	12	8	10	5	7	8	10	**8**	**10**
North & West (n)	(355)		(417)		(487)		(642)		(879)		(2,797)	
One year	21	23	21	22	19	20	18	19	20	22	**22**	**23**
Five years	7	10	6	8	8	10	5	7	7	9	**7**	**10**
WALES (n)	(136)		(156)		(272)		(266)		(168)		(1,047)	
One year	24	25	17	17	16	16	21	21	23	24	**23**	**24**
Five years	14	18	6	6	7	8	9	11	11	14	**11**	**14**

[1] Including those for whom no deprivation category was assigned. C - crude survival, R - relative survival (see Chapter 3). Number of patients contributing to each analysis in parentheses.

Table 13.2 Crude and relative survival (%) at one, five and ten years since diagnosis: NHS Region, deprivation category and calendar period of diagnosis

ADULTS	1971-75						1976-80					
	Affluent C R	2 C R	3 C R	4 C R	Deprived C R	All[1] C R	Affluent C R	2 C R	3 C R	4 C R	Deprived C R	All[1] C R
ENGLAND & WALES	*(6,014)*	*(7,492)*	*(8,759)*	*(9,712)*	*(8,999)*	*(51,549)*	*(7,255)*	*(8,667)*	*(10,295)*	*(11,448)*	*(10,816)*	*(50,633)*
One year	17 18	16 17	16 16	16 17	17 18	**16 17**	18 19	17 18	17 18	17 18	17 18	**17 18**
Five years	4 5	4 5	4 5	4 6	5 6	**4 5**	5 7	5 6	5 6	5 7	5 7	**5 7**
Ten years	3 4	3 4	3 4	3 5	3 5	**3 4**	3 6	3 6	3 5	3 6	3 6	**3 5**
ENGLAND	*(5,716)*	*(7,069)*	*(8,045)*	*(8,753)*	*(8,495)*	*(47,862)*	*(6,944)*	*(8,190)*	*(9,486)*	*(10,453)*	*(10,254)*	*(47,014)*
One year	17 17	16 17	16 17	16 17	17 18	**16 17**	18 19	17 18	16 17	17 18	17 18	**17 18**
Five years	4 5	4 5	4 5	4 5	5 6	**4 5**	5 6	5 6	5 6	5 6	5 7	**5 6**
Ten years	3 4	3 4	3 4	3 5	3 5	**3 4**	3 5	3 5	3 5	3 5	3 5	**3 5**
Northern & Yorkshire	*(468)*	*(693)*	*(961)*	*(1,294)*	*(1,803)*	*(7,490)*	*(651)*	*(977)*	*(1,298)*	*(1,809)*	*(2,573)*	*(7,491)*
One year	14 15	13 14	13 14	12 12	14 15	**16 17**	16 17	14 15	13 14	15 15	14 15	**16 17**
Five years	3 3	3 4	3 4	3 4	3 4	**4 5**	5 6	3 4	4 5	4 5	4 5	**5 6**
Ten years	2 3	2 3	2 4	2 3	2 3	**2 4**	3 5	2 4	3 4	2 4	3 4	**3 5**
Trent	*(392)*	*(701)*	*(914)*	*(1,196)*	*(1,079)*	*(4,717)*	*(543)*	*(818)*	*(1,068)*	*(1,385)*	*(1,168)*	*(5,113)*
One year	14 15	14 15	13 14	13 14	16 17	**17 18**	15 16	14 15	16 17	14 15	14 14	**17 17**
Five years	3 3	4 5	4 5	3 4	5 6	**5 6**	5 6	4 5	5 7	5 6	4 6	**5 7**
Ten years	1 2	2 4	2 4	2 3	3 5	**3 5**	3 5	3 5	4 6	3 5	2 4	**4 6**
Anglia & Oxford	*(765)*	*(839)*	*(869)*	*(700)*	*(380)*	*(4,038)*	*(1,004)*	*(997)*	*(994)*	*(832)*	*(408)*	*(4,404)*
One year	15 16	18 19	14 15	17 18	15 16	**18 19**	15 16	17 18	14 15	17 18	17 18	**18 19**
Five years	4 5	5 7	4 5	6 7	5 6	**5 7**	5 6	4 6	4 5	5 6	6 7	**5 7**
Ten years	3 4	3 5	3 5	4 6	3 5	**4 6**	3 6	3 4	3 4	3 5	4 7	**3 6**
North Thames	*(813)*	*(910)*	*(993)*	*(1,124)*	*(997)*	*(5,524)*	*(711)*	*(797)*	*(876)*	*(956)*	*(825)*	*(4,344)*
One year	17 18	17 17	16 17	18 19	17 18	**19 20**	18 19	19 20	18 19	19 20	22 23	**21 22**
Five years	6 7	5 6	4 5	6 7	6 8	**6 8**	6 7	6 8	6 7	6 8	8 10	**7 9**
Ten years	3 5	3 5	3 4	4 6	4 7	**4 6**	4 6	4 6	4 7	4 6	5 9	**5 8**
South Thames	*(1,405)*	*(1,362)*	*(1,485)*	*(1,480)*	*(829)*	*(7,222)*	*(1,461)*	*(1,290)*	*(1,333)*	*(1,289)*	*(645)*	*(6,364)*
One year	14 14	12 12	13 14	14 15	15 16	**16 16**	15 16	13 14	12 13	13 14	16 17	**16 17**
Five years	3 3	3 3	3 4	3 4	4 5	**3 4**	4 6	4 6	3 3	3 4	6 9	**5 6**
Ten years	1 2	1 2	2 3	2 3	3 4	**2 4**	3 4	3 5	2 3	2 4	4 7	**3 5**
South & West	*(946)*	*(1,429)*	*(1,366)*	*(820)*	*(326)*	*(5,392)*	*(1,124)*	*(1,566)*	*(1,574)*	*(843)*	*(340)*	*(5,656)*
One year	14 15	14 15	14 15	16 17	15 15	**17 17**	16 16	15 16	16 17	16 17	15 16	**18 19**
Five years	4 4	4 5	4 5	5 7	4 5	**5 6**	5 6	5 7	4 5	6 7	5 7	**6 7**
Ten years	3 4	2 4	3 5	4 6	3 4	**3 5**	3 5	3 6	3 4	4 7	2 4	**4 6**
West Midlands	*(201)*	*(265)*	*(380)*	*(554)*	*(577)*	*(5,140)*	*(606)*	*(735)*	*(1,025)*	*(1,348)*	*(1,556)*	*(5,342)*
One year	16 17	14 14	14 15	17 18	14 15	**17 18**	17 18	16 17	16 17	15 15	14 15	**17 18**
Five years	5 7	3 4	4 5	3 4	4 5	**4 6**	4 5	4 5	5 6	4 5	4 5	**5 6**
Ten years	4 6	2 3	3 4	2 4	3 4	**3 5**	2 4	3 5	3 5	2 4	2 4	**3 5**
North & West	*(726)*	*(870)*	*(1,077)*	*(1,585)*	*(2,504)*	*(8,339)*	*(844)*	*(1,010)*	*(1,318)*	*(1,991)*	*(2,739)*	*(8,300)*
One year	13 14	13 14	10 11	11 12	13 14	**14 15**	14 15	14 15	12 13	11 12	12 13	**15 16**
Five years	4 5	2 3	2 2	3 4	3 4	**3 4**	3 4	4 5	3 4	3 5	3 4	**4 5**
Ten years	2 3	2 3	1 2	2 3	2 3	**2 3**	2 3	3 4	2 4	2 4	2 4	**3 4**
WALES	*(298)*	*(423)*	*(714)*	*(959)*	*(504)*	*(3,687)*	*(311)*	*(477)*	*(809)*	*(995)*	*(562)*	*(3,619)*
One year	16 17	13 13	11 12	14 15	14 14	**16 17**	17 18	13 14	17 18	15 16	17 18	**18 19**
Five years	4 6	4 5	3 4	4 5	5 6	**5 7**	7 9	5 6	7 8	6 8	6 7	**7 9**
Ten years	3 5	3 5	2 3	2 5	3 6	**3 5**	6 9	4 6	4 7	4 7	3 6	**5 8**

[1] Including those for whom no deprivation category was assigned. C - crude survival, R - relative survival (see Chapter 3). Number of patients contributing to each analysis in parentheses.

	1981-85						1986-90						
	Affluent C R	*2* C R	*3* C R	*4* C R	*Deprived* C R	*All¹* C R	*Affluent* C R	*2* C R	*3* C R	*4* C R	*Deprived* C R	*All¹* C R	***ADULTS***
	(7,286)	(8,797)	(10,054)	(10,980)	(10,035)	(48,292)	(6,664)	(8,151)	(9,280)	(10,058)	(8,947)	(43,585)	**ENGLAND & WALES**
One year	24 25	23 24	22 23	21 22	21 22	**22 23**	26 27	25 26	24 25	23 25	24 25	**24 26**	
Five years	8 9	8 10	7 9	7 9	7 9	**7 9**	8 10	8 10	8 10	8 10	8 11	**8 10**	
	5 8	5 8	4 7	4 7	4 8	**5 8**							
	(6,991)	(8,322)	(9,267)	(10,044)	(9,478)	(44,921)	(6,335)	(7,721)	(8,615)	(9,353)	(8,526)	(40,936)	**ENGLAND**
One year	23 25	23 24	22 23	21 22	21 22	**22 23**	26 27	25 26	24 26	24 25	24 25	**24 26**	
Five years	7 9	8 10	7 9	6 8	7 9	**7 9**	8 10	8 10	8 10	8 10	8 11	**8 10**	
	5 8	5 8	4 7	4 7	4 8	**4 8**							
	(677)	(990)	(1,314)	(1,814)	(2,328)	(7,234)	(703)	(950)	(1,258)	(1,601)	(2,010)	(6,572)	**Northern & Yorkshire**
One year	22 24	21 22	18 19	19 20	19 20	**21 23**	20 21	21 22	23 24	22 24	21 22	**23 24**	
Five years	7 9	7 9	5 6	6 8	6 9	**7 9**	6 8	6 8	7 10	7 10	8 10	**8 10**	
	4 7	5 8	3 5	4 7	4 8	**5 8**							
	(597)	(853)	(1,073)	(1,481)	(1,079)	(5,128)	(542)	(747)	(1,012)	(1,279)	(1,202)	(4,796)	**Trent**
One year	22 23	22 23	18 19	20 21	18 19	**21 23**	30 32	22 24	22 23	20 22	21 22	**24 25**	
Five years	7 8	5 7	6 8	6 8	5 7	**6 8**	11 13	7 9	7 9	7 10	7 9	**8 10**	
	5 8	4 7	4 6	3 6	3 5	**4 7**							
	(1,039)	(1,041)	(991)	(758)	(364)	(4,271)	(968)	(1,109)	(978)	(714)	(353)	(4,158)	**Anglia & Oxford**
One year	21 22	21 22	19 21	20 22	19 20	**22 23**	22 24	22 24	22 23	23 24	16 17	**23 24**	
Five years	7 8	7 9	6 8	5 6	3 4	**6 8**	7 9	5 7	6 8	7 10	6 8	**7 8**	
	4 7	4 7	4 7	3 6	2 4	**4 7**							
	(611)	(707)	(735)	(740)	(653)	(3,503)	(541)	(736)	(820)	(942)	(804)	(3,945)	**North Thames**
One year	25 26	26 27	26 28	23 25	24 25	**27 28**	32 34	29 31	28 30	29 30	31 33	**31 33**	
Five years	9 11	9 12	11 14	9 12	9 12	**10 13**	9 12	10 13	8 10	9 12	10 14	**10 13**	
	6 9	6 10	8 13	6 11	7 11	**7 12**							
	(1,320)	(1,211)	(1,129)	(973)	(442)	(5,328)	(989)	(1,007)	(925)	(828)	(471)	(4,292)	**South Thames**
One year	23 24	20 22	22 23	21 23	24 26	**24 25**	28 30	25 27	23 25	24 26	29 31	**27 29**	
Five years	8 10	7 9	6 8	7 9	8 11	**8 10**	9 11	7 10	7 9	7 10	11 15	**8 11**	
	5 8	5 8	4 8	4 8	6 10	**5 9**							
	(1,058)	(1,526)	(1,508)	(804)	(325)	(5,366)	(967)	(1,309)	(1,356)	(954)	(365)	(4,983)	**South & West**
One year	23 24	22 23	20 22	20 22	21 22	**23 25**	25 26	24 26	23 25	25 27	25 27	**26 27**	
Five years	8 10	9 11	6 8	7 9	8 11	**8 11**	9 11	9 12	8 10	9 13	7 10	**9 12**	
	5 8	5 9	3 6	5 8	6 10	**5 9**							
	(706)	(854)	(1,174)	(1,545)	(1,599)	(5,929)	(681)	(834)	(1,099)	(1,405)	(1,177)	(5,235)	**West Midlands**
One year	20 22	21 23	20 21	19 20	19 20	**22 23**	22 23	23 25	23 25	19 20	20 22	**23 24**	
Five years	6 7	7 8	6 8	5 7	7 8	**7 9**	7 8	8 11	8 10	7 9	8 11	**8 10**	
	4 6	4 6	5 8	3 6	4 7	**4 7**							
	(983)	(1,140)	(1,343)	(1,929)	(2,688)	(8,162)	(944)	(1,029)	(1,167)	(1,630)	(2,144)	(6,955)	**North & West**
One year	17 18	17 18	16 17	15 16	16 17	**18 19**	21 23	21 22	18 19	18 20	21 22	**21 23**	
Five years	5 6	5 7	4 6	4 6	5 7	**5 7**	7 9	6 8	7 9	5 7	7 9	**7 9**	
	2 4	3 5	2 5	3 5	3 6	**3 6**							
	(295)	(475)	(787)	(936)	(557)	(3,371)	(329)	(430)	(665)	(705)	(421)	(2,649)	**WALES**
One year	24 26	20 21	20 22	21 22	19 20	**23 24**	21 22	19 20	19 20	18 19	21 22	**22 23**	
Five years	10 13	7 10	6 8	7 10	7 9	**8 10**	8 12	7 8	7 9	7 10	7 9	**9 11**	
	7 11	5 9	4 7	5 9	4 7	**5 9**							

¹ Including those for whom no deprivation category was assigned. C - crude survival, R - relative survival (see Chapter 3).
Number of patients contributing to each analysis in parentheses.

Fig. 13.1   Relative survival up to ten years: trends by calendar period of diagnosis (1971-90) and NHS Region

ADULTS

England & Wales

Relative survival (%)

Years since diagnosis

Period of diagnosis
- 1986-90
- 1981-85
- 1976-80
- 1971-75

Northern & Yorkshire

Relative survival (%)

Years since diagnosis

Trent

Relative survival (%)

Years since diagnosis

Anglia & Oxford

Relative survival (%)

Years since diagnosis

North Thames

Relative survival (%)

Years since diagnosis

South Thames

Relative survival (%)

Years since diagnosis

South & West

Relative survival (%)

Years since diagnosis

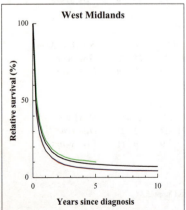

West Midlands

Relative survival (%)

Years since diagnosis

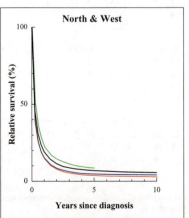

North & West

Relative survival (%)

Years since diagnosis

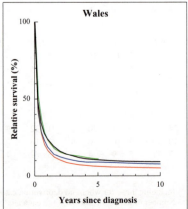

Wales

Relative survival (%)

Years since diagnosis

Fig. 13.2   Relative survival up to five years, by deprivation category and NHS Region: patients diagnosed 1986-90

ADULTS

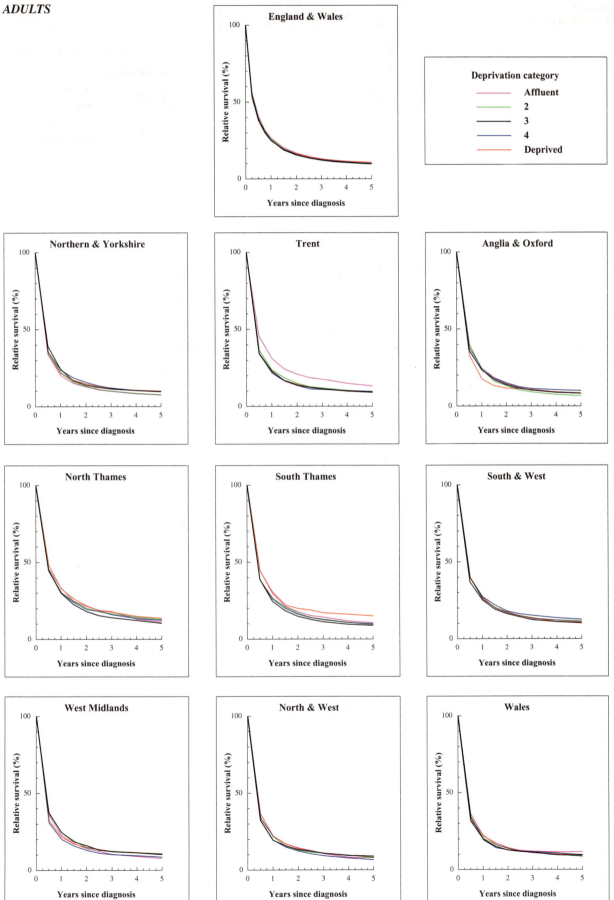

Fig. 13.3   Relative survival at five years by deprivation category, period of diagnosis (1981-90) and NHS Region

ADULTS
(Note vertical scale)

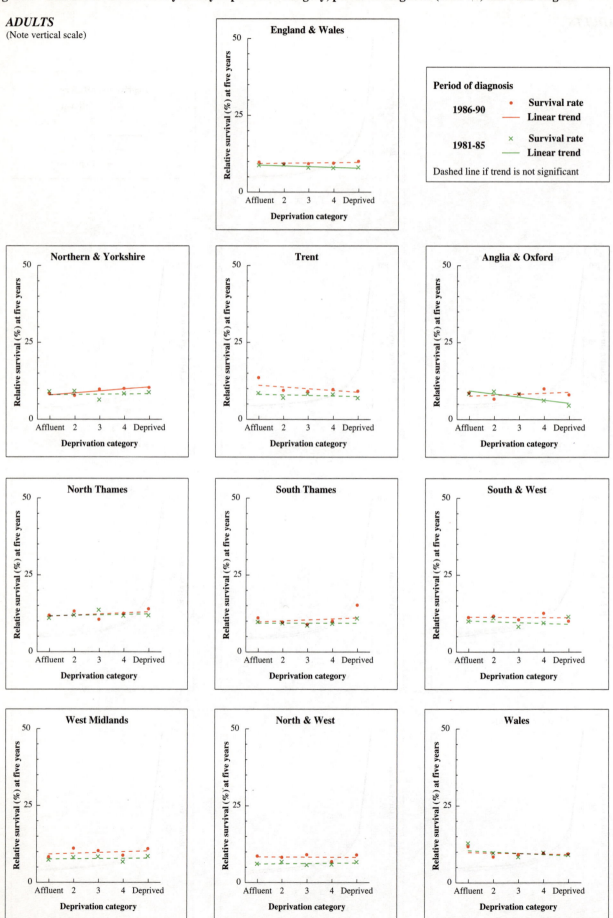

Fig. 13.4 Relative survival (%) by NHS Region and sex:
one-year and five-year survival for patients diagnosed 1986-90, with 95% confidence intervals

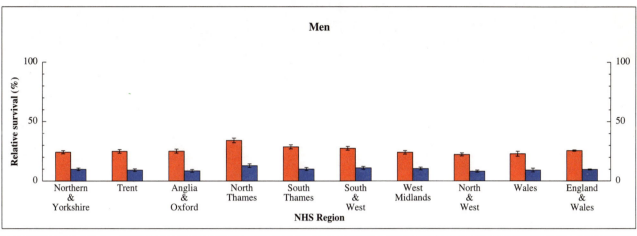

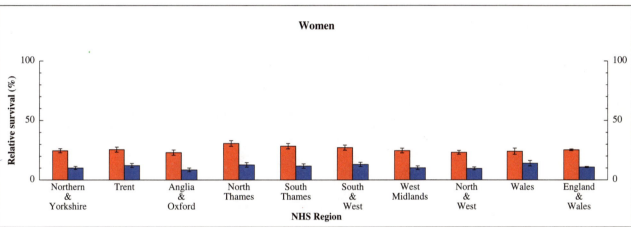

Fig. 13.5 Relative survival (%) at five years, with 95% confidence interval, by age at diagnosis and sex:
England and Wales, patients diagnosed 1986-90

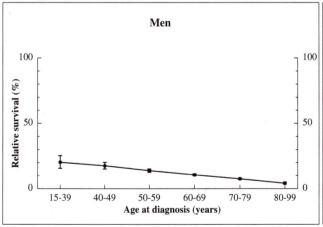

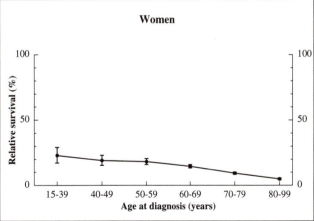

Chapter 14

SMALL INTESTINE

In sharp contrast with other parts of the intestinal tract (oesophagus, stomach, large bowel), cancer of the small intestine is rare, with only about 400 new cases each year in England and Wales. In 1990, the age-standardised (European) incidence in England and Wales was 0.7 per 100,000 in men and 0.5 in women, about half the levels in Japan, Sweden and the USA. The male-to-female sex ratio in incidence is about 1.5 to 1. Small bowel cancer is very rare under the age of 40. Mortality rates have been about 0.5 per 100,000 in men and 0.3 in women over the past 25 years.

Adenocarcinomas of the intestinal mucosa are the most common type: they frequently spread to regional lymph nodes and the liver by the time of diagnosis. Carcinoid tumours, leiomyosarcomas of the muscle wall and primary lymphomas are the other main types. Almost 100 primary lymphomas were re-assigned for analysis as extra-nodal lymphomas (Chapter 41).

The survival analyses here are based on 5,600 patients, about 86% of those eligible, who were diagnosed between 1971 and 1990 and followed up to the end of 1995. Some 4% of patients were excluded because their vital status was unknown at 6 July 1997, and a further 9% because they had zero or unknown survival (Table 14.1).

About 26% of cases occurred in the duodenum and 22% in the jejunum and ileum, while in a third of cases no anatomic detail was specified (Table 14A). Almost 60% were specified as adenocarcinomas, including mucinous adenocarcinoma (4%). Leiomyosarcomas accounted for 8% of cases, and 21% were unspecified epithelial neoplasms.

Survival trends

Relative survival from cancer of the small intestine in adults diagnosed during 1986-90 in England and Wales was only moderate: about 50% at one year and 25% at five years (Table 14.2). The corresponding figures for patients diagnosed in the early 1970s were about 40% and 20%. Ten-year survival has remained at around 20% in most regions (Table 14.2, Figure 14.1).

One-year survival differs by up to 20% between regions and by up to 10% at five years: survival in Anglia & Oxford has generally been highest. Survival among women diagnosed in 1986-90 was lower in North & West than the average for England and Wales (Table 14.2, Figure 14.4).

Survival declines with age, more sharply in men than women (Figure 14.5). Five-year relative survival for men diagnosed during 1986-90 was about 40% under age 50 at diagnosis but less than 10% for those aged 80-99, whereas survival in elderly women remained at around 20% (see Table 4.7).

After adjustment for age, one-year relative survival improved slightly more in men than women, and it is now the same in both sexes (50%). The average gain in successive quinquennia was 3% for women and almost 5% for men (see Table 4.4). The small (1%) average improvements in five-year survival in successive calendar periods were not significant in either men or women.

Deprivation

There was little difference in survival between deprivation categories during the 1970s. The gap in one-year survival widened to 10% by the early 1980s. For some 1,500 patients of known deprivation status who were diagnosed during 1986-90, the gap in one-year survival between the most affluent and most deprived groups was 20%, one of

Table 14A Sub-site distribution of cancers of the small intestine

ICD-9 code	Site description	1971-75 No.	%	1976-80 No.	%	1981-85 No.	%	1986-90 No.	%	Total No.	%
152.0	Duodenum	327	25.3	340	24.9	373	25.5	426	28.6	**1,466**	**26.1**
152.1	Jejunum[1]	-	-	76	5.6	182	12.4	158	10.6	**416**	**7.4**
152.2	Ileum[1]	-	-	139	10.2	362	24.7	314	21.1	**815**	**14.5**
152.3	Meckel's diverticulum[1]	-	-	3	0.2	9	0.6	8	0.5	**20**	**0.4**
152.8	Other	564	43.6	359	26.3	31	2.1	21	1.4	**975**	**17.4**
152.9	Small intestine, unspecified	403	31.1	448	32.8	508	34.7	561	37.7	**1,920**	**34.2**
	Total	**1,294**		**1,365**		**1,465**		**1,488**		**5,612**	

[1] From 1971 to 1978, jejunum, ileum and Meckel's diverticulum were included in the ICD-8 rubric 152.8.

the largest survival gradients for any adult cancer (see Table 4.9). The gap in five-year survival between deprivation groups was less pronounced (6-7%), but it was significant for patients diagnosed during 1986-90 (Figure 14.3).

International comparison

Five-year survival for patients diagnosed with cancer of the small bowel in England and Wales during 1986-90 (26% for men and 23% for women) was some 14% lower than the average for men in Europe, and 9% lower for women. The highest regional survival rates were similar to the rates in Scotland and to the European average. Survival rates in the USA were more than twice as high as in England and Wales (see Table 4.12).

Table 14.1 Ineligible and excluded records, and patients included in analyses, by calendar period of diagnosis

	Calendar period of diagnosis																	
	1971-75			1976-80			1981-85			1986-90			Overall		
Total registered	**1,641**			**1,715**			**1,762**			**1,854**			**6,972**		
Ineligible	Records	Patients	%	Records	Patients	%	Records	Patients	%	Records	Patients	%	Records	Patients	%
Incomplete data[1]	4	4		11	11		34	34		34	34		83	**83**	
Resident outside England and Wales	12	12		28	28		11	11		8	6		59	**57**	
In situ neoplasm[2]	3	3		0	0		0	0		0	0		3	**3**	
Benign or uncertain behaviour[2]	93	87		70	67		1	1		0	0		164	**155**	
Metastatic[2]	25	24		12	12		0	0		1	1		38	**37**	
Otherwise ineligible[3]	-	-		-	-		-	-		-	-		-	**-**	
Lymphoma[4]	67	66		39	33		11	0		13	0		130	**99**	
Leukaemia or myeloma[4]	0	0		1	1		0	0		0	0		1	**1**	
		196			152			46			41			435	
Total eligible		**1,445**	**100**		**1,563**	**100**		**1,716**	**100**		**1,813**	**100**		**6,537**	**100**
Aged under 15 or 100 years or over at diagnosis	2	0	0.0	8	3	0.2	1	0	0.0	5	5	0.3	16	**8**	0.1
Vital status unknown[5]	73	58	4.0	89	64	4.1	79	70	4.1	108	103	5.7	349	**295**	4.5
Duplicate registration[6]	0	0	0.0	0	0	0.0	0	0	0.0	0	0	0.0	0	**0**	0.0
Multiple primary[7]	0	0	0.0	1	1	<0.1	6	6	0.3	5	4	0.2	12	**11**	0.2
Synchronous tumours[8]	3	1	<0.1	4	3	0.2	11	10	0.6	8	5	0.3	26	**19**	0.3
Sex not known	0	0	0.0	0	0	0.0	0	0	0.0	0	0	0.0	0	**0**	0.0
Sex incompatible with site code[9]	-	-	-	-	-	-	-	-	-	-	-	-	-	**-**	-
Invalid dates[10]	0	0	0.0	0	0	0.0	0	0	0.0	2	1	<0.1	2	**1**	<0.1
Zero survival[11]	113	92	6.4	147	127	8.1	191	165	9.6	219	207	11.4	670	**591**	9.0
Total exclusions		**151**	**10.4**		**198**	**12.7**		**251**	**14.6**		**325**	**17.9**		**925**	**14.2**
Patients included in analysis		**1,294**	**89.6**		**1,365**	**87.3**		**1,465**	**85.4**		**1,488**	**82.1**		**5,612**	**85.8**

[1] Main data item(s) invalid or incompatible with one another: name, sex, date of birth, date of diagnosis and (if present) date of death, postcode, site, morphology and behaviour

[2] Tumour either *in situ* (behaviour code 2), benign (0), uncertain if benign or malignant (1) or metastatic (6 or 9)

[3] Not applicable for this malignancy

[4] Morphology that of lymphoma (included with non-Hodgkin lymphoma, Ch. 41) or myeloma (included with multiple myeloma, Ch. 43), or leukaemia (excluded)

[5] Tracing of vital status at National Health Service Central Register incomplete at 6 July 1997

[6] Same site code, sex, cancer registry and cancer registry number as an earlier registration

[7] Same site code and person number as an earlier registration(s): mostly confirmed multiple primary tumours at this site, some unresolved duplicate registrations

[8] Same site code, sex, date of birth and date of diagnosis as another registration(s): mostly synchronous or (in paired organs) bilateral tumours in same anatomic site in one individual, not previously linked: also some duplicate registrations

[9] Not applicable for this malignancy

[10] Impossible sequence of dates (birth, diagnosis and, if present, emigration or death); or date of death after 6 July 1997

[11] Date of diagnosis same as date of death: some are patients who did die on the day of diagnosis, but most were registered solely from a death certificate (death certificate only, or DCO), and their survival time is thus unknown

Table 14.2   Crude and relative survival (%) at one, five and ten years since diagnosis: NHS Region, deprivation category and calendar period of diagnosis

	1971-75												1976-80											
	Affluent		2		3		4		Deprived		All[1]		Affluent		2		3		4		Deprived		All[1]	
	C	R	C	R	C	R	C	R	C	R	C	R	C	R	C	R	C	R	C	R	C	R	C	R
ENGLAND & WALES																								
Men	*(84)*		*(125)*		*(121)*		*(125)*		*(90)*		*(684)*		*(121)*		*(156)*		*(136)*		*(139)*		*(103)*		*(679)*	
One year	35	37	39	41	40	41	34	35	35	37	37	38	47	49	45	47	41	42	36	38	35	37	41	43
Five years	13	15	22	27	17	22	24	28	14	17	18	22	20	23	22	27	20	24	17	22	11	14	18	23
Ten years	8	13	16	27	8	15	17	23	8	14	12	19	13	18	15	24	11	18	10	20	8	12	12	20
Women	*(102)*		*(103)*		*(105)*		*(86)*		*(90)*		*(610)*		*(124)*		*(141)*		*(147)*		*(130)*		*(113)*		*(686)*	
One year	46	47	46	48	49	51	31	32	40	42	42	44	40	41	39	40	41	43	44	46	44	46	41	42
Five years	16	18	20	24	20	24	16	19	15	18	18	21	18	20	20	23	24	28	22	25	21	24	20	23
Ten years	15	17	14	21	15	23	12	16	12	15	13	19	12	16	14	20	15	19	15	20	16	23	14	19
Adults	*(186)*		*(228)*		*(226)*		*(211)*		*(180)*		*(1,294)*		*(245)*		*(297)*		*(283)*		*(269)*		*(216)*		*(1,365)*	
One year	41	43	42	44	44	46	33	34	38	39	39	41	43	45	42	43	41	43	40	42	40	42	41	43
Five years	15	17	21	26	19	23	21	24	15	18	18	22	19	22	21	25	22	26	19	23	16	19	19	23
Ten years	12	16	15	25	11	19	15	20	10	14	13	19	13	17	15	21	13	18	12	20	12	19	13	19

	1971-75						1976-80					
	Men		Women		Adults		Men		Women		Adults	
	C	R	C	R	C	R	C	R	C	R	C	R
Northern & Yorkshire	*(69)*		*(65)*		*(134)*		*(97)*		*(77)*		*(174)*	
One year	40	42	39	41	40	41	36	38	36	38	36	38
Five years	19	23	15	18	17	21	18	23	24	27	20	25
Ten years	11	15	7	12	9	14	11	20	18	25	14	22
Trent	*(86)*		*(50)*		*(136)*		*(65)*		*(61)*		*(126)*	
One year	37	39	57	58	44	46	48	50	43	44	46	47
Five years	16	20	24	28	19	24	19	24	24	26	21	26
Ten years	9	14	18	24	12	19	13	19	14	19	13	20
Anglia & Oxford	*(65)*		*(59)*		*(124)*		*(61)*		*(66)*		*(127)*	
One year	46	48	54	56	50	52	49	51	39	40	44	45
Five years	18	21	22	25	20	23	25	28	19	21	22	24
Ten years	11	17	18	24	14	20	17	24	11	14	14	19
North Thames	*(64)*		*(61)*		*(125)*		*(72)*		*(51)*		*(123)*	
One year	39	41	39	40	39	40	42	44	31	32	37	39
Five years	23	27	14	16	18	21	18	21	12	14	15	18
Ten years	-	-	9	14	15	21	11	20	7	10	9	16
South Thames	*(99)*		*(112)*		*(211)*		*(97)*		*(134)*		*(231)*	
One year	40	42	36	38	38	40	36	38	43	44	40	42
Five years	19	23	12	15	15	18	15	18	24	27	20	24
Ten years	13	22	12	15	12	18	7	12	15	20	11	17
South & West	*(83)*		*(67)*		*(150)*		*(95)*		*(92)*		*(187)*	
One year	30	32	43	45	36	38	47	49	44	46	46	48
Five years	21	25	15	18	19	22	26	33	21	24	23	28
Ten years	13	20	11	16	12	18	18	32	16	23	17	27
West Midlands	*(67)*		*(56)*		*(123)*		*(70)*		*(67)*		*(137)*	
One year	42	44	42	43	42	43	45	47	51	53	48	50
Five years	19	23	23	28	21	25	21	25	25	28	23	27
Ten years	16	23	9	13	13	18	15	23	20	27	18	26
North & West	*(103)*		*(104)*		*(207)*		*(82)*		*(94)*		*(176)*	
One year	32	33	38	39	35	36	25	27	35	36	30	32
Five years	15	19	22	25	19	22	5	7	12	15	9	11
Ten years	9	15	18	22	13	19	2	4	9	13	6	9
WALES	*(48)*		*(36)*		*(84)*		*(40)*		*(44)*		*(84)*	
One year	24	26	45	47	33	34	45	48	44	46	45	47
Five years	13	16	23	26	17	21	24	32	18	22	21	26
Ten years	7	15	21	26	13	21	17	32	9	11	13	19

[1] Including those for whom no deprivation category was assigned. C - crude survival, R - relative survival (see Chapter 3).
Number of patients contributing to each analysis in parentheses.

1981-85						1986-90						
Affluent	2	3	4	Deprived	All[1]	Affluent	2	3	4	Deprived	All[1]	
C R	C R	C R	C R	C R	C R	C R	C R	C R	C R	C R	C R	

ENGLAND & WALES

Men

(147)	(144)	(159)	(147)	(118)	(739)	(160)	(156)	(137)	(145)	(127)	(732)	
48 50	42 44	40 42	38 40	43 46	42 45	58 60	57 59	48 50	43 46	36 38	49 52	*One year*
23 27	16 20	16 21	13 16	19 24	18 22	25 31	20 24	19 24	20 25	19 25	21 26	*Five years*
15 21	13 18	8 13	8 14	15 23	12 18							

Women

(141)	(165)	(156)	(130)	(114)	(726)	(152)	(157)	(176)	(140)	(125)	(756)	
56 59	53 55	46 48	48 51	42 44	49 51	48 50	54 56	56 58	43 44	31 33	48 49	*One year*
27 32	23 27	25 29	23 27	20 23	23 27	19 22	23 27	28 33	14 17	10 14	20 23	*Five years*
17 24	13 18	12 19	11 16	15 21	13 19							

Adults

(288)	(309)	(315)	(277)	(232)	(1,465)	(312)	(313)	(313)	(285)	(252)	(1,488)	
52 54	48 50	43 45	43 45	43 45	45 48	54 56	55 57	53 55	43 45	34 35	48 50	*One year*
25 29	20 24	21 25	17 21	19 23	20 24	22 26	22 26	24 29	17 21	15 19	20 25	*Five years*
16 22	13 18	10 16	10 14	15 22	13 19							

Men	Women	Adults				Men	Women	Adults				

Northern & Yorkshire

(95)	(87)	(182)				(93)	(86)	(179)				
42 44	41 44	42 44				50 52	48 50	49 51				*One year*
21 26	29 33	25 30				24 32	20 23	22 28				*Five years*
14 23	11 16	13 20										

Trent

(63)	(48)	(111)				(71)	(83)	(154)				
41 44	*39 41*	*41 43*				*48 51*	*42 44*	*45 47*				*One year*
18 21	*19 21*	*18 22*				*21 27*	*16 20*	*18 23*				*Five years*
15 20	*13 20*	*14 21*										

Anglia & Oxford

(77)	(77)	(154)				(89)	(88)	(177)				
43 45	63 64	53 55				58 60	57 59	58 60				*One year*
21 26	26 32	24 29				24 29	23 28	24 28				*Five years*
17 23	13 19	15 22										

North Thames

(35)	(58)	(93)				(61)	(65)	(126)				
26 28	*52 54*	*42 44*				*47 49*	*52 53*	*49 51*				*One year*
9 10	*25 32*	*19 24*				*16 20*	*22 25*	*19 23*				*Five years*
6 7	*19 30*	*14 23*										

South Thames

(88)	(79)	(167)				(65)	(85)	(150)				
44 45	42 44	43 45				*53 56*	*56 59*	*55 57*				*One year*
18 23	10 12	14 17				*22 27*	*20 24*	*21 25*				*Five years*
10 17	- -	*10 15*										

South & West

(86)	(104)	(190)				(110)	(105)	(215)				
43 46	55 58	50 53				47 49	49 50	48 50				*One year*
17 22	28 33	23 28				21 26	22 25	22 26				*Five years*
13 18	16 23	15 21										

West Midlands

(107)	(113)	(220)				(110)	(98)	(208)				
51 54	48 50	50 52				49 51	46 48	47 50				*One year*
23 29	28 31	26 30				22 27	21 25	21 26				*Five years*
15 24	18 24	17 24										

North & West

(120)	(101)	(221)				(100)	(103)	(203)				
40 42	49 51	44 46				45 47	31 32	38 39				*One year*
14 17	24 29	18 23				18 23	12 15	15 19				*Five years*
6 9	11 17	9 13										

WALES

(68)	(59)	(127)				(33)	(43)	(76)				
39 42	*41 43*	*40 43*				*45 48*	*54 57*	*50 53*				*One year*
12 16	*8 9*	*10 13*				*18 23*	*25 30*	*22 27*				*Five years*
12 16	*5 6*	*9 11*										

[1] Including those for whom no deprivation category was assigned. C - crude survival, R - relative survival (see Chapter 3).
Number of patients contributing to each analysis in parentheses.

Fig. 14.1   Relative survival up to ten years: trends by calendar period of diagnosis (1971-90) and NHS Region

ADULTS

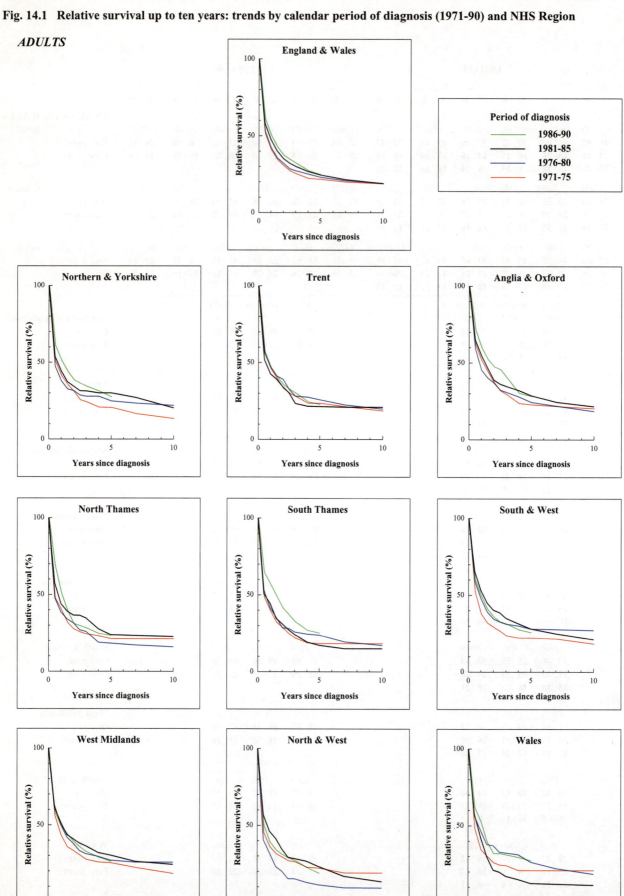

Fig. 14.2 Relative survival up to five years, by deprivation category: England and Wales, patients diagnosed 1986-90

ADULTS

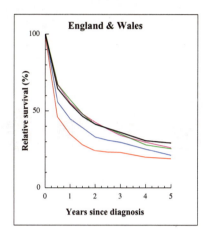

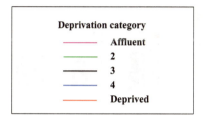

Fig. 14.3 Relative survival at five years, by deprivation category and period of diagnosis (1981-90): England and Wales

ADULTS
(Note vertical scale)

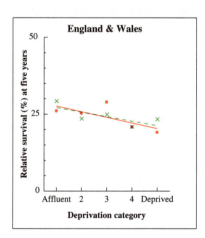

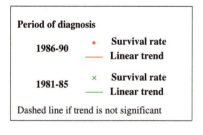

Fig. 14.4 Relative survival (%) by NHS Region and sex:
  one-year and five-year survival for patients diagnosed 1986-90, with 95% confidence intervals

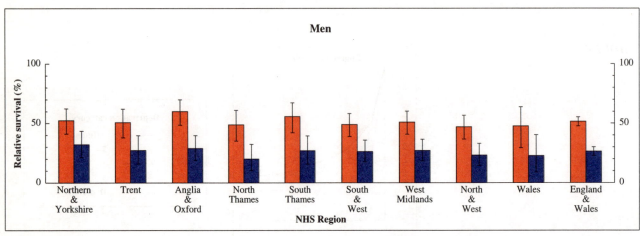

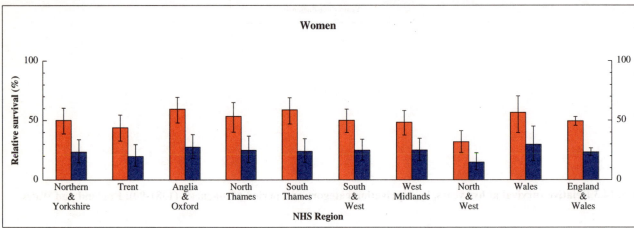

Fig. 14.5 Relative survival (%) at five years, with 95% confidence interval, by age at diagnosis and sex:
  England and Wales, patients diagnosed 1986-90

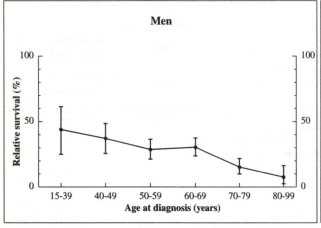

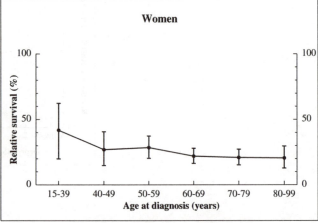

Chapter 15

COLON

Colon cancer accounts for about 9% of all malignant tumours in adults, almost one in 11 cancers. Some 17,000 new cases are diagnosed each year in England and Wales, and it is the third most common cancer in both sexes combined. Incidence in men has been increasing at about 1% a year for the last 25 years, less quickly than in other countries in western Europe, but rates in women have changed very little. Incidence in men is about 50% greater than in women. Mortality has been falling at about 2% every five years in men and 5% in women. The increase in incidence is more marked in the elderly, while the decline in mortality is more marked at younger ages. The consistency of these patterns with those in other developed countries strongly suggests the influence of earlier diagnosis and more successful treatment[1].

Over 300,000 patients were registered with cancer of the colon in England and Wales during the period 1971-90, the average number each year rising from 13,000 to 17,000 (Table 15.1). Some 3,700 patients were considered ineligible for analysis because the tumour had been assigned to the intestinal tract without further specification (see Chapter 2). In a further 2,200 patients, mainly diagnosed in the 1970s, the tumour was *in situ*, benign, of uncertain behaviour, or apparently metastatic in the colon from another primary site.

The survival analyses here relate to almost 250,000 patients with colon cancer diagnosed in England and Wales during the period 1971-90, and followed up to the end of 1995. They represent about 85% of those eligible (Table 15.1). About 4% of patients were excluded because their vital status was unknown at 6 July 1977, and 11% because their survival was zero or unknown. The proportion of patients in this category increased from 9% to 13% in successive five-year periods from 1971-75 to 1986-90. Some patients with zero survival will in fact have died on the day of diagnosis, for example after presenting with an abdominal emergency; others will have been recorded solely from a death certificate. The two categories cannot be distinguished from the data available in the national cancer registry for patients diagnosed before 1993. Among patients with colon cancer diagnosed in 1993, however, 36% of the 1,400 patients for whom the dates of diagnosis and death were the same were *not* recorded solely from a death certificate, and they must be assumed to have died on the day of diagnosis (see Table 2.9). Even so, most patients in this category would still have been recorded solely from a death certificate, and their true survival is thus unknown. There is evidence that patients recorded solely from a death certificate tend to have shorter than average survival[2,3]. The exclusion of

some patients with survival that is either shorter than average or actually zero would tend to increase the estimate of survival.

Almost a third of the tumours arose in the sigmoid colon, and about a quarter in the transverse colon and flexures (Table 15A). About 60% were adenocarcinomas and another 30% were classified as unspecified carcinoma.

Survival trends

Relative survival for patients diagnosed in England and Wales during 1986-90 was 60% at one year and 39% at five years. For those diagnosed during 1971-75, an average of 15 years earlier, the corresponding survival rates were 45% and 26% (Table 15.2).

There has been a clear and substantial improvement in national survival rates for patients diagnosed in each successive five-year period (Figure 15.1). Nationally, and in most regions, the improvement over time was more marked between the late 1970s and the early 1980s. Gains have occurred in all the NHS Regions and Wales, although in several regions there was no apparent improvement in five-year survival for patients diagnosed during 1986-90 compared with those diagnosed during 1981-85.

The average improvement in one-year survival between successive quinquennia was 7%. The corresponding average gain in five-year survival was 5% (see Table 4.4). These estimates of the trend in survival are adjusted for changes in the age distribution of colon cancer patients over the 20 years 1971-90, and for the precision of survival estimates in each period. The effect of age-adjustment on the estimate of survival trends can be seen by comparison with the data in Table 15.2. The unadjusted net gain in five-year survival between 1971-75 and 1986-90 was 13% for men and 14% for women, compared to 16% after adjustment.

The trends in survival differ quite markedly between NHS Regions. One-year survival has improved much less since the early 1970s in Wales (about 4% increase every five years) than in England (7%) (see Table-Figure 4.5C). Trends have been higher than average in South Thames (8-9%). Five-year survival in South & West and the Thames regions was about 5% higher than the national average for patients diagnosed during 1986-90 (Figure 15.4).

The effect of age on survival is less marked than for many major tumours (Figure 15.5). Five-year survival for patients diagnosed during 1986-90 is in the range 40% to

Table 15A Sub-site distribution of colon cancer[1]

ICD-9 code	Site description	1971-75 No.	%	1976-80 No.	%	1981-85 No.	%	1986-90 No.	%	Total No.	%
153.0	Hepatic flexure	14,165	25.4	10,441	17.3	1,744	2.7	1,654	2.4	**28,004**	**11.2**
153.1	Transverse colon	8,254	14.8	7,213	12.0	4,823	7.4	4,568	6.7	**24,858**	**10.0**
153.2	Descending colon	4,094	7.3	3,753	6.2	3,474	5.3	2,921	4.3	**14,242**	**5.7**
153.3	Sigmoid colon	16,413	29.5	18,303	30.4	19,858	30.5	20,136	29.4	**74,710**	**29.9**
153.4	Caecum	-	-	4,604	7.6	12,989	19.9	13,451	19.6	**31,044**	**12.4**
153.5	Appendix	-	-	143	0.2	348	0.5	318	0.5	**809**	**0.3**
153.6	Ascending colon	-	-	1,949	3.2	5,105	7.8	5,052	7.4	**12,106**	**4.8**
153.7	Splenic flexure	-	-	1,069	1.8	2,359	3.6	1,996	2.9	**5,424**	**2.2**
153.8	Other	70	0.1	192	0.3	500	0.8	601	0.9	**1,363**	**0.5**
153.9	Unspecified	12,727	22.8	12,621	20.9	13,990	21.5	17,784	26.0	**57,122**	**22.9**
	Total	**55,723**		**60,288**		**65,190**		**68,481**		**249,682**	

[1] The 70 tumours coded to NCR special code 153.7 (multiple parts) during 1971-75 have been assigned to ICD-9 153.8 (other) for the purposes of this table (see Chapter 2). Tumours of the splenic flexure were coded with tumours of the transverse and hepatic flexures (ICD-8 153.1) during 1971-78; tumours of caecum, appendix and ascending colon were coded together in ICD-8 153.0.

50% at all ages under 80 years at diagnosis, and about 30% for patients aged 80-99 (see Table 4.7).

Deprivation

Survival has been lower for patients in the more deprived groups since the early 1970s, although the gap in five-year survival does not usually exceed 5% (Figure 15.2).

There was a significant deprivation gradient in five-year survival for patients diagnosed in England and Wales during the 1980s (Figure 15.3). Among more than 60,000 patients of known deprivation status who were diagnosed during 1981-85, the survival deficit for patients in the most deprived group, compared with the most affluent, was about 7% at one year and 5% at five years. For a similar number of patients diagnosed during 1986-90, the deprivation gaps were 6% and 4% (see Table 4.9). The gradient is echoed in most regions and is statistically significant in several of them (Figure 15.3). In Wales, the difference for patients diagnosed in 1986-90 was about 10% in both sexes (Table 15.2). Similar deprivation gradients have been reported from Finland[4] and France[5].

International comparison

Five-year survival is similar to that in Scotland, but 8% lower than the average for Europe (see Table 4.12). Detailed comparisons have been published[6]. The highest regional survival rates (South & West) are similar to the European average, but still almost 20% lower than in the USA. In Sweden, survival for patients diagnosed during 1988-90 was 74% at one year and 55% at five years, more than 10% higher than in England and Wales[7].

1.   Coleman MP, Estève J, Damiecki P, Arslan A, Renard H. *Trends in cancer incidence and mortality (IARC Scientific Publications No. 121).* Lyon: International Agency for Research on Cancer, 1993

2.   Pollock AM, Vickers N. The impact on colorectal cancer survival of cases registered by 'death certificate only': implications for national survival rates. Br J Cancer 1994; 70: 1229-1231

3.   Berrino F, Estève J, Coleman MP. Basic issues in the estimation and comparison of cancer patient survival. In: Berrino F, Sant M, Verdecchia A, Capocaccia R, Hakulinen T, Estève J, (eds.) *Survival of cancer patients in Europe: the EUROCARE study. (IARC Scientific Publications No. 132).* Lyon: International Agency for Research on Cancer, 1995, pp1-14

4.   Auvinen A. Social class and colon cancer survival in Finland. Cancer 1992; 70: 402-409

5.   Monnet E, Boutron M-C, Faivre J, Milan C. Influence of socioeconomic status on prognosis of colorectal cancer. A population-based study in Côte d'Or. Cancer 1993; 72: 1165-1170

6.   Sant M, Capocaccia R, Verdecchia A, Gatta G, Micheli A, Mariotto A, Hakulinen T, Berrino F, EUROCARE Working Group. Comparisons of colon cancer survival among European countries: the EUROCARE study. Int J Cancer 1995; 63: 43-48

7.   Blomqvist P, Ekbom A, Nyrén O, Krusemo UB, Bergström R, Adami H-O. Survival after colon cancer 1973-1990 in Sweden. Convergence between catchment areas. Ann Surg 1997; 225: 208-216

Table 15.1 Ineligible and excluded records, and patients included in analyses, by calendar period of diagnosis

	Calendar period of diagnosis														
	1971-75			**1976-80**			**1981-85**			**1986-90**			**Overall**		
Total registered	**66,139**			**71,819**			**79,510**			**86,489**			**303,957**		
Ineligible	*Records*	*Patients*	*%*	*Records*	*Patients*	*%*	*Records*	*Patients*	*%*	*Records*	*Patients*	*%*	*Records*	*Patients*	*%*
Incomplete data[1]	110	110		194	194		198	198		199	199		701	**701**	
Resident outside England and Wales	564	562		425	416		326	282		348	266		1,663	**1,526**	
In situ neoplasm[2]	42	40		9	9		0	0		0	0		51	**49**	
Benign or uncertain behaviour[2]	818	556		519	364		6	6		2	2		1,345	**928**	
Metastatic[2]	821	816		368	366		1	1		14	14		1,204	**1,197**	
Otherwise ineligible[3]	632	569		825	767		976	964		1,410	1,403		3,843	**3,703**	
Lymphoma[4]	36	36		26	26		12	0		10	0		84	**62**	
Leukaemia or myeloma[4]	0	0		0	0		1	0		0	0		1	**0**	
		2,689			**2,142**			**1,451**			**1,884**			**8,166**	
Total eligible		**63,450**	**100**		**69,677**	**100**		**78,059**	**100**		**84,605**	**100**		**295,791**	**100**
Aged under 15 or 100 years or over at diagnosis	66	49	<0.1	47	35	<0.1	57	54	<0.1	82	78	<0.1	252	**216**	<0.1
Vital status unknown[5]	2,089	1,803	2.8	2,866	2,606	3.7	3,254	3,039	3.9	4,590	4,366	5.2	12,799	**11,814**	4.0
Duplicate registration[6]	0	0	0.0	2	0	0.0	0	0	0.0	3	3	<0.1	5	**3**	<0.1
Multiple primary[7]	6	1	<0.1	43	36	<0.1	268	225	0.3	403	346	0.4	720	**608**	0.2
Synchronous tumours[8]	309	220	0.3	271	178	0.3	377	297	0.4	552	397	0.5	1,509	**1,092**	0.4
Sex not known	0	0	0.0	0	0	0.0	0	0	0.0	0	0	0.0	0	**0**	0.0
Sex incompatible with site code[9]	-	-	-	-	-	-	-	-	-	-	-	-	-	**-**	-
Invalid dates[10]	5	3	<0.1	10	4	<0.1	24	13	<0.1	43	26	<0.1	82	**46**	<0.1
Zero survival[11]	6,300	5,651	8.9	7,147	6,530	9.4	10,325	9,241	11.8	12,201	10,908	12.9	35,973	**32,330**	10.9
Total exclusions		**7,727**	**12.2**		**9,389**	**13.5**		**12,869**	**16.5**		**16,124**	**19.1**		**46,109**	**15.6**
Patients included in analysis		**55,723**	**87.8**		**60,288**	**86.5**		**65,190**	**83.5**		**68,481**	**80.9**		**249,682**	**84.4**

[1] Main data item(s) invalid or incompatible with one another: name, sex, date of birth, date of diagnosis and (if present) date of death, postcode, site, morphology and behaviour

[2] Tumour either *in situ* (behaviour code 2), benign (0), uncertain if benign or malignant (1) or metastatic (6 or 9)

[3] Intestinal tract, part unspecified (ICD-8 153.9 for period 1971-78; ICD-9 159.0 for period 1979-90)

[4] Morphology that of lymphoma (included with non-Hodgkin lymphoma, Ch. 41) or myeloma (included with multiple myeloma, Ch. 43), or leukaemia (excluded)

[5] Tracing of vital status at National Health Service Central Register incomplete at 6 July 1997

[6] Same site code, sex, cancer registry and cancer registry number as an earlier registration

[7] Same site code and person number as an earlier registration(s): mostly confirmed multiple primary tumours at this site, some unresolved duplicate registrations

[8] Same site code, sex, date of birth and date of diagnosis as another registration(s): mostly synchronous or (in paired organs) bilateral tumours in same anatomic site in one individual, not previously linked: also some duplicate registrations

[9] Not applicable for this malignancy

[10] Impossible sequence of dates (birth, diagnosis and, if present, emigration or death); or date of death after 6 July 1997

[11] Date of diagnosis same as date of death: some are patients who did die on the day of diagnosis, but most were registered solely from a death certificate (death certificate only, or DCO), and their survival time is thus unknown

**Table 15.2   Crude and relative survival (%) at one, five and ten years since diagnosis:
NHS Region, deprivation category and calendar period of diagnosis**

	1971-75												1976-80											
MEN	*Affluent*		*2*		*3*		*4*		*Deprived*		*All[1]*		*Affluent*		*2*		*3*		*4*		*Deprived*		*All[1]*	
	C	R	C	R	C	R	C	R	C	R	C	R	C	R	C	R	C	R	C	R	C	R	C	R
ENGLAND & WALES	*(3,707)*		*(4,038)*		*(4,188)*		*(3,933)*		*(3,203)*		*(23,747)*		*(5,051)*		*(5,404)*		*(5,437)*		*(5,129)*		*(4,234)*		*(26,356)*	
One year	46	48	44	47	42	45	42	44	39	42	**43**	**45**	50	53	51	54	46	48	46	49	42	45	**47**	**50**
Five years	22	28	22	29	20	26	18	24	17	22	**20**	**26**	26	33	25	33	22	30	22	30	20	27	**23**	**31**
Ten years	14	24	14	26	12	23	12	22	10	20	**13**	**23**	18	30	16	29	14	27	15	27	12	24	**15**	**27**
ENGLAND	*(3,553)*		*(3,834)*		*(3,860)*		*(3,565)*		*(3,012)*		*(22,155)*		*(4,889)*		*(5,136)*		*(5,038)*		*(4,759)*		*(4,025)*		*(24,751)*	
One year	46	48	44	47	42	45	41	44	39	42	**43**	**45**	50	53	51	54	45	48	46	49	43	45	**47**	**50**
Five years	22	28	22	29	20	26	18	24	17	23	**20**	**26**	26	33	25	33	22	30	22	30	20	27	**23**	**31**
Ten years	14	23	15	26	12	23	12	22	11	20	**13**	**23**	18	30	16	29	14	27	15	27	12	24	**15**	**27**
Northern & Yorkshire	*(254)*		*(338)*		*(404)*		*(500)*		*(611)*		*(2,987)*		*(428)*		*(576)*		*(681)*		*(839)*		*(973)*		*(3,587)*	
One year	46	49	42	45	35	37	34	36	34	36	**40**	**43**	46	49	45	48	38	40	40	43	36	39	**42**	**44**
Five years	23	28	17	23	16	23	14	19	13	18	**18**	**24**	25	33	24	32	17	24	20	27	16	22	**20**	**28**
Ten years	13	21	12	21	9	19	8	16	7	14	**11**	**20**	16	28	15	29	12	22	13	24	10	21	**13**	**25**
Trent	*(257)*		*(319)*		*(477)*		*(535)*		*(432)*		*(2,242)*		*(320)*		*(474)*		*(544)*		*(650)*		*(522)*		*(2,579)*	
One year	41	43	40	42	44	47	39	42	34	36	**41**	**44**	45	48	49	52	44	47	44	47	44	48	**46**	**49**
Five years	21	27	19	25	20	25	19	24	15	20	**19**	**25**	21	27	20	27	24	32	22	29	19	26	**22**	**29**
Ten years	14	23	13	21	11	21	12	21	9	18	**12**	**21**	14	25	14	25	14	26	15	27	13	25	**14**	**25**
Anglia & Oxford	*(477)*		*(508)*		*(477)*		*(329)*		*(145)*		*(2,208)*		*(687)*		*(667)*		*(574)*		*(379)*		*(171)*		*(2,590)*	
One year	49	51	45	47	42	45	40	42	49	52	**46**	**49**	50	52	52	55	47	50	49	52	39	42	**50**	**53**
Five years	25	31	23	29	20	26	17	22	22	28	**22**	**28**	25	32	26	34	22	30	25	33	17	23	**25**	**33**
Ten years	16	27	14	24	12	23	11	20	13	21	**14**	**24**	17	30	17	30	13	25	17	29	10	20	**16**	**29**
North Thames	*(516)*		*(503)*		*(484)*		*(455)*		*(350)*		*(2,618)*		*(509)*		*(475)*		*(447)*		*(458)*		*(349)*		*(2,329)*	
One year	49	52	48	51	39	41	47	50	45	48	**48**	**50**	54	57	55	58	51	54	54	58	49	52	**54**	**57**
Five years	25	30	26	34	19	24	22	29	22	29	**24**	**31**	24	30	29	37	26	36	28	36	26	35	**27**	**36**
Ten years	18	28	17	29	12	22	15	27	15	26	**16**	**28**	18	28	19	33	17	33	19	35	15	31	**18**	**32**
South Thames	*(884)*		*(662)*		*(626)*		*(526)*		*(274)*		*(3,231)*		*(1,061)*		*(804)*		*(667)*		*(574)*		*(210)*		*(3,518)*	
One year	41	43	39	42	39	42	41	44	40	43	**42**	**45**	47	50	47	50	43	46	43	46	45	48	**47**	**49**
Five years	17	21	20	26	17	23	17	24	17	23	**19**	**25**	25	31	24	31	20	26	22	30	23	32	**23**	**31**
Ten years	12	19	13	23	9	18	12	22	12	21	**12**	**22**	17	28	14	27	12	25	13	26	12	24	**14**	**27**
South & West	*(647)*		*(885)*		*(753)*		*(383)*		*(137)*		*(3,097)*		*(894)*		*(1,128)*		*(932)*		*(452)*		*(150)*		*(3,685)*	
One year	49	52	45	47	44	47	44	47	36	38	**46**	**49**	53	56	52	55	47	50	53	56	50	54	**52**	**55**
Five years	24	31	22	30	22	29	22	29	17	24	**23**	**31**	29	38	26	35	25	33	28	37	26	36	**27**	**36**
Ten years	14	24	15	28	14	27	14	25	9	18	**15**	**27**	20	36	16	32	16	31	20	35	18	33	**18**	**34**
West Midlands	*(143)*		*(149)*		*(160)*		*(196)*		*(211)*		*(2,291)*		*(446)*		*(439)*		*(558)*		*(594)*		*(626)*		*(2,697)*	
One year	47	50	50	53	52	56	41	44	38	40	**45**	**48**	53	55	50	53	48	52	45	48	47	50	**49**	**52**
Five years	17	24	30	37	26	33	17	23	18	24	**21**	**28**	30	38	26	34	24	32	20	26	23	32	**25**	**33**
Ten years	12	22	21	34	18	33	13	23	13	23	**15**	**26**	20	36	18	31	16	29	12	23	14	27	**16**	**29**
North & West	*(375)*		*(470)*		*(479)*		*(641)*		*(852)*		*(3,481)*		*(544)*		*(573)*		*(635)*		*(813)*		*(1,024)*		*(3,766)*	
One year	35	37	37	39	35	37	33	36	35	38	**37**	**39**	42	44	43	46	36	38	37	40	36	38	**40**	**42**
Five years	17	21	18	23	14	18	12	17	15	19	**16**	**20**	19	24	19	25	17	23	16	22	17	23	**18**	**24**
Ten years	10	17	11	21	9	16	6	14	9	19	**9**	**18**	13	21	13	24	11	21	11	20	10	21	**12**	**22**
WALES	*(154)*		*(204)*		*(328)*		*(368)*		*(191)*		*(1,592)*		*(162)*		*(268)*		*(399)*		*(370)*		*(209)*		*(1,605)*	
One year	44	46	43	45	41	43	43	45	34	37	**42**	**45**	45	48	52	55	47	50	42	44	40	43	**47**	**50**
Five years	27	34	16	21	19	25	20	27	12	16	**19**	**26**	26	32	27	35	22	32	22	30	17	23	**23**	**32**
Ten years	16	27	11	21	14	24	14	26	8	15	**14**	**25**	16	31	19	35	13	30	15	29	12	21	**15**	**30**

[1]   Including those for whom no deprivation category was assigned. C - crude survival, R - relative survival (see Chapter 3).
    Number of patients contributing to each analysis in parentheses.

1981-85

MEN		Affluent C	R	2 C	R	3 C	R	4 C	R	Deprived C	R	All[1] C	R
ENGLAND & WALES	(n)	(6,007)		(6,225)		(6,118)		(5,585)		(4,761)		(29,386)	
	One year	58	61	56	59	54	57	53	56	51	55	54	58
	Five years	30	39	28	37	26	35	26	36	25	35	27	36
		20	34	18	33	17	32	17	33	15	31	18	33
ENGLAND	(n)	(5,812)		(5,977)		(5,730)		(5,184)		(4,515)		(27,755)	
	One year	58	61	56	60	54	57	53	56	51	55	54	58
	Five years	31	39	28	37	27	36	26	35	25	34	27	36
		20	34	18	33	18	33	17	33	15	31	18	33
Northern & Yorkshire	(n)	(554)		(669)		(770)		(900)		(1,104)		(4,061)	
	One year	57	61	56	60	52	56	50	54	49	53	53	57
	Five years	33	43	28	39	27	37	25	35	24	34	27	37
		21	39	19	37	18	36	16	33	15	31	18	34
Trent	(n)	(458)		(568)		(683)		(794)		(520)		(3,043)	
	One year	58	61	57	61	53	57	52	56	51	55	55	58
	Five years	28	38	26	35	26	35	23	33	25	37	26	35
		19	32	16	32	17	34	16	32	13	32	17	32
Anglia & Oxford	(n)	(841)		(820)		(697)		(413)		(181)		(3,014)	
	One year	59	63	54	57	54	58	48	51	51	55	55	59
	Five years	31	40	28	37	28	38	27	36	29	40	29	38
		21	36	18	34	18	35	17	33	20	38	19	34
North Thames	(n)	(511)		(471)		(426)		(413)		(338)		(2,206)	
	One year	58	61	60	64	56	60	60	64	62	67	60	63
	Five years	30	39	28	38	28	37	33	44	35	48	31	41
		20	35	18	33	19	35	21	40	23	48	20	37
South Thames	(n)	(1,126)		(880)		(676)		(541)		(233)		(3,601)	
	One year	59	63	55	58	54	58	58	62	48	52	57	61
	Five years	32	42	27	38	26	36	27	39	25	36	29	39
		21	37	18	36	17	33	15	35	17	33	18	34
South & West	(n)	(1,100)		(1,308)		(1,152)		(492)		(183)		(4,352)	
	One year	59	62	57	61	57	61	57	61	53	56	58	62
	Five years	31	41	29	40	29	40	30	42	24	36	30	40
		21	38	20	39	19	39	20	42	18	36	20	38
West Midlands	(n)	(533)		(560)		(621)		(745)		(743)		(3,235)	
	One year	58	61	53	57	51	54	49	52	50	54	53	56
	Five years	29	37	28	37	24	32	23	32	24	35	26	35
		19	35	17	32	16	30	16	31	15	33	17	31
North & West	(n)	(689)		(701)		(705)		(886)		(1,213)		(4,243)	
	One year	47	49	51	54	45	48	46	49	45	49	48	51
	Five years	23	31	24	32	21	29	22	31	20	29	22	30
		14	26	16	29	13	26	14	29	12	25	14	26
WALES	(n)	(195)		(248)		(388)		(401)		(246)		(1,631)	
	One year	52	55	52	56	49	52	56	60	50	54	53	57
	Five years	23	32	25	32	19	27	30	42	27	38	26	36
		16	32	17	30	13	27	22	40	16	35	17	33

1986-90

MEN		Affluent C	R	2 C	R	3 C	R	4 C	R	Deprived C	R	All[1] C	R
ENGLAND & WALES	(n)	(6,384)		(7,009)		(6,847)		(6,194)		(4,904)		(31,651)	
	One year	60	63	60	63	56	60	56	59	55	59	58	61
	Five years	31	40	30	39	28	37	28	38	28	38	29	39
ENGLAND	(n)	(6,190)		(6,770)		(6,532)		(5,860)		(4,701)		(30,296)	
	One year	60	63	60	63	57	60	56	60	56	59	58	61
	Five years	31	40	30	39	28	38	28	38	28	38	29	39
Northern & Yorkshire	(n)	(598)		(769)		(774)		(923)		(1,078)		(4,164)	
	One year	58	62	55	59	54	58	54	58	53	58	55	59
	Five years	30	41	28	38	25	36	27	39	27	39	28	37
Trent	(n)	(463)		(673)		(800)		(798)		(688)		(3,432)	
	One year	58	62	61	65	52	57	56	61	54	59	57	61
	Five years	29	39	31	43	26	36	28	40	27	39	29	38
Anglia & Oxford	(n)	(948)		(987)		(750)		(496)		(200)		(3,409)	
	One year	55	59	55	58	56	60	54	59	55	60	56	59
	Five years	30	40	27	37	28	40	25	36	30	39	28	38
North Thames	(n)	(577)		(611)		(638)		(561)		(454)		(2,902)	
	One year	63	67	66	70	60	64	62	67	60	66	63	66
	Five years	34	45	32	44	30	40	30	43	29	42	31	41
South Thames	(n)	(1,086)		(930)		(760)		(596)		(257)		(3,679)	
	One year	65	68	63	67	58	62	58	63	60	64	62	65
	Five years	32	42	31	43	29	41	28	40	31	46	31	41
South & West	(n)	(1,072)		(1,411)		(1,198)		(689)		(232)		(4,622)	
	One year	61	65	63	68	62	67	56	61	58	62	62	65
	Five years	33	44	32	44	31	45	31	45	28	41	32	43
West Midlands	(n)	(681)		(631)		(826)		(885)		(672)		(3,717)	
	One year	60	64	57	61	55	59	56	60	55	59	57	60
	Five years	32	43	29	39	27	38	29	43	27	39	29	39
North & West	(n)	(765)		(758)		(786)		(912)		(1,120)		(4,371)	
	One year	54	58	54	58	50	54	52	56	52	56	53	56
	Five years	26	36	26	36	24	34	25	35	25	37	26	35
WALES	(n)	(194)		(239)		(315)		(334)		(203)		(1,355)	
	One year	53	56	53	56	45	50	47	50	44	48	50	53
	Five years	30	42	32	44	21	31	22	33	25	34	28	37

[1] Including those for whom no deprivation category was assigned. C - crude survival, R - relative survival (see Chapter 3).
Number of patients contributing to each analysis in parentheses.

Table 15.2   Crude and relative survival (%) at one, five and ten years since diagnosis: NHS Region, deprivation category and calendar period of diagnosis

WOMEN		1971-75						1976-80					
		Affluent	2	3	4	Deprived	All[1]	Affluent	2	3	4	Deprived	All[1]
		C R	C R	C R	C R	C R	C R	C R	C R	C R	C R	C R	C R
ENGLAND & WALES		*(4,958)*	*(5,408)*	*(5,565)*	*(5,299)*	*(4,407)*	*(31,976)*	*(6,067)*	*(6,936)*	*(7,075)*	*(6,689)*	*(5,558)*	*(33,932)*
	One year	45 47	44 46	43 45	40 42	39 40	42 44	49 51	46 48	46 48	44 46	43 45	45 48
	Five years	23 27	22 27	22 27	19 24	19 24	21 25	27 32	24 29	24 30	23 28	22 28	24 30
	Ten years	17 25	16 24	16 24	13 21	13 22	15 23	19 30	17 27	17 28	16 26	15 25	17 27
ENGLAND		*(4,756)*	*(5,163)*	*(5,132)*	*(4,860)*	*(4,172)*	*(29,999)*	*(5,839)*	*(6,632)*	*(6,659)*	*(6,211)*	*(5,305)*	*(32,001)*
	One year	45 47	44 46	43 45	40 42	38 40	42 44	49 51	46 49	45 48	44 46	43 45	45 48
	Five years	23 27	23 27	22 26	19 24	19 24	21 25	27 32	24 29	24 29	23 28	22 28	24 29
	Ten years	17 25	16 25	16 24	13 21	13 22	15 23	19 30	17 27	17 27	16 26	15 26	17 27
Northern & Yorkshire		*(340)*	*(461)*	*(553)*	*(640)*	*(819)*	*(3,973)*	*(528)*	*(713)*	*(817)*	*(1,032)*	*(1,335)*	*(4,541)*
	One year	41 43	44 46	37 39	36 38	33 34	39 41	50 53	40 42	43 45	42 44	41 43	44 46
	Five years	17 21	21 25	17 22	16 20	15 18	18 22	26 32	19 23	23 28	22 27	20 26	22 28
	Ten years	12 19	17 24	13 19	10 17	10 16	13 20	18 30	14 23	18 27	15 24	15 24	16 26
Trent		*(322)*	*(420)*	*(550)*	*(725)*	*(592)*	*(2,889)*	*(386)*	*(564)*	*(725)*	*(765)*	*(583)*	*(3,140)*
	One year	40 42	39 41	41 43	37 39	40 42	41 43	49 51	45 47	43 45	45 47	40 42	45 48
	Five years	17 21	20 23	20 24	19 24	19 23	20 24	27 33	24 29	23 28	24 31	22 28	25 31
	Ten years	13 19	14 20	14 21	14 21	14 23	14 22	18 29	18 27	17 26	16 27	16 26	18 28
Anglia & Oxford		*(646)*	*(637)*	*(598)*	*(408)*	*(178)*	*(2,842)*	*(822)*	*(823)*	*(722)*	*(510)*	*(219)*	*(3,248)*
	One year	43 45	46 48	43 46	37 39	42 44	44 46	48 50	44 47	44 47	44 47	43 46	47 49
	Five years	22 27	24 30	21 26	19 23	22 26	22 27	25 31	22 27	22 28	24 28	23 30	24 30
	Ten years	17 25	18 27	16 24	13 20	15 26	16 25	18 29	17 27	17 25	19 26	17 27	18 28
North Thames		*(779)*	*(698)*	*(637)*	*(653)*	*(498)*	*(3,725)*	*(604)*	*(579)*	*(638)*	*(606)*	*(477)*	*(3,030)*
	One year	49 51	48 50	46 48	43 45	42 45	48 50	53 55	50 52	46 48	46 49	49 51	50 52
	Five years	26 31	25 30	25 30	22 27	23 29	25 31	28 34	27 32	25 30	24 30	27 34	27 33
	Ten years	21 31	18 27	19 28	17 26	17 26	19 29	19 30	19 30	17 28	17 28	18 32	19 31
South Thames		*(1,185)*	*(1,038)*	*(935)*	*(784)*	*(399)*	*(4,826)*	*(1,344)*	*(1,280)*	*(992)*	*(862)*	*(336)*	*(5,141)*
	One year	41 43	40 41	38 40	40 43	35 37	41 43	46 48	43 45	40 43	40 43	41 43	44 46
	Five years	19 23	19 23	17 22	19 24	16 21	19 24	25 31	20 25	19 24	21 27	22 28	22 28
	Ten years	14 22	14 20	12 20	13 20	11 19	13 21	18 29	14 22	13 22	14 23	14 25	15 25
South & West		*(787)*	*(1,130)*	*(1,012)*	*(513)*	*(184)*	*(3,999)*	*(966)*	*(1,411)*	*(1,276)*	*(638)*	*(214)*	*(4,711)*
	One year	46 48	45 47	46 48	43 45	39 41	46 48	53 55	49 51	48 50	50 52	47 49	50 53
	Five years	26 30	24 29	25 31	21 27	20 25	25 30	29 36	28 34	25 32	27 34	24 31	28 34
	Ten years	19 28	18 27	18 28	15 23	14 23	17 28	22 34	20 31	19 30	17 29	17 29	20 32
West Midlands		*(165)*	*(213)*	*(229)*	*(286)*	*(291)*	*(3,015)*	*(448)*	*(538)*	*(671)*	*(743)*	*(738)*	*(3,199)*
	One year	52 54	43 46	45 48	40 42	43 45	43 46	46 49	46 49	46 48	41 43	44 46	46 48
	Five years	22 28	18 22	21 26	20 24	21 26	22 27	24 29	24 29	27 32	22 26	22 28	24 30
	Ten years	17 26	15 22	16 25	15 22	16 25	17 25	17 26	19 28	19 29	16 23	15 25	18 27
North & West		*(532)*	*(566)*	*(618)*	*(851)*	*(1,211)*	*(4,730)*	*(741)*	*(724)*	*(818)*	*(1,055)*	*(1,403)*	*(4,991)*
	One year	40 42	35 36	36 38	32 33	33 35	37 38	41 43	42 44	42 44	35 37	37 39	40 43
	Five years	21 25	19 23	18 21	13 17	17 22	18 22	23 28	21 26	21 26	17 22	18 24	21 26
	Ten years	14 21	13 19	13 19	9 16	12 19	12 20	16 25	14 22	14 24	13 21	13 21	14 24
WALES		*(202)*	*(245)*	*(433)*	*(439)*	*(235)*	*(1,977)*	*(228)*	*(304)*	*(416)*	*(478)*	*(253)*	*(1,931)*
	One year	51 53	39 41	43 45	39 41	38 39	43 45	42 44	44 46	47 50	44 47	41 43	47 50
	Five years	23 29	19 23	22 27	21 26	18 22	21 26	24 30	26 32	27 32	23 28	22 25	26 32
	Ten years	14 23	13 20	16 25	13 22	13 20	14 24	19 28	18 27	20 31	16 25	12 19	18 29

[1] Including those for whom no deprivation category was assigned. C - crude survival, R - relative survival (see Chapter 3). Number of patients contributing to each analysis in parentheses.

WOMEN

Survival (%), by deprivation category — C = crude survival, R = relative survival. Number of patients contributing to each analysis in parentheses.

1981-85

Region	Measure	Affluent (C R)	2 (C R)	3 (C R)	4 (C R)	Deprived (C R)	All¹ (C R)
ENGLAND & WALES	(n)	(6,965)	(7,507)	(7,733)	(7,053)	(5,645)	(35,804)
	One year	56 59	54 57	54 57	51 53	49 52	**53 56**
	Five years	31 38	29 36	29 36	27 34	25 33	**28 35**
		23 35	21 33	21 33	19 31	18 30	**20 33**
ENGLAND	(n)	(6,760)	(7,226)	(7,270)*	(6,630)	(5,400)	(34,002)
	One year	56 59	54 57	54 57	50 53	49 52	**53 56**
	Five years	31 38	29 36	29 36	27 34	25 33	**28 36**
		23 35	21 33	21 33	19 31	18 30	**20 33**
Northern & Yorkshire	(n)	(629)	(773)	(906)	(1,138)	(1,305)	(4,839)
	One year	52 56	49 53	51 54	48 51	47 50	**50 54**
	Five years	28 35	26 33	27 34	26 33	23 30	**26 34**
		19 30	20 31	19 31	17 30	16 28	**18 31**
Trent	(n)	(504)	(648)	(808)	(854)	(620)	(3,476)
	One year	55 58	52 55	51 53	45 48	52 55	**51 54**
	Five years	30 37	27 34	25 31	24 29	26 34	**27 33**
		22 34	20 31	18 29	17 27	19 31	**19 31**
Anglia & Oxford	(n)	(963)	(936)	(877)	(540)	(201)	(3,595)
	One year	58 61	55 58	54 57	49 51	57 60	**56 59**
	Five years	32 40	31 38	30 38	29 37	33 41	**31 39**
		23 36	22 36	22 35	18 31	26 40	**22 36**
North Thames	(n)	(634)	(635)	(582)	(536)	(385)	(2,833)
	One year	62 65	57 60	61 64	58 61	62 66	**61 64**
	Five years	35 43	31 37	34 42	31 38	34 45	**33 42**
		25 38	21 33	24 36	20 33	22 39	**23 37**
South Thames	(n)	(1,385)	(1,158)	(1,023)	(760)	(294)	(4,825)
	One year	54 57	55 58	52 55	52 55	49 52	**54 57**
	Five years	30 37	31 38	26 34	29 37	28 36	**29 37**
		22 34	22 36	18 31	21 35	20 34	**21 35**
South & West	(n)	(1,213)	(1,550)	(1,344)	(680)	(215)	(5,156)
	One year	59 62	56 59	55 59	56 59	56 59	**57 60**
	Five years	34 42	29 36	31 39	31 39	33 41	**32 40**
		24 39	20 33	23 37	22 35	23 38	**23 37**
West Midlands	(n)	(615)	(654)	(775)	(875)	(866)	(3,818)
	One year	57 60	50 53	53 57	48 51	46 49	**52 55**
	Five years	30 36	27 33	29 36	25 31	24 31	**27 34**
		23 34	18 30	20 32	19 30	18 29	**20 32**
North & West	(n)	(817)	(872)	(955)	(1,247)	(1,514)	(5,460)
	One year	47 50	50 53	46 49	44 47	43 46	**47 50**
	Five years	26 31	25 31	25 31	23 29	21 27	**24 30**
		18 28	17 28	18 29	16 27	14 24	**17 28**
WALES	(n)	(205)	(281)	(463)	(423)	(245)	(1,802)
	One year	52 55	55 59	52 55	53 56	43 45	**53 56**
	Five years	27 32	30 38	29 36	29 35	23 28	**28 36**
		21 32	20 34	20 33	20 34	16 26	**20 34**

1986-90

Region	Measure	Affluent (C R)	2 (C R)	3 (C R)	4 (C R)	Deprived (C R)	All¹ (C R)
ENGLAND & WALES	(n)	(7,189)	(8,168)	(8,230)	(7,305)	(5,511)	(36,830)
	One year	58 61	58 61	57 60	55 58	53 56	**56 59**
	Five years	33 41	33 41	31 39	29 37	28 36	**31 39**
ENGLAND	(n)	(6,949)	(7,871)	(7,820)	(6,905)	(5,295)	(35,183)
	One year	58 61	58 61	58 61	55 58	53 56	**57 60**
	Five years	33 41	33 41	31 39	29 37	28 36	**31 39**
Northern & Yorkshire	(n)	(717)	(910)	(1,014)	(1,032)	(1,193)	(4,891)
	One year	54 58	53 56	51 54	52 55	52 55	**53 56**
	Five years	32 40	28 35	28 36	27 35	27 35	**29 37**
Trent	(n)	(515)	(735)	(870)	(867)	(734)	(3,732)
	One year	56 59	60 63	56 59	54 57	52 55	**56 60**
	Five years	30 36	34 42	30 38	29 37	27 35	**31 39**
Anglia & Oxford	(n)	(1,073)	(1,125)	(915)	(624)	(215)	(3,984)
	One year	56 59	51 54	54 57	53 56	41 44	**54 57**
	Five years	31 38	31 38	30 38	27 34	23 29	**30 38**
North Thames	(n)	(665)	(783)	(762)	(702)	(572)	(3,584)
	One year	60 62	65 68	64 67	60 63	63 66	**63 66**
	Five years	33 40	35 42	33 41	34 41	34 42	**34 42**
South Thames	(n)	(1,227)	(1,137)	(1,025)	(756)	(322)	(4,554)
	One year	61 64	62 66	60 63	61 64	60 64	**61 65**
	Five years	33 41	34 43	32 41	32 41	32 41	**33 42**
South & West	(n)	(1,146)	(1,568)	(1,456)	(893)	(291)	(5,385)
	One year	64 67	59 61	61 64	59 62	60 64	**61 64**
	Five years	37 46	35 44	35 44	32 41	35 43	**35 45**
West Midlands	(n)	(667)	(703)	(818)	(962)	(718)	(3,899)
	One year	54 57	60 63	57 59	52 55	51 54	**56 59**
	Five years	32 40	35 44	31 39	28 34	29 37	**31 40**
North & West	(n)	(939)	(910)	(960)	(1,069)	(1,250)	(5,154)
	One year	53 57	52 55	53 56	49 51	45 48	**51 54**
	Five years	30 38	28 36	27 35	24 31	24 31	**27 35**
WALES	(n)	(240)	(297)	(410)	(400)	(216)	(1,647)
	One year	55 57	52 54	51 53	46 48	44 46	**52 54**
	Five years	33 41	31 38	26 34	25 31	24 30	**30 37**

¹ Including those for whom no deprivation category was assigned. C - crude survival, R - relative survival (see Chapter 3).
Number of patients contributing to each analysis in parentheses.

Table 15.2   Crude and relative survival (%) at one, five and ten years since diagnosis: NHS Region, deprivation category and calendar period of diagnosis

ADULTS	1971-75 Affluent C	R	2 C	R	3 C	R	4 C	R	Deprived C	R	All[1] C	R	1976-80 Affluent C	R	2 C	R	3 C	R	4 C	R	Deprived C	R	All[1] C	R
ENGLAND & WALES	(8,665)		(9,446)		(9,753)		(9,232)		(7,610)		(55,723)		(11,118)		(12,340)		(12,512)		(11,818)		(9,792)		(60,288)	
One year	45	48	44	46	43	45	41	43	39	41	42	45	49	52	48	51	46	48	45	47	43	45	46	49
Five years	22	28	22	28	21	26	19	24	18	23	20	26	26	33	25	31	23	30	23	29	21	28	24	30
Ten years	16	25	16	25	14	24	13	22	12	21	14	23	18	30	17	28	16	27	15	26	14	25	16	27
ENGLAND	(8,309)		(8,997)		(8,992)		(8,425)		(7,184)		(52,154)		(10,728)		(11,768)		(11,697)		(10,970)		(9,330)		(56,752)	
One year	45	47	44	46	43	45	40	43	39	41	42	45	50	52	48	51	45	48	45	47	43	45	46	49
Five years	22	27	22	28	21	26	19	24	18	23	20	26	26	33	24	31	23	30	23	29	21	28	23	30
Ten years	16	25	16	25	14	24	13	21	12	21	14	23	18	30	17	28	16	27	15	26	14	25	16	27
Northern & Yorkshire	(594)		(799)		(957)		(1,140)		(1,430)		(6,960)		(956)		(1,289)		(1,498)		(1,871)		(2,308)		(8,128)	
One year	43	45	43	45	36	38	35	37	33	35	40	42	48	51	42	45	40	43	41	44	39	41	43	45
Five years	19	24	19	24	17	22	15	19	14	18	18	23	25	32	21	27	20	26	21	27	18	25	21	28
Ten years	12	20	15	22	11	19	9	17	9	15	12	20	17	29	15	25	15	25	14	24	13	23	15	26
Trent	(579)		(739)		(1,027)		(1,260)		(1,024)		(5,131)		(706)		(1,038)		(1,269)		(1,415)		(1,105)		(5,719)	
One year	41	42	40	41	43	45	38	40	37	40	41	43	47	50	46	49	44	46	44	47	42	45	46	48
Five years	19	23	20	24	20	24	19	24	17	22	19	24	24	30	22	28	23	30	23	30	21	27	23	30
Ten years	13	21	13	21	13	21	13	21	12	21	13	22	16	27	16	26	16	26	16	26	15	25	16	27
Anglia & Oxford	(1,123)		(1,145)		(1,075)		(737)		(323)		(5,050)		(1,509)		(1,490)		(1,296)		(889)		(390)		(5,838)	
One year	45	47	45	48	43	45	38	40	45	48	45	47	49	51	48	50	45	48	46	49	42	44	48	51
Five years	24	29	24	29	21	26	18	22	22	27	22	28	25	32	24	30	22	28	24	30	20	26	24	31
Ten years	17	26	16	26	14	24	12	20	14	24	15	25	18	29	17	28	15	25	18	28	14	24	17	29
North Thames	(1,295)		(1,201)		(1,121)		(1,108)		(848)		(6,343)		(1,113)		(1,054)		(1,085)		(1,064)		(826)		(5,359)	
One year	49	51	48	51	43	45	45	47	44	46	48	50	53	56	52	55	48	50	50	52	49	52	51	54
Five years	26	31	26	31	22	27	22	28	23	29	24	31	26	32	28	35	25	33	26	33	26	34	27	34
Ten years	20	30	18	28	16	25	16	26	16	27	18	29	18	30	19	32	17	29	18	31	17	32	18	32
South Thames	(2,069)		(1,700)		(1,561)		(1,310)		(673)		(8,057)		(2,405)		(2,084)		(1,659)		(1,436)		(546)		(8,659)	
One year	41	43	39	42	38	41	41	43	37	39	41	44	46	49	44	47	42	44	41	44	42	45	45	47
Five years	18	23	19	24	17	22	18	24	17	22	19	24	25	31	22	28	19	25	21	28	22	30	22	29
Ten years	13	21	13	21	11	19	12	21	11	20	13	21	18	29	14	24	13	23	13	25	14	24	15	26
South & West	(1,434)		(2,015)		(1,765)		(896)		(321)		(7,096)		(1,860)		(2,539)		(2,208)		(1,090)		(364)		(8,396)	
One year	48	50	45	47	45	47	43	46	37	40	46	48	53	56	50	53	47	50	51	54	48	51	51	54
Five years	25	31	23	29	24	30	22	28	19	25	24	30	29	37	27	34	25	32	27	35	25	33	27	35
Ten years	17	26	16	27	16	28	14	25	12	22	16	28	21	35	18	31	18	31	18	31	18	31	19	33
West Midlands	(308)		(362)		(389)		(482)		(502)		(5,306)		(894)		(977)		(1,229)		(1,337)		(1,364)		(5,896)	
One year	50	52	46	49	48	51	40	43	41	43	44	47	49	52	48	51	47	49	42	45	45	48	47	50
Five years	20	26	23	28	23	29	19	24	20	25	22	27	27	33	25	31	25	32	21	26	22	30	24	31
Ten years	14	24	18	27	17	28	14	23	14	24	16	26	19	30	18	30	18	29	14	23	14	26	17	28
North & West	(907)		(1,036)		(1,097)		(1,492)		(2,063)		(8,211)		(1,285)		(1,297)		(1,453)		(1,868)		(2,427)		(8,757)	
One year	38	40	36	38	36	38	32	34	34	36	37	39	41	44	43	45	39	42	36	38	36	38	40	43
Five years	19	24	18	23	16	20	13	17	16	21	17	21	21	26	20	26	19	25	17	22	18	23	20	25
Ten years	12	19	12	20	11	18	8	15	11	19	11	19	15	23	14	22	13	23	12	21	12	21	13	23
WALES	(356)		(449)		(761)		(807)		(426)		(3,569)		(390)		(572)		(815)		(848)		(462)		(3,536)	
One year	48	50	41	43	42	44	41	43	36	38	42	45	43	46	48	50	47	50	43	46	41	43	47	50
Five years	25	31	18	22	21	26	20	26	15	19	20	26	25	31	26	33	25	32	23	29	19	24	25	32
Ten years	15	24	12	20	15	25	14	24	11	18	14	24	18	29	19	30	17	31	16	27	12	20	17	29

[1] Including those for whom no deprivation category was assigned. C - crude survival, R - relative survival (see Chapter 3).
Number of patients contributing to each analysis in parentheses.

ADULTS

	1981-85 Affluent C	R	2 C	R	3 C	R	4 C	R	Deprived C	R	All[1] C	R	1986-90 Affluent C	R	2 C	R	3 C	R	4 C	R	Deprived C	R	All[1] C	R
ENGLAND & WALES	*(12,972)*		*(13,732)*		*(13,851)*		*(12,638)*		*(10,406)*		*(65,190)*		*(13,573)*		*(15,177)*		*(15,077)*		*(13,499)*		*(10,415)*		*(68,481)*	
One year	57	60	55	58	54	57	52	55	50	53	54	57	59	62	59	62	57	60	55	58	54	57	57	60
Five years	31	38	29	36	28	36	27	35	25	33	28	36	32	40	32	40	30	38	29	37	28	37	30	39
	21	34	20	33	19	33	18	32	17	30	19	33												
ENGLAND	*(12,572)*		*(13,203)*		*(13,000)*		*(11,814)*		*(9,915)*		*(61,757)*		*(13,139)*		*(14,641)*		*(14,352)*		*(12,765)*		*(9,996)*		*(65,479)*	
One year	57	60	55	58	54	57	51	54	50	53	54	57	59	62	59	62	57	60	56	59	54	58	57	60
Five years	31	39	29	36	28	36	27	35	25	33	28	36	32	40	32	40	30	39	29	38	28	37	30	39
	21	35	20	33	19	33	18	32	17	30	19	33												
Northern & Yorkshire	*(1,183)*		*(1,442)*		*(1,676)*		*(2,038)*		*(2,409)*		*(8,900)*		*(1,315)*		*(1,679)*		*(1,788)*		*(1,955)*		*(2,271)*		*(9,055)*	
One year	55	58	52	56	52	55	49	52	48	52	52	55	56	59	54	58	52	56	53	56	53	56	54	58
Five years	30	38	27	36	27	36	25	34	24	32	27	35	31	41	28	36	27	36	27	37	27	37	28	37
	20	34	19	34	18	33	17	31	15	29	18	32												
Trent	*(962)*		*(1,216)*		*(1,491)*		*(1,648)*		*(1,140)*		*(6,519)*		*(978)*		*(1,408)*		*(1,670)*		*(1,665)*		*(1,422)*		*(7,164)*	
One year	56	59	54	57	52	55	49	52	51	55	53	56	57	61	60	64	54	58	55	59	53	57	57	60
Five years	29	37	27	34	25	33	23	31	26	35	26	34	30	37	33	42	28	37	29	39	27	37	30	38
	21	33	18	32	18	31	17	29	16	32	18	31												
Anglia & Oxford	*(1,804)*		*(1,756)*		*(1,574)*		*(953)*		*(382)*		*(6,609)*		*(2,021)*		*(2,112)*		*(1,665)*		*(1,120)*		*(415)*		*(7,393)*	
One year	58	62	54	58	54	58	48	51	54	58	55	59	56	59	53	56	55	58	54	57	48	51	55	58
Five years	31	40	29	37	29	38	28	37	31	41	30	39	31	39	29	38	29	39	26	35	26	34	29	38
	22	36	20	35	20	35	18	32	23	39	21	35												
North Thames	*(1,145)*		*(1,106)*		*(1,008)*		*(949)*		*(723)*		*(5,039)*		*(1,242)*		*(1,394)*		*(1,400)*		*(1,263)*		*(1,026)*		*(6,486)*	
One year	60	63	58	62	59	62	59	62	62	66	60	64	61	64	65	69	62	66	61	65	62	66	63	66
Five years	33	41	29	38	32	40	32	41	35	47	32	41	33	42	34	43	31	41	33	42	32	42	33	42
	23	37	20	33	22	36	21	37	23	43	22	37												
South Thames	*(2,511)*		*(2,038)*		*(1,699)*		*(1,301)*		*(527)*		*(8,426)*		*(2,313)*		*(2,067)*		*(1,785)*		*(1,352)*		*(579)*		*(8,233)*	
One year	57	60	55	58	53	56	54	58	49	52	55	59	63	66	62	66	59	63	59	63	60	64	62	65
Five years	31	39	29	38	26	35	28	38	27	36	29	38	33	41	33	43	31	41	30	40	31	43	32	42
	22	36	21	36	18	32	18	36	19	34	20	35												
South & West	*(2,313)*		*(2,858)*		*(2,496)*		*(1,172)*		*(398)*		*(9,508)*		*(2,218)*		*(2,979)*		*(2,654)*		*(1,582)*		*(523)*		*(10,007)*	
One year	59	62	56	60	56	60	56	60	54	58	57	61	62	66	61	64	62	65	58	61	59	63	61	65
Five years	33	42	29	38	30	39	30	40	29	39	31	40	35	45	34	44	33	44	32	42	32	42	34	44
	23	39	20	35	21	38	21	37	21	37	22	38												
West Midlands	*(1,148)*		*(1,214)*		*(1,396)*		*(1,620)*		*(1,609)*		*(7,053)*		*(1,348)*		*(1,334)*		*(1,644)*		*(1,847)*		*(1,390)*		*(7,616)*	
One year	57	60	52	55	52	56	49	52	48	51	52	55	57	60	58	62	56	59	54	57	53	57	56	60
Five years	30	37	27	35	27	35	24	32	24	33	27	34	32	42	32	42	29	39	29	38	28	38	30	40
	21	35	18	31	18	31	18	30	17	31	18	32												
North & West	*(1,506)*		*(1,573)*		*(1,660)*		*(2,133)*		*(2,727)*		*(9,703)*		*(1,704)*		*(1,668)*		*(1,746)*		*(1,981)*		*(2,370)*		*(9,525)*	
One year	47	50	50	53	45	49	45	48	44	47	47	50	54	57	53	57	51	55	50	53	48	52	52	55
Five years	25	31	25	32	23	30	23	30	20	28	23	30	28	37	27	36	26	34	24	33	25	34	26	35
	16	27	17	29	16	28	15	28	13	25	15	27												
WALES	*(400)*		*(529)*		*(851)*		*(824)*		*(491)*		*(3,433)*		*(434)*		*(536)*		*(725)*		*(734)*		*(419)*		*(3,002)*	
One year	52	55	54	57	51	54	54	58	46	49	53	56	54	57	52	55	48	52	47	49	44	47	51	54
Five years	25	32	27	36	25	33	30	38	25	33	27	36	32	41	32	40	24	32	24	32	24	32	29	37
	18	32	19	32	17	31	21	37	16	31	19	34												

[1] Including those for whom no deprivation category was assigned. C - crude survival, R - relative survival (see Chapter 3). Number of patients contributing to each analysis in parentheses.

Fig. 15.1   Relative survival up to ten years: trends by calendar period of diagnosis (1971-90) and NHS Region

ADULTS

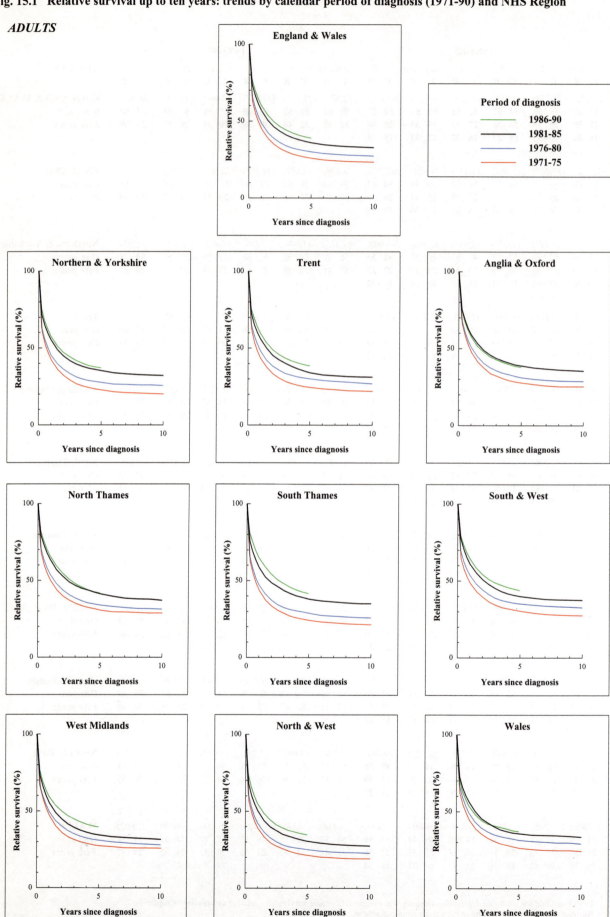

Fig. 15.2   Relative survival up to five years, by deprivation category and NHS Region: patients diagnosed 1986-90

ADULTS

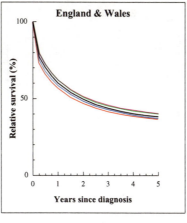

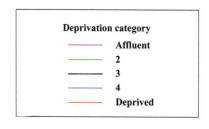

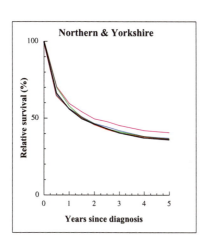

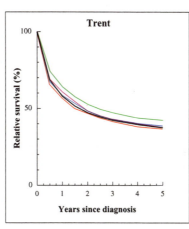

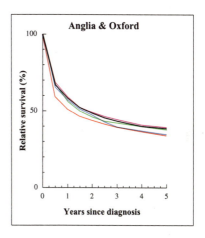

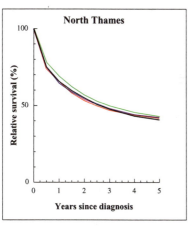

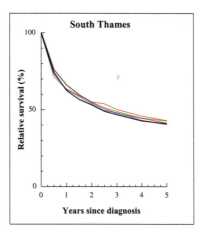

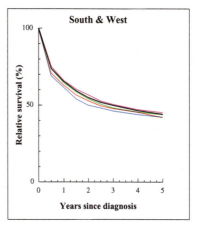

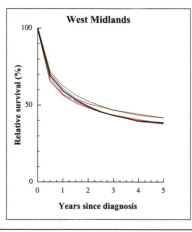

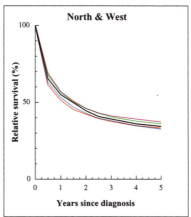

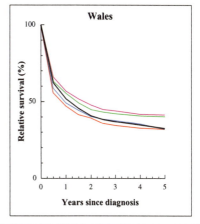

Fig. 15.3 Relative survival at five years by deprivation category, period of diagnosis (1981-90) and NHS Region

ADULTS

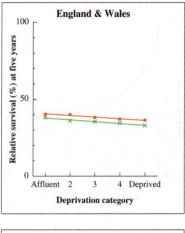

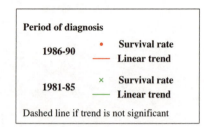

Period of diagnosis

1986-90   ● Survival rate
           — Linear trend

1981-85   × Survival rate
           — Linear trend

Dashed line if trend is not significant

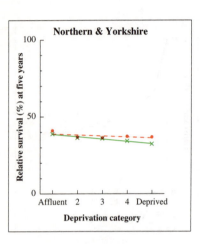

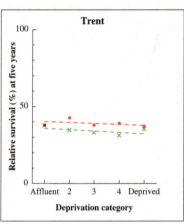

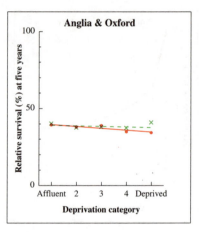

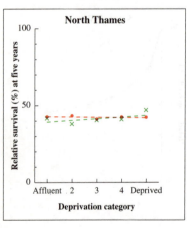

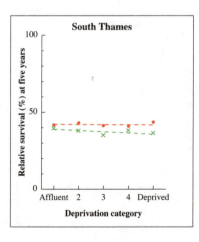

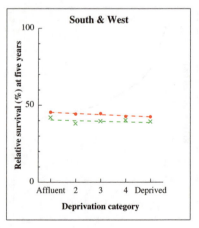

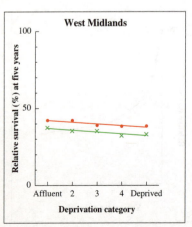

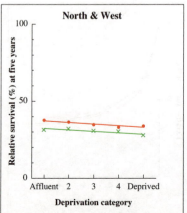

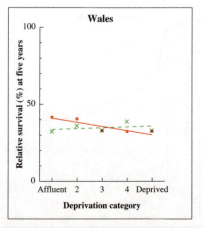

Fig. 15.4 Relative survival (%) by NHS Region and sex:
   one-year and five-year survival for patients diagnosed 1986-90, with 95% confidence intervals

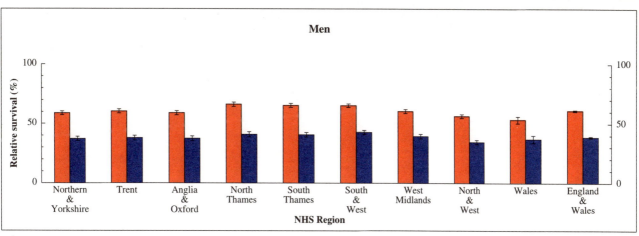

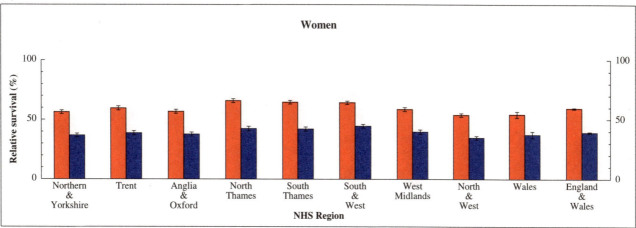

Fig. 15.5 Relative survival (%) at five years, with 95% confidence interval, by age at diagnosis and sex:
   England and Wales, patients diagnosed 1986-90

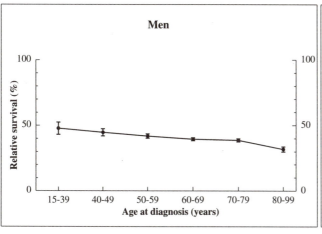

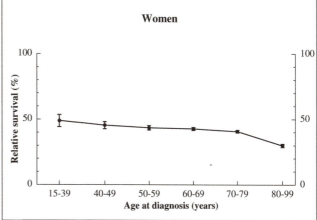

Chapter 16

RECTUM

Cancer of the rectum is less frequent than colon cancer, accounting for 6% of malignant tumours, and it ranks as the fifth most common cancer in adults. About 10,500 new cases are diagnosed in England and Wales each year, with about 20% more cases in men than women. The annual number of new cases has risen by about 20% since 1971. Recent trends in incidence and mortality are similar to those for colon cancer (Chapter 15).

The survival analyses reported here relate to 168,000 adults registered with cancer of the rectum in England and Wales during the period 1971-90, and followed up to the end of 1995. Almost 1,200 patients registered in the 1970s were considered ineligible for analysis because the tumour was either in situ, benign, of uncertain behaviour, or apparently metastatic in the rectum from another primary site (Table 16.1). About 11% of the 190,000 patients eligible for analysis were excluded, 4% because their vital status was unknown at 6 July 1997, and 7% because their survival was zero or unknown. The proportion of patients excluded for unknown vital status in each five-year period increased from 3% to 5%, and the proportion with zero or unknown survival from 6% to 8%. Almost 1,000 patients (0.5%) had a synchronous or subsequent neoplasm, but survival is reckoned from the date of the first primary cancer.

Overall, 80% of the cancers arose in the rectum, but the proportion arising at the junction with the sigmoid colon increased from 11% to 18% between the early 1970s and the late 1980s (Table 16A). Almost two-thirds (64%) of the tumours were adenocarcinomas, and 3% were squamous carcinomas, rising from 2% to 4% by the late 1980s. A further 24% were poorly specified carcinomas.

Survival trends

Relative survival for adults diagnosed in England and Wales during 1986-90 was 66% at one year and 38% at five years. These survival rates are 10-12% higher than for patients diagnosed some 15 years earlier (Table 16.2).

There has been consistent improvement over time in survival from rectal cancer. As for colon cancer, the gains in survival were most clear-cut between the late 1970s and early 1980s, both nationally and in many regions (Figure 16.1). The survival curves for rectal cancer differ in shape from those for colon cancer (Figure 15.1). One-year survival from rectal cancer is higher than for colon cancer, but five-year survival is similar, because substantial excess mortality persists well beyond the first year after

diagnosis: the survival curves for colon cancer approach a plateau earlier.

The pattern of survival by age and sex for rectal cancer resembles that for colon cancer (Figure 16.5). For patients diagnosed during 1986-90, five-year survival was in the range 40% to 48% for both men and women up to age 80 at diagnosis, and about 10% lower for very elderly patients (see Table 4.7).

The average gain in one-year survival between successive five-year periods was 5%, and almost as much (4%) for survival five years after diagnosis (see Table 4.4). The age-standardised trends have been very similar for men and women, with an improvement of 5-7% in one-year and five-year survival rates around 1980, and smaller gains for patients diagnosed during the late 1980s. The improvement for rectal cancer has been slightly less than for cancer of the colon, and five-year survival is now very similar for both cancers.

Regional differences in survival and in survival trends over time appear less marked than for colon cancer (Figure 16.4; see also Table-Figure 4.5D).

A small but consistent survival advantage of 1-2% has been evident for women at all ages and throughout the 1970s and 1980s.

Deprivation

One-year and five-year survival from rectal cancer is lower in the more deprived groups for patients diagnosed in England and Wales during 1986-90 (Figure 16.2). Similar patterns occurred in six of the NHS Regions, and the gap is particularly marked in South Thames, where the difference in five-year survival between the most affluent and most deprived categories reaches 14% (Table 16.2): this can be clearly seen from the survival curves.

The fact that overall survival in South Thames is similar to the national average despite this marked deprivation gradient is partly due to the fact that there are three times as many patients in the most affluent group as in the most deprived group in this region.

In striking contrast, however, overall survival rates in Anglia & Oxford and South & West are almost identical to those in South Thames, and the distribution of patients between deprivation groups is very similar, but there is no difference in survival between the deprivation categories in these regions (Table 16.2, Figure 16.2).

Table 16A Sub-site distribution of cancer of the rectum

ICD-9 code	Site description	1971-75		1976-80		1981-85		1986-90		Total	
		No.	%	No.	%	No.	%	No.	%	No.	%
154.0	Rectosigmoid junction	4,100	10.7	5,350	13.0	7,303	16.4	8,195	18.5	24,948	4.8
154.1	Rectum	33,413	87.0	34,563	84.1	35,604	79.7	34,083	76.8	137,663	81.7
154.2	Anal canal	899	2.3	858	2.1	866	1.9	956	2.2	3,579	2.1
154.3	Anus, unspecified	-	-	242	0.6	627	1.4	859	1.9	1,728	1.0
154.8	Other	-	-	107	0.3	257	0.6	295	0.7	659	0.4
	Total	**38,412**		**41,120**		**44,657**		**44,388**		**168,577**	

The deprivation gradient in five-year survival for patients diagnosed during 1981-85 and 1986-90 is shown in Figure 16.3. Nationally, the gradient is significant in both periods. There is variation between the regions in the deprivation gradient for survival. While there was no appreciable difference between deprivation groups in three of the regions, the gradient is statistically significant in four (Figure 16.3).

For 87,000 patients of known deprivation status diagnosed in England and Wales during the 1980s, the gap in survival between the most affluent and the most deprived groups was 7-8% at one year and 5-7% at five years (see Table 4.9).

International comparison

The pattern is broadly similar to that for colon cancer. Average survival in England and Wales – 37% and 39% for men and women, respectively, at five years after diagnosis – is very similar to that in Scotland, but the 4% to 6% deficit compared with the European average is less than for cancer of the colon. The highest regional survival rates (South & West, as for colon cancer) are similar to the European average, but still 20% less than in the USA (see Table 4.12).

Table 16.1 Ineligible and excluded records, and patients included in analyses, by calendar period of diagnosis

	Calendar period of diagnosis														
	1971-75			1976-80			1981-85			1986-90			Overall		
Total registered	43,230			46,733			51,016			52,063			193,042		
Ineligible	*Records*	*Patients*	*%*	*Records*	*Patients*	*%*	*Records*	*Patients*	*%*	*Records*	*Patients*	*%*	*Records*	*Patients*	*%*
Incomplete data[1]	35	35		92	92		149	149		146	146		422	**422**	
Resident outside England and Wales	422	422		342	336		243	219		242	207		1,249	**1,184**	
In situ neoplasm[2]	162	162		39	38		0	0		1	1		202	**201**	
Benign or uncertain behaviour[2]	443	328		357	308		1	1		2	2		803	**639**	
Metastatic[2]	236	233		115	114		2	2		3	3		356	**352**	
Otherwise ineligible[3]	-	-		-	-		-	-		-	-		-	**-**	
Lymphoma[4]	9	9		4	4		1	0		2	0		16	**13**	
Leukaemia or myeloma[4]	1	0		1	0		2	0		0	0		4	**0**	
		1,189			892			371			359			2,811	
Total eligible		42,041	100		45,841	100		50,645	100		51,704	100		190,231	100
Aged under 15 or 100 years or over at diagnosis	13	13	<0.1	22	22	<0.1	19	18	<0.1	25	24	<0.1	79	**77**	<0.1
Vital status unknown[5]	1,368	1,178	2.8	1,975	1,789	3.9	2,009	1,869	3.7	2,917	2,818	5.5	8,269	**7,654**	4.0
Duplicate registration[6]	1	1	<0.1	1	0	0.0	0	0	0.0	0	0	0.0	2	**1**	<0.1
Multiple primary[7]	2	0	0.0	31	23	<0.1	209	192	0.4	252	232	0.4	494	**447**	0.2
Synchronous tumours[8]	160	105	0.2	137	85	0.2	216	165	0.3	275	176	0.3	788	**531**	0.3
Sex not known	0	0	0.0	0	0	0.0	0	0	0.0	0	0	0.0	0	**0**	0.0
Sex incompatible with site code[9]	-	-	-	-	-	-	-	-	-	-	-	-	-	**-**	-
Invalid dates[10]	3	3	<0.1	5	2	<0.1	9	6	<0.1	34	13	<0.1	51	**24**	<0.1
Zero survival[11]	2,535	2,329	5.5	3,004	2,800	6.1	4,052	3,738	7.4	4,342	4,053	7.8	13,933	**12,920**	6.8
Total exclusions		3,629	8.6		4,721	10.3		5,988	11.8		7,316	14.1		21,654	11.4
Patients included in analysis		38,412	91.4		41,120	89.7		44,657	88.2		44,388	85.9		168,577	88.6

[1] Main data item(s) invalid or incompatible with one another: name, sex, date of birth, date of diagnosis and (if present) date of death, postcode, site, morphology and behaviour

[2] Tumour either *in situ* (behaviour code 2), benign (0), uncertain if benign or malignant (1) or metastatic (6 or 9)

[3] Not applicable for this malignancy

[4] Morphology that of lymphoma (included with non-Hodgkin lymphoma, Ch. 41) or myeloma (included with multiple myeloma, Ch. 43), or leukaemia (excluded)

[5] Tracing of vital status at National Health Service Central Register incomplete at 6 July 1997

[6] Same site code, sex, cancer registry and cancer registry number as an earlier registration

[7] Same site code and person number as an earlier registration(s): mostly confirmed multiple primary tumours at this site, some unresolved duplicate registrations

[8] Same site code, sex, date of birth and date of diagnosis as another registration(s): mostly synchronous or (in paired organs) bilateral tumours in same anatomic site in one individual, not previously linked: also some duplicate registrations

[9] Not applicable for this malignancy

[10] Impossible sequence of dates (birth, diagnosis and, if present, emigration or death); or date of death after 6 July 1997

[11] Date of diagnosis same as date of death: some are patients who did die on the day of diagnosis, but most were registered solely from a death certificate (death certificate only, or DCO), and their survival time is thus unknown

Table 16.2   Crude and relative survival (%) at one, five and ten years since diagnosis: NHS Region, deprivation category and calendar period of diagnosis

MEN	1971-75												1976-80											
	Affluent		*2*		*3*		*4*		*Deprived*		*All[1]*		*Affluent*		*2*		*3*		*4*		*Deprived*		*All[1]*	
	C	R	C	R	C	R	C	R	C	R	C	R	C	R	C	R	C	R	C	R	C	R	C	R
ENGLAND & WALES	(2,799)		(3,369)		(3,660)		(3,730)		(3,191)		(21,203)		(3,787)		(4,258)		(4,735)		(4,627)		(4,235)		(22,599)	
One year	55	58	54	57	52	56	51	54	48	51	52	55	57	60	57	60	55	59	54	58	52	55	55	58
Five years	24	30	22	29	21	27	21	27	19	26	21	27	26	33	26	34	23	31	23	30	21	29	24	31
Ten years	15	25	14	25	13	23	12	23	11	21	13	23	16	28	16	29	14	26	14	26	13	25	14	26
ENGLAND	(2,683)		(3,201)		(3,393)		(3,395)		(3,024)		(19,871)		(3,646)		(4,057)		(4,414)		(4,250)		(4,025)		(21,153)	
One year	55	58	54	57	52	56	51	54	48	51	52	55	57	60	57	60	55	59	55	58	52	55	55	58
Five years	24	30	22	29	20	27	21	28	19	26	21	27	26	34	26	34	23	31	23	30	21	29	24	31
Ten years	15	26	15	26	12	22	13	24	11	21	13	23	16	28	16	29	14	26	13	25	13	25	15	26
Northern & Yorkshire	(181)		(300)		(371)		(489)		(631)		(2,849)		(345)		(445)		(515)		(754)		(963)		(3,092)	
One year	52	55	54	57	44	48	53	56	42	46	50	53	57	60	57	61	56	60	52	56	51	55	54	58
Five years	19	24	24	31	17	22	21	28	18	24	19	26	29	36	27	35	22	31	21	28	20	27	23	31
Ten years	10	17	16	27	11	19	12	23	9	17	12	22	19	32	16	29	13	25	11	23	12	24	14	25
Trent	(219)		(337)		(434)		(599)		(479)		(2,282)		(281)		(444)		(560)		(643)		(525)		(2,533)	
One year	52	55	57	60	53	56	48	51	49	53	52	55	56	59	53	56	50	54	53	56	54	58	54	57
Five years	21	26	23	30	20	28	21	29	18	25	21	28	22	28	23	30	21	28	21	27	22	31	22	29
Ten years	12	21	16	26	12	21	13	24	11	22	13	23	15	25	14	25	14	26	12	22	11	26	13	25
Anglia & Oxford	(367)		(407)		(385)		(314)		(153)		(1,884)		(516)		(496)		(470)		(302)		(143)		(2,004)	
One year	57	60	60	63	52	55	52	55	53	57	56	59	54	58	55	59	55	58	56	60	49	53	55	58
Five years	25	32	23	30	21	28	21	28	20	27	23	30	26	34	25	33	22	29	26	35	18	26	25	33
Ten years	16	26	13	24	13	23	11	23	11	19	13	24	16	30	17	28	14	24	14	27	11	25	15	27
North Thames	(413)		(418)		(405)		(377)		(314)		(2,188)		(396)		(355)		(437)		(377)		(315)		(1,948)	
One year	62	65	52	55	60	63	52	56	54	57	57	60	58	61	60	64	54	58	60	63	57	61	58	62
Five years	29	36	27	34	24	31	27	35	24	31	26	34	28	35	33	41	25	33	28	37	25	33	28	36
Ten years	18	29	18	32	16	26	17	31	16	29	17	30	18	30	21	35	15	27	18	34	19	33	18	32
South Thames	(648)		(588)		(578)		(508)		(262)		(2,834)		(710)		(643)		(623)		(456)		(224)		(2,814)	
One year	53	56	47	50	52	55	49	53	48	51	51	54	54	58	57	60	55	58	53	57	55	59	55	59
Five years	22	29	17	23	19	25	20	26	17	23	20	26	26	32	26	33	24	32	21	28	24	33	25	32
Ten years	14	25	11	19	9	18	12	23	9	18	12	22	16	27	15	27	14	27	12	25	15	28	15	27
South & West	(451)		(653)		(631)		(349)		(122)		(2,424)		(625)		(847)		(767)		(379)		(161)		(2,893)	
One year	52	55	56	59	53	56	52	56	44	47	54	57	62	65	58	62	56	60	58	63	48	51	58	62
Five years	25	31	23	30	22	29	22	29	20	26	23	30	28	36	25	34	25	34	28	39	29	37	27	36
Ten years	18	30	14	25	12	23	15	26	10	18	14	26	17	31	14	28	16	30	17	34	21	36	16	30
West Midlands	(129)		(141)		(194)		(213)		(292)		(2,492)		(364)		(368)		(516)		(665)		(745)		(2,701)	
One year	51	54	50	53	54	57	56	59	46	50	53	56	56	60	56	60	58	61	56	59	51	55	56	59
Five years	24	30	22	29	20	26	26	33	15	20	21	27	24	31	26	32	23	30	23	31	21	28	23	30
Ten years	14	25	17	28	14	26	17	30	9	17	13	24	16	28	18	29	14	25	13	25	12	23	14	26
North & West	(275)		(357)		(395)		(546)		(771)		(2,918)		(409)		(459)		(526)		(674)		(949)		(3,168)	
One year	52	55	49	52	44	47	42	44	45	48	46	49	54	56	50	53	50	53	49	53	45	49	50	53
Five years	18	23	19	26	16	22	13	17	19	26	17	22	24	31	24	33	18	25	18	25	18	25	20	27
Ten years	10	17	13	23	10	20	7	13	10	19	10	18	13	25	18	29	12	22	11	21	11	21	12	23
WALES	(116)		(168)		(267)		(335)		(167)		(1,332)		(141)		(201)		(321)		(377)		(210)		(1,446)	
One year	43	46	44	46	52	55	49	52	51	55	49	53	51	54	55	59	54	57	47	50	50	53	53	56
Five years	22	28	14	19	21	28	17	22	20	25	19	25	22	28	27	37	24	32	21	29	19	25	23	31
Ten years	12	21	9	17	16	27	10	19	13	21	12	23	14	25	16	32	14	25	15	28	9	17	14	26

[1] Including those for whom no deprivation category was assigned. C - crude survival, R - relative survival (see Chapter 3). Number of patients contributing to each analysis in parentheses.

	1981-85											1986-90												MEN
	Affluent		*2*		*3*		*4*		*Deprived*		*All*[1]		*Affluent*		*2*		*3*		*4*		*Deprived*		*All*[1]		
	C	R	C	R	C	R	C	R	C	R	C	R	C	R	C	R	C	R	C	R	C	R	C	R	
	(4,396)		*(4,768)*		*(5,319)*		*(5,279)*		*(4,475)*		*(24,867)*		*(4,543)*		*(5,353)*		*(5,376)*		*(5,253)*		*(4,258)*		*(25,064)*		**ENGLAND & WALES**
	65	69	62	66	60	64	58	61	57	61	**60**	**64**	67	70	65	68	64	68	59	63	60	64	**63**	**67**	*One year*
	31	39	29	38	27	35	25	34	23	32	**27**	**36**	31	40	30	38	29	38	26	34	26	35	**28**	**37**	*Five years*
	20	34	18	32	16	30	15	29	14	26	**17**	**30**													
	(4,190)		*(4,531)*		*(4,853)*		*(4,800)*		*(4,237)*		*(23,062)*		*(4,336)*		*(5,072)*		*(4,987)*		*(4,772)*		*(4,019)*		*(23,389)*		**ENGLAND**
	65	69	63	67	61	65	58	62	56	61	**60**	**64**	67	70	65	69	65	69	60	63	60	64	**63**	**67**	*One year*
	31	40	29	38	27	35	25	34	23	31	**27**	**36**	32	40	30	38	29	38	26	34	26	35	**28**	**37**	*Five years*
	20	33	18	32	16	30	15	29	14	26	**17**	**30**													
	(417)		*(527)*		*(712)*		*(866)*		*(1,068)*		*(3,649)*		*(466)*		*(636)*		*(657)*		*(781)*		*(937)*		*(3,504)*		**Northern & Yorkshire**
	69	73	65	70	57	61	58	61	54	59	**59**	**64**	66	70	63	68	62	67	56	60	61	65	**62**	**65**	*One year*
	30	39	31	41	25	34	23	32	22	31	**25**	**34**	28	37	28	39	29	40	25	35	26	36	**27**	**36**	*Five years*
	18	34	19	37	16	30	14	28	14	28	**16**	**29**													
	(368)		*(436)*		*(646)*		*(767)*		*(534)*		*(2,765)*		*(411)*		*(537)*		*(622)*		*(723)*		*(552)*		*(2,850)*		**Trent**
	66	70	57	61	58	62	58	62	57	61	**59**	**63**	65	70	66	70	64	69	58	62	58	63	**62**	**66**	*One year*
	32	41	26	34	26	36	26	36	24	33	**27**	**35**	30	41	32	44	26	36	24	33	27	38	**28**	**36**	*Five years*
	20	34	17	32	16	30	15	30	13	26	**16**	**29**													
	(614)		*(623)*		*(513)*		*(312)*		*(164)*		*(2,278)*		*(640)*		*(697)*		*(584)*		*(350)*		*(171)*		*(2,451)*		**Anglia & Oxford**
	65	69	60	64	61	65	60	64	52	56	**61**	**66**	64	68	65	70	63	68	59	64	58	62	**64**	**67**	*One year*
	32	42	26	34	29	38	31	42	23	31	**29**	**38**	32	41	28	38	29	41	29	40	28	39	**30**	**39**	*Five years*
	22	39	16	31	19	35	17	35	15	28	**18**	**33**													
	(397)		*(384)*		*(386)*		*(379)*		*(320)*		*(1,910)*		*(421)*		*(471)*		*(511)*		*(467)*		*(365)*		*(2,293)*		**North Thames**
	67	71	67	72	66	70	64	68	61	66	**66**	**70**	69	74	69	74	69	74	65	70	67	73	**68**	**72**	*One year*
	32	40	34	44	28	38	33	44	29	40	**31**	**41**	31	40	31	42	30	41	25	35	29	40	**30**	**38**	*Five years*
	21	34	22	38	18	34	22	39	19	36	**20**	**34**													
	(831)		*(638)*		*(603)*		*(458)*		*(160)*		*(2,811)*		*(738)*		*(687)*		*(541)*		*(455)*		*(233)*		*(2,699)*		**South Thames**
	62	67	65	70	62	67	59	63	62	67	**63**	**67**	72	76	63	69	68	73	61	66	60	65	**66**	**70**	*One year*
	27	36	28	39	28	39	24	34	28	37	**27**	**37**	38	50	27	36	28	41	25	36	25	37	**30**	**39**	*Five years*
	18	32	15	31	17	34	16	31	15	30	**17**	**31**													
	(668)		*(906)*		*(790)*		*(415)*		*(122)*		*(3,011)*		*(690)*		*(952)*		*(810)*		*(512)*		*(172)*		*(3,158)*		**South & West**
	68	72	62	67	62	66	60	65	55	60	**63**	**67**	64	68	65	69	65	70	63	67	61	67	**65**	**68**	*One year*
	34	45	31	41	28	38	28	38	22	30	**30**	**40**	29	38	31	42	31	44	29	41	32	46	**31**	**40**	*Five years*
	21	39	19	36	16	33	20	38	*11*	*25*	**19**	**34**													
	(421)		*(493)*		*(623)*		*(759)*		*(756)*		*(3,069)*		*(457)*		*(511)*		*(639)*		*(760)*		*(589)*		*(2,977)*		**West Midlands**
	66	71	60	64	63	67	57	60	58	62	**61**	**65**	65	70	67	71	64	68	60	65	64	68	**64**	**68**	*One year*
	36	46	27	36	28	37	22	30	24	33	**27**	**35**	32	41	33	44	30	42	26	36	27	39	**29**	**38**	*Five years*
	25	43	16	30	17	34	13	25	14	29	**17**	**30**													
	(474)		*(524)*		*(580)*		*(844)*		*(1,113)*		*(3,569)*		*(513)*		*(581)*		*(623)*		*(724)*		*(1,000)*		*(3,457)*		**North & West**
	58	62	60	64	53	57	51	56	53	57	**55**	**59**	63	67	61	65	59	64	56	60	55	59	**58**	**62**	*One year*
	26	34	26	36	19	28	21	30	21	29	**22**	**30**	31	39	25	33	27	37	24	34	22	32	**25**	**33**	*Five years*
	15	28	18	35	12	26	13	26	13	25	**14**	**26**													
	(206)		*(237)*		*(466)*		*(479)*		*(238)*		*(1,805)*		*(207)*		*(281)*		*(389)*		*(481)*		*(239)*		*(1,675)*		**WALES**
	60	65	54	58	56	59	52	56	59	63	**56**	**60**	65	69	56	60	56	59	55	59	55	59	**58**	**61**	*One year*
	28	36	27	37	26	36	25	34	27	36	**26**	**35**	27	37	28	37	27	39	24	35	26	35	**27**	**36**	*Five years*
	20	36	16	32	16	31	15	28	16	33	**16**	**30**													

[1] Including those for whom no deprivation category was assigned. C - crude survival, R - relative survival (see Chapter 3).
Number of patients contributing to each analysis in parentheses.

Table 16.2 Crude and relative survival (%) at one, five and ten years since diagnosis: NHS Region, deprivation category and calendar period of diagnosis

		1971-75						1976-80					
WOMEN		*Affluent*	*2*	*3*	*4*	*Deprived*	*All[1]*	*Affluent*	*2*	*3*	*4*	*Deprived*	*All[1]*
		C R	C R	C R	C R	C R	C R	C R	C R	C R	C R	C R	C R
ENGLAND & WALES		*(2,539)*	*(2,811)*	*(3,061)*	*(2,838)*	*(2,496)*	*(17,209)*	*(3,169)*	*(3,586)*	*(3,877)*	*(3,789)*	*(3,249)*	*(18,521)*
	One year	56 59	53 55	51 54	48 51	48 50	**51 54**	57 60	54 57	53 56	53 56	51 54	**53 56**
	Five years	26 32	24 29	25 30	22 27	21 26	**23 29**	29 35	26 32	25 31	25 31	23 29	**25 32**
	Ten years	18 26	17 26	17 27	15 24	15 23	**16 25**	21 31	18 29	17 27	17 27	15 25	**18 28**
ENGLAND		*(2,447)*	*(2,688)*	*(2,854)*	*(2,618)*	*(2,388)*	*(16,227)*	*(3,053)*	*(3,389)*	*(3,595)*	*(3,501)*	*(3,094)*	*(17,321)*
	One year	56 59	53 55	51 54	49 52	48 50	**51 54**	57 60	54 57	53 56	53 56	51 54	**54 56**
	Five years	26 31	24 29	25 30	22 28	21 26	**23 29**	28 35	26 32	25 31	25 31	23 29	**25 32**
	Ten years	18 26	17 26	17 27	15 25	15 23	**16 25**	21 31	18 28	17 27	17 27	16 25	**17 27**
Northern & Yorkshire		*(171)*	*(237)*	*(300)*	*(376)*	*(482)*	*(2,131)*	*(269)*	*(373)*	*(520)*	*(585)*	*(741)*	*(2,554)*
	One year	56 58	53 56	50 52	44 47	47 49	**50 53**	56 60	51 54	53 56	51 54	49 52	**52 55**
	Five years	27 32	20 25	24 29	23 29	19 24	**23 28**	29 36	25 31	23 31	21 28	21 26	**23 30**
	Ten years	19 27	14 22	15 25	15 24	13 20	**16 25**	22 33	17 27	15 25	13 20	14 22	**15 25**
Trent		*(186)*	*(263)*	*(336)*	*(378)*	*(313)*	*(1,618)*	*(218)*	*(320)*	*(369)*	*(450)*	*(385)*	*(1,801)*
	One year	57 59	48 51	55 58	43 45	42 44	**49 52**	53 55	53 56	48 50	51 53	51 55	**52 55**
	Five years	25 30	23 28	27 33	18 23	20 24	**22 28**	28 34	23 30	21 25	24 30	23 29	**24 30**
	Ten years	14 22	15 23	19 28	13 21	15 23	**15 24**	18 29	17 29	15 23	15 24	15 25	**16 26**
Anglia & Oxford		*(324)*	*(317)*	*(315)*	*(226)*	*(96)*	*(1,457)*	*(415)*	*(414)*	*(357)*	*(288)*	*(120)*	*(1,659)*
	One year	60 63	54 56	49 52	48 51	45 47	**54 57**	58 62	52 55	54 56	48 51	51 55	**54 58**
	Five years	30 35	25 31	23 28	23 28	16 20	**25 30**	27 33	27 34	24 30	29 34	23 30	**27 34**
	Ten years	22 31	17 27	16 25	17 26	10 16	**18 27**	19 27	17 27	16 25	21 32	14 26	**18 29**
North Thames		*(404)*	*(375)*	*(330)*	*(328)*	*(275)*	*(1,963)*	*(333)*	*(320)*	*(347)*	*(369)*	*(268)*	*(1,706)*
	One year	59 62	58 61	52 55	51 54	52 55	**56 59**	59 62	56 59	57 60	60 63	58 62	**59 62**
	Five years	27 32	27 33	29 35	23 28	21 26	**27 33**	29 36	29 35	29 36	30 37	28 36	**29 37**
	Ten years	20 29	20 29	19 32	16 25	14 24	**19 29**	23 33	22 32	20 31	22 32	20 30	**22 33**
South Thames		*(615)*	*(537)*	*(557)*	*(448)*	*(257)*	*(2,664)*	*(698)*	*(601)*	*(558)*	*(469)*	*(212)*	*(2,701)*
	One year	51 55	50 53	48 51	50 53	49 52	**51 54**	55 58	52 54	56 59	48 51	59 62	**54 57**
	Five years	21 26	21 26	22 27	20 25	22 29	**22 28**	26 32	25 31	27 34	22 27	28 36	**25 32**
	Ten years	13 20	14 21	16 25	13 20	13 23	**15 23**	19 28	17 26	18 31	15 24	20 32	**17 29**
South & West		*(385)*	*(560)*	*(521)*	*(253)*	*(100)*	*(2,027)*	*(512)*	*(704)*	*(609)*	*(300)*	*(98)*	*(2,331)*
	One year	56 58	54 56	50 52	56 58	47 49	**54 56**	57 60	56 59	53 56	62 65	53 56	**57 60**
	Five years	30 35	27 33	22 28	28 34	18 22	**26 32**	28 34	27 33	27 34	28 35	23 29	**27 34**
	Ten years	20 29	19 29	15 24	21 33	11 18	**18 29**	22 32	19 29	18 28	20 30	18 28	**19 30**
West Midlands		*(107)*	*(107)*	*(150)*	*(161)*	*(211)*	*(1,872)*	*(260)*	*(270)*	*(386)*	*(445)*	*(499)*	*(1,893)*
	One year	54 57	55 57	46 49	52 55	45 48	**51 54**	56 59	55 58	52 55	55 58	49 52	**54 56**
	Five years	30 35	26 32	21 28	26 31	23 29	**25 30**	31 39	23 28	26 32	24 30	24 30	**26 32**
	Ten years	21 29	20 30	18 27	19 28	20 29	**18 28**	21 32	17 26	20 29	16 26	17 28	**18 29**
North & West		*(255)*	*(292)*	*(345)*	*(448)*	*(654)*	*(2,495)*	*(348)*	*(387)*	*(449)*	*(595)*	*(771)*	*(2,676)*
	One year	48 50	42 44	49 51	45 48	46 49	**46 49**	53 56	47 50	47 49	48 51	46 48	**49 52**
	Five years	23 27	19 23	24 29	21 26	20 25	**20 25**	29 34	21 26	19 24	23 30	20 26	**23 29**
	Ten years	13 19	15 22	19 29	14 23	14 23	**14 23**	21 32	15 23	14 21	15 26	13 22	**16 25**
WALES		*(92)*	*(123)*	*(207)*	*(220)*	*(108)*	*(982)*	*(116)*	*(197)*	*(282)*	*(288)*	*(155)*	*(1,200)*
	One year	54 56	51 54	51 54	36 38	54 56	**48 51**	57 60	54 57	51 53	54 56	51 53	**53 56**
	Five years	28 33	24 30	27 32	14 17	24 30	**22 28**	35 44	30 39	26 32	26 33	22 27	**27 34**
	Ten years	20 31	19 29	19 29	8 13	16 22	**16 25**	23 38	24 37	18 28	19 30	12 19	**19 30**

[1] Including those for whom no deprivation category was assigned. C - crude survival, R - relative survival (see Chapter 3). Number of patients contributing to each analysis in parentheses.

Values below are given as **C R** (C - crude survival, R - relative survival).

	1981-85						1986-90						**WOMEN**
	Affluent	*2*	*3*	*4*	*Deprived*	*All*[1]	*Affluent*	*2*	*3*	*4*	*Deprived*	*All*[1]	
	C R	C R	C R	C R	C R	C R	C R	C R	C R	C R	C R	C R	
	(3,638)	*(3,990)*	*(4,265)*	*(4,075)*	*(3,339)*	*(19,790)*	*(3,670)*	*(4,167)*	*(4,323)*	*(3,945)*	*(2,982)*	*(19,324)*	**ENGLAND & WALES**
	62 66	61 64	59 62	57 61	56 59	59 62	65 69	63 66	62 66	61 65	59 62	62 66	*One year*
	30 37	31 39	29 36	29 36	25 32	29 36	34 41	32 39	31 38	31 39	28 35	31 39	*Five years*
	22 33	21 34	20 31	19 31	17 28	20 31							
	(3,486)	*(3,769)*	*(3,933)*	*(3,693)*	*(3,134)*	*(18,381)*	*(3,491)*	*(3,944)*	*(3,999)*	*(3,618)*	*(2,787)*	*(18,003)*	**ENGLAND**
	63 66	61 64	59 62	58 61	56 60	59 63	66 69	63 66	63 66	62 65	59 62	63 66	*One year*
	30 37	31 39	29 36	28 36	26 32	29 36	34 42	32 39	32 39	31 39	27 35	31 39	*Five years*
	22 33	21 33	20 32	19 30	17 29	20 31							
	(325)	*(436)*	*(542)*	*(644)*	*(765)*	*(2,750)*	*(365)*	*(438)*	*(524)*	*(574)*	*(621)*	*(2,537)*	**Northern & Yorkshire**
	59 62	60 63	59 63	56 60	53 57	57 61	70 74	61 65	60 64	57 60	56 60	61 64	*One year*
	29 37	31 37	31 38	27 35	25 32	28 36	38 47	33 42	27 36	28 35	27 34	30 39	*Five years*
	19 30	20 31	22 34	18 30	16 27	19 31							
	(300)	*(333)*	*(486)*	*(544)*	*(365)*	*(2,033)*	*(280)*	*(395)*	*(423)*	*(508)*	*(384)*	*(1,998)*	**Trent**
	62 65	53 56	55 58	57 60	51 54	56 60	65 68	55 58	58 61	61 64	56 59	59 63	*One year*
	30 37	26 32	28 35	29 36	25 32	28 35	33 42	28 36	30 38	30 39	26 33	30 38	*Five years*
	20 32	17 27	20 31	20 32	19 30	20 31							
	(471)	*(461)*	*(412)*	*(283)*	*(119)*	*(1,788)*	*(490)*	*(513)*	*(466)*	*(305)*	*(136)*	*(1,925)*	**Anglia & Oxford**
	60 64	64 68	58 62	60 64	59 64	61 65	63 66	61 64	62 66	63 67	56 59	62 66	*One year*
	31 39	33 42	30 38	32 40	27 33	31 40	34 42	31 40	34 42	36 44	30 38	34 42	*Five years*
	23 34	24 38	21 34	20 34	18 27	22 35							
	(357)	*(348)*	*(325)*	*(299)*	*(233)*	*(1,595)*	*(365)*	*(410)*	*(433)*	*(405)*	*(273)*	*(1,934)*	**North Thames**
	63 66	68 72	64 68	64 67	72 75	67 70	70 73	69 73	68 72	68 71	65 68	69 72	*One year*
	30 36	34 41	37 45	36 43	36 46	35 43	36 43	34 42	33 40	36 44	29 36	34 42	*Five years*
	22 33	24 36	25 37	24 36	21 37	23 37							
	(721)	*(580)*	*(531)*	*(445)*	*(171)*	*(2,553)*	*(632)*	*(590)*	*(529)*	*(381)*	*(175)*	*(2,339)*	**South Thames**
	63 67	59 62	61 65	57 60	55 58	60 64	68 72	65 69	66 69	62 66	60 64	65 69	*One year*
	29 35	31 38	29 36	26 32	22 28	29 36	34 42	31 39	32 40	30 37	22 27	31 39	*Five years*
	21 32	20 34	20 32	18 28	15 28	20 32							
	(562)	*(786)*	*(695)*	*(340)*	*(105)*	*(2,570)*	*(549)*	*(781)*	*(705)*	*(412)*	*(171)*	*(2,633)*	**South & West**
	68 72	62 65	61 64	63 67	70 75	64 67	66 70	64 67	64 68	65 68	64 67	65 69	*One year*
	34 42	35 43	29 36	32 41	31 37	33 41	35 43	33 41	34 42	33 41	33 40	34 42	*Five years*
	24 36	22 35	20 31	17 30	19 31	21 34							
	(310)	*(339)*	*(425)*	*(488)*	*(514)*	*(2,098)*	*(340)*	*(359)*	*(428)*	*(436)*	*(364)*	*(1,941)*	**West Midlands**
	61 65	62 65	57 60	55 58	59 63	59 62	65 69	58 62	62 66	62 65	62 66	63 66	*One year*
	31 39	29 36	29 37	25 32	25 31	28 35	33 40	32 39	30 36	30 37	30 39	31 39	*Five years*
	24 34	21 33	20 30	16 26	17 29	19 31							
	(440)	*(486)*	*(517)*	*(650)*	*(862)*	*(2,994)*	*(470)*	*(458)*	*(491)*	*(597)*	*(663)*	*(2,696)*	**North & West**
	57 61	55 59	51 55	51 55	50 54	53 57	59 62	62 65	59 62	55 58	54 57	58 61	*One year*
	25 32	27 34	22 28	24 31	23 29	24 31	28 34	27 33	29 36	26 34	24 32	27 34	*Five years*
	18 28	18 28	13 23	16 27	16 27	16 27							
	(152)	*(221)*	*(332)*	*(382)*	*(205)*	*(1,409)*	*(179)*	*(223)*	*(324)*	*(327)*	*(195)*	*(1,321)*	**WALES**
	54 58	56 59	53 56	55 58	52 55	55 58	57 59	61 64	54 57	59 61	65 67	60 63	*One year*
	30 36	32 40	27 34	31 40	22 26	29 37	27 33	32 39	25 31	30 38	30 38	31 39	*Five years*
	23 30	23 39	20 30	22 38	12 18	20 33							

[1] Including those for whom no deprivation category was assigned. C - crude survival, R - relative survival (see Chapter 3).
Number of patients contributing to each analysis in parentheses.

Table 16.2   Crude and relative survival (%) at one, five and ten years since diagnosis: NHS Region, deprivation category and calendar period of diagnosis

	1971-75												1976-80											
ADULTS	Affluent		2		3		4		Deprived		All[1]		Affluent		2		3		4		Deprived		All[1]	
	C	R	C	R	C	R	C	R	C	R	C	R	C	R	C	R	C	R	C	R	C	R	C	R
ENGLAND & WALES	(5,338)		(6,180)		(6,721)		(6,568)		(5,687)		(38,412)		(6,956)		(7,844)		(8,612)		(8,416)		(7,484)		(41,120)	
One year	55	58	53	56	52	55	50	52	48	51	51	54	57	60	55	59	54	57	54	57	51	55	54	57
Five years	25	31	23	29	22	29	21	27	20	26	22	28	27	34	26	33	24	31	24	31	22	29	24	31
Ten years	16	26	16	26	15	25	13	24	13	22	14	24	18	30	17	29	15	27	15	26	14	25	16	27
ENGLAND	(5,130)		(5,889)		(6,247)		(6,013)		(5,412)		(36,098)		(6,699)		(7,446)		(8,009)		(7,751)		(7,119)		(38,474)	
One year	56	58	53	56	52	55	50	53	48	51	52	54	57	60	55	59	54	57	54	57	51	55	54	57
Five years	25	31	23	29	22	29	22	28	20	26	22	28	27	34	26	33	24	31	24	31	22	29	24	31
Ten years	16	26	16	26	15	25	14	24	12	22	14	24	18	30	17	29	15	27	15	26	14	25	16	27
Northern & Yorkshire	(352)		(537)		(671)		(865)		(1,113)		(4,980)		(614)		(818)		(1,035)		(1,339)		(1,704)		(5,646)	
One year	54	57	54	57	47	50	49	52	44	47	50	53	57	60	55	58	54	58	52	55	50	54	53	57
Five years	23	28	22	28	20	25	22	28	19	24	21	27	29	36	26	33	23	31	21	28	20	27	23	30
Ten years	14	23	15	25	13	22	13	23	11	19	13	23	20	33	16	28	14	25	12	22	13	22	14	25
Trent	(405)		(600)		(770)		(977)		(792)		(3,900)		(499)		(764)		(929)		(1,093)		(910)		(4,334)	
One year	54	57	53	56	54	57	46	49	46	49	51	54	55	58	53	56	49	52	52	55	53	56	53	56
Five years	22	28	23	29	23	30	20	26	19	25	22	28	24	31	23	30	21	27	22	28	22	30	23	30
Ten years	13	22	15	25	15	25	13	23	13	23	14	24	16	27	15	26	14	24	13	23	13	25	14	25
Anglia & Oxford	(691)		(724)		(700)		(540)		(249)		(3,341)		(931)		(910)		(827)		(590)		(263)		(3,663)	
One year	58	61	57	60	51	53	50	53	50	53	55	58	56	59	54	57	54	57	52	55	50	53	55	58
Five years	27	33	24	31	22	28	22	28	19	24	24	30	27	34	26	34	23	29	27	35	20	28	26	33
Ten years	19	29	15	25	14	24	14	25	10	18	15	26	17	28	17	28	15	25	18	31	13	26	17	28
North Thames	(817)		(793)		(735)		(705)		(589)		(4,151)		(729)		(675)		(784)		(746)		(583)		(3,654)	
One year	61	64	55	58	56	59	52	55	53	56	56	60	59	62	58	61	56	59	60	63	57	61	58	62
Five years	28	34	27	33	26	33	25	32	22	29	26	34	28	35	31	38	26	35	29	37	27	35	29	37
Ten years	19	29	19	30	18	29	16	28	15	26	18	30	21	32	21	34	17	29	20	33	19	32	20	32
South Thames	(1,263)		(1,125)		(1,135)		(956)		(519)		(5,498)		(1,408)		(1,244)		(1,181)		(925)		(436)		(5,515)	
One year	52	55	49	51	50	53	50	53	48	51	51	54	55	58	54	57	55	59	50	54	57	61	55	58
Five years	22	27	19	24	20	26	20	26	19	26	21	27	26	32	25	32	25	33	21	28	26	34	25	32
Ten years	14	22	12	20	12	22	12	21	11	21	13	23	17	28	16	27	16	29	14	25	17	30	16	28
South & West	(836)		(1,213)		(1,152)		(602)		(222)		(4,451)		(1,137)		(1,551)		(1,376)		(679)		(259)		(5,224)	
One year	54	56	55	58	52	55	54	57	45	48	54	57	60	63	57	61	55	58	60	64	50	53	58	61
Five years	27	33	25	31	22	29	24	31	19	24	24	31	28	35	26	34	26	34	28	37	27	34	27	35
Ten years	19	30	17	28	14	24	17	30	10	18	16	27	19	32	16	29	17	29	18	32	20	33	18	31
West Midlands	(236)		(248)		(344)		(374)		(503)		(4,364)		(624)		(638)		(902)		(1,110)		(1,244)		(4,594)	
One year	52	55	52	55	50	54	54	57	46	49	52	55	56	59	56	59	55	59	55	59	51	54	55	58
Five years	27	33	24	30	21	27	26	32	19	24	23	29	27	34	25	30	24	31	23	30	22	29	24	31
Ten years	17	27	18	29	16	26	18	29	14	23	15	26	18	30	17	28	16	27	14	25	14	25	16	27
North & West	(530)		(649)		(740)		(994)		(1,425)		(5,413)		(757)		(846)		(975)		(1,269)		(1,720)		(5,844)	
One year	50	53	46	48	46	49	43	46	46	49	46	49	54	56	48	52	48	52	49	52	45	48	49	52
Five years	21	25	19	25	20	25	16	21	19	25	18	24	26	33	23	29	19	24	21	28	19	25	21	28
Ten years	11	18	14	23	14	24	10	17	12	21	12	21	17	29	16	26	13	21	13	24	12	22	14	24
WALES	(208)		(291)		(474)		(555)		(275)		(2,314)		(257)		(398)		(603)		(665)		(365)		(2,646)	
One year	48	51	47	49	52	55	44	46	52	55	49	52	54	57	54	58	52	55	50	53	50	53	53	56
Five years	24	30	18	24	24	30	16	20	21	27	21	27	28	35	29	38	25	32	23	31	20	26	25	33
Ten years	15	27	13	22	17	28	9	17	14	22	14	24	18	31	20	35	16	27	17	29	10	18	16	28

[1] Including those for whom no deprivation category was assigned. C - crude survival, R - relative survival (see Chapter 3).
Number of patients contributing to each analysis in parentheses.

	1981-85						1986-90						ADULTS
	Affluent	*2*	*3*	*4*	*Deprived*	*All*[1]	*Affluent*	*2*	*3*	*4*	*Deprived*	*All*[1]	
	C R	C R	C R	C R	C R	C R	C R	C R	C R	C R	C R	C R	
	(8,034)	*(8,758)*	*(9,584)*	*(9,354)*	*(7,814)*	*(44,657)*	*(8,213)*	*(9,520)*	*(9,699)*	*(9,198)*	*(7,240)*	*(44,388)*	**ENGLAND & WALES**
One year	64 67	62 65	59 63	58 61	56 60	60 63	66 69	64 67	63 67	60 64	60 63	63 66	*One year*
Five years	31 38	30 38	28 36	27 35	24 32	28 36	32 40	31 39	30 38	28 36	27 35	30 38	*Five years*
	21 33	19 33	18 31	17 30	15 27	18 31							
	(7,676)	*(8,300)*	*(8,786)*	*(8,493)*	*(7,371)*	*(41,443)*	*(7,827)*	*(9,016)*	*(8,986)*	*(8,390)*	*(6,806)*	*(41,392)*	**ENGLAND**
One year	64 68	62 66	60 64	58 62	56 60	60 64	66 70	64 68	64 68	61 64	60 63	63 66	*One year*
Five years	31 39	30 38	28 36	27 35	24 32	28 36	33 41	31 39	30 39	28 36	27 35	30 38	*Five years*
	21 33	19 33	18 31	17 29	15 27	18 31							
	(742)	*(963)*	*(1,254)*	*(1,510)*	*(1,833)*	*(6,399)*	*(831)*	*(1,074)*	*(1,181)*	*(1,355)*	*(1,558)*	*(6,041)*	**Northern & Yorkshire**
One year	64 68	63 67	58 62	57 61	54 58	59 63	68 72	62 67	61 66	56 60	59 63	61 65	*One year*
Five years	29 38	31 39	28 36	25 34	23 31	27 35	32 42	30 40	28 38	26 35	26 36	28 37	*Five years*
	18 32	20 34	18 32	16 29	15 28	17 30							
	(668)	*(769)*	*(1,132)*	*(1,311)*	*(899)*	*(4,798)*	*(691)*	*(932)*	*(1,045)*	*(1,231)*	*(936)*	*(4,848)*	**Trent**
One year	64 68	55 59	57 60	58 61	54 58	58 62	65 69	61 65	62 66	59 63	57 61	61 65	*One year*
Five years	31 39	26 33	27 35	27 36	24 32	27 35	31 41	31 40	28 37	26 35	27 36	29 37	*Five years*
	20 33	17 30	18 31	17 31	15 28	18 30							
	(1,085)	*(1,084)*	*(925)*	*(595)*	*(283)*	*(4,066)*	*(1,130)*	*(1,210)*	*(1,050)*	*(655)*	*(307)*	*(4,376)*	**Anglia & Oxford**
One year	63 66	62 66	60 64	60 64	55 59	61 65	64 67	63 67	63 67	.61 65	57 61	63 67	*One year*
Five years	31 41	29 38	30 38	32 41	25 32	30 39	33 42	29 39	31 41	32 42	29 39	31 40	*Five years*
	22 37	20 34	20 35	18 35	16 27	20 34							
	(754)	*(732)*	*(711)*	*(678)*	*(553)*	*(3,505)*	*(786)*	*(881)*	*(944)*	*(872)*	*(638)*	*(4,227)*	**North Thames**
One year	65 69	68 72	65 69	64 68	66 70	66 70	69 73	69 73	69 73	67 71	66 71	68 72	*One year*
Five years	31 38	34 43	32 42	34 44	32 43	33 42	33 42	33 42	31 41	30 40	29 38	32 40	*Five years*
	21 33	23 37	21 35	23 38	20 37	22 36							
	(1,552)	*(1,218)*	*(1,134)*	*(903)*	*(331)*	*(5,364)*	*(1,370)*	*(1,277)*	*(1,070)*	*(836)*	*(408)*	*(5,038)*	**South Thames**
One year	63 67	62 66	62 66	58 62	58 62	62 66	70 74	64 69	67 71	62 66	60 65	66 70	*One year*
Five years	28 35	30 39	28 37	25 33	24 33	28 36	36 46	29 38	30 40	27 37	23 32	30 39	*Five years*
	19 32	17 32	18 33	17 30	15 29	18 32							
	(1,230)	*(1,692)*	*(1,485)*	*(755)*	*(227)*	*(5,581)*	*(1,239)*	*(1,733)*	*(1,515)*	*(924)*	*(343)*	*(5,791)*	**South & West**
One year	68 72	62 66	61 65	62 66	62 67	63 67	65 69	64 68	65 69	64 68	63 67	65 69	*One year*
Five years	34 44	32 42	28 37	30 40	26 34	31 40	32 41	32 42	33 43	31 41	33 43	32 41	*Five years*
	22 38	21 36	18 32	19 34	14 28	20 34							
	(731)	*(832)*	*(1,048)*	*(1,247)*	*(1,270)*	*(5,167)*	*(797)*	*(870)*	*(1,067)*	*(1,196)*	*(953)*	*(4,918)*	**West Midlands**
One year	64 68	61 65	61 64	56 60	58 62	60 64	65 69	63 67	63 67	61 65	63 67	63 67	*One year*
Five years	34 43	28 36	29 37	23 31	24 32	27 35	32 41	32 42	30 39	28 36	28 39	30 39	*Five years*
	25 39	18 31	18 32	14 25	15 29	18 30							
	(914)	*(1,010)*	*(1,097)*	*(1,494)*	*(1,975)*	*(6,563)*	*(983)*	*(1,039)*	*(1,114)*	*(1,321)*	*(1,663)*	*(6,153)*	**North & West**
One year	57 61	58 62	52 56	51 55	52 56	54 58	61 64	61 65	59 63	55 59	54 58	58 62	*One year*
Five years	25 33	27 35	21 28	22 30	22 29	23 31	29 37	26 33	28 37	25 34	23 32	26 34	*Five years*
	17 28	18 31	13 24	15 27	14 26	15 27							
	(358)	*(458)*	*(798)*	*(861)*	*(443)*	*(3,214)*	*(386)*	*(504)*	*(713)*	*(808)*	*(434)*	*(2,996)*	**WALES**
One year	58 62	55 58	54 58	53 57	56 59	55 59	61 64	58 62	55 58	57 60	60 63	59 62	*One year*
Five years	29 36	29 38	26 35	28 37	24 31	28 36	27 35	30 38	26 35	27 36	28 36	29 37	*Five years*
	21 33	20 36	17 31	18 33	14 25	18 31							

[1] Including those for whom no deprivation category was assigned. C - crude survival, R - relative survival (see Chapter 3). Number of patients contributing to each analysis in parentheses.

Fig. 16.1   Relative survival up to ten years: trends by calendar period of diagnosis (1971-90) and NHS Region

ADULTS

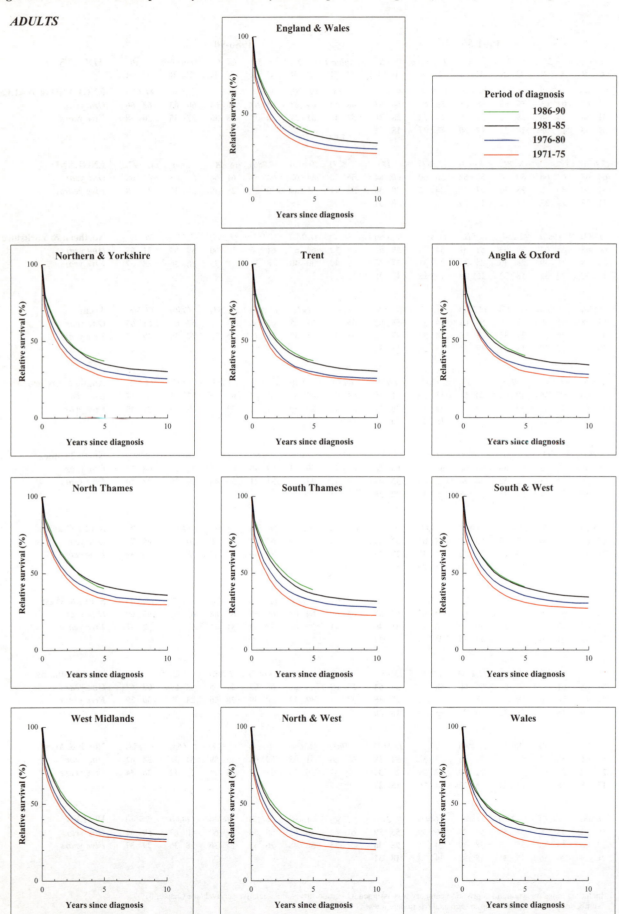

Fig. 16.2   Relative survival up to five years, by deprivation category and NHS Region: patients diagnosed 1986-90

ADULTS

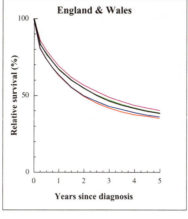

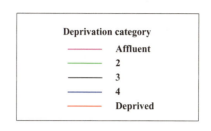

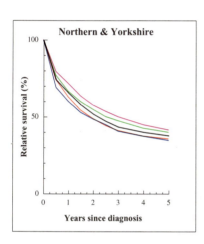

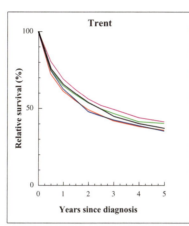

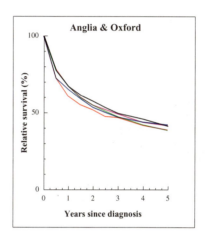

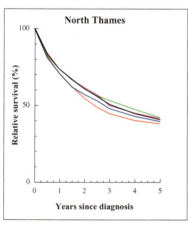

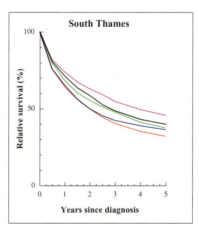

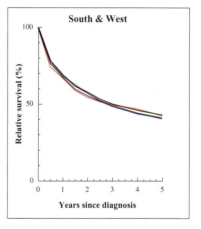

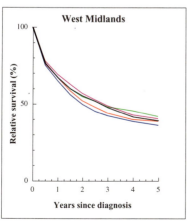

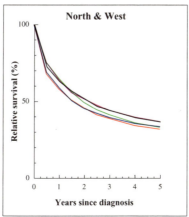

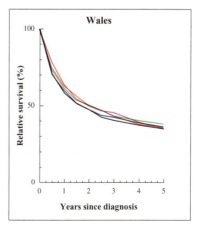

Fig. 16.3 Relative survival at five years by deprivation category, period of diagnosis (1981-90) and NHS Region

ADULTS

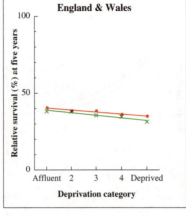

England & Wales

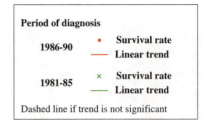

Period of diagnosis

1986-90   ● Survival rate
           ── Linear trend

1981-85   ✕ Survival rate
           ── Linear trend

Dashed line if trend is not significant

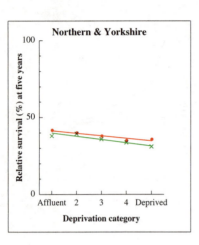

Northern & Yorkshire

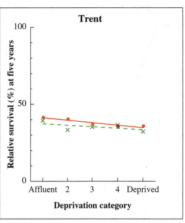

Trent

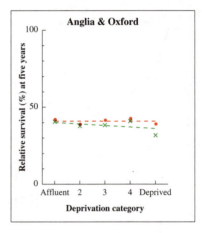

Anglia & Oxford

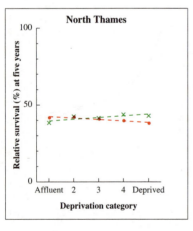

North Thames

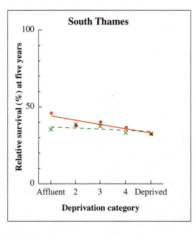

South Thames

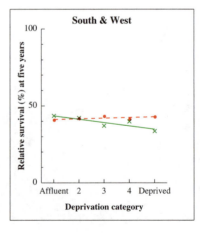

South & West

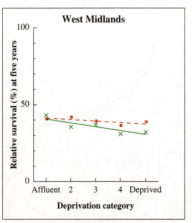

West Midlands

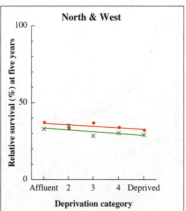

North & West

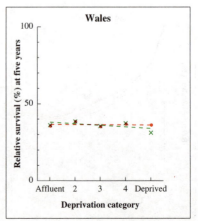

Wales

Fig. 16.4 **Relative survival (%) by NHS Region and sex:**
           one-year and five-year survival for patients diagnosed 1986-90, with 95% confidence intervals

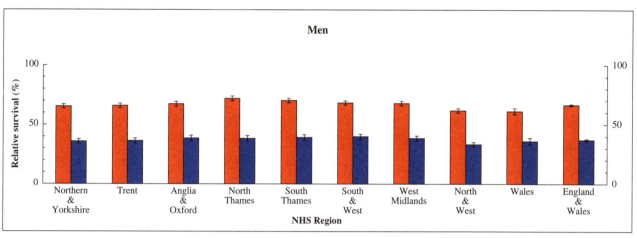

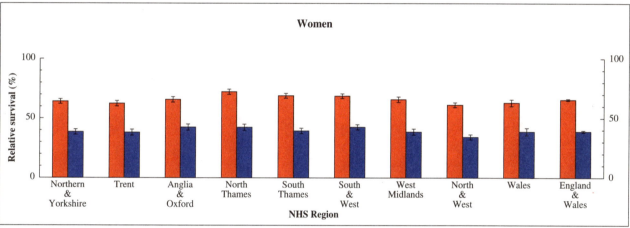

Fig. 16.5 **Relative survival (%) at five years, with 95% confidence interval, by age at diagnosis and sex:**
           England and Wales, patients diagnosed 1986-90

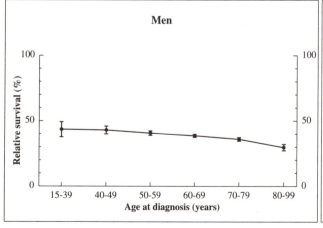
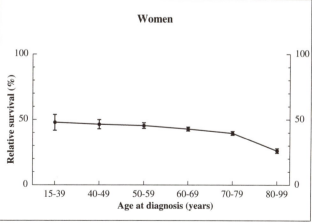

Chapter 17

LIVER

Primary malignancies of the liver are uncommon in England and Wales, but the liver is frequently the site of metastatic spread from primary cancers arising in other organs. The incidence of primary carcinoma of the liver in England and Wales is less than 3 per 100,000 per year in men and about half that in women: primary liver cancers account for fewer than one in 200 malignancies in adults. Incidence has been increasing in the UK[1], particularly in older men[2]. Liver tumours usually present in an advanced stage, and in most clinical series the proportion considered surgically removable has been low, even in the absence of cirrhosis[3]. Until non-invasive imaging techniques became widely available, the clinical condition of the patient when first seen often precluded a precise (or morphological) diagnosis, and some tumours included in this series, particularly those registered during the 1970s, are likely to have been metastatic from other organs rather than primary tumours of the liver.

Survival from hepatocellular carcinoma has been low even in clinical series, with 10-15% peri-operative mortality rates from bleeding or liver failure considered acceptable. Surgical resection and arterial embolisation of the tumour have been the main therapeutic options. Transplantation has been used more recently in selected patients.

The availability of such a large population-based series of liver cancers diagnosed over a 20-year period represents an opportunity to examine survival trends and geographical differences in survival. The well-known difficulties in registering liver cancer were taken into account in preparing the data for analysis, but the quality of the data cannot be as high as in small clinical series. The results presented here must be interpreted in this light.

About 100 tumours a year are described at registration simply as 'cancer of the liver', or else are not specified as primary cancers. Such tumours were not considered eligible for analysis (Table 17.1, "otherwise ineligible"). Since 1980, most of these tumours have been explicitly coded as metastatic (e.g. all but 23 of 480 registered during 1986-90).

The survival analyses cover 12,000 adults registered with a primary malignancy of the liver in England and Wales during the period 1971-90, and followed up to the end of 1995. Among almost 18,000 eligible patients, 5% were excluded because their vital status was unknown at 6 July 1997 (Table 17.1). A further 25% were excluded because their recorded survival was zero or unknown. An appreciable proportion of these patients would be expected to have true zero survival, since operative

mortality for liver cancer has been high, but many cases in this category would have been registered solely from a death certificate, and the two groups cannot be distinguished in these data (see Chapter 2). The proportion in this category rose from 22% in 1971-75 to 28% in 1986-90, even though operative mortality has fallen.

The possible influence of these exclusions on patterns of survival needs comment. There were marked regional differences. During the 1980s, 10% to 30% of liver cancer patients eligible for analysis in most regions had a recorded survival of zero, but for patients diagnosed during 1981-85 this proportion reached 40% in South Thames and 60% in North Thames. Because the actual survival of these patients, if known, would be expected to have been shorter than for other patients (see Chapter 2), survival estimates for liver cancer in these regions would be biassed upward by their exclusion.

By contrast, patients with zero recorded survival had the same distribution across deprivation groups as all other patients throughout the 1970s and 1980s (data not shown): any deprivation gradient in survival should therefore be unaffected.

Most cases were primary carcinomas of the liver. Overall, about one in seven were carcinomas of the intra-hepatic bile ducts, but this proportion rose gradually to one in five by the late 1980s (Table 17A). Two-thirds (62%) of the tumours were adenocarcinomas. This proportion increased from 50% in the early 1970s, and during 1986-90 almost three-quarters (71%) were recorded as hepatocellular carcinoma (41%), cholangiocarcinoma (26%) or simply adenocarcinoma (5%). During this period, only 0.3% were classified as squamous carcinomas, but 25% were coded as carcinoma without further specification.

Survival trends

Relative survival for adults diagnosed in England and Wales during 1986-90 was 9% at one year and just 2% at five years. The corresponding figures were 3% and 1% for patients diagnosed in 1971-75. There were no major differences between men and women (Table 17.2).

This improvement of about 6% in one-year survival is visible as a lifting of the heel of the survival curves for patients diagnosed in the 1980s compared to the curves for earlier periods, both nationally and in most regions (Figure 17.1). The apparently larger increase in survival at one and five years for patients diagnosed in North Thames during the 1980s is anomalous (see Figure 17.4). This increase

Table 17A Sub-site distribution of liver cancer

ICD-9 code	Site description	1971-75		1976-80		1981-85		1986-90		Total	
		No.	%	No.	%	No.	%	No.	%	No.	%
155.0	Liver, primary	2,356	90.7	2,519	90.4	2,763	85.2	2,970	81.0	**10,608**	**86.3**
155.1	Intrahepatic bile ducts	241	9.3	269	9.6	479	14.8	696	19.0	**1,685**	**13.7**
	Total	**2,597**		**2,788**		**3,242**		**3,666**		**12,293**	

was similar in all deprivation groups, mirroring the distribution of patients excluded with zero or unknown survival, so the larger increase in survival in North Thames may well be an artefact (see above). Even so, only 41 of the 644 patients with liver cancer included in the analyses for this region during the 1980s had a recorded survival greater than five years.

Survival for patients diagnosed during 1986-90 was extremely low for men and women at all ages (Figure 17.5). Under the age of 40, five-year survival was 10% to 15%. At 40 or over, survival was less than 5% (see Table 4.7).

The average gain in one-year survival, estimated from the age-standardised relative survival rates in successive five-year periods, was about 2% for men and women (see Table 4.4). For five-year survival, the corresponding average improvement between successive quinquennia was only 0.5%. These improvements are fairly regular, and with more than 1,000 patients contributing to the survival estimate for men and women in each five-year period, even these small gains are statistically significant. The gains in survival are extremely limited in clinical or public health terms.

Deprivation

Despite the poor overall survival for liver cancer, the one-year survival of patients diagnosed with liver cancer in the 1980s in England and Wales is consistently lower for the more deprived groups (Table 17.2, Figure 17.2). This gradient is apparent in all regions except Anglia & Oxford and North & West, where there is no difference between deprivation groups (data not shown).

For 3,200 patients of known deprivation status who were diagnosed in England and Wales during 1981-85, the gap in one-year survival between the most deprived and the most affluent groups was 2% (see Table 4.9). For 3,600 patients diagnosed in 1986-90, the gap was 4%. Survival is very low for all deprivation groups, so while these differences are numerically small, they are relatively large. The gap of 4% implies an average survival of 6.5% for the most deprived group compared to 10.6% for the most affluent. Although the survival estimates themselves may be somewhat inflated (see above and Chapter 2), exclusions for zero or unknown survival were evenly distributed between deprivation categories, and the deprivation gradient in one-year survival is unlikely to be biassed.

Differences between the deprivation categories in survival at five years were less than 1%, and were not significant (Figure 17.3 – note expanded y-axis scale).

International comparison

Five-year survival from liver cancer for patients diagnosed in England and Wales during 1986-90 was 2%, about half the corresponding average for Europe in 1985-89 (see Table 4.12). If the highest regional survival rate of 8% (North Thames) is discounted, there is very little variation between regions. Survival in the USA is higher than in Europe.

1.   Taylor-Robinson SD, Foster GR, Arora S, Hargreaves S, Thomas HC. Increase in primary liver cancer in the UK, 1979-94. Lancet 1997; 350: 1142-1143

2.   de vos Irvine H, Goldberg D, Hole D, McMenamin J. Trends in primary liver cancer. Lancet 1998; 351: 215-216

3.   Gennari L, Doci R, Bozzetti F. Liver tumours. In: Peckham M, Pinedo HM, Veronesi U, (eds.) Oxford textbook of oncology. Oxford: Oxford Medical Publications, 1995

Table 17.1 Ineligible and excluded records, and patients included in analyses, by calendar period of diagnosis

Calendar period of diagnosis

	1971-75			1976-80			1981-85			1986-90			Overall		
Total registered	**4,372**			**4,638**			**5,367**			**6,101**			**20,478**		
Ineligible	*Records*	*Patients*	*%*	*Records*	*Patients*	*%*	*Records*	*Patients*	*%*	*Records*	*Patients*	*%*	*Records*	*Patients*	*%*
Incomplete data[1]	3	3		11	11		21	21		13	13		48	**48**	
Resident outside England and Wales	86	86		90	90		46	44		28	26		250	**246**	
In situ neoplasm[2]	4	4		0	0		0	0		0	0		4	**4**	
Benign or uncertain behaviour[2]	39	11		21	18		0	0		0	0		60	**29**	
Metastatic[2]	227	226		254	253		470	465		461	459		1,412	**1,403**	
Otherwise ineligible[3]	489	457		501	276		472	3		480	23		1942	**759**	
Lymphoma[4]	11	9		2	2		0	0		0	0		13	**11**	
Leukaemia or myeloma[4]	0	0		0	0		0	0		0	0		0	**0**	
		796			650			533			521			2,500	
Total eligible		**3,576**	**100**		**3,988**	**100**		**4,834**	**100**		**5,580**	**100**		**17,978**	**100**
Aged under 15 or 100 years or over at diagnosis	55	46	1.3	58	51	1.3	58	49	1.0	73	72	1.3	244	**218**	1.2
Vital status unknown[5]	194	132	3.7	281	207	5.2	300	234	4.8	326	283	5.1	1,101	**856**	4.8
Duplicate registration[6]	0	0	0.0	0	0	0.0	0	0	0.0	0	0	0.0	0	**0**	0.0
Multiple primary[7]	0	0	0.0	2	2	<0.1	12	10	0.2	10	7	0.1	24	**19**	0.1
Synchronous tumours[8]	26	24	0.7	25	16	0.4	28	8	0.2	21	17	0.3	100	**65**	0.4
Sex not known	0	0	0.0	0	0	0.0	0	0	0.0	0	0	0.0	0	**0**	0.0
Sex incompatible with site code[9]	-	-	-	-	-	-	-	-	-	-	-	-	-	**-**	-
Invalid dates[10]	1	1	<0.1	0	0	0.0	1	1	<0.1	0	0	0.0	2	**2**	<0.1
Zero survival[11]	1,043	776	21.7	1,152	924	23.2	1,492	1,290	26.7	1,750	1,535	27.5	5,437	**4,525**	25.2
Total exclusions		**979**	**27.4**		**1,200**	**30.1**		**1,592**	**32.9**		**1,914**	**34.3**		**5,685**	**31.6**
Patients included in analysis		**2,597**	**72.6**		**2,788**	**69.9**		**3,242**	**67.1**		**3,666**	**65.7**		**12,293**	**68.4**

[1] Main data item(s) invalid or incompatible with one another: name, sex, date of birth, date of diagnosis and (if present) date of death, postcode, site, morphology and behaviour

[2] Tumour either *in situ* (behaviour code 2), benign (0), uncertain if benign or malignant (1) or metastatic (6 or 9)

[3] Liver, unspecified (ICD-8 197.8, 1971-78) or not specified as primary or secondary (ICD-9 155.2, 1979-90)

[4] Morphology that of lymphoma (included with non-Hodgkin lymphoma, Ch. 41) or myeloma (included with multiple myeloma, Ch. 43), or leukaemia (excluded)

[5] Tracing of vital status at National Health Service Central Register incomplete at 6 July 1997

[6] Same site code, sex, cancer registry and cancer registry number as an earlier registration

[7] Same site code and person number as an earlier registration(s): mostly confirmed multiple primary tumours at this site, some unresolved duplicate registrations

[8] Same site code, sex, date of birth and date of diagnosis as another registration(s): mostly synchronous or (in paired organs) bilateral tumours in same anatomic site in one individual, not previously linked: also some duplicate registrations

[9] Not applicable for this malignancy

[10] Impossible sequence of dates (birth, diagnosis and, if present, emigration or death); or date of death after 6 July 1997

[11] Date of diagnosis same as date of death: some are patients who did die on the day of diagnosis, but most were registered solely from a death certificate (death certificate only, or DCO), and their survival time is thus unknown

Table 17.2   Crude and relative survival (%) at one, five and ten years since diagnosis: NHS Region, deprivation category and calendar period of diagnosis

ENGLAND & WALES

	1971-75 Affluent C R	2 C R	3 C R	4 C R	Deprived C R	All[1] C R	1976-80 Affluent C R	2 C R	3 C R	4 C R	Deprived C R	All[1] C R
Men	(219)	(250)	(293)	(287)	(296)	(1,588)	(271)	(292)	(341)	(384)	(425)	(1,749)
One year	3 3	3 3	4 4	2 2	1 1	2 3	5 5	3 4	3 3	2 2	3 3	3 3
Five years	0 0	1 1	1 1	0 0	0 1	1 1	2 2	1 1	0 0	1 1	1 1	1 1
Ten years	0 0	1 1	1 1	0 0	0 0	0 1	1 2	1 1	0 0	1 1	1 1	1 1
Women	(144)	(177)	(192)	(205)	(149)	(1,009)	(168)	(217)	(226)	(210)	(192)	(1,039)
One year	5 5	5 5	2 2	4 4	4 5	4 4	11 11	5 5	5 5	6 6	4 4	6 6
Five years	1 1	2 2	0 0	1 1	1 1	1 1	2 3	1 1	2 2	1 1	0 0	1 1
Ten years	0 0	1 2	0 0	1 1	1 1	1 1	2 3	0 0	1 1	0 1	0 0	1 1
Adults	(363)	(427)	(485)	(492)	(445)	(2,597)	(439)	(509)	(567)	(594)	(617)	(2,788)
One year	3 4	4 4	3 3	3 3	2 2	3 3	7 7	4 4	4 4	3 3	4 4	4 4
Five years	0 1	1 1	1 1	1 1	1 1	1 1	2 2	1 1	1 1	1 1	1 1	1 1
Ten years	0 0	1 1	0 1	1 1	0 1	1 1	2 2	1 1	0 1	1 1	1 1	1 1

	1971-75 Men C R	Women C R	Adults C R	1976-80 Men C R	Women C R	Adults C R
Northern & Yorkshire	(212)	(123)	(335)	(189)	(118)	(307)
One year	1 1	6 6	3 3	2 2	5 5	3 3
Five years	0 0	1 1	0 0	1 1	0 0	1 1
Ten years	0 0	1 1	0 0	1 1	0 0	0 1
Trent	(152)	(121)	(273)	(170)	(109)	(279)
One year	2 2	2 2	2 2	2 2	3 3	2 2
Five years	1 1	1 1	1 1	0 0	0 0	0 0
Ten years	0 0	1 1	0 1	0 0	0 0	0 0
Anglia & Oxford	(134)	(66)	(200)	(140)	(66)	(206)
One year	2 2	2 3	2 2	3 3	11 11	5 5
Five years	1 1	1 1	1 1	1 2	3 3	2 2
Ten years	0 0	1 1	0 0	1 1	2 2	1 2
North Thames	(246)	(165)	(411)	(179)	(102)	(281)
One year	3 3	2 2	2 2	3 3	6 6	4 4
Five years	1 1	0 0	1 1	0 1	1 1	1 1
Ten years	1 1	0 0	1 1	0 1	0 0	0 0
South Thames	(196)	(113)	(309)	(238)	(128)	(366)
One year	2 2	7 7	4 4	7 7	10 11	8 8
Five years	1 1	2 2	1 1	2 3	2 2	2 2
Ten years	1 1	2 2	1 1	2 2	1 1	1 2
South & West	(184)	(142)	(326)	(215)	(147)	(362)
One year	5 5	2 2	3 4	3 3	6 6	4 4
Five years	1 2	1 1	1 2	0 0	2 2	1 1
Ten years	1 2	1 1	1 1	0 0	1 1	0 1
West Midlands	(146)	(63)	(209)	(181)	(100)	(281)
One year	3 3	3 3	3 3	5 5	7 7	5 6
Five years	0 0	0 0	0 0	2 2	2 2	2 2
Ten years	0 0	0 0	0 0	2 2	1 1	1 2
North & West	(206)	(128)	(334)	(273)	(152)	(425)
One year	2 2	6 6	3 3	2 2	3 3	3 3
Five years	0 0	1 1	0 1	0 0	1 1	1 1
Ten years	0 0	1 1	0 0	0 0	1 1	0 1
WALES	(112)	(88)	(200)	(164)	(117)	(281)
One year	3 3	5 5	4 4	3 3	6 7	4 5
Five years	1 1	2 2	1 1	1 1	2 3	1 2
Ten years	1 1	1 1	1 1	1 1	2 3	- -

[1] Including those for whom no deprivation category was assigned. C - crude survival, R - relative survival (see Chapter 3). Number of patients contributing to each analysis in parentheses.

	1981-85						1986-90						
	Affluent C R	*2* C R	*3* C R	*4* C R	*Deprived* C R	*All*[1] C R	*Affluent* C R	*2* C R	*3* C R	*4* C R	*Deprived* C R	*All*[1] C R	
													ENGLAND & WALES
	(313)	(372)	(376)	(400)	(460)	(1,945)	(378)	(435)	(430)	(505)	(479)	(2,253)	**Men**
One year	5 5	4 5	4 4	6 6	5 5	5 5	11 12	9 9	8 8	7 7	7 7	8 8	*One year*
Five years	1 1	1 1	1 1	2 3	1 1	1 1	3 3	2 3	1 2	2 2	2 2	2 2	*Five years*
	1 1	0 0	1 1	2 2	1 1	1 1							
	(218)	(245)	(266)	(285)	(260)	(1,297)	(257)	(266)	(304)	(309)	(261)	(1,413)	**Women**
	14 14	9 9	7 7	7 7	5 6	8 8	11 12	8 8	8 9	8 8	6 6	8 9	*One year*
	4 4	2 2	2 3	3 3	1 2	2 3	2 2	2 2	2 2	1 2	1 1	2 2	*Five years*
	3 3	1 1	2 3	1 2	1 1	2 2							
	(531)	(617)	(642)	(685)	(720)	(3,242)	(635)	(701)	(734)	(814)	(740)	(3,666)	**Adults**
	8 8	6 6	5 5	6 6	5 5	6 6	11 12	8 9	8 8	7 8	7 7	8 9	*One year*
	2 3	1 1	1 2	2 3	1 1	2 2	2 3	2 3	2 2	2 2	2 2	2 2	*Five years*
	1 2	1 1	1 2	2 2	1 1	1 2							

	Men	Women	Adults				Men	Women	Adults				
	(249)	(154)	(403)				(301)	(166)	(467)				**Northern & Yorkshire**
	4 4	8 8	5 6				7 7	8 9	7 8				*One year*
	1 2	3 3	2 2				1 2	1 1	1 2				*Five years*
	1 1	2 2	1 2										
	(214)	(178)	(392)				(211)	(175)	(386)				**Trent**
	3 3	5 5	4 4				4 4	11 12	7 7				*One year*
	1 1	2 2	1 1				0 1	2 2	1 1				*Five years*
	0 0	1 1	1 1										
	(163)	(105)	(268)				(194)	(118)	(312)				**Anglia & Oxford**
	5 5	12 13	7 8				8 8	8 9	8 8				*One year*
	0 0	4 4	1 1				2 2	2 2	2 2				*Five years*
	0 0	2 2	1 1										
	(186)	(95)	(281)				(244)	(119)	(363)				**North Thames**
	7 7	20 21	11 11				19 20	21 22	19 20				*One year*
	2 2	8 8	4 4				6 8	6 8	6 8				*Five years*
	2 2	5 7	3 3										
	(225)	(158)	(383)				(236)	(153)	(389)				**South Thames**
	7 7	11 12	9 9				13 13	10 10	11 12				*One year*
	2 2	3 3	2 3				3 3	2 3	3 3				*Five years*
	2 2	2 3	- -										
	(244)	(155)	(399)				(279)	(195)	(474)				**South & West**
	7 7	7 7	7 7				6 7	6 6	6 7				*One year*
	1 2	2 3	2 2				1 1	1 2	1 1				*Five years*
	1 1	1 2	1 2										
	(173)	(105)	(278)				(205)	(103)	(308)				**West Midlands**
	3 3	4 4	3 3				7 7	5 5	6 6				*One year*
	1 1	1 1	1 1				3 3	1 1	2 2				*Five years*
	1 1	0 0	1 1										
	(365)	(254)	(619)				(398)	(240)	(638)				**North & West**
	4 4	5 6	4 5				6 6	5 5	5 6				*One year*
	0 0	1 1	1 1				1 1	1 1	1 1				*Five years*
	0 0	1 1	1 1										
	(126)	(93)	(219)				(185)	(144)	(329)				**WALES**
	5 6	10 11	7 8				8 8	8 9	8 8				*One year*
	3 3	4 5	3 4				3 4	1 1	2 3				*Five years*
	2 3	2 3	2 3										

[1] Including those for whom no deprivation category was assigned. C - crude survival, R - relative survival (see Chapter 3).
Number of patients contributing to each analysis in parentheses.

Fig. 17.1   Relative survival up to ten years: trends by calendar period of diagnosis (1971-90) and NHS Region

ADULTS

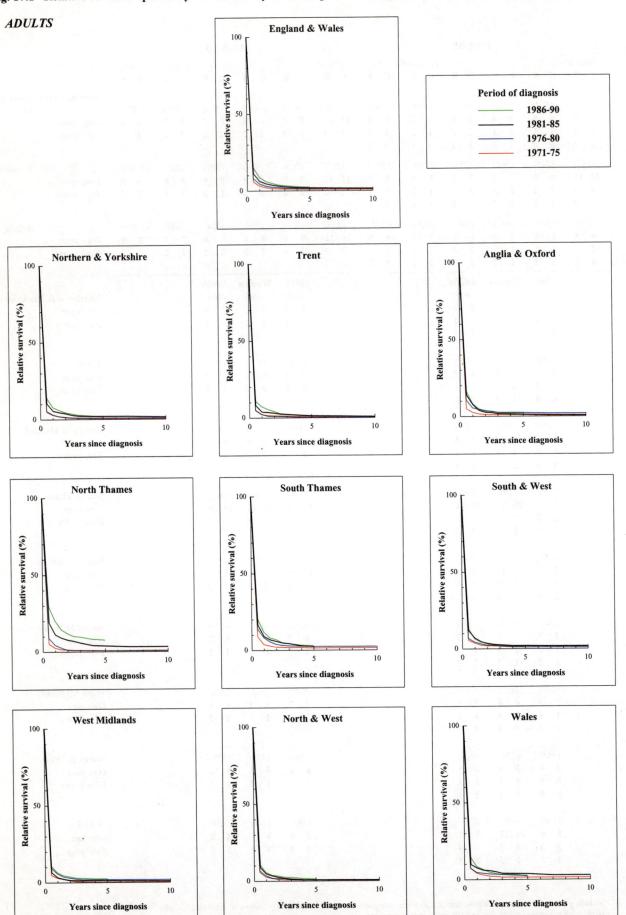

Fig. 17.2 Relative survival up to five years, by deprivation category: England and Wales, patients diagnosed 1986-90

ADULTS

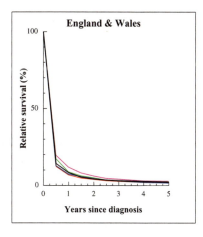

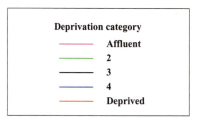

Fig. 17.3 Relative survival at five years, by deprivation category and period of diagnosis (1981-90): England and Wales

ADULTS
(Note vertical scale)

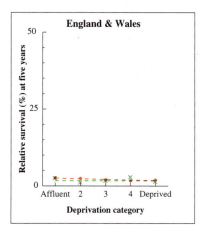

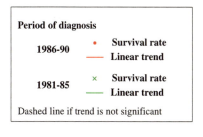

Fig. 17.4 **Relative survival (%) by NHS Region and sex:**
one-year and five-year survival for patients diagnosed 1986-90, with 95% confidence intervals

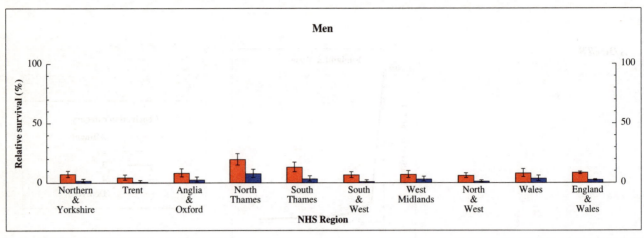

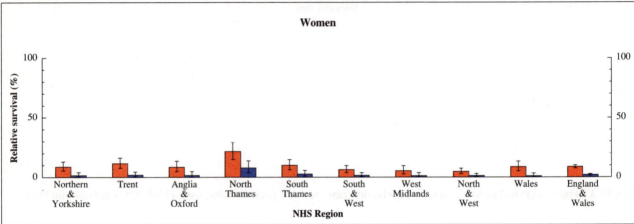

Fig. 17.5 **Relative survival (%) at five years, with 95% confidence interval, by age at diagnosis and sex:**
England and Wales, patients diagnosed 1986-90

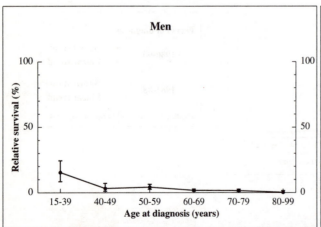
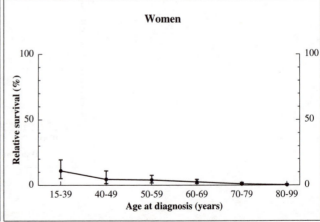

Chapter 18

GALLBLADDER

Primary cancer of the gallbladder and bile ducts is uncommon in England and Wales, accounting for 0.7% of malignant tumours in adults. There has been little change in age-standardised incidence rates since 1971, and about 1,200 new cases are registered each year in England and Wales. Cancer of the gallbladder is one of the few cancers for which there is a female excess, with three cases in women for every two in men. Cancers of the extra-hepatic bile ducts and the ampulla of Vater are more common in men.

Cholesterol gallstones, obesity and increasing parity are the main risk factors for gallbladder cancer in Western populations. Recorded mortality from cancer of the gallbladder in women fell in several Anglo-Saxon populations up to 1985. In England and Wales, mortality fell by about 1% a year[1]. The decline in mortality has been partly attributed to increasing rates of cholecystectomy, with fewer gallbladders at risk[2]. Age-standardised incidence in women has remained between 1.8 and 2.1 per 100,000 per year in the period 1979-90[3], however, and trends in cholecystectomy appear unrelated to gallstone prevalence[4]; there may be other reasons for the observed mortality trends. Patients with longstanding ulcerative colitis have a higher risk of tumours of the extra-hepatic bile ducts.

The survival analyses reported here cover 20,100 adults registered with cancer of the gallbladder in England and Wales between 1971 and 1990, and followed up to the end of 1995, some 83% of those eligible (Table 18.1). Of the 24,000 eligible patients, about 4% were excluded because their vital status was unknown at 6 July 1997, and 12% because their recorded survival was zero or unknown. About 80% of tumours arose in the gallbladder itself or in the extra-hepatic bile ducts, and 14% in the ampulla of Vater (Table 18A). Among cases diagnosed during 1986-90, 13% were classified as cholangiocarcinoma and another 34% simply as adenocarcinoma; 40% were unspecified carcinomas.

Survival trends

Relative survival for adults diagnosed in England and Wales during 1986-90 was 22% at one year and 9% at five years, compared to 12% and 4%, respectively, for those diagnosed an average of 15 years earlier, in 1971-75 (Table 18.2). The improvement in survival is visible in the successive survival curves in Figure 18.1, both nationally and in most regions.

Unusually, survival has been higher for men than for women since the 1970s, mainly among younger patients (Table 18.2, data for men at end of chapter; Figure 18.5). Among more than 700 patients aged under 60 at diagnosis during 1986-90, for example, five-year survival for men in each age group was at least 5% higher than for women (see Table 4.7); at older ages there was little difference.

The average improvement in age-standardised one-year survival between successive five-year periods was 3-4%. The corresponding gain in five-year survival was about 2% (see Table 4.4).

Deprivation

Survival by deprivation category is only presented at national level for men, because of the lack of precision of regional estimates (see Chapter 3).

Differences in survival between deprivation groups for patients diagnosed during 1986-90 are rather variable between regions (Figure 18.2). This variability is not due to regional differences in the numbers of patients with unknown deprivation category (see Table 4.3), or to differences between deprivation categories in the proportion of patients excluded with zero survival (data not shown).

Survival is often higher in the second group than in the most affluent, particularly at one year after diagnosis, but patients in the most deprived group generally have the lowest survival (Table 18.2). Among 5,000 patients of known deprivation status who were diagnosed during 1986-90, the regression estimate of one-year survival for patients in the most affluent group was about 24%, with a significant deficit of almost 6% for patients in the most deprived group (see Table 4.9). The gap was wider for patients diagnosed during 1986-90 than in the early 1980s. Much of the limited overall improvement in one-year survival from gallbladder cancer in England and Wales during the 1980s is attributable to gains for patients in the more affluent categories.

The gradient in five-year survival was 1% in the early 1980s and 2% for patients diagnosed during 1986-90, but the gradient is not statistically significant in either period (Figure 18.3).

International comparison

Five-year survival from gallbladder cancer in England and Wales is about 9%, similar to the levels in Scotland but some 3% lower than the average for Europe (see Table

Table 18A Sub-site distribution of gallbladder cancer

ICD-9 code	Site description	1971-75		1976-80		1981-85		1986-90		Total	
		No.	%	No.	%	No.	%	No.	%	No.	%
156.0	Gallbladder	2,261	46.5	2,268	44.0	2,227	43.7	2,062	41.4	**8,818**	**43.9**
156.1	Extrahepatic bile ducts	1,729	35.6	2,053	39.8	1,943	38.1	1,900	38.1	**7,625**	**37.9**
156.2	Ampulla of Vater	667	13.7	630	12.2	717	14.1	821	16.5	**2,835**	**14.1**
156.8	Other	-	-	11	0.2	31	0.6	17	0.3	**59**	**0.3**
156.9	Biliary tract, part unspecified	202	4.2	194	3.8	182	3.6	185	3.7	**763**	**3.8**
	Total	**4,859**		**5,156**		**5,100**		**4,985**		**20,100**	

4.12). Survival for the US population covered by the SEER programme is 18% for men and 14% for women.

1.   Nectoux J, Coleman MP. Trends in biliary tract cancer. Rev épidémiol santé publ 1993; 41: 113-122

2.   Diehl AK, Beral V. Cholecystectomy and changing mortality from gallbladder cancer. Lancet 1981; 2: 187-189

3.   Office for National Statistics. *Cancer statistics, registrations: England and Wales, 1991. Series MB1 no. 24.* London: The Stationery Office, 1997

4.   Bateson MC. Gallbladder disease and cholecystectomy rate are independently variable. Lancet 1984; ii: 621-624

Table 18.1 Ineligible and excluded records, and patients included in analyses, by calendar period of diagnosis

		Calendar period of diagnosis														
	1971-75			1976-80			1981-85			1986-90			Overall		
Total registered	**5,828**			**6,193**			**6,236**			**6,259**			**24,516**		
Ineligible	*Records*	*Patients*	*%*	*Records*	*Patients*	*%*	*Records*	*Patients*	*%*	*Records*	*Patients*	*%*	*Records*	*Patients*	*%*
Incomplete data[1]	2	2		4	4		12	12		9	9		27	**27**	
Resident outside England and Wales	46	46		43	42		27	24		18	14		134	**126**	
In situ neoplasm[2]	7	7		1	1		0	0		0	0		8	**8**	
Benign or uncertain behaviour[2]	18	8		11	8		0	0		1	1		30	**17**	
Metastatic[2]	168	167		60	58		0	0		1	1		229	**226**	
Otherwise ineligible[3]	-	-		-	-		-	-		-	-		-	**-**	
Lymphoma[4]	0	0		0	0		0	0		0	0		0	**0**	
Leukaemia or myeloma[4]	0	0		0	0		0	0		0	0		0	**0**	
		230			113			36			25			404	
Total eligible	**5,598**	**100**		**6,080**	**100**		**6,200**	**100**		**6,234**	**100**		**24,112**	**100**	
Aged under 15 or 100 years or over at diagnosis	2	2	<0.1	5	5	<0.1	1	1	<0.1	8	8	0.1	16	**16**	<0.1
Vital status unknown[5]	178	150	2.7	239	217	3.6	246	233	3.8	313	305	4.9	976	**905**	3.8
Duplicate registration[6]	0	0	0.0	0	0	0.0	0	0	0.0	0	0	0.0	0	**0**	0.0
Multiple primary[7]	0	0	0.0	5	5	<0.1	18	14	0.2	22	20	0.3	45	**39**	0.2
Synchronous tumours[8]	8	6	0.1	15	9	0.1	21	20	0.3	20	16	0.3	64	**51**	0.2
Sex not known	0	0	0.0	0	0	0.0	0	0	0.0	0	0	0.0	0	**0**	0.0
Sex incompatible with site code[9]	-	-	-	-	-	-	-	-	-	-	-	-	-	**-**	-
Invalid dates[10]	2	2	<0.1	1	0	0.0	2	2	<0.1	3	2	<0.1	8	**6**	<0.1
Zero survival[11]	625	579	10.3	738	688	11.3	885	830	13.4	948	898	14.4	3,196	**2,995**	12.4
Total exclusions		**739**	**13.2**		**924**	**15.2**		**1,100**	**17.7**		**1,249**	**20.0**		**4,012**	**16.6**
Patients included in analysis		**4,859**	**86.8**		**5,156**	**84.8**		**5,100**	**82.3**		**4,985**	**80.0**		**20,100**	**83.4**

[1] Main data item(s) invalid or incompatible with one another: name, sex, date of birth, date of diagnosis and (if present) date of death, postcode, site, morphology and behaviour

[2] Tumour either *in situ* (behaviour code 2), benign (0), uncertain if benign or malignant (1) or metastatic (6 or 9)

[3] Not applicable for this malignancy

[4] Morphology that of lymphoma (included with non-Hodgkin lymphoma, Ch. 41) or myeloma (included with multiple myeloma, Ch. 43), or leukaemia (excluded)

[5] Tracing of vital status at National Health Service Central Register incomplete at 6 July 1997

[6] Same site code, sex, cancer registry and cancer registry number as an earlier registration

[7] Same site code and person number as an earlier registration(s): mostly confirmed multiple primary tumours at this site, some unresolved duplicate registrations

[8] Same site code, sex, date of birth and date of diagnosis as another registration(s): mostly synchronous or (in paired organs) bilateral tumours in same anatomic site in one individual, not previously linked: also some duplicate registrations

[9] Not applicable for this malignancy

[10] Impossible sequence of dates (birth, diagnosis and, if present, emigration or death); or date of death after 6 July 1997

[11] Date of diagnosis same as date of death: some are patients who did die on the day of diagnosis, but most were registered solely from a death certificate (death certificate only, or DCO), and their survival time is thus unknown

Table 18.2   Crude and relative survival (%) at one, five and ten years since diagnosis:
NHS Region, deprivation category and calendar period of diagnosis

	1971-75						1976-80					
WOMEN	*Affluent*	*2*	*3*	*4*	*Deprived*	*All*[1]	*Affluent*	*2*	*3*	*4*	*Deprived*	*All*[1]
	C R	C R	C R	C R	C R	C R	C R	C R	C R	C R	C R	C R
ENGLAND & WALES	*(419)*	*(513)*	*(526)*	*(557)*	*(459)*	*(3,004)*	*(507)*	*(613)*	*(694)*	*(633)*	*(583)*	*(3,100)*
One year	11 11	10 10	10 11	11 11	10 10	**11 11**	14 14	11 12	11 12	12 12	9 10	**11 12**
Five years	4 4	3 4	2 3	3 3	2 3	**3 3**	4 5	3 4	3 4	3 4	3 4	**3 4**
Ten years	3 4	2 2	2 2	2 3	2 3	**2 3**	3 4	2 3	2 4	2 3	2 4	**2 4**
ENGLAND	*(406)*	*(490)*	*(489)*	*(520)*	*(438)*	*(2,842)*	*(488)*	*(575)*	*(645)*	*(589)*	*(560)*	*(2,910)*
One year	11 11	10 10	11 11	11 11	9 10	**11 11**	14 15	11 12	11 11	11 12	9 10	**11 12**
Five years	4 4	3 4	2 3	3 3	2 2	**3 3**	4 5	3 4	3 4	3 3	3 4	**3 4**
Ten years	2 4	2 3	2 2	2 3	1 2	**2 3**	3 4	2 3	2 4	2 2	2 4	**2 3**
Northern & Yorkshire	*(24)*	*(43)*	*(57)*	*(65)*	*(72)*	*(353)*	*(51)*	*(75)*	*(84)*	*(105)*	*(140)*	*(463)*
One year	9 9	9 9	11 11	11 11	10 10	**12 12**	8 9	11 12	13 13	11 11	10 11	**11 11**
Five years	3 4	0 0	4 5	2 2	2 2	**3 3**	*1 1*	1 2	8 8	4 5	3 3	**3 4**
Ten years	0 0	0 0	4 5	*1 1*	1 2	**2 3**	*1 1*	*1 1*	6 8	2 4	2 3	**2 4**
Trent	*(36)*	*(52)*	*(49)*	*(84)*	*(54)*	*(304)*	*(28)*	*(58)*	*(81)*	*(94)*	*(68)*	*(336)*
One year	25 25	11 12	7 7	8 8	12 13	**11 12**	3 3	8 8	7 8	10 11	5 5	**7 8**
Five years	15 17	5 7	4 5	2 2	2 2	**4 5**	0 0	3 4	1 1	3 4	2 2	**2 3**
Ten years	15 17	4 7	4 5	*1 2*	*1 1*	**3 5**	0 0	2 3	0 0	2 2	2 2	**1 2**
Anglia & Oxford	*(46)*	*(72)*	*(69)*	*(49)*	*(22)*	*(297)*	*(74)*	*(80)*	*(78)*	*(44)*	*(18)*	*(302)*
One year	20 21	15 16	5 5	6 6	13 14	**11 12**	20 21	12 12	7 7	16 17	5 5	**12 13**
Five years	6 7	3 4	1 1	1 1	0 0	**3 3**	5 6	6 6	3 4	4 5	0 0	**4 5**
Ten years	4 5	2 3	0 0	*1 1*	0 0	**2 2**	1 2	4 5	3 4	3 3	0 0	**3 4**
North Thames	*(59)*	*(54)*	*(60)*	*(74)*	*(54)*	*(339)*	*(58)*	*(57)*	*(53)*	*(67)*	*(40)*	*(277)*
One year	8 8	9 9	9 9	18 19	16 16	**10 11**	22 23	22 23	11 12	17 18	*18 19*	**18 19**
Five years	2 2	3 3	2 2	8 10	6 7	**4 4**	*14 17*	3 5	0 0	2 2	3 2	**4 5**
Ten years	0 0	*1 3*	2 2	7 9	3 5	**2 4**	*11 14*	3 5	0 0	2 2	3 2	**4 5**
South Thames	*(98)*	*(94)*	*(98)*	*(89)*	*(44)*	*(460)*	*(112)*	*(77)*	*(85)*	*(66)*	*(40)*	*(385)*
One year	8 8	12 13	12 12	13 14	5 5	**11 11**	10 10	8 8	11 12	7 7	*16 16*	**10 10**
Five years	4 4	3 4	2 2	1 1	*1 1*	**2 3**	3 3	4 5	3 4	1 1	5 5	**3 3**
Ten years	3 4	1 1	1 1	*1 1*	*1 1*	**1 3**	2 2	4 5	3 4	0 0	5 5	**2 3**
South & West	*(67)*	*(90)*	*(77)*	*(51)*	*(20)*	*(332)*	*(66)*	*(94)*	*(113)*	*(48)*	*(21)*	*(356)*
One year	13 13	6 6	12 13	5 5	6 6	**9 9**	21 22	14 15	10 10	5 6	15 16	**12 13**
Five years	2 3	2 3	3 3	2 3	4 4	**3 3**	6 8	6 7	3 3	1 1	7 8	**4 5**
Ten years	*1 3*	2 3	1 2	2 2	4 4	**2 3**	6 8	3 4	1 2	0 0	7 8	**3 4**
West Midlands	*(18)*	*(21)*	*(16)*	*(22)*	*(48)*	*(304)*	*(36)*	*(60)*	*(67)*	*(78)*	*(82)*	*(327)*
One year	5 5	4 5	27 27	8 8	9 9	**10 11**	9 10	5 5	17 18	16 17	13 14	**13 13**
Five years	0 0	2 2	0 0	2 2	*1 1*	**2 2**	2 2	1 1	3 4	4 5	4 4	**3 4**
Ten years	0 0	0 0	0 0	2 2	*1 1*	**1 2**	0 0	0 0	3 4	2 3	4 4	**2 3**
North & West	*(58)*	*(64)*	*(63)*	*(86)*	*(124)*	*(453)*	*(63)*	*(74)*	*(84)*	*(87)*	*(151)*	*(464)*
One year	6 6	10 11	16 17	12 13	7 7	**10 11**	15 15	9 10	11 11	10 10	4 4	**9 9**
Five years	1 1	3 3	4 4	3 4	1 1	**2 3**	4 4	1 1	3 4	2 2	2 2	**2 3**
Ten years	0 0	*1 1*	1 1	2 4	*1 1*	**1 2**	2 4	0 0	2 4	2 2	1 1	**1 2**
WALES	*(13)*	*(23)*	*(37)*	*(37)*	*(21)*	*(162)*	*(19)*	*(38)*	*(49)*	*(44)*	*(23)*	*(190)*
One year	15 15	4 4	3 4	*11 11*	15 16	**10 10**	9 9	16 17	20 21	15 16	17 18	**17 17**
Five years	5 5	2 2	2 2	0 0	12 12	**3 4**	5 5	5 6	6 8	7 8	4 6	**6 8**
Ten years	5 5	0 0	2 2	0 0	12 12	**3 4**	0 0	4 5	3 5	7 8	4 6	**5 7**

---

[1] Including those for whom no deprivation category was assigned. C - crude survival, R - relative survival (see Chapter 3).
Number of patients contributing to each analysis in parentheses.

		1981-85												1986-90											
Region / WOMEN		*Affluent* C R		*2* C R		*3* C R		*4* C R		*Deprived* C R		*All¹* C R		*Affluent* C R		*2* C R		*3* C R		*4* C R		*Deprived* C R		*All¹* C R	
ENGLAND & WALES	(n)	(548)		(617)		(660)		(634)		(620)		(3,116)		(520)		(546)		(642)		(683)		(558)		(2,969)	
	One year	18	19	16	16	14	15	16	16	15	15	16	17	20	20	20	21	19	20	21	22	16	17	19	20
	Five years	6	7	6	7	3	4	5	6	4	6	5	6	6	7	7	9	6	7	8	10	5	6	7	8
		3	5	4	6	2	3	3	5	3	5	3	5												
ENGLAND	(n)	(527)		(588)		(616)		(581)		(591)		(2,930)		(500)		(530)		(595)		(639)		(529)		(2,805)	
	One year	18	19	16	17	15	15	15	16	15	16	16	16	20	21	20	20	19	20	20	21	16	16	19	20
	Five years	5	7	6	7	4	4	5	6	5	6	5	6	6	7	7	8	6	8	8	11	4	6	6	8
		3	5	4	6	2	3	3	5	3	5	3	5												
Northern & Yorkshire	(n)	(64)		(83)		(73)		(105)		(129)		(457)		(60)		(76)		(85)		(105)		(136)		(463)	
	One year	18	19	14	15	16	18	14	15	19	20	16	17	22	23	22	23	19	20	24	25	15	16	20	21
	Five years	8	9	4	5	5	6	3	4	8	12	6	7	7	7	7	10	6	8	8	10	4	5	6	8
		5	6	3	5	4	6	1	2	4	8	3	6												
Trent	(n)	(49)		(65)		(72)		(96)		(90)		(373)		(50)		(65)		(86)		(100)		(73)		(374)	
	One year	21	24	13	14	13	14	13	13	12	13	14	15	10	11	10	11	16	17	18	19	17	18	15	16
	Five years	7	8	5	5	1	1	5	6	4	5	4	5	2	2	5	5	6	7	7	8	5	6	5	6
		2	2	5	5	1	1	3	5	3	4	3	5												
Anglia & Oxford	(n)	(61)		(74)		(79)		(35)		(31)		(284)		(77)		(72)		(69)		(59)		(18)		(297)	
	One year	17	18	8	9	11	11	17	18	6	6	11	12	30	31	18	19	10	10	19	20	13	14	19	20
	Five years	3	4	2	3	3	3	10	11	2	3	4	4	12	16	7	9	3	4	4	5	0	0	6	8
		0	0	1	3	3	3	6	10	2	3	2	4												
North Thames	(n)	(48)		(37)		(46)		(42)		(46)		(222)		(35)		(47)		(58)		(51)		(42)		(237)	
	One year	18	19	20	21	21	22	16	17	20	21	19	20	24	26	28	29	29	30	34	36	24	25	29	30
	Five years	10	11	13	14	6	7	6	8	7	8	8	10	8	8	12	16	10	11	15	20	9	10	11	14
		6	9	13	14	1	2	4	5	7	8	6	9												
South Thames	(n)	(99)		(80)		(88)		(59)		(25)		(354)		(79)		(61)		(68)		(54)		(23)		(287)	
	One year	19	20	20	21	14	15	17	18	10	11	17	18	22	23	28	29	23	25	28	29	24	25	25	26
	Five years	7	9	8	10	4	5	7	8	3	3	6	7	5	6	5	6	5	6	7	9	0	0	5	6
		6	9	5	6	3	3	4	4	3	3	4	6												
South & West	(n)	(94)		(116)		(95)		(64)		(15)		(390)		(68)		(85)		(86)		(70)		(26)		(336)	
	One year	15	16	20	21	15	16	13	14	0	0	16	16	21	22	20	21	18	19	17	18	10	11	18	19
	Five years	4	5	8	9	4	5	6	7	0	0	5	6	7	8	9	11	7	8	7	8	3	4	7	9
		2	5	4	6	1	2	5	7	0	0	3	5												
West Midlands	(n)	(53)		(55)		(83)		(77)		(110)		(380)		(59)		(60)		(73)		(106)		(89)		(388)	
	One year	21	22	14	14	13	14	20	21	19	20	18	18	14	15	17	18	28	29	18	19	21	22	20	21
	Five years	6	7	5	5	2	2	4	5	4	4	4	5	4	5	6	7	11	12	11	13	8	10	8	10
		4	6	2	3	2	2	4	5	3	4	3	5												
North & West	(n)	(59)		(78)		(80)		(103)		(145)		(470)		(72)		(64)		(70)		(94)		(122)		(423)	
	One year	15	16	16	17	15	16	12	12	13	14	14	15	13	13	15	16	15	16	13	14	9	10	12	13
	Five years	0	0	8	9	4	5	2	2	3	4	3	4	3	4	3	3	2	3	8	10	2	3	4	5
		0	0	5	7	1	2	1	1	2	3	2	3												
WALES	(n)	(21)		(29)		(44)		(53)		(29)		(186)		(20)		(16)		(47)		(44)		(29)		(164)	
	One year	21	22	15	15	7	7	23	24	14	14	17	18	16	16	45	46	13	14	25	26	31	32	25	26
	Five years	12	15	3	3	1	1	9	10	3	3	6	7	8	9	16	17	3	3	7	7	8	9	9	11
		0	0	3	3	0	0	7	8	3	3	3	5												

¹ Including those for whom no deprivation category was assigned. C - crude survival, R - relative survival (see Chapter 3).
Number of patients contributing to each analysis in parentheses.

Table 18.2   Crude and relative survival (%) at one, five and ten years since diagnosis: NHS Region, deprivation category and calendar period of diagnosis

ADULTS	1971-75 Affluent C R	2 C R	3 C R	4 C R	Deprived C R	All[1] C R	1976-80 Affluent C R	2 C R	3 C R	4 C R	Deprived C R	All[1] C R
ENGLAND & WALES	*(689)*	*(814)*	*(859)*	*(873)*	*(741)*	*(4,859)*	*(916)*	*(1,007)*	*(1,119)*	*(1,028)*	*(967)*	*(5,156)*
One year	**13 14**	**10 11**	**12 13**	**12 13**	**10 11**	**12 12**	**16 17**	**12 13**	**11 12**	**12 13**	**11 11**	**12 13**
Five years	3 4	3 3	3 4	3 4	3 3	3 4	5 6	4 5	4 5	3 4	3 4	4 5
Ten years	2 3	1 2	2 3	2 3	2 3	2 3	3 5	2 4	3 5	2 3	2 4	3 4
ENGLAND	*(666)*	*(780)*	*(805)*	*(807)*	*(700)*	*(4,587)*	*(880)*	*(956)*	*(1,045)*	*(952)*	*(927)*	*(4,853)*
One year	13 13	11 11	12 13	12 12	10 11	**12 12**	16 17	12 13	11 12	12 13	10 11	**12 13**
Five years	3 4	3 3	3 4	3 4	2 3	**3 3**	5 6	3 4	4 5	3 4	3 4	**4 4**
Ten years	2 3	1 2	1 2	2 3	2 3	**2 3**	3 5	2 4	3 5	2 3	2 4	**2 4**
Northern & Yorkshire	*(44)*	*(73)*	*(94)*	*(110)*	*(120)*	*(609)*	*(97)*	*(116)*	*(136)*	*(191)*	*(236)*	*(786)*
One year	*10 10*	14 14	15 15	12 12	8 9	**13 13**	15 16	11 12	13 14	12 12	13 13	**13 13**
Five years	*1 2*	*0 0*	6 8	3 3	2 2	**3 4**	5 5	4 4	7 8	4 5	5 6	**5 6**
Ten years	*0 0*	*0 0*	4 6	2 2	1 2	**2 3**	5 5	2 3	5 8	2 5	3 6	**3 5**
Trent	*(55)*	*(84)*	*(85)*	*(124)*	*(95)*	*(484)*	*(66)*	*(89)*	*(121)*	*(142)*	*(118)*	*(548)*
One year	25 26	13 13	9 9	9 9	14 15	**12 13**	15 15	9 10	8 8	9 10	6 6	**9 9**
Five years	*11 13*	4 5	4 5	2 3	3 4	**4 5**	2 3	3 3	1 1	3 3	2 2	**2 3**
Ten years	*11 13*	3 4	4 5	1 2	*3 4*	**3 5**	2 3	*1 2*	0 0	2 2	2 2	**1 2**
Anglia & Oxford	*(80)*	*(109)*	*(113)*	*(81)*	*(35)*	*(483)*	*(116)*	*(134)*	*(121)*	*(72)*	*(37)*	*(490)*
One year	25 27	14 14	8 8	12 13	22 22	**14 15**	20 21	12 13	10 10	20 21	4 5	**14 14**
Five years	6 7	3 3	2 2	4 4	5 5	**3 4**	4 5	5 6	4 6	7 9	3 3	**5 6**
Ten years	5 6	2 3	0 0	2 4	5 5	**2 3**	2 2	3 5	4 6	5 7	0 0	**3 4**
North Thames	*(97)*	*(103)*	*(112)*	*(109)*	*(80)*	*(567)*	*(98)*	*(98)*	*(97)*	*(98)*	*(63)*	*(462)*
One year	10 10	12 13	14 15	15 15	17 18	**13 13**	19 19	20 21	10 11	16 17	24 25	**17 18**
Five years	1 1	4 4	2 2	6 7	6 7	**3 4**	11 12	2 4	3 3	2 3	*3 3*	**4 6**
Ten years	0 0	3 4	*1 1*	5 7	4 6	**2 4**	7 10	2 4	2 3	2 3	3 3	**4 6**
South Thames	*(160)*	*(137)*	*(149)*	*(137)*	*(65)*	*(702)*	*(191)*	*(137)*	*(136)*	*(102)*	*(68)*	*(646)*
One year	10 10	11 11	13 13	14 15	5 5	**11 12**	14 15	9 10	9 10	11 12	12 12	**11 12**
Five years	2 3	2 3	2 3	2 3	*1 1*	**2 3**	4 4	4 6	3 3	2 2	3 4	**3 4**
Ten years	2 2	0 1	1 2	2 2	*1 1*	**1 2**	2 3	4 6	*3 3*	*1 1*	3 4	**3 4**
South & West	*(121)*	*(144)*	*(123)*	*(71)*	*(30)*	*(536)*	*(129)*	*(168)*	*(186)*	*(78)*	*(35)*	*(620)*
One year	14 14	8 8	11 11	7 7	*12 12*	**10 10**	19 20	15 15	12 13	11 12	*13 14*	**14 15**
Five years	3 3	3 4	3 4	3 3	2 3	**3 3**	5 7	5 6	.5 6	3 4	4 5	**5 6**
Ten years	1 3	2 4	1 2	1 2	2 3	**2 3**	5 7	3 4	3 6	3 4	4 5	**3 5**
West Midlands	*(28)*	*(25)*	*(28)*	*(42)*	*(66)*	*(477)*	*(72)*	*(96)*	*(109)*	*(123)*	*(123)*	*(528)*
One year	7 7	4 4	22 22	*11 11*	8 9	**10 11**	9 9	11 12	16 17	14 14	13 14	**13 13**
Five years	0 0	2 2	2 2	2 3	1 1	**2 3**	2 2	1 2	3 4	3 4	4 5	**3 3**
Ten years	0 0	0 0	*1 1*	1 2	*1 1*	**1 2**	*1 1*	0 0	3 4	2 2	3 4	**2 2**
North & West	*(81)*	*(105)*	*(101)*	*(133)*	*(209)*	*(729)*	*(111)*	*(118)*	*(139)*	*(146)*	*(247)*	*(773)*
One year	6 6	8 8	16 16	13 14	8 8	**11 11**	16 17	10 11	10 10	9 9	7 7	**9 10**
Five years	2 2	2 2	2 3	3 4	2 2	**2 3**	6 7	1 2	3 4	1 2	2 2	**2 3**
Ten years	*1 1*	1 2	0 1	2 4	*1 1*	**1 2**	2 4	0 1	2 3	*1 2*	1 1	**1 2**
WALES	*(23)*	*(34)*	*(54)*	*(66)*	*(41)*	*(272)*	*(36)*	*(51)*	*(74)*	*(76)*	*(40)*	*(303)*
One year	*19 19*	2 2	9 10	18 19	*10 10*	**11 12**	*13 14*	17 18	18 19	17 18	*12 13*	**16 17**
Five years	6 7	1 1	3 4	4 5	5 5	**3 4**	8 9	8 8	5 7	5 6	4 5	**6 7**
Ten years	2 3	0 0	3 4	3 3	5 5	**3 3**	5 6	4 6	3 4	4 6	3 4	**4 6**

---

[1] Including those for whom no deprivation category was assigned. C - crude survival, R - relative survival (see Chapter 3).
Number of patients contributing to each analysis in parentheses.

	1981-85												1986-90												ADULTS
	Affluent		*2*		*3*		*4*		*Deprived*		*All*[1]		*Affluent*		*2*		*3*		*4*		*Deprived*		*All*[1]		
	C	R	C	R	C	R	C	R	C	R	C	R	C	R	C	R	C	R	C	R	C	R	C	R	
	(928)		*(997)*		*(1,090)*		*(993)*		*(1,022)*		*(5,100)*		*(878)*		*(958)*		*(1,083)*		*(1,127)*		*(900)*		*(4,985)*		**ENGLAND & WALES**
	19	**20**	**17**	**18**	**15**	**16**	**16**	**17**	**16**	**17**	**17**	**18**	**22**	**24**	**23**	**25**	**19**	**20**	**20**	**21**	**17**	**18**	**21**	**22**	*One year*
	6	**8**	**6**	**7**	**4**	**5**	**5**	**7**	**5**	**6**	**5**	**6**	**7**	**8**	**9**	**11**	**6**	**7**	**8**	**10**	**5**	**6**	**7**	**9**	*Five years*
	4	**6**	**4**	**6**	**2**	**4**	**4**	**6**	**3**	**5**	**3**	**5**													
	(890)		*(954)*		*(1,020)*		*(920)*		*(979)*		*(4,814)*		*(845)*		*(928)*		*(1,010)*		*(1,064)*		*(856)*		*(4,727)*		**ENGLAND**
	19	20	18	19	16	17	15	16	16	17	**17**	**18**	23	24	23	24	19	20	20	21	17	18	**20**	**22**	*One year*
	6	7	6	7	4	5	5	6	5	6	**5**	**6**	7	8	8	11	6	7	8	10	5	6	**7**	**9**	*Five years*
	4	6	4	6	3	4	3	5	3	5	**3**	**5**													
	(106)		*(125)*		*(126)*		*(160)*		*(226)*		*(749)*		*(98)*		*(121)*		*(147)*		*(176)*		*(214)*		*(757)*		**Northern & Yorkshire**
	18	19	16	17	13	14	16	16	21	22	**17**	**18**	21	22	24	25	20	21	23	25	17	18	**21**	**22**	*One year*
	6	8	3	4	3	4	3	4	7	10	**5**	**6**	6	7	12	16	7	9	7	9	4	6	**7**	**9**	*Five years*
	5	7	*3*	*4*	*2*	*3*	2	3	4	7	**3**	**5**													
	(82)		*(105)*		*(136)*		*(158)*		*(154)*		*(637)*		*(87)*		*(116)*		*(153)*		*(164)*		*(117)*		*(637)*		**Trent**
	20	21	17	18	16	17	11	12	14	14	**15**	**16**	17	18	19	21	17	18	19	20	14	15	**17**	**18**	*One year*
	8	9	8	10	3	3	4	5	6	7	**5**	**6**	5	5	8	10	6	7	6	8	4	6	**6**	**7**	*Five years*
	2	3	6	8	2	3	2	4	4	6	**3**	**5**													
	(110)		*(119)*		*(131)*		*(62)*		*(45)*		*(477)*		*(138)*		*(129)*		*(126)*		*(107)*		*(23)*		*(526)*		**Anglia & Oxford**
	20	21	11	12	14	15	13	14	*11*	*12*	**14**	**15**	29	30	20	21	16	17	20	21	*10*	*10*	**21**	**22**	*One year*
	4	5	2	2	3	4	7	7	2	3	**4**	**5**	11	14	7	8	2	3	5	6	0	0	**6**	**8**	*Five years*
	2	3	1	2	*3*	*4*	4	7	2	3	**3**	**4**													
	(79)		*(64)*		*(83)*		*(59)*		*(81)*		*(369)*		*(63)*		*(80)*		*(93)*		*(90)*		*(69)*		*(403)*		**North Thames**
	21	23	25	26	23	24	17	19	26	28	**23**	**24**	25	27	32	34	28	29	29	31	29	31	**29**	**31**	*One year*
	8	9	9	10	8	9	9	*12*	10	13	**9**	**11**	*4*	*5*	15	21	10	12	12	16	13	15	**11**	**14**	*Five years*
	4	7	7	9	5	8	*8*	*10*	9	13	**7**	**10**													
	(168)		*(131)*		*(138)*		*(89)*		*(38)*		*(573)*		*(117)*		*(124)*		*(106)*		*(97)*		*(38)*		*(486)*		**South Thames**
	21	22	26	27	16	17	19	20	*15*	*16*	**20**	**22**	25	26	36	38	22	24	27	28	20	21	**27**	**28**	*One year*
	8	10	10	12	5	6	9	11	5	5	**8**	**9**	7	9	8	10	4	5	10	13	*3*	*3*	**7**	**9**	*Five years*
	7	*10*	6	10	*4*	*5*	5	7	5	5	**5**	**8**													
	(152)		*(197)*		*(153)*		*(88)*		*(24)*		*(622)*		*(113)*		*(147)*		*(144)*		*(112)*		*(39)*		*(559)*		**South & West**
	18	19	21	22	19	20	15	15	*4*	*4*	**18**	**19**	26	27	23	24	18	19	15	16	*16*	*17*	**20**	**22**	*One year*
	5	6	7	9	6	8	7	9	*0*	*0*	**6**	**7**	10	12	10	14	6	8	7	8	6	7	**8**	**11**	*Five years*
	3	5	3	7	4	6	6	9	*0*	*0*	**4**	**6**													
	(86)		*(86)*		*(131)*		*(139)*		*(177)*		*(624)*		*(99)*		*(106)*		*(130)*		*(160)*		*(150)*		*(647)*		**West Midlands**
	21	22	12	12	12	12	19	20	17	18	**16**	**17**	20	21	20	21	26	27	20	21	20	21	**21**	**22**	*One year*
	7	8	3	4	2	3	6	8	4	4	**4**	**5**	6	8	8	10	9	11	11	13	6	8	**8**	**10**	*Five years*
	6	7	2	2	*1*	*1*	5	7	3	4	**3**	**4**													
	(107)		*(127)*		*(122)*		*(165)*		*(234)*		*(763)*		*(130)*		*(105)*		*(111)*		*(158)*		*(206)*		*(712)*		**North & West**
	15	16	13	14	15	15	14	15	12	13	**14**	**15**	17	18	11	12	12	13	15	16	13	14	**14**	**14**	*One year*
	2	2	6	7	3	4	3	3	2	3	**3**	**4**	4	5	2	3	2	2	7	10	4	5	**4**	**5**	*Five years*
	1	2	4	6	*1*	*2*	2	3	1	2	**2**	**3**													
	(38)		*(43)*		*(70)*		*(73)*		*(43)*		*(286)*		*(33)*		*(30)*		*(73)*		*(63)*		*(44)*		*(258)*		**WALES**
	21	*22*	*11*	*11*	*13*	*14*	*24*	*26*	*20*	*22*	**18**	**20**	*16*	*17*	*40*	*42*	*19*	*20*	*19*	*20*	*24*	*26*	**24**	**25**	*One year*
	12	*17*	*2*	*2*	*2*	*3*	*9*	*11*	*5*	*7*	**6**	**8**	*5*	*7*	*12*	*16*	*5*	*6*	*6*	*6*	*6*	*6*	**8**	**11**	*Five years*
	1	*2*	*2*	*2*	*1*	*2*	*6*	*8*	*2*	*3*	**3**	**5**													

[1] Including those for whom no deprivation category was assigned. C - crude survival, R - relative survival (see Chapter 3).
Number of patients contributing to each analysis in parentheses.

Fig. 18.1   Relative survival up to ten years: trends by calendar period of diagnosis (1971-90) and NHS Region

ADULTS

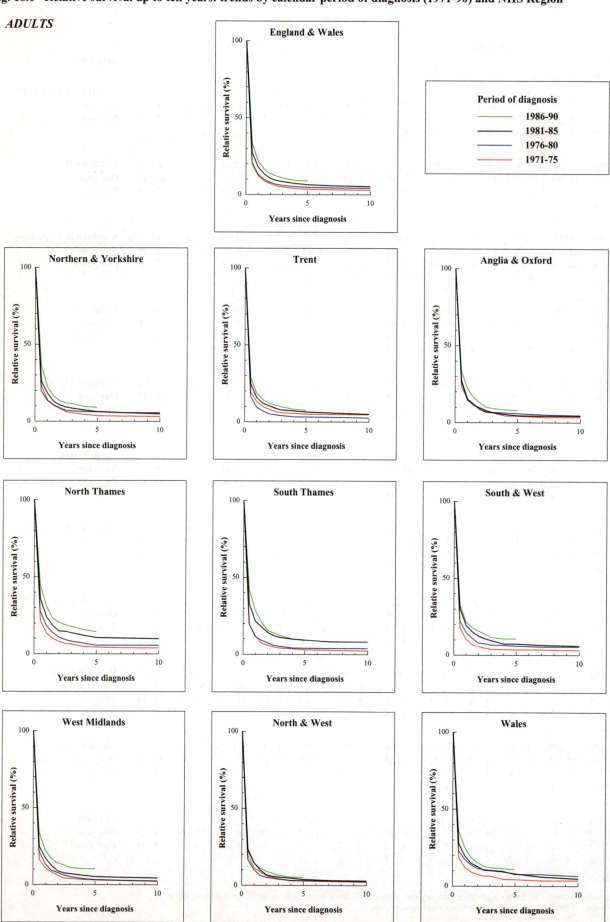

Fig. 18.2   Relative survival up to five years, by deprivation category and NHS Region: patients diagnosed 1986-90

ADULTS

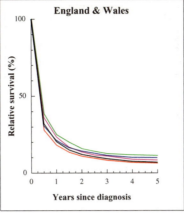

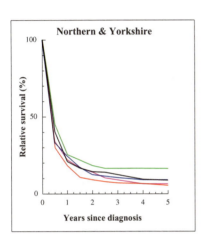

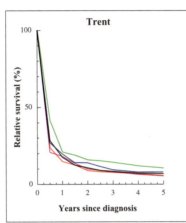

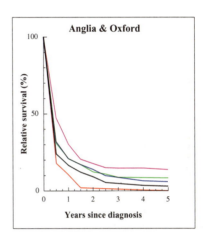

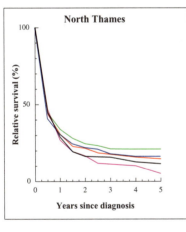

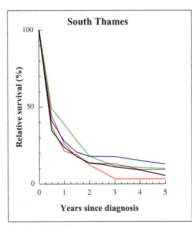

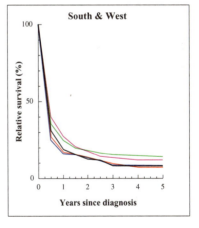

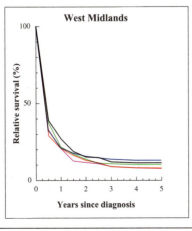

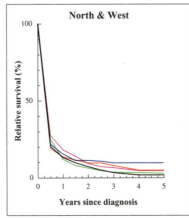

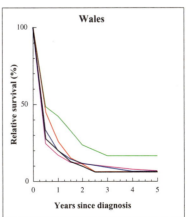

Fig. 18.3   Relative survival at five years by deprivation category, period of diagnosis (1981-90) and NHS Region

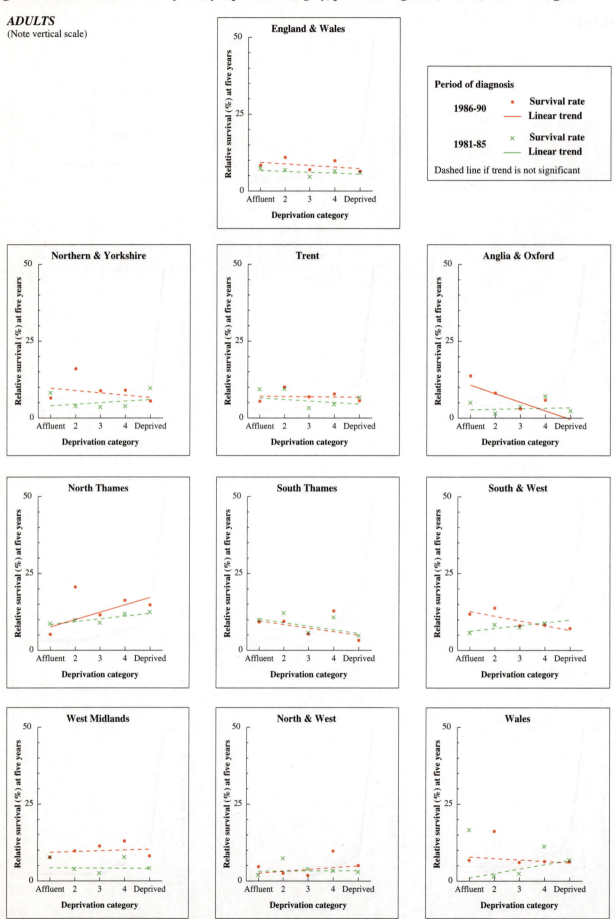

Fig. 18.4 Relative survival (%) by NHS Region and sex:

  one-year and five-year survival for patients diagnosed 1986-90, with 95% confidence intervals

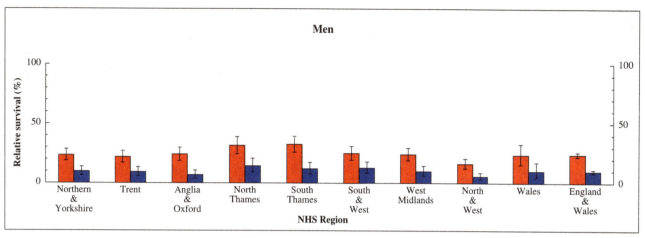

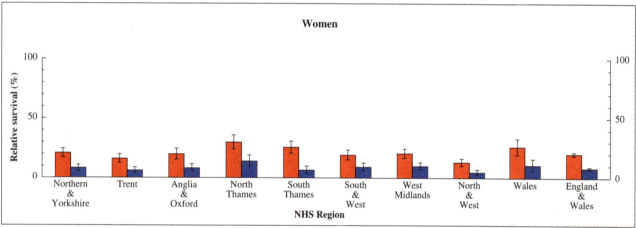

Fig. 18.5 Relative survival (%) at five years, with 95% confidence interval, by age at diagnosis and sex:

  England and Wales, patients diagnosed 1986-90

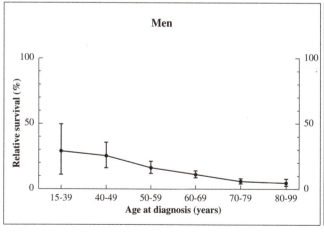

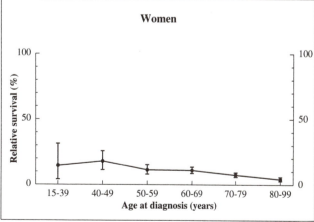

Table 18.2   Crude and relative survival (%) at one, five and ten years since diagnosis: NHS Region, deprivation category and calendar period of diagnosis

MEN	1971-75 Affluent		2		3		4		Deprived		All[1]		1976-80 Affluent		2		3		4		Deprived		All[1]	
	C	R	C	R	C	R	C	R	C	R	C	R	C	R	C	R	C	R	C	R	C	R	C	R
ENGLAND & WALES	(270)		(301)		(333)		(316)		(282)		(1,855)		(409)		(394)		(425)		(395)		(384)		(2,056)	
One year	17	17	11	12	16	16	15	16	11	12	14	14	19	20	14	15	12	12	13	14	12	13	14	15
Five years	2	3	2	3	4	5	4	6	3	3	3	4	6	7	4	5	4	5	3	4	4	4	4	5
Ten years	1	2	1	2	2	3	3	4	2	3	2	3	3	5	3	5	3	5	2	4	2	3	3	5

	1981-85 Affluent		2		3		4		Deprived		All[1]		1986-90 Affluent		2		3		4		Deprived		All[1]	
	C	R	C	R	C	R	C	R	C	R	C	R	C	R	C	R	C	R	C	R	C	R	C	R
	(380)		(380)		(430)		(359)		(402)		(1,984)		(358)		(412)		(441)		(444)		(342)		(2,016)	
One year	21	22	20	21	18	19	16	17	19	21	19	20	27	28	28	29	20	21	20	21	19	20	23	24
Five years	7	8	5	7	5	6	6	8	5	7	6	7	8	10	11	14	5	7	7	9	6	7	7	10
Ten years	5	7	3	5	4	6	4	7	3	5	4	6												

	1971-75		1976-80		1981-85		1986-90	
Northern & Yorkshire	(256)		(323)		(292)		(294)	
One year	14	15	16	17	18	19	22	23
Five years	3	4	7	9	3	5	7	10
Ten years	2	3	5	8	2	4		
Trent	(180)		(212)		(264)		(263)	
One year	15	16	12	12	16	17	21	22
Five years	4	5	2	3	7	9	7	9
Ten years	3	4	1	2	4	6		
Anglia & Oxford	(186)		(188)		(193)		(229)	
One year	20	21	16	17	19	20	23	24
Five years	5	6	6	7	4	5	6	7
Ten years	3	5	3	6	4	5		
North Thames	(228)		(185)		(147)		(166)	
One year	16	17	16	17	28	30	30	32
Five years	3	4	5	6	9	11	12	14
Ten years	2	4	3	5	7	10		
South Thames	(242)		(261)		(219)		(199)	
One year	12	13	13	14	25	27	31	33
Five years	2	3	4	5	10	13	10	12
Ten years	1	2	3	4	7	12		
South & West	(204)		(264)		(232)		(223)	
One year	12	12	16	17	23	25	24	25
Five years	3	4	5	7	7	9	9	13
Ten years	1	2	4	7	5	8		
West Midlands	(173)		(201)		(244)		(259)	
One year	10	11	13	14	15	16	23	24
Five years	3	4	2	3	5	7	7	10
Ten years	2	3	1	2	3	5		
North & West	(276)		(309)		(293)		(289)	
One year	10	11	11	11	13	14	15	16
Five years	2	3	2	3	3	4	5	6
Ten years	1	2	1	2	2	3		
WALES	(110)		(113)		(100)		(94)	
One year	14	15	16	17	21	23	22	24
Five years	4	5	5	6	7	9	7	10
Ten years	2	3	3	5	2	5		

[1] Including those for whom no deprivation category was assigned. C - crude survival, R - relative survival (see Chapter 3). Number of patients contributing to each analysis in parentheses.

Chapter 19

PANCREAS

Cancer of the pancreas is the eighth most frequent cancer in England and Wales, accounting for some 3% of cancers in both sexes combined. The average number of new cases rose steadily to 6,000 a year by 1990. Incidence rates are 20-50% higher in men in each age group, but the overall number of cases is similar in men and women, because incidence increases steeply with age and elderly women greatly outnumber elderly men in the population. Incidence in men reached a peak in the mid-1970s, and it has been declining at 3-4% every five years since then. For women, incidence was still increasing slightly at the end of the 1980s, but the patterns by age suggest an imminent plateau and decline. Mortality trends in most countries have closely followed those for incidence, since survival has remained extremely low everywhere.

The causes of cancer of the pancreas are unknown. Diagnosis almost invariably occurs at an advanced stage of disease, because symptoms such as pain and weight loss are non-specific. In a population series of almost 6,000 patients registered during 1977-86, curative resection was possible in less than 3% of patients, and peri-operative mortality was over 20%[1]. Mortality was lower in 1,000 patients operated in specialist units over the 20 years to 1996[2].

The survival analyses here include over 90,000 adults registered with cancer of the pancreas in England and Wales during the period 1971-90, and followed up to the end of 1995. Over 1,000 cases diagnosed during the 1970s were excluded because the tumour was coded as a metastatic cancer in the pancreas from a primary cancer elsewhere (Table 19.1). Given the frequently occult early clinical course of cancer of the pancreas, it is likely that most of these would have been primary pancreatic cancers, metastatic to other organs when diagnosed, and that they were incorrectly coded (behaviour code 6, rather than 3). For consistency with the rules applied to other cancers, however, these patients were not included in the analyses (see Chapter 2).

Of the 114,000 patients eligible for analysis, 4% were excluded because the patient's vital status was not known at 6 July 1997, and a further 17% because the duration of survival was zero or unknown. The proportion of patients in this category increased from 12% in 1971-75 to 22% for patients diagnosed in 1986-90.

Almost half the tumours arose in the head of the pancreas, and only 3% in the body of the pancreas, but for more than 40% there was no anatomic detail (Table 19A). Adenocarcinomas of various types accounted for 20%,

increasing slightly from 17% in the early 1970s to 21% by the late 1980s (data not shown), but 70% of the tumours had been assigned a morphology code for unspecified carcinoma, and for 9% no morphology was available. This reflects the difficulty throughout much of the period 1971-90 in making the diagnosis early enough for surgery of curative intent – or even biopsy for definitive morphological diagnosis – to be considered clinically justifiable.

Survival trends

Relative survival for adults diagnosed in England and Wales during 1986-90 was 11% at one year and only 3% at five years. Patients diagnosed during 1971-75 had 9% survival at one year and 2% at five years (Table 19.2). There was little difference between men and women.

The survival curves are strikingly uniform in all regions and time periods (Figure 19.1), with the minor and more recent exception of the Thames regions (see below).

Survival has improved over time, but the gains are extremely small. Survival at one year has improved slightly more than survival at five years. Trends in age-standardised relative survival rates show that one-year survival has increased by about 1% in England and Wales between successive five-year periods in both men and women (see Table 4.4). The corresponding gain in five-year survival was just 0.3% for men and 0.2% for women. The very large numbers of patients contributing to these analyses (about 11,000 for each sex and in each time period) mean that even these very small gains in survival are statistically significant.

This is of little comfort: the age-adjusted five-year survival estimates for England and Wales, rounded to the nearest whole number, have remained at 2% for both men and women since 1971-75.

The small trends in survival are similar in most regions (see Table-Figure 4.5E). One-year survival estimates for North Thames and South Thames increased by 2-3% during the 1980s: more than in other regions (see also Table 19.2). In North Thames, in particular, much of even this small increase is likely to be an artefact, due to the very high proportion (52%) of pancreatic cancers registered solely from death certificates during the period 1981-85, shortly before cancer registration for the then NW Thames and NE Thames Regional Health Authorities was transferred elsewhere. The higher survival rate for patients diagnosed during 1986-90 in Wales may also be misleading, as 30% of eligible patients there were also

Table 19A Sub-site distribution of cancer of the pancreas

ICD-9 code	Site description	1971-75 No.	%	1976-80 No.	%	1981-85 No.	%	1986-90 No.	%	Total No.	%
157.0	Head of pancreas	10,440	49.3	11,787	50.7	11,488	49.7	10,728	47.0	**44,443**	**49.2**
157.1	Body of pancreas	-	-	571	2.5	1,075	4.7	735	3.2	**2,381**	**2.6**
157.2	Tail of pancreas	-	-	163	0.7	424	1.8	362	1.6	**949**	**1.1**
157.3	Pancreatic duct	-	-	13	0.1	21	0.1	31	0.1	**65**	**0.1**
157.4	Islets of Langerhans	-	-	10	0.0	26	0.1	33	0.1	**69**	**0.1**
157.8	Other	2,014	9.5	1,440	6.2	378	1.6	238	1.0	**4,070**	**4.5**
157.9	Part unspecified	8,707	41.1	9,277	39.9	9,696	42.0	10,690	46.9	**38,370**	**42.5**
	Total	**21,161**		**23,261**		**23,108**		**22,817**		**90,347**	

registered solely from death certificates (data not shown). These patients were excluded from analysis: they can be expected to have had survival that was even shorter than the average (see Chapter 2). In West Midlands and North & West Regions, the gains in one-year survival in men were significantly less than the national average gain for England and Wales, but the average gain is inflated by the unduly large trends in the Thames regions. Finally, if the exceptionally high proportion of patients excluded for zero or unknown survival has caused bias in the estimation of survival, the true five-year survival rate is likely to be even lower than 2-3%.

Survival declines with age (Figure 19.5). Pancreatic cancer is extremely rare under the age of 40, but five-year survival among some 200 patients diagnosed in this age range during 1986-90 was about 12%: above this age, survival was 5% or less (see Table 4.7).

Deprivation

Differences in survival between patients in the various deprivation categories are small (Figure 19.2), but they occur in both sexes and they are consistent over time (Table 19.2).

There is regional variation in the survival differences between deprivation categories (Figure 19.3 – note expanded y-axis scale), but the overall national pattern is clear. For the 22,600 patients of known deprivation status who were diagnosed during 1981-85, the fitted estimate of one-year relative survival for the most affluent group of patients was 9.7%, and the deficit in survival for the most deprived patients was 2%. The survival gradient was the same for a further 22,600 patients of known deprivation status who were diagnosed in 1986-90 (see Table 4.9).

Survival at five years was extremely low, and differences between deprivation categories were correspondingly small. The fitted estimate of relative survival for patients in the most affluent category diagnosed during 1986-90

was 2.6%, and the deficit for the most deprived group was small (0.6%) but just significant. The deprivation gradient for England and Wales (solid red line) in Figure 19.3 reflects this.

In contrast to regional survival rates, bias in the estimation of deprivation gradients in survival is unlikely. Whilst there are large regional differences in the proportion of patients excluded from analysis because of zero or unknown survival, there are no important differences between deprivation groups excluded for this reason. Both crude and relative survival show gradients, which cannot therefore be due to the life tables used to adjust for expected mortality: the use of deprivation-specific life tables would also produce a smaller deprivation gradient than conventional analysis with a single national life table (see Chapter 3).

International comparison

Survival is uniformly very low in Scotland, in the other European countries included in the EUROCARE study, and in the USA. Survival in England and Wales is no exception (see Table 4.12).

1.   Bramhall SR, Allum WH, Jones AG, Allwood A, Cummins C, Neoptolemos JP. Treatment and survival in 13,560 patients with pancreatic cancer, and incidence of the disease, in the West Midlands: an epidemiological study. Br J Surg 1995; 82: 111-115

2.   Neoptolemos JP, Russell RCG, Bramhall S, Theis B. Low mortality following resection for pancreatic and periampullary tumours in 1026 patients: UK survey of specialist pancreatic units. Br J Surg 1997; 84: 1370-1376

Table 19.1 Ineligible and excluded records, and patients included in analyses, by calendar period of diagnosis

	Calendar period of diagnosis														
	1971-75			**1976-80**			**1981-85**			**1986-90**			**Overall**		
Total registered	**26,008**			**28,733**			**30,253**			**31,042**			**116,036**		
Ineligible	*Records*	*Patients*	*%*	*Records*	*Patients*	*%*	*Records*	*Patients*	*%*	*Records*	*Patients*	*%*	*Records*	*Patients*	*%*
Incomplete data[1]	6	6		30	30		58	58		114	114		208	**208**	
Resident outside England and Wales	248	248		171	169		119	111		104	70		642	**598**	
In situ neoplasm[2]	4	4		1	1		0	0		0	0		5	**5**	
Benign or uncertain behaviour[2]	118	29		42	34		0	0		0	0		160	**63**	
Metastatic[2]	704	696		343	336		0	0		2	2		1,049	**1,034**	
Otherwise ineligible[3]	-	-		-	-		-	-		-	-		-	**-**	
Lymphoma[4]	7	7		7	7		2	0		0	0		16	**14**	
Leukaemia or myeloma[4]	0	0		0	0		0	0		1	0		1	**0**	
		990			577			169			186			1,922	
Total eligible		**25,018**	**100**		**28,156**	**100**		**30,084**	**100**		**30,856**	**100**		**114,114**	**100**
Aged under 15 or 100 years or over at diagnosis	8	7	<0.1	12	12	<0.1	22	21	<0.1	22	20	<0.1	64	**60**	<0.1
Vital status unknown[5]	813	703	2.8	1,065	949	3.4	1,249	1,185	3.9	1,258	1,207	3.9	4,385	**4,044**	3.5
Duplicate registration[6]	0	0	0.0	0	0	0.0	0	0	0.0	1	1	<0.1	1	**1**	<0.1
Multiple primary[7]	0	0	0.0	10	9	<0.1	63	54	0.2	68	56	0.2	141	**119**	0.1
Synchronous tumours[8]	101	81	0.3	55	35	0.1	72	65	0.2	73	59	0.2	301	**240**	0.2
Sex not known	0	0	0.0	0	0	0.0	0	0	0.0	0	0	0.0	0	**0**	0.0
Sex incompatible with site code[9]	-	-	-	-	-	-	-	-	-	-	-	-	-	**-**	-
Invalid dates[10]	0	0	0.0	3	3	<0.1	11	9	<0.1	16	11	<0.1	30	**23**	<0.1
Zero survival[11]	3,364	3,066	12.3	4,153	3,887	13.8	5,991	5,642	18.8	6,973	6,685	21.7	20,481	**19,280**	16.9
Total exclusions		**3,857**	**15.4**		**4,895**	**17.4**		**6,976**	**23.2**		**8,039**	**26.1**		**23,767**	**20.8**
Patients included in analysis		**21,161**	**84.6**		**23,261**	**82.6**		**23,108**	**76.8**		**22,817**	**73.9**		**90,347**	**79.2**

[1] Main data item(s) invalid or incompatible with one another: name, sex, date of birth, date of diagnosis and (if present) date of death, postcode, site, morphology and behaviour

[2] Tumour either *in situ* (behaviour code 2), benign (0), uncertain if benign or malignant (1) or metastatic (6 or 9)

[3] Not applicable for this malignancy

[4] Morphology that of lymphoma (included with non-Hodgkin lymphoma, Ch. 41) or myeloma (included with multiple myeloma, Ch. 43), or leukaemia (excluded)

[5] Tracing of vital status at National Health Service Central Register incomplete at 6 July 1997

[6] Same site code, sex, cancer registry and cancer registry number as an earlier registration

[7] Same site code and person number as an earlier registration(s): mostly confirmed multiple primary tumours at this site, some unresolved duplicate registrations

[8] Same site code, sex, date of birth and date of diagnosis as another registration(s): mostly synchronous or (in paired organs) bilateral tumours in same anatomic site in one individual, not previously linked: also some duplicate registrations

[9] Not applicable for this malignancy

[10] Impossible sequence of dates (birth, diagnosis and, if present, emigration or death); or date of death after 6 July 1997

[11] Date of diagnosis same as date of death: some are patients who did die on the day of diagnosis, but most were registered solely from a death certificate (death certificate only, or DCO), and their survival time is thus unknown

Table 19.2 Crude and relative survival (%) at one, five and ten years since diagnosis:
NHS Region, deprivation category and calendar period of diagnosis

MEN		1971-75											1976-80											
		Affluent		*2*		*3*		*4*		*Deprived*		*All*[1]		*Affluent*		*2*		*3*		*4*		*Deprived*		*All*[1]
		C	R	C	R	C	R	C	R	C	R	C	R	C	R	C	R	C	R	C	R	C	R	C	R
ENGLAND & WALES		*(1,677)*		*(1,863)*		*(1,952)*		*(1,900)*		*(1,739)*		*(11,317)*		*(2,160)*		*(2,350)*		*(2,566)*		*(2,403)*		*(2,224)*		*(12,238)*	
	One year	9	9	9	9	8	9	8	8	7	8	**8**	**9**	9	10	9	10	8	9	9	9	8	9	**9**	**9**
	Five years	2	2	2	2	1	2	2	3	1	2	**2**	**2**	2	2	1	2	2	2	2	3	2	3	**2**	**2**
	Ten years	1	2	1	2	1	1	1	3	1	2	**1**	**2**	1	2	1	2	1	2	1	2	1	2	**1**	**2**
ENGLAND		*(1,607)*		*(1,777)*		*(1,791)*		*(1,718)*		*(1,626)*		*(10,538)*		*(2,075)*		*(2,228)*		*(2,409)*		*(2,220)*		*(2,101)*		*(11,456)*	
	One year	9	9	9	9	8	9	8	8	7	7	**8**	**8**	9	10	9	10	8	9	9	9	8	8	**9**	**9**
	Five years	2	2	2	2	1	2	2	3	1	1	**2**	**2**	2	2	1	2	2	2	2	3	2	2	**2**	**2**
	Ten years	1	2	1	2	1	1	1	3	1	1	**1**	**2**	1	2	1	2	1	2	1	2	1	2	**1**	**2**
Northern & Yorkshire		*(133)*		*(156)*		*(211)*		*(237)*		*(368)*		*(1,530)*		*(201)*		*(250)*		*(299)*		*(407)*		*(519)*		*(1,718)*	
	One year	10	10	3	3	6	7	7	7	5	5	**8**	**8**	4	4	8	9	5	6	9	9	4	5	**8**	**9**
	Five years	1	1	1	1	1	2	2	2	1	1	**2**	**2**	1	1	2	2	2	2	2	3	1	2	**2**	**3**
	Ten years	*1*	*1*	*1*	*1*	0	1	*1*	*2*	*1*	*1*	**1**	**2**	1	1	1	2	1	1	1	2	1	1	**1**	**2**
Trent		*(116)*		*(155)*		*(217)*		*(274)*		*(232)*		*(1,090)*		*(133)*		*(194)*		*(270)*		*(293)*		*(263)*		*(1,187)*	
	One year	7	8	10	10	8	8	9	9	5	5	**9**	**10**	8	8	8	9	9	9	7	8	7	8	**10**	**10**
	Five years	2	3	2	2	1	1	3	3	1	1	**2**	**2**	2	2	2	2	2	2	2	2	3	4	**3**	**3**
	Ten years	0	1	1	2	0	0	2	3	*0*	*0*	**1**	**2**	2	2	1	1	1	2	1	1	2	3	**2**	**3**
Anglia & Oxford		*(234)*		*(226)*		*(214)*		*(145)*		*(68)*		*(1,021)*		*(300)*		*(284)*		*(281)*		*(173)*		*(86)*		*(1,173)*	
	One year	4	4	8	8	6	6	4	4	6	6	**7**	**7**	8	8	9	9	6	7	3	3	6	7	**9**	**9**
	Five years	1·	1	2	2	0	0	1	1	1	1	**1**	**2**	2	2	1	1	2	2	1	1	0	0	**1**	**2**
	Ten years	*1*	*1*	*1*	*2*	*0*	*0*	*1*	*1*	*1*	*1*	**1**	**1**	1	1	*1*	*1*	1	2	*1*	*1*	*0*	*0*	**1**	**1**
North Thames		*(230)*		*(236)*		*(226)*		*(222)*		*(177)*		*(1,249)*		*(218)*		*(194)*		*(241)*		*(221)*		*(187)*		*(1,098)*	
	One year	8	8	9	9	6	7	7	7	11	12	**10**	**10**	10	10	6	7	7	7	8	8	9	10	**10**	**10**
	Five years	2	2	1	2	1	2	3	4	2	3	**2**	**3**	3	3	1	1	1	2	1	2	-	-	**2**	**3**
	Ten years	2	2	*1*	*2*	*1*	*1*	2	3	1	3	**2**	**3**	2	3	*1*	*1*	1	2	1	2	2	3	**2**	**3**
South Thames		*(413)*		*(355)*		*(310)*		*(293)*		*(148)*		*(1,658)*		*(528)*		*(449)*		*(366)*		*(293)*		*(153)*		*(1,892)*	
	One year	7	7	7	7	5	6	7	7	5	6	**8**	**8**	5	5	8	8	4	4	10	10	5	6	**8**	**9**
	Five years	1	1	1	1	1	1	3	3	1	1	**2**	**2**	1	1	1	2	0	1	2	2	0	0	**1**	**2**
	Ten years	1	1	*1*	*1*	0	1	1	3	*1*	*1*	**1**	**2**	*0*	*1*	1	1	0	1	1	2	0	0	**1**	**1**
South & West		*(276)*		*(416)*		*(321)*		*(165)*		*(66)*		*(1,379)*		*(294)*		*(456)*		*(419)*		*(192)*		*(73)*		*(1,487)*	
	One year	6	7	6	7	9	9	6	7	1	1	**8**	**9**	9	10	6	6	7	7	9	10	5	6	**9**	**10**
	Five years	2	2	2	2	2	2	1	2	0	0	**2**	**3**	2	3	1	1	1	2	3	4	1	2	**2**	**3**
	Ten years	1	2	*1*	*2*	1	1	*1*	*1*	0	0	**1**	**2**	1	3	1	1	1	1	2	4	*1*	*2*	**1**	**3**
West Midlands		*(50)*		*(47)*		*(86)*		*(86)*		*(120)*		*(981)*		*(170)*		*(162)*		*(220)*		*(256)*		*(296)*		*(1,118)*	
	One year	13	14	16	17	6	7	5	5	7	7	**8**	**8**	8	8	7	7	7	7	5	5	6	7	**8**	**9**
	Five years	3	4	3	4	0	0	1	1	1	1	**1**	**1**	1	1	0	1	1	2	0	1	0	1	**1**	**1**
	Ten years	3	4	3	4	0	0	0	0	*1*	*1*	**1**	**1**	1	1	0	*1*	1	1	0	*1*	0	0	**1**	**1**
North & West		*(155)*		*(186)*		*(206)*		*(296)*		*(447)*		*(1,630)*		*(231)*		*(239)*		*(313)*		*(385)*		*(524)*		*(1,783)*	
	One year	8	8	5	5	7	7	5	5	5	5	**7**	**7**	10	10	9	9	6	6	6	6	6	6	**8**	**9**
	Five years	1	1	1	1	2	2	1	2	0	1	**1**	**2**	2	2	1	1	1	1	1	2	1	1	**1**	**2**
	Ten years	*1*	*1*	0	0	1	2	0	1	*0*	*1*	**1**	**1**	*1*	*1*	0	0	0	0	1	1	*1*	*1*	**1**	**1**
WALES		*(70)*		*(86)*		*(161)*		*(182)*		*(113)*		*(779)*		*(85)*		*(122)*		*(157)*		*(183)*		*(123)*		*(782)*	
	One year	8	9	9	9	6	6	5	5	7	8	**10**	**10**	11	11	6	6	6	6	7	7	12	12	**10**	**10**
	Five years	3	4	3	4	1	1	1	2	3	3	**3**	**4**	2	2	1	1	2	2	-	-	4	5	**3**	**3**
	Ten years	3	4	2	2	*1*	*1*	*1*	*2*	2	3	**2**	**4**	2	2	*1*	*1*	1	1	1	3	4	5	**2**	**3**

[1] Including those for whom no deprivation category was assigned. C - crude survival, R - relative survival (see Chapter 3).
Number of patients contributing to each analysis in parentheses.

	1981-85												1986-90												
	Affluent		*2*		*3*		*4*		*Deprived*		*All*[1]		*Affluent*		*2*		*3*		*4*		*Deprived*		*All*[1]		**MEN**
	C	R	C	R	C	R	C	R	C	R	C	R	C	R	C	R	C	R	C	R	C	R	C	R	
	(2,201)		*(2,379)*		*(2,422)*		*(2,397)*		*(2,158)*		*(11,835)*		*(2,139)*		*(2,350)*		*(2,395)*		*(2,303)*		*(1,917)*		*(11,214)*		**ENGLAND & WALES**
	11	**12**	**10**	**10**	**9**	**10**	**10**	**10**	**9**	**10**	**10**	**11**	**11**	**12**	**12**	**12**	**11**	**12**	**11**	**11**	**9**	**10**	**11**	**12**	*One year*
	2	**3**	**2**	**2**	**2**	**3**	**2**	**3**	**2**	**3**	**2**	**3**	2	3	2	3	3	3	2	3	2	2	2	3	*Five years*
	1	**2**	**1**	**2**	**2**	**3**	**2**	**3**	**2**	**3**	**2**	**2**													
	(2,117)		*(2,282)*		*(2,270)*		*(2,195)*		*(2,040)*		*(11,124)*		*(2,061)*		*(2,258)*		*(2,258)*		*(2,130)*		*(1,822)*		*(10,617)*		**ENGLAND**
	11	12	10	10	9	10	10	10	9	10	**10**	**10**	11	11	12	12	11	12	11	12	9	10	**11**	**11**	*One year*
	2	2	2	2	2	3	2	3	2	3	**2**	**3**	2	3	2	3	3	3	2	3	2	2	**2**	**3**	*Five years*
	1	2	1	2	1	2	1	2	2	3	**1**	**2**													
	(188)		*(267)*		*(311)*		*(372)*		*(536)*		*(1,700)*		*(195)*		*(257)*		*(302)*		*(329)*		*(416)*		*(1,508)*		**Northern & Yorkshire**
	8	8	9	9	8	8	9	9	9	9	**10**	**11**	9	10	11	11	11	12	9	9	8	8	**11**	**12**	*One year*
	2	2	2	2	2	2	2	2	2	3	**2**	**3**	2	2	2	3	1	2	1	2	2	2	**2**	**2**	*Five years*
	1	1	1	1	1	2	1	2	1	2	**1**	**2**													
	(173)		*(205)*		*(257)*		*(336)*		*(261)*		*(1,241)*		*(182)*		*(244)*		*(296)*		*(293)*		*(273)*		*(1,289)*		**Trent**
	8	8	11	11	7	8	8	8	9	10	**10**	**11**	8	9	9	9	11	11	9	9	6	6	**10**	**11**	*One year*
	1	1	1	1	2	2	2	2	2	2	**2**	**2**	2	3	2	3	3	5	2	2	1	1	**2**	**3**	*Five years*
	0	1	*1*	*1*	1	1	1	2	1	2	**1**	**2**													
	(333)		*(329)*		*(270)*		*(146)*		*(74)*		*(1,186)*		*(355)*		*(337)*		*(261)*		*(182)*		*(66)*		*(1,207)*		**Anglia & Oxford**
	8	9	9	9	5	6	5	6	6	7	**9**	**9**	11	12	7	8	9	10	9	10	6	7	**10**	**11**	*One year*
	2	3	1	2	1	1	1	2	0	0	**2**	**2**	2	3	1	2	1	2	1	1	1	2	**2**	**2**	*Five years*
	2	3	1	1	*1*	*1*	1	1	0	0	**1**	**2**													
	(211)		*(167)*		*(183)*		*(172)*		*(154)*		*(907)*		*(207)*		*(196)*		*(219)*		*(254)*		*(175)*		*(1,079)*		**North Thames**
	12	12	9	9	14	14	12	12	11	12	**13**	**14**	14	15	18	19	13	13	11	12	17	18	**16**	**17**	*One year*
	2	2	4	4	3	4	3	4	6	7	**4**	**5**	4	5	4	5	3	4	3	4	6	7	**4**	**6**	*Five years*
	2	2	2	4	*3*	*4*	2	4	5	7	**3**	**5**													
	(451)		*(357)*		*(307)*		*(226)*		*(125)*		*(1,521)*		*(383)*		*(332)*		*(252)*		*(207)*		*(104)*		*(1,296)*		**South Thames**
	13	13	9	10	8	8	9	10	11	12	**12**	**13**	10	11	13	14	11	12	13	14	14	15	**13**	**14**	*One year*
	3	4	2	3	1	1	3	4	2	3	**3**	**3**	3	4	2	3	2	2	3	4	2	3	**3**	**3**	*Five years*
	3	3	1	2	1	1	*3*	*4*	2	3	**2**	**3**													
	(314)		*(446)*		*(389)*		*(210)*		*(51)*		*(1,455)*		*(294)*		*(400)*		*(413)*		*(237)*		*(81)*		*(1,432)*		**South & West**
	9	9	8	9	8	9	9	9	8	9	**11**	**11**	11	12	10	10	8	8	13	14	8	8	**12**	**12**	*One year*
	1	2	-	-	3	4	2	2	3	3	**2**	**3**	2	3	2	3	2	3	3	4	0	0	**2**	**3**	*Five years*
	1	1	1	2	2	4	*1*	*2*	*1*	*2*	**1**	**3**													
	(180)		*(206)*		*(265)*		*(318)*		*(304)*		*(1,280)*		*(166)*		*(210)*		*(241)*		*(286)*		*(234)*		*(1,141)*		**West Midlands**
	8	8	4	4	7	7	6	7	6	6	**8**	**8**	4	4	9	10	8	9	6	6	6	6	**8**	**8**	*One year*
	1	1	0	0	1	2	0	1	1	1	**1**	**1**	2	2	1	2	3	4	0	0	1	1	**2**	**2**	*Five years*
	1	*1*	0	0	1	2	*0*	*1*	0	*1*	**1**	**1**													
	(267)		*(305)*		*(288)*		*(415)*		*(535)*		*(1,834)*		*(279)*		*(282)*		*(274)*		*(342)*		*(473)*		*(1,665)*		**North & West**
	8	9	6	7	6	6	6	6	5	6	**7**	**8**	8	8	6	6	8	8	8	8	6	6	**8**	**9**	*One year*
	0	0	1	1	2	2	2	2	1	2	**2**	**2**	0	1	0	0	1	2	2	2	0	1	**1**	**1**	*Five years*
	0	*0*	1	1	1	1	1	2	1	2	**1**	**1**													
	(84)		*(97)*		*(152)*		*(202)*		*(118)*		*(711)*		*(78)*		*(92)*		*(137)*		*(173)*		*(95)*		*(597)*		**WALES**
	16	17	9	9	9	9	7	8	10	11	**12**	**13**	17	18	11	11	6	7	8	8	11	11	**12**	**13**	*One year*
	3	5	3	3	3	3	3	3	4	5	**4**	**5**	5	5	4	4	2	2	3	3	5	6	**5**	**6**	*Five years*
	3	*5*	*3*	*3*	*3*	*3*	2	3	4	5	**3**	**5**													

[1] Including those for whom no deprivation category was assigned. C - crude survival, R - relative survival (see Chapter 3). Number of patients contributing to each analysis in parentheses.

Table 19.2 Crude and relative survival (%) at one, five and ten years since diagnosis: NHS Region, deprivation category and calendar period of diagnosis

WOMEN	1971-75												1976-80											
	Affluent		2		3		4		Deprived		All[1]		Affluent		2		3		4		Deprived		All[1]	
	C	R	C	R	C	R	C	R	C	R	C	R	C	R	C	R	C	R	C	R	C	R	C	R
ENGLAND & WALES	*(1,490)*		*(1,668)*		*(1,764)*		*(1,647)*		*(1,439)*		*(9,844)*		*(1,941)*		*(2,105)*		*(2,257)*		*(2,281)*		*(1,941)*		*(11,023)*	
One year	11	11	10	10	8	9	8	8	6	7	9	9	9	10	9	9	9	9	7	8	8	8	8	9
Five years	2	2	2	3	2	2	2	2	1	2	2	2	2	2	1	1	2	2	1	2	1	2	2	2
Ten years	1	2	2	2	1	2	1	2	1	2	1	2	1	2	1	1	1	2	1	2	1	1	1	2
ENGLAND	*(1,441)*		*(1,593)*		*(1,645)*		*(1,517)*		*(1,371)*		*(9,297)*		*(1,875)*		*(2,007)*		*(2,107)*		*(2,122)*		*(1,851)*		*(10,383)*	
One year	11	11	10	10	8	9	8	8	6	7	9	9	9	10	9	9	9	9	7	8	8	8	8	9
Five years	2	2	2	2	2	2	2	2	1	2	2	2	2	2	1	1	2	2	1	2	1	2	2	2
Ten years	1	2	1	2	1	2	1	2	1	1	1	2	1	2	1	1	1	2	1	2	1	1	1	2
Northern & Yorkshire	*(102)*		*(158)*		*(164)*		*(200)*		*(330)*		*(1,285)*		*(189)*		*(214)*		*(306)*		*(384)*		*(483)*		*(1,618)*	
One year	8	9	5	5	13	14	7	7	3	3	8	8	8	8	5	5	7	7	8	9	4	4	8	8
Five years	0	0	0	0	3	3	2	2	1	1	2	2	1	1	1	1	1	1	1	2	1	1	1	1
Ten years	0	0	0	0	2	3	1	2	0	0	1	2	1	1	1	1	0	1	1	2	0	1	1	1
Trent	*(102)*		*(139)*		*(150)*		*(197)*		*(152)*		*(827)*		*(134)*		*(189)*		*(221)*		*(267)*		*(208)*		*(1,057)*	
One year	12	12	8	8	5	5	10	10	5	5	9	9	7	8	6	6	9	9	5	5	9	9	9	9
Five years	4	5	1	1	2	3	2	3	1	2	2	3	2	2	1	1	2	2	2	2	2	2	2	2
Ten years	2	3	0	1	2	3	2	3	1	2	2	3	1	2	1	1	1	1	2	2	1	2	1	2
Anglia & Oxford	*(216)*		*(208)*		*(203)*		*(150)*		*(55)*		*(960)*		*(280)*		*(259)*		*(249)*		*(168)*		*(66)*		*(1,067)*	
One year	9	10	7	8	8	8	3	4	7	8	9	9	7	7	9	10	7	7	6	6	2	2	9	9
Five years	2	2	2	3	2	2	0	0	3	3	2	2	1	1	1	1	2	2	1	1	0	0	1	2
Ten years	1	1	2	3	1	2	0	0	1	2	1	2	1	1	1	1	1	1	0	0	0	0	1	1
North Thames	*(201)*		*(208)*		*(211)*		*(216)*		*(159)*		*(1,113)*		*(191)*		*(201)*		*(184)*		*(174)*		*(171)*		*(968)*	
One year	9	9	9	9	8	8	9	10	6	6	10	10	6	6	8	9	7	8	7	7	10	10	10	10
Five years	1	2	2	2	2	3	3	3	2	2	2	3	1	1	1	1	1	1	1	1	2	2	2	2
Ten years	1	2	2	2	2	3	2	3	1	1	2	2	1	1	1	1	1	1	0	0	1	2	1	2
South Thames	*(414)*		*(353)*		*(344)*		*(251)*		*(155)*		*(1,667)*		*(443)*		*(396)*		*(338)*		*(295)*		*(137)*		*(1,709)*	
One year	10	10	7	8	6	6	5	6	5	5	9	10	7	8	8	8	7	7	6	6	6	6	8	9
Five years	1	1	2	2	1	1	1	1	2	2	2	2	1	1	1	1	2	3	1	1	2	2	2	2
Ten years	1	1	1	2	0	0	1	1	1	2	1	2	1	1	1	1	1	2	0	1	1	1	1	2
South & West	*(213)*		*(324)*		*(290)*		*(163)*		*(49)*		*(1,156)*		*(295)*		*(384)*		*(363)*		*(189)*		*(54)*		*(1,340)*	
One year	9	10	10	10	5	5	5	5	2	2	9	9	9	9	7	7	8	8	8	8	16	17	10	11
Five years	3	4	2	3	1	1	2	2	0	0	2	3	3	4	1	2	3	3	2	2	2	2	2	3
Ten years	2	3	2	2	1	1	1	1	0	0	2	2	1	3	1	1	1	2	1	2	0	0	1	2
West Midlands	*(52)*		*(51)*		*(65)*		*(67)*		*(100)*		*(805)*		*(129)*		*(160)*		*(173)*		*(239)*		*(234)*		*(955)*	
One year	5	5	9	10	10	10	4	4	5	5	7	7	10	10	7	8	5	5	4	4	7	8	8	8
Five years	2	2	0	0	0	0	2	2	1	1	1	1	1	1	1	1	0	1	0	0	1	1	1	1
Ten years	2	2	0	0	0	0	0	0	1	1	1	1	1	1	0	1	0	1	0	0	0	0	1	1
North & West	*(141)*		*(152)*		*(218)*		*(273)*		*(371)*		*(1,484)*		*(214)*		*(204)*		*(273)*		*(406)*		*(498)*		*(1,669)*	
One year	6	7	8	8	5	5	5	6	6	6	8	8	7	7	8	8	7	7	5	5	5	5	7	8
Five years	1	1	2	2	1	1	1	1	1	1	1	1	2	2	1	1	0	1	1	1	1	1	1	1
Ten years	1	1	1	2	0	1	0	0	1	1	1	1	1	2	1	1	0	1	1	1	0	1	1	1
WALES	*(49)*		*(75)*		*(119)*		*(130)*		*(68)*		*(547)*		*(66)*		*(98)*		*(150)*		*(159)*		*(90)*		*(640)*	
One year	12	12	8	8	7	8	9	10	8	8	10	11	8	8	8	8	6	6	7	7	9	10	9	9
Five years	2	2	3	3	2	3	3	3	1	1	3	4	2	2	2	2	2	2	2	3	2	3	3	3
Ten years	2	2	2	3	1	3	2	2	1	1	2	3	2	2	1	2	2	2	2	2	0	0	2	3

---

[1] Including those for whom no deprivation category was assigned. C - crude survival, R - relative survival (see Chapter 3). Number of patients contributing to each analysis in parentheses.

Each cell shows paired values C R (C - crude survival, R - relative survival). Number of patients contributing to each analysis in parentheses.

	|	1981-85					|	1986-90					WOMEN
	Affluent	2	3	4	Deprived	All[1]	Affluent	2	3	4	Deprived	All[1]	
	C R	C R	C R	C R	C R	C R	C R	C R	C R	C R	C R	C R	
(counts)	(2,011)	(2,258)	(2,357)	(2,385)	(1,998)	(11,273)	(2,053)	(2,420)	(2,568)	(2,486)	(1,935)	(11,603)	**ENGLAND & WALES**
	11 11	12 12	9 10	9 9	9 10	10 10	11 11	10 10	10 10	9 9	10 10	10 10	*One year*
	2 2	2 3	2 2	2 2	1 2	2 2	3 3	2 2	2 2	2 2	2 3	2 2	*Five years*
	1 2	1 2	1 1	1 2	1 1	1 2							
(counts)	(1,924)	(2,174)	(2,220)	(2,191)	(1,894)	(10,596)	(1,977)	(2,322)	(2,415)	(2,301)	(1,858)	(10,993)	**ENGLAND**
	11 11	11 12	9 10	9 9	9 10	10 10	11 11	10 10	10 10	9 9	10 10	10 10	*One year*
	2 2	2 3	1 2	2 2	1 2	2 2	2 3	2 2	2 2	2 2	2 3	2 2	*Five years*
	1 2	1 2	1 1	1 2	1 1	1 2							
(counts)	(178)	(228)	(291)	(375)	(448)	(1,539)	(193)	(292)	(341)	(363)	(462)	(1,663)	**Northern & Yorkshire**
	8 8	12 12	7 7	9 9	8 9	10 11	10 11	9 9	9 9	6 6	6 6	9 9	*One year*
	1 1	2 2	2 2	1 1	1 1	1 2	2 3	2 2	1 1	0 0	1 2	1 2	*Five years*
	1 1	1 1	1 1	1 1	0 1	1 1							
(counts)	(134)	(198)	(259)	(304)	(241)	(1,143)	(159)	(211)	(270)	(300)	(240)	(1,185)	**Trent**
	7 7	8 8	5 6	6 6	6 6	7 7	8 8	6 6	8 8	8 8	8 8	9 10	*One year*
	0 0	- -	1 1	1 1	1 1	1 1	3 3	1 1	1 1	1 1	2 2	2 2	*Five years*
	0 0	1 2	0 1	1 1	1 1	1 1							
(counts)	(315)	(296)	(263)	(166)	(68)	(1,129)	(338)	(345)	(284)	(226)	(75)	(1,274)	**Anglia & Oxford**
	8 8	9 10	5 6	8 9	7 8	9 10	8 8	7 7	7 7	7 8	6 6	8 9	*One year*
	2 2	2 2	1 1	2 2	1 1	2 2	2 3	1 1	1 1	2 2	1 1	1 2	*Five years*
	1 2	1 2	0 0	2 2	0 0	1 2							
(counts)	(172)	(180)	(161)	(153)	(108)	(791)	(192)	(228)	(213)	(245)	(167)	(1,084)	**North Thames**
	9 10	11 12	10 11	10 10	14 15	12 13	13 14	9 9	9 9	14 14	16 16	13 14	*One year*
	1 2	2 2	3 3	2 2	4 5	3 3	4 5	2 2	1 2	1 2	5 6	3 3	*Five years*
	1 2	2 2	2 2	2 2	3 4	2 3							
(counts)	(429)	(381)	(307)	(284)	(110)	(1,568)	(338)	(359)	(287)	(225)	(106)	(1,343)	**South Thames**
	13 14	11 11	12 12	7 8	8 8	12 13	12 13	12 13	10 11	11 12	9 10	13 13	*One year*
	2 3	2 3	1 2	2 2	2 2	2 3	2 3	3 4	2 2	2 3	1 2	3 3	*Five years*
	1 2	1 2	1 1	1 2	1 1	1 2							
(counts)	(300)	(447)	(425)	(216)	(71)	(1,501)	(308)	(415)	(426)	(262)	(87)	(1,511)	**South & West**
	11 11	12 12	7 8	8 9	8 9	11 12	8 9	9 10	11 11	6 7	6 6	10 11	*One year*
	2 2	3 4	1 2	3 4	2 3	3 4	1 1	2 2	3 3	2 2	1 2	2 3	*Five years*
	1 2	2 3	0 1	2 3	2 3	2 3							
(counts)	(145)	(168)	(232)	(258)	(280)	(1,093)	(189)	(200)	(262)	(284)	(240)	(1,183)	**West Midlands**
	7 8	8 8	7 7	6 6	8 8	8 9	9 9	6 6	9 9	6 6	7 8	8 9	*One year*
	0 0	1 2	0 0	1 1	0 1	1 1	2 2	1 1	1 1	2 3	2 2	2 2	*Five years*
	0 0	0 0	0 0	1 1	0 0	0 0							
(counts)	(251)	(276)	(282)	(435)	(568)	(1,832)	(260)	(272)	(332)	(396)	(481)	(1,750)	**North & West**
	7 7	5 6	8 9	7 7	6 7	8 9	9 9	9 9	7 7	4 4	8 8	8 9	*One year*
	1 1	0 0	2 2	2 2	1 1	1 2	2 2	1 1	2 2	1 1	2 2	2 2	*Five years*
	0 0	0 0	1 1	1 1	0 1	1 1							
(counts)	(87)	(84)	(137)	(194)	(104)	(677)	(76)	(98)	(153)	(185)	(77)	(610)	**WALES**
	8 9	11 12	10 10	10 10	9 10	11 12	13 13	9 10	7 8	11 11	7 7	11 12	*One year*
	1 1	5 5	4 4	2 3	2 2	3 3	4 4	2 2	3 3	2 3	1 1	3 4	*Five years*
	1 1	3 4	2 4	1 2	2 2	2 3							

[1] Including those for whom no deprivation category was assigned. C - crude survival, R - relative survival (see Chapter 3).
Number of patients contributing to each analysis in parentheses.

Table 19.2　Crude and relative survival (%) at one, five and ten years since diagnosis: NHS Region, deprivation category and calendar period of diagnosis

ADULTS		1971-75 Affluent C	R	2 C	R	3 C	R	4 C	R	Deprived C	R	All[1] C	R	1976-80 Affluent C	R	2 C	R	3 C	R	4 C	R	Deprived C	R	All[1] C	R
ENGLAND & WALES		(3,167)		(3,531)		(3,716)		(3,547)		(3,178)		(21,161)		(4,101)		(4,455)		(4,823)		(4,684)		(4,165)		(23,261)	
	One year	10	10	9	10	8	9	8	8	7	7	**8**	**9**	9	10	9	10	8	9	8	9	8	9	**9**	**9**
	Five years	2	2	2	2	1	2	2	3	1	2	**2**	**2**	2	2	1	2	2	2	2	2	2	2	**2**	**2**
	Ten years	1	2	1	2	1	2	1	2	1	2	**1**	**2**	1	2	1	2	1	2	1	2	1	2	**1**	**2**
ENGLAND		(3,048)		(3,370)		(3,436)		(3,235)		(2,997)		(19,835)		(3,950)		(4,235)		(4,516)		(4,342)		(3,952)		(21,839)	
	One year	10	10	9	10	8	9	8	8	7	7	**8**	**9**	9	10	9	10	9	9	8	9	8	8	**9**	**9**
	Five years	2	2	2	2	1	2	2	3	1	2	**2**	**2**	2	2	1	2	2	2	2	2	2	2	**2**	**2**
	Ten years	1	2	1	2	1	2	1	2	1	1	**1**	**2**	1	2	1	2	1	2	1	2	1	2	**1**	**2**
Northern & Yorkshire		(235)		(314)		(375)		(437)		(698)		(2,815)		(390)		(464)		(605)		(791)		(1,002)		(3,336)	
	One year	9	9	4	4	9	10	7	7	4	4	**8**	**8**	6	6	7	7	6	6	8	9	4	5	**8**	**8**
	Five years	1	1	1	1	2	2	2	2	1	1	**2**	**2**	1	1	1	1	1	2	2	2	1	1	**2**	**2**
	Ten years	1	1	0	1	1	2	1	2	1	1	**1**	**2**	1	1	1	1	0	1	1	2	1	1	**1**	**2**
Trent		(218)		(294)		(367)		(471)		(384)		(1,917)		(267)		(383)		(491)		(560)		(471)		(2,244)	
	One year	9	10	9	9	7	7	9	9	5	5	**9**	**9**	8	8	7	8	9	9	6	7	8	8	**9**	**10**
	Five years	3	4	1	2	1	2	2	3	1	1	**2**	**3**	2	2	1	1	2	2	2	2	2	3	**2**	**3**
	Ten years	1	2	1	2	1	2	2	3	1	1	**1**	**2**	1	2	1	1	1	2	1	2	2	3	**1**	**3**
Anglia & Oxford		(450)		(434)		(417)		(295)		(123)		(1,981)		(580)		(543)		(530)		(341)		(152)		(2,240)	
	One year	7	7	8	8	7	7	4	4	6	7	**8**	**8**	7	8	9	9	7	7	4	5	4	4	**9**	**9**
	Five years	1	2	2	2	1	1	1	1	2	2	**2**	**2**	1	1	1	1	2	2	1	1	0	0	**1**	**2**
	Ten years	1	1	2	2	1	1	0	1	1	1	**1**	**2**	1	1	1	1	1	1	1	1	0	0	**1**	**1**
North Thames		(431)		(444)		(437)		(438)		(336)		(2,362)		(409)		(395)		(425)		(395)		(358)		(2,066)	
	One year	8	9	9	9	7	7	8	8	8	9	**10**	**10**	8	8	7	8	7	8	7	8	9	10	**10**	**10**
	Five years	2	2	1	2	2	2	3	3	2	2	**2**	**3**	2	2	1	1	1	2	1	1	2	3	**2**	**3**
	Ten years	1	2	1	2	1	2	2	3	1	2	**2**	**3**	1	2	1	1	1	2	1	1	2	3	**1**	**2**
South Thames		(827)		(708)		(654)		(544)		(303)		(3,325)		(971)		(845)		(704)		(588)		(290)		(3,601)	
	One year	8	9	7	7	6	6	6	7	5	5	**9**	**9**	6	6	8	8	5	6	7	8	6	6	**8**	**9**
	Five years	1	1	2	2	1	1	2	3	-	-	**2**	**2**	1	1	1	1	1	2	1	2	1	2	**1**	**2**
	Ten years	1	1	1	2	0	1	1	2	1	1	**1**	**2**	1	1	1	1	1	1	1	2	1	1	**1**	**2**
South & West		(489)		(740)		(611)		(328)		(115)		(2,535)		(589)		(840)		(782)		(381)		(127)		(2,827)	
	One year	8	8	8	8	7	7	6	6	1	1	**9**	**9**	9	9	6	7	7	8	9	9	10	10	**10**	**10**
	Five years	2	3	2	2	1	2	1	2	0	0	**2**	**3**	2	3	1	2	2	2	2	3	2	2	**2**	**3**
	Ten years	1	3	1	2	1	1	1	1	0	0	**1**	**2**	1	3	1	1	1	2	2	3	1	1	**1**	**2**
West Midlands		(102)		(98)		(151)		(153)		(220)		(1,786)		(299)		(322)		(393)		(495)		(530)		(2,073)	
	One year	8	9	13	13	8	8	4	5	6	6	**7**	**8**	9	9	7	7	6	6	5	5	7	7	**8**	**8**
	Five years	2	3	2	2	0	0	1	1	1	1	**1**	**1**	-	-	1	1	1	1	0	0	1	1	**1**	**1**
	Ten years	2	3	2	2	0	0	0	0	1	1	**1**	**1**	1	1	0	1	1	1	0	0	0	0	**1**	**1**
North & West		(296)		(338)		(424)		(569)		(818)		(3,114)		(445)		(443)		(586)		(791)		(1,022)		(3,452)	
	One year	7	7	6	6	6	6	5	5	5	6	**7**	**8**	8	9	9	9	6	7	5	5	5	6	**8**	**8**
	Five years	1	1	1	1	1	1	1	1	1	1	**1**	**1**	2	2	1	1	1	1	1	1	1	1	**1**	**2**
	Ten years	1	1	0	0	1	1	0	1	1	1	**1**	**1**	1	2	0	1	0	0	1	1	1	1	**1**	**1**
WALES		(119)		(161)		(280)		(312)		(181)		(1,326)		(151)		(220)		(307)		(342)		(213)		(1,422)	
	One year	10	10	8	9	7	7	7	7	7	8	**10**	**11**	9	10	7	7	6	6	7	7	11	11	**9**	**10**
	Five years	3	3	3	4	1	2	2	3	2	2	**3**	**4**	2	2	2	2	2	2	-	-	3	4	**3**	**3**
	Ten years	3	3	2	3	1	2	1	2	2	2	**2**	**4**	2	2	1	2	1	2	2	3	2	3	**2**	**3**

[1]　Including those for whom no deprivation category was assigned. C - crude survival, R - relative survival (see Chapter 3).
　　Number of patients contributing to each analysis in parentheses.

Each data cell shows two figures: **C** – crude survival and **R** – relative survival. Counts (number of patients) are given in parentheses.

	1981-85						**1986-90**						
ADULTS	Affluent (C R)	2 (C R)	3 (C R)	4 (C R)	Deprived (C R)	All¹ (C R)	Affluent (C R)	2 (C R)	3 (C R)	4 (C R)	Deprived (C R)	All¹ (C R)	
ENGLAND & WALES	(4,212)	(4,637)	(4,779)	(4,782)	(4,156)	(23,108)	(4,192)	(4,770)	(4,963)	(4,789)	(3,852)	(22,817)	
	11 12	11 11	9 10	9 10	9 10	10 10	11 12	11 11	11 11	10 10	10 10	10 11	One year
	2 2	2 3	2 2	2 3	2 2	2 2	2 3	2 3	2 3	2 2	2 3	2 3	Five years
	1 2	1 2	1 2	1 2	1 2	1 2							
ENGLAND	(4,041)	(4,456)	(4,490)	(4,386)	(3,934)	(21,720)	(4,038)	(4,580)	(4,673)	(4,431)	(3,680)	(21,610)	
	11 11	11 11	9 10	9 10	9 10	10 10	11 11	11 11	11 11	10 10	9 10	10 11	One year
	2 2	2 3	2 2	2 2	2 2	2 2	2 3	2 3	2 3	2 2	2 2	2 3	Five years
	1 2	1 2	1 2	1 2	1 2	1 2							
Northern & Yorkshire	(366)	(495)	(602)	(747)	(984)	(3,239)	(388)	(549)	(643)	(692)	(878)	(3,171)	
	8 8	10 11	7 8	9 9	8 9	10 11	10 10	10 10	10 11	7 8	7 7	10 10	One year
	1 1	2 2	2 2	1 2	1 2	2 2	2 2	2 3	1 2	1 1	1 2	2 2	Five years
	1 1	1 1	1 2	1 2	1 1	1 2							
Trent	(307)	(403)	(516)	(640)	(502)	(2,384)	(341)	(455)	(566)	(593)	(513)	(2,474)	
	7 8	9 10	6 7	7 7	7 8	9 9	8 8	7 8	9 10	8 9	7 7	10 10	One year
	1 1	- -	1 2	2 2	1 2	2 2	2 3	2 2	2 3	1 2	1 2	2 3	Five years
	0 1	1 2	1 1	1 2	1 2	1 2							
Anglia & Oxford	(648)	(625)	(533)	(312)	(142)	(2,315)	(693)	(682)	(545)	(408)	(141)	(2,481)	
	8 9	9 9	5 6	7 7	7 7	9 10	9 10	7 7	8 8	8 9	6 7	9 10	One year
	2 3	2 2	1 1	2 2	1 1	2 2	2 3	1 1	1 2	1 1	1 1	2 2	Five years
	1 2	1 1	0 1	1 2	0 0	1 2							
North Thames	(383)	(347)	(344)	(325)	(262)	(1,698)	(399)	(424)	(432)	(499)	(342)	(2,163)	
	11 11	10 10	12 13	11 11	12 13	13 13	14 14	13 14	11 11	12 13	16 17	15 15	One year
	2 2	3 4	3 3	- -	5 6	3 4	4 5	3 4	2 3	2 3	5 6	4 5	Five years
	1 2	2 3	2 3	2 3	4 6	3 4							
South Thames	(880)	(738)	(614)	(510)	(235)	(3,089)	(721)	(691)	(539)	(432)	(210)	(2,639)	
	13 14	10 11	10 10	8 9	9 10	12 13	11 12	12 13	11 11	12 13	12 12	13 14	One year
	3 3	2 3	1 1	- -	2 2	2 3	3 3	2 3	2 2	3 3	2 3	3 3	Five years
	2 3	1 2	1 1	2 3	2 2	2 2							
South & West	(614)	(893)	(814)	(426)	(122)	(2,956)	(602)	(815)	(839)	(499)	(168)	(2,943)	
	10 10	10 10	8 8	8 9	8 9	11 12	10 10	10 10	9 10	10 10	7 7	11 11	One year
	2 2	2 3	2 3	2 3	2 3	3 3	1 2	2 3	2 3	2 3	1 1	2 3	Five years
	1 2	2 2	1 2	1 2	2 3	2 3							
West Midlands	(325)	(374)	(497)	(576)	(584)	(2,373)	(355)	(410)	(503)	(570)	(474)	(2,324)	
	8 8	6 6	7 7	6 7	7 7	8 8	7 7	7 8	9 9	6 6	6 7	8 9	One year
	0 0	1 1	1 1	1 1	1 1	1 1	2 2	1 1	2 3	1 1	1 2	2 2	Five years
	0 0	0 0	1 1	1 1	0 0	1 1							
North & West	(518)	(581)	(570)	(850)	(1,103)	(3,666)	(539)	(554)	(606)	(738)	(954)	(3,415)	
	7 8	6 6	7 8	6 7	6 6	8 8	8 9	7 8	7 8	6 6	7 7	8 9	One year
	0 1	1 1	2 2	2 2	1 1	1 2	1 1	1 1	2 2	1 2	1 1	1 2	Five years
	0 0	0 1	1 1	1 1	1 1	1 1							
WALES	(171)	(181)	(289)	(396)	(222)	(1,388)	(154)	(190)	(290)	(358)	(172)	(1,207)	
	12 13	10 11	9 10	9 9	10 10	12 12	15 16	10 10	7 7	9 10	9 9	12 12	One year
	2 2	4 4	3 4	2 3	3 4	3 4	4 5	3 3	2 3	2 3	3 4	4 5	Five years
	2 2	3 4	3 4	1 3	3 4	3 4							

¹ Including those for whom no deprivation category was assigned. C - crude survival, R - relative survival (see Chapter 3).
Number of patients contributing to each analysis in parentheses.

Fig. 19.1 Relative survival up to ten years: trends by calendar period of diagnosis (1971-90) and NHS Region

ADULTS

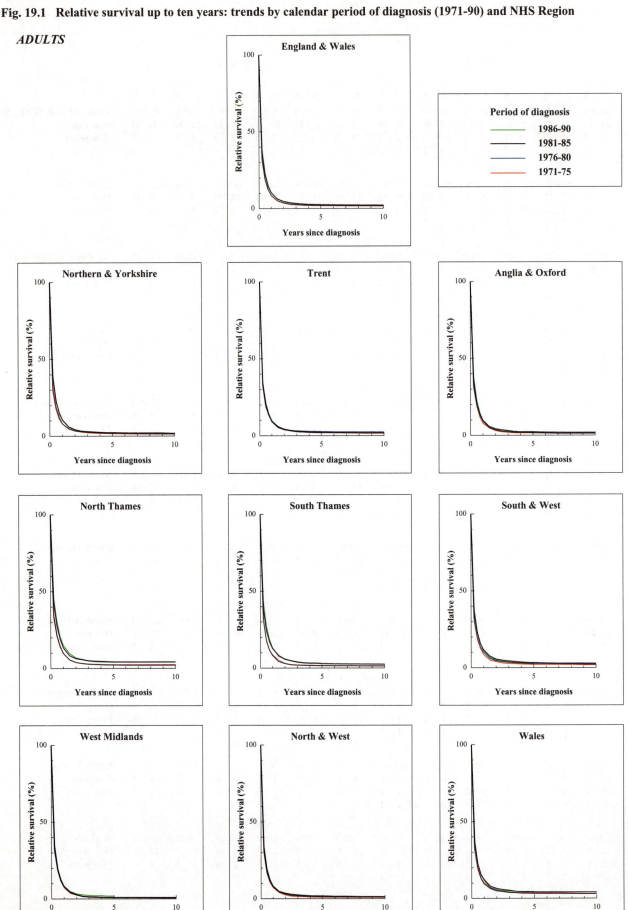

Fig. 19.2   Relative survival up to five years, by deprivation category and NHS Region: patients diagnosed 1986-90

ADULTS

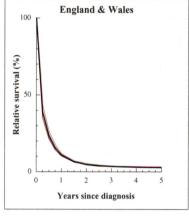

England & Wales

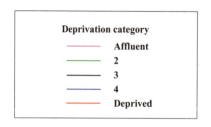

Deprivation category
- Affluent
- 2
- 3
- 4
- Deprived

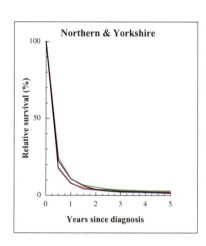

Northern & Yorkshire

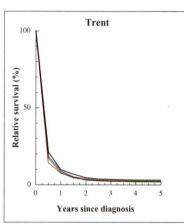

Trent

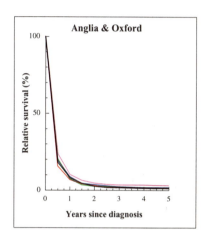

Anglia & Oxford

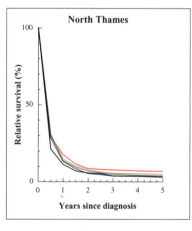

North Thames

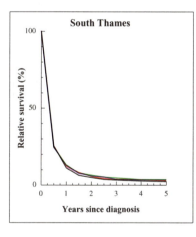

South Thames

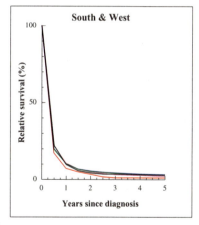

South & West

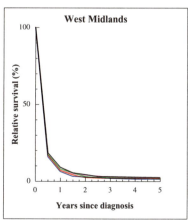

West Midlands

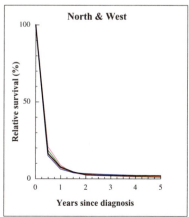

North & West

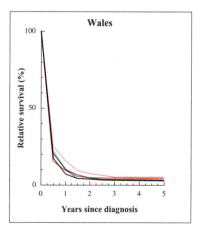

Wales

Fig. 19.3   Relative survival at five years by deprivation category, period of diagnosis (1981-90) and NHS Region

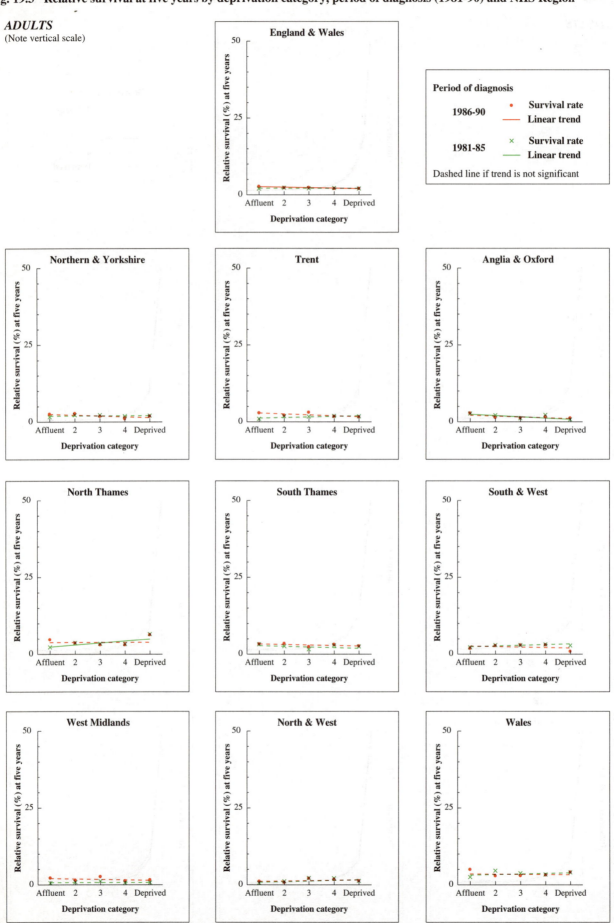

ADULTS
(Note vertical scale)

Fig. 19.4 Relative survival (%) by NHS Region and sex:
one-year and five-year survival for patients diagnosed 1986-90, with 95% confidence intervals

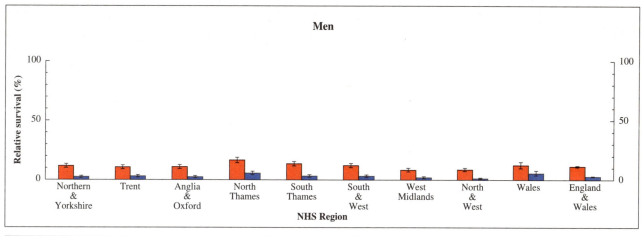

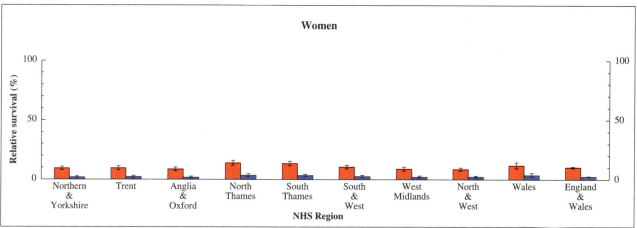

Fig. 19.5 Relative survival (%) at five years, with 95% confidence interval, by age at diagnosis and sex:
England and Wales, patients diagnosed 1986-90

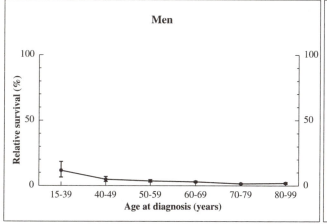

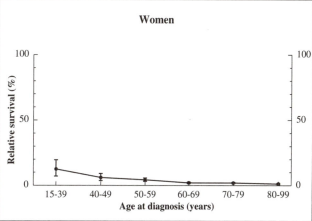

Chapter 20

NASAL CAVITIES AND PARANASAL SINUSES

Cancers of the nasal cavity and paranasal sinuses are uncommon. There are about 400 new cases a year in England and Wales, with incidence rates of 1 per 100,000 in men and 0.5 in women. They represent about one in 400 (0.25%) of all cancers in adults. This anatomic location includes the nasal cavity, cartilage and septum, the maxillary antrum, and the frontal, sphenoidal and ethmoid sinuses which develop from the nasal cavity and communicate with it. Occupations known to increase cancer risk include those involving exposure to nickel compounds and to leather and wood dusts[1].

Diagnosis is often late, because there may be no symptoms until the tumour has invaded adjacent structures, such as the palate, maxilla or orbit of the eye. Surgery and radiotherapy are used, according to the location and extent of spread of the tumour.

The survival analyses here include 7,300 adults registered with cancer of the nasal cavities or paranasal sinuses in England and Wales during the period 1971-90, and followed up to the end of 1995. Of the 8,000 eligible patients, 5% were excluded because their vital status was unknown at 6 July 1997, and 3% because their survival was zero or unknown. Among patients diagnosed during 1986-90, the proportions excluded for these reasons were 6% and 5%, respectively (Table 20.1).

The tissues of the maxillary antrum and nasal cavity itself were most often involved (40% and 37%, respectively): these proportions have not changed much since 1971 (Table 20A). Since 1979, when tumours of the ethmoid, frontal and sphenoid sinuses were separately coded for the first time (in ICD-9), about 10% of tumours have been assigned to the ethmoid sinus. A similar proportion arose in the middle ear and in the mastoid cavity. Almost half (48%) of the tumours were squamous carcinomas, and

12% were adenocarcinomas. There were 287 melanomas (4%), about 15 each year. A quarter were poorly specified carcinomas.

Survival trends

Relative survival for adults diagnosed in England and Wales during 1986-90 was 66% at one year and 41% at five years. The corresponding figures for those diagnosed in 1971-75 were 68% and 39% (Table 20.2). Regional estimates of five-year survival for patients diagnosed in the late 1980s, each based on 150-250 patients, ranged from 34% in North & West to 51% in Wales (Figure 20.4).

The lack of substantial improvement in national average survival since the early 1970s is clear from Figure 20.1, which also reflects regional differences in survival trends. There appears to have been no appreciable gain in relative survival. Estimates of the average gain in one-year survival in successive five-year periods from the early 1970s up to 1986-90 were close to zero, after adjustment for age (see Table 4.4). The national average gain in five-year survival between successive five-year periods was also very low (0.5%) in both sexes, and not statistically significant.

Relative survival at five years is about 50% in both men and women up to age 60 at diagnosis (Figure 20.5), but is in the range 20% to 40% at older ages (see Table 4.7).

Deprivation

Differences in survival between deprivation groups are quite marked, both for men and women. With the exception of patients diagnosed during the late 1970s, the gap in relative survival between the most affluent and most deprived groups in England and Wales was 10% or more

Table 20A Sub-site distribution of cancers of the nasal cavities and paranasal sinuses

ICD-9 code	Site description	1971-75		1976-80		1981-85		1986-90		Total	
		No.	%	No.	%	No.	%	No.	%	No.	%
160.0	Nasal cavity	808	37.8	654	36.6	564	34.1	660	38.6	2,686	36.8
160.1	Middle ear and mastoid	191	8.9	180	10.1	174	10.5	144	8.4	689	9.4
160.2	Maxillary antrum	894	41.8	704	39.4	668	40.4	670	39.2	2,936	40.3
160.3	Ethmoid sinus	-	-	66	3.7	184	11.1	167	9.8	417	5.7
160.4	Frontal sinus	-	-	5	0.3	9	0.5	9	0.5	23	0.3
160.5	Sphenoidal sinus	-	-	5	0.3	17	1.0	13	0.8	35	0.5
160.8	Other	228	10.7	130	7.3	20	1.2	14	0.8	392	5.4
160.9	Unspecified	17	0.8	44	2.5	19	1.1	34	2.0	114	1.6
	Total	**2,138**		**1,788**		**1,655**		**1,711**		**7,292**	

(Table 20.2). The most deprived group had the poorest survival in each period. Differences between other deprivation categories are less marked. For example, relative survival was 6% to 12% lower among the 309 adults in the most deprived category who were diagnosed in England and Wales during 1986-90 than in all other deprivation categories. This is also clear from Figure 20.2.

The deprivation gradient in survival nevertheless flattened slightly between the early and late 1980s (Figure 20.3). For over 1,600 patients of known deprivation status who were diagnosed during 1981-85, the gap in one-year and five-year survival between the most affluent and the most deprived groups was 11% (see Table 4.9). For the 1,700 patients of known status diagnosed in the five years up to 1990, this gap had fallen to some 6-7%, and it was no longer statistically significant.

International comparison

Five-year survival was 42% for men and 39% for women diagnosed with a cancer of the nasal cavity or sinuses during 1986-90 in England and Wales, some 3% to 5% below the average for Europe (see Table 4.12). The national five-year survival rate in the affluent group was comparable with the rate in Scotland and with the European average. The highest regional survival rates exceed 50% at five years, and are comparable with those in the USA.

1. Malker HS, Weiner JA, McLaughlin JK. Register epidemiology studies of recent cancer trends in selected workers. Ann N Y Acad Sci 1990; 609: 322-333

Table 20.1 Ineligible and excluded records, and patients included in analyses, by calendar period of diagnosis

	Calendar period of diagnosis														
	1971-75			1976-80			1981-85			1986-90			Overall		
Total registered	**2,408**			**2,051**			**1,926**			**2,024**			**8,409**		
Ineligible	*Records*	*Patients*	*%*	*Records*	*Patients*	*%*	*Records*	*Patients*	*%*	*Records*	*Patients*	*%*	*Records*	*Patients*	*%*
Incomplete data[1]	3	3		11	11		52	52		39	39		105	**105**	
Resident outside England and Wales	36	36		33	33		10	9		15	14		94	**92**	
In situ neoplasm[2]	30	30		3	3		0	0		0	0		33	**33**	
Benign or uncertain behaviour[2]	44	37		26	24		3	2		1	1		74	**64**	
Metastatic[2]	6	6		2	2		0	0		0	0		8	**8**	
Otherwise ineligible[3]	-	-		-	-		-	-		-	-		-	**-**	
Lymphoma[4]	26	25		15	14		10	0		2	0		53	**39**	
Leukaemia or myeloma[4]	0	0		1	1		0	0		1	0		2	**1**	
		137			88			63			54			342	
Total eligible		**2,271**	**100**		**1,963**	**100**		**1,863**	**100**		**1,970**	**100**		**8,067**	**100**
Aged under 15 or 100 years or over at diagnosis	20	16	0.7	18	14	0.7	23	20	1.1	22	21	1.1	83	**71**	0.9
Vital status unknown[5]	109	81	3.6	152	119	6.1	97	79	4.2	133	127	6.4	491	**406**	5.0
Duplicate registration[6]	0	0	0.0	0	0	0.0	0	0	0.0	1	1	<0.1	1	**1**	<0.1
Multiple primary[7]	0	0	0.0	0	0	0.0	8	8	0.4	11	9	0.5	19	**17**	0.2
Synchronous tumours[8]	6	3	0.1	3	2	0.1	9	9	0.5	7	5	0.3	25	**19**	0.2
Sex not known	0	0	0.0	0	0	0.0	0	0	0.0	0	0	0.0	0	**0**	0.0
Sex incompatible with site code[9]	-	-	-	-	-	-	-	-	-	-	-	-	-	**-**	-
Invalid dates[10]	1	1	<0.1	0	0	0.0	0	0	0.0	2	1	<0.1	3	**2**	<0.1
Zero survival[11]	35	32	1.4	51	40	2.0	104	92	4.9	103	95	4.8	293	**259**	3.2
Total exclusions		**133**	**5.9**		**175**	**8.9**		**208**	**11.2**		**259**	**13.1**		**775**	**9.6**
Patients included in analysis		**2,138**	**94.1**		**1,788**	**91.1**		**1,655**	**88.8**		**1,711**	**86.9**		**7,292**	**90.4**

[1] Main data item(s) invalid or incompatible with one another: name, sex, date of birth, date of diagnosis and (if present) date of death, postcode, site, morphology and behaviour

[2] Tumour either *in situ* (behaviour code 2), benign (0), uncertain if benign or malignant (1) or metastatic (6 or 9)

[3] Not applicable for this malignancy

[4] Morphology that of lymphoma (included with non-Hodgkin lymphoma, Ch. 41) or myeloma (included with multiple myeloma, Ch. 43), or leukaemia (excluded)

[5] Tracing of vital status at National Health Service Central Register incomplete at 6 July 1997

[6] Same site code, sex, cancer registry and cancer registry number as an earlier registration

[7] Same site code and person number as an earlier registration(s): mostly confirmed multiple primary tumours at this site, some unresolved duplicate registrations

[8] Same site code, sex, date of birth and date of diagnosis as another registration(s): mostly synchronous or (in paired organs) bilateral tumours in same anatomic site in one individual, not previously linked: also some duplicate registrations

[9] Not applicable for this malignancy

[10] Impossible sequence of dates (birth, diagnosis and, if present, emigration or death); or date of death after 6 July 1997

[11] Date of diagnosis same as date of death: some are patients who did die on the day of diagnosis, but most were registered solely from a death certificate (death certificate only, or DCO), and their survival time is thus unknown

Table 20.2 Crude and relative survival (%) at one, five and ten years since diagnosis: NHS Region, deprivation category and calendar period of diagnosis

	1971-75						1976-80					
	Affluent C R	2 C R	3 C R	4 C R	Deprived C R	All[1] C R	Affluent C R	2 C R	3 C R	4 C R	Deprived C R	All[1] C R
ENGLAND & WALES												
Men	*(173)*	*(180)*	*(233)*	*(180)*	*(174)*	*(1,177)*	*(178)*	*(178)*	*(220)*	*(210)*	*(205)*	*(1,038)*
One year	68 71	71 75	63 67	67 71	68 72	65 69	65 68	69 72	60 63	64 68	65 69	64 68
Five years	36 45	31 40	33 42	33 42	25 33	31 39	34 43	35 42	30 38	30 39	29 39	31 40
Ten years	29 42	22 38	22 37	20 33	17 29	21 35	25 38	22 38	16 27	19 33	19 32	20 33
Women	*(127)*	*(150)*	*(161)*	*(166)*	*(144)*	*(961)*	*(125)*	*(138)*	*(147)*	*(167)*	*(145)*	*(750)*
One year	69 72	64 67	67 69	67 69	59 62	65 67	62 64	61 63	60 63	57 59	59 62	60 63
Five years	36 44	32 39	34 42	33 40	31 39	32 39	33 40	34 40	29 35	29 35	29 36	31 37
Ten years	24 37	24 37	22 36	21 33	19 30	22 34	19 27	23 31	19 31	15 23	15 24	18 27
Adults	*(300)*	*(330)*	*(394)*	*(346)*	*(318)*	*(2,138)*	*(303)*	*(316)*	*(367)*	*(377)*	*(350)*	*(1,788)*
One year	69 72	68 71	65 68	67 70	64 68	65 68	64 66	65 68	60 63	61 64	62 66	62 66
Five years	36 45	31 40	34 42	33 41	28 36	32 39	34 41	34 41	29 37	29 38	29 38	31 38
Ten years	27 39	23 37	22 37	21 33	18 29	22 34	23 33	22 34	17 28	17 29	17 29	19 30

	Men	Women	Adults				Men	Women	Adults			
Northern & Yorkshire	*(130)*	*(105)*	*(235)*				*(131)*	*(88)*	*(219)*			
One year	61 64	58 61	60 63				60 63	60 63	60 63			
Five years	26 34	24 29	25 32				30 38	29 34	30 36			
Ten years	19 28	*12 24*	16 28				21 32	21 28	21 31			
Trent	*(120)*	*(106)*	*(226)*				*(119)*	*(85)*	*(204)*			
One year	66 71	67 71	67 71				54 56	63 65	58 60			
Five years	30 37	29 35	30 36				23 29	35 40	28 34			
Ten years	17 30	20 29	19 30				16 24	*11 18*	14 22			
Anglia & Oxford	*(125)*	*(82)*	*(207)*				*(106)*	*(69)*	*(175)*			
One year	70 75	68 72	69 74				66 69	*64 67*	65 68			
Five years	37 46	34 42	36 44				29 36	*35 44*	31 39			
Ten years	26 41	*24 38*	26 39				15 27	*25 38*	18 32			
North Thames	*(174)*	*(133)*	*(307)*				*(121)*	*(96)*	*(217)*			
One year	69 73	71 74	70 73				71 73	63 67	67 71			
Five years	35 46	43 52	38 49				42 52	31 40	37 47			
Ten years	25 42	27 43	26 43				25 40	21 34	24 37			
South Thames	*(146)*	*(120)*	*(266)*				*(132)*	*(99)*	*(231)*			
One year	62 65	64 68	63 66				66 70	58 61	62 66			
Five years	31 39	33 42	32 40				35 44	32 40	34 42			
Ten years	18 32	23 36	21 33				22 39	16 24	19 31			
South & West	*(152)*	*(130)*	*(282)*				*(127)*	*(91)*	*(218)*			
One year	71 74	63 65	67 70				64 69	54 57	60 64			
Five years	34 45	33 42	34 44				33 44	26 33	30 39			
Ten years	27 41	22 39	25 41				*21 41*	15 22	19 31			
West Midlands	*(117)*	*(87)*	*(204)*				*(108)*	*(74)*	*(182)*			
One year	59 63	59 61	59 62				62 66	*63 66*	63 66			
Five years	28 34	27 33	27 33				24 31	*32 39*	27 35			
Ten years	18 30	*21 33*	19 32				17 27	*23 36*	19 31			
North & West	*(147)*	*(142)*	*(289)*				*(144)*	*(107)*	*(251)*			
One year	66 69	65 68	65 68				71 75	58 61	65 69			
Five years	28 34	28 34	28 34				33 43	25 32	30 38			
Ten years	18 27	21 30	19 29				20 33	15 22	18 28			
WALES	*(66)*	*(56)*	*(122)*				*(50)*	*(41)*	*(91)*			
One year	*59 63*	*63 67*	60 65				*66 70*	*52 55*	59 63			
Five years	*29 38*	*41 54*	34 45				*26 33*	*35 41*	30 39			
Ten years	*23 36*	*28 49*	25 43				*16 24*	*25 35*	20 31			

[1] Including those for whom no deprivation category was assigned. C - crude survival, R - relative survival (see Chapter 3).
Number of patients contributing to each analysis in parentheses.

1981-85 / 1986-90

1981-85 Affluent C R	2 C R	3 C R	4 C R	Deprived C R	All[1] C R	1986-90 Affluent C R	2 C R	3 C R	4 C R	Deprived C R	All[1] C R		
												ENGLAND & WALES	
(184)	*(167)*	*(188)*	*(213)*	*(147)*	*(915)*	*(199)*	*(185)*	*(212)*	*(220)*	*(185)*	*(1,018)*		*Men*
66 69	61 64	60 64	59 63	58 61	61 65	74 78	61 64	63 67	65 69	56 59	64 67	*One year*	
34 42	31 39	30 38	26 35	24 32	29 38	37 45	30 38	31 40	36 47	29 36	33 42	*Five years*	
22 37	18 27	19 31	14 28	14 24	18 30								
(128)	*(133)*	*(165)*	*(151)*	*(140)*	*(740)*	*(123)*	*(139)*	*(140)*	*(154)*	*(124)*	*(693)*		*Women*
72 76	61 65	66 69	59 62	55 58	62 66	57 59	62 66	67 71	61 65	61 64	62 65	*One year*	
40 47	29 36	33 41	32 41	26 32	32 39	34 41	38 46	32 39	33 43	25 30	32 39	*Five years*	
32 43	18 29	20 32	21 35	22 30	22 34								
(312)	*(300)*	*(353)*	*(364)*	*(287)*	*(1,655)*	*(322)*	*(324)*	*(352)*	*(374)*	*(309)*	*(1,711)*		*Adults*
69 72	61 65	63 67	59 63	57 60	62 65	67 71	62 64	65 68	64 67	58 61	63 66	*One year*	
36 44	30 38	31 40	28 37	25 32	30 39	36 44	34 41	32 39	35 45	28 33	33 41	*Five years*	
26 40	18 27	19 32	17 31	18 28	20 32								

1981-85 Men C R	Women C R	Adults C R	1986-90 Men C R	Women C R	Adults C R		
						Northern & Yorkshire	
(128)	*(108)*	*(236)*	*(110)*	*(71)*	*(181)*		
54 57	67 70	60 63	57 59	58 61	57 60	*One year*	
23 31	39 45	30 38	31 40	24 28	28 35	*Five years*	
9 18	*26 43*	17 31					
(84)	*(70)*	*(154)*	*(84)*	*(65)*	*(149)*	**Trent**	
64 69	*61 66*	63 67	58 60	*70 74*	63 66	*One year*	
28 36	*24 32*	26 34	28 35	*29 38*	28 36	*Five years*	
18 29	*14 24*	*16 27*					
(110)	*(69)*	*(179)*	*(124)*	*(53)*	*(177)*	**Anglia & Oxford**	
70 75	*65 69*	68 72	66 69	*61 65*	65 68	*One year*	
40 51	*31 40*	36 47	31 39	*33 42*	31 39	*Five years*	
27 45	*20 33*	24 40					
(87)	*(66)*	*(153)*	*(132)*	*(91)*	*(223)*	**North Thames**	
60 63	*63 66*	61 64	69 73	70 75	69 74	*One year*	
31 42	*34 43*	32 42	39 50	43 51	41 50	*Five years*	
17 29	*23 38*	20 33					
(122)	*(119)*	*(241)*	*(131)*	*(76)*	*(207)*	**South Thames**	
64 68	*63 66*	63 67	68 71	*60 63*	65 68	*One year*	
36 45	*32 40*	34 43	37 45	*39 46*	38 46	*Five years*	
25 41	*20 34*	23 38					
(115)	*(88)*	*(203)*	*(119)*	*(84)*	*(203)*	**South & West**	
60 63	*57 61*	59 62	69 72	62 65	66 69	*One year*	
27 34	*30 37*	28 35	36 43	28 37	33 41	*Five years*	
18 28	*25 37*	21 32					
(98)	*(71)*	*(169)*	*(84)*	*(68)*	*(152)*	**West Midlands**	
66 70	*68 72*	67 71	55 57	*54 58*	55 58	*One year*	
23 30	*32 38*	27 33	28 36	*29 37*	29 36	*Five years*	
10 18	*23 32*	16 25					
(115)	*(106)*	*(221)*	*(139)*	*(113)*	*(252)*	**North & West**	
50 53	*57 61*	53 57	61 63	57 61	59 62	*One year*	
22 30	*29 38*	26 34	25 32	27 34	26 34	*Five years*	
14 24	*22 33*	18 29					
(56)	*(43)*	*(99)*	*(95)*	*(72)*	*(167)*	**WALES**	
71 76	*59 63*	*66 70*	*71 75*	*64 67*	*68 72*	*One year*	
37 50	*37 44*	*37 48*	*43 56*	*39 46*	*41 51*	*Five years*	
25 44	*25 32*	*25 39*					

[1] Including those for whom no deprivation category was assigned. C - crude survival, R - relative survival (see Chapter 3). Number of patients contributing to each analysis in parentheses.

Fig. 20.1   Relative survival up to ten years: trends by calendar period of diagnosis (1971-90) and NHS Region

ADULTS

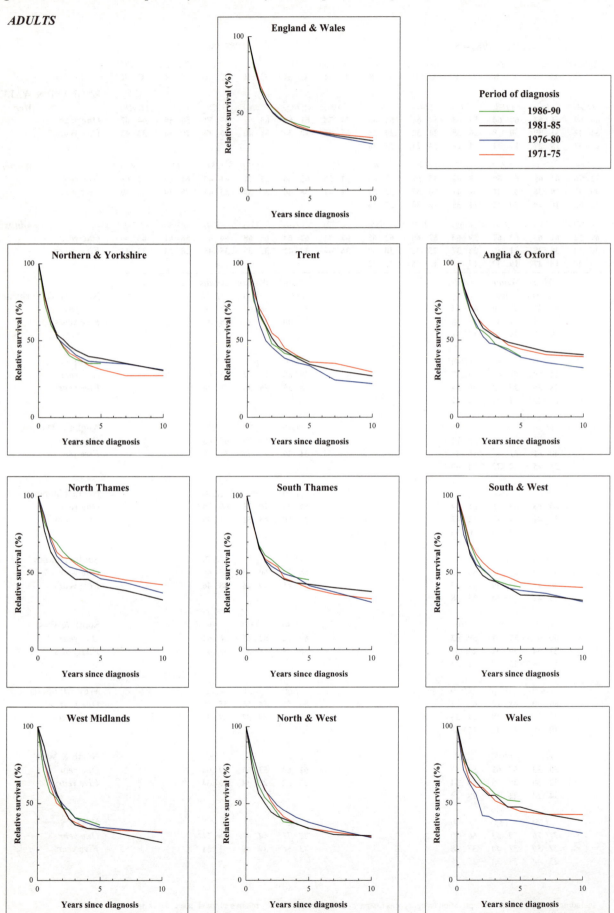

Fig. 20.2 Relative survival up to five years, by deprivation category: England and Wales, patients diagnosed 1986-90

ADULTS

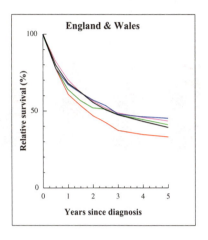

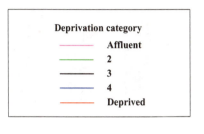

Fig. 20.3 Relative survival at five years, by deprivation category and period of diagnosis (1981-90): England and Wales

ADULTS

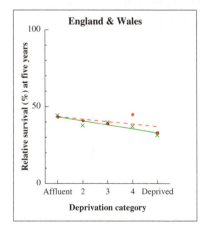

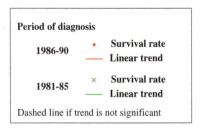

Fig. 20.4 Relative survival (%) by NHS Region and sex:
   one-year and five-year survival for patients diagnosed 1986-90, with 95% confidence intervals

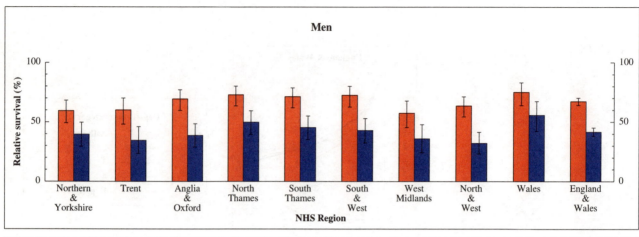

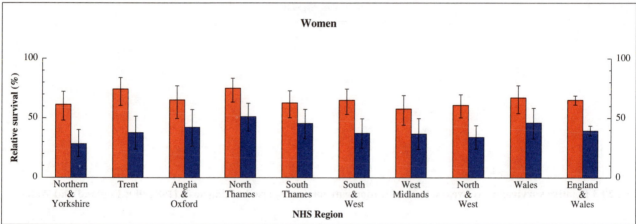

Fig. 20.5 Relative survival (%) at five years, with 95% confidence interval, by age at diagnosis and sex:
   England and Wales, patients diagnosed 1986-90

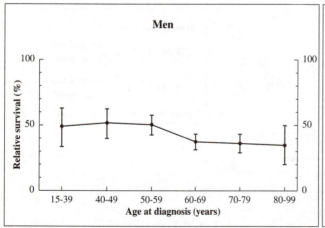

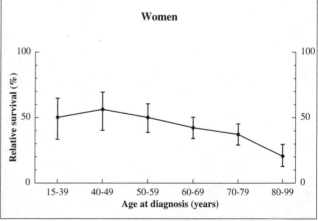

Chapter 21

LARYNX

Cancer of the larynx is one of the more common malignancies in England and Wales, with over 2,000 new cases a year, of which 80% occur in men. Incidence rates are about 6.6 and 1.4 per 100,000 per year in men and women, respectively. The rates have been broadly stable in both sexes over the last 20 years, but the patterns by age suggest a declining risk for successive generations of men in England and Wales, although the risk has been increasing in Scotland.

Cancer of the larynx is caused principally by tobacco smoking and alcohol consumption. Tobacco dominates the risk for cancers of the vocal cords and glottis, while alcohol is more prominent for cancers of the supraglottis[1,2]. This is relevant for interpretation of survival differences between men and women in England and Wales, and for international comparison. Glottal cancers are more common in men, give rise to hoarseness when the tumour is still small, can often be treated surgically, and are responsive to radiotherapy. They tend to have higher survival than supraglottic tumours. Cancers of the supraglottis are more common in women and do not give rise to early symptoms of hoarseness. Diagnosis is often later than for cancers of the glottis, surgery may be less successful and survival is lower. In the Latin countries of western Europe, supraglottic carcinoma is more common than cancer of the glottis, and this has an impact on overall survival from laryngeal cancer[3].

The survival analyses reported here include 32,000 patients diagnosed with cancer of the larynx in England and Wales during the period 1971-90 and followed up to the end of 1995, about 90% of those eligible. Some 5% of patients were excluded because their vital status was unknown at 6 July 1997, and another 4% because their survival was zero or unknown (Table 21.1). About 45% of laryngeal tumours diagnosed during the 1980s arose in the glottis (endolarynx), including the vocal cords, and about 16% in the supraglottis (epilarynx). It can be difficult to distinguish between tumours of the supraglottis and those of the hypopharynx (see Chapter 11), and as with other tumours in the head and neck, the precise anatomic origin within the larynx can also be difficult to determine. Such information was unavailable for 38% of patients, although this proportion had fallen from 42% to 35% by 1986-90 (Table 21A). Tumours of the larynx below the cords (subglottis) were rare (1-2% of the total). Four in five (79%) laryngeal tumours were squamous carcinomas, and a further 15% were coded as carcinoma without further specification. About 1% were papillary or verrucous carcinomas, and less than 0.5% were classified as adenocarcinoma.

Survival trends

Men diagnosed during 1986-90 in England and Wales had relative survival of 84% at one year and 64% at five years, 5-6% higher than for women (Table 21.2; data for women at end of chapter). The corresponding survival rates for men diagnosed during the early 1970s were 80% and 60%, with a 5-10% deficit for women. Survival increased between the late 1970s and the early 1980s, but there has been little further improvement for patients diagnosed in the late 1980s (Figure 21.1). Such gains in survival as there have been occurred during the 1970s in several regions, but about five years later in other regions.

There is a continuing decline in relative survival up to at least ten years after diagnosis.

The average gain in one-year survival was 2% every five years, and the corresponding gain in five-year survival was 2-3%, but most of the overall improvement occurred around 1980 (see Table 4.4). The difference in five-year survival between men and women fell from 9% in the early 1970s to 6% in the late 1980s, after adjustment for the age distribution of patients between the sexes and over time.

Survival falls with age, but the decline is much steeper for women (Figure 21.5). Survival at five years was 65% or more for patients diagnosed under age 60 during 1986-90, but at older ages survival for women was significantly lower than for men by 6% to 16%, and for women aged 80-99 at diagnosis the five-year survival rate was only 37% (see Table 4.7).

Regional five-year survival rates for patients diagnosed during the late 1980s were in the range 59-70% for men (Figure 21.4). For women the range was wider (45-66%), although the confidence intervals around the rates for women are relatively large. Five-year survival in Wales appeared to fall during the 1980s (Table 21.2).

Deprivation

Survival at both one and five years has generally been higher in the more affluent groups than in the more deprived groups since the early 1970s (Table 21.2), and for patients diagnosed during 1986-90 this is broadly consistent between regions (Figure 21.2). Almost all the deprivation gradient in survival is due to the differences between men. This is only partly because men account for about 80% of the cases. The five-year survival gap for men

Table 21A Sub-site distribution of cancer of the larynx

ICD-9 code	Site description	1971-75 No.	%	1976-80 No.	%	1981-85 No.	%	1986-90 No.	%	Total No.	%
161.0	Glottis	2,805	38.7	3,236	41.4	3,681	44.3	3,959	45.1	**13,681**	**42.5**
161.1	Supraglottis	-	-	495	6.3	1,315	15.8	1,493	17.0	**3,303**	**10.3**
161.2	Subglottis	-	-	79	1.0	181	2.2	161	1.8	**421**	**1.3**
161.3	Laryngeal cartilages	-	-	31	0.4	52	0.6	39	0.4	**122**	**0.4**
161.8	Other	1,378	19.0	941	12.0	83	1.0	47	0.5	**2,449**	**7.6**
161.9	Unspecified	3,062	42.3	3,035	38.8	2,996	36.1	3,088	35.1	**12,181**	**37.9**
	Total	**7,245**		**7,817**		**8,308**		**8,787**		**32,157**	

diagnosed during 1986-90 was 13%, for example, whereas survival for women in the deprived group was higher than in the affluent group, although there was no regular pattern.

The gap in survival between the most affluent and the most deprived groups was 5% at one year and 9% at five years throughout the 1980s (see Table 4.9). These estimates are based on more than 8,000 patients diagnosed during each period (1981-85 and 1986-90) and with a known deprivation status. With the exception of Anglia & Oxford, the deprivation gradient in survival is systematic throughout the regions, although not statistically significant in each region (Figure 21.3). Survival for patients in Wales diagnosed during 1986-90 was lower than for those diagnosed during 1981-85 in four of the five deprivation categories.

International comparison

Five-year survival from laryngeal cancer was 64% for men and 58% for women who were diagnosed during 1986-90 in England and Wales, the same as in Scotland (see Table 4.12). Survival for men is similar to the average survival in Europe, but for women it is 5% lower than the average. The differences in parity with Europe between men and women in England and Wales may be due to different sub-types of laryngeal cancer between the sexes. Separate comparisons of survival between the sub-sites of the larynx would be needed to clarify this.

Survival for men and women is 4% lower than in the USA.

The highest regional survival rates at five years were 70% for men (Anglia & Oxford) and 66% for women (Northern & Yorkshire, West Midlands). These are higher than the European average and similar to survival rates in the USA.

1   Tuyns A, Audigier JC. Double wave cohort increase for oesophageal and laryngeal cancer in France in relation to reduced alcohol consumption during the Second World War. Digestion 1976; 14: 197-208

2.   Tuyns A, Estève J, Raymond L, Berrino F, Benhamou E, Blanchet F, Boffetta P, Crosignani P, del Moral Aldaz A, Lehmann W, Merletti F, Péquignot G, Riboli E, Sancho-Garnier H, Terracini B, Zubiri A, Zubiri L. Cancer of the larynx/hypopharynx, tobacco and alcohol: IARC international case-control study in Turin and Varese (Italy), Zaragoza and Navarra (Spain), Geneva (Switzerland) and Calvados (France). Int J Cancer 1988; 41: 483-491

3.   Berrino F, Sant M, Verdecchia A, Capocaccia R, Hakulinen T, Estève J, (eds.) *Survival of cancer patients in Europe: the EUROCARE study. (IARC Scientific Publications No. 132).* Lyon: International Agency for Research on Cancer, 1995

Table 21.1 Ineligible and excluded records, and patients included in analyses, by calendar period of diagnosis

							Calendar period of diagnosis									
	1971-75			1976-80			1981-85			1986-90			Overall		
Total registered	**8,464**			**8,914**			**9,281**			**10,106**			**36,765**		
Ineligible	*Records*	*Patients*	*%*	*Records*	*Patients*	*%*	*Records*	*Patients*	*%*	*Records*	*Patients*	*%*	*Records*	*Patients*	*%*
Incomplete data[1]	10	10		31	31		37	37		40	40		118	**118**	
Resident outside England and Wales	267	267		174	173		75	70		40	26		556	**536**	
In situ neoplasm[2]	320	303		62	62		1	1		3	2		386	**368**	
Benign or uncertain behaviour[2]	99	80		95	78		0	0		0	0		194	**158**	
Metastatic[2]	12	12		7	6		0	0		0	0		19	**18**	
Otherwise ineligible[3]	-	-		-	-		-	-		-	-		-	**-**	
Lymphoma[4]	5	5		2	1		1	0		2	0		10	**6**	
Leukaemia or myeloma[4]	3	3		0	0		0	0		0	0		3	**3**	
		680			351			108			68			1,207	
Total eligible		7,784	100		8,563	100		9,173	100		10,038	100		35,558	100
Aged under 15 or 100 years or over at diagnosis	1	1	<0.1	2	2	<0.1	9	8	<0.1	11	11	0.1	23	**22**	<0.1
Vital status unknown[5]	539	304	3.9	603	460	5.4	481	418	4.6	722	696	6.9	2,345	**1,878**	5.3
Duplicate registration[6]	0	0	0.0	1	1	<0.1	1	0	0.0	0	0	0.0	2	**1**	<0.1
Multiple primary[7]	0	0	0.0	15	11	0.1	43	42	0.5	67	61	0.6	125	**114**	0.3
Synchronous tumours[8]	28	14	0.2	38	13	0.2	36	26	0.3	35	14	0.1	137	**67**	0.2
Sex not known	0	0	0.0	0	0	0.0	0	0	0.0	0	0	0.0	0	**0**	0.0
Sex incompatible with site code[9]	-	-	-	-	-	-	-	-	-	-	-	-	-	**-**	-
Invalid dates[10]	0	0	0.0	1	0	0.0	3	1	<0.1	13	8	<0.1	17	**9**	<0.1
Zero survival[11]	247	220	2.8	300	259	3.0	422	370	4.0	512	461	4.6	1,481	**1,310**	3.7
Total exclusions		539	6.9		746	8.7		865	9.4		1,251	12.5		3,401	9.6
Patients included in analysis		7,245	93.1		7,817	91.3		8,308	90.6		8,787	87.5		32,157	90.4

[1] Main data item(s) invalid or incompatible with one another: name, sex, date of birth, date of diagnosis and (if present) date of death, postcode, site, morphology and behaviour

[2] Tumour either *in situ* (behaviour code 2), benign (0), uncertain if benign or malignant (1) or metastatic (6 or 9)

[3] Not applicable for this malignancy

[4] Morphology that of lymphoma (included with non-Hodgkin lymphoma, Ch. 41) or myeloma (included with multiple myeloma, Ch. 43), or leukaemia (excluded)

[5] Tracing of vital status at National Health Service Central Register incomplete at 6 July 1997

[6] Same site code, sex, cancer registry and cancer registry number as an earlier registration

[7] Same site code and person number as an earlier registration(s): mostly confirmed multiple primary tumours at this site, some unresolved duplicate registrations

[8] Same site code, sex, date of birth and date of diagnosis as another registration(s): mostly synchronous or (in paired organs) bilateral tumours in same anatomic site in one individual, not previously linked: also some duplicate registrations

[9] Not applicable for this malignancy

[10] Impossible sequence of dates (birth, diagnosis and, if present, emigration or death); or date of death after 6 July 1997

[11] Date of diagnosis same as date of death: some are patients who did die on the day of diagnosis, but most were registered solely from a death certificate (death certificate only, or DCO), and their survival time is thus unknown

Table 21.2   Crude and relative survival (%) at one, five and ten years since diagnosis: NHS Region, deprivation category and calendar period of diagnosis

Each cell gives **C R** (C - crude survival, R - relative survival). First six data columns are **1971-75**; last six are **1976-80**.

MEN	Affluent C R	2 C R	3 C R	4 C R	Deprived C R	All¹ C R	Affluent C R	2 C R	3 C R	4 C R	Deprived C R	All¹ C R
ENGLAND & WALES	*(773)*	*(875)*	*(1,024)*	*(1,067)*	*(1,133)*	*(6,054)*	*(910)*	*(1,115)*	*(1,215)*	*(1,449)*	*(1,457)*	*(6,438)*
One year	78 82	79 84	76 80	75 80	73 77	76 80	79 83	79 84	77 82	75 80	73 77	76 80
Five years	51 64	48 62	47 62	47 61	45 58	47 60	52 66	51 66	48 62	46 60	43 55	47 61
Ten years	34 55	31 56	29 52	30 53	27 48	30 51	36 58	36 60	33 56	32 54	28 49	32 54
ENGLAND	*(736)*	*(824)*	*(948)*	*(963)*	*(1,062)*	*(5,646)*	*(882)*	*(1,057)*	*(1,137)*	*(1,342)*	*(1,380)*	*(6,039)*
One year	78 82	79 84	76 80	75 79	73 77	**76 80**	79 83	80 84	77 81	76 80	73 77	**76 80**
Five years	51 64	48 62	48 63	48 62	45 58	**47 60**	52 66	52 66	48 62	47 60	43 55	**48 61**
Ten years	34 54	31 55	30 53	30 54	27 48	**30 51**	36 58	36 60	34 56	32 55	28 49	**33 55**
Northern & Yorkshire	*(59)*	*(74)*	*(96)*	*(147)*	*(214)*	*(815)*	*(76)*	*(117)*	*(144)*	*(219)*	*(352)*	*(932)*
One year	*76 81*	*84 88*	*73 78*	*76 80*	*66 70*	**74 78**	*75 79*	*86 90*	*74 78*	*76 80*	*70 73*	**75 78**
Five years	*38 49*	*53 63*	*49 64*	*42 56*	*39 52*	**44 57**	*55 68*	*58 71*	*44 59*	*50 63*	*42 55*	**47 61**
Ten years	*25 39*	*36 62*	*30 54*	*24 45*	*24 42*	**28 49**	*41 62*	*37 63*	*31 54*	*33 56*	*27 50*	**32 54**
Trent	*(46)*	*(98)*	*(103)*	*(133)*	*(134)*	*(566)*	*(74)*	*(97)*	*(123)*	*(180)*	*(148)*	*(642)*
One year	*85 87*	*84 87*	*76 79*	*74 77*	*78 82*	**78 81**	*78 81*	*70 76*	*80 83*	*78 83*	*70 73*	**75 79**
Five years	*63 70*	*49 62*	*52 67*	*43 55*	*49 64*	**49 63**	*54 69*	*43 57*	*50 65*	*48 62*	*34 44*	**45 59**
Ten years	*48 63*	*28 48*	*33 55*	*29 52*	*24 43*	**30 51**	*33 52*	*32 57*	*33 56*	*34 60*	*23 41*	**31 53**
Anglia & Oxford	*(83)*	*(87)*	*(130)*	*(82)*	*(44)*	*(477)*	*(129)*	*(139)*	*(112)*	*(98)*	*(52)*	*(563)*
One year	*78 81*	*89 92*	*79 84*	*72 76*	*73 77*	**78 82**	*81 85*	*79 85*	*71 76*	*81 86*	*88 93*	**79 84**
Five years	*47 62*	*55 70*	*48 60*	*47 59*	*36 46*	**47 59**	*52 69*	*58 73*	*49 64*	*53 69*	*65 79*	**54 70**
Ten years	*29 50*	*33 59*	*26 49*	*27 43*	*18 39*	**27 49**	*40 68*	*42 67*	*36 59*	*38 66*	*38 60*	**39 64**
North Thames	*(122)*	*(127)*	*(141)*	*(140)*	*(148)*	*(777)*	*(103)*	*(141)*	*(147)*	*(179)*	*(132)*	*(725)*
One year	*85 89*	*83 87*	*77 80*	*77 82*	*79 84*	**79 83**	*85 91*	*83 87*	*85 88*	*79 84*	*79 84*	**82 87**
Five years	*63 77*	*58 73*	*50 64*	*55 71*	*52 70*	**54 69**	*55 71*	*54 71*	*49 63*	*52 67*	*44 57*	**51 66**
Ten years	*44 68*	*43 68*	*37 60*	*37 64*	*32 59*	**37 62**	*45 71*	*43 70*	*34 58*	*35 63*	*27 49*	**36 61**
South Thames	*(186)*	*(153)*	*(150)*	*(143)*	*(93)*	*(800)*	*(193)*	*(153)*	*(160)*	*(145)*	*(59)*	*(750)*
One year	*75 79*	*74 79*	*74 79*	*77 81*	*75 81*	**74 79**	*77 81*	*80 85*	*75 80*	*74 78*	*79 83*	**76 81**
Five years	*41 53*	*38 49*	*43 57*	*53 67*	*49 64*	**44 57**	*53 67*	*51 66*	*46 60*	*43 56*	*46 60*	**48 63**
Ten years	*27 46*	*27 48*	*25 45*	*35 60*	*27 49*	**27 48**	*35 57*	*30 54*	*35 60*	*30 50*	*36 60*	**33 56**
South & West	*(139)*	*(163)*	*(168)*	*(96)*	*(58)*	*(688)*	*(148)*	*(204)*	*(199)*	*(123)*	*(61)*	*(776)*
One year	*76 80*	*73 78*	*74 77*	*67 72*	*65 69*	**73 77**	*72 75*	*78 83*	*75 80*	*74 78*	*77 81*	**75 80**
Five years	*53 66*	*45 58*	*42 56*	*35 48*	*32 38*	**42 55**	*46 58*	*47 60*	*45 57*	*43 56*	*34 46*	**45 57**
Ten years	*34 54*	*30 54*	*26 51*	*24 45*	*16 27*	**27 49**	*31 53*	*31 53*	*30 51*	*30 51*	*23 44*	**30 51**
West Midlands	*(27)*	*(32)*	*(39)*	*(62)*	*(91)*	*(600)*	*(56)*	*(85)*	*(92)*	*(170)*	*(197)*	*(607)*
One year	*78 80*	*78 86*	*97 100*	*81 85*	*77 80*	**79 83**	*91 93*	*82 87*	*83 86*	*76 80*	*78 83*	**80 84**
Five years	*52 56*	*40 56*	*67 78*	*50 63*	*45 54*	**46 58**	*53 64*	*54 68*	*56 67*	*43 56*	*49 62*	**49 62**
Ten years	*44 54*	*22 45*	*38 54*	*36 56*	*34 53*	**31 50**	*43 61*	*40 63*	*38 65*	*33 52*	*29 51*	**35 56**
North & West	*(74)*	*(90)*	*(121)*	*(160)*	*(280)*	*(923)*	*(103)*	*(121)*	*(160)*	*(228)*	*(379)*	*(1,044)*
One year	*75 80*	*73 78*	*72 76*	*74 78*	*72 75*	**72 76**	*78 83*	*77 80*	*75 79*	*71 75*	*69 73*	**72 76**
Five years	*56 67*	*47 61*	*50 65*	*52 66*	*47 60*	**48 62**	*47 59*	*49 63*	*51 66*	*43 56*	*42 54*	**45 57**
Ten years	*32 50*	*28 50*	*32 56*	*31 56*	*30 52*	**31 54**	*25 46*	*37 63*	*34 59*	*28 49*	*29 52*	**30 52**
WALES	*(37)*	*(51)*	*(76)*	*(104)*	*(71)*	*(408)*	*(28)*	*(58)*	*(78)*	*(107)*	*(77)*	*(399)*
One year	*81 84*	*84 88*	*72 77*	*80 85*	*72 76*	**77 81**	*82 90*	*70 75*	*81 86*	*71 74*	*64 68*	**72 77**
Five years	*64 75*	*53 65*	*42 58*	*43 56*	*38 48*	**46 59**	*53 78*	*41 51*	*46 62*	*45 59*	*45 57*	**45 59**
Ten years	*46 63*	*31 60*	*22 42*	*26 46*	*24 41*	**28 51**	*31 67*	*31 50*	*25 54*	*30 51*	*32 55*	**28 51**

¹ Including those for whom no deprivation category was assigned. C - crude survival, R - relative survival (see Chapter 3). Number of patients contributing to each analysis in parentheses.

	1981-85												**1986-90**												
	Affluent		*2*		*3*		*4*		*Deprived*		*All*[1]		*Affluent*		*2*		*3*		*4*		*Deprived*		*All*[1]		**MEN**
	C	R	C	R	C	R	C	R	C	R	C	R	C	R	C	R	C	R	C	R	C	R	C	R	
	(1,047)		*(1,148)*		*(1,347)*		*(1,534)*		*(1,687)*		*(6,933)*		*(1,025)*		*(1,218)*		*(1,474)*		*(1,633)*		*(1,746)*		*(7,195)*		**ENGLAND & WALES**
	84	89	80	85	77	81	79	84	78	82	79	83	84	88	80	85	81	85	79	83	77	81	80	84	*One year*
	58	71	53	67	51	65	50	64	48	61	51	64	58	71	54	68	52	65	49	63	46	58	51	64	*Five years*
	40	63	38	63	35	58	32	54	31	52	34	56													
	(1,015)		*(1,097)*		*(1,249)*		*(1,414)*		*(1,598)*		*(6,502)*		*(975)*		*(1,148)*		*(1,382)*		*(1,498)*		*(1,657)*		*(6,733)*		**ENGLAND**
	85	89	81	85	77	81	79	84	77	82	79	83	84	88	81	85	81	85	80	84	77	81	80	84	*One year*
	58	71	53	67	50	64	50	63	47	60	51	64	58	71	54	69	51	64	49	63	46	58	51	64	*Five years*
	40	63	38	63	35	58	32	54	31	52	34	56													
	(85)		*(121)*		*(174)*		*(245)*		*(409)*		*(1,048)*		*(92)*		*(155)*		*(180)*		*(254)*		*(361)*		*(1,049)*		**Northern & Yorkshire**
	92	96	80	85	79	84	77	81	76	81	79	83	84	87	78	84	78	82	78	83	72	76	77	80	*One year*
	61	*78*	54	69	52	67	51	66	48	64	51	65	*57*	*69*	55	69	44	58	49	66	42	56	47	59	*Five years*
	44	*68*	37	64	35	64	31	56	33	58	34	56													
	(80)		*(113)*		*(153)*		*(195)*		*(196)*		*(747)*		*(67)*		*(102)*		*(155)*		*(186)*		*(237)*		*(748)*		**Trent**
	82	88	87	91	80	83	83	87	75	79	81	85	82	87	88	92	83	88	86	91	79	82	83	87	*One year*
	56	72	51	65	51	66	48	61	45	58	49	62	57	72	65	80	57	74	52	71	48	62	54	68	*Five years*
	40	70	38	62	37	65	30	52	31	56	34	57													
	(136)		*(129)*		*(120)*		*(104)*		*(59)*		*(560)*		*(106)*		*(129)*		*(156)*		*(111)*		*(53)*		*(558)*		**Anglia & Oxford**
	86	90	82	86	77	82	81	86	87	89	82	86	77	83	79	85	84	90	83	86	79	84	81	86	*One year*
	58	71	57	70	54	66	*56*	*71*	70	84	57	70	55	73	52	71	56	71	*56*	*74*	62	76	56	70	*Five years*
	39	*60*	*42*	*67*	*40*	*64*	*35*	*67*	*47*	*73*	40	63													
	(125)		*(146)*		*(137)*		*(179)*		*(150)*		*(755)*		*(130)*		*(133)*		*(177)*		*(190)*		*(220)*		*(876)*		**North Thames**
	91	*96*	81	86	83	88	87	91	79	83	84	89	88	92	84	88	81	85	84	89	82	87	83	87	*One year*
	65	*83*	64	79	61	79	55	71	48	63	58	74	63	77	57	74	55	70	49	63	46	58	53	65	*Five years*
	52	*83*	*43*	*72*	*44*	*76*	38	65	29	54	41	68													
	(227)		*(176)*		*(186)*		*(166)*		*(104)*		*(892)*		*(184)*		*(161)*		*(172)*		*(168)*		*(90)*		*(788)*		**South Thames**
	84	89	85	89	73	78	76	82	81	85	79	84	89	94	85	91	81	87	77	81	*86*	*91*	83	88	*One year*
	56	70	53	70	49	65	53	69	*49*	*63*	51	65	61	78	50	69	48	63	51	65	*56*	*73*	53	67	*Five years*
	37	63	36	67	39	64	*35*	*62*	30	54	35	59													
	(179)		*(198)*		*(180)*		*(117)*		*(51)*		*(747)*		*(136)*		*(214)*		*(209)*		*(145)*		*(64)*		*(778)*		**South & West**
	79	83	74	79	76	80	76	81	*80*	*84*	76	81	79	84	79	84	79	83	76	82	*74*	*77*	78	82	*One year*
	53	67	45	59	46	61	*46*	*61*	*39*	*49*	47	60	45	61	51	67	53	68	44	62	*44*	*55*	48	61	*Five years*
	33	56	32	55	27	47	25	46	27	44	29	48													
	(70)		*(78)*		*(134)*		*(171)*		*(208)*		*(668)*		*(106)*		*(124)*		*(152)*		*(195)*		*(207)*		*(787)*		**West Midlands**
	89	93	87	89	73	76	81	86	82	87	81	86	*92*	*95*	80	86	83	87	76	81	79	83	81	84	*One year*
	59	72	*60*	*74*	42	55	51	66	48	62	50	63	*68*	*85*	*56*	*73*	45	61	48	62	46	59	51	63	*Five years*
	37	70	*47*	*74*	25	47	37	62	25	46	32	53													
	(113)		*(136)*		*(165)*		*(237)*		*(421)*		*(1,085)*		*(154)*		*(130)*		*(181)*		*(249)*		*(425)*		*(1,149)*		**North & West**
	80	83	76	81	74	79	72	77	74	79	74	79	78	83	74	79	81	85	77	83	76	80	77	81	*One year*
	56	*71*	47	62	49	64	43	56	44	57	46	58	58	74	*51*	*70*	53	69	49	68	45	60	49	63	*Five years*
	43	*69*	36	62	34	61	28	49	31	54	32	53													
	(32)		*(51)*		*(98)*		*(120)*		*(89)*		*(431)*		*(50)*		*(70)*		*(92)*		*(135)*		*(89)*		*(462)*		**WALES**
	75	78	71	78	79	84	80	84	*82*	*86*	78	83	90	95	76	80	85	90	78	82	71	76	80	83	*One year*
	56	69	49	62	*54*	*72*	*56*	*71*	*57*	*72*	53	67	52	68	54	68	*54*	*72*	49	64	40	54	51	63	*Five years*
	35	*62*	*37*	*62*	*35*	*66*	*35*	*63*	*39*	*66*	35	57													

[1] Including those for whom no deprivation category was assigned. C - crude survival, R - relative survival (see Chapter 3).
Number of patients contributing to each analysis in parentheses.

**Table 21.2   Crude and relative survival (%) at one, five and ten years since diagnosis:
NHS Region, deprivation category and calendar period of diagnosis**

	1971-75												1976-80											
ADULTS	*Affluent*		*2*		*3*		*4*		*Deprived*		*All*[1]		*Affluent*		*2*		*3*		*4*		*Deprived*		*All*[1]	
	C	R	C	R	C	R	C	R	C	R	C	R	C	R	C	R	C	R	C	R	C	R	C	R
ENGLAND & WALES	*(914)*		*(1,034)*		*(1,214)*		*(1,292)*		*(1,371)*		*(7,245)*		*(1,084)*		*(1,361)*		*(1,456)*		*(1,757)*		*(1,812)*		*(7,817)*	
One year	78	81	79	83	75	79	75	79	73	76	75	79	77	81	79	83	76	80	76	80	72	76	76	79
Five years	51	63	48	60	47	59	47	59	44	56	46	58	51	64	50	63	48	60	47	59	44	55	47	59
Ten years	34	53	31	53	30	51	31	53	28	48	30	50	35	56	35	57	33	53	32	52	29	49	32	52
ENGLAND	*(870)*		*(976)*		*(1,131)*		*(1,162)*		*(1,286)*		*(6,757)*		*(1,046)*		*(1,288)*		*(1,355)*		*(1,623)*		*(1,709)*		*(7,309)*	
One year	78	81	79	83	75	79	75	79	73	77	75	79	77	81	79	84	76	80	76	80	73	77	76	80
Five years	51	62	47	59	47	59	47	60	45	57	46	58	50	63	51	64	48	61	47	60	44	55	48	60
Ten years	34	53	31	53	30	51	31	53	28	48	30	50	35	56	35	57	33	54	33	53	29	49	33	53
Northern & Yorkshire	*(68)*		*(88)*		*(117)*		*(182)*		*(265)*		*(1,018)*		*(94)*		*(142)*		*(165)*		*(260)*		*(436)*		*(1,126)*	
One year	76	80	82	86	72	76	76	80	66	70	73	77	72	77	87	91	75	79	78	81	71	74	75	79
Five years	38	48	52	62	45	57	46	60	38	49	43	55	51	63	58	69	46	61	52	64	43	55	48	61
Ten years	24	36	37	59	29	48	29	52	25	43	29	48	37	57	38	60	34	56	33	53	29	48	33	52
Trent	*(58)*		*(114)*		*(124)*		*(158)*		*(157)*		*(672)*		*(82)*		*(124)*		*(139)*		*(220)*		*(189)*		*(774)*	
One year	84	86	82	86	72	75	75	78	78	82	77	80	78	81	72	77	79	83	80	84	72	76	76	80
Five years	60	68	49	61	52	65	44	55	51	65	50	62	55	69	43	56	49	64	50	63	41	52	47	60
Ten years	46	61	29	48	36	58	31	53	26	45	32	53	33	52	34	56	31	51	37	60	28	46	33	54
Anglia & Oxford	*(98)*		*(97)*		*(147)*		*(97)*		*(48)*		*(546)*		*(153)*		*(169)*		*(130)*		*(111)*		*(63)*		*(662)*	
One year	74	77	89	92	80	85	73	77	75	80	78	82	79	83	79	85	74	80	82	87	87	92	79	84
Five years	44	57	52	65	50	62	46	56	37	47	46	57	52	68	56	69	51	66	53	67	62	74	54	68
Ten years	26	44	33	57	29	53	27	41	18	35	28	47	40	66	40	60	36	58	38	61	34	53	38	60
North Thames	*(148)*		*(151)*		*(177)*		*(170)*		*(175)*		*(935)*		*(118)*		*(160)*		*(173)*		*(210)*		*(166)*		*(858)*	
One year	84	88	83	86	77	80	78	83	79	83	79	83	80	85	82	85	83	87	78	82	77	82	80	84
Five years	62	74	56	69	47	58	53	65	50	65	52	65	52	66	52	68	47	57	49	62	46	58	49	62
Ten years	43	64	42	66	37	54	37	61	33	58	37	60	41	66	41	67	33	52	33	55	28	51	34	56
South Thames	*(217)*		*(185)*		*(178)*		*(174)*		*(109)*		*(953)*		*(227)*		*(190)*		*(200)*		*(179)*		*(79)*		*(925)*	
One year	75	79	76	79	74	79	76	79	76	81	75	79	76	79	81	85	73	78	73	77	78	82	76	80
Five years	43	54	41	51	41	54	50	63	48	62	44	56	51	65	50	64	47	60	44	57	47	59	48	62
Ten years	29	48	26	45	23	42	33	57	26	46	27	47	35	56	29	52	35	55	30	48	37	59	33	54
South & West	*(157)*		*(200)*		*(192)*		*(109)*		*(69)*		*(799)*		*(180)*		*(248)*		*(238)*		*(154)*		*(72)*		*(943)*	
One year	76	80	74	78	74	77	69	74	65	69	73	77	72	75	77	81	74	78	74	79	76	80	74	78
Five years	53	66	43	53	40	51	37	50	29	37	42	53	44	54	46	58	44	56	45	57	33	42	44	55
Ten years	36	55	29	48	25	46	25	44	16	28	27	46	29	48	30	50	30	50	29	47	22	39	29	48
West Midlands	*(33)*		*(35)*		*(48)*		*(68)*		*(104)*		*(689)*		*(66)*		*(98)*		*(121)*		*(196)*		*(237)*		*(726)*	
One year	79	80	77	84	94	96	80	84	78	81	80	84	89	91	83	87	80	83	77	81	78	83	80	83
Five years	55	59	40	54	67	77	49	60	47	56	47	58	54	66	52	64	53	63	45	57	49	62	49	61
Ten years	46	56	20	38	39	54	36	54	35	53	33	50	46	64	40	60	37	62	35	54	32	54	36	56
North & West	*(91)*		*(106)*		*(148)*		*(204)*		*(359)*		*(1,145)*		*(126)*		*(157)*		*(189)*		*(293)*		*(467)*		*(1,295)*	
One year	76	81	72	77	71	75	73	77	71	75	71	75	78	83	77	80	75	79	71	75	68	71	72	76
Five years	57	70	48	61	49	63	50	62	47	59	47	60	49	60	52	66	50	64	43	54	41	51	45	56
Ten years	32	51	29	50	31	52	31	54	31	50	31	51	27	47	36	58	33	54	28	48	28	49	30	49
WALES	*(44)*		*(58)*		*(83)*		*(130)*		*(85)*		*(488)*		*(38)*		*(73)*		*(101)*		*(134)*		*(103)*		*(508)*	
One year	77	80	83	86	72	77	74	79	68	72	74	79	87	95	68	71	75	78	71	74	63	66	70	74
Five years	58	69	50	62	44	59	41	52	37	46	45	58	60	83	44	53	40	51	42	54	41	51	43	55
Ten years	41	57	31	59	24	43	28	47	25	41	29	50	36	71	34	53	22	46	30	49	29	47	28	48

[1] Including those for whom no deprivation category was assigned. C - crude survival, R - relative survival (see Chapter 3).
Number of patients contributing to each analysis in parentheses.

	| 1981-85						| 1986-90					
ADULTS	Affluent (C R)	2 (C R)	3 (C R)	4 (C R)	Deprived (C R)	All¹ (C R)	Affluent (C R)	2 (C R)	3 (C R)	4 (C R)	Deprived (C R)	All¹ (C R)
ENGLAND & WALES	(1,217)	(1,375)	(1,595)	(1,880)	(2,046)	(8,308)	(1,215)	(1,452)	(1,798)	(2,019)	(2,187)	(8,787)
One year	83 87	80 84	77 81	79 83	77 81	79 83	83 87	80 84	80 83	78 82	78 81	79 83
Five years	57 70	52 65	51 64	50 63	48 60	51 63	56 69	53 67	50 62	50 63	47 59	51 63
	40 61	37 59	35 57	33 54	32 52	35 55						
ENGLAND	(1,175)	(1,310)	(1,473)	(1,724)	(1,936)	(7,762)	(1,156)	(1,364)	(1,680)	(1,849)	(2,076)	(8,211)
One year	83 87	80 84	77 81	79 83	77 81	**79 83**	83 87	80 84	79 83	79 82	78 82	**80 83**
Five years	57 70	52 65	51 64	50 62	47 60	**51 63**	57 69	53 67	50 62	50 63	48 59	**51 63**
	40 61	38 60	36 57	33 53	32 52	**35 55**						
Northern & Yorkshire	(94)	(137)	(209)	(299)	(495)	(1,250)	(115)	(184)	(230)	(303)	(463)	(1,305)
One year	91 95	78 82	79 84	77 81	77 81	**78 83**	82 86	77 82	78 82	79 83	75 79	**77 81**
Five years	64 80	52 67	52 66	52 65	50 64	**52 65**	57 69	53 66	45 60	51 68	46 59	**49 61**
	45 68	34 59	35 61	32 55	35 59	**35 56**						
Trent	(97)	(134)	(183)	(233)	(234)	(892)	(79)	(116)	(192)	(243)	(296)	(928)
One year	79 83	86 90	81 84	84 88	73 77	**80 84**	81 85	86 90	81 85	86 89	80 83	**82 86**
Five years	53 68	50 63	53 66	46 59	43 57	**48 60**	54 68	61 75	55 70	52 68	49 62	**53 65**
	37 62	38 58	39 64	30 49	30 54	**34 54**						
Anglia & Oxford	(157)	(149)	(147)	(125)	(72)	(664)	(131)	(158)	(189)	(138)	(70)	(689)
One year	86 90	80 84	77 82	83 88	85 87	**82 86**	74 80	80 85	81 87	80 83	77 81	**79 83**
Five years	60 73	55 68	56 69	56 70	67 80	**57 70**	53 70	54 72	54 68	57 75	54 67	**55 68**
	40 62	42 66	39 63	37 69	44 66	**40 62**						
North Thames	(143)	(180)	(161)	(216)	(170)	(890)	(152)	(157)	(205)	(225)	(260)	(1,027)
One year	90 95	82 86	83 88	86 90	79 82	**84 88**	89 93	85 88	79 82	83 87	82 86	**83 86**
Five years	62 78	61 74	61 77	56 70	48 62	**57 72**	62 75	57 72	53 67	48 61	47 60	**52 64**
	50 77	42 68	44 72	41 66	29 50	**41 65**						
South Thames	(265)	(219)	(206)	(207)	(123)	(1,058)	(220)	(188)	(228)	(206)	(110)	(969)
One year	83 87	85 89	73 77	76 81	79 83	**78 83**	87 92	84 89	78 83	76 80	83 88	**81 86**
Five years	55 67	52 67	46 61	52 66	48 61	**50 63**	58 73	48 66	46 58	51 65	53 69	**51 64**
	36 57	37 64	37 60	35 58	31 54	**35 56**						
South & West	(199)	(236)	(211)	(142)	(65)	(876)	(156)	(254)	(254)	(172)	(78)	(926)
One year	79 83	74 78	76 80	78 83	82 85	**77 81**	78 82	78 83	77 81	75 80	77 80	**77 81**
Five years	54 67	46 59	47 61	48 63	44 54	**48 61**	45 59	51 65	50 64	45 60	45 55	**48 60**
	35 59	32 52	28 47	27 49	34 53	**31 50**						
West Midlands	(83)	(88)	(153)	(199)	(257)	(787)	(124)	(152)	(177)	(246)	(247)	(950)
One year	87 90	85 87	74 78	81 85	82 87	**81 85**	92 95	78 83	83 87	75 79	80 84	**80 84**
Five years	54 68	60 74	44 56	53 67	48 60	**50 63**	68 84	54 69	49 65	48 61	48 60	**52 64**
	36 66	49 74	27 47	39 62	26 46	**33 53**						
North & West	(137)	(167)	(203)	(303)	(520)	(1,345)	(179)	(155)	(205)	(316)	(552)	(1,417)
One year	76 79	74 78	73 78	71 76	73 77	**73 77**	78 82	76 81	80 83	76 81	77 81	**77 81**
Five years	55 69	45 58	49 64	42 54	44 56	**46 57**	57 71	51 69	51 66	49 67	47 60	**50 63**
	43 68	35 56	35 60	28 46	32 53	**33 52**						
WALES	(42)	(65)	(122)	(156)	(110)	(546)	(59)	(88)	(118)	(170)	(111)	(576)
One year	78 82	72 78	77 81	79 84	79 83	**77 82**	86 90	73 77	80 85	76 79	74 79	**78 81**
Five years	64 76	49 63	49 64	57 71	52 65	**52 66**	49 61	51 65	49 63	46 57	45 57	**49 59**
	46 72	34 52	33 57	37 63	36 56	**35 56**						

¹ Including those for whom no deprivation category was assigned. C - crude survival, R - relative survival (see Chapter 3). Number of patients contributing to each analysis in parentheses.

Fig. 21.1 Relative survival up to ten years: trends by calendar period of diagnosis (1971-90) and NHS Region

ADULTS

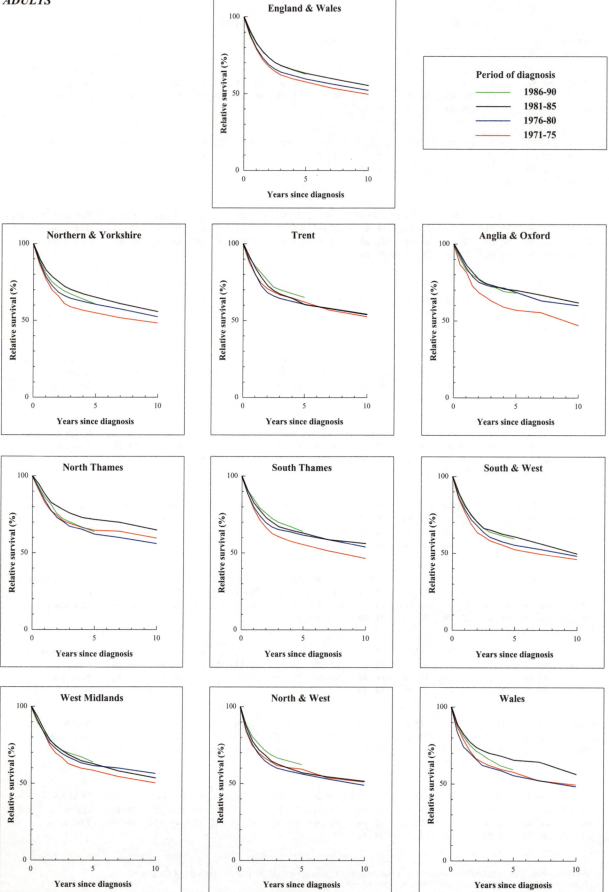

Fig. 21.2   Relative survival up to five years, by deprivation category and NHS Region: patients diagnosed 1986-90

ADULTS

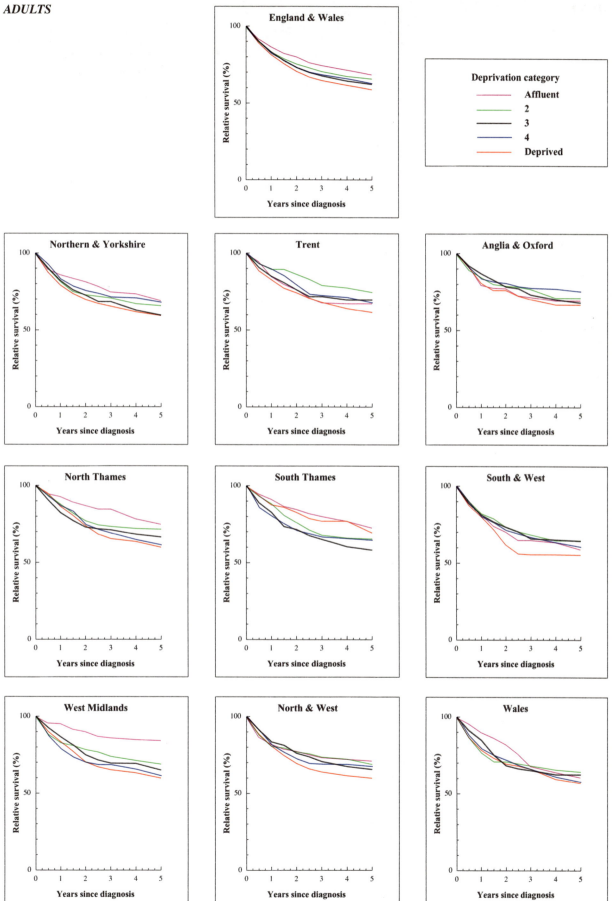

Fig. 21.3 Relative survival at five years by deprivation category, period of diagnosis (1981-90) and NHS Region

ADULTS

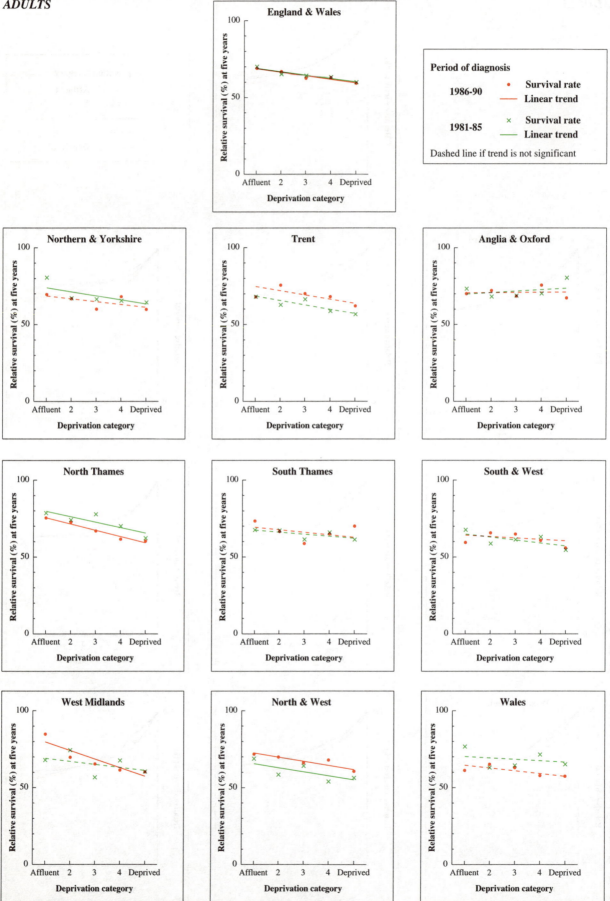

Fig. 21.4   Relative survival (%) by NHS Region and sex:
         one-year and five-year survival for patients diagnosed 1986-90, with 95% confidence intervals

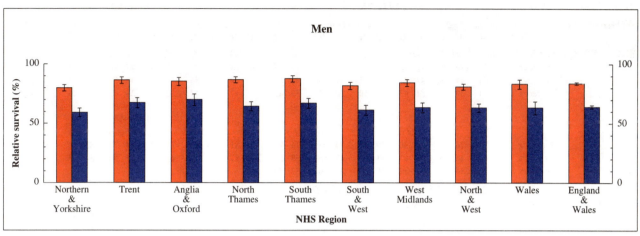

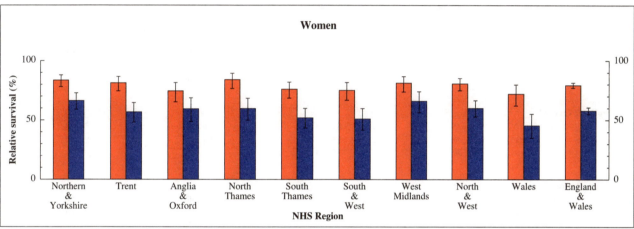

Fig. 21.5   Relative survival (%) at five years, with 95% confidence interval, by age at diagnosis and sex:
         England and Wales, patients diagnosed 1986-90

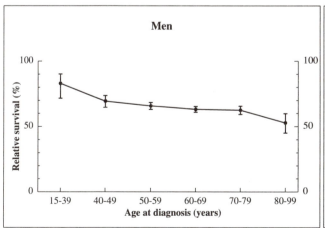

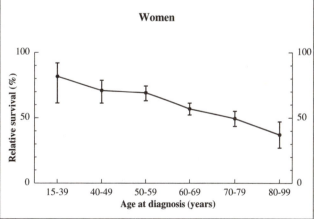

Table 21.2 Crude and relative survival (%) at one, five and ten years since diagnosis: NHS Region, deprivation category and calendar period of diagnosis

	1971-75						1976-80					
WOMEN	Affluent	2	3	4	Deprived	All[1]	Affluent	2	3	4	Deprived	All[1]
	C R	C R	C R	C R	C R	C R	C R	C R	C R	C R	C R	C R
ENGLAND & WALES	(141)	(159)	(190)	(225)	(238)	(1,191)	(174)	(246)	(241)	(308)	(355)	(1,379)
One year	75 77	76 78	72 75	74 76	71 74	73 75	70 73	77 79	72 74	77 80	71 74	74 76
Five years	50 56	44 50	41 47	44 51	41 50	43 50	46 54	47 54	45 52	48 56	46 54	46 54
Ten years	33 44	30 42	31 43	36 50	32 45	33 45	33 47	32 44	30 41	34 45	33 46	32 44

	1981-85						1986-90					
	(170)	(227)	(248)	(346)	(359)	(1,375)	(190)	(234)	(324)	(386)	(441)	(1,592)
One year	75 78	76 78	77 80	79 83	74 77	76 79	78 81	76 80	71 74	74 77	82 85	76 79
Five years	56 64	46 55	52 61	51 60	47 55	50 58	48 57	48 57	44 51	51 60	54 63	49 58
Ten years	40 54	34 46	37 52	38 53	36 50	37 50						

	1971-75	1976-80	1981-85	1986-90
Northern & Yorkshire	(203)	(194)	(202)	(256)
One year	71 74	78 80	78 82	80 84
Five years	41 49	53 60	57 65	56 66
Ten years	34 46	37 48	39 56	
Trent	(106)	(132)	(145)	(180)
One year	72 74	81 84	77 79	79 81
Five years	51 60	56 68	43 52	49 57
Ten years	41 56	41 58	33 43	
Anglia & Oxford	(69)	(99)	(104)	(131)
One year	75 79	81 85	81 85	71 74
Five years	40 44	53 61	58 67	49 59
Ten years	30 38	35 44	40 58	
North Thames	(158)	(133)	(135)	(151)
One year	79 82	69 72	82 85	81 84
Five years	43 49	39 45	53 61	50 60
Ten years	36 49	25 35	42 55	
South Thames	(153)	(175)	(166)	(181)
One year	75 78	73 77	75 78	73 76
Five years	43 50	49 58	44 52	43 52
Ten years	26 40	32 48	32 45	
South & West	(111)	(167)	(129)	(148)
One year	72 74	71 74	80 83	73 75
Five years	36 41	39 47	54 65	44 51
Ten years	26 35	25 37	39 57	
West Midlands	(89)	(119)	(119)	(163)
One year	84 87	77 80	80 83	78 81
Five years	55 63	49 56	52 60	55 66
Ten years	42 54	44 55	39 53	
North & West	(222)	(251)	(260)	(268)
One year	67 70	71 73	68 71	77 81
Five years	43 50	44 52	44 52	51 60
Ten years	30 42	28 40	34 46	
WALES	(80)	(109)	(115)	(114)
One year	61 64	63 65	75 79	69 72
Five years	40 49	37 43	47 59	40 45
Ten years	33 44	27 38	36 52	

[1] Including those for whom no deprivation category was assigned. C - crude survival, R - relative survival (see Chapter 3). Number of patients contributing to each analysis in parentheses.

Chapter 22

LUNG

Lung cancer is the most common cancer in the world. In England and Wales, it is the most common cancer among men, accounting for about one in four cancers (25,000 cases a year), and the third most common among women (one in 9; 12,000 cases a year). Lung cancer alone accounts for 6% of deaths from all causes. Incidence rates are typical for developed countries, about 100 per 100,000 per year for men, and 45 for women. The male-to-female ratio is about 3 to 1 after adjustment for age. Lung cancer is more common among lower social class groups and in the north of England (Northern & Yorkshire, North & West).

Incidence in men in the UK has been the highest in the world, but rates have been falling by 7-10% every five years since the mid-1970s. In sharp contrast, incidence among UK women was increasing by 10% every five years until the mid 1970s, and in 1990 lung cancer overtook breast cancer in Glasgow as the most common cancer in women[1]. The rate of increase in women slowed down by the end of the 1980s, and the pattern by age suggests a progressive decline in life-time risk for women born after 1930. These incidence trends mirror earlier trends in tobacco smoking, which causes 90% of lung cancers. Lung cancer mortality trends mimic those for incidence, because survival has been poor for decades. Screening with sputum cytology or chest X-rays does not reduce mortality.

Most patients when diagnosed have advanced disease for which no curative treatment is available. Small cell carcinoma accounts for about a quarter of cases. It is aggressive: more than 60% of patients have regional lymph node involvement and metastases in liver, bone or brain when diagnosed, and although it is sensitive to radiotherapy and chemotherapy, long-term survival remains poor. About 75% of lung tumours are non-small cell cancers (squamous carcinoma, adenocarcinoma and large cell carcinoma), for which surgery is potentially curative, but less than 20% of patients are operable.

The survival analyses here relate to almost 600,000 patients diagnosed with primary cancer of the trachea, bronchus and lung in England and Wales during 1971-90 and followed up to the end of 1995. Of the 714,000 patients eligible for inclusion in the analyses, 4% were excluded because their vital status was unknown at 6 July 1997, and a further 13% because their survival was zero or unknown (Table 22.1). More than 1,500 patients had multiple primary cancers, but survival was only analysed from the diagnosis of the first tumour.

Cancers of the bronchus and lung predominate: cancers of the trachea accounted for only 0.3% of cases (Table 22A). About a quarter of cases were coded to the upper lobe, 15% to the middle and lower lobes and 10% to the main bronchus, but for half the cases there was no anatomic detail. About 25% were papillary or squamous carcinomas and 6% adenocarcinomas, but 55% were poorly specified carcinomas, and for 14% the histology was either not specified or not known.

Survival trends

Survival from lung cancer among men diagnosed in England and Wales during 1986-90 was 21% at one year and 6% after five years. One-year survival was 2% higher than for men diagnosed during 1971-75, but five-year survival has not changed (Table 22.2). For women, one-year survival increased from 16% to 20% over the same period. One-year survival in most regions was in the range 18-21%; survival in the Thames regions has been 2-3% higher than average in both sexes since the early 1970s, and for patients diagnosed during 1986-90 the rates were 24-28%.

The survival curves up to ten years show very little change since the early 1970s, either nationally or within any of the regions (Figure 22.1). There has nevertheless been a small increase in survival in all regions, after adjustment for gradual changes in the age distribution of patients. Age-standardised relative survival at one year after diagnosis has increased by an average of 1.5% to 2% in successive five-year periods to reach 19% for those diagnosed during 1986-90, and five-year survival rose by about 0.5% every five years (see Table 4.4). One-year survival was about 2% lower in women than in men during the 1970s, but the increase in one-year survival has been greater for women than for men in every region, and survival rates are now similar in both sexes.

The same general pattern of small improvements is seen in most regions, with some variation. Regional differences in survival trends are often statistically significant because of the large numbers of patients, but they are small, and five-year survival rates for men and women diagnosed during 1986-90 in every region are in the range 4% to 7% (Figure 22.4; see also Table-Figure 4.5F).

Survival falls with age at diagnosis (Figure 22.5). Patients aged less than 50 at diagnosis during 1986-90 had five-year relative survival rates of 10-20%, but survival was 3% or less for those aged 70 or over (see Table 4.7).

Table 22A Sub-site distribution of lung cancer

ICD-9 code	Site description	1971-75 No.	%	1976-80 No.	%	1981-85 No.	%	1986-90 No.	%	Total No.	%
162.0	Trachea	380	0.3	453	0.3	404	0.3	399	0.3	1,636	0.3
162.1	Bronchus and lung (1971-78)[1]	142,019	99.7	89,900	59.4	-	-	-	-	231,919	39.2
162.2	Main bronchus	-	-	5,000	3.3	13,232	8.7	12,115	8.3	30,347	5.1
162.3	Upper lobe, bronchus or lung	-	-	13,515	8.9	36,407	23.9	33,076	22.6	82,998	14.0
162.4	Middle lobe, bronchus or lung	-	-	1,362	0.9	3,981	2.6	3,806	2.6	9,149	1.5
162.5	Lower lobe, bronchus or lung	-	-	7,100	4.7	18,535	12.1	16,093	11.0	41,728	7.0
162.8	Other	-	-	636	0.4	1,421	0.9	1,433	1.0	3,490	0.6
162.9	Bronchus and lung, unspecified	-	-	33,322	22.0	78,572	51.5	79,153	54.2	191,047	32.3
	Total	**142,399**		**151,288**		**152,552**		**146,075**		**592,314**	

[1]Cancers of the bronchus and lung were coded to 162.1 in ICD-8 (1971-78) and to 162.2-162.9 in ICD-9 (1979-90)

Deprivation

In England and Wales as a whole, and in most regions, both one-year and five-year survival were slightly lower in the deprived group than in the affluent (Table 22.2; Figure 22.2). The deprivation gradient in five-year survival was similar in most regions and statistically significant in four of the nine: there was little change in this pattern during the 1980s (Figure 22.3; note expanded y-axis scale). The gap between the affluent and the most deprived groups was about 3% for one-year survival, and 1% for five-year survival (see Table 4.9).

The deprivation gradient in lung cancer survival is unlikely to be due to differences in the stage of disease at diagnosis between deprivation groups, or to bias from the exclusion of cases registered only from a death certificate. A similar gradient has been reported for 40,000 patients diagnosed during 1980-89 in South Thames Region, about 15% of the national caseload for that period[2]. The proportion of patients with non-local disease was similar in all deprivation groups, and adjustment for stage did not affect the gradient. The proportion of death-certificate-only cases excluded from analysis was slightly *higher* in the more deprived groups, and this was shown to have reduced the gradient.

International comparison

For both men and women, five-year survival in England and Wales was similar to that in Scotland (also to the rates in Denmark, Slovenia, Poland and Estonia), but about 5% lower than the European average, and 7% to 10% lower than in the USA (see Table 4.12). Even the best regional survival rates, or survival in England and Wales for the most affluent group, were significantly below the average rates in Europe and the USA.

1.   Gillis CR, Hole D, Lamont DW, Graham AC, Ramage S. The incidences of lung cancer and breast cancer in women in Glasgow. Br Med J 1992; 305: 1331

2.   Schrijvers CTM, Mackenbach J, Lutz J-M, Quinn MJ, Coleman MP. Deprivation, stage at diagnosis and cancer survival. Int J Cancer 1995; 63: 324-329

Table 22.1 Ineligible and excluded records, and patients included in analyses, by calendar period of diagnosis

	Calendar period of diagnosis															
	1971-75			1976-80			1981-85			1986-90			Overall		
Total registered	165,125			179,008			189,835			187,728			721,696		
Ineligible	*Records*	*Patients*	*%*	*Records*	*Patients*	*%*	*Records*	*Patients*	*%*	*Records*	*Patients*	*%*	*Records*	*Patients*	*%*
Incomplete data[1]	135	135		189	189		232	232		370	370		926	**926**	
Resident outside England and Wales	1,305	1,301		834	827		686	607		587	402		3,412	**3,137**	
In situ neoplasm[2]	117	114		27	23		0	0		1	1		145	**138**	
Benign or uncertain behaviour[2]	564	200		321	183		0	0		0	0		885	**383**	
Metastatic[2]	2,315	2,298		995	981		6	6		38	38		3,354	**3,323**	
Otherwise ineligible[3]	-	-		-	-		-	-		-	-		-	**-**	
Lymphoma[4]	32	28		19	15		6	0		8	0		65	**43**	
Leukaemia or myeloma[4]	1	1		0	0		2	0		2	0		5	**1**	
		4,077			2,218			845			811			7,951	
Total eligible		161,048	100		176,790	100		188,990	100		186,917	100		713,745	100
Aged under 15 or 100 years or over at diagnosis	46	43	<0.1	29	26	<0.1	40	35	<0.1	62	59	<0.1	177	**163**	<0.1
Vital status unknown[5]	5,224	4,500	2.8	7,044	6,455	3.7	7,980	7,611	4.0	8,142	7,840	4.2	28,390	**26,406**	3.7
Duplicate registration[6]	0	0	0.0	3	2	<0.1	0	0	0.0	9	9	<0.1	12	**11**	<0.1
Multiple primary[7]	0	0	0.0	160	138	<0.1	720	619	0.3	884	774	0.4	1,764	**1,531**	0.2
Synchronous tumours[8]	945	847	0.5	854	714	0.4	1,054	900	0.5	1,098	915	0.5	3,951	**3,376**	0.5
Sex not known	0	0	0.0	0	0	0.0	0	0	0.0	2	2	<0.1	2	**2**	<0.1
Sex incompatible with site code[9]	-	-	-	-	-	-	-	-	-	-	-	-	-	**-**	-
Invalid dates[10]	5	4	<0.1	17	13	<0.1	67	55	<0.1	116	76	<0.1	205	**148**	<0.1
Zero survival[11]	14,357	13,255	8.2	19,383	18,154	10.3	29,142	27,218	14.4	32,929	31,167	16.7	95,811	**89,794**	12.6
Total exclusions		18,649	11.6		25,502	14.4		36,438	19.3		40,842	21.9		**121,431**	17.0
Patients included in analysis		142,399	88.4		151,288	85.6		152,552	80.7		146,075	78.1		**592,314**	83.0

[1] Main data item(s) invalid or incompatible with one another: name, sex, date of birth, date of diagnosis and (if present) date of death, postcode, site, morphology and behaviour

[2] Tumour either *in situ* (behaviour code 2), benign (0), uncertain if benign or malignant (1) or metastatic (6 or 9)

[3] Not applicable for this malignancy

[4] Morphology that of lymphoma (included with non-Hodgkin lymphoma, Ch. 41) or myeloma (included with multiple myeloma, Ch. 43), or leukaemia (excluded)

[5] Tracing of vital status at National Health Service Central Register incomplete at 6 July 1997

[6] Same site code, sex, cancer registry and cancer registry number as an earlier registration

[7] Same site code and person number as an earlier registration(s): mostly confirmed multiple primary tumours at this site, some unresolved duplicate registrations

[8] Same site code, sex, date of birth and date of diagnosis as another registration(s): mostly synchronous or (in paired organs) bilateral tumours in same anatomic site in one individual, not previously linked: also some duplicate registrations

[9] Not applicable for this malignancy

[10] Impossible sequence of dates (birth, diagnosis and, if present, emigration or death); or date of death after 6 July 1997

[11] Date of diagnosis same as date of death: some are patients who did die on the day of diagnosis, but most were registered solely from a death certificate (death certificate only, or DCO), and their survival time is thus unknown

Table 22.2 Crude and relative survival (%) at one, five and ten years since diagnosis: NHS Region, deprivation category and calendar period of diagnosis

MEN	1971-75												1976-80											
	Affluent		*2*		*3*		*4*		*Deprived*		*All[1]*		*Affluent*		*2*		*3*		*4*		*Deprived*		*All[1]*	
	C	R	C	R	C	R	C	R	C	R	C	R	C	R	C	R	C	R	C	R	C	R	C	R
ENGLAND & WALES	*(13,564)*		*(16,582)*		*(19,361)*		*(21,562)*		*(21,723)*		*(115,065)*		*(16,302)*		*(20,098)*		*(23,181)*		*(26,137)*		*(26,954)*		*(117,556)*	
One year	19	20	19	20	18	19	17	18	17	17	**18**	**19**	20	21	18	19	17	18	16	17	16	17	**17**	**18**
Five years	5	6	5	6	5	6	5	6	4	6	**5**	**6**	5	7	5	6	4	6	4	5	4	5	**4**	**6**
Ten years	3	5	3	4	3	4	3	5	3	4	**3**	**4**	3	5	3	5	3	4	3	4	3	5	**3**	**4**
ENGLAND	*(13,101)*		*(15,938)*		*(18,102)*		*(19,939)*		*(20,629)*		*(108,780)*		*(15,769)*		*(19,235)*		*(21,779)*		*(24,410)*		*(25,811)*		*(111,085)*	
One year	19	20	19	20	18	19	17	18	17	17	**18**	**19**	20	21	18	19	17	18	17	17	16	17	**17**	**18**
Five years	5	6	5	6	5	6	5	6	4	6	**5**	**6**	5	7	5	6	4	6	4	5	4	5	**4**	**6**
Ten years	3	5	3	4	3	4	3	5	3	4	**3**	**4**	3	5	3	5	3	4	2	4	3	4	**3**	**4**
Northern & Yorkshire	*(905)*		*(1,332)*		*(1,978)*		*(2,720)*		*(4,302)*		*(15,837)*		*(1,443)*		*(2,081)*		*(2,962)*		*(4,102)*		*(6,445)*		*(17,482)*	
One year	17	18	15	16	14	15	13	13	13	14	**16**	**17**	17	18	14	14	13	13	13	14	13	14	**15**	**16**
Five years	4	5	4	4	3	4	3	4	3	4	**4**	**5**	4	5	3	4	3	4	3	4	3	4	**3**	**4**
Ten years	2	3	2	3	2	3	2	3	2	3	**2**	**3**	2	4	2	3	2	3	2	3	2	3	**2**	**3**
Trent	*(987)*		*(1,409)*		*(1,912)*		*(2,728)*		*(2,564)*		*(10,506)*		*(1,147)*		*(1,687)*		*(2,303)*		*(3,075)*		*(2,819)*		*(11,360)*	
One year	17	18	16	17	14	15	15	16	16	16	**17**	**18**	16	16	14	15	15	16	14	15	14	14	**16**	**17**
Five years	4	5	3	4	3	4	4	5	4	5	**4**	**5**	4	5	4	5	3	4	3	4	3	4	**4**	**5**
Ten years	2	3	2	3	2	3	3	4	3	4	**3**	**4**	3	4	3	5	2	3	2	3	2	4	**3**	**4**
Anglia & Oxford	*(1,760)*		*(1,976)*		*(2,098)*		*(1,602)*		*(878)*		*(9,496)*		*(2,262)*		*(2,404)*		*(2,354)*		*(1,794)*		*(903)*		*(10,108)*	
One year	18	18	18	18	17	18	19	20	18	18	**19**	**20**	20	21	17	18	17	18	16	17	15	16	**19**	**20**
Five years	4	5	5	6	5	6	6	8	4	5	**5**	**6**	5	6	4	5	4	5	3	4	4	5	**4**	**5**
Ten years	3	4	3	5	3	4	4	6	1	2	**3**	**5**	3	5	2	3	2	4	2	3	2	4	**2**	**4**
North Thames	*(2,223)*		*(2,606)*		*(2,680)*		*(3,187)*		*(2,800)*		*(15,289)*		*(1,865)*		*(2,219)*		*(2,388)*		*(2,826)*		*(2,509)*		*(12,306)*	
One year	19	20	20	21	20	21	20	20	19	20	**21**	**22**	23	24	21	22	21	22	22	23	21	22	**23**	**24**
Five years	6	7	5	6	6	7	5	7	6	8	**6**	**7**	6	7	5	7	5	7	6	7	6	7	**6**	**7**
Ten years	4	6	3	5	4	6	4	6	4	6	**4**	**6**	4	6	3	5	3	5	4	6	4	7	**4**	**6**
South Thames	*(3,476)*		*(3,335)*		*(3,542)*		*(3,444)*		*(2,095)*		*(17,252)*		*(3,612)*		*(3,494)*		*(3,422)*		*(3,293)*		*(1,671)*		*(16,309)*	
One year	19	20	18	19	17	18	18	19	16	17	**19**	**20**	19	20	19	19	17	18	15	16	16	17	**19**	**20**
Five years	5	6	4	5	4	5	5	6	4	5	**5**	**6**	5	7	5	6	4	5	4	5	4	6	**5**	**6**
Ten years	2	4	2	4	2	4	3	4	2	4	**3**	**4**	3	5	2	4	2	4	2	4	3	5	**3**	**4**
South & West	*(2,038)*		*(3,100)*		*(3,057)*		*(1,794)*		*(827)*		*(11,918)*		*(2,373)*		*(3,617)*		*(3,459)*		*(2,079)*		*(911)*		*(12,954)*	
One year	18	18	18	19	16	17	16	17	17	17	**18**	**19**	18	18	17	18	16	17	15	16	14	15	**18**	**19**
Five years	5	6	5	6	4	5	5	6	6	7	**5**	**6**	4	5	5	7	5	6	5	6	4	6	**5**	**7**
Ten years	3	4	3	5	3	4	3	4	3	6	**3**	**5**	3	5	4	6	3	5	3	5	3	5	**3**	**6**
West Midlands	*(406)*		*(584)*		*(798)*		*(1,115)*		*(1,506)*		*(11,036)*		*(1,235)*		*(1,526)*		*(2,083)*		*(3,017)*		*(3,827)*		*(11,839)*	
One year	16	17	18	19	19	20	13	14	17	17	**18**	**19**	18	19	17	17	17	18	15	15	15	16	**18**	**19**
Five years	3	3	4	5	4	5	4	5	5	6	**4**	**5**	5	6	4	5	4	5	3	4	4	5	**4**	**5**
Ten years	2	2	2	3	3	4	2	4	3	4	**3**	**4**	4	6	2	4	2	4	2	3	2	4	**3**	**4**
North & West	*(1,306)*		*(1,596)*		*(2,037)*		*(3,349)*		*(5,657)*		*(17,446)*		*(1,832)*		*(2,207)*		*(2,808)*		*(4,224)*		*(6,726)*		*(18,727)*	
One year	15	15	13	14	14	14	12	13	13	13	**15**	**16**	16	17	13	14	13	13	12	13	13	14	**15**	**16**
Five years	4	5	4	4	4	5	4	4	4	5	**4**	**5**	4	5	4	4	3	4	3	4	4	5	**4**	**5**
Ten years	3	5	2	3	2	4	2	3	2	3	**3**	**4**	3	4	2	3	2	3	2	3	2	4	**2**	**4**
WALES	*(463)*		*(644)*		*(1,259)*		*(1,623)*		*(1,094)*		*(6,285)*		*(533)*		*(863)*		*(1,402)*		*(1,727)*		*(1,143)*		*(6,471)*	
One year	17	18	14	15	15	16	14	15	15	16	**16**	**17**	20	21	15	16	14	15	14	15	15	16	**17**	**18**
Five years	4	5	4	5	4	5	4	5	4	4	**4**	**5**	7	8	4	5	5	6	5	7	5	6	**6**	**7**
Ten years	3	5	2	4	2	4	2	4	2	4	**3**	**4**	5	7	3	4	3	5	3	6	3	5	**4**	**6**

[1] Including those for whom no deprivation category was assigned. C - crude survival, R - relative survival (see Chapter 3).
Number of patients contributing to each analysis in parentheses.

1981-85

MEN

Region		Affluent C	Affluent R	2 C	2 R	3 C	3 R	4 C	4 R	Deprived C	Deprived R	All[1] C	All[1] R
ENGLAND & WALES	(n)	(16,436)		(19,365)		(22,664)		(25,766)		(25,721)		(112,291)	
	One year	21	22	20	21	19	20	19	20	18	19	19	20
	Five years	5	6	5	6	5	6	5	6	4	6	5	6
		3	5	3	5	3	5	3	5	3	5	3	5
ENGLAND	(n)	(15,896)		(18,589)		(21,314)		(24,075)		(24,541)		(106,232)	
	One year	21	22	20	21	19	20	19	20	18	19	19	20
	Five years	5	6	5	6	4	6	5	6	4	6	5	6
		3	5	3	5	3	4	3	5	3	4	3	5
Northern & Yorkshire	(n)	(1,492)		(2,043)		(2,825)		(4,065)		(6,201)		(16,866)	
	One year	16	17	16	17	15	16	15	16	15	16	17	18
	Five years	4	5	4	5	3	5	4	5	3	4	4	5
		2	4	2	4	2	3	2	4	2	3	2	4
Trent	(n)	(1,223)		(1,745)		(2,486)		(3,402)		(2,852)		(11,783)	
	One year	18	19	17	17	15	15	16	17	14	15	17	18
	Five years	4	5	4	5	3	3	4	5	3	4	4	5
		2	4	2	4	2	3	2	4	2	3	2	4
Anglia & Oxford	(n)	(2,354)		(2,428)		(2,465)		(1,821)		(785)		(10,043)	
	One year	17	18	17	18	17	18	15	16	16	17	18	19
	Five years	4	5	4	5	4	6	3	4	4	6	4	5
		3	4	2	4	3	5	2	3	3	5	3	4
North Thames	(n)	(1,643)		(1,853)		(1,963)		(2,241)		(1,835)		(9,696)	
	One year	26	28	24	26	22	23	24	25	27	28	26	27
	Five years	8	10	6	8	6	8	7	10	8	11	7	9
		5	8	4	7	4	7	5	9	5	9	5	8
South Thames	(n)	(3,532)		(3,144)		(3,016)		(2,809)		(1,373)		(14,455)	
	One year	22	24	20	21	20	21	20	22	19	20	22	23
	Five years	5	6	5	6	5	6	5	7	4	5	5	6
		3	5	3	5	3	5	3	6	2	5	3	5
South & West	(n)	(2,367)		(3,430)		(3,256)		(1,982)		(849)		(12,204)	
	One year	18	19	18	19	17	18	18	19	16	17	19	20
	Five years	5	6	4	5	4	5	5	7	5	6	5	6
		3	5	2	4	3	5	3	5	3	5	3	5
West Midlands	(n)	(1,286)		(1,617)		(2,404)		(3,299)		(3,741)		(12,429)	
	One year	17	18	18	19	17	18	16	17	15	16	18	19
	Five years	3	4	4	5	4	5	3	5	4	5	4	5
		2	4	2	4	2	4	2	3	2	3	2	4
North & West	(n)	(1,999)		(2,329)		(2,899)		(4,456)		(6,905)		(18,756)	
	One year	18	19	16	17	14	15	15	16	14	15	17	18
	Five years	4	6	4	5	4	5	4	5	4	5	4	5
		3	5	3	5	2	4	2	4	2	4	3	4
WALES	(n)	(540)		(776)		(1,350)		(1,691)		(1,180)		(6,059)	
	One year	20	21	20	21	17	18	17	18	17	18	20	21
	Five years	8	10	6	7	5	7	5	7	5	7	6	8
		5	8	4	7	3	6	3	6	4	7	4	7

1986-90

MEN

Region		Affluent C	Affluent R	2 C	2 R	3 C	3 R	4 C	4 R	Deprived C	Deprived R	All[1] C	All[1] R
ENGLAND & WALES	(n)	(14,463)		(18,152)		(21,193)		(23,905)		(22,955)		(101,688)	
	One year	22	23	20	21	20	21	19	20	19	20	20	21
	Five years	5	6	5	6	4	6	4	6	4	6	5	6
ENGLAND	(n)	(13,844)		(17,339)		(19,922)		(22,346)		(21,942)		(96,258)	
	One year	22	23	20	21	20	21	19	20	18	19	20	21
	Five years	5	6	5	6	4	6	4	5	4	6	5	6
Northern & Yorkshire	(n)	(1,443)		(2,159)		(2,883)		(3,920)		(5,394)		(15,910)	
	One year	18	19	17	18	16	17	16	18	15	16	18	19
	Five years	4	6	5	6	3	5	4	5	3	5	4	5
Trent	(n)	(1,168)		(1,758)		(2,368)		(3,019)		(2,872)		(11,226)	
	One year	18	19	18	19	17	18	15	16	13	14	17	18
	Five years	4	5	4	6	4	6	3	5	3	4	4	5
Anglia & Oxford	(n)	(2,185)		(2,435)		(2,403)		(1,764)		(751)		(9,599)	
	One year	19	20	17	18	17	18	17	19	17	18	19	20
	Five years	5	6	4	5	3	5	4	5	5	6	4	5
North Thames	(n)	(1,277)		(1,763)		(2,073)		(2,404)		(2,242)		(10,039)	
	One year	28	29	26	27	25	27	24	26	27	29	26	28
	Five years	6	7	5	7	5	6	6	7	6	8	6	7
South Thames	(n)	(2,450)		(2,589)		(2,455)		(2,212)		(1,259)		(11,154)	
	One year	25	26	22	24	22	23	22	23	23	25	24	25
	Five years	6	7	5	6	4	6	4	5	4	6	5	6
South & West	(n)	(1,935)		(2,870)		(2,897)		(2,123)		(922)		(10,818)	
	One year	19	20	18	19	18	19	17	18	17	18	19	20
	Five years	4	5	4	6	4	6	4	6	4	5	5	6
West Midlands	(n)	(1,367)		(1,519)		(2,268)		(2,973)		(2,716)		(10,888)	
	One year	19	21	18	19	18	20	16	17	15	16	19	20
	Five years	5	6	4	6	4	5	3	5	4	5	4	5
North & West	(n)	(2,019)		(2,246)		(2,575)		(3,931)		(5,786)		(16,624)	
	One year	19	20	17	18	15	16	15	16	16	18	18	19
	Five years	6	7	4	5	4	5	3	5	4	6	5	6
WALES	(n)	(619)		(813)		(1,271)		(1,559)		(1,013)		(5,430)	
	One year	17	18	17	18	18	19	16	17	19	20	19	20
	Five years	4	6	5	7	5	7	6	7	6	8	6	8

[1] Including those for whom no deprivation category was assigned. C - crude survival, R - relative survival (see Chapter 3). Number of patients contributing to each analysis in parentheses.

Table 22.2   Crude and relative survival (%) at one, five and ten years since diagnosis: NHS Region, deprivation category and calendar period of diagnosis

WOMEN	1971-75 Affluent C	R	2 C	R	3 C	R	4 C	R	Deprived C	R	All[1] C	R	1976-80 Affluent C	R	2 C	R	3 C	R	4 C	R	Deprived C	R	All[1] C	R
ENGLAND & WALES	(3,578)		(4,156)		(4,786)		(4,993)		(4,985)		(27,334)		(4,999)		(5,887)		(6,726)		(7,185)		(7,489)		(33,732)	
One year	18	19	15	16	15	16	16	16	16	16	16	16	18	19	18	18	17	18	15	16	15	16	17	17
Five years	5	6	4	5	4	5	4	5	4	5	5	5	5	6	5	5	5	5	4	5	4	5	5	5
Ten years	4	5	3	3	3	4	3	4	3	4	3	4	4	5	3	4	3	4	3	4	3	4	3	4
ENGLAND	(3,471)		(3,994)		(4,556)		(4,668)		(4,785)		(26,106)		(4,852)		(5,676)		(6,391)		(6,767)		(7,196)		(32,135)	
One year	18	18	15	16	15	16	16	16	16	16	16	16	18	18	18	18	17	17	15	15	15	15	16	17
Five years	5	6	4	5	4	5	4	5	4	5	5	5	5	6	5	5	5	5	4	5	4	5	4	5
Ten years	4	5	2	3	3	4	3	4	3	4	3	4	4	5	3	4	3	4	3	3	3	4	3	4
Northern & Yorkshire	(242)		(359)		(468)		(620)		(957)		(3,608)		(422)		(618)		(853)		(1,155)		(1,836)		(5,002)	
One year	15	15	14	14	12	12	13	13	13	13	14	15	14	14	15	16	13	13	13	14	13	14	15	15
Five years	5	5	4	5	3	3	3	3	3	4	4	4	4	4	4	4	3	3	3	4	4	4	4	4
Ten years	3	3	3	4	2	2	2	3	2	3	2	3	2	3	3	4	2	3	2	2	2	3	2	3
Trent	(201)		(303)		(435)		(568)		(518)		(2,202)		(308)		(417)		(538)		(735)		(677)		(2,756)	
One year	17	17	12	13	15	15	16	16	14	14	17	17	16	17	16	16	16	17	13	14	14	15	16	17
Five years	5	6	4	5	4	5	5	6	3	3	5	6	5	6	6	6	5	5	5	5	4	4	5	6
Ten years	3	4	2	3	3	3	4	5	2	3	3	4	3	4	3	4	3	4	3	4	2	2	3	4
Anglia & Oxford	(445)		(423)		(482)		(323)		(176)		(2,118)		(680)		(704)		(716)		(542)		(265)		(3,023)	
One year	16	17	14	15	15	15	15	16	12	12	17	17	17	17	16	17	16	16	11	12	13	14	16	17
Five years	5	5	3	3	5	6	3	3	5	6	4	5	5	5	5	5	4	4	3	3	4	4	4	5
Ten years	3	4	2	2	3	5	2	3	4	5	3	4	3	4	3	4	3	3	2	2	2	3	3	4
North Thames	(610)		(680)		(696)		(770)		(679)		(3,889)		(605)		(639)		(745)		(895)		(776)		(3,825)	
One year	18	18	16	17	16	16	15	15	19	20	18	19	18	18	17	18	18	18	18	19	17	18	19	19
Five years	6	7	4	5	5	6	4	5	7	7	5	6	6	7	4	4	5	6	6	6	5	6	5	6
Ten years	5	6	2	3	3	4	3	4	5	6	4	5	5	6	3	3	4	5	4	5	3	4	4	5
South Thames	(1,013)		(942)		(1,022)		(962)		(603)		(5,007)		(1,259)		(1,139)		(1,137)		(1,045)		(514)		(5,419)	
One year	17	17	14	15	14	15	16	16	15	16	17	18	17	18	17	17	16	16	14	15	13	13	17	18
Five years	4	5	2	3	3	4	4	5	4	5	4	5	5	5	4	5	4	5	4	4	4	4	5	5
Ten years	3	4	1	2	2	3	3	4	3	4	3	4	3	4	3	3	3	4	2	3	2	3	3	4
South & West	(505)		(753)		(792)		(464)		(149)		(2,957)		(741)		(1,132)		(1,071)		(577)		(261)		(3,919)	
One year	18	19	14	14	17	17	14	14	10	10	17	17	18	19	16	17	16	16	14	15	15	15	18	18
Five years	5	6	5	5	5	6	5	6	3	3	5	6	6	7	5	5	5	5	4	5	4	4	5	6
Ten years	3	4	3	4	3	4	3	4	1	1	3	4	4	6	3	4	3	4	3	4	2	2	3	5
West Midlands	(112)		(128)		(164)		(198)		(306)		(2,116)		(284)		(398)		(500)		(653)		(841)		(2,728)	
One year	10	10	11	12	10	10	15	15	12	12	15	15	14	15	20	20	17	18	13	13	14	15	17	17
Five years	2	2	3	3	3	3	3	3	3	4	4	4	3	3	3	3	5	6	2	3	4	4	4	4
Ten years	2	2	1	2	2	2	2	3	2	2	2	3	2	2	2	2	4	5	1	1	2	3	2	3
North & West	(343)		(406)		(497)		(763)		(1,397)		(4,209)		(553)		(629)		(831)		(1,165)		(2,026)		(5,463)	
One year	15	16	12	12	10	10	12	12	12	13	14	14	16	16	13	14	13	13	11	12	12	12	14	15
Five years	5	6	4	5	4	4	3	3	4	4	4	5	5	5	4	5	3	4	3	4	3	4	4	5
Ten years	3	4	3	3	2	3	2	3	3	3	3	4	3	4	3	4	2	3	2	3	3	3	3	4
WALES	(107)		(162)		(230)		(325)		(200)		(1,228)		(147)		(211)		(335)		(418)		(293)		(1,597)	
One year	19	20	17	18	13	13	17	17	15	15	17	17	21	22	17	17	16	17	19	20	16	16	20	20
Five years	6	7	6	7	3	3	5	5	5	5	5	5	8	8	7	8	5	6	5	5	6	7	6	7
Ten years	6	7	4	7	2	2	3	4	4	4	3	5	6	8	5	6	3	4	3	5	4	7	4	6

[1] Including those for whom no deprivation category was assigned. C - crude survival, R - relative survival (see Chapter 3).
Number of patients contributing to each analysis in parentheses.

1981-85						1986-90						
Affluent	*2*	*3*	*4*	*Deprived*	*All*[1]	*Affluent*	*2*	*3*	*4*	*Deprived*	*All*[1]	*WOMEN*
C R	C R	C R	C R	C R	C R	C R	C R	C R	C R	C R	C R	
(6,049)	(6,949)	(7,983)	(9,115)	(9,328)	(40,261)	(6,167)	(7,839)	(9,090)	(10,480)	(10,360)	(44,387)	**ENGLAND & WALES**
21 22	19 20	19 19	19 19	18 18	19 20	21 22	21 21	20 20	19 19	19 20	20 20	*One year*
6 7	5 6	5 5	5 6	5 6	5 6	6 7	5 6	5 6	5 5	5 5	5 6	*Five years*
4 6	4 5	3 4	3 5	3 5	3 5							
(5,860)	(6,665)	(7,495)	(8,481)	(8,941)	(38,126)	(5,917)	(7,475)	(8,535)	(9,736)	(9,886)	(41,921)	**ENGLAND**
21 22	19 20	19 19	19 19	18 18	19 20	21 22	21 21	19 20	18 19	19 20	20 20	*One year*
6 7	5 6	4 5	5 6	5 5	5 6	6 7	5 6	5 6	4 5	5 5	5 6	*Five years*
4 6	4 5	3 4	3 4	3 4	3 5							
(556)	(737)	(986)	(1,487)	(2,329)	(6,172)	(686)	(1,046)	(1,275)	(1,892)	(2,555)	(7,501)	**Northern & Yorkshire**
18 18	18 19	16 17	15 15	15 15	17 18	19 20	16 17	17 17	15 15	15 16	17 18	*One year*
6 7	5 6	5 5	4 5	3 4	5 6	5 6	4 5	4 5	3 4	3 4	4 5	*Five years*
4 5	4 5	3 3	3 4	3 3	3 4							
(494)	(604)	(752)	(1,104)	(943)	(3,914)	(458)	(647)	(906)	(1,105)	(1,190)	(4,318)	**Trent**
16 17	20 20	14 15	16 17	18 18	18 19	15 16	19 20	16 17	12 13	15 16	17 17	*One year*
5 6	5 6	4 4	4 4	4 4	5 5	3 4	4 5	4 4	3 3	4 5	4 5	*Five years*
3 4	3 5	2 3	3 3	2 3	3 4							
(874)	(837)	(832)	(611)	(298)	(3,512)	(905)	(1,018)	(919)	(757)	(329)	(3,945)	**Anglia & Oxford**
19 19	14 15	16 16	14 15	12 13	17 18	20 21	17 17	18 19	15 15	15 15	19 19	*One year*
5 6	3 3	2 3	3 4	1 1	4 4	5 5	4 5	5 6	4 5	4 4	5 6	*Five years*
4 5	2 3	2 2	2 3	1 1	3 4							
(618)	(674)	(755)	(868)	(713)	(3,709)	(577)	(790)	(981)	(1,095)	(979)	(4,534)	**North Thames**
26 27	23 24	23 23	24 25	25 25	25 26	26 27	25 26	24 25	27 28	30 31	27 28	*One year*
7 8	7 8	6 6	7 7	8 9	7 8	8 9	7 9	5 6	6 7	6 7	7 8	*Five years*
5 7	5 7	3 5	4 6	6 8	5 7							
(1,317)	(1,205)	(1,208)	(1,136)	(572)	(5,671)	(1,088)	(1,177)	(1,158)	(1,059)	(589)	(5,165)	**South Thames**
20 21	21 22	21 22	21 22	20 21	22 23	23 24	23 24	19 20	20 21	22 23	23 24	*One year*
6 7	6 7	5 6	5 6	5 5	6 7	7 8	5 5	5 6	4 5	5 6	5 6	*Five years*
5 6	4 6	3 4	4 5	3 4	4 5							
(834)	(1,244)	(1,219)	(717)	(297)	(4,442)	(803)	(1,237)	(1,283)	(910)	(412)	(4,677)	**South & West**
20 20	15 15	17 18	18 19	17 18	19 19	20 21	18 19	16 16	16 17	18 18	19 19	*One year*
6 7	4 4	4 5	5 6	6 7	5 6	6 6	5 5	5 6	4 5	5 5	5 6	*Five years*
4 5	3 3	3 4	3 5	4 5	4 5							
(423)	(475)	(687)	(854)	(1,045)	(3,516)	(489)	(538)	(789)	(1,097)	(1,020)	(3,951)	**West Midlands**
21 21	16 17	16 16	15 16	15 16	17 18	19 20	20 21	16 16	16 17	17 17	19 19	*One year*
4 5	4 5	3 3	4 4	4 4	4 4	4 5	6 6	5 6	3 4	4 5	5 6	*Five years*
3 3	2 3	2 2	2 3	3 4	2 3							
(744)	(889)	(1,056)	(1,704)	(2,744)	(7,190)	(911)	(1,022)	(1,224)	(1,821)	(2,812)	(7,830)	**North & West**
17 18	16 17	14 15	14 15	15 15	17 17	17 18	16 16	18 18	17 18	16 17	19 19	*One year*
5 6	4 4	4 4	4 5	4 5	5 5	5 5	5 5	5 6	5 6	4 5	5 6	*Five years*
3 4	3 4	2 3	2 3	3 4	3 4							
(189)	(284)	(488)	(634)	(387)	(2,135)	(250)	(364)	(555)	(744)	(474)	(2,466)	**WALES**
21 22	19 20	19 20	18 19	17 18	21 21	22 23	20 21	20 20	21 22	18 19	22 23	*One year*
7 8	7 8	7 9	5 6	7 8	7 8	7 7	6 7	8 9	6 7	7 8	8 9	*Five years*
6 7	5 6	5 7	4 5	4 6	5 7							

[1] Including those for whom no deprivation category was assigned. C - crude survival, R - relative survival (see Chapter 3). Number of patients contributing to each analysis in parentheses.

Table 22.2 Crude and relative survival (%) at one, five and ten years since diagnosis: NHS Region, deprivation category and calendar period of diagnosis

ADULTS	1971-75 Affluent C	R	2 C	R	3 C	R	4 C	R	Deprived C	R	All[1] C	R	1976-80 Affluent C	R	2 C	R	3 C	R	4 C	R	Deprived C	R	All[1] C	R
ENGLAND & WALES	*(17,142)*		*(20,738)*		*(24,147)*		*(26,555)*		*(26,708)*		*(142,399)*		*(21,301)*		*(25,985)*		*(29,907)*		*(33,322)*		*(34,443)*		*(151,288)*	
One year	19	20	18	19	17	18	17	18	16	17	17	18	20	20	18	19	17	18	16	17	16	17	17	18
Five years	5	6	5	6	4	5	5	6	4	5	5	6	5	7	5	6	4	6	4	5	4	5	5	6
Ten years	3	5	3	4	3	4	3	5	3	4	3	4	3	5	3	4	3	4	3	4	3	4	3	4
ENGLAND	*(16,572)*		*(19,932)*		*(22,658)*		*(24,607)*		*(25,414)*		*(134,886)*		*(20,621)*		*(24,911)*		*(28,170)*		*(31,177)*		*(33,007)*		*(143,220)*	
One year	19	20	18	19	18	18	17	18	16	17	18	18	20	20	18	19	17	18	16	17	16	17	17	18
Five years	5	6	5	5	5	6	5	6	4	6	5	6	5	6	5	6	4	5	4	5	4	5	4	5
Ten years	3	5	3	4	3	4	3	5	3	4	3	4	3	5	3	4	3	4	2	4	3	4	3	4
Northern & Yorkshire	*(1,147)*		*(1,691)*		*(2,446)*		*(3,340)*		*(5,259)*		*(19,445)*		*(1,865)*		*(2,699)*		*(3,815)*		*(5,257)*		*(8,281)*		*(22,484)*	
One year	16	17	15	15	14	14	13	13	13	14	16	16	16	17	14	15	13	13	13	14	13	14	15	16
Five years	4	5	4	5	3	4	3	3	3	4	4	5	4	5	3	4	3	4	3	4	3	4	4	4
Ten years	2	3	2	3	2	3	2	3	2	3	2	3	2	4	2	3	2	3	2	3	2	3	2	3
Trent	*(1,188)*		*(1,712)*		*(2,347)*		*(3,296)*		*(3,082)*		*(12,708)*		*(1,455)*		*(2,104)*		*(2,841)*		*(3,810)*		*(3,496)*		*(14,116)*	
One year	17	18	16	16	14	15	15	16	15	16	17	18	16	16	14	15	15	16	14	14	14	14	16	17
Five years	4	5	4	4	4	4	4	5	4	5	4	5	5	5	4	5	4	4	3	4	3	4	4	5
Ten years	2	4	2	3	2	3	3	4	2	4	3	4	3	4	3	4	2	4	2	3	2	3	3	4
Anglia & Oxford	*(2,205)*		*(2,399)*		*(2,580)*		*(1,925)*		*(1,054)*		*(11,614)*		*(2,942)*		*(3,108)*		*(3,070)*		*(2,336)*		*(1,168)*		*(13,131)*	
One year	17	18	17	18	17	17	19	19	17	17	19	20	19	20	17	18	17	17	15	16	15	16	18	19
Five years	4	5	4	5	5	6	6	7	4	5	5	6	5	6	4	5	4	4	3	4	4	5	4	5
Ten years	3	4	3	4	3	4	4	5	2	3	3	5	3	5	2	3	2	4	2	3	2	4	3	4
North Thames	*(2,833)*		*(3,286)*		*(3,376)*		*(3,957)*		*(3,479)*		*(19,178)*		*(2,470)*		*(2,858)*		*(3,133)*		*(3,721)*		*(3,285)*		*(16,131)*	
One year	19	20	19	20	19	20	19	19	19	20	20	21	22	22	20	21	20	21	21	22	20	21	22	23
Five years	6	7	5	6	5	7	5	6	6	8	6	7	6	7	5	6	5	7	6	7	6	7	6	7
Ten years	4	6	3	4	4	5	4	6	4	6	4	6	4	6	3	5	3	5	4	6	4	6	4	6
South Thames	*(4,489)*		*(4,277)*		*(4,564)*		*(4,406)*		*(2,698)*		*(22,259)*		*(4,871)*		*(4,633)*		*(4,559)*		*(4,338)*		*(2,185)*		*(21,728)*	
One year	18	19	17	18	17	17	18	19	16	17	19	20	18	19	18	19	17	17	15	16	15	16	18	19
Five years	5	5	4	4	4	5	5	6	4	5	5	6	5	6	5	6	4	5	4	5	4	5	5	6
Ten years	3	4	2	3	2	3	3	4	2	4	3	4	3	5	2	4	2	4	2	3	3	5	3	4
South & West	*(2,543)*		*(3,853)*		*(3,849)*		*(2,258)*		*(976)*		*(14,875)*		*(3,114)*		*(4,749)*		*(4,530)*		*(2,656)*		*(1,172)*		*(16,873)*	
One year	18	18	17	18	16	17	16	16	16	16	18	19	18	18	17	18	16	17	15	16	14	15	18	19
Five years	5	6	5	6	4	5	5	6	5	6	5	6	5	6	5	6	5	6	5	6	4	5	5	7
Ten years	3	4	3	5	3	4	3	4	3	5	3	5	3	5	4	5	3	5	3	5	3	4	3	5
West Midlands	*(518)*		*(712)*		*(962)*		*(1,313)*		*(1,812)*		*(13,152)*		*(1,519)*		*(1,924)*		*(2,583)*		*(3,670)*		*(4,668)*		*(14,567)*	
One year	15	15	17	18	18	18	14	14	16	16	18	18	18	18	17	18	17	18	14	15	15	16	17	18
Five years	3	3	4	5	4	5	4	4	5	6	4	5	5	6	4	4	4	5	3	4	4	5	4	5
Ten years	2	2	2	3	3	4	2	3	3	4	3	4	3	5	2	3	3	4	2	3	2	4	3	4
North & West	*(1,649)*		*(2,002)*		*(2,534)*		*(4,112)*		*(7,054)*		*(21,655)*		*(2,385)*		*(2,836)*		*(3,639)*		*(5,389)*		*(8,752)*		*(24,190)*	
One year	15	15	13	14	13	14	12	13	13	13	15	15	16	16	13	14	13	13	12	13	13	14	15	15
Five years	5	5	4	4	4	5	3	4	4	4	4	5	4	5	4	5	3	4	3	4	4	4	4	5
Ten years	3	4	2	3	2	4	2	3	2	3	3	4	3	4	2	3	2	3	2	3	2	4	3	4
WALES	*(570)*		*(806)*		*(1,489)*		*(1,948)*		*(1,294)*		*(7,513)*		*(680)*		*(1,074)*		*(1,737)*		*(2,145)*		*(1,436)*		*(8,068)*	
One year	17	18	15	16	15	16	15	15	15	16	16	17	20	21	16	16	14	15	15	16	15	16	17	18
Five years	5	6	5	6	4	5	4	5	4	5	4	5	7	8	5	6	5	6	5	6	5	6	6	7
Ten years	4	5	3	4	2	4	3	4	3	4	3	4	5	8	3	5	3	5	3	6	3	6	4	6

[1] Including those for whom no deprivation category was assigned. C - crude survival, R - relative survival (see Chapter 3).
Number of patients contributing to each analysis in parentheses.

Measure	1981-85 Affluent C	R	2 C	R	3 C	R	4 C	R	Deprived C	R	All[1] C	R	1986-90 Affluent C	R	2 C	R	3 C	R	4 C	R	Deprived C	R	All[1] C	R	ADULTS
(n)	(22,485)		(26,314)		(30,647)		(34,881)		(35,049)		(152,552)		(20,630)		(25,991)		(30,283)		(34,385)		(33,315)		(146,075)		**ENGLAND & WALES**
One year	21	22	20	21	19	20	19	20	18	19	**19**	**20**	22	23	20	21	20	21	19	20	19	20	**20**	**21**	*One year*
Five years	5	7	5	6	5	6	5	6	4	6	**5**	**6**	5	6	5	6	5	6	4	5	4	6	**5**	**6**	*Five years*
	4	5	3	5	3	4	3	5	3	5	**3**	**5**													
(n)	(21,756)		(25,254)		(28,809)		(32,556)		(33,482)		(144,358)		(19,761)		(24,814)		(28,457)		(32,082)		(31,828)		(138,179)		**ENGLAND**
One year	21	22	20	20	19	19	19	19	18	19	**19**	**20**	22	23	20	21	20	21	19	20	19	20	**20**	**21**	*One year*
Five years	5	7	5	6	4	5	5	6	4	6	**5**	**6**	5	6	5	6	5	6	4	5	4	5	**5**	**6**	*Five years*
	4	5	3	5	3	4	3	5	3	4	**3**	**5**													
(n)	(2,048)		(2,780)		(3,811)		(5,552)		(8,530)		(23,038)		(2,129)		(3,205)		(4,158)		(5,812)		(7,949)		(23,411)		**Northern & Yorkshire**
One year	17	17	17	18	15	16	15	16	15	16	**17**	**18**	18	19	17	17	16	17	16	17	15	16	**17**	**18**	*One year*
Five years	5	6	4	5	4	5	4	5	3	4	**4**	**5**	5	6	4	5	4	5	4	5	3	4	**4**	**5**	*Five years*
	3	4	3	4	2	3	3	4	2	4	**3**	**4**													
(n)	(1,717)		(2,349)		(3,238)		(4,506)		(3,795)		(15,697)		(1,626)		(2,405)		(3,274)		(4,124)		(4,062)		(15,544)		**Trent**
One year	18	19	17	18	14	15	16	17	15	16	**17**	**18**	17	18	18	19	17	17	14	15	14	15	**17**	**18**	*One year*
Five years	4	5	4	5	3	4	4	5	3	4	**4**	**5**	4	5	4	6	4	5	3	4	3	4	**4**	**5**	*Five years*
	3	4	3	4	2	3	2	4	2	3	**2**	**4**													
(n)	(3,228)		(3,265)		(3,297)		(2,432)		(1,083)		(13,555)		(3,090)		(3,453)		(3,322)		(2,521)		(1,080)		(13,544)		**Anglia & Oxford**
One year	17	18	17	17	17	18	15	16	15	16	**18**	**19**	19	20	17	18	18	18	17	17	16	17	**19**	**20**	*One year*
Five years	4	5	3	4	4	5	3	4	4	4	**4**	**5**	5	6	4	5	4	5	4	5	5	6	**4**	**5**	*Five years*
	3	5	2	4	3	4	2	3	2	4	**3**	**4**													
(n)	(2,261)		(2,527)		(2,718)		(3,109)		(2,548)		(13,405)		(1,854)		(2,553)		(3,054)		(3,499)		(3,221)		(14,573)		**North Thames**
One year	26	27	24	25	22	23	24	25	26	28	**26**	**27**	27	29	25	27	25	26	25	27	28	29	**27**	**28**	*One year*
Five years	7	9	6	8	6	7	7	9	8	10	**7**	**9**	6	8	6	8	5	6	6	7	6	8	**6**	**7**	*Five years*
	5	8	4	7	4	6	5	8	6	9	**5**	**8**													
(n)	(4,849)		(4,349)		(4,224)		(3,945)		(1,945)		(20,126)		(3,538)		(3,766)		(3,613)		(3,271)		(1,848)		(16,319)		**South Thames**
One year	22	23	20	21	20	21	21	22	19	20	**22**	**23**	25	26	23	24	21	22	21	22	23	24	**23**	**24**	*One year*
Five years	5	7	5	6	5	6	5	6	4	5	**5**	**7**	6	7	5	6	5	6	4	5	5	6	**5**	**6**	*Five years*
	4	5	3	5	3	5	3	5	3	4	**3**	**5**													
(n)	(3,201)		(4,674)		(4,475)		(2,699)		(1,146)		(16,646)		(2,738)		(4,107)		(4,180)		(3,033)		(1,334)		(15,495)		**South & West**
One year	19	20	17	18	17	18	18	19	16	17	**19**	**20**	19	20	18	19	17	18	17	18	17	18	**19**	**20**	*One year*
Five years	5	7	4	5	4	5	5	6	5	6	**5**	**6**	5	6	4	6	4	6	4	5	4	5	**5**	**6**	*Five years*
	3	5	2	4	3	5	3	5	3	5	**3**	**5**													
(n)	(1,709)		(2,092)		(3,091)		(4,153)		(4,786)		(15,945)		(1,856)		(2,057)		(3,057)		(4,070)		(3,736)		(14,839)		**West Midlands**
One year	18	19	18	18	17	18	16	17	15	16	**18**	**19**	19	20	18	19	18	19	16	17	16	17	**19**	**19**	*One year*
Five years	3	4	4	5	4	5	3	4	4	5	**4**	**5**	5	6	5	6	4	6	3	4	4	5	**4**	**6**	*Five years*
	2	3	2	4	2	3	2	3	2	4	**2**	**4**													
(n)	(2,743)		(3,218)		(3,955)		(6,160)		(9,649)		(25,946)		(2,930)		(3,268)		(3,799)		(5,752)		(8,598)		(24,454)		**North & West**
One year	18	18	16	17	14	15	15	16	14	15	**17**	**18**	18	19	17	17	16	17	16	17	16	17	**18**	**19**	*One year*
Five years	5	6	4	5	4	5	4	5	4	5	**4**	**5**	5	7	4	5	4	5	4	5	4	5	**5**	**6**	*Five years*
	3	4	3	5	2	3	2	4	2	4	**3**	**4**													
(n)	(729)		(1,060)		(1,838)		(2,325)		(1,567)		(8,194)		(869)		(1,177)		(1,826)		(2,303)		(1,487)		(7,896)		**WALES**
One year	20	21	20	20	18	19	17	18	17	18	**20**	**21**	18	19	18	19	19	20	18	19	19	20	**20**	**21**	*One year*
Five years	8	9	6	7	6	8	5	7	6	7	**6**	**8**	5	6	6	7	6	8	6	7	6	8	**7**	**8**	*Five years*
	5	8	4	7	4	6	4	6	4	6	**4**	**7**													

[1] Including those for whom no deprivation category was assigned. C - crude survival, R - relative survival (see Chapter 3). Number of patients contributing to each analysis in parentheses.

Fig. 22.1   Relative survival up to ten years: trends by calendar period of diagnosis (1971-90) and NHS Region

ADULTS

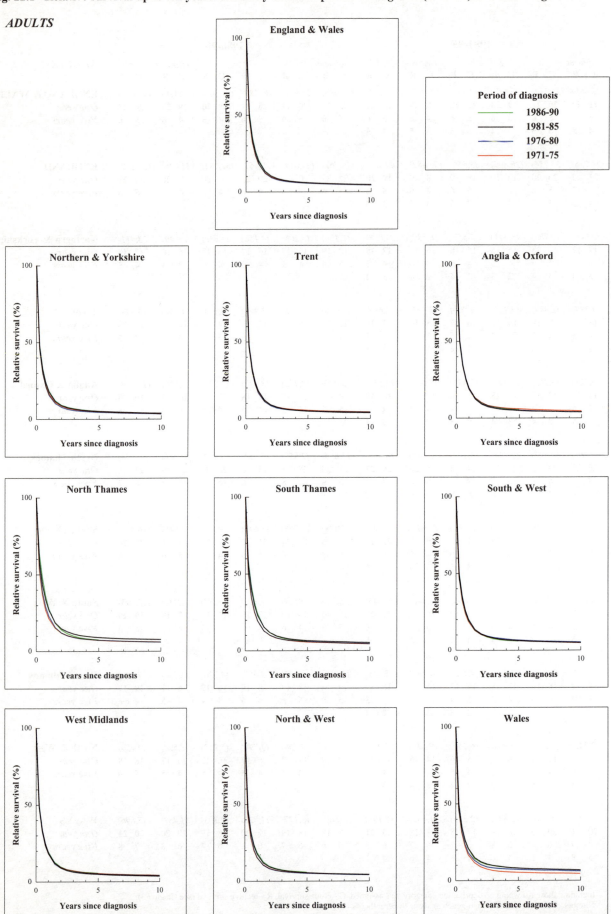

Fig. 22.2   Relative survival up to five years, by deprivation category and NHS Region: patients diagnosed 1986-90

ADULTS

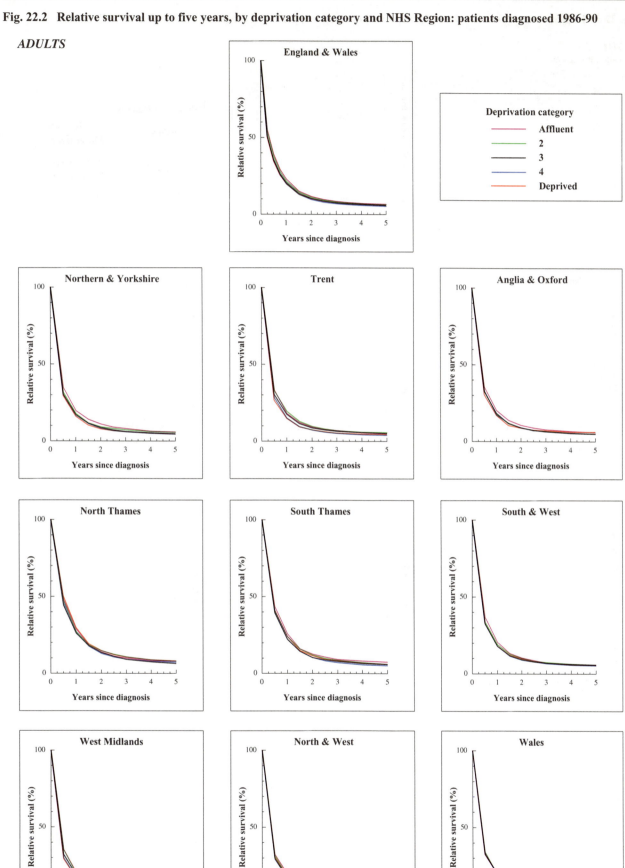

Fig. 22.3   Relative survival at five years by deprivation category, period of diagnosis (1981-90) and NHS Region

ADULTS
(Note vertical scale)

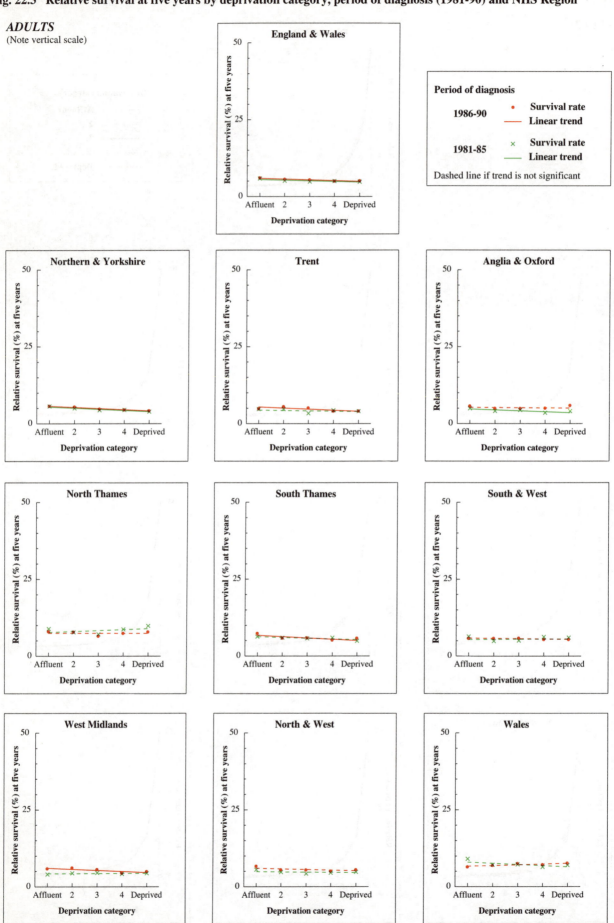

Fig. 22.4 Relative survival (%) by NHS Region and sex:
      one-year and five-year survival for patients diagnosed 1986-90, with 95% confidence intervals

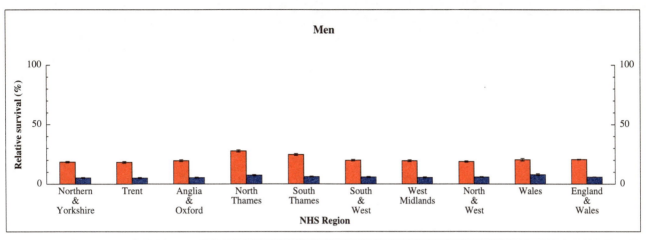

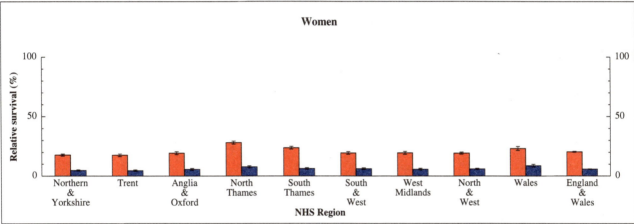

Fig. 22.5 Relative survival (%) at five years, with 95% confidence interval, by age at diagnosis and sex:
      England and Wales, patients diagnosed 1986-90

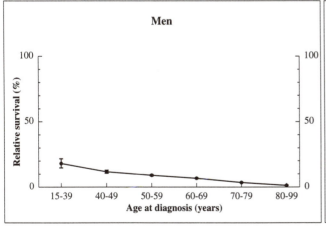

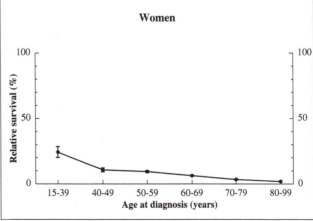

Chapter 23

PLEURA

Malignant neoplasms of the pleural lining of the lungs and thorax are almost all mesotheliomas. The great majority are induced by occupational or occupation-related exposure to asbestos, mainly in the shipbuilding and insulation industries. Men with mesothelioma outnumber women by 5 to 1.

These malignancies have been uncommon, but there has been a four-fold increase in incidence since 1971, and almost 1,000 new cases were diagnosed in England and Wales in 1990. Incidence will continue to increase for 10 years or more until the full effect of industrial asbestos exposure controls introduced during the 1960s becomes apparent.

Age-standardised incidence in 1990 was 3.0 per 100,000 per year in men and 0.6 in women. Mesothelioma is extremely rare before age 40, because the latency between asbestos exposure and disease is usually 30 years or more. Mesotheliomas also occur in the ovary, testis, bowel and peritoneum, but these are rare. Mortality from pleural malignancies in men also increased four-fold between 1971 and 1995.

The analyses presented here are based on almost 7,500 patients diagnosed with malignant pleural tumours in England and Wales between 1971 and 1990 and followed up to the end of 1995, about 82% of those eligible (Table 23.1). About 4% were excluded because their vital status was unknown at 6 July 1997, and a further 14% because their survival was zero or unknown, most having been registered from death certificate information alone. The proportions of cases registered solely from a death certificate were particularly high in the Thames and South Western regions in the 1980s. The great majority (98%) of cases were simply coded to the pleura, unspecified (Table 23A). Overall about 77% of cases were specified as mesotheliomas, but this proportion rose from about 50% in the 1970s to over 80% by the late 1980s (data not shown). About 7% of the tumours were specified as

adenocarcinomas, and the remainder were poorly specified epithelial tumours. Mesotheliomas represented more than 90% of pleural malignancies to which a specific morphology had been assigned, and pleural tumours will be referred to here as mesotheliomas.

Survival trends

Survival from mesothelioma is very low. One-year survival for adults diagnosed during 1986-90 was about 27%, and five-year survival was only 4%. The corresponding figures for those diagnosed during 1971-75 were 24% and 3% (Table 23.2). The lack of any marked improvement in survival is clear from Figure 23.1, although there was a 2-4% improvement in five-year survival in West Midlands, North & West and Wales during the 1980s, also visible in the survival curves.

The age-standardised trends in survival from mesothelioma for men are small, and very similar to those for cancer of the lung. The improvement between successive five-year periods approaches 2% for one-year survival, and 1% for five-year survival. Women show a small average gain in survival at one year, and none at all for survival at five years (see Table 4.4).

There is very little variation with age in survival from mesothelioma (Figure 23.5), and relative survival is generally less than 10% in every age group for both men and women (see Table 4.7).

Relative survival was slightly higher for women than for men throughout the 1970s and 1980s (Table 23.2), and age-adjusted survival rates were also 2% to 3% higher, but age-specific survival rates for patients diagnosed during 1986-90 do not all show an advantage for women (see Tables 4.4 and 4.7).

Regional variation in survival during the 1970s is partly due to statistical instability from the small number of

Table 23A Sub-site distribution of malignant neoplasms of the pleura

ICD-9 code	Site description	1971-75		1976-80		1981-85		1986-90		Total	
		No.	%	No.	%	No.	%	No.	%	No.	%
163.0	Parietal	-	-	46	3.3	19	1.0	25	0.8	90	1.2
163.1	Visceral	-	-	0	0.0	2	0.1	2	0.1	4	0.1
163.8	Other	-	-	1	0.1	19	1.0	32	1.0	52	0.7
163.9	Pleura, unspecified[1]	872	100.0	1,330	96.6	1,952	98.0	3,145	98.2	7,299	98.0
	Total	**872**		**1,377**		**1,992**		**3,204**		**7,445**	

[1] All pleural neoplasms were coded to 163.0 in ICD-8 (1971-78): they are assigned to the corresponding ICD-9 category (163.9) in this table.

cases. The number of cases had at least doubled by the late 1980s in every region, and it increased five-fold or more in South Thames and South & West. For patients diagnosed during 1986-90, relative survival in each region was broadly similar to that for England and Wales as a whole (Figure 23.4).

Deprivation

Differences in five-year survival between deprivation categories were very small for patients diagnosed during 1986-90 (Figure 23.2), and the gradient was not significant (Figure 23.3). Survival at one year was consistently 1-2% lower in the deprived group for patients diagnosed during the 1980s, although the gap was not statistically significant (see Table 4.9).

International comparison

Five years after diagnosis, relative survival for patients diagnosed with mesothelioma in England and Wales during 1986-90 was slightly higher than in Scotland but 2% to 3% lower than the average for Europe (see Table 4.12). The estimates for England and Wales depend on over 3,000 patients: although these differences are small, they are statistically significant. The highest regional estimates of five-year survival (7% for men and 17% for women) are for Wales. The estimate for women depends on only 29 cases, and is not precise. If this estimate is excluded, none of the regional survival estimates is greater than the European average.

Table 23.1 Ineligible and excluded records, and patients included in analyses, by calendar period of diagnosis

	Calendar period of diagnosis														
	1971-75			**1976-80**			**1981-85**			**1986-90**			**Overall**		
Total registered	**1,023**			**1,632**			**2,480**			**4,048**			**9,183**		
Ineligible	*Records*	*Patients*	*%*	*Records*	*Patients*	*%*	*Records*	*Patients*	*%*	*Records*	*Patients*	*%*	*Records*	*Patients*	*%*
Incomplete data[1]	0	0		4	4		12	12		17	17		33	**33**	
Resident outside England and Wales	8	8		5	5		4	4		7	5		24	**22**	
In situ neoplasm[2]	0	0		0	0		0	0		0	0		0	**0**	
Benign or uncertain behaviour[2]	10	10		6	6		0	0		0	0		16	**16**	
Metastatic[2]	7	7		6	6		0	0		1	1		14	**14**	
Otherwise ineligible[3]	-	-		-	-		-	-		-	-		-	**-**	
Lymphoma[4]	3	3		4	4		0	0		0	0		7	**7**	
Leukaemia or myeloma[4]	0	0		0	0		0	0		0	0		0	**0**	
		28			25			16			23			92	
Total eligible		**995**	**100**		**1,607**	**100**		**2,464**	**100**		**4,025**	**100**		**9,091**	**100**
Aged under 15 or 100 years or over at diagnosis	1	1	0.1	1	1	<0.1	2	2	<0.1	3	3	<0.1	7	**7**	<0.1
Vital status unknown[5]	33	28	2.8	64	60	3.7	105	103	4.2	181	174	4.3	383	**365**	4.0
Duplicate registration[6]	0	0	0.0	0	0	0.0	0	0	0.0	0	0	0.0	0	**0**	0.0
Multiple primary[7]	0	0	0.0	2	2	0.1	1	1	<0.1	7	7	0.2	10	**10**	0.1
Synchronous tumours[8]	3	2	0.2	3	2	0.1	5	5	0.2	10	5	0.1	21	**14**	0.2
Sex not known	0	0	0.0	0	0	0.0	0	0	0.0	0	0	0.0	0	**0**	0.0
Sex incompatible with site code[9]	-	-	-	-	-	-	-	-	-	-	-	-	-	**-**	-
Invalid dates[10]	0	0	0.0	0	0	0.0	2	2	<0.1	1	0	0.0	3	**2**	<0.1
Zero survival[11]	95	92	9.2	177	165	10.3	378	359	14.6	656	632	15.7	1,306	**1,248**	13.7
Total exclusions		**123**	**12.4**		**230**	**14.3**		**472**	**19.2**		**821**	**20.4**		**1,646**	**18.1**
Patients included in analysis		**872**	**87.6**		**1,377**	**85.7**		**1,992**	**80.8**		**3,204**	**79.6**		**7,445**	**81.9**

[1] Main data item(s) invalid or incompatible with one another: name, sex, date of birth, date of diagnosis and (if present) date of death, postcode, site, morphology and behaviour

[2] Tumour either *in situ* (behaviour code 2), benign (0), uncertain if benign or malignant (1) or metastatic (6 or 9)

[3] Not applicable for this malignancy

[4] Morphology that of lymphoma (included with non-Hodgkin lymphoma, Ch. 41) or myeloma (included with multiple myeloma, Ch. 43), or leukaemia (excluded)

[5] Tracing of vital status at National Health Service Central Register incomplete at 6 July 1997

[6] Same site code, sex, cancer registry and cancer registry number as an earlier registration

[7] Same site code and person number as an earlier registration(s): mostly confirmed multiple primary tumours at this site, some unresolved duplicate registrations

[8] Same site code, sex, date of birth and date of diagnosis as another registration(s): mostly synchronous or (in paired organs) bilateral tumours in same anatomic site in one individual, not previously linked: also some duplicate registrations

[9] Not applicable for this malignancy

[10] Impossible sequence of dates (birth, diagnosis and, if present, emigration or death); or date of death after 6 July 1997

[11] Date of diagnosis same as date of death: some are patients who did die on the day of diagnosis, but most were registered solely from a death certificate (death certificate only, or DCO), and their survival time is thus unknown

Table 23.2   Crude and relative survival (%) at one, five and ten years since diagnosis: NHS Region, deprivation category and calendar period of diagnosis

1971-75 / 1976-80 — deprivation category

	Affluent C R	2 C R	3 C R	4 C R	Deprived C R	All[1] C R	Affluent C R	2 C R	3 C R	4 C R	Deprived C R	All[1] C R
ENGLAND & WALES												
Men	(83)	(100)	(98)	(126)	(124)	(647)	(164)	(158)	(206)	(237)	(219)	(1,021)
One year	29 29	22 23	15 15	20 21	26 27	23 23	24 24	24 25	20 21	21 22	18 19	21 21
Five years	4 4	2 3	1 1	2 2	3 3	2 3	2 3	2 2	1 1	2 2	1 1	1 2
Ten years	2 3	0 0	1 1	2 2	2 3	1 2	2 3	1 1	1 1	1 2	0 0	1 1
Women	(27)	(33)	(43)	(41)	(48)	(225)	(63)	(53)	(59)	(86)	(79)	(356)
One year	35 36	22 22	17 17	31 31	30 31	27 27	14 14	31 32	27 28	19 20	18 19	22 22
Five years	3 4	8 10	2 2	5 6	3 3	5 5	5 5	4 4	3 3	1 1	2 3	3 4
Ten years	3 4	8 10	2 2	0 1	3 3	3 4	1 2	2 2	3 3	1 1	2 3	2 3
Adults	(110)	(133)	(141)	(167)	(172)	(872)	(227)	(211)	(265)	(323)	(298)	(1,377)
One year	30 31	22 23	15 16	23 23	27 28	24 24	21 22	26 27	22 22	20 21	18 19	21 22
Five years	4 4	4 4	1 2	2 3	3 3	3 3	3 4	2 2	1 1	2 2	1 1	2 2
Ten years	3 3	2 3	1 2	1 2	2 3	2 3	2 3	1 2	1 1	1 1	1 1	1 2

NHS Region — 1971-75 / 1976-80

	Men C R	Women C R	Adults C R	Men C R	Women C R	Adults C R
Northern & Yorkshire	(111)	(46)	(157)	(189)	(73)	(262)
One year	22 23	34 35	26 27	25 26	17 17	22 23
Five years	2 3	6 7	3 4	1 2	1 1	1 2
Ten years	0 0	2 2	1 1	0 1	1 1	0 1
Trent	(29)	(33)	(62)	(61)	(27)	(88)
One year	12 13	32 33	23 24	31 32	22 23	28 29
Five years	4 4	6 7	5 5	2 2	0 0	1 1
Ten years	4 4	6 7	5 5	0 0	0 0	0 0
Anglia & Oxford	(39)	(18)	(57)	(89)	(19)	(108)
One year	26 26	43 44	31 32	15 16	55 56	22 23
Five years	0 0	16 17	5 6	1 1	11 11	3 3
Ten years	0 0	11 17	4 6	1 1	4 5	2 2
North Thames	(102)	(36)	(138)	(96)	(50)	(146)
One year	23 24	18 18	22 22	22 23	19 19	21 22
Five years	4 5	0 0	3 3	1 1	7 7	3 3
Ten years	2 3	0 0	2 2	1 1	7 7	3 3
South Thames	(46)	(20)	(66)	(124)	(29)	(153)
One year	39 40	24 26	35 36	28 30	30 31	29 30
Five years	3 3	0 0	2 2	3 3	3 3	3 3
Ten years	3 3	0 0	2 2	2 2	3 3	2 3
South & West	(93)	(10)	(103)	(148)	(28)	(176)
One year	14 15	16 17	15 15	18 19	14 14	18 18
Five years	2 2	0 0	2 2	3 4	6 6	3 4
Ten years	1 1	0 0	1 1	3 4	0 0	2 3
West Midlands	(49)	(9)	(58)	(66)	(24)	(90)
One year	19 20	- -	20 20	17 18	33 34	21 22
Five years	0 0	- -	0 0	1 1	0 0	1 1
Ten years	0 0	- -	0 0	1 1	0 0	1 1
North & West	(142)	(38)	(180)	(204)	(84)	(288)
One year	25 26	21 22	25 25	17 17	13 13	16 16
Five years	3 3	2 3	3 3	1 1	0 0	1 1
Ten years	2 3	2 3	2 3	0 1	0 0	0 0
WALES	(36)	(15)	(51)	(44)	(22)	(66)
One year	21 22	21 22	21 22	10 11	36 37	18 19
Five years	3 3	11 11	5 6	0 0	10 10	2 3
Ten years	3 3	11 11	5 6	0 0	5 8	1 2

[1] Including those for whom no deprivation category was assigned. C - crude survival, R - relative survival (see Chapter 3).
Number of patients contributing to each analysis in parentheses.

ENGLAND & WALES

	1981-85 Affluent C	R	2 C	R	3 C	R	4 C	R	Deprived C	R	All[1] C	R	1986-90 Affluent C	R	2 C	R	3 C	R	4 C	R	Deprived C	R	All[1] C	R	
Men (n)	(304)		(317)		(327)		(329)		(286)		(1,594)		(515)		(553)		(569)		(535)		(410)		(2,604)		
One year	26	27	22	23	26	27	25	26	25	26	25	26	25	25	25	26	24	25	27	29	22	23	25	26	
Five years	3	3	3	4	3	4	4	4	4	4	3	4	2	3	4	4	3	4	3	4	3	4	3	4	
	1	1	3	4	1	2	2	3	2	3	2	2													
Women (n)	(76)		(84)		(61)		(72)		(92)		(398)		(109)		(123)		(136)		(113)		(114)		(600)		
One year	37	38	29	30	23	24	28	29	22	22	27	28	26	26	31	32	32	33	21	21	23	24	27	28	
Five years	4	5	5	6	2	2	2	2	0	0	2	3	3	3	2	2	6	6	7	8	3	4	4	5	
	2	4	4	5	2	2	2	2	0	0	2	3													
Adults (n)	(380)		(401)		(388)		(401)		(378)		(1,992)		(624)		(676)		(705)		(648)		(524)		(3,204)		
One year	28	29	24	25	25	26	25	26	24	25	25	26	25	25	26	27	26	27	26	27	23	24	25	26	
Five years	3	3	4	4	3	3	3	4	3	3	3	4	2	3	3	4	4	4	4	5	3	4	3	4	
	1	2	3	4	1	2	2	3	2	2	2	3													

	1981-85 Men C	R	Women C	R	Adults C	R	1986-90 Men C	R	Women C	R	Adults C	R	Region
(n)	(301)		(82)		(383)		(418)		(102)		(520)		**Northern & Yorkshire**
One year	24	25	31	32	26	27	24	25	22	23	23	24	
Five years	3	3	1	2	2	3	3	3	1	1	2	3	
	2	2	1	2	2	2							
(n)	(95)		(25)		(120)		(163)		(40)		(203)		**Trent**
One year	13	14	12	12	13	13	23	24	16	16	22	22	
Five years	2	2	4	4	2	2	2	3	0	0	2	2	
	1	2	4	4	1	2							
(n)	(131)		(42)		(173)		(234)		(47)		(281)		**Anglia & Oxford**
One year	25	26	30	30	26	27	25	26	39	40	27	28	
Five years	4	4	0	0	3	3	4	5	6	7	5	5	
	4	4	0	0	3	3							
(n)	(138)		(32)		(170)		(324)		(99)		(423)		**North Thames**
One year	33	35	36	37	34	35	28	29	28	29	28	29	
Five years	5	6	6	7	5	6	5	6	5	6	5	6	
	2	2	6	7	3	3							
(n)	(178)		(55)		(233)		(310)		(84)		(394)		**South Thames**
One year	26	27	30	31	27	28	25	26	34	35	27	28	
Five years	4	5	7	8	5	6	2	2	5	5	2	3	
	2	3	5	7	3	4							
(n)	(309)		(32)		(341)		(523)		(78)		(601)		**South & West**
One year	23	24	36	37	24	25	24	25	19	19	23	24	
Five years	3	4	0	0	3	4	2	2	2	3	2	2	
	2	3	0	0	2	2							
(n)	(92)		(21)		(113)		(141)		(32)		(173)		**West Midlands**
One year	28	29	21	21	26	27	30	31	24	25	29	30	
Five years	1	1	0	0	1	1	4	5	6	7	4	5	
	0	0	0	0	0	0							
(n)	(315)		(83)		(398)		(429)		(89)		(518)		**North & West**
One year	24	25	19	20	23	24	25	26	31	32	26	27	
Five years	2	3	3	3	2	3	3	4	6	6	4	5	
	1	2	1	1	1	2							
(n)	(35)		(26)		(61)		(62)		(29)		(91)		**WALES**
One year	33	35	25	26	30	31	23	24	33	34	26	27	
Five years	10	11	0	0	6	7	6	7	16	17	9	10	
	5	6	0	0	3	3							

[1] Including those for whom no deprivation category was assigned. C - crude survival, R - relative survival (see Chapter 3). Number of patients contributing to each analysis in parentheses.

Fig. 23.1   Relative survival up to ten years: trends by calendar period of diagnosis (1971-90) and NHS Region

ADULTS

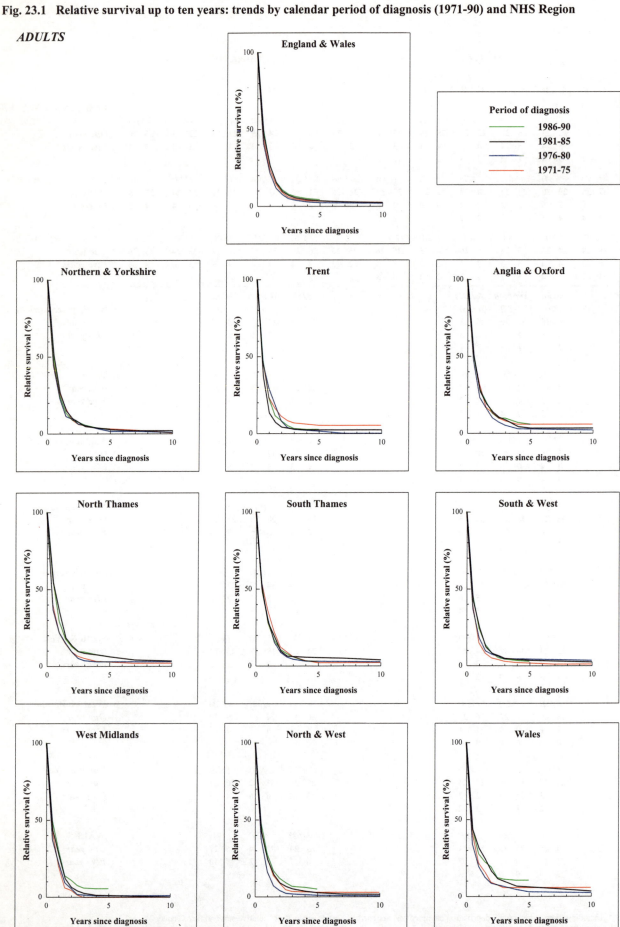

Fig. 23.2 Relative survival up to five years, by deprivation category: England and Wales, patients diagnosed 1986-90

ADULTS

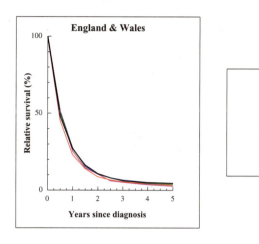

Fig. 23.3 Relative survival at five years, by deprivation category and period of diagnosis (1981-90): England and Wales

ADULTS
(Note vertical scale)

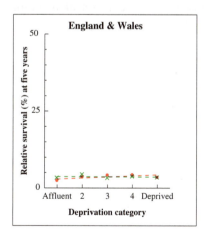
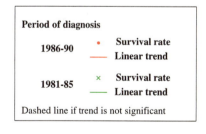

Fig. 23.4 Relative survival (%) by NHS Region and sex:
one-year and five-year survival for patients diagnosed 1986-90, with 95% confidence intervals

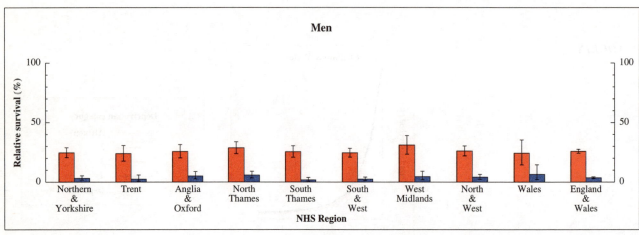

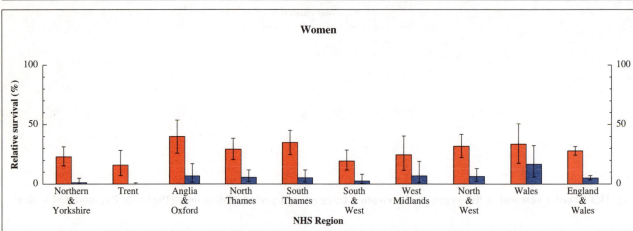

Fig. 23.5 Relative survival (%) at five years, with 95% confidence interval, by age at diagnosis and sex:
England and Wales, patients diagnosed 1986-90

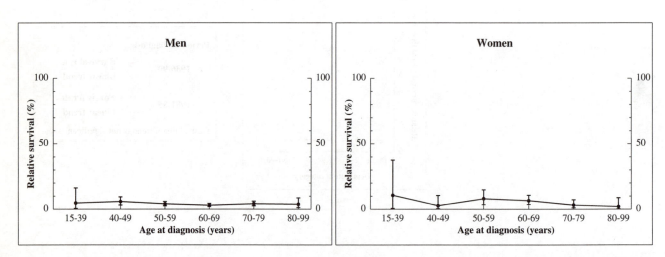

Chapter 24

THYMUS

The thymus is an organ of epithelial tissue in the upper mediastinum which has a complex role in the development of cellular immunity during childhood and adolescence. T-lymphocytes mature in the thymus, which tends to atrophy after adolescence. The thymus was classified as an endocrine gland in the International Classification of Diseases until 1978. Malignancy of the thymus is extremely rare: there are on average just 40 cases a year in England and Wales. Most of these are thymomas, although germ cell tumours, lymphomas and carcinoids also arise in the thymus. Thymoma is slightly more frequent in men than in women, with a sex ratio of around 1.4 to 1.

Between 1979 and 1990, the recorded incidence of malignant thymoma in men in England and Wales doubled to 0.1 per 100,000, but there was little change among women, the age-standardised (Europe) rate remaining at around 0.07 per 100,000. Mortality has fluctuated around 0.05 per 100,000 in both sexes over the same period.

The causes of thymoma are unknown. Tumours of the thymus are associated with myasthenia gravis and red cell aplasia, an indication of its immunological role. Benign or malignant thymomas occur in 10-30% of patients with myasthenia gravis[1], the symptoms of which may lead to diagnosis of the tumour. Thymectomy has been used to treat severe myasthenia that is unresponsive to immunosuppressive agents, even in the absence of thymoma. Thymectomy was carried out for benign enlargement of the thymus in adolescents during the 1930s and 1940s: this increases cancer risk in neonatal rodents, but there is no convincing evidence of increased cancer risk in man[2]. Thymomas are usually of low-grade malignancy, but occasionally metastasise to the lungs. Treatment is by surgical resection.

The survival analyses reported here relate only to malignant thymoma. They include almost 600 patients diagnosed from 1971 to 1990, about 85% of eligible adults: some 3% of eligible cases were excluded only because they arose in children under the age of 15 years (Table 24.1). Another 6% of those eligible were excluded because their vital status was unknown at 6 July 1997, and a further 9% because their survival was zero or unknown. The thymus has a single fourth-digit rubric in the ICD, and all the tumours analysed were coded to 194.2 in ICD-8 (1971-78) or to 164.0 in ICD-9 (1979-90). Three-quarters (77%) of the tumours were histologically coded as thymomas, 3% were squamous cell carcinomas, 2% germ cell neoplasms and the remainder were poorly specified carcinomas.

Survival trends

Survival from malignant thymoma for patients diagnosed in England and Wales during 1986-90 was 70% at one year and 46% at five years (Table 24.2). There has been substantial improvement in national survival rates since the early 1970s, but the regional pattern is very irregular (Figure 24.1). In Wales and in several regions, there were too few cases to enable survival to be estimated at all for men or for women in one or more five-year periods, and all the regional survival estimates are imprecise (see Chapter 3).

Survival for women has increased more than for men. The age-standardised trends show an average increase of 3-5% in successive five-year periods for women, compared to an average 1% gain every five years for men (to 66%), although neither rate of increase is statistically significant (see Table 4.4). The gains in five-year survival are modest, reaching 44% for men and 48% for women diagnosed during 1986-90.

Even though all the age-specific estimates of relative survival for men and women are imprecise, it is clear that five-year survival from thymoma falls sharply with age (Figure 24.5). Five-year survival exceeded 50% for those aged less than 50 at diagnosis during 1986-90, but it was 20% or less for those aged 70 or more (see Table 4.7).

Deprivation

The survival curves for patients diagnosed during 1986-90 show wide differences between deprivation categories, with the most affluent having the highest survival (Figure 24.2). These differences broadly echo the patterns seen for many other cancers, but they are imprecise because of the small numbers of patients.

The estimates of the deprivation gradient in survival for the periods 1981-85 and 1986-90 each depend on some 160 patients of known deprivation status who were included in the analyses (see Table 4.9). The gap in survival between the most affluent and the most deprived groups for patients diagnosed during the late 1980s was 10% at one year and 15% at five years after diagnosis, but neither deprivation gradient is statistically significant (Figure 24.3).

International comparison

We have not found other population-based survival estimates for malignant thymoma. Crude five-year survival of 52% has been reported among 53 patients diagnosed with thymoma in Geneva during 1966-90, of whom 14 also had myasthenia gravis[3]. In a hospital series of over 500 patients with myasthenia treated in Rome during the period 1971-1991, of whom 111 also had a thymoma, five-year survival of about 85% was reported[1]. The presence of myasthenia did not appear to alter the prognosis.

Crude survival for patients diagnosed in England and Wales during 1986-90 was 42% at five years (Table 24.2), although selection bias could be the main reason for the discrepancies.

1.   Palmisani MT, Evoli A, Batocchi AP, Provenzano C, Tonali P. Myasthenia gravis associated with thymoma: clinical characteristics and long-term outcome. Eur Neurol 1994; 34: 78-82

2.   Vessey MP, Doll R, Norman-Smith B, Hill ID. Thymectomy and cancer: a further report. Br J Cancer 1979; 39: 193-195

3.   Etienne T, Deleaval PJ, Spiliopoulos A, Megevand R. Thymoma: prognostic factors. Eur J Cardiothorac Surg 1993; 7: 449-452

Table 24.1 Ineligible and excluded records, and patients included in analyses, by calendar period of diagnosis

Calendar period of diagnosis

	1971-75			1976-80			1981-85			1986-90			Overall		
Total registered	**160**			**190**			**210**			**214**			**774**		
Ineligible	*Records*	*Patients*	*%*	*Records*	*Patients*	*%*	*Records*	*Patients*	*%*	*Records*	*Patients*	*%*	*Records*	*Patients*	*%*
Incomplete data[1]	0	0		3	3		7	7		4	4		14	**14**	
Resident outside England and Wales	4	4		4	4		2	1		3	2		13	**11**	
In situ neoplasm[2]	0	0		0	0		0	0		0	0		0	**0**	
Benign or uncertain behaviour[2]	23	21		9	8		0	0		0	0		32	**29**	
Metastatic[2]	1	1		0	0		0	0		0	0		1	**1**	
Otherwise ineligible[3]	-	-		-	-		-	-		-	-		-	**-**	
Lymphoma[4]	4	4		0	0		0	0		0	0		4	**4**	
Leukaemia or myeloma[4]	0	0		0	0		0	0		0	0		0	**0**	
		30			**15**			**8**			**6**			**59**	
Total eligible		**130**	**100**		**175**	**100**		**202**	**100**		**208**	**100**		**715**	**100**
Aged under 15 or 100 years or over at diagnosis	10	6	4.6	3	2	1.1	8	8	4.0	3	3	1.4	24	**19**	2.7
Vital status unknown[5]	7	4	3.1	17	14	8.0	14	13	6.4	15	13	6.3	53	**44**	6.2
Duplicate registration[6]	0	0	0.0	0	0	0.0	0	0	0.0	0	0	0.0	0	**0**	0.0
Multiple primary[7]	0	0	0.0	0	0	0.0	2	2	1.0	1	1	0.5	3	**3**	0.4
Synchronous tumours[8]	0	0	0.0	0	0	0.0	1	0	0.0	1	0	0.0	2	**0**	0.0
Sex not known	0	0	0.0	0	0	0.0	0	0	0.0	0	0	0.0	0	**0**	0.0
Sex incompatible with site code[9]	-	-	-	-	-	-	-	-	-	-	-	-	-	**-**	-
Invalid dates[10]	0	0	0.0	0	0	0.0	0	0	0.0	0	0	0.0	0	**0**	0.0
Zero survival[11]	17	10	7.7	22	18	10.3	19	14	6.9	24	24	11.5	82	**66**	9.2
Total exclusions		**20**	**15.4**		**34**	**19.4**		**37**	**18.3**		**41**	**19.7**		**132**	**18.5**
Patients included in analysis		**110**	**84.6**		**141**	**80.6**		**165**	**81.7**		**167**	**80.3**		**583**	**81.5**

[1]　Main data item(s) invalid or incompatible with one another: name, sex, date of birth, date of diagnosis and (if present) date of death, postcode, site, morphology and behaviour

[2]　Tumour either *in situ* (behaviour code 2), benign (0), uncertain if benign or malignant (1) or metastatic (6 or 9)

[3]　Not applicable for this malignancy

[4]　Morphology that of lymphoma (included with non-Hodgkin lymphoma, Ch. 41) or myeloma (included with multiple myeloma, Ch. 43), or leukaemia (excluded)

[5]　Tracing of vital status at National Health Service Central Register incomplete at 6 July 1997

[6]　Same site code, sex, cancer registry and cancer registry number as an earlier registration

[7]　Same site code and person number as an earlier registration(s): mostly confirmed multiple primary tumours at this site, some unresolved duplicate registrations

[8]　Same site code, sex, date of birth and date of diagnosis as another registration(s): mostly synchronous or (in paired organs) bilateral tumours in same anatomic site in one individual, not previously linked: also some duplicate registrations

[9]　Not applicable for this malignancy

[10]　Impossible sequence of dates (birth, diagnosis and, if present, emigration or death); or date of death after 6 July 1997

[11]　Date of diagnosis same as date of death: some are patients who did die on the day of diagnosis, but most were registered solely from a death certificate (death certificate only, or DCO), and their survival time is thus unknown

Table 24.2 Crude and relative survival (%) at one, five and ten years since diagnosis: NHS Region, deprivation category and calendar period of diagnosis

	1971-75						1976-80					
	Affluent	2	3	4	Deprived	All[1]	Affluent	2	3	4	Deprived	All[1]
	C R	C R	C R	C R	C R	C R	C R	C R	C R	C R	C R	C R
ENGLAND & WALES												
Men	(5)	(19)	(11)	(6)	(8)	(59)	(26)	(15)	(16)	(13)	(11)	(84)
One year	- -	41 42	72 74	- -	- -	54 55	63 64	73 74	69 71	70 71	52 53	67 69
Five years	- -	26 28	44 45	- -	- -	30 33	40 42	60 61	43 44	31 33	24 24	42 46
Ten years	- -	16 23	23 25	- -	- -	18 24	30 32	46 50	26 33	31 33	24 24	32 39
Women	(12)	(11)	(6)	(7)	(5)	(51)	(15)	(11)	(10)	(14)	(6)	(57)
One year	58 58	60 61	- -	- -	- -	60 60	58 59	81 81	45 46	86 86	- -	68 68
Five years	32 33	26 27	- -	- -	- -	33 35	26 27	63 64	21 22	57 58	- -	43 46
Ten years	25 25	18 19	- -	- -	- -	25 27	26 27	53 56	21 22	23 25	- -	30 34
Adults	(17)	(30)	(17)	(13)	(13)	(110)	(41)	(26)	(26)	(27)	(17)	(141)
One year	40 40	49 50	76 78	48 49	68 68	56 57	62 62	77 77	60 61	78 79	58 59	67 68
Five years	22 23	26 28	38 43	33 36	38 39	31 34	35 37	61 62	34 35	45 47	40 41	43 45
Ten years	17 18	17 20	26 31	24 26	38 39	21 26	28 31	49 53	23 29	26 30	29 31	31 36

	Men	Women	Adults				Men	Women	Adults
Northern & Yorkshire	(8)	(6)	(14)				(11)	(6)	(17)
One year	- -	- -	42 42				43 44	- -	58 60
Five years	- -	- -	27 28				34 35	- -	52 54
Ten years	- -	- -	14 26				25 28	- -	29 32
Trent	(5)	(3)	(8)				(9)	(4)	(13)
One year	- -	- -	- -				- -	- -	69 71
Five years	- -	- -	- -				- -	- -	62 65
Ten years	- -	- -	- -				- -	- -	54 59
Anglia & Oxford	(6)	(10)	(16)				(12)	(9)	(21)
One year	- -	68 69	60 62				74 76	- -	70 71
Five years	- -	39 41	36 39				39 41	- -	41 43
Ten years	- -	20 21	13 15				31 33	- -	32 35
North Thames	(9)	(10)	(19)				(16)	(7)	(23)
One year	- -	60 60	52 53				87 87	- -	77 78
Five years	- -	30 31	35 37				56 58	- -	48 49
Ten years	- -	18 19	23 26				44 55	- -	39 48
South Thames	(8)	(7)	(15)				(8)	(4)	(12)
One year	- -	- -	59 61				- -	- -	38 38
Five years	- -	- -	40 44				- -	- -	31 31
Ten years	- -	- -	40 44				- -	- -	16 20
South & West	(5)	(5)	(10)				(9)	(9)	(18)
One year	- -	- -	79 79				- -	- -	61 62
Five years	- -	- -	30 32				- -	- -	44 45
Ten years	- -	- -	30 32				- -	- -	33 36
West Midlands	(7)	(5)	(12)				(9)	(7)	(16)
One year	- -	- -	59 60				- -	- -	69 70
Five years	- -	- -	26 26				- -	- -	24 25
Ten years	- -	- -	26 26				- -	- -	6 7
North & West	(7)	(4)	(11)				(10)	(8)	(18)
One year	- -	- -	52 52				79 80	- -	77 77
Five years	- -	- -	33 35				38 40	- -	32 33
Ten years	- -	- -	11 17				28 38	- -	26 33
WALES	(4)	(1)	(5)				(0)	(3)	(3)
One year	- -	- -	- -				- -	- -	- -
Five years	- -	- -	- -				- -	- -	- -
Ten years	- -	- -	- -				- -	- -	- -

[1] Including those for whom no deprivation category was assigned. C - crude survival, R - relative survival (see Chapter 3). Number of patients contributing to each analysis in parentheses.

1981-85 / 1986-90 — ENGLAND & WALES

	Affluent C R	2 C R	3 C R	4 C R	Deprived C R	All[1] C R	Affluent C R	2 C R	3 C R	4 C R	Deprived C R	All[1] C R	
Men	*(15)*	*(18)*	*(20)*	*(22)*	*(10)*	*(87)*	*(32)*	*(9)*	*(22)*	*(20)*	*(8)*	*(92)*	
One year	68 69	70 71	64 66	77 78	80 82	72 73	71 73	51 52	68 70	79 80	40 41	67 69	*One year*
Five years	28 29	41 44	45 49	49 56	70 72	45 51	53 55	38 39	41 47	37 41	25 28	42 46	*Five years*
	12 13	31 36	16 18	18 26	60 72	24 31							
Women	*(18)*	*(16)*	*(17)*	*(14)*	*(13)*	*(78)*	*(20)*	*(15)*	*(18)*	*(11)*	*(7)*	*(75)*	
One year	77 80	74 76	62 63	58 60	56 57	67 69	85 87	72 74	63 64	64 66	- -	70 72	*One year*
Five years	72 74	33 34	28 29	36 37	35 39	42 45	54 57	39 41	34 38	46 48	- -	42 47	*Five years*
	50 55	33 34	11 16	29 30	15 20	28 33							
Adults	*(33)*	*(34)*	*(37)*	*(36)*	*(23)*	*(165)*	*(52)*	*(24)*	*(40)*	*(31)*	*(15)*	*(167)*	
One year	73 74	72 73	63 65	70 71	68 69	69 71	76 78	63 64	66 67	74 75	60 61	69 70	*One year*
Five years	51 53	38 42	37 40	44 50	51 58	44 48	53 56	37 39	38 43	39 45	33 39	42 46	*Five years*
	33 37	32 38	13 16	22 30	34 48	26 32							

	Men	Women	Adults	Men	Women	Adults	
Northern & Yorkshire	*(23)*	*(10)*	*(33)*	*(16)*	*(8)*	*(24)*	
One year	83 83	79 87	82 83	58 58	75 77	63 64	*One year*
Five years	57 61	49 55	54 59	38 39	37 40	38 40	*Five years*
	39 50	40 48	39 50				
Trent	*(4)*	*(9)*	*(13)*	*(5)*	*(12)*	*(17)*	
One year	- -	- -	65 67	- -	66 67	71 72	*One year*
Five years	- -	- -	43 48	- -	19 21	31 34	*Five years*
	- -	- -	29 35				
Anglia & Oxford	*(8)*	*(9)*	*(17)*	*(14)*	*(11)*	*(25)*	
One year	- -	- -	63 63	64 66	64 65	64 65	*One year*
Five years	- -	- -	24 25	57 62	45 46	52 57	*Five years*
	- -	- -	7 7				
North Thames	*(5)*	*(6)*	*(11)*	*(14)*	*(8)*	*(22)*	
One year	- -	- -	73 76	93 93	72 75	86 88	*One year*
Five years	- -	- -	54 56	50 53	59 63	54 60	*Five years*
	- -	- -	18 23				
South Thames	*(10)*	*(8)*	*(18)*	*(14)*	*(16)*	*(30)*	
One year	90 94	- -	89 96	86 89	75 77	80 82	*One year*
Five years	58 66	- -	60 70	58 65	56 59	57 62	*Five years*
	41 47	- -	50 59				
South & West	*(11)*	*(11)*	*(22)*	*(8)*	*(8)*	*(16)*	
One year	52 53	73 74	63 63	39 41	62 65	51 53	*One year*
Five years	34 35	47 48	41 42	25 27	25 26	25 26	*Five years*
	9 11	28 29	18 20				
West Midlands	*(11)*	*(6)*	*(17)*	*(8)*	*(5)*	*(13)*	
One year	45 46	- -	52 53	75 75	- -	85 85	*One year*
Five years	21 22	- -	24 25	39 39	- -	47 48	*Five years*
	8 13	- -	11 14				
North & West	*(6)*	*(14)*	*(20)*	*(11)*	*(2)*	*(13)*	
One year	- -	48 50	58 60	47 48	- -	56 57	*One year*
Five years	- -	42 44	54 56	15 16	- -	28 30	*Five years*
	- -	21 25	29 35				
WALES	*(9)*	*(5)*	*(14)*	*(2)*	*(5)*	*(7)*	
One year	- -	- -	70 71	- -	- -	- -	*One year*
Five years	- -	- -	27 32	- -	- -	- -	*Five years*
	- -	- -	15 18				

[1] Including those for whom no deprivation category was assigned. C - crude survival, R - relative survival (see Chapter 3). Number of patients contributing to each analysis in parentheses.

Fig. 24.1   Relative survival up to ten years: trends by calendar period of diagnosis (1971-90) and NHS Region

ADULTS

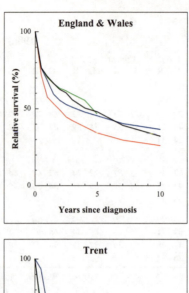

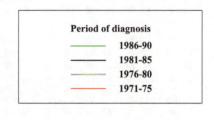

Period of diagnosis
- 1986-90
- 1981-85
- 1976-80
- 1971-75

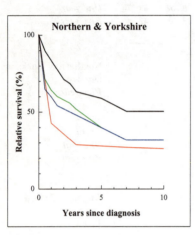

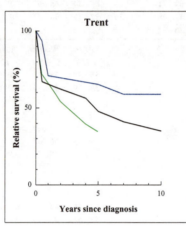

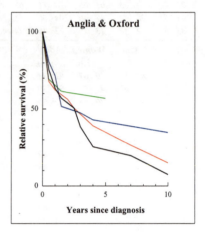

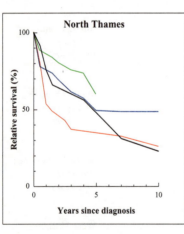

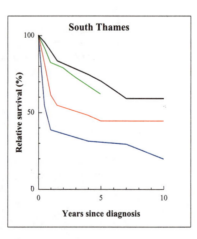

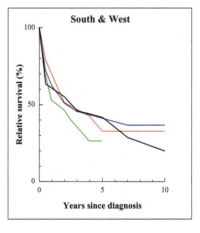

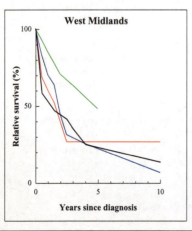

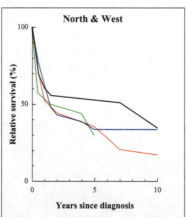

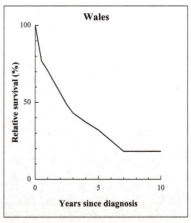

Fig. 24.2 Relative survival up to five years, by deprivation category: England and Wales, patients diagnosed 1986-90

ADULTS

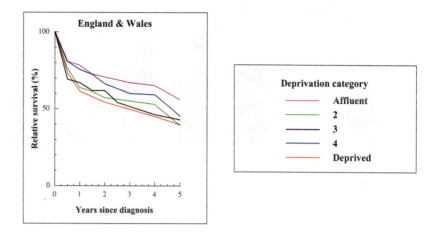

Fig. 24.3 Relative survival at five years, by deprivation category and period of diagnosis (1981-90): England and Wales

ADULTS

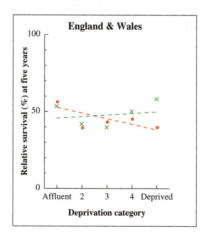
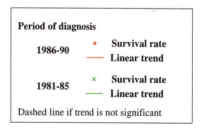

Fig. 24.4 Relative survival (%) by NHS Region and sex:
one-year and five-year survival for patients diagnosed 1986-90, with 95% confidence intervals

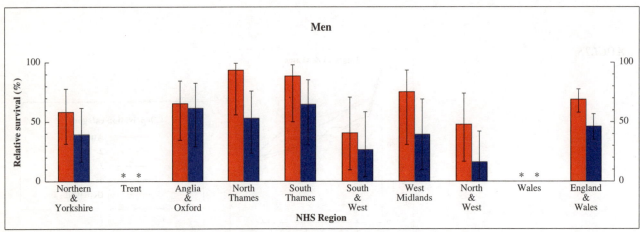

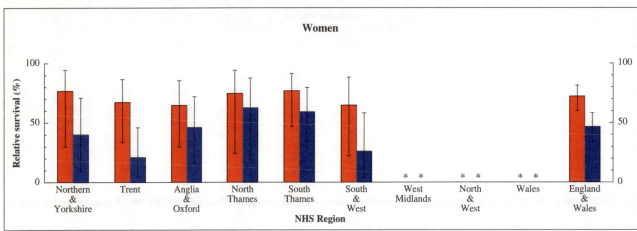

Fig. 24.5 Relative survival (%) at five years, with 95% confidence interval, by age at diagnosis and sex:
England and Wales, patients diagnosed 1986-90

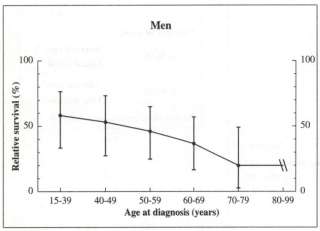 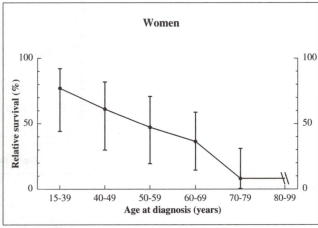

Chapter 25

BONE

Primary malignant neoplasms of bone and articular cartilage in adults are uncommon, with about 450 new cases each year in England and Wales. They are slightly more common in men than women, with a sex ratio of 1.4 to 1. Bone tumours in England and Wales have a bimodal age distribution, with a first peak in adolescence (10-14 years), principally due to osteosarcoma; incidence increases again after age 45. The incidence of bone cancer has remained at around 1 per 100,000 in men and 0.7 in women since the early 1970s.

International comparison of incidence and mortality is difficult for bone tumours, partly because of their rarity, but also because of variable distinction (at cancer registration) of primary bone tumours from tumours that are metastatic to bone. Recorded mortality from bone tumours in England and Wales and in several other countries fell by around 30% every five years in men and women up to 1985. By 1995, mortality rates in England and Wales were 0.4 per 100,000 in men and 0.2 in women. This substantial decline in national mortality rates in the face of relatively stable incidence is likely to be due in part to improved diagnosis during life (and a consequent decline in incorrect certification of the underlying cause of death as a primary bone tumour in patients with cancer metastatic to bone), but also to more effective treatment of bone tumours[1]. If national mortality trends are due to better treatment, this should be reflected in national survival trends.

The causes of primary malignancy in bone are poorly understood. Ionising radiation is the only known environmental agent, and no occupational exposure has been implicated. Paget's disease predisposes to osteosarcoma at older ages, more frequently in men. The adolescent peak in osteosarcoma has been related to the growth spurt in puberty. Osteosarcoma of the long bones and chondrosarcoma are more common in adults, while Ewing's sarcoma is more common in children and young adults. Surgery, radiotherapy and chemotherapy are all used to treat these tumours. While bone is an uncommon site for primary tumours, the bones are frequently the site of metastases from common cancers such as those of breast, prostate, and lung, and the morphologic basis of diagnosis is important in interpreting trends.

The survival analyses reported here are based on some 6,400 adults with a primary malignant neoplasm of bone diagnosed in England and Wales in the period 1971 to 1990 and followed up to the end of 1995. These patients represent about 74% of those eligible, but almost 15% of eligible cases arose in children under 15, and the patients included in the analyses represent about 89% of the eligible adults (Table 25.1). Survival from osteosarcoma and Ewing's sarcoma in children is presented in Chapters 58 and 59, respectively. About 5% of patients were excluded because their vital status was unknown at 6 July 1997, and a further 5% because their survival was zero or unknown: the proportion in this last category increased from 4% to 8% by 1986-90.

About a third (32%) of bone tumours arose in the long bones of the lower limb, almost a fifth (19%) in the pelvis and sacrum, and a tenth in the long bones of the upper limb (Table 25A). Another 10% arose in the bones of the skull, face and jaw. Three-quarters had been assigned a typical morphology for primary malignancy of bone or articular cartilage. There were 2,000 osteosarcomas, accounting for almost a third (31%) of the tumours: among these were 86 arising in Paget's disease. Chondrosarcomas comprised a fifth (22%) of tumours, and 7% were Ewing's sarcoma in adults (aged 15-99): this proportion rose from 4% in the early 1970s to 8-10% by the late 1980s. A further 7% were coded as unspecified sarcomas, and 5% as fibrosarcomas.

Survival trends

Survival rates for adults diagnosed with malignant tumours of bone in England and Wales during 1986-90 were about 70% at one year and 50% at five years, some 15% to 20% higher than for patients diagnosed during 1971-75 (Table 25.2). The survival curves show two main phases of improvement, both nationally and in most regions (Figure 25.1). Five-year survival improved substantially between the early and late 1970s, and survival at both one year and five years improved between the early and late 1980s. The largest improvements appeared to occur about five years later than elsewhere in Northern & Yorkshire and ten years later in Wales[a].

After adjustment of the results in each period to the age structure of all adults included in the analyses for 1986-90, there was an average gain of about 5% between successive five-year periods in the national one-year survival rate (see Table 4.4). The corresponding gain in five-year survival was around 6%. Survival is still higher for women than for men in most age groups, and an overall survival advantage of 2% to 6% for women has persisted since the early 1970s, after adjustment (see Chapter 3).

Survival from malignancies of bone falls with age (Figure 25.5). For bone tumours diagnosed during 1986-90 in England and Wales, five-year survival was about 60-70%

Table 25A Sub-site distribution for malignant neoplasms of bone

ICD-9 code	Site description	1971-75 No.	%	1976-80 No.	%	1981-85 No.	%	1986-90 No.	%	Total No.	%
170.0	Bones of skull and face	96	6.1	105	6.8	97	5.8	123	7.5	**421**	**6.5**
170.1	Lower jaw bone	60	3.8	55	3.6	72	4.3	57	3.5	**244**	**3.8**
170.2	Vertebral column, excluding sacrum and coccyx	130	8.2	128	8.3	167	9.9	131	8.0	**556**	**8.6**
170.3	Ribs, sternum and clavicle	102	6.5	106	6.9	140	8.3	118	7.2	**466**	**7.2**
170.4	Long bones of upper limb and scapula	201	12.7	181	11.8	187	11.1	139	8.5	**708**	**11.0**
170.5	Upper limb, short bones	25	1.6	32	2.1	43	2.6	40	2.5	**140**	**2.2**
170.6	Pelvic bones, sacrum and coccyx	325	20.6	314	20.4	285	16.9	269	16.5	**1,193**	**18.5**
170.7	Lower limb, long bones	538	34.1	504	32.7	543	32.2	474	29.1	**2,059**	**32.0**
170.8	Lower limb, short bones	50	3.2	45	2.9	39	2.3	39	2.4	**173**	**2.7**
170.9	Unspecified	51	3.2	69	4.5	113	6.7	240	14.7	**473**	**7.4**
	Total	**1,578**		**1,539**		**1,686**		**1,630**		**6,433**	

for patients aged under 50 at diagnosis, falling sharply to around 20% for patients aged 80-99 (see Table 4.7). Of all the bone tumours in adults, about 40% in men and 33% in women occurred under the age of 40. Bone tumours in young adults are clinically as well as morphologically different from those arising in elderly patients, and separate analyses will be necessary to show any differences in survival trends for these broad categories of bone malignancy.

Survival for both men and women diagnosed with bone tumours in North Thames during 1986-90 is higher than in England and Wales as a whole (Figure 25.4). This difference may be partly due to the exclusion of a higher than average proportion of cases registered solely from a death certificate (data not shown), but artefact is a most unlikely explanation for this regional difference. Survival from bone tumours in North Thames was also higher than the national average during the 1970s (Table 25.2, Figure 25.2), when the proportion of cases registered solely from a death certificate (and thus excluded from analysis) was similar to the national average, and the large increase in survival during the late 1980s in North Thames has occurred without any change in the proportion of cases excluded for this reason.

Deprivation

There were large differences in one-year and five-year survival between deprivation groups in the late 1970s, particularly for women; the differences are much smaller at ten years after diagnosis (Table 25.2). For patients diagnosed during 1981-85, there is still a substantial (9%) gap in one-year survival between the most affluent and the most deprived groups (see Table 4.9), but the difference in five-year survival is 3%, and is not statistically significant. For patients diagnosed during 1986-90, there is no difference in one-year survival between deprivation groups, and five-year survival is slightly higher in the most

deprived group than the affluent group, although none of the differences is significant (Figures 25.2 and 25.3). These patterns suggest an improvement in equity of survival.

International comparison

Five-year survival from bone tumours in adults diagnosed during 1986-90 in England and Wales was 50% in men and 53% in women, similar to the average survival for Europe (1985-89) and about 8% higher than in Scotland. Survival at five years in the USA (1986-90) was much higher, at 60% for men and 71% for women; these rates are similar to the highest regional survival rates in England and Wales (see Table 4.12).

a The survival curve for patients diagnosed during 1971-75 in Wales (Figure 25.2) stops at five years because of non-convergence (see Chapter 3), but the ten-year survival rate is clearly very close to 23% (Table 25.2).

1. Coleman MP, Estève J, Damiecki P, Arslan A, Renard H. *Trends in cancer incidence and mortality (IARC Scientific Publications No. 121)*. Lyon: International Agency for Research on Cancer, 1993

Table 25.1 Ineligible and excluded records, and patients included in analyses, by calendar period of diagnosis

Calendar period of diagnosis

	1971-75			1976-80			1981-85			1986-90			Overall		
Total registered	**2,301**			**2,237**			**2,577**			**2,623**			**9,738**		
Ineligible	Records	Patients	%	Records	Patients	%	Records	Patients	%	Records	Patients	%	Records	Patients	%
Incomplete data[1]	3	3		60	60		266	266		307	307		636	**636**	
Resident outside England and Wales	62	62		42	41		45	43		27	17		176	**163**	
In situ neoplasm[2]	3	2		1	1		0	0		0	0		4	**3**	
Benign or uncertain behaviour[2]	115	107		83	75		0	0		0	0		198	**182**	
Metastatic[2]	73	72		9	9		2	2		1	1		85	**84**	
Otherwise ineligible[3]	-	-		-	-		-	-		-	-		-	**-**	
Lymphoma[4]	17	16		8	4		8	0		1	0		34	**20**	
Leukaemia or myeloma[4]	7	7		2	0		9	1		4	0		22	**8**	
		269			190			312			325			1,096	
Total eligible		**2,032**	**100**		**2,047**	**100**		**2,265**	**100**		**2,298**	**100**		**8,642**	**100**
Aged under 15 or 100 years or over at diagnosis	342	308	15.2	334	323	15.8	359	328	14.5	314	296	12.9	1,349	**1,255**	14.5
Vital status unknown[5]	126	63	3.1	172	111	5.4	159	87	3.8	270	187	8.1	727	**448**	5.2
Duplicate registration[6]	0	0	0.0	0	0	0.0	0	0	0.0	0	0	0.0	0	**0**	0.0
Multiple primary[7]	0	0	0.0	1	1	<0.1	16	12	0.5	7	3	0.1	24	**16**	0.2
Synchronous tumours[8]	5	4	0.2	7	3	0.1	26	15	0.7	8	6	0.3	46	**28**	0.3
Sex not known	0	0	0.0	0	0	0.0	0	0	0.0	0	0	0.0	0	**0**	0.0
Sex incompatible with site code[9]	-	-	-	-	-	-	-	-	-	-	-	-	-	**-**	-
Invalid dates[10]	0	0	0.0	0	0	0.0	1	0	0.0	1	0	0.0	2	**0**	0.0
Zero survival[11]	97	79	3.9	91	70	3.4	168	137	6.0	208	176	7.7	564	**462**	5.3
Total exclusions		**454**	**22.3**		**508**	**24.8**		**579**	**25.6**		**668**	**29.1**		**2,209**	**25.6**
Patients included in analysis		**1,578**	**77.7**		**1,539**	**75.2**		**1,686**	**74.4**		**1,630**	**70.9**		**6,433**	**74.4**

[1] Main data item(s) invalid or incompatible with one another: name, sex, date of birth, date of diagnosis and (if present) date of death, postcode, site, morphology and behaviour

[2] Tumour either *in situ* (behaviour code 2), benign (0), uncertain if benign or malignant (1) or metastatic (6 or 9)

[3] Not applicable for this malignancy

[4] Morphology that of lymphoma (included with non-Hodgkin lymphoma, Ch. 41) or myeloma (included with multiple myeloma, Ch. 43), or leukaemia (excluded)

[5] Tracing of vital status at National Health Service Central Register incomplete at 6 July 1997

[6] Same site code, sex, cancer registry and cancer registry number as an earlier registration

[7] Same site code and person number as an earlier registration(s): mostly confirmed multiple primary tumours at this site, some unresolved duplicate registrations

[8] Same site code, sex, date of birth and date of diagnosis as another registration(s): mostly synchronous or (in paired organs) bilateral tumours in same anatomic site in one individual, not previously linked: also some duplicate registrations

[9] Not applicable for this malignancy

[10] Impossible sequence of dates (birth, diagnosis and, if present, emigration or death); or date of death after 6 July 1997

[11] Date of diagnosis same as date of death: some are patients who did die on the day of diagnosis, but most were registered solely from a death certificate (death certificate only, or DCO), and their survival time is thus unknown

Table 25.2 Crude and relative survival (%) at one, five and ten years since diagnosis: NHS Region, deprivation category and calendar period of diagnosis

	1971-75						1976-80					
	Affluent	*2*	*3*	*4*	*Deprived*	*All[1]*	*Affluent*	*2*	*3*	*4*	*Deprived*	*All[1]*
	C R	C R	C R	C R	C R	C R	C R	C R	C R	C R	C R	C R
ENGLAND & WALES												
Men	*(113)*	*(155)*	*(168)*	*(138)*	*(155)*	*(884)*	*(188)*	*(171)*	*(158)*	*(166)*	*(154)*	*(867)*
One year	51 53	48 51	51 53	50 53	55 58	52 54	59 61	60 62	44 47	53 56	50 52	54 56
Five years	25 28	25 28	23 26	24 29	29 33	24 28	32 37	32 35	25 28	34 37	33 37	31 35
Ten years	20 27	20 25	20 25	19 26	24 30	19 25	23 30	26 32	21 26	25 31	30 37	26 32
Women	*(102)*	*(126)*	*(109)*	*(121)*	*(104)*	*(694)*	*(133)*	*(129)*	*(134)*	*(136)*	*(111)*	*(672)*
One year	61 64	49 51	52 54	48 50	47 49	52 54	64 67	64 66	55 57	53 55	52 54	57 59
Five years	31 35	23 25	30 34	19 22	26 29	27 30	37 43	41 46	34 38	28 31	31 34	34 38
Ten years	28 34	18 21	21 27	14 20	18 23	22 26	33 40	33 43	27 34	19 26	26 34	27 35
Adults	*(215)*	*(281)*	*(277)*	*(259)*	*(259)*	*(1,578)*	*(321)*	*(300)*	*(292)*	*(302)*	*(265)*	*(1,539)*
One year	56 58	49 51	51 53	49 51	52 54	52 54	61 63	61 64	49 52	53 55	51 53	55 58
Five years	28 31	24 27	26 29	22 26	28 32	25 29	34 39	36 40	29 33	31 34	32 36	32 36
Ten years	24 30	19 23	20 26	17 23	22 27	20 26	27 34	29 37	24 30	23 29	29 35	26 33

	Men	Women	Adults				Men	Women	Adults
Northern & Yorkshire	*(119)*	*(67)*	*(186)*				*(106)*	*(66)*	*(172)*
One year	47 50	62 63	53 55				51 53	46 49	49 52
Five years	16 19	32 35	22 25				24 27	25 28	24 28
Ten years	15 19	25 31	18 24				- -	18 24	20 26
Trent	*(93)*	*(65)*	*(158)*				*(95)*	*(65)*	*(160)*
One year	49 52	52 53	50 52				57 60	57 59	57 60
Five years	21 24	28 30	23 27				38 43	31 36	35 40
Ten years	16 22	24 30	20 26				28 36	28 34	28 35
Anglia & Oxford	*(62)*	*(53)*	*(115)*				*(77)*	*(55)*	*(132)*
One year	59 61	56 58	58 59				54 55	67 69	59 61
Five years	33 37	26 29	30 33				36 39	49 55	41 46
Ten years	27 34	24 26	26 31				28 34	36 46	31 40
North Thames	*(115)*	*(98)*	*(213)*				*(84)*	*(73)*	*(157)*
One year	59 61	52 54	55 58				59 62	62 64	60 63
Five years	33 38	31 34	32 36				37 39	38 44	37 42
Ten years	28 36	24 29	26 33				27 34	35 43	31 39
South Thames	*(112)*	*(94)*	*(206)*				*(95)*	*(89)*	*(184)*
One year	51 53	48 50	50 52				63 66	57 60	60 63
Five years	21 24	26 31	23 27				39 46	32 37	36 41
Ten years	17 22	19 25	18 23				33 44	22 31	28 37
South & West	*(94)*	*(99)*	*(193)*				*(116)*	*(93)*	*(209)*
One year	52 55	55 58	54 57				45 47	56 59	50 53
Five years	23 26	26 30	25 29				26 29	36 40	30 35
Ten years	20 26	21 27	20 26				20 25	28 38	24 31
West Midlands	*(76)*	*(66)*	*(142)*				*(83)*	*(78)*	*(161)*
One year	58 60	53 54	55 57				64 66	66 68	65 67
Five years	33 37	27 30	30 33				37 40	37 41	37 41
Ten years	25 34	19 22	22 28				32 36	32 37	32 37
North & West	*(157)*	*(109)*	*(266)*				*(131)*	*(92)*	*(223)*
One year	48 51	44 46	47 49				45 47	56 58	49 52
Five years	21 25	26 30	23 27				24 28	31 34	27 30
Ten years	14 19	19 24	16 21				21 25	25 33	23 28
WALES	*(56)*	*(43)*	*(99)*				*(80)*	*(61)*	*(141)*
One year	43 45	51 54	46 49				54 57	47 49	51 53
Five years	22 24	20 22	21 23				30 35	24 26	28 31
Ten years	19 24	18 22	- -				23 34	21 26	22 31

[1] Including those for whom no deprivation category was assigned. C - crude survival, R - relative survival (see Chapter 3). Number of patients contributing to each analysis in parentheses.

ENGLAND & WALES

	1981-85						1986-90						
	Affluent C R	*2* C R	*3* C R	*4* C R	*Deprived* C R	*All*[1] C R	*Affluent* C R	*2* C R	*3* C R	*4* C R	*Deprived* C R	*All*[1] C R	
Men													
	(196)	*(168)*	*(194)*	*(187)*	*(171)*	*(945)*	*(189)*	*(175)*	*(195)*	*(199)*	*(144)*	*(918)*	
One year	67 69	56 58	66 69	62 65	58 61	62 65	74 76	68 71	66 69	70 72	75 76	71 73	
Five years	39 43	34 38	41 46	33 38	36 41	37 41	46 50	46 51	41 45	44 51	49 53	45 50	
	32 39	28 34	32 41	24 30	27 36	29 36							
Women													
	(151)	*(144)*	*(145)*	*(145)*	*(125)*	*(741)*	*(136)*	*(160)*	*(153)*	*(154)*	*(98)*	*(712)*	
One year	68 71	65 67	56 58	58 61	51 53	60 62	68 70	69 72	68 71	65 68	68 71	67 70	
Five years	36 40	37 41	31 34	37 41	29 33	34 38	44 48	48 54	47 54	47 52	53 57	47 53	
	26 34	29 35	21 26	30 37	24 31	26 33							
Adults													
	(347)	*(312)*	*(339)*	*(332)*	*(296)*	*(1,686)*	*(325)*	*(335)*	*(348)*	*(353)*	*(242)*	*(1,630)*	
One year	67 70	60 62	62 64	60 63	55 58	61 64	71 74	69 71	67 70	68 70	72 74	69 72	
Five years	38 41	35 39	37 41	35 39	33 38	36 40	45 49	47 52	43 49	45 52	51 55	46 51	
	29 37	28 35	27 34	27 33	26 34	27 34							

	1981-85			1986-90			
	Men C R	*Women* C R	*Adults* C R	*Men* C R	*Women* C R	*Adults* C R	
Northern & Yorkshire							
	(108)	*(83)*	*(191)*	*(103)*	*(68)*	*(171)*	
One year	65 68	64 67	64 67	69 71	68 71	69 71	
Five years	43 45	39 45	41 45	43 46	52 56	46 50	
	35 40	30 38	33 40				
Trent							
	(87)	*(86)*	*(173)*	*(86)*	*(79)*	*(165)*	
One year	64 69	56 57	60 63	73 75	60 63	67 69	
Five years	42 48	27 29	35 39	48 52	43 46	45 49	
	30 39	25 29	27 34				
Anglia & Oxford							
	(101)	*(66)*	*(167)*	*(98)*	*(49)*	*(147)*	
One year	67 69	*66 69*	66 69	68 70	*67 69*	68 70	
Five years	41 45	*39 46*	40 45	41 44	*43 47*	41 45	
	33 41	*30 37*	32 39				
North Thames							
	(91)	*(57)*	*(148)*	*(135)*	*(76)*	*(211)*	
One year	63 65	*70 74*	65 68	84 87	75 77	81 83	
Five years	39 42	*38 43*	39 42	55 61	*64 71*	59 65	
	31 38	*31 40*	31 40				
South Thames							
	(105)	*(88)*	*(193)*	*(131)*	*(111)*	*(242)*	
One year	60 62	68 71	63 66	73 76	73 76	73 76	
Five years	36 40	41 47	38 43	43 50	50 55	46 53	
	26 34	27 36	26 35				
South & West							
	(98)	*(83)*	*(181)*	*(124)*	*(98)*	*(222)*	
One year	71 73	*68 73*	70 73	64 66	73 75	68 70	
Five years	41 45	35 40	38 43	38 43	44 51	41 47	
	35 42	24 32	30 38				
West Midlands							
	(93)	*(71)*	*(164)*	*(71)*	*(65)*	*(136)*	
One year	74 77	*65 68*	70 73	*73 76*	*77 81*	75 78	
Five years	39 42	*42 47*	40 44	*53 56*	*53 61*	53 58	
	27 34	*34 44*	30 38				
North & West							
	(132)	*(84)*	*(216)*	*(85)*	*(78)*	*(163)*	
One year	59 62	54 56	57 59	63 66	53 57	58 61	
Five years	36 40	30 33	34 37	39 43	29 36	34 40	
	27 34	26 31	27 33				
WALES							
	(130)	*(123)*	*(253)*	*(85)*	*(88)*	*(173)*	
One year	46 48	40 41	43 45	63 66	59 62	61 64	
Five years	21 26	23 25	22 26	47 50	44 48	45 50	
	18 24	15 18	16 21				

[1] Including those for whom no deprivation category was assigned. C - crude survival, R - relative survival (see Chapter 3). Number of patients contributing to each analysis in parentheses.

Fig. 25.1   Relative survival up to ten years: trends by calendar period of diagnosis (1971-90) and NHS Region

ADULTS

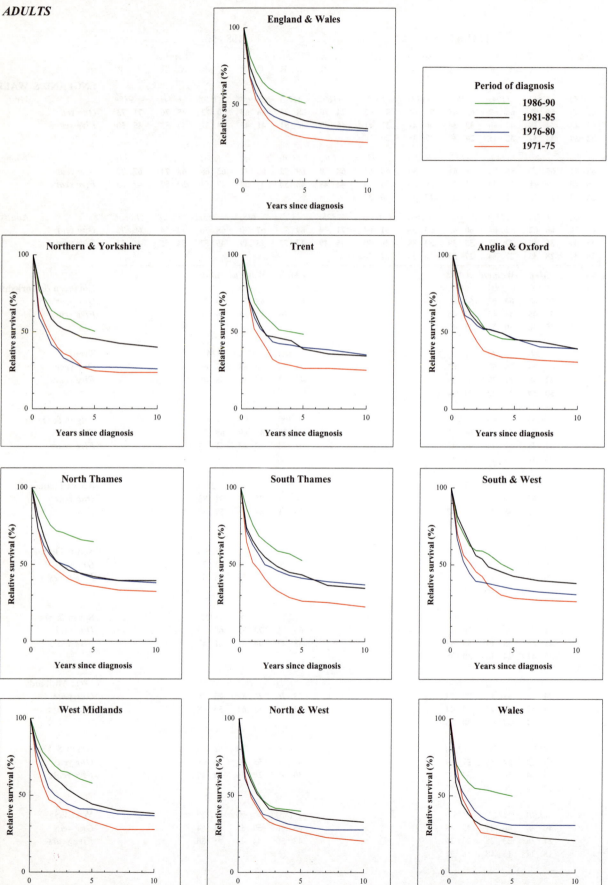

Fig. 25.2 Relative survival up to five years, by deprivation category: England and Wales, patients diagnosed 1986-90

ADULTS

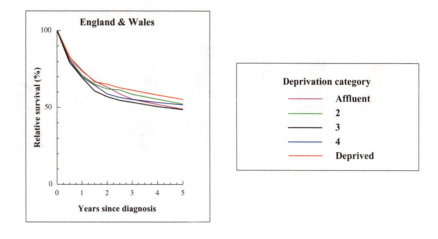

Fig. 25.3 Relative survival at five years, by deprivation category and period of diagnosis (1981-90): England and Wales

ADULTS

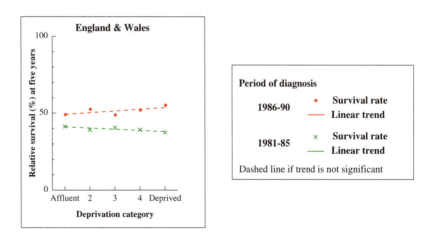

Fig. 25.4 Relative survival (%) by NHS Region and sex:
one-year and five-year survival for patients diagnosed 1986-90, with 95% confidence intervals

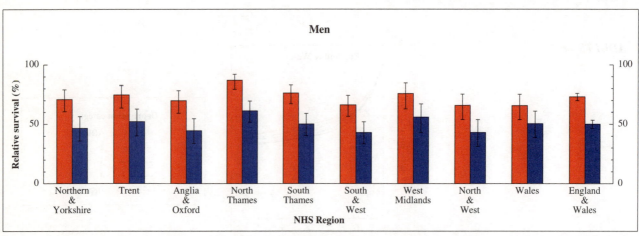

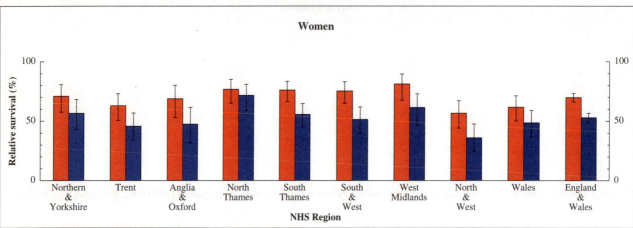

Fig. 25.5 Relative survival (%) at five years, with 95% confidence interval, by age at diagnosis and sex:
England and Wales, patients diagnosed 1986-90

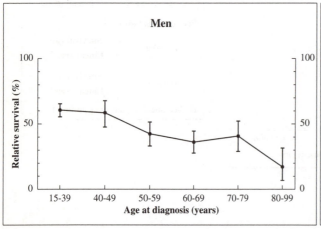

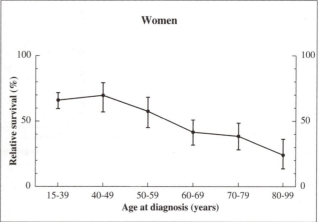

318

Chapter 26

CONNECTIVE TISSUE

Sarcomas of connective and other soft tissues are uncommon. About 1,100 new patients are registered with one of these malignancies in England and Wales each year. Sarcomas arise from cells in blood and lymphatic vessel walls, striated and smooth muscle, cartilage, and fibrous, fat and other connective tissues: as such they can occur in almost any part of the body. They differ considerably in age distribution and prognosis, as well as in location and morphologic type. The International Classification of Diseases has largely been based on the anatomic location of tumours, and – in contrast with childhood tumours – tumour registries have not generally published information on soft tissue tumours classified by their morphology. Incidence is probably under-estimated because sarcomas may be assigned to the organ or part of the body in which they arise, rather than to connective tissue, although sarcomas with an unspecified site are included in this rubric. Tumours of connective and soft tissue are therefore more heterogeneous than most, and the survival results presented here may be less clinically relevant, although they should be comparable to other population-based data.

Soft tissue sarcomas may be caused by medical treatment with ionising radiation for malignant tumours and other conditions, and by the radiological contrast medium Thorotrast (thorium dioxide), now disused. Occupational exposure to phenoxy herbicides and dioxin may also increase risk, and vinyl chloride can cause angiosarcoma of liver. Epidemic Kaposi sarcoma is linked to the human immunodeficiency virus and forms part of the clinical definition of the acquired immune deficiency syndrome[1].

The incidence of connective tissue neoplasms for men in England and Wales increased steadily after 1971 to reach 2.3 per 100,000 in 1990, but incidence in women has remained around 1.7. Incidence is slightly higher in men than women: the sex ratio in 1990 was 1.3 to 1. Recorded mortality has nearly doubled since 1971, and in 1995 the age-standardised rates were about 1 per 100,000 in both sexes (600 deaths).

The survival analyses reported here are based on 14,500 patients registered with a malignant neoplasm of connective tissue in England and Wales during the period 1971-90 and followed up to the end of 1995. They represent about 85% of all eligible patients, although 6% were excluded solely because they arose in children under 15, and the analyses include over 90% of eligible adults (Table 26.1). A further 5% were excluded because their vital status was unknown at 6 July 1997, and 4% because their survival was zero or unknown.

Just over one-third (37%) of the tumours arose in the lower limb, and one-seventh (14%) in the upper limb (Table 26A). Almost a third again occurred in the soft tissues of the chest, abdomen and pelvis. The number of sarcomas of unspecified site increased markedly, accounting for more than one in six soft tissue tumours during 1986-90. This suggests more imprecise anatomic specification of sarcomas that would previously have been assigned to other organs, and it may account for part of the recorded increase in incidence.

Almost a third (30%) of the tumours were fibrosarcomas. Myosarcomas accounted for one-sixth (17%), most of them (11%) leiomyosarcomas, and liposarcomas accounted for another 16%. Haemangiosarcomas represented 5%, including Kaposi sarcoma (1%), which was more commonly recorded in the late 1980s. About 14% of the tumours were coded to sarcoma without further detail. It is worth noting that although the records for this analysis were not selected on the basis of morphology, no records were coded to nephroblastoma, hepatoblastoma or germ cell neoplasm, and only 7% had no morphology or had been assigned to non-specific codes for epithelial tumours. This gives some confidence in the quality of the data for tumours assigned to this anatomic rubric, even though some sarcomas are assigned to other anatomic sites.

Survival trends

Relative survival from connective tissue tumours for adults diagnosed in England and Wales during 1986-90 was 73% at one year and 51% at five years (Table 26.2). Survival at ten years is often only about 3% to 5% lower than survival at five years. Survival in England and Wales as a whole has improved steadily in successive five-year periods (Figure 26.1), but the improvements are variable between regions, and there has been little change during the 1980s.

Survival trends since the early 1970s show a national average increase of 2% in one-year survival between successive five-year periods, after adjustment for differences in the age distribution of patients (see Table 4.4). These steady gains are similar in men and women. Five-year survival has increased at an average of 2-3% every five years. For men the largest increase in five-year survival occurred in the late 1970s, and for women in the early 1980s.

The range of regional five-year survival rates for patients diagnosed during 1986-90 is slightly wider for women

Table 26A Sub-site distribution for malignant neoplasms of connective and other soft tissue

ICD-9 code	Site description	1971-75 No.	%	1976-80 No.	%	1981-85 No.	%	1986-90 No.	%	Total No.	%
171.0	Head, face and neck	250	7.5	290	8.4	304	8.4	295	7.1	**1,139**	**7.8**
171.1	Trunk (1971-78 only)[1]	1,263	37.7	585	17.0	-	-	-	-	**1,848**	**12.7**
171.2	Upper limb, including shoulder	450	13.4	507	14.8	515	14.3	550	13.3	**2,022**	**13.9**
171.3	Lower limb, including hip	1,277	38.1	1,335	38.9	1,355	37.5	1,333	32.3	**5,300**	**36.5**
171.4	Thorax	-	-	115	3.3	274	7.6	251	6.1	**640**	**4.4**
171.5	Abdomen	-	-	94	2.7	295	8.2	307	7.4	**696**	**4.8**
171.6	Pelvis	-	-	137	4.0	410	11.4	397	9.6	**944**	**6.5**
171.7	Trunk, unspecified	-	-	55	1.6	142	3.9	175	4.2	**372**	**2.6**
171.8	Other	-	-	9	0.3	46	1.3	97	2.4	**152**	**1.0**
171.9	Site unspecified[2]	109	3.3	307	8.9	270	7.5	721	17.5	**1,407**	**9.7**
	Total	**3,349**		**3,434**		**3,611**		**4,126**		**14,520**	

[1] Tumours of connective and other soft tissues of the trunk were coded to 171.1 in ICD-8 (1971-78) and to 171.4-171.8 in ICD-9 (1979-90)
[2] Includes most types of sarcoma when the anatomic site is unspecified

(39% to 65%) than for men (37% to 57%), although survival is lowest in the North & West for both sexes (Figure 26.4). Women in North Thames had higher survival at one year (87%) and five years (65%) than the average for England and Wales (Table 26.2), but survival for men is similar to that in South Thames and Anglia & Oxford. One-year survival for men in Wales was 58%, much lower than the national average. Regional survival rates for men and women in each period are based on 200-300 patients, and are statistically precise (see Chapter 3).

Survival falls with age, but the decline is notably steeper for women than for men (Figure 26.5). For men aged 50 or more at diagnosis, there is very little difference in relative survival with age: it remains at 45% to 50%. For women, there is a progressive decline in relative survival with age, from 65% for those under 40 years of age at diagnosis to 33% for those aged 80-99 (see Table 4.7). Each age-specific rate is based on 200-500 patients. These striking differences between men and women in the patterns of survival by age could be partly due to variation between the sexes in the range of tumour types in each age group.

Deprivation

Survival patterns by deprivation are variable, but there is generally a difference in favour of the more affluent group (Table 26.2). Regional variation in the survival patterns by deprivation for patients diagnosed during 1986-90 can be seen in Figure 26.2, although the pattern is more consistent nationally (Figure 26.3).

When there is a large deprivation gradient in survival, regional differences in survival could arise simply from a higher proportion of patients living in deprived areas in that region. In fact, 24% of patients diagnosed with connective tissue tumours in North & West Region during 1986-90 were in the deprived group (compared with 28% for all tumours combined: see Table 4.3), and five-year

survival in each deprivation category in North & West was 10% or more below the national average for that category (Figure 26.3). Survival at five years in this region was about 10% lower in each deprivation category for patients who were diagnosed during 1986-90 than for those diagnosed during 1981-85, even though survival rates at one year are broadly similar in the two periods (Table 26.2, see also Figure 26.3). This odd pattern may relate to changes in treatment policy: it appears difficult to explain otherwise.

The national gap in survival at one and five years between the most affluent and most deprived groups was 9% for patients diagnosed in 1981-85 (see Table 4.9). For those diagnosed during 1986-90, the gap in one-year survival was similar, but the gap in five-year survival was 5% (Figure 26.3).

International comparison

Five-year survival for patients diagnosed in England and Wales during 1986-90 (about 50%) was similar to that in Scotland, but 4-10% below the average for patients diagnosed about the same time in Europe and the USA (see Table 4.12). The highest regional survival rates in England and Wales are comparable with those in the USA.

1.   Zahm SH, Tucker MA, Fraumeni JF. Soft tissue sarcomas. In: Schottenfeld D, Fraumeni JF, (eds.) *Cancer epidemiology and prevention.* 2nd edn. Oxford: Oxford University Press, 1996, pp984-999

Table 26.1 Ineligible and excluded records, and patients included in analyses, by calendar period of diagnosis

	Calendar period of diagnosis																	
	1971-75			**1976-80**			**1981-85**			**1986-90**			**Overall**		
Total registered	**4,059**			**4,241**			**4,568**			**5,305**			**18,173**		
Ineligible	*Records*	*Patients*	*%*	*Records*	*Patients*	*%*	*Records*	*Patients*	*%*	*Records*	*Patients*	*%*	*Records*	*Patients*	*%*
Incomplete data[1]	11	11		86	86		212	212		221	221		530	**530**	
Resident outside England and Wales	75	74		46	46		41	38		27	15		189	**173**	
In situ neoplasm[2]	7	7		3	3		0	0		0	0		10	**10**	
Benign or uncertain behaviour[2]	121	103		97	88		4	3		1	0		223	**194**	
Metastatic[2]	17	17		8	7		0	0		0	0		25	**24**	
Otherwise ineligible[3]	-	-		-	-		-	-		-	-		-	**-**	
Lymphoma[4]	69	68		46	41		7	0		3	0		125	**109**	
Leukaemia or myeloma[4]	0	0		1	0		4	0		1	0		6	**0**	
		280			**271**			**253**			**236**			**1,040**	
Total eligible		**3,779**	**100**		**3,970**	**100**		**4,315**	**100**		**5,069**	**100**		**17,133**	**100**
Aged under 15 or 100 years or over at diagnosis	234	201	5.3	245	220	5.5	291	273	6.3	315	309	6.1	1,085	**1,003**	5.9
Vital status unknown[5]	194	118	3.1	252	203	5.1	259	199	4.6	419	345	6.8	1,124	**865**	5.0
Duplicate registration[6]	0	0	0.0	0	0	0.0	0	0	0.0	0	0	0.0	0	**0**	0.0
Multiple primary[7]	0	0	0.0	6	3	<0.1	15	13	0.3	21	19	0.4	42	**35**	0.2
Synchronous tumours[8]	17	9	0.2	21	2	<0.1	19	13	0.3	24	10	0.2	81	**34**	0.2
Sex not known	0	0	0.0	0	0	0.0	0	0	0.0	0	0	0.0	0	**0**	0.0
Sex incompatible with site code[9]	-	-	-	-	-	-	-	-	-	-	-	-	-	**-**	-
Invalid dates[10]	1	0	0.0	2	2	<0.1	3	2	<0.1	5	1	<0.1	11	**5**	<0.1
Zero survival[11]	118	102	2.7	132	106	2.7	241	204	4.7	298	259	5.1	789	**671**	3.9
Total exclusions		**430**	**11.4**		**536**	**13.5**		**704**	**16.3**		**943**	**18.6**		**2,613**	**15.3**
Patients included in analysis		**3,349**	**88.6**		**3,434**	**86.5**		**3,611**	**83.7**		**4,126**	**81.4**		**14,520**	**84.7**

[1] Main data item(s) invalid or incompatible with one another: name, sex, date of birth, date of diagnosis and (if present) date of death, postcode, site, morphology and behaviour

[2] Tumour either *in situ* (behaviour code 2), benign (0), uncertain if benign or malignant (1) or metastatic (6 or 9)

[3] Not applicable for this malignancy

[4] Morphology that of lymphoma (included with non-Hodgkin lymphoma, Ch. 41) or myeloma (included with multiple myeloma, Ch. 43), or leukaemia (excluded)

[5] Tracing of vital status at National Health Service Central Register incomplete at 6 July 1997

[6] Same site code, sex, cancer registry and cancer registry number as an earlier registration

[7] Same site code and person number as an earlier registration(s): mostly confirmed multiple primary tumours at this site, some unresolved duplicate registrations

[8] Same site code, sex, date of birth and date of diagnosis as another registration(s): mostly synchronous or (in paired organs) bilateral tumours in same anatomic site in one individual, not previously linked: also some duplicate registrations

[9] Not applicable for this malignancy

[10] Impossible sequence of dates (birth, diagnosis and, if present, emigration or death); or date of death after 6 July 1997

[11] Date of diagnosis same as date of death: some are patients who did die on the day of diagnosis, but most were registered solely from a death certificate (death certificate only, or DCO), and their survival time is thus unknown

Table 26.2   Crude and relative survival (%) at one, five and ten years since diagnosis: NHS Region, deprivation category and calendar period of diagnosis

MEN

	1971-75						1976-80					
	Affluent	*2*	*3*	*4*	*Deprived*	*All[1]*	*Affluent*	*2*	*3*	*4*	*Deprived*	*All[1]*
	C R	C R	C R	C R	C R	C R	C R	C R	C R	C R	C R	C R
ENGLAND & WALES	*(254)*	*(291)*	*(238)*	*(261)*	*(217)*	*(1,592)*	*(358)*	*(365)*	*(361)*	*(357)*	*(299)*	*(1,815)*
One year	70 72	68 70	58 60	60 62	63 66	65 67	70 73	65 68	67 70	65 68	63 67	66 69
Five years	40 47	37 44	32 38	34 40	31 38	36 43	45 53	42 49	41 49	39 47	36 44	40 48
Ten years	30 45	30 40	22 31	24 36	23 36	27 38	35 48	32 44	32 45	31 45	29 42	31 45

	1981-85						1986-90					
	(404)	*(386)*	*(372)*	*(360)*	*(318)*	*(1,881)*	*(474)*	*(455)*	*(445)*	*(427)*	*(350)*	*(2,178)*
One year	74 77	67 70	68 71	69 72	59 62	68 71	77 80	70 73	71 74	65 68	67 70	70 73
Five years	49 57	39 46	39 48	41 50	35 42	41 49	45 54	40 47	42 50	42 49	42 50	42 50
Ten years	38 52	31 43	29 42	32 46	26 39	31 45						

	1971-75	1976-80	1981-85	1986-90
Northern & Yorkshire	*(221)*	*(240)*	*(301)*	*(327)*
One year	62 65	61 65	66 69	71 74
Five years	33 41	35 44	36 44	43 50
Ten years	24 35	28 43	28 42	
Trent	*(164)*	*(189)*	*(167)*	*(228)*
One year	66 68	68 71	64 67	68 72
Five years	39 45	42 50	42 52	41 48
Ten years	28 41	30 43	37 50	
Anglia & Oxford	*(158)*	*(178)*	*(200)*	*(233)*
One year	64 67	69 72	69 72	71 75
Five years	37 43	42 51	47 56	46 53
Ten years	29 38	34 47	35 48	
North Thames	*(197)*	*(175)*	*(171)*	*(261)*
One year	65 66	66 69	74 78	76 79
Five years	36 43	44 53	47 58	48 55
Ten years	30 42	33 47	35 49	
South Thames	*(199)*	*(194)*	*(178)*	*(223)*
One year	66 69	67 70	71 75	77 79
Five years	32 38	38 46	43 52	49 57
Ten years	23 33	29 39	32 45	
South & West	*(203)*	*(269)*	*(295)*	*(309)*
One year	71 73	71 74	70 72	70 74
Five years	37 46	50 60	44 50	43 51
Ten years	27 42	42 59	32 45	
West Midlands	*(162)*	*(180)*	*(187)*	*(241)*
One year	65 68	66 68	72 75	73 76
Five years	36 45	38 45	44 53	41 51
Ten years	27 38	28 41	34 51	
North & West	*(224)*	*(251)*	*(237)*	*(229)*
One year	59 62	63 66	63 66	64 67
Five years	35 43	36 42	35 42	31 37
Ten years	25 39	29 41	26 39	
WALES	*(64)*	*(139)*	*(145)*	*(127)*
One year	62 66	65 67	59 62	55 58
Five years	39 46	38 45	34 41	36 44
Ten years	31 42	26 36	26 40	

[1] Including those for whom no deprivation category was assigned. C - crude survival, R - relative survival (see Chapter 3). Number of patients contributing to each analysis in parentheses.

Table 26.2   Crude and relative survival (%) at one, five and ten years since diagnosis: NHS Region, deprivation category and calendar period of diagnosis

WOMEN

	1971-75												1976-80											
	Affluent		2		3		4		Deprived		All[1]		Affluent		2		3		4		Deprived		All[1]	
	C	R	C	R	C	R	C	R	C	R	C	R	C	R	C	R	C	R	C	R	C	R	C	R
ENGLAND & WALES	(266)		(312)		(305)		(305)		(234)		(1,757)		(301)		(321)		(312)		(341)		(276)		(1,619)	
One year	64	66	67	69	65	67	60	63	64	66	65	67	68	70	66	69	64	67	65	67	61	63	65	68
Five years	39	44	42	50	41	46	39	45	40	47	40	46	40	45	39	44	37	41	39	46	39	44	39	44
Ten years	29	38	35	46	31	41	31	41	30	43	31	41	32	40	29	38	28	38	30	43	32	42	30	41

	1981-85												1986-90											
	Affluent		2		3		4		Deprived		All[1]		Affluent		2		3		4		Deprived		All[1]	
	(375)		(348)		(345)		(343)		(279)		(1,730)		(368)		(444)		(423)		(376)		(299)		(1,948)	
One year	74	76	70	73	68	71	70	73	65	68	69	72	72	74	70	72	72	75	68	70	65	67	70	72
Five years	50	57	43	49	43	50	44	52	40	47	44	51	49	55	42	48	48	56	39	46	40	46	44	51
Ten years	39	50	33	43	34	46	35	46	33	45	35	46												

	1971-75		1976-80		1981-85		1986-90	
Northern & Yorkshire	(213)		(213)		(293)		(310)	
One year	70	72	61	63	69	72	65	67
Five years	45	52	37	42	41	48	35	42
Ten years	35	48	28	37	32	42		
Trent	(206)		(134)		(151)		(214)	
One year	70	72	64	67	65	67	67	69
Five years	46	52	39	47	41	47	38	44
Ten years	35	45	36	46	35	44		
Anglia & Oxford	(196)		(158)		(177)		(214)	
One year	66	68	74	78	75	78	69	73
Five years	42	46	47	55	55	62	45	54
Ten years	33	41	36	48	42	54		
North Thames	(239)		(173)		(140)		(211)	
One year	58	60	64	67	77	80	83	87
Five years	34	38	42	48	49	56	56	65
Ten years	25	35	32	44	38	53		
South Thames	(173)		(200)		(158)		(194)	
One year	59	62	64	66	68	71	73	76
Five years	34	41	37	44	45	53	51	58
Ten years	25	37	32	42	36	48		
South & West	(253)		(233)		(279)		(260)	
One year	64	67	65	68	72	75	69	72
Five years	44	52	36	42	45	54	48	56
Ten years	36	49	30	40	35	50		
West Midlands	(153)		(168)		(177)		(228)	
One year	70	72	65	68	75	78	72	74
Five years	36	44	39	45	44	53	46	55
Ten years	29	39	30	41	34	48		
North & West	(230)		(217)		(234)		(196)	
One year	62	65	66	68	65	68	63	66
Five years	37	45	39	45	43	52	34	39
Ten years	26	37	27	38	35	48		
WALES	(94)		(123)		(121)		(121)	
One year	68	70	63	64	56	58	66	68
Five years	44	48	31	37	31	37	46	53
Ten years	36	45	22	32	20	28		

[1] Including those for whom no deprivation category was assigned. C - crude survival, R - relative survival (see Chapter 3). Number of patients contributing to each analysis in parentheses.

Table 26.2 Crude and relative survival (%) at one, five and ten years since diagnosis: NHS Region, deprivation category and calendar period of diagnosis

ADULTS	1971-75												1976-80											
	Affluent		2		3		4		Deprived		All[1]		Affluent		2		3		4		Deprived		All[1]	
	C	R	C	R	C	R	C	R	C	R	C	R	C	R	C	R	C	R	C	R	C	R	C	R
ENGLAND & WALES	(520)		(603)		(543)		(566)		(451)		(3,349)		(659)		(686)		(673)		(698)		(575)		(3,434)	
One year	67	69	67	70	62	64	60	62	64	66	65	67	69	71	66	68	66	68	65	68	62	65	66	68
Five years	39	45	39	47	37	43	37	43	36	43	38	44	43	49	40	47	39	46	39	47	38	44	40	47
Ten years	30	41	32	43	27	37	28	39	27	40	29	40	33	44	31	41	30	42	30	45	31	42	31	43
ENGLAND	(501)		(583)		(513)		(535)		(432)		(3,191)		(629)		(651)		(610)		(636)		(539)		(3,172)	
One year	67	69	67	69	62	64	59	62	64	67	65	67	68	71	65	68	68	71	64	67	62	65	66	69
Five years	39	46	39	46	37	43	36	43	36	43	38	44	43	49	40	47	41	48	39	47	37	44	40	47
Ten years	30	41	32	42	27	36	27	38	27	40	29	40	33	44	31	42	32	44	31	45	31	42	31	43
Northern & Yorkshire	(47)		(65)		(57)		(73)		(68)		(434)		(58)		(78)		(75)		(110)		(120)		(453)	
One year	58	60	69	72	50	52	65	68	73	78	66	69	71	74	70	72	57	59	58	61	56	58	61	64
Five years	36	40	43	51	26	29	46	55	44	52	39	46	40	46	43	50	43	50	31	38	30	35	36	43
Ten years	21	26	34	45	17	22	37	52	38	52	30	42	28	38	35	46	32	46	23	35	26	34	28	39
Trent	(39)		(55)		(79)		(92)		(66)		(370)		(41)		(65)		(60)		(87)		(59)		(323)	
One year	77	80	69	71	61	63	69	72	62	64	68	71	73	75	70	73	58	61	70	72	62	64	67	69
Five years	49	54	42	46	38	45	50	57	36	41	43	49	46	53	46	55	27	31	54	65	32	39	41	49
Ten years	33	42	29	35	30	42	37	51	30	40	32	43	32	43	38	55	20	27	44	62	28	37	33	45
Anglia & Oxford	(75)		(82)		(67)		(64)		(20)		(354)		(92)		(84)		(65)		(59)		(26)		(336)	
One year	63	65	66	69	73	75	52	54	85	88	65	68	71	74	69	72	83	87	61	64	68	70	72	75
Five years	41	46	38	44	37	42	33	37	60	66	39	45	42	47	45	55	52	64	35	43	49	52	45	53
Ten years	34	43	36	43	25	32	26	31	40	56	31	40	32	40	33	49	37	52	28	43	49	52	35	47
North Thames	(89)		(84)		(71)		(79)		(66)		(436)		(81)		(61)		(57)		(80)		(60)		(348)	
One year	67	69	69	70	59	61	47	48	61	63	61	63	63	66	57	59	69	72	65	68	70	74	65	68
Five years	39	45	35	40	34	40	28	31	33	40	35	40	44	50	36	41	38	46	44	52	53	58	43	50
Ten years	31	41	34	39	23	33	20	29	25	38	27	38	36	49	21	28	32	43	32	47	43	52	33	46
South Thames	(87)		(82)		(57)		(66)		(32)		(372)		(109)		(86)		(84)		(71)		(23)		(394)	
One year	64	66	63	64	61	64	52	54	71	77	63	66	75	78	59	61	63	65	63	65	65	67	65	68
Five years	35	43	31	36	40	45	19	23	31	37	33	39	45	54	36	42	34	40	35	41	44	49	38	45
Ten years	26	39	23	30	24	34	16	22	19	33	24	35	36	47	26	35	29	37	31	40	35	47	30	40
South & West	(99)		(134)		(104)		(59)		(22)		(456)		(126)		(147)		(127)		(60)		(27)		(502)	
One year	68	70	69	71	62	64	79	83	51	53	67	70	64	67	69	71	72	74	66	72	70	72	69	71
Five years	44	52	42	51	39	45	40	51	30	33	41	49	41	48	40	46	50	56	43	56	40	48	43	51
Ten years	36	50	33	47	30	41	32	46	21	33	32	46	34	44	34	45	41	56	36	56	33	47	36	50
West Midlands	(23)		(30)		(15)		(24)		(25)		(315)		(56)		(63)		(70)		(76)		(79)		(348)	
One year	78	81	62	65	73	83	63	64	56	59	67	70	62	64	65	67	62	64	68	70	68	71	65	68
Five years	35	44	39	53	46	63	29	31	24	29	36	44	42	50	43	48	31	36	38	45	39	45	39	45
Ten years	31	44	32	46	33	57	29	31	20	24	28	39	30	43	35	45	24	34	27	42	29	41	29	41
North & West	(42)		(51)		(63)		(78)		(133)		(454)		(66)		(67)		(72)		(93)		(145)		(468)	
One year	67	69	66	70	61	63	51	54	61	64	60	63	70	71	61	63	77	81	64	66	58	61	65	67
Five years	33	37	45	54	39	43	35	40	34	44	36	44	44	51	33	38	46	52	35	43	35	41	37	43
Ten years	19	32	35	51	31	37	23	32	22	37	26	38	35	51	19	28	33	44	28	41	28	40	28	40
WALES	(19)		(20)		(30)		(31)		(19)		(158)		(30)		(35)		(63)		(62)		(36)		(262)	
One year	69	70	75	78	63	65	71	74	48	49	66	68	76	80	72	74	43	45	72	74	61	62	64	66
Five years	32	34	49	56	40	44	42	48	31	34	42	47	43	51	40	49	18	22	35	42	41	46	35	41
Ten years	22	31	43	54	36	44	39	45	26	28	34	44	-	-	28	35	12	19	24	32	25	45	24	35

[1] Including those for whom no deprivation category was assigned. C - crude survival, R - relative survival (see Chapter 3).
Number of patients contributing to each analysis in parentheses.

	1981-85						1986-90						
	Affluent C R	2 C R	3 C R	4 C R	Deprived C R	All[1] C R	Affluent C R	2 C R	3 C R	4 C R	Deprived C R	All[1] C R	*ADULTS*
	(779)	*(734)*	*(717)*	*(703)*	*(597)*	*(3,611)*	*(842)*	*(899)*	*(868)*	*(803)*	*(649)*	*(4,126)*	**ENGLAND & WALES**
	74 77	69 71	68 71	69 73	62 65	69 71	75 77	70 72	71 74	66 69	66 69	70 73	*One year*
	49 57	41 48	41 49	43 51	37 45	42 50	47 54	41 48	45 53	40 48	41 48	43 51	*Five years*
	39 51	32 43	31 44	33 45	29 42	33 45							
	(741)	*(696)*	*(654)*	*(632)*	*(562)*	*(3,345)*	*(804)*	*(853)*	*(806)*	*(735)*	*(625)*	*(3,878)*	**ENGLAND**
	75 78	69 72	69 72	70 73	63 66	**69 72**	76 78	71 73	72 75	67 70	66 69	**71 73**	*One year*
	50 58	41 48	42 51	44 52	37 45	**43 51**	47 54	41 48	45 53	41 48	41 48	**43 51**	*Five years*
	40 52	33 43	32 46	34 47	29 42	**34 46**							
	(89)	*(105)*	*(107)*	*(134)*	*(151)*	*(594)*	*(89)*	*(112)*	*(127)*	*(121)*	*(184)*	*(637)*	**Northern & Yorkshire**
	71 75	77 81	68 71	69 72	58 60	**68 71**	81 82	69 71	68 71	66 69	64 66	**68 71**	*One year*
	49 59	44 51	39 45	38 45	29 37	**39 46**	46 50	38 44	41 49	39 46	36 44	**39 46**	*Five years*
	37 49	36 47	30 40	31 43	21 34	**30 42**							
	(53)	*(62)*	*(56)*	*(79)*	*(67)*	*(318)*	*(70)*	*(90)*	*(100)*	*(99)*	*(81)*	*(442)*	**Trent**
	70 73	59 61	60 63	67 69	65 68	**64 67**	67 70	72 75	68 70	65 69	64 66	**67 70**	*One year*
	48 55	32 38	30 38	48 54	48 58	**42 49**	38 45	45 53	41 48	35 43	38 43	**40 46**	*Five years*
	- -	23 31	21 34	- -	43 55	**36 47**							
	(110)	*(96)*	*(75)*	*(54)*	*(28)*	*(377)*	*(146)*	*(108)*	*(97)*	*(78)*	*(15)*	*(447)*	**Anglia & Oxford**
	72 76	80 84	65 69	65 68	74 77	**71 75**	71 76	67 71	74 79	65 69	74 77	**70 74**	*One year*
	57 65	54 62	42 50	50 56	42 46	**51 58**	45 53	39 46	49 61	47 57	47 53	**45 53**	*Five years*
	44 58	*39 53*	33 46	39 49	23 36	**38 51**							
	(75)	*(60)*	*(55)*	*(71)*	*(47)*	*(311)*	*(94)*	*(88)*	*(92)*	*(95)*	*(85)*	*(472)*	**North Thames**
	75 78	*73 76*	76 79	80 85	72 75	**76 79**	76 80	*49 58*	84 87	*79 82*	*84 88*	**79 82**	*One year*
	46 55	*45 53*	54 61	47 56	47 54	**48 57**	53 63	*49 58*	53 61	46 53	56 66	**52 60**	*Five years*
	30 44	35 45	41 54	39 53	40 51	**37 50**							
	(110)	*(81)*	*(63)*	*(56)*	*(20)*	*(336)*	*(108)*	*(105)*	*(86)*	*(73)*	*(36)*	*(417)*	**South Thames**
	76 79	70 73	65 67	72 75	49 51	**70 73**	83 86	75 78	72 74	*66 69*	*78 84*	**75 78**	*One year*
	50 58	38 45	*44 57*	52 61	15 22	**44 53**	66 74	48 57	47 54	*39 46*	*46 55*	**50 58**	*Five years*
	40 49	*33 44*	31 50	37 50	11 17	**34 47**							
	(149)	*(166)*	*(152)*	*(62)*	*(24)*	*(574)*	*(120)*	*(181)*	*(138)*	*(86)*	*(39)*	*(569)*	**South & West**
	77 80	63 65	*75 78*	*76 79*	57 61	**71 74**	71 74	71 73	74 78	*63 66*	*64 66*	**70 73**	*One year*
	51 59	36 42	45 53	48 59	45 49	**44 52**	45 53	41 48	48 61	*44 57*	*51 59*	**45 54**	*Five years*
	40 56	30 39	33 49	37 51	24 39	**34 47**							
	(75)	*(62)*	*(62)*	*(77)*	*(87)*	*(364)*	*(91)*	*(92)*	*(93)*	*(95)*	*(89)*	*(469)*	**West Midlands**
	81 83	72 75	78 82	66 70	70 72	**73 76**	80 82	71 73	77 81	71 74	61 63	**72 75**	*One year*
	50 56	48 59	43 52	42 52	39 49	**44 53**	*45 55*	39 50	46 57	46 55	39 47	**44 53**	*Five years*
	39 52	35 53	32 48	31 49	33 48	**34 49**							
	(80)	*(64)*	*(84)*	*(99)*	*(138)*	*(471)*	*(86)*	*(77)*	*(73)*	*(88)*	*(96)*	*(425)*	**North & West**
	74 77	56 58	59 63	66 70	63 65	**64 67**	75 77	68 70	*61 64*	59 62	56 59	**64 66**	*One year*
	48 57	34 41	40 52	35 44	37 46	**39 47**	36 43	30 35	*35 43*	28 32	32 38	**32 38**	*Five years*
	39 55	26 41	35 52	25 38	29 42	**30 43**							
	(38)	*(38)*	*(63)*	*(71)*	*(35)*	*(266)*	*(38)*	*(46)*	*(62)*	*(68)*	*(24)*	*(248)*	**WALES**
	55 57	59 62	56 58	65 68	44 47	**58 60**	63 64	52 54	59 62	60 62	75 79	**61 63**	*One year*
	26 32	33 43	28 32	35 42	33 35	**33 39**	44 52	33 42	38 45	39 45	49 58	**41 48**	*Five years*
	19 27	23 39	20 29	24 35	31 35	**24 34**							

[1] Including those for whom no deprivation category was assigned. C - crude survival, R - relative survival (see Chapter 3).
Number of patients contributing to each analysis in parentheses.

Fig. 26.1   Relative survival up to ten years: trends by calendar period of diagnosis (1971-90) and NHS Region

ADULTS

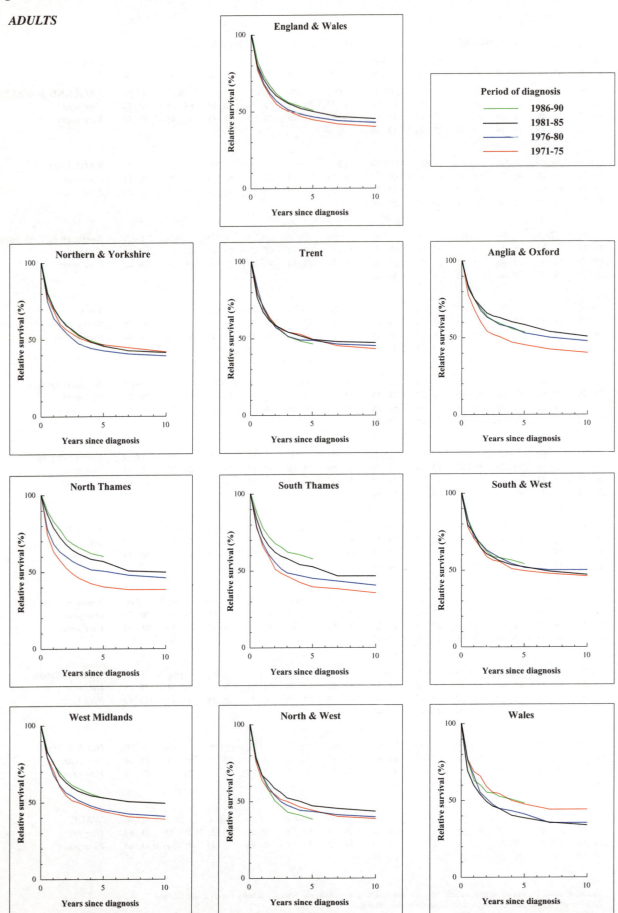

Fig. 26.2 Relative survival up to five years, by deprivation category and NHS Region: patients diagnosed 1986-90

ADULTS

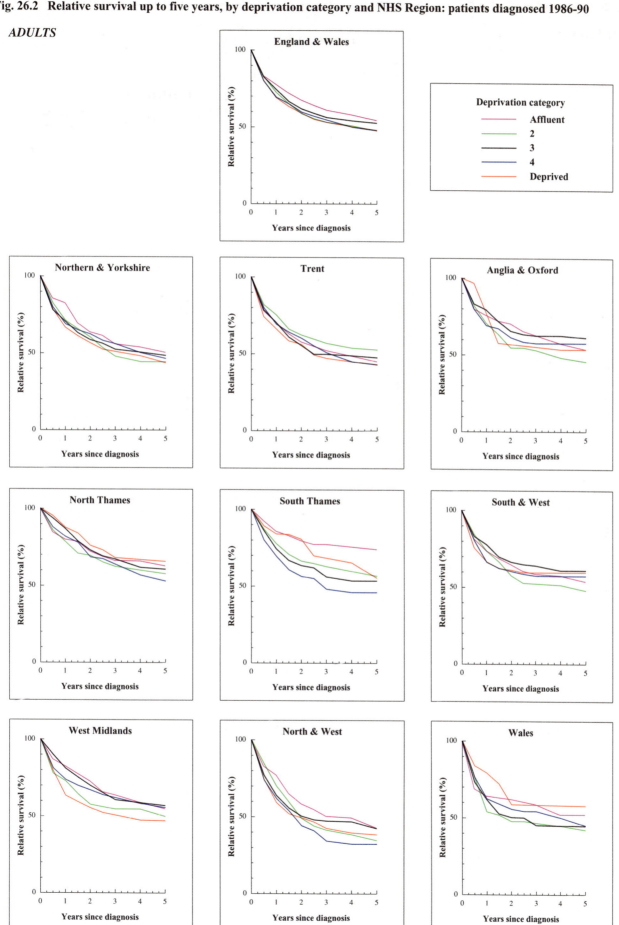

Fig. 26.3 Relative survival at five years by deprivation category, period of diagnosis (1981-90) and NHS Region

ADULTS

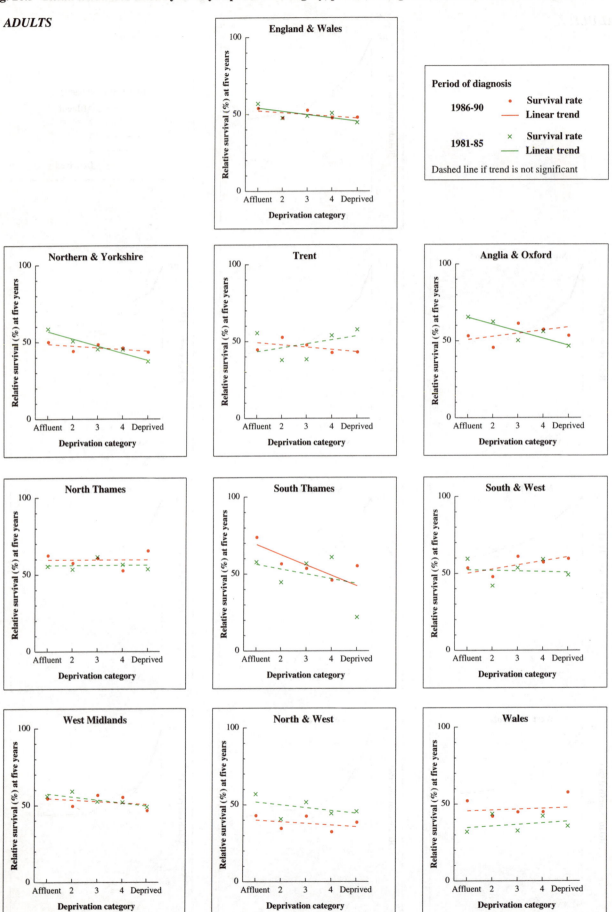

Fig. 26.4 Relative survival (%) by NHS Region and sex:
one-year and five-year survival for patients diagnosed 1986-90, with 95% confidence intervals

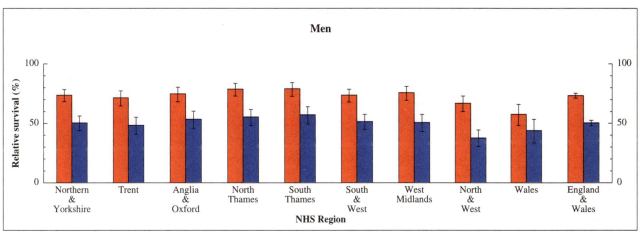

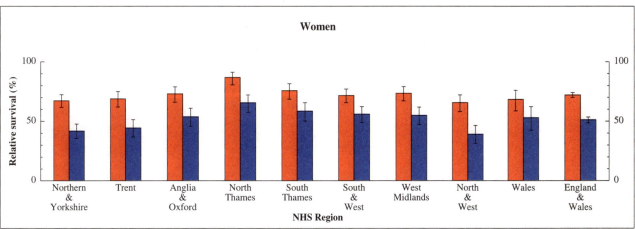

Fig. 26.5 Relative survival (%) at five years, with 95% confidence interval, by age at diagnosis and sex:
England and Wales, patients diagnosed 1986-90

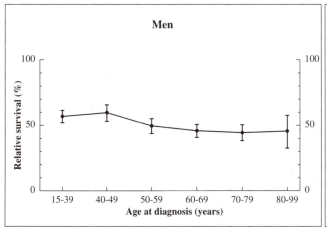

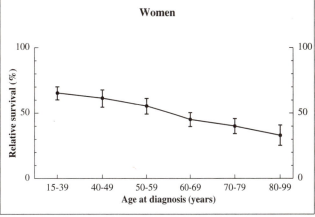

Chapter 27

MELANOMA OF SKIN

Melanoma arises from the pigment-producing melanocytes, found mainly in the skin, but also in the choroid of the eye, the bowel and the genital tract. Basal and squamous cell carcinomas of the skin develop from keratinocytes: they are much more common than melanoma, but are often diagnosed and treated without pathological confirmation, are incompletely registered, and are very rarely fatal. Only melanoma of skin is considered here; melanomas arising elsewhere are included in the analyses for the relevant organ or tissue.

About 3,700 new cases of melanoma of the skin were registered in England and Wales in 1990, some 2% of all cancers. Incidence is higher in women than men, at about 7 and 4 per 100,000 per year, respectively, and higher in South & West Region than in the north and the midlands. Incidence is three times higher among more affluent groups[1], an uncommon pattern.

Melanoma of the skin has been increasing rapidly in the UK since the 1960s, as in many other countries. Incidence has been rising by 30-50% every five years, more rapidly in women than in men; the numbers of patients more than doubled between 1971 and 1990. The increases affect men and women in all age groups, and they are not due to changes in diagnostic criteria[2]. Mortality has been increasing by 12-15% every five years – rapidly, but much less so than for incidence – and the rate of increase has slowed down further in the 1990s. This difference between incidence and mortality trends suggests substantial and long-term improvements in survival. Higher survival in women and increasing survival for men and women have been reported from Sweden[3] and from Scotland[4], where the trends have been ascribed to greater public awareness, earlier diagnosis and thinner lesions[5]. Melanoma is more common in fair-skinned Caucasian populations in sunny climates.

Survival from melanoma is much higher with thin tumours (<1.5 mm), when surgery may be curative; the prognosis for disseminated disease remains poor. Earlier diagnosis contributes to socioeconomic differences in survival. In a population study of 3,000 cases diagnosed in the West of Scotland during the period 1979-93, thin melanomas were more common in the affluent, although survival was still worse in the deprived group after adjustment for tumour thickness[1]. The proportion of thin tumours increased more in deprived groups during the study period.

The survival analyses reported here relate to almost 43,000 patients diagnosed with malignant melanoma of the skin in England and Wales during the period 1971-90

and followed up to the end of 1995, about 90% of those eligible (Table 27.1). About 6% were excluded because their vital status was not known at 6 July 1997, and a further 3% because their survival was zero or unknown. The proportion excluded with unknown vital status was unusually high (8%) for those diagnosed during 1986-90. About 40% of melanomas diagnosed during the 1980s arose on the leg or hip, a slight fall since the early 1970s (Table 27A). Another 21% arose on the skin of the trunk, and about 16% on the face, head and neck. More than 98% of the tumours were assigned to morphology codes for malignant melanoma or naevus: during the period 1986-90, 64% were classified as malignant melanoma, 18% as superficial spreading melanoma and 10% as nodular melanoma.

Survival trends

Survival from melanoma is high. For patients diagnosed during 1986-90, relative survival in men was 91% at one year and 70% at five years; the corresponding survival rates for women were 95% and 84% (Table 27.2). There have been large improvements in survival since the early 1970s in both sexes and in all regions of the country (Figure 27.1).

In England and Wales as a whole, one-year survival rose by 12% for men and 5% for women, and the corresponding gains in five-year survival were 22% and 17%, after adjustment for age (see Table 4.4). Improvement was significantly more rapid for men than for women. For one-year survival, the average gain between successive five-year periods was 4% for men and 2% for women; the corresponding gains in five-year survival were 8% and 6%, respectively. The *rate of increase* in survival is constrained, however (since survival cannot exceed 100%), and men have been catching up: the five-year survival advantage for women was no less than 21% for patients diagnosed in the early 1970s; this had fallen to 14% for those diagnosed in the late 1980s.

In West Midlands, the average gain in five-year survival was high (10% every five years for men, and 7% for women), while in Wales the overall increase was smaller than average, and survival for patients diagnosed during 1986-90 was about 10% below the national average. Survival is otherwise similar in all regions (Figure 27.4; see also Table-Figure 4.5G).

Survival falls with age (Figure 27.5). For men diagnosed in 1986-90, five-year survival was 60-80% for those aged under 70 at diagnosis, but only 44% for those aged 80-99

Table 27A Sub-site distribution of melanoma

ICD-9 code	Site description	1971-75 No.	%	1976-80 No.	%	1981-85 No.	%	1986-90 No.	%	Total No.	%
172.0	Lip	21	0.3	20	0.2	23	0.2	34	0.2	**98**	**0.2**
172.1	Eyelid, including canthus	75	1.1	77	0.9	78	0.7	94	0.6	**324**	**0.8**
172.2	Ear and external auricular canal	78	1.2	94	1.1	142	1.2	185	1.2	**499**	**1.2**
172.3	Other & unspecified parts of face	757	11.6	953	11.4	1,188	10.3	1,671	10.5	**4,569**	**10.8**
172.4	Scalp and neck	254	3.9	318	3.8	364	3.1	546	3.4	**1,482**	**3.5**
172.5	Trunk except scrotum[1]	1,213	18.5	*748*	*8.9*	2,308	19.9	3,507	22.0	*see footnote*	
172.6	Upper limb including shoulder[1]	945	14.4	*1,483*	*17.7*	1,942	16.8	2,705	17.0	*see footnote*	
172.7	Lower limb including hip[1]	2,953	45.1	*2,206*	*26.3*	4,902	42.3	6,052	38.0	*see footnote*	
172.8	Other (scrotum for 1971-75)[1]	34	0.5	*1,992*	*23.7*	51	0.4	98	0.6	*see footnote*	
172.9	Site unspecified	224	3.4	505	6.0	591	5.1	1,048	6.6	**2,368**	**5.6**
	Total	**6,554**		**8,396**		**11,589**		**15,940**		**42,479**	

[1] In ICD-8 (1971-78), the trunk except scrotum was coded to 172.6, upper limb to 172.7 and lower limb to 172.8. Scrotum was coded to 172.5 (it is coded to 187.7 in ICD-9). For 1971-75, melanomas at these sites were reallocated to the correct ICD-9 code in the table (34 scrotal melanomas are shown in 172.8, not used in ICD-8), but for 1976-80, the italicised counts for 172.5-172.8 include melanomas coded under both revisions. Row totals for these codes are thus misleading, and have been suppressed.

(see Table 4.7). The pattern with age was broadly similar for women, but at higher levels: 80-90% for those aged under 70, and just over 60% in the elderly.

Deprivation

In England and Wales as a whole, and in most regions, there are substantial differences in survival between deprivation groups (Table 27.2, Figure 27.2).

Quantitative estimates of the deprivation gradient are given in Table 4.9. For more than 11,000 patients of known deprivation status diagnosed during 1981-85, the gap between the most affluent and most deprived groups was about 6% for one-year survival, and 13% for survival at five years. Among more than 15,000 patients diagnosed during 1986-90, survival had improved by 10% for the deprived group and 4% for the affluent, reducing the deprivation gap in five-year survival to 8%. This pattern of variation in survival by deprivation and over time is broadly similar across the regions (Figure 27.3). The explanation may be similar to that in Scotland (see above)[1].

International comparison

Five-year survival from melanoma of the skin in England and Wales (70% for men and 84% for women) is 4% lower than in Scotland, but similar to the average for Europe (see Table 4.12). Five-year survival rates for melanoma across Europe in the late 1980s varied widely: from less than 50% in Slovenia, Poland and Estonia to over 80% in Sweden and Switzerland for men, and from about 70% to 90% for women[6]. The highest regional survival rates in England and Wales, and national survival rates for the affluent group, are similar to those in Scotland, but still 5% to 10% lower than in the USA (see Figure 4.12).

1. MacKie R, Hole DJ. Incidence and thickness of primary tumours and survival of patients with cutaneous malignant melanoma in relation to socioeconomic status. Br Med J 1996; 312: 1125-1128

2. van der Esch EP, Muir CS, Nectoux J, Macfarlane GJ, Maisonneuve P, Bharucha H, Briggs J, Cooke RA, Dempster AG, Essex WB, Hofer PA, Hood AF, Ironside P, Larsen TE, Little JH, Philipp R, Pfau RS, Prade M, Pozharisski KM, Rilke F, Schafler K. Temporal change in diagnostic criteria as a cause of the increase of malignant melanoma over time is unlikely. Int J Cancer 1991; 47: 483-490

3. Thörn M, Adami H-O, Bergström R, Ringborg U, Krusemo UB. Trends in survival from malignant melanoma: remarkable improvement in 23 years. J Natl Cancer Inst 1989; 81: 611-617

4. Black RJ, Clarke JA, Warner JM. Malignant melanoma in Scotland: trends in incidence and survival, 1968-87. Health Bull (Edinb) 1991; 49: 97-105

5. MacKie R, Hunter JA, Aitchison TC, Hole D, McLaren K, Rankin R, Blessing K, Evans AT, Hutcheon AW, Jones DH, Soutar DS, Watson ACH, Cornbleet MA, Smyth JF. Cutaneous malignant melanoma, Scotland, 1979-89. The Scottish Melanoma Group. Lancet 1992; 339: 971-975

6. Berrino F, Capocaccia R, Estève J, Gatta G, Hakulinen T, Micheli M, Sant M, Verdecchia A, (eds.) *Survival of cancer patients in Europe: the EUROCARE study, II. (IARC Scientific Publications No. 151).* Lyon: International Agency for Research on Cancer, 1999

Table 27.1 Ineligible and excluded records, and patients included in analyses, by calendar period of diagnosis

	1971-75			**1976-80**			**1981-85**			**1986-90**			**Overall**		
Total registered	**7,205**			**9,324**			**12,882**			**18,533**			**47,944**		
Ineligible	*Records*	*Patients*	*%*	*Records*	*Patients*	*%*	*Records*	*Patients*	*%*	*Records*	*Patients*	*%*	*Records*	*Patients*	*%*
Incomplete data[1]	4	4		9	9		60	60		57	57		130	**130**	
Resident outside England and Wales	73	72		99	96		78	67		86	67		336	**302**	
In situ neoplasm[2]	10	10		4	4		0	0		2	2		16	**16**	
Benign or uncertain behaviour[2]	91	80		70	62		0	0		0	0		161	**142**	
Metastatic[2]	64	60		34	34		50	49		76	76		224	**219**	
Otherwise ineligible[3]	-	-		-	-		-	-		-	-		-	**-**	
Lymphoma[4]	3	3		1	1		4	0		3	0		11	**4**	
Leukaemia or myeloma[4]	0	0		0	0		0	0		0	0		0	**0**	
		229			**206**			**176**			**202**			**813**	
Total eligible		**6,976**	**100**		**9,118**	**100**		**12,706**	**100**		**18,331**	**100**		**47,131**	**100**
Aged under 15 or 100 years or over at diagnosis	40	39	0.6	59	55	0.6	58	57	0.4	71	70	0.4	228	**221**	0.5
Vital status unknown[5]	306	264	3.8	490	440	4.8	576	525	4.1	1,558	1,512	8.2	2,930	**2,741**	5.8
Duplicate registration[6]	0	0	0.0	2	1	<0.1	1	0	0.0	0	0	0.0	3	**1**	<0.1
Multiple primary[7]	0	0	0.0	4	2	<0.1	46	42	0.3	98	91	0.5	148	**135**	0.3
Synchronous tumours[8]	24	12	0.2	35	17	0.2	57	39	0.3	157	76	0.4	273	**144**	0.3
Sex not known	0	0	0.0	0	0	0.0	0	0	0.0	0	0	0.0	0	**0**	0.0
Sex incompatible with site code[9]	1	0	0.0	4	0	0.0	0	0	0.0	0	0	0.0	5	**0**	0.0
Invalid dates[10]	0	0	0.0	0	0	0.0	2	0	0.0	13	6	<0.1	15	**6**	<0.1
Zero survival[11]	122	107	1.5	229	207	2.3	503	454	3.6	703	636	3.5	1,557	**1,404**	3.0
Total exclusions		**422**	**6.0**		**722**	**7.9**		**1,117**	**8.8**		**2,391**	**13.0**		**4,652**	**9.9**
Patients included in analysis		**6,554**	**94.0**		**8,396**	**92.1**		**11,589**	**91.2**		**15,940**	**87.0**		**42,479**	**90.1**

[1] Main data item(s) invalid or incompatible with one another: name, sex, date of birth, date of diagnosis and (if present) date of death, postcode, site, morphology and behaviour

[2] Tumour either *in situ* (behaviour code 2), benign (0), uncertain if benign or malignant (1) or metastatic (6 or 9)

[3] Not applicable for this malignancy

[4] Morphology that of lymphoma (included with non-Hodgkin lymphoma, Ch. 41) or myeloma (included with multiple myeloma, Ch. 43), or leukaemia (excluded)

[5] Tracing of vital status at National Health Service Central Register incomplete at 6 July 1997

[6] Same site code, sex, cancer registry and cancer registry number as an earlier registration

[7] Same site code and person number as an earlier registration(s): mostly confirmed multiple primary tumours at this site, some unresolved duplicate registrations

[8] Same site code, sex, date of birth and date of diagnosis as another registration(s): mostly synchronous or (in paired organs) bilateral tumours in same anatomic site in one individual, not previously linked: also some duplicate registrations

[9] Site code specific to one sex; tumour registered to opposite sex: scrotum (172.5 in ICD-8, 1971-78)

[10] Impossible sequence of dates (birth, diagnosis and, if present, emigration or death); or date of death after 6 July 1997

[11] Date of diagnosis same as date of death: some are patients who did die on the day of diagnosis, but most were registered solely from a death certificate (death certificate only, or DCO), and their survival time is thus unknown

Table 27.2   **Crude and relative survival (%) at one, five and ten years since diagnosis: NHS Region, deprivation category and calendar period of diagnosis**

MEN		1971-75						1976-80					
		Affluent	2	3	4	Deprived	All[1]	Affluent	2	3	4	Deprived	All[1]
		C R	C R	C R	C R	C R	C R	C R	C R	C R	C R	C R	C R
ENGLAND & WALES		*(430)*	*(394)*	*(377)*	*(341)*	*(226)*	*(2,161)*	*(749)*	*(610)*	*(578)*	*(464)*	*(313)*	*(2,849)*
	One year	81 83	80 83	77 79	74 77	70 73	**76 79**	85 87	81 84	81 83	75 78	71 75	**80 82**
	Five years	48 54	42 48	42 49	35 42	28 35	**41 47**	53 59	45 53	42 50	39 48	37 44	**44 52**
	Ten years	36 45	30 41	32 43	26 36	18 27	**30 40**	40 49	36 46	31 41	29 41	26 40	**33 45**
ENGLAND		*(411)*	*(379)*	*(356)*	*(319)*	*(217)*	*(2,050)*	*(712)*	*(581)*	*(537)*	*(418)*	*(296)*	*(2,659)*
	One year	81 83	80 83	78 81	74 77	69 72	**77 79**	85 87	81 83	81 84	76 79	72 75	**80 83**
	Five years	48 54	42 48	43 50	36 42	28 34	**41 48**	53 59	45 52	42 50	39 48	37 45	**44 52**
	Ten years	36 45	31 42	32 43	26 36	18 26	**30 41**	39 49	35 46	31 42	30 42	25 39	**33 45**
Northern & Yorkshire		*(23)*	*(34)*	*(25)*	*(31)*	*(37)*	*(228)*	*(56)*	*(39)*	*(56)*	*(45)*	*(78)*	*(286)*
	One year	82 84	88 89	81 84	68 72	59 61	**77 79**	78 80	74 76	82 89	70 73	68 73	**74 77**
	Five years	48 54	53 61	40 50	29 33	26 31	**42 49**	54 58	49 59	44 54	32 42	38 46	**43 52**
	Ten years	30 36	35 45	32 45	17 23	8 13	**28 38**	32 39	36 47	28 47	27 37	30 46	**30 42**
Trent		*(27)*	*(34)*	*(43)*	*(48)*	*(34)*	*(203)*	*(49)*	*(54)*	*(64)*	*(68)*	*(37)*	*(285)*
	One year	74 79	74 77	86 88	69 72	56 60	**72 76**	81 83	69 71	76 79	67 69	76 78	**72 75**
	Five years	49 57	44 48	54 62	33 36	20 27	**40 47**	47 50	35 42	34 43	33 42	43 54	**37 45**
	Ten years	33 45	41 48	27 39	24 31	12 17	**27 38**	28 32	28 36	28 43	26 39	27 44	**27 38**
Anglia & Oxford		*(69)*	*(56)*	*(44)*	*(37)*	*(18)*	*(251)*	*(106)*	*(63)*	*(58)*	*(54)*	*(20)*	*(315)*
	One year	77 79	80 82	65 70	87 90	71 74	**77 79**	89 92	81 84	79 82	76 80	80 87	**82 85**
	Five years	54 59	38 42	43 49	40 48	34 37	**43 50**	56 63	46 53	43 51	45 53	31 37	**47 55**
	Ten years	42 49	27 37	38 46	27 35	28 34	**34 43**	43 53	31 45	31 45	31 41	7 11	**34 45**
North Thames		*(68)*	*(55)*	*(44)*	*(52)*	*(34)*	*(275)*	*(72)*	*(66)*	*(58)*	*(43)*	*(30)*	*(279)*
	One year	72 74	80 82	77 78	71 74	80 83	**76 78**	90 92	85 87	79 81	75 76	77 80	**82 84**
	Five years	38 43	45 50	43 50	42 45	31 38	**40 45**	58 62	36 43	36 42	32 37	44 50	**41 47**
	Ten years	30 38	37 46	41 49	34 41	16 23	**31 40**	40 49	30 38	27 36	25 34	30 50	**31 41**
South Thames		*(103)*	*(73)*	*(58)*	*(45)*	*(23)*	*(336)*	*(166)*	*(107)*	*(64)*	*(46)*	*(18)*	*(427)*
	One year	81 84	82 87	79 81	82 86	78 80	**81 84**	89 90	83 86	80 83	91 94	84 87	**86 89**
	Five years	44 50	41 50	41 47	38 46	38 44	**42 50**	55 62	47 56	49 55	47 58	45 51	**50 58**
	Ten years	33 42	24 41	30 38	25 37	20 26	**29 40**	44 55	32 46	43 50	31 48	21 31	**38 51**
South & West		*(79)*	*(73)*	*(97)*	*(52)*	*(10)*	*(342)*	*(141)*	*(152)*	*(134)*	*(62)*	*(24)*	*(532)*
	One year	90 92	82 85	78 81	68 69	80 83	**80 82**	80 82	84 86	85 88	74 76	63 65	**82 84**
	Five years	53 60	40 48	43 51	- -	40 44	**43 51**	49 56	49 56	49 59	40 52	33 36	**48 56**
	Ten years	37 48	34 45	36 48	29 38	40 44	**34 46**	42 53	41 51	35 46	35 52	29 31	**39 50**
West Midlands		*(19)*	*(18)*	*(11)*	*(14)*	*(17)*	*(186)*	*(53)*	*(48)*	*(50)*	*(43)*	*(34)*	*(237)*
	One year	95 95	61 63	81 83	71 75	83 85	**73 76**	83 85	79 81	82 85	82 86	79 84	**81 84**
	Five years	63 64	- -	46 49	30 33	23 29	**40 47**	49 52	41 45	34 42	32 37	36 42	**39 45**
	Ten years	47 55	20 29	28 35	8 14	23 29	**31 42**	32 40	35 40	24 35	25 37	24 33	**28 38**
North & West		*(23)*	*(36)*	*(34)*	*(40)*	*(44)*	*(229)*	*(69)*	*(52)*	*(53)*	*(57)*	*(55)*	*(298)*
	One year	87 87	75 83	80 81	77 81	66 69	**75 78**	83 84	81 84	77 80	79 83	66 68	**76 79**
	Five years	48 51	41 51	32 38	38 44	23 29	**36 43**	50 57	48 53	39 44	47 62	32 42	**43 51**
	Ten years	48 51	24 33	14 20	30 38	18 29	**26 34**	35 52	38 48	21 28	33 49	24 41	**30 44**
WALES		*(19)*	*(15)*	*(21)*	*(22)*	*(9)*	*(111)*	*(37)*	*(29)*	*(41)*	*(46)*	*(17)*	*(190)*
	One year	74 75	79 82	56 58	68 72	- -	**70 73**	81 85	90 92	81 82	63 66	58 60	**74 78**
	Five years	47 49	33 37	32 35	27 35	- -	**33 39**	57 66	55 60	36 44	39 46	34 36	**44 52**
	Ten years	32 36	12 15	23 29	18 31	- -	**21 29**	52 63	51 57	29·36	17 24	28 36	**35 45**

[1] Including those for whom no deprivation category was assigned. C - crude survival, R - relative survival (see Chapter 3).
Number of patients contributing to each analysis in parentheses.

MEN

	1981-85												1986-90											
	Affluent C	R	2 C	R	3 C	R	4 C	R	Deprived C	R	All[1] C	R	Affluent C	R	2 C	R	3 C	R	4 C	R	Deprived C	R	All[1] C	R
ENGLAND & WALES	*(1,067)*		*(936)*		*(734)*		*(639)*		*(424)*		*(3,896)*		*(1,651)*		*(1,477)*		*(1,303)*		*(899)*		*(557)*		*(5,964)*	
One year	86	89	82	86	83	86	81	84	75	77	82	85	89	92	89	92	87	90	83	87	85	89	87	91
Five years	59	65	48	56	51	60	45	53	42	50	50	59	63	71	62	72	61	71	53	63	53	65	60	70
	45	57	38	51	39	52	34	48	31	45	39	52												
ENGLAND	*(1,035)*		*(896)*		*(704)*		*(587)*		*(396)*		*(3,687)*		*(1,605)*		*(1,425)*		*(1,242)*		*(833)*		*(525)*		*(5,689)*	
One year	87	89	82	86	84	87	83	85	75	78	83	86	89	92	89	92	88	91	85	88	86	91	88	91
Five years	59	65	48	55	51	60	46	55	41	50	51	59	63	71	63	72	61	71	54	65	55	68	60	71
	45	56	38	50	38	52	35	48	31	45	39	52												
Northern & Yorkshire	*(86)*		*(92)*		*(69)*		*(73)*		*(80)*		*(407)*		*(151)*		*(134)*		*(140)*		*(101)*		*(106)*		*(635)*	
One year	87	89	88	91	78	82	85	87	80	84	84	86	90	93	93	96	86	89	83	87	82	86	87	91
Five years	65	71	48	55	42	52	49	57	45	57	50	58	62	74	70	82	60	72	49	61	54	68	60	71
	49	66	40	55	37	46	34	48	36	57	39	54												
Trent	*(71)*		*(78)*		*(71)*		*(81)*		*(52)*		*(353)*		*(119)*		*(146)*		*(134)*		*(104)*		*(78)*		*(582)*	
One year	85	87	80	85	75	77	82	85	71	74	79	82	86	89	86	91	82	85	87	92	89	93	86	89
Five years	58	65	35	42	42	54	43	52	42	51	44	52	57	65	63	76	60	71	58	73	55	68	59	69
	39	52	25	37	28	46	32	44	32	51	31	44												
Anglia & Oxford	*(156)*		*(132)*		*(85)*		*(66)*		*(13)*		*(460)*		*(230)*		*(184)*		*(171)*		*(72)*		*(25)*		*(694)*	
One year	91	94	79	83	89	94	83	86	77	79	86	89	93	96	89	93	90	93	83	89	88	97	90	93
Five years	62	71	51	59	58	67	42	51	55	57	55	63	66	76	65	76	68	79	56	68	48	63	65	74
	49	64	40	50	43	58	35	48	55	57	43	56												
North Thames	*(102)*		*(75)*		*(61)*		*(67)*		*(49)*		*(357)*		*(184)*		*(154)*		*(148)*		*(117)*		*(82)*		*(701)*	
One year	76	77	81	84	80	83	85	87	79	81	80	82	92	95	89	92	88	91	93	95	88	92	90	93
Five years	45	49	48	56	42	51	48	54	39	45	45	51	65	75	59	70	60	71	58	67	56	69	60	70
	36	43	39	50	34	43	37	50	33	42	36	45												
South Thames	*(212)*		*(153)*		*(84)*		*(70)*		*(32)*		*(576)*		*(286)*		*(199)*		*(152)*		*(99)*		*(44)*		*(792)*	
One year	90	92	84	87	87	90	84	88	78	80	86	89	87	90	88	92	85	89	82	87	80	83	86	89
Five years	60	66	45	54	56	67	51	66	50	57	53	61	63	73	59	69	58	69	51	61	54	66	59	68
	44	55	32	47	37	53	40	59	31	52	38	51												
South & West	*(206)*		*(226)*		*(191)*		*(82)*		*(30)*		*(752)*		*(278)*		*(359)*		*(257)*		*(144)*		*(40)*		*(1,083)*	
One year	85	89	83	87	83	85	83	85	76	80	83	87	86	89	90	94	87	90	85	88	95	98	88	90
Five years	59	67	54	64	57	68	51	60	39	46	56	65	63	71	63	74	58	71	62	76	63	77	62	72
	49	62	43	61	44	61	36	53	36	46	44	59												
West Midlands	*(85)*		*(68)*		*(76)*		*(63)*		*(55)*		*(352)*		*(175)*		*(127)*		*(121)*		*(105)*		*(47)*		*(581)*	
One year	89	94	80	83	90	94	82	86	80	83	85	89	91	94	90	97	94	97	82	87	87	90	90	94
Five years	65	75	40	45	51	60	46	55	34	43	48	56	65	74	65	79	67	78	51	63	64	80	63	74
	48	63	28	39	36	48	35	55	24	34	35	48												
North & West	*(117)*		*(72)*		*(67)*		*(85)*		*(85)*		*(430)*		*(182)*		*(122)*		*(119)*		*(91)*		*(103)*		*(621)*	
One year	85	87	79	83	85	88	77	80	65	68	78	81	86	91	86	89	91	97	80	85	87	95	86	91
Five years	56	62	50	58	43	53	40	48	38	51	46	55	60	72	58	69	58	71	45	56	51	65	56	67
	42	52	46	58	34	50	30	42	22	37	35	48												
WALES	*(32)*		*(40)*		*(30)*		*(52)*		*(28)*		*(209)*		*(46)*		*(52)*		*(61)*		*(66)*		*(32)*		*(275)*	
One year	81	83	85	87	73	75	66	69	64	66	75	77	87	89	84	85	77	79	66	68	55	57	76	78
Five years	59	65	58	65	46	54	29	37	46	51	45	54	59	64	42	50	60	66	38	45	22	26	47	54
	49	65	43	60	46	54	25	34	36	44	38	51												

---

[1] Including those for whom no deprivation category was assigned. C - crude survival, R - relative survival (see Chapter 3).
Number of patients contributing to each analysis in parentheses.

Table 27.2   Crude and relative survival (%) at one, five and ten years since diagnosis: NHS Region, deprivation category and calendar period of diagnosis

WOMEN	1971-75 Affluent		2		3		4		Deprived		All[1]		1976-80 Affluent		2		3		4		Deprived		All[1]	
	C	R	C	R	C	R	C	R	C	R	C	R	C	R	C	R	C	R	C	R	C	R	C	R
ENGLAND & WALES	*(918)*		*(845)*		*(766)*		*(658)*		*(413)*		*(4,393)*		*(1,464)*		*(1,290)*		*(1,146)*		*(884)*		*(555)*		*(5,547)*	
One year	89	91	87	89	87	89	88	91	84	87	87	90	90	92	87	90	89	92	84	87	84	87	87	90
Five years	66	73	60	68	59	67	58	66	52	59	59	67	70	78	64	72	61	70	58	67	55	65	63	72
Ten years	56	68	50	62	48	61	46	58	39	53	48	61	62	73	56	70	51	66	44	58	43	59	53	67
ENGLAND	*(878)*		*(802)*		*(714)*		*(586)*		*(392)*		*(4,113)*		*(1,411)*		*(1,230)*		*(1,074)*		*(811)*		*(519)*		*(5,215)*	
One year	89	91	88	90	87	89	89	91	84	87	**87**	**90**	90	93	88	90	90	92	85	88	84	88	**88**	**91**
Five years	66	73	60	68	58	66	58	66	52	60	**59**	**67**	71	78	65	73	62	71	58	67	55	64	**64**	**72**
Ten years	56	68	51	63	47	60	47	59	39	53	**48**	**61**	62	73	57	70	51	66	45	59	43	59	**54**	**68**
Northern & Yorkshire	*(45)*		*(65)*		*(73)*		*(71)*		*(77)*		*(465)*		*(95)*		*(103)*		*(124)*		*(120)*		*(103)*		*(563)*	
One year	89	90	85	86	84	86	90	93	79	82	**85**	**87**	90	92	85	87	86	88	87	88	82	85	**86**	**88**
Five years	58	68	57	63	48	54	52	61	48	58	**53**	**61**	68	74	63	72	60	70	63	71	61	73	**63**	**72**
Ten years	51	63	49	59	40	51	41	55	34	49	**44**	**55**	59	73	56	72	43	60	47	63	46	65	**50**	**67**
Trent	*(42)*		*(51)*		*(80)*		*(84)*		*(57)*		*(340)*		*(82)*		*(92)*		*(80)*		*(92)*		*(73)*		*(430)*	
One year	81	85	82	83	80	82	86	87	77	79	**82**	**84**	91	95	87	88	90	94	81	83	79	82	**86**	**88**
Five years	60	67	61	67	56	62	52	60	47	54	**55**	**64**	68	75	65	73	61	73	54	61	55	64	**61**	**69**
Ten years	50	58	47	62	49	58	41	51	40	51	**46**	**58**	57	67	52	66	53	68	42	52	47	63	**51**	**63**
Anglia & Oxford	*(121)*		*(106)*		*(87)*		*(64)*		*(22)*		*(452)*		*(213)*		*(166)*		*(114)*		*(69)*		*(29)*		*(613)*	
One year	91	92	90	91	93	97	95	97	87	90	**91**	**93**	93	96	91	94	86	88	87	90	90	91	**90**	**93**
Five years	70	77	59	67	60	69	74	81	54	60	**64**	**72**	76	85	69	78	61	67	52	59	49	56	**67**	**75**
Ten years	58	69	55	65	45	57	59	74	45	56	**54**	**66**	69	83	66	78	49	60	42	55	36	47	**59**	**73**
North Thames	*(124)*		*(95)*		*(71)*		*(89)*		*(57)*		*(496)*		*(120)*		*(114)*		*(127)*		*(78)*		*(44)*		*(502)*	
One year	89	90	83	86	87	91	88	89	90	91	**87**	**89**	89	92	86	88	88	91	86	88	89	95	**87**	**90**
Five years	57	64	60	66	59	68	62	67	53	61	**59**	**66**	78	86	60	66	57	64	60	69	46	56	**62**	**70**
Ten years	49	60	51	59	49	63	46	54	42	55	**48**	**59**	67	81	53	63	48	61	48	63	41	54	**53**	**66**
South Thames	*(242)*		*(166)*		*(126)*		*(96)*		*(56)*		*(751)*		*(327)*		*(209)*		*(162)*		*(119)*		*(39)*		*(905)*	
One year	89	91	92	94	87	89	87	88	87	90	**89**	**91**	91	93	91	94	92	95	85	88	90	95	**90**	**93**
Five years	69	75	62	70	59	66	57	68	57	65	**62**	**70**	66	75	66	75	66	75	63	75	77	83	**66**	**76**
Ten years	58	69	50	63	45	58	51	65	43	65	**51**	**64**	59	71	57	72	55	71	49	68	53	76	**56**	**71**
South & West	*(183)*		*(221)*		*(177)*		*(87)*		*(21)*		*(772)*		*(323)*		*(336)*		*(276)*		*(122)*		*(43)*		*(1,130)*	
One year	91	92	88	91	90	92	91	94	76	77	**89**	**92**	89	91	85	88	93	95	85	89	79	82	**88**	**90**
Five years	71	78	64	72	67	75	62	68	48	55	**65**	**74**	72	79	64	73	65	73	64	73	49	57	**66**	**75**
Ten years	63	76	53	65	57	72	49	62	33	41	**55**	**69**	63	74	57	71	56	71	48	62	32	40	**57**	**70**
West Midlands	*(40)*		*(28)*		*(35)*		*(22)*		*(29)*		*(388)*		*(114)*		*(93)*		*(87)*		*(79)*		*(78)*		*(458)*	
One year	88	91	89	93	91	92	81	89	90	92	**88**	**91**	88	89	91	93	92	95	87	90	86	89	**89**	**91**
Five years	60	64	71	76	60	64	54	60	55	62	**59**	**66**	68	74	67	73	62	70	58	66	59	67	**63**	**71**
Ten years	50	61	60	75	48	60	49	56	44	62	**45**	**59**	57	66	52	64	48	66	44	58	46	64	**50**	**65**
North & West	*(81)*		*(70)*		*(65)*		*(73)*		*(73)*		*(449)*		*(137)*		*(117)*		*(104)*		*(132)*		*(110)*		*(614)*	
One year	91	94	84	87	82	84	89	93	88	91	**87**	**90**	91	93	87	91	86	88	84	88	85	89	**87**	**90**
Five years	69	78	-	-	41	46	51	60	55	65	**52**	**60**	69	77	61	70	56	68	50	57	46	56	**57**	**66**
Ten years	52	69	41	60	36	46	41	55	36	49	**39**	**54**	59	70	53	66	49	66	37	48	38	53	**48**	**62**
WALES	*(40)*		*(43)*		*(52)*		*(72)*		*(21)*		*(280)*		*(53)*		*(60)*		*(72)*		*(73)*		*(36)*		*(332)*	
One year	85	87	72	73	89	91	85	87	80	83	**82**	**85**	79	81	78	80	85	88	75	79	72	75	**79**	**82**
Five years	68	74	51	56	65	74	54	65	46	49	**57**	**66**	60	70	55	61	56	67	56	65	58	67	**58**	**68**
Ten years	53	67	37	44	51	68	42	54	41	45	**44**	**57**	56	70	43	54	47	66	41	53	41	56	**48**	**63**

[1] Including those for whom no deprivation category was assigned. C - crude survival, R - relative survival (see Chapter 3). Number of patients contributing to each analysis in parentheses.

WOMEN

1981-85

	Affluent C R	2 C R	3 C R	4 C R	Deprived C R	All[1] C R	
	(2,097)	*(1,819)*	*(1,550)*	*(1,248)*	*(795)*	*(7,693)*	**ENGLAND & WALES**
One year	**94 95**	**92 95**	**90 93**	**90 93**	**86 89**	**91 94**	
Five years	**75 82**	**70 79**	**68 77**	**65 74**	**61 70**	**69 78**	
	65 77	**59 73**	**58 73**	**51 66**	**49 65**	**58 73**	
	(2,031)	*(1,756)*	*(1,463)*	*(1,154)*	*(746)*	*(7,305)*	**ENGLAND**
One year	94 96	92 95	90 93	91 94	87 90	**91 94**	
Five years	75 82	71 79	68 77	64 74	61 71	**70 78**	
	65 77	60 73	58 72	50 66	49 65	**58 73**	
	(169)	*(176)*	*(190)*	*(158)*	*(183)*	*(888)*	**Northern & Yorkshire**
One year	94 96	93 96	95 97	89 92	86 89	**91 94**	
Five years	77 84	75 83	72 80	62 71	58 70	**69 78**	
	66 79	64 80	61 75	44 59	47 65	**57 73**	
	(137)	*(130)*	*(137)*	*(155)*	*(81)*	*(641)*	**Trent**
One year	91 94	88 92	82 84	92 96	85 89	**88 91**	
Five years	74 81	68 80	61 70	64 73	60 68	**66 75**	
	68 77	58 72	55 69	49 64	45 62	**56 70**	
	(324)	*(215)*	*(148)*	*(117)*	*(40)*	*(863)*	**Anglia & Oxford**
One year	97 98	91 94	90 92	96 96	80 86	**93 95**	
Five years	76 82	68 77	68 75	*74 84*	*67 84*	**72 80**	
	66 76	58 71	50 62	*64 83*	*54 77*	**60 74**	
	(185)	*(178)*	*(112)*	*(98)*	*(58)*	*(645)*	**North Thames**
One year	91 93	92 94	91 92	*91 93*	*91 92*	**91 93**	
Five years	72 78	72 79	73 80	65 74	*72 83*	**70 78**	
	58 69	61 73	65 79	*46 63*	*53 68*	**58 71**	
	(439)	*(251)*	*(192)*	*(153)*	*(54)*	*(1,126)*	**South Thames**
One year	93 95	94 97	93 96	90 93	*91 94*	**93 96**	
Five years	74 81	66 75	68 76	65 76	*57 65*	**69 78**	
	60 72	54 68	59 74	49 65	*48 65*	**57 71**	
	(397)	*(476)*	*(368)*	*(167)*	*(52)*	*(1,508)*	**South & West**
One year	94 96	93 96	89 92	92 95	82 85	**92 95**	
Five years	80 89	74 83	69 78	71 79	*65 76*	**74 83**	
	70 84	62 77	57 74	60 74	*48 63*	**62 78**	
	(171)	*(149)*	*(172)*	*(154)*	*(120)*	*(779)*	**West Midlands**
One year	95 99	91 94	93 95	88 91	93 95	**92 95**	
Five years	77 84	73 81	70 77	60 67	67 76	**70 78**	
	71 80	62 74	61 74	46 60	56 76	**60 73**	
	(209)	*(181)*	*(144)*	*(152)*	*(158)*	*(855)*	**North & West**
One year	91 93	91 95	90 93	90 92	83 86	**89 92**	
Five years	71 78	66 73	65 74	57 66	55 63	**64 72**	
	64 73	58 70	56 68	47 59	45 59	**55 67**	
	(66)	*(63)*	*(87)*	*(94)*	*(49)*	*(388)*	**WALES**
One year	89 91	81 84	88 91	83 85	81 84	**85 88**	
Five years	74 82	59 71	67 77	67 76	55 62	**65 75**	
	66 81	51 64	63 77	53 68	45 62	**56 72**	

1986-90

	Affluent C R	2 C R	3 C R	4 C R	Deprived C R	All[1] C R	
	(2,649)	*(2,534)*	*(2,114)*	*(1,617)*	*(902)*	*(9,976)*	**ENGLAND & WALES**
One year	**94 97**	**93 96**	**92 95**	**91 94**	**90 93**	**92 95**	
Five years	**78 87**	**76 86**	**73 83**	**70 80**	**67 79**	**74 84**	
	(2,572)	*(2,445)*	*(2,021)*	*(1,516)*	*(859)*	*(9,533)*	**ENGLAND**
One year	95 97	93 96	92 95	91 95	90 94	**93 96**	
Five years	79 87	76 86	74 84	70 81	67 80	**75 85**	
	(226)	*(222)*	*(206)*	*(209)*	*(170)*	*(1,048)*	**Northern & Yorkshire**
One year	96 98	94 96	93 97	89 93	87 92	**92 96**	
Five years	79 88	80 91	80 90	69 82	62 76	**74 87**	
	(182)	*(238)*	*(213)*	*(182)*	*(127)*	*(948)*	**Trent**
One year	94 97	89 92	88 91	91 93	90 93	**90 93**	
Five years	80 89	77 86	72 82	72 83	70 83	**75 85**	
	(405)	*(338)*	*(257)*	*(121)*	*(52)*	*(1,195)*	**Anglia & Oxford**
One year	96 97	94 97	91 94	89 93	92 96	**93 96**	
Five years	79 87	76 86	69 81	71 80	*71 85*	**75 85**	
	(250)	*(249)*	*(237)*	*(196)*	*(101)*	*(1,060)*	**North Thames**
One year	96 98	94 95	95 97	90 93	94 95	**94 96**	
Five years	78 87	76 84	78 87	70 80	76 86	**76 85**	
	(489)	*(365)*	*(269)*	*(197)*	*(77)*	*(1,425)*	**South Thames**
One year	95 98	93 96	93 96	95 98	86 90	**94 96**	
Five years	78 87	75 84	73 81	73 82	65 77	**75 84**	
	(476)	*(598)*	*(412)*	*(236)*	*(81)*	*(1,810)*	**South & West**
One year	93 95	93 95	92 95	93 97	*91 95*	**93 95**	
Five years	79 86	74 84	69 80	70 80	*67 74*	**74 83**	
	(252)	*(203)*	*(223)*	*(167)*	*(102)*	*(956)*	**West Midlands**
One year	94 97	95 98	95 98	91 94	- -	**94 98**	
Five years	79 87	76 85	79 89	72 82	*74 89*	**77 87**	
	(292)	*(232)*	*(204)*	*(208)*	*(149)*	*(1,091)*	**North & West**
One year	93 96	94 97	93 95	91 95	89 93	**92 95**	
Five years	80 87	79 88	76 88	64 76	61 72	**73 84**	
	(77)	*(89)*	*(93)*	*(101)*	*(43)*	*(443)*	**WALES**
One year	84 86	*91 92*	80 82	82 84	84 86	**85 88**	
Five years	66 77	*78 88*	58 68	64 73	53 66	**67 78**	

[1] Including those for whom no deprivation category was assigned. C - crude survival, R - relative survival (see Chapter 3). Number of patients contributing to each analysis in parentheses.

Table 27.2 Crude and relative survival (%) at one, five and ten years since diagnosis: NHS Region, deprivation category and calendar period of diagnosis

				1971-75								1976-80				
ADULTS	Affluent	2	3	4	Deprived	All[1]	Affluent	2	3	4	Deprived	All[1]
	C R	C R	C R	C R	C R	C R	C R	C R	C R	C R	C R	C R
ENGLAND & WALES	*(1,348)*	*(1,239)*	*(1,143)*	*(999)*	*(639)*	*(6,554)*	*(2,213)*	*(1,900)*	*(1,724)*	*(1,348)*	*(868)*	*(8,396)*
One year	87 89	85 87	84 86	83 86	79 82	84 86	88 90	85 88	86 89	81 84	79 83	85 87
Five years	61 67	54 61	53 61	50 58	43 51	53 61	64 71	58 66	55 64	52 61	49 58	57 65
Ten years	50 61	44 56	43 55	39 51	31 44	42 54	54 65	50 62	44 58	39 53	37 52	46 60
ENGLAND	*(1,289)*	*(1,181)*	*(1,070)*	*(905)*	*(609)*	*(6,163)*	*(2,123)*	*(1,811)*	*(1,611)*	*(1,229)*	*(815)*	*(7,874)*
One year	87 89	85 88	84 87	84 86	79 82	84 86	88 91	86 88	87 89	82 85	80 83	85 88
Five years	60 67	54 62	53 61	50 58	43 52	53 61	65 72	58 66	55 64	52 61	49 58	57 66
Ten years	50 61	45 57	42 55	40 51	31 44	42 55	54 65	50 63	44 58	40 54	37 52	47 60
Northern & Yorkshire	*(68)*	*(99)*	*(98)*	*(102)*	*(114)*	*(693)*	*(151)*	*(142)*	*(180)*	*(165)*	*(181)*	*(849)*
One year	87 88	86 87	83 85	83 87	72 75	82 84	86 88	82 84	84 88	82 84	76 80	82 85
Five years	55 65	56 62	46 53	45 53	41 50	49 57	63 69	59 69	56 65	55 64	51 61	56 65
Ten years	44 56	44 54	38 50	33 46	25 39	38 50	49 62	51 65	38 55	42 57	39 56	43 59
Trent	*(69)*	*(85)*	*(123)*	*(132)*	*(91)*	*(543)*	*(131)*	*(146)*	*(144)*	*(160)*	*(110)*	*(715)*
One year	78 82	79 81	82 84	80 82	69 72	78 81	88 90	80 82	84 87	75 77	78 81	80 83
Five years	55 63	54 59	55 62	45 52	37 45	50 57	60 65	54 62	49 60	45 53	51 61	52 60
Ten years	43 52	45 56	41 52	35 44	29 40	39 51	46 53	43 56	41 57	35 46	40 59	41 53
Anglia & Oxford	*(190)*	*(162)*	*(131)*	*(101)*	*(40)*	*(703)*	*(319)*	*(229)*	*(172)*	*(123)*	*(49)*	*(928)*
One year	86 88	86 88	84 89	92 94	80 83	86 88	92 95	88 91	84 86	82 86	86 91	87 90
Five years	64 71	52 58	54 63	61 69	45 50	57 64	70 78	63 71	55 62	49 56	42 50	60 69
Ten years	53 62	45 55	42 53	47 60	37 47	47 58	60 73	57 69	43 55	37 49	23 36	50 64
North Thames	*(192)*	*(150)*	*(115)*	*(141)*	*(91)*	*(771)*	*(192)*	*(180)*	*(185)*	*(121)*	*(74)*	*(781)*
One year	83 85	82 84	83 86	82 84	86 89	83 85	90 92	85 88	85 87	82 84	84 88	86 88
Five years	50 57	54 60	53 61	55 60	45 53	52 59	71 77	51 58	51 57	50 58	45 53	55 62
Ten years	42 52	46 53	46 58	42 49	32 44	42 52	57 69	44 55	41 53	40 54	37 52	45 57
South Thames	*(345)*	*(239)*	*(184)*	*(141)*	*(79)*	*(1,087)*	*(493)*	*(316)*	*(226)*	*(165)*	*(57)*	*(1,332)*
One year	87 89	89 92	85 86	85 87	85 87	86 89	90 92	89 91	89 92	87 90	88 92	89 92
Five years	61 68	56 64	54 60	51 61	52 59	56 64	62 70	60 69	61 69	59 70	67 75	61 70
Ten years	50 61	42 57	40 51	43 57	37 53	44 57	54 66	48 64	52 64	45 63	43 65	50 65
South & West	*(262)*	*(294)*	*(274)*	*(139)*	*(31)*	*(1,114)*	*(464)*	*(488)*	*(410)*	*(184)*	*(67)*	*(1,662)*
One year	90 92	87 89	86 89	82 84	77 79	86 89	86 88	85 87	90 93	82 85	73 76	86 88
Five years	66 72	58 67	58 68	51 58	45 52	58 67	65 72	60 68	59 69	56 66	43 49	60 69
Ten years	55 67	48 61	49 64	42 55	35 43	48 62	57 68	52 65	49 63	43 59	31 37	51 64
West Midlands	*(59)*	*(46)*	*(46)*	*(36)*	*(46)*	*(574)*	*(167)*	*(141)*	*(137)*	*(122)*	*(112)*	*(695)*
One year	90 92	78 81	89 90	77 83	87 89	83 86	86 88	87 89	88 91	85 89	84 87	86 89
Five years	61 65	54 60	56 61	44 50	44 52	53 60	62 67	58 64	52 61	49 57	52 60	55 63
Ten years	49 59	45 60	44 55	34 44	37 52	41 53	49 58	46 56	39 55	38 51	39 54	43 56
North & West	*(104)*	*(106)*	*(99)*	*(113)*	*(117)*	*(678)*	*(206)*	*(169)*	*(157)*	*(189)*	*(165)*	*(912)*
One year	90 92	81 86	81 83	85 89	80 83	83 86	88 90	85 89	83 86	83 87	79 82	83 87
Five years	64 72	45 55	38 43	46 54	43 52	46 55	62 70	57 64	51 59	49 58	42 51	52 61
Ten years	51 66	36 50	29 39	37 50	29 41	35 48	51 64	48 60	39 53	36 48	33 48	42 56
WALES	*(59)*	*(58)*	*(73)*	*(94)*	*(30)*	*(391)*	*(90)*	*(89)*	*(113)*	*(119)*	*(53)*	*(522)*
One year	81 83	74 76	79 82	81 84	83 86	79 81	80 83	82 84	83 86	70 74	68 70	78 81
Five years	61 68	47 51	56 65	48 60	42 48	50 58	59 68	55 61	49 58	49 58	51 56	53 62
Ten years	46 59	31 37	43 60	36 49	35 45	38 50	54 66	46 56	41 54	32 43	37 50	43 56

[1] Including those for whom no deprivation category was assigned. C - crude survival, R - relative survival (see Chapter 3).
Number of patients contributing to each analysis in parentheses.

	1981-85 Affluent (C R)	2 (C R)	3 (C R)	4 (C R)	Deprived (C R)	All¹ (C R)	1986-90 Affluent (C R)	2 (C R)	3 (C R)	4 (C R)	Deprived (C R)	All¹ (C R)	ADULTS
counts	(3,164)	(2,755)	(2,284)	(1,887)	(1,219)	(11,589)	(4,300)	(4,011)	(3,417)	(2,516)	(1,459)	(15,940)	**ENGLAND & WALES**
One year	91 93	89 92	88 91	87 90	82 85	88 91	92 95	92 94	90 93	88 91	88 92	91 94	*One year*
Five years	70 77	63 71	63 72	58 67	54 64	63 72	73 81	71 81	69 79	64 74	62 74	69 79	*Five years*
	59 70	52 66	52 66	45 60	43 58	52 66							
counts	(3,066)	(2,652)	(2,167)	(1,741)	(1,142)	(10,992)	(4,177)	(3,870)	(3,263)	(2,349)	(1,384)	(15,222)	**ENGLAND**
One year	91 93	89 92	88 91	88 91	83 86	89 91	92 95	92 95	91 94	89 92	89 93	91 94	*One year*
Five years	70 77	63 72	63 72	58 67	54 64	63 72	73 81	71 81	69 79	65 75	63 75	69 80	*Five years*
	59 70	52 66	52 66	45 60	43 59	52 66							
counts	(255)	(268)	(259)	(231)	(263)	(1,295)	(377)	(356)	(346)	(310)	(276)	(1,683)	**Northern & Yorkshire**
One year	91 93	91 94	90 94	88 90	84 87	89 92	94 96	94 96	90 94	87 91	85 90	90 94	*One year*
Five years	73 80	66 73	64 73	58 67	54 66	63 72	72 83	76 87	72 83	63 75	59 73	69 81	*Five years*
	60 75	56 71	54 68	41 56	44 62	51 67							
counts	(208)	(208)	(208)	(236)	(133)	(994)	(301)	(384)	(347)	(286)	(205)	(1,530)	**Trent**
One year	89 91	85 90	79 82	89 92	80 83	85 88	91 94	88 92	86 89	89 93	89 93	89 92	*One year*
Five years	69 76	56 66	55 65	57 66	53 62	58 67	71 79	72 82	67 78	67 79	64 78	69 79	*Five years*
	58 70	45 59	46 62	43 58	40 58	47 61							
counts	(480)	(347)	(233)	(183)	(53)	(1,323)	(635)	(522)	(428)	(193)	(77)	(1,889)	**Anglia & Oxford**
One year	95 97	87 90	90 93	91 93	79 84	91 93	95 97	92 96	91 94	87 91	91 96	92 95	*One year*
Five years	71 78	62 71	65 73	63 73	64 75	66 74	74 83	72 83	69 80	65 77	64 78	71 81	*Five years*
	61 73	51 63	47 62	53 71	54 71	54 68							
counts	(287)	(253)	(173)	(165)	(107)	(1,002)	(434)	(403)	(385)	(313)	(183)	(1,761)	**North Thames**
One year	86 87	88 91	87 89	88 91	86 87	87 89	94 97	92 94	92 94	91 94	91 94	92 95	*One year*
Five years	62 68	65 72	62 71	58 66	57 66	61 69	72 82	69 79	71 81	66 75	67 79	70 79	*Five years*
	50 60	55 67	54 67	42 57	44 56	50 62							
counts	(651)	(404)	(276)	(223)	(86)	(1,702)	(775)	(564)	(421)	(296)	(121)	(2,217)	**South Thames**
One year	92 94	91 93	91 94	88 92	86 89	91 93	92 95	91 94	90 94	91 94	83 87	91 94	*One year*
Five years	69 76	58 67	64 73	60 72	54 63	63 72	72 82	69 79	68 77	66 75	61 72	69 79	*Five years*
	55 67	46 60	52 68	46 63	42 60	51 64							
counts	(603)	(702)	(559)	(249)	(82)	(2,260)	(754)	(957)	(669)	(380)	(121)	(2,893)	**South & West**
One year	91 94	90 93	87 90	89 91	80 83	89 92	91 93	92 95	90 93	90 94	93 95	91 93	*One year*
Five years	73 82	68 77	65 75	64 73	56 65	68 77	73 81	70 80	65 77	67 79	65 75	69 79	*Five years*
	63 77	56 72	53 69	52 68	44 58	56 72							
counts	(256)	(217)	(248)	(217)	(175)	(1,131)	(427)	(330)	(344)	(272)	(149)	(1,537)	**West Midlands**
One year	93 97	88 91	92 95	87 90	89 91	90 93	93 96	93 98	94 98	88 92	93 97	93 96	*One year*
Five years	73 81	63 71	64 72	56 64	57 67	63 72	74 82	72 83	75 85	64 76	71 86	71 82	*Five years*
	63 75	51 64	53 67	43 59	46 63	52 66							
counts	(326)	(253)	(211)	(237)	(243)	(1,285)	(474)	(354)	(323)	(299)	(252)	(1,712)	**North & West**
One year	89 91	87 91	89 91	86 88	77 80	86 89	90 94	91 94	92 95	88 92	88 94	90 94	*One year*
Five years	66 73	62 69	58 68	51 60	49 58	58 66	72 82	72 82	70 82	58 70	57 69	67 78	*Five years*
	56 66	55 67	49 63	41 53	37 52	48 61							
counts	(98)	(103)	(117)	(146)	(77)	(597)	(123)	(141)	(154)	(167)	(75)	(718)	**WALES**
One year	87 88	82 85	84 87	77 79	75 78	81 84	85 88	88 90	78 81	76 78	71 74	82 84	*One year*
Five years	69 77	58 69	61 72	54 65	52 59	58 68	64 73	65 77	59 67	54 62	40 48	59 69	*Five years*
	61 76	48 62	59 72	43 60	42 55	50 65							

¹ Including those for whom no deprivation category was assigned. C - crude survival, R - relative survival (see Chapter 3).
Number of patients contributing to each analysis in parentheses.

Fig. 27.1 Relative survival up to ten years: trends by calendar period of diagnosis (1971-90) and NHS Region

ADULTS

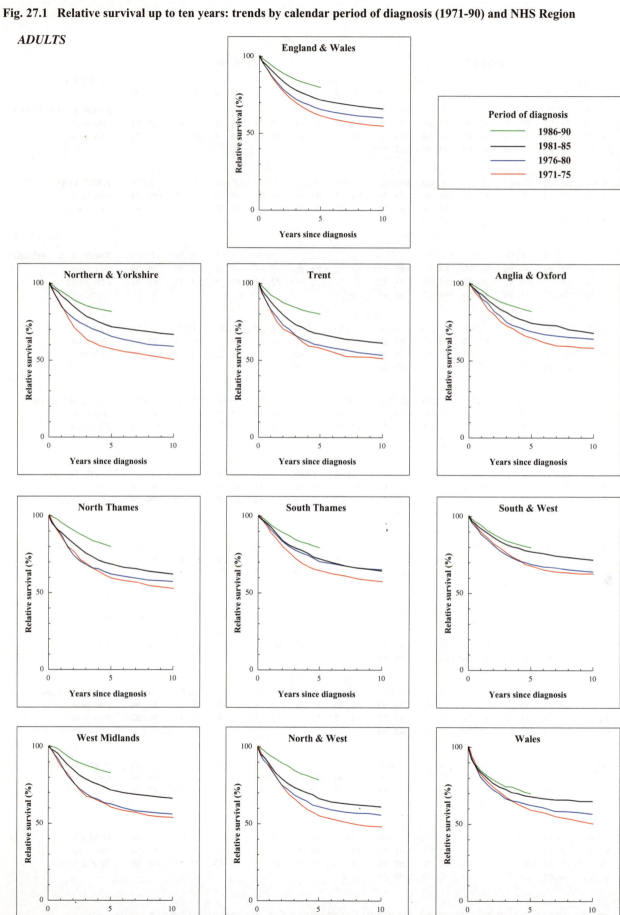

Fig. 27.2   Relative survival up to five years, by deprivation category and NHS Region: patients diagnosed 1986-90

ADULTS

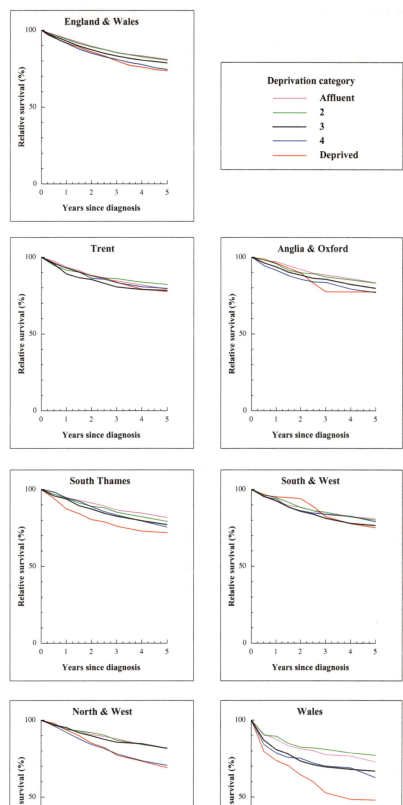

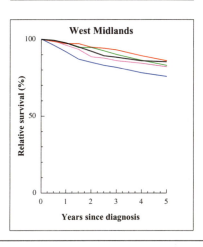

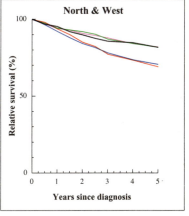

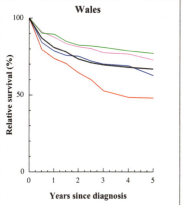

Fig. 27.3   Relative survival at five years by deprivation category, period of diagnosis (1981-90) and NHS Region

ADULTS

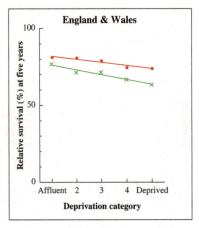

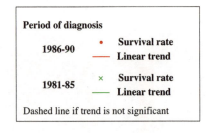

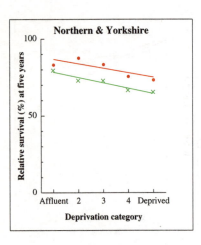

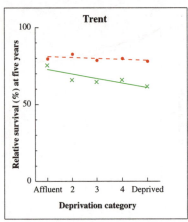

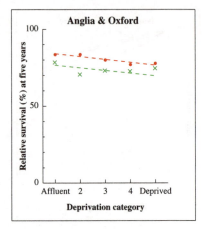

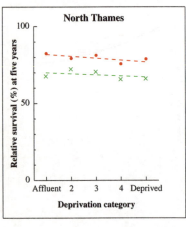

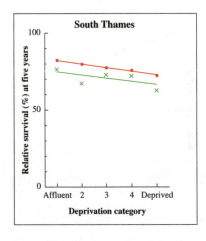

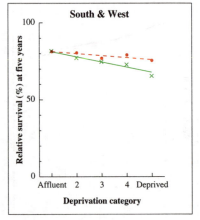

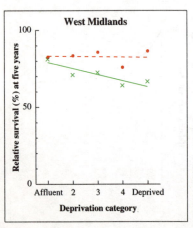

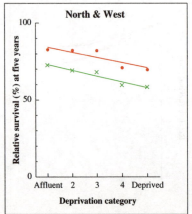

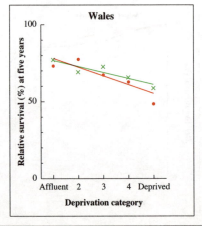

Fig. 27.4 Relative survival (%) by NHS Region and sex:
       one-year and five-year survival for patients diagnosed 1986-90, with 95% confidence intervals

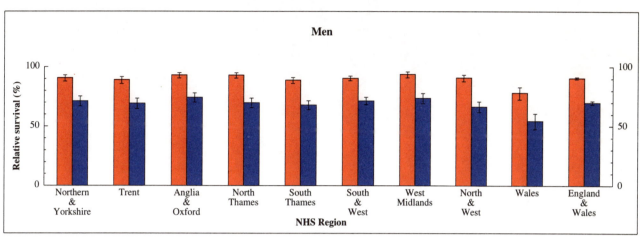

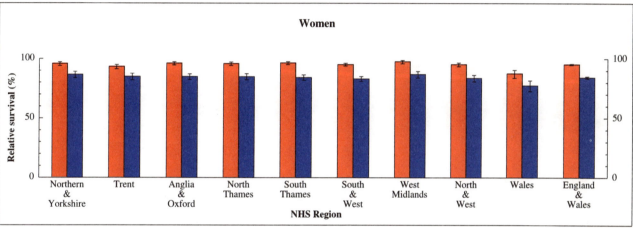

Fig. 27.5 Relative survival (%) at five years, with 95% confidence interval, by age at diagnosis and sex:
       England and Wales, patients diagnosed 1986-90

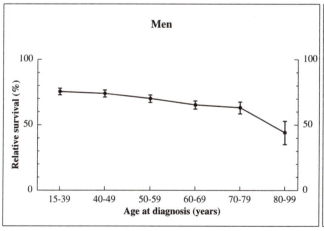
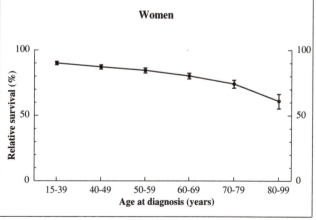

Chapter 28

BREAST

Breast cancer is the most common cancer in women in the UK and worldwide, representing about 20% of all cancers. There were 29,000 new cases in England and Wales in 1990, more than twice as many as for bowel cancer. Breast cancer in men is rare. Incidence in women is similar to that in other European countries, and was increasing at 5% every five years until the late 1980s. Mortality was the highest in Europe, but increasing more slowly (3% every five years), and the patterns by age in the late 1980s suggested an imminent peak and decline[1]: mortality has fallen sharply since then[2]. A national breast cancer screening programme began in 1988, with women aged 50-64 being invited for mammography every three years, but the decline in mortality is too early to be attributable to screening: changes in the number of children and age at first birth[3] and improved treatment, including widespread use of tamoxifen[4], are more likely explanations.

Relative survival at five years ranges from 85% for stage I disease to 20% for stage IV[5], and differences in stage distribution may explain some international differences in survival[6].

The survival analyses reported here include 411,000 patients diagnosed with a primary malignant breast cancer in England and Wales during the period 1971-90 and followed up to the end of 1995, about 89% of those eligible. Some 17,000 patients were ineligible because the breast tumour was either benign, *in situ*, of uncertain malignancy, or metastatic to the breast from a primary tumour in another organ (Table 28.1). Of the 461,000 eligible patients, 5% were excluded because their vital status was unknown at 6 July 1997, and another 5% because their survival was either zero or unknown. Almost 1% had multiple or synchronous cancers: each patient is included only once, and survival is analysed from the date of the first cancer.

More than 99% of the patients were women (Table 28A). During the 1980s, about 16% of tumours arose in the upper outer quadrant of the breast, but for 60% there was no anatomic detail. The proportion of tumours described as adenocarcinoma fell from 36% to 15% by the late 1980s, while those described as ductal, lobular or medullary carcinoma rose from less than 10% to more than 50%: almost all the increase was in intraduct carcinoma (data not shown). The proportion of tumours with non-specific or no morphology fell from 47% to 30%.

Survival trends

For patients diagnosed in England and Wales during

1986-90, one-year relative survival was 90% for men and women, and five-year survival was 68% for women and 72% for men. Survival in the early 1970s was 82-84% at one year and 54-56% at five years (Table 28.2: data for men at end of chapter).

Survival has improved steadily over time and in all regions (Figure 28.1). The relative survival curves still show excess mortality (declining curve) for at least ten years after diagnosis, but they have flattened slightly since the 1970s. This is reflected in the trends for one-year and five-year survival. One-year survival rose by an average of 2% every five years between 1971-75 and 1986-90, but the average improvement in five-year survival was 4% every five years, after adjustment for differences in the age distribution of patients between the sexes and over time (see Table 4.4).

For women, regional improvements in one-year survival range from 1% to 3% every five years (see Table-Figure 4.5H). The differences in trend are small, but the numbers of patients are large: one-year and five-year survival increased significantly faster than the national average in South Thames, at 3.0% and 5.5% every five years, respectively. Regional differences in survival at one and five years are not marked for women or men diagnosed during 1986-90[a] (Figure 28.4).

The well-known pattern of survival with age for women with breast cancer is seen in Figure 28.5. Among 7,000 women diagnosed under age 40 during 1986-90, relative survival at five years was 66% (95% confidence interval 65% to 67%). This is 7% lower than for women aged 40-49, and similar to the survival rates in women aged 70-79 at diagnosis. Relative survival for men is around 70-80% up to age 79, but only 41% in men aged 80-99 (see Table 4.7).

Deprivation

Survival for women is 5% to 10% higher for affluent groups than for deprived groups. The differences are remarkably consistent at one, five and ten years since diagnosis, in all regions, throughout the 1970s and 1980s. National and regional survival curves for women diagnosed during 1986-90 are shown in Figure 28.2.

The deprivation gradients in five-year survival are similar in most regions for the early 1980s and the late 1980s, although improvements in survival occurred for all groups (Figure 28.3). Among more than 100,000 patients of known deprivation status diagnosed in the early 1980s, the gap in survival between affluent and deprived groups was

Table 28A Sub-site distribution of breast cancer

ICD-9 code	Site description[1]	1971-75 No.	%	1976-80 No.	%	1981-85 No.	%	1986-90 No.	%	Total No.	%
174.0	Nipple and areola	-	-	-	-	1,519	1.4	1,604	1.4	**3,123**	-
174.1	Central portion	-	-	-	-	3,367	3.2	4,037	3.4	**7,404**	-
174.2	Upper-inner quadrant	-	-	-	-	4,848	4.6	5,864	5.0	**10,712**	-
174.3	Lower-inner quadrant	-	-	-	-	1,954	1.9	2,389	2.0	**4,343**	-
174.4	Upper-outer quadrant	-	-	*5,355*	-	16,013	15.2	21,270	18.1	**42,638**	-
174.5	Lower-outer quadrant	-	-	*1,067*	-	3,085	2.9	3,824	3.2	**7,976**	-
174.6	Axillary tail	-	-	*277*	-	789	0.7	793	0.7	**1,859**	-
174.8	Other	-	-	*2,184*	-	6,180	5.9	7,706	6.5	**16,070**	-
174.9	Breast unspecified	90,454	99.1	86,881	-	66,761	63.4	69,396	58.9	**313,492**	-
174	**Total for female breast**	**90,454**	**99.1**	**95,764**	**99.1**	**104,516**	**99.2**	**116,883**	**99.3**	**407,617**	**99.2**
175	**Male breast**	**796**	**0.9**	**892**	**0.9**	**840**	**0.8**	**856**	**0.7**	**3,384**	**0.8**
	Total	**91,250**		**96,656**		**105,356**		**117,739**		**411,001**	

[1] Breast was coded simply 174, without subdivision, in ICD-8 (1971-78). Special codes 174.0-174.3 used by the national cancer registry during that period (see Chapter 2) are incompatible with ICD-9, and tumours with those codes have been assigned to 174.9 in this table. Tumours of the skin of the breast are not included. Italicised data in column 2 relate to 1979-80 only. Misleading percentages have been suppressed.

5% at one year and 9% at five years; for 116,000 patients diagnosed in the late 1980s, the gap was similar: 5% at one year and 8% at five years (see Table 4.9).

In a previous study of 30,000 women diagnosed in South Thames during the 1980s, the deprivation gap in five-year survival was 11%: this was not due to differences in stage at diagnosis[7]. The slightly smaller 9% gap seen here may be due to the use of deprivation-specific life tables, which minimise bias in estimating the deprivation gradient, compared to estimations made with a single life table (see Chapter 3).

International comparison

Survival for women diagnosed in England and Wales during 1986-90 was similar to that in Scotland (68% vs. 66%) but 5% lower than the average for patients diagnosed during 1985-89 in the European populations included in the EUROCARE study (see Table 4.12): Slovenia, Slovakia, Poland and Estonia had similar or lower survival. Both the highest regional survival rate in England and Wales and the national survival rate for women in the affluent group were still lower than the European average. Survival for men was similar to the European average. For both women and men, survival was more than 10% lower than in the USA.

European differences in breast cancer survival have been explored in detail[8,9]. For women diagnosed during 1978-85, the hazard of death in the UK was higher at all ages than for other European countries in the EUROCARE study, particularly in the first six months after diagnosis (based on 8,700 deaths in the first six months): this suggests that later diagnosis and or adherence to treatment protocols may be important.

[a] For a few Health Authorities in Anglia & Oxford Region, doubt was raised over linkage of 1990 cancer registrations to death data. Analyses were repeated after exclusion of all patients registered in Oxford cancer registry in 1990 (9.4% of regional total for 1986-90). Crude and relative one-year and five-year regional survival rates for 1986-90 were either the same or 1-2% higher in each deprivation category.

1.   Coleman MP, Estève J, Damiecki P, Arslan A, Renard H. *Trends in cancer incidence and mortality (IARC Scientific Publications No. 121).* Lyon: International Agency for Research on Cancer, 1993

2.   Beral V, Hermon C, Reeves G, Peto R. Sudden fall in breast cancer death rates in England and Wales. Lancet 1995; 345: 1642-1643

3.   Hermon C, Beral V. Breast cancer mortality rates are levelling off or beginning to decline in many western countries: analysis of time trends, age-cohort and age-period models of breast cancer mortality in 20 countries. Br J Cancer 1996; 73: 955-960

4.   Early Breast Cancer Trialists' Collaborative Group. Systemic treatment of early breast cancer by hormonal, cytotoxic or immune therapy. Lancet 1992; 339: 1-15, 71-85

5.   Yorkshire Cancer Organisation. *Cancer in Yorkshire. Cancer registry report series, 3: breast cancer.* Leeds: YCO, 1995

6.   Karjalainen S, Aareleid T, Hakulinen T, Pukkala E, Rahu M, Tekkel M. Survival of female breast cancer patients in Finland and Estonia: stage at diagnosis an important determinant of the difference between countries. Soc Sci Med 1989; 28: 233-238

7.   Schrijvers CTM, Mackenbach J, Lutz J-M, Quinn MJ, Coleman MP. Deprivation and survival from breast cancer. Br J Cancer 1995; 72: 738-743

8.   Sant M, Capocaccia R, Verdecchia A, Estève J, Gatta G, Micheli A, Coleman MP, Berrino F, EUROCARE Working Group. Survival of women with breast cancer in Europe: variation with age, year of diagnosis and country. Int J Cancer 1998; 77: 679-683

9.   Quinn MJ, Martinez-Garcia C, Berrino F. Variations in survival from breast cancer in Europe by age and country, 1978-89. Eur J Cancer 1998; 34: 2204-2211

Table 28.1 Ineligible and excluded records, and patients included in analyses, by calendar period of diagnosis

	Calendar period of diagnosis														
	1971-75			**1976-80**			**1981-85**			**1986-90**			**Overall**		
Total registered	**105,642**			**112,345**			**123,082**			**143,383**			**484,452**		
Ineligible	*Records*	*Patients*	*%*	*Records*	*Patients*	*%*	*Records*	*Patients*	*%*	*Records*	*Patients*	*%*	*Records*	*Patients*	*%*
Incomplete data[1]	927	927		1,114	1,114		224	224		220	220		2,485	**2,485**	
Resident outside England and Wales	1,102	1,084		741	718		597	509		622	511		3,062	**2,822**	
In situ neoplasm[2]	3,882	3,715		3,221	3,167		3,739	3,721		4,980	4,955		15,822	**15,558**	
Benign or uncertain behaviour[2]	1,369	692		1,006	381		2	2		1	1		2,378	**1,076**	
Metastatic[2]	533	528		215	211		11	11		32	32		791	**782**	
Otherwise ineligible[3]	190	110		182	121		0	0		0	0		372	**231**	
Lymphoma[4]	16	16		9	9		4	0		2	0		31	**25**	
Leukaemia or myeloma[4]	1	0		0	0		1	0		2	0		4	**0**	
		7,072			5,721			4,467			5,719			22,979	
Total eligible		**98,570**	**100**		**106,624**	**100**		**118,615**	**100**		**137,664**	**100**		**461,473**	**100**
Aged under 15 or 100 years or over at diagnosis	33	32	<0.1	54	52	<0.1	83	79	<0.1	120	117	<0.1	290	**280**	<0.1
Vital status unknown[5]	4,254	3,422	3.5	5,938	5,193	4.9	5,453	4,955	4.2	8,615	8,083	5.9	24,260	**21,653**	4.7
Duplicate registration[6]	0	0	0.0	6	5	<0.1	3	1	<0.1	3	3	<0.1	12	**9**	<0.1
Multiple primary[7]	9	5	<0.1	217	94	<0.1	751	427	0.4	1,715	1,372	1.0	2,692	**1,898**	0.4
Synchronous tumours[8]	1,044	535	0.5	825	413	0.4	790	471	0.4	1,495	800	0.6	4,154	**2,219**	0.5
Sex not known	0	0	0.0	0	0	0.0	0	0	0.0	0	0	0.0	0	**0**	0.0
Sex incompatible with site code[9]	0	0	0.0	2	2	<0.1	9	4	<0.1	43	22	<0.1	54	**28**	<0.1
Invalid dates[10]	3	2	<0.1	24	15	<0.1	42	20	<0.1	108	44	<0.1	177	**81**	<0.1
Zero survival[11]	3,780	3,324	3.4	4,544	4,194	3.9	7,909	7,302	6.2	10,308	9,484	6.9	26,541	**24,304**	5.3
Total exclusions		**7,320**	**7.4**		**9,968**	**9.3**		**13,259**	**11.2**		**19,925**	**14.5**		**50,472**	**10.9**
Patients included in analysis		**91,250**	**92.6**		**96,656**	**90.7**		**105,356**	**88.8**		**117,739**	**85.5**		**411,001**	**89.1**

[1] Main data item(s) invalid or incompatible with one another: name, sex, date of birth, date of diagnosis and (if present) date of death, postcode, site, morphology and behaviour

[2] Tumour either *in situ* (behaviour code 2), benign (0), uncertain if benign or malignant (1) or metastatic (6 or 9)

[3] Skin of breast (ICD-8 site code (ONS modification) 174.2, 1971-78)

[4] Morphology that of lymphoma (included with non-Hodgkin lymphoma, Ch. 41) or myeloma (included with multiple myeloma, Ch. 43), or leukaemia (excluded)

[5] Tracing of vital status at National Health Service Central Register incomplete at 6 July 1997

[6] Same site code, sex, cancer registry and cancer registry number as an earlier registration

[7] Same site code and person number as an earlier registration(s): mostly confirmed multiple primary tumours at this site, some unresolved duplicate registrations

[8] Same site code, sex, date of birth and date of diagnosis as another registration(s): mostly synchronous or (in paired organs) bilateral tumours in same anatomic site in one individual, not previously linked: also some duplicate registrations

[9] Site code specific to one sex; tumour registered to opposite sex: in ICD-9 (1979-90), the breast is coded 174 (women) or 175 (men)

[10] Impossible sequence of dates (birth, diagnosis and, if present, emigration or death); or date of death after 6 July 1997

[11] Date of diagnosis same as date of death: some are patients who did die on the day of diagnosis, but most were registered solely from a death certificate (death certificate only, or DCO), and their survival time is thus unknown

Table 28.2 **Crude and relative survival (%) at one, five and ten years since diagnosis: NHS Region, deprivation category and calendar period of diagnosis**

		1971-75											1976-80											
WOMEN		*Affluent*		*2*		*3*		*4*		*Deprived*		*All[1]*		*Affluent*		*2*		*3*		*4*		*Deprived*		*All[1]*	
		C	R	C	R	C	R	C	R	C	R	C	R	C	R	C	R	C	R	C	R	C	R	C	R
ENGLAND & WALES		*(15,197)*		*(15,862)*		*(15,722)*		*(14,574)*		*(11,443)*		*(90,454)*		*(19,270)*		*(19,729)*		*(19,915)*		*(18,117)*		*(14,567)*		*(95,764)*	
	One year	84	87	82	85	81	84	79	82	78	81	81	84	86	89	85	88	82	86	81	84	80	83	83	86
	Five years	52	59	49	56	47	54	45	52	42	49	47	54	57	64	54	62	51	60	50	58	47	55	52	60
	Ten years	35	45	32	42	30	41	28	39	27	37	31	41	39	50	36	48	34	47	33	45	30	42	35	47
ENGLAND		*(14,638)*		*(15,048)*		*(14,647)*		*(13,331)*		*(10,747)*		*(84,822)*		*(18,555)*		*(18,803)*		*(18,567)*		*(16,674)*		*(13,839)*		*(89,835)*	
	One year	84	87	82	85	81	84	79	82	78	81	**81**	**84**	86	89	85	88	83	86	81	85	80	84	**83**	**86**
	Five years	52	59	49	56	47	54	45	52	42	50	**47**	**54**	57	64	54	62	51	60	50	58	47	55	**52**	**60**
	Ten years	35	45	32	43	30	41	28	39	27	37	**31**	**41**	39	51	36	48	34	47	33	45	30	42	**35**	**47**
Northern & Yorkshire		*(1,041)*		*(1,262)*		*(1,546)*		*(1,728)*		*(2,037)*		*(10,496)*		*(1,554)*		*(1,874)*		*(2,340)*		*(2,642)*		*(3,008)*		*(11,658)*	
	One year	84	87	83	86	80	83	78	81	76	79	**81**	**84**	88	90	85	89	82	86	79	82	81	84	**82**	**86**
	Five years	48	54	49	56	48	56	47	54	43	50	**46**	**54**	58	66	54	62	52	60	47	55	47	55	**51**	**59**
	Ten years	32	41	31	42	31	43	29	40	28	39	**30**	**42**	40	51	37	49	35	48	31	43	31	44	**34**	**47**
Trent		*(1,115)*		*(1,460)*		*(1,750)*		*(2,020)*		*(1,526)*		*(8,672)*		*(1,476)*		*(1,739)*		*(2,126)*		*(2,239)*		*(1,729)*		*(9,649)*	
	One year	84	87	83	85	83	86	80	83	80	82	**81**	**84**	86	88	85	88	81	85	82	85	79	83	**83**	**86**
	Five years	53	59	48	55	46	52	46	52	43	51	**47**	**54**	57	64	55	62	50	57	52	60	46	55	**52**	**60**
	Ten years	36	45	31	40	30	40	30	41	26	36	**30**	**41**	39	50	37	49	34	45	34	47	29	41	**34**	**47**
Anglia & Oxford		*(1,886)*		*(1,769)*		*(1,621)*		*(1,181)*		*(481)*		*(7,927)*		*(2,458)*		*(2,348)*		*(1,981)*		*(1,277)*		*(568)*		*(9,053)*	
	One year	86	89	83	85	82	85	80	83	80	83	**82**	**86**	87	90	86	89	84	88	82	86	81	84	**85**	**88**
	Five years	53	60	49	55	46	54	45	52	41	48	**48**	**56**	55	63	53	61	53	62	49	57	45	53	**52**	**61**
	Ten years	35	46	32	43	29	40	28	38	27	36	**31**	**42**	38	49	35	48	36	49	32	44	31	44	**35**	**48**
North Thames		*(2,473)*		*(2,224)*		*(1,951)*		*(1,912)*		*(1,454)*		*(11,279)*		*(2,158)*		*(2,050)*		*(1,936)*		*(1,921)*		*(1,540)*		*(9,980)*	
	One year	84	87	83	86	83	86	81	83	78	81	**82**	**85**	87	89	86	89	85	88	83	86	82	86	**85**	**88**
	Five years	54	61	51	57	49	56	47	54	42	49	**49**	**56**	59	66	55	62	54	61	50	59	50	58	**54**	**62**
	Ten years	37	46	32	42	33	43	30	40	28	38	**32**	**43**	41	52	37	49	36	47	33	45	33	46	**36**	**48**
South Thames		*(3,355)*		*(2,651)*		*(2,454)*		*(2,097)*		*(1,024)*		*(12,801)*		*(3,877)*		*(3,118)*		*(2,611)*		*(2,028)*		*(880)*		*(13,293)*	
	One year	84	87	81	84	80	83	78	81	75	78	**81**	**84**	86	90	84	87	82	85	80	83	77	81	**83**	**86**
	Five years	50	57	47	54	44	51	41	49	40	47	**46**	**54**	57	65	52	61	49	57	49	58	44	54	**52**	**61**
	Ten years	34	44	31	41	28	38	25	34	24	35	**29**	**40**	39	51	34	46	31	43	30	43	27	38	**33**	**46**
South & West		*(2,519)*		*(3,303)*		*(2,879)*		*(1,419)*		*(449)*		*(11,577)*		*(3,091)*		*(3,980)*		*(3,301)*		*(1,670)*		*(519)*		*(13,062)*	
	One year	83	85	82	84	81	84	80	83	78	80	**81**	**84**	85	89	83	86	80	84	83	87	78	82	**83**	**86**
	Five years	54	60	51	58	48	55	45	53	41	47	**49**	**57**	57	65	54	62	51	59	51	60	42	51	**53**	**62**
	Ten years	36	46	34	46	32	43	29	40	27	38	**32**	**44**	40	51	35	48	33	46	34	48	27	39	**35**	**48**
West Midlands		*(631)*		*(656)*		*(718)*		*(872)*		*(958)*		*(9,567)*		*(1,710)*		*(1,590)*		*(2,006)*		*(2,305)*		*(2,289)*		*(10,077)*	
	One year	90	92	85	88	84	88	82	85	80	83	**83**	**86**	88	91	85	89	85	88	84	87	80	84	**84**	**87**
	Five years	53	59	51	59	47	54	46	53	45	53	**48**	**55**	58	64	54	61	54	61	53	61	48	56	**53**	**61**
	Ten years	38	47	35	47	30	39	28	39	29	41	**31**	**42**	41	51	37	48	36	48	34	46	30	41	**35**	**47**
North & West		*(1,618)*		*(1,723)*		*(1,728)*		*(2,102)*		*(2,818)*		*(12,503)*		*(2,231)*		*(2,104)*		*(2,266)*		*(2,592)*		*(3,306)*		*(13,063)*	
	One year	83	85	81	84	80	83	78	81	77	80	**79**	**82**	84	87	84	87	82	85	79	83	79	83	**81**	**85**
	Five years	52	58	48	55	45	52	44	52	42	49	**45**	**53**	56	63	55	64	51	60	48	57	45	54	**50**	**59**
	Ten years	34	44	31	42	29	40	28	39	26	36	**29**	**40**	39	49	38	50	35	49	32	45	29	41	**34**	**47**
WALES		*(559)*		*(814)*		*(1,075)*		*(1,243)*		*(696)*		*(5,632)*		*(715)*		*(926)*		*(1,348)*		*(1,443)*		*(728)*		*(5,929)*	
	One year	84	86	77	79	79	82	78	80	79	81	**78**	**81**	82	85	82	85	81	83	80	83	78	81	**81**	**84**
	Five years	52	60	43	49	46	53	43	49	44	49	**45**	**51**	52	59	52	58	49	57	48	56	42	49	**49**	**57**
	Ten years	38	50	27	36	31	41	28	38	28	37	**29**	**39**	35	45	37	48	34	47	34	47	29	38	**34**	**46**

[1] Including those for whom no deprivation category was assigned. C - crude survival, R - relative survival (see Chapter 3). Number of patients contributing to each analysis in parentheses.

	1981-85 Affluent C	R	2 C	R	3 C	R	4 C	R	Deprived C	R	All[1] C	R	1986-90 Affluent C	R	2 C	R	3 C	R	4 C	R	Deprived C	R	All[1] C	R	WOMEN
	(22,925)		*(22,230)*		*(21,755)*		*(19,860)*		*(15,207)*		*(104,516)*		*(25,829)*		*(26,620)*		*(25,137)*		*(21,929)*		*(15,816)*		*(116,883)*		**ENGLAND & WALES**
One year	88	91	86	89	85	88	84	87	81	85	**85**	**88**	89	92	88	91	87	90	85	89	83	86	**87**	**90**	*One year*
Five years	60	67	56	64	54	63	52	61	49	58	**54**	**63**	63	70	60	69	59	68	56	65	53	63	**59**	**68**	*Five years*
	43	55	39	52	37	51	35	49	32	46	**37**	**51**													
	(22,163)		*(21,257)*		*(20,369)*		*(18,410)*		*(14,425)*		*(98,557)*		*(24,785)*		*(25,484)*		*(23,700)*		*(20,433)*		*(15,050)*		*(110,666)*		**ENGLAND**
	88	91	86	89	85	88	84	88	81	85	**85**	**88**	89	92	88	91	87	90	86	89	83	87	**87**	**90**	*One year*
	60	68	56	64	54	63	52	61	49	58	**54**	**63**	63	71	60	69	59	68	56	66	53	63	**59**	**68**	*Five years*
	43	55	39	52	37	51	35	49	32	45	**37**	**51**													
	(1,891)		*(2,214)*		*(2,524)*		*(2,899)*		*(3,249)*		*(12,940)*		*(2,299)*		*(2,798)*		*(2,800)*		*(2,855)*		*(3,048)*		*(13,899)*		**Northern & Yorkshire**
	88	91	86	89	87	90	85	89	80	84	**85**	**88**	88	91	88	91	87	91	85	89	83	87	**86**	**90**	*One year*
	60	67	56	64	58	67	53	63	48	57	**54**	**63**	61	69	61	71	58	67	55	65	54	64	**58**	**67**	*Five years*
	43	54	40	54	39	54	36	51	31	45	**37**	**52**													
	(1,746)		*(2,032)*		*(2,269)*		*(2,561)*		*(1,714)*		*(10,392)*		*(1,876)*		*(2,431)*		*(2,682)*		*(2,512)*		*(1,929)*		*(11,474)*		**Trent**
	87	90	85	88	83	87	83	86	81	85	**84**	**87**	87	90	87	91	86	89	83	86	81	85	**85**	**88**	*One year*
	56	64	55	63	52	60	49	57	48	58	**52**	**61**	62	69	60	67	58	66	53	63	52	61	**57**	**66**	*Five years*
	39	50	37	49	35	47	33	45	30	44	**35**	**48**													
	(3,074)		*(2,692)*		*(2,252)*		*(1,441)*		*(577)*		*(10,322)*		*(3,672)*		*(3,463)*		*(2,721)*		*(1,715)*		*(636)*		*(12,332)*		**Anglia & Oxford**
	87	91	86	90	84	88	83	87	83	86	**86**	**89**	88	92	87	90	86	90	85	89	84	87	**87**	**90**	*One year*
	60	68	56	65	55	64	55	65	51	60	**56**	**66**	62	70	59	68	60	70	57	66	50	59	**59**	**69**	*Five years*
	44	56	39	53	37	52	37	52	36	49	**39**	**54**													
	(2,477)		*(2,185)*		*(1,996)*		*(1,982)*		*(1,441)*		*(10,274)*		*(2,876)*		*(2,984)*		*(2,864)*		*(2,750)*		*(2,228)*		*(14,111)*		**North Thames**
	89	92	89	92	87	90	87	90	86	89	**88**	**91**	92	95	90	92	89	92	89	92	86	90	**89**	**92**	*One year*
	61	69	58	66	55	64	56	64	54	64	**57**	**66**	67	75	63	72	60	68	60	69	55	65	**62**	**70**	*Five years*
	45	57	43	55	38	51	38	51	38	53	**41**	**54**													
	(4,835)		*(3,459)*		*(2,924)*		*(2,196)*		*(883)*		*(14,839)*		*(4,745)*		*(3,917)*		*(3,088)*		*(2,377)*		*(1,130)*		*(15,511)*		**South Thames**
	89	92	87	90	86	89	86	89	80	83	**86**	**90**	90	93	90	93	88	92	88	91	86	89	**89**	**92**	*One year*
	61	70	56	66	55	64	54	63	50	59	**57**	**66**	65	73	62	71	60	70	55	65	55	64	**61**	**70**	*Five years*
	43	57	39	53	37	52	37	51	34	47	**39**	**54**													
	(3,481)		*(4,190)*		*(3,608)*		*(1,743)*		*(552)*		*(13,983)*		*(3,737)*		*(4,737)*		*(4,150)*		*(2,502)*		*(866)*		*(16,098)*		**South & West**
	87	90	84	87	84	88	83	87	83	86	**85**	**88**	87	90	86	89	85	88	86	89	86	89	**86**	**89**	*One year*
	60	68	54	63	54	63	53	62	53	62	**55**	**64**	61	69	59	67	58	67	57	66	57	67	**59**	**68**	*Five years*
	42	54	37	50	37	51	35	49	32	46	**38**	**52**													
	(2,053)		*(1,935)*		*(2,328)*		*(2,550)*		*(2,495)*		*(11,474)*		*(2,514)*		*(2,335)*		*(2,660)*		*(2,695)*		*(2,036)*		*(12,326)*		**West Midlands**
	89	92	88	91	85	88	84	87	82	85	**85**	**89**	90	93	89	92	89	92	85	89	83	86	**88**	**91**	*One year*
	59	67	56	65	54	62	51	60	49	57	**53**	**62**	63	70	62	70	62	71	57	66	52	62	**59**	**68**	*Five years*
	43	55	39	51	34	46	33	46	31	44	**35**	**49**													
	(2,606)		*(2,550)*		*(2,468)*		*(3,038)*		*(3,514)*		*(14,333)*		*(3,066)*		*(2,819)*		*(2,735)*		*(3,027)*		*(3,177)*		*(14,915)*		**North & West**
	87	90	84	87	83	86	82	86	80	84	**83**	**87**	87	91	87	91	85	89	84	88	81	85	**85**	**89**	*One year*
	58	66	53	61	51	60	49	59	46	55	**51**	**60**	60	69	59	69	55	64	54	64	50	60	**56**	**65**	*Five years*
	41	52	37	49	35	49	33	47	30	43	**35**	**48**													
	(762)		*(973)*		*(1,386)*		*(1,450)*		*(782)*		*(5,959)*		*(1,044)*		*(1,136)*		*(1,437)*		*(1,496)*		*(766)*		*(6,217)*		**WALES**
	82	85	83	86	80	83	77	80	82	85	**80**	**83**	84	86	80	83	81	84	80	83	79	81	**81**	**84**	*One year*
	56	63	57	65	52	60	50	57	52	60	**53**	**61**	59	66	56	64	55	63	54	63	54	62	**56**	**65**	*Five years*
	38	50	41	53	37	50	35	49	35	48	**37**	**51**													

[1] Including those for whom no deprivation category was assigned. C - crude survival, R - relative survival (see Chapter 3).
Number of patients contributing to each analysis in parentheses.

Fig. 28.1   Relative survival up to ten years: trends by calendar period of diagnosis (1971-90) and NHS Region

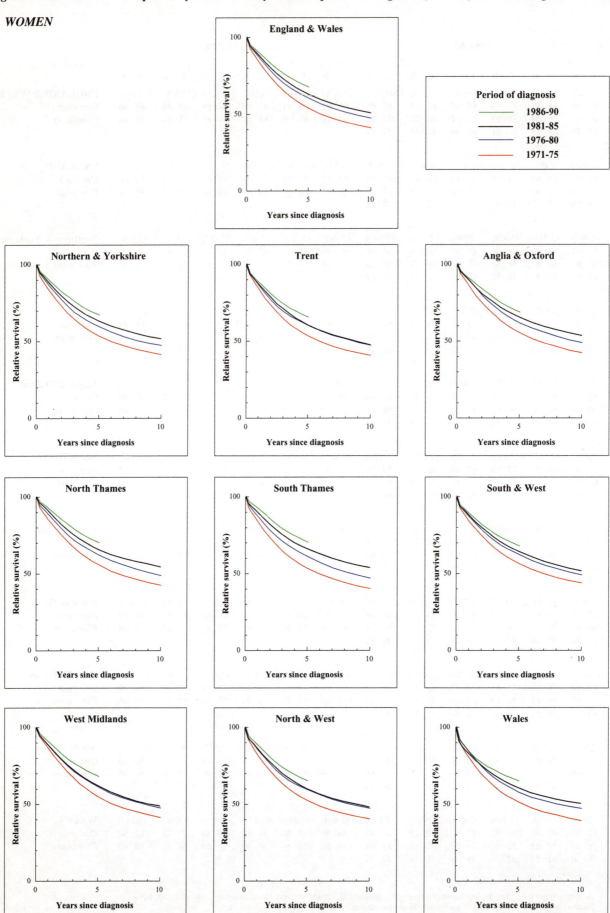

WOMEN

Period of diagnosis
- 1986-90
- 1981-85
- 1976-80
- 1971-75

Fig. 28.2   Relative survival up to five years, by deprivation category and NHS Region: patients diagnosed 1986-90

WOMEN

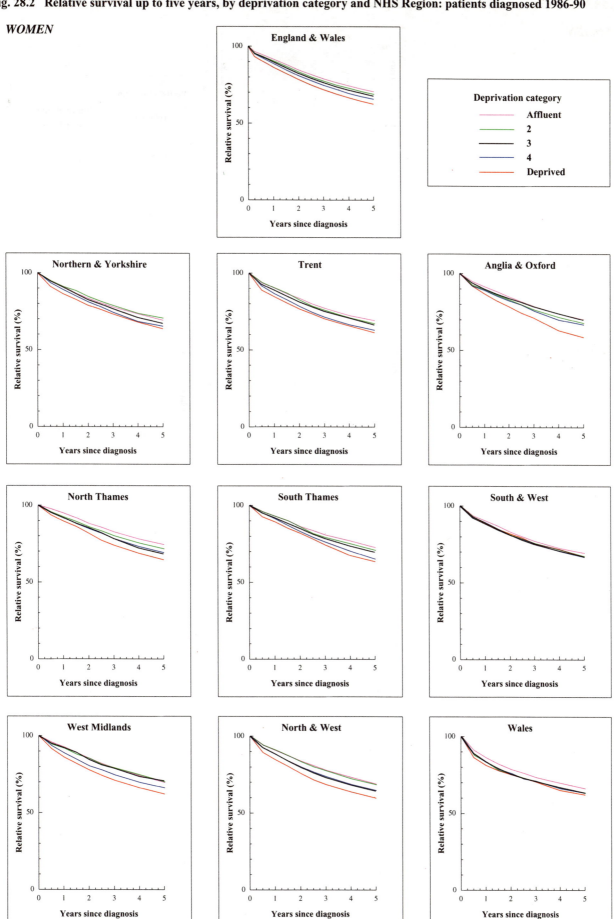

Fig. 28.3 Relative survival at five years by deprivation category, period of diagnosis (1981-90) and NHS Region

WOMEN

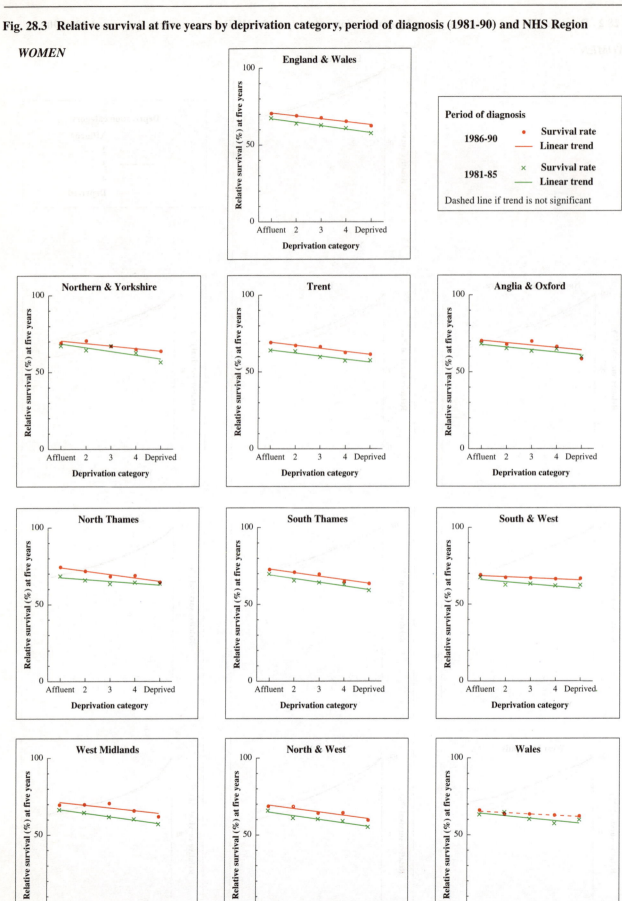

Fig. 28.4 Relative survival (%) by NHS Region and sex:
one-year and five-year survival for patients diagnosed 1986-90, with 95% confidence intervals

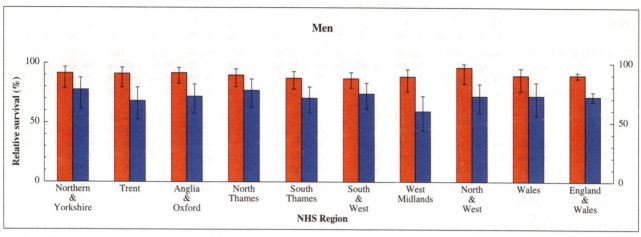

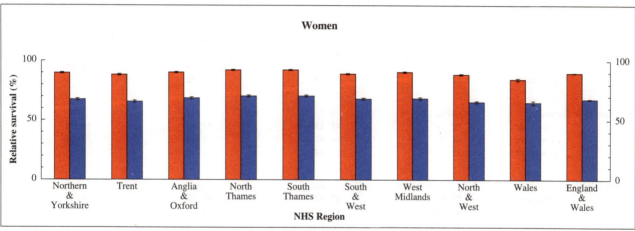

Fig. 28.5 Relative survival (%) at five years, with 95% confidence interval, by age at diagnosis and sex:
England and Wales, patients diagnosed 1986-90

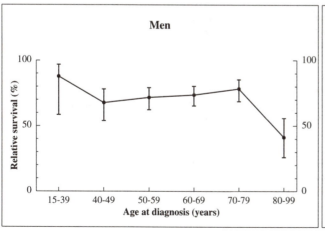

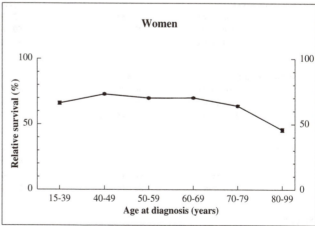

Table 28.2   Crude and relative survival (%) at one, five and ten years since diagnosis:
NHS Region, deprivation category and calendar period of diagnosis

MEN	1971-75						1976-80					
	Affluent	*2*	*3*	*4*	*Deprived*	*All*[1]	*Affluent*	*2*	*3*	*4*	*Deprived*	*All*[1]
	C R	C R	C R	C R	C R	C R	C R	C R	C R	C R	C R	C R
ENGLAND & WALES	*(119)*	*(131)*	*(140)*	*(144)*	*(106)*	*(796)*	*(180)*	*(169)*	*(180)*	*(169)*	*(144)*	*(892)*
One year	85 88	81 86	74 78	74 80	74 80	77 82	81 85	82 89	78 85	80 86	80 85	80 85
Five years	54 67	44 58	40 52	38 51	41 57	43 56	57 70	51 69	42 56	44 58	43 56	48 62
Ten years	37 59	30 51	25 46	20 37	22 38	26 45	41 62	29 54	24 42	29 48	24 44	30 52
	1981-85						**1986-90**					
	(168)	*(171)*	*(181)*	*(163)*	*(140)*	*(840)*	*(185)*	*(225)*	*(177)*	*(160)*	*(91)*	*(856)*
One year	83 89	82 88	75 80	78 84	74 78	79 84	90 95	83 89	85 90	80 85	81 88	84 90
Five years	56 73	43 59	44 59	40 55	30 43	43 59	60 75	52 68	55 73	52 72	44 61	54 72
Ten years	29 50	26 49	24 44	20 38	15 29	23 43						

	1971-75	1976-80	1981-85	1986-90
Northern & Yorkshire	*(87)*	*(87)*	*(104)*	*(86)*
One year	76 81	71 76	74 80	85 91
Five years	39 55	34 43	40 59	58 78
Ten years	16 34	19 30	21 42	
Trent	*(76)*	*(74)*	*(92)*	*(82)*
One year	78 84	81 86	78 82	86 91
Five years	40 53	43 57	38 53	51 68
Ten years	22 37	26 49	17 32	
Anglia & Oxford	*(68)*	*(72)*	*(74)*	*(96)*
One year	72 77	81 85	77 84	85 92
Five years	41 54	48 65	38 52	54 72
Ten years	29 45	33 54	21 37	
North Thames	*(97)*	*(100)*	*(88)*	*(101)*
One year	77 81	87 92	80 84	85 90
Five years	47 62	53 68	45 62	59 77
Ten years	32 49	31 50	23 43	
South Thames	*(117)*	*(196)*	*(105)*	*(126)*
One year	72 77	85 92	86 92	81 87
Five years	39 52	59 76	52 70	52 71
Ten years	30 48	40 66	29 54	
South & West	*(126)*	*(117)*	*(122)*	*(133)*
One year	81 84	73 79	88 93	82 87
Five years	47 57	50 67	51 67	54 74
Ten years	25 43	30 57	33 61	
West Midlands	*(81)*	*(75)*	*(83)*	*(70)*
One year	80 86	83 87	72 77	84 89
Five years	40 53	40 51	36 49	46 59
Ten years	24 45	32 50	19 38	
North & West	*(94)*	*(103)*	*(105)*	*(95)*
One year	82 87	85 89	74 80	89 96
Five years	46 59	52 66	39 52	55 72
Ten years	28 48	27 48	19 36	
WALES	*(50)*	*(68)*	*(67)*	*(67)*
One year	75 82	63 68	77 83	85 90
Five years	52 69	32 42	47 63	58 73
Ten years	34 60	17 35	25 43	

[1] Including those for whom no deprivation category was assigned. C - crude survival, R - relative survival (see Chapter 3).
Number of patients contributing to each analysis in parentheses.

Chapter 29

CERVIX

Cervical cancer is the second most common cancer in women world-wide (about 15% of all cancers), but it ranks sixth (4%) among women in the UK. About 4,300 new cases are diagnosed in England and Wales each year, and there are some 1,300 deaths.

A national screening programme has been in operation since the 1960s. Women aged 20-64 are invited for a cervical smear every five years with the aim of detecting pre-malignant disease (cervical dysplasia and *in situ* carcinoma). A key indicator of effectiveness would be a decline in the incidence of invasive cancer. Cancer of the cervix is largely caused by the human papilloma virus, transmitted sexually[1,2].

Incidence trends in the UK have been unusual. A gradual decline in overall incidence in the 1970s was followed by a rise in the late 1980s. The same pattern of age-specific trend was observed in three large populations (South Thames, West Midlands and Scotland): an increase of more than 6% a year among women under 45, a smaller rise for elderly women (65-74), but a fall among women aged 45-64. Mortality fell during the same period by about 1% a year, with a similar relativity between age groups[3]. These trends have been attributed to the ineffectiveness of the cervical screening programme[4], although subsequent improvements have been reported[5] and pre-malignant lesions rose substantially during the late 1980s (see below). Mortality trends for cervical cancer are more difficult to interpret, because some deaths are classified to 'uterus, not otherwise specified', but the discrepancy between incidence and mortality trends is likely to be attributable at least partly to improvements in survival.

The survival analyses here include almost 74,000 women diagnosed with invasive cancer of the cervix in England and Wales during the period 1971-90 and followed up to the end of 1995, about 91% of those eligible (Table 29.1).

More than 160,000 women registered with *in situ* carcinoma of the cervix were ineligible for these analyses: the numbers rose exponentially to reach more than 20,000 a year by the end of the 1980s. Tumours of uncertain malignancy were excluded, together with those coded as *in situ* carcinoma (ICD-9 233.1), but with the behaviour code 3 (malignant). Of the 81,000 women with invasive carcinoma, 5% in all were excluded because their vital status was unknown at 6 July 1997, and a total of 3% because their survival was zero or unknown: but the proportion excluded for these reasons rose to 7% and 4% respectively by 1986-90.

During the 1980s, about 6% of cervical cancers were assigned to the endocervix (cervical canal), ten times as many as to the exocervix, but more than 90% of tumours were simply coded to the uterine cervix, without anatomical detail (Table 29A). During the 1980s, squamous carcinoma accounted for two-thirds of the tumours, and 10% were adenocarcinomas (7% in the 1970s), but almost a quarter (23%) were coded as poorly specified carcinomas; for 1% there was no information on morphology.

Survival trends

Relative survival for women diagnosed with cervical cancer in England and Wales during 1986-90 was 83% at one year and 64% at five years, a 10-15% increase since the early 1970s (Table 29.2). Improvements in survival since the early 1970s are clearly shown in Figure 29.1.

One-year survival has increased by an average of 2% between successive five-year periods of diagnosis, and the corresponding average gain in five-year survival was just over 3%, after adjustment for changes in the age distribution of patients over time (see Table 4.4).

Survival improved earlier in some regions than others (Figure 29.1). For women diagnosed during 1986-90, regional differences in survival are generally small, and a 5% range around the national average five-year survival includes all regions (Figure 29.4). The speed of improvement in survival with time has nonetheless differed significantly between regions. Thus one-year and five-year survival rates from cervical cancer in Northern & Yorkshire were the lowest of all regions for women diagnosed in the early 1970s, but amongst the highest for women diagnosed in the late 1980s: age-standardised five-year survival was similar to that in Anglia & Oxford (63-64%), compared to an 8% deficit for the early 1970s (see Table-Figure 4.5J). The speed of improvement in five-year survival in Northern & Yorkshire (5.5% every five years) was significantly greater than in all other regions.

Survival falls steeply with age at diagnosis, even after adjustment for other causes of death (Figure 29.5). Under the age of 50, five-year relative survival is 70% or higher, but this falls to about 40% for women diagnosed in their seventies, and only 20% for older women (see Table 4.7).

Deprivation

Survival in England and Wales has consistently been lower for women in deprived groups than for those in the most

Table 29A Sub-site distribution of cancer of the cervix

ICD-9 code	Site description	1971-75 No.	%	1976-80 No.	%	1981-85 No.	%	1986-90 No.	%	Total No.	%
180.0	Endocervix	-	-	*479*	-	1,264	6.7	1,144	6.0	**2,887**	-
180.1	Exocervix	-	-	-	-	74	0.4	63	0.3	**137**	-
180.8	Other	-	-	*25*	-	99	0.5	247	1.3	**371**	-
180.9	Unspecified[1]	18,152	100.0	17,391	-	17,389	92.4	17,654	92.4	**70,586**	-
	Total	**18,152**		**17,895**		**18,826**		**19,108**		**73,981**	

[1] Cervix was coded simply 180, without subdivision, in ICD-8 (1971-78). Special codes 180.1-180.2 used by the national cancer registry during that period (see Chapter 2) are incompatible with ICD-9, and tumours with those codes have been assigned to 180.9 in this table. Italicised data in column 2 relate to 1979-80 only. Misleading percentages have been suppressed.

affluent groups (Table 29.2). The differences in survival were still evident in most regions in the late 1980s (Figure 29.2), but the deprivation gradient in five-year survival was less steep than for women diagnosed during the early 1980s (Figure 29.3).

For more than 18,000 women of known deprivation status who were diagnosed during 1981-85, the deprivation gap in survival was 5% one year after diagnosis and 8% five years after diagnosis, but for a similar number of women diagnosed in the period 1986-90, the gap in five-year survival between the most affluent and the most deprived groups was 4%.

The striking gains in survival in Northern & Yorkshire during the 1980s were most marked for women in the more deprived groups; the significant gradient seen in the first half of the decade is absent for women diagnosed during 1986-90 (Table 29.2, Figure 29.3).

The deprivation gap in five-year survival among 3,800 women diagnosed with cervical cancer in South Thames during 1980-89 was 8%, similar to the national gradient observed here for 1981-85: it was not accounted for by stage at diagnosis[6].

International comparison

Five-year survival in England and Wales for women diagnosed during 1986-90 (64%) was slightly higher than in Scotland or the average for Europe (62%). The highest regional survival rate (68%, West Midlands) was almost as high as in the USA (see Table 4.12).

1. Beral V. Cancer of the cervix: a sexually transmitted infection? Lancet 1974; i: 1037-1040
2. Muñoz N, Bosch FX, Shah KV, Meheus A, (eds.) *The epidemiology of cervical cancer and human papillomavirus. (IARC Scientific Publications No. 119).* Lyon: International Agency for Research on Cancer, 1992
3. Coleman MP, Estève J, Damiecki P, Arslan A, Renard H. *Trends in cancer incidence and mortality (IARC Scientific Publications No. 121).* Lyon: International Agency for Research on Cancer, 1993
4. Murphy M, Campbell MJ, Goldblatt P. Twenty years screening for cancer of the uterine cervix in Great Britain, 1964-84: further evidence for its ineffectiveness. J Epidemiol Comm Hlth 1988; 42: 49-53
5. Sasieni P, Cuzick J, Farmery E. Accelerated decline in cervical cancer mortality in England and Wales. Lancet 1995; 346: 1566-1567
6. Schrijvers CTM, Mackenbach J, Lutz J-M, Quinn MJ, Coleman MP. Deprivation, stage at diagnosis and cancer survival. Int J Cancer 1995; 63: 324-329

Table 29.1 Ineligible and excluded records, and patients included in analyses, by calendar period of diagnosis

Calendar period of diagnosis

	1971-75			1976-80			1981-85			1986-90			Overall		
Total registered	**33,790**			**40,932**			**61,285**			**112,637**			**248,644**		
Ineligible	*Records*	*Patients*	*%*	*Records*	*Patients*	*%*	*Records*	*Patients*	*%*	*Records*	*Patients*	*%*	*Records*	*Patients*	*%*
Incomplete data[1]	318	318		282	282		212	212		3,351	3,351		4,163	**4,163**	
Resident outside England and Wales	285	281		267	261		266	249		650	524		1,468	**1,315**	
In situ neoplasm[2]	13,717	13,677		20,900	20,784		40,469	40,147		90,831	87,110		165,917	**161,718**	
Benign or uncertain behaviour[2]	174	55		117	58		1	1		0	0		292	**114**	
Metastatic[2]	39	38		12	12		0	0		7	7		58	**57**	
Otherwise ineligible[3]	37	20		73	65		89	86		28	28		227	**199**	
Lymphoma[4]	1	1		1	1		1	0		0	0		3	**2**	
Leukaemia or myeloma[4]	0	0		0	0		0	0		0	0		0	**0**	
	14,390			**21,463**			**40,695**			**91,020**			**167,568**		
Total eligible	**19,400**	**100**		**19,469**	**100**		**20,590**	**100**		**21,617**	**100**		**81,076**	**100**	
Aged under 15 or 100 years or over at diagnosis	7	4	<0.1	7	5	<0.1	11	5	<0.1	22	6	<0.1	47	**20**	<0.1
Vital status unknown[5]	1,816	840	4.3	2,311	1,078	5.5	2,738	947	4.6	7,955	1,493	6.9	14,820	**4,358**	5.4
Duplicate registration[6]	0	0	0.0	3	1	<0.1	0	0	0.0	0	0	0.0	3	**1**	<0.1
Multiple primary[7]	67	1	<0.1	134	17	<0.1	321	96	0.5	396	134	0.6	918	**248**	0.3
Synchronous tumours[8]	91	26	0.1	149	29	0.1	318	40	0.2	922	38	0.2	1,480	**133**	0.2
Sex not known	0	0	0.0	0	0	0.0	0	0	0.0	0	0	0.0	0	**0**	0.0
Sex incompatible with site code[9]	0	0	0.0	3	1	<0.1	1	1	<0.1	2	2	<0.1	6	**4**	<0.1
Invalid dates[10]	2	1	<0.1	4	2	<0.1	11	5	<0.1	275	6	<0.1	292	**14**	<0.1
Zero survival[11]	413	376	1.9	500	441	2.3	759	670	3.3	931	830	3.8	2,603	**2,317**	2.9
Total exclusions		**1,248**	6.4		**1,574**	8.1		**1,764**	8.6		**2,509**	11.6		**7,095**	8.8
Patients included in analysis		**18,152**	93.6		**17,895**	91.9		**18,826**	91.4		**19,108**	88.4		**73,981**	91.2

[1] Main data item(s) invalid or incompatible with one another: name, sex, date of birth, date of diagnosis and (if present) date of death, postcode, site, morphology and behaviour

[2] Tumour either *in situ* (behaviour code 2), benign (0), uncertain if benign or malignant (1) or metastatic (6 or 9)

[3] Site code for *in situ* neoplasm of cervix (ICD-9 233.1, ICD-8 234.0), but behaviour code *not* equal to 2 (*in situ)*

[4] Morphology that of lymphoma (included with non-Hodgkin lymphoma, Ch. 41) or myeloma (included with multiple myeloma, Ch. 43), or leukaemia (excluded)

[5] Tracing of vital status at National Health Service Central Register incomplete at 6 July 1997

[6] Same site code, sex, cancer registry and cancer registry number as an earlier registration

[7] Same site code and person number as an earlier registration(s): mostly confirmed multiple primary tumours at this site, some unresolved duplicate registrations

[8] Same site code, sex, date of birth and date of diagnosis as another registration(s): mostly synchronous or (in paired organs) bilateral tumours in same anatomic site in one individual, not previously linked: also some duplicate registrations

[9] Site code specific to one sex; tumour registered to opposite sex

[10] Impossible sequence of dates (birth, diagnosis and, if present, emigration or death); or date of death after 6 July 1997

[11] Date of diagnosis same as date of death: some are patients who did die on the day of diagnosis, but most were registered solely from a death certificate (death certificate only, or DCO), and their survival time is thus unknown

Table 29.2   Crude and relative survival (%) at one, five and ten years since diagnosis: NHS Region, deprivation category and calendar period of diagnosis

WOMEN	1971-75 Affluent C	R	2 C	R	3 C	R	4 C	R	Deprived C	R	All[1] C	R	1976-80 Affluent C	R	2 C	R	3 C	R	4 C	R	Deprived C	R	All[1] C	R
ENGLAND & WALES	*(1,865)*		*(2,399)*		*(2,926)*		*(3,504)*		*(3,720)*		*(18,152)*		*(2,175)*		*(2,790)*		*(3,533)*		*(4,058)*		*(4,596)*		*(17,895)*	
One year	76	77	74	76	73	75	74	76	72	73	**73**	**75**	77	79	77	79	73	75	75	77	73	75	**75**	**77**
Five years	51	56	45	50	46	51	45	50	44	48	**46**	**50**	52	57	51	56	48	53	49	54	48	53	**49**	**54**
Ten years	43	52	37	46	37	46	36	45	36	44	**37**	**46**	44	52	42	50	39	48	41	50	40	49	**41**	**49**
ENGLAND	*(1,779)*		*(2,267)*		*(2,678)*		*(3,186)*		*(3,481)*		*(16,897)*		*(2,081)*		*(2,624)*		*(3,234)*		*(3,696)*		*(4,284)*		*(16,488)*	
One year	75	77	74	76	73	75	74	75	71	73	**73**	**75**	77	79	77	79	73	75	75	76	73	75	**74**	**76**
Five years	50	55	45	50	46	51	45	49	44	48	**45**	**50**	52	57	50	55	47	52	48	53	47	53	**49**	**54**
Ten years	42	51	37	45	38	46	36	44	36	44	**37**	**45**	43	52	42	50	39	48	40	49	39	48	**40**	**49**
Northern & Yorkshire	*(137)*		*(223)*		*(284)*		*(475)*		*(769)*		*(2,756)*		*(168)*		*(307)*		*(430)*		*(624)*		*(966)*		*(2,561)*	
One year	74	76	65	67	69	70	70	72	67	69	**69**	**71**	71	73	75	77	71	73	71	73	69	71	**71**	**73**
Five years	41	47	39	44	40	44	41	45	41	45	**41**	**45**	46	51	49	54	42	46	43	48	45	50	**44**	**50**
Ten years	35	43	33	40	30	38	29	38	34	42	**33**	**40**	40	49	38	47	33	42	35	45	36	46	**36**	**45**
Trent	*(183)*		*(205)*		*(329)*		*(472)*		*(453)*		*(1,789)*		*(171)*		*(281)*		*(382)*		*(544)*		*(503)*		*(1,938)*	
One year	77	78	70	72	77	79	76	78	68	70	**73**	**75**	76	78	74	76	74	76	73	75	73	75	**74**	**76**
Five years	58	64	39	44	49	54	45	50	44	48	**46**	**51**	53	58	51	56	49	54	48	53	46	52	**49**	**54**
Ten years	50	60	31	39	39	47	35	43	36	44	**37**	**46**	42	49	44	55	41	52	40	48	38	47	**41**	**49**
Anglia & Oxford	*(239)*		*(265)*		*(312)*		*(246)*		*(137)*		*(1,337)*		*(293)*		*(293)*		*(325)*		*(222)*		*(132)*		*(1,309)*	
One year	74	76	77	79	72	75	73	75	78	80	**75**	**77**	78	81	82	84	73	75	79	81	73	76	**77**	**80**
Five years	54	58	50	54	49	54	45	50	49	53	**49**	**54**	50	56	51	55	52	58	52	57	51	56	**51**	**57**
Ten years	48	55	41	51	44	53	39	47	43	51	**43**	**51**	45	54	43	50	42	52	44	53	41	51	**43**	**52**
North Thames	*(264)*		*(323)*		*(359)*		*(450)*		*(366)*		*(1,986)*		*(213)*		*(237)*		*(310)*		*(369)*		*(381)*		*(1,568)*	
One year	76	79	77	80	76	77	73	75	74	75	**75**	**77**	74	76	77	80	75	77	75	77	77	78	**76**	**78**
Five years	52	57	49	54	49	54	46	50	46	51	**48**	**53**	49	53	49	54	47	52	49	53	48	52	**48**	**53**
Ten years	44	53	41	50	40	49	39	48	35	44	**39**	**49**	39	47	42	52	38	46	39	47	40	48	**39**	**48**
South Thames	*(382)*		*(382)*		*(407)*		*(427)*		*(279)*		*(2,065)*		*(406)*		*(361)*		*(408)*		*(370)*		*(214)*		*(1,869)*	
One year	71	73	72	74	68	70	73	75	76	78	**73**	**75**	81	83	75	77	69	71	72	74	77	79	**75**	**77**
Five years	43	47	42	46	42	46	45	50	43	48	**43**	**48**	55	61	47	52	41	46	47	51	50	56	**48**	**53**
Ten years	34	42	33	41	33	41	36	45	34	43	**35**	**43**	45	55	37	46	33	42	37	45	43	53	**39**	**48**
South & West	*(311)*		*(500)*		*(510)*		*(354)*		*(175)*		*(2,029)*		*(353)*		*(535)*		*(561)*		*(374)*		*(155)*		*(2,061)*	
One year	77	79	73	75	75	77	76	78	69	72	**74**	**76**	73	75	75	77	71	72	79	81	77	79	**75**	**77**
Five years	50	55	45	50	47	51	47	51	40	45	**46**	**51**	48	53	51	56	48	54	55	60	50	55	**50**	**56**
Ten years	43	53	37	45	40	48	37	45	36	43	**38**	**46**	40	48	42	50	41	50	46	56	39	47	**42**	**51**
West Midlands	*(64)*		*(95)*		*(120)*		*(188)*		*(246)*		*(1,837)*		*(217)*		*(250)*		*(365)*		*(489)*		*(661)*		*(2,008)*	
One year	78	80	80	82	77	79	76	77	76	78	**75**	**77**	84	86	81	83	75	77	80	81	77	79	**78**	**80**
Five years	56	62	57	62	48	54	45	48	47	53	**47**	**51**	60	65	58	64	51	55	55	61	53	58	**54**	**59**
Ten years	45	57	48	61	39	49	34	42	43	50	**38**	**47**	52	59	49	57	41	49	45	57	45	54	**45**	**55**
North & West	*(199)*		*(274)*		*(357)*		*(574)*		*(1,056)*		*(3,098)*		*(260)*		*(360)*		*(453)*		*(704)*		*(1,272)*		*(3,174)*	
One year	79	80	77	79	73	75	72	74	72	73	**74**	**76**	75	77	76	78	77	79	72	74	71	73	**73**	**75**
Five years	54	59	42	47	46	51	44	50	44	49	**46**	**51**	51	56	47	53	49	54	45	50	45	50	**47**	**51**
Ten years	45	55	37	44	36	45	36	47	36	44	**38**	**46**	43	52	39	48	39	48	37	46	37	45	**38**	**46**
WALES	*(86)*		*(132)*		*(248)*		*(318)*		*(239)*		*(1,255)*		*(94)*		*(166)*		*(299)*		*(362)*		*(312)*		*(1,407)*	
One year	80	82	72	73	74	76	75	77	74	76	**74**	**76**	85	86	75	77	73	75	79	81	79	80	**78**	**79**
Five years	64	66	49	53	43	49	48	52	47	52	**48**	**53**	61	65	57	63	52	57	56	62	59	64	**56**	**61**
Ten years	52	60	41	47	33	42	39	48	40	48	**40**	**48**	55	62	46	55	43	52	47	58	54	63	**48**	**57**

[1]   Including those for whom no deprivation category was assigned. C - crude survival, R - relative survival (see Chapter 3). Number of patients contributing to each analysis in parentheses.

		1981-85						1986-90					
WOMEN		Affluent (C R)	2 (C R)	3 (C R)	4 (C R)	Deprived (C R)	All[1] (C R)	Affluent (C R)	2 (C R)	3 (C R)	4 (C R)	Deprived (C R)	All[1] (C R)
ENGLAND & WALES	(n)	(2,694)	(3,118)	(3,771)	(4,319)	(4,519)	(18,826)	(2,805)	(3,352)	(3,851)	(4,342)	(4,518)	(19,108)
	One year	81 83	80 82	79 81	78 80	77 79	**79 81**	83 85	82 84	82 84	80 82	79 81	**81 83**
	Five years	59 64	56 61	55 60	52 57	51 56	**54 59**	61 66	59 65	60 65	57 63	56 62	**58 64**
		52 61	48 57	46 55	43 53	42 51	**45 55**						
ENGLAND	(n)	(2,590)	(2,937)	(3,468)	(3,947)	(4,236)	(17,471)	(2,640)	(3,123)	(3,542)	(3,938)	(4,236)	(17,666)
	One year	81 83	80 83	80 82	78 80	77 79	**79 81**	83 85	82 84	82 84	80 82	79 81	**81 83**
	Five years	59 64	56 61	55 60	52 57	51 56	**54 59**	61 66	60 65	60 65	58 63	56 62	**59 64**
		52 60	48 57	46 55	43 52	41 51	**45 54**						
Northern & Yorkshire	(n)	(232)	(354)	(474)	(673)	(1,027)	(2,801)	(270)	(373)	(526)	(635)	(918)	(2,740)
	One year	77 79	80 83	81 83	74 76	75 77	**77 79**	86 88	79 81	82 85	82 85	81 83	**82 84**
	Five years	62 67	53 59	57 63	47 52	50 55	**52 57**	64 70	60 64	60 68	62 68	61 67	**61 67**
		55 64	45 54	50 60	39 48	40 50	**43 53**						
Trent	(n)	(219)	(301)	(434)	(567)	(504)	(2,038)	(226)	(326)	(413)	(510)	(612)	(2,091)
	One year	83 85	82 84	77 79	77 79	76 78	**78 80**	79 81	81 83	82 85	79 81	79 81	**80 82**
	Five years	59 63	56 61	55 60	51 56	52 57	**54 59**	60 65	58 64	58 64	55 61	55 62	**57 63**
		53 60	49 59	47 56	43 52	42 52	**46 55**						
Anglia & Oxford	(n)	(390)	(375)	(323)	(281)	(175)	(1,571)	(392)	(396)	(336)	(299)	(162)	(1,597)
	One year	81 83	79 81	83 85	81 83	83 84	**81 84**	84 87	82 84	80 82	79 81	83 84	**82 84**
	Five years	60 64	57 63	61 68	58 64	57 62	**59 65**	62 68	63 69	58 64	57 63	65 71	**61 67**
		52 60	50 60	51 65	48 58	43 54	**50 60**						
North Thames	(n)	(267)	(280)	(333)	(373)	(342)	(1,626)	(215)	(284)	(316)	(423)	(373)	(1,673)
	One year	84 87	78 80	82 84	82 84	81 83	**81 84**	84 87	83 86	83 85	80 82	78 80	**81 84**
	Five years	63 68	53 58	58 62	55 61	59 65	**57 63**	59 65	55 62	55 60	56 61	54 59	**56 61**
		56 66	46 54	48 58	44 55	48 60	**48 58**						
South Thames	(n)	(483)	(396)	(438)	(403)	(223)	(2,027)	(478)	(410)	(423)	(372)	(264)	(1,997)
	One year	79 82	83 86	78 81	79 82	79 81	**79 82**	83 85	83 85	82 83	83 85	80 82	**82 85**
	Five years	56 60	53 59	51 56	53 59	54 59	**53 59**	58 62	55 61	59 64	58 63	50 55	**57 62**
		47 55	44 55	43 52	45 56	44 53	**45 54**						
South & West	(n)	(444)	(598)	(627)	(374)	(172)	(2,273)	(438)	(612)	(636)	(466)	(265)	(2,431)
	One year	79 81	81 83	79 81	80 82	75 77	**79 81**	84 86	84 87	81 83	79 82	80 82	**82 84**
	Five years	59 64	60 65	51 56	55 60	51 54	**56 61**	65 69	60 65	59 64	55 61	56 61	**59 64**
		51 59	50 59	43 52	44 55	41 48	**47 56**						
West Midlands	(n)	(248)	(270)	(376)	(511)	(612)	(2,031)	(287)	(280)	(448)	(523)	(565)	(2,115)
	One year	84 86	80 82	80 81	79 81	80 82	**80 82**	83 84	83 85	85 87	82 84	78 80	**82 84**
	Five years	65 69	57 63	57 62	54 59	55 60	**57 61**	64 70	66 70	66 71	63 69	57 62	**63 68**
		58 67	53 61	49 56	47 56	48 57	**49 58**						
North & West	(n)	(307)	(363)	(463)	(765)	(1,181)	(3,104)	(334)	(442)	(444)	(710)	(1,077)	(3,022)
	One year	81 83	79 81	79 81	75 77	74 76	**76 78**	78 81	80 82	83 85	77 79	78 80	**79 81**
	Five years	55 60	53 60	52 57	48 53	45 50	**49 54**	57 62	59 65	60 66	53 59	53 59	**56 61**
		48 58	44 53	39 47	40 49	36 45	**40 49**						
WALES	(n)	(104)	(181)	(303)	(372)	(283)	(1,355)	(165)	(229)	(309)	(404)	(282)	(1,442)
	One year	83 84	77 79	73 75	77 79	76 77	**77 79**	77 79	80 82	79 80	81 83	77 79	**79 81**
	Five years	66 69	56 61	52 57	56 61	56 61	**56 61**	62 68	55 60	58 63	55 62	54 59	**57 63**
		58 68	51 60	45 54	47 57	48 59	**48 58**						

[1] Including those for whom no deprivation category was assigned. C - crude survival, R - relative survival (see Chapter 3). Number of patients contributing to each analysis in parentheses.

Fig. 29.1   Relative survival up to ten years: trends by calendar period of diagnosis (1971-90) and NHS Region

WOMEN

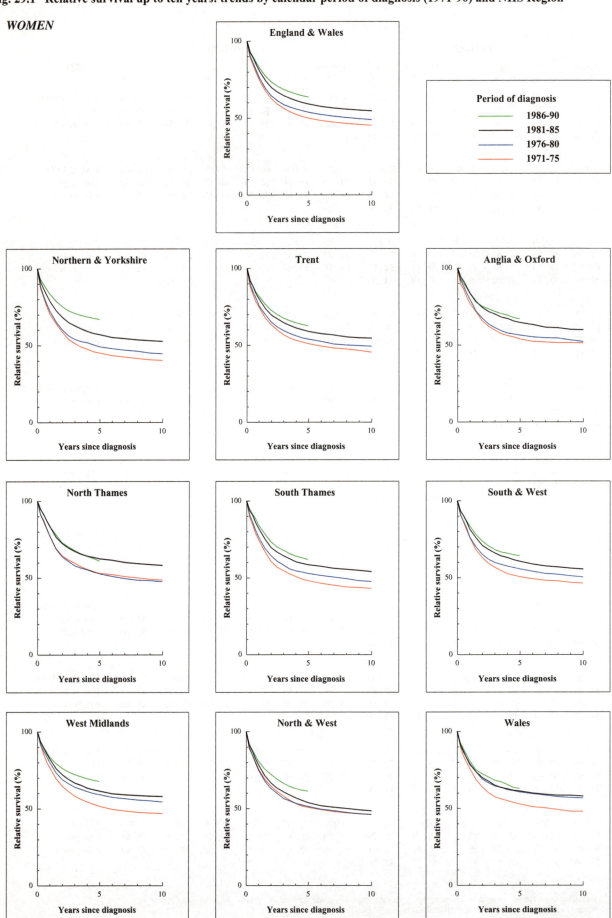

Fig. 29.2   Relative survival up to five years, by deprivation category and NHS Region: patients diagnosed 1986-90

WOMEN

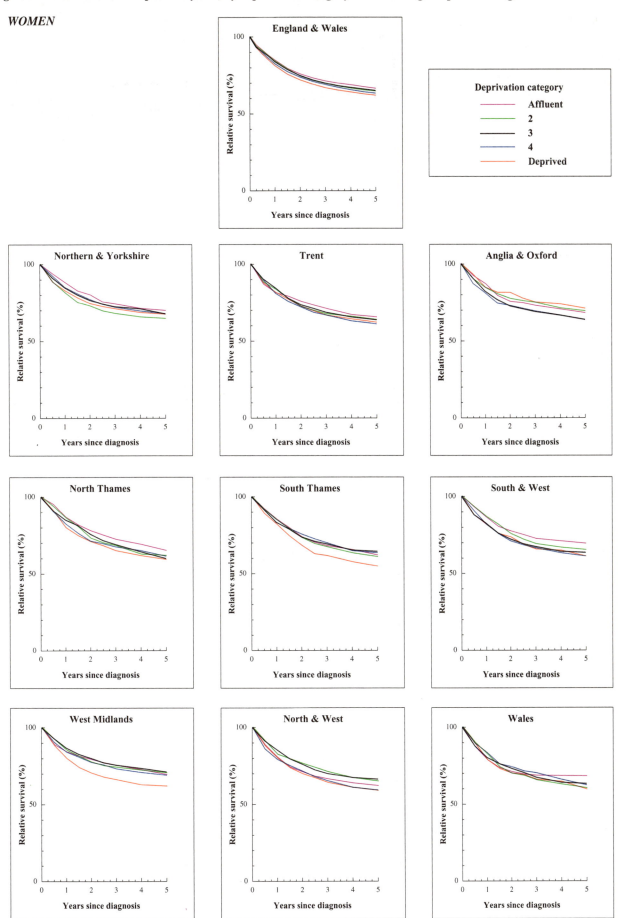

Fig. 29.3   Relative survival at five years by deprivation category, period of diagnosis (1981-90) and NHS Region

WOMEN

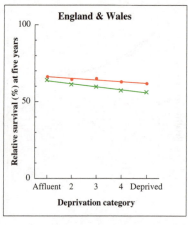

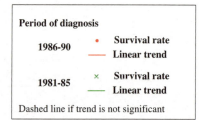

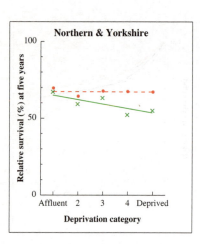

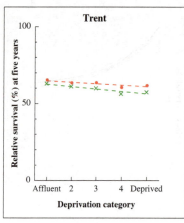

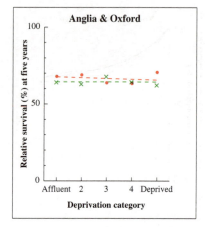

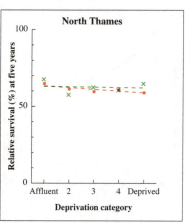

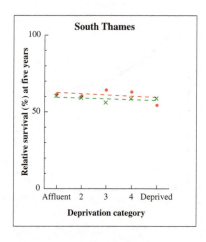

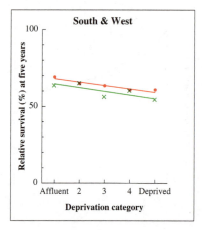

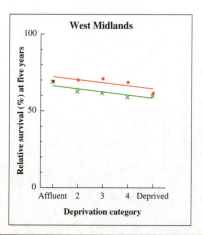

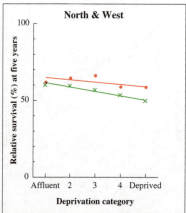

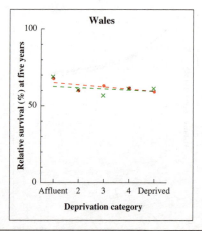

Fig. 29.4 Relative survival (%) by NHS Region:

          one-year and five-year survival for women diagnosed 1986-90, with 95% confidence intervals

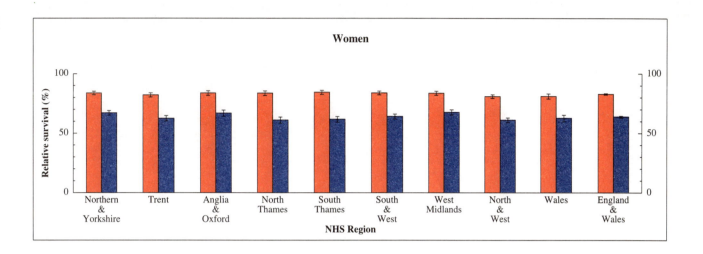

Fig. 29.5 Relative survival (%) at five years, with 95% confidence interval, by age at diagnosis:

          England and Wales, women diagnosed 1986-90

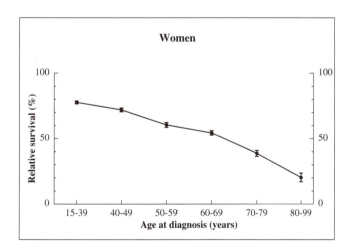

Chapter 30

UTERUS

Cancer of the body of the uterus arises in about 3,800 women every year in England and Wales. It ranks fifth, accounting for 3.5% of all cancers in women. Incidence in England and Wales was broadly stable during the 1980s, but incidence fell in some regions. The protective effect of combined oral contraceptives[1] and more recent post-menopausal oestrogen replacement formulations[2], and hysterectomy for benign conditions are all likely to be involved. Incidence is higher in the north than the south. Mortality fell by about 10% every five years during the 1970s and 1980s; there were some 700 deaths in 1995.

Most uterine tumours are adenocarcinomas of the endometrial lining, with some leiomyosarcomas of the smooth muscle wall.

The survival analyses reported here include almost 65,000 women diagnosed with cancer of the uterus during the period 1986-90 and followed up to the end of 1995, about 92% of those eligible. Cancers of the uterus should not normally be registered without sufficient information being available to classify them to the cervix or the body of the uterus[3]. In practice, about 10% of uterine tumours registered in England and Wales during 1971-90 were assigned to a non-specific code for uterus (182.9 in ICD-8 or 179 in ICD-9: see Table 2.3). The proportion ranged from under 1% to 28% between cancer registries during 1986-90 (data not shown); it fell in Northern & Yorkshire, and rose in South & West and Trent. These 8,000 tumours were ineligible for analysis because they could not be reliably assigned to the cervix or body of uterus (Table 30.1).

About 4% of the 71,000 women eligible for analysis were excluded because their vital status was unknown at 6 July 1997, and a further 4% because their survival was either zero or unknown. Almost all tumours were coded to the body of the uterus: only about 100 were coded to the lower segment (isthmus) or involved both parts (Table 30A). Three-quarters (75%) were adenocarcinomas; 1.4% were leiomyosarcoma and another 1% poorly specified sarcoma, and 16% were poorly specified carcinomas. These proportions were stable over time.

Survival trends

For women diagnosed with uterine cancer in England and Wales during 1986-90, one-year survival was 85% and five-year survival 73%, compared with 81% and 66%, respectively, for women diagnosed in the early 1970s (Table 30.2). Improvements in survival during the 1970s were larger than in the 1980s (Figure 30.1).

After adjustment for differences in age distribution over time, one-year survival in England and Wales as a whole increased by an average of 2% between successive five-year periods, whilst the corresponding gain in five-year survival was 3% (see Table 4.4). These patterns of improvement in survival are closely similar to those for cancer of the cervix (Chapter 29), and one-year survival rates are also similar, but five-year survival for cancer of the uterus is 10% higher than for the cervix.

Five-year survival rates in all regions are within a 9% range around the national average for women diagnosed during 1986-90 (Figure 30.4), but regional trends have differed markedly over time. Survival was high in West Midlands in the 1970s but has not changed much, whilst there have been large gains in North & West and in South Thames, where survival was low. Five-year survival for women diagnosed during the late 1980s in Anglia & Oxford appears slightly lower than previously, and there has been little change in Wales (Figure 30.1).

Quantitative estimates of the average gain in survival in each region, adjusted to the age distribution of all women with uterine cancer diagnosed in 1986-90, closely reflect the visual impression of the differences in (unadjusted) regional survival curves in Figure 30.1. Survival at one and five years after diagnosis was higher than average in South & West in the early 1970s, improved as much as the average, and remains 3-4% higher than average (see Table-Figure 4.5K). Five-year survival has improved by 4-5% every five years in South Thames, more than the national average rate of improvement.

Despite these regional differences in survival trends since the early 1970s, it should be stressed that regional differences in survival for women diagnosed during 1986-90 are not large, and may be partly accounted for by survival differences between deprivation groups (see below) and regional differences in the proportion of women in each deprivation category (see Table 4.3).

Survival from cancer of the body of the uterus falls after age 60 at diagnosis (Figure 30.5), but in a pattern quite different from that for the uterine cervix (see Figure 29.5). Under age 60, relative survival from uterine cancer at five years was 80% or more for women diagnosed during 1986-90, some 10% higher than for cervical cancer (see Table 4.7). For women in their sixties, five-year survival was 74%, falling to 42% for women aged 80-99, but at all ages over 60, survival was 20% higher than for cervical cancer.

Table 30A Sub-site distribution of cancer of the uterus

ICD-9 code	Site description	1971-75 No.	1971-75 %	1976-80 No.	1976-80 %	1981-85 No.	1981-85 %	1986-90 No.	1986-90 %	Total No.	Total %
182.0	Corpus uteri, except isthmus	15,390	100.0	16,303	99.9	16,609	99.7	16,471	99.8	**64,773**	**99.8**
182.1	Isthmus (lower segment)	-	-	1	0.0	10	0.1	6	0.0	**17**	**0.0**
182.8	Other	-	-	13	0.1	47	0.3	31	0.2	**91**	**0.1**
		15,390		**16,317**		**16,666**		**16,508**		**64,881**	

Deprivation

Differences in survival between deprivation categories are rarely more than 6% at five years (Table 30.2), but survival is generally lower for women in more deprived groups (Figure 30.2). At a national level, the deprivation gradient has been small but statistically significant during the 1980s (Figure 30.3). At regional level, the deprivation gradient is negative in five regions, but significantly so only in West Midlands for patients diagnosed in the late 1980s, and in four regions (Northern & Yorkshire, Anglia & Oxford, South Thames, South & West), there is no systematic difference in survival between deprivation categories (Table 30.2).

At a national level, the deprivation gradient among more than 16,000 women diagnosed during the early 1980s was 5% one year after diagnosis and 6% at five years, and for a similar number of women diagnosed during 1986-90, the gap between the most affluent and most deprived groups was 4% (see Table 4.9).

The deprivation gap of 9% in relative survival between affluent and deprived groups among 5,000 women diagnosed in South Thames during 1980-89 was unaffected by adjustment for stage of disease at diagnosis[4]. Although the deprivation gradient seen here is thus unlikely to be accounted for by differences in stage, deprivation-specific life tables allow estimation of the gradient with less bias (see Chapter 3), and the national estimate of 4% is probably more realistic.

International comparison

Five-year relative survival for women aged 15-99 who were diagnosed in England and Wales during 1986-90 was similar to that in Scotland and to the average survival rate in Europe (72-75%). The highest regional survival rate (76%, South & West) was 9% lower than the survival rate in the USA (see Table 4.12).

1.   Villard-Mackintosh L, Murphy M. Endometrial cancer trends in England and Wales: a possible protective effect of oral contraception. Int J Epidemiol 1990; 19: 255-258

2.   Persky V, Davis F, Barrett R, Ruby E, Sailer C, Levy P. Recent time trends in uterine cancer. Am J Public Health 1990; 80: 935-939

3.   Jensen OM, Parkin DM, MacLennan R, Muir CS, Skeet RG, (eds.) *Cancer registration: principles and methods. (IARC Scientific Publications No. 95).* Lyon: International Agency for Research on Cancer, 1991

4.   Schrijvers CTM, Mackenbach J, Lutz J-M, Quinn MJ, Coleman MP. Deprivation, stage at diagnosis and cancer survival. Int J Cancer 1995; 63: 324-329

Table 30.1 Ineligible and excluded records, and patients included in analyses, by calendar period of diagnosis

	Calendar period of diagnosis														
	1971-75			1976-80			1981-85			1986-90			Overall		
Total registered	18,616			20,082			20,564			20,792			80,054		
Ineligible	Records	Patients	%	Records	Patients	%	Records	Patients	%	Records	Patients	%	Records	Patients	%
Incomplete data[1]	19	19		37	37		73	73		87	87		216	**216**	
Resident outside England and Wales	154	154		120	119		72	63		85	65		431	**401**	
In situ neoplasm[2]	111	111		22	22		0	0		1	1		134	**134**	
Benign or uncertain behaviour[2]	240	191		194	175		2	1		4	2		440	**369**	
Metastatic[2]	56	55		34	32		0	0		1	1		91	**88**	
Otherwise ineligible[3]	1813	1750		2203	2148		2273	2245		1983	1959		8272	**8102**	
Lymphoma[4]	2	2		0	0		0	0		0	0		2	**2**	
Leukaemia or myeloma[4]	0	0		1	0		0	0		0	0		1	**0**	
		2,282			2,533			2,382			2,115			9,312	
Total eligible		16,334	100		17,549	100		18,182	100		18,677	100		70,742	100
Aged under 15 or 100 years or over at diagnosis	4	4	<0.1	6	1	<0.1	8	5	<0.1	10	8	<0.1	28	**18**	<0.1
Vital status unknown[5]	675	533	3.3	884	725	4.1	757	644	3.5	1,254	1,085	5.8	3,570	**2,987**	4.2
Duplicate registration[6]	0	0	0.0	0	0	0.0	0	0	0.0	1	0	0.0	1	**0**	0.0
Multiple primary[7]	2	1	<0.1	21	16	<0.1	78	43	0.2	76	61	0.3	177	**121**	0.2
Synchronous tumours[8]	67	28	0.2	73	23	0.1	113	57	0.3	116	39	0.2	369	**147**	0.2
Sex not known	0	0	0.0	0	0	0.0	0	0	0.0	0	0	0.0	0	**0**	0.0
Sex incompatible with site code[9]	0	0	0.0	1	1	<0.1	0	0	0.0	3	1	<0.1	4	**2**	<0.1
Invalid dates[10]	0	0	0.0	4	3	<0.1	3	3	<0.1	16	4	<0.1	23	**10**	<0.1
Zero survival[11]	678	378	2.3	774	463	2.6	1,124	764	4.2	1,368	971	5.2	3,944	**2,576**	3.6
Total exclusions		944	5.8		1,232	7.0		1,516	8.3		2,169	11.6		**5,861**	8.3
Patients included in analysis		15,390	94.2		16,317	93.0		16,666	91.7		16,508	88.4		64,881	91.7

[1] Main data item(s) invalid or incompatible with one another: name, sex, date of birth, date of diagnosis and (if present) date of death, postcode, site, morphology and behaviour

[2] Tumour either *in situ* (behaviour code 2), benign (0), uncertain if benign or malignant (1) or metastatic (6 or 9)

[3] Uterus, not otherwise specified (i.e. either cervix *or* body of uterus) (ICD-8 182.9, 1971-78; ICD-9 179-, 1979-90)

[4] Morphology that of lymphoma (included with non-Hodgkin lymphoma, Ch. 41) or myeloma (included with multiple myeloma, Ch. 43), or leukaemia (excluded)

[5] Tracing of vital status at National Health Service Central Register incomplete at 6 July 1997

[6] Same site code, sex, cancer registry and cancer registry number as an earlier registration

[7] Same site code and person number as an earlier registration(s): mostly confirmed multiple primary tumours at this site, some unresolved duplicate registrations

[8] Same site code, sex, date of birth and date of diagnosis as another registration(s): mostly synchronous or (in paired organs) bilateral tumours in same anatomic site in one individual, not previously linked: also some duplicate registrations

[9] Site code specific to one sex; tumour registered to opposite sex

[10] Impossible sequence of dates (birth, diagnosis and, if present, emigration or death); or date of death after 6 July 1997

[11] Date of diagnosis same as date of death: some are patients who did die on the day of diagnosis, but most were registered solely from a death certificate (death certificate only, or DCO), and their survival time is thus unknown

Table 30.2 Crude and relative survival (%) at one, five and ten years since diagnosis: NHS Region, deprivation category and calendar period of diagnosis

	1971-75						1976-80					
WOMEN	*Affluent* C R	*2* C R	*3* C R	*4* C R	*Deprived* C R	*All*[1] C R	*Affluent* C R	*2* C R	*3* C R	*4* C R	*Deprived* C R	*All*[1] C R
ENGLAND & WALES	*(2,494)*	*(2,725)*	*(2,815)*	*(2,560)*	*(1,891)*	*(15,390)*	*(3,184)*	*(3,435)*	*(3,432)*	*(3,245)*	*(2,315)*	*(16,317)*
One year	80 83	78 81	78 81	78 81	77 80	78 81	81 84	80 83	82 85	78 82	77 80	80 83
Five years	60 69	57 65	56 65	55 64	56 66	57 66	62 72	61 71	60 71	56 67	55 66	59 69
Ten years	49 66	46 62	45 60	44 62	46 64	46 63	52 69	51 67	48 67	46 63	44 62	48 66
ENGLAND	*(2,407)*	*(2,611)*	*(2,639)*	*(2,342)*	*(1,784)*	*(14,483)*	*(3,108)*	*(3,305)*	*(3,264)*	*(3,043)*	*(2,218)*	*(15,534)*
One year	80 83	78 81	78 81	78 81	77 80	**79 81**	81 84	80 83	82 85	78 82	77 80	**80 83**
Five years	59 68	57 65	56 66	55 65	56 66	**57 66**	62 71	61 71	60 71	56 67	55 66	**59 69**
Ten years	49 65	46 62	45 60	45 62	46 64	**46 63**	52 69	51 67	48 67	46 63	43 62	**48 66**
Northern & Yorkshire	*(128)*	*(151)*	*(229)*	*(264)*	*(317)*	*(1,525)*	*(217)*	*(266)*	*(372)*	*(430)*	*(460)*	*(1,796)*
One year	80 82	79 82	78 82	78 81	78 82	**78 81**	83 85	79 82	80 83	76 79	77 80	**78 81**
Five years	62 72	65 73	54 64	57 67	58 68	**58 68**	65 75	58 68	60 71	52 61	56 68	**57 68**
Ten years	49 66	56 72	44 64	46 66	48 67	**47 65**	56 73	52 66	49 69	40 56	41 62	**46 64**
Trent	*(200)*	*(285)*	*(332)*	*(377)*	*(274)*	*(1,600)*	*(242)*	*(302)*	*(368)*	*(445)*	*(259)*	*(1,673)*
One year	81 84	77 80	78 81	75 78	77 79	**77 80**	82 84	81 85	81 84	79 82	80 83	**80 83**
Five years	59 69	56 64	52 61	56 64	58 68	**56 65**	61 70	65 75	63 74	60 71	56 66	**61 72**
Ten years	48 65	47 62	39 51	47 62	48 68	**46 62**	53 70	54 71	49 68	50 68	46 61	**50 68**
Anglia & Oxford	*(351)*	*(357)*	*(364)*	*(192)*	*(104)*	*(1,550)*	*(490)*	*(488)*	*(363)*	*(274)*	*(123)*	*(1,814)*
One year	82 85	80 84	82 85	79 81	82 84	**82 84**	83 86	81 85	84 88	79 82	75 77	**82 85**
Five years	59 69	59 68	58 67	58 66	*60 71*	**60 69**	64 75	63 73	66 76	61 72	52 64	**63 74**
Ten years	46 63	50 65	47 62	48 65	*48 65*	**48 65**	55 74	54 72	51 71	50 69	38 55	**52 71**
North Thames	*(389)*	*(358)*	*(340)*	*(347)*	*(255)*	*(1,918)*	*(360)*	*(338)*	*(340)*	*(356)*	*(230)*	*(1,697)*
One year	83 85	79 82	85 87	79 82	81 85	**81 84**	81 84	81 84	84 87	81 84	81 83	**82 85**
Five years	63 71	55 62	60 70	57 66	59 68	**59 68**	64 73	61 69	59 69	55 66	58 67	**60 70**
Ten years	53 70	47 61	49 64	48 64	49 67	**49 66**	54 69	50 66	46 65	46 63	48 65	**49 67**
South Thames	*(562)*	*(513)*	*(434)*	*(407)*	*(183)*	*(2,320)*	*(726)*	*(556)*	*(530)*	*(377)*	*(165)*	*(2,506)*
One year	78 80	76 78	69 72	73 76	75 78	**75 78**	79 82	78 80	81 84	77 80	70 73	**78 82**
Five years	56 65	52 61	49 57	49 58	50 59	**52 61**	60 70	56 65	57 69	52 63	53 62	**56 67**
Ten years	46 62	39 54	38 52	39 56	38 53	**41 57**	51 67	42 59	44 63	43 59	43 59	**45 63**
South & West	*(474)*	*(629)*	*(562)*	*(321)*	*(120)*	*(2,332)*	*(545)*	*(782)*	*(652)*	*(383)*	*(109)*	*(2,565)*
One year	81 84	79 82	79 82	86 89	73 76	**80 83**	80 82	79 82	83 87	78 81	74 77	**80 83**
Five years	61 70	59 68	63 72	63 75	55 66	**60 70**	62 71	63 72	60 72	59 69	56 66	**61 71**
Ten years	53 70	47 64	49 64	52 71	*45 62*	**48 66**	51 68	51 68	49 68	45 62	*40 57*	**49 67**
West Midlands	*(103)*	*(100)*	*(147)*	*(141)*	*(159)*	*(1,608)*	*(275)*	*(294)*	*(375)*	*(396)*	*(440)*	*(1,809)*
One year	86 90	79 81	85 88	79 83	76 79	**83 86**	83 87	84 87	80 83	81 85	80 83	**81 85**
Five years	65 75	60 66	63 73	52 64	61 71	**61 71**	61 70	65 76	63 73	59 69	60 71	**61 72**
Ten years	*59 75*	*47 59*	52 69	44 60	49 68	**51 68**	51 69	55 73	53 71	49 66	47 68	**50 69**
North & West	*(200)*	*(218)*	*(231)*	*(293)*	*(372)*	*(1,630)*	*(253)*	*(279)*	*(264)*	*(382)*	*(432)*	*(1,674)*
One year	74 77	77 80	72 75	76 79	74 77	**74 77**	80 83	78 81	77 81	76 80	71 74	**76 79**
Five years	52 60	53 62	49 59	48 57	52 62	**50 60**	58 67	59 69	52 62	52 63	48 59	**53 64**
Ten years	40 53	41 56	39 56	34 52	42 60	**39 56**	48 63	50 65	46 62	42 60	39 57	**44 61**
WALES	*(87)*	*(114)*	*(176)*	*(218)*	*(107)*	*(907)*	*(76)*	*(130)*	*(168)*	*(202)*	*(97)*	*(783)*
One year	78 80	78 81	73 75	73 76	78 80	**76 78**	*91 94*	84 87	79 82	79 82	84 87	**82 85**
Five years	*64 75*	56 63	52 60	53 62	57 66	**55 64**	65 75	65 73	55 66	58 68	*63 72*	**59 70**
Ten years	*57 71*	50 60	45 59	43 59	47 62	**46 61**	*57 74*	47 65	41 59	48 66	49 68	**47 66**

[1] Including those for whom no deprivation category was assigned. C - crude survival, R - relative survival (see Chapter 3).
Number of patients contributing to each analysis in parentheses.

	1981-85												1986-90												WOMEN
	Affluent		*2*		*3*		*4*		*Deprived*		*All[1]*		*Affluent*		*2*		*3*		*4*		*Deprived*		*All[1]*		
	C	R	C	R	C	R	C	R	C	R	C	R	C	R	C	R	C	R	C	R	C	R	C	R	
	(3,487)		*(3,566)*		*(3,629)*		*(3,258)*		*(2,349)*		*(16,666)*		*(3,416)*		*(3,647)*		*(3,765)*		*(3,198)*		*(2,235)*		*(16,508)*		**ENGLAND & WALES**
One year	83	86	82	85	81	85	79	82	79	82	**81**	**84**	84	87	83	86	82	86	81	84	80	83	**82**	**85**	*One year*
Five years	65	74	63	73	61	71	58	69	58	70	**61**	**72**	64	74	63	74	62	73	60	71	59	71	**62**	**73**	*Five years*
	54	71	52	70	50	69	47	65	46	65	**50**	**68**													
	(3,390)		*(3,433)*		*(3,445)*		*(3,019)*		*(2,215)*		*(15,793)*		*(3,291)*		*(3,504)*		*(3,587)*		*(2,976)*		*(2,119)*		*(15,680)*		**ENGLAND**
One year	83	86	82	85	82	85	79	82	78	82	**81**	**84**	84	87	83	86	83	86	81	84	80	84	**82**	**85**	*One year*
Five years	65	74	63	74	61	72	58	69	58	70	**61**	**72**	64	74	63	74	62	73	60	72	59	71	**62**	**73**	*Five years*
	54	71	52	71	50	69	47	65	46	66	**50**	**69**													
	(284)		*(315)*		*(393)*		*(405)*		*(463)*		*(1,883)*		*(286)*		*(352)*		*(430)*		*(391)*		*(380)*		*(1,851)*		**Northern & Yorkshire**
One year	83	86	80	83	84	88	80	84	78	82	**81**	**85**	85	88	80	83	83	86	79	84	80	85	**81**	**85**	*One year*
Five years	65	74	61	71	59	71	59	71	58	72	**60**	**72**	64	75	62	73	61	73	61	74	60	72	**62**	**74**	*Five years*
	53	71	52	69	48	70	46	65	45	68	**48**	**69**													
	(252)		*(363)*		*(421)*		*(428)*		*(278)*		*(1,757)*		*(248)*		*(350)*		*(367)*		*(376)*		*(284)*		*(1,634)*		**Trent**
One year	81	84	84	87	80	83	78	81	81	84	**81**	**84**	81	83	83	86	84	88	78	81	79	82	**81**	**84**	*One year*
Five years	63	73	63	74	59	69	58	67	59	70	**60**	**71**	66	74	61	71	63	72	59	69	55	67	**61**	**71**	*Five years*
	52	67	49	69	49	66	49	66	45	65	**48**	**67**													
	(536)		*(494)*		*(450)*		*(264)*		*(115)*		*(1,887)*		*(494)*		*(517)*		*(432)*		*(289)*		*(116)*		*(1,878)*		**Anglia & Oxford**
One year	83	87	83	86	79	83	79	83	71	75	**81**	**85**	81	84	81	85	79	82	79	82	78	82	**80**	**83**	*One year*
Five years	66	75	64	74	60	70	64	74	56	69	**63**	**73**	60	70	60	71	61	73	59	69	60	71	**60**	**71**	*Five years*
	53	69	55	72	52	69	53	69	*44*	*63*	**52**	**70**													
	(361)		*(348)*		*(317)*		*(376)*		*(217)*		*(1,651)*		*(405)*		*(393)*		*(418)*		*(379)*		*(293)*		*(1,955)*		**North Thames**
One year	82	85	85	88	83	87	81	84	79	83	**82**	**86**	87	90	86	89	85	87	84	88	83	86	**85**	**88**	*One year*
Five years	62	70	62	72	65	76	61	71	60	71	**62**	**72**	65	74	69	79	59	68	61	72	60	71	**63**	**73**	*Five years*
	55	70	52	69	56	75	49	67	49	67	**52**	**70**													
	(779)		*(604)*		*(505)*		*(388)*		*(164)*		*(2,526)*		*(593)*		*(530)*		*(523)*		*(395)*		*(171)*		*(2,259)*		**South Thames**
One year	84	87	81	84	80	83	78	81	81	84	**81**	**85**	84	88	83	86	83	87	83	86	83	86	**83**	**87**	*One year*
Five years	67	77	63	73	59	70	56	66	58	69	**61**	**72**	64	74	65	75	63	76	60	72	63	73	**63**	**75**	*Five years*
	56	73	53	70	49	69	47	63	47	65	**51**	**70**													
	(536)		*(684)*		*(615)*		*(298)*		*(94)*		*(2,296)*		*(522)*		*(705)*		*(664)*		*(389)*		*(146)*		*(2,443)*		**South & West**
One year	84	87	84	88	82	85	79	82	*84*	*88*	**83**	**86**	84	86	85	88	86	89	83	86	85	88	**85**	**88**	*One year*
Five years	66	75	66	77	61	71	56	66	*61*	*71*	**63**	**74**	65	74	65	76	65	76	65	76	70	82	**65**	**76**	*Five years*
	55	72	52	72	48	65	45	63	*50*	*66*	**50**	**69**													
	(339)		*(317)*		*(401)*		*(457)*		*(429)*		*(1,963)*		*(373)*		*(340)*		*(408)*		*(402)*		*(309)*		*(1,840)*		**West Midlands**
One year	85	89	80	83	82	85	78	81	81	84	**81**	**84**	86	89	83	86	79	82	81	84	80	84	**82**	**85**	*One year*
Five years	66	78	63	73	63	74	58	68	62	73	**62**	**73**	69	79	63	74	59	70	60	71	61	73	**62**	**74**	*Five years*
	55	74	52	69	52	69	47	66	49	68	**51**	**69**													
	(303)		*(308)*		*(343)*		*(403)*		*(455)*		*(1,830)*		*(370)*		*(317)*		*(345)*		*(355)*		*(420)*		*(1,820)*		**North & West**
One year	79	82	79	81	83	85	79	82	73	77	**78**	**82**	83	87	80	84	79	83	78	81	76	80	**79**	**83**	*One year*
Five years	59	69	60	70	64	73	56	67	54	65	**58**	**69**	61	71	59	70	61	72	58	68	54	65	**58**	**69**	*Five years*
	50	67	50	68	54	73	44	62	43	61	**48**	**67**													
	(97)		*(133)*		*(184)*		*(239)*		*(134)*		*(873)*		*(125)*		*(143)*		*(178)*		*(222)*		*(116)*		*(828)*		**WALES**
One year	82	85	80	82	80	82	79	82	83	86	**80**	**83**	85	87	79	82	80	83	79	82	76	79	**80**	**83**	*One year*
Five years	*59*	*68*	55	64	56	64	59	69	56	64	**58**	**67**	58	67	60	73	62	71	53	62	53	62	**57**	**68**	*Five years*
	48	*62*	46	64	44	62	46	64	41	54	**46**	**63**													

[1] Including those for whom no deprivation category was assigned. C - crude survival, R - relative survival (see Chapter 3).
Number of patients contributing to each analysis in parentheses.

Fig. 30.1 Relative survival up to ten years: trends by calendar period of diagnosis (1971-90) and NHS Region

WOMEN

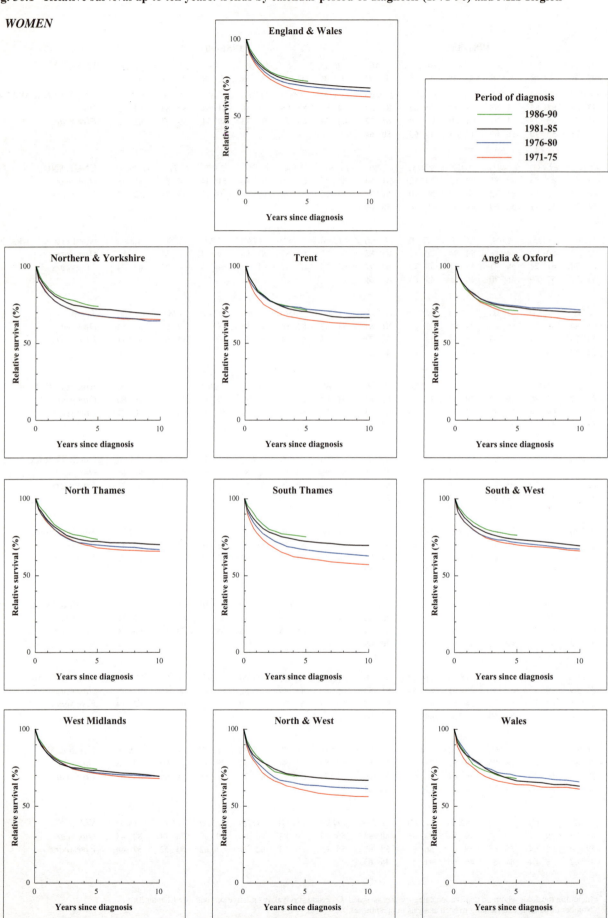

Fig. 30.2 Relative survival up to five years, by deprivation category and NHS Region: patients diagnosed 1986-90

WOMEN

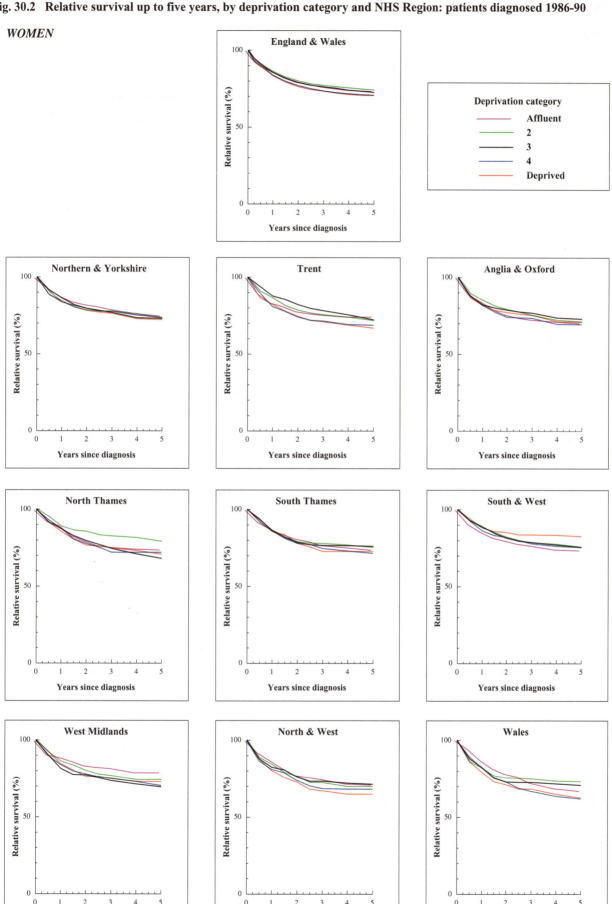

Fig. 30.3 Relative survival at five years by deprivation category, period of diagnosis (1981-90) and NHS Region

WOMEN

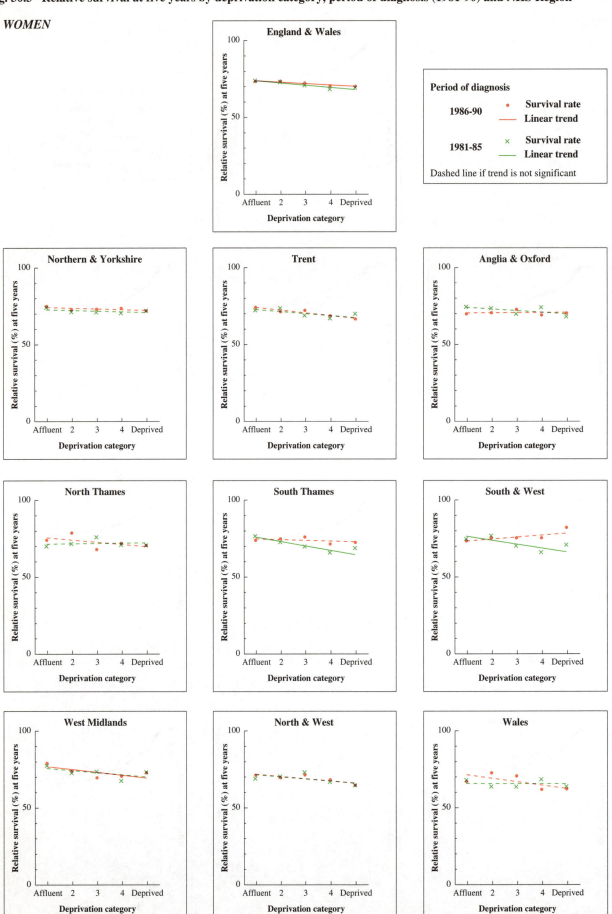

Fig. 30.4 Relative survival (%) by NHS Region:
      one-year and five-year survival for women diagnosed 1986-90, with 95% confidence intervals

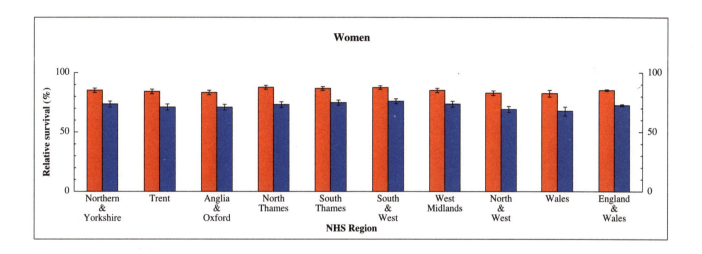

Fig. 30.5 Relative survival (%) at five years, with 95% confidence interval, by age at diagnosis:
      England and Wales, women diagnosed 1986-90

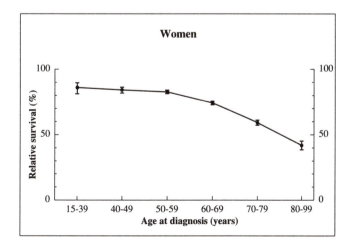

Chapter 31

OVARY

Ovarian cancer is the fourth commonest cancer in women in England and Wales, with about 5,000 new cases each year (5% of all cancers). The age-standardised rate (Europe) is around 17 per 100,000 per year. Successive generations have had slightly higher risk, but there has been a decline for women born after 1930. Late menarche, high parity, earlier menopause and particularly the use of combined oral contraceptives all reduce the risk, probably by reducing the number of ovulatory cycles. Overall mortality increased for many years, but was stable in the late 1980s. The earlier increase may have been partly artefactual, due to improved death certification, but may also reflect a true increase in risk with declining parity. Mortality in women under 45 was falling by more than 4% a year in the late 1980s, and mortality has been falling for successive generations born after 1930. This decline has not been considered attributable to treatment or to prior oophorectomy for benign disease[1].

Ovarian cancer is rare under the age of 40. Almost all the tumours are epithelial: incidence rises steeply with age. Germ cell and follicular tumours account for only 1-2%, but they are relatively more common in young women. Survival is generally low, because symptoms of abdominal pain or distension are non-specific, and disease is usually advanced at diagnosis. Five-year survival in hospital series may vary from less than 10% to 70% or more according to stage at diagnosis. Ovarian cancer accounts for 6% of all cancer deaths in women, and for more deaths than all the other gynaecological cancers combined. There is no evidence that screening reduces mortality, but several large randomised controlled trials are in progress.

The survival analyses reported here include more than 78,000 women diagnosed with ovarian cancer during 1971-90 and followed up to the end of 1995, about 87% of those eligible (Table 31.1). Tumours of uncertain malignancy were not included. Of over 90,000 women eligible for analysis, 4% were excluded because their vital status was unknown at 6 July 1997, and a further 9% because their survival was zero or unknown: the proportion excluded for this reason rose from 9% during 1971-75 to 17% in 1986-90.

Almost all tumours occurred in the ovary, and only 1% in the Fallopian tubes or other uterine adnexa (Table 31A). About a third (32%) were serous, papillary or mucinous cystadenocarcinomas, another 35% were coded to other types of adenocarcinoma, and 25% were other types of epithelial tumours, poorly specified. Granulosa cell tumours accounted for 1%, and a similar proportion were dysgerminoma or malignant teratoma. For 7% the morphology was either non-specific or unavailable.

Survival trends

Survival for women diagnosed with cancer of the ovary in England and Wales during 1986-90 was 56% at one year and 29% at five years, compared to 46% and 23% for women diagnosed during 1971-75 (Table 31.2). Excess mortality five years after diagnosis is low, and relative survival at ten years is only 3-5% lower than five-year survival.

Nationally, survival for women diagnosed during the 1980s was 5-6% higher than for women diagnosed during the 1970s, but trends differ widely between NHS Regions (Figure 31.1).

One-year survival in England and Wales improved by an average of 4% between successive five-year periods. The corresponding average gain in five-year survival was 3% every five years, after adjustment for changes in the age distribution of patients over time. Most of the improvement occurred by the early 1980s, and there has been little subsequent change (see Table 4.4). It seems likely that part of the fall in mortality during the 1980s is related to improved survival, as well as to changes in incidence.

Regional survival rates for women diagnosed during 1986-90 vary within a 11% range around the national average for one-year survival and a 9% range for five-year survival (Figure 31.4). Regional trends vary between an average gain every five years of 2% and 7% for one-year survival, and from 1% to 4% for five-year survival. Gains have been highest in Northern & Yorkshire and the Thames regions (Figure 31.1; see Table-Figure 4.5L).

Survival falls sharply with age after adjustment for other causes of death (Figure 31.5). For more than 1,200 women diagnosed under age 40 during the period 1986-90, relative survival at five years was 69%, but for women aged 40-49 the rate was only 43%, and this fell to less than 20% for women aged 70 or more (see Table 4.7).

Deprivation

Differences in survival between affluent and deprived groups are more evident at one year after diagnosis than at five years (Figure 31.2). In Anglia & Oxford, for example, the 7% advantage in one-year survival for women in the affluent group diagnosed during 1986-90 is absent at five years (Table 31.2). Survival estimates for each deprivation

Table 31A Sub-site distribution of cancer of the ovary and uterine adnexa

ICD-9 code	Site description	1971-75 No.	1971-75 %	1976-80 No.	1976-80 %	1981-85 No.	1981-85 %	1986-90 No.	1986-90 %	Total No.	Total %
183.0	Ovary	17,592	98.9	18,796	98.9	20,199	98.6	20,968	98.7	77,555	98.8
183.1	*Fallopian tube & broad ligament*[1]	184	1.0	100	0.5	-	-	-	-	**284**	**0.4**
183.2	Fallopian tube	-	-	61	0.3	175	0.9	190	0.9	**426**	**0.5**
183.3	Broad ligament	-	-	8	0.0	21	0.1	6	0.0	**35**	**0.0**
183.4	Parametrium	-	-	0	0.0	19	0.1	10	0.0	**29**	**0.0**
183.5	Round ligament	-	-	0	0.0	2	0.0	1	0.0	**3**	**0.0**
183.8	Other	-	-	24	0.1	27	0.1	41	0.2	**92**	**0.1**
183.9	Uterine adnexa, unspecified	6	0.0	20	0.1	37	0.2	25	0.1	**88**	**0.1**
	Total	**17,782**		**19,009**		**20,480**		**21,241**		**78,512**	

[1] Tumours of the Fallopian tube and broad ligament were classified together in ICD-8 (1971-78); the code 183.1 was not used in ICD-9.

group in each region are generally based on 200-800 patients, and they are precise.

Nationally, the deprivation gradient in five-year survival is significant throughout the 1980s, but it was less steep for women diagnosed during 1986-90, and it is variable between regions (Figure 31.3). Among more than 20,000 women diagnosed in England and Wales during the early 1980s, there was a 9% gap in one-year survival between affluent and deprived groups: this was 7% for women diagnosed in the late 1980s. The gap in five-year survival was 2-3% (see Table 4.9). The much flatter deprivation gradient for five-year survival suggests that diagnosis among women in more affluent groups tends to occur earlier.

There was no consistent difference in survival between deprivation categories among 6,000 women diagnosed with ovarian cancer in South Thames during 1980-89[2]: this is reflected in the results here (Table 31.2).

International comparison

Five-year survival for women diagnosed with ovarian cancer in England and Wales during 1986-90 was similar to that in Scotland (29-30%), and higher than in Slovenia, Poland and Estonia. It was 6% lower than the average for Europe (see Table 4.12). The highest regional five-year survival rate (34%, North Thames) is similar to the European average, but 11% below the survival rate in the USA.

1.   Villard-Mackintosh L, Vessey MP, Jones L. The effects of oral contraceptives and parity on ovarian cancer trends in women under 55 years of age. Br J Obstet Gynaecol 1989; 96: 783-788

2.   Schrijvers CTM, Mackenbach J, Lutz J-M, Quinn MJ, Coleman MP. Deprivation, stage at diagnosis and cancer survival. Int J Cancer 1995; 63: 324-329

Table 31.1 Ineligible and excluded records, and patients included in analyses, by calendar period of diagnosis

	1971-75			1976-80			1981-85			1986-90			Overall		
Total registered	**20,648**			**22,029**			**24,145**			**25,814**			**92,636**		
Ineligible	*Records*	*Patients*	*%*	*Records*	*Patients*	*%*	*Records*	*Patients*	*%*	*Records*	*Patients*	*%*	*Records*	*Patients*	*%*
Incomplete data[1]	5	5		26	26		81	81		68	68		180	**180**	
Resident outside England and Wales	157	156		134	133		105	90		78	58		474	**437**	
In situ neoplasm[2]	22	22		17	17		0	0		0	0		39	**39**	
Benign or uncertain behaviour[2]	323	286		198	188		10	10		8	7		539	**491**	
Metastatic[2]	636	632		256	252		4	4		12	12		908	**900**	
Otherwise ineligible[3]	-	-		-	-		-	-		-	-		-	**-**	
Lymphoma[4]	5	4		5	5		1	0		0	0		11	**9**	
Leukaemia or myeloma[4]	0	0		0	0		0	0		1	0		1	**0**	
		1,105			**621**			**185**			**145**			**2,056**	
Total eligible		**19,543**	**100**		**21,408**	**100**		**23,960**	**100**		**25,669**	**100**		**90,580**	**100**
Aged under 15 or 100 years or over at diagnosis	43	38	0.2	57	50	0.2	57	57	0.2	71	69	0.3	228	**214**	0.2
Vital status unknown[5]	670	581	3.0	897	792	3.7	1,012	949	4.0	1,371	1,317	5.1	3,950	**3,639**	4.0
Duplicate registration[6]	0	0	0.0	0	0	0.0	0	0	0.0	2	0	0.0	2	**0**	0.0
Multiple primary[7]	2	1	<0.1	17	15	<0.1	100	85	0.4	149	136	0.5	268	**237**	0.3
Synchronous tumours[8]	66	46	0.2	62	41	0.2	113	88	0.4	131	78	0.3	372	**253**	0.3
Sex not known	0	0	0.0	0	0	0.0	1	0	0.0	2	0	0.0	3	**0**	0.0
Sex incompatible with site code[9]	0	0	0.0	0	0	0.0	1	1	<0.1	2	2	<0.1	3	**3**	<0.1
Invalid dates[10]	0	0	0.0	2	0	0.0	4	3	<0.1	13	5	<0.1	19	**8**	<0.1
Zero survival[11]	1,203	1,095	5.6	1,600	1,501	7.0	2,484	2,297	9.6	3,009	2,821	11.0	8,296	**7,714**	8.5
Total exclusions		**1,761**	**9.0**		**2,399**	**11.2**		**3,480**	**14.5**		**4,428**	**17.3**		**12,068**	**13.3**
Patients included in analysis		**17,782**	**91.0**		**19,009**	**88.8**		**20,480**	**85.5**		**21,241**	**82.7**		**78,512**	**86.7**

[1] Main data item(s) invalid or incompatible with one another: name, sex, date of birth, date of diagnosis and (if present) date of death, postcode, site, morphology and behaviour

[2] Tumour either *in situ* (behaviour code 2), benign (0), uncertain if benign or malignant (1) or metastatic (6 or 9)

[3] Not applicable for this malignancy

[4] Morphology that of lymphoma (included with non-Hodgkin lymphoma, Ch. 41) or myeloma (included with multiple myeloma, Ch. 43), or leukaemia (excluded)

[5] Tracing of vital status at National Health Service Central Register incomplete at 6 July 1997

[6] Same site code, sex, cancer registry and cancer registry number as an earlier registration

[7] Same site code and person number as an earlier registration(s): mostly confirmed multiple primary tumours at this site, some unresolved duplicate registrations

[8] Same site code, sex, date of birth and date of diagnosis as another registration(s): mostly synchronous or (in paired organs) bilateral tumours in same anatomic site in one individual, not previously linked: also some duplicate registrations

[9] Site code specific to one sex; tumour registered to opposite sex

[10] Impossible sequence of dates (birth, diagnosis and, if present, emigration or death); or date of death after 6 July 1997

[11] Date of diagnosis same as date of death: some are patients who did die on the day of diagnosis, but most were registered solely from a death certificate (death certificate only, or DCO), and their survival time is thus unknown

Table 31.2 Crude and relative survival (%) at one, five and ten years since diagnosis: NHS Region, deprivation category and calendar period of diagnosis

WOMEN		1971-75												1976-80											
		Affluent		2		3		4		Deprived		All[1]		Affluent		2		3		4		Deprived		All[1]	
		C	R	C	R	C	R	C	R	C	R	C	R	C	R	C	R	C	R	C	R	C	R	C	R
ENGLAND & WALES		(2,930)		(3,094)		(3,251)		(2,888)		(2,227)		(17,782)		(3,779)		(3,923)		(3,988)		(3,696)		(2,820)		(19,009)	
	One year	46	47	46	47	45	46	44	45	44	45	45	46	50	51	47	48	45	46	44	45	44	45	46	47
	Five years	21	22	21	23	20	23	20	22	20	22	21	23	24	26	22	24	21	24	21	23	20	23	22	24
	Ten years	16	19	17	20	16	19	16	19	16	20	16	20	19	23	17	20	16	20	16	20	15	19	17	21
ENGLAND		(2,828)		(2,951)		(3,063)		(2,675)		(2,108)		(16,815)		(3,674)		(3,749)		(3,773)		(3,421)		(2,701)		(17,999)	
	One year	46	47	45	46	45	46	43	44	43	44	45	46	50	51	47	48	45	46	44	45	43	44	46	47
	Five years	21	23	21	23	20	22	19	21	20	22	21	23	23	26	22	24	21	24	20	23	20	22	21	24
	Ten years	16	20	17	20	15	19	15	19	15	20	16	20	19	23	16	20	16	20	16	19	15	19	17	21
Northern & Yorkshire		(211)		(270)		(342)		(358)		(414)		(2,246)		(323)		(399)		(460)		(550)		(652)		(2,454)	
	One year	41	42	34	35	41	42	42	43	37	38	41	42	47	48	46	47	41	42	39	40	39	40	42	43
	Five years	13	14	13	14	16	17	18	21	16	19	17	19	21	22	21	23	18	20	16	18	18	20	19	21
	Ten years	9	11	9	11	11	14	14	18	14	18	13	16	16	19	16	20	14	17	13	16	13	17	14	18
Trent		(215)		(281)		(340)		(414)		(291)		(1,690)		(307)		(338)		(444)		(444)		(345)		(1,932)	
	One year	46	47	51	52	49	49	48	49	49	50	48	49	51	53	49	50	47	48	43	44	42	43	47	48
	Five years	23	24	25	26	22	25	21	24	26	28	23	26	24	26	23	26	22	25	23	26	22	25	23	26
	Ten years	19	23	21	24	18	21	17	21	18	23	19	22	21	24	16	21	18	22	19	23	16	21	18	22
Anglia & Oxford		(358)		(351)		(376)		(214)		(110)		(1,621)		(508)		(465)		(432)		(257)		(105)		(1,847)	
	One year	48	49	51	52	45	46	44	45	34	35	47	48	48	49	46	47	44	45	41	43	45	46	46	47
	Five years	23	26	25	27	22	25	22	25	15	16	23	25	24	26	23	26	23	26	19	21	18	20	22	25
	Ten years	18	22	22	25	17	22	18	22	11	14	18	22	20	23	19	23	18	23	15	19	10	13	18	22
North Thames		(412)		(408)		(363)		(376)		(277)		(2,067)		(371)		(387)		(367)		(366)		(255)		(1,822)	
	One year	49	50	47	48	47	48	47	48	52	53	48	49	51	52	53	54	48	49	54	55	46	47	51	52
	Five years	23	25	21	23	20	22	21	23	26	29	22	25	23	25	22	24	22	24	24	26	19	21	22	25
	Ten years	19	22	16	20	15	18	17	21	23	29	18	22	17	21	17	21	17	21	17	20	13	18	17	21
South Thames		(755)		(604)		(605)		(466)		(243)		(2,926)		(859)		(701)		(568)		(455)		(180)		(2,920)	
	One year	43	44	43	44	40	41	39	40	39	40	42	43	49	50	42	43	40	41	41	43	40	41	44	45
	Five years	18	20	18	20	15	18	18	20	15	17	18	20	23	26	18	20	17	19	17	19	22	25	20	22
	Ten years	14	17	13	16	13	16	15	18	12	15	14	17	19	23	13	16	13	16	14	17	17	21	15	19
South & West		(496)		(637)		(536)		(281)		(95)		(2,284)		(638)		(777)		(687)		(343)		(121)		(2,671)	
	One year	45	46	44	45	43	44	40	41	38	39	43	44	50	51	47	48	46	47	45	46	57	58	49	50
	Five years	22	24	23	25	22	24	18	20	22	23	22	24	24	26	23	25	22	25	19	22	27	32	23	26
	Ten years	17	22	19	23	16	21	16	19	18	23	17	22	19	22	16	21	17	22	16	20	20	28	18	22
West Midlands		(84)		(92)		(125)		(133)		(137)		(1,563)		(273)		(288)		(359)		(411)		(406)		(1,767)	
	One year	57	58	54	55	56	58	51	52	51	52	52	53	59	60	52	53	48	49	48	49	48	50	51	52
	Five years	26	28	22	24	29	32	21	24	21	24	26	28	30	33	27	30	27	29	26	29	25	28	27	30
	Ten years	22	25	15	18	20	24	15	18	13	17	20	24	25	31	21	26	21	25	19	22	21	26	22	26
North & West		(297)		(308)		(376)		(433)		(541)		(2,418)		(395)		(394)		(456)		(595)		(637)		(2,586)	
	One year	43	44	39	40	41	42	35	36	40	41	41	42	41	42	43	45	42	43	37	38	37	38	40	42
	Five years	19	20	18	20	18	20	14	16	17	19	18	20	18	20	18	20	16	18	17	20	13	15	17	19
	Ten years	15	17	13	16	13	16	10	13	12	16	13	16	15	17	14	18	13	16	13	17	10	13	13	17
WALES		(102)		(143)		(188)		(213)		(119)		(967)		(105)		(174)		(215)		(275)		(119)		(1,010)	
	One year	35	35	49	50	49	51	50	51	51	52	48	49	57	59	45	46	43	44	45	46	55	56	47	48
	Five years	14	15	21	23	27	30	25	28	24	26	23	26	27	29	24	27	21	22	23	25	31	35	25	28
	Ten years	12	14	18	23	20	25	20	25	20	24	19	24	19	24	17	21	15	19	19	23	19	26	18	24

[1] Including those for whom no deprivation category was assigned. C - crude survival, R - relative survival (see Chapter 3). Number of patients contributing to each analysis in parentheses.

	1981-85												1986-90												WOMEN
	Affluent		*2*		*3*		*4*		*Deprived*		*All¹*		*Affluent*		*2*		*3*		*4*		*Deprived*		*All¹*		
	C	R	C	R	C	R	C	R	C	R	C	R	C	R	C	R	C	R	C	R	C	R	C	R	
	(4,319)		*(4,359)*		*(4,269)*		*(3,953)*		*(3,102)*		*(20,480)*		*(4,545)*		*(4,752)*		*(4,532)*		*(4,187)*		*(2,983)*		*(21,241)*		**ENGLAND & WALES**
	56	58	52	54	50	51	50	51	48	50	51	53	58	59	55	56	53	55	51	53	51	53	54	56	*One year*
	27	30	25	28	24	27	25	28	23	26	25	28	28	31	26	29	26	29	25	29	25	29	26	29	*Five years*
	21	26	19	24	19	23	19	24	18	23	19	24													
	(4,149)		*(4,150)*		*(3,983)*		*(3,610)*		*(2,908)*		*(19,176)*		*(4,370)*		*(4,514)*		*(4,219)*		*(3,892)*		*(2,830)*		*(20,032)*		**ENGLAND**
	57	58	52	54	50	51	50	51	48	49	51	53	58	60	55	56	53	54	52	53	51	53	54	56	*One year*
	27	30	25	28	24	27	25	28	23	26	25	28	28	31	26	29	25	29	26	29	25	28	26	29	*Five years*
	21	26	19	24	19	23	19	24	18	23	19	24													
	(340)		*(426)*		*(449)*		*(523)*		*(657)*		*(2,432)*		*(408)*		*(487)*		*(522)*		*(616)*		*(620)*		*(2,664)*		**Northern & Yorkshire**
	54	55	53	54	47	49	47	48	47	48	50	51	55	56	54	55	49	51	48	50	48	50	51	53	*One year*
	26	28	27	31	21	23	20	23	24	27	24	27	27	29	27	30	21	24	22	25	23	26	24	27	*Five years*
	19	23	21	26	16	20	15	20	18	24	18	23													
	(323)		*(407)*		*(455)*		*(506)*		*(317)*		*(2,023)*		*(346)*		*(479)*		*(465)*		*(495)*		*(354)*		*(2,148)*		**Trent**
	60	62	53	54	53	55	51	52	44	46	53	55	59	60	60	61	50	51	52	53	53	55	55	57	*One year*
	32	35	25	28	24	27	26	29	22	25	26	29	31	33	31	35	24	27	26	29	25	28	28	31	*Five years*
	26	30	21	25	19	24	19	25	16	21	20	25													
	(585)		*(553)*		*(465)*		*(305)*		*(109)*		*(2,063)*		*(609)*		*(636)*		*(516)*		*(324)*		*(120)*		*(2,225)*		**Anglia & Oxford**
	56	58	51	53	50	51	52	54	48	50	53	55	57	58	52	54	53	55	51	52	49	51	54	56	*One year*
	29	32	26	29	28	31	26	29	28	31	28	31	28	30	29	32	26	30	29	33	28	30	28	31	*Five years*
	22	27	19	24	22	26	22	28	21	27	22	27													
	(418)		*(353)*		*(335)*		*(321)*		*(250)*		*(1,708)*		*(467)*		*(470)*		*(479)*		*(449)*		*(395)*		*(2,333)*		**North Thames**
	60	61	58	59	58	59	57	58	48	49	57	58	67	69	61	62	57	58	59	61	57	59	61	63	*One year*
	28	30	26	28	27	30	28	31	28	31	28	30	32	35	31	34	28	31	30	33	27	31	30	34	*Five years*
	22	26	19	23	20	25	21	26	24	30	21	26													
	(913)		*(687)*		*(563)*		*(444)*		*(196)*		*(2,921)*		*(849)*		*(707)*		*(565)*		*(438)*		*(180)*		*(2,782)*		**South Thames**
	58	59	53	54	49	51	51	53	57	59	54	56	63	65	56	58	54	56	54	55	60	63	58	60	*One year*
	27	29	24	26	26	29	26	30	25	28	26	29	30	33	24	27	28	31	24	27	30	34	28	31	*Five years*
	21	25	18	22	20	25	19	24	17	23	20	25													
	(679)		*(815)*		*(717)*		*(361)*		*(120)*		*(2,775)*		*(703)*		*(800)*		*(676)*		*(457)*		*(163)*		*(2,818)*		**South & West**
	56	58	50	52	46	47	47	48	47	48	50	52	49	51	47	48	50	51	54	55	50	52	50	52	*One year*
	27	30	25	27	22	25	26	29	23	26	25	28	22	25	20	22	21	24	26	29	25	27	22	25	*Five years*
	21	25	20	24	17	21	22	28	*21*	*26*	19	24													
	(407)		*(405)*		*(480)*		*(504)*		*(492)*		*(2,311)*		*(458)*		*(453)*		*(481)*		*(515)*		*(383)*		*(2,301)*		**West Midlands**
	58	59	53	55	49	51	49	51	54	55	53	55	58	59	54	55	54	56	52	54	51	53	54	56	*One year*
	27	30	27	29	27	29	26	29	24	27	26	29	29	32	26	29	29	33	27	30	26	30	28	31	*Five years*
	23	28	20	25	20	24	20	25	19	25	21	26													
	(484)		*(504)*		*(519)*		*(646)*		*(767)*		*(2,943)*		*(530)*		*(482)*		*(515)*		*(598)*		*(615)*		*(2,761)*		**North & West**
	48	50	46	48	43	45	43	44	39	40	45	46	53	55	56	58	51	53	43	45	44	46	50	52	*One year*
	21	23	20	23	19	21	22	24	17	20	20	23	26	29	23	25	24	28	21	24	21	24	24	27	*Five years*
	17	21	15	19	14	18	14	18	13	17	15	19													
	(170)		*(209)*		*(286)*		*(343)*		*(194)*		*(1,304)*		*(175)*		*(238)*		*(313)*		*(295)*		*(153)*		*(1,209)*		**WALES**
	52	54	48	49	50	52	52	53	54	56	52	53	47	48	53	54	55	56	44	45	52	53	51	53	*One year*
	25	27	24	27	26	29	27	30	26	29	26	29	24	26	29	31	27	31	23	25	28	32	27	30	*Five years*
	19	22	19	24	19	24	21	27	20	25	20	25													

[1] Including those for whom no deprivation category was assigned. C - crude survival, R - relative survival (see Chapter 3).
Number of patients contributing to each analysis in parentheses.

Fig. 31.1 Relative survival up to ten years: trends by calendar period of diagnosis (1971-90) and NHS Region

WOMEN

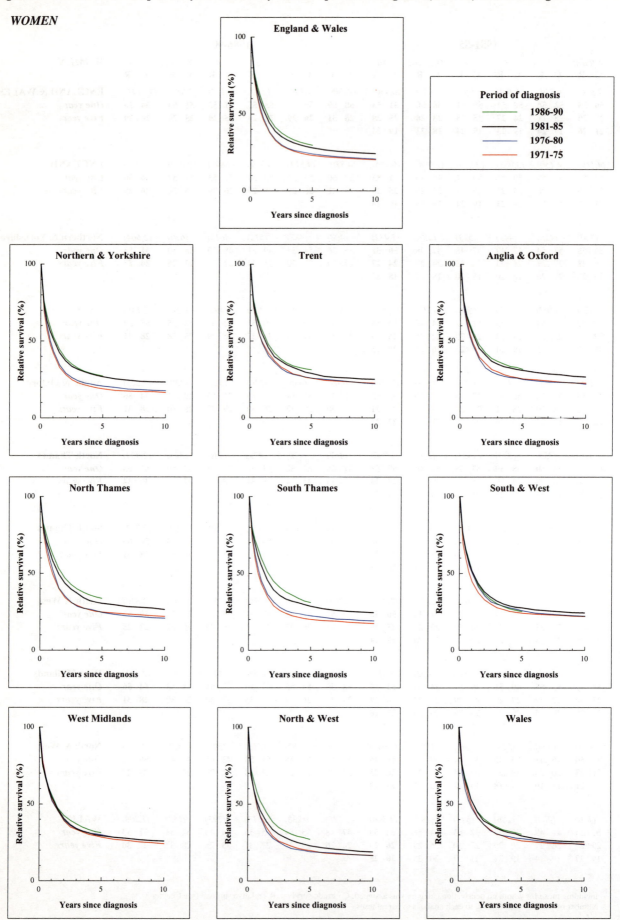

Fig. 31.2 Relative survival up to five years, by deprivation category and NHS Region: patients diagnosed 1986-90

WOMEN

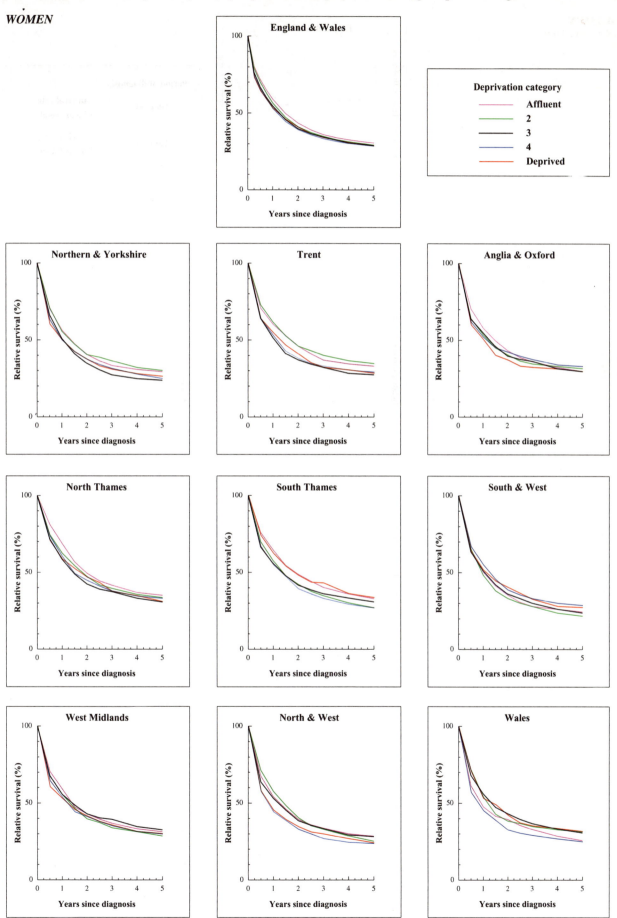

Fig. 31.3   Relative survival at five years by deprivation category, period of diagnosis (1981-90) and NHS Region

WOMEN
(Note vertical scale)

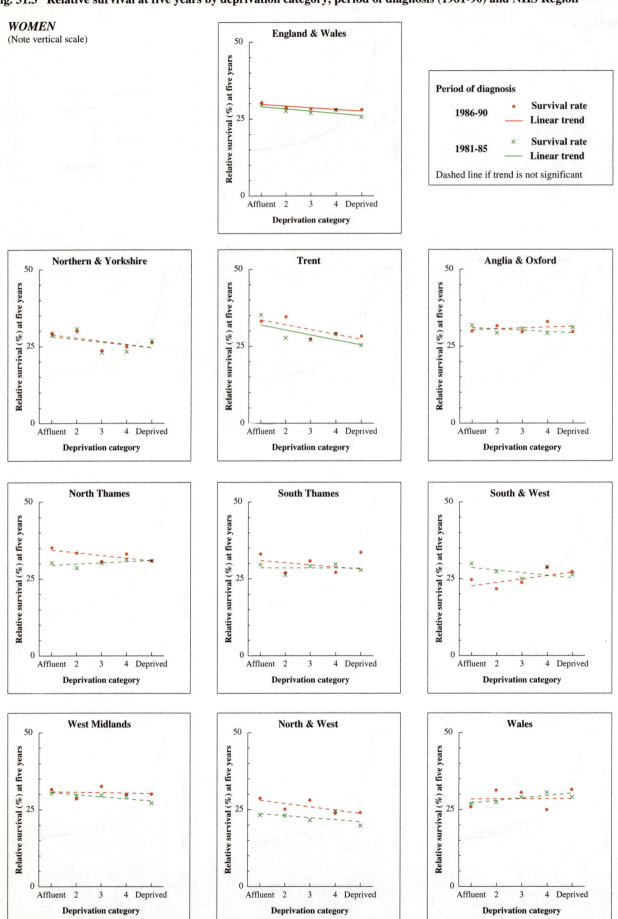

Fig. 31.4 Relative survival (%) by NHS Region:
one-year and five-year survival for women diagnosed 1986-90, with 95% confidence intervals

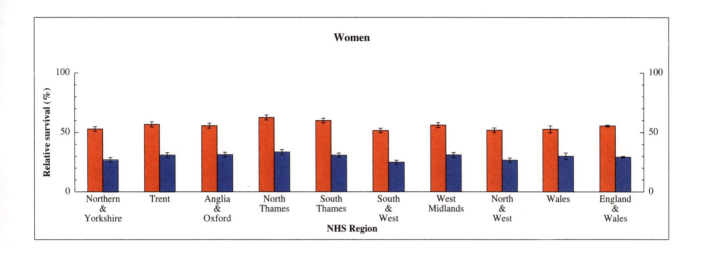

Fig. 31.5 Relative survival (%) at five years, with 95% confidence interval, by age at diagnosis:
England and Wales, women diagnosed 1986-90

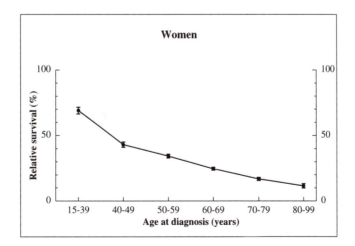

Chapter 32

VAGINA AND VULVA

Cancers of the vagina and vulva are uncommon. About 1,100 new cases are diagnosed in England and Wales each year, of which 900 are cancers of the external genitalia: vaginal cancer is rare. Incidence in England and Wales is 1.5 per 100,000 per year for vulvar cancer and 0.4 for vaginal cancer. Cancers at both sites are rare before the age of 50. Incidence has been stable since the 1970s. Mortality fell from 16.4 per 100,000 per year in 1971 to 9.9 by 1995.

Little is known about the causes of cancer of the vulva and vagina. Risk is higher in women who have had other anogenital cancers, suggesting shared risk factors. Infection with human papilloma virus is likely to be important, as with cervical cancer. Tobacco smoking is associated with increased risk, but no link has been shown with oral contraceptives. Vaginal cancer risk is high in women irradiated for cervical cancer. Treatment is by surgery or radiotherapy.

The survival analyses reported here include 18,700 women who were diagnosed with cancer of the vagina or vulva during the period 1971-90 and followed up to the end of 1995, about 90% of those eligible (Table 32.1). About 4% of women were excluded because their vital status was unknown at 6 July 1997, and a further 5% because their survival was either zero or unknown: women excluded in this category comprised 7% of those eligible for the period 1986-90.

About three-quarters of the tumours arose in the external genitalia (Table 32A). Tumours of the labia majora were more commonly specified than those of the labia minora, but most tumours were coded to the vulva, without further anatomic detail. Two-thirds (68%) of the tumours were squamous carcinomas. Melanoma accounted for 3%, basal cell carcinoma 4% and adenocarcinoma a further 3%, but 20% were other, poorly specified epithelial tumours.

Survival trends

Relative survival for women diagnosed in England and Wales during 1986-90 was 72% at one year and 53% at five years (Table 32.2; at end of chapter). Compared with the survival rates for women diagnosed in the early 1970s, this represents a 7% increase in one-year survival and a 9% rise in survival at five years. There has been improvement in each five-year period since the early 1970s (Figure 32.1). Survival has improved in all regions: the increase in regional five-year survival rates ranges from 5% to 15%.

Survival at both one year and five years has improved consistently in England and Wales by about 3% every five years since the early 1970s, after adjustment to the age distribution of women diagnosed during 1986-90 (see Table 4.4).

For women diagnosed during 1986-90, regional survival rates at one year and five years vary within a range of about 10% around the national average (Figure 32.4). Apart from those for Wales, the regional survival rates each depend on some 400-700 patients.

Relative survival falls steadily with age (Figure 32.5). For 360 women diagnosed under the age of 50 in England and Wales during 1986-90, five-year survival was about 80%. Survival was 66% for women aged 50-59 at diagnosis, and about 10% lower for each further decade of age, reaching 38% for women aged 80-99 at diagnosis (see Table 4.7).

Table 32A Sub-site distribution of cancers of the vagina and vulva

ICD-9 code	Site description	1971-75 No.	%	1976-80 No.	%	1981-85 No.	%	1986-90 No.	%	Total No.	%
184.0	Vagina	857	18.7	902	19.0	921	19.4	855	18.3	**3,535**	**18.9**
184.1	Labia majora[1]	-	-	72	1.5	190	4.0	173	3.7	**435**	**2.3**
184.2	Labia minora	-	-	38	0.8	80	1.7	52	1.1	**170**	**0.9**
184.3	Clitoris	-	-	29	0.6	86	1.8	62	1.3	**177**	**0.9**
184.4	Vulva, unspecified[1]	3,634	79.3	3,584	75.6	3,333	70.3	3,415	73.3	**13,966**	**74.6**
184.8	Other	31	0.7	39	0.8	32	0.7	28	0.6	**130**	**0.7**
184.9	Site unspecified	62	1.4	76	1.6	102	2.2	75	1.6	**315**	**1.7**
	Total	**4,584**		**4,740**		**4,744**		**4,660**		**18,728**	

[1] Tumours of the vulva were coded to 184.1 during 1971-78 (ICD-8): they have been assigned to the corresponding ICD-9 code in this table.

Deprivation

For women diagnosed during the 1980s, one-year survival has consistently been lowest in the most deprived groups (Table 32.2).

Among more than 4,600 women of known deprivation status who were diagnosed in England and Wales during the early 1980s, the deprivation gap in one-year survival was 8% (95% confidence interval 3% to 12%). For the 4,600 women diagnosed during the late 1980s, the deprivation gap was 5% (see Table 4.9).

Differences in five-year survival for women have been smaller (Figure 32.2). Regional patterns of survival between deprivation groups are not consistent (data not shown). The 5% deprivation gap in five-year survival for women diagnosed in England and Wales during the early 1980s is not statistically significant; the difference between deprivation categories for women diagnosed during the late 1980s was 2% (Figure 32.3).

International comparison

Five-year survival for women diagnosed with cancer of the vagina or vulva in England and Wales during 1986-90 was 53%, similar to the survival rates in Scotland and to the average for Europe (see Table 4.12). The highest regional five-year survival rate was 59% (West Midlands), still 7% below the average for the states covered by the SEER programme in the USA, 66%.

Table 32.1 Ineligible and excluded records, and patients included in analyses, by calendar period of diagnosis

		1971-75			1976-80			1981-85			1986-90			Overall	
Total registered		**5,169**			**5,275**			**5,322**			**5,499**			**21,265**	
Ineligible	*Records*	*Patients*	*%*	*Records*	*Patients*	*%*	*Records*	*Patients*	*%*	*Records*	*Patients*	*%*	*Records*	*Patients*	*%*
Incomplete data[1]	9	9		14	14		30	30		25	25		78	**78**	
Resident outside England and Wales	45	45		32	32		21	15		20	18		118	**110**	
In situ neoplasm[2]	142	141		62	62		0	0		0	0		204	**203**	
Benign or uncertain behaviour[2]	71	56		30	24		0	0		0	0		101	**80**	
Metastatic[2]	24	24		3	3		1	1		1	1		29	**29**	
Otherwise ineligible[3]	-	-		-	-		-	-		-	-		-	**-**	
Lymphoma[4]	8	8		1	1		0	0		0	0		9	**9**	
Leukaemia or myeloma[4]	0	0		0	0		0	0		0	0		0	**0**	
		283			136			46			44			509	
Total eligible		**4,886**	**100**		**5,139**	**100**		**5,276**	**100**		**5,455**	**100**		**20,756**	**100**
Aged under 15 or 100 years or over at diagnosis	12	12	0.2	14	14	0.3	10	10	0.2	20	18	0.3	56	**54**	0.3
Vital status unknown[5]	200	167	3.4	243	215	4.2	202	188	3.6	368	357	6.5	1,013	**927**	4.5
Duplicate registration[6]	0	0	0.0	0	0	0.0	0	0	0.0	0	0	0.0	0	**0**	0.0
Multiple primary[7]	0	0	0.0	6	3	<0.1	24	21	0.4	35	31	0.6	65	**55**	0.3
Synchronous tumours[8]	22	8	0.2	23	12	0.2	33	24	0.5	26	15	0.3	104	**59**	0.3
Sex not known	0	0	0.0	0	0	0.0	0	0	0.0	0	0	0.0	0	**0**	0.0
Sex incompatible with site code[9]	1	1	<0.1	1	1	<0.1	1	1	<0.1	0	0	0.0	3	**3**	<0.1
Invalid dates[10]	0	0	0.0	0	0	0.0	1	0	0.0	7	1	<0.1	8	**1**	<0.1
Zero survival[11]	139	114	2.3	164	154	3.0	316	288	5.5	414	373	6.8	1,033	**929**	4.5
Total exclusions		**302**	**6.2**		**399**	**7.8**		**532**	**10.1**		**795**	**14.6**		**2,028**	**9.8**
Patients included in analysis		**4,584**	**93.8**		**4,740**	**92.2**		**4,744**	**89.9**		**4,660**	**85.4**		**18,728**	**90.2**

[1]   Main data item(s) invalid or incompatible with one another: name, sex, date of birth, date of diagnosis and (if present) date of death, postcode, site, morphology and behaviour

[2]   Tumour either *in situ* (behaviour code 2), benign (0), uncertain if benign or malignant (1) or metastatic (6 or 9)

[3]   Not applicable for this malignancy

[4]   Morphology that of lymphoma (included with non-Hodgkin lymphoma, Ch. 41) or myeloma (included with multiple myeloma, Ch. 43), or leukaemia (excluded)

[5]   Tracing of vital status at National Health Service Central Register incomplete at 6 July 1997

[6]   Same site code, sex, cancer registry and cancer registry number as an earlier registration

[7]   Same site code and person number as an earlier registration(s): mostly confirmed multiple primary tumours at this site, some unresolved duplicate registrations

[8]   Same site code, sex, date of birth and date of diagnosis as another registration(s): mostly synchronous or (in paired organs) bilateral tumours in same anatomic site in one individual, not previously linked: also some duplicate registrations

[9]   Site code specific to one sex; tumour registered to opposite sex

[10]   Impossible sequence of dates (birth, diagnosis and, if present, emigration or death); or date of death after 6 July 1997

[11]   Date of diagnosis same as date of death: some are patients who did die on the day of diagnosis, but most were registered solely from a death certificate (death certificate only, or DCO), and their survival time is thus unknown

Fig. 32.1   Relative survival up to ten years: trends by calendar period of diagnosis (1971-90) and NHS Region

WOMEN

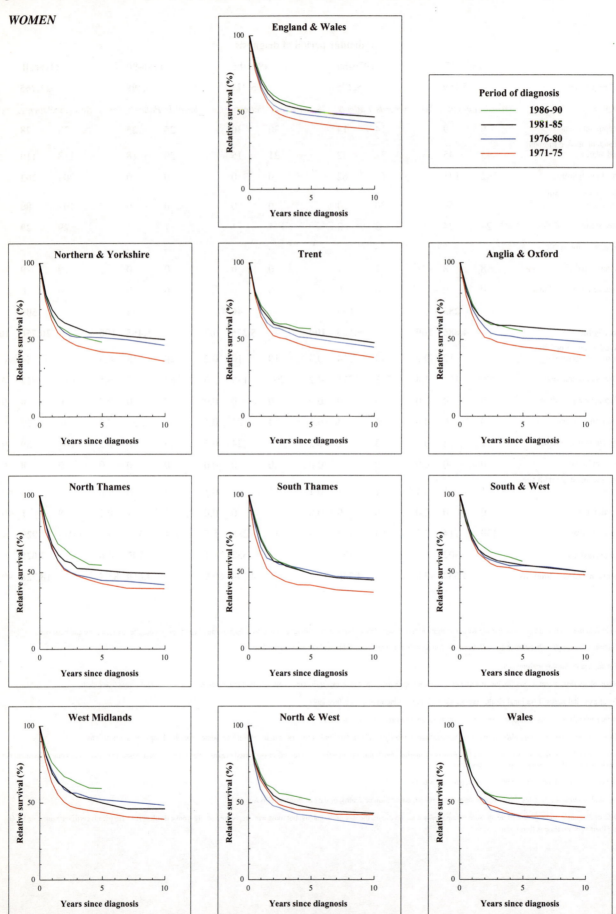

Fig. 32.2 Relative survival up to five years, by deprivation category: England and Wales, patients diagnosed 1986-90

WOMEN

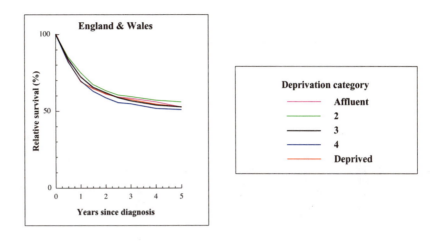

Fig. 32.3 Relative survival at five years, by deprivation category and period of diagnosis (1981-90): England and Wales

WOMEN

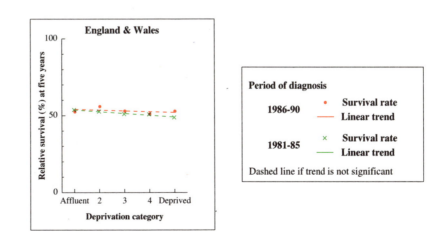

Fig. 32.4 Relative survival (%) by NHS Region:
   one-year and five-year survival for women diagnosed 1986-90, with 95% confidence intervals

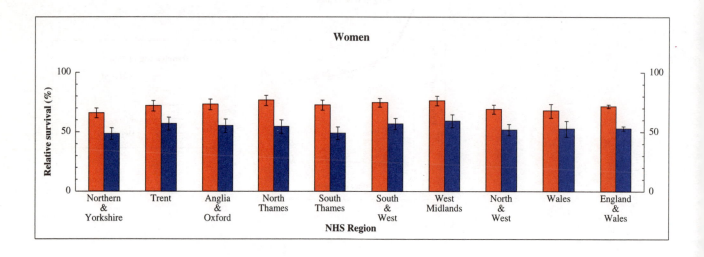

Fig. 32.5 Relative survival (%) at five years, with 95% confidence interval, by age at diagnosis:
   England and Wales, women diagnosed 1986-90

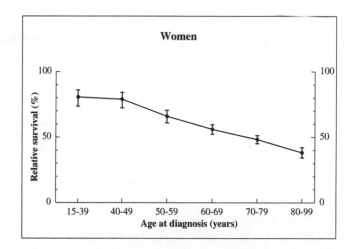

Table 32.2 Crude and relative survival (%) at one, five and ten years since diagnosis: NHS Region, deprivation category and calendar period of diagnosis

WOMEN	1971-75												1976-80											
	Affluent		*2*		*3*		*4*		*Deprived*		*All[1]*		*Affluent*		*2*		*3*		*4*		*Deprived*		*All[1]*	
	C	R	C	R	C	R	C	R	C	R	C	R	C	R	C	R	C	R	C	R	C	R	C	R
ENGLAND & WALES	*(597)*		*(769)*		*(835)*		*(794)*		*(657)*		*(4,584)*		*(773)*		*(964)*		*(993)*		*(973)*		*(808)*		*(4,740)*	
One year	62	65	65	68	58	61	61	64	60	63	61	65	64	67	64	68	64	69	61	65	60	64	62	66
Five years	36	45	37	48	32	40	34	43	32	40	34	44	38	49	39	50	38	49	37	48	37	48	37	48
Ten years	26	43	27	43	21	36	23	39	23	38	24	39	26	44	28	46	26	42	25	43	25	42	26	43
	1981-85												1986-90											
	(799)		*(957)*		*(1,009)*		*(1,015)*		*(841)*		*(4,744)*		*(772)*		*(951)*		*(1,060)*		*(1,023)*		*(794)*		*(4,660)*	
One year	68	73	67	71	64	69	63	67	61	66	65	69	67	72	71	75	68	72	65	70	65	69	67	72
Five years	41	53	40	52	39	51	38	50	37	48	39	51	40	52	42	56	40	53	38	51	40	53	40	53
Ten years	29	48	29	49	28	48	28	47	26	44	28	47												

	1971-75		1976-80		1981-85		1986-90	
Northern & Yorkshire	*(645)*		*(637)*		*(653)*		*(619)*	
One year	60	63	63	67	65	70	61	66
Five years	33	42	39	51	41	55	35	48
Ten years	21	36	25	46	29	50		
Trent	*(507)*		*(519)*		*(489)*		*(470)*	
One year	61	65	63	68	65	70	68	72
Five years	34	45	39	51	40	53	44	57
Ten years	23	38	27	45	27	48		
Anglia & Oxford	*(432)*		*(442)*		*(430)*		*(467)*	
One year	62	66	65	70	66	71	68	73
Five years	34	45	38	51	44	58	40	55
Ten years	23	39	27	48	34	55		
North Thames	*(469)*		*(394)*		*(367)*		*(443)*	
One year	63	67	62	66	63	68	72	77
Five years	34	43	35	45	38	51	41	55
Ten years	24	39	25	42	28	49		
South Thames	*(581)*		*(572)*		*(576)*		*(521)*	
One year	58	62	62	66	65	70	68	73
Five years	32	41	38	51	36	49	37	49
Ten years	22	37	26	46	26	45		
South & West	*(555)*		*(666)*		*(668)*		*(654)*	
One year	66	71	67	71	67	72	70	75
Five years	39	50	42	54	41	54	43	57
Ten years	28	48	29	50	29	50		
West Midlands	*(479)*		*(467)*		*(465)*		*(520)*	
One year	60	64	67	72	64	69	72	76
Five years	35	44	41	52	37	50	44	59
Ten years	24	39	30	48	25	46		
North & West	*(639)*		*(737)*		*(700)*		*(680)*	
One year	61	65	55	59	61	66	64	69
Five years	35	45	31	41	35	47	37	52
Ten years	25	42	20	36	24	43		
WALES	*(277)*		*(306)*		*(396)*		*(286)*	
One year	60	63	59	63	63	68	64	68
Five years	32	41	31	41	39	49	42	53
Ten years	22	40	21	33	32	47		

[1] Including those for whom no deprivation category was assigned. C - crude survival, R - relative survival (see Chapter 3).
Number of patients contributing to each analysis in parentheses.

Chapter 33

PROSTATE

Cancer of the prostate is the second most frequent cancer among men in England and Wales. More than 13,000 new cases occur each year (13% of all cancers in men), and the incidence rate in 1990 was 53 per 100,000. Incidence has been increasing at 10-15% every five years in England and Wales, as in many developed countries, where the lifetime risk up to age 74 for men born around 1940 is already an estimated 3%. Mortality has not been increasing quite as quickly; the corresponding risk of death is 1.3%. There are about 9,500 deaths a year from prostate cancer.

The causes of prostate cancer are essentially unknown, although hormonal factors are involved, and diet may exert an indirect influence. The proportion of cases attributable to genetic factors, ionising radiation, occupational exposure to cadmium or, possibly, vasectomy appears low. Geographic variations of up to 30% in incidence occur in England and Wales[1], but worldwide variation is at least 70-fold. Prostate cancer is rare under the age of 50, and more than half the cases arise over the age of 75.

Latent carcinoma of the prostate is common, particularly in old age, and such cancers are often diagnosed incidentally on histological examination of a prostate gland removed for benign prostatic hypertrophy, or discovered at autopsy. Most cancer registries record such tumours, and the recorded incidence of prostate cancer can thus be influenced by the frequency of prostatectomy carried out for benign disease[2]. Since these cancers may have a better prognosis than cancers diagnosed in men with symptoms due to prostate cancer, their influence on overall trends in population-based survival rates must also be borne in mind[3].

Urinary symptoms or pain from bone metastases may lead to the diagnosis. Five-year relative survival varies with stage at diagnosis from 70% or more with malignancy confined to the prostate to about 20% with bony metastases[4]. Surgery, radiotherapy and hormone therapy all have a role in treatment.

The survival analyses reported here include more than 162,000 men who were diagnosed with a malignant prostate tumour in England and Wales during the period 1971-90 and followed up to the end of 1995, representing about 87% of those eligible (Table 33.1). Some 4% of eligible patients were excluded because their vital status was unknown at 6 July 1997, and a further 8% because their survival was zero or unknown. About 70 men were diagnosed over age 100: these men were not included in analyses.

Over half (56%) the tumours were recorded as adenocarcinomas, but 37% were coded as epithelial malignancy without further detail. Less than 1% were transitional cell carcinomas, and sarcomas were extremely rare. There was no anatomic subdivision of the prostate in the eighth or ninth revisions of the International Classification of Diseases.

Survival trends

Relative survival from prostate cancer among men diagnosed in England and Wales during 1986-90 was 77% at one year and 42% at five years, an overall increase of some 10% compared with the early 1970s (Table 33.2). Excess mortality continues well beyond five years after diagnosis, and the relative survival curves are still falling quickly at ten years (Figure 33.1).

No change in survival occurred during the 1980s, either nationally or in any of the regions. The survival curves for men diagnosed in the early and late 1980s are almost exactly superposed (Figure 33.1). The national average improvement in survival between successive five-year periods of diagnosis from 1971-75 up to 1986-90 was 4% for one-year survival and almost 4% for five-year survival, after adjustment for differences in the age distribution of patients over time (see Table 4.4). Survival actually improved by about 5-6% every five years between 1971-75 and 1981-85.

The average rate of increase in age-adjusted survival between the early 1970s and the late 1980s varied widely between the regions, from 2% to 6% every five years for one-year survival and from 1% to 5% for five-year survival (see Table-Figure 4.5M). The broad national pattern of increase was strikingly uniform in all regions, however: a large increase in survival in the 1970s, and little further change between men diagnosed in the early 1980s and those diagnosed during 1986-90. Age-adjusted survival in the Thames regions, which showed the highest average rate of improvement in survival, actually declined slightly for men diagnosed in the late 1980s. Regional survival rates for men diagnosed during the late 1980s varied between 70% and 82% at one year, but less widely at five years (40-45%).

The age-dependence of prostate cancer survival is unusual, and it resembles the pattern for breast cancer, although shifted to a higher age (Figure 33.5, cf. 28.5). Prostate cancer is rare under age 50, but the five-year relative survival rate among 170 men aged 40-49 at diagnosis during 1986-90 was 31% (95% confidence interval 24% to 38%), compared to 46% (45-47%) among more than 12,000 men diagnosed in the age range 60-69 (see Table 4.7). Survival in men aged 80-99 was similar to that for men forty years younger.

Deprivation

Survival has generally been 3-6% lower for men in deprived groups than for men in affluent groups (Table 33.2), and the patterns for men diagnosed in 1986-90 are typical (Figure 33.2), but the pattern has changed over time and it differs between regions.

A significant deprivation gradient in survival was seen for men diagnosed in England and Wales during the 1980s (Figure 33.3). Among some 43,000 men diagnosed in the period 1981-85, the deprivation gap was 7% at one year and 6% at five years, but for more than 51,000 men diagnosed during 1986-90, the corresponding figures were 4% and 3%, a considerable improvement (see Table 4.9).

For men diagnosed during 1986-90, there was a significant deprivation gap of 7% in North and West; there was little or no difference in survival between deprivation groups in South & West, North Thames or West Midlands regions (Table 33.2, Figure 33.3). Regional survival estimates are generally based on 400 to 1,000 or more patients in each deprivation group, and are precise.

International comparison

Five-year relative survival among men diagnosed with prostate cancer in England and Wales during 1986-90 (42%) was 6% lower than in Scotland and 14% less than the average for Europe (see Table 4.12). Even the highest regional five-year survival rate in England and Wales (46%, North Thames) is 10% lower than the European average. The range of five-year survival rates among men diagnosed during 1985-89 in 14 countries contributing to the EUROCARE project was wide, from 39% (Estonia) to 72% (Switzerland).

European differences in survival pale into insignificance, however, compared with the extraordinary 86% five-year survival rate for men diagnosed during 1986-90 in the states covered by the SEER programme in the USA. The populations involved are extremely large, and the 95% confidence intervals around the survival estimates are vanishingly small by comparison with these differences (see Figure 4.12). Regardless of any difference in the efficacy of treatment, there can be little doubt that the prostate cancers included in the survival estimates for these large population series are biologically quite different.

The recorded incidence of prostate cancer in the USA increased by 82% in the five years 1986-91, and by 20% in 1990 alone, a "dramatic epidemic" attributable mainly to widespread testing with prostate-specific antigen[5]. This huge increase in incidence alone suggests that the SEER data must now include a much higher proportion of less advanced or less aggressive tumours than in Europe, and this conclusion is supported by the very rapid increase in survival in the area covered by the SEER programme. Among white men diagnosed with prostate cancer in the USA during 1983-89, five-year relative survival was (only) 79%[6], compared to 86% for men diagnosed during 1986-90[7] (see Table 4.12): the mid-points of these overlapping periods are only two years apart. This sharp increase in five-year relative survival to 86% contrasts with the unchanged survival of 42% for men diagnosed during the 1980s in England and Wales. Reliable comparison of prostate cancer survival between Europe and the USA will require adjustment for stage of disease and modality of diagnosis, including whether disease was symptomatic, or whether initial diagnosis was based on trans-rectal ultrasound, prostate-specific antigen or rectal examination[3].

1.   Swerdlow AJ, dos Santos Silva I. *Atlas of cancer incidence in England and Wales 1968-85.* Oxford: Oxford University Press, 1993

2.   Potosky AL, Kessler LG, Gridley G, Brown CC, Horm JW. Rise in prostatic cancer incidence associated with increased use of transurethral resection. J Natl Cancer Inst 1990; 82: 1624-1628

3. Berrino F, Estève J, Coleman MP. Basic issues in the estimation and comparison of cancer patient survival. In: Berrino F, Sant M, Verdecchia A, Capocaccia R, Hakulinen T, Estève J, (eds.) *Survival of cancer patients in Europe: the EUROCARE study. (IARC Scientific Publications No. 132).* Lyon: International Agency for Research on Cancer, 1995, pp1-14

4.   Thames Cancer Registry. *Cancer in South East England 1992: cancer incidence, prevalence, treatment and survival in residents of the District Health Authorities in South East England.* Sutton: Thames Cancer Registry, 1995

5.   Potosky AL, Miller BA, Albertsen PC, Kramer BS. The role of increasing detection in the rising incidence of prostate cancer. J Am Med Assoc 1995; 273: 548-552

6.   Schottenfeld D. Prostate cancer. In: Schottenfeld D, Fraumeni JF, (eds.) *Cancer epidemiology and prevention.* 2nd edn. Oxford: Oxford University Press, 1996, pp1180-1206

7.   National Cancer Institute. SEER Stat – cancer incidence public use database 1973-95. Release 1.1. Bethesda, MD: National Cancer Institute, 1998

Table 33.1 Ineligible and excluded records, and patients included in analyses, by calendar period of diagnosis

	\multicolumn{15}{c}{Calendar period of diagnosis}														
	1971-75			1976-80			1981-85			1986-90			Overall		
Total registered	34,409			41,087			50,898			62,257			188,651		
Ineligible	*Records*	*Patients*	*%*	*Records*	*Patients*	*%*	*Records*	*Patients*	*%*	*Records*	*Patients*	*%*	*Records*	*Patients*	*%*
Incomplete data[1]	100	100		171	171		70	70		96	96		437	**437**	
Resident outside England and Wales	303	300		270	258		234	208		271	210		1,078	**976**	
In situ neoplasm[2]	58	58		6	6		0	0		1	1		65	**65**	
Benign or uncertain behaviour[2]	100	16		148	14		0	0		0	0		248	**30**	
Metastatic[2]	243	240		113	113		0	0		4	4		360	**357**	
Otherwise ineligible[3]	-	-		-	-		-	-		-	-		-	**-**	
Lymphoma[4]	0	0		1	1		0	0		1	0		2	**1**	
Leukaemia or myeloma[4]	0	0		0	0		1	0		0	0		1	**0**	
		714			563			278			311			1,866	
Total eligible		33,695	100		40,524	100		50,620	100		61,946	100		186,785	100
Aged under 15 or 100 years or over at diagnosis	22	21	<0.1	17	16	<0.1	12	11	<0.1	26	23	<0.1	77	**71**	<0.1
Vital status unknown[5]	1,120	1,018	3.0	1,992	1,855	4.6	1,993	1,887	3.7	3,673	3,546	5.7	8,778	**8,306**	4.4
Duplicate registration[6]	1	1	<0.1	0	0	0.0	0	0	0.0	0	0	0.0	1	**1**	<0.1
Multiple primary[7]	0	0	0.0	39	38	<0.1	225	204	0.4	437	397	0.6	701	**639**	0.3
Synchronous tumours[8]	97	86	0.3	131	99	0.2	208	163	0.3	350	241	0.4	786	**589**	0.3
Sex not known	0	0	0.0	0	0	0.0	0	0	0.0	0	0	0.0	0	**0**	0.0
Sex incompatible with site code[9]	0	0	0.0	0	0	0.0	3	3	<0.1	14	14	<0.1	17	**17**	<0.1
Invalid dates[10]	2	2	<0.1	2	2	<0.1	9	6	<0.1	51	23	<0.1	64	**33**	<0.1
Zero survival[11]	2,251	2,139	6.3	2,866	2,729	6.7	4,688	4,359	8.6	6,243	5,792	9.4	16,048	**15,019**	8.0
Total exclusions		3,267	9.7		4,739	11.7		6,633	13.1		10,036	16.2		24,675	13.2
Patients included in analysis		30,428	90.3		35,785	88.3		43,987	86.9		51,910	83.8		162,110	86.8

[1] Main data item(s) invalid or incompatible with one another: name, sex, date of birth, date of diagnosis and (if present) date of death, postcode, site, morphology and behaviour

[2] Tumour either *in situ* (behaviour code 2), benign (0), uncertain if benign or malignant (1) or metastatic (6 or 9)

[3] Not applicable for this malignancy

[4] Morphology that of lymphoma (included with non-Hodgkin lymphoma, Ch. 41) or myeloma (included with multiple myeloma, Ch. 43), or leukaemia (excluded)

[5] Tracing of vital status at National Health Service Central Register incomplete at 6 July 1997

[6] Same site code, sex, cancer registry and cancer registry number as an earlier registration

[7] Same site code and person number as an earlier registration(s): mostly confirmed multiple primary tumours at this site, some unresolved duplicate registrations

[8] Same site code, sex, date of birth and date of diagnosis as another registration(s): mostly synchronous or (in paired organs) bilateral tumours in same anatomic site in one individual, not previously linked: also some duplicate registrations

[9] Site code specific to one sex; tumour registered to opposite sex

[10] Impossible sequence of dates (birth, diagnosis and, if present, emigration or death); or date of death after 6 July 1997

[11] Date of diagnosis same as date of death: some are patients who did die on the day of diagnosis, but most were registered solely from a death certificate (death certificate only, or DCO), and their survival time is thus unknown

Table 33.2   Crude and relative survival (%) at one, five and ten years since diagnosis: NHS Region, deprivation category and calendar period of diagnosis

MEN		1971-75						1976-80					
		Affluent	*2*	*3*	*4*	*Deprived*	*All[1]*	*Affluent*	*2*	*3*	*4*	*Deprived*	*All[1]*
		C R	C R	C R	C R	C R	C R	C R	C R	C R	C R	C R	C R
ENGLAND & WALES		*(4,995)*	*(5,447)*	*(5,634)*	*(4,866)*	*(3,702)*	*(30,428)*	*(6,977)*	*(7,744)*	*(7,624)*	*(6,676)*	*(5,097)*	*(35,785)*
	One year	64 70	63 69	61 67	60 66	60 66	62 67	69 74	66 71	65 71	64 70	62 68	65 71
	Five years	24 36	22 33	22 33	20 32	20 31	22 33	28 41	25 38	26 40	23 37	23 37	25 39
	Ten years	10 23	9 21	8 20	8 21	8 21	8 21	11 25	10 23	9 23	9 24	9 24	10 24
ENGLAND		*(4,798)*	*(5,179)*	*(5,257)*	*(4,446)*	*(3,523)*	*(28,589)*	*(6,768)*	*(7,434)*	*(7,147)*	*(6,166)*	*(4,860)*	*(33,734)*
	One year	64 70	63 69	62 67	60 66	60 66	62 68	69 74	66 72	66 72	64 70	62 68	65 71
	Five years	24 36	22 34	21 33	20 32	20 31	22 33	28 41	26 38	26 40	23 37	23 37	25 39
	Ten years	10 23	9 21	8 20	8 21	8 21	8 21	11 25	10 23	9 23	9 24	9 24	10 24
Northern & Yorkshire		*(351)*	*(451)*	*(519)*	*(590)*	*(691)*	*(3,571)*	*(574)*	*(724)*	*(833)*	*(951)*	*(1,038)*	*(4,232)*
	One year	66 72	61 66	63 69	60 66	58 64	62 68	67 74	69 75	66 73	63 69	62 68	65 72
	Five years	23 36	21 31	19 30	19 30	17 27	20 31	28 41	27 41	25 40	22 35	24 39	25 40
	Ten years	9 25	6 15	5 11	7 17	7 19	7 18	8 20	11 27	9 23	8 22	9 27	9 24
Trent		*(316)*	*(445)*	*(565)*	*(648)*	*(492)*	*(2,728)*	*(478)*	*(668)*	*(742)*	*(829)*	*(607)*	*(3,428)*
	One year	59 64	55 60	57 62	56 61	55 61	57 63	62 67	60 65	58 63	61 67	60 66	60 66
	Five years	20 30	20 30	23 35	18 28	17 28	20 32	27 40	23 34	23 35	24 37	24 37	24 37
	Ten years	7 15	8 18	8 21	7 17	6 18	7 19	11 27	8 20	8 22	10 27	12 32	10 25
Anglia & Oxford		*(696)*	*(733)*	*(693)*	*(439)*	*(204)*	*(3,214)*	*(999)*	*(998)*	*(882)*	*(568)*	*(241)*	*(3,869)*
	One year	67 73	69 75	65 71	62 68	70 76	68 74	70 76	67 73	68 74	65 71	66 72	68 75
	Five years	29 43	26 38	22 34	23 35	24 37	25 38	25 37	24 36	29 44	26 40	28 43	26 40
	Ten years	10 23	11 25	8 19	8 19	11 28	9 23	11 25	9 21	11 26	9 23	11 29	10 25
North Thames		*(713)*	*(666)*	*(591)*	*(611)*	*(455)*	*(3,404)*	*(699)*	*(688)*	*(657)*	*(632)*	*(505)*	*(3,312)*
	One year	66 73	65 70	64 70	64 70	63 69	65 71	73 79	68 74	71 78	72 79	69 75	71 78
	Five years	28 40	25 36	23 35	23 35	24 38	25 38	34 48	24 36	30 45	29 44	26 41	29 44
	Ten years	12 25	9 21	11 26	11 27	10 28	11 26	14 31	11 25	11 29	12 31	12 31	12 29
South Thames		*(1,173)*	*(911)*	*(929)*	*(739)*	*(334)*	*(4,514)*	*(1,530)*	*(1,262)*	*(1,003)*	*(745)*	*(329)*	*(5,177)*
	One year	64 70	60 66	60 66	59 65	62 68	62 68	69 75	63 69	63 70	63 70	61 68	65 72
	Five years	21 31	18 26	19 29	19 30	22 35	20 31	28 41	27 40	25 38	23 37	24 37	26 40
	Ten years	10 23	6 15	7 19	8 23	8 22	9 22	12 27	9 23	8 21	8 21	9 25	10 24
South & West		*(963)*	*(1,319)*	*(1,187)*	*(521)*	*(159)*	*(4,587)*	*(1,317)*	*(1,811)*	*(1,609)*	*(716)*	*(221)*	*(5,920)*
	One year	62 67	65 71	62 68	60 66	71 78	63 70	68 73	67 73	65 72	63 69	57 64	66 72
	Five years	24 35	24 37	24 37	22 34	17 26	24 37	27 40	27 40	26 40	24 38	22 36	26 40
	Ten years	9 21	10 25	9 23	10 24	7 18	9 24	12 27	10 25	9 25	10 27	7 19	10 25
West Midlands		*(162)*	*(190)*	*(242)*	*(229)*	*(304)*	*(2,811)*	*(473)*	*(539)*	*(656)*	*(692)*	*(747)*	*(3,169)*
	One year	68 73	67 74	61 67	65 71	60 66	62 68	72 79	67 72	67 73	67 73	64 70	67 73
	Five years	24 35	23 34	24 37	26 41	19 29	22 34	31 45	25 37	24 38	22 34	21 34	24 38
	Ten years	11 25	10 24	8 20	10 28	7 19	9 22	12 28	10 25	9 24	10 28	7 19	9 24
North & West		*(424)*	*(464)*	*(531)*	*(669)*	*(884)*	*(3,760)*	*(698)*	*(744)*	*(765)*	*(1,033)*	*(1,172)*	*(4,627)*
	One year	58 64	57 62	54 60	55 60	55 61	56 62	63 70	62 68	63 69	59 65	57 64	61 67
	Five years	23 35	18 28	15 24	16 26	19 30	18 29	23 35	22 34	22 35	18 30	20 33	21 34
	Ten years	9 21	7 19	4 11	4 12	7 19	6 17	8 20	10 25	9 22	6 19	8 23	8 22
WALES		*(197)*	*(268)*	*(377)*	*(420)*	*(179)*	*(1,839)*	*(209)*	*(310)*	*(477)*	*(510)*	*(237)*	*(2,051)*
	One year	52 57	56 61	55 60	58 64	55 60	56 62	67 75	63 69	60 65	60 66	54 59	61 67
	Five years	16 25	16 24	22 33	21 33	21 32	20 31	29 44	20 32	23 37	25 39	18 29	25 39
	Ten years	6 15	8 20	10 24	8 20	7 18	8 21	13 29	11 25	9 23	10 30	8 22	11 27

[1] Including those for whom no deprivation category was assigned. C - crude survival, R - relative survival (see Chapter 3).
Number of patients contributing to each analysis in parentheses.

| | 1981-85 | | | | | | 1986-90 | | | | | | MEN |
	Affluent C R	*2* C R	*3* C R	*4* C R	*Deprived* C R	*All*[1] C R	*Affluent* C R	*2* C R	*3* C R	*4* C R	*Deprived* C R	*All*[1] C R	
	(9,185)	*(9,792)*	*(9,622)*	*(8,155)*	*(6,110)*	*(43,987)*	*(11,157)*	*(12,171)*	*(11,371)*	*(9,743)*	*(6,912)*	*(51,910)*	**ENGLAND & WALES**
	74 81	72 79	70 77	68 75	67 74	71 77	73 79	72 78	72 78	70 76	69 75	71 77	*One year*
	31 45	28 42	27 41	26 40	25 40	28 42	30 43	29 42	28 42	27 42	26 40	28 42	*Five years*
	14 30	12 27	11 27	10 26	10 27	11 28							
	(8,896)	*(9,388)*	*(9,000)*	*(7,545)*	*(5,791)*	*(41,485)*	*(10,659)*	*(11,680)*	*(10,722)*	*(9,075)*	*(6,586)*	*(49,166)*	**ENGLAND**
	75 81	72 79	70 77	69 75	67 74	**71 78**	74 79	72 78	72 78	71 77	69 75	**72 78**	*One year*
	31 45	29 43	27 41	26 40	26 40	**28 42**	30 43	29 42	28 42	27 42	26 40	**28 42**	*Five years*
	14 30	12 27	11 26	10 27	10 27	**11 28**							
	(821)	*(989)*	*(1,150)*	*(1,210)*	*(1,254)*	*(5,506)*	*(1,085)*	*(1,226)*	*(1,350)*	*(1,363)*	*(1,377)*	*(6,438)*	**Northern & Yorkshire**
	73 80	72 79	72 79	68 74	67 74	**70 77**	72 79	72 80	73 81	70 77	69 77	**71 78**	*One year*
	29 44	28 44	26 42	26 43	25 41	**27 42**	30 47	28 45	28 48	28 46	25 42	**28 43**	*Five years*
	13 32	12 33	10 29	10 29	9 27	**11 28**							
	(704)	*(900)*	*(1,027)*	*(1,137)*	*(779)*	*(4,576)*	*(825)*	*(1,105)*	*(1,200)*	*(1,178)*	*(945)*	*(5,262)*	**Trent**
	74 80	70 76	67 73	68 75	69 76	**70 76**	72 79	69 76	72 80	68 75	66 73	**70 76**	*One year*
	32 48	28 43	25 40	27 43	25 42	**28 42**	30 48	28 43	27 43	29 46	23 39	**27 42**	*Five years*
	14 37	10 25	10 28	11 31	11 32	**11 28**							
	(1,222)	*(1,223)*	*(1,098)*	*(637)*	*(270)*	*(4,565)*	*(1,627)*	*(1,808)*	*(1,355)*	*(883)*	*(301)*	*(6,016)*	**Anglia & Oxford**
	76 82	74 80	71 78	72 79	67 73	**73 80**	72 79	73 80	73 80	69 76	64 72	**72 78**	*One year*
	30 45	30 47	29 44	27 42	28 44	**29 44**	30 46	30 47	28 46	25 41	23 40	**29 43**	*Five years*
	14 33	13 32	12 33	11 30	12 36	**13 30**							
	(863)	*(804)*	*(752)*	*(704)*	*(565)*	*(3,764)*	*(1,108)*	*(1,169)*	*(1,194)*	*(1,082)*	*(841)*	*(5,533)*	**North Thames**
	79 85	73 79	74 81	72 79	74 81	**75 81**	77 84	76 84	77 85	77 84	78 86	**77 83**	*One year*
	36 53	33 50	31 47	30 47	31 48	**33 48**	32 48	31 48	30 47	32 50	31 51	**31 46**	*Five years*
	17 40	14 33	14 35	14 39	14 39	**15 35**							
	(1,892)	*(1,517)*	*(1,208)*	*(882)*	*(331)*	*(6,052)*	*(1,978)*	*(1,758)*	*(1,357)*	*(953)*	*(463)*	*(6,585)*	**South Thames**
	76 83	76 83	73 80	73 81	66 73	**74 81**	77 84	77 84	75 83	77 86	73 81	**77 83**	*One year*
	33 49	30 45	29 45	25 40	28 44	**30 44**	31 47	30 48	27 44	28 47	24 40	**29 43**	*Five years*
	14 35	13 32	12 32	11 30	13 36	**13 30**							
	(1,687)	*(2,202)*	*(1,781)*	*(794)*	*(260)*	*(6,959)*	*(1,825)*	*(2,512)*	*(2,025)*	*(1,247)*	*(394)*	*(8,076)*	**South & West**
	72 79	71 78	69 76	67 73	63 70	**70 77**	70 76	69 75	68 75	69 76	68 76	**69 75**	*One year*
	30 45	27 42	25 39	24 39	23 36	**27 41**	28 43	27 43	26 42	27 44	27 44	**27 40**	*Five years*
	11 28	10 26	10 28	9 25	10 27	**10 25**							
	(787)	*(758)*	*(923)*	*(928)*	*(921)*	*(4,364)*	*(885)*	*(912)*	*(1,103)*	*(1,068)*	*(819)*	*(4,815)*	**West Midlands**
	74 81	72 79	70 77	68 75	69 76	**71 78**	73 79	71 79	72 80	71 78	67 74	**71 77**	*One year*
	30 46	26 41	28 44	25 39	26 43	**27 42**	29 44	27 43	28 46	27 44	27 44	**28 42**	*Five years*
	13 33	10 28	11 30	9 25	10 30	**11 27**							
	(920)	*(995)*	*(1,061)*	*(1,253)*	*(1,411)*	*(5,699)*	*(1,326)*	*(1,190)*	*(1,138)*	*(1,301)*	*(1,446)*	*(6,441)*	**North & West**
	71 78	69 75	66 73	62 69	63 70	**66 73**	73 81	69 76	68 75	65 73	65 72	**68 75**	*One year*
	29 44	27 43	24 38	24 41	23 37	**25 39**	31 49	27 43	27 45	22 39	25 42	**26 41**	*Five years*
	12 31	12 31	8 24	10 31	9 28	**10 27**							
	(289)	*(404)*	*(622)*	*(610)*	*(319)*	*(2,502)*	*(498)*	*(491)*	*(649)*	*(668)*	*(326)*	*(2,744)*	**WALES**
	63 70	64 70	64 71	63 70	63 70	**64 71**	64 70	69 76	67 73	61 68	63 68	**65 71**	*One year*
	29 43	23 36	28 43	23 38	23 38	**26 40**	28 44	29 47	29 46	26 44	25 41	**28 43**	*Five years*
	14 34	10 27	13 38	9 27	10 30	**12 29**							

[1] Including those for whom no deprivation category was assigned. C - crude survival, R - relative survival (see Chapter 3). Number of patients contributing to each analysis in parentheses.

Fig. 33.1   Relative survival up to ten years: trends by calendar period of diagnosis (1971-90) and NHS Region

MEN

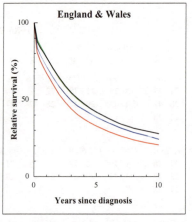

England & Wales

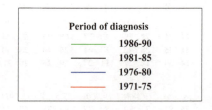

Period of diagnosis
— 1986-90
— 1981-85
— 1976-80
— 1971-75

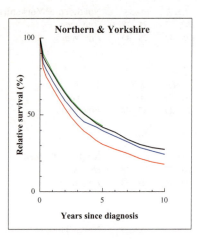

Northern & Yorkshire

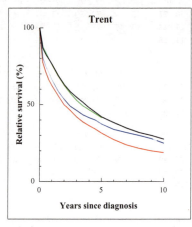

Trent

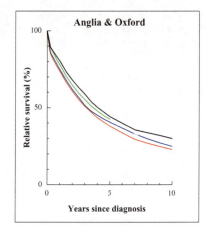

Anglia & Oxford

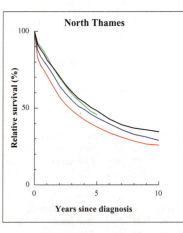

North Thames

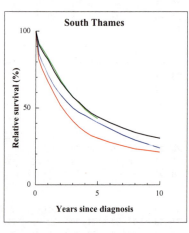

South Thames

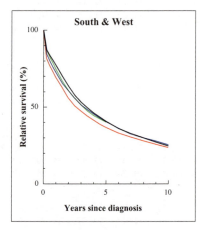

South & West

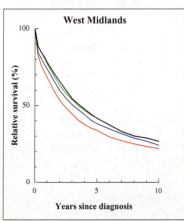

West Midlands

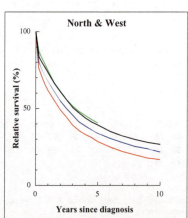

North & West

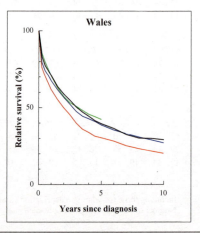

Wales

Fig. 33.2 Relative survival up to five years, by deprivation category and NHS Region: patients diagnosed 1986-90

MEN

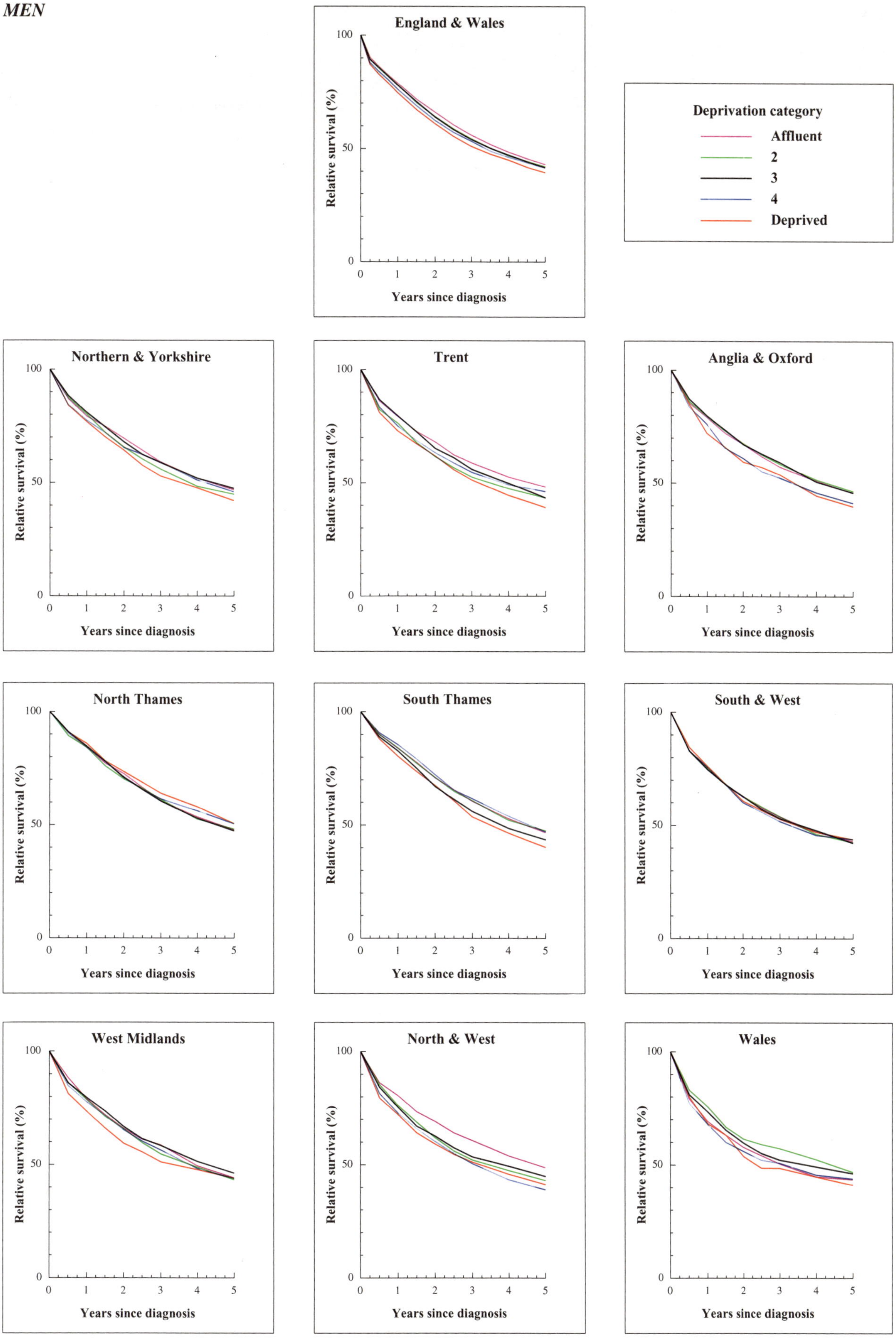
England & Wales
Northern & Yorkshire
Trent
Anglia & Oxford
North Thames
South Thames
South & West
West Midlands
North & West
Wales
Relative survival (%)
Years since diagnosis
0
50
100
1
2
3
4
5
Deprivation category
Affluent
2
3
4
Deprived

Fig. 33.3   Relative survival at five years by deprivation category, period of diagnosis (1981-90) and NHS Region

MEN

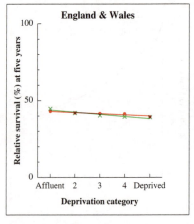

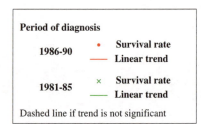

Period of diagnosis

1986-90 ● Survival rate
— Linear trend

1981-85 ✕ Survival rate
— Linear trend

Dashed line if trend is not significant

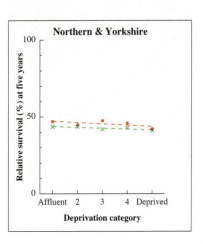

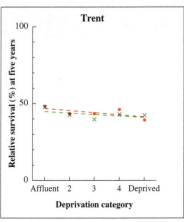

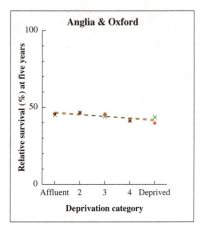

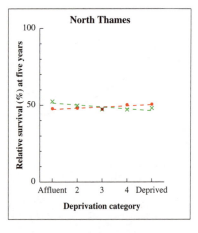

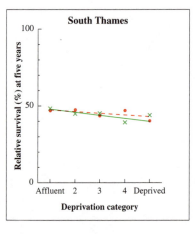

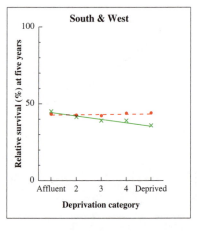

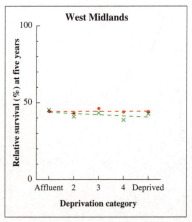

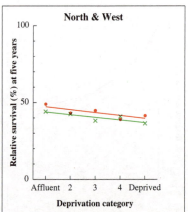

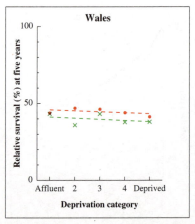

Fig. 33.4 Relative survival (%) by NHS Region:

  one-year and five-year survival for men diagnosed 1986-90, with 95% confidence intervals

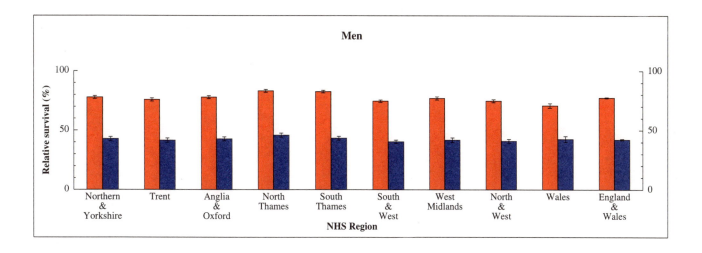

Fig. 33.5 Relative survival (%) at five years, with 95% confidence interval, by age at diagnosis:

  England and Wales, men diagnosed 1986-90

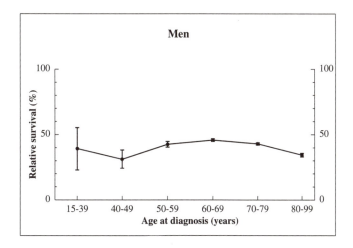

Chapter 34

TESTIS

Testicular cancer accounts for about 1% of all cancers in men, and about 1,300 new cases are diagnosed each year in England and Wales. It is the commonest cancer in men under 35, accounting for a third of all cancers below that age; more than 80% of testicular cancers occur under the age of 45. Incidence has been increasing rapidly in England and Wales and in many developed countries, mainly in younger men, and often by more than 10% every five years. Mortality fell by 70% between 1971 and 1995, mainly between about 1975 and the early 1980s.

The peak incidence for teratoma occurs among men aged 20-29, and the peak for seminoma about 10 years later. The risk is high in an undescended testis, and maldescent has become more common, but this can account for little of the rapid increase in incidence[1]. In men who have had testicular cancer, risk for the contralateral testis is high. Genetic traits account for a small proportion of cases. Risk is higher in single men and in higher socio-economic groups.

Cis-platinum drugs transformed the treatment of testicular tumours in the 1970s. Surgical removal of the testis (orchidectomy) and chemotherapy are the mainstay of treatment, and testicular tumours are radiosensitive. Chemotherapy may be effective in disseminated seminoma, but less so with teratoma. Survival is sharply dependent on stage at diagnosis: over 95% for localised disease, 85% for stage II and only 70% for stage IV[2].

The trends in mortality reflect a major therapeutic advance, and this can now be seen in national survival trends. The improvements in survival and mortality tend to overshadow the rapid and unexplained increase in the number of new cases of testicular cancer.

The survival analyses reported here include 16,000 men who were diagnosed with testicular cancer in England and Wales during 1971-90 and followed up to the end of 1995, about 91% of those eligible. More than 600 tumours diagnosed in the 1970s were judged ineligible because their behaviour was recorded as benign or of uncertain malignancy (Table 34.1). About 6% of men were excluded because their vital status was unknown at 6 July 1997, and a further 1% because their survival was zero or unknown. Less than 1% of testicular tumours were recorded as having arisen in an undescended testis (Table 34A). Overall, 53% of the tumours were seminomas and 33% teratomas. The proportions have fluctuated, but seminomas have become less common (60% in the early 1970s to 50% in the late 1980s), while the proportion of teratomas rose from 30% in the early 1970s to 40% around

1980, then fell to 30% by the late 1980s. Choriocarcinoma and trophoblastic teratoma accounted for 1%, but Leydig cell (0.2%) and Sertoli cell tumours and mesotheliomas were extremely rare.

Survival trends

Relative survival from testicular cancer for men diagnosed in England and Wales during 1986-90 was 96% at one year and 91% at five years, the highest survival from any tumour in men (Table 34.2). These rates are 15-20% higher than for men diagnosed in the early 1970s. Rapid improvement occurred between the early 1970s and the late 1970s, and again up to the early 1980s, but with national average five-year survival already around 90% for men diagnosed during 1981-85, subsequent improvements have been less marked (Figure 34.1)[a].

Three to five years after diagnosis, excess mortality is very low in all regions, and the relative survival curves are almost flat.

Age-standardised trends show a 5-7% improvement in one-year survival during the 1970s and a similar further gain by the early 1980s (see Table 4.4). Five-year survival increased by no less than 20% during the same ten years, between 1971-75 and 1981-85. Further gains in survival for men diagnosed during the late 1980s were small, of the order of 1-2%.

With the striking overall improvement in survival, the difference between one-year and five-year survival has shrunk markedly (see Table-Figure 4.5N). The pattern of improvement since the 1970s has been very similar in all regions, with rapid early gains in the 1970s and early 1980s, and small further improvements by the late 1980s.

Overall survival is high, and five-year survival in some regions is now higher than one-year survival in others (Table 34.2), although these differences are not statistically significant after adjustment for age (see Table 4.5N). Regional one-year survival rates for men diagnosed during 1986-90 ranged between 91% and 98%, and the regional range in five-year survival was 87% to 95% (Figure 34.4).

Relative survival falls sharply with age (Figure 34.5). Five-year survival for men diagnosed under the age of 60 during 1986-90 was about 90%, falling to around 70% for men aged 60-79 at diagnosis, and although testicular cancer is rare over the age of 80, relative survival is less than 40% (see Table 4.7).

Table 34A Sub-site distribution of testicular cancer

ICD-9 code	Site description	1971-75 No.	%	1976-80 No.	%	1981-85 No.	%	1986-90 No.	%	Total No.	%
186.0	Undescended	-	-	14	0.4	43	1.0	39	0.7	**96**	**0.6**
186.9	Other and unspecified	2,901	100.0	3,420	99.6	4,378	99.0	5,542	99.3	**16,241**	**99.4**
	Total	**2,901**		**3,434**		**4,421**		**5,581**		**16,337**	

Deprivation

The deprivation gap in survival was 13% or more for men diagnosed in England and Wales during the early 1970s (Table 34.2). Differences became less marked with the rapid gains in overall survival during the next ten years. Even so, survival curves for the five deprivation groups in most regions are still clearly separate for men diagnosed during 1986-90 (Figure 34.2). There was a significant deprivation gradient in five-year survival for men diagnosed during the 1980s in England and Wales. The gradient is significant in either the early or the late 1980s in six of the nine regions (Figure 34.3). Since five-year relative survival for all deprivation groups combined now approaches 100% in the regions with highest survival, it is not surprising that the deprivation gradient is less steep in these regions.

Among 4,300 men of known deprivation status diagnosed during 1981-85, the deprivation gap was 4% for one-year survival and 7% for survival at five years (see Table 4.9). For almost 5,500 men diagnosed during 1986-90, survival among men in the affluent group had risen by a further 1.5% or so, to about 94% at five years, but the deprivation gap in survival between affluent and deprived groups had barely altered.

International comparison

Five-year survival for men diagnosed with testicular cancer in England and Wales during 1986-90 was virtually the same as in Scotland and to the average survival rate in Europe (90-92%; see Table 4.12). Survival trends since the 1970s and the pattern by age have also been similar to those in Scotland.

The highest regional five-year survival rate, and survival for men in the affluent group in England and Wales as a whole (93-95%), were similar to the survival rate in the USA.

[a] The survival curve for men diagnosed in Wales during 1981-85 stops at five years (non-convergence, see Chapter 3), but there were six deaths between five and ten years after diagnosis, and the ten-year survival rate was clearly about 81% (Table 34.2).

1.   Morris Brown L, Pottern LM, Hoover RN. Testicular cancer in young men: the search for causes of the epidemic increase in the United States. J Epidemiol Comm Hlth 1987; 41: 349-354

2.   Thames Cancer Registry. *Cancer in South East England 1992: cancer incidence, prevalence, treatment and survival in residents of the District Health Authorities in South East England.* Sutton: Thames Cancer Registry, 1995

Table 34.1 Ineligible and excluded records, and patients included in analyses, by calendar period of diagnosis

	Calendar period of diagnosis														
	1971-75			1976-80			1981-85			1986-90			Overall		
Total registered	**3,586**			**4,091**			**4,887**			**6,247**			**18,811**		
Ineligible	*Records*	*Patients*	*%*	*Records*	*Patients*	*%*	*Records*	*Patients*	*%*	*Records*	*Patients*	*%*	*Records*	*Patients*	*%*
Incomplete data[1]	10	10		13	13		28	28		19	19		70	**70**	
Resident outside England and Wales	52	48		65	63		53	46		25	13		195	**170**	
In situ neoplasm[2]	0	0		0	0		0	0		0	0		0	**0**	
Benign or uncertain behaviour[2]	393	379		249	232		0	0		1	1		643	**612**	
Metastatic[2]	11	11		5	4		0	0		0	0		16	**15**	
Otherwise ineligible[3]	-	-		-	-		-	-		-	-		-	**-**	
Lymphoma[4]	18	18		21	21		3	0		4	0		46	**39**	
Leukaemia or myeloma[4]	0	0		2	1		0	0		0	0		2	**1**	
		466			334			74			33			907	
Total eligible		**3,120**	**100**		**3,757**	**100**		**4,813**	**100**		**6,214**	**100**		**17,904**	**100**
Aged under 15 or 100 years or over at diagnosis	64	54	1.7	73	66	1.8	73	69	1.4	73	71	1.1	283	**260**	1.5
Vital status unknown[5]	169	120	3.8	248	185	4.9	280	246	5.1	480	454	7.3	1,177	**1,005**	5.6
Duplicate registration[6]	0	0	0.0	0	0	0.0	0	0	0.0	0	0	0.0	0	**0**	0.0
Multiple primary[7]	0	0	0.0	3	3	<0.1	8	7	0.1	25	25	0.4	36	**35**	0.2
Synchronous tumours[8]	20	9	0.3	19	7	0.2	25	7	0.1	54	15	0.2	118	**38**	0.2
Sex not known	0	0	0.0	0	0	0.0	0	0	0.0	0	0	0.0	0	**0**	0.0
Sex incompatible with site code[9]	0	0	0.0	1	1	<0.1	0	0	0.0	1	1	<0.1	2	**2**	<0.1
Invalid dates[10]	1	1	<0.1	0	0	0.0	1	0	0.0	5	3	<0.1	7	**4**	<0.1
Zero survival[11]	45	35	1.1	80	61	1.6	73	63	1.3	73	64	1.0	271	**223**	1.2
Total exclusions		219	7.0		323	8.6		392	8.1		633	10.2		**1,567**	8.8
Patients included in analysis		2,901	93.0		3,434	91.4		4,421	91.9		5,581	89.8		**16,337**	91.2

[1] Main data item(s) invalid or incompatible with one another: name, sex, date of birth, date of diagnosis and (if present) date of death, postcode, site, morphology and behaviour

[2] Tumour either *in situ* (behaviour code 2), benign (0), uncertain if benign or malignant (1) or metastatic (6 or 9)

[3] Not applicable for this malignancy

[4] Morphology that of lymphoma (included with non-Hodgkin lymphoma, Ch. 41) or myeloma (included with multiple myeloma, Ch. 43), or leukaemia (excluded)

[5] Tracing of vital status at National Health Service Central Register incomplete at 6 July 1997

[6] Same site code, sex, cancer registry and cancer registry number as an earlier registration

[7] Same site code and person number as an earlier registration(s): mostly confirmed multiple primary tumours at this site, some unresolved duplicate registrations

[8] Same site code, sex, date of birth and date of diagnosis as another registration(s): mostly synchronous or (in paired organs) bilateral tumours in same anatomic site in one individual, not previously linked: also some duplicate registrations

[9] Site code specific to one sex; tumour registered to opposite sex

[10] Impossible sequence of dates (birth, diagnosis and, if present, emigration or death); or date of death after 6 July 1997

[11] Date of diagnosis same as date of death: some are patients who did die on the day of diagnosis, but most were registered solely from a death certificate (death certificate only, or DCO), and their survival time is thus unknown

Table 34.2   Crude and relative survival (%) at one, five and ten years since diagnosis: NHS Region, deprivation category and calendar period of diagnosis

MEN	1971-75 Affluent C R	2 C R	3 C R	4 C R	Deprived C R	All[1] C R	1976-80 Affluent C R	2 C R	3 C R	4 C R	Deprived C R	All[1] C R
ENGLAND & WALES	(622)	(525)	(488)	(430)	(290)	(2,901)	(794)	(709)	(695)	(613)	(479)	(3,434)
One year	85 86	82 83	81 82	81 82	72 73	81 82	89 90	86 87	88 89	84 85	84 86	86 87
Five years	74 76	65 69	67 70	63 66	57 61	66 69	80 82	75 77	78 81	73 76	72 75	76 78
Ten years	70 75	62 67	62 69	59 65	52 59	62 67	77 82	71 76	74 80	70 75	67 74	72 78
ENGLAND	(603)	(498)	(459)	(393)	(269)	(2,745)	(767)	(688)	(661)	(564)	(451)	(3,249)
One year	86 86	82 83	81 82	80 81	72 73	81 82	89 90	86 87	88 89	84 86	84 85	86 87
Five years	74 76	65 69	67 70	63 66	57 60	66 69	80 82	75 76	78 81	73 76	71 75	76 78
Ten years	70 75	62 68	63 69	59 65	51 59	62 68	- -	71 75	74 80	69 76	67 74	72 78
Northern & Yorkshire	(39)	(43)	(52)	(63)	(47)	(335)	(67)	(76)	(87)	(84)	(105)	(425)
One year	84 85	74 76	79 80	84 85	59 62	77 79	88 90	87 87	93 94	84 87	91 92	89 90
Five years	66 70	56 59	63 68	68 71	47 50	62 65	84 85	74 75	77 79	70 74	70 74	75 77
Ten years	58 65	56 59	59 68	66 70	42 46	58 63	- -	74 75	76 78	- -	65 74	72 77
Trent	(60)	(46)	(54)	(56)	(45)	(281)	(61)	(60)	(67)	(67)	(53)	(320)
One year	85 86	74 75	87 88	77 78	75 76	80 82	79 81	78 79	88 89	77 78	85 86	82 83
Five years	78 80	61 64	70 75	64 67	58 59	67 70	71 73	67 67	76 79	64 65	77 81	71 73
Ten years	75 79	61 64	- -	63 67	56 58	65 70	71 73	65 66	75 77	62 65	70 77	68 72
Anglia & Oxford	(73)	(57)	(40)	(33)	(18)	(245)	(132)	(87)	(75)	(50)	(15)	(372)
One year	88 88	91 93	82 84	85 87	94 95	87 88	91 92	91 91	85 86	82 83	87 87	88 89
Five years	79 80	74 79	72 74	61 65	66 67	72 75	83 84	79 81	68 73	70 75	87 87	77 80
Ten years	77 80	68 76	67 72	48 57	55 67	67 73	- -	75 78	63 71	64 71	87 87	72 78
North Thames	(71)	(53)	(52)	(48)	(32)	(291)	(82)	(62)	(60)	(71)	(47)	(331)
One year	79 79	91 91	85 85	79 81	68 69	81 82	90 90	85 87	85 86	93 93	85 87	88 89
Five years	70 71	74 74	71 74	65 67	59 60	68 70	85 88	72 75	78 81	85 86	74 78	79 82
Ten years	66 69	68 70	63 71	54 64	59 60	63 69	84 88	69 74	73 79	85 86	74 78	78 82
South Thames	(162)	(101)	(80)	(61)	(25)	(461)	(166)	(121)	(91)	(80)	(30)	(520)
One year	88 89	81 83	73 73	77 77	72 72	80 81	92 92	89 91	89 90	86 88	80 82	88 89
Five years	70 73	69 74	63 63	61 62	56 58	66 68	83 85	78 79	86 87	77 81	73 78	79 82
Ten years	67 72	- -	60 61	58 59	48 58	63 67	80 84	- -	78 85	74 81	- -	76 81
South & West	(105)	(105)	(80)	(38)	(8)	(362)	(107)	(128)	(114)	(56)	(20)	(440)
One year	88 88	81 82	85 86	74 74	- -	83 84	88 89	84 85	86 87	89 90	75 76	86 86
Five years	81 81	57 61	66 71	58 59	- -	67 70	73 76	74 76	75 80	73 76	70 71	74 77
Ten years	75 79	53 60	60 67	55 59	- -	62 68	70 74	69 73	70 78	70 74	70 71	70 75
West Midlands	(24)	(30)	(36)	(26)	(31)	(361)	(63)	(83)	(74)	(65)	(64)	(353)
One year	88 89	87 87	86 87	88 90	65 65	80 81	84 85	87 87	84 85	88 88	84 85	85 86
Five years	79 81	73 75	75 77	58 59	48 52	65 68	71 73	77 78	78 80	80 82	73 74	76 78
Ten years	75 76	67 69	75 77	58 59	32 40	58 63	- -	69 72	77 80	72 78	69 74	71 76
North & West	(69)	(63)	(65)	(68)	(63)	(409)	(89)	(71)	(93)	(91)	(117)	(488)
One year	82 83	78 80	77 77	82 83	78 80	80 81	92 93	83 84	95 96	77 79	78 79	85 86
Five years	72 74	63 69	65 65	64 68	62 67	65 69	84 86	75 75	83 86	66 68	66 69	74 77
Ten years	- -	62 68	60 62	- -	- -	63 69	80 84	72 75	82 86	61 65	61 67	71 77
WALES	(19)	(27)	(29)	(37)	(21)	(156)	(27)	(21)	(34)	(49)	(28)	(185)
One year	79 79	85 87	83 83	86 87	71 71	83 84	96 96	95 95	88 88	80 80	93 93	88 88
Five years	69 70	62 63	65 70	62 68	66 67	67 70	81 82	76 84	79 83	73 74	75 76	77 80
Ten years	69 70	55 56	55 65	- -	61 62	62 67	74 78	71 83	74 77	73 74	68 70	73 79

[1]   Including those for whom no deprivation category was assigned. C - crude survival, R - relative survival (see Chapter 3).
Number of patients contributing to each analysis in parentheses.

	1981-85												1986-90												MEN
	Affluent		2		3		4		Deprived		All[1]		Affluent		2		3		4		Deprived		All[1]		
	C	R	C	R	C	R	C	R	C	R	C	R	C	R	C	R	C	R	C	R	C	R	C	R	
ENGLAND & WALES	*(1,032)*		*(1,018)*		*(897)*		*(795)*		*(566)*		*(4,421)*		*(1,353)*		*(1,179)*		*(1,177)*		*(1,015)*		*(736)*		*(5,581)*		**ENGLAND & WALES**
One year	95	96	95	96	92	94	93	94	91	93	94	95	96	97	95	96	95	96	93	94	91	93	95	96	*One year*
Five years	90	92	88	91	85	89	82	86	82	87	86	89	91	93	90	93	89	92	85	89	83	87	89	91	*Five years*
	86	92	84	90	82	88	79	86	79	85	82	89													
ENGLAND	*(1,006)*		*(981)*		*(845)*		*(739)*		*(530)*		*(4,200)*		*(1,301)*		*(1,132)*		*(1,100)*		*(947)*		*(694)*		*(5,286)*		**ENGLAND**
One year	95	96	95	96	92	94	93	94	91	92	94	95	96	97	95	96	96	97	94	94	92	93	95	96	*One year*
Five years	90	92	88	91	85	89	82	86	82	87	86	89	91	93	90	93	90	92	86	89	84	87	89	91	*Five years*
	86	92	85	90	82	88	78	85	78	85	83	89													
Northern & Yorkshire	*(83)*		*(122)*		*(106)*		*(124)*		*(116)*		*(561)*		*(120)*		*(121)*		*(122)*		*(110)*		*(147)*		*(630)*		**Northern & Yorkshire**
One year	94	95	95	95	92	94	91	92	89	91	92	93	100	100	96	97	93	94	89	90	88	90	93	94	*One year*
Five years	87	89	87	89	86	88	77	80	82	86	83	86	95	96	95	96	89	91	78	83	77	80	86	89	*Five years*
	84	87	-	-	82	87	73	78	79	86	80	86													
Trent	*(82)*		*(96)*		*(97)*		*(92)*		*(68)*		*(441)*		*(98)*		*(111)*		*(142)*		*(111)*		*(91)*		*(554)*		**Trent**
One year	89	91	92	92	93	95	91	93	94	95	92	93	97	97	93	95	96	97	92	93	91	92	94	95	*One year*
Five years	79	81	83	85	86	90	84	89	81	83	83	86	93	97	85	88	87	90	84	86	81	83	86	89	*Five years*
	79	81	79	82	78	85	79	87	78	80	79	83													
Anglia & Oxford	*(193)*		*(153)*		*(100)*		*(75)*		*(30)*		*(573)*		*(221)*		*(145)*		*(114)*		*(96)*		*(37)*		*(631)*		**Anglia & Oxford**
One year	97	98	98	99	95	96	93	95	93	96	97	97	95	95	96	96	93	93	94	95	86	91	94	95	*One year*
Five years	95	97	92	94	86	90	88	91	87	89	91	94	91	92	90	91	84	87	81	84	81	88	88	90	*Five years*
	92	97	-	-	83	87	-	-	87	89	89	94													
North Thames	*(109)*		*(86)*		*(93)*		*(75)*		*(54)*		*(423)*		*(164)*		*(132)*		*(145)*		*(151)*		*(107)*		*(723)*		**North Thames**
One year	96	97	94	94	93	95	93	94	93	93	94	95	96	97	97	98	97	98	99	99	98	99	97	98	*One year*
Five years	90	93	92	92	92	94	84	87	91	91	90	93	93	95	94	96	89	92	93	96	91	93	92	95	*Five years*
	-	-	-	-	-	-	81	85	87	89	-	-													
South Thames	*(152)*		*(128)*		*(92)*		*(70)*		*(32)*		*(494)*		*(235)*		*(186)*		*(154)*		*(120)*		*(47)*		*(783)*		**South Thames**
One year	97	98	96	99	91	94	96	96	100	100	96	97	96	97	96	96	98	98	94	94	91	93	96	96	*One year*
Five years	91	95	86	90	80	85	87	90	88	97	87	91	90	92	92	93	94	96	89	90	83	85	91	93	*Five years*
	89	95	82	88	76	83	-	-	84	94	83	90													
South & West	*(166)*		*(192)*		*(147)*		*(84)*		*(13)*		*(621)*		*(198)*		*(216)*		*(199)*		*(123)*		*(39)*		*(784)*		**South & West**
One year	95	96	94	94	95	95	88	91	92	95	94	95	92	94	94	95	97	98	93	94	100	100	95	96	*One year*
Five years	89	91	87	90	90	94	74	80	85	88	86	89	86	88	88	94	90	92	85	89	97	100	88	91	*Five years*
	84	89	80	88	88	94	71	80	77	83	81	89													
West Midlands	*(85)*		*(79)*		*(105)*		*(89)*		*(73)*		*(440)*		*(96)*		*(92)*		*(107)*		*(109)*		*(85)*		*(491)*		**West Midlands**
One year	93	93	99	99	87	90	97	97	92	93	93	94	99	99	95	96	95	97	91	92	86	88	93	94	*One year*
Five years	88	89	91	92	86	88	79	83	78	82	84	87	97	98	89	92	90	93	87	90	78	81	88	91	*Five years*
	82	89	89	90	-	-	73	80	-	-	81	87													
North & West	*(136)*		*(125)*		*(105)*		*(130)*		*(144)*		*(647)*		*(169)*		*(129)*		*(117)*		*(127)*		*(141)*		*(690)*		**North & West**
One year	96	96	95	96	87	89	92	94	87	89	92	93	96	97	95	96	97	98	94	96	93	95	95	96	*One year*
Five years	90	93	91	93	76	79	84	88	79	85	84	88	91	93	88	91	94	95	87	92	87	91	90	93	*Five years*
	86	93	88	91	72	79	-	-	73	82	80	87													
WALES	*(26)*		*(37)*		*(52)*		*(56)*		*(36)*		*(221)*		*(52)*		*(47)*		*(77)*		*(68)*		*(42)*		*(295)*		**WALES**
One year	96	96	92	93	94	95	93	93	92	95	93	94	96	97	89	90	88	91	89	92	83	85	90	91	*One year*
Five years	96	96	78	83	81	84	89	91	83	89	84	88	94	95	81	88	83	86	76	83	76	79	83	87	*Five years*
	92	96	73	80	79	82	-	-	83	89	-	-													

[1] Including those for whom no deprivation category was assigned. C - crude survival, R - relative survival (see Chapter 3).
Number of patients contributing to each analysis in parentheses.

Fig. 34.1 Relative survival up to ten years: trends by calendar period of diagnosis (1971-90) and NHS Region

MEN

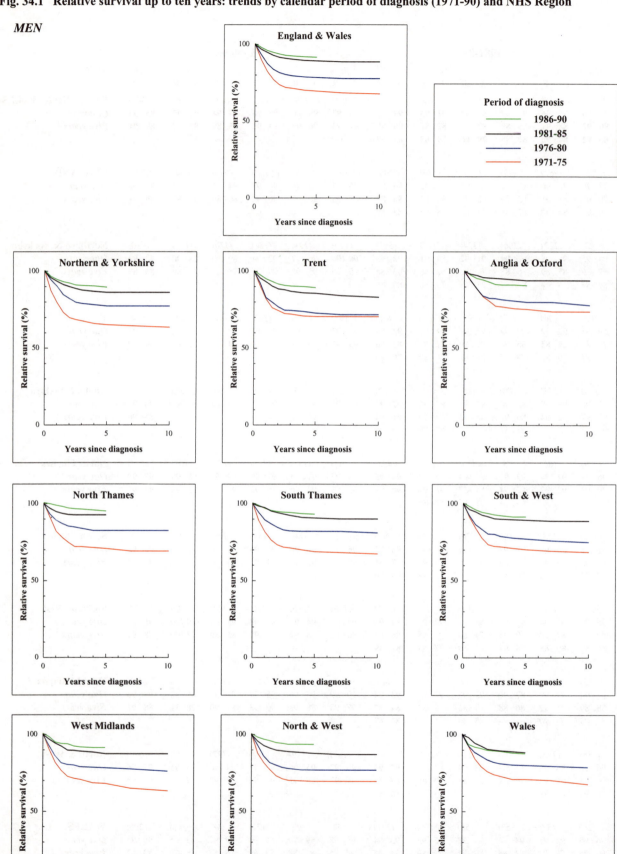

Fig. 34.2   Relative survival up to five years, by deprivation category and NHS Region: patients diagnosed 1986-90

MEN

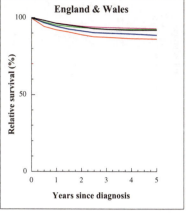

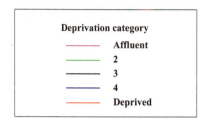

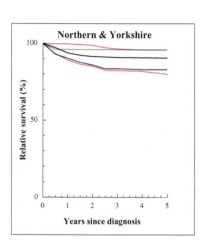

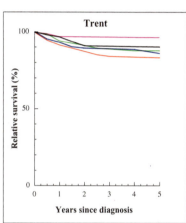

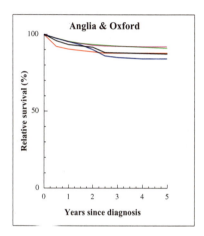

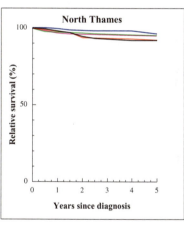

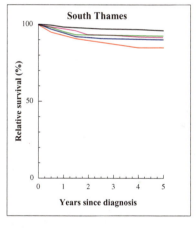

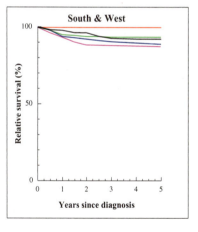

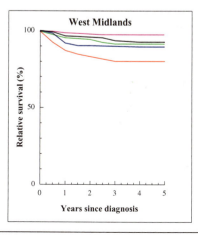

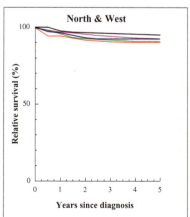

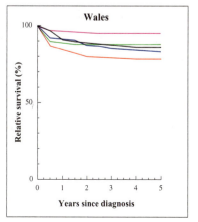

Fig. 34.3 Relative survival at five years by deprivation category, period of diagnosis (1981-90) and NHS Region

MEN

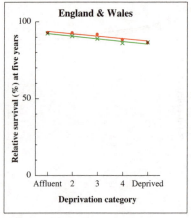

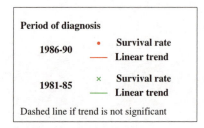

Period of diagnosis

1986-90 • Survival rate
         — Linear trend

1981-85 × Survival rate
         — Linear trend

Dashed line if trend is not significant

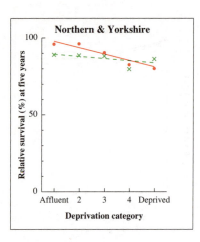

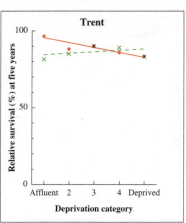

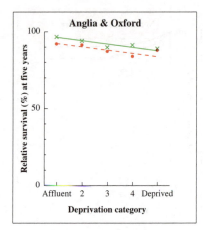

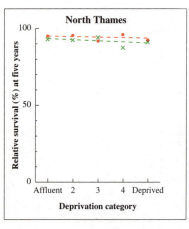

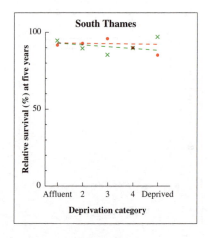

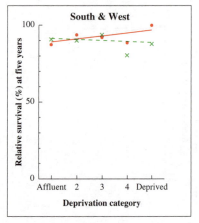

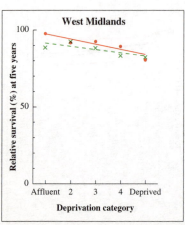

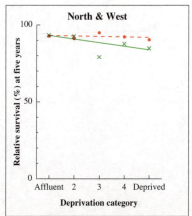

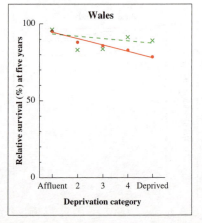

Fig. 34.4 Relative survival (%) by NHS Region:
one-year and five-year survival for men diagnosed 1986-90, with 95% confidence intervals

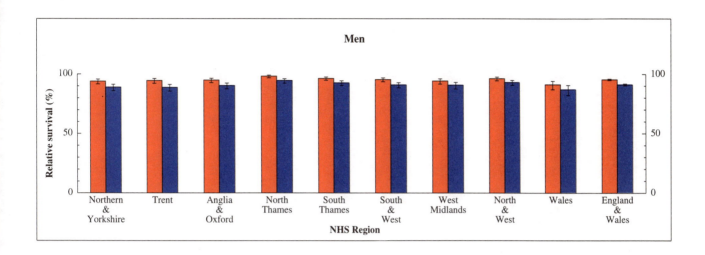

Fig. 34.5 Relative survival (%) at five years, with 95% confidence interval, by age at diagnosis:
England and Wales, men diagnosed 1986-90

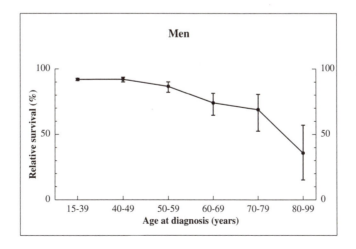

Chapter 35

PENIS

Cancer of the penis is rare in England and Wales. About 300 cases are diagnosed each year, less than 0.5% of all cancers. Incidence during the 1980s was stable, at about 1.5 per 100,000 per year, although mortality declined steadily by about 3% each year up to 1995. Three-quarters of the cases arise over the age of 60. Penile cancer is defined here as cancer of the foreskin, glans or shaft of the penis, excluding cancers of other male genital organs (epididymis, spermatic cord) which are also rare, but are usually included in analyses with tumours of the penis itself.

Carcinoma of the penis is more common in uncircumcised men, and in men with a history of sexually transmitted disease. Human papilloma virus is a risk factor, and cervical cancer is more common in the partners of men with penile cancer. Tumours are mostly squamous carcinomas, and more rarely melanoma, Kaposi sarcoma, adenocarcinoma of urethral glands and transitional cell carcinoma of the urethra.

The survival analyses reported here include 5,000 men diagnosed with a malignant tumour of the penis during the period 1971-90 and followed up to the end of 1995, about 92% of those eligible (Table 35.1). About 5% of the men were excluded because their vital status was unknown at 6 July 1997 and a further 3% because their survival was zero or unknown. About 16% of tumours arose on the glans, and 10% on the foreskin, but for more than 70% the part of the penis in which the tumour arose was not specified (Table 35A). Four-fifths (79%) of the tumours were squamous carcinomas, including verrucous carcinoma (2%). Melanoma accounted for 1.2%, and 16% were poorly specified carcinomas. Leiomyosarcoma and Kaposi sarcoma were extremely rare (under 0.2%).

Survival trends

For men diagnosed with penile cancer in England and Wales during 1986-90, relative survival at one year was 84%, and five-year survival was 69% (Table 35.2; at end of chapter). The corresponding survival rates for men diagnosed in the early 1970s were 78% at one year and 62% at five years. In England and Wales as a whole, survival improved during the 1970s and again during the 1980s, but for men diagnosed in the early 1980s there was little improvement compared with men diagnosed in the late 1970s (Figure 35.1). Ten-year survival is typically no more than 5% lower than five-year survival: this reduction in excess mortality five years and more after diagnosis is reflected in the rather flat survival curves.

The national average improvement between successive five-year periods has been 2% for both one-year and five-year survival, after adjustment of the relative survival rates for changes in the age distribution of men with cancer of the penis diagnosed during 1986-90 (see Table 4.4). Regional survival estimates are based on 150-200 men in each five-year period, and are not all precise (see Chapter 3); regional trends are correspondingly more variable (Figure 35.1). Survival in Anglia & Oxford Region was well above the national average for men diagnosed during the period 1986-90 (Figure 35.4).

Relative survival declines only slightly with age (Figure 35.5). For men under the age of 60 at diagnosis in England and Wales during 1986-90, five-year survival was in the range 70% to 80%, while for men aged 70 or more when diagnosed, the range of five-year survival rates was 60% to 70%, only 10% lower (see Table 4.7).

Deprivation

Survival has not always differed between deprivation groups, but when it has, survival has usually been lowest in the most deprived group. The deprivation gap between most affluent and deprived groups in England and Wales in survival at one year and five years was 10% in the early 1970s, for example, but there was no consistent pattern for men diagnosed in the late 1970s (Table 35.2).

Table 35A Sub-site distribution of cancer of the penis

ICD-9 code	Site description	1971-75		1976-80		1981-85		1986-90		Total	
		No.	%	No.	%	No.	%	No.	%	No.	%
187.1	Prepuce (foreskin)	-	-	39	3.1	101	7.7	141	10.6	**281**	**5.5**
187.2	Glans penis	-	-	86	6.8	214	16.3	212	16.0	**512**	**10.1**
187.3	Body of penis	-	-	4	0.3	10	0.8	21	1.6	**35**	**0.7**
187.4	Penis, part unspecified[1]	1,165	100.0	1,143	89.9	989	75.3	951	71.8	**4,248**	**83.7**
	Total	**1,165**		**1,272**		**1,314**		**1,325**		**5,076**	

[1] Penile tumours were coded to 187.0 in ICD-8 (1971-78): they have been assigned to the corresponding code in ICD-9 in this table.

Among 1,300 men of known deprivation status who were diagnosed with penile cancer in England and Wales during the early 1980s, the estimated survival rates in the affluent group were 82% at one year and 65% at five years, but the deprivation gap was only 1%, not statistically significant (see Table 4.9). For a further 1,300 men diagnosed during 1986-90, the gap was 5% at one year and 3% at five years, but the gradient was not significant in either period (Figure 35.3).

International comparison

Five-year survival from cancer of the penis for men diagnosed in England and Wales during 1986-90 was 69%, within 3% of the corresponding survival rate for Scotland and the USA, as well as the average for Europe, based on data from 17 countries (see Table 4.12). The highest precise regional five-year survival rate was 76% (West Midlands, based on 136 cases), among the highest survival rates in Europe, along with parts of France (80%, based on 45 cases), Spain (77%, 98 cases) and Italy (75%, 126 cases)[1].

1.   Berrino F, Capocaccia R, Estève J, Gatta G, Hakulinen T, Micheli M, Sant M, Verdecchia A, (eds.) *Survival of cancer patients in Europe: the EUROCARE study, II. (IARC Scientific Publications No. 151).* Lyon: International Agency for Research on Cancer, 1999

Table 35.1 Ineligible and excluded records, and patients included in analyses, by calendar period of diagnosis

| | | | | | | | Calendar period of diagnosis | | | | | | | | |
	1971-75			1976-80			1981-85			1986-90			Overall		
Total registered	**1,328**			**1,424**			**1,429**			**1,486**			**5,667**		
Ineligible	*Records*	*Patients*	*%*	*Records*	*Patients*	*%*	*Records*	*Patients*	*%*	*Records*	*Patients*	*%*	*Records*	*Patients*	*%*
Incomplete data[1]	8	8		10	10		5	5		7	7		30	**30**	
Resident outside England and Wales	12	12		12	11		7	7		4	3		35	**33**	
In situ neoplasm[2]	48	48		24	23		1	1		0	0		73	**72**	
Benign or uncertain behaviour[2]	16	8		16	7		0	0		0	0		32	**15**	
Metastatic[2]	6	6		1	1		0	0		0	0		7	**7**	
Otherwise ineligible[3]	-	-		-	-		-	-		-	-		-	**-**	
Lymphoma[4]	0	0		1	1		1	0		0	0		2	**1**	
Leukaemia or myeloma[4]	0	0		0	0		0	0		0	0		0	**0**	
		82			53			13			10			158	
Total eligible		**1,246**	**100**		**1,371**	**100**		**1,416**	**100**		**1,476**	**100**		**5,509**	**100**
Aged under 15 or 100 years or over at diagnosis	2	2	0.2	2	2	0.1	0	0	0.0	1	1	<0.1	5	**5**	<0.1
Vital status unknown[5]	50	43	3.5	76	66	4.8	63	58	4.1	90	86	5.8	279	**253**	4.6
Duplicate registration[6]	0	0	0.0	0	0	0.0	0	0	0.0	0	0	0.0	0	**0**	0.0
Multiple primary[7]	0	0	0.0	1	1	<0.1	8	6	0.4	9	9	0.6	18	**16**	0.3
Synchronous tumours[8]	2	0	0.0	3	2	0.1	4	2	0.1	3	2	0.1	12	**6**	0.1
Sex not known	0	0	0.0	0	0	0.0	0	0	0.0	0	0	0.0	0	**0**	0.0
Sex incompatible with site code[9]	0	0	0.0	1	1	<0.1	0	0	0.0	0	0	0.0	1	**1**	<0.1
Invalid dates[10]	0	0	0.0	0	0	0.0	1	0	0.0	1	0	0.0	2	**0**	0.0
Zero survival[11]	38	36	2.9	33	27	2.0	42	36	2.5	61	53	3.6	174	**152**	2.8
Total exclusions		**81**	**6.5**		**99**	**7.2**		**102**	**7.2**		**151**	**10.2**		**433**	**7.9**
Patients included in analysis		**1,165**	**93.5**		**1,272**	**92.8**		**1,314**	**92.8**		**1,325**	**89.8**		**5,076**	**92.1**

---

[1] Main data item(s) invalid or incompatible with one another: name, sex, date of birth, date of diagnosis and (if present) date of death, postcode, site, morphology and behaviour

[2] Tumour either *in situ* (behaviour code 2), benign (0), uncertain if benign or malignant (1) or metastatic (6 or 9)

[3] Not applicable for this malignancy

[4] Morphology that of lymphoma (included with non-Hodgkin lymphoma, Ch. 41) or myeloma (included with multiple myeloma, Ch. 43), or leukaemia (excluded)

[5] Tracing of vital status at National Health Service Central Register incomplete at 6 July 1997

[6] Same site code, sex, cancer registry and cancer registry number as an earlier registration

[7] Same site code and person number as an earlier registration(s): mostly confirmed multiple primary tumours at this site, some unresolved duplicate registrations

[8] Same site code, sex, date of birth and date of diagnosis as another registration(s): mostly synchronous or (in paired organs) bilateral tumours in same anatomic site in one individual, not previously linked: also some duplicate registrations

[9] Site code specific to one sex; tumour registered to opposite sex

[10] Impossible sequence of dates (birth, diagnosis and, if present, emigration or death); or date of death after 6 July 1997

[11] Date of diagnosis same as date of death: some are patients who did die on the day of diagnosis, but most were registered solely from a death certificate (death certificate only, or DCO), and their survival time is thus unknown

Fig. 35.1 Relative survival up to ten years: trends by calendar period of diagnosis (1971-90) and NHS Region

MEN

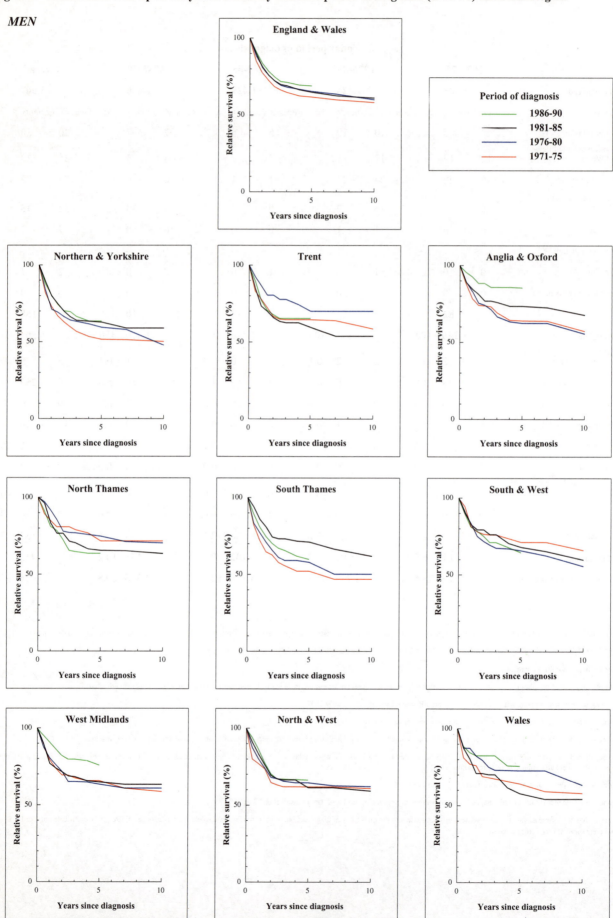

Fig. 35.2 Relative survival up to five years, by deprivation category: England and Wales, patients diagnosed 1986-90

MEN

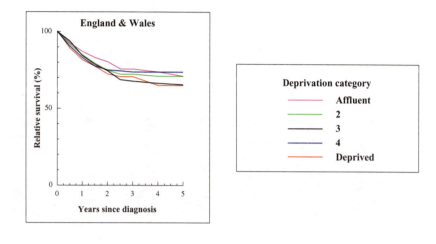

Fig. 35.3 Relative survival at five years, by deprivation category and period of diagnosis (1981-90): England and Wales

MEN

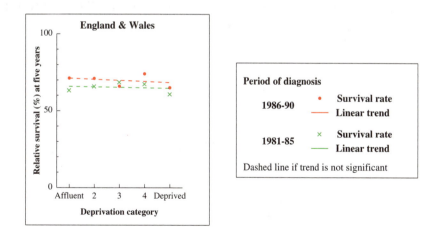

Fig. 35.4 Relative survival (%) by NHS Region:
one-year and five-year survival for men diagnosed 1986-90, with 95% confidence intervals

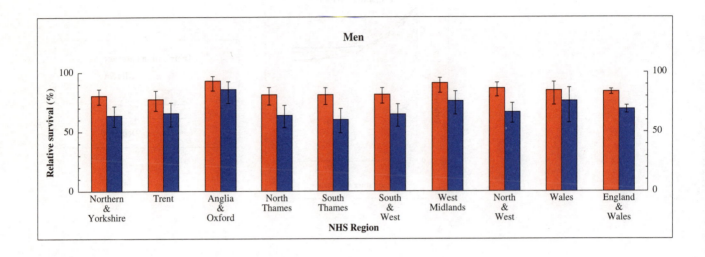

Fig. 35.5 Relative survival (%) at five years, with 95% confidence interval, by age at diagnosis:
England and Wales, men diagnosed 1986-90

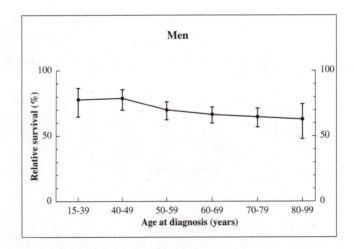

Table 35.2   Crude and relative survival (%) at one, five and ten years since diagnosis: NHS Region, deprivation category and calendar period of diagnosis

ENGLAND & WALES

MEN	1971-75						1976-80					
	Affluent	2	3	4	Deprived	All[1]	Affluent	2	3	4	Deprived	All[1]
	C R	C R	C R	C R	C R	C R	C R	C R	C R	C R	C R	C R
ENGLAND & WALES	(167)	(167)	(199)	(206)	(191)	(1,165)	(165)	(242)	(272)	(271)	(257)	(1,272)
One year	77 81	76 81	75 81	73 78	68 73	73 78	77 82	77 83	69 75	78 84	78 84	76 81
Five years	52 69	53 69	51 71	46 58	40 52	47 62	52 68	49 64	44 60	46 63	52 70	48 65
Ten years	36 67	37 64	34 63	33 57	27 50	32 58	40 67	33 58	29 53	33 60	35 61	33 60

	1981-85						1986-90					
	(232)	(272)	(277)	(257)	(244)	(1,314)	(227)	(283)	(267)	(298)	(230)	(1,325)
One year	73 79	77 82	78 84	75 81	72 78	75 81	82 87	79 85	80 85	78 83	77 81	79 84
Five years	48 63	47 65	50 68	50 67	44 60	48 65	55 71	52 71	48 65	55 74	48 65	52 69
Ten years	32 57	33 61	35 64	32 63	30 56	33 61						

	1971-75	1976-80	1981-85	1986-90
Northern & Yorkshire	(182)	(180)	(199)	(201)
One year	69 73	66 71	74 80	75 80
Five years	40 52	40 59	46 63	47 64
Ten years	28 50	25 48	29 59	
Trent	(138)	(153)	(143)	(118)
One year	74 78	80 86	67 73	73 78
Five years	47 65	50 70	43 60	52 66
Ten years	31 59	41 70	28 54	
Anglia & Oxford	(84)	(112)	(139)	(136)
One year	74 79	78 83	79 85	87 93
Five years	46 64	48 62	55 73	67 86
Ten years	27 57	32 55	36 68	
North Thames	(113)	(117)	(110)	(140)
One year	80 85	85 92	78 83	77 81
Five years	56 72	57 75	50 65	49 64
Ten years	40 72	38 70	34 63	
South Thames	(139)	(124)	(142)	(152)
One year	68 73	72 77	79 86	76 81
Five years	40 52	42 58	52 71	43 60
Ten years	28 47	28 50	34 62	
South & West	(145)	(155)	(152)	(174)
One year	75 82	77 83	77 83	77 81
Five years	48 72	50 66	51 68	47 65
Ten years	31 66	31 56	30 60	
West Midlands	(124)	(145)	(143)	(136)
One year	74 80	76 81	72 77	85 91
Five years	52 66	49 63	50 65	59 76
Ten years	33 59	36 61	39 63	
North & West	(172)	(200)	(202)	(189)
One year	72 77	73 80	78 83	82 87
Five years	49 62	47 65	44 61	51 67
Ten years	37 61	34 62	34 59	
WALES	(68)	(86)	(84)	(79)
One year	73 77	82 87	75 80	80 85
Five years	52 64	56 73	42 58	55 76
Ten years	37 58	39 63	28 54	

[1] Including those for whom no deprivation category was assigned. C - crude survival, R - relative survival (see Chapter 3).
Number of patients contributing to each analysis in parentheses.

Chapter 36

BLADDER

Cancer of the urinary bladder occurs in 8,000 men and 3,000 women in England and Wales each year. It ranks fourth among men, accounting for 8% of all cancers, and ninth among women (3%). Incidence in men was increasing by about 2% every five years up to the end of the 1980s, and more quickly in women (6-7%); by 1990, annual incidence rates were 33 per 100,000 in men and 12 in women. Men have a three-fold higher risk than women at each age. Mortality from bladder cancer in men in England and Wales fell by about 1% a year during the 1980s, while the death rate in women did not change. In 1995, there were about 5,000 deaths from bladder cancer.

Tobacco smoking is responsible for 30-40% of bladder cancers in developed countries. Occupational exposure to industrial carcinogens in rubber, organic dye and metal refining can cause bladder cancer. Cyclophosphamide and some other chemotherapy drugs also increase the risk.

Most bladder cancers are transitional cell carcinomas. These tumours often arise as small papillomas of the urothelial lining of the bladder, and small biopsies may not enable the pathologist to determine unequivocally if the tumour is invasive or *in situ*. Many cancer registries include urothelial papillomas described as benign or of uncertain malignancy when reporting cancer incidence[1]. Where papillomas comprise a large proportion of all bladder cancers, this complicates international comparison of incidence[2].

Bladder papillomas described as benign, *in situ* or of uncertain behaviour are excluded from the analyses here, but the increasing tendency for papillomas to be described and treated as malignant, rather than pre-malignant, has led to their inclusion by cancer registries that would not include 'benign' papillomas. Some of these tumours are of low invasive potential[3]. The biological spectrum of bladder cancers recorded as malignant has shifted, and this in turn complicates the international comparison of cancer survival[4], much as for prostate cancer (see Chapter 33). Time trends and geographic differences in population-based survival rates are therefore more difficult to interpret for bladder cancer than for many other cancers.

The survival analyses reported here include 168,000 adults diagnosed with a malignant bladder tumour in England and Wales during the period 1971-90 and followed up to the end of 1995, about 91% of those eligible. Only bladder tumours explicitly described as malignant were included. More than 21,000 tumours registered as non-malignant had been registered during 1971-90, but these were not included in the initial data set (Table 36A). Another 1,400 tumours registered during the

1970s, included in the initial data set because they had the site code for a primary malignant neoplasm of bladder (ICD-8 188), were later excluded as ineligible because the behaviour was coded as benign, *in situ*, or of uncertain malignancy. About 5% of the 185,000 eligible adults were excluded from analysis because their vital status was unknown at 6 July 1997, and another 4% because their survival was either zero or unknown.

More than 1,000 patients had synchronous or multiple primary cancers, but each patient was included only once in the analyses, from the date of the first malignancy. More than 90% of the tumours diagnosed since 1979 were assigned to the bladder without further anatomic detail (data not shown). The proportion specified as transitional cell carcinoma rose from 50% to 75% between the 1970s and the 1980s. Papillary transitional cell carcinoma accounted for much of this increase, the proportion rising steadily from less than 20% in the early 1970s to over 30% during 1986-90. Five per cent were squamous carcinomas and 2% adenocarcinoma, mainly solid or papillary adenocarcinomas; leiomyosarcoma and other sarcomas were very rare. Overall, about 20% of tumours were poorly specified carcinomas: the proportion fell steadily from 25% in the early 1970s to 12% by the mid-1980s, but had risen again to 20% by 1990.

Survival trends

Relative survival for men diagnosed with bladder cancer in England and Wales during 1986-90 was 81% at one year and 65% at five years; the corresponding survival rates for women were 72% and 58% (Table 36.2). Survival improved similarly in both sexes between the early and late 1970s and again by the early 1980s, but there was little further improvement for patients diagnosed during 1986-90 (Figure 36.1). This pattern is largely consistent in the NHS Regions of England.

The survival advantage for men has been consistent for years: it is greater for survival at one year than at five years. After adjustment to the age distribution of all adults with bladder cancer in 1986-90, one-year relative survival increased from 66% to 80% for men, and from 59% to 72% for women, an average gain of almost 5% every five years, although there was less improvement in the 1980s (see Table 4.4). Five-year survival rose from 44% to 62% for men and from 42% to 57% for women, an average gain of 5% to 6% every five years: again, improvement was most marked between patients diagnosed in the early and late 1970s. The average improvement in five-year survival was slightly faster for men than for women, so the difference in survival between the sexes had increased by

Table 36A   Bladder tumours *not* included, by behaviour[1] and NHS Region

NHS Region	Benign No.	Benign %	*In situ* No.	*In situ* %	Uncertain No.	Uncertain %	Unspecified No.	Unspecified %	Total
Northern & Yorkshire	874	26.3	1,778	53.5	474	14.3	200	6.0	**3,326**
Trent	813	47.1	130	7.5	632	36.6	151	8.7	**1,726**
Anglia & Oxford	649	38.4	160	9.5	796	47.1	86	5.1	**1,691**
North Thames	704	47.0	215	14.4	385	25.7	193	12.9	**1,497**
South Thames	863	70.2	184	15.0	156	12.7	27	2.2	**1,230**
South & West	1,542	46.3	136	4.1	713	21.4	936	28.1	**3,327**
West Midlands	1,081	46.8	133	5.8	535	23.2	562	24.3	**2,311**
North & West	1,871	58.5	229	7.2	965	30.2	134	4.2	**3,199**
Wales	1,836	53.3	68	2.0	1,434	41.6	105	3.0	**3,443**
	10,233	**48.2**	**3,033**	**13.2**	**6,090**	**28.1**	**2,394**	**10.5**	**21,750**

[1]Bladder tumours coded as benign (ICD-9 223.3), *in situ* (233.7), or of uncertain (236.7) or unspecified behaviour (239.4).

the late 1980s. This is unusual: among the common malignancies, men have a survival advantage only for cancers of the larynx, breast and bladder (see Table 4.8).

Quantitative estimates of regional trends in survival from bladder cancer confirm the visual impression from Figure 36.1. After adjustment to a common age standard, improvement in survival has been quickest in Wales and North & West Region, where survival was amongst the lowest for patients diagnosed in the early 1970s (see Table-Figure 4.5P). Regional differences in survival are no longer as marked as they were (Figure 36.4). In Wales, one-year and five-year survival improved significantly more quickly than the national average for both men and women, by 7% to 8% every five years. Five-year survival in Wales was significantly higher than the national average for patients diagnosed during 1986-90. The distinctive trend in Wales could be due to improved treatment, but changes in the type of tumour registered should also be considered as a possible explanation.

Survival falls with age in both sexes, but less rapidly for younger women (Figure 36.5). For men under 50 at diagnosis, relative survival at five years is 83-88%, compared with about 75% for women aged less than 60 at diagnosis (see Table 4.7). Survival falls to 74% for men aged 50-59, similar to that for women. Above the age of 60 at diagnosis, relative survival at five years falls by about 1% for every year in both sexes, and it is significantly higher for men.

Differences in the proportion of small papillomas could underlie at least part of the observed differences in survival between men and women, by age, between regions and over time: more detailed studies would be needed to resolve this.

Deprivation

Survival is higher for men and women in affluent groups (Figure 36.2). This pattern is largely consistent at one, five and ten years after diagnosis, in most regions and for men and women diagnosed in the 1970s and the 1980s (Table

36.2; see Table 4.9). The deprivation gradient in five-year survival in England and Wales is significant and it did not flatten during the 1980s (Figure 36.3). The gradient was similar and statistically significant in six regions.

There was no deprivation gradient in five-year survival in Wales or the Thames regions during the 1980s, however, and the large survival gains in Wales during the 1980s have been experienced by men and women in all deprivation categories (Table 36.2, Figure 36.3).

International comparison

Five-year survival in England and Wales for adults diagnosed during 1986-90 was the same as in Scotland and within 2-3% of the average for Europe (see Table 4.12). The survival advantage for men is seen in most countries. The highest regional survival rates are above the European average, but still 10-14% lower than in the USA. It is likely that part of the survival advantage in the USA is due to differences in the definition of malignancy and in the proportion of papillary tumours, but this remains to be confirmed.

1.   Parkin DM, Muir CS, Whelan SL, Gao Y-T, Ferlay J, Powell J, (eds.) *Cancer incidence in five continents, volume VI (IARC Scientific Publications No. 120).* Lyon: International Agency for Research on Cancer, 1993

2.   Coleman MP, Estève J, Damiecki P, Arslan A, Renard H. *Trends in cancer incidence and mortality (IARC Scientific Publications No. 121).* Lyon: International Agency for Research on Cancer, 1993

3.   Newling DWW, Denis L, Gerard J-P, Scalliet PGM, Stoter G. Bladder cancer. In: Peckham M, Pinedo HM, Veronesi U, (eds.) *Oxford textbook of oncology.* Oxford: Oxford Medical Publications, 1995

4.   Berrino F, Estève J, Coleman MP. Basic issues in the estimation and comparison of cancer patient survival. In: Berrino F, Sant M, Verdecchia A, Capocaccia R, Hakulinen T, Estève J, (eds.) *Survival of cancer patients in Europe: the EUROCARE study. (IARC Scientific Publications No. 132).* Lyon: International Agency for Research on Cancer, 1995, pp1-14

Table 36.1 Ineligible and excluded records, and patients included in analyses, by calendar period of diagnosis

	Calendar period of diagnosis															
	1971-75			**1976-80**			**1981-85**			**1986-90**			**Overall**		
Total registered	**38,221**			**42,989**			**50,716**			**56,718**			**188,644**		
Ineligible	*Records*	*Patients*	*%*	*Records*	*Patients*	*%*	*Records*	*Patients*	*%*	*Records*	*Patients*	*%*	*Records*	*Patients*	*%*
Incomplete data[1]	153	153		160	160		56	56		101	101		470	**470**	
Resident outside England and Wales	358	350		249	238		225	204		217	154		1,049	**946**	
In situ neoplasm[2]	126	121		38	38		0	0		2	2		166	**161**	
Benign or uncertain behaviour[2]	1,222	1,061		323	170		8	8		0	0		1,553	**1,239**	
Metastatic[2]	54	53		27	27		0	0		1	1		82	**81**	
Otherwise ineligible[3]	-	-		-	-		-	-		-	-		-	-	
Lymphoma[4]	5	5		2	2		1	0		4	0		12	**7**	
Leukaemia or myeloma[4]	0	0		0	0		0	0		0	0		0	**0**	
		1,743			635			268			258			2,904	
Total eligible		**36,478**	**100**		**42,354**	**100**		**50,448**	**100**		**56,460**	**100**		**185,740**	**100**
Aged under 15 or 100 years or over at diagnosis	19	18	<0.1	32	31	<0.1	27	24	<0.1	52	50	<0.1	130	**123**	<0.1
Vital status unknown[5]	1,341	1,136	3.1	2,030	1,846	4.4	1,974	1,857	3.7	3,568	3,453	6.1	8,913	**8,292**	4.5
Duplicate registration[6]	0	0	0.0	4	3	<0.1	1	1	<0.1	2	1	<0.1	7	**5**	<0.1
Multiple primary[7]	0	0	0.0	45	41	<0.1	249	233	0.5	443	415	0.7	737	**689**	0.4
Synchronous tumours[8]	107	81	0.2	131	86	0.2	267	194	0.4	321	184	0.3	826	**545**	0.3
Sex not known	0	0	0.0	0	0	0.0	0	0	0.0	0	0	0.0	0	**0**	0.0
Sex incompatible with site code[9]	-	-	-	-	-	-	-	-	-	-	-	-	-	-	-
Invalid dates[10]	3	2	<0.1	5	5	<0.1	15	7	<0.1	60	27	<0.1	83	**41**	<0.1
Zero survival[11]	1,050	974	2.7	1,522	1,403	3.3	2,667	2,450	4.9	3,299	3,012	5.3	8,538	**7,839**	4.2
Total exclusions		**2,211**	6.1		**3,415**	8.1		**4,766**	9.4		**7,142**	12.6		**17,534**	9.4
Patients included in analysis		**34,267**	93.9		**38,939**	91.9		**45,682**	90.6		**49,318**	87.4		**168,206**	90.6

[1] Main data item(s) invalid or incompatible with one another: name, sex, date of birth, date of diagnosis and (if present) date of death, postcode, site, morphology and behaviour

[2] Tumour either *in situ* (behaviour code 2), benign (0), uncertain if benign or malignant (1) or metastatic (6 or 9)

[3] Not applicable for this malignancy

[4] Morphology that of lymphoma (included with non-Hodgkin lymphoma, Ch. 41) or myeloma (included with multiple myeloma, Ch. 43), or leukaemia (excluded)

[5] Tracing of vital status at National Health Service Central Register incomplete at 6 July 1997

[6] Same site code, sex, cancer registry and cancer registry number as an earlier registration

[7] Same site code and person number as an earlier registration(s): mostly confirmed multiple primary tumours at this site, some unresolved duplicate registrations

[8] Same site code, sex, date of birth and date of diagnosis as another registration(s): mostly synchronous or (in paired organs) bilateral tumours in same anatomic site in one individual, not previously linked: also some duplicate registrations

[9] Not applicable for this malignancy

[10] Impossible sequence of dates (birth, diagnosis and, if present, emigration or death); or date of death after 6 July 1997

[11] Date of diagnosis same as date of death: some are patients who did die on the day of diagnosis, but most were registered solely from a death certificate (death certificate only, or DCO), and their survival time is thus unknown

Table 36.2 Crude and relative survival (%) at one, five and ten years since diagnosis: NHS Region, deprivation category and calendar period of diagnosis

MEN	1971-75 Affluent		2		3		4		Deprived		All[1]		1976-80 Affluent		2		3		4		Deprived		All[1]	
	C	R	C	R	C	R	C	R	C	R	C	R	C	R	C	R	C	R	C	R	C	R	C	R
ENGLAND & WALES	(3,722)		(4,190)		(4,521)		(4,450)		(3,866)		(25,315)		(5,016)		(5,459)		(5,951)		(5,781)		(5,165)		(28,555)	
One year	71	75	67	71	67	71	65	70	63	67	66	70	73	78	72	77	70	75	69	74	68	73	70	75
Five years	41	55	38	51	38	51	35	48	34	46	37	50	46	61	43	59	42	58	41	57	40	56	42	58
Ten years	26	47	25	46	24	45	22	43	21	41	23	43	31	54	28	51	27	51	26	51	25	51	27	51
ENGLAND	(3,609)		(4,035)		(4,290)		(4,160)		(3,674)		(24,104)		(4,820)		(5,254)		(5,638)		(5,428)		(4,952)		(27,084)	
One year	71	76	67	72	67	72	65	70	63	67	66	71	73	78	72	77	70	75	69	74	68	73	70	75
Five years	42	55	39	52	38	51	35	49	34	47	37	50	47	62	44	59	42	58	41	57	40	56	43	58
Ten years	27	47	25	46	24	45	22	43	21	41	23	43	31	55	28	52	27	52	26	51	25	50	27	52
Northern & Yorkshire	(277)		(374)		(474)		(536)		(789)		(3,414)		(414)		(526)		(737)		(921)		(1,248)		(3,947)	
One year	68	72	67	72	69	74	63	67	59	63	64	69	75	80	73	77	69	74	68	73	68	73	70	75
Five years	38	50	40	53	38	53	33	46	32	44	36	50	50	66	46	61	40	55	38	54	41	58	42	58
Ten years	25	44	26	46	22	44	20	38	20	41	23	43	31	57	30	56	24	49	24	50	26	54	26	52
Trent	(238)		(344)		(462)		(596)		(518)		(2,391)		(391)		(500)		(609)		(689)		(614)		(2,895)	
One year	69	73	66	70	66	71	70	74	63	67	67	71	73	77	71	76	73	79	72	77	69	75	72	77
Five years	42	54	38	50	39	52	37	51	31	43	37	50	46	60	42	59	47	64	43	60	41	58	44	61
Ten years	26	44	26	43	25	45	23	45	20	38	24	44	28	49	26	51	30	57	28	57	24	50	27	53
Anglia & Oxford	(485)		(523)		(496)		(365)		(146)		(2,277)		(664)		(670)		(577)		(364)		(193)		(2,580)	
One year	71	76	69	73	67	71	67	72	68	73	68	73	72	77	74	79	71	76	69	74	74	80	72	77
Five years	45	61	42	56	39	53	39	53	40	54	41	55	46	61	46	60	45	61	43	59	47	65	45	61
Ten years	29	54	28	50	26	48	24	47	22	41	26	49	32	56	30	54	32	58	28	54	29	54	31	56
North Thames	(582)		(608)		(613)		(668)		(550)		(3,410)		(545)		(625)		(608)		(692)		(565)		(3,156)	
One year	71	75	67	72	70	74	69	73	69	74	69	73	75	79	73	78	71	76	71	76	72	78	72	77
Five years	39	50	37	49	43	57	39	53	39	53	39	53	52	68	43	58	42	56	43	60	44	61	44	61
Ten years	24	41	25	45	27	48	26	49	25	47	25	47	38	67	28	52	26	49	27	55	28	60	29	55
South Thames	(895)		(790)		(760)		(735)		(405)		(3,917)		(1,107)		(919)		(874)		(739)		(329)		(4,215)	
One year	71	75	66	71	65	70	63	68	62	67	66	71	72	77	71	76	69	75	70	75	72	78	71	76
Five years	39	52	36	49	33	45	35	49	34	48	36	50	44	59	41	57	42	59	43	60	44	64	43	60
Ten years	25	45	22	41	20	41	19	38	18	37	22	42	27	51	25	50	26	53	27	55	29	62	26	52
South & West	(722)		(923)		(905)		(440)		(157)		(3,431)		(866)		(1,132)		(1,102)		(548)		(214)		(4,006)	
One year	76	80	70	75	69	74	65	69	67	71	70	75	75	81	74	79	70	75	72	78	74	80	73	78
Five years	48	63	42	56	40	54	34	46	42	55	41	56	48	64	47	65	43	62	46	63	46	63	46	64
Ten years	30	53	28	51	26	50	22	45	29	55	27	50	33	59	29	58	30	59	31	59	31	61	31	59
West Midlands	(91)		(93)		(150)		(193)		(207)		(2,022)		(327)		(338)		(479)		(564)		(617)		(2,356)	
One year	64	68	66	70	72	76	69	74	57	61	64	69	71	76	66	70	69	74	66	70	68	73	68	73
Five years	33	44	39	50	40	52	39	52	28	41	34	47	45	60	40	52	38	53	41	55	39	54	40	55
Ten years	22	39	17	35	22	40	24	44	22	39	21	40	31	54	25	45	23	45	26	49	25	49	26	48
North & West	(319)		(380)		(430)		(627)		(902)		(3,242)		(506)		(544)		(652)		(911)		(1,172)		(3,929)	
One year	67	72	60	64	57	61	59	64	60	65	60	64	68	72	68	73	65	70	63	69	62	66	65	70
Five years	37	48	32	44	28	39	28	39	32	43	30	41	43	58	39	53	35	50	34	49	34	47	36	51
Ten years	24	43	21	41	16	33	18	36	18	36	18	36	29	52	26	47	24	48	20	43	20	42	23	45
WALES	(113)		(155)		(231)		(290)		(192)		(1,211)		(196)		(205)		(313)		(353)		(213)		(1,471)	
One year	61	65	56	60	60	65	58	62	58	63	60	64	70	74	64	68	68	73	64	69	61	66	66	70
Five years	37	51	29	40	37	50	34	43	31	41	34	46	35	49	34	47	38	53	39	55	37	51	37	53
Ten years	24	45	19	35	25	46	21	38	20	36	22	41	21	43	20	39	21	42	22	45	27	51	23	45

[1] Including those for whom no deprivation category was assigned. C - crude survival, R - relative survival (see Chapter 3). Number of patients contributing to each analysis in parentheses.

MEN

	1981-85 Affluent C	R	2 C	R	3 C	R	4 C	R	Deprived C	R	All[1] C	R	1986-90 Affluent C	R	2 C	R	3 C	R	4 C	R	Deprived C	R	All[1] C	R
ENGLAND & WALES	*(6,344)*		*(6,729)*		*(6,887)*		*(6,912)*		*(5,607)*		*(33,180)*		*(6,659)*		*(7,599)*		*(7,699)*		*(7,430)*		*(5,726)*		*(35,539)*	
One year	78	83	75	81	74	80	73	78	71	77	74	80	78	83	78	84	76	82	74	80	73	78	76	81
Five years	50	66	47	64	47	64	45	62	44	62	46	63	51	67	50	68	48	65	45	63	44	62	48	65
	33	59	30	56	29	56	28	55	28	56	30	56												
ENGLAND	*(6,110)*		*(6,478)*		*(6,443)*		*(6,423)*		*(5,332)*		*(31,334)*		*(6,333)*		*(7,249)*		*(7,187)*		*(6,799)*		*(5,399)*		*(33,306)*	
One year	78	83	76	81	74	80	73	78	71	77	74	80	78	83	78	84	76	82	74	79	73	78	76	81
Five years	50	66	47	64	47	65	45	63	44	62	46	64	51	67	50	67	47	65	45	62	44	61	47	65
	33	59	30	56	30	57	28	56	27	56	30	57												
Northern & Yorkshire	*(574)*		*(758)*		*(905)*		*(1,114)*		*(1,300)*		*(4,713)*		*(583)*		*(747)*		*(859)*		*(993)*		*(1,143)*		*(4,348)*	
One year	77	83	76	81	74	81	73	79	72	79	74	80	76	82	75	81	73	80	70	77	70	76	72	78
Five years	49	67	48	66	45	65	44	62	44	64	46	63	50	68	46	65	47	71	40	61	42	61	44	62
	35	64	31	60	28	58	27	57	27	59	29	56												
Trent	*(429)*		*(647)*		*(781)*		*(870)*		*(619)*		*(3,371)*		*(508)*		*(699)*		*(823)*		*(868)*		*(739)*		*(3,651)*	
One year	75	81	72	77	69	74	71	77	70	76	71	77	79	86	80	86	74	81	73	79	73	79	76	81
Five years	47	65	45	63	41	58	45	64	45	67	45	62	50	70	50	71	47	67	44	64	45	66	47	64
	32	62	29	57	27	54	29	62	28	63	29	56												
Anglia & Oxford	*(794)*		*(770)*		*(704)*		*(488)*		*(216)*		*(3,033)*		*(850)*		*(874)*		*(766)*		*(518)*		*(207)*		*(3,236)*	
One year	77	82	76	81	73	79	71	77	74	80	75	80	76	83	76	82	77	84	73	79	70	76	75	81
Five years	53	70	50	68	47	67	44	65	48	69	49	67	49	69	51	71	45	65	46	65	44	63	48	65
	36	65	33	63	29	60	28	60	30	61	32	59												
North Thames	*(667)*		*(682)*		*(663)*		*(642)*		*(542)*		*(3,249)*		*(684)*		*(872)*		*(891)*		*(864)*		*(704)*		*(4,120)*	
One year	80	85	75	81	76	82	74	80	75	82	76	82	77	84	82	88	80	86	78	85	79	85	79	85
Five years	51	68	48	67	49	68	47	67	47	69	49	67	50	69	52	72	49	70	50	72	47	69	50	67
	34	63	30	62	30	62	32	65	31	66	32	60												
South Thames	*(1,302)*		*(1,003)*		*(889)*		*(805)*		*(324)*		*(4,478)*		*(1,262)*		*(1,196)*		*(1,004)*		*(854)*		*(353)*		*(4,748)*	
One year	79	85	75	81	77	83	76	82	74	81	77	83	81	87	79	85	78	84	75	82	78	86	79	84
Five years	48	66	46	65	50	72	48	69	44	66	47	66	50	69	48	71	49	71	44	65	45	67	48	66
	30	59	29	57	33	67	29	64	29	65	30	58												
South & West	*(1,109)*		*(1,363)*		*(1,100)*		*(644)*		*(203)*		*(4,535)*		*(1,062)*		*(1,449)*		*(1,261)*		*(839)*		*(274)*		*(4,932)*	
One year	77	83	79	84	75	81	75	81	74	79	77	82	80	86	78	84	76	83	75	81	72	78	77	83
Five years	51	69	48	67	46	65	48	67	46	65	48	66	51	71	51	73	48	71	48	69	43	64	49	67
	34	64	31	62	28	58	31	63	34	63	31	59												
West Midlands	*(509)*		*(540)*		*(612)*		*(778)*		*(740)*		*(3,201)*		*(586)*		*(622)*		*(723)*		*(834)*		*(659)*		*(3,443)*	
One year	76	81	75	81	76	81	71	77	69	74	73	79	75	81	81	87	76	82	76	82	70	76	76	81
Five years	47	63	48	66	46	64	42	61	39	57	44	61	51	71	52	71	47	68	44	64	44	64	47	64
	32	58	33	62	32	61	27	55	25	55	29	55												
North & West	*(726)*		*(715)*		*(789)*		*(1,082)*		*(1,388)*		*(4,754)*		*(798)*		*(790)*		*(860)*		*(1,029)*		*(1,320)*		*(4,828)*	
One year	76	82	72	78	74	81	69	75	69	75	72	78	77	84	76	82	76	82	73	79	71	77	74	79
Five years	51	69	43	61	50	71	42	60	42	60	45	62	53	73	48	71	47	67	43	65	41	62	46	64
	35	64	27	56	32	66	25	54	26	56	28	55												
WALES	*(234)*		*(251)*		*(444)*		*(489)*		*(275)*		*(1,846)*		*(326)*		*(350)*		*(512)*		*(631)*		*(327)*		*(2,233)*	
One year	74	80	70	75	70	75	70	76	69	73	70	76	77	84	78	83	76	82	75	82	79	85	77	82
Five years	44	61	43	58	43	62	41	59	47	65	43	60	54	78	52	72	51	72	50	73	53	75	52	70
	28	59	29	56	26	58	26	54	30	61	27	53												

[1] Including those for whom no deprivation category was assigned. C - crude survival, R - relative survival (see Chapter 3). Number of patients contributing to each analysis in parentheses.

Table 36.2   Crude and relative survival (%) at one, five and ten years since diagnosis: NHS Region, deprivation category and calendar period of diagnosis

WOMEN	1971-75 Affluent C R	2 C R	3 C R	4 C R	Deprived C R	All¹ C R	1976-80 Affluent C R	2 C R	3 C R	4 C R	Deprived C R	All¹ C R
ENGLAND & WALES	*(1,236)*	*(1,431)*	*(1,650)*	*(1,640)*	*(1,364)*	*(8,952)*	*(1,788)*	*(1,967)*	*(2,154)*	*(2,164)*	*(1,831)*	*(10,384)*
One year	60 64	58 62	58 61	57 60	54 57	**57 60**	66 69	63 67	63 66	58 62	58 61	**61 65**
Five years	36 46	36 45	36 45	34 43	34 43	**35 43**	44 55	43 53	40 52	36 46	37 47	**39 50**
Ten years	26 41	24 38	25 40	24 39	23 38	**24 38**	32 51	31 49	28 46	26 43	25 41	**28 46**
ENGLAND	*(1,197)*	*(1,382)*	*(1,572)*	*(1,541)*	*(1,314)*	*(8,540)*	*(1,746)*	*(1,908)*	*(2,042)*	*(2,039)*	*(1,755)*	*(9,882)*
One year	60 64	58 62	58 61	57 60	54 57	**57 60**	66 70	64 67	63 67	59 63	58 62	**62 65**
Five years	36 46	36 44	35 45	35 44	34 42	**35 44**	44 56	43 54	40 52	37 47	37 47	**40 50**
Ten years	26 41	23 37	25 39	24 39	23 38	**24 38**	33 51	31 49	28 47	27 43	25 41	**28 46**
Northern & Yorkshire	*(85)*	*(113)*	*(187)*	*(214)*	*(286)*	*(1,197)*	*(164)*	*(216)*	*(311)*	*(346)*	*(467)*	*(1,551)*
One year	61 64	54 57	56 59	54 56	52 55	**55 58**	65 68	59 62	62 65	60 62	58 62	**61 64**
Five years	*40 49*	36 43	33 41	34 41	34 41	**34 42**	45 54	35 46	38 50	36 46	39 50	**38 50**
Ten years	*22 35*	24 37	23 37	24 36	25 39	**23 38**	34 53	27 43	27 47	27 42	26 45	**28 46**
Trent	*(83)*	*(126)*	*(167)*	*(191)*	*(164)*	*(810)*	*(130)*	*(165)*	*(201)*	*(257)*	*(212)*	*(998)*
One year	58 61	53 56	56 59	58 61	56 59	**57 61**	67 71	54 57	69 72	63 67	58 61	**62 66**
Five years	23 29	27 32	35 44	33 41	36 45	**33 41**	43 56	38 45	47 60	44 53	37 47	**41 53**
Ten years	*15 26*	19 29	23 35	21 33	26 42	**22 36**	*30 53*	28 43	33 53	30 48	27 45	**30 49**
Anglia & Oxford	*(144)*	*(162)*	*(143)*	*(119)*	*(53)*	*(714)*	*(233)*	*(231)*	*(181)*	*(139)*	*(56)*	*(878)*
One year	57 60	61 65	61 65	61 64	*39 41*	**58 62**	63 67	65 68	59 63	61 64	52 56	**62 66**
Five years	32 40	38 49	38 50	40 48	*32 38*	**37 47**	42 55	42 54	38 49	33 43	35 46	**39 51**
Ten years	24 38	28 43	28 46	26 42	*23 36*	**26 43**	30 49	30 50	29 47	25 43	28 43	**29 48**
North Thames	*(214)*	*(208)*	*(218)*	*(276)*	*(202)*	*(1,260)*	*(208)*	*(235)*	*(219)*	*(276)*	*(171)*	*(1,149)*
One year	53 55	66 69	57 60	58 61	56 60	**59 62**	64 68	67 71	62 65	63 67	61 65	**64 68**
Five years	31 38	43 53	35 46	39 49	35 45	**36 46**	44 54	48 58	42 55	40 52	39 51	**43 55**
Ten years	21 35	28 44	24 39	32 49	21 37	**25 43**	33 51	34 53	28 44	27 48	26 44	**30 49**
South Thames	*(313)*	*(269)*	*(290)*	*(249)*	*(157)*	*(1,411)*	*(390)*	*(327)*	*(328)*	*(267)*	*(116)*	*(1,528)*
One year	61 65	53 56	55 59	60 64	60 63	**58 62**	66 70	63 66	60 64	53 57	68 71	**61 65**
Five years	37 47	30 36	31 40	35 46	38 48	**34 44**	43 54	40 51	37 50	31 40	49 61	**39 51**
Ten years	26 41	18 29	22 35	21 38	25 44	**22 38**	31 49	27 44	28 46	24 40	*29 50*	**27 47**
South & West	*(196)*	*(301)*	*(301)*	*(148)*	*(55)*	*(1,102)*	*(310)*	*(396)*	*(390)*	*(192)*	*(81)*	*(1,423)*
One year	64 67	61 64	59 62	60 64	*72 76*	**62 66**	70 74	67 71	64 68	65 70	64 68	**66 71**
Five years	45 55	40 49	40 49	42 53	*43 53*	**42 53**	50 62	48 60	44 55	40 51	*43 53*	**45 58**
Ten years	34 53	25 41	30 45	31 50	*25 41*	**29 48**	39 61	34 54	29 49	30 49	27 47	**32 54**
West Midlands	*(30)*	*(43)*	*(72)*	*(58)*	*(81)*	*(720)*	*(115)*	*(125)*	*(148)*	*(190)*	*(203)*	*(788)*
One year	*80 86*	*45 48*	*69 72*	*49 52*	52 55	**58 61**	63 67	63 67	58 62	58 62	57 61	**60 64**
Five years	*59 67*	*31 34*	*44 55*	*30 38*	34 41	**36 44**	42 53	41 51	34 43	37 45	30 38	**36 46**
Ten years	*39 53*	*18 25*	*23 38*	*24 37*	*19 30*	**24 38**	29 43	32 50	23 38	25 39	20 31	**25 40**
North & West	*(132)*	*(160)*	*(194)*	*(286)*	*(316)*	*(1,326)*	*(196)*	*(213)*	*(264)*	*(372)*	*(449)*	*(1,567)*
One year	63 66	55 59	53 56	50 52	47 49	**53 56**	64 67	59 63	63 66	52 55	53 56	**57 61**
Five years	36 45	33 42	28 36	26 33	28 35	**29 37**	40 50	41 53	39 49	32 39	31 39	**36 45**
Ten years	29 42	23 36	19 31	17 27	20 32	**20 32**	29 43	29 49	25 42	22 36	21 34	**24 40**
WALES	*(39)*	*(49)*	*(78)*	*(99)*	*(50)*	*(412)*	*(42)*	*(59)*	*(112)*	*(125)*	*(76)*	*(502)*
One year	*61 63*	*64 65*	54 57	54 57	49 51	**56 59**	*52 56*	*57 60*	56 59	45 48	48 50	**52 55**
Five years	*38 46*	*47 57*	*38 47*	25 32	*38 41*	**34 43**	*36 46*	*40 49*	34 46	30 36	*34 42*	**34 44**
Ten years	*26 46*	*35 51*	*30 45*	19 29	22 30	**25 39**	*22 39*	*32 49*	20 36	22 33	*30 40*	**25 40**

¹ Including those for whom no deprivation category was assigned. C - crude survival, R - relative survival (see Chapter 3). Number of patients contributing to each analysis in parentheses.

	1981-85						1986-90						WOMEN
	Affluent	*2*	*3*	*4*	*Deprived*	*All¹*	*Affluent*	*2*	*3*	*4*	*Deprived*	*All¹*	
	C R	C R	C R	C R	C R	C R	C R	C R	C R	C R	C R	C R	
	(2,207)	*(2,472)*	*(2,636)*	*(2,614)*	*(2,302)*	*(12,502)*	*(2,439)*	*(2,990)*	*(3,007)*	*(2,897)*	*(2,276)*	*(13,779)*	**ENGLAND & WALES**
One year	70 74	70 74	68 72	65 69	61 65	67 71	71 75	70 73	67 71	67 71	65 68	68 72	*One year*
Five years	48 60	48 61	46 58	44 55	41 52	45 57	48 62	47 59	46 58	45 57	41 52	45 58	*Five years*
	35 55	34 55	31 52	30 48	28 47	32 52							
	(2,134)	*(2,371)*	*(2,502)*	*(2,420)*	*(2,207)*	*(11,845)*	*(2,328)*	*(2,820)*	*(2,811)*	*(2,655)*	*(2,155)*	*(12,897)*	**ENGLAND**
One year	70 74	70 74	68 72	65 69	62 65	67 71	70 74	70 74	66 70	67 70	64 68	68 71	*One year*
Five years	48 60	48 61	46 58	44 55	41 53	45 57	48 61	46 59	45 58	45 57	40 52	45 57	*Five years*
	35 56	34 55	32 53	29 48	28 47	32 52							
	(209)	*(284)*	*(369)*	*(431)*	*(547)*	*(1,863)*	*(235)*	*(287)*	*(398)*	*(431)*	*(517)*	*(1,873)*	**Northern & Yorkshire**
One year	72 76	69 73	67 71	64 68	63 67	66 71	69 73	69 73	65 68	65 69	62 65	66 70	*One year*
Five years	50 64	50 64	44 57	45 57	43 55	46 59	51 62	42 53	42 54	42 53	33 43	41 53	*Five years*
	35 58	38 62	28 48	28 46	30 50	31 52							
	(187)	*(218)*	*(289)*	*(332)*	*(258)*	*(1,291)*	*(164)*	*(234)*	*(292)*	*(335)*	*(276)*	*(1,301)*	**Trent**
One year	65 69	70 74	67 71	62 66	64 68	66 70	69 73	65 70	68 72	68 72	67 71	68 72	*One year*
Five years	46 57	47 60	47 59	41 52	41 52	45 57	46 61	42 53	48 62	47 59	41 52	45 58	*Five years*
	35 53	35 56	33 55	27 46	28 48	31 52							
	(250)	*(283)*	*(238)*	*(170)*	*(73)*	*(1,032)*	*(313)*	*(324)*	*(275)*	*(190)*	*(76)*	*(1,186)*	**Anglia & Oxford**
One year	70 74	70 74	72 76	60 63	*54 57*	68 72	68 73	69 73	56 60	61 65	*63 66*	64 69	*One year*
Five years	47 58	47 58	46 60	43 53	*42 50*	46 58	46 62	46 61	38 50	40 52	*38 49*	43 57	*Five years*
	33 52	35 52	35 59	31 47	*33 48*	34 54							
	(221)	*(239)*	*(224)*	*(244)*	*(220)*	*(1,161)*	*(235)*	*(306)*	*(338)*	*(343)*	*(256)*	*(1,525)*	**North Thames**
One year	76 80	70 73	74 78	69 74	67 70	71 76	73 77	72 76	74 78	75 78	66 70	73 77	*One year*
Five years	50 62	43 53	51 63	44 55	47 59	47 60	50 64	49 62	52 66	51 66	46 60	50 65	*Five years*
	34 54	31 50	37 59	28 48	33 53	33 55							
	(490)	*(423)*	*(374)*	*(282)*	*(121)*	*(1,756)*	*(410)*	*(486)*	*(365)*	*(309)*	*(151)*	*(1,757)*	**South Thames**
One year	68 73	70 74	70 74	70 74	69 73	69 74	75 79	72 76	70 74	68 72	70 74	71 76	*One year*
Five years	46 58	48 62	49 61	47 60	39 52	46 60	46 59	48 61	43 56	46 59	46 60	46 60	*Five years*
	35 56	34 56	31 53	34 57	*30 49*	33 56							
	(375)	*(465)*	*(436)*	*(216)*	*(78)*	*(1,623)*	*(371)*	*(548)*	*(474)*	*(287)*	*(102)*	*(1,795)*	**South & West**
One year	70 74	72 76	67 71	64 68	*63 66*	69 73	69 73	69 73	66 69	65 68	67 70	68 72	*One year*
Five years	52 64	51 65	44 57	44 54	*44 54*	48 62	50 62	48 60	47 59	41 52	42 51	47 59	*Five years*
	38 60	36 58	32 51	29 46	*38 53*	34 56							
	(156)	*(162)*	*(227)*	*(291)*	*(283)*	*(1,126)*	*(224)*	*(240)*	*(294)*	*(322)*	*(221)*	*(1,309)*	**West Midlands**
One year	68 72	66 69	63 67	68 72	57 61	65 68	63 67	68 72	67 71	65 68	66 70	66 70	*One year*
Five years	48 59	48 60	44 54	47 58	39 50	45 57	46 58	48 61	46 59	44 55	43 54	45 58	*Five years*
	36 53	34 51	30 46	34 52	26 44	32 50							
	(246)	*(297)*	*(345)*	*(454)*	*(627)*	*(1,993)*	*(376)*	*(395)*	*(375)*	*(438)*	*(556)*	*(2,151)*	**North & West**
One year	66 70	69 73	64 68	62 66	57 61	63 67	69 73	67 72	62 66	63 67	61 65	65 69	*One year*
Five years	45 56	48 59	44 55	40 51	36 48	41 53	47 60	45 58	43 55	42 53	41 52	44 56	*Five years*
	33 51	29 49	30 51	26 44	23 40	27 47							
	(73)	*(101)*	*(134)*	*(194)*	*(95)*	*(657)*	*(111)*	*(170)*	*(196)*	*(242)*	*(121)*	*(882)*	**WALES**
One year	*71 76*	*64 66*	*65 69*	*61 64*	*57 60*	63 66	*75 79*	*69 72*	*77 80*	*71 75*	*69 73*	73 77	*One year*
Five years	*48 59*	*46 56*	*42 55*	*44 56*	- -	43 54	*54 68*	*49 59*	*52 63*	*50 62*	*51 65*	51 64	*Five years*
	37 54	*37 54*	*23 43*	*31 52*	*28 40*	30 50							

¹   Including those for whom no deprivation category was assigned. C - crude survival, R - relative survival (see Chapter 3).
    Number of patients contributing to each analysis in parentheses.

Table 36.2 Crude and relative survival (%) at one, five and ten years since diagnosis: NHS Region, deprivation category and calendar period of diagnosis

ADULTS	1971-75												1976-80											
	Affluent		*2*		*3*		*4*		*Deprived*		*All*[1]		*Affluent*		*2*		*3*		*4*		*Deprived*		*All*[1]	
	C	R	C	R	C	R	C	R	C	R	C	R	C	R	C	R	C	R	C	R	C	R	C	R
ENGLAND & WALES	*(4,958)*		*(5,621)*		*(6,171)*		*(6,090)*		*(5,230)*		*(34,267)*		*(6,804)*		*(7,426)*		*(8,105)*		*(7,945)*		*(6,996)*		*(38,939)*	
One year	68	72	65	69	64	69	63	67	60	64	64	68	71	75	69	74	68	73	66	70	65	70	68	72
Five years	40	52	38	50	37	49	35	47	34	45	36	48	46	60	43	57	41	56	40	54	39	54	42	56
Ten years	26	46	25	44	24	44	22	42	21	40	23	42	31	53	28	51	27	50	26	48	25	48	27	50
ENGLAND	*(4,806)*		*(5,417)*		*(5,862)*		*(5,701)*		*(4,988)*		*(32,644)*		*(6,566)*		*(7,162)*		*(7,680)*		*(7,467)*		*(6,707)*		*(36,966)*	
One year	68	73	65	69	65	69	63	67	60	64	64	68	71	76	70	74	68	73	66	71	66	70	68	72
Five years	40	52	38	50	37	49	35	47	34	46	36	48	46	60	43	58	42	56	40	54	39	54	42	56
Ten years	26	45	25	44	24	43	23	42	21	40	23	42	31	54	29	51	27	51	26	49	25	48	27	50
Northern & Yorkshire	*(362)*		*(487)*		*(661)*		*(750)*		*(1,075)*		*(4,611)*		*(578)*		*(742)*		*(1,048)*		*(1,267)*		*(1,715)*		*(5,498)*	
One year	66	70	64	68	65	69	60	64	57	61	62	66	72	77	69	73	67	72	66	70	65	70	67	72
Five years	39	50	39	51	36	50	34	44	32	44	36	47	48	63	43	57	39	54	38	51	41	56	41	56
Ten years	24	42	26	44	22	42	21	37	21	41	23	42	32	56	29	52	25	49	25	47	26	51	27	50
Trent	*(321)*		*(470)*		*(629)*		*(787)*		*(682)*		*(3,201)*		*(521)*		*(665)*		*(810)*		*(946)*		*(826)*		*(3,893)*	
One year	66	70	63	67	63	68	67	71	61	65	64	68	71	76	67	71	72	77	70	74	66	71	69	74
Five years	37	47	35	46	38	50	36	48	32	44	36	48	45	59	41	56	47	63	43	58	40	55	43	59
Ten years	23	39	24	39	24	42	23	42	21	39	24	42	28	50	27	49	30	56	28	54	25	49	28	52
Anglia & Oxford	*(629)*		*(685)*		*(639)*		*(484)*		*(199)*		*(2,991)*		*(897)*		*(901)*		*(758)*		*(503)*		*(249)*		*(3,458)*	
One year	68	72	67	71	66	70	66	70	60	64	66	71	70	74	72	76	68	73	66	71	69	75	70	74
Five years	42	56	41	54	39	52	39	51	38	51	40	53	45	60	45	58	44	58	40	54	44	61	44	59
Ten years	28	50	28	48	26	47	25	47	23	41	26	48	32	55	30	53	31	55	28	51	28	52	30	54
North Thames	*(796)*		*(816)*		*(831)*		*(944)*		*(752)*		*(4,670)*		*(753)*		*(860)*		*(827)*		*(968)*		*(736)*		*(4,305)*	
One year	66	70	67	71	66	70	65	69	66	70	66	70	72	76	72	76	68	73	68	73	69	75	70	75
Five years	37	47	39	50	41	54	39	52	38	50	38	51	50	64	44	59	42	56	42	58	43	59	44	59
Ten years	24	39	26	45	26	46	28	51	23	44	25	46	37	62	29	53	26	47	27	53	27	56	29	54
South Thames	*(1,208)*		*(1,059)*		*(1,050)*		*(984)*		*(562)*		*(5,328)*		*(1,497)*		*(1,246)*		*(1,202)*		*(1,006)*		*(445)*		*(5,743)*	
One year	68	73	63	67	62	67	63	67	61	66	64	69	70	75	69	73	67	72	66	70	71	76	68	73
Five years	39	51	34	46	32	44	35	48	35	48	36	48	44	58	41	55	41	56	40	55	46	63	42	57
Ten years	25	44	21	38	20	39	19	38	20	40	22	41	28	51	26	48	26	51	26	52	29	58	27	51
South & West	*(918)*		*(1,224)*		*(1,206)*		*(588)*		*(212)*		*(4,533)*		*(1,176)*		*(1,528)*		*(1,492)*		*(740)*		*(295)*		*(5,429)*	
One year	73	77	68	72	67	71	64	68	68	72	68	73	74	79	72	77	68	73	70	76	72	76	71	76
Five years	47	61	42	54	40	53	36	48	42	55	41	55	49	64	47	64	43	60	45	60	45	60	46	62
Ten years	31	53	27	48	27	49	24	46	28	51	27	50	34	60	31	57	30	56	31	57	30	58	31	58
West Midlands	*(121)*		*(136)*		*(222)*		*(251)*		*(288)*		*(2,742)*		*(442)*		*(463)*		*(627)*		*(754)*		*(820)*		*(3,144)*	
One year	68	72	59	63	71	75	64	69	56	59	63	67	69	74	65	69	67	71	64	68	65	70	66	71
Five years	39	51	37	45	41	53	37	50	30	41	35	46	44	58	40	52	37	51	40	52	36	50	39	53
Ten years	26	43	18	32	23	39	24	44	21	37	22	40	30	51	27	47	23	44	26	46	24	44	26	46
North & West	*(451)*		*(540)*		*(624)*		*(913)*		*(1,218)*		*(4,568)*		*(702)*		*(757)*		*(916)*		*(1,283)*		*(1,621)*		*(5,496)*	
One year	66	70	59	63	56	60	56	60	57	61	58	62	67	71	65	70	65	69	60	65	59	63	63	67
Five years	37	47	33	43	28	38	27	37	31	41	30	40	43	55	40	53	37	50	34	46	33	45	36	49
Ten years	25	43	22	39	17	32	18	33	19	35	19	35	29	49	27	48	24	46	21	41	20	40	23	43
WALES	*(152)*		*(204)*		*(309)*		*(389)*		*(242)*		*(1,623)*		*(238)*		*(264)*		*(425)*		*(478)*		*(289)*		*(1,973)*	
One year	61	65	58	62	59	63	57	61	56	60	59	63	67	71	62	66	65	69	59	63	58	61	62	66
Five years	37	50	33	45	37	49	31	40	33	42	34	45	36	49	36	48	37	51	37	49	36	48	36	50
Ten years	25	46	23	41	26	46	21	36	20	35	23	41	21	42	23	42	21	40	22	42	28	48	23	44

---

[1] Including those for whom no deprivation category was assigned. C - crude survival, R - relative survival (see Chapter 3). Number of patients contributing to each analysis in parentheses.

	1981-85						**1986-90**						**ADULTS**
	Affluent	*2*	*3*	*4*	*Deprived*	*All*[1]	*Affluent*	*2*	*3*	*4*	*Deprived*	*All*[1]	
	C R	C R	C R	C R	C R	C R	C R	C R	C R	C R	C R	C R	
	(8,551)	(9,201)	(9,523)	(9,526)	(7,909)	(45,682)	(9,098)	(10,589)	(10,706)	(10,327)	(8,002)	(49,318)	**ENGLAND & WALES**
One year	76 80	74 79	72 78	70 76	68 73	72 77	76 81	76 81	74 79	72 77	71 75	74 79	*One year*
Five years	49 64	47 63	46 63	44 60	43 59	46 62	50 66	49 65	47 63	45 61	43 59	47 63	*Five years*
	34 58	31 56	30 55	29 53	28 53	30 55							
	(8,244)	(8,849)	(8,945)	(8,843)	(7,539)	(43,179)	(8,661)	(10,069)	(9,998)	(9,454)	(7,554)	(46,203)	**ENGLAND**
One year	76 81	74 79	73 78	71 76	69 74	72 77	76 81	76 81	74 78	72 77	70 75	74 79	*One year*
Five years	49 65	48 63	47 63	45 60	43 59	46 62	50 65	49 65	47 63	45 61	43 58	47 63	*Five years*
	34 58	31 56	30 56	29 53	28 53	30 55							
	(783)	(1,042)	(1,274)	(1,545)	(1,847)	(6,576)	(818)	(1,034)	(1,257)	(1,424)	(1,660)	(6,221)	**Northern & Yorkshire**
One year	76 81	74 79	72 78	71 76	69 75	72 77	74 79	73 78	70 76	68 74	68 73	70 75	*One year*
Five years	49 66	49 66	45 63	44 61	44 61	46 62	51 66	45 61	46 65	41 58	39 55	43 59	*Five years*
	35 62	33 61	28 55	27 54	28 56	29 55							
	(616)	(865)	(1,070)	(1,202)	(877)	(4,662)	(672)	(933)	(1,115)	(1,203)	(1,015)	(4,952)	**Trent**
One year	72 77	72 76	68 73	69 74	68 73	70 75	77 83	77 82	73 78	71 77	71 77	73 78	*One year*
Five years	47 63	46 62	43 58	44 60	44 62	45 60	49 67	48 66	47 66	45 63	44 62	46 63	*Five years*
	33 59	30 57	29 54	29 57	28 58	30 55							
	(1,044)	(1,053)	(942)	(658)	(289)	(4,065)	(1,163)	(1,198)	(1,041)	(708)	(283)	(4,422)	**Anglia & Oxford**
One year	75 80	74 79	73 79	68 74	69 74	73 78	74 80	74 80	72 77	69 75	68 74	72 78	*One year*
Five years	51 67	49 65	47 65	44 62	47 64	48 64	48 67	50 68	43 61	44 62	42 59	47 63	*Five years*
	35 61	33 60	31 60	29 56	31 60	32 58							
	(888)	(921)	(887)	(886)	(762)	(4,410)	(919)	(1,178)	(1,229)	(1,207)	(960)	(5,645)	**North Thames**
One year	79 84	74 79	75 81	73 78	73 78	75 80	76 82	80 85	78 84	77 83	75 81	78 83	*One year*
Five years	51 67	46 63	50 67	46 63	47 66	48 65	50 68	51 69	50 69	50 70	47 67	50 67	*Five years*
	34 61	30 59	32 62	31 60	32 62	32 59							
	(1,792)	(1,426)	(1,263)	(1,087)	(445)	(6,234)	(1,672)	(1,682)	(1,369)	(1,163)	(504)	(6,505)	**South Thames**
One year	76 81	74 79	75 80	74 80	73 79	75 80	80 85	77 83	76 81	73 79	76 82	77 82	*One year*
Five years	47 64	47 64	50 69	47 67	43 61	47 64	49 67	48 68	47 67	45 63	46 65	47 64	*Five years*
	31 59	31 57	33 63	31 62	29 60	31 58							
	(1,484)	(1,828)	(1,536)	(860)	(281)	(6,158)	(1,433)	(1,997)	(1,735)	(1,126)	(376)	(6,727)	**South & West**
One year	75 80	77 82	73 78	72 78	71 75	74 80	77 82	76 81	73 79	73 78	71 76	75 80	*One year*
Five years	51 68	49 67	45 63	47 64	46 62	48 65	51 68	50 69	48 67	46 64	43 60	49 65	*Five years*
	35 63	32 61	29 56	30 58	35 61	32 58							
	(665)	(702)	(839)	(1,069)	(1,023)	(4,327)	(810)	(862)	(1,017)	(1,156)	(880)	(4,752)	**West Midlands**
One year	74 79	73 78	72 77	70 75	65 70	71 76	72 77	78 83	73 79	73 78	69 75	73 78	*One year*
Five years	47 62	48 64	45 61	44 60	39 55	44 60	50 67	51 68	46 65	44 61	44 61	47 63	*Five years*
	33 57	33 59	31 56	29 54	25 52	30 54							
	(972)	(1,012)	(1,134)	(1,536)	(2,015)	(6,747)	(1,174)	(1,185)	(1,235)	(1,467)	(1,876)	(6,979)	**North & West**
One year	73 79	71 77	71 77	67 72	65 70	69 74	75 80	73 79	72 77	70 75	68 73	71 76	*One year*
Five years	49 65	44 61	48 66	41 57	40 56	44 60	51 69	47 66	45 63	43 61	41 59	45 61	*Five years*
	35 61	28 54	31 61	25 51	25 50	28 53							
	(307)	(352)	(578)	(683)	(370)	(2,503)	(437)	(520)	(708)	(873)	(448)	(3,115)	**WALES**
One year	73 79	68 72	69 74	68 73	66 70	68 73	76 82	75 79	76 81	74 80	76 82	76 80	*One year*
Five years	45 61	44 57	42 60	42 59	43 57	43 58	54 74	51 67	51 69	50 69	52 72	52 68	*Five years*
	30 57	31 56	25 54	28 53	30 56	28 53							

[1] Including those for whom no deprivation category was assigned. C - crude survival, R - relative survival (see Chapter 3). Number of patients contributing to each analysis in parentheses.

Fig. 36.1   Relative survival up to ten years: trends by calendar period of diagnosis (1971-90) and NHS Region

ADULTS

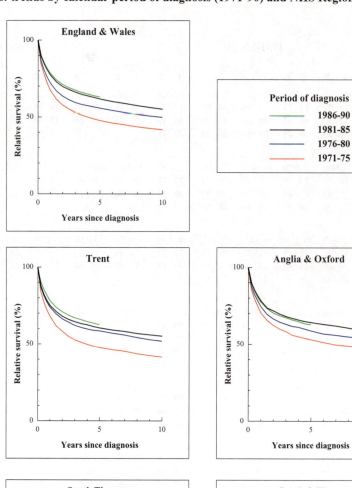

Period of diagnosis
- 1986-90
- 1981-85
- 1976-80
- 1971-75

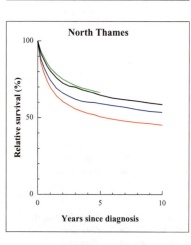

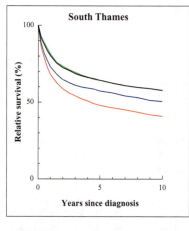

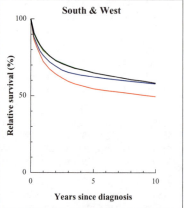

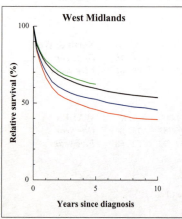

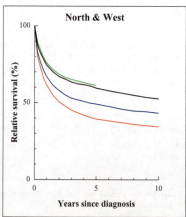

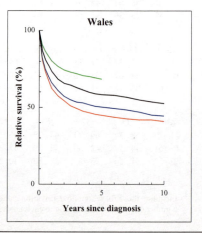

Fig. 36.2   Relative survival up to five years, by deprivation category and NHS Region: patients diagnosed 1986-90

ADULTS

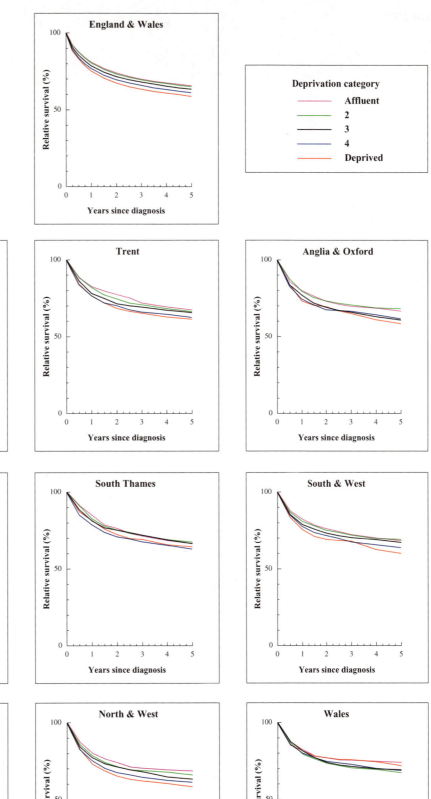

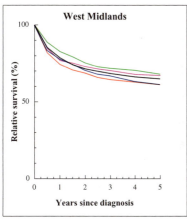

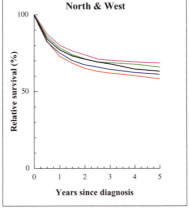

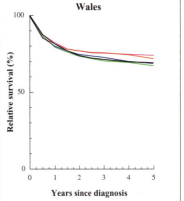

Fig. 36.3 Relative survival at five years by deprivation category, period of diagnosis (1981-90) and NHS Region

ADULTS

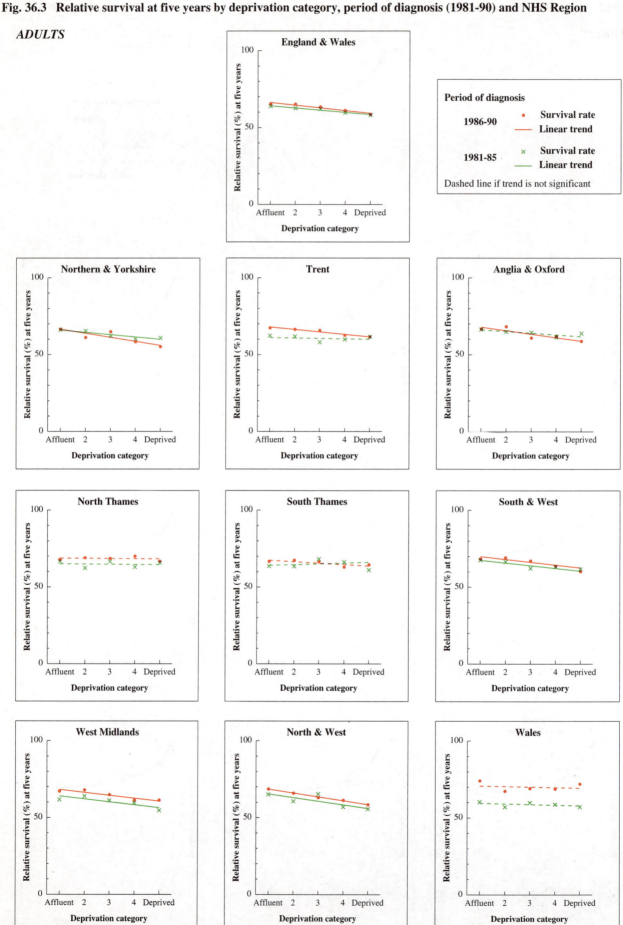

Fig. 36.4 Relative survival (%) by NHS Region and sex:
one-year and five-year survival for patients diagnosed 1986-90, with 95% confidence intervals

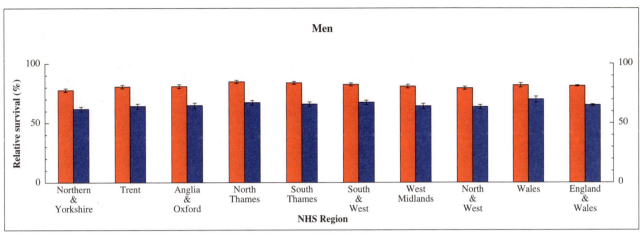

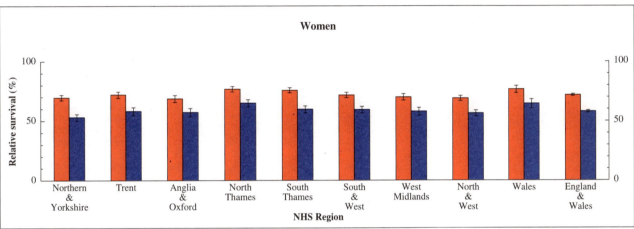

Fig. 36.5 Relative survival (%) at five years, with 95% confidence interval, by age at diagnosis and sex:
England and Wales, patients diagnosed 1986-90

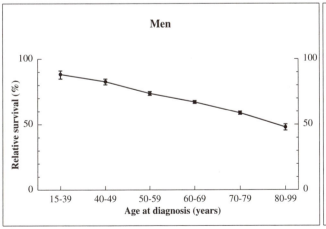

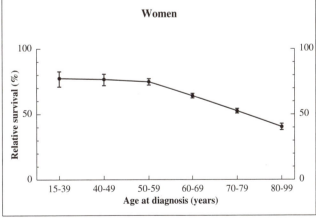

Chapter 37

KIDNEY

Cancers of the kidney, renal pelvis or ureter rank as the tenth most common malignancy in men in England and Wales, among whom there are 2,600 new cases each year (2.5% of all cancers). Among women, there are 1,600 cases a year (fifteenth, 1.5%). Incidence in both sexes increased by more than 10% every five years in the 20 years to 1990, reaching 10 per 100,000 per year in men and about 5 in women. Kidney cancer is rare before age 40, and twice as common in men at all ages above 40. Mortality has also been increasing, but less rapidly than for incidence: the difference in the trends probably reflects improvements in survival, although improved registration may account for part of the increase in incidence. New cases and deaths have been increasing more rapidly for women than for men.

Tobacco smoking doubles the risk of renal cancer, and about a third of cases are attributable to it. Phenacetin-containing analgesics, cadmium exposure and occupational exposure to aromatic hydrocarbons are also known risk factors. Clear cell adenocarcinomas comprise 80% of malignant kidney tumours, and transitional cell carcinomas of the renal pelvis and ureter account for most of the remainder. Renal tumours spread through the lymphatics, and metastasise to lung, liver and bone through the bloodstream, often many years after nephrectomy. Diagnosis usually follows symptoms of pain or haematuria, but improvements in survival may arise from the detection of smaller, clinically silent tumours with abdominal ultrasound.

The survival analyses reported here include 49,000 adults with primary malignant tumours of the kidney diagnosed in England and Wales during 1971-90 and followed up to the end of 1995, about 82% of those eligible (Table 37.1). Over 1,400 tumours arising mainly in children were excluded. For 4% of the eligible adults, the vital status at 6 July 1997 was unknown, and 11% were excluded because their survival was either zero or unknown: this group accounted for 13% of adults diagnosed during 1986-90. Over 80% were renal tumours, and tumours of the renal pelvis and ureter represented about 8% and 6% respectively (Table 37A). Over 40% of the tumours were clear cell or renal cell adenocarcinomas, and a further 12% papillary, tubular or unspecified adenocarcinomas: less than 0.2% were described as nephroblastoma. Transitional cell carcinoma accounted for 17%, and 25% had non-specific morphology codes.

Survival trends

Relative survival for men diagnosed in England and Wales

during 1986-90 was 58% at one year and 40% at five years, about 5% higher than for women (Table 37.2). Five-year survival improved during the 1970s in England and Wales as a whole, but there was considerable regional variation in the pattern of improvement, and gains were smaller during the 1980s (Figure 37.1).

National average improvements in survival between successive five-year periods up to 1986-90 have been slightly larger for men than for women. One-year survival increased by an average of 3.8% every five years for men and 2.9% for women, while the corresponding gains in five-year survival were 3.3% and 2.6%, after adjustment to the age distribution of men and women diagnosed during 1986-90 (see Table 4.4). Survival was the same for men and women diagnosed in the early 1970s, but 3% higher for men diagnosed during 1986-90.

Survival trends in the NHS Regions have been diverse. Average gains every five years in one-year survival have ranged from 1% to 6% for men, and from 2% to 4% for women. Age-standardised one-year survival rates for patients diagnosed during 1986-90 vary across the range 52-64% for men and 45-60% for women (see Table-Figure 4.5Q). The regional range of improvement in five-year survival is also wide. In Anglia & Oxford Region, where survival for patients diagnosed during 1971-75 was amongst the highest in both sexes, five-year survival has fluctuated, and is now among the lowest; improvements have also been smaller than average in Wales.

Survival falls steadily with advancing age at diagnosis in both men and women (Figure 37.5). Relative survival at five years is 50% or more for men aged less than 50 at diagnosis, about 5% higher than for women. Survival is lower for patients aged 50 or more, but the differences between men and women are generally smaller.

Deprivation

The deprivation gradient for survival from kidney cancer in England and Wales has been largely consistent since the early 1970s for men and for women, although not in all regions, and it has generally been greater for one-year survival than for survival at five years after diagnosis (Table 37.2, Figure 37.2). There has been no consistent difference in survival between deprivation groups in Trent or South & West since the late 1970s. The significant gradient in five-year survival in England and Wales as a whole was similar for patients diagnosed in the early 1980s and those diagnosed during 1986-90 (Figure 37.3).

Table 37A   Sub-site distribution of cancer of the kidney and other urinary organs

ICD-9 code	Site description	1971-75 No.	1971-75 %	1976-80 No.	1976-80 %	1981-85 No.	1981-85 %	1986-90 No.	1986-90 %	Total No.	Total %
189.0	Kidney, except pelvis	8,210	82.9	9,113	82.1	10,692	80.8	12,803	84.4	**40,818**	**82.6**
189.1	Renal pelvis	786	7.9	973	8.8	1,158	8.8	1,020	6.7	**3,937**	**8.0**
189.2	Ureter[1]	615	6.2	698	6.3	876	6.6	893	5.9	**3,082**	**6.2**
189.3	Urethra[2]	–	–	102	0.9	246	1.9	224	1.5	**572**	**1.2**
189.4	Paraurethral glands	–	–	1	0.0	4	0.0	1	0.0	**6**	**0.0**
189.8	Other	–	–	17	0.2	62	0.5	114	0.8	**193**	**0.4**
189.9	Site unspecified	293	3.0	198	1.8	194	1.5	115	0.8	**800**	**1.6**
	Total	**9,904**		**11,102**		**13,232**		**15,170**		**49,408**	

[1] Excludes ureteric orifice of bladder.
[2] Excludes urethral orifice of bladder.

Among almost 13,000 patients of known deprivation status who were diagnosed in England and Wales during 1981-85, there was a significant 10% deficit in one-year survival for the most deprived group of patients relative to the most affluent group, and a 5% gap for survival at five years (see Table 4.9). For the 15,000 patients diagnosed during 1986-90, the deprivation gap was 7% for one-year survival and 4% for five-year survival.

Relative survival among affluent and deprived groups varies widely between the NHS Regions, and is higher for deprived patients in some regions than for affluent patients in others. Thus for patients diagnosed during 1986-90, five-year survival for the affluent group varied from 47% in West Midlands to 32% in Wales, while for the deprived group the range was from 42% in North Thames to 22% in Anglia & Oxford. Regional survival estimates for each deprivation category depend on 100-500 cases, and are generally precise. The regional comparisons of relative survival are adjusted for differences in background mortality between the regions, the deprivation categories and the sexes (see Chapter 3).

Regardless of the level of survival, it is difficult to explain the presence of a gradient in some regions and its absence in others as the result of regional differences in the definition or registration of renal cancer, all the more so because for most of the NHS Regions the data for patients diagnosed during 1971-90 come from more than one cancer registry (see Chapter 2). The explanation seems more likely to involve regional differences in the stage of disease at diagnosis and in the management of patients with cancer of the kidney in the various deprivation groups.

International comparison

Survival for patients diagnosed in England and Wales during 1986-90 was similar to that in Scotland, but 10% below the average for Europe, and 20% below survival rates in the USA (see Table 4.12).

Table 37.1 Ineligible and excluded records, and patients included in analyses, by calendar period of diagnosis

	Calendar period of diagnosis															
	1971-75			**1976-80**			**1981-85**			**1986-90**			**Overall**		
Total registered	**12,045**			**13,595**			**16,382**			**19,274**			**61,296**		
Ineligible	*Records*	*Patients*	*%*	*Records*	*Patients*	*%*	*Records*	*Patients*	*%*	*Records*	*Patients*	*%*	*Records*	*Patients*	*%*
Incomplete data[1]	13	13		33	33		64	64		60	60		170	**170**	
Resident outside England and Wales	105	104		107	105		70	63		75	63		357	**335**	
In situ neoplasm[2]	21	20		6	6		0	0		0	0		27	**26**	
Benign or uncertain behaviour[2]	95	68		89	77		1	1		1	1		186	**147**	
Metastatic[2]	165	164		53	53		0	0		1	1		219	**218**	
Otherwise ineligible[3]	-	-		-	-		-	-		-	-		-	-	
Lymphoma[4]	6	6		4	4		3	0		3	0		16	**10**	
Leukaemia or myeloma[4]	0	0		0	0		3	1		3	2		6	**3**	
		375			278			129			127			909	
Total eligible		**11,670**	**100**		**13,317**	**100**		**16,253**	**100**		**19,147**	**100**		**60,387**	**100**
Aged under 15 or 100 years or over at diagnosis	402	391	3.4	384	365	2.7	350	326	2.0	378	359	1.9	1,514	**1,441**	2.4
Vital status unknown[5]	404	332	2.8	582	494	3.7	619	569	3.5	1,101	1,013	5.3	2,706	**2,408**	4.0
Duplicate registration[6]	0	0	0.0	0	0	0.0	0	0	0.0	0	0	0.0	0	**0**	0.0
Multiple primary[7]	1	0	0.0	17	14	0.1	66	56	0.3	93	84	0.4	177	**154**	0.3
Synchronous tumours[8]	39	24	0.2	43	24	0.2	61	46	0.3	86	40	0.2	229	**134**	0.2
Sex not known	0	0	0.0	0	0	0.0	0	0	0.0	0	0	0.0	0	**0**	0.0
Sex incompatible with site code[9]	-	-	-	-	-	-	-	-	-	-	-	-	-	**-**	-
Invalid dates[10]	0	0	0.0	1	1	<0.1	6	5	<0.1	13	6	<0.1	20	**12**	<0.1
Zero survival[11]	1,116	1,019	8.7	1,419	1,317	9.9	2,159	2,019	12.4	2,624	2,475	12.9	7,318	**6,830**	11.3
Total exclusions		**1,766**	**15.1**		**2,215**	**16.6**		**3,021**	**18.6**		**3,977**	**20.8**		**10,979**	**18.2**
Patients included in analysis		**9,904**	**84.9**		**11,102**	**83.4**		**13,232**	**81.4**		**15,170**	**79.2**		**49,408**	**81.8**

[1] Main data item(s) invalid or incompatible with one another: name, sex, date of birth, date of diagnosis and (if present) date of death, postcode, site, morphology and behaviour

[2] Tumour either *in situ* (behaviour code 2), benign (0), uncertain if benign or malignant (1) or metastatic (6 or 9)

[3] Not applicable for this malignancy

[4] Morphology that of lymphoma (included with non-Hodgkin lymphoma, Ch. 41) or myeloma (included with multiple myeloma, Ch. 43), or leukaemia (excluded)

[5] Tracing of vital status at National Health Service Central Register incomplete at 6 July 1997

[6] Same site code, sex, cancer registry and cancer registry number as an earlier registration

[7] Same site code and person number as an earlier registration(s): mostly confirmed multiple primary tumours at this site, some unresolved duplicate registrations

[8] Same site code, sex, date of birth and date of diagnosis as another registration(s): mostly synchronous or (in paired organs) bilateral tumours in same anatomic site in one individual, not previously linked: also some duplicate registrations

[9] Not applicable for this malignancy

[10] Impossible sequence of dates (birth, diagnosis and, if present, emigration or death); or date of death after 6 July 1997

[11] Date of diagnosis same as date of death: some are patients who did die on the day of diagnosis, but most were registered solely from a death certificate (death certificate only, or DCO), and their survival time is thus unknown

Table 37.2   Crude and relative survival (%) at one, five and ten years since diagnosis: NHS Region, deprivation category and calendar period of diagnosis

MEN	1971-75 Affluent C R	2 C R	3 C R	4 C R	Deprived C R	All[1] C R	1976-80 Affluent C R	2 C R	3 C R	4 C R	Deprived C R	All[1] C R
ENGLAND & WALES	*(1,009)*	*(1,075)*	*(1,110)*	*(1,011)*	*(867)*	*(6,225)*	*(1,388)*	*(1,366)*	*(1,425)*	*(1,341)*	*(1,169)*	*(6,990)*
One year	52 54	48 50	46 48	48 51	39 41	48 50	52 54	50 52	48 50	47 49	46 49	49 51
Five years	28 33	26 31	25 30	27 33	19 25	26 31	30 36	29 35	26 32	26 33	25 32	27 34
Ten years	18 25	16 24	17 25	16 26	13 21	16 25	20 30	19 28	16 26	17 28	16 26	18 28
ENGLAND	*(966)*	*(1,020)*	*(1,028)*	*(923)*	*(824)*	*(5,849)*	*(1,335)*	*(1,312)*	*(1,332)*	*(1,230)*	*(1,116)*	*(6,575)*
One year	52 54	48 50	47 49	48 50	40 42	48 50	52 54	50 52	48 50	47 49	47 49	49 51
Five years	28 33	26 31	25 30	27 33	19 24	26 31	30 36	29 35	26 32	27 33	26 33	27 34
Ten years	18 25	16 25	17 25	16 27	12 20	16 25	20 30	19 28	16 26	17 28	16 27	18 28
Northern & Yorkshire	*(70)*	*(101)*	*(115)*	*(122)*	*(174)*	*(824)*	*(114)*	*(140)*	*(171)*	*(191)*	*(285)*	*(924)*
One year	54 56	46 48	44 46	47 49	38 39	46 48	55 58	50 52	46 48	45 47	45 48	48 50
Five years	24 30	24 29	21 26	22 28	14 17	21 27	34 42	30 36	23 28	23 30	22 29	26 33
Ten years	17 28	15 21	12 22	14 25	8 15	14 23	*23 39*	18 27	14 22	13 23	14 25	16 26
Trent	*(76)*	*(90)*	*(103)*	*(111)*	*(96)*	*(538)*	*(100)*	*(126)*	*(144)*	*(163)*	*(125)*	*(682)*
One year	44 45	49 51	39 41	48 50	33 34	44 46	51 52	48 50	49 51	47 48	42 43	48 50
Five years	24 26	23 28	25 31	26 32	19 24	25 30	31 36	30 34	23 28	24 30	23 30	26 32
Ten years	15 21	17 23	13 21	14 22	11 18	15 23	20 28	21 29	15 23	15 25	17 27	17 27
Anglia & Oxford	*(121)*	*(133)*	*(136)*	*(100)*	*(43)*	*(604)*	*(172)*	*(182)*	*(141)*	*(104)*	*(37)*	*(661)*
One year	54 56	50 52	52 54	56 58	*38 40*	53 55	59 61	48 51	49 52	44 46	53 57	52 54
Five years	36 42	32 40	31 35	31 38	21 27	32 39	34 39	30 36	32 40	28 35	*18 23*	31 38
Ten years	25 36	23 40	18 25	*18 31*	17 25	21 32	20 30	21 32	18 30	18 28	*15 20*	20 30
North Thames	*(168)*	*(164)*	*(163)*	*(148)*	*(118)*	*(841)*	*(172)*	*(122)*	*(159)*	*(107)*	*(118)*	*(700)*
One year	56 57	45 46	49 50	49 51	42 44	50 52	48 50	47 49	40 42	54 57	55 58	49 52
Five years	34 39	22 27	26 31	25 31	20 26	27 32	30 34	25 32	23 29	27 35	29 38	27 34
Ten years	23 33	13 20	17 27	17 27	16 26	18 28	23 33	21 31	14 21	*16 30*	17 32	19 30
South Thames	*(220)*	*(168)*	*(176)*	*(150)*	*(84)*	*(860)*	*(279)*	*(205)*	*(198)*	*(136)*	*(64)*	*(932)*
One year	48 50	44 46	43 45	49 51	44 46	47 49	50 52	46 48	48 50	44 46	*40 43*	47 50
Five years	20 23	21 25	27 32	27 34	23 28	24 28	28 34	27 33	26 33	25 31	22 28	27 33
Ten years	14 18	11 17	18 28	13 22	*10 17*	14 22	19 29	16 27	15 26	17 28	*14 21*	17 27
South & West	*(178)*	*(230)*	*(173)*	*(92)*	*(34)*	*(788)*	*(220)*	*(290)*	*(234)*	*(146)*	*(50)*	*(975)*
One year	54 57	48 50	48 50	47 49	*36 37*	50 52	56 58	50 52	48 51	48 51	*47 49*	51 54
Five years	33 39	27 33	24 30	30 37	*16 19*	29 35	29 36	30 36	29 36	28 35	*28 35*	30 37
Ten years	15 22	18 29	19 29	*22 34*	14 16	18 29	20 30	18 27	18 31	17 27	*15 27*	18 29
West Midlands	*(44)*	*(33)*	*(48)*	*(65)*	*(75)*	*(602)*	*(104)*	*(110)*	*(137)*	*(164)*	*(165)*	*(687)*
One year	*53 55*	*61 63*	*47 49*	36 38	37 38	49 51	43 45	60 62	48 50	52 54	48 50	50 52
Five years	*37 40*	*37 44*	*18 23*	26 30	23 30	28 35	31 38	30 36	24 30	31 38	28 35	29 36
Ten years	*29 35*	*14 19*	*16 23*	*18 28*	16 25	20 29	24 33	18 27	17 28	23 35	17 27	20 30
North & West	*(89)*	*(101)*	*(114)*	*(135)*	*(200)*	*(792)*	*(174)*	*(137)*	*(148)*	*(219)*	*(272)*	*(1,014)*
One year	45 47	48 50	40 42	43 45	37 39	44 45	48 50	43 45	46 48	39 41	41 44	45 47
Five years	17 19	24 28	15 18	25 31	16 21	21 26	23 28	26 32	19 23	23 28	26 34	24 31
Ten years	5 7	16 21	11 15	14 24	10 16	13 20	13 22	17 24	12 21	15 26	15 27	15 25
WALES	*(43)*	*(55)*	*(82)*	*(88)*	*(43)*	*(376)*	*(53)*	*(54)*	*(93)*	*(111)*	*(53)*	*(415)*
One year	45 46	51 52	44 46	48 51	*33 35*	47 49	49 51	53 54	40 42	42 44	*40 42*	46 48
Five years	23 27	24 30	29 34	24 30	29 31	27 34	27 31	23 28	24 30	21 26	*18 21*	25 30
Ten years	*17 23*	*5 7*	17 25	12 20	22 28	16 25	*21 27*	*14 21*	18 27	17 25	*12 21*	18 26

[1] Including those for whom no deprivation category was assigned. C - crude survival, R - relative survival (see Chapter 3).
Number of patients contributing to each analysis in parentheses.

MEN

1981-85 | C = crude survival, R = relative survival. Values listed as C R per cell. Column headings: Affluent, 2, 3, 4, Deprived, All[1].

	Affluent C R	2 C R	3 C R	4 C R	Deprived C R	All[1] C R
ENGLAND & WALES (n)	*(1,770)*	*(1,737)*	*(1,765)*	*(1,647)*	*(1,272)*	*(8,374)*
One year	56 58	56 59	53 55	51 53	46 48	**53 55**
Five years	33 39	32 39	29 36	30 37	26 34	**30 37**
	21 31	21 32	20 31	19 31	17 28	**20 31**
ENGLAND (n)	*(1,724)*	*(1,665)*	*(1,639)*	*(1,536)*	*(1,210)*	*(7,922)*
One year	56 58	56 59	54 56	51 54	45 48	**53 56**
Five years	33 39	32 39	30 37	30 37	26 34	**30 38**
	21 31	21 32	20 32	19 32	17 28	**20 31**
Northern & Yorkshire (n)	*(165)*	*(184)*	*(215)*	*(243)*	*(313)*	*(1,138)*
One year	60 62	54 57	47 50	49 52	43 46	**50 53**
Five years	35 41	32 40	28 35	25 33	24 32	**29 36**
	27 41	21 34	19 33	18 31	16 28	**20 32**
Trent (n)	*(155)*	*(188)*	*(210)*	*(219)*	*(139)*	*(918)*
One year	53 55	53 56	49 51	47 49	47 50	**51 53**
Five years	33 38	24 30	26 33	29 37	27 34	**28 35**
	19 28	16 26	18 28	15 27	16 26	**17 27**
Anglia & Oxford (n)	*(241)*	*(203)*	*(203)*	*(138)*	*(55)*	*(862)*
One year	55 57	56 58	58 60	58 62	*51 54*	**57 59**
Five years	34 41	34 40	38 49	38 47	*33 42*	**36 44**
	25 36	23 35	26 43	25 41	*27 42*	**25 38**
North Thames (n)	*(166)*	*(149)*	*(151)*	*(129)*	*(87)*	*(693)*
One year	63 65	55 58	59 61	51 55	55 58	**58 61**
Five years	34 42	27 33	31 38	35 47	29 40	**32 40**
	19 31	19 28	22 33	24 41	*20 38*	**21 33**
South Thames (n)	*(348)*	*(242)*	*(198)*	*(157)*	*(58)*	*(1,043)*
One year	58 61	57 60	59 62	58 61	47 49	**58 61**
Five years	34 42	31 39	37 46	31 39	29 35	**34 42**
	20 33	20 33	26 43	21 35	*25 35*	**22 35**
South & West (n)	*(287)*	*(341)*	*(290)*	*(156)*	*(52)*	*(1,154)*
One year	55 56	58 60	51 53	48 50	*56 58*	**55 57**
Five years	31 37	36 45	24 31	30 37	*32 38*	**31 39**
	19 30	24 37	15 26	19 36	*26 37*	**20 32**
West Midlands (n)	*(157)*	*(141)*	*(170)*	*(228)*	*(189)*	*(896)*
One year	58 59	61 64	54 56	51 53	36 38	**52 55**
Five years	32 38	36 46	28 36	30 38	22 28	**30 37**
	22 32	26 42	18 32	18 32	14 22	**20 31**
North & West (n)	*(205)*	*(217)*	*(202)*	*(266)*	*(317)*	*(1,218)*
One year	45 47	51 54	49 52	45 47	40 43	**47 49**
Five years	25 31	29 37	26 34	23 31	23 32	**26 33**
	17 26	19 32	16 28	17 27	13 23	**16 26**
WALES (n)	*(46)*	*(72)*	*(126)*	*(111)*	*(62)*	*(452)*
One year	*51 53*	*55 58*	*39 41*	*44 46*	*54 56*	**48 51**
Five years	*24 30*	*27 37*	*20 25*	*27 34*	*31 38*	**26 32**
	17 30	*19 32*	*12 23*	*15 27*	*16 29*	**16 26**

1986-90 | Column headings: Affluent, 2, 3, 4, Deprived, All[1]; values C R.

	Affluent C R	2 C R	3 C R	4 C R	Deprived C R	All[1] C R
ENGLAND & WALES (n)	*(1,973)*	*(2,046)*	*(2,053)*	*(1,861)*	*(1,504)*	*(9,527)*
One year	59 61	56 58	55 58	53 56	50 53	**55 58**
Five years	34 41	33 40	32 40	30 38	29 37	**32 40**
ENGLAND (n)	*(1,886)*	*(1,936)*	*(1,925)*	*(1,719)*	*(1,448)*	*(8,990)*
One year	60 62	56 58	55 57	54 57	50 53	**55 58**
Five years	34 41	32 39	32 40	31 38	29 37	**32 40**
Northern & Yorkshire (n)	*(195)*	*(215)*	*(257)*	*(263)*	*(310)*	*(1,252)*
One year	57 59	53 55	48 51	54 58	47 50	**52 55**
Five years	34 41	31 41	28 38	30 38	28 37	**30 38**
Trent (n)	*(158)*	*(189)*	*(262)*	*(233)*	*(220)*	*(1,064)*
One year	59 62	57 60	52 55	56 59	47 50	**54 57**
Five years	34 42	38 49	28 38	32 43	29 38	**32 41**
Anglia & Oxford (n)	*(312)*	*(283)*	*(220)*	*(141)*	*(62)*	*(1,030)*
One year	54 57	51 54	52 54	43 45	*41 42*	**52 54**
Five years	29 36	29 36	31 41	28 36	*24 32*	**30 37**
North Thames (n)	*(232)*	*(204)*	*(220)*	*(210)*	*(187)*	*(1,068)*
One year	64 67	59 62	60 63	64 67	60 63	**62 64**
Five years	32 40	39 48	33 44	39 49	37 47	**36 45**
South Thames (n)	*(299)*	*(279)*	*(204)*	*(193)*	*(89)*	*(1,082)*
One year	60 63	61 64	63 67	52 54	61 65	**60 63**
Five years	35 44	31 39	37 47	27 35	*32 43*	**33 40**
South & West (n)	*(300)*	*(355)*	*(299)*	*(200)*	*(80)*	*(1,243)*
One year	57 60	54 57	58 61	54 57	50 54	**56 59**
Five years	36 43	29 37	36 47	32 43	*36 44*	**34 41**
West Midlands (n)	*(164)*	*(172)*	*(213)*	*(222)*	*(167)*	*(940)*
One year	63 67	51 53	50 53	49 52	49 52	**53 55**
Five years	38 49	28 35	30 40	26 32	29 37	**31 38**
North & West (n)	*(226)*	*(239)*	*(250)*	*(257)*	*(333)*	*(1,311)*
One year	59 62	57 60	53 56	50 54	46 49	**54 56**
Five years	35 44	34 44	29 37	28 38	23 32	**30 38**
WALES (n)	*(87)*	*(110)*	*(128)*	*(142)*	*(56)*	*(537)*
One year	*50 52*	*52 54*	*55 57*	*43 45*	*50 53*	**51 53**
Five years	*29 35*	*37 49*	*30 39*	*22 31*	*28 35*	**30 38**

[1] Including those for whom no deprivation category was assigned. C - crude survival, R - relative survival (see Chapter 3).
Number of patients contributing to each analysis in parentheses.

Table 37.2   Crude and relative survival (%) at one, five and ten years since diagnosis:
NHS Region, deprivation category and calendar period of diagnosis

WOMEN		1971-75							1976-80					
		Affluent C R	2 C R	3 C R	4 C R	Deprived C R	All[1] C R	Affluent C R	2 C R	3 C R	4 C R	Deprived C R	All[1] C R
ENGLAND & WALES		*(540)*	*(617)*	*(654)*	*(596)*	*(545)*	*(3,679)*	*(771)*	*(788)*	*(889)*	*(809)*	*(672)*	*(4,112)*
	One year	47 49	47 49	45 46	46 48	42 43	45 47	48 50	45 46	47 49	42 44	45 47	45 47
	Five years	28 32	26 29	24 28	27 32	21 25	25 30	27 31	26 31	28 34	24 28	26 32	26 31
	Ten years	21 28	17 24	18 25	20 27	15 21	18 25	19 25	19 26	20 29	17 24	19 28	19 26
ENGLAND		*(520)*	*(588)*	*(616)*	*(556)*	*(519)*	*(3,491)*	*(751)*	*(747)*	*(838)*	*(757)*	*(644)*	*(3,887)*
	One year	48 50	47 49	44 46	47 48	42 43	46 47	48 49	44 46	47 48	43 44	45 47	45 47
	Five years	29 33	26 30	24 28	28 33	21 25	26 30	26 30	26 30	28 33	24 28	26 32	26 31
	Ten years	21 28	18 25	18 25	20 27	14 22	18 26	18 25	19 26	20 29	17 24	19 28	19 26
Northern & Yorkshire		*(44)*	*(62)*	*(88)*	*(87)*	*(119)*	*(571)*	*(71)*	*(124)*	*(112)*	*(129)*	*(165)*	*(619)*
	One year	49 50	39 40	31 32	48 50	35 36	43 45	43 45	40 42	48 50	41 43	38 40	43 45
	Five years	25 27	16 18	10 13	24 30	19 22	21 25	29 35	22 27	31 38	23 28	25 30	26 32
	Ten years	*11 14*	*10 15*	*7 12*	*16 22*	15 20	14 20	*21 28*	18 25	21 31	17 25	16 25	19 27
Trent		*(29)*	*(41)*	*(61)*	*(69)*	*(58)*	*(291)*	*(46)*	*(61)*	*(74)*	*(105)*	*(93)*	*(388)*
	One year	59 60	27 28	49 51	44 46	*35 36*	44 45	*43 44*	46 48	51 53	51 52	50 52	50 52
	Five years	*31 37*	*15 16*	32 35	30 35	20 23	28 32	*30 33*	21 24	30 35	24 29	31 36	28 33
	Ten years	*24 33*	*11 13*	24 29	22 31	12 23	21 29	*20 26*	*12 18*	24 34	14 22	23 32	20 29
Anglia & Oxford		*(74)*	*(64)*	*(76)*	*(40)*	*(24)*	*(321)*	*(100)*	*(70)*	*(87)*	*(47)*	*(20)*	*(346)*
	One year	52 54	55 58	39 40	*42 43*	*34 35*	46 48	46 47	42 43	53 56	*47 48*	*38 39*	48 50
	Five years	36 39	32 36	25 29	*26 30*	*30 32*	29 34	24 27	25 27	24 29	*31 37*	*15 16*	26 30
	Ten years	*27 38*	*22 33*	20 24	*18 24*	*23 29*	23 30	18 24	15 19	13 19	*18 27*	*15 16*	17 23
North Thames		*(86)*	*(78)*	*(87)*	*(91)*	*(79)*	*(463)*	*(92)*	*(80)*	*(99)*	*(87)*	*(52)*	*(420)*
	One year	62 64	56 57	44 46	57 59	40 42	53 54	53 54	54 55	29 30	49 51	*45 46*	47 48
	Five years	38 43	34 40	24 28	40 46	17 20	31 37	21 25	38 43	19 23	28 33	26 29	27 32
	Ten years	27 36	29 39	17 25	28 39	*13 20*	24 33	16 23	*31 42*	13 20	23 31	*17 23*	20 29
South Thames		*(128)*	*(107)*	*(92)*	*(102)*	*(48)*	*(516)*	*(174)*	*(118)*	*(121)*	*(94)*	*(40)*	*(576)*
	One year	36 38	52 54	47 49	50 52	*55 56*	47 49	42 43	36 37	51 52	32 33	*41 43*	42 43
	Five years	21 24	25 30	25 29	25 29	*26 32*	24 29	22 25	24 27	29 36	21 23	*18 22*	24 28
	Ten years	18 23	16 22	16 23	20 28	*14 19*	17 24	15 21	16 22	19 28	13 17	*12 16*	16 22
South & West		*(88)*	*(145)*	*(114)*	*(45)*	*(26)*	*(468)*	*(119)*	*(163)*	*(148)*	*(66)*	*(19)*	*(533)*
	One year	51 52	40 42	42 44	36 37	*51 53*	45 46	58 59	42 43	47 49	*45 46*	*37 38*	48 50
	Five years	35 39	28 31	26 32	21 24	*40 44*	30 34	31 35	23 27	30 34	*31 37*	26 29	28 33
	Ten years	26 33	17 23	*25 31*	*8 11*	*21 27*	20 28	22 32	16 23	24 31	*27 33*	21 25	21 29
West Midlands		*(20)*	*(20)*	*(28)*	*(28)*	*(35)*	*(343)*	*(52)*	*(60)*	*(77)*	*(80)*	*(77)*	*(354)*
	One year	*39 40*	*48 49*	*62 64*	*26 27*	*41 43*	46 47	*50 52*	*41 42*	49 51	41 42	55 57	49 51
	Five years	*14 17*	*29 33*	*28 33*	*22 23*	*16 20*	25 30	*28 34*	*25 28*	29 33	22 28	32 39	29 34
	Ten years	*9 12*	*14 24*	*11 15*	*15 20*	*8 14*	17 25	*17 24*	*16 24*	19 26	12 20	29 38	20 28
North & West		*(51)*	*(71)*	*(70)*	*(94)*	*(130)*	*(518)*	*(97)*	*(71)*	*(120)*	*(149)*	*(178)*	*(651)*
	One year	*32 33*	43 44	41 43	40 41	40 42	42 43	40 42	*54 55*	39 40	33 34	42 43	42 43
	Five years	*14 16*	19 22	17 21	22 26	18 21	20 24	26 30	*32 37*	26 32	14 17	22 28	24 29
	Ten years	*11 16*	*12 17*	*12 18*	16 22	13 20	14 21	15 22	*24 32*	21 30	12 17	16 23	18 25
WALES		*(20)*	*(29)*	*(38)*	*(40)*	*(26)*	*(188)*	*(20)*	*(41)*	*(51)*	*(52)*	*(28)*	*(225)*
	One year	*32 33*	*44 46*	*46 47*	*39 40*	*40 41*	40 41	*60 61*	*51 54*	*45 48*	*33 33*	*41 42*	45 46
	Five years	*18 19*	*16 17*	*23 25*	*24 27*	*19 20*	21 23	*46 47*	*30 35*	*31 36*	*22 27*	*27 33*	28 34
	Ten years	*4 7*	*10 12*	*23 25*	*19 26*	*15 18*	16 21	*35 38*	*19 28*	*24 36*	*11 15*	*24 31*	20 29

[1] Including those for whom no deprivation category was assigned. C - crude survival, R - relative survival (see Chapter 3).
Number of patients contributing to each analysis in parentheses.

	1981-85												1986-90													
	Affluent		*2*		*3*		*4*		*Deprived*		*All*[1]		*Affluent*		*2*		*3*		*4*		*Deprived*		*All*[1]		WOMEN	
	C	R	C	R	C	R	C	R	C	R	C	R	C	R	C	R	C	R	C	R	C	R	C	R		
	(926)		(983)		(979)		(1,041)		(801)		(4,858)		(1,075)		(1,179)		(1,201)		(1,158)		(970)		(5,643)		**ENGLAND & WALES**	
One year	52	53	53	55	53	55	48	50	45	46	50	52	54	56	51	52	52	54	49	51	50	52	51	53	*One year*	
Five years	32	36	31	36	29	34	28	34	27	33	29	34	33	37	31	36	31	36	29	34	28	33	30	36	*Five years*	
	23	31	22	30	21	29	20	28	19	27	21	29														
	(890)		(943)		(926)		(974)		(760)		(4,596)		(1,035)		(1,114)		(1,140)		(1,075)		(914)		(5,325)		**ENGLAND**	
One year	52	54	53	55	53	55	48	50	44	46	50	52	54	56	51	53	53	55	49	51	51	53	52	54	*One year*	
Five years	31	36	31	36	28	33	28	34	27	32	29	34	33	38	31	37	31	37	30	35	28	33	31	36	*Five years*	
	23	31	22	29	20	28	20	28	18	26	21	28														
	(101)		(116)		(127)		(212)		(209)		(776)		(111)		(135)		(174)		(191)		(214)		(831)		**Northern & Yorkshire**	
One year	57	59	51	52	48	50	44	46	39	41	47	49	42	43	41	43	54	56	48	50	54	56	50	52	*One year*	
Five years	34	41	27	30	25	30	25	30	23	27	26	32	30	35	29	33	32	37	29	33	31	37	31	36	*Five years*	
	25	35	18	25	18	27	20	26	17	24	19	28														
	(78)		(102)		(117)		(122)		(109)		(532)		(90)		(128)		(152)		(144)		(134)		(650)		**Trent**	
One year	40	41	53	55	47	49	48	50	41	43	48	49	63	66	50	53	51	53	45	47	49	51	52	54	*One year*	
Five years	23	25	28	32	19	22	31	37	28	33	27	31	37	41	23	29	30	36	26	31	34	42	30	36	*Five years*	
	14	20	20	28	15	18	22	31	17	24	18	25														
	(129)		(118)		(112)		(72)		(21)		(464)		(144)		(150)		(116)		(88)		(34)		(535)		**Anglia & Oxford**	
One year	51	53	54	56	57	59	47	50	44	45	53	55	53	55	49	51	51	52	54	56	*19*	*20*	51	53	*One year*	
Five years	34	39	36	41	27	32	34	38	26	35	32	38	33	36	28	32	23	29	30	34	*7*	*7*	28	33	*Five years*	
	32	37	30	39	19	27	20	26	*17*	*27*	26	33														
	(85)		(79)		(69)		(71)		(47)		(358)		(107)		(91)		(110)		(113)		(96)		(532)		**North Thames**	
One year	44	45	63	65	*57*	*58*	53	55	62	64	56	58	54	56	63	65	61	63	56	59	57	60	58	60	*One year*	
Five years	29	32	30	35	*39*	*47*	37	44	37	50	34	41	32	37	40	46	34	41	35	40	27	34	33	39	*Five years*	
	20	26	23	31	*29*	*43*	25	34	*31*	*50*	25	36														
	(156)		(153)		(129)		(107)		(43)		(615)		(171)		(167)		(129)		(109)		(49)		(634)		**South Thames**	
One year	54	56	50	51	54	56	54	56	*51*	*54*	53	55	62	64	49	51	53	55	54	56	*52*	*53*	55	57	*One year*	
Five years	36	42	32	37	31	34	33	38	*36*	*43*	33	38	32	37	30	34	34	39	32	35	22	27	31	36	*Five years*	
	23	32	23	31	20	25	28	36	*23*	*32*	23	31														
	(149)		(189)		(141)		(92)		(29)		(627)		(167)		(224)		(189)		(120)		(40)		(744)		**South & West**	
One year	54	56	52	55	51	53	49	52	40	42	52	54	49	50	51	53	52	54	49	51	*50*	*52*	51	53	*One year*	
Five years	29	33	32	38	29	35	30	37	*28*	*30*	30	36	34	38	31	38	32	38	28	33	*28*	*32*	32	38	*Five years*	
	20	27	22	31	21	28	23	32	*18*	*22*	21	30														
	(83)		(71)		(93)		(135)		(99)		(488)		(102)		(91)		(127)		(123)		(116)		(561)		**West Midlands**	
One year	59	61	*54*	*56*	54	56	40	42	41	43	49	51	58	60	52	53	45	47	50	52	49	51	52	53	*One year*	
Five years	30	35	*32*	*38*	30	34	18	22	27	32	27	32	37	42	33	40	32	37	34	39	25	29	33	39	*Five years*	
	19	27	*15*	*20*	21	27	11	17	19	27	17	24														
	(109)		(115)		(138)		(163)		(203)		(736)		(143)		(128)		(143)		(187)		(231)		(838)		**North & West**	
One year	47	48	44	46	50	52	41	43	42	44	46	48	48	50	46	48	52	54	38	40	46	48	48	50	*One year*	
Five years	27	31	23	27	26	31	23	29	22	26	25	30	28	34	32	36	25	30	23	28	25	29	27	32	*Five years*	
	22	30	15	20	21	30	14	21	14	21	17	25														
	(36)		(40)		(53)		(67)		(41)		(262)		(40)		(65)		(61)		(83)		(56)		(318)		**WALES**	
One year	*45*	*46*	*46*	*48*	*52*	*54*	*50*	*52*	*45*	*47*	49	51	*50*	*51*	*48*	*49*	*36*	*37*	*41*	*42*	*36*	*37*	44	45	*One year*	
Five years	*37*	*45*	*33*	*37*	*41*	*47*	*30*	*34*	*34*	*39*	35	41	*19*	*23*	*27*	*31*	*25*	*30*	*23*	*27*	*23*	*26*	26	31	*Five years*	
	27	*36*	*33*	*37*	*32*	*41*	*20*	*26*	*20*	*32*	26	35														

[1] Including those for whom no deprivation category was assigned. C - crude survival, R - relative survival (see Chapter 3).
Number of patients contributing to each analysis in parentheses.

Table 37.2   Crude and relative survival (%) at one, five and ten years since diagnosis: NHS Region, deprivation category and calendar period of diagnosis

ADULTS		1971-75												1976-80											
		Affluent		*2*		*3*		*4*		*Deprived*		*All[1]*		*Affluent*		*2*		*3*		*4*		*Deprived*		*All[1]*	
		C	R	C	R	C	R	C	R	C	R	C	R	C	R	C	R	C	R	C	R	C	R	C	R
ENGLAND & WALES		*(1,549)*		*(1,692)*		*(1,764)*		*(1,607)*		*(1,412)*		*(9,904)*		*(2,159)*		*(2,154)*		*(2,314)*		*(2,150)*		*(1,841)*		*(11,102)*	
	One year	51	52	48	50	46	48	48	50	40	42	47	49	51	53	48	50	47	49	45	47	46	48	47	49
	Five years	28	33	26	30	25	29	27	33	20	25	26	31	29	34	28	34	27	33	25	31	26	32	27	33
	Ten years	19	26	16	24	17	25	17	27	13	21	17	25	20	28	19	27	18	27	17	26	17	27	18	27
ENGLAND		*(1,486)*		*(1,608)*		*(1,644)*		*(1,479)*		*(1,343)*		*(9,340)*		*(2,086)*		*(2,059)*		*(2,170)*		*(1,987)*		*(1,760)*		*(10,462)*	
	One year	51	53	48	50	46	48	48	50	40	42	47	49	51	53	48	50	48	50	46	47	46	48	47	49
	Five years	29	33	26	31	24	29	27	33	20	25	26	31	29	34	28	33	27	33	26	31	26	33	27	33
	Ten years	19	27	17	25	17	25	18	27	13	21	17	25	19	28	19	27	17	27	17	26	17	27	18	27
Northern & Yorkshire		*(114)*		*(163)*		*(203)*		*(209)*		*(293)*		*(1,395)*		*(185)*		*(264)*		*(283)*		*(320)*		*(450)*		*(1,543)*	
	One year	52	53	43	45	39	40	47	49	37	38	45	47	51	53	46	47	47	49	44	45	43	45	46	48
	Five years	24	29	21	25	16	20	23	29	16	19	21	26	32	40	26	32	26	32	23	29	23	30	26	32
	Ten years	15	21	13	18	10	17	15	24	10	17	14	21	22	34	18	26	17	26	15	24	15	25	17	27
Trent		*(105)*		*(131)*		*(164)*		*(180)*		*(154)*		*(829)*		*(146)*		*(187)*		*(218)*		*(268)*		*(218)*		*(1,070)*	
	One year	48	49	42	43	43	45	47	48	33	35	44	46	48	50	48	49	50	52	48	50	45	47	49	50
	Five years	26	29	21	24	27	32	28	33	19	24	26	31	31	35	27	32	26	30	24	30	26	32	27	32
	Ten years	17	25	15	20	17	24	17	26	12	20	17	25	20	27	18	26	18	27	15	24	20	29	18	27
Anglia & Oxford		*(195)*		*(197)*		*(212)*		*(140)*		*(67)*		*(925)*		*(272)*		*(252)*		*(228)*		*(151)*		*(57)*		*(1,007)*	
	One year	53	55	51	54	47	49	52	53	37	38	51	53	54	56	47	49	51	53	45	47	48	50	51	53
	Five years	36	41	32	38	29	33	30	35	25	30	31	37	30	35	29	34	29	35	29	36	17	22	30	35
	Ten years	26	37	23	37	19	26	18	29	19	27	22	32	19	28	19	28	16	25	18	27	15	20	19	28
North Thames		*(254)*		*(242)*		*(250)*		*(239)*		*(197)*		*(1,304)*		*(264)*		*(202)*		*(258)*		*(194)*		*(170)*		*(1,120)*	
	One year	58	60	48	50	47	49	52	54	41	43	51	53	50	51	50	52	36	37	52	54	52	55	49	50
	Five years	35	41	26	31	26	30	31	37	19	23	28	34	27	31	30	37	22	26	28	34	28	36	27	33
	Ten years	24	34	19	27	17	26	21	32	15	23	20	30	20	30	25	35	13	21	19	31	17	30	19	30
South Thames		*(348)*		*(275)*		*(268)*		*(252)*		*(132)*		*(1,376)*		*(453)*		*(323)*		*(319)*		*(230)*		*(104)*		*(1,508)*	
	One year	44	46	47	49	44	46	49	51	48	50	47	49	47	49	43	44	49	51	39	41	41	43	45	47
	Five years	21	24	23	27	26	31	26	32	24	29	24	28	26	30	26	31	27	34	23	28	21	26	26	31
	Ten years	15	20	13	19	18	26	16	25	12	18	15	23	17	26	16	25	17	27	16	23	13	20	16	25
South & West		*(266)*		*(375)*		*(287)*		*(137)*		*(60)*		*(1,256)*		*(339)*		*(453)*		*(382)*		*(212)*		*(69)*		*(1,508)*	
	One year	53	55	45	47	46	47	43	45	42	44	48	50	57	59	47	49	48	50	47	49	44	46	50	52
	Five years	34	39	28	32	25	31	27	33	27	31	29	35	30	36	27	33	29	35	29	37	27	35	29	36
	Ten years	19	27	18	26	22	30	17	25	17	23	19	28	21	31	17	26	20	32	20	31	16	28	19	29
West Midlands		*(64)*		*(53)*		*(76)*		*(93)*		*(110)*		*(945)*		*(156)*		*(170)*		*(214)*		*(244)*		*(242)*		*(1,041)*	
	One year	48	50	56	57	52	55	33	34	38	40	48	49	45	47	53	55	48	50	48	50	50	52	50	52
	Five years	29	34	34	40	22	27	24	29	21	27	27	33	30	36	28	33	26	31	28	35	29	36	29	35
	Ten years	23	29	14	20	14	21	17	27	13	22	19	28	21	29	17	26	18	27	20	30	20	32	20	29
North & West		*(140)*		*(172)*		*(184)*		*(229)*		*(330)*		*(1,310)*		*(271)*		*(208)*		*(268)*		*(368)*		*(450)*		*(1,665)*	
	One year	40	41	46	48	40	42	41	43	38	40	43	45	45	47	47	49	42	44	37	38	42	44	43	45
	Five years	16	18	22	26	15	19	23	29	16	21	21	25	24	29	28	34	22	27	19	23	25	31	24	30
	Ten years	7	11	14	20	11	16	15	23	11	18	13	20	14	22	19	27	16	25	14	23	15	25	16	25
WALES		*(63)*		*(84)*		*(120)*		*(128)*		*(69)*		*(564)*		*(73)*		*(95)*		*(144)*		*(163)*		*(81)*		*(640)*	
	One year	41	42	48	50	44	47	45	48	36	37	45	47	52	54	52	54	42	44	39	40	41	42	45	47
	Five years	21	25	21	26	27	31	24	30	25	27	25	30	32	36	27	33	27	33	21	27	22	25	26	32
	Ten years	13	20	6	9	19	26	14	22	19	24	16	24	25	31	16	25	20	30	15	22	16	24	19	27

[1] Including those for whom no deprivation category was assigned. C - crude survival, R - relative survival (see Chapter 3).
Number of patients contributing to each analysis in parentheses.

1981-85

	Affluent C	R	2 C	R	3 C	R	4 C	R	Deprived C	R	All[1] C	R	ADULTS
(2,696)		*(2,720)*		*(2,744)*		*(2,688)*		*(2,073)*		*(13,232)*			**ENGLAND & WALES**
55	57	55	57	53	55	50	52	45	48	**52**	**54**		*One year*
32	38	31	38	29	35	29	36	27	34	**30**	**36**		*Five years*
22	31	22	32	20	30	19	30	18	27	**20**	**30**		
(2,614)		*(2,608)*		*(2,565)*		*(2,510)*		*(1,970)*		*(12,518)*			**ENGLAND**
55	57	55	57	53	56	50	52	45	47	**52**	**54**		*One year*
32	38	32	38	29	36	29	36	26	33	**30**	**36**		*Five years*
22	31	21	31	20	30	20	30	17	27	**20**	**30**		
(266)		*(300)*		*(342)*		*(455)*		*(522)*		*(1,914)*			**Northern & Yorkshire**
59	61	53	55	48	50	47	49	42	44	**49**	**51**		*One year*
35	42	30	36	27	33	25	31	23	30	**28**	**34**		*Five years*
26	40	20	30	19	30	19	28	16	27	**20**	**30**		
(233)		*(290)*		*(327)*		*(341)*		*(248)*		*(1,450)*			**Trent**
48	50	53	56	48	50	47	50	45	47	**50**	**52**		*One year*
29	34	26	31	24	29	30	37	27	34	**28**	**33**		*Five years*
18	25	17	27	17	24	17	30	16	25	**18**	**26**		
(370)		*(321)*		*(315)*		*(210)*		*(76)*		*(1,326)*			**Anglia & Oxford**
53	55	55	57	58	60	54	58	49	52	**56**	**58**		*One year*
34	41	35	41	34	43	37	44	*31*	*40*	**35**	**42**		*Five years*
27	37	26	37	24	37	23	35	*25*	*37*	**25**	**36**		
(251)		*(228)*		*(220)*		*(200)*		*(134)*		*(1,051)*			**North Thames**
56	58	58	60	58	60	52	55	57	60	**57**	**60**		*One year*
32	39	28	34	34	41	36	45	32	44	**33**	**40**		*Five years*
19	30	20	29	24	36	24	39	24	42	**23**	**34**		
(504)		*(395)*		*(327)*		*(264)*		*(101)*		*(1,658)*			**South Thames**
57	59	54	56	57	60	57	59	49	51	**56**	**58**		*One year*
35	42	31	38	34	41	32	39	32	39	**33**	**40**		*Five years*
21	33	21	32	24	35	24	35	24	35	**22**	**33**		
(436)		*(530)*		*(431)*		*(248)*		*(81)*		*(1,781)*			**South & West**
54	56	56	58	51	53	48	51	50	52	**54**	**56**		*One year*
30	36	35	42	26	32	30	37	30	36	**31**	**38**		*Five years*
19	29	23	35	17	27	20	34	*23*	*34*	**21**	**31**		
(240)		*(212)*		*(263)*		*(363)*		*(288)*		*(1,384)*			**West Midlands**
58	60	59	61	54	56	47	49	38	40	**51**	**53**		*One year*
31	37	35	43	29	35	25	32	24	30	**29**	**35**		*Five years*
21	30	23	34	19	29	15	26	16	24	**19**	**28**		
(314)		*(332)*		*(340)*		*(429)*		*(520)*		*(1,954)*			**North & West**
45	47	49	51	50	52	43	46	41	43	**46**	**49**		*One year*
26	31	27	34	26	33	23	30	23	29	**25**	**32**		*Five years*
19	28	17	27	18	29	16	25	13	22	**17**	**26**		
(82)		*(112)*		*(179)*		*(178)*		*(103)*		*(714)*			**WALES**
48	50	52	55	43	45	46	48	50	53	**49**	**51**		*One year*
30	36	30	38	26	32	28	34	32	39	**29**	**36**		*Five years*
21	32	24	37	18	29	17	26	*18*	*31*	**20**	**30**		

1986-90

	Affluent C	R	2 C	R	3 C	R	4 C	R	Deprived C	R	All[1] C	R
(3,048)		*(3,225)*		*(3,254)*		*(3,019)*		*(2,474)*		*(15,170)*			**ENGLAND & WALES**
57	59	54	56	54	56	52	54	50	53	**54**	**56**	*One year*
34	40	32	39	31	39	30	36	29	36	**31**	**38**	*Five years*
(2,921)		*(3,050)*		*(3,065)*		*(2,794)*		*(2,362)*		*(14,315)*			**ENGLAND**
58	60	54	56	54	57	52	55	51	53	**54**	**56**	*One year*
34	40	32	38	32	39	30	37	29	36	**32**	**38**	*Five years*
(306)		*(350)*		*(431)*		*(454)*		*(524)*		*(2,083)*			**Northern & Yorkshire**
52	53	48	50	50	53	52	54	50	52	**51**	**54**	*One year*
33	39	30	38	30	37	29	36	29	36	**31**	**37**	*Five years*
(248)		*(317)*		*(414)*		*(377)*		*(354)*		*(1,714)*			**Trent**
60	63	54	57	52	54	52	54	48	50	**53**	**56**	*One year*
35	42	32	40	29	37	30	38	31	40	**32**	**39**	*Five years*
(456)		*(433)*		*(336)*		*(229)*		*(96)*		*(1,565)*			**Anglia & Oxford**
54	56	50	53	51	53	47	49	33	35	**51**	**53**	*One year*
30	36	29	35	28	36	29	35	17	22	**29**	**35**	*Five years*
(339)		*(295)*		*(330)*		*(323)*		*(283)*		*(1,600)*			**North Thames**
61	63	60	63	60	63	61	64	59	62	**61**	**63**	*One year*
32	39	39	47	33	43	37	46	34	42	**35**	**43**	*Five years*
(470)		*(446)*		*(333)*		*(302)*		*(138)*		*(1,716)*			**South Thames**
61	64	56	59	59	62	53	55	58	60	**58**	**61**	*One year*
34	41	30	37	36	44	28	36	28	36	**32**	**39**	*Five years*
(467)		*(579)*		*(488)*		*(320)*		*(120)*		*(1,987)*			**South & West**
54	56	53	55	55	58	52	55	50	53	**54**	**57**	*One year*
35	41	30	38	35	43	30	39	33	41	**33**	**40**	*Five years*
(266)		*(263)*		*(340)*		*(345)*		*(283)*		*(1,501)*			**West Midlands**
61	64	51	53	48	50	49	52	49	52	**53**	**55**	*One year*
38	47	30	36	31	39	29	34	28	34	**32**	**38**	*Five years*
(369)		*(367)*		*(393)*		*(444)*		*(564)*		*(2,149)*			**North & West**
55	57	53	56	53	55	45	48	46	48	**51**	**54**	*One year*
32	40	33	42	28	34	26	33	24	31	**29**	**36**	*Five years*
(127)		*(175)*		*(189)*		*(225)*		*(112)*		*(855)*			**WALES**
50	52	50	52	49	51	42	44	43	45	**49**	**50**	*One year*
26	32	33	41	28	37	22	30	25	32	**28**	**35**	*Five years*

[1] Including those for whom no deprivation category was assigned. C - crude survival, R - relative survival (see Chapter 3).
Number of patients contributing to each analysis in parentheses.

Fig. 37.1 Relative survival up to ten years: trends by calendar period of diagnosis (1971-90) and NHS Region

ADULTS

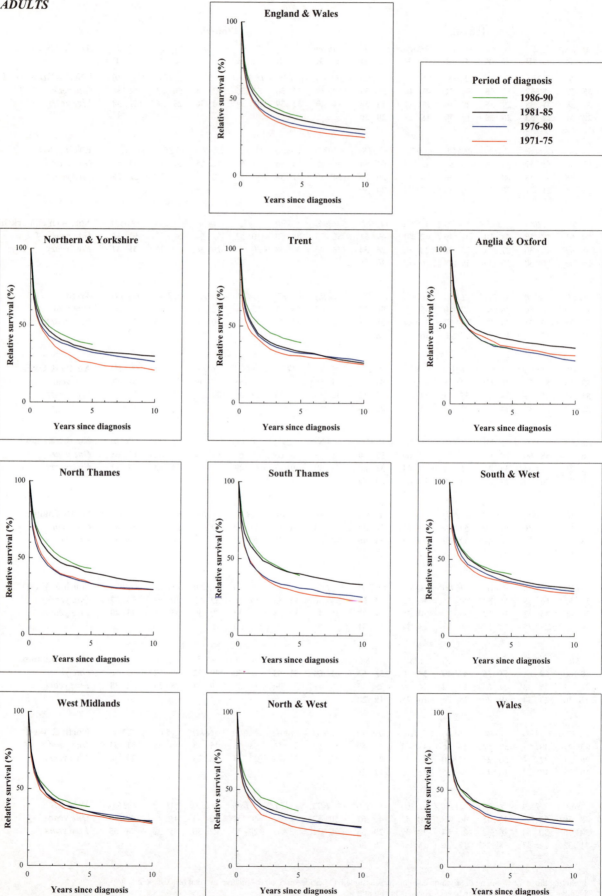

Fig. 37.2 Relative survival up to five years, by deprivation category and NHS Region: patients diagnosed 1986-90

ADULTS

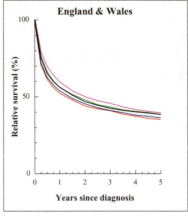

England & Wales

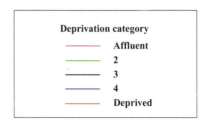

Deprivation category
- Affluent
- 2
- 3
- 4
- Deprived

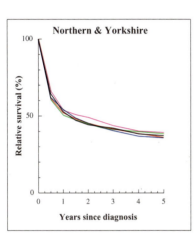

Northern & Yorkshire

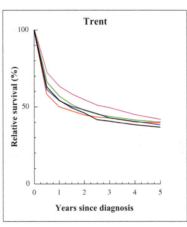

Trent

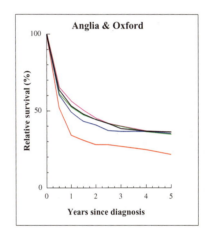

Anglia & Oxford

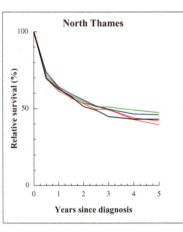

North Thames

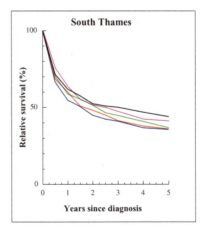

South Thames

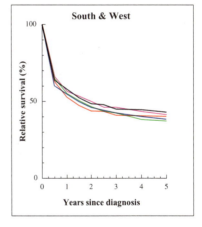

South & West

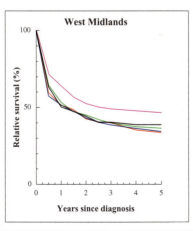

West Midlands

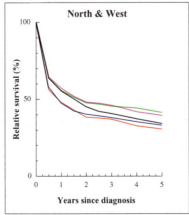

North & West

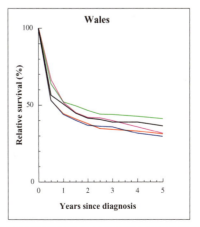

Wales

Fig. 37.3 Relative survival at five years by deprivation category, period of diagnosis (1981-90) and NHS Region

ADULTS

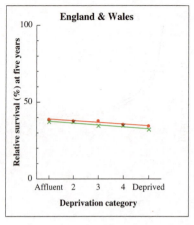

England & Wales

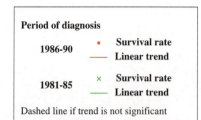

Period of diagnosis

1986-90   •   Survival rate
       ──   Linear trend

1981-85   ×   Survival rate
       ──   Linear trend

Dashed line if trend is not significant

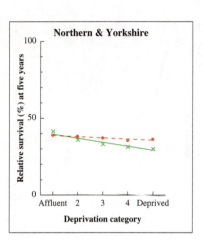

Northern & Yorkshire

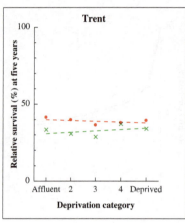

Trent

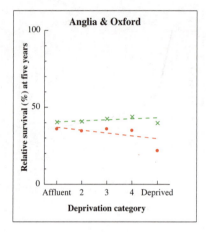

Anglia & Oxford

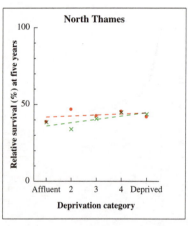

North Thames

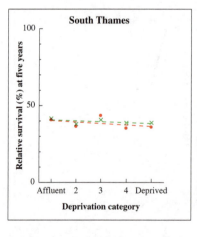

South Thames

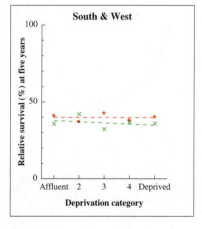

South & West

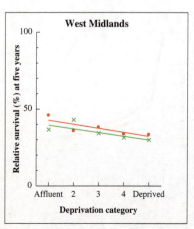

West Midlands

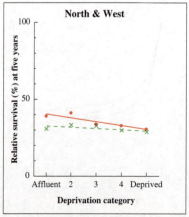

North & West

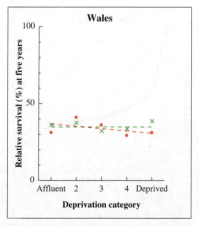

Wales

Fig. 37.4 Relative survival (%) by NHS Region and sex:
   one-year and five-year survival for patients diagnosed 1986-90, with 95% confidence intervals

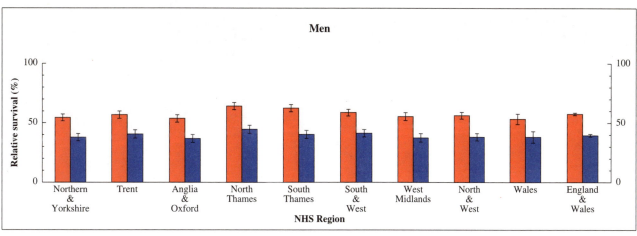

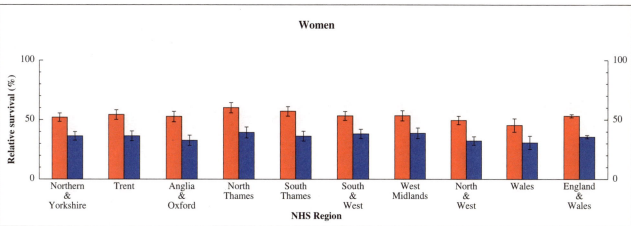

Fig. 37.5 Relative survival (%) at five years, with 95% confidence interval, by age at diagnosis and sex:
   England and Wales, patients diagnosed 1986-90

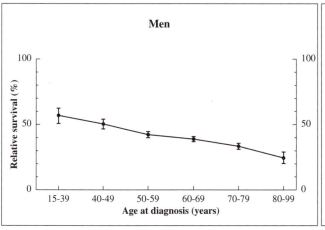
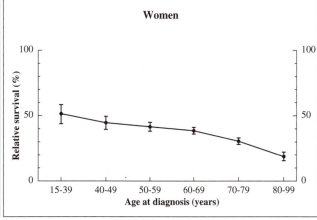

Malignant tumours of the eye are rare in England and Wales, with about 500 new cases each year in adults. Incidence is slightly higher in men. Age-standardised (Europe) incidence rates have risen by about 4% every year, from 0.7 per 100,000 in 1980 to 1.0 in 1990 in men, and from 0.6 to 0.9 in women. The increase began in the early 1960s, and is not an artefact of improved diagnosis or registration[1]. Mortality rose from the 1950s to the early 1970s[2], but has fallen steadily since then, from a peak of 0.4 per 100,000 in 1971 to 0.2 in 1995 in men, and from 0.4 to 0.1 in women.

In developed countries, malignant melanomas of the choroid are the most common tumours of the eye in adults. Squamous carcinoma of the eye is more common in Africa. Retinoblastoma, the most common malignancy of the eye in children, is discussed in Chapter 55. The causes of ocular melanoma are largely unknown. Incidence is higher in rural than urban areas, but there is no association with latitude, as there is for melanoma of the skin, and solar ultraviolet radiation is probably a minor risk factor. Incidence may be higher in higher socioeconomic groups[1]. Developments in radiotherapy and surgical techniques are not thought to have improved survival[2].

The survival analyses reported here include 6,100 adults diagnosed with a malignant tumour of the eye in England and Wales during the period 1971-90 and followed up to the end of 1995, After the exclusion of about 10% of tumours diagnosed in children under 15, about 92% of eligible adults were included in the analyses (Table 38.1). Malignant tumours of the choroid accounted for around 45% of cases, with a further 10% in the eyeball, 5% in the conjunctiva, and just under a third unspecified. The majority of tumours (88%) were malignant melanomas: for most of the remainder the morphology was poorly specified.

Survival trends

Survival from malignant tumours of the eye for adults diagnosed in England and Wales during 1986-90 was 93% at one year and 71% at five years (Table 38.2). Survival is similar in men and women. There has been only gradual improvement in survival nationally, and regional patterns of survival with time are variable: the survival curve for a five-year period in each region generally includes 100-250 patients. Trends in West Midlands are particularly irregular (Figure 38.2).

After standardisation for age, one-year survival increased from about 88% for patients diagnosed during 1971-75 to 92-93% for those diagnosed in 1986-90. Five-year survival increased from 60% to 67-70%. The increases were small but consistent between successive periods. The national average gain in survival between successive five-year periods of diagnosis has been 1-2% for one-year survival and 2-3% for survival at five years (see Table 4.4). Ten-year survival showed little change, remaining at around 60% (Table 38.2).

Five-year survival for patients diagnosed in 1986-90 falls with age (Figure 38.5), but the decline is less steep than for many epithelial tumours, and resembles that for melanoma of the skin (see Chapter 27). Five-year survival is 70% or more for those aged less than 60 at diagnosis, and around 67% for those aged 60-79. Survival is broadly similar for men and women at all ages (see Table 4.7).

Survival from tumours of the eye showed little variation by region (Table 38.2, Figure 38.4).

Deprivation

Survival at one year was broadly similar in all deprivation groups, and differences in survival up to five years for patients diagnosed in 1986-90 were not systematic (Figure 38.2).

There was a significant deprivation gradient in five-year survival among patients diagnosed in 1981-85, but the gradient was less steep, and not significant, for those diagnosed during 1986-90 (Figure 38.3).

Among 1,500 patients of known deprivation status who were diagnosed during the early 1980s, the gap in one-year survival between the most affluent and the most deprived groups was an estimated 4%. For five-year survival the deprivation gap was 9%. For the 1,700 patients diagnosed during 1986-90, the differences between deprivation groups were smaller, and not statistically significant (see Table 4.9).

International comparison

Five-year survival for patients diagnosed with a malignant tumour of the eye in England and Wales during 1986-90 was 70-71%, about 6% lower than in the USA for the same period (see Table 4.12). The highest regional survival rates (77-78%) were similar to those in the USA. Survival in the USA did not improve during the 1980s: five-year relative survival was 79% for patients diagnosed during 1983-87[3].

The survival rates for Scotland and the rest of Europe in

Table 38A Sub-site distribution for malignant tumours of the eye[1]

ICD-9 code	Site description	1971-75 No.	%	1976-80 No.	%	1981-85 No.	%	1986-90 No.	%	Total No.	%
190.0	Eyeball	-	-	*60*	*4.3*	148	9.5	169	9.7	**377**	-
190.1	Orbit	-	-	-	-	78	5.0	79	4.5	**157**	-
190.2	Lacrimal gland	-	-	-	-	13	0.8	6	0.3	**19**	-
190.3	Conjunctiva	-	-	-	-	69	4.4	81	4.7	**150**	-
190.4	Cornea	-	-	*4*	*0.3*	11	0.7	12	0.7	**27**	-
190.5	Retina	-	-	*8*	*0.6*	26	1.7	32	1.8	**66**	-
190.6	Choroid	-	-	*189*	*13.7*	675	43.4	789	45.4	**1,653**	-
190.7	Lacrimal duct	-	-	-	-	7	0.4	14	0.8	**21**	-
190.8	Other	-	-	-	-	11	0.7	7	0.4	**18**	-
190.9	Part unspecified	1,431		1,121	81.1	518	33.3	548	31.5	**3,618**	-
	Total	**1,431**		**1,382**		**1,556**		**1,737**		**6,106**	

[1] Eye was coded simply 190, without subdivision, in ICD-8 (1971-78). Special codes 190.1-190.3 and 190.7-190.9 used by the national cancer registry during that period (see Chapter 2) are incompatible with ICD-9, and tumours with those codes have been assigned to 190.9 in this table. Italicised data in column 2 relate to 1979-80 only. Misleading percentages have been suppressed.

Table 4.12 are not strictly comparable with those for England and Wales reported here, because they only include melanoma of the choroid: results from the EUROCARE study show that five-year survival from melanoma of the choroid in England and Wales for 1985-89 was 81% for men (based on 203 cases) and 70% for women (175 cases)[4]. These rates were based on data from about half the population of England; the numbers of cases also reflect the fact that melanoma of the choroid accounted for 45% of all eye tumours in England and Wales during 1986-90 (Table 38A).

1. Swerdlow AJ. Epidemiology of eye cancer in adults in England and Wales, 1962-1977. Am J Epidemiol 1983; 118: 294-300

2. Foss AJE, Dolin PJ. Trends in eye cancer mortality among adults in the USA and England and Wales. Br J Cancer 1996; 74: 1687-1689

3. Polednak AP, Flannery JT. Brain, other central nervous system, and eye cancer. Cancer 1995; 75: 330-337

4. Berrino F, Capocaccia R, Estève J, Gatta G, Hakulinen T, Micheli M, Sant M, Verdecchia A, (eds.) *Survival of cancer patients in Europe: the EUROCARE study, II. (IARC Scientific Publications No. 151)*. Lyon: International Agency for Research on Cancer, 1999

Table 38.1 Ineligible and excluded records, and patients included in analyses, by calendar period of diagnosis

	Calendar period of diagnosis																
	1971-75			1976-80			1981-85			1986-90			Overall		
Total registered	1,909			1,810			1,976			2,267			7,962		
Ineligible	*Records*	*Patients*	*%*	*Records*	*Patients*	*%*	*Records*	*Patients*	*%*	*Records*	*Patients*	*%*	*Records*	*Patients*	*%*
Incomplete data[1]	12	12		38	38		90	90		78	78		218	**218**	
Resident outside England and Wales	73	73		48	45		31	23		16	14		168	**155**	
In situ neoplasm[2]	14	14		1	1		0	0		1	1		16	**16**	
Benign or uncertain behaviour[2]	44	29		47	29		0	0		0	0		91	**58**	
Metastatic[2]	11	11		2	2		0	0		0	0		13	**13**	
Otherwise ineligible[3]	-	-		-	-		-	-		-	-		-	**-**	
Lymphoma[4]	23	21		27	20		8	0		7	0		65	**41**	
Leukaemia or myeloma[4]	0	0		0	0		1	0		0	0		1	**0**	
		160			135			113			93			501	
Total eligible	1,749	100		1,675	100		1,863	100		2,174	100		7,461	100	
Aged under 15 or 100 years or over at diagnosis	254	217	12.4	229	187	11.2	210	155	8.3	193	154	7.1	886	713	9.6
Vital status unknown[5]	134	57	3.3	135	72	4.3	124	75	4.0	250	209	9.6	643	413	5.5
Duplicate registration[6]	0	0	0.0	0	0	0.0	0	0	0.0	0	0	0.0	0	0	0.0
Multiple primary[7]	0	0	0.0	1	1	<0.1	7	7	0.4	2	1	<0.1	10	9	0.1
Synchronous tumours[8]	21	9	0.5	13	2	0.1	15	6	0.3	18	2	<0.1	67	19	0.3
Sex not known	0	0	0.0	0	0	0.0	0	0	0.0	0	0	0.0	0	0	0.0
Sex incompatible with site code[9]	-	-	-	-	-	-	-	-	-	-	-	-	-	-	-
Invalid dates[10]	0	0	0.0	0	0	0.0	0	0	0.0	6	0	0.0	6	0	0.0
Zero survival[11]	43	35	2.0	40	31	1.9	73	64	3.4	76	71	3.3	232	201	2.7
Total exclusions		318	18.2		293	17.5		307	16.5		437	20.1		1,355	18.2
Patients included in analysis		1,431	81.8		1,382	82.5		1,556	83.5		1,737	79.9		6,106	81.8

[1] Main data item(s) invalid or incompatible with one another: name, sex, date of birth, date of diagnosis and (if present) date of death, postcode, site, morphology and behaviour

[2] Tumour either *in situ* (behaviour code 2), benign (0), uncertain if benign or malignant (1) or metastatic (6 or 9)

[3] Not applicable for this malignancy

[4] Morphology that of lymphoma (included with non-Hodgkin lymphoma, Ch. 41) or myeloma (included with multiple myeloma, Ch. 43), or leukaemia (excluded)

[5] Tracing of vital status at National Health Service Central Register incomplete at 6 July 1997

[6] Same site code, sex, cancer registry and cancer registry number as an earlier registration

[7] Same site code and person number as an earlier registration(s): mostly confirmed multiple primary tumours at this site, some unresolved duplicate registrations

[8] Same site code, sex, date of birth and date of diagnosis as another registration(s): mostly synchronous or (in paired organs) bilateral tumours in same anatomic site in one individual, not previously linked: also some duplicate registrations

[9] Not applicable for this malignancy

[10] Impossible sequence of dates (birth, diagnosis and, if present, emigration or death); or date of death after 6 July 1997

[11] Date of diagnosis same as date of death: some are patients who did die on the day of diagnosis, but most were registered solely from a death certificate (death certificate only, or DCO), and their survival time is thus unknown

Table 38.2   Crude and relative survival (%) at one, five and ten years since diagnosis: NHS Region, deprivation category and calendar period of diagnosis

		1971-75						1976-80					
		Affluent	2	3	4	Deprived	All[1]	Affluent	2	3	4	Deprived	All[1]
		C R	C R	C R	C R	C R	C R	C R	C R	C R	C R	C R	C R
ENGLAND & WALES													
Men		*(117)*	*(111)*	*(119)*	*(127)*	*(74)*	*(696)*	*(156)*	*(142)*	*(154)*	*(145)*	*(100)*	*(728)*
	One year	87 93	89 93	81 85	85 89	86 90	86 90	83 87	91 95	86 90	92 96	83 88	87 92
	Five years	59 72	53 62	46 55	54 67	59 73	55 67	56 67	62 74	50 64	55 69	44 57	54 67
	Ten years	45 62	37 51	33 50	41 61	45 67	40 60	44 60	46 72	38 59	41 65	31 50	41 62
Women		*(115)*	*(132)*	*(130)*	*(118)*	*(78)*	*(735)*	*(113)*	*(131)*	*(145)*	*(137)*	*(95)*	*(654)*
	One year	86 90	86 88	88 90	85 88	87 91	87 90	92 94	92 95	92 95	88 92	83 89	89 93
	Five years	57 66	57 67	55 65	47 55	51 60	54 64	62 70	59 68	62 73	52 62	46 59	56 66
	Ten years	45 59	42 57	35 51	37 49	37 48	41 56	52 64	47 65	48 65	37 55	34 52	43 60
Adults		*(232)*	*(243)*	*(249)*	*(245)*	*(152)*	*(1,431)*	*(269)*	*(273)*	*(299)*	*(282)*	*(195)*	*(1,382)*
	One year	87 91	87 90	85 88	85 89	87 90	86 90	87 91	91 95	89 93	90 94	83 89	88 93
	Five years	58 69	55 65	50 60	51 61	55 67	55 65	59 68	61 71	56 69	53 66	45 58	55 67
	Ten years	45 60	40 55	34 50	39 55	41 58	41 58	47 62	46 68	43 63	39 59	32 50	42 61

		Men	Women	Adults				Men	Women	Adults			
Northern & Yorkshire		*(99)*	*(94)*	*(193)*				*(105)*	*(104)*	*(209)*			
	One year	83 87	85 88	84 87				84 88	92 94	88 92			
	Five years	52 64	51 62	52 63				52 64	61 70	57 68			
	Ten years	38 54	39 55	38 54				37 56	41 60	39 59			
Trent		*(64)*	*(64)*	*(128)*				*(75)*	*(76)*	*(151)*			
	One year	79 83	89 94	84 89				84 89	91 94	87 92			
	Five years	51 63	55 66	53 65				52 66	58 66	55 66			
	Ten years	36 56	39 56	37 56				41 66	51 63	46 64			
Anglia & Oxford		*(71)*	*(60)*	*(131)*				*(62)*	*(76)*	*(138)*			
	One year	86 91	88 93	87 92				87 91	92 97	90 94			
	Five years	65 80	50 61	58 71				48 56	55 69	52 63			
	Ten years	52 73	42 54	48 64				37 51	45 67	41 58			
North Thames		*(85)*	*(95)*	*(180)*				*(99)*	*(76)*	*(175)*			
	One year	92 96	92 96	92 96				92 96	93 98	93 98			
	Five years	62 72	63 75	63 74				58 74	63 76	60 75			
	Ten years	46 63	43 59	44 61				47 70	51 72	49 71			
South Thames		*(98)*	*(104)*	*(202)*				*(104)*	*(74)*	*(178)*			
	One year	88 92	82 85	85 88				92 98	88 95	90 97			
	Five years	48 58	50 58	49 58				63 77	57 68	60 73			
	Ten years	36 52	43 58	39 56				45 71	41 57	43 65			
South & West		*(74)*	*(102)*	*(176)*				*(109)*	*(94)*	*(203)*			
	One year	88 92	86 89	87 90				88 93	89 93	89 93			
	Five years	53 64	61 71	57 69				58 72	60 73	59 72			
	Ten years	42 62	45 60	44 61				43 67	46 64	44 65			
West Midlands		*(88)*	*(86)*	*(174)*				*(54)*	*(50)*	*(104)*			
	One year	92 97	95 98	94 97				91 93	90 95	90 95			
	Five years	66 80	62 72	64 76				46 55	48 56	47 57			
	Ten years	49 77	48 68	48 73				33 49	40 52	36 52			
North & West		*(86)*	*(86)*	*(172)*				*(63)*	*(54)*	*(117)*			
	One year	82 87	79 82	81 85				82 87	83 87	83 87			
	Five years	43 56	49 58	46 58				52 66	46 55	49 61			
	Ten years	27 49	34 46	30 47				36 57	33 45	35 51			
WALES		*(31)*	*(44)*	*(75)*				*(57)*	*(50)*	*(107)*			
	One year	77 80	84 86	81 84				80 85	80 82	80 83			
	Five years	54 63	43 50	48 57				51 66	44 51	48 58			
	Ten years	39 50	33 42	36 48				37 58	36 51	36 54			

[1] Including those for whom no deprivation category was assigned. C - crude survival, R - relative survival (see Chapter 3). Number of patients contributing to each analysis in parentheses.

1981-85

	Affluent C R	2 C R	3 C R	4 C R	Deprived C R	All[1] C R		
								ENGLAND & WALES
	(169)	*(159)*	*(162)*	*(144)*	*(113)*	*(769)*		*Men*
	89 92	91 94	88 93	83 88	88 91	88 91		*One year*
	64 73	56 69	54 66	53 65	54 68	56 68		*Five years*
	48 63	45 63	41 60	*34 56*	*39 59*	42 61		
	(148)	*(176)*	*(165)*	*(150)*	*(128)*	*(787)*		*Women*
	95 96	90 92	86 92	87 90	88 91	89 92		*One year*
	64 72	57 65	64 75	50 59	54 63	58 67		*Five years*
	48 63	37 50	45 64	*34 48*	37 52	40 55		
	(317)	*(335)*	*(327)*	*(294)*	*(241)*	*(1,556)*		*Adults*
	91 94	90 93	87 92	85 89	88 91	88 92		*One year*
	64 73	56 66	59 71	51 62	54 65	57 68		*Five years*
	48 63	41 57	43 63	*34 52*	*38 55*	41 58		

1986-90

	Affluent C R	2 C R	3 C R	4 C R	Deprived C R	All[1] C R	
(191)	*(199)*	*(198)*	*(150)*	*(120)*	*(872)*		*Men*
91 94	90 94	86 91	94 96	*88 93*	90 94		*One year*
65 75	59 71	52 63	62 73	51 65	58 70		*Five years*
(197)	*(183)*	*(174)*	*(167)*	*(133)*	*(865)*		*Women*
90 92	90 94	92 96	89 93	88 92	90 93		*One year*
59 68	60 70	63 75	63 77	56 66	60 71		*Five years*
(388)	*(382)*	*(372)*	*(317)*	*(253)*	*(1,737)*		*Adults*
90 93	90 94	89 93	91 95	88 92	90 93		*One year*
62 71	59 71	57 69	62 75	53 65	59 71		*Five years*

1981-85 Men C R	Women C R	Adults C R		1986-90 Men C R	Women C R	Adults C R	Region
(109)	*(100)*	*(209)*		*(113)*	*(136)*	*(249)*	**Northern & Yorkshire**
85 90	92 95	88 93		88 92	90 95	89 94	*One year*
54 66	57 64	55 65		52 63	53 65	53 64	*Five years*
40 58	*46 59*	*43 59*					
(79)	*(95)*	*(174)*		*(77)*	*(82)*	*(159)*	**Trent**
87 93	89 93	88 93		*83 88*	*94 96*	*89 93*	*One year*
58 73	54 64	56 68		*47 59*	*60 69*	*53 65*	*Five years*
42 64	*41 55*	*41 58*					
(77)	*(72)*	*(149)*		*(73)*	*(70)*	*(143)*	**Anglia & Oxford**
94 96	*86 91*	*90 93*		*93 100*	*89 92*	*91 95*	*One year*
65 80	*54 63*	*59 72*		*60 71*	*58 71*	*59 70*	*Five years*
49 71	*39 54*	*44 63*					
(90)	*(92)*	*(182)*		*(142)*	*(116)*	*(258)*	**North Thames**
96 100	*92 97*	*94 98*		93 95	95 97	94 96	*One year*
62 76	*70 84*	*66 80*		63 75	67 75	65 76	*Five years*
52 74	*53 76*	*53 76*					
(106)	*(99)*	*(205)*		*(137)*	*(141)*	*(278)*	**South Thames**
90 96	96 98	93 97		93 97	89 93	91 95	*One year*
64 77	60 69	62 73		60 71	61 72	61 72	*Five years*
50 71	*41 54*	*45 63*					
(125)	*(119)*	*(244)*		*(134)*	*(129)*	*(263)*	**South & West**
85 88	92 94	89 91		89 93	91 93	90 93	*One year*
56 66	61 72	59 69		60 73	65 77	62 75	*Five years*
40 56	*38 54*	*39 55*					
(45)	*(56)*	*(101)*		*(56)*	*(53)*	*(109)*	**West Midlands**
87 93	86 88	86 90		88 91	89 93	88 92	*One year*
34 42	*54 61*	*45 53*		55 69	59 69	57 68	*Five years*
22 37	*32 42*	*28 39*					
(82)	*(100)*	*(182)*		*(91)*	*(80)*	*(171)*	**North & West**
80 83	82 86	81 85		*86 90*	*86 90*	*86 90*	*One year*
46 56	*50 60*	*49 58*		*53 67*	*62 74*	*57 71*	*Five years*
28 43	*31 47*	*30 45*					
(56)	*(54)*	*(110)*		*(49)*	*(58)*	*(107)*	**WALES**
82 86	82 84	82 84		*90 93*	*83 85*	*86 89*	*One year*
57 70	54 63	55 67		*69 78*	*55 64*	*62 72*	*Five years*
43 65	*39 58*	*41 64*					

[1] Including those for whom no deprivation category was assigned. C - crude survival, R - relative survival (see Chapter 3). Number of patients contributing to each analysis in parentheses.

Fig. 38.1   Relative survival up to ten years: trends by calendar period of diagnosis (1971-90) and NHS Region

ADULTS

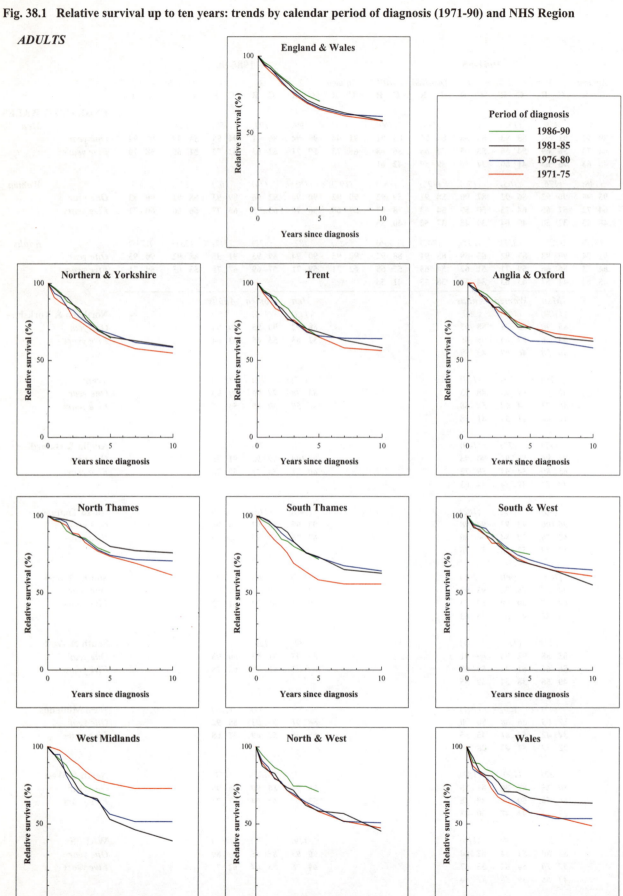

Fig. 38.2 Relative survival up to five years, by deprivation category: England and Wales, patients diagnosed 1986-90

ADULTS

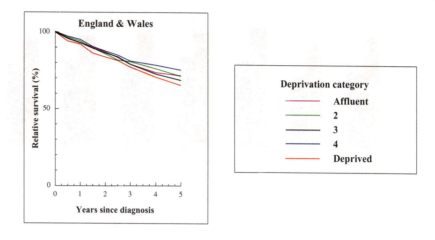

Fig. 38.3 Relative survival at five years, by deprivation categoryand period of diagnosis (1981-90): England and Wales

ADULTS

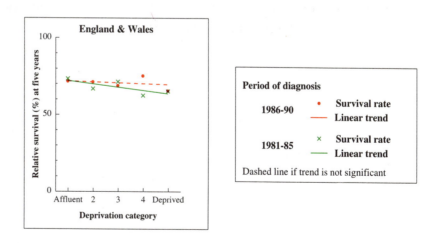

Fig. 38.4   Relative survival (%) by NHS Region and sex:
one-year and five-year survival for patients diagnosed 1986-90, with 95% confidence intervals

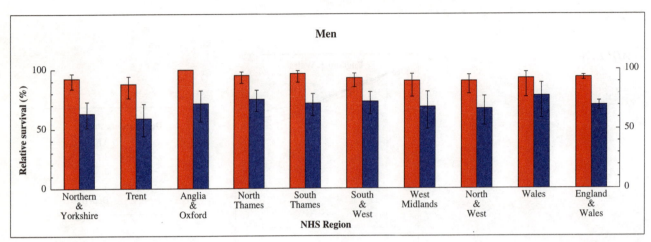

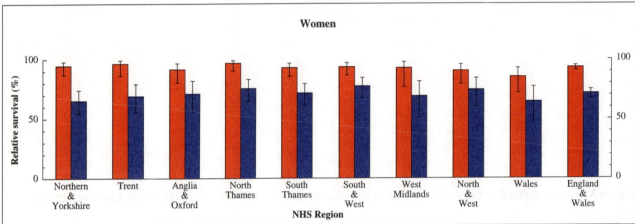

Fig. 38.5   Relative survival (%) at five years, with 95% confidence interval, by age at diagnosis and sex:
England and Wales, patients diagnosed 1986-90

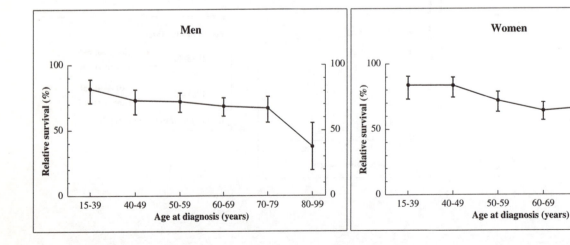

Chapter 39

BRAIN

Brain tumours comprise 1.5% of all malignant neoplasms in adults in England and Wales, with over 3,000 new cases and 2,500 deaths each year. Most are gliomas, including astrocytoma, oligodendroglioma, and ependymoma. The other common types, meningioma and neuroma, are predominantly benign. This chapter covers only malignant tumours of the brain. Malignant tumours of other parts of the central nervous system (cranial nerves and spinal cord) were excluded, and tumours of the brain itself were only included if explicitly specified as malignant primary tumours. Survival from tumours of the spinal cord and the pituitary gland is addressed in Chapter 62.

Brain tumours are 1.5 times more common in men. In 1990, the age-standardised (Europe) incidence in England and Wales was 7.3 per 100,000 per year in men and 4.8 in women, an increase of about 25% since 1971. Recorded incidence is highest in developed countries, particularly Ireland, the Nordic countries, New Zealand and the USA, in part because tumours of benign or unspecified behaviour are often included with malignant tumours[1].

Cancers in other organs frequently metastasise to the brain, especially those of lung, breast, kidney, gastrointestinal tract and prostate, and malignant melanoma of skin. These are also excluded, insofar as is possible: tumours with the behaviour code for metastatic disease (/6) were not included, and the distribution of morphologic types was examined.

The accuracy of diagnosis of cerebral tumours has improved. Computed tomography was introduced in the 1970s, magnetic resonance imaging became widespread during the 1980s, and stereotactic biopsy can enable histological diagnosis of unresectable tumours. These tools provide better information on the location, size and invasion of the tumour. Better discrimination between primary brain tumours and metastases from cancers in other organs probably accounts for part of the increase in incidence of brain tumours observed over the past 20 years. The impact on trends in survival is likely to be smaller, because prognosis in adults is very poor for many primary brain tumours, as it is for cerebral metastases, and has not changed greatly. Five-year survival for low-grade astrocytomas is around 60% in clinical series, while high-grade tumours have only 30% survival at one year[2].

The survival analyses reported here are based on 41,000 adults diagnosed with a primary malignant tumour of the brain in England and Wales during the period 1971-90 and followed up to the end of 1995. After the exclusion of some 8% of tumours arising mainly in children aged under

15, these represent about 86% of tumours in eligible adults (Table 39.1). Survival from brain tumours in children is discussed in Chapter 54. About 5% of eligible adults were excluded because their vital status was unknown at 6 July 1997, and another 8% because their survival was zero or unknown. During 1986-90, over 40% of brain tumours arose in the frontal, parietal and temporal lobes, but the site was poorly specified or unspecified for almost as many again (Table 39A). Only 3% arose in the cerebellum or brain stem. Most tumours were gliomas (86%): 28% astrocytoma, 18% glioblastoma, 3% oligodendroblastoma and 37% other or unspecified glioma. Less than 100 tumours (0.2%) had an epithelial morphology, although for 10% the morphology was poorly specified or absent. Almost 26,000 brain tumours recorded in the national cancer registry for the period 1971-90, a third of all brain tumours, were of benign (65%), uncertain (1%) or unspecified (34%) behaviour: these tumours were *not* included in the initial data set (see Chapter 2) and thus do not appear in Table 39.1.

Survival trends

Relative survival for adults diagnosed with a malignant tumour of the brain in England and Wales during 1986-90 was about 30% at one year and 15% at five years (Table 39.2). One-year and five-year survival have improved since the early 1970s by about 10% and 7%, respectively. Ten-year survival was just 8% for patients diagnosed in 1981-85. Survival in England and Wales as a whole has improved gradually since the early 1970s, but the timing and extent of improvement in survival varies between NHS Regions (Figure 39.1).

For patients diagnosed during 1986-90, survival was 2% higher for women than for men, after adjustment to a common age standard. The national trends in survival since the early 1970s have been closely similar in both sexes (see Table 4.4). The increase in survival with time has been fairly regular: the average increase between successive five-year periods of diagnosis was 3% for one-year survival and 2% for five-year survival.

Survival improved in most regions, but the regional trends in age-adjusted survival show some unusual differences between men and women. Relative survival at both one year and five years for men in North & West Region has apparently improved much less than for women in the same region, and five-year survival for men diagnosed during 1986-90 was 10%, significantly lower than the national average, compared with 19% for women, significantly higher than the average (see Table-Figure

Table 39A Sub-site distribution for malignant tumours of brain

ICD-9 code	Site description	1971-75 No.	%	1976-80 No.	%	1981-85 No.	%	1986-90 No.	%	Total No.	%
191.0	Cerebrum (except lobes, ventricles)	-	-	357	3.8	1,094	10.1	1,102	9.2	**2,553**	**6.2**
191.1	Frontal lobe	-	-	626	6.7	1,891	17.5	1,804	15.0	**4,321**	**10.5**
191.2	Temporal lobe	-	-	464	4.9	1,187	11.0	1,309	10.9	**2,960**	**7.2**
191.3	Parietal lobe	-	-	613	6.5	1,821	16.9	1,863	15.5	**4,297**	**10.5**
191.4	Occipital lobe	-	-	89	0.9	263	2.4	306	2.5	**658**	**1.6**
191.5	Ventricle	-	-	58	0.6	130	1.2	148	1.2	**336**	**0.8**
191.6	Cerebellum	-	-	104	1.1	271	2.5	210	1.7	**585**	**1.4**
191.7	Brain stem	-	-	54	0.6	148	1.4	157	1.3	**359**	**0.9**
191.8	Other	-	-	434	4.6	1,124	10.4	1,344	11.2	**2,902**	**7.1**
191.9	Unspecified	8,928	100	6,581	70.2	2,874	26.6	3,758	31.3	**22,141**	**53.9**
	Total	**8,928**		**9,380**		**10,803**		**12,001**		**41,112**	

4.5R). Survival for women diagnosed in Anglia & Oxford during 1971-75 was 27% at one year and 14% at five years, much higher than the national average, but since then there has apparently been no improvement at all. Five-year survival for women in Wales was also among the highest in 1971-75, and again has apparently improved very little. Some anomalous trends in survival may be due to shifts over time in the accuracy of diagnosis and coding in a given region, but it is difficult to explain differences in survival trends between men and women within a region on this basis.

Survival five years after the diagnosis of a malignant brain tumour falls rapidly with age at diagnosis (Figure 39.5). For adults under 40 at diagnosis during 1986-90, five-year relative survival was above 40%, while for those aged 40-49 at diagnosis, it was only 22% for men and 27% for women (see Table 4.7). Five-year survival was 3% or less in those aged 60 or over at diagnosis.

Deprivation

One-year survival has generally been higher among affluent groups (Table 39.2).

Survival curves for adults diagnosed in England and Wales during 1986-90 are similar for the various deprivation categories (Figure 39.2). There was no difference in five-year survival from malignant brain tumours in England and Wales as a whole in either 1981-85 or 1986-90 (Figure 39.3). One-year survival was 4% higher for the affluent group than the most deprived group, however, based on the survival of about 11,000 patients of known deprivation status in each period (see Table 4.7). For patients diagnosed in Northern & Yorkshire during 1986-90, there was a 9% gap in one-year survival, and the deprivation gradient in five-year survival (15% to 9%) was significant (Figure 39.3). Regional estimates of survival within each deprivation category are precise, generally depending on 200 or more cases.

International comparison

Five-year survival from malignant tumours for adults diagnosed in England and Wales during 1986-90 – 13% in men and 16% in women – was similar to that in Scotland, but 4-5% below the average for Europe, and about 10% lower than in the USA (see Table 4.12). The highest regional survival rates in England and Wales (18% and 20%, North Thames) were the same as the average for Europe.

1.   Doll R, Fraumeni JF, Muir CS (eds.) Trends in cancer incidence and mortality. Cancer Surv 1994; 19/20

2.   Neal AJ, Hoskin PJ. *Clinical oncology.* London: Edward Arnold, 1994

Table 39.1 Ineligible and excluded records, and patients included in analyses, by calendar period of diagnosis

	\multicolumn{15}{c}{Calendar period of diagnosis}													
	1971-75			1976-80			1981-85			1986-90			Overall		
Total registered	11,397			12,045			14,279			15,723			53,444		
Ineligible	*Records*	*Patients*	*%*	*Records*	*Patients*	*%*	*Records*	*Patients*	*%*	*Records*	*Patients*	*%*	*Records*	*Patients*	*%*
Incomplete data[1]	61	61		97	97		246	246		156	156		560	**560**	
Resident outside England and Wales	220	210		162	159		146	136		88	72		616	**577**	
In situ neoplasm[2]	1	0		1	1		0	0		0	0		2	**1**	
Uncertain behaviour[2]	93	83		22	22		0	0		0	0		115	**105**	
Metastatic[2]	38	33		5	5		0	0		1	1		44	**39**	
Otherwise ineligible[3]	-	-		-	-		-	-		-	-		-	**-**	
Lymphoma[4]	68	64		60	57		62	57		28	19		218	**197**	
Leukaemia or myeloma[4]	0	0		6	1		6	0		8	0		20	**1**	
		451			342			439			248			1,480	
Total eligible		10,946	100		11,703	100		13,840	100		15,475	100		51,964	100
Aged under 15 or 100 years or over at diagnosis	1,074	982	9.0	1,008	959	8.2	1,037	972	7.0	1,087	1,052	6.8	4,206	**3,965**	7.6
Vital status unknown[5]	605	338	3.1	782	542	4.6	873	687	5.0	1,046	866	5.6	3,306	**2,433**	4.7
Duplicate registration[6]	0	0	0.0	0	0	0.0	0	0	0.0	0	0	0.0	0	**0**	0.0
Multiple primary[7]	0	0	0.0	13	9	<0.1	91	79	0.6	80	65	0.4	184	**153**	0.3
Synchronous tumours[8]	25	18	0.2	37	28	0.2	63	44	0.3	123	64	0.4	248	**154**	0.3
Sex not known	0	0	0.0	0	0	0.0	0	0	0.0	0	0	0.0	0	**0**	0.0
Sex incompatible with site code[9]	-	-	-	-	-	-	-	-	-	-	-	-	-	**-**	-
Invalid dates[10]	1	1	<0.1	0	0	0.0	6	3	<0.1	15	4	<0.1	22	**8**	<0.1
Zero survival[11]	772	679	6.2	915	785	6.7	1,456	1,252	9.0	1,566	1,423	9.2	4,709	**4,139**	8.0
Total exclusions		2,018	18.4		2,323	19.8		3,037	21.9		3,474	22.4		10,852	20.9
Patients included in analysis		8,928	81.6		9,380	80.2		10,803	78.1		12,001	77.6		41,112	79.1

---

[1] Main data item(s) invalid or incompatible with one another: name, sex, date of birth, date of diagnosis and (if present) date of death, postcode, site, morphology and behaviour

[2] Tumour either *in situ* (behaviour code 2), uncertain if malignant (1) or metastatic (6 or 9). A few benign tumours were retained for analysis (0.4%) (see Ch.2)

[3] Not applicable for this malignancy

[4] Morphology that of lymphoma (included with non-Hodgkin lymphoma, Ch. 41) or myeloma (included with multiple myeloma, Ch. 43), or leukaemia (excluded)

[5] Tracing of vital status at National Health Service Central Register incomplete at 6 July 1997

[6] Same site code, sex, cancer registry and cancer registry number as an earlier registration

[7] Same site code and person number as an earlier registration(s): mostly confirmed multiple primary tumours at this site, some unresolved duplicate registrations

[8] Same site code, sex, date of birth and date of diagnosis as another registration(s): mostly synchronous or (in paired organs) bilateral tumours in same anatomic site in one individual, not previously linked: also some duplicate registrations

[9] Not applicable for this malignancy

[10] Impossible sequence of dates (birth, diagnosis and, if present, emigration or death); or date of death after 6 July 1997

[11] Date of diagnosis same as date of death: some are patients who did die on the day of diagnosis, but most were registered solely from a death certificate (death certificate only, or DCO), and their survival time is thus unknown

Table 39.2 Crude and relative survival (%) at one, five and ten years since diagnosis: NHS Region, deprivation category and calendar period of diagnosis

MEN	1971-75 Affluent C R	2 C R	3 C R	4 C R	Deprived C R	All¹ C R	1976-80 Affluent C R	2 C R	3 C R	4 C R	Deprived C R	All¹ C R
ENGLAND & WALES	(904)	(908)	(847)	(859)	(736)	(5,228)	(1,194)	(1,120)	(1,105)	(971)	(803)	(5,417)
One year	22 23	20 21	20 20	20 21	21 21	21 21	24 25	22 23	24 25	23 24	22 23	23 24
Five years	7 8	6 7	7 8	7 8	7 7	7 8	9 10	9 9	9 10	9 10	10 11	9 10
Ten years	5 6	4 5	5 6	5 6	5 5	5 6	6 6	5 6	6 7	5 6	6 7	6 7
ENGLAND	(874)	(873)	(799)	(780)	(691)	(4,933)	(1,162)	(1,073)	(1,042)	(908)	(762)	(5,141)
One year	22 22	20 20	20 20	20 20	20 21	21 21	24 25	22 23	23 24	23 24	23 23	23 24
Five years	7 8	6 6	7 8	7 8	7 7	7 7	9 10	9 9	9 9	9 9	10 11	9 10
Ten years	5 6	4 5	5 6	5 6	5 5	5 5	6 7	5 6	6 6	5 6	6 7	6 6
Northern & Yorkshire	(61)	(91)	(83)	(82)	(121)	(571)	(109)	(114)	(147)	(153)	(167)	(712)
One year	20 20	11 12	16 16	14 14	22 22	20 20	21 22	16 16	20 21	22 23	23 23	22 23
Five years	3 3	4 4	5 5	5 5	6 7	6 6	8 8	5 6	10 11	10 11	10 10	9 10
Ten years	1 1	2 2	4 4	4 4	4 5	4 4	7 7	3 3	4 5	6 7	5 6	6 6
Trent	(66)	(82)	(101)	(98)	(81)	(483)	(70)	(93)	(99)	(93)	(99)	(462)
One year	15 15	18 18	12 12	14 15	20 20	18 19	22 22	29 29	19 20	23 23	21 22	25 26
Five years	7 7	7 7	6 6	8 8	4 5	8 8	3 3	13 14	4 4	12 12	10 11	10 10
Ten years	7 7	5 5	4 4	6 6	3 3	6 6	1 1	9 9	2 2	6 6	8 10	6 7
Anglia & Oxford	(148)	(121)	(104)	(71)	(37)	(548)	(168)	(150)	(103)	(105)	(29)	(576)
One year	21 22	21 21	25 25	25 25	26 27	25 26	28 28	22 23	24 25	28 29	14 14	27 27
Five years	8 9	6 6	11 11	8 9	10 10	9 10	12 13	8 8	6 7	10 11	0 0	9 10
Ten years	6 7	4 5	8 9	6 7	10 10	7 8	7 8	5 5	4 4	5 6	0 0	6 6
North Thames	(139)	(118)	(108)	(115)	(101)	(669)	(126)	(102)	(112)	(119)	(96)	(584)
One year	22 22	27 27	30 31	29 30	23 24	27 28	26 26	25 26	31 31	23 24	33 34	28 28
Five years	8 8	10 11	10 11	8 8	9 9	9 10	10 11	9 9	12 12	11 11	14 15	11 12
Ten years	2 2	9 9	8 8	6 7	7 7	6 7	6 7	6 6	10 11	6 7	10 11	8 8
South Thames	(193)	(149)	(131)	(135)	(79)	(764)	(325)	(220)	(197)	(122)	(51)	(963)
One year	24 25	20 21	16 17	17 18	22 23	21 22	24 25	19 20	23 23	19 20	13 13	23 23
Five years	6 7	4 4	4 5	6 6	8 9	6 7	9 10	8 8	8 9	5 6	9 9	9 9
Ten years	6 6	2 2	3 3	3 3	6 7	5 5	6 7	5 6	6 7	4 5	5 6	6 6
South & West	(132)	(180)	(139)	(90)	(26)	(618)	(160)	(198)	(171)	(89)	(27)	(672)
One year	17 17	13 13	13 13	13 13	7 7	15 15	18 18	21 22	19 19	23 23	20 20	22 22
Five years	5 5	5 5	7 7	6 7	4 4	6 6	9 10	8 9	6 7	9 9	16 17	9 10
Ten years	3 3	3 4	5 5	5 6	4 4	4 4	6 6	6 7	4 5	4 4	11 12	6 7
West Midlands	(32)	(43)	(38)	(56)	(58)	(517)	(68)	(80)	(90)	(87)	(88)	(416)
One year	34 34	20 20	25 25	13 14	12 13	19 20	25 26	26 27	25 26	21 21	18 18	24 25
Five years	12 12	6 4	10 10	3 3	4 4	6 6	9 10	13 14	13 14	4 5	7 8	10 10
Ten years	9 10	3 3	8 8	1 1	1 2	4 4	5 6	6 6	8 9	1 1	4 5	5 6
North & West	(103)	(89)	(95)	(133)	(188)	(763)	(136)	(116)	(123)	(140)	(205)	(756)
One year	16 16	21 21	15 15	16 16	14 14	18 19	19 19	17 18	17 18	15 15	19 19	19 20
Five years	7 8	6 7	4 4	7 7	5 5	6 7	7 7	7 7	6 7	5 5	8 8	7 8
Ten years	7 8	4 5	3 3	4 5	3 3	5 5	5 5	2 2	5 6	4 4	5 5	5 5
WALES	(30)	(35)	(48)	(79)	(45)	(295)	(32)	(47)	(63)	(63)	(41)	(276)
One year	39 40	35 36	16 17	23 24	23 23	28 29	20 21	20 21	36 36	22 23	15 15	25 26
Five years	6 6	12 13	5 5	10 10	6 6	9 10	2 2	5 6	14 15	15 16	8 9	11 11
Ten years	1 1	12 13	5 5	6 7	6 6	7 8	0 0	5 6	13 14	8 9	4 5	7 8

¹ Including those for whom no deprivation category was assigned. C - crude survival, R - relative survival (see Chapter 3).
Number of patients contributing to each analysis in parentheses.

	1981-85												1986-90												MEN
	Affluent		*2*		*3*		*4*		*Deprived*		*All¹*		*Affluent*		*2*		*3*		*4*		*Deprived*		*All¹*		
	C	R	C	R	C	R	C	R	C	R	C	R	C	R	C	R	C	R	C	R	C	R	C	R	
	(1,437)		*(1,311)*		*(1,257)*		*(1,171)*		*(902)*		*(6,232)*		*(1,650)*		*(1,576)*		*(1,417)*		*(1,258)*		*(941)*		*(6,942)*		**ENGLAND & WALES**
One year	28	28	26	27	27	28	25	26	24	25	26	27	30	30	31	32	28	29	28	29	25	26	29	30	*One year*
Five years	11	11	11	11	11	12	11	12	11	12	11	12	13	13	14	15	12	13	12	13	11	12	13	13	*Five years*
	7	8	7	8	6	7	7	7	7	8	7	8													
	(1,394)		*(1,244)*		*(1,181)*		*(1,066)*		*(849)*		*(5,849)*		*(1,586)*		*(1,507)*		*(1,313)*		*(1,155)*		*(884)*		*(6,527)*		**ENGLAND**
	28	28	26	26	27	28	24	25	23	24	26	26	29	30	31	32	29	29	29	30	25	26	29	30	*One year*
	11	11	11	11	11	11	10	11	11	12	10	11	13	14	14	15	12	13	12	13	12	13	13	14	*Five years*
	7	8	7	8	6	7	6	7	7	8	7	7													
	(134)		*(139)*		*(157)*		*(197)*		*(190)*		*(825)*		*(149)*		*(161)*		*(166)*		*(170)*		*(203)*		*(856)*		**Northern & Yorkshire**
	31	32	22	22	26	27	20	21	18	18	24	24	28	28	29	30	28	28	27	28	22	23	27	28	*One year*
	12	12	11	11	13	14	10	11	7	7	11	11	13	14	17	19	11	13	14	15	9	10	13	14	*Five years*
	8	9	9	9	8	8	6	7	5	5	7	8													
	(99)		*(124)*		*(136)*		*(145)*		*(86)*		*(597)*		*(114)*		*(147)*		*(164)*		*(162)*		*(121)*		*(711)*		**Trent**
	27	28	25	25	23	23	21	22	15	16	24	25	21	21	23	24	28	29	26	26	20	20	25	26	*One year*
	14	14	11	11	9	9	7	7	10	10	10	11	9	10	9	10	11	12	10	11	10	11	11	11	*Five years*
	10	11	5	5	5	6	3	3	8	8	6	7													
	(201)		*(178)*		*(136)*		*(100)*		*(51)*		*(681)*		*(288)*		*(241)*		*(156)*		*(125)*		*(32)*		*(848)*		**Anglia & Oxford**
	19	20	23	23	25	25	27	28	22	22	24	24	29	29	28	29	23	24	28	29	23	24	29	29	*One year*
	6	7	9	9	7	8	12	13	8	8	9	9	12	13	11	12	12	13	12	13	12	12	12	13	*Five years*
	5	5	6	7	4	4	8	8	4	5	6	6													
	(147)		*(122)*		*(109)*		*(110)*		*(100)*		*(602)*		*(163)*		*(163)*		*(150)*		*(147)*		*(101)*		*(750)*		**North Thames**
	38	38	36	37	31	31	33	34	33	34	35	36	32	33	46	47	34	35	41	43	38	39	38	39	*One year*
	15	16	15	16	8	9	20	21	19	21	16	17	15	16	19	20	13	14	15	16	19	20	17	18	*Five years*
	7	8	13	14	7	7	13	15	14	16	11	12													
	(311)		*(184)*		*(174)*		*(121)*		*(47)*		*(866)*		*(271)*		*(222)*		*(165)*		*(118)*		*(46)*		*(844)*		**South Thames**
	29	29	26	26	27	28	21	22	38	39	28	29	34	34	33	34	32	33	29	30	38	39	33	34	*One year*
	9	10	12	12	8	9	7	7	25	26	11	11	16	17	15	15	14	15	12	14	17	18	15	16	*Five years*
	6	7	9	10	6	7	6	6	17	18	8	9													
	(231)		*(249)*		*(216)*		*(103)*		*(28)*		*(849)*		*(237)*		*(282)*		*(214)*		*(148)*		*(50)*		*(936)*		**South & West**
	27	27	20	20	24	25	24	25	9	9	25	25	22	22	27	28	25	26	26	27	27	28	26	27	*One year*
	10	11	7	7	11	11	12	13	6	6	10	11	7	7	13	14	11	11	10	11	10	11	11	11	*Five years*
	7	7	5	5	5	5	8	9	6	6	6	7													
	(114)		*(117)*		*(135)*		*(125)*		*(135)*		*(631)*		*(148)*		*(126)*		*(131)*		*(144)*		*(119)*		*(675)*		**West Midlands**
	24	25	33	34	30	31	18	19	20	21	26	27	30	31	37	38	29	29	27	28	21	22	31	31	*One year*
	12	13	9	9	13	14	6	6	9	9	10	11	15	17	16	17	13	14	11	11	14	15	15	16	*Five years*
	8	9	6	6	6	6	2	2	5	6	6	6													
	(157)		*(131)*		*(118)*		*(165)*		*(212)*		*(798)*		*(216)*		*(165)*		*(167)*		*(141)*		*(212)*		*(907)*		**North & West**
	18	19	18	19	22	22	25	25	20	21	22	23	32	33	21	21	22	23	21	22	19	20	25	25	*One year*
	8	8	10	10	10	11	8	8	7	8	9	9	13	14	9	10	8	9	8	8	6	7	10	10	*Five years*
	6	6	5	6	5	6	4	4	4	4	5	5													
	(43)		*(67)*		*(76)*		*(105)*		*(53)*		*(383)*		*(64)*		*(69)*		*(104)*		*(103)*		*(57)*		*(415)*		**WALES**
	31	32	34	36	28	29	33	34	36	37	34	35	30	31	32	33	23	24	15	15	18	18	23	24	*One year*
	19	20	16	17	17	18	20	22	13	14	18	19	9	9	19	20	11	12	10	11	5	6	11	12	*Five years*
	19	20	7	8	13	15	11	13	6	7	11	12													

¹ Including those for whom no deprivation category was assigned. C - crude survival, R - relative survival (see Chapter 3).
Number of patients contributing to each analysis in parentheses.

Table 39.2   Crude and relative survival (%) at one, five and ten years since diagnosis: NHS Region, deprivation category and calendar period of diagnosis

WOMEN		1971-75						1976-80					
		Affluent C R	2 C R	3 C R	4 C R	Deprived C R	All¹ C R	Affluent C R	2 C R	3 C R	4 C R	Deprived C R	All¹ C R
ENGLAND & WALES		(639)	(643)	(631)	(612)	(505)	(3,700)	(847)	(783)	(797)	(725)	(622)	(3,963)
	One year	23 23	21 22	23 23	22 22	21 22	22 22	28 29	22 23	24 25	22 22	21 21	24 24
	Five years	10 10	9 9	9 9	9 10	11 11	10 10	13 13	8 8	10 10	10 11	9 10	10 11
	Ten years	6 7	6 6	6 7	7 8	7 8	7 7	8 8	5 5	7 7	6 7	6 7	7 7
ENGLAND		(615)	(611)	(586)	(555)	(468)	(3,468)	(821)	(742)	(754)	(667)	(595)	(3,740)
	One year	22 23	21 21	22 22	22 22	21 21	22 22	28 29	22 23	24 24	21 21	21 21	23 24
	Five years	9 10	9 9	8 8	9 10	10 11	9 9	13 13	8 8	10 10	10 10	9 9	10 10
	Ten years	6 7	5 6	6 6	7 8	7 8	6 7	8 8	5 6	7 7	6 6	6 7	6 7
Northern & Yorkshire		(50)	(66)	(66)	(65)	(84)	(434)	(67)	(88)	(97)	(128)	(154)	(548)
	One year	20 21	16 16	13 13	20 21	26 27	20 21	27 27	24 24	22 22	22 23	15 15	23 23
	Five years	5 5	3 3	3 3	6 7	15 15	7 7	9 10	9 10	8 8	11 11	4 5	9 9
	Ten years	5 5	3 3	3 3	6 7	11 11	6 6	8 8	4 5	5 6	9 9	2 2	6 7
Trent		(32)	(59)	(63)	(76)	(75)	(357)	(63)	(60)	(69)	(64)	(52)	(321)
	One year	23 23	18 18	17 17	27 27	11 11	19 19	34 34	29 30	34 34	20 20	26 26	29 30
	Five years	10 10	8 8	10 10	13 13	6 7	10 11	17 17	15 15	15 15	8 9	15 16	14 14
	Ten years	7 7	5 5	8 9	9 10	5 6	7 8	10 11	10 11	7 7	5 5	14 14	9 9
Anglia & Oxford		(86)	(82)	(77)	(59)	(25)	(372)	(123)	(97)	(103)	(66)	(22)	(438)
	One year	27 27	24 24	27 28	28 28	23 23	29 29	25 26	15 15	22 22	18 18	37 38	23 23
	Five years	13 13	10 10	10 10	10 10	12 12	12 13	11 11	4 4	7 7	4 5	9 9	8 8
	Ten years	8 8	8 8	9 9	7 8	8 8	10 10	7 7	2 2	6 6	1 1	9 9	5 5
North Thames		(90)	(84)	(86)	(82)	(56)	(466)	(90)	(83)	(78)	(71)	(66)	(416)
	One year	26 26	31 32	20 20	20 20	23 24	27 28	38 39	26 26	32 32	26 26	18 19	30 31
	Five years	11 12	12 13	5 5	7 7	8 8	11 11	21 22	8 8	13 13	10 11	8 8	13 14
	Ten years	6 6	7 8	3 4	6 6	8 8	7 8	16 17	3 4	8 8	6 7	6 7	9 9
South Thames		(163)	(109)	(99)	(112)	(49)	(584)	(194)	(127)	(136)	(103)	(46)	(641)
	One year	22 23	12 12	17 18	16 17	16 16	19 19	25 25	18 18	16 16	18 18	24 24	22 22
	Five years	7 8	5 5	4 4	4 4	8 8	7 7	12 12	4 4	9 9	8 8	15 16	10 10
	Ten years	5 5	4 4	3 4	3 3	3 4	4 5	6 7	3 4	6 6	3 4	15 16	6 7
South & West		(104)	(110)	(99)	(51)	(21)	(411)	(141)	(145)	(107)	(62)	(23)	(492)
	One year	15 15	17 17	16 16	21 21	18 18	18 18	19 20	18 19	18 18	18 18	25 26	21 21
	Five years	9 9	7 7	6 6	14 15	18 18	9 9	8 8	6 6	8 8	12 13	7 8	9 9
	Ten years	6 6	3 3	4 5	12 13	14 17	6 7	5 5	4 4	6 6	7 7	4 5	6 6
West Midlands		(25)	(30)	(37)	(23)	(35)	(330)	(40)	(60)	(62)	(65)	(65)	(296)
	One year	7 8	16 16	33 34	18 18	13 13	21 21	39 39	26 26	19 19	13 14	18 18	22 23
	Five years	7 8	11 11	18 18	18 18	0 0	10 11	19 19	14 14	6 7	5 5	5 5	9 10
	Ten years	7 8	8 9	12 12	14 14	0 0	7 8	10 11	9 9	6 7	5 5	4 4	7 7
North & West		(65)	(71)	(59)	(87)	(123)	(514)	(103)	(82)	(102)	(108)	(167)	(588)
	One year	16 16	18 18	24 25	11 11	18 19	20 20	23 23	17 17	21 21	20 20	17 17	20 20
	Five years	4 4	8 8	8 8	6 6	10 10	9 9	11 11	7 7	9 10	12 13	8 8	10 10
	Ten years	3 3	3 3	2 2	4 4	6 7	4 5	4 4	6 6	8 9	7 7	4 4	6 6
WALES		(24)	(32)	(45)	(57)	(37)	(232)	(26)	(41)	(43)	(58)	(27)	(223)
	One year	37 38	22 23	31 32	20 20	26 26	28 28	36 37	23 23	33 34	28 28	23 23	30 30
	Five years	21 21	10 10	21 21	11 11	11 12	16 16	14 14	4 5	10 11	13 14	20 20	14 14
	Ten years	12 12	7 8	12 13	8 8	7 7	10 10	4 4	3 3	6 6	10 11	14 14	8 9

¹ Including those for whom no deprivation category was assigned. C - crude survival, R - relative survival (see Chapter 3).
  Number of patients contributing to each analysis in parentheses.

	1981-85												1986-90												WOMEN
---	Affluent		2		3		4		Deprived		All[1]		Affluent		2		3		4		Deprived		All[1]		
	C	R	C	R	C	R	C	R	C	R	C	R	C	R	C	R	C	R	C	R	C	R	C	R	
	(1,015)		*(1,005)*		*(889)*		*(873)*		*(686)*		*(4,571)*		*(1,190)*		*(1,147)*		*(1,083)*		*(920)*		*(671)*		*(5,059)*		**ENGLAND & WALES**
One year	28	28	26	27	26	27	25	26	24	25	26	27	32	32	30	31	27	28	31	31	29	29	30	31	*One year*
Five years	12	12	12	13	12	12	12	13	11	12	12	13	17	17	14	15	13	14	16	17	15	15	15	16	*Five years*
	8	8	7	8	8	9	7	8	8	9	8	8													
	(980)		*(961)*		*(818)*		*(806)*		*(654)*		*(4,296)*		*(1,125)*		*(1,075)*		*(1,010)*		*(834)*		*(619)*		*(4,699)*		**ENGLAND**
One year	28	28	26	27	26	26	25	26	24	24	**26**	**26**	32	33	30	31	28	28	31	31	29	30	**30**	**31**	*One year*
Five years	12	13	12	13	11	12	13	13	11	12	**12**	**13**	17	17	14	15	13	14	16	17	14	15	**15**	**16**	*Five years*
	8	9	7	8	8	8	7	8	8	9	**8**	**8**													
	(94)		*(123)*		*(120)*		*(155)*		*(169)*		*(672)*		*(116)*		*(125)*		*(125)*		*(127)*		*(139)*		*(634)*		**Northern & Yorkshire**
One year	24	25	18	19	21	21	18	18	24	24	**23**	**24**	32	33	30	31	21	21	20	20	19	20	**26**	**27**	*One year*
Five years	9	9	8	9	6	7	6	6	10	10	**9**	**9**	15	16	9	10	7	8	8	8	7	8	**10**	**11**	*Five years*
	5	5	5	6	6	7	3	3	8	8	**6**	**7**													
	(70)		*(100)*		*(102)*		*(108)*		*(86)*		*(472)*		*(77)*		*(99)*		*(110)*		*(107)*		*(83)*		*(477)*		**Trent**
One year	30	30	23	24	26	26	24	25	26	26	**27**	**27**	32	33	28	28	20	20	29	30	23	24	**28**	**29**	*One year*
Five years	-	-	10	10	15	16	8	8	18	19	**13**	**14**	15	15	16	17	11	11	13	14	9	10	**14**	**15**	*Five years*
	10	11	6	7	9	10	5	6	13	14	**9**	**10**													
	(150)		*(123)*		*(92)*		*(60)*		*(31)*		*(465)*		*(195)*		*(165)*		*(126)*		*(89)*		*(40)*		*(617)*		**Anglia & Oxford**
One year	26	26	30	30	26	26	27	28	*13*	*13*	**29**	**29**	24	25	23	24	18	18	27	28	*19*	*19*	**24**	**24**	*One year*
Five years	10	10	12	13	11	12	22	23	5	5	**13**	**13**	14	15	12	12	8	8	14	15	*11*	*12*	**13**	**13**	*Five years*
	7	7	7	7	9	10	*8*	*8*	*0*	*0*	**7**	**8**													
	(116)		*(86)*		*(79)*		*(94)*		*(68)*		*(453)*		*(125)*		*(127)*		*(100)*		*(105)*		*(73)*		*(540)*		**North Thames**
One year	30	30	30	31	31	31	33	33	25	25	**31**	**32**	39	39	29	29	41	42	33	33	29	29	**35**	**36**	*One year*
Five years	12	13	14	15	15	16	20	21	15	16	**16**	**17**	21	21	14	15	21	22	19	20	17	18	**19**	**20**	*Five years*
	10	10	12	13	12	13	16	18	14	14	**13**	**14**													
	(211)		*(149)*		*(106)*		*(90)*		*(35)*		*(615)*		*(219)*		*(143)*		*(138)*		*(86)*		*(33)*		*(636)*		**South Thames**
One year	29	29	28	29	27	28	24	24	29	30	**29**	**30**	32	32	26	27	31	31	40	41	*13*	*13*	**32**	**32**	*One year*
Five years	12	13	12	13	14	15	12	12	*14*	*15*	**14**	**14**	16	16	16	17	16	16	24	25	*7*	*7*	**17**	**18**	*Five years*
	7	8	7	7	9	10	5	6	*4*	*5*	**8**	**8**													
	(158)		*(195)*		*(151)*		*(76)*		*(21)*		*(613)*		*(174)*		*(195)*		*(184)*		*(106)*		*(31)*		*(692)*		**South & West**
One year	26	26	22	22	20	20	32	33	*3*	*3*	**25**	**25**	30	31	32	32	19	19	25	25	*33*	*33*	**29**	**29**	*One year*
Five years	10	11	11	12	10	10	15	15	*1*	*1*	**12**	**12**	17	17	15	15	8	8	13	13	*9*	*9*	**14**	**15**	*Five years*
	8	8	6	6	6	6	8	9	*0*	*0*	**7**	**7**													
	(70)		*(77)*		*(75)*		*(105)*		*(73)*		*(403)*		*(99)*		*(89)*		*(100)*		*(96)*		*(65)*		*(449)*		**West Midlands**
One year	24	24	26	26	29	29	23	23	17	17	**26**	**26**	34	35	38	38	30	31	29	29	38	39	**35**	**35**	*One year*
Five years	15	16	14	14	10	10	8	8	4	4	**10**	**11**	19	20	17	17	12	13	14	15	19	20	**17**	**17**	*Five years*
	8	8	6	6	6	6	3	3	4	4	**6**	**6**													
	(111)		*(108)*		*(93)*		*(118)*		*(171)*		*(603)*		*(120)*		*(132)*		*(127)*		*(118)*		*(155)*		*(654)*		**North & West**
One year	20	20	20	20	18	18	17	18	21	22	**21**	**21**	29	29	30	30	32	33	36	36	35	36	**34**	**34**	*One year*
Five years	12	13	10	11	6	6	12	13	10	10	**11**	**12**	13	14	11	11	18	19	18	19	21	22	**17**	**18**	*Five years*
	7	8	5	6	2	2	9	10	7	8	**7**	**7**													
	(35)		*(44)*		*(71)*		*(67)*		*(32)*		*(275)*		*(65)*		*(72)*		*(73)*		*(86)*		*(52)*		*(360)*		**WALES**
One year	24	24	29	29	27	27	29	29	27	27	**27**	**27**	26	26	28	29	24	24	26	26	24	25	**29**	**29**	*One year*
Five years	3	3	16	17	14	15	11	11	*12*	*12*	**12**	**12**	13	14	12	12	15	16	15	15	*15*	*15*	**15**	**16**	*Five years*
	0	0	14	15	9	9	6	7	12	12	**8**	**9**													

[1] Including those for whom no deprivation category was assigned. C - crude survival, R - relative survival (see Chapter 3).
Number of patients contributing to each analysis in parentheses.

Table 39.2   Crude and relative survival (%) at one, five and ten years since diagnosis: NHS Region, deprivation category and calendar period of diagnosis

	1971-75						1976-80					
ADULTS	*Affluent*	*2*	*3*	*4*	*Deprived*	*All[1]*	*Affluent*	*2*	*3*	*4*	*Deprived*	*All[1]*
	C R	C R	C R	C R	C R	C R	C R	C R	C R	C R	C R	C R
ENGLAND & WALES	*(1,543)*	*(1,551)*	*(1,478)*	*(1,471)*	*(1,241)*	*(8,928)*	*(2,041)*	*(1,903)*	*(1,902)*	*(1,696)*	*(1,425)*	*(9,380)*
One year	23 23	21 21	21 21	21 21	21 21	21 22	26 26	22 23	24 25	23 23	22 22	24 24
Five years	8 9	7 8	8 8	8 9	8 9	8 9	11 11	8 9	9 10	10 10	10 10	10 10
Ten years	6 6	5 5	6 6	6 7	6 6	6 6	7 7	5 6	6 7	6 6	6 7	6 7
ENGLAND	*(1,489)*	*(1,484)*	*(1,385)*	*(1,335)*	*(1,159)*	*(8,401)*	*(1,983)*	*(1,815)*	*(1,796)*	*(1,575)*	*(1,357)*	*(8,881)*
One year	22 22	20 21	21 21	20 21	21 21	**21 21**	26 26	22 23	23 24	22 23	22 22	**23 24**
Five years	8 8	7 8	8 8	8 8	8 9	**8 8**	11 11	8 9	9 10	9 10	9 10	**10 10**
Ten years	6 6	5 5	5 6	6 6	6 6	**5 6**	7 7	5 6	6 7	6 6	6 7	**6 7**
Northern & Yorkshire	*(111)*	*(157)*	*(149)*	*(147)*	*(205)*	*(1,005)*	*(176)*	*(202)*	*(244)*	*(281)*	*(321)*	*(1,260)*
One year	20 20	13 14	15 15	17 17	23 24	**20 20**	23 24	19 20	21 22	22 23	19 19	**23 23**
Five years	4 4	3 3	4 4	6 6	10 10	**6 7**	8 9	7 7	9 10	11 11	7 7	**9 10**
Ten years	3 3	2 2	3 4	5 5	7 7	**5 5**	7 8	4 4	5 5	7 8	4 4	**6 6**
Trent	*(98)*	*(141)*	*(164)*	*(174)*	*(156)*	*(840)*	*(133)*	*(153)*	*(168)*	*(157)*	*(151)*	*(783)*
One year	18 18	18 18	14 14	19 20	15 16	**19 19**	27 28	29 30	25 26	21 22	23 23	**27 27**
Five years	8 8	7 8	7 8	10 10	5 6	**9 9**	9 10	14 14	8 8	10 11	12 13	**11 12**
Ten years	7 7	5 5	6 6	7 8	4 5	**7 7**	5 6	9 10	4 4	5 6	10 11	**7 8**
Anglia & Oxford	*(234)*	*(203)*	*(181)*	*(130)*	*(62)*	*(920)*	*(291)*	*(247)*	*(206)*	*(171)*	*(51)*	*(1,014)*
One year	24 24	22 23	26 26	26 26	25 26	**27 27**	27 27	19 20	23 23	24 25	*24 25*	**25 25**
Five years	10 10	8 8	10 11	9 10	10 11	**11 11**	11 12	6 6	7 7	8 8	*4 4*	**9 9**
Ten years	7 7	6 6	8 10	7 7	9 9	**8 9**	7 7	3 4	5 5	4 4	*4 4*	**5 6**
North Thames	*(229)*	*(202)*	*(194)*	*(197)*	*(157)*	*(1,135)*	*(216)*	*(185)*	*(190)*	*(190)*	*(162)*	*(1,000)*
One year	23 24	29 29	26 26	25 26	23 24	**27 28**	31 31	26 26	31 32	24 25	27 27	**29 29**
Five years	9 10	11 12	8 8	7 8	8 9	**10 10**	15 15	8 9	12 13	11 11	11 12	**12 13**
Ten years	3 4	8 9	6 6	6 7	7 8	**7 7**	10 11	5 5	9 10	6 7	8 9	**8 9**
South Thames	*(356)*	*(258)*	*(230)*	*(247)*	*(128)*	*(1,348)*	*(519)*	*(347)*	*(333)*	*(225)*	*(97)*	*(1,604)*
One year	23 24	17 17	17 17	17 17	20 20	**20 21**	24 25	19 19	20 20	19 19	17 18	**23 23**
Five years	7 7	4 5	4 4	5 5	8 9	**7 7**	10 11	6 7	8 9	6 7	12 12	**9 10**
Ten years	5 6	3 3	3 4	3 3	5 5	**4 5**	6 7	4 5	6 7	4 4	*10 10*	**6 7**
South & West	*(236)*	*(290)*	*(238)*	*(141)*	*(47)*	*(1,029)*	*(301)*	*(343)*	*(278)*	*(151)*	*(50)*	*(1,164)*
One year	16 16	14 15	14 14	16 16	11 11	**16 16**	19 19	20 20	18 19	21 21	22 22	**21 22**
Five years	6 7	6 6	6 7	9 10	9 9	**7 7**	9 9	7 8	7 7	10 11	*12 12*	**9 9**
Ten years	4 4	3 4	4 5	7 9	7 9	**5 5**	5 6	5 6	5 5	5 6	*7 9*	**6 7**
West Midlands	*(57)*	*(73)*	*(75)*	*(79)*	*(93)*	*(847)*	*(108)*	*(140)*	*(152)*	*(152)*	*(153)*	*(712)*
One year	22 22	19 19	29 29	15 15	13 13	**20 20**	30 31	26 27	22 23	18 18	18 18	**24 24**
Five years	10 10	9 8	14 15	7 7	3 3	**8 8**	13 13	13 14	11 11	4 5	6 7	**10 10**
Ten years	8 9	6 5	*10 11*	5 5	1 1	**5 5**	7 8	7 8	8 8	2 2	4 4	**6 6**
North & West	*(168)*	*(160)*	*(154)*	*(220)*	*(311)*	*(1,277)*	*(239)*	*(198)*	*(225)*	*(248)*	*(372)*	*(1,344)*
One year	16 16	20 20	18 19	14 14	16 16	**19 19**	21 21	17 17	19 19	17 17	18 18	**20 20**
Five years	6 6	7 7	5 5	7 7	7 7	**7 8**	9 9	7 7	8 8	8 8	8 8	**8 9**
Ten years	5 6	4 4	2 2	4 5	4 4	**5 5**	5 5	3 4	6 8	5 6	4 5	**5 6**
WALES	*(54)*	*(67)*	*(93)*	*(136)*	*(82)*	*(527)*	*(58)*	*(88)*	*(106)*	*(121)*	*(68)*	*(499)*
One year	38 39	30 30	24 24	21 22	24 25	**28 28**	27 28	21 22	35 35	25 25	18 18	**27 28**
Five years	12 12	11 12	12 13	10 11	9 9	**12 13**	8 8	5 5	13 13	14 15	13 13	**12 13**
Ten years	5 5	*10 11*	8 9	7 8	7 7	**8 9**	2 2	4 4	10 12	9 10	8 9	**8 9**

[1] Including those for whom no deprivation category was assigned. C - crude survival, R - relative survival (see Chapter 3).
   Number of patients contributing to each analysis in parentheses.

1981-85 and **1986-90** — *ADULTS*

Each period has columns: Affluent, 2, 3, 4, Deprived, All[1]; with C - crude survival, R - relative survival. First six data columns = 1981-85; next six = 1986-90.

ADULTS	Affluent (C R)	2 (C R)	3 (C R)	4 (C R)	Deprived (C R)	All[1] (C R)	Affluent (C R)	2 (C R)	3 (C R)	4 (C R)	Deprived (C R)	All[1] (C R)
ENGLAND & WALES (n)	(2,452)	(2,316)	(2,146)	(2,044)	(1,588)	(10,803)	(2,840)	(2,723)	(2,500)	(2,178)	(1,612)	(12,001)
One year	28 28	26 27	27 27	25 26	24 25	26 27	31 31	31 31	28 28	29 30	27 27	29 30
Five years	11 12	11 12	11 12	12 12	11 12	11 12	14 15	14 15	13 13	14 14	13 14	14 14
	8 8	7 8	7 8	7 8	8 9	7 8						
ENGLAND (n)	(2,374)	(2,205)	(1,999)	(1,872)	(1,503)	(10,145)	(2,711)	(2,582)	(2,323)	(1,989)	(1,503)	(11,226)
One year	28 28	26 26	27 27	25 25	23 24	26 26	31 31	31 31	28 29	30 31	27 28	30 30
Five years	11 12	11 12	11 12	11 12	11 12	11 12	15 15	14 15	12 13	14 14	13 14	14 14
	7 8	7 8	7 7	7 7	8 8	7 8						
Northern & Yorkshire (n)	(228)	(262)	(277)	(352)	(359)	(1,497)	(265)	(286)	(291)	(297)	(342)	(1,490)
One year	28 29	20 21	24 24	19 20	21 21	24 24	30 30	29 30	25 26	24 24	21 21	27 27
Five years	11 11	10 10	10 11	8 9	8 9	10 10	14 15	14 14	9 10	11 12	9 9	12 13
	7 8	7 8	7 8	5 5	6 7	7 7						
Trent (n)	(169)	(224)	(238)	(253)	(172)	(1,069)	(191)	(246)	(274)	(269)	(204)	(1,188)
One year	28 29	24 24	24 25	23 23	21 21	25 26	25 26	25 26	25 25	27 28	21 22	27 27
Five years	13 13	10 11	11 12	7 8	14 14	11 12	11 12	12 13	11 12	11 12	10 11	12 13
	10 11	6 6	7 8	4 4	10 11	7 8						
Anglia & Oxford (n)	(351)	(301)	(228)	(160)	(82)	(1,146)	(483)	(406)	(282)	(214)	(72)	(1,465)
One year	22 23	26 26	25 25	27 28	18 19	26 26	27 27	26 27	21 21	28 29	21 21	27 27
Five years	8 8	10 11	9 9	16 17	7 7	10 11	13 14	11 12	10 11	13 14	12 12	12 13
	6 6	7 7	6 6	8 8	2 2	6 7						
North Thames (n)	(263)	(208)	(188)	(204)	(168)	(1,055)	(288)	(290)	(250)	(252)	(174)	(1,290)
One year	34 35	34 35	31 31	33 34	30 30	33 34	35 36	38 39	37 38	37 39	34 35	37 38
Five years	14 15	15 16	11 12	20 22	17 19	16 17	18 18	17 18	16 17	17 18	18 19	18 19
	8 9	13 14	9 10	15 17	14 16	12 13						
South Thames (n)	(522)	(333)	(280)	(211)	(82)	(1,481)	(490)	(365)	(303)	(204)	(79)	(1,480)
One year	29 29	27 27	27 28	22 23	34 35	29 29	33 33	30 31	32 32	34 35	27 27	33 33
Five years	10 11	12 12	10 11	9 9	20 22	12 13	16 17	15 16	15 16	17 19	13 13	16 17
	7 7	8 9	7 8	6 6	11 13	8 9						
South & West (n)	(389)	(444)	(367)	(179)	(49)	(1,462)	(411)	(477)	(398)	(254)	(81)	(1,628)
One year	26 27	21 21	22 23	28 28	6 6	25 25	25 26	29 30	22 23	26 26	29 30	27 28
Five years	10 11	9 9	10 11	13 14	3 3	11 11	11 11	14 14	9 10	11 12	10 11	12 13
	7 8	5 6	5 6	8 9	3 3	6 7						
West Midlands (n)	(184)	(194)	(210)	(230)	(208)	(1,034)	(247)	(215)	(231)	(240)	(184)	(1,124)
One year	24 24	30 31	30 30	20 21	19 20	26 26	32 33	37 38	29 30	28 28	27 28	32 33
Five years	13 14	11 11	12 12	7 7	7 7	10 11	17 18	16 17	13 13	12 13	15 17	15 16
	8 9	6 6	6 6	2 2	5 5	6 6						
North & West (n)	(268)	(239)	(211)	(283)	(383)	(1,401)	(336)	(297)	(294)	(259)	(367)	(1,561)
One year	19 20	19 20	20 21	22 22	21 21	22 22	31 32	25 25	27 27	28 28	26 27	29 29
Five years	9 10	10 11	8 9	10 10	9 9	10 10	13 14	10 10	13 14	12 13	12 13	13 13
	6 7	5 6	4 4	6 7	5 6	6 6						
WALES (n)	(78)	(111)	(147)	(172)	(85)	(658)	(129)	(141)	(177)	(189)	(109)	(775)
One year	28 29	32 33	27 28	31 32	32 33	31 32	28 29	30 31	24 24	20 20	21 21	26 26
Five years	12 12	16 17	15 17	16 17	13 13	15 16	11 12	15 16	12 14	12 13	10 10	13 14
	10 12	9 11	11 12	9 10	8 9	10 11						

[1] Including those for whom no deprivation category was assigned. C - crude survival, R - relative survival (see Chapter 3).
Number of patients contributing to each analysis in parentheses.

Fig. 39.1 Relative survival up to ten years: trends by calendar period of diagnosis (1971-90) and NHS Region

ADULTS

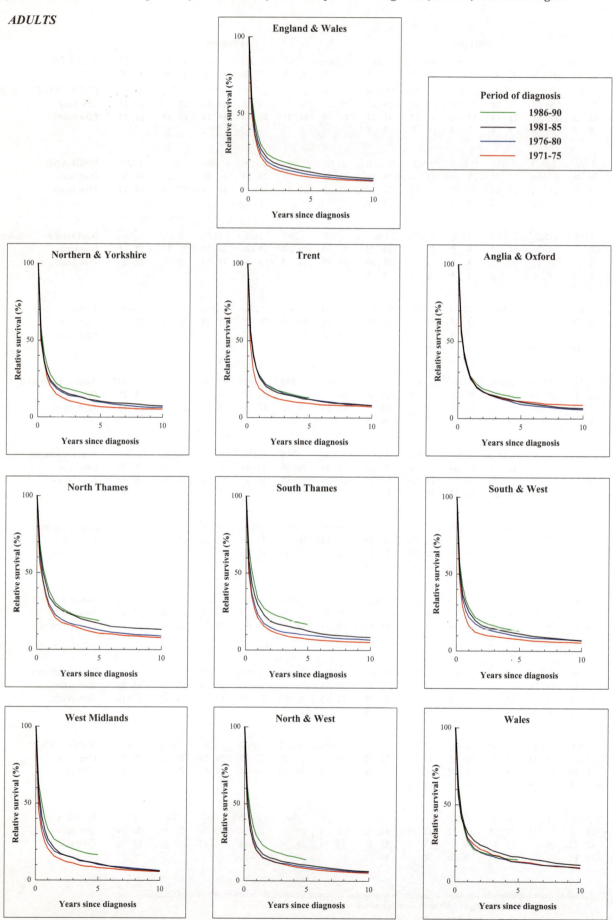

Fig. 39.2   Relative survival up to five years, by deprivation category and NHS Region: patients diagnosed 1986-90

ADULTS

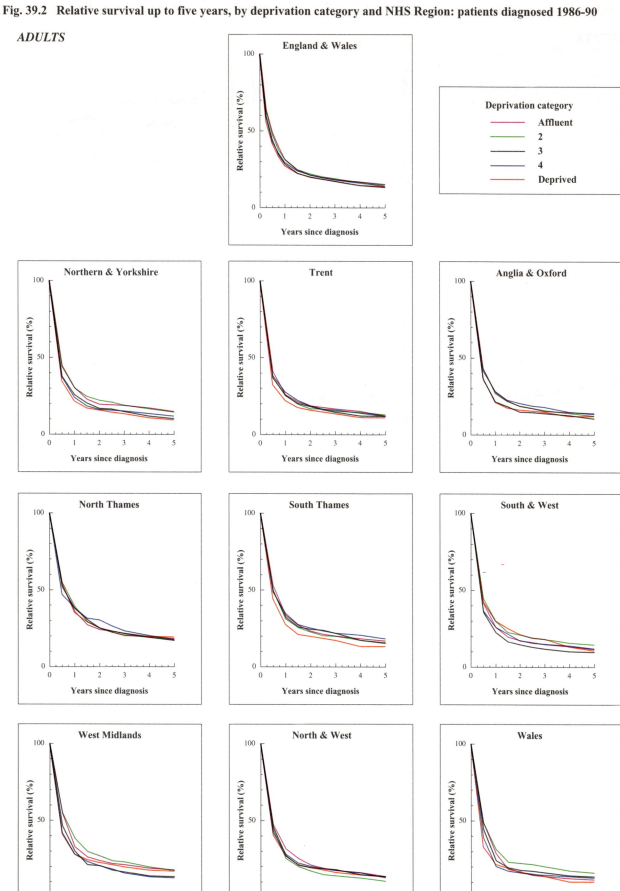

Fig. 39.3   Relative survival at five years by deprivation category, period of diagnosis (1981-90) and NHS Region

ADULTS
(Note vertical scale)

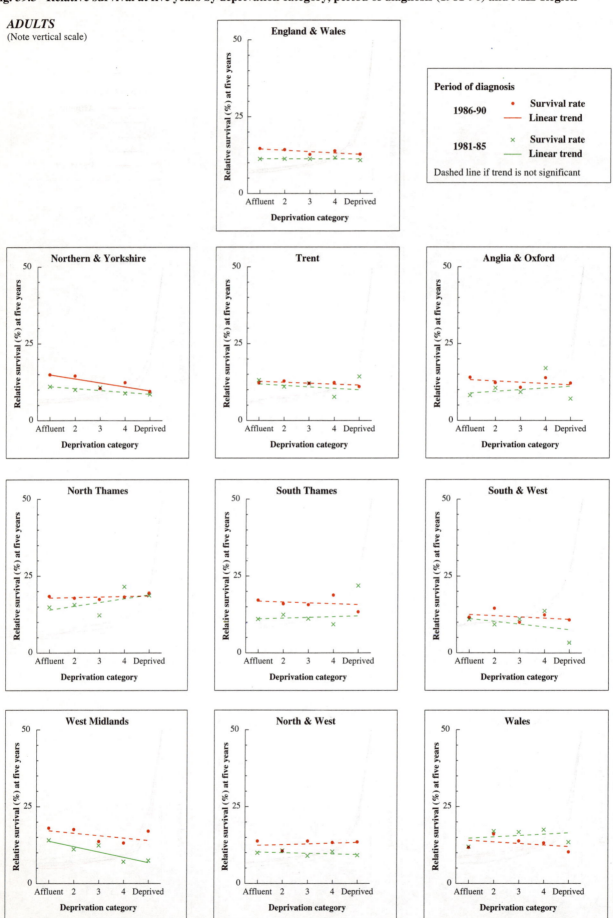

Fig. 39.4 Relative survival (%) by NHS Region and sex:
one-year and five-year survival for patients diagnosed 1986-90, with 95% confidence intervals

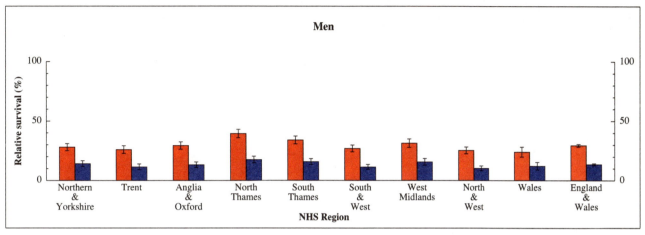

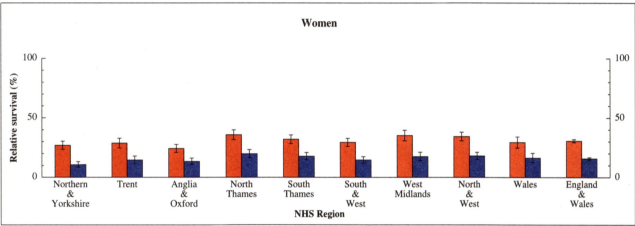

Fig. 39.5 Relative survival (%) at five years, with 95% confidence interval, by age at diagnosis and sex:
England and Wales, patients diagnosed 1986-90

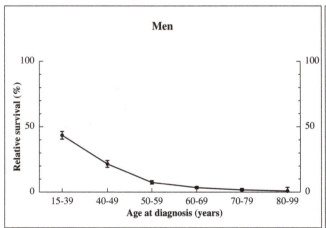

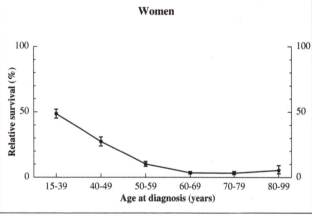

Chapter 40

THYROID

Cancer of the thyroid is generally uncommon, representing less than 0.5% of all malignant neoplasms among adults in England and Wales. In 1990, there were almost 900 new cases of thyroid cancer, and age-standardised (Europe) incidence rates were 2.3 per 100,000 per year in women and 0.9 in men. The sex ratio declines after middle age, but three out of every four cases arise in women. Incidence in England and Wales increased by about 10% every five years in both sexes, but mortality has declined by 8% every five years for men and by 17% for women. Similar patterns have been observed in a number of developed countries. There are about 250 deaths a year from thyroid cancer.

Several morphologically distinct types of thyroid cancer exist, each with a very different clinical picture and prognosis. Papillary and follicular carcinomas are the most common, in young women and in men and women of middle age, respectively, and they can often be treated effectively with total or sub-total thyroidectomy and radio-iodine. Anaplastic carcinoma is less common, but aggressive: it occurs in the elderly, and is often metastatic at diagnosis. Medullary carcinoma is also uncommon, but can be treated with surgery and radiotherapy. Incidence patterns by morphologic type are available[1], but the proportion of unspecified type varies by region, and mortality data are not available by morphologic type. The analyses reported here include all types of thyroid cancer.

Survival was analysed for some 13,500 adults diagnosed with cancer of the thyroid in England and Wales during the period 1971-90 and followed up to the end of 1995, some 87% of those eligible (Table 40.1). Almost 6% of patients were excluded because their vital status was unknown at 6 July 1997, and a further 6% because their survival was zero or unknown. Papillary adenocarcinoma accounted for about 9% of the tumours throughout, but mixed papillary and follicular adenocarcinoma rose from 10% in the early 1970s to about 20% by the late 1980s, and those described as papillary carcinoma increased sharply from 12% to almost 30% over the same period. Follicular and medullary adenocarcinoma accounted for 18% and 5% of the tumours, respectively, and 5% were anaplastic carcinomas: these proportions were stable over time.

Survival trends

Relative survival for adults diagnosed with thyroid cancer in England and Wales during 1986-90 was 78% at one year and 73% at five years. The survival advantage for women has increased steadily: five-year survival was 6% higher for women diagnosed in the early 1970s, and it rose to 11% for women diagnosed during the late 1980s (Table 40.2).

Survival curves for thyroid cancer reach a plateau much sooner after diagnosis than for most cancers, indicating very low excess mortality (relative to the general population) within three years of diagnosis. Survival improved substantially in all regions between the early 1970s and the late 1980s, although in several regions there was little or no further improvement during the 1980s (Figure 40.1).

One-year and five-year survival in England and Wales improved by 12-15% between the early 1970s and the late 1980s. This represented an average gain of 4-5% every five years, after adjustment for changes in the age distribution of patients over time. Improvements in age-standardised survival over time were similar in men and women (see Table 4.4): but see below.

Survival falls steeply with age at diagnosis, even after adjustment for background mortality (Figure 40.5). Thus for more than 1,000 women under the age of 40 at diagnosis during 1986-90, five-year survival was higher than 95%. Men aged 50-69 at diagnosis have five-year relative survival rates around 60%, but survival falls sharply in both sexes at older ages (see Table 4.7). This pattern is largely due to differences in incidence by age for the main types of thyroid cancer: papillary carcinoma is dominant in younger patients, and has high survival, while anaplastic carcinoma occurs mainly in the elderly, and has very poor survival.

The apparent discrepancy between the trends in relative survival before and after standardisation for age merits comment here. The data in Table 40.2 are summary survival rates for all adults in the age range 15-99 at diagnosis. The increase in five-year survival for women since 1971-75 represents in part a shift in the biological spectrum of malignancy with age. In the 1980s, papillary carcinoma of the thyroid, with five-year survival rates of 95% or more, has become more common in young women. Overall survival rates for women reflect this change. Age-specific survival rates are higher in women under age 60 and higher in men aged 80-99 (see Table 4.7). For thyroid cancer, therefore, age-standardisation provides an incomplete picture of the trends in survival with time (see Table 4.4).

The appropriate description of trends in thyroid cancer survival is that of a substantial increase in survival for men and women, with part of the larger increase in women being due to a substantial shift towards a less aggressive type of thyroid cancer at younger ages. Trends in age-specific survival will need to be examined separately for each of the main morphological types of thyroid cancer in order to clarify this issue.

Deprivation

Survival from thyroid cancer has been lower in the most deprived group than in the most affluent group since the early 1970s for both men and women (Table 40.2, Figure 40.2). The deprivation gap in survival for men and women diagnosed during the period 1981-85 diagnosis was about 10% at one, five and ten years after diagnosis.

Among 3,500 patients of known deprivation status who were diagnosed with thyroid cancer in England and Wales during the period 1986-90, the deprivation gap had fallen to about 6% at both one and five years after diagnosis (see Table 4.9). On average, therefore, survival improved slightly more for the deprived groups than for the affluent groups during the 1980s, but the deprivation gradient in survival was statistically significant for patients diagnosed from 1981 up to 1990, with follow-up to 1995 (Figure 40.3).

International comparison

Five-year survival for adults diagnosed with thyroid cancer in England and Wales during 1986-90 was 65% in men and 76% in women, some 4% lower than the average for Europe. Survival in England and Wales was similar to that in Scotland for women, but lower for men. The highest regional survival rates, 83% for both men and women (North Thames), were well above the average for Europe, but still more than 10% lower than five-year survival in the USA: about 95% in both sexes (see Table 4.12).

As for prostate cancer (Chapter 33), these very large discrepancies in population-based survival between Europe and the USA for what is ostensibly the same cancer strongly suggest a different biological spectrum of malignancy in the two populations of cancer patients. It seems unlikely that survival differences of 20-30% between the UK and the USA for thyroid cancer could be solely due to differences in treatment.

1.   dos Santos Silva I, Swerdlow AJ. Thyroid cancer epidemiology in England and Wales: time trends and geographical distribution. Br J Cancer 1993; 67: 330-340

Table 40.1 Ineligible and excluded records, and patients included in analyses, by calendar period of diagnosis

	Calendar period of diagnosis														
	1971-75			1976-80			1981-85			1986-90			Overall		
Total registered	3,541			3,931			4,195			4,309			15,976		
Ineligible	*Records*	*Patients*	*%*	*Records*	*Patients*	*%*	*Records*	*Patients*	*%*	*Records*	*Patients*	*%*	*Records*	*Patients*	*%*
Incomplete data[1]	26	26		37	37		65	65		66	66		194	**194**	
Resident outside England and Wales	58	51		55	55		33	26		23	15		169	**147**	
In situ neoplasm[2]	1	1		0	0		0	0		0	0		1	**1**	
Benign or uncertain behaviour[2]	58	32		32	5		1	1		0	0		91	**38**	
Metastatic[2]	59	58		27	27		0	0		1	1		87	**86**	
Otherwise ineligible[3]	-	-		-	-		-	-		-	-		-	**-**	
Lymphoma[4]	37	36		19	17		13	0		8	0		77	**53**	
Leukaemia or myeloma[4]	0	0		1	1		0	0		0	0		1	**1**	
		204			142			92			82			520	
Total eligible		3,337	100		3,789	100		4,103	100		4,227	100		15,456	100
Aged under 15 or 100 years or over at diagnosis	28	23	0.7	25	23	0.6	31	29	0.7	36	34	0.8	120	**109**	0.7
Vital status unknown[5]	206	149	4.5	261	205	5.4	276	223	5.4	359	310	7.3	1,102	**887**	5.7
Duplicate registration[6]	0	0	0.0	0	0	0.0	0	0	0.0	0	0	0.0	0	**0**	0.0
Multiple primary[7]	0	0	0.0	3	3	<0.1	14	11	0.3	18	17	0.4	35	**31**	0.2
Synchronous tumours[8]	11	9	0.3	5	2	<0.1	15	5	0.1	31	10	0.2	62	**26**	0.2
Sex not known	0	0	0.0	0	0	0.0	0	0	0.0	0	0	0.0	0	**0**	0.0
Sex incompatible with site code[9]	-	-	-	-	-	-	-	-	-	-	-	-	-	**-**	-
Invalid dates[10]	0	0	0.0	0	0	0.0	3	3	<0.1	2	0	0.0	5	**3**	<0.1
Zero survival[11]	214	187	5.6	246	221	5.8	295	264	6.4	289	273	6.5	1,044	**945**	6.1
Total exclusions		368	11.0		454	12.0		535	13.0		644	15.2		2,001	12.9
Patients included in analysis		2,969	89.0		3,335	88.0		3,568	87.0		3,583	84.8		13,455	87.1

[1] Main data item(s) invalid or incompatible with one another: name, sex, date of birth, date of diagnosis and (if present) date of death, postcode, site, morphology and behaviour

[2] Tumour either *in situ* (behaviour code 2), benign (0), uncertain if benign or malignant (1) or metastatic (6 or 9)

[3] Not applicable for this malignancy

[4] Morphology that of lymphoma (included with non-Hodgkin lymphoma, Ch. 41) or myeloma (included with multiple myeloma, Ch. 43), or leukaemia (excluded)

[5] Tracing of vital status at National Health Service Central Register incomplete at 6 July 1997

[6] Same site code, sex, cancer registry and cancer registry number as an earlier registration

[7] Same site code and person number as an earlier registration(s): mostly confirmed multiple primary tumours at this site, some unresolved duplicate registrations

[8] Same site code, sex, date of birth and date of diagnosis as another registration(s): mostly synchronous or (in paired organs) bilateral tumours in same anatomic site in one individual, not previously linked: also some duplicate registrations

[9] Not applicable for this malignancy

[10] Impossible sequence of dates (birth, diagnosis and, if present, emigration or death); or date of death after 6 July 1997

[11] Date of diagnosis same as date of death: some are patients who did die on the day of diagnosis, but most were registered solely from a death certificate (death certificate only, or DCO), and their survival time is thus unknown

Table 40.2　Crude and relative survival (%) at one, five and ten years since diagnosis: NHS Region, deprivation category and calendar period of diagnosis

	1971-75						1976-80					
	Affluent	2	3	4	Deprived	All[1]	Affluent	2	3	4	Deprived	All[1]
	C R	C R	C R	C R	C R	C R	C R	C R	C R	C R	C R	C R
ENGLAND & WALES												
Men	(137)	(138)	(129)	(139)	(100)	(780)	(180)	(151)	(168)	(174)	(145)	(849)
One year	59 61	60 63	53 55	56 59	54 57	57 60	58 60	55 58	61 64	57 60	54 56	57 60
Five years	39 47	43 50	36 43	40 49	37 44	39 47	41 48	40 48	45 56	44 53	38 46	42 51
Ten years	32 46	37 50	29 42	32 45	30 42	32 45	34 45	31 43	34 51	36 52	30 44	34 48
Women	(332)	(381)	(371)	(406)	(306)	(2,189)	(480)	(511)	(498)	(492)	(412)	(2,486)
One year	62 64	61 63	62 64	53 56	57 59	58 60	67 71	65 68	66 69	58 61	56 59	63 66
Five years	50 57	49 57	50 61	39 47	42 49	45 53	56 64	55 63	52 62	46 56	45 52	51 60
Ten years	44 54	39 52	43 58	33 45	35 48	38 50	50 62	49 62	46 61	40 53	35 48	45 58
Adults	(469)	(519)	(500)	(545)	(406)	(2,969)	(660)	(662)	(666)	(666)	(557)	(3,335)
One year	61 63	60 63	60 62	54 57	56 59	57 60	65 68	63 66	65 68	58 61	56 58	61 64
Five years	47 55	47 55	46 56	39 47	40 48	43 51	52 59	52 59	50 60	46 55	43 51	49 58
Ten years	40 52	38 51	39 54	33 45	34 47	36 49	45 58	45 58	43 59	39 53	34 47	42 56

	Men	Women	Adults				Men	Women	Adults			
Northern & Yorkshire	(86)	(251)	(337)				(97)	(301)	(398)			
One year	49 50	53 55	52 54				51 54	55 58	54 57			
Five years	31 37	39 47	37 44				36 44	42 51	41 49			
Ten years	23 35	31 47	29 44				30 44	36 50	35 49			
Trent	(73)	(184)	(257)				(95)	(275)	(370)			
One year	54 57	61 64	59 62				55 58	63 67	61 65			
Five years	38 46	44 53	42 51				44 50	53 64	50 60			
Ten years	29 41	35 50	34 47				38 47	45 61	43 57			
Anglia & Oxford	(70)	(203)	(273)				(66)	(242)	(308)			
One year	65 68	61 63	62 64				66 69	63 66	63 67			
Five years	52 58	48 56	49 56				54 63	53 60	53 61			
Ten years	39 50	42 54	41 53				48 62	49 58	49 60			
North Thames	(112)	(288)	(400)				(82)	(256)	(338)			
One year	58 62	66 68	64 67				63 68	77 80	74 78			
Five years	41 51	54 63	51 60				43 53	67 76	61 72			
Ten years	37 51	46 60	43 57				32 46	61 75	54 70			
South Thames	(103)	(305)	(408)				(133)	(300)	(433)			
One year	54 56	57 60	56 59				62 65	66 68	65 67			
Five years	31 36	46 55	42 51				47 56	53 63	51 61			
Ten years	29 36	40 53	37 49				36 50	49 62	45 60			
South & West	(116)	(303)	(419)				(123)	(354)	(477)			
One year	53 56	62 64	59 62				53 56	68 71	64 67			
Five years	38 45	47 56	45 53				39 47	56 64	52 60			
Ten years	27 38	39 52	35 48				33 45	48 62	44 58			
West Midlands	(81)	(185)	(266)				(71)	(236)	(307)			
One year	70 72	54 57	59 62				56 58	63 66	62 65			
Five years	45 52	43 50	44 51				41 47	50 59	48 57			
Ten years	37 47	40 49	39 49				30 42	44 56	40 54			
North & West	(96)	(315)	(411)				(125)	(378)	(503)			
One year	59 62	51 53	53 55				55 58	54 57	54 57			
Five years	38 48	37 44	37 45				40 51	43 52	42 51			
Ten years	- -	30 42	31 44				30 50	35 48	34 48			
WALES	(43)	(155)	(198)				(57)	(144)	(201)			
One year	51 52	53 56	52 55				55 58	59 62	58 61			
Five years	42 48	42 51	42 51				35 42	42 52	40 49			
Ten years	31 40	39 49	37 48				26 38	35 50	32 46			

[1]　Including those for whom no deprivation category was assigned. C - crude survival, R - relative survival (see Chapter 3). Number of patients contributing to each analysis in parentheses.

1981-85 / 1986-90 — ENGLAND & WALES

	Affluent C R	2 C R	3 C R	4 C R	Deprived C R	All[1] C R	Affluent C R	2 C R	3 C R	4 C R	Deprived C R	All[1] C R	
													ENGLAND & WALES
	(224)	(179)	(209)	(174)	(141)	(954)	(201)	(221)	(212)	(198)	(150)	(992)	**Men**
One year	76 79	69 72	71 74	69 74	60 63	70 74	75 77	68 71	74 78	68 72	71 75	71 74	
Five years	59 69	54 62	53 63	49 62	44 53	52 63	59 66	57 67	53 65	53 65	53 63	55 65	
	47 63	43 56	41 57	41 60	35 51	42 58							
	(552)	(549)	(514)	(502)	(437)	(2,614)	(571)	(560)	(551)	(463)	(405)	(2,591)	**Women**
One year	75 78	74 78	72 76	69 73	66 69	72 75	81 84	78 82	73 76	77 80	73 76	77 80	
Five years	65 74	61 71	62 73	57 67	52 62	60 70	72 81	70 77	62 71	66 75	64 74	67 76	
	58 71	56 70	53 69	50 65	44 60	53 68							
	(776)	(728)	(723)	(676)	(578)	(3,568)	(772)	(781)	(763)	(661)	(555)	(3,583)	**Adults**
One year	75 78	73 76	72 76	69 73	64 68	71 75	79 82	76 79	73 77	74 77	72 76	75 78	
Five years	63 73	59 69	59 70	55 66	50 60	58 68	69 77	66 74	60 70	62 73	61 72	64 73	
	55 70	53 67	50 66	48 64	42 57	50 65							

Regions

	Men C R	Women C R	Adults C R	Men C R	Women C R	Adults C R	
	(108)	(323)	(431)	(115)	(333)	(448)	**Northern & Yorkshire**
One year	68 71	70 74	69 73	78 83	75 78	76 79	
Five years	49 60	57 67	55 66	63 74	65 75	64 74	
	38 55	50 64	47 62				
	(95)	(244)	(339)	(88)	(249)	(337)	**Trent**
One year	65 69	68 72	67 71	71 73	74 77	73 76	
Five years	52 57	58 68	56 65	54 62	61 70	59 68	
	42 53	51 66	48 61				
	(90)	(246)	(336)	(104)	(307)	(411)	**Anglia & Oxford**
One year	79 84	76 79	77 80	75 80	78 81	77 81	
Five years	60 70	64 74	63 74	56 65	69 79	65 76	
	47 65	55 71	53 70				
	(108)	(234)	(342)	(126)	(314)	(440)	**North Thames**
One year	73 76	77 81	76 80	85 88	83 87	84 87	
Five years	54 64	65 73	61 70	71 83	72 83	71 83	
	45 60	57 70	53 68				
	(122)	(372)	(494)	(131)	(319)	(450)	**South Thames**
One year	73 78	74 77	74 77	72 75	78 81	76 79	
Five years	53 65	62 73	60 71	51 59	67 76	62 71	
	45 61	55 71	52 69				
	(135)	(431)	(566)	(135)	(361)	(496)	**South & West**
One year	71 75	76 80	75 79	65 67	77 80	74 76	
Five years	53 68	65 76	63 74	53 60	69 78	65 73	
	40 57	60 75	55 71				
	(104)	(277)	(381)	(126)	(255)	(381)	**West Midlands**
One year	63 66	64 68	64 68	66 69	77 80	73 76	
Five years	47 56	55 65	52 63	52 61	69 77	64 72	
	33 48	46 60	43 57				
	(142)	(345)	(487)	(126)	(285)	(411)	**North & West**
One year	69 72	65 69	66 70	60 63	73 77	69 73	
Five years	53 64	54 65	54 64	44 54	63 73	57 68	
	44 64	48 64	47 64				
	(50)	(142)	(192)	(41)	(168)	(209)	**WALES**
One year	73 77	74 77	73 77	65 67	76 78	74 76	
Five years	47 57	56 67	54 64	50 57	66 73	63 71	
	41 55	46 63	45 60				

[1] Including those for whom no deprivation category was assigned. C - crude survival, R - relative survival (see Chapter 3).
Number of patients contributing to each analysis in parentheses.

Fig. 40.1   Relative survival up to ten years: trends by calendar period of diagnosis (1971-90) and NHS Region

ADULTS

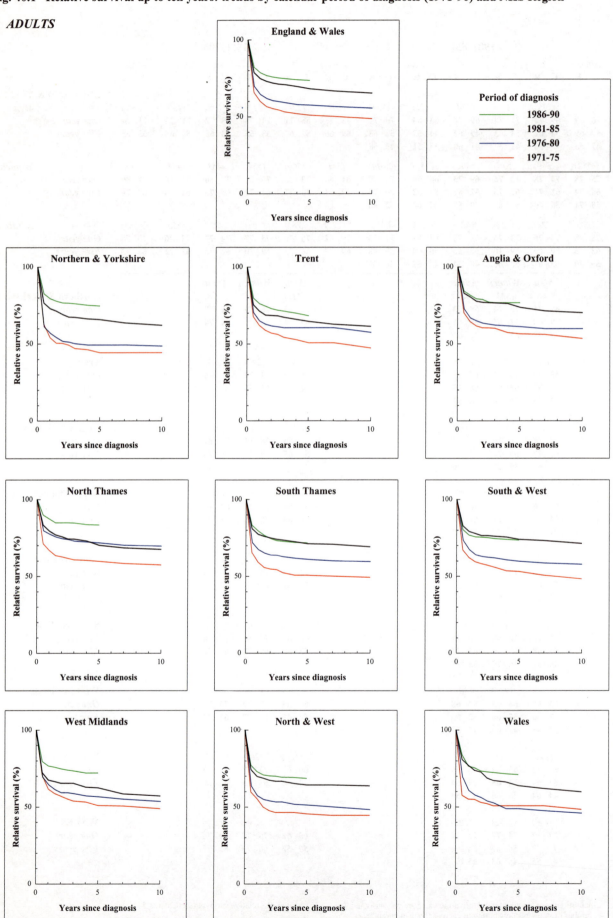

Fig. 40.2 Relative survival up to five years, by deprivation category: England and Wales, patients diagnosed 1986-90

ADULTS

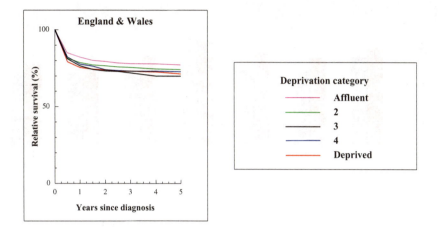

Fig. 40.3 Relative survival at five years, by deprivation category and period of diagnosis (1981-90): England and Wales

ADULTS

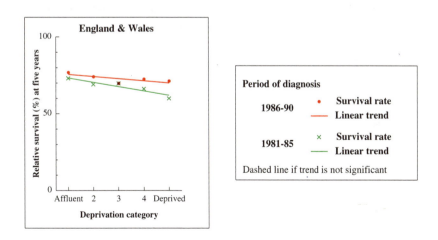

Fig. 40.4 Relative survival (%) by NHS Region and sex:
           one-year and five-year survival for patients diagnosed 1986-90, with 95% confidence intervals

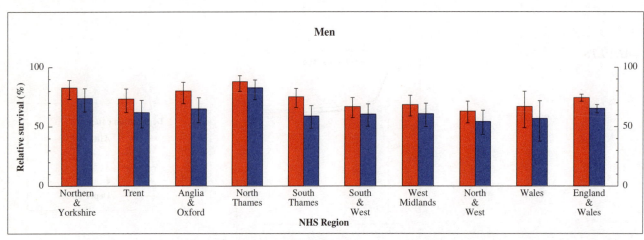

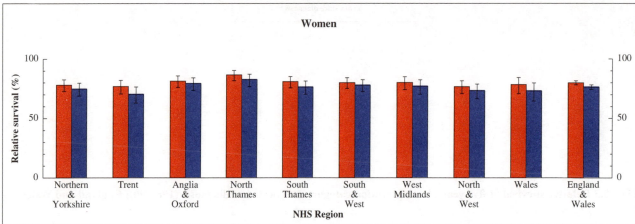

Fig. 40.5 Relative survival (%) at five years, with 95% confidence interval, by age at diagnosis and sex:
           England and Wales, patients diagnosed 1986-90

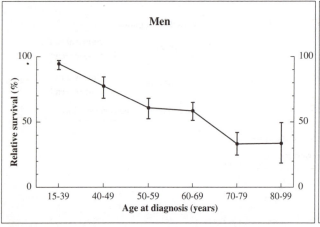

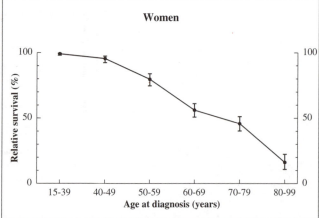

Chapter 41

NON-HODGKIN LYMPHOMA

Non-Hodgkin lymphoma is the most common malignancy among the leukaemias and lymphomas, with 6,000 new cases each year in England and Wales. The male-to-female sex ratio is about 1.6 to 1. Lymphomas other than Hodgkin's disease comprise a very disparate group of malignancies with different clinical behaviour and prognosis. Trends in aetiology and survival for the different types of lymphomas may not become clear when such broad lymphoma groups are adopted for epidemiological study[1]. Unfortunately, it appears impossible to re-classify population lymphoma series from the 1970s to reflect subsequent changes in pathological classification and coding, and thus to define a coherent population series of more precisely defined lymphomas over such a long period of time. Lymphoma incidence and mortality are usually reported for the same broad groups, and this facilitates the interpretation of population survival trends. Incidence and mortality have changed rapidly, and the overall trends in survival are also unmistakable.

Incidence of the malignant lymphomas has more than doubled since 1971, and age-standardised (Europe) incidence in 1990 was 13.2 per 100,000 for men and 8.4 for women. Despite the difficulties of definition and registration, the increase is likely to be real[2]. As elsewhere in Europe, mortality has been rising less rapidly than incidence[3]; death rates in England and Wales in 1995 were 7.7 and 4.8 per 100,000 in men and women, respectively. Until about 30 years ago, most lymphomas were fatal, but effective combination chemotherapy, initially developed for Hodgkin's disease, has resulted in the cure of some advanced tumours. The consistency of trends in many countries suggests that improvements in treatment for the lymphomas have prevented mortality from rising as quickly as incidence.

The survival analyses reported here are based on almost 68,000 patients diagnosed with a malignant lymphoma in England and Wales between 1971 and 1990 and followed up to the end of 1995. After exclusion of some 2% of eligible lymphomas, diagnosed in children under 15, these patients represent about 87% of the eligible adults (Table 41.1). Five per cent were excluded because their vital status was unknown at 6 July 1997, and 8% because their survival was either zero or unknown. Survival from malignant lymphoma in childhood is discussed in Chapter 53.

Information on the different types of lymphoma included in the analyses is given in Table 41A. The shift in classification and coding of the lymphomas is clear, with a huge decline in neoplasms coded as reticulosarcoma and lymphosarcoma (ICD-9 200) and a corresponding increase in other types of lymphoma (202). Separate analysis of trends in survival from lymphomas assigned to these two rubrics would be misleading.

Most of the extra-nodal lymphomas diagnosed in the 1980s had already been reassigned to this category in the original data. A further 1,860 lymphomas, mainly diagnosed in the 1970s, were reassigned from solid tumour sites (see Chapter 2). The brain, the stomach and the oropharynx were the sites most frequently involved. After application of the same exclusion criteria as for other lymphomas, some 1,200 of these lymphomas were included in the analyses (Table 41.1).

Almost 80% of the neoplasms were histologically coded as lymphomas, diffuse or not otherwise specified; the proportion increased from 53% in 1971 to 87% in 1990. Reticulosarcomas accounted for 11% of cases overall (38% in 1971, but less than 1% in 1990) and 9% were nodular or follicular lymphomas. Morphology and site codes were largely consistent.

Survival trends

Relative survival for patients with a malignant lymphoma diagnosed in England and Wales during 1986-90 was 65% at one year and 45% at five years. Ten-year survival for patients diagnosed during 1981-85 was 34% (Table 41.2). Survival in England and Wales as a whole has improved substantially in each successive five-year period. In most regions, the largest increase in survival occurred between patients diagnosed in the late 1970s and those diagnosed in the early 1980s, although there has been a smaller increase in Anglia & Oxford, where survival was already high in the 1970s (Figure 41.1).

The overall improvement in national survival for non-Hodgkin lymphoma since the early 1970s has been about 14% for both men and women. The average improvement between successive five-year periods of diagnosis was almost 5% for both one-year and five-year survival, after adjustment for changes in the age distribution of patients over time (see Table 4.4). Women diagnosed during 1986-90 had a 3-4% survival advantage over men at all times up to five years after diagnosis.

Regional differences in five-year survival are quite marked. For patients diagnosed during 1986-90, survival was above average for men and women in North Thames, South Thames and South & West (Figure 41.4). These

Table 41A Sub-type distribution of non-Hodgkin lymphoma

ICD-9 code	Description	1971-75 No.	%	1976-80 No.	%	1981-85 No.	%	1986-90 No.	%	Total No.	%
200.0	Reticulosarcoma	3,504	29.3	1,985	14.0	855	4.7	214	0.9	**6,558**	**9.7**
200.1	Lymphosarcoma	4,703	39.3	4,145	29.2	2,301	12.8	1,224	5.2	**12,373**	**18.2**
200.2	Burkitt's tumour	-	-	13	0.1	48	0.3	47	0.2	**108**	**0.2**
200.8	Other	-	-	184	1.3	512	2.8	662	2.8	**1,358**	**2.0**
200	**Lymphosarcoma and reticulosarcoma**	**8,207**	**68.6**	**6,327**	**44.5**	**3,716**	**20.6**	**2,147**	**9.1**	**20,397**	**30.0**
202.0	Nodular lymphoma	464	3.9	1,080	7.6	1,805	10.0	1,945	8.2	**5,294**	**7.8**
202.1	Mycosis fungoides	278	2.3	371	2.6	393	2.2	454	1.9	**1,496**	**2.2**
202.2	Sézary's disease[1]	-	-	-	-	19	0.1	25	0.1	**44**	**0.1**
202.3	Malignant histiocytosis	-	-	58	0.4	138	0.8	51	0.2	**247**	**0.4**
202.4	Leukaemic reticuloendotheliosis	-	-	92	0.6	296	1.6	403	1.7	**791**	**1.2**
202.5	Letterer-Siwe disease	-	-	3	0.0	3	0.0	8	0.0	**14**	**0.0**
202.6	Malignant mast-cell tumours	-	-	0	0.0	3	0.0	6	0.0	**9**	**0.0**
202.8	Other lymphomas	-	-	2,930	20.6	11,573	64.2	18,552	78.2	**33,055**	**48.7**
202.9	Other and unspecified	2,320	19.4	2,920	20.5	26	0.1	113	0.5	**5,379**	**7.9**
202	**Other malignant neoplasm of lymphoid and histiocytic tissue**	**3,062**	**25.6**	**7,454**	**52.4**	**14,256**	**79.1**	**21,557**	**90.9**	**46,329**	**68.2**
140-199	**Extra-nodal**	**695**	**5.8**	**433**	**3.0**	**49**	**0.3**	**15**	**0.1**	**1,192**	**1.8**
	Total	**11,964**		**14,214**		**18,021**		**23,719**		**67,918**	

[1] The ICD-8 code 202.2 (other primary malignant forms of lymphoma) is not compatible with the ICD-9 code 202.2: the 1,852 neoplasms so coded in 1971-75 and 2,274 in 1976-80 have been reassigned to 202.9 for this table.

regions have seen the largest overall increase in survival: there was less variation between the regions for patients diagnosed during the early 1970s (see Table-Figure 4.5S). Survival improved in all regions, but in Anglia & Oxford, where survival was among the highest in 1971-75, the average increase in survival was about 3% every five years, although survival in 1986-90 is still close to the national average. Gains in survival for women in Wales were also less than average, and for women diagnosed during 1986-90, five-year survival was 6% lower than in England and Wales as a whole.

Relative survival from malignant lymphoma falls with age at diagnosis for both men and women (Figure 41.5). For adults under the age of 50, five-year relative survival was about 60-70%, but this falls steadily with increasing age to about 20% for those aged 80-99 at diagnosis (see Table 4.4). There is a small survival advantage for younger women, but this also declines with age.

Deprivation

Survival from malignant lymphomas in England and Wales has consistently been lower among deprived groups than for the most affluent group (Table 41.2). The survival curves for patients diagnosed during 1986-90 are typical (Figure 41.2). A deprivation gradient in five-year survival was present in most regions during the early and late 1980s, and statistically significant in one or both periods in six of the eight NHS Regions in England (Figure 41.3). There was no systematic difference in survival between deprivation groups in Wales.

Among more than 17,000 patients diagnosed with a malignant lymphoma in England and Wales during the early 1980s, the deprivation gap in survival between the most affluent and the most deprived groups was 9% at one year and 7% at five years (see Table 4.9). For more than 23,000 patients diagnosed during the next five years, the deprivation gap was 7%.

International comparison

Relative survival at five years in England and Wales (45%) is similar to that in Scotland, and 3% less than the average for Europe. The highest regional survival rates in England and Wales for the period 1986-90 (52%) are the same as those for 1983-89 in the USA[4], and still similar to the survival rates in the USA for the period 1986-90 (49% for men, 56% for women: see Table 4.12).

1.   Scherr PA, Mueller NE. Non-Hodgkin's lymphomas. In: Schottenfeld D, Fraumeni JF, (eds.) *Cancer epidemiology and prevention*. 2nd edn. Oxford: Oxford University Press, 1996, pp920-945

2.   Barnes N, Cartwright RA, O'Brien C, Richards IDG, Roberts B, Bird CC. Rising incidence of lymphoid malignancies – true or false? Br J Cancer 1986; 53: 393-398

3.   Doll R. Progress against cancer: an epidemiologic assessment. Am J Epidemiol 1991; 134: 675-688

4. National Cancer Institute. *SEER Cancer statistics review 1973-1990. Publication no. 93-2789*. Bethesda: National Institutes of Health, 1993

Table 41.1 Ineligible and excluded records, and patients included in analyses, by calendar period of diagnosis

	\multicolumn{15}{c}{Calendar period of diagnosis}													
	1971-75			**1976-80**			**1981-85**			**1986-90**			**Overall**		
Total registered	**13,751**			**16,957**			**22,547**			**30,009**			**83,264**		
Ineligible	*Records*	*Patients*	*%*	*Records*	*Patients*	*%*	*Records*	*Patients*	*%*	*Records*	*Patients*	*%*	*Records*	*Patients*	*%*
Incomplete data[1]	59	59		280	280		1,087	1,087		1,130	1,130		2,556	**2,556**	
Resident outside England and Wales	235	233		219	211		157	141		144	87		755	**672**	
In situ neoplasm[2]	3	3		0	0		0	0		0	0		3	**3**	
Benign or uncertain behaviour[2]	89	32		106	50		0	0		0	0		195	**82**	
Metastatic[2]	7	7		16	16		3	0		2	0		28	**23**	
Otherwise ineligible[3]	-	-		-	-		-	-		-	-		-	**-**	
Lymphoma[4]	-	-		-	-		-	-		-	-		-	**-**	
Leukaemia or myeloma[4]	-	-		-	-		-	-		-	-		-	**-**	
		334			**557**			**1,228**			**1,217**			**3,336**	
Total eligible		**13,417**	**100**		**16,400**	**100**		**21,319**	**100**		**28,792**	**100**		**79,928**	**100**
Aged under 15 or 100 years or over at diagnosis	390	365	2.7	481	443	2.7	400	370	1.7	428	406	1.4	1,699	**1,584**	2.0
Vital status unknown[5]	630	427	3.2	904	666	4.1	1,058	856	4.0	1,988	1,700	5.9	4,580	**3,649**	4.6
Duplicate registration[6]	0	0	0.0	2	1	<0.1	0	0	0.0	0	0	0.0	2	**1**	<0.1
Multiple primary[7]	51	48	0.4	75	70	0.4	232	204	1.0	189	168	0.6	547	**490**	0.6
Synchronous tumours[8]	50	24	0.2	73	39	0.2	75	37	0.2	120	70	0.2	318	**170**	0.2
Sex not known	0	0	0.0	0	0	0.0	0	0	0.0	0	0	0.0	0	**0**	0.0
Sex incompatible with site code[9]	-	-	-	-	-	-	-	-	-	-	-	-	-	**-**	-
Invalid dates[10]	0	0	0.0	1	0	0.0	4	4	<0.1	27	16	<0.1	32	**20**	<0.1
Zero survival[11]	667	589	4.4	1,096	967	5.9	2,068	1,827	8.6	2,954	2,713	9.4	6,785	**6,096**	7.6
Total exclusions		**1,453**	**10.8**		**2,186**	**13.3**		**3,298**	**15.5**		**5,073**	**17.6**		**12,010**	**15.0**
Patients included in analysis		**11,964**	**89.2**		**14,214**	**86.7**		**18,021**	**84.5**		**23,719**	**82.4**		**67,918**	**85.0**
Of which, extra-nodal		695			433			49			15			**1,192**	

[1] Main data item(s) invalid or incompatible with one another: name, sex, date of birth, date of diagnosis and (if present) date of death, postcode, site, morphology and behaviour

[2] Tumour either *in situ* (behaviour code 2), benign (0), uncertain if benign or malignant (1) or metastatic (6 or 9)

[3] Not applicable for this malignancy

[4] Not applicable for this malignancy

[5] Tracing of vital status at National Health Service Central Register incomplete at 6 July 1997

[6] Same site code, sex, cancer registry and cancer registry number as an earlier registration

[7] Same site code and person number as an earlier registration(s): mostly confirmed multiple primary tumours at this site, some unresolved duplicate registrations

[8] Same site code, sex, date of birth and date of diagnosis as another registration(s): mostly synchronous or (in paired organs) bilateral tumours in same anatomic site in one individual, not previously linked: also some duplicate registrations

[9] Not applicable for this malignancy

[10] Impossible sequence of dates (birth, diagnosis and, if present, emigration or death); or date of death after 6 July 1997

[11] Date of diagnosis same as date of death: some are patients who did die on the day of diagnosis, but most were registered solely from a death certificate (death certificate only, or DCO), and their survival time is thus unknown

Table 41.2 Crude and relative survival (%) at one, five and ten years since diagnosis: NHS Region, deprivation category and calendar period of diagnosis

	1971-75												1976-80											
MEN	Affluent		2		3		4		Deprived		All[1]		Affluent		2		3		4		Deprived		All[1]	
	C	R	C	R	C	R	C	R	C	R	C	R	C	R	C	R	C	R	C	R	C	R	C	R
ENGLAND & WALES	*(1,092)*		*(1,067)*		*(1,073)*		*(1,028)*		*(853)*		*(6,258)*		*(1,509)*		*(1,663)*		*(1,562)*		*(1,388)*		*(1,134)*		*(7,545)*	
One year	56	58	50	52	52	55	49	51	50	52	52	54	58	61	56	58	55	58	52	55	50	52	55	57
Five years	28	34	26	31	27	33	26	31	28	34	27	32	33	40	33	39	30	37	28	35	26	33	31	37
Ten years	19	27	17	25	18	26	17	25	18	27	18	26	21	29	21	30	20	30	18	27	17	26	20	29
ENGLAND	*(1,050)*		*(1,031)*		*(1,008)*		*(952)*		*(808)*		*(5,908)*		*(1,454)*		*(1,599)*		*(1,472)*		*(1,316)*		*(1,080)*		*(7,163)*	
One year	56	58	50	52	53	55	49	51	50	52	52	54	58	60	56	58	55	58	52	55	50	52	55	57
Five years	29	34	26	31	27	33	26	31	27	34	27	33	33	40	33	39	30	37	29	35	26	33	31	37
Ten years	20	28	17	25	18	26	17	25	18	28	18	26	21	30	21	30	20	29	18	27	17	26	20	29
Northern & Yorkshire	*(57)*		*(73)*		*(92)*		*(100)*		*(152)*		*(710)*		*(121)*		*(132)*		*(171)*		*(191)*		*(230)*		*(864)*	
One year	46	47	54	56	50	52	40	42	44	46	49	51	59	61	49	51	49	52	45	48	46	48	50	52
Five years	27	33	26	32	29	36	20	25	25	32	25	31	31	35	35	42	25	30	21	26	22	28	26	32
Ten years	19	25	15	22	18	28	11	19	17	26	16	24	15	21	22	34	17	24	12	17	14	22	16	23
Trent	*(81)*		*(97)*		*(115)*		*(148)*		*(115)*		*(605)*		*(97)*		*(151)*		*(179)*		*(173)*		*(112)*		*(739)*	
One year	51	54	46	48	52	54	39	41	49	51	49	50	52	55	50	52	59	61	53	55	48	51	55	57
Five years	27	32	26	30	28	35	24	31	25	30	27	32	27	31	25	29	36	47	25	29	25	32	30	36
Ten years	18	26	17	22	16	25	18	28	18	26	18	27	16	22	15	22	26	39	17	23	20	31	20	28
Anglia & Oxford	*(148)*		*(142)*		*(123)*		*(77)*		*(35)*		*(597)*		*(227)*		*(211)*		*(162)*		*(113)*		*(65)*		*(801)*	
One year	55	57	51	53	59	61	50	52	70	73	56	58	54	56	58	60	53	56	59	62	64	68	57	59
Five years	30	35	25	28	38	45	32	36	37	45	32	37	29	35	30	35	29	36	38	45	36	46	31	37
Ten years	23	32	18	25	28	39	24	31	24	35	24	33	20	28	16	24	22	31	24	34	25	40	20	29
North Thames	*(172)*		*(166)*		*(152)*		*(174)*		*(119)*		*(860)*		*(189)*		*(189)*		*(183)*		*(180)*		*(153)*		*(921)*	
One year	62	63	52	54	57	59	49	50	59	61	56	58	67	69	62	64	59	61	56	58	57	59	60	63
Five years	33	38	28	33	31	37	28	33	32	39	31	37	40	47	39	45	32	38	30	37	31	38	35	42
Ten years	24	33	18	26	20	28	17	23	26	35	21	30	27	37	24	35	23	32	24	35	20	29	24	34
South Thames	*(247)*		*(183)*		*(178)*		*(153)*		*(78)*		*(899)*		*(291)*		*(286)*		*(212)*		*(164)*		*(64)*		*(1,065)*	
One year	57	60	45	47	47	49	52	54	51	53	52	54	57	59	54	56	52	55	51	54	52	54	55	57
Five years	27	32	20	23	21	26	25	30	28	32	26	30	30	35	34	41	24	31	34	42	28	36	31	38
Ten years	17	23	16	23	12	18	18	25	15	24	17	24	19	26	22	32	15	22	21	32	21	32	19	28
South & West	*(200)*		*(236)*		*(211)*		*(109)*		*(49)*		*(885)*		*(268)*		*(371)*		*(277)*		*(156)*		*(58)*		*(1,170)*	
One year	50	52	56	58	51	53	51	54	35	36	51	53	57	58	52	54	56	58	54	57	38	40	55	57
Five years	27	32	31	37	25	30	24	29	22	27	26	32	35	43	29	35	32	40	27	34	18	23	31	38
Ten years	19	28	19	29	16	24	17	24	15	21	18	26	23	33	18	25	21	33	16	24	9	14	19	28
West Midlands	*(53)*		*(46)*		*(47)*		*(54)*		*(78)*		*(618)*		*(101)*		*(126)*		*(143)*		*(155)*		*(161)*		*(689)*	
One year	60	62	43	44	58	60	50	51	43	45	50	53	66	68	61	64	55	57	47	50	41	43	53	56
Five years	32	36	13	15	23	28	28	33	26	31	25	31	45	50	39	45	25	32	28	33	22	26	31	37
Ten years	17	23	9	12	19	28	16	21	17	23	17	24	30	41	29	40	13	23	18	28	12	17	19	29
North & West	*(92)*		*(88)*		*(90)*		*(137)*		*(182)*		*(734)*		*(160)*		*(133)*		*(145)*		*(184)*		*(237)*		*(914)*	
One year	60	62	38	39	45	47	50	52	46	48	49	51	50	53	60	62	54	57	48	50	49	52	53	56
Five years	23	27	24	28	18	22	23	28	25	31	25	30	30	37	34	39	31	38	26	33	28	34	29	36
Ten years	15	22	13	17	13	20	16	22	15	24	15	22	18	26	27	36	21	31	17	26	19	29	20	29
WALES	*(42)*		*(36)*		*(65)*		*(76)*		*(45)*		*(350)*		*(55)*		*(64)*		*(90)*		*(72)*		*(54)*		*(382)*	
One year	57	59	46	47	46	48	51	53	48	50	51	53	65	67	49	51	53	55	49	51	46	49	53	55
Five years	25	30	27	31	27	31	26	33	30	35	26	32	34	41	31	36	32	39	24	33	25	29	29	36
Ten years	15	21	16	23	14	21	17	29	13	17	15	23	17	26	25	33	22	33	16	26	16	21	19	28

[1] Including those for whom no deprivation category was assigned. C - crude survival, R - relative survival (see Chapter 3).
Number of patients contributing to each analysis in parentheses.

1981-85 | 1986-90 — MEN

Values shown as "C R" (C - crude survival, R - relative survival). Number of patients contributing to each analysis in parentheses.

	Row	Affluent	2	3	4	Deprived	All[1]	Affluent	2	3	4	Deprived	All[1]
		1981-85						*1986-90*					
ENGLAND & WALES	N	(2,109)	(2,041)	(1,928)	(1,780)	(1,423)	(9,489)	(2,836)	(2,895)	(2,570)	(2,396)	(1,800)	(12,639)
	One year	65 68	61 64	59 62	58 61	55 58	60 63	65 68	64 67	61 64	62 65	59 62	62 65
	Five years	39 46	36 43	33 41	31 39	31 38	34 42	40 48	38 46	36 44	36 44	34 43	37 45
		27 37	25 36	23 34	20 31	21 32	23 34						
ENGLAND	N	(2,044)	(1,970)	(1,825)	(1,684)	(1,361)	(9,055)	(2,732)	(2,783)	(2,433)	(2,252)	(1,727)	(12,053)
	One year	65 68	61 64	59 63	57 60	55 58	60 63	65 68	64 67	61 64	62 65	59 62	63 66
	Five years	39 46	36 43	34 42	30 38	30 37	34 42	40 48	38 46	36 44	36 45	34 43	37 46
		27 37	25 36	23 34	19 30	21 31	23 34						
Northern & Yorkshire	N	(151)	(183)	(235)	(228)	(272)	(1,085)	(232)	(239)	(255)	(305)	(353)	(1,388)
	One year	68 70	57 60	61 65	51 53	54 57	58 61	62 65	57 61	53 56	56 60	53 57	57 60
	Five years	38 46	37 46	32 40	29 37	31 39	33 41	40 48	28 36	26 34	36 47	30 41	33 41
		26 36	25 36	19 30	17 27	21 33	21 31						
Trent	N	(153)	(220)	(222)	(264)	(189)	(1,054)	(188)	(279)	(311)	(286)	(251)	(1,320)
	One year	60 63	58 60	56 59	52 55	52 55	56 59	59 63	60 64	58 62	57 61	51 54	58 61
	Five years	38 45	35 41	29 37	27 34	28 34	31 38	37 46	35 45	34 43	32 43	28 35	34 41
		26 35	23 32	21 32	15 25	15 25	20 29						
Anglia & Oxford	N	(279)	(213)	(187)	(120)	(56)	(876)	(374)	(381)	(245)	(171)	(77)	(1,263)
	One year	62 64	61 63	60 63	51 53	47 48	59 62	60 63	61 65	62 65	55 57	62 66	61 63
	Five years	35 43	37 44	34 45	29 34	27 33	35 42	36 45	37 46	36 46	33 40	37 46	36 44
		24 34	29 37	25 39	16 24	22 30	25 34						
North Thames	N	(238)	(217)	(199)	(219)	(165)	(1,058)	(344)	(325)	(320)	(299)	(241)	(1,563)
	One year	67 69	60 63	65 68	67 71	67 70	65 68	69 73	69 72	67 71	65 69	70 74	68 71
	Five years	43 52	40 47	42 53	38 47	44 52	42 50	47 57	46 57	40 52	38 49	43 54	43 52
		27 40	26 36	31 46	22 36	33 48	28 40						
South Thames	N	(496)	(339)	(265)	(211)	(96)	(1,446)	(504)	(425)	(314)	(306)	(155)	(1,732)
	One year	68 71	65 68	61 64	64 68	55 59	65 68	70 74	68 73	64 68	67 70	67 70	68 71
	Five years	41 50	37 46	36 44	32 41	28 34	37 45	42 52	40 49	39 49	37 47	40 53	40 49
		26 39	27 41	26 40	25 38	23 33	26 39						
South & West	N	(358)	(445)	(352)	(175)	(57)	(1,434)	(545)	(672)	(511)	(325)	(103)	(2,177)
	One year	59 62	58 62	59 62	63 66	58 61	60 63	67 71	66 69	61 65	66 70	63 67	65 69
	Five years	33 41	34 42	30 39	31 39	36 45	33 41	40 50	38 48	37 46	36 48	33 42	38 47
		25 38	25 38	20 33	20 33	29 45	23 35						
West Midlands	N	(159)	(148)	(187)	(170)	(194)	(866)	(235)	(180)	(210)	(225)	(207)	(1,064)
	One year	59 63	63 66	54 56	55 58	54 57	58 60	65 68	59 62	57 61	61 65	57 60	61 63
	Five years	34 41	37 45	32 39	29 38	25 32	31 39	41 50	33 41	36 46	34 43	31 41	35 43
		24 34	29 45	23 34	21 36	15 26	22 34						
North & West	N	(210)	(205)	(178)	(297)	(332)	(1,236)	(310)	(282)	(267)	(335)	(340)	(1,546)
	One year	74 77	58 62	54 57	51 54	48 51	57 60	57 60	62 66	58 62	59 63	56 59	59 62
	Five years	44 52	30 38	34 41	26 33	26 33	32 39	34 42	39 49	34 45	35 46	33 44	35 44
		34 47	19 30	19 30	17 27	17 27	21 32						
WALES	N	(65)	(71)	(103)	(96)	(62)	(434)	(104)	(112)	(137)	(144)	(73)	(586)
	One year	52 54	61 64	59 61	59 62	50 53	58 61	59 62	61 65	55 58	57 59	60 64	59 61
	Five years	31 39	28 35	26 33	37 46	38 44	34 42	35 41	38 47	34 43	31 36	33 43	34 41
		24 38	19 28	19 28	26 40	33 44	25 36						

[1] Including those for whom no deprivation category was assigned. C - crude survival, R - relative survival (see Chapter 3). Number of patients contributing to each analysis in parentheses.

Table 41.2 Crude and relative survival (%) at one, five and ten years since diagnosis: NHS Region, deprivation category and calendar period of diagnosis

		1971-75													1976-80											
WOMEN		Affluent		2		3		4		Deprived		All[1]		Affluent		2		3		4		Deprived		All[1]	
		C	R	C	R	C	R	C	R	C	R	C	R	C	R	C	R	C	R	C	R	C	R	C	R
ENGLAND & WALES		(983)		(940)		(1,033)		(929)		(772)		(5,706)		(1,380)		(1,354)		(1,374)		(1,265)		(1,013)		(6,669)	
	One year	56	58	53	55	52	54	50	52	45	46	51	53	59	61	56	58	54	56	54	56	52	54	55	57
	Five years	33	37	31	36	27	32	28	32	24	28	28	33	35	41	33	38	32	38	32	37	32	38	33	38
	Ten years	22	28	20	28	18	25	17	23	16	23	19	26	23	31	22	30	23	32	21	29	20	29	22	30
ENGLAND		(948)		(889)		(976)		(872)		(747)		(5,427)		(1,341)		(1,302)		(1,292)		(1,179)		(961)		(6,321)	
	One year	56	58	52	54	52	53	50	51	45	46	51	53	59	61	55	57	54	57	54	56	52	54	55	57
	Five years	32	37	31	36	27	32	27	32	24	28	28	33	35	40	33	38	32	38	33	38	32	38	33	38
	Ten years	22	28	20	27	18	25	16	22	16	23	19	26	23	30	22	30	23	32	21	29	20	29	22	30
Northern & Yorkshire		(58)		(62)		(105)		(112)		(133)		(677)		(118)		(120)		(137)		(165)		(192)		(765)	
	One year	53	54	46	48	47	50	49	51	39	41	48	50	56	58	50	52	53	56	61	64	44	46	53	56
	Five years	38	43	28	34	30	36	25	30	20	23	27	32	26	31	34	40	31	37	37	44	23	28	30	36
	Ten years	20	26	17	24	18	29	16	23	15	20	18	26	17	22	26	36	21	30	24	33	17	23	21	29
Trent		(58)		(80)		(97)		(136)		(91)		(503)		(78)		(120)		(142)		(165)		(111)		(638)	
	One year	46	47	53	55	59	61	49	51	48	49	53	55	58	60	59	61	49	51	49	50	50	52	53	55
	Five years	26	28	35	43	34	40	23	26	24	29	30	35	42	45	33	38	30	35	30	34	41	46	34	40
	Ten years	15	18	19	28	23	31	14	19	17	24	18	25	28	33	24	32	20	29	22	28	24	31	23	31
Anglia & Oxford		(152)		(105)		(104)		(64)		(29)		(516)		(216)		(177)		(150)		(89)		(46)		(703)	
	One year	61	62	54	57	52	54	56	57	43	44	56	58	60	62	49	51	51	54	58	62	50	52	55	58
	Five years	39	45	42	50	28	33	30	35	30	35	35	41	36	42	31	36	32	38	33	39	34	36	34	40
	Ten years	25	33	29	47	19	26	16	21	13	17	23	33	22	29	20	28	24	34	20	28	19	25	22	31
North Thames		(176)		(140)		(140)		(151)		(145)		(870)		(172)		(154)		(144)		(158)		(127)		(794)	
	One year	59	60	51	53	53	55	55	57	51	53	54	56	64	66	65	67	59	61	59	61	63	67	62	64
	Five years	35	39	28	31	31	35	32	38	31	36	31	36	39	44	41	48	34	39	37	43	42	51	39	45
	Ten years	27	35	19	24	23	29	17	24	25	34	21	29	27	34	27	36	25	34	21	29	29	42	25	35
South Thames		(223)		(160)		(174)		(159)		(76)		(881)		(309)		(221)		(221)		(171)		(84)		(1,068)	
	One year	56	58	50	52	45	46	50	51	40	42	51	53	54	56	50	52	49	51	52	54	53	56	52	54
	Five years	31	35	23	27	21	26	31	35	18	22	27	32	31	36	26	31	30	36	32	38	28	36	29	35
	Ten years	21	27	14	18	13	19	17	23	12	17	17	23	22	30	19	26	22	31	22	30	20	29	21	30
South & West		(157)		(220)		(230)		(83)		(26)		(792)		(238)		(306)		(263)		(132)		(46)		(1,014)	
	One year	51	53	48	50	51	53	35	36	26	27	49	51	60	61	55	57	58	60	52	54	48	51	57	59
	Five years	28	31	26	31	25	30	22	26	11	12	26	31	37	43	31	36	32	37	34	40	27	33	33	39
	Ten years	20	26	18	25	16	23	12	17	11	12	18	25	22	31	21	28	22	30	21	29	19	28	21	30
West Midlands		(48)		(35)		(42)		(63)		(84)		(535)		(82)		(82)		(106)		(139)		(137)		(546)	
	One year	53	54	56	59	50	51	49	51	33	35	47	48	55	56	53	56	58	61	46	48	48	50	52	55
	Five years	31	34	34	41	24	27	20	24	15	18	25	29	31	35	28	32	33	39	28	32	27	33	30	35
	Ten years	22	29	28	37	19	23	10	17	9	13	17	24	20	26	17	23	22	33	19	26	18	26	20	28
North & West		(76)		(87)		(84)		(104)		(163)		(653)		(128)		(122)		(129)		(160)		(218)		(793)	
	One year	54	56	61	62	50	52	42	44	47	49	51	53	59	61	59	61	49	51	48	50	51	53	53	56
	Five years	22	25	42	47	17	20	25	29	25	30	27	33	35	42	38	45	32	37	25	29	30	35	32	38
	Ten years	10	13	22	29	12	16	18	26	13	20	17	24	19	26	23	32	23	30	16	23	16	23	19	27
WALES		(35)		(51)		(57)		(57)		(25)		(279)		(39)		(52)		(82)		(86)		(52)		(348)	
	One year	63	64	60	62	51	52	49	51	33	34	53	55	59	61	62	65	51	53	46	48	52	54	54	56
	Five years	35	37	32	38	32	34	32	39	18	19	31	36	54	56	36	41	33	40	21	25	34	40	33	39
	Ten years	16	21	22	31	22	26	25	34	6	9	21	30	38	45	19	28	25	33	17	24	19	26	23	32

[1] Including those for whom no deprivation category was assigned. C - crude survival, R - relative survival (see Chapter 3).
Number of patients contributing to each analysis in parentheses.

1981-85 / 1986-90 — WOMEN

Region	Period	Measure	Affluent C	Affluent R	2 C	2 R	3 C	3 R	4 C	4 R	Deprived C	Deprived R	All[1] C	All[1] R
ENGLAND & WALES	1981-85	(n)	(1,848)		(1,811)		(1,754)		(1,677)		(1,269)		(8,532)	
	1981-85	One year	64	67	61	63	59	62	57	60	56	59	60	62
	1981-85	Five years	38	45	36	43	35	42	34	41	33	41	36	42
	1981-85		26	35	25	35	24	33	22	32	22	32	24	34
	1986-90	(n)	(2,305)		(2,448)		(2,364)		(2,211)		(1,652)		(11,080)	
	1986-90	One year	66	69	63	65	63	66	60	63	58	60	62	65
	1986-90	Five years	43	50	38	45	38	46	34	41	35	42	38	45
ENGLAND	1981-85	(n)	(1,796)		(1,734)		(1,666)		(1,563)		(1,200)		(8,104)	
	1981-85	One year	64	66	61	63	59	62	57	60	56	59	60	62
	1981-85	Five years	38	44	37	43	35	42	34	41	34	41	36	42
	1981-85		26	35	25	35	24	33	22	32	22	32	24	34
	1986-90	(n)	(2,216)		(2,341)		(2,229)		(2,067)		(1,569)		(10,504)	
	1986-90	One year	67	69	63	66	63	66	61	64	58	61	63	65
	1986-90	Five years	43	51	38	45	38	46	34	41	35	42	38	45
Northern & Yorkshire	1981-85	(n)	(153)		(159)		(203)		(255)		(241)		(1,022)	
	1981-85	One year	59	61	63	65	58	61	55	57	53	56	58	61
	1981-85	Five years	36	43	40	47	35	42	34	41	28	35	35	42
	1981-85		27	36	29	43	24	34	23	35	16	25	23	35
	1986-90	(n)	(188)		(230)		(225)		(269)		(285)		(1,204)	
	1986-90	One year	59	61	55	58	57	59	56	59	54	57	57	59
	1986-90	Five years	41	48	34	40	33	41	32	39	32	39	34	42
Trent	1981-85	(n)	(122)		(166)		(194)		(228)		(162)		(881)	
	1981-85	One year	65	68	60	63	62	64	60	62	44	46	59	62
	1981-85	Five years	35	41	35	40	36	41	34	41	27	33	34	40
	1981-85		23	33	27	36	24	32	20	30	19	27	23	33
	1986-90	(n)	(172)		(214)		(240)		(254)		(223)		(1,104)	
	1986-90	One year	64	66	61	63	57	59	56	58	47	49	57	60
	1986-90	Five years	42	49	34	39	35	43	33	39	28	35	35	41
Anglia & Oxford	1981-85	(n)	(238)		(209)		(169)		(101)		(50)		(780)	
	1981-85	One year	57	60	54	56	51	53	54	57	45	47	55	57
	1981-85	Five years	34	40	33	39	28	34	39	48	27	32	33	40
	1981-85		24	31	24	35	19	27	25	36	21	28	23	32
	1986-90	(n)	(281)		(290)		(212)		(147)		(61)		(1,000)	
	1986-90	One year	63	65	61	63	61	63	54	57	61	65	61	64
	1986-90	Five years	42	49	32	37	39	46	27	33	32	38	36	43
North Thames	1981-85	(n)	(236)		(224)		(176)		(169)		(138)		(952)	
	1981-85	One year	69	71	68	71	67	70	63	65	66	69	67	70
	1981-85	Five years	39	44	47	54	45	55	39	45	40	47	42	49
	1981-85		30	37	34	44	34	48	26	35	33	44	31	41
	1986-90	(n)	(275)		(281)		(298)		(307)		(240)		(1,424)	
	1986-90	One year	73	76	67	70	67	69	64	67	71	74	68	71
	1986-90	Five years	53	62	46	53	40	48	36	43	47	55	44	52
South Thames	1981-85	(n)	(411)		(287)		(265)		(200)		(76)		(1,276)	
	1981-85	One year	68	71	64	67	57	60	56	59	73	76	63	66
	1981-85	Five years	43	51	36	44	35	43	27	32	44	53	37	45
	1981-85		31	42	23	34	23	33	16	23	33	45	25	36
	1986-90	(n)	(452)		(381)		(305)		(239)		(129)		(1,526)	
	1986-90	One year	72	75	67	69	72	75	65	67	67	70	69	72
	1986-90	Five years	45	53	39	46	44	52	34	41	46	55	42	50
South & West	1981-85	(n)	(338)		(403)		(319)		(186)		(52)		(1,337)	
	1981-85	One year	58	60	57	60	59	61	58	61	46	48	58	61
	1981-85	Five years	32	38	30	35	33	38	38	45	31	37	33	39
	1981-85		22	29	20	28	21	29	25	35	24	31	21	30
	1986-90	(n)	(419)		(519)		(501)		(309)		(98)		(1,859)	
	1986-90	One year	69	71	63	65	64	67	67	70	64	68	66	69
	1986-90	Five years	45	53	38	46	37	44	38	45	45	54	40	48
West Midlands	1981-85	(n)	(119)		(126)		(137)		(177)		(159)		(728)	
	1981-85	One year	67	70	55	57	58	60	49	51	51	54	56	59
	1981-85	Five years	38	47	36	40	35	40	22	27	35	42	33	40
	1981-85		22	32	23	30	24	32	13	19	22	33	21	30
	1986-90	(n)	(171)		(190)		(213)		(231)		(167)		(975)	
	1986-90	One year	61	64	64	67	58	60	56	58	49	51	59	61
	1986-90	Five years	37	45	42	49	35	42	35	40	25	29	35	42
North & West	1981-85	(n)	(179)		(160)		(203)		(247)		(322)		(1,128)	
	1981-85	One year	63	66	56	59	59	62	55	57	59	62	59	62
	1981-85	Five years	40	47	36	45	33	39	36	45	34	42	37	45
	1981-85		24	34	27	39	23	33	25	35	20	29	24	35
	1986-90	(n)	(258)		(236)		(235)		(311)		(366)		(1,412)	
	1986-90	One year	59	62	63	65	60	63	61	63	55	57	60	63
	1986-90	Five years	36	42	39	47	40	48	32	40	31	37	35	43
WALES	1981-85	(n)	(52)		(77)		(88)		(114)		(69)		(428)	
	1981-85	One year	79	81	58	60	61	63	57	59	54	56	61	64
	1981-85	Five years	40	45	31	36	37	43	35	42	30	34	35	41
	1981-85		31	37	21	30	20	27	23	32	23	30	24	32
	1986-90	(n)	(89)		(107)		(135)		(144)		(83)		(576)	
	1986-90	One year	50	53	53	55	57	59	49	51	54	56	54	56
	1986-90	Five years	28	31	29	34	38	45	28	33	35	43	33	39

[1] Including those for whom no deprivation category was assigned. C - crude survival, R - relative survival (see Chapter 3).
Number of patients contributing to each analysis in parentheses.

Table 41.2 Crude and relative survival (%) at one, five and ten years since diagnosis: NHS Region, deprivation category and calendar period of diagnosis

ADULTS		1971-75											1976-80											
		Affluent		2		3		4		Deprived		All[1]		Affluent		2		3		4		Deprived		All[1]	
		C	R	C	R	C	R	C	R	C	R	C	R	C	R	C	R	C	R	C	R	C	R	C	R
ENGLAND & WALES		*(2,075)*		*(2,007)*		*(2,106)*		*(1,957)*		*(1,625)*		*(11,964)*		*(2,889)*		*(3,017)*		*(2,936)*		*(2,653)*		*(2,147)*		*(14,214)*	
	One year	56	58	52	53	52	54	49	51	47	49	51	53	59	61	56	58	55	57	53	55	51	53	55	57
	Five years	30	35	28	33	27	33	27	32	26	31	28	33	34	40	33	39	31	38	30	36	29	36	31	38
	Ten years	20	27	18	26	18	26	17	24	17	25	18	26	22	30	22	30	22	31	20	28	19	27	21	29
ENGLAND		*(1,998)*		*(1,920)*		*(1,984)*		*(1,824)*		*(1,555)*		*(11,335)*		*(2,795)*		*(2,901)*		*(2,764)*		*(2,495)*		*(2,041)*		*(13,484)*	
	One year	56	58	51	53	52	54	49	51	47	49	51	53	58	61	56	58	55	57	53	56	51	53	55	57
	Five years	30	35	28	33	27	33	27	32	26	31	28	33	34	40	33	39	31	37	30	36	29	36	32	38
	Ten years	21	28	18	26	18	26	17	24	17	25	18	26	22	30	22	30	21	31	20	28	19	28	21	29
Northern & Yorkshire		*(115)*		*(135)*		*(197)*		*(212)*		*(285)*		*(1,387)*		*(239)*		*(252)*		*(308)*		*(356)*		*(422)*		*(1,629)*	
	One year	49	51	51	52	49	51	45	47	42	43	49	51	57	59	50	52	51	53	53	56	45	47	51	54
	Five years	33	38	27	34	30	36	23	27	23	28	26	31	29	33	35	41	28	33	28	34	23	28	28	34
	Ten years	19	26	16	24	18	28	14	21	16	23	17	25	16	22	24	35	19	27	17	25	15	23	18	26
Trent		*(139)*		*(177)*		*(212)*		*(284)*		*(206)*		*(1,108)*		*(175)*		*(271)*		*(321)*		*(338)*		*(223)*		*(1,377)*	
	One year	49	51	49	51	55	57	44	46	49	50	51	53	55	57	54	56	55	57	51	53	49	52	54	56
	Five years	27	31	30	36	31	37	24	28	25	30	28	33	34	38	29	33	34	41	27	31	33	41	32	38
	Ten years	17	23	18	24	19	28	16	23	17	25	18	26	22	27	19	27	23	34	19	25	22	32	21	30
Anglia & Oxford		*(300)*		*(247)*		*(227)*		*(141)*		*(64)*		*(1,113)*		*(443)*		*(388)*		*(312)*		*(202)*		*(111)*		*(1,504)*	
	One year	58	60	53	55	55	58	53	54	58	59	56	58	57	59	54	56	52	55	59	62	58	61	56	59
	Five years	34	40	32	37	33	39	31	36	33	41	33	39	32	38	30	35	31	37	36	42	35	43	32	38
	Ten years	24	32	23	34	24	33	20	26	19	28	24	33	21	29	18	25	23	33	22	31	23	34	21	30
North Thames		*(348)*		*(306)*		*(292)*		*(325)*		*(264)*		*(1,730)*		*(361)*		*(343)*		*(327)*		*(338)*		*(280)*		*(1,715)*	
	One year	60	62	52	54	55	57	52	53	55	56	55	57	66	68	63	66	59	61	57	60	60	63	61	63
	Five years	34	39	28	32	31	36	30	35	32	37	31	36	40	46	40	46	33	38	33	40	36	44	37	43
	Ten years	25	34	18	25	21	29	17	23	26	34	21	30	27	35	25	36	24	33	22	32	24	35	25	35
South Thames		*(470)*		*(343)*		*(352)*		*(312)*		*(154)*		*(1,780)*		*(600)*		*(507)*		*(433)*		*(335)*		*(148)*		*(2,133)*	
	One year	57	59	47	49	46	48	51	53	46	48	51	54	55	58	52	54	50	53	52	54	53	55	53	56
	Five years	29	33	21	25	21	26	28	33	23	27	26	31	30	35	31	37	27	34	33	40	28	36	30	36
	Ten years	19	25	15	20	13	19	17	24	14	20	17	23	20	28	21	29	19	27	21	31	21	31	20	29
South & West		*(357)*		*(456)*		*(441)*		*(192)*		*(75)*		*(1,677)*		*(506)*		*(677)*		*(540)*		*(288)*		*(104)*		*(2,184)*	
	One year	51	52	52	54	51	53	44	46	32	33	50	52	58	60	54	55	57	59	53	56	43	45	56	58
	Five years	27	32	28	34	25	30	23	28	18	22	26	32	36	43	30	35	32	39	30	37	22	27	32	38
	Ten years	20	27	19	27	16	23	15	21	14	19	18	25	23	32	19	27	22	31	18	27	13	21	20	29
West Midlands		*(101)*		*(81)*		*(89)*		*(117)*		*(162)*		*(1,153)*		*(183)*		*(208)*		*(249)*		*(294)*		*(298)*		*(1,235)*	
	One year	56	58	49	50	54	56	49	51	38	40	49	51	61	63	58	61	56	59	47	49	44	46	53	55
	Five years	31	35	21	26	23	28	24	29	20	24	25	30	39	44	35	40	29	35	28	32	24	30	30	36
	Ten years	20	26	17	23	19	25	13	19	13	18	17	24	26	34	24	33	17	27	19	26	14	21	20	28
North & West		*(168)*		*(175)*		*(174)*		*(241)*		*(345)*		*(1,387)*		*(288)*		*(255)*		*(274)*		*(344)*		*(455)*		*(1,707)*	
	One year	57	59	49	51	48	50	46	49	47	49	50	52	54	57	59	61	52	54	48	50	50	53	53	56
	Five years	23	26	33	38	17	21	24	29	25	31	26	31	32	39	36	42	31	37	25	32	29	35	31	37
	Ten years	13	17	17	23	12	18	17	23	14	22	16	23	19	26	25	34	22	31	16	25	17	26	20	28
WALES		*(77)*		*(87)*		*(122)*		*(133)*		*(70)*		*(629)*		*(94)*		*(116)*		*(172)*		*(158)*		*(106)*		*(730)*	
	One year	60	61	54	56	48	50	50	52	43	44	52	54	63	65	55	57	52	54	47	49	49	52	53	56
	Five years	29	35	30	37	29	33	29	36	26	30	28	34	42	50	33	38	33	39	22	28	29	34	31	38
	Ten years	15	22	20	29	17	23	21	33	10	15	18	26	26	37	22	31	24	33	16	25	18	24	20	30

[1] Including those for whom no deprivation category was assigned. C - crude survival, R - relative survival (see Chapter 3). Number of patients contributing to each analysis in parentheses.

	1981-85						1986-90						ADULTS
	Affluent	2	3	4	Deprived	All[1]	Affluent	2	3	4	Deprived	All[1]	
	C R	C R	C R	C R	C R	C R	C R	C R	C R	C R	C R	C R	
	(3,957)	*(3,852)*	*(3,682)*	*(3,457)*	*(2,692)*	*(18,021)*	*(5,141)*	*(5,343)*	*(4,934)*	*(4,607)*	*(3,452)*	*(23,719)*	**ENGLAND & WALES**
	65 67	61 63	59 62	57 60	55 58	60 63	65 68	64 66	62 65	61 64	59 61	62 65	*One year*
	38 45	36 43	34 41	32 40	32 39	35 42	41 49	38 45	37 45	35 42	34 42	37 45	*Five years*
	26 36	25 35	23 33	21 31	22 32	24 34							
	(3,840)	*(3,704)*	*(3,491)*	*(3,247)*	*(2,561)*	*(17,159)*	*(4,948)*	*(5,124)*	*(4,662)*	*(4,319)*	*(3,296)*	*(22,557)*	**ENGLAND**
	65 67	61 63	59 62	57 60	55 58	60 63	66 69	64 67	62 65	61 64	59 61	63 66	*One year*
	38 45	36 43	34 42	32 39	32 39	35 42	42 49	38 46	37 45	35 43	34 42	38 45	*Five years*
	26 36	25 36	23 34	21 31	21 32	24 34							
	(304)	*(342)*	*(438)*	*(483)*	*(513)*	*(2,107)*	*(420)*	*(469)*	*(480)*	*(574)*	*(638)*	*(2,592)*	**Northern & Yorkshire**
	63 65	60 62	59 63	53 55	53 57	58 61	60 63	56 59	55 57	56 59	54 57	57 60	*One year*
	37 44	38 46	33 41	32 39	29 37	34 41	41 48	31 38	30 37	34 43	31 40	33 41	*Five years*
	27 36	27 39	21 32	20 31	18 30	22 33							
	(275)	*(386)*	*(416)*	*(492)*	*(351)*	*(1,935)*	*(360)*	*(493)*	*(551)*	*(540)*	*(474)*	*(2,424)*	**Trent**
	62 65	59 61	59 62	56 59	48 51	58 60	62 65	61 63	58 61	57 59	49 52	58 60	*One year*
	37 43	35 41	32 39	30 37	27 34	33 39	39 47	35 42	35 43	33 41	28 35	34 41	*Five years*
	25 34	25 34	22 32	18 28	17 26	21 31							
	(517)	*(422)*	*(356)*	*(221)*	*(106)*	*(1,656)*	*(655)*	*(671)*	*(457)*	*(318)*	*(138)*	*(2,263)*	**Anglia & Oxford**
	60 62	57 60	56 58	53 55	46 48	57 60	61 64	61 64	61 64	54 57	62 65	61 64	*One year*
	35 41	35 41	31 39	34 41	27 33	34 41	39 47	35 42	37 46	30 37	35 42	36 44	*Five years*
	24 33	27 36	22 33	20 30	21 29	24 33							
	(474)	*(441)*	*(375)*	*(388)*	*(303)*	*(2,010)*	*(619)*	*(606)*	*(618)*	*(606)*	*(481)*	*(2,987)*	**North Thames**
	68 70	64 67	66 69	65 68	67 70	66 69	71 74	68 71	67 70	65 68	71 74	68 71	*One year*
	41 48	44 51	44 54	38 46	42 50	42 49	50 59	46 55	40 50	37 46	45 54	44 52	*Five years*
	29 39	30 40	32 46	24 35	33 46	29 40							
	(907)	*(626)*	*(530)*	*(411)*	*(172)*	*(2,722)*	*(956)*	*(806)*	*(619)*	*(545)*	*(284)*	*(3,258)*	**South Thames**
	68 71	65 68	59 62	60 63	63 67	64 67	71 74	68 71	68 72	66 69	67 70	69 72	*One year*
	42 50	37 45	36 44	30 37	35 42	37 45	43 53	39 48	42 51	36 44	43 54	41 49	*Five years*
	28 40	25 38	24 36	21 31	28 39	26 37							
	(696)	*(848)*	*(671)*	*(361)*	*(109)*	*(2,771)*	*(964)*	*(1,191)*	*(1,012)*	*(634)*	*(201)*	*(4,036)*	**South & West**
	59 61	58 61	59 62	61 63	52 55	59 62	68 71	64 67	63 66	66 70	64 67	66 69	*One year*
	33 40	32 38	31 38	35 43	34 41	33 40	42 51	38 47	37 45	37 46	38 48	39 47	*Five years*
	23 33	23 33	20 31	22 34	27 37	22 32							
	(278)	*(274)*	*(324)*	*(347)*	*(353)*	*(1,594)*	*(406)*	*(370)*	*(423)*	*(456)*	*(374)*	*(2,039)*	**West Midlands**
	63 66	60 62	56 58	52 54	53 56	57 60	64 67	61 64	58 60	59 62	54 56	60 62	*One year*
	36 44	36 42	33 39	25 32	29 37	32 39	39 48	38 45	35 43	35 42	28 35	35 43	*Five years*
	23 33	27 36	24 33	17 27	18 29	22 32							
	(389)	*(365)*	*(381)*	*(544)*	*(654)*	*(2,364)*	*(568)*	*(518)*	*(502)*	*(646)*	*(706)*	*(2,958)*	**North & West**
	69 72	57 61	57 60	53 55	54 56	58 61	58 61	62 66	59 62	60 63	55 58	59 62	*One year*
	42 50	33 41	33 40	31 39	30 38	34 42	35 42	39 48	37 46	34 43	32 40	35 43	*Five years*
	29 41	22 34	21 32	21 31	18 28	22 33							
	(117)	*(148)*	*(191)*	*(210)*	*(131)*	*(862)*	*(193)*	*(219)*	*(272)*	*(288)*	*(156)*	*(1,162)*	**WALES**
	64 67	59 62	60 62	58 60	53 55	60 62	55 58	57 60	56 58	53 55	57 60	56 59	*One year*
	35 42	29 36	31 38	36 44	34 39	34 41	32 37	33 40	36 45	29 35	34 43	34 40	*Five years*
	27 37	20 30	20 28	25 35	28 38	24 34							

[1] Including those for whom no deprivation category was assigned. C - crude survival, R - relative survival (see Chapter 3).
Number of patients contributing to each analysis in parentheses.

Fig. 41.1 Relative survival up to ten years: trends by calendar period of diagnosis (1971-90) and NHS Region

ADULTS

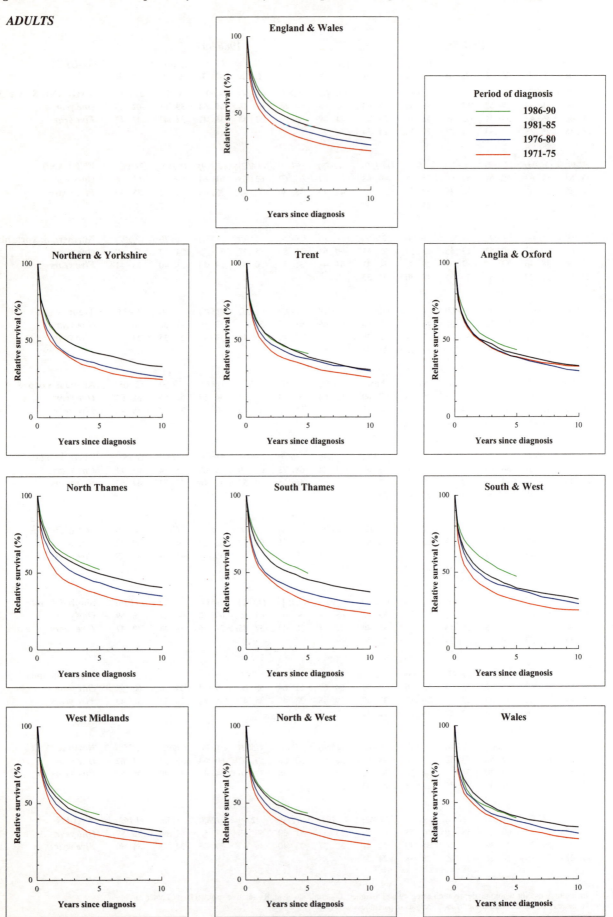

Fig. 41.2 Relative survival up to five years, by deprivation category and NHS Region: patients diagnosed 1986-90

ADULTS

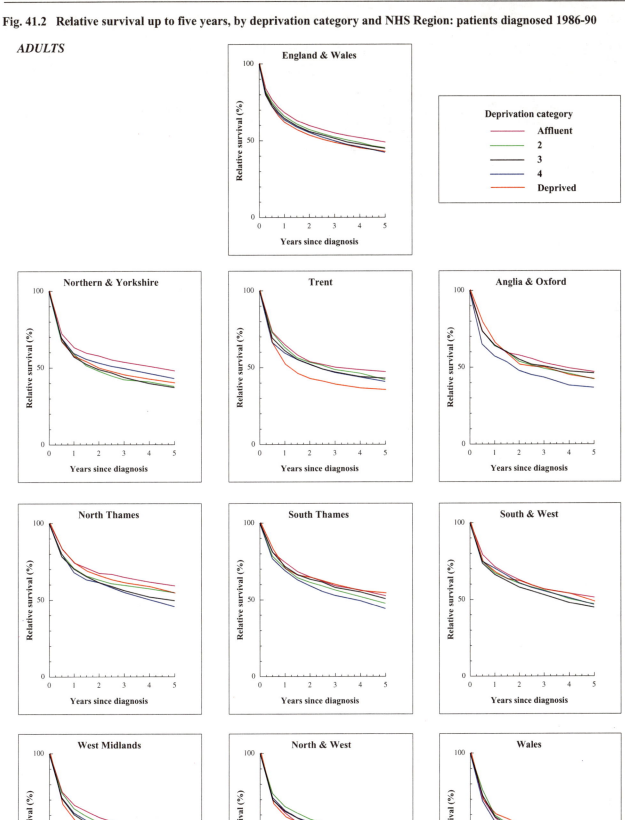

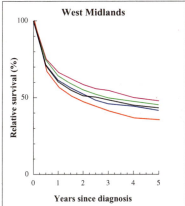

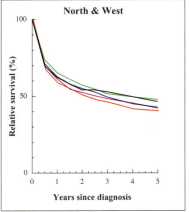

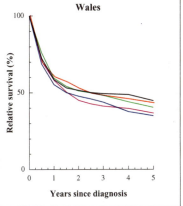

Fig. 41.3   Relative survival at five years by deprivation category, period of diagnosis (1981-90) and NHS Region

ADULTS

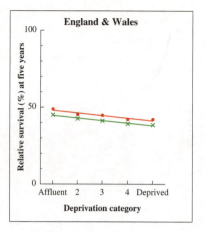

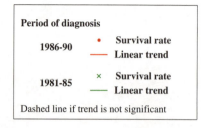

Period of diagnosis

1986-90   ● Survival rate
          — Linear trend

1981-85   × Survival rate
          — Linear trend

Dashed line if trend is not significant

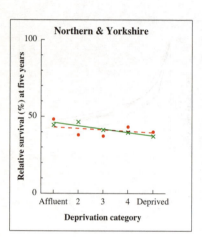

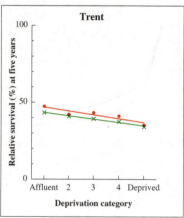

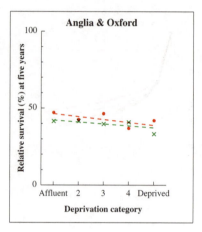

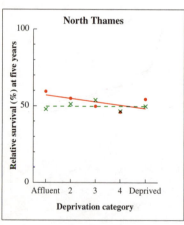

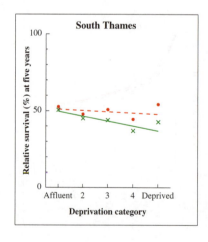

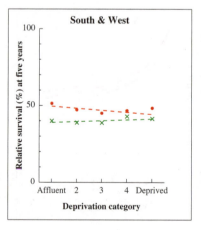

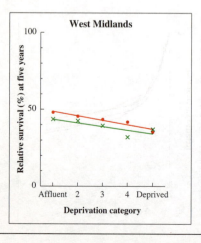

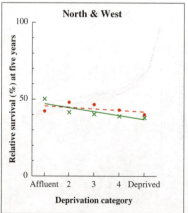

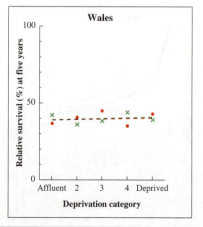

Fig. 41.4   Relative survival (%) by NHS Region and sex:
      one-year and five-year survival for patients diagnosed 1986-90, with 95% confidence intervals

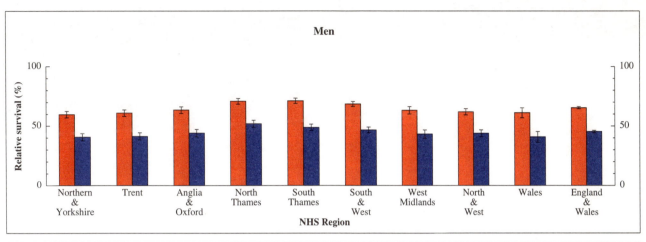

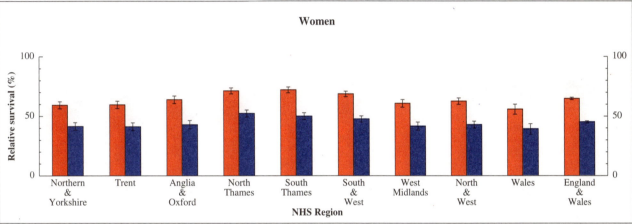

Fig. 41.5   Relative survival (%) at five years, with 95% confidence interval, by age at diagnosis and sex:
      England and Wales, patients diagnosed 1986-90

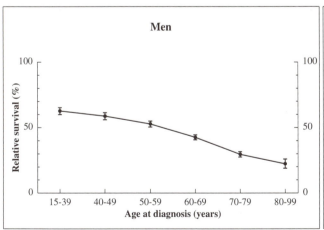

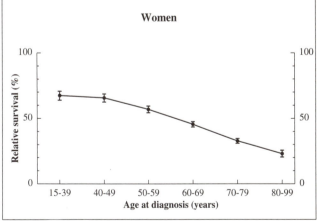

Chapter 42

HODGKIN'S DISEASE

Hodgkin's disease is an uncommon malignant lymphoma, with around 1,200 new cases a year in England and Wales, about 0.5% of all cancers. The characteristic bi-nuclear Reed-Sternberg cell enables it to be distinguished morphologically from other lymphomas, which are much more common (see Chapter 41). Hodgkin's disease also differs from other lymphomas in its patterns of incidence by age, recent trends in incidence, and in its response to treatment. In developed countries, Hodgkin's disease in adults typically has a bimodal age distribution, with a peak in young adults aged 15-30, followed by a decline and plateau in middle age and a further peak above age 60. Incidence in England and Wales has fallen slowly but steadily since the early 1970s, slightly more in men than women, although the increasing trend in Yorkshire up to 1982 was divergent[1]. In 1990, the incidence rates in England and Wales were 2.7 per 100,000 for men and 1.8 for women. Hodgkin's disease is more common in men than women at all ages, with an overall sex ratio of 1.5 to 1.

In common with other countries, mortality in England and Wales has fallen sharply since the early 1970s, by about 20% every five years. Between 1971 and 1995, mortality fell from 22 to 6 per 100,000 among men and from 11 to 4 in women. This steep decline is undoubtedly due to intensive combined radiotherapy and chemotherapy regimes that greatly improved survival[2].

The causes of Hodgkin's disease are not clearly defined. In young adults it may be a rare consequence of infection by the Epstein-Barr herpesvirus, which is also associated with infectious mononucleosis, Burkitt's lymphoma and nasopharyngeal carcinoma. Occupational risk factors for Hodgkin's disease may also include wood dust, solvents, and agricultural chemicals. Different histological subtypes occur in the different age groups: young adults predominantly show nodular sclerosis of the lymph nodes, while older adults show a mixed cellularity[3].

The survival analyses reported here are based on 22,600 patients diagnosed with Hodgkin's disease in England and Wales during the period 1971-90 and followed up to the end of 1995. After the exclusion of 1,000 children under 15 at diagnosis, about 4% of all eligible patients, those included in the analyses represent about 91% of eligible adults (Table 42.1). Survival from Hodgkin's disease in childhood is discussed in Chapter 52. About 5% of adults were excluded because their vital status was unknown at 6 July 1997, and 3% because their survival was either zero or unknown. Among patients diagnosed during 1986-90, nodular sclerosis accounted for 38% (Table 42A): the increase from 26% in the early 1980s appears to be largely due to more specific coding, since Hodgkin's disease of unspecified type fell by a similar amount during the same period. Hodgkin's disease of mixed cellularity accounted for 16% of cases during the 1980s. Two-thirds (59%) were assigned the morphology code for Hodgkin's disease without further specification, but the distribution of cases to which a more specific morphology code had been assigned was broadly similar to that of the sub-types identified by ICD-9 site code in Table 42A (data not shown).

Survival trends

Relative survival for adults diagnosed with Hodgkin's disease in England and Wales during 1986-90 was 88% at one year and 75% at five years, much higher than the corresponding figures for patients diagnosed during 1971-75, 73% at one year and 53% at five years (Table 42.2).

These large improvements in survival are apparent from Figure 42.1. In most regions, substantial gains in one-year and five-year survival occurred in each successive five-year period of diagnosis from the early 1970s to the late 1980s. The largest increase generally occurred around 1980, and the pace of improvement slowed down in several regions in the late 1980s.

Ten-year survival from Hodgkin's disease for patients diagnosed during 1981-85 was 63% (Table 42.2), but excess mortality relative to the general population continues for at least ten years after diagnosis, and the survival curves do not reach a plateau (Figure 42.1).

Nationally, the improvements in survival have been more marked for men, and survival is now very similar in both sexes. One-year survival for men increased by an average of 4% between successive five-year periods of diagnosis, compared to 3% for women, after adjustment for differences in the age distribution of patients between the sexes and over time (see Table 4.4). The corresponding average gains in five-year survival were about 5% every five years, but the difference in speed of improvement between men and women was less marked.

The regional patterns of survival suggest a north-south divide, with higher survival in the south of the country. In Anglia & Oxford, North Thames, South Thames, and South & West, five-year survival for men and women combined was higher than the national average (53%) for patients diagnosed in 1971-75, and it remained at or above the national average (75%) for patients diagnosed in those regions during 1986-90.

Table 42A Sub-site distribution of Hodgkin's disease[1]

ICD-9 code	Site description	1971-75 No.	1971-75 %	1976-80 No.	1976-80 %	1981-85 No.	1981-85 %	1986-90 No.	1986-90 %	Total No.	Total %
201.0	Hodgkin's paragranuloma	-	-	2	0.0	5	0.1	5	0.1	**12**	**0.1**
201.1	Hodgkin's granuloma	-	-	5	0.1	6	0.1	3	0.1	**14**	**0.1**
201.2	Hodgkin's sarcoma	-	-	12	0.2	9	0.2	3	0.1	**24**	**0.1**
201.4	Lymphocytic-histiocytic	-	-	161	2.7	425	7.8	278	5.5	**864**	**3.8**
201.5	Nodular sclerosis	-	-	406	6.9	1,397	25.8	1,929	38.4	**3,732**	**16.5**
201.6	Mixed cellularity	-	-	314	5.3	891	16.4	801	16.0	**2,006**	**8.9**
201.7	Lymphocytic depletion	-	-	180	3.0	279	5.1	171	3.4	**630**	**2.8**
201.9	Unspecified	6,278	100.0	4,825	81.7	2,409	44.4	1,831	36.5	**15,343**	**67.8**
	Total	**6,278**		**5,905**		**5,421**		**5,021**		**22,625**	

[1] Hodgkin's disease was first classified separately from other lymphomas only in 1968 (ICD-8), but was not subdivided until 1979, when ICD-9 codes allowed the alternative Parker-Jackson (201.0-201.2) and Rye (Lukes-Butler) classifications (201.4-201.7).

There was variation between regions in the timing and extent of improvements in survival. The largest overall increases in five-year survival occurred in Northern & Yorkshire (27%) and West Midlands (25%), two of the regions with the lowest survival rates in the early 1970s. Five-year survival in West Midlands Region rose by 14% during the 1970s alone, while in North Thames and North & West, no further gains in survival occurred for patients diagnosed during 1986-90, compared with those diagnosed in the early 1980s. The regional range in five-year survival rates for patients diagnosed during the late 1980s was still about 10% (Table 42.2, Figure 42.4).

Regional trends in five-year survival varied widely for women. In Northern & Yorkshire, the five-year survival rate of 77% for women diagnosed during 1986-90 was 30% higher than the corresponding figure for women diagnosed during 1971-75, while in Trent the increase was only 8%, from 60% to 68% (Table 42.2). Regional trends in survival were less variable for men. With the exception of Wales, regional survival rates for men and women in each five-year period of diagnosis are based on 200 or more patients, and are precise (see Chapter 3).

Survival depends heavily on age at diagnosis in both men and women (Figure 42.5). For patients diagnosed during 1986-90 under the age of 40 – fortunately the great majority in both sexes – relative survival at five years was 86%. Relative survival at five years falls by 10% or more with each successive decade of age at diagnosis, however, to 20% or less for those aged 80-99 (see Table 4.7).

Deprivation

Survival in England and Wales since the early 1970s has been lowest in the two most deprived groups (Table 42.2). The gap varies between regions (Figure 42.2), and in several regions there was no systematic difference in five-year survival between deprivation groups for patients diagnosed during the 1980s (Figure 42.3).

In England and Wales as a whole, the deprivation gradient in five-year survival was consistent and statistically significant throughout the 1980s (Figure 42.3). Thus among 5,300 patients of known deprivation status who were diagnosed during 1981-85, the deprivation gap between the most affluent and most deprived groups was 6% for both one-year and five-year survival. Survival improved for all deprivation groups during the 1980s, but the corresponding gap in survival among 5,000 patients diagnosed during 1986-90 was still 6% at one year and 8% at five years (see Table 4.7).

International comparison

Five-year survival rates for men and women diagnosed in England and Wales during 1986-90 were 75%, similar to the average for Europe and about 7% higher than in Scotland. The highest regional survival rates (81-82%, Thames regions) were similar to those for patients diagnosed during 1986-90 in the regions covered by the SEER programme in the USA (see Table 4.12).

1.   Barnes N, Cartwright RA, O'Brien C, Richards IDG, Roberts B, Bird CC. Rising incidence of lymphoid malignancies – true or false? Br J Cancer 1986; 53: 393-398

2.   Cancer Research Campaign. *Trends in cancer survival in Great Britain: cases registered between 1960 and 1974.* London: CRC, 1982

3.   Alexander FE, McKinney PA, Williams J, Ricketts TJ, Cartwright RA. Epidemiological evidence for the 'two-disease' hypothesis in Hodgkin's disease. Int J Epidemiol 1991; 20: 354-361

Table 42.1 Ineligible and excluded records, and patients included in analyses, by calendar period of diagnosis

	Calendar period of diagnosis														
	1971-75			**1976-80**			**1981-85**			**1986-90**			**Overall**		
Total registered	**7,240**			**6,951**			**6,554**			**6,221**			**26,966**		
Ineligible	*Records*	*Patients*	*%*	*Records*	*Patients*	*%*	*Records*	*Patients*	*%*	*Records*	*Patients*	*%*	*Records*	*Patients*	*%*
Incomplete data[1]	47	47		92	92		256	256		280	280		675	**675**	
Resident outside England and Wales	153	153		118	118		71	59		31	23		373	**353**	
In situ neoplasm[2]	0	0		0	0		0	0		0	0		0	**0**	
Benign or uncertain behaviour[2]	0	0		0	0		0	0		0	0		0	**0**	
Metastatic[2]	0	0		0	0		0	0		0	0		0	**0**	
Otherwise ineligible[3]	-	-		-	-		-	-		-	-		-	**-**	
Lymphoma[4]	-	-		-	-		-	-		-	-		-	**-**	
Leukaemia or myeloma[4]	-	-		-	-		-	-		-	-		-	**-**	
	200			**210**			**315**			**303**			**1,028**		
Total eligible	**7,040**	**100**		**6,741**	**100**		**6,239**	**100**		**5,918**	**100**		**25,938**	**100**	
Aged under 15 or 100 years or over at diagnosis	266	247	3.5	302	286	4.2	282	260	4.2	228	214	3.6	1,078	**1,007**	3.9
Vital status unknown[5]	438	297	4.2	485	354	5.3	394	306	4.9	489	428	7.2	1,806	**1,385**	5.3
Duplicate registration[6]	0	0	0.0	1	0	0.0	0	0	0.0	0	0	0.0	1	**0**	0.0
Multiple primary[7]	0	0	0.0	2	1	<0.1	15	14	0.2	30	28	0.5	47	**43**	0.2
Synchronous tumours[8]	14	9	0.1	20	11	0.2	24	12	0.2	28	8	0.1	86	**40**	0.2
Sex not known	0	0	0.0	0	0	0.0	0	0	0.0	0	0	0.0	0	**0**	0.0
Sex incompatible with site code[9]	-	-		-	-		-	-		-	-		-	**-**	
Invalid dates[10]	1	1	<0.1	1	1	<0.1	1	0	0.0	3	0	0.0	6	**2**	<0.1
Zero survival[11]	226	208	3.0	211	183	2.7	258	226	3.6	241	219	3.7	936	**836**	3.2
Total exclusions		**762**	**10.8**		**836**	**12.4**		**818**	**13.1**		**897**	**15.2**		**3,313**	**12.8**
Patients included in analysis		**6,278**	**89.2**		**5,905**	**87.6**		**5,421**	**86.9**		**5,021**	**84.8**		**22,625**	**87.2**

[1] Main data item(s) invalid or incompatible with one another: name, sex, date of birth, date of diagnosis and (if present) date of death, postcode, site, morphology and behaviour

[2] Tumour either *in situ* (behaviour code 2), benign (0), uncertain if benign or malignant (1) or metastatic (6 or 9)

[3] Not applicable for this malignancy

[4] Not applicable for this malignancy

[5] Tracing of vital status at National Health Service Central Register incomplete at 6 July 1997

[6] Same site code, sex, cancer registry and cancer registry number as an earlier registration

[7] Same site code and person number as an earlier registration(s): mostly confirmed multiple primary tumours at this site, some unresolved duplicate registrations

[8] Same site code, sex, date of birth and date of diagnosis as another registration(s): mostly synchronous or (in paired organs) bilateral tumours in same anatomic site in one individual, not previously linked: also some duplicate registrations

[9] Not applicable for this malignancy

[10] Impossible sequence of dates (birth, diagnosis and, if present, emigration or death); or date of death after 6 July 1997

[11] Date of diagnosis same as date of death: some are patients who did die on the day of diagnosis, but most were registered solely from a death certificate (death certificate only, or DCO), and their survival time is thus unknown

Table 42.2 Crude and relative survival (%) at one, five and ten years since diagnosis: NHS Region, deprivation category and calendar period of diagnosis

MEN	1971-75						1976-80					
	Affluent	*2*	*3*	*4*	*Deprived*	*All[1]*	*Affluent*	*2*	*3*	*4*	*Deprived*	*All[1]*
	C R	C R	C R	C R	C R	C R	C R	C R	C R	C R	C R	C R
ENGLAND & WALES	*(613)*	*(635)*	*(629)*	*(634)*	*(526)*	*(3,835)*	*(713)*	*(727)*	*(691)*	*(676)*	*(565)*	*(3,543)*
One year	76 78	72 74	72 74	69 71	68 70	71 73	78 80	77 79	74 76	74 76	72 74	75 77
Five years	53 58	52 57	51 55	45 49	44 49	49 53	61 66	57 61	56 62	51 57	51 56	55 61
Ten years	45 51	40 46	40 47	35 41	37 45	39 46	50 57	47 54	45 53	39 48	40 49	45 52
	1981-85						**1986-90**					
	(676)	*(656)*	*(707)*	*(572)*	*(532)*	*(3,219)*	*(661)*	*(600)*	*(624)*	*(581)*	*(427)*	*(2,931)*
One year	85 88	84 86	84 87	80 84	81 84	83 86	91 93	85 88	85 88	85 88	85 87	87 89
Five years	67 72	64 69	63 69	61 67	63 69	64 69	74 79	70 75	70 76	68 73	65 71	70 75
Ten years	58 64	53 61	52 60	51 61	54 64	54 62						

	1971-75	1976-80	1981-85	1986-90
Northern & Yorkshire	*(472)*	*(465)*	*(389)*	*(272)*
One year	70 72	70 72	78 81	83 86
Five years	45 49	51 56	60 66	68 73
Ten years	33 38	41 48	49 57	
Trent	*(416)*	*(399)*	*(356)*	*(308)*
One year	66 68	69 72	82 84	81 83
Five years	44 49	48 53	62 67	67 71
Ten years	37 43	40 46	49 58	
Anglia & Oxford	*(335)*	*(354)*	*(330)*	*(265)*
One year	78 80	81 83	84 87	90 91
Five years	54 58	60 65	65 71	72 77
Ten years	44 50	50 56	57 65	
North Thames	*(514)*	*(377)*	*(344)*	*(402)*
One year	74 76	79 81	92 95	93 94
Five years	51 55	59 63	74 79	75 80
Ten years	41 47	46 54	64 72	
South Thames	*(424)*	*(433)*	*(441)*	*(405)*
One year	73 75	75 78	86 89	88 91
Five years	54 59	56 62	67 73	74 81
Ten years	43 49	43 53	59 66	
South & West	*(472)*	*(461)*	*(373)*	*(376)*
One year	72 75	81 83	84 87	90 92
Five years	53 59	61 67	62 67	71 76
Ten years	44 52	49 57	52 58	
West Midlands	*(366)*	*(334)*	*(355)*	*(314)*
One year	69 72	75 77	80 83	86 89
Five years	42 47	55 60	60 66	65 70
Ten years	33 40	46 53	51 59	
North & West	*(560)*	*(467)*	*(465)*	*(424)*
One year	69 71	73 75	82 85	84 86
Five years	49 53	55 61	63 69	68 73
Ten years	40 47	46 55	54 62	
WALES	*(276)*	*(253)*	*(166)*	*(165)*
One year	63 65	71 73	79 83	82 85
Five years	46 50	51 57	60 67	65 72
Ten years	34 41	38 46	49 61	

[1] Including those for whom no deprivation category was assigned. C - crude survival, R - relative survival (see Chapter 3).
Number of patients contributing to each analysis in parentheses.

Table 42.2   Crude and relative survival (%) at one, five and ten years since diagnosis: NHS Region, deprivation category and calendar period of diagnosis

WOMEN	1971-75						1976-80					
	Affluent	*2*	*3*	*4*	*Deprived*	*All*[1]	*Affluent*	*2*	*3*	*4*	*Deprived*	*All*[1]
	C R	C R	C R	C R	C R	C R	C R	C R	C R	C R	C R	C R
ENGLAND & WALES	*(372)*	*(413)*	*(436)*	*(387)*	*(299)*	*(2,443)*	*(482)*	*(468)*	*(470)*	*(468)*	*(359)*	*(2,362)*
One year	71 73	72 74	74 76	70 71	70 72	71 73	78 80	78 80	74 77	73 75	70 72	75 77
Five years	51 56	51 56	52 57	48 53	49 53	49 54	63 67	59 65	54 59	53 58	53 58	56 61
Ten years	38 44	43 48	41 49	37 43	42 50	39 45	53 58	52 60	45 53	45 51	42 50	48 55
	1981-85						**1986-90**					
	(479)	*(493)*	*(420)*	*(413)*	*(344)*	*(2,202)*	*(441)*	*(436)*	*(462)*	*(404)*	*(314)*	*(2,090)*
One year	85 87	82 84	83 86	77 79	76 79	81 83	88 89	85 87	86 89	81 83	81 84	85 87
Five years	70 75	66 73	68 73	61 66	62 67	65 71	75 79	72 77	71 76	62 67	67 72	70 75
Ten years	61 68	55 63	60 68	50 59	52 61	56 64						

	1971-75	1976-80	1981-85	1986-90
Northern & Yorkshire	*(291)*	*(270)*	*(264)*	*(205)*
One year	64 66	68 70	76 79	84 87
Five years	43 47	49 54	59 66	72 77
Ten years	34 40	40 47	49 59	
Trent	*(246)*	*(238)*	*(234)*	*(233)*
One year	77 79	75 77	76 79	79 82
Five years	56 60	55 60	60 65	63 68
Ten years	42 48	46 54	51 59	
Anglia & Oxford	*(213)*	*(232)*	*(223)*	*(213)*
One year	69 71	79 81	83 85	83 85
Five years	49 53	57 62	68 73	67 73
Ten years	40 48	50 57	57 66	
North Thames	*(341)*	*(267)*	*(266)*	*(264)*
One year	76 78	83 86	87 89	88 91
Five years	54 58	65 70	74 80	76 81
Ten years	44 51	57 63	66 73	
South Thames	*(267)*	*(308)*	*(296)*	*(299)*
One year	75 78	75 78	87 89	90 92
Five years	52 57	58 64	72 78	77 82
Ten years	42 49	49 56	60 70	
South & West	*(332)*	*(337)*	*(258)*	*(268)*
One year	71 73	78 81	82 84	90 92
Five years	51 56	58 64	69 74	73 78
Ten years	40 47	48 57	62 69	
West Midlands	*(259)*	*(219)*	*(213)*	*(208)*
One year	71 73	81 82	79 82	87 89
Five years	50 53	63 67	63 68	74 79
Ten years	38 42	54 59	52 60	
North & West	*(337)*	*(330)*	*(301)*	*(296)*
One year	68 71	68 71	80 82	78 80
Five years	46 51	53 58	64 71	59 65
Ten years	39 45	43 51	55 64	
WALES	*(157)*	*(161)*	*(147)*	*(104)*
One year	59 61	61 63	74 76	79 81
Five years	42 45	43 48	54 59	66 70
Ten years	30 34		42 48	

[1] Including those for whom no deprivation category was assigned. C - crude survival, R - relative survival (see Chapter 3).
Number of patients contributing to each analysis in parentheses.

Table 42.2 Crude and relative survival (%) at one, five and ten years since diagnosis: NHS Region, deprivation category and calendar period of diagnosis

		1971-75								1976-80						
ADULTS		Affluent	2	3	4	Deprived	All[1]		Affluent	2	3	4	Deprived	All[1]
		C R	C R	C R	C R	C R	C R		C R	C R	C R	C R	C R	C R
ENGLAND & WALES		(985)	(1,048)	(1,065)	(1,021)	(825)	(6,278)		(1,195)	(1,195)	(1,161)	(1,144)	(924)	(5,905)
	One year	74 76	72 74	73 75	69 71	68 71	71 73		78 80	77 79	74 76	73 76	71 74	75 77
	Five years	53 57	52 56	51 56	46 51	46 51	49 53		62 66	58 63	55 61	52 57	52 57	56 61
	Ten years	42 48	41 47	40 47	36 42	39 47	39 45		51 58	49 56	45 53	42 49	41 49	46 53
ENGLAND		(957)	(987)	(988)	(929)	(757)	(5,845)		(1,150)	(1,143)	(1,062)	(1,043)	(873)	(5,491)
	One year	75 77	73 75	73 75	70 72	69 71	71 73		78 80	78 80	75 77	74 76	71 74	75 78
	Five years	53 57	53 57	51 56	47 51	46 50	49 54		62 67	58 63	56 62	52 58	52 57	56 61
	Ten years	43 49	42 48	40 47	37 43	39 47	40 46		52 58	50 57	45 54	42 50	41 49	46 54
Northern & Yorkshire		(51)	(80)	(111)	(128)	(149)	(763)		(112)	(129)	(120)	(170)	(185)	(735)
	One year	76 78	75 76	72 73	63 65	68 69	68 69		73 75	65 67	70 72	70 72	69 71	69 72
	Five years	57 62	52 57	43 47	42 46	45 49	44 48		56 60	48 52	55 60	45 51	50 55	50 55
	Ten years	47 54	41 46	29 33	27 34	38 46	33 39		45 50	43 49	48 55	33 40	40 49	41 48
Trent		(87)	(113)	(128)	(159)	(108)	(662)		(88)	(117)	(155)	(139)	(117)	(637)
	One year	68 70	76 78	72 74	67 68	65 67	70 72		86 87	71 72	67 69	71 73	70 72	72 74
	Five years	50 55	57 62	47 52	45 49	46 49	49 53		67 69	47 51	47 52	51 56	49 56	51 56
	Ten years	35 42	44 50	41 48	37 42	36 42	39 45		56 62	41 47	38 45	43 52	38 49	42 50
Anglia & Oxford		(120)	(121)	(111)	(85)	(31)	(548)		(174)	(144)	(117)	(86)	(45)	(586)
	One year	74 76	80 82	74 76	78 80	57 59	75 77		84 85	88 90	71 73	77 79	80 81	80 82
	Five years	50 53	61 64	51 56	47 52	38 46	52 56		66 71	66 70	54 58	49 52	49 53	59 63
	Ten years	41 47	52 55	40 50	36 43	- -	42 49		55 62	57 63	47 53	37 42	44 53	50 56
North Thames		(204)	(152)	(122)	(159)	(110)	(855)		(138)	(118)	(115)	(136)	(105)	(644)
	One year	77 78	77 79	73 75	77 79	72 75	75 77		76 78	75 77	88 92	81 83	83 86	81 83
	Five years	51 54	50 53	56 60	55 60	51 58	52 57		63 66	54 56	67 74	59 65	63 70	61 66
	Ten years	42 47	38 42	44 52	44 50	44 55	42 49		55 60	48 52	53 67	47 54	49 59	51 58
South Thames		(179)	(149)	(133)	(104)	(60)	(691)		(227)	(168)	(125)	(120)	(61)	(741)
	One year	79 81	66 68	74 76	73 75	75 77	74 76		78 80	76 79	76 79	68 70	72 74	75 78
	Five years	55 59	54 58	51 55	50 57	53 57	53 58		63 69	56 61	52 60	48 55	52 57	57 63
	Ten years	43 48	44 50	40 46	37 45	45 50	42 49		53 62	46 53	39 49	41 49	33 39	45 54
South & West		(174)	(221)	(210)	(89)	(40)	(804)		(183)	(242)	(199)	(107)	(33)	(798)
	One year	77 79	74 75	72 74	64 67	74 77	72 74		78 80	82 84	76 78	85 88	78 81	80 82
	Five years	58 63	50 55	55 59	51 55	49 56	52 57		57 62	64 69	57 64	62 69	54 60	60 66
	Ten years	47 54	41 47	41 49	46 52	42 51	43 50		50 56	54 63	41 49	49 58	51 57	48 57
West Midlands		(35)	(30)	(52)	(62)	(55)	(625)		(94)	(93)	(111)	(112)	(138)	(553)
	One year	74 75	67 70	68 69	71 72	59 60	70 72		84 86	81 84	80 82	69 71	73 76	77 79
	Five years	37 39	40 47	52 54	48 51	33 37	45 49		70 75	63 69	58 64	52 56	52 56	58 63
	Ten years	31 36	30 36	41 44	36 41	29 37	35 41		52 58	57 63	55 60	45 51	41 48	49 56
North & West		(107)	(121)	(121)	(143)	(204)	(897)		(134)	(132)	(120)	(173)	(189)	(797)
	One year	68 71	64 66	76 78	68 69	70 73	69 71		71 72	79 81	75 77	72 75	64 66	71 73
	Five years	56 60	51 58	53 58	39 42	45 51	48 52		58 62	60 68	57 65	53 59	47 52	54 60
	Ten years	44 50	41 50	45 55	31 36	41 48	39 46		47 54	50 61	48 58	44 54	38 47	45 53
WALES		(28)	(61)	(77)	(92)	(68)	(433)		(45)	(52)	(99)	(101)	(51)	(414)
	One year	60 61	55 57	69 71	61 62	66 68	61 63		75 76	64 66	65 67	71 74	66 67	67 69
	Five years	42 46	40 44	52 56	40 42	47 53	44 48		51 54	53 61	45 49	46 53	54 60	48 54
	Ten years	32 36	30 34	39 46	28 33	35 45	32 38		44 52	41 49	41 47	35 46	44 52	39 47

[1] Including those for whom no deprivation category was assigned. C - crude survival, R - relative survival (see Chapter 3).
Number of patients contributing to each analysis in parentheses.

ADULTS

Columns for each period: Affluent, 2, 3, 4, Deprived, All[1]. C – crude survival, R – relative survival.

Region	Measure	1981-85 Aff C	R	2 C	R	3 C	R	4 C	R	Dep C	R	All C	R	1986-90 Aff C	R	2 C	R	3 C	R	4 C	R	Dep C	R	All C	R
ENGLAND & WALES	(n)	(1,155)		(1,149)		(1,127)		(985)		(876)		(5,421)		(1,102)		(1,036)		(1,086)		(985)		(741)		(5,021)	
	One year	85	88	83	85	84	86	79	82	79	82	**82**	**85**	90	92	85	88	86	89	84	86	83	86	**86**	**88**
	Five years	69	73	65	70	65	70	61	67	62	68	**65**	**70**	74	79	71	76	70	76	65	71	66	71	**70**	**75**
		59	66	54	62	55	63	50	60	54	63	**55**	**63**												
ENGLAND	(n)	(1,118)		(1,096)		(1,063)		(905)		(823)		(5,108)		(1,057)		(984)		(1,028)		(924)		(700)		(4,752)	
	One year	86	88	83	85	84	87	80	83	79	82	**83**	**85**	90	92	86	88	86	89	84	86	84	86	**86**	**89**
	Five years	69	73	65	70	65	71	61	67	63	69	**65**	**71**	74	79	71	77	71	76	66	71	66	72	**70**	**76**
		59	66	54	62	56	64	51	61	55	64	**55**	**63**												
Northern & Yorkshire	(n)	(80)		(105)		(124)		(138)		(197)		(653)		(67)		(72)		(109)		(116)		(107)		(477)	
	One year	82	86	75	77	85	87	75	78	73	76	**77**	**80**	90	93	85	88	82	87	81	84	83	86	**84**	**87**
	Five years	67	73	60	67	70	75	54	60	54	60	**60**	**66**	73	78	75	80	72	78	63	69	65	71	**69**	**75**
		59	65	48	57	56	64	43	53	46	57	**49**	**58**												
Trent	(n)	(92)		(111)		(140)		(144)		(97)		(590)		(77)		(101)		(137)		(128)		(94)		(541)	
	One year	86	87	78	80	87	90	73	77	78	80	**80**	**82**	88	90	79	82	84	88	77	79	74	77	**80**	**83**
	Five years	68	73	61	67	68	73	53	59	61	66	**61**	**66**	74	77	69	74	68	74	60	66	58	64	**65**	**70**
		57	64	51	60	54	64	43	54	48	57	**50**	**58**												
Anglia & Oxford	(n)	(171)		(143)		(125)		(67)		(32)		(553)		(162)		(119)		(94)		(67)		(32)		(478)	
	One year	82	84	84	86	86	89	85	87	78	80	**84**	**86**	86	88	86	89	87	89	91	93	85	85	**87**	**89**
	Five years	69	73	64	71	65	74	70	75	47	51	**66**	**72**	67	72	65	74	78	83	74	77	63	67	**69**	**75**
		61	67	53	62	57	69	60	68	44	48	**57**	**66**												
North Thames	(n)	(163)		(132)		(115)		(106)		(76)		(610)		(141)		(127)		(135)		(124)		(125)		(666)	
	One year	91	94	92	93	90	93	87	90	85	87	**90**	**92**	92	93	92	94	84	87	92	94	94	96	**91**	**93**
	Five years	73	78	74	79	74	81	72	79	72	77	**74**	**80**	79	84	81	85	67	74	77	82	71	77	**75**	**81**
		65	71	62	69	70	79	58	69	67	74	**65**	**73**												
South Thames	(n)	(225)		(175)		(148)		(109)		(54)		(737)		(223)		(161)		(136)		(118)		(48)		(704)	
	One year	86	88	86	89	88	90	85	89	81	85	**86**	**89**	94	97	88	90	84	88	90	93	81	85	**89**	**92**
	Five years	71	75	69	76	67	70	71	78	68	74	**69**	**75**	83	90	76	82	70	76	66	72	71	76	**75**	**81**
		59	66	61	69	55	60	63	75	65	74	**59**	**67**												
South & West	(n)	(164)		(206)		(149)		(64)		(29)		(631)		(144)		(193)		(150)		(107)		(43)		(644)	
	One year	88	90	85	87	78	80	79	81	79	80	**83**	**86**	93	95	90	92	89	91	87	88	88	92	**90**	**92**
	Five years	70	75	64	69	62	67	59	63	65	69	**65**	**70**	72	78	70	76	73	79	71	77	79	85	**72**	**77**
		64	69	53	60	53	61	48	54	62	68	**56**	**63**												
West Midlands	(n)	(94)		(100)		(116)		(126)		(129)		(568)		(102)		(79)		(130)		(113)		(94)		(522)	
	One year	83	85	83	84	80	82	78	81	77	80	**80**	**82**	89	90	86	89	87	90	85	88	86	89	**87**	**89**
	Five years	62	67	68	71	54	60	59	65	63	68	**61**	**67**	71	75	71	76	66	71	66	73	70	77	**69**	**74**
		50	56	55	62	47	55	51	61	54	62	**51**	**59**												
North & West	(n)	(129)		(124)		(146)		(151)		(209)		(766)		(141)		(132)		(137)		(151)		(157)		(720)	
	One year	84	87	72	74	81	83	81	83	85	88	**81**	**84**	86	88	77	80	88	91	76	79	79	82	**81**	**84**
	Five years	64	70	57	63	63	70	60	66	69	76	**64**	**70**	69	76	63	67	73	81	55	61	61	67	**64**	**70**
		56	65	46	54	53	64	51	59	60	69	**54**	**63**												
WALES	(n)	(37)		(53)		(64)		(80)		(53)		(313)		(45)		(52)		(58)		(61)		(41)		(269)	
	One year	73	75	87	89	77	82	69	72	73	76	**77**	**79**	84	87	75	77	86	88	78	81	76	79	**81**	**83**
	Five years	65	68	66	73	55	63	53	58	53	57	**57**	**63**	73	78	60	67	67	76	60	66	56	62	**65**	**71**
		52	57	51	62	47	55	43	54	36	43	**46**	**54**												

[1] Including those for whom no deprivation category was assigned. C - crude survival, R - relative survival (see Chapter 3). Number of patients contributing to each analysis in parentheses.

Fig. 42.1   Relative survival up to ten years: trends by calendar period of diagnosis (1971-90) and NHS Region

ADULTS

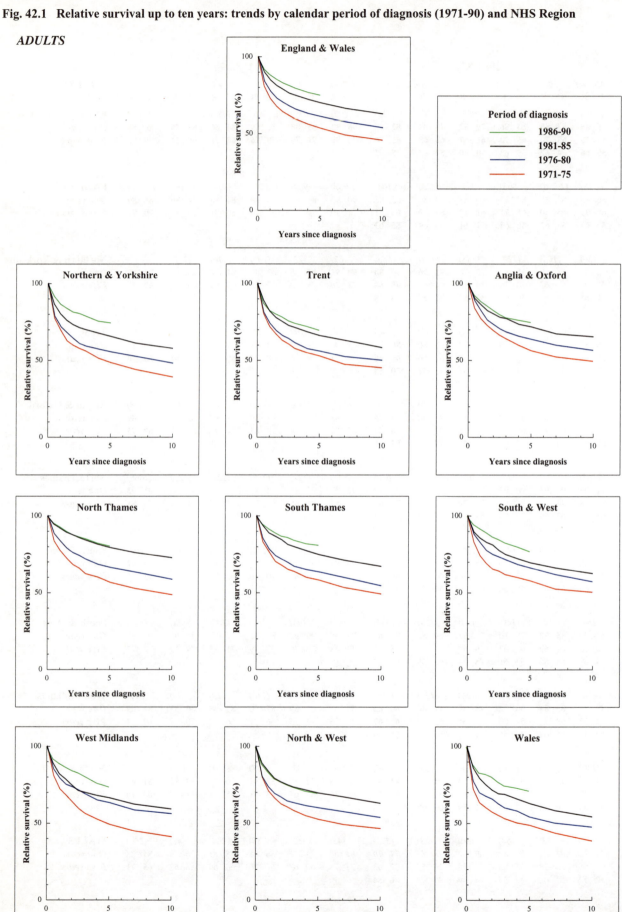

Fig. 42.2   Relative survival up to five years, by deprivation category and NHS Region: patients diagnosed 1986-90

ADULTS

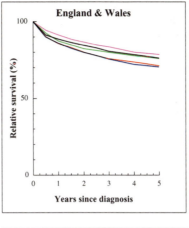

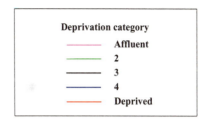

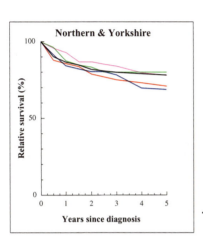

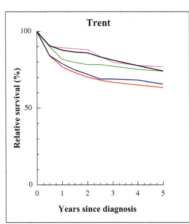

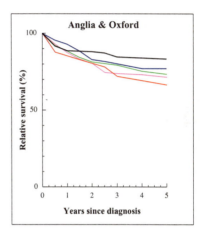

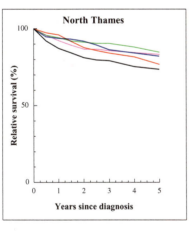

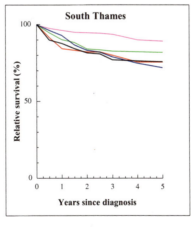

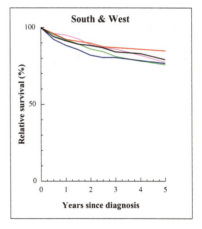

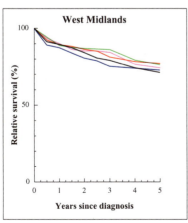

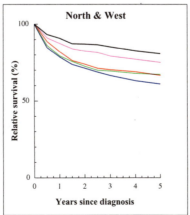

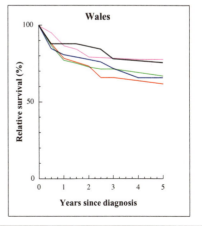

Fig. 42.3   Relative survival at five years by deprivation category, period of diagnosis (1981-90) and NHS Region

ADULTS

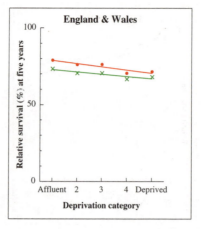

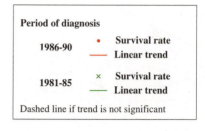

Period of diagnosis

1986-90   • Survival rate   — Linear trend

1981-85   × Survival rate   — Linear trend

Dashed line if trend is not significant

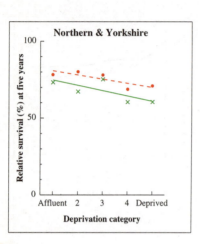

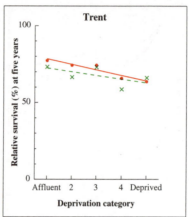

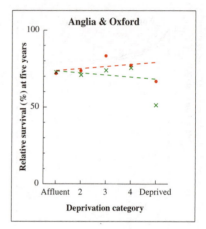

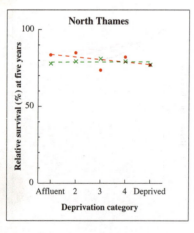

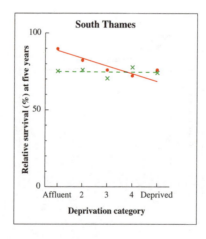

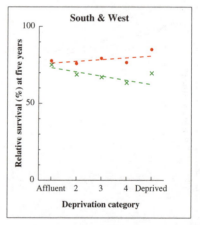

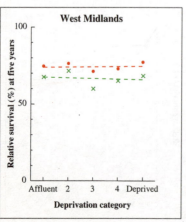

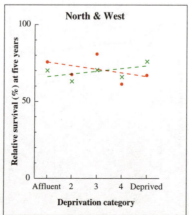

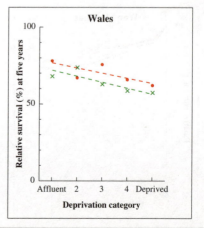

Fig. 42.4 Relative survival (%) by NHS Region and sex:
one-year and five-year survival for patients diagnosed 1986-90, with 95% confidence intervals

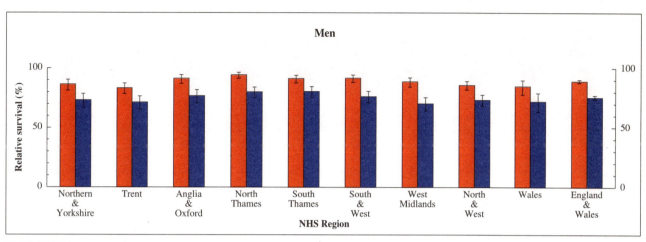

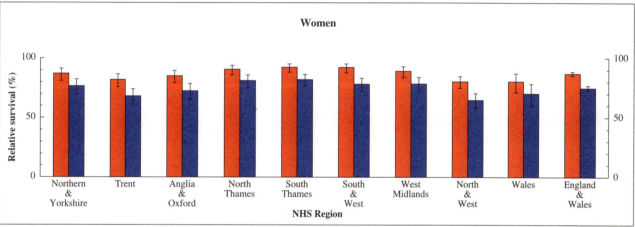

Fig. 42.5 Relative survival (%) at five years, with 95% confidence interval, by age at diagnosis and sex:
England and Wales, patients diagnosed 1986-90

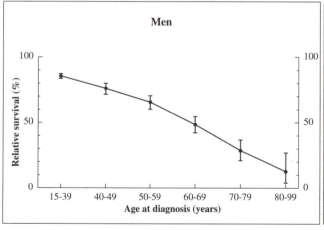

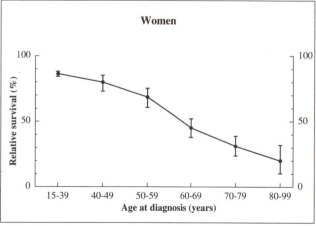

Chapter 43

MULTIPLE MYELOMA

Multiple myeloma is an uncommon haematological neoplasm, giving rise to some 2,600 new cases in England and Wales each year, about 1% of all cancers. It is a malignant clonal proliferation of plasma cells in the bone marrow. Less than 1% of patients with myeloma are diagnosed under age 40: incidence increases steeply with age, and rates in men are 1.5- to 2-fold higher than in women at all ages from 40. Age-standardised (Europe) incidence in England and Wales in 1990 was 5.1 per 100,000 per year in men and 3.5 in women.

Risk is increased by ionising radiation, whether from nuclear weapons, low-dose exposure in the nuclear industry, or radiotherapy (e.g. for cervical cancer or ankylosing spondylitis). Occupational exposures to pesticides and solvents have also been associated with higher risk.

Over the period 1971-90, the incidence of myeloma in England and Wales increased by almost 90% in men (3.5% each year on average) and by 70% in women (3% each year), although the rates of increase slowed down in the late 1980s. Mortality also increased up to the late 1980s, but less rapidly than incidence (2.5% each year in men, 2% in women), and mortality rates levelled off in both men and women in the early 1990s. The increases in mortality mainly affected adults over age 65, and trends were small or negative at younger ages. Increases in mortality among the elderly were probably due in part to better diagnosis and improved accuracy of death certification.

Most patients present with bone pain or pathological fractures, and diagnosis is based on X-rays showing multiple destructive bone lesions, atypical plasma cells on microscopic examination of bone marrow aspirate, and characteristic M-proteins in urine or blood. The diagnosis may be made earlier in the course of disease with recent and more sensitive diagnostic procedures, or on follow-up of asymptomatic patients with monoclonal gammopathy of uncertain significance, a known precursor of myeloma with overlapping clinical features. The date of diagnosis of myeloma (and thus the duration of survival) would be influenced by the frequency of follow-up of such patients, and this may have influenced trends in both incidence and survival[1].

The survival analyses here are based on over 35,000 adults diagnosed with multiple myeloma in England and Wales during the period 1971-90 and followed up to the end of 1995, about 85% of those eligible (Table 43.1). Some 4% of patients were excluded because their vital status was unknown at 6 July 1997, and 11% because their survival was zero or unknown: the proportion in the latter category rose from 6% for patients diagnosed during 1971-75 to 14% for those diagnosed during 1986-90. It is likely that most patients in this category were diagnosed solely from a death certificate, which may lead to some artefactual increase in survival (see Chapter 2). Since 1979, about 10 cases of plasma cell leukaemia have been recorded each year (Table 43A) with the corresponding morphology (data not shown).

Survival trends

The patterns of survival from multiple myeloma are unusual. Relative survival one year after diagnosis was about 55% for adults diagnosed in England and Wales during 1986-90, but five-year survival was only 20%. For patients diagnosed during 1981-85, followed up likewise to the end of 1995, ten-year survival was only 7% (Table 43.2). Substantial excess mortality, relative to that of the general population, persists well beyond the first year after diagnosis, and the relative survival curves continue to decline up to ten years after diagnosis (Figure 43.1). Although survival has increased, most of the gains have been in one-year and five-year survival. Ten-year survival has risen very little since the 1970s: from about 5% to 7%.

After adjustment for differences in the age distribution of patients between the sexes and over time, one-year survival in England and Wales as a whole improved by about 18%, an average gain of 6.5% between successive five-year periods of diagnosis for men and 5.9% for women (see Table 4.4). The corresponding improvement in survival at five years between the early 1970s and the late 1980s was 7-8%, a national average gain of about 2.5% between successive five-year periods. Women have had a survival advantage of about 5% at one year and about 2% at five years since the early 1970s.

Regional variation in survival is fairly wide (Figure 43.4). For patients diagnosed during 1986-90, regional one-year survival rates show a 20% absolute range around the national average of 55-56%, both for men (48-67%) and for women (47-67%; Table 43.2). Regional five-year survival rates cover a 10% range for men (14-24%), and for women the range is two-fold, from 14% to 28%. Survival is lower than the national average in North & West, and improvements since the early 1970s were smaller than in other regions (Table 43.2, Figure 43.1). The regional estimates are each based on 300-700 patients, and are statistically precise (see Chapter 3). Exclusion of cases registered solely from a death

Table 43A Sub-site distribution of multiple myeloma

ICD-9 code	Site description	1971-75		1976-80		1981-85		1986-90		Total	
		No.	%	No.	%	No.	%	No.	%	No.	%
203.0	Multiple myeloma	6,727	100.0	8,260	99.8	9,727	99.6	10,470	99.4	**35,184**	**99.7**
203.1	Plasma cell leukaemia	-	-	16	0.2	40	0.4	54	0.5	**110**	**0.3**
203.8	Other immunoproliferative neoplasms	-	-	4	0.0	1	0.0	6	0.1	**11**	**0.0**
	Total	**6,727**		**8,280**		**9,768**		**10,530**		**35,305**	

certificate may partly account for the higher survival rate in North Thames during the 1980s, but examination of regional patterns in such exclusions over time does not suggest that any resulting bias could account for the regional differences in survival trends since the 1970s (data not shown).

Survival declines steadily with increasing age at diagnosis (Figure 43.5). For patients diagnosed during 1986-90, five-year survival was 30% or more in those under 60 when diagnosed, falling to less than 10% for patients aged 80-99 (see Table 4.7).

Deprivation

For patients diagnosed in England and Wales during 1986-90, one-year survival was higher in the affluent group than in more deprived groups. The patterns were less consistent in the various regions, and there was much less difference in five-year survival (Table 43.2, Figure 43.2).

Among some 9,500 patients of known deprivation status who were diagnosed during 1981-85, the deprivation gap in one-year survival between the most affluent and the most deprived groups was 7% (see Table 4.7). For more than 10,000 patients diagnosed during the next five years, the deprivation gap was also about 7%.

Differences in five-year survival were smaller (2% or less), and were not statistically significant. This is reflected in Figure 43.3 (note expanded y-axis scale), which shows the lack of systematic differences in survival at five years after diagnosis in most regions.

International comparison

One-year survival for patients diagnosed with multiple myeloma in England and Wales during 1986-90 (55%) was about 10% lower than the European average. It was similar to that in Denmark, but 15-20% lower than in Sweden, Finland, France or Italy[2].

Five-year survival for patients diagnosed 1986-90 was about 20%, similar to that in Scotland, but also 8-10% lower than the average survival rate for Europe for the period 1985-89, or the survival rates for patients diagnosed during 1986-90 in the USA (28-30%; see Table 4.12).

1.   Berrino F, Estève J, Coleman MP. Basic issues in the estimation and comparison of cancer patient survival. In: Berrino F, Sant M, Verdecchia A, Capocaccia R, Hakulinen T, Estève J, (eds.) *Survival of cancer patients in Europe: the EUROCARE study. (IARC Scientific Publications No. 132).* Lyon: International Agency for Research on Cancer, 1995, pp1-14

2.   Berrino F, Capocaccia R, Estève J, Gatta G, Hakulinen T, Micheli M, Sant M, Verdecchia A, (eds.) *Survival of cancer patients in Europe: the EUROCARE study, II. (IARC Scientific Publications No. 151).* Lyon: International Agency for Research on Cancer, 1999

Table 43.1 Ineligible and excluded records, and patients included in analyses, by calendar period of diagnosis

	\multicolumn{15}{c}{Calendar period of diagnosis}														
	1971-75			1976-80			1981-85			1986-90			Overall		
Total registered	7,593			9,581			11,791			13,339			42,304		
Ineligible	Records	Patients	%	Records	Patients	%	Records	Patients	%	Records	Patients	%	Records	Patients	%
Incomplete data[1]	9	9		46	46		93	93		90	90		238	**238**	
Resident outside England and Wales	41	41		40	39		48	41		52	37		181	**158**	
In situ neoplasm[2]	0	0		0	0		0	0		0	0		0	**0**	
Benign or uncertain behaviour[2]	150	140		125	113		0	0		0	0		275	**253**	
Metastatic[2]	0	0		0	0		0	0		0	0		0	**0**	
Otherwise ineligible[3]	-	-		-	-		-	-		-	-		-	**-**	
Lymphoma[4]	-	-		-	-		-	-		-	-		-	**-**	
Leukaemia or myeloma[4]	-	-		-	-		-	-		-	-		-	**-**	
		190			198			134			127			649	
Total eligible		7,403	100		9,383	100		11,657	100		13,212	100		41,655	100
Aged under 15 or 100 years or over at diagnosis	0	0	0.0	7	5	<0.1	7	6	<0.1	6	6	<0.1	20	17	<0.1
Vital status unknown[5]	263	226	3.1	430	391	4.2	478	448	3.8	716	690	5.2	1,887	1,755	4.2
Duplicate registration[6]	0	0	0.0	1	1	<0.1	0	0	0.0	1	1	<0.1	2	2	<0.1
Multiple primary[7]	0	0	0.0	4	4	<0.1	38	36	0.3	54	44	0.3	96	84	0.2
Synchronous tumours[8]	13	11	0.1	10	9	<0.1	28	23	0.2	42	31	0.2	93	74	0.2
Sex not known	0	0	0.0	0	0	0.0	0	0	0.0	0	0	0.0	0	0	0.0
Sex incompatible with site code[9]	-	-	-	-	-	-	-	-	-	-	-	-	-	-	-
Invalid dates[10]	0	0	0.0	0	0	0.0	5	2	<0.1	13	6	<0.1	18	8	<0.1
Zero survival[11]	463	439	5.9	741	693	7.4	1,468	1,374	11.8	2,010	1,904	14.4	4,682	4,410	10.6
Total exclusions		676	9.1		1,103	11.8		1,889	16.2		2,682	20.3		6,350	15.2
Patients included in analysis		6,727	90.9		8,280	88.2		9,768	83.8		10,530	79.7		35,305	84.8
Of which, extra-medullary		7			0			0			0			7	

[1] Main data item(s) invalid or incompatible with one another: name, sex, date of birth, date of diagnosis and (if present) date of death, postcode, site, morphology and behaviour

[2] Tumour either *in situ* (behaviour code 2), benign (0), uncertain if benign or malignant (1) or metastatic (6 or 9)

[3] Not applicable for this malignancy

[4] Not applicable for this malignancy

[5] Tracing of vital status at National Health Service Central Register incomplete at 6 July 1997

[6] Same site code, sex, cancer registry and cancer registry number as an earlier registration

[7] Same site code and person number as an earlier registration(s): mostly confirmed multiple primary tumours at this site, some unresolved duplicate registrations

[8] Same site code, sex, date of birth and date of diagnosis as another registration(s): mostly synchronous or (in paired organs) bilateral tumours in same anatomic site in one individual, not previously linked: also some duplicate registrations

[9] Not applicable for this malignancy

[10] Impossible sequence of dates (birth, diagnosis and, if present, emigration or death); or date of death after 6 July 1997

[11] Date of diagnosis same as date of death: some are patients who did die on the day of diagnosis, but most were registered solely from a death certificate (death certificate only, or DCO), and their survival time is thus unknown

Table 43.2 Crude and relative survival (%) at one, five and ten years since diagnosis: NHS Region, deprivation category and calendar period of diagnosis

		1971-75												1976-80											
MEN		Affluent		2		3		4		Deprived		All[1]		Affluent		2		3		4		Deprived		All[1]	
		C	R	C	R	C	R	C	R	C	R	C	R	C	R	C	R	C	R	C	R	C	R	C	R
ENGLAND & WALES		(541)		(572)		(550)		(537)		(437)		(3,305)		(849)		(850)		(885)		(793)		(646)		(4,189)	
	One year	39	40	39	41	38	40	40	42	41	44	39	41	45	47	40	42	42	44	39	42	40	43	42	44
	Five years	11	13	11	13	11	14	12	15	12	15	11	13	13	16	11	14	12	15	12	15	10	14	12	15
	Ten years	4	6	3	5	4	6	4	7	4	7	4	6	4	7	3	5	4	6	5	8	4	8	4	7
ENGLAND		(511)		(551)		(512)		(488)		(408)		(3,093)		(828)		(817)		(826)		(712)		(601)		(3,920)	
	One year	39	41	38	40	38	40	40	42	41	44	39	41	45	47	40	42	42	44	39	41	40	43	42	44
	Five years	11	13	11	13	11	14	12	15	12	15	11	13	13	16	11	14	12	15	11	15	10	14	12	15
	Ten years	4	6	3	5	4	6	4	6	4	7	4	6	4	7	3	5	3	6	5	8	4	8	4	7
Northern & Yorkshire		(40)		(47)		(61)		(69)		(81)		(424)		(69)		(76)		(112)		(115)		(136)		(520)	
	One year	29	30	40	42	22	23	34	35	27	28	32	34	44	48	25	27	34	36	39	41	37	39	39	41
	Five years	10	11	8	9	6	7	8	9	5	6	8	10	14	17	9	10	9	14	8	10	11	16	11	15
	Ten years	2	3	0	0	4	5	0	1	1	2	2	3	7	9	2	3	2	5	4	5	4	8	4	7
Trent		(37)		(41)		(64)		(87)		(61)		(322)		(56)		(72)		(104)		(111)		(88)		(442)	
	One year	43	46	44	46	43	46	28	30	43	44	39	41	42	44	37	39	43	46	43	45	52	55	46	49
	Five years	14	15	15	18	16	19	10	13	14	19	13	17	16	20	8	9	12	15	10	13	15	21	13	17
	Ten years	4	7	4	6	2	3	5	7	9	19	5	8	5	7	2	3	5	8	1	2	5	10	4	7
Anglia & Oxford		(69)		(79)		(56)		(44)		(25)		(333)		(131)		(100)		(110)		(55)		(26)		(433)	
	One year	34	35	46	48	24	26	48	51	51	53	41	43	48	51	38	40	41	44	36	38	40	42	44	47
	Five years	12	14	15	18	7	9	16	19	20	22	13	16	10	12	9	11	10	13	13	17	7	8	11	14
	Ten years	2	3	4	6	2	3	7	10	11	14	4	7	4	6	4	5	3	5	3	6	0	0	4	6
North Thames		(78)		(79)		(57)		(76)		(54)		(392)		(82)		(80)		(71)		(80)		(70)		(396)	
	One year	55	58	37	39	49	52	44	47	53	57	49	51	41	44	41	43	47	49	40	42	55	59	46	49
	Five years	10	12	5	6	15	16	19	23	12	15	13	16	16	22	7	9	12	15	11	14	10	14	13	16
	Ten years	5	6	2	3	7	10	6	10	5	10	6	9	6	9	3	6	3	4	6	9	6	9	5	9
South Thames		(113)		(88)		(86)		(84)		(39)		(450)		(182)		(162)		(116)		(86)		(33)		(610)	
	One year	40	42	31	32	37	39	45	48	32	35	38	40	42	45	40	42	42	44	32	35	31	33	42	44
	Five years	11	12	11	14	6	8	5	7	10	12	9	11	11	15	14	18	16	21	13	16	8	10	14	18
	Ten years	5	7	3	5	1	1	2	3	2	3	3	5	3	5	3	5	3	6	7	11	3	4	4	6
South & West		(101)		(132)		(96)		(44)		(13)		(443)		(175)		(188)		(156)		(63)		(32)		(651)	
	One year	33	34	34	36	45	47	37	39	50	53	40	42	44	46	41	43	38	40	38	40	45	48	43	46
	Five years	9	11	13	15	15	19	18	21	28	32	14	18	14	18	11	14	11	14	9	11	18	23	12	16
	Ten years	3	5	4	6	4	6	2	4	2	3	5	7	4	8	4	5	3	6	6	9	9	18	4	7
West Midlands		(19)		(22)		(26)		(20)		(31)		(302)		(58)		(66)		(82)		(80)		(72)		(363)	
	One year	44	45	23	25	34	36	23	24	32	33	41	43	33	35	36	38	41	42	44	48	32	34	40	43
	Five years	8	8	3	3	11	14	8	10	6	7	8	10	10	12	5	7	9	12	15	19	5	6	9	12
	Ten years	0	0	3	3	3	4	5	6	2	4	3	5	3	4	1	2	3	5	7	13	4	6	4	7
North & West		(54)		(63)		(66)		(64)		(104)		(427)		(75)		(73)		(75)		(122)		(144)		(505)	
	One year	16	17	31	32	27	28	39	41	40	42	35	36	43	45	40	43	38	40	25	27	27	29	35	37
	Five years	4	5	5	7	7	9	10	13	12	16	9	11	10	13	16	19	7	9	8	11	7	10	10	12
	Ten years	3	4	1	1	3	4	1	2	2	3	2	3	4	6	4	7	1	2	4	6	3	8	3	6
WALES		(30)		(21)		(38)		(49)		(29)		(212)		(21)		(33)		(59)		(81)		(45)		(269)	
	One year	33	34	50	52	38	39	34	37	41	43	41	43	50	52	40	42	41	43	38	40	38	39	44	46
	Five years	11	12	9	10	14	16	16	20	9	11	12	14	7	8	22	26	17	22	13	18	3	4	15	20
	Ten years	4	5	0	0	7	8	11	16	6	8	6	8	1	2	6	9	11	17	4	8	1	2	6	10

---

[1] Including those for whom no deprivation category was assigned. C - crude survival, R - relative survival (see Chapter 3).
Number of patients contributing to each analysis in parentheses.

	1981-85						1986-90						MEN
	Affluent C R	*2* C R	*3* C R	*4* C R	*Deprived* C R	*All¹* C R	*Affluent* C R	*2* C R	*3* C R	*4* C R	*Deprived* C R	*All¹* C R	
ENGLAND & WALES	*(1,049)*	*(1,068)*	*(1,000)*	*(923)*	*(750)*	*(4,902)*	*(1,119)*	*(1,285)*	*(1,171)*	*(1,000)*	*(773)*	*(5,392)*	ENGLAND & WALES
One year	54 57	47 50	47 50	46 49	48 52	**48 52**	57 61	53 56	50 53	47 50	50 54	**52 55**	*One year*
Five years	14 18	14 18	12 16	12 15	15 21	**13 17**	16 20	14 18	14 18	14 18	14 19	**14 19**	*Five years*
	5 8	4 6	4 7	4 7	6 10	**4 7**							
ENGLAND	*(1,000)*	*(1,024)*	*(922)*	*(844)*	*(705)*	*(4,574)*	*(1,062)*	*(1,215)*	*(1,103)*	*(934)*	*(731)*	*(5,078)*	ENGLAND
One year	53 56	47 50	46 49	46 49	49 53	**48 51**	57 61	53 56	49 52	48 51	51 54	**52 55**	*One year*
Five years	14 18	14 18	12 16	11 15	15 21	**13 17**	16 20	15 19	14 18	14 18	14 19	**14 19**	*Five years*
	5 8	4 6	3 6	4 7	6 10	**4 7**							
Northern & Yorkshire	*(100)*	*(141)*	*(122)*	*(136)*	*(177)*	*(685)*	*(114)*	*(146)*	*(145)*	*(156)*	*(160)*	*(727)*	Northern & Yorkshire
One year	45 48	43 47	46 49	38 41	42 46	**44 48**	59 64	52 57	42 45	39 43	45 48	**48 52**	*One year*
Five years	9 11	16 21	12 16	8 11	11 16	**12 16**	20 26	16 21	12 16	11 17	14 19	**15 19**	*Five years*
	4 6	3 5	3 6	2 5	6 13	**4 7**							
Trent	*(89)*	*(98)*	*(124)*	*(136)*	*(103)*	*(552)*	*(83)*	*(138)*	*(144)*	*(127)*	*(97)*	*(589)*	Trent
One year	60 63	36 38	45 48	43 47	53 57	**49 52**	53 58	51 56	50 54	50 54	43 46	**51 55**	*One year*
Five years	20 26	9 12	11 16	12 17	13 18	**13 18**	13 18	14 19	17 23	16 22	11 15	**15 20**	*Five years*
	8 12	1 2	4 8	5 10	6 10	**5 8**							
Anglia & Oxford	*(143)*	*(136)*	*(112)*	*(58)*	*(23)*	*(476)*	*(165)*	*(184)*	*(133)*	*(70)*	*(35)*	*(589)*	Anglia & Oxford
One year	46 49	49 52	40 43	49 52	53 58	**48 51**	48 51	50 53	46 49	31 33	42 45	**47 50**	*One year*
Five years	15 18	13 16	14 19	*13 16*	21 31	**15 19**	16 22	11 15	14 19	9 12	*15 19*	**14 17**	*Five years*
	3 5	1 2	4 7	*3 4*	8 16	**3 5**							
North Thames	*(87)*	*(79)*	*(81)*	*(71)*	*(66)*	*(392)*	*(108)*	*(115)*	*(110)*	*(115)*	*(97)*	*(548)*	North Thames
One year	65 68	46 50	50 53	*63 65*	73 79	**60 63**	69 74	58 62	59 65	54 58	74 77	**63 67**	*One year*
Five years	12 15	17 21	15 20	17 21	*31 40*	**18 22**	18 24	17 22	13 19	19 27	21 30	**18 23**	*Five years*
	2 4	6 8	*5 9*	*10 17*	*11 19*	**7 10**							
South Thames	*(198)*	*(144)*	*(120)*	*(104)*	*(31)*	*(624)*	*(185)*	*(174)*	*(132)*	*(84)*	*(45)*	*(628)*	South Thames
One year	55 58	43 46	49 53	52 56	*42 44*	**51 54**	64 69	53 57	51 55	54 59	*50 54*	**57 60**	*One year*
Five years	19 25	13 16	17 22	13 19	*23 29*	**16 21**	18 23	19 28	15 22	19 28	*11 17*	**18 24**	*Five years*
	9 14	3 6	2 4	3 5	*12 16*	**5 9**							
South & West	*(210)*	*(251)*	*(183)*	*(93)*	*(36)*	*(790)*	*(203)*	*(254)*	*(217)*	*(144)*	*(39)*	*(864)*	South & West
One year	48 51	50 53	38 41	46 49	*27 29*	**47 50**	58 62	54 58	52 55	51 55	*43 46*	**55 58**	*One year*
Five years	9 12	16 21	4 6	17 23	*4 6*	**11 15**	15 19	12 15	16 21	12 17	*14 18*	**14 18**	*Five years*
	3 5	4 8	2 5	*5 10*	*2 4*	**4 6**							
West Midlands	*(79)*	*(75)*	*(94)*	*(112)*	*(126)*	*(490)*	*(86)*	*(99)*	*(117)*	*(121)*	*(100)*	*(526)*	West Midlands
One year	49 52	40 42	42 45	45 49	44 48	**46 49**	57 59	44 47	39 42	50 53	53 58	**49 52**	*One year*
Five years	12 16	3 4	9 13	7 10	14 18	**10 13**	12 17	11 14	9 13	13 18	13 19	**12 15**	*Five years*
	6 9	*1 2*	*2 4*	*2 4*	4 7	**3 5**							
North & West	*(94)*	*(100)*	*(86)*	*(134)*	*(143)*	*(565)*	*(118)*	*(105)*	*(105)*	*(117)*	*(158)*	*(607)*	North & West
One year	52 55	47 50	47 50	31 34	46 49	**45 48**	44 47	46 49	44 48	38 42	46 49	**45 48**	*One year*
Five years	13 16	15 20	12 17	6 9	14 20	**12 16**	11 15	16 23	8 10	7 9	10 15	**11 14**	*Five years*
	6 9	7 11	5 8	2 4	4 8	**5 8**							
WALES	*(49)*	*(44)*	*(78)*	*(79)*	*(45)*	*(328)*	*(57)*	*(70)*	*(68)*	*(66)*	*(42)*	*(314)*	WALES
One year	72 75	*38 40*	*57 62*	46 50	*30 32*	**52 55**	58 61	*52 56*	*52 56*	38 41	38 41	**49 52**	*One year*
Five years	21 28	*13 15*	*17 24*	13 17	*14 17*	**16 21**	14 20	*12 16*	*17 22*	*17 23*	15 20	**15 20**	*Five years*
	7 14	*8 10*	*10 21*	*3 6*	*2 4*	**6 11**							

¹ Including those for whom no deprivation category was assigned. C - crude survival, R - relative survival (see Chapter 3).
   Number of patients contributing to each analysis in parentheses.

MULTIPLE MYELOMA *(ICD-9 203)*

Table 43.2 Crude and relative survival (%) at one, five and ten years since diagnosis: NHS Region, deprivation category and calendar period of diagnosis

	1971-75												**1976-80**											
WOMEN	*Affluent*		*2*		*3*		*4*		*Deprived*		*All[1]*		*Affluent*		*2*		*3*		*4*		*Deprived*		*All[1]*	
	C	R	C	R	C	R	C	R	C	R	C	R	C	R	C	R	C	R	C	R	C	R	C	R
ENGLAND & WALES	*(561)*		*(604)*		*(615)*		*(587)*		*(435)*		*(3,422)*		*(785)*		*(842)*		*(842)*		*(794)*		*(639)*		*(4,091)*	
One year	40	41	42	43	43	45	40	42	35	37	40	42	48	50	44	46	45	47	43	45	43	45	44	46
Five years	11	13	14	16	13	15	12	14	7	9	11	13	14	16	13	16	13	16	12	15	12	15	13	16
Ten years	5	6	5	7	4	6	3	5	2	4	4	5	5	7	5	8	4	6	4	7	3	5	4	7
ENGLAND	*(546)*		*(570)*		*(576)*		*(541)*		*(409)*		*(3,209)*		*(758)*		*(800)*		*(777)*		*(706)*		*(599)*		*(3,786)*	
One year	39	41	43	44	43	45	41	43	35	37	41	42	48	50	43	45	44	46	43	45	42	44	44	46
Five years	11	13	14	17	13	15	11	13	7	9	11	13	13	16	13	16	13	16	11	13	11	14	12	15
Ten years	5	7	5	7	4	6	3	4	3	5	4	6	5	6	5	8	4	6	4	6	3	5	4	6
Northern & Yorkshire	*(44)*		*(51)*		*(69)*		*(77)*		*(77)*		*(438)*		*(66)*		*(95)*		*(104)*		*(123)*		*(134)*		*(536)*	
One year	37	38	41	43	38	40	32	33	39	41	38	40	45	47	44	46	30	31	39	41	46	48	43	46
Five years	7	8	10	12	10	13	8	9	5	6	9	12	10	13	13	15	7	10	15	18	7	9	11	14
Ten years	4	5	5	7	4	6	1	2	2	5	3	5	4	7	6	10	1	2	8	12	1	2	4	7
Trent	*(50)*		*(47)*		*(66)*		*(85)*		*(59)*		*(338)*		*(64)*		*(61)*		*(92)*		*(103)*		*(78)*		*(412)*	
One year	42	44	39	40	47	49	33	34	31	33	41	42	55	57	38	40	48	50	44	46	44	46	47	49
Five years	9	12	12	14	11	13	11	13	4	5	10	12	10	12	7	8	20	24	11	15	17	20	14	17
Ten years	3	4	4	5	4	6	3	4	2	3	3	4	5	7	3	5	9	13	3	5	5	8	5	8
Anglia & Oxford	*(53)*		*(70)*		*(65)*		*(40)*		*(15)*		*(293)*		*(118)*		*(99)*		*(85)*		*(62)*		*(29)*		*(417)*	
One year	40	43	51	54	43	44	37	39	30	31	43	45	45	48	47	50	48	49	43	46	37	38	47	49
Five years	15	17	20	24	14	16	17	19	10	10	15	18	10	12	18	21	17	21	7	9	12	13	14	17
Ten years	3	4	6	8	3	4	3	4	0	0	3	5	2	3	6	8	5	7	1	2	0	0	4	6
North Thames	*(101)*		*(83)*		*(72)*		*(73)*		*(60)*		*(423)*		*(84)*		*(69)*		*(67)*		*(87)*		*(65)*		*(385)*	
One year	44	45	43	44	43	44	46	48	40	42	46	48	55	56	50	53	37	39	46	47	40	42	47	49
Five years	11	13	12	14	15	18	16	18	9	12	13	16	15	17	15	18	12	16	11	13	11	14	13	16
Ten years	7	10	1	2	1	2	7	11	1	3	4	6	5	7	7	11	2	4	5	8	3	7	5	7
South Thames	*(129)*		*(102)*		*(101)*		*(104)*		*(40)*		*(519)*		*(159)*		*(148)*		*(115)*		*(88)*		*(37)*		*(575)*	
One year	27	28	33	35	44	46	46	48	27	29	39	41	45	46	43	45	43	45	42	44	33	35	44	46
Five years	9	10	11	13	8	9	9	11	4	4	9	11	13	15	12	15	10	12	12	14	0	0	11	14
Ten years	3	4	6	7	3	4	2	2	2	3	4	5	7	9	6	9	6	7	4	5	0	0	5	8
South & West	*(101)*		*(137)*		*(115)*		*(69)*		*(28)*		*(492)*		*(132)*		*(192)*		*(162)*		*(82)*		*(25)*		*(618)*	
One year	44	46	45	47	38	40	43	44	38	40	45	46	47	49	40	42	40	42	51	54	54	57	45	47
Five years	13	15	17	21	13	15	12	15	12	13	15	18	16	19	16	20	12	15	11	15	23	26	15	19
Ten years	5	7	6	9	8	11	7	10	2	3	6	9	4	7	6	8	4	5	3	5	8	12	5	7
West Midlands	*(11)*		*(21)*		*(32)*		*(25)*		*(31)*		*(280)*		*(57)*		*(46)*		*(72)*		*(65)*		*(81)*		*(322)*	
One year	41	43	33	33	41	43	27	28	35	36	41	43	42	43	36	37	56	59	29	30	26	27	39	41
Five years	17	18	10	11	11	13	7	7	6	7	9	10	11	13	6	7	11	13	5	7	7	9	9	10
Ten years	17	18	5	6	6	10	0	0	6	7	3	5	5	6	0	0	4	6	0	0	2	3	2	3
North & West	*(57)*		*(59)*		*(56)*		*(68)*		*(99)*		*(426)*		*(78)*		*(90)*		*(80)*		*(96)*		*(150)*		*(521)*	
One year	32	33	35	37	36	38	36	37	26	27	35	37	46	48	32	34	40	42	32	34	42	44	41	43
Five years	9	10	11	13	13	15	8	9	7	8	10	11	16	18	7	9	10	12	6	7	13	15	10	13
Ten years	5	6	5	6	3	4	0	0	3	4	3	4	2	3	3	4	2	2	2	2	3	5	3	4
WALES	*(15)*		*(34)*		*(39)*		*(46)*		*(26)*		*(213)*		*(27)*		*(42)*		*(65)*		*(88)*		*(40)*		*(305)*	
One year	50	51	22	23	39	41	30	31	34	34	33	34	35	35	51	53	48	50	43	46	50	52	48	50
Five years	13	14	2	2	14	16	15	17	8	9	10	11	16	20	14	15	16	18	23	28	22	26	20	24
Ten years	0	0	2	2	6	10	4	6	0	0	3	5	12	18	2	4	4	5	8	14	10	14	7	12

[1] Including those for whom no deprivation category was assigned. C - crude survival, R - relative survival (see Chapter 3).
Number of patients contributing to each analysis in parentheses.

WOMEN

		1981-85											1986-90										
		Affluent		*2*		*3*		*4*		*Deprived*		*All*[1]		*Affluent*		*2*		*3*		*4*		*Deprived*		*All*[1]
		C	R	C	R	C	R	C	R	C	R	C	R	C	R	C	R	C	R	C	R	C	R	C	R
ENGLAND & WALES	(n)	(961)		(1,019)		(1,082)		(988)		(697)		(4,866)		(1,041)		(1,116)		(1,139)		(1,022)		(760)		(5,138)	
	One year	52	54	53	56	48	51	49	52	45	47	50	52	55	58	54	56	54	56	51	53	50	53	53	56
	Five years	18	22	14	18	14	17	14	17	12	16	14	18	16	19	16	20	15	19	16	19	16	20	16	20
		7	10	4	5	4	7	4	6	4	7	4	7												
ENGLAND	(n)	(918)		(977)		(1,011)		(911)		(656)		(4,568)		(993)		(1,065)		(1,079)		(940)		(718)		(4,837)	
	One year	53	55	53	55	48	51	49	51	45	47	49	52	55	58	54	57	53	56	51	53	50	53	53	56
	Five years	18	22	14	17	13	16	14	17	12	15	14	18	16	19	16	20	15	18	15	19	15	19	15	19
		7	10	3	5	4	6	3	5	4	7	4	7												
Northern & Yorkshire	(n)	(80)		(114)		(144)		(155)		(148)		(649)		(101)		(130)		(153)		(156)		(148)		(696)	
	One year	56	59	32	34	47	50	50	53	36	38	45	48	58	62	55	59	50	53	40	42	49	52	51	54
	Five years	19	25	9	11	13	17	11	15	11	15	13	17	17	21	25	31	13	16	11	14	9	12	15	19
		5	7	2	4	5	8	2	3	3	6	3	6												
Trent	(n)	(74)		(96)		(108)		(145)		(71)		(495)		(101)		(78)		(151)		(130)		(106)		(568)	
	One year	52	55	50	52	42	45	48	50	49	53	49	52	55	58	54	56	52	55	48	51	43	45	52	55
	Five years	22	27	14	17	8	9	17	20	15	21	15	19	15	17	15	19	16	20	15	18	9	11	14	18
		8	12	3	5	4	7	4	6	1	2	4	7												
Anglia & Oxford	(n)	(120)		(117)		(121)		(68)		(28)		(460)		(137)		(145)		(110)		(73)		(38)		(507)	
	One year	45	48	45	48	42	44	41	43	30	32	45	48	43	45	42	44	47	49	44	46	27	29	44	47
	Five years	18	22	13	15	17	22	12	16	3	4	15	19	13	16	10	12	15	19	17	19	9	12	13	17
		5	7	3	4	4	6	2	3	3	4	4	6												
North Thames	(n)	(107)		(66)		(71)		(86)		(67)		(403)		(107)		(102)		(94)		(93)		(100)		(502)	
	One year	52	55	64	67	54	57	44	46	46	48	53	55	65	67	64	66	51	54	69	75	68	71	64	67
	Five years	21	26	16	19	13	15	17	20	15	19	18	21	23	28	22	26	14	17	29	38	26	31	23	28
		10	15	5	6	3	4	4	6	8	13	7	10												
South Thames	(n)	(181)		(165)		(134)		(85)		(35)		(635)		(157)		(156)		(124)		(100)		(36)		(581)	
	One year	55	58	62	65	47	50	52	55	47	49	55	58	62	65	59	62	63	66	60	64	48	50	61	64
	Five years	19	22	16	21	15	18	22	27	8	11	16	21	19	24	14	17	21	25	20	25	25	33	19	23
		6	8	3	4	3	5	11	17	2	4	4	7												
South & West	(n)	(187)		(244)		(213)		(109)		(38)		(818)		(191)		(258)		(233)		(130)		(38)		(858)	
	One year	54	56	53	56	46	49	47	50	42	45	51	54	53	56	56	59	52	54	53	56	47	49	54	57
	Five years	15	18	17	21	14	18	14	17	13	16	15	19	15	18	18	21	13	17	14	17	10	12	15	19
		8	12	7	10	5	8	4	6	8	11	6	10												
West Midlands	(n)	(82)		(72)		(85)		(111)		(111)		(465)		(87)		(88)		(91)		(116)		(87)		(471)	
	One year	47	49	51	53	57	60	47	49	53	56	52	55	54	57	46	47	59	61	51	54	44	47	52	54
	Five years	11	14	13	15	15	18	10	12	11	14	12	15	14	17	10	12	14	16	14	17	12	15	13	16
		4	7	2	2	4	5	1	1	4	5	3	4												
North & West	(n)	(87)		(103)		(135)		(152)		(158)		(643)		(112)		(108)		(123)		(142)		(165)		(654)	
	One year	45	48	51	54	40	43	48	51	41	43	46	49	46	49	44	46	42	45	38	40	48	51	45	48
	Five years	14	17	9	11	6	8	9	11	10	13	10	12	6	8	12	14	11	13	7	8	18	22	11	14
		6	8	2	3	2	3	2	4	5	8	3	5												
WALES	(n)	(43)		(42)		(71)		(77)		(41)		(298)		(48)		(51)		(60)		(82)		(42)		(301)	
	One year	39	40	64	67	56	59	51	54	43	46	52	56	44	46	44	47	62	65	51	54	50	53	54	57
	Five years	16	19	18	22	19	26	13	17	14	23	16	22	18	22	15	17	23	30	19	24	23	27	22	28
		5	9	7	10	6	11	5	8	2	4	5	10												

[1] Including those for whom no deprivation category was assigned. C - crude survival, R - relative survival (see Chapter 3).
Number of patients contributing to each analysis in parentheses.

Table 43.2   Crude and relative survival (%) at one, five and ten years since diagnosis: NHS Region, deprivation category and calendar period of diagnosis

ADULTS	1971-75												1976-80											
	Affluent		2		3		4		Deprived		All[1]		Affluent		2		3		4		Deprived		All[1]	
	C	R	C	R	C	R	C	R	C	R	C	R	C	R	C	R	C	R	C	R	C	R	C	R
ENGLAND & WALES	*(1,102)*		*(1,176)*		*(1,165)*		*(1,124)*		*(872)*		*(6,727)*		*(1,634)*		*(1,692)*		*(1,727)*		*(1,587)*		*(1,285)*		*(8,280)*	
One year	39	41	40	42	41	42	40	42	38	40	**40**	**42**	46	49	42	44	43	46	41	43	41	44	**43**	**45**
Five years	11	13	12	15	12	14	12	15	10	12	**11**	**13**	13	16	12	15	13	16	12	15	11	14	**12**	**15**
Ten years	4	6	4	6	4	6	4	5	3	5	**4**	**6**	5	7	4	6	4	6	5	7	4	6	**4**	**7**
ENGLAND	*(1,057)*		*(1,121)*		*(1,088)*		*(1,029)*		*(817)*		*(6,302)*		*(1,586)*		*(1,617)*		*(1,603)*		*(1,418)*		*(1,200)*		*(7,706)*	
One year	39	41	41	42	41	43	41	42	38	40	**40**	**42**	46	49	42	44	43	45	41	43	41	44	**43**	**45**
Five years	11	13	13	15	12	14	12	14	10	12	**11**	**13**	13	16	12	15	12	15	11	14	11	14	**12**	**15**
Ten years	4	6	4	6	4	6	3	5	3	5	**4**	**6**	4	7	4	6	4	6	4	7	4	6	**4**	**6**
Northern & Yorkshire	*(84)*		*(98)*		*(130)*		*(146)*		*(158)*		*(862)*		*(135)*		*(171)*		*(216)*		*(238)*		*(270)*		*(1,056)*	
One year	33	34	40	42	31	32	33	34	33	34	**35**	**37**	45	48	36	37	32	34	39	41	41	44	**41**	**43**
Five years	8	9	9	11	8	10	8	10	5	6	**9**	**11**	12	15	11	13	9	12	12	15	9	12	**11**	**14**
Ten years	3	5	3	4	4	6	1	1	2	3	**3**	**4**	6	9	4	7	2	3	6	9	3	5	**4**	**7**
Trent	*(87)*		*(88)*		*(130)*		*(172)*		*(120)*		*(660)*		*(120)*		*(133)*		*(196)*		*(214)*		*(166)*		*(854)*	
One year	43	45	41	43	45	47	31	32	37	39	**40**	**42**	49	52	37	39	46	48	44	46	49	51	**46**	**49**
Five years	11	14	14	16	13	16	10	13	9	12	**11**	**14**	13	16	7	9	16	20	11	14	16	21	**13**	**17**
Ten years	*3*	*5*	4	6	3	5	4	5	6	11	**4**	**6**	5	7	3	4	7	11	2	3	5	9	**5**	**7**
Anglia & Oxford	*(122)*		*(149)*		*(121)*		*(84)*		*(40)*		*(626)*		*(249)*		*(199)*		*(195)*		*(117)*		*(55)*		*(850)*	
One year	37	39	48	51	34	36	43	45	*43*	*45*	**42**	**44**	47	49	43	45	44	46	40	42	*39*	*40*	**46**	**48**
Five years	13	15	17	21	11	13	16	19	*17*	*19*	**14**	**17**	10	12	13	16	13	16	10	13	*10*	*12*	**13**	**15**
Ten years	3	4	5	7	2	3	5	7	7	9	**4**	**6**	3	5	5	7	4	6	2	4	0	0	**4**	**6**
North Thames	*(179)*		*(162)*		*(129)*		*(149)*		*(114)*		*(815)*		*(166)*		*(149)*		*(138)*		*(167)*		*(135)*		*(781)*	
One year	49	51	40	41	46	47	45	47	46	49	**47**	**49**	48	50	45	48	42	44	43	45	48	50	**47**	**49**
Five years	11	13	9	10	15	17	17	21	11	14	**13**	**16**	16	19	11	13	12	15	11	13	11	14	**13**	**16**
Ten years	6	8	2	2	4	5	7	11	3	6	**5**	**8**	5	8	5	8	3	4	6	9	5	9	**5**	**8**
South Thames	*(242)*		*(190)*		*(187)*		*(188)*		*(79)*		*(969)*		*(341)*		*(310)*		*(231)*		*(174)*		*(70)*		*(1,185)*	
One year	33	34	32	34	40	42	46	48	30	32	**38**	**40**	43	46	41	43	42	45	37	40	32	34	**43**	**45**
Five years	10	11	11	13	7	9	7	9	7	9	**9**	**11**	12	15	13	16	13	16	12	15	3	4	**13**	**16**
Ten years	4	5	4	6	2	3	2	3	2	3	**3**	**5**	5	7	4	7	4	7	5	8	*1*	*2*	**4**	**7**
South & West	*(202)*		*(269)*		*(211)*		*(113)*		*(41)*		*(935)*		*(307)*		*(380)*		*(318)*		*(145)*		*(57)*		*(1,269)*	
One year	39	40	40	42	42	43	40	42	*42*	*44*	**43**	**44**	45	47	41	43	39	41	45	48	49	52	**44**	**47**
Five years	11	13	15	18	14	17	14	17	*17*	*19*	**15**	**18**	15	19	14	17	12	15	10	13	20	25	**14**	**17**
Ten years	4	6	5	8	6	9	5	8	2	3	**6**	**8**	4	7	4	7	3	6	4	7	9	14	**4**	**7**
West Midlands	*(30)*		*(43)*		*(58)*		*(45)*		*(62)*		*(582)*		*(115)*		*(112)*		*(154)*		*(145)*		*(153)*		*(685)*	
One year	*43*	*45*	28	29	*38*	*40*	25	26	33	35	**41**	**43**	38	39	36	38	48	51	37	40	29	30	**40**	**42**
Five years	*11*	*12*	7	7	*11*	*14*	7	8	6	7	**8**	**10**	11	12	5	7	10	13	10	13	6	8	**9**	**11**
Ten years	5	8	5	5	4	7	2	3	4	7	**3**	**5**	4	6	1	1	3	5	4	6	3	4	**3**	**5**
North & West	*(111)*		*(122)*		*(122)*		*(132)*		*(203)*		*(853)*		*(153)*		*(163)*		*(155)*		*(218)*		*(294)*		*(1,026)*	
One year	24	25	33	34	31	33	38	39	33	35	**35**	**36**	45	46	36	38	39	41	28	30	34	36	**38**	**40**
Five years	7	8	8	9	10	12	9	11	9	12	**9**	**11**	13	16	11	14	8	10	7	9	10	13	**10**	**12**
Ten years	4	5	3	4	3	4	1	1	2	3	**3**	**4**	3	4	3	6	2	2	3	5	3	6	**3**	**5**
WALES	*(45)*		*(55)*		*(77)*		*(95)*		*(55)*		*(425)*		*(48)*		*(75)*		*(124)*		*(169)*		*(85)*		*(574)*	
One year	38	40	*33*	*34*	39	40	32	34	38	39	**37**	**39**	*41*	*43*	46	48	45	47	41	43	43	46	**46**	**48**
Five years	12	13	4	5	14	16	16	19	9	10	**11**	**13**	*13*	*16*	17	20	16	19	18	24	12	15	**18**	**22**
Ten years	3	3	*1*	*2*	7	*11*	7	11	3	6	**4**	**6**	7	12	4	7	7	10	6	12	5	9	**7**	**11**

[1]   Including those for whom no deprivation category was assigned. C - crude survival, R - relative survival (see Chapter 3). Number of patients contributing to each analysis in parentheses.

	1981-85						1986-90						ADULTS
	Affluent	*2*	*3*	*4*	*Deprived*	*All[1]*	*Affluent*	*2*	*3*	*4*	*Deprived*	*All[1]*	
	C R	C R	C R	C R	C R	C R	C R	C R	C R	C R	C R	C R	
	(2,010)	*(2,087)*	*(2,082)*	*(1,911)*	*(1,447)*	*(9,768)*	*(2,160)*	*(2,401)*	*(2,310)*	*(2,022)*	*(1,533)*	*(10,530)*	**ENGLAND & WALES**
One year	**53 56**	**50 53**	**48 51**	**48 50**	47 50	**49 52**	**56 59**	**53 56**	**52 55**	**49 52**	**50 53**	**52 55**	*One year*
Five years	**16 20**	**14 18**	**13 16**	**13 16**	**14 18**	**14 18**	**16 20**	**15 19**	**15 19**	**15 19**	**15 19**	**15 19**	*Five years*
	6 9	**4 6**	**4 7**	**4 6**	**5 9**	**4 7**							
	(1,918)	*(2,001)*	*(1,933)*	*(1,755)*	*(1,361)*	*(9,142)*	*(2,055)*	*(2,280)*	*(2,182)*	*(1,874)*	*(1,449)*	*(9,915)*	**ENGLAND**
One year	53 56	50 53	47 50	47 50	47 50	**49 52**	56 59	53 56	51 54	49 52	51 54	**52 55**	*One year*
Five years	16 20	14 17	12 16	13 16	14 18	**14 17**	16 20	15 19	14 18	14 19	15 19	**15 19**	*Five years*
	6 9	4 6	4 6	4 6	5 9	**4 7**							
	(180)	*(255)*	*(266)*	*(291)*	*(325)*	*(1,334)*	*(215)*	*(276)*	*(298)*	*(312)*	*(308)*	*(1,423)*	**Northern & Yorkshire**
One year	50 53	38 41	47 50	44 47	39 42	**45 48**	59 63	54 58	46 49	39 42	47 50	**50 53**	*One year*
Five years	13 17	13 16	12 17	9 13	11 15	**12 16**	19 24	20 26	12 16	11 15	12 16	**15 19**	*Five years*
	4 6	3 5	4 7	2 4	5 9	**4 6**							
	(163)	*(194)*	*(232)*	*(281)*	*(174)*	*(1,047)*	*(184)*	*(216)*	*(295)*	*(257)*	*(203)*	*(1,157)*	**Trent**
One year	56 59	43 46	44 47	46 49	52 55	**49 52**	54 57	52 56	51 54	49 52	43 46	**51 55**	*One year*
Five years	21 26	11 14	10 13	14 19	14 19	**14 18**	14 17	14 19	17 22	15 20	10 13	**15 19**	*Five years*
	8 12	2 4	4 8	5 7	4 7	**5 7**							
	(263)	*(253)*	*(233)*	*(126)*	*(51)*	*(936)*	*(302)*	*(329)*	*(243)*	*(143)*	*(73)*	*(1,096)*	**Anglia & Oxford**
One year	46 49	47 50	41 44	44 47	*40 43*	**47 49**	45 48	46 49	46 49	38 40	34 37	**46 48**	*One year*
Five years	16 20	13 16	15 20	13 16	*11 15*	**15 19**	15 19	11 14	14 19	13 16	12 15	**13 17**	*Five years*
	4 6	2 3	4 6	2 3	*6 10*	**3 5**							
	(194)	*(145)*	*(152)*	*(157)*	*(133)*	*(795)*	*(215)*	*(217)*	*(204)*	*(208)*	*(197)*	*(1,050)*	**North Thames**
One year	58 61	55 58	52 55	52 55	59 63	**56 59**	67 71	61 64	55 59	61 65	71 75	**63 67**	*One year*
Five years	18 21	16 20	14 17	17 21	23 29	**18 22**	21 26	19 24	14 18	24 32	24 31	**20 26**	*Five years*
	7 10	5 7	4 6	7 10	10 16	**7 10**							
	(379)	*(309)*	*(254)*	*(189)*	*(66)*	*(1,259)*	*(342)*	*(330)*	*(256)*	*(184)*	*(81)*	*(1,209)*	**South Thames**
One year	55 58	53 56	48 51	52 56	*45 47*	**53 56**	63 67	56 59	57 60	58 62	49 52	**59 62**	*One year*
Five years	19 24	15 19	16 20	17 22	*15 20*	**16 21**	19 23	17 22	18 23	20 26	18 25	**18 24**	*Five years*
	8 11	3 5	3 5	6 11	*7 10*	**5 8**							
	(397)	*(495)*	*(396)*	*(202)*	*(74)*	*(1,608)*	*(394)*	*(512)*	*(450)*	*(274)*	*(77)*	*(1,722)*	**South & West**
One year	50 53	52 55	43 46	47 50	35 37	**49 52**	56 59	55 58	52 55	52 55	45 48	**55 58**	*One year*
Five years	12 15	16 21	10 12	15 20	9 11	**13 17**	15 19	15 18	15 19	13 17	12 15	**15 18**	*Five years*
	5 9	5 9	4 6	4 8	5 8	**5 8**							
	(161)	*(147)*	*(179)*	*(223)*	*(237)*	*(955)*	*(173)*	*(187)*	*(208)*	*(237)*	*(187)*	*(997)*	**West Midlands**
One year	48 51	46 48	49 52	46 49	48 52	**49 52**	55 58	45 47	48 51	50 53	49 52	**51 53**	*One year*
Five years	11 15	8 10	12 15	8 11	13 16	**11 14**	13 16	10 13	11 14	13 18	13 17	**12 16**	*Five years*
	5 8	1 2	3 5	1 2	4 6	**3 5**							
	(181)	*(203)*	*(221)*	*(286)*	*(301)*	*(1,208)*	*(230)*	*(213)*	*(228)*	*(259)*	*(323)*	*(1,261)*	**North & West**
One year	49 52	49 52	43 46	40 43	43 46	**45 48**	45 48	45 48	43 47	38 41	47 50	**45 48**	*One year*
Five years	14 16	12 15	8 11	8 10	12 16	**11 14**	9 11	14 18	9 12	7 9	14 20	**11 14**	*Five years*
	5 8	4 6	3 5	2 4	5 8	**4 6**							
	(92)	*(86)*	*(149)*	*(156)*	*(86)*	*(626)*	*(105)*	*(121)*	*(128)*	*(148)*	*(84)*	*(615)*	**WALES**
One year	56 58	51 53	57 61	48 52	36 39	**52 55**	51 54	49 52	57 60	45 48	44 47	**51 54**	*One year*
Five years	18 24	16 19	18 24	13 17	14 20	**16 21**	16 22	13 17	20 26	18 24	19 23	**18 24**	*Five years*
	6 12	7 10	8 16	4 7	2 4	**6 10**							

[1] Including those for whom no deprivation category was assigned. C - crude survival, R - relative survival (see Chapter 3).
Number of patients contributing to each analysis in parentheses.

Fig. 43.1   Relative survival up to ten years: trends by calendar period of diagnosis (1971-90) and NHS Region

ADULTS

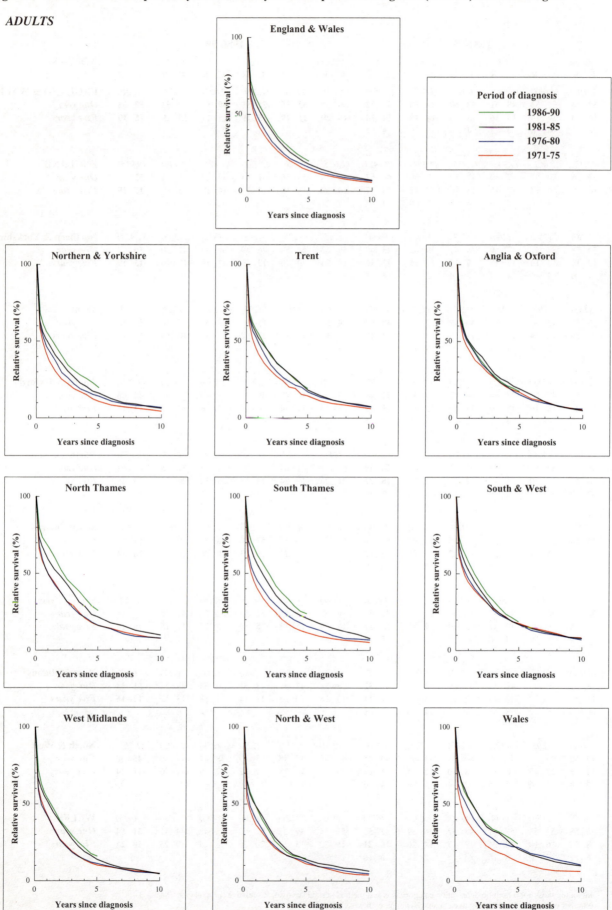

Fig. 43.2   Relative survival up to five years, by deprivation category and NHS Region: patients diagnosed 1986-90

ADULTS

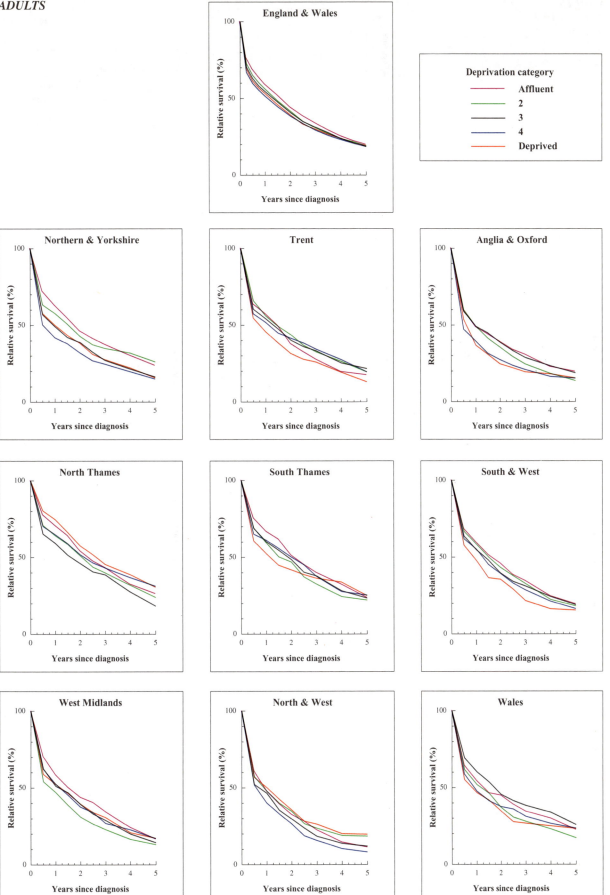

Fig. 43.3   Relative survival at five years by deprivation category, period of diagnosis (1981-90) and NHS Region

ADULTS
(Note vertical scale)

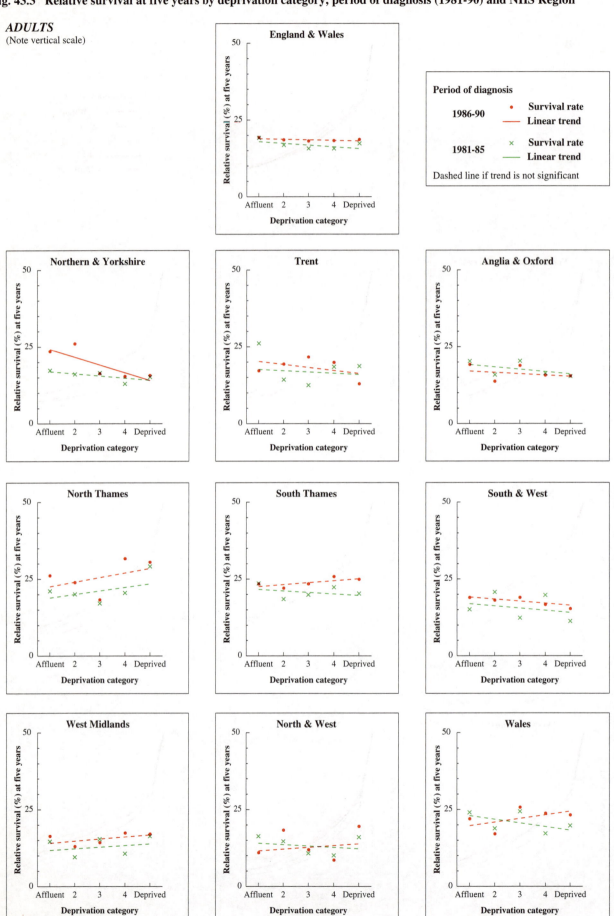

Fig. 43.4 Relative survival (%) by NHS Region and sex:
  one-year and five-year survival for patients diagnosed 1986-90, with 95% confidence intervals

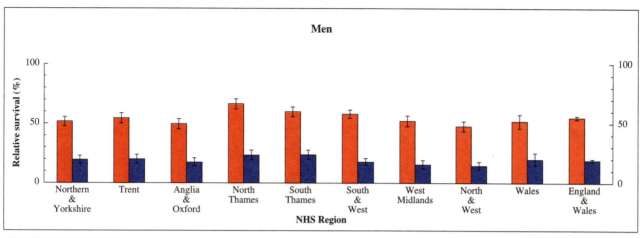

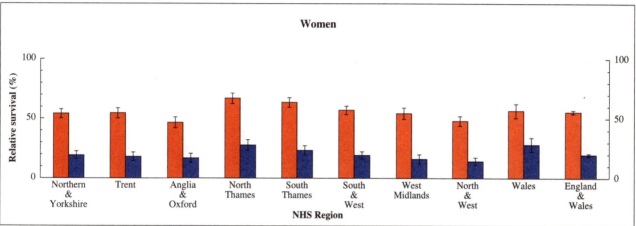

Fig. 43.5 Relative survival (%) at five years, with 95% confidence interval, by age at diagnosis and sex:
  England and Wales, patients diagnosed 1986-90

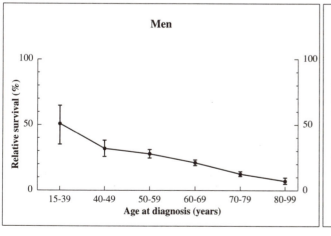

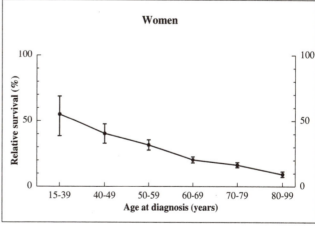

Chapter 44

ACUTE LYMPHOID LEUKAEMIA

Acute lymphoid leukaemia is uncommon in adults, comprising some 6% of all leukaemias and about one in 700 of all cancers in adults. More than half of all acute lymphoid leukaemias occur in children under 15, among whom it accounts for more than three-quarters of all leukaemias, and is the most common malignancy (Chapter 52). About 350 new cases are diagnosed in adults in England and Wales each year. In adults, acute lymphoid leukaemia represents about 15% of all lymphoid leukaemias: the chronic form is five times more common. Survival from all leukaemias combined is discussed in Chapter 49.

Acute lymphoid leukaemia is more common in men than women, with a sex ratio of 1.6 to 1; age-standardised (Europe) incidence rates in England and Wales in 1990 were 0.8 per 100,000 in men and 0.5 in women. In adults aged 15 and over, more than half the cases occur under age 50, and acute lymphoid leukaemia is rare over the age of 70. Overall incidence has risen by 20% in men and by 32% in women since 1971, while mortality remained unchanged between 1971 and 1995.

The survival analyses reported here are based on 4,000 adults who were diagnosed with acute lymphoid leukaemia in England and Wales between 1971 and 1990 and followed up to the end of 1995. After the exclusion of children under 15, who accounted for 55% of all eligible cases, the patients included in the analyses represented 85% of eligible adults (Table 44.1). About 3% of patients (6% of eligible adults) were excluded because their vital status was unknown at 6 July 1997, and 4% (8% of adults) because their survival was zero or unknown.

Survival trends

Relative survival for adults diagnosed with acute lymphoid leukaemia in England and Wales during the period 1986-90 was 53% at one year and 26% at five years, some 20% higher than for adults diagnosed during 1971-75 (Table 44.2). The progressive improvements in both short-term and long-term survival in England and Wales as a whole are clear from Figure 44.1.

Regional differences in the pattern of improvement are striking: one-year and five-year survival rates were extremely low in West Midlands and Wales during the early 1970s, but rose by 20% within the next five years. In Anglia & Oxford, by contrast, survival for patients diagnosed in the early 1970s was already 46% at one year and 21% at five years, some 10-20% higher than the national average for that period; survival did not improve

further until the late 1980s. Survival in North Thames was also high in the early 1970s (Table 44.2, Figure 44.1). Given the major impact of chemotherapy regimes introduced in the early 1970s on survival, these regional differences in survival trends strongly suggest that widespread use of effective chemotherapy regimes occurred earlier in the (then) Oxford and East Anglia Regional Health Authorities, and in NE and NW Thames, than in other regions.

Quantitative estimates of the national survival trends show an overall improvement of 20% or more in one-year survival to 50% in men and 54% in women, an average gain of 6-7% between successive quinquennia (see Table 4.4). More than half of the overall increase in one-year survival (11-14%) occurred between the early and late 1970s. These estimates of relative survival are adjusted for differences in the age distribution of patients over time and between the sexes; the estimates of trend are also adjusted for the precision of the relative survival estimate for each five-year period (see Chapter 3).

Five-year survival doubled in men between the early 1970s and the late 1980s, from 10% to 22%, while the increase for women was from 11% to 29%. The corresponding average gain between successive five-year periods of diagnosis was 4.4% for men and 5.5% for women. There was little difference between the sexes in the early 1970s, but women diagnosed during 1986-90 had a survival advantage over men of 7% at five years (see Table 4.4).

Relative survival from acute lymphoid leukaemia in adults falls steadily with age (Figure 44.5). For adults diagnosed in England and Wales during 1986-90, five-year survival was about 40% in those aged 15-39 at diagnosis, but it falls to less than 10% in those aged 70 or over (see Table 4.7). Survival is slightly higher in women than in men at all ages.

Deprivation

For patients diagnosed with acute lymphoid leukaemia in England and Wales during 1986-90, there were no systematic differences between deprivation categories in relative survival up to five years (Figure 44.2).

Among 1,000 adults of known deprivation status who were diagnosed during 1981-85, the difference between the most affluent and the most deprived group in one-year relative survival was estimated at 7%, but this was not statistically significant (see Table 4.9). The corresponding estimate for 1,100 adults of known status diagnosed during 1986-90 was only 2%. Differences between deprivation groups in five-year survival were negligible: this is confirmed in Figure 44.3 (note expanded vertical scale), which shows no consistent difference in relative survival at five years in either the early or late 1980s.

International comparison

Five-year survival for adults diagnosed with acute lymphoid leukemia in England and Wales during 1986-90 (25-27%) was broadly similar to the average for Europe and the USA (see Table 4.12). The highest regional survival rates in England and Wales were imprecise, and the high survival rate for men in Scotland (34%) also depends on only 70 cases[1].

1.   Berrino F, Capocaccia R, Estève J, Gatta G, Hakulinen T, Micheli M, Sant M, Verdecchia A, (eds.) *Survival of cancer patients in Europe: the EUROCARE study, II. (IARC Scientific Publications No. 151).* Lyon: International Agency for Research on Cancer, 1999

Table 44.1 Ineligible and excluded records, and patients included in analyses, by calendar period of diagnosis

Calendar period of diagnosis

	1971-75			1976-80			1981-85			1986-90			Overall		
Total registered	**2,396**			**2,654**			**2,897**			**2,850**			**10,797**		
Ineligible	Records	Patients	%	Records	Patients	%	Records	Patients	%	Records	Patients	%	Records	Patients	%
Incomplete data[1]	8	8		5	5		4	4		5	5		22	**22**	
Resident outside England and Wales	43	43		53	53		48	45		12	8		156	**149**	
In situ neoplasm[2]	0	0		0	0		0	0		0	0		0	**0**	
Benign or uncertain behaviour[2]	0	0		0	0		0	0		0	0		0	**0**	
Metastatic[3]	-	-		-	-		-	-		-	-		-	**-**	
Otherwise ineligible[3]	-	-		-	-		-	-		-	-		-	**-**	
Lymphoma[4]	-	-		-	-		-	-		-	-		-	**-**	
Leukaemia or myeloma[4]	-	-		-	-		-	-		-	-		-	**-**	
		51			**58**			**49**			**13**			**171**	
Total eligible		**2,345**	**100**		**2,596**	**100**		**2,848**	**100**		**2,837**	**100**		**10,626**	**100**
Aged under 15 or 100 years or over at diagnosis	1,440	1,403	59.8	1,508	1,471	56.7	1,486	1,463	51.4	1,471	1,463	51.6	5,905	**5,800**	54.6
Vital status unknown[5]	129	36	1.5	179	58	2.2	243	106	3.7	225	86	3.0	776	**286**	2.7
Duplicate registration[6]	0	0	0.0	0	0	0.0	1	0	0.0	0	0	0.0	1	**0**	0.0
Multiple primary[7]	0	0	0.0	1	0	0.0	10	1	<0.1	10	5	0.2	21	**6**	<0.1
Synchronous tumours[8]	1	0	0.0	7	1	<0.1	4	0	0.0	16	5	0.2	28	**6**	<0.1
Sex not known	0	0	0.0	0	0	0.0	0	0	0.0	0	0	0.0	0	**0**	0.0
Sex incompatible with site code[9]	-	-	-	-	-	-	-	-	-	-	-	-	-	**-**	-
Invalid dates[10]	0	0	0.0	0	0	0.0	0	0	0.0	1	0	0.0	1	**0**	0.0
Zero survival[11]	93	55	2.3	120	76	2.9	230	161	5.7	198	158	5.6	641	**450**	4.2
Total exclusions		**1,494**	**63.7**		**1,606**	**61.9**		**1,731**	**60.8**		**1,717**	**60.5**		**6,548**	**61.6**
Patients included in analysis		**851**	**36.3**		**990**	**38.1**		**1,117**	**39.2**		**1,120**	**39.5**		**4,078**	**38.4**
Percentage of eligible adults			**90.3**			**88.0**			**80.6**			**81.5**			**84.5**

[1] Main data item(s) invalid or incompatible with one another: name, sex, date of birth, date of diagnosis and (if present) date of death, postcode, site, morphology and behaviour

[2] Tumour either *in situ* (behaviour code 2), benign (0), uncertain if benign or malignant (1) or metastatic (6 or 9)

[3] Not applicable for this malignancy

[4] Not applicable for this malignancy

[5] Tracing of vital status at National Health Service Central Register incomplete at 6 July 1997

[6] Same site code, sex, cancer registry and cancer registry number as an earlier registration

[7] Same site code and person number as an earlier registration(s): mostly confirmed multiple primary tumours at this site, some unresolved duplicate registrations

[8] Same site code, sex, date of birth and date of diagnosis as another registration(s): mostly synchronous or (in paired organs) bilateral tumours in same anatomic site in one individual, not previously linked: also some duplicate registrations

[9] Not applicable for this malignancy

[10] Impossible sequence of dates (birth, diagnosis and, if present, emigration or death); or date of death after 6 July 1997

[11] Date of diagnosis same as date of death: some are patients who did die on the day of diagnosis, but most were registered solely from a death certificate (death certificate only, or DCO), and their survival time is thus unknown

Table 44.2   Crude and relative survival (%) at one, five and ten years since diagnosis: NHS Region, deprivation category and calendar period of diagnosis

		1971-75									1976-80								
		Affluent		2		3		4		Deprived		All[1]		Affluent		2		3		4		Deprived		All[1]	
		C	R	C	R	C	R	C	R	C	R	C	R	C	R	C	R	C	R	C	R	C	R	C	R
ENGLAND & WALES																										
Men		*(74)*		*(84)*		*(99)*		*(80)*		*(61)*		*(498)*		*(128)*		*(105)*		*(104)*		*(103)*		*(114)*		*(584)*	
	One year	37	37	22	23	29	30	21	22	31	32	27	28	43	44	42	43	33	34	44	46	40	41	41	42
	Five years	7	8	5	5	9	11	10	12	10	11	8	9	12	13	17	18	14	15	17	20	9	10	13	14
	Ten years	2	3	4	4	7	8	8	9	7	7	5	6	9	10	12	13	13	14	12	14	6	7	10	11
Women		*(69)*		*(60)*		*(72)*		*(49)*		*(44)*		*(353)*		*(83)*		*(97)*		*(81)*		*(64)*		*(66)*		*(406)*	
	One year	31	32	34	35	15	15	25	25	21	21	25	25	31	32	35	37	46	48	38	39	41	42	38	39
	Five years	11	12	16	17	3	3	10	10	4	4	8	9	14	15	14	15	20	20	16	18	19	20	16	17
	Ten years	8	9	12	13	1	1	8	9	2	2	6	7	11	13	13	15	15	17	13	15	14	17	13	15
Adults		*(143)*		*(144)*		*(171)*		*(129)*		*(105)*		*(851)*		*(211)*		*(202)*		*(185)*		*(167)*		*(180)*		*(990)*	
	One year	34	35	27	28	23	23	22	23	26	27	26	27	39	40	39	40	39	40	42	43	40	41	40	41
	Five years	9	10	9	10	7	7	10	11	7	8	8	9	13	14	15	16	17	18	17	19	12	14	14	15
	Ten years	5	6	7	8	4	5	8	9	5	5	6	6	10	11	13	14	13	16	12	14	9	11	11	13

		Men		Women		Adults								Men		Women		Adults	
Northern & Yorkshire		*(53)*		*(52)*		*(105)*								*(71)*		*(60)*		*(131)*	
	One year	20	20	31	31	25	25							36	38	37	38	36	38
	Five years	4	5	5	5	5	5							3	3	12	13	7	8
	Ten years	4	5	3	4	4	5							3	3	12	13	7	8
Trent		*(57)*		*(36)*		*(93)*								*(48)*		*(27)*		*(75)*	
	One year	16	17	11	11	14	14							42	43	43	43	42	43
	Five years	7	8	2	2	5	6							9	9	21	21	13	14
	Ten years	4	4	2	2	3	3							7	7	14	15	9	10
Anglia & Oxford		*(57)*		*(29)*		*(86)*								*(73)*		*(36)*		*(109)*	
	One year	39	40	55	57	45	46							43	45	33	34	40	41
	Five years	16	17	27	29	20	21							18	19	14	15	17	18
	Ten years	11	11	20	23	14	15							10	11	9	10	10	11
North Thames		*(62)*		*(30)*		*(92)*								*(67)*		*(52)*		*(119)*	
	One year	49	49	23	23	40	41							57	59	44	46	52	53
	Five years	17	19	12	12	15	16							21	22	18	19	19	21
	Ten years	15	17	8	9	13	14							16	18	13	15	15	17
South Thames		*(67)*		*(67)*		*(134)*								*(67)*		*(55)*		*(122)*	
	One year	21	21	28	29	25	25							38	39	32	34	35	37
	Five years	5	6	10	11	7	8							11	12	12	13	11	12
	Ten years	4	4	8	9	5	7							11	12	12	13	11	12
South & West		*(55)*		*(46)*		*(101)*								*(83)*		*(66)*		*(149)*	
	One year	25	26	16	17	22	22							31	32	37	38	34	35
	Five years	4	4	10	10	7	7							15	16	17	19	16	17
	Ten years	0	0	5	7	2	4							12	13	14	16	13	14
West Midlands		*(43)*		*(22)*		*(65)*								*(50)*		*(35)*		*(85)*	
	One year	26	27	13	13	22	23							41	42	42	43	41	43
	Five years	2	2	0	0	1	1							17	18	19	20	18	19
	Ten years	2	2	0	0	1	1							15	16	17	17	16	16
North & West		*(73)*		*(51)*		*(124)*								*(93)*		*(59)*		*(152)*	
	One year	31	32	24	24	27	28							43	44	38	39	41	42
	Five years	9	10	8	8	9	9							10	11	17	17	13	14
	Ten years	6	7	3	3	5	6							8	9	12	14	10	11
WALES		*(31)*		*(20)*		*(51)*								*(32)*		*(16)*		*(48)*	
	One year	11	11	17	17	14	14							40	41	34	35	38	39
	Five years	4	4	0	0	3	3							19	24	23	23	20	24
	Ten years	2	2	0	0	2	2							7	9	23	23	12	15

[1] Including those for whom no deprivation category was assigned. C - crude survival, R - relative survival (see Chapter 3). Number of patients contributing to each analysis in parentheses.

1981-85 / 1986-90 — ENGLAND & WALES

	Affluent C R	2 C R	3 C R	4 C R	Deprived C R	All[1] C R	Affluent C R	2 C R	3 C R	4 C R	Deprived C R	All[1] C R	
	(142)	*(132)*	*(110)*	*(125)*	*(93)*	*(621)*	*(149)*	*(155)*	*(135)*	*(133)*	*(95)*	*(674)*	**Men**
One year	46 48	44 45	39 41	42 44	46 47	44 45	48 49	57 59	49 51	54 55	57 59	53 54	
Five years	18 19	19 20	15 16	20 21	24 26	19 21	22 23	26 27	21 23	22 24	25 27	23 25	
	14 15	15 16	12 15	16 18	20 22	15 17							
	(98)	*(103)*	*(100)*	*(102)*	*(87)*	*(496)*	*(83)*	*(108)*	*(96)*	*(96)*	*(59)*	*(446)*	**Women**
One year	49 51	44 44	40 41	36 38	37 38	41 42	52 54	53 55	49 50	54 55	33 34	50 51	
Five years	22 24	19 21	13 15	18 19	17 17	18 19	25 27	29 31	22 23	30 32	20 21	25 27	
	19 21	15 17	9 11	14 16	15 16	14 16							
	(240)	*(235)*	*(210)*	*(227)*	*(180)*	*(1,117)*	*(232)*	*(263)*	*(231)*	*(229)*	*(154)*	*(1,120)*	**Adults**
One year	47 49	44 45	39 41	40 41	42 43	43 44	50 51	56 57	49 51	54 55	48 49	51 53	
Five years	20 21	19 20	14 16	19 20	21 22	19 20	23 24	27 29	21 23	26 28	23 25	24 26	
	16 18	15 17	11 12	15 17	17 19	15 17							

Regional

	Men	Women	Adults	Men	Women	Adults	
	(102)	*(68)*	*(170)*	*(91)*	*(38)*	*(129)*	**Northern & Yorkshire**
One year	42 43	40 41	41 42	49 50	*52 53*	50 51	
Five years	17 18	16 18	17 18	23 24	*28 30*	25 27	
	17 18	*15 17*	*16 18*				
	(55)	*(59)*	*(114)*	*(51)*	*(46)*	*(97)*	**Trent**
One year	*44 46*	*38 40*	41 43	*63 65*	*42 43*	53 54	
Five years	15 16	13 14	14 15	*27 28*	*27 28*	27 28	
	8 9	10 11	9 10				
	(53)	*(34)*	*(87)*	*(50)*	*(42)*	*(92)*	**Anglia & Oxford**
One year	*40 41*	*49 50*	44 45	*52 53*	*59 61*	55 57	
Five years	16 17	25 26	19 20	*21 22*	*22 23*	21 22	
	12 14	*22 23*	16 17				
	(70)	*(40)*	*(110)*	*(64)*	*(55)*	*(119)*	**North Thames**
One year	*49 50*	*46 48*	48 49	*60 61*	*44 45*	53 54	
Five years	25 28	28 29	26 28	*23 24*	*20 22*	22 23	
	22 25	23 25	23 25				
	(76)	*(61)*	*(137)*	*(71)*	*(57)*	*(128)*	**South Thames**
One year	41 45	47 49	44 47	*60 61*	*53 55*	57 59	
Five years	16 17	17 19	16 18	*27 28*	*34 36*	30 32	
	12 16	*11 15*	12 16				
	(92)	*(77)*	*(169)*	*(122)*	*(59)*	*(181)*	**South & West**
One year	46 48	39 39	42 44	55 57	*56 58*	56 57	
Five years	25 26	16 17	21 22	23 24	*33 35*	26 28	
	18 20	15 15	17 18				
	(73)	*(54)*	*(127)*	*(90)*	*(48)*	*(138)*	**West Midlands**
One year	37 38	*48 49*	42 43	52 53	*49 50*	51 52	
Five years	17 18	*21 22*	19 20	19 21	*19 20*	19 20	
	13 14	*17 19*	15 17				
	(77)	*(75)*	*(152)*	*(92)*	*(76)*	*(168)*	**North & West**
One year	48 51	33 34	41 43	40 41	47 48	43 45	
Five years	17 19	10 11	14 15	20 22	23 25	21 23	
	12 15	7 8	10 11				
	(23)	*(28)*	*(51)*	*(43)*	*(25)*	*(68)*	**WALES**
One year	*46 47*	*35 36*	40 41	*46 48*	*42 43*	45 46	
Five years	*29 31*	*25 26*	27 29	*30 35*	*18 20*	26 29	
	25 28	*19 20*	22 24				

[1] Including those for whom no deprivation category was assigned. C - crude survival, R - relative survival (see Chapter 3). Number of patients contributing to each analysis in parentheses.

Fig. 44.1   Relative survival up to ten years: trends by calendar period of diagnosis (1971-90) and NHS Region

ADULTS

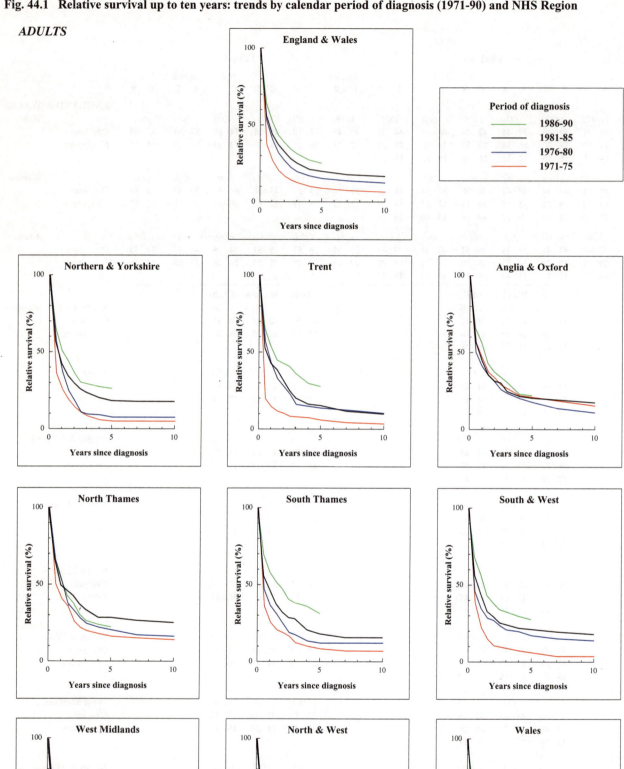

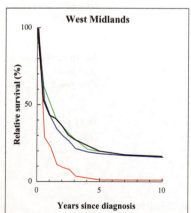

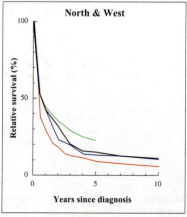

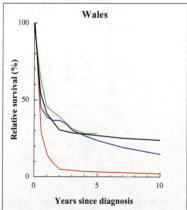

Fig. 44.2 Relative survival up to five years, by deprivation category: England and Wales, patients diagnosed 1986-90

ADULTS

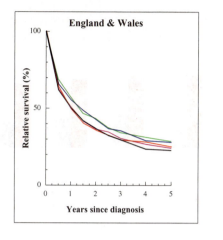

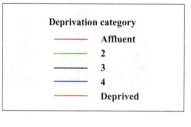

Fig. 44.3 Relative survival at five years, by deprivation category and period of diagnosis (1981-90): England and Wales

ADULTS
(Note vertical scale)

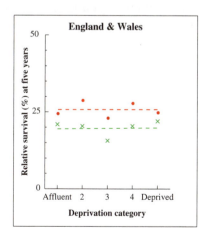

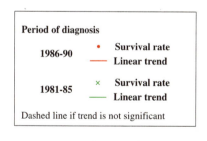

Fig. 44.4 Relative survival (%) by NHS Region and sex:
one-year and five-year survival for patients diagnosed 1986-90, with 95% confidence intervals

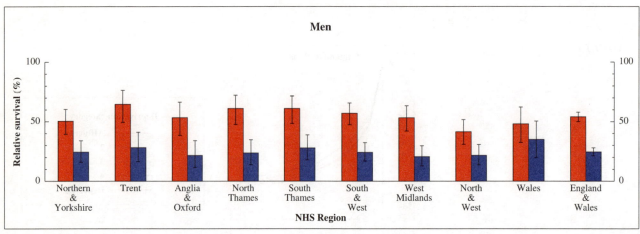

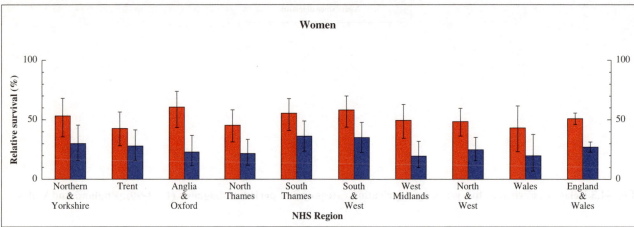

Fig. 44.5 Relative survival (%) at five years, with 95% confidence interval, by age at diagnosis and sex:
England and Wales, patients diagnosed 1986-90

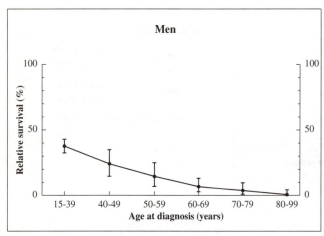

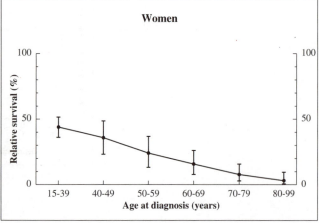

Chapter 45

CHRONIC LYMPHOID LEUKAEMIA

Chronic lymphoid leukaemia accounts for a third of all leukaemias in adults, and some 1,600 new cases are diagnosed in England and Wales each year, about 0.8% of all cancers. In adults, it is five times more common than acute lymphoid leukaemia. Survival from all leukaemias combined is discussed in Chapter 49.

In direct contrast with the adult form of acute lymphoid leukaemia, chronic lymphoid leukaemia is very rare below the age of 50 (less than 5% of cases), but incidence increases rapidly above that age. It is more common in men, with a sex ratio of almost 2.5 to 1. Age-standardised (Europe) incidence in men in England and Wales increased by more than 50% between 1971 and 1990 to 4.7 per 100,000, and remained around 2 in women. Mortality rates were stable at about 2 per 100,000 in men and 1 in women between 1971 and 1995.

The survival analyses reported here are based on almost 22,000 adults diagnosed with chronic lymphoid leukaemia in England and Wales during the period 1971-90 and followed up to the end of 1995. They represent about 84% of eligible patients (Table 45.1). Some 4% of patients were excluded because their vital status was unknown at 6 July 1997, and 11% because they had zero or unknown survival: many of the latter would have been registered solely from a death certificate.

Survival trends

Relative survival for adults diagnosed with chronic lymphoid leukaemia in England and Wales during 1986-90 was 77% at one year and 51% at five years, about 15-18% higher than for patients diagnosed during the early 1970s (Table 45.2). Both long-term and short-term survival improved steadily between successive five-year periods in England and Wales, but the largest increase occurred for patients diagnosed around 1980, when national average survival at one, five and ten years after diagnosis rose by 6-9% in the space of five years. Survival also improved in most regions, although the timing and extent of the gains in survival varied between regions (Figure 45.1).

The shape of the survival curves is quite different from those for acute lymphoid leukaemia in adults (see Chapter 44). One-year survival from chronic lymphoid leukaemia is more than 20% higher than for the acute form (77% compared to 53%), but excess mortality relative to that of the general population continues for at least 10 years, and the relative survival curves decline steadily, with little evidence of a plateau. Ten years after diagnosis, relative

survival from chronic lymphoid leukaemia is only about 10% higher than for the acute form (28% compared to 17%).

In a further contrast with acute lymphoid leukaemia, improvements in longer-term survival from chronic lymphoid leukaemia have been at least as great as the gains in short-term survival. Thus the national average increase in one-year survival of 16-17% between patients diagnosed during the early 1970s and those diagnosed during 1986-90 was matched by a gain of 18-19% in five-year survival. These increases are adjusted for differences in the age distribution of patients over time and between the sexes (see Table 4.4). Both short-term and long-term survival improved by an average of 6-7% every five years over this period. The national average five-year survival rate rose by 10% between the periods 1976-80 and 1981-85 alone, after adjustment for age.

Substantial improvements occurred in most regions, but regional variation in survival remains wide (Figure 45.4). One-year survival for patients diagnosed during 1986-90 ranged from 65% to 85%, and five-year survival from 37% to 62%. Both one-year and five-year survival in North & West Region were significantly lower than the national average in men and women. Regional estimates are each based on 500-1,000 patients.

Survival in North Thames and Anglia & Oxford was about 10% higher than the national average for patients diagnosed during 1971-75 (Table 45.2). Overall improvements in survival for adults diagnosed in Anglia & Oxford have been very small, however (Figure 45.1), because the 7% increase in survival for men is largely balanced by a decline of 8% in overall one-year and five-year survival for women. This anomaly needs explanation, but appears more likely to be due to change in the age distribution of patients or the definition of disease than to any decline in the efficacy of treatment for women as opposed to men. Survival estimates in this region are based on some 200 women diagnosed in each five-year period; very few cases in either Oxford or East Anglia were registered solely from a death certificate.

Since the early 1970s, women have consistently had a 4-8% survival advantage up to five years after diagnosis, after adjustment to the joint age distribution of cases among men and women.

Relative survival falls steadily with age (Figure 45.5). Five-year survival for adults diagnosed during 1986-90 under the age of 50 was around 80%, while for those aged

80-99, it was around 30%, after adjustment for mortality from other causes (see Table 4.4).

Deprivation

Survival from chronic lymphoid leukaemia in England and Wales has consistently been higher among more affluent groups than among deprived groups (Table 45.2). Survival up to five years for adults in the different deprivation groups diagnosed during 1986-90 is clearly demarcated within a year after diagnosis, although the curves are broadly parallel thereafter (Figure 45.2). The pattern is more variable within regions, in part because regional survival curves for each deprivation group are based on about 100 patients.

Among almost 6,000 patients of known deprivation status diagnosed in England and Wales during 1981-85, the gap in relative survival between the most affluent and the most deprived groups was 7% at one year and 10% at five years. The deprivation gap among a further 6,000 patients of known status who were diagnosed during 1986-90 was 10% at one year and 11% at five years (see Table 4.9). These patterns are reflected in the significant deprivation gradients for patients diagnosed in England and Wales in both the early and late 1980s (Figure 45.3).

International comparison

Five-year survival rates for adults diagnosed with chronic lymphoid leukaemia in England and Wales during 1986-90 (around 50%) were similar to those in Scotland, but 13-14% lower than the average for Europe (1985-89). The highest regional survival rates (around 60%) were closer to the European average, but still about 10% lower than survival rates in the USA (Table-Figure 4.12).

Table 45.1 Ineligible and excluded records, and patients included in analyses, by calendar period of diagnosis

	Calendar period of diagnosis														
	1971-75			1976-80			1981-85			1986-90			Overall		
Total registered	5,016			5,695			7,302			8,040			26,053		
Ineligible	*Records*	*Patients*	*%*	*Records*	*Patients*	*%*	*Records*	*Patients*	*%*	*Records*	*Patients*	*%*	*Records*	*Patients*	*%*
Incomplete data[1]	0	0		5	5		11	11		15	15		31	**31**	
Resident outside England and Wales	28	28		25	22		35	24		27	14		115	**88**	
In situ neoplasm[2]	0	0		0	0		0	0		0	0		0	**0**	
Benign or uncertain behaviour[2]	0	0		0	0		0	0		0	0		0	**0**	
Metastatic[3]	-	-		-	-		-	-		-	-		-	**-**	
Otherwise ineligible[3]	-	-		-	-		-	-		-	-		-	**-**	
Lymphoma[4]	-	-		-	-		-	-		-	-		-	**-**	
Leukaemia or myeloma[4]	-	-		-	-		-	-		-	-		-	**-**	
		28			27			35			29			119	
Total eligible		**4,988**	**100**		**5,668**	**100**		**7,267**	**100**		**8,011**	**100**		**25,934**	**100**
Aged under 15 or 100 years or over at diagnosis	8	7	0.1	5	5	<0.1	6	6	<0.1	9	9	0.1	28	**27**	0.1
Vital status unknown[5]	178	158	3.2	245	233	4.1	260	239	3.3	441	423	5.3	1,124	**1,053**	4.1
Duplicate registration[6]	0	0	0.0	2	2	<0.1	0	0	0.0	0	0	0.0	2	**2**	<0.1
Multiple primary[7]	0	0	0.0	1	1	<0.1	12	10	0.1	31	28	0.3	44	**39**	0.2
Synchronous tumours[8]	6	6	0.1	11	8	0.1	13	11	0.2	28	20	0.2	58	**45**	0.2
Sex not known	0	0	0.0	0	0	0.0	0	0	0.0	0	0	0.0	0	**0**	0.0
Sex incompatible with site code[9]	-	-	-	-	-	-	-	-	-	-	-	-	-	**-**	-
Invalid dates[10]	1	1	<0.1	2	2	<0.1	7	6	<0.1	7	5	<0.1	17	**14**	<0.1
Zero survival[11]	332	323	6.5	467	441	7.8	1,004	957	13.2	1,259	1,200	15.0	3,062	**2,921**	11.3
Total exclusions		**495**	**9.9**		**692**	**12.2**		**1,229**	**16.9**		**1,685**	**21.0**		**4,101**	**15.8**
Patients included in analysis		**4,493**	**90.1**		**4,976**	**87.8**		**6,038**	**83.1**		**6,326**	**79.0**		**21,833**	**84.2**

[1]   Main data item(s) invalid or incompatible with one another: name, sex, date of birth, date of diagnosis and (if present) date of death, postcode, site, morphology and behaviour

[2]   Tumour either *in situ* (behaviour code 2), benign (0), uncertain if benign or malignant (1) or metastatic (6 or 9)

[3]   Not applicable for this malignancy

[4]   Not applicable for this malignancy

[5]   Tracing of vital status at National Health Service Central Register incomplete at 6 July 1997

[6]   Same site code, sex, cancer registry and cancer registry number as an earlier registration

[7]   Same site code and person number as an earlier registration(s): mostly confirmed multiple primary tumours at this site, some unresolved duplicate registrations

[8]   Same site code, sex, date of birth and date of diagnosis as another registration(s): mostly synchronous or (in paired organs) bilateral tumours in same anatomic site in one individual, not previously linked: also some duplicate registrations

[9]   Not applicable for this malignancy

[10]   Impossible sequence of dates (birth, diagnosis and, if present, emigration or death); or date of death after 6 July 1997

[11]   Date of diagnosis same as date of death: some are patients who did die on the day of diagnosis, but most were registered solely from a death certificate (death certificate only, or DCO), and their survival time is thus unknown

Table 45.2   Crude and relative survival (%) at one, five and ten years since diagnosis: NHS Region, deprivation category and calendar period of diagnosis

MEN

	1971-75						1976-80					
	Affluent	*2*	*3*	*4*	*Deprived*	*All*[1]	*Affluent*	*2*	*3*	*4*	*Deprived*	*All*[1]
	C R	C R	C R	C R	C R	C R	C R	C R	C R	C R	C R	C R
ENGLAND & WALES	(423)	(485)	(460)	(420)	(346)	(2,649)	(537)	(646)	(616)	(571)	(437)	(2,920)
One year	58 62	60 64	56 60	58 62	58 63	58 62	64 69	60 65	55 60	57 61	59 64	59 63
Five years	26 33	23 31	21 29	25 35	23 31	23 32	30 40	27 37	25 35	27 37	26 38	27 37
Ten years	9 14	9 16	7 14	10 20	9 17	8 15	12 20	11 20	11 20	11 21	8 17	10 20

	1981-85						1986-90					
	(699)	(756)	(731)	(686)	(550)	(3,491)	(759)	(856)	(850)	(654)	(547)	(3,692)
One year	71 76	67 73	65 71	66 72	64 70	67 73	75 80	76 82	69 75	68 73	66 72	71 77
Five years	39 53	33 46	31 43	31 43	29 41	33 45	41 54	37 51	33 46	31 45	32 45	35 49
Ten years	19 33	13 24	11 22	12 24	9 18	13 25						

	1971-75	1976-80	1981-85	1986-90
Northern & Yorkshire	(345)	(439)	(493)	(538)
One year	56 60	59 64	67 73	70 76
Five years	21 30	25 36	32 44	36 50
Ten years	6 11	10 20	10 18	
Trent	(290)	(343)	(441)	(413)
One year	62 67	55 60	66 73	71 77
Five years	28 37	27 36	33 47	35 49
Ten years	10 19	10 18	11 21	
Anglia & Oxford	(267)	(323)	(324)	(350)
One year	59 63	61 65	61 66	65 70
Five years	27 35	26 35	30 42	26 37
Ten years	9 17	9 17	12 23	
North Thames	(296)	(233)	(238)	(349)
One year	66 71	71 76	75 80	78 84
Five years	31 43	38 53	36 49	40 54
Ten years	12 22	17 33	17 31	
South Thames	(375)	(338)	(400)	(449)
One year	57 62	61 67	71 77	80 86
Five years	24 34	27 39	41 58	43 58
Ten years	10 17	10 19	19 38	
South & West	(365)	(453)	(500)	(611)
One year	57 60	59 64	63 69	74 80
Five years	23 31	30 42	30 43	43 59
Ten years	9 17	14 25	14 26	
West Midlands	(236)	(229)	(405)	(359)
One year	58 63	57 61	73 79	72 78
Five years	21 28	26 37	35 48	33 47
Ten years	5 10	9 19	12 22	
North & West	(340)	(408)	(508)	(416)
One year	57 61	54 58	62 68	63 68
Five years	19 25	20 28	28 40	25 37
Ten years	7 13	6 11	11 22	
WALES	(135)	(154)	(182)	(207)
One year	48 52	55 59	65 71	61 67
Five years	11 15	27 40	31 44	25 37
Ten years	6 11	12 23	14 29	

[1] Including those for whom no deprivation category was assigned. C - crude survival, R - relative survival (see Chapter 3). Number of patients contributing to each analysis in parentheses.

Table 45.2   Crude and relative survival (%) at one, five and ten years since diagnosis: NHS Region, deprivation category and calendar period of diagnosis

WOMEN	1971-75						1976-80					
	Affluent	*2*	*3*	*4*	*Deprived*	*All[1]*	*Affluent*	*2*	*3*	*4*	*Deprived*	*All[1]*
	C R	C R	C R	C R	C R	C R	C R	C R	C R	C R	C R	C R
ENGLAND & WALES	*(290)*	*(352)*	*(312)*	*(303)*	*(234)*	*(1,844)*	*(382)*	*(419)*	*(432)*	*(411)*	*(311)*	*(2,056)*
One year	60 64	57 61	62 66	56 61	55 60	59 63	65 70	60 65	62 67	59 63	58 62	61 65
Five years	30 39	27 35	30 38	22 30	23 32	27 35	33 43	29 39	33 43	31 41	27 38	30 41
Ten years	15 25	10 17	12 21	11 21	10 18	12 20	15 25	14 24	13 22	16 28	14 25	14 24
	1981-85						**1986-90**					
	(523)	*(547)*	*(554)*	*(477)*	*(401)*	*(2,547)*	*(525)*	*(606)*	*(596)*	*(502)*	*(372)*	*(2,634)*
One year	72 77	68 74	68 73	67 73	61 66	67 73	74 80	73 78	73 78	69 74	64 68	71 76
Five years	42 55	38 51	37 50	33 46	34 47	37 50	42 58	41 55	39 53	36 50	33 45	39 53
Ten years	21 37	18 33	17 31	19 34	15 27	18 32						

	1971-75	1976-80	1981-85	1986-90
Northern & Yorkshire	*(241)*	*(294)*	*(409)*	*(373)*
One year	60 65	59 65	66 72	77 84
Five years	30 39	26 37	36 50	40 57
Ten years	10 19	12 22	15 29	
Trent	*(169)*	*(211)*	*(298)*	*(265)*
One year	62 66	62 65	70 77	66 71
Five years	30 39	37 49	37 53	34 48
Ten years	13 22	19 30	21 39	
Anglia & Oxford	*(198)*	*(219)*	*(219)*	*(219)*
One year	71 77	68 74	66 73	64 69
Five years	36 50	30 43	30 43	30 42
Ten years	17 31	15 28	15 30	
North Thames	*(180)*	*(156)*	*(154)*	*(260)*
One year	70 76	62 67	75 81	77 83
Five years	36 47	33 46	47 62	45 60
Ten years	17 28	12 23	27 44	
South Thames	*(285)*	*(321)*	*(332)*	*(325)*
One year	54 58	63 69	74 81	78 84
Five years	27 36	31 41	44 60	46 62
Ten years	14 25	13 24	24 43	
South & West	*(258)*	*(328)*	*(364)*	*(543)*
One year	59 64	65 71	64 70	76 82
Five years	26 36	32 45	36 49	47 64
Ten years	12 23	16 29	15 28	
West Midlands	*(180)*	*(126)*	*(261)*	*(208)*
One year	62 66	59 64	74 81	67 72
Five years	21 27	29 39	40 54	33 45
Ten years	7 12	15 26	18 32	
North & West	*(227)*	*(274)*	*(376)*	*(317)*
One year	40 43	48 51	57 62	56 60
Five years	14 18	25 34	30 43	25 37
Ten years	3 6	11 19	14 29	
WALES	*(106)*	*(127)*	*(134)*	*(124)*
One year	50 52	62 67	65 71	67 73
Five years	18 24	33 44	36 52	34 50
Ten years	10 20	15 27	19 37	

[1] Including those for whom no deprivation category was assigned. C - crude survival, R - relative survival (see Chapter 3).
Number of patients contributing to each analysis in parentheses.

Table 45.2 Crude and relative survival (%) at one, five and ten years since diagnosis: NHS Region, deprivation category and calendar period of diagnosis

ADULTS	1971-75 Affluent C	R	2 C	R	3 C	R	4 C	R	Deprived C	R	All[1] C	R	1976-80 Affluent C	R	2 C	R	3 C	R	4 C	R	Deprived C	R	All[1] C	R
ENGLAND & WALES	(713)		(837)		(772)		(723)		(580)		(4,493)		(919)		(1,065)		(1,048)		(982)		(748)		(4,976)	
One year	59	63	59	63	58	63	57	61	57	62	58	62	65	69	60	65	58	63	58	62	58	63	60	64
Five years	28	36	25	33	24	33	24	33	23	31	25	33	31	41	28	38	28	39	28	39	26	38	28	39
Ten years	11	18	9	16	9	17	11	20	9	17	10	17	13	22	12	22	12	21	13	24	10	21	12	22
ENGLAND	(683)		(812)		(721)		(671)		(550)		(4,252)		(881)		(1,021)		(984)		(904)		(722)		(4,695)	
One year	60	64	59	63	60	64	58	62	58	63	59	63	65	69	61	65	59	63	57	61	59	63	60	64
Five years	28	36	25	34	26	34	25	33	23	33	25	34	31	41	28	38	28	39	28	38	26	38	28	39
Ten years	11	19	10	17	9	17	11	20	9	18	10	17	13	21	12	22	11	20	13	23	11	21	12	22
Northern & Yorkshire	(56)		(92)		(87)		(87)		(119)		(586)		(89)		(115)		(145)		(169)		(189)		(733)	
One year	65	68	60	65	70	74	58	63	53	57	58	62	75	80	60	66	56	61	50	54	61	67	59	65
Five years	25	31	26	36	30	40	27	36	16	24	25	34	36	47	24	33	28	39	23	31	24	36	26	37
Ten years	8	13	13	22	7	13	6	11	5	10	8	15	17	28	9	19	7	14	12	20	11	24	11	21
Trent	(56)		(79)		(70)		(113)		(92)		(459)		(75)		(104)		(112)		(147)		(108)		(554)	
One year	69	73	64	69	62	65	63	68	57	63	62	66	70	75	61	64	61	65	52	56	51	56	58	62
Five years	26	33	35	44	34	44	25	32	26	37	28	37	36	44	33	42	34	46	27	37	25	36	30	41
Ten years	11	20	7	12	17	28	14	27	10	19	11	20	15	22	15	24	14	25	14	25	10	20	13	23
Anglia & Oxford	(93)		(116)		(100)		(55)		(20)		(465)		(130)		(145)		(133)		(85)		(28)		(542)	
One year	59	64	67	72	62	66	56	60	69	77	64	69	62	67	67	72	56	60	73	79	67	71	64	69
Five years	32	43	29	39	26	35	33	47	29	42	31	41	32	44	24	33	23	31	29	40	42	56	27	38
Ten years	15	24	11	21	11	23	18	36	9	25	12	23	14	26	9	17	13	21	11	21	10	18	12	22
North Thames	(99)		(97)		(76)		(84)		(62)		(476)		(84)		(84)		(81)		(67)		(57)		(389)	
One year	70	76	70	75	67	73	67	72	65	70	67	73	65	69	65	70	70	74	69	74	66	71	67	72
Five years	42	56	32	41	26	36	34	47	36	47	33	44	33	42	42	57	29	41	31	44	47	62	36	50
Ten years	16	26	10	19	14	26	18	32	14	24	14	25	16	27	21	36	9	19	12	25	19	39	15	29
South Thames	(169)		(138)		(124)		(109)		(52)		(660)		(184)		(147)		(138)		(99)		(45)		(659)	
One year	55	59	52	55	59	64	51	55	61	66	56	60	67	73	62	66	60	65	56	60	63	69	62	68
Five years	25	33	24	32	24	33	21	31	28	38	25	35	30	41	26	37	30	41	30	40	24	36	29	40
Ten years	10	16	10	15	9	18	9	19	15	27	12	20	11	18	11	24	13	25	15	27	7	14	11	22
South & West	(117)		(184)		(141)		(88)		(29)		(623)		(168)		(241)		(205)		(101)		(34)		(781)	
One year	61	65	53	56	58	63	52	56	65	70	57	62	64	68	61	66	61	65	67	73	55	58	62	67
Five years	27	36	21	29	27	36	21	29	34	48	24	33	30	41	29	41	31	42	41	54	30	41	31	43
Ten years	12	21	10	19	11	18	8	16	17	38	10	19	14	25	15	26	14	25	18	33	14	24	15	27
West Midlands	(23)		(28)		(32)		(40)		(45)		(416)		(53)		(65)		(59)		(94)		(81)		(355)	
One year	52	55	59	63	58	63	62	66	61	65	60	64	60	63	48	52	57	61	59	63	61	66	58	62
Five years	26	30	20	30	24	30	23	28	18	25	21	28	35	44	23	30	34	44	23	32	23	34	27	37
Ten years	8	11	10	19	0	0	7	12	6	12	6	11	11	19	10	20	12	21	13	28	8	19	11	22
North & West	(70)		(78)		(91)		(95)		(131)		(567)		(98)		(120)		(111)		(142)		(180)		(682)	
One year	45	47	49	54	42	45	56	61	53	58	50	54	54	58	55	59	49	53	47	50	54	58	52	56
Five years	16	20	15	21	14	18	18	25	20	27	17	23	22	29	24	33	17	25	25	34	23	33	22	31
Ten years	7	10	3	6	2	4	7	13	7	13	5	10	5	9	8	16	7	13	9	17	9	17	8	14
WALES	(30)		(25)		(51)		(52)		(30)		(241)		(38)		(44)		(64)		(78)		(26)		(281)	
One year	48	50	53	57	38	41	45	48	42	44	49	52	61	65	54	57	53	57	64	68	55	59	58	62
Five years	22	32	7	10	9	11	16	22	8	10	14	19	31	39	25	33	29	39	35	49	18	27	30	42
Ten years	9	16	0	0	6	11	11	18	6	7	8	15	21	34	7	11	16	26	15	28	7	11	13	25

[1] Including those for whom no deprivation category was assigned. C - crude survival, R - relative survival (see Chapter 3).
Number of patients contributing to each analysis in parentheses.

	1981-85 Affluent (C R)	2 (C R)	3 (C R)	4 (C R)	Deprived (C R)	All¹ (C R)	1986-90 Affluent (C R)	2 (C R)	3 (C R)	4 (C R)	Deprived (C R)	All¹ (C R)	ADULTS
	(1,222)	*(1,303)*	*(1,285)*	*(1,163)*	*(951)*	*(6,038)*	*(1,284)*	*(1,462)*	*(1,446)*	*(1,156)*	*(919)*	*(6,326)*	**ENGLAND & WALES**
One year	**71 77**	**67 73**	**66 72**	**67 73**	**63 68**	**67 73**	**74 80**	**75 80**	**71 76**	**68 74**	**65 70**	**71 77**	*One year*
Five years	**40 54**	**35 48**	**33 46**	**32 44**	**31 44**	**34 47**	**41 56**	**39 53**	**36 49**	**33 47**	**32 45**	**37 51**	*Five years*
	20 35	**15 28**	**14 26**	**15 28**	**12 22**	**15 28**							
	(1,188)	*(1,249)*	*(1,209)*	*(1,085)*	*(905)*	*(5,722)*	*(1,234)*	*(1,406)*	*(1,370)*	*(1,066)*	*(873)*	*(5,995)*	**ENGLAND**
	71 77	67 73	66 72	67 73	63 69	**67 73**	75 81	75 80	71 77	68 73	66 71	**72 77**	*One year*
	41 54	35 48	33 46	32 44	31 44	**34 47**	42 57	39 53	36 49	34 48	33 45	**37 51**	*Five years*
	20 35	15 27	14 26	15 27	12 22	**15 28**							
	(117)	*(161)*	*(172)*	*(209)*	*(231)*	*(902)*	*(135)*	*(190)*	*(188)*	*(188)*	*(206)*	*(911)*	**Northern & Yorkshire**
	70 76	66 71	68 74	65 71	67 72	**67 73**	77 83	78 85	74 82	72 77	67 72	**73 79**	*One year*
	41 56	32 43	36 52	31 45	32 44	**34 47**	48 68	41 58	36 51	33 47	33 48	**38 53**	*Five years*
	20 36	*12 22*	*10 20*	*13 26*	*11 20*	**12 23**							
	(100)	*(160)*	*(161)*	*(200)*	*(116)*	*(739)*	*(88)*	*(139)*	*(150)*	*(149)*	*(150)*	*(678)*	**Trent**
	76 83	61 66	68 74	69 77	68 73	**68 74**	69 75	74 80	69 75	68 74	65 71	**69 75**	*One year*
	43 57	36 51	37 53	31 45	30 44	**35 49**	*31 45*	41 60	32 48	36 51	32 44	**35 49**	*Five years*
	27 48	*12 24*	*16 30*	*14 28*	*9 19*	**15 28**							
	(166)	*(155)*	*(132)*	*(59)*	*(24)*	*(543)*	*(149)*	*(187)*	*(129)*	*(82)*	*(15)*	*(569)*	**Anglia & Oxford**
	65 70	66 71	63 69	57 61	61 67	**63 69**	66 73	64 69	68 73	58 63	67 73	**65 70**	*One year*
	33 44	32 47	28 39	21 30	25 34	**30 42**	32 48	28 40	23 33	27 36	20 27	**28 39**	*Five years*
	17 33	12 25	10 21	*15 29*	*4 8*	**13 26**							
	(95)	*(95)*	*(85)*	*(59)*	*(52)*	*(392)*	*(122)*	*(133)*	*(149)*	*(111)*	*(82)*	*(609)*	**North Thames**
	76 82	78 83	*67 72*	78 83	75 81	**75 81**	78 84	81 86	80 86	74 80	77 82	**78 83**	*One year*
	49 65	*38 52*	*34 45*	*43 58*	*38 57*	**40 54**	*43 61*	39 53	42 57	47 65	40 54	**42 57**	*Five years*
	24 38	*20 36*	*22 36*	*24 49*	*16 30*	**21 36**							
	(214)	*(189)*	*(159)*	*(108)*	*(39)*	*(732)*	*(243)*	*(199)*	*(159)*	*(112)*	*(46)*	*(774)*	**South Thames**
	76 82	77 84	69 74	68 75	65 71	**72 79**	83 89	81 88	75 82	77 83	73 80	**79 85**	*One year*
	45 60	42 58	39 54	*44 62*	*52 66*	**43 59**	53 69	43 59	43 59	*35 50*	38 53	**44 60**	*Five years*
	25 45	20 38	15 31	*25 46*	*30 54*	**21 41**							
	(204)	*(246)*	*(224)*	*(122)*	*(46)*	*(864)*	*(271)*	*(325)*	*(336)*	*(157)*	*(60)*	*(1,154)*	**South & West**
	72 78	62 67	61 67	60 66	59 63	**64 69**	76 82	81 86	73 80	67 74	72 80	**75 81**	*One year*
	40 54	32 45	30 42	27 38	32 45	**33 45**	44 61	49 68	42 59	43 62	48 68	**45 62**	*Five years*
	17 31	14 27	15 29	*11 23*	14 26	**14 27**							
	(131)	*(108)*	*(138)*	*(157)*	*(125)*	*(666)*	*(110)*	*(107)*	*(125)*	*(113)*	*(112)*	*(567)*	**West Midlands**
	77 83	74 80	79 85	76 83	59 64	**73 80**	77 83	69 75	67 73	66 73	71 76	**70 76**	*One year*
	43 60	*40 54*	36 50	35 49	28 41	**36 50**	37 49	*29 43*	36 52	26 40	*37 55*	**33 46**	*Five years*
	16 31	*17 31*	13 26	12 23	14 25	**14 26**							
	(161)	*(135)*	*(138)*	*(171)*	*(272)*	*(884)*	*(116)*	*(126)*	*(134)*	*(154)*	*(202)*	*(733)*	**North & West**
	61 66	64 68	57 62	61 66	58 64	**60 66**	68 74	65 70	57 61	60 65	54 59	**60 65**	*One year*
	35 48	29 41	26 38	27 39	28 41	**29 41**	*35 50*	27 40	22 32	23 38	22 33	**25 37**	*Five years*
	17 33	*14 33*	12 23	13 26	10 21	**13 25**							
	(34)	*(54)*	*(76)*	*(78)*	*(46)*	*(316)*	*(50)*	*(56)*	*(76)*	*(90)*	*(46)*	*(331)*	**WALES**
	64 71	*67 72*	*63 69*	*64 71*	*56 58*	**65 71**	*52 56*	*64 70*	*61 67*	*70 77*	*57 63*	**63 69**	*One year*
	29 39	*39 52*	*32 49*	*30 44*	*32 42*	**33 47**	*19 30*	*27 39*	*31 47*	*30 45*	*27 42*	**29 42**	*Five years*
	14 29	*22 38*	*12 32*	*18 37*	*11 20*	**16 32**							

¹ Including those for whom no deprivation category was assigned. C - crude survival, R - relative survival (see Chapter 3).
Number of patients contributing to each analysis in parentheses.

Fig. 45.1   Relative survival up to ten years: trends by calendar period of diagnosis (1971-90) and NHS Region

ADULTS

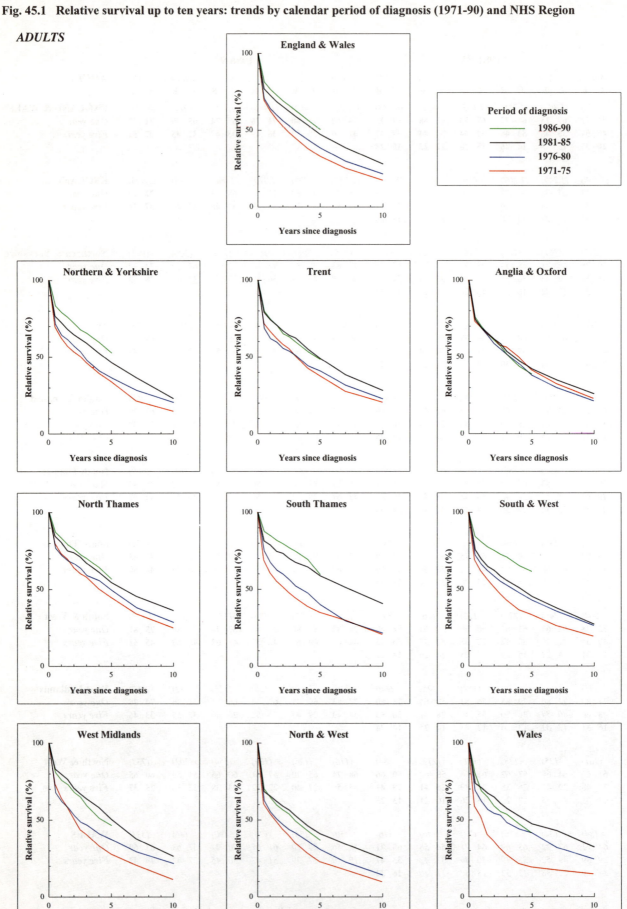

Fig. 45.2   Relative survival up to five years, by deprivation category and NHS Region: patients diagnosed 1986-90

ADULTS

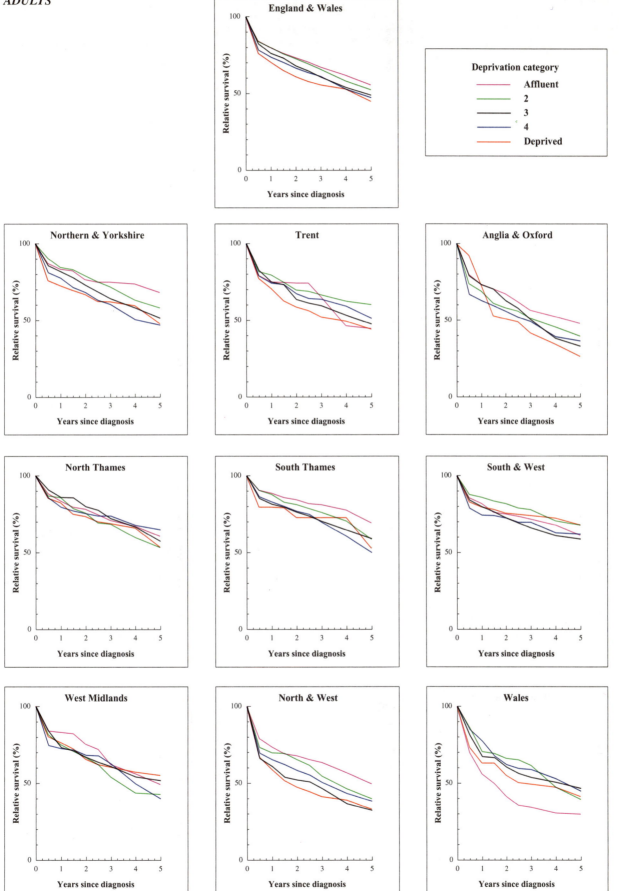

Fig. 45.3 Relative survival at five years by deprivation category, period of diagnosis (1981-90) and NHS Region

ADULTS

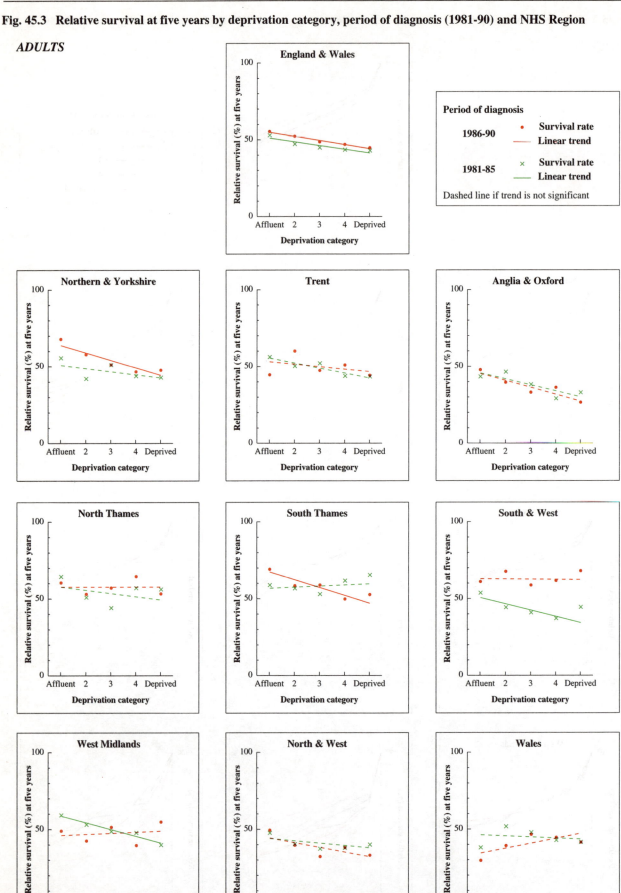

Fig. 45.4 Relative survival (%) by NHS Region and sex:
         one-year and five-year survival for patients diagnosed 1986-90, with 95% confidence intervals

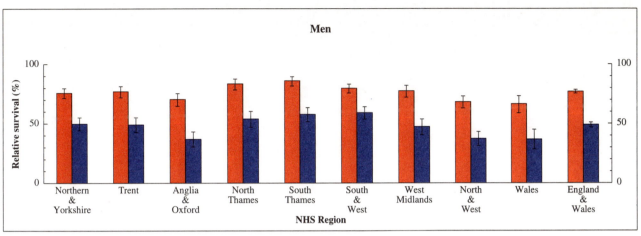

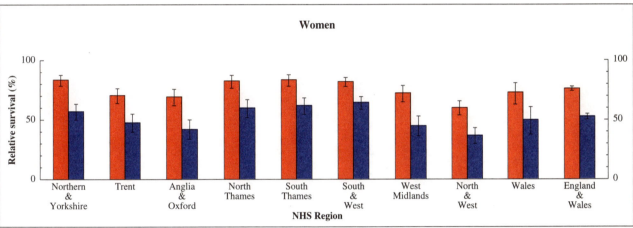

Fig. 45.5 Relative survival (%) at five years, with 95% confidence interval, by age at diagnosis and sex:
         England and Wales, patients diagnosed 1986-90

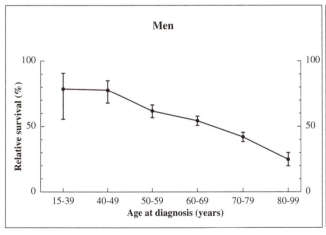
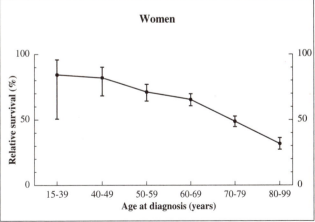

Chapter 46

ACUTE MYELOID LEUKAEMIA

Acute myeloid leukaemia accounts for about 30% of all leukaemias in adults, and about 1,500 new cases are diagnosed in England and Wales each year, about 0.7% of all cancers. Overall, acute myeloid leukaemia is marginally less common than chronic lymphoid leukaemia, the most common leukaemia in adults, but it is five times more frequent under the age of 50. It is about twice as frequent as chronic myeloid leukaemia at all ages.

Less than 4% of cases occur in children. Acute myeloid leukaemia is more common in men than women, with a sex ratio of 1.4 to 1; age-standardised (Europe) incidence rates in England and Wales were 3.8 per 100,000 in men and 2.7 in women in 1990. Incidence has risen by about 70% in both sexes since 1971. Mortality also rose between 1971 and 1995, but to a lesser degree, by 40% for men and 26% for women. In 1995 there were nearly 1,500 deaths from acute myeloid leukaemia. Survival from all leukaemias combined is discussed in Chapter 49.

The survival analyses reported here are based on 19,000 adults diagnosed with acute lymphoid leukaemia in England and Wales between 1971 and 1990 and followed up to the end of 1995. After the exclusion of children under 15 at diagnosis, some 4% of all eligible cases, these patients represented about 83% of eligible adults (Table 46.1). About 4% of adults were excluded because their vital status was not known at 6 July 1997, and a further 12% because their survival was zero or unknown. Since 1979, when a separate category became available in the International Classification of Diseases, about 2.5% of these leukaemias have been coded as acute promyelocytic leukaemia.

Survival trends

Relative survival for adults diagnosed with acute myeloid leukaemia in England and Wales during 1986-90 was 24% at one year and 8% at five years, more than double the extremely low survival rates for patients diagnosed in the early 1970s: 11% at one year and 2% at five years (Table 46.2). The gains in survival are shown graphically in Figure 46.1.

The relative survival curves reach a plateau two to three years after diagnosis, and subsequent mortality is not much higher than that of the general population from which the leukaemia patients are drawn. Relative survival ten years after diagnosis is on average only 1-2% lower than at five years.

One-year survival improved steadily by about 5% between successive quinquennia in both men and women, rising from 10% for patients diagnosed in the early 1970s to 25% for those diagnosed up to 1990, after adjustment for differences in the age distribution of patients over time and between the sexes (see Table 4.4). Five-year survival also improved substantially, but less quickly than for survival at one year, and the average gain was 2-3% every five years. Most (4%) of the overall 6-7% increase in the national average five-year survival rate occurred between 1976-80 and 1981-85. One-year survival continued to increase during the 1980s, but five-year survival barely changed.

In North Thames, five-year survival fell in men between the early and late 1980s, and did not improve in women (Table 46.2). The regional estimates of survival were based on 200 or more cases for each sex in each period, but a fall in the proportion of cases registered solely from a death certificate in this region may partly explain this apparent anomaly. Five-year survival in other regions ranges from 5% to 10% (Figure 46.4).

Survival declines with age to very low levels (Figure 46.5). Overall survival at five years differs very little between men and women, but women diagnosed during 1986-90 under the age of 40 had a significant survival advantage over men (37% compared to 25%). Relative survival at five years is the same in men and women at age 50 or over, and it is only 2% or less for those aged 70 or more when diagnosed (see Table 4.7).

Deprivation

Relative survival at one year has generally been highest in the most affluent group of patients, but there are no consistent differences in five-year survival (Table 46.2). Survival curves for patients in the various deprivation groups who were diagnosed during 1986-90 are not generally distinct, except for Anglia & Oxford (Figure 46.2).

Among 5,000 adults of known deprivation status who were diagnosed with acute myeloid leukaemia in England and Wales during 1981-85, there was a significant deprivation gap of 7% in one-year survival between the most affluent and most deprived groups (see Table 4.9). The deprivation gap in one-year survival was still 7% among more than 5,000 adults diagnosed during 1986-90.

The small deprivation gradients in five-year survival in England and Wales as a whole are not statistically significant in either the early or late 1980s (Figure 46.3 – note expanded vertical scale). The regional differences in five-year survival between deprivation groups are not generally significant either.

International comparison

Five-year survival in England and Wales (8-9%) was 2-4% lower than the average for Europe and the USA (see Table 4.12).

Table 46.1 Ineligible and excluded records, and patients included in analyses, by calendar period of diagnosis

	Calendar period of diagnosis														
	1971-75			1976-80			1981-85			1986-90			Overall		
Total registered	**4,732**			**5,528**			**6,492**			**7,400**			**24,152**		
Ineligible	*Records*	*Patients*	*%*	*Records*	*Patients*	*%*	*Records*	*Patients*	*%*	*Records*	*Patients*	*%*	*Records*	*Patients*	*%*
Incomplete data[1]	3	3		4	4		10	10		11	11		28	**28**	
Resident outside England and Wales	52	51		67	67		59	52		29	19		207	**189**	
In situ neoplasm[2]	0	0		0	0		0	0		0	0		0	**0**	
Benign or uncertain behaviour[2]	0	0		0	0		0	0		0	0		0	**0**	
Metastatic[3]	-	-		-	-		-	-		-	-		-	**-**	
Otherwise ineligible[3]	-	-		-	-		-	-		-	-		-	**-**	
Lymphoma[4]	-	-		-	-		-	-		-	-		-	**-**	
Leukaemia or myeloma[4]	-	-		-	-		-	-		-	-		-	**-**	
		54			**71**			**62**			**30**			**217**	
Total eligible		**4,678**	**100**		**5,457**	**100**		**6,430**	**100**		**7,370**	**100**		**23,935**	**100**
Aged under 15 or 100 years or over at diagnosis	281	270	5.8	238	228	4.2	254	242	3.8	255	255	3.5	1,028	**995**	4.2
Vital status unknown[5]	180	142	3.0	237	187	3.4	285	231	3.6	351	310	4.2	1,053	**870**	3.6
Duplicate registration[6]	0	0	0.0	0	0	0.0	0	0	0.0	0	0	0.0	0	**0**	0.0
Multiple primary[7]	0	0	0.0	1	0	0.0	6	6	<0.1	10	8	0.1	17	**14**	<0.1
Synchronous tumours[8]	6	5	0.1	5	4	<0.1	9	6	<0.1	20	14	0.2	40	**29**	0.1
Sex not known	0	0	0.0	0	0	0.0	0	0	0.0	0	0	0.0	0	**0**	0.0
Sex incompatible with site code[9]	-	-	-	-	-	-	-	-	-	-	-	-	-	**-**	-
Invalid dates[10]	0	0	0.0	0	0	0.0	3	2	<0.1	4	2	<0.1	7	**4**	<0.1
Zero survival[11]	361	323	6.9	536	484	8.9	1,003	914	14.2	1,231	1,155	15.7	3,131	**2,876**	12.0
Total exclusions		**740**	**15.8**		**903**	**16.5**		**1,401**	**21.8**		**1,744**	**23.7**		**4,788**	**20.0**
Patients included in analysis		**3,938**	**84.2**		**4,554**	**83.5**		**5,029**	**78.2**		**5,626**	**76.3**		**19,147**	**80.0**

[1] Main data item(s) invalid or incompatible with one another: name, sex, date of birth, date of diagnosis and (if present) date of death, postcode, site, morphology and behaviour

[2] Tumour either *in situ* (behaviour code 2), benign (0), uncertain if benign or malignant (1) or metastatic (6 or 9)

[3] Not applicable for this malignancy

[4] Not applicable for this malignancy

[5] Tracing of vital status at National Health Service Central Register incomplete at 6 July 1997

[6] Same site code, sex, cancer registry and cancer registry number as an earlier registration

[7] Same site code and person number as an earlier registration(s): mostly confirmed multiple primary tumours at this site, some unresolved duplicate registrations

[8] Same site code, sex, date of birth and date of diagnosis as another registration(s): mostly synchronous or (in paired organs) bilateral tumours in same anatomic site in one individual, not previously linked: also some duplicate registrations

[9] Not applicable for this malignancy

[10] Impossible sequence of dates (birth, diagnosis and, if present, emigration or death); or date of death after 6 July 1997

[11] Date of diagnosis same as date of death: some are patients who did die on the day of diagnosis, but most were registered solely from a death certificate (death certificate only, or DCO), and their survival time is thus unknown

Table 46.2   Crude and relative survival (%) at one, five and ten years since diagnosis: NHS Region, deprivation category and calendar period of diagnosis

MEN	1971-75						1976-80					
	Affluent	2	3	4	Deprived	All[1]	Affluent	2	3	4	Deprived	All[1]
	C R	C R	C R	C R	C R	C R	C R	C R	C R	C R	C R	C R
ENGLAND & WALES	*(359)*	*(328)*	*(382)*	*(338)*	*(276)*	*(2,044)*	*(473)*	*(468)*	*(484)*	*(435)*	*(365)*	*(2,312)*
One year	14 15	9 9	12 12	12 13	10 10	11 12	15 16	16 16	14 15	14 14	14 15	15 16
Five years	2 2	2 2	2 2	2 2	2 2	2 2	2 3	2 2	4 5	3 3	4 4	3 3
Ten years	2 2	1 2	*1 2*	1 2	1 1	1 2	2 2	2 2	2 3	2 3	3 3	2 3

	1981-85						1986-90					
	(583)	*(546)*	*(528)*	*(509)*	*(375)*	*(2,603)*	*(614)*	*(649)*	*(675)*	*(522)*	*(422)*	*(2,915)*
One year	20 21	20 21	21 23	21 22	16 17	20 21	28 29	26 27	22 23	18 19	24 25	24 25
Five years	5 6	6 7	7 7	7 9	5 7	6 7	7 8	7 8	7 8	5 6	8 9	7 8
Ten years	4 4	4 5	5 7	5 7	4 5	4 6						

	1971-75	1976-80	1981-85	1986-90
Northern & Yorkshire	*(225)*	*(281)*	*(322)*	*(370)*
One year	9 9	13 13	16 17	18 18
Five years	1 2	3 3	6 7	6 7
Ten years	0 0	3 3	4 5	
Trent	*(170)*	*(214)*	*(332)*	*(319)*
One year	9 9	13 14	21 23	22 23
Five years	2 2	4 4	8 10	6 7
Ten years	2 2	2 3	6 8	
Anglia & Oxford	*(193)*	*(263)*	*(252)*	*(309)*
One year	13 14	16 17	13 14	22 23
Five years	3 3	2 2	3 3	6 7
Ten years	2 2	1 2	2 3	
North Thames	*(276)*	*(210)*	*(211)*	*(329)*
One year	16 16	22 23	34 36	34 36
Five years	3 4	5 6	16 19	9 11
Ten years	2 3	5 6	11 15	
South Thames	*(290)*	*(356)*	*(376)*	*(300)*
One year	12 12	13 14	21 22	30 32
Five years	1 1	2 2	5 6	7 8
Ten years	1 1	1 1	4 5	
South & West	*(249)*	*(344)*	*(350)*	*(436)*
One year	9 9	16 16	20 22	26 27
Five years	2 2	3 3	3 4	7 8
Ten years	2 2	2 2	2 3	
West Midlands	*(228)*	*(223)*	*(286)*	*(238)*
One year	12 13	12 13	19 20	18 19
Five years	1 1	3 4	5 6	5 6
Ten years	*1 1*	2 2	3 4	
North & West	*(293)*	*(299)*	*(356)*	*(419)*
One year	9 9	13 14	19 20	23 24
Five years	1 1	3 3	5 7	6 8
Ten years	0 1	2 2	4 6	
WALES	*(120)*	*(122)*	*(118)*	*(195)*
One year	11 12	20 21	21 22	21 22
Five years	2 2	5 6	6 8	10 11
Ten years	1 2	5 5	6 8	

[1] Including those for whom no deprivation category was assigned. C - crude survival, R - relative survival (see Chapter 3).
Number of patients contributing to each analysis in parentheses.

Table 46.2   Crude and relative survival (%) at one, five and ten years since diagnosis: NHS Region, deprivation category and calendar period of diagnosis

WOMEN	1971-75						1976-80					
	Affluent	2	3	4	Deprived	All[1]	Affluent	2	3	4	Deprived	All[1]
	C R	C R	C R	C R	C R	C R	C R	C R	C R	C R	C R	C R
ENGLAND & WALES	(317)	(340)	(310)	(321)	(264)	(1,894)	(417)	(489)	(470)	(407)	(369)	(2,242)
One year	14 14	11 11	8 8	10 10	8 8	10 10	17 17	16 16	13 13	16 16	10 10	14 15
Five years	2 3	2 2	2 2	2 2	1 1	2 2	4 4	3 4	3 3	4 5	2 3	3 4
Ten years	2 2	1 1	1 1	1 2	*1 1*	1 1	3 4	3 3	2 3	3 4	2 2	3 3

	1981-85						1986-90					
	(492)	(485)	(540)	(493)	(357)	(2,426)	(542)	(575)	(605)	(505)	(460)	(2,711)
One year	25 26	24 24	19 20	19 20	14 14	20 21	28 28	26 27	18 19	21 22	22 23	23 24
Five years	8 9	7 8	7 8	6 6	6 6	7 8	11 13	8 9	6 6	6 7	9 10	8 9
Ten years	6 7	6 7	6 7	4 5	4 5	5 6						

	1971-75	1976-80	1981-85	1986-90
Northern & Yorkshire	(198)	(271)	(301)	(386)
One year	6 7	9 9	15 16	15 15
Five years	0 0	1 1	5 6	6 6
Ten years	*0 0*	1 1	4 5	
Trent	(142)	(241)	(252)	(319)
One year	10 10	13 13	19 20	18 18
Five years	2 2	4 4	8 9	8 8
Ten years	2 2	2 3	6 8	
Anglia & Oxford	(198)	(248)	(247)	(299)
One year	11 11	18 19	13 14	23 24
Five years	1 2	3 4	4 5	8 9
Ten years	1 1	3 3	3 4	
North Thames	(238)	(247)	(210)	(272)
One year	18 19	23 24	32 33	29 30
Five years	3 4	8 9	12 13	12 13
Ten years	2 4	7 8	9 10	
South Thames	(318)	(311)	(302)	(284)
One year	8 8	13 13	22 23	32 33
Five years	2 2	3 4	7 8	11 12
Ten years	1 2	*3 3*	5 6	
South & West	(222)	(306)	(386)	(375)
One year	9 9	18 19	25 26	23 24
Five years	0 1	4 4	9 9	6 7
Ten years	0 0	3 3	6 8	
West Midlands	(175)	(200)	(287)	(231)
One year	8 9	10 10	18 19	22 23
Five years	1 1	2 3	4 4	5 5
Ten years	0 0	1 2	2 3	
North & West	(291)	(307)	(330)	(392)
One year	7 7	12 13	18 18	26 27
Five years	1 1	3 4	6 7	10 11
Ten years	0 0	2 3	5 5	
WALES	(112)	(111)	(111)	(153)
One year	20 20	12 12	23 24	20 21
Five years	4 4	2 2	10 11	8 9
Ten years	3 4	1 2	*8 10*	

[1]   Including those for whom no deprivation category was assigned. C - crude survival, R - relative survival (see Chapter 3). Number of patients contributing to each analysis in parentheses.

Table 46.2   Crude and relative survival (%) at one, five and ten years since diagnosis: NHS Region, deprivation category and calendar period of diagnosis

ADULTS		1971-75										1976-80										
		Affluent		2		3		4		Deprived		All[1]		Affluent		2		3		4		Deprived		All[1]
		C R		C R		C R		C R		C R		C R		C R		C R		C R		C R		C R
ENGLAND & WALES		*(676)*		*(668)*		*(692)*		*(659)*		*(540)*		*(3,938)*		*(890)*		*(957)*		*(954)*		*(842)*		*(734)*		*(4,554)*
	One year	14 14		10 10		10 10		11 12		9 9		11 11		16 16		16 16		13 14		15 15		12 12		15 15
	Five years	2 2		2 2		2 2		2 2		1 2		2 2		3 4		3 3		3 4		3 4		3 3		3 4
	Ten years	2 2		1 1		1 2		1 2		1 1		1 1		2 3		2 3		2 3		3 3		2 3		2 3
ENGLAND		*(664)*		*(632)*		*(639)*		*(605)*		*(506)*		*(3,706)*		*(868)*		*(914)*		*(904)*		*(796)*		*(692)*		*(4,321)*
	One year	14 14		10 10		10 10		10 11		9 9		**10 11**		16 17		16 16		13 14		14 15		11 12		**14 15**
	Five years	2 2		1 1		2 2		2 2		1 2		**2 2**		3 4		3 3		3 4		3 4		3 3		**3 4**
	Ten years	2 2		1 1		1 2		1 2		1 1		**1 1**		2 3		2 3		2 3		3 3		2 2		**2 3**
Northern & Yorkshire		*(35)*		*(44)*		*(50)*		*(73)*		*(81)*		*(423)*		*(67)*		*(94)*		*(100)*		*(120)*		*(162)*		*(552)*
	One year	3 3		12 12		5 6		13 13		8 8		**8 8**		9 9		11 12		13 13		13 14		8 9		**11 11**
	Five years	0 0		2 2		0 0		3 3		1 1		**1 1**		2 2		1 1		1 2		2 2		2 2		**2 2**
	Ten years	0 0		2 2		0 0		0 0		*1 1*		**0 0**		2 2		*1 1*		1 1		2 2		2 2		**2 2**
Trent		*(44)*		*(44)*		*(76)*		*(73)*		*(53)*		*(312)*		*(57)*		*(85)*		*(103)*		*(116)*		*(86)*		*(455)*
	One year	12 12		12 12		6 6		14 15		6 7		**9 10**		14 15		10 10		15 15		13 14		14 14		**13 14**
	Five years	3 3		0 0		2 2		4 4		1 1		**2 2**		3 4		1 2		4 6		5 6		4 5		**4 4**
	Ten years	3 3		0 0		2 2		4 4		0 0		**2 2**		1 2		1 2		3 5		3 4		*3 4*		**2 3**
Anglia & Oxford		*(105)*		*(93)*		*(64)*		*(52)*		*(24)*		*(391)*		*(142)*		*(116)*		*(129)*		*(74)*		*(28)*		*(511)*
	One year	11 11		12 12		11 12		6 7		6 6		**12 12**		21 22		18 19		12 12		17 18		*13 13*		**17 18**
	Five years	2 2		2 2		1 1		1 1		2 2		**2 2**		3 3		3 3		3 4		2 2		0 0		**3 3**
	Ten years	1 1		1 1		*1 1*		*1 1*		2 2		**1 1**		2 3		1 2		*3 3*		2 2		0 0		**2 3**
North Thames		*(113)*		*(91)*		*(83)*		*(102)*		*(74)*		*(514)*		*(95)*		*(107)*		*(96)*		*(82)*		*(62)*		*(457)*
	One year	21 22		14 14		18 19		16 17		13 14		**17 17**		27 28		26 26		17 18		22 22		19 20		**23 23**
	Five years	3 4		4 5		1 1		4 4		3 3		**3 4**		8 9		7 8		6 6		8 8		4 4		**7 8**
	Ten years	3 4		4 5		0 0		3 3		1 1		**2 3**		8 9		6 7		5 5		8 8		1 1		**6 7**
South Thames		*(173)*		*(126)*		*(121)*		*(94)*		*(46)*		*(608)*		*(192)*		*(154)*		*(140)*		*(104)*		*(45)*		*(667)*
	One year	16 16		7 7		7 7		10 10		8 8		**10 10**		14 14		15 16		8 8		9 10		22 24		**13 14**
	Five years	2 2		1 1		1 2		1 1		4 4		**2 2**		3 3		0 1		2 2		2 2		*11 13*		**3 3**
	Ten years	1 1		1 1		*1 2*		*1 1*		2 3		**1 1**		2 2		0 1		*1 1*		2 2		7 9		**2 2**
South & West		*(103)*		*(132)*		*(117)*		*(59)*		*(22)*		*(471)*		*(159)*		*(183)*		*(159)*		*(84)*		*(34)*		*(650)*
	One year	10 11		8 9		12 13		4 5		1 1		**9 9**		16 17		14 15		19 20		25 26		8 8		**17 17**
	Five years	1 1		0 0		4 4		0 0		0 0		**1 2**		3 4		4 4		4 4		4 4		0 0		**3 4**
	Ten years	*1 1*		0 0		2 3		0 0		0 0		**1 1**		2 3		3 4		1 2		2 3		0 0		**2 3**
West Midlands		*(24)*		*(31)*		*(43)*		*(42)*		*(71)*		*(403)*		*(68)*		*(76)*		*(76)*		*(97)*		*(104)*		*(423)*
	One year	*16 17*		*14 15*		*13 14*		*14 15*		11 12		**10 11**		8 8		16 17		10 11		11 12		9 10		**11 11**
	Five years	3 3		0 0		2 2		0 0		1 1		**1 1**		0 0		3 4		6 7		2 3		1 1		**3 3**
	Ten years	3 3		0 0		2 2		0 0		*1 1*		**1 1**		0 0		3 3		4 5		1 2		0 0		**1 2**
North & West		*(67)*		*(71)*		*(85)*		*(110)*		*(135)*		*(584)*		*(88)*		*(99)*		*(101)*		*(119)*		*(171)*		*(606)*
	One year	13 14		3 4		8 8		7 7		8 8		**8 8**		18 18		16 16		13 13		11 11		10 10		**13 13**
	Five years	2 2		1 1		1 2		1 1		1 1		**1 1**		4 4		5 6		1 1		3 3		3 3		**3 3**
	Ten years	2 2		0 0		*1 1*		0 1		0 0		**0 0**		3 3		4 5		1 1		2 3		2 2		**2 3**
WALES		*(12)*		*(36)*		*(53)*		*(54)*		*(34)*		*(232)*		*(22)*		*(43)*		*(50)*		*(46)*		*(42)*		*(233)*
	One year	19 20		21 21		8 8		21 21		*14 15*		**15 15**		6 6		*14 15*		17 17		17 18		21 22		**16 17**
	Five years	8 9		9 10		0 0		3 3		0 0		**3 3**		2 2		0 0		5 5		2 3		8 8		**4 4**
	Ten years	8 9		7 8		0 0		*1 2*		0 0		**2 3**		0 0		0 0		3 3		2 3		8 8		**3 3**

[1] Including those for whom no deprivation category was assigned. C - crude survival, R - relative survival (see Chapter 3).
Number of patients contributing to each analysis in parentheses.

ADULTS

Values shown as C R (C – crude survival, R – relative survival). Numbers of patients in parentheses.

Region		1981-85 Affluent	2	3	4	Deprived	All[1]	1986-90 Affluent	2	3	4	Deprived	All[1]
ENGLAND & WALES	(n)	(1,075)	(1,031)	(1,068)	(1,002)	(732)	(5,029)	(1,156)	(1,224)	(1,280)	(1,027)	(882)	(5,626)
	One year	23 23	22 23	20 21	20 21	15 16	20 21	28 29	26 27	20 21	20 21	23 24	23 24
	Five years	7 7	6 7	7 8	6 7	6 7	6 7	9 10	7 8	6 7	6 6	9 10	7 8
		5 5	5 6	5 7	4 6	4 5	5 6						
ENGLAND	(n)	(1,047)	(999)	(1,016)	(935)	(703)	(4,800)	(1,112)	(1,162)	(1,199)	(932)	(828)	(5,278)
	One year	23 24	22 23	20 21	20 20	15 15	20 21	28 29	26 27	21 21	20 21	23 24	24 25
	Five years	7 7	6 7	7 7	6 7	6 6	6 7	9 10	7 8	6 7	6 7	9 10	7 8
		5 5	5 6	5 7	4 6	4 5	5 6						
Northern & Yorkshire	(n)	(77)	(119)	(121)	(143)	(152)	(623)	(121)	(112)	(178)	(164)	(180)	(756)
	One year	16 16	24 25	16 17	14 15	10 10	15 16	17 18	15 16	11 12	15 16	22 23	16 17
	Five years	5 6	7 8	5 5	5 6	5 6	5 6	8 9	8 10	3 4	4 4	7 9	6 7
		3 3	5 6	4 5	4 6	5 5	4 5						
Trent	(n)	(116)	(92)	(141)	(135)	(93)	(584)	(92)	(129)	(152)	(132)	(131)	(638)
	One year	25 26	26 27	20 21	20 21	12 13	20 21	23 24	28 29	18 19	16 17	17 18	20 21
	Five years	10 11	11 12	7 8	11 13	3 4	8 10	8 9	10 11	7 8	6 7	4 5	7 8
		7 9	10 12	5 7	5 7	3 4	6 8						
Anglia & Oxford	(n)	(134)	(125)	(128)	(80)	(23)	(499)	(177)	(174)	(145)	(75)	(33)	(608)
	One year	20 21	11 12	11 12	8 8	17 18	13 14	29 31	23 24	21 21	13 14	14 14	22 23
	Five years	6 7	3 3	2 2	2 2	6 7	3 4	11 13	5 5	7 8	3 4	4 4	7 8
		4 5	3 3	2 2	2 2	6 7	3 3						
North Thames	(n)	(105)	(92)	(87)	(80)	(53)	(421)	(113)	(130)	(125)	(112)	(108)	(601)
	One year	34 35	36 38	29 30	32 33	33 34	33 34	34 35	32 33	33 34	28 29	32 33	32 33
	Five years	17 18	15 17	11 12	11 13	16 17	14 16	10 12	10 11	7 9	13 16	12 14	10 12
		13 15	10 12	8 10	8 11	8 11	10 12						
South Thames	(n)	(205)	(157)	(128)	(115)	(41)	(678)	(174)	(142)	(110)	(88)	(57)	(584)
	One year	23 24	17 18	21 22	30 32	16 17	21 22	37 38	33 35	18 19	30 32	39 40	31 32
	Five years	4 5	5 6	6 7	9 10	7 7	6 7	12 14	10 12	5 6	6 7	11 13	9 10
		3 3	5 5	5 6	7 8	4 6	4 5						
South & West	(n)	(188)	(202)	(189)	(101)	(31)	(736)	(196)	(231)	(231)	(114)	(35)	(811)
	One year	26 27	23 24	20 21	21 22	26 28	23 24	27 28	25 26	23 24	23 24	20 20	24 25
	Five years	6 6	5 5	9 10	3 4	8 9	6 7	7 7	4 5	8 9	9 11	11 12	7 8
		4 5	3 4	7 9	3 4	2 3	4 5						
West Midlands	(n)	(85)	(112)	(124)	(128)	(118)	(573)	(87)	(90)	(113)	(92)	(83)	(469)
	One year	15 16	19 20	23 25	17 18	16 17	18 19	27 28	18 18	22 23	17 17	16 17	20 21
	Five years	2 3	3 4	6 6	5 5	4 5	4 5	5 6	2 3	7 8	1 2	8 8	5 5
		2 2	2 2	6 6	2 2	3 3	3 4						
North & West	(n)	(137)	(100)	(98)	(153)	(192)	(686)	(152)	(154)	(145)	(155)	(201)	(811)
	One year	19 20	24 25	25 26	18 19	12 13	18 19	27 28	31 32	21 22	20 21	25 26	25 26
	Five years	6 6	7 9	9 10	5 7	4 5	6 7	9 10	10 11	5 7	4 5	11 13	8 9
		2 3	7 8	7 10	4 7	3 5	4 6						
WALES	(n)	(28)	(32)	(52)	(67)	(29)	(229)	(44)	(62)	(81)	(95)	(54)	(348)
	One year	15 16	24 25	24 25	25 26	25 26	22 23	25 27	24 26	16 16	19 20	21 22	21 22
	Five years	5 5	11 11	9 9	8 9	6 9	8 10	15 16	9 10	10 10	5 6	6 7	9 10
		5 5	4 5	9 9	6 8	6 9	7 9						

[1] Including those for whom no deprivation category was assigned. C - crude survival, R - relative survival (see Chapter 3).
Number of patients contributing to each analysis in parentheses.

Fig. 46.1   Relative survival up to ten years: trends by calendar period of diagnosis (1971-90) and NHS Region

ADULTS

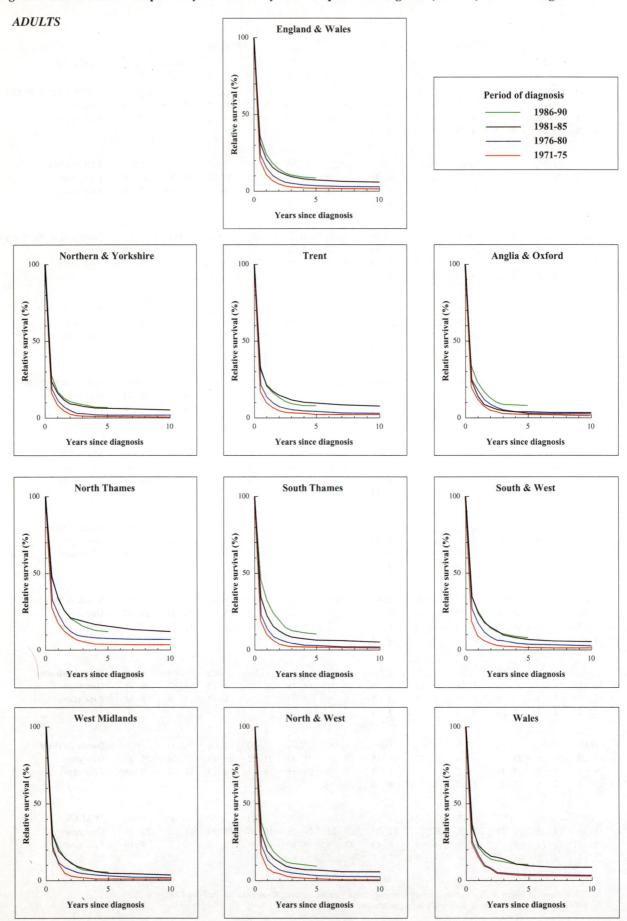

Fig. 46.2 Relative survival up to five years, by deprivation category and NHS Region: patients diagnosed 1986-90

ADULTS

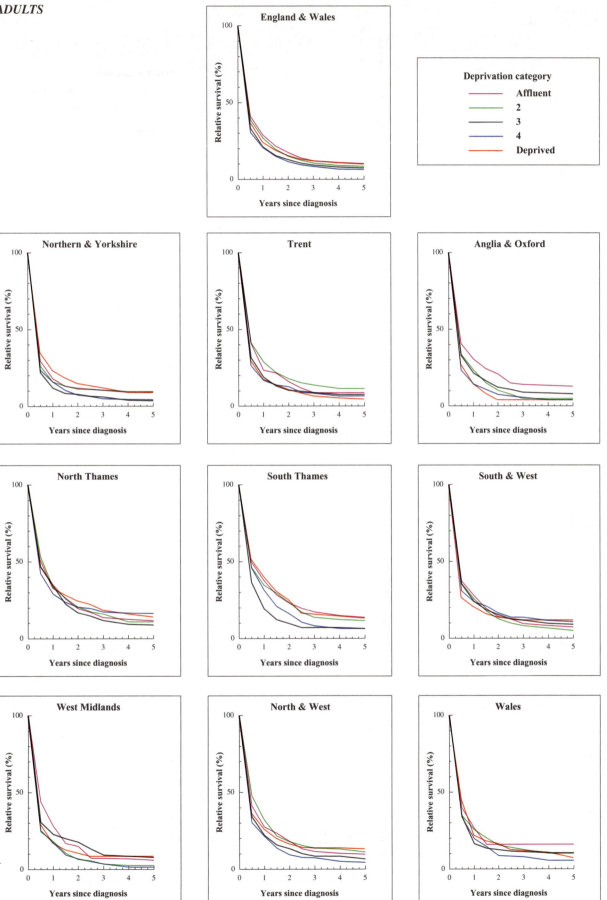

Fig. 46.3   Relative survival at five years by deprivation category, period of diagnosis (1981-90) and NHS Region

ADULTS
(Note vertical scale)

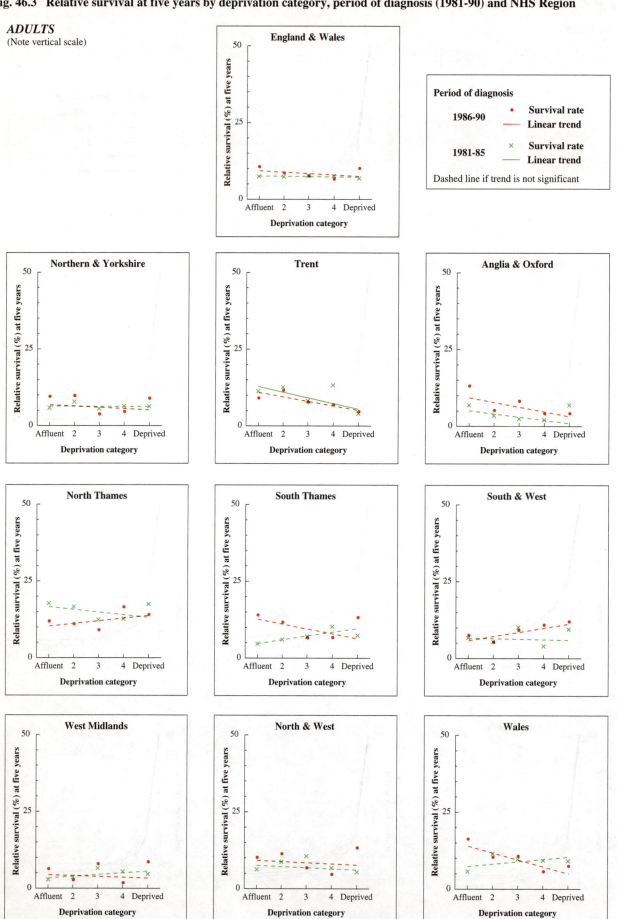

Fig. 46.4 Relative survival (%) by NHS Region and sex:
one-year and five-year survival for patients diagnosed 1986-90, with 95% confidence intervals

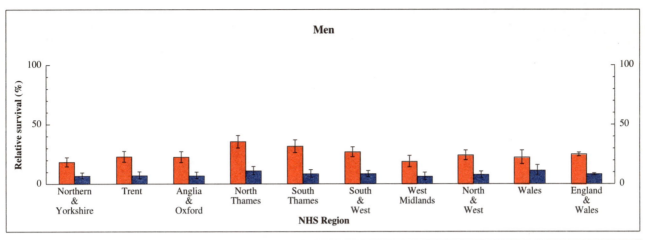

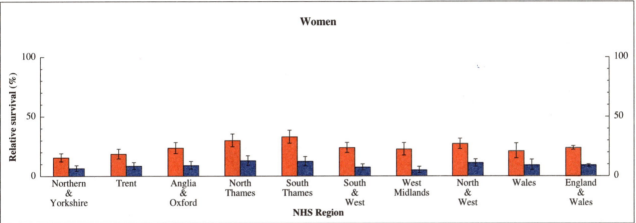

Fig. 46.5 Relative survival (%) at five years, with 95% confidence interval, by age at diagnosis and sex:
England and Wales, patients diagnosed 1986-90

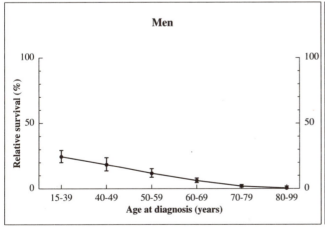

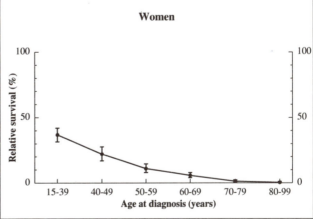

Chapter 47

CHRONIC MYELOID LEUKAEMIA

Chronic myeloid leukaemia accounts for about one in six leukaemias in adults, and nearly 800 new cases are diagnosed in England and Wales each year, somewhat less than 0.5% of all malignancies.

Chronic myeloid leukaemia is very rare in children. It is more common in men than in women, with a sex ratio of 1.5 to 1. The age-standardised (Europe) incidence rates in England and Wales in 1990 were 1.9 per 100,000 in men and 1.3 in women. Incidence has risen by about 50% in men and 30% in women since 1971. Mortality, however, has declined by around 30% in both sexes from a peak in the late 1970s. In 1995, there were just over 500 deaths from chronic myeloid leukaemia in England and Wales. Survival from all leukaemias combined is discussed in Chapter 49.

The survival analyses reported here are based on 10,000 adults diagnosed with chronic myeloid leukaemia in England and Wales between 1971 and 1990 and followed up to the end of 1995, about 84% of eligible patients (Table 47.1). About 4% of patients were excluded because their vital status was not known at 6 July 1997, and 12% because their survival was either zero or unknown.

Survival trends

Survival from chronic myeloid leukaemia for adults diagnosed in England and Wales during 1986-90 was 61% at one year, but only 22% at five years. Ten-year survival for those diagnosed during 1981-85 and followed up to the end of 1995 was only 6% (Table 47.2).

Both short-term and long-term survival have improved in England and Wales as a whole, but there was virtually no change between the early and late 1970s, and the sharpest increase occurred around 1980 (Figure 47.1). After adjustment for differences in age distribution of the patients over time and between the sexes, one-year survival rose by about 15% from the early 1970s to 60% for those diagnosed during 1986-90, an average gain of some 5-6% every five years (see Table 4.4). Five-year survival doubled from about 10% to 20-23% over the same elapsed time, representing a smaller gain of some 3-4% between successive quinquennia.

It is perhaps worth noting that while much the largest increase (8-10%) in survival one year after diagnosis occurred between 1976-80 and 1981-85 alone, the largest increase in five-year survival (5-7%) occurred some five years later, between patients diagnosed during the early 1980s and those diagnosed during 1986-90 (see Table 4.4).

This is clearly visible in Figure 47.1. The national survival curves for the early and late 1970s are virtually superposed; the curve for 1981-85 is clearly higher at all time intervals after diagnosis, while the curve for patients diagnosed during 1986-90 only becomes higher some three years after diagnosis.

The shape of the survival curves for chronic myeloid leukaemia resembles those for chronic lymphoid leukaemia (Chapter 45) and multiple myeloma (Chapter 43). One-year survival from chronic myeloid leukaemia is more than 30% higher than for acute myeloid leukaemia (61% compared to 24%), but for chronic myeloid leukaemia, excess mortality relative to that of the general population continues for up to 10 years after diagnosis. The difference in ten-year survival between acute and chronic myeloid leukaemia is negligible.

For patients diagnosed during 1986-90, one-year survival rates ranged from 49% in North & West to 72% in North Thames (Table 47.2). Five-year survival rates ranged from 18% (Trent) to 33% (Wales) (Figure 47.4).

Relative survival declines steadily with age (Figure 47.5). For patients diagnosed during 1986-90 below the age of 40, five-year survival is above 40%, but it falls steadily to below 20% among patients aged 70 or more at diagnosis (see Table 4.7).

Deprivation

Relative survival one and five years after diagnosis of chronic myeloid leukaemia in England and Wales is generally some 4-8% lower in the most deprived group of patients than in the most affluent group, although intermediate deprivation groups sometimes have higher survival (Table 47.2). For patients diagnosed during 1986-90, the survival curves for the most affluent and the most deprived groups are clearly separate, while the curves for the intermediate groups overlap (Figure 47.2).

Among 2,400 patients of known deprivation status who were diagnosed in England and Wales during 1981-85, the deprivation gap in one-year survival was 4%, but among 2,800 patients of known status who were diagnosed during 1986-90, the deprivation gap was 7%, and was statistically significant (see Table 4.9). The estimated differences between deprivation groups in survival at five years were smaller (about 3%) and were not statistically significant for either the early or late 1980s: this is reflected in Figure 47.3 (note expanded vertical scale).

International comparison

Five-year survival rates for adults diagnosed with chronic myeloid leukaemia in England and Wales during 1986-90 (22-23%) were similar to those in Scotland, but 7-9% lower than the average for Europe and the USA (see Table 4.12). The highest regional survival rates (34%; Wales, North Thames) are similar to those in the USA.

Table 47.1 Ineligible and excluded records, and patients included in analyses, by calendar period of diagnosis

	Calendar period of diagnosis														
	1971-75			**1976-80**			**1981-85**			**1986-90**			**Overall**		
Total registered	2,556			2,872			3,153			3,804			12,385		
Ineligible	*Records*	*Patients*	*%*	*Records*	*Patients*	*%*	*Records*	*Patients*	*%*	*Records*	*Patients*	*%*	*Records*	*Patients*	*%*
Incomplete data[1]	0	0		0	0		3	3		9	9		12	**12**	
Resident outside England and Wales	25	25		34	34		31	29		17	10		107	**98**	
In situ neoplasm[2]	0	0		0	0		0	0		0	0		0	**0**	
Benign or uncertain behaviour[2]	0	0		0	0		0	0		0	0		0	**0**	
Metastatic[3]	-	-		-	-		-	-		-	-		-	**-**	
Otherwise ineligible[3]	-	-		-	-		-	-		-	-		-	**-**	
Lymphoma[4]	-	-		-	-		-	-		-	-		-	**-**	
Leukaemia or myeloma[4]	-	-		-	-		-	-		-	-		-	**-**	
		25			34			32			19			110	
Total eligible		2,531	100		2,838	100		3,121	100		3,785	100		12,275	100
Aged under 15 or 100 years or over at diagnosis	51	51	2.0	39	38	1.3	39	39	1.2	50	50	1.3	179	**178**	1.5
Vital status unknown[5]	102	90	3.6	132	118	4.2	139	116	3.7	231	215	5.7	604	**539**	4.4
Duplicate registration[6]	1	1	<0.1	0	0	0.0	0	0	0.0	1	1	<0.1	2	**2**	<0.1
Multiple primary[7]	0	0	0.0	0	0	0.0	4	4	0.1	16	15	0.4	20	**19**	0.2
Synchronous tumours[8]	4	3	0.1	3	2	<0.1	7	5	0.2	13	10	0.3	27	**20**	0.2
Sex not known	0	0	0.0	0	0	0.0	0	0	0.0	0	0	0.0	0	**0**	0.0
Sex incompatible with site code[9]	-	-	-	-	-	-	-	-	-	-	-	-	-	**-**	-
Invalid dates[10]	0	0	0.0	0	0	0.0	1	0	0.0	3	2	<0.1	4	**2**	<0.1
Zero survival[11]	186	176	7.0	313	290	10.2	469	440	14.1	631	596	15.7	1,599	**1,502**	12.2
Total exclusions		321	12.7		448	15.8		604	19.4		889	23.5		2,262	18.4
Patients included in analysis		2,210	87.3		2,390	84.2		2,517	80.6		2,896	76.5		10,013	81.6

[1] Main data item(s) invalid or incompatible with one another: name, sex, date of birth, date of diagnosis and (if present) date of death, postcode, site, morphology and behaviour

[2] Tumour either *in situ* (behaviour code 2), benign (0), uncertain if benign or malignant (1) or metastatic (6 or 9)

[3] Not applicable for this malignancy

[4] Not applicable for this malignancy

[5] Tracing of vital status at National Health Service Central Register incomplete at 6 July 1997

[6] Same site code, sex, cancer registry and cancer registry number as an earlier registration

[7] Same site code and person number as an earlier registration(s): mostly confirmed multiple primary tumours at this site, some unresolved duplicate registrations

[8] Same site code, sex, date of birth and date of diagnosis as another registration(s): mostly synchronous or (in paired organs) bilateral tumours in same anatomic site in one individual, not previously linked: also some duplicate registrations

[9] Not applicable for this malignancy

[10] Impossible sequence of dates (birth, diagnosis and, if present, emigration or death); or date of death after 6 July 1997

[11] Date of diagnosis same as date of death: some are patients who did die on the day of diagnosis, but most were registered solely from a death certificate (death certificate only, or DCO), and their survival time is thus unknown

Table 47.2 Crude and relative survival (%) at one, five and ten years since diagnosis: NHS Region, deprivation category and calendar period of diagnosis

	1971-75						1976-80					
	Affluent C R	*2* C R	*3* C R	*4* C R	*Deprived* C R	*All[1]* C R	*Affluent* C R	*2* C R	*3* C R	*4* C R	*Deprived* C R	*All[1]* C R
ENGLAND & WALES												
Men	*(190)*	*(180)*	*(219)*	*(199)*	*(135)*	*(1,138)*	*(226)*	*(276)*	*(261)*	*(244)*	*(208)*	*(1,270)*
One year	52 55	51 54	47 49	48 51	47 50	49 52	55 57	51 55	47 50	48 51	42 45	49 52
Five years	11 12	8 10	11 12	9 11	10 12	10 12	12 14	10 12	14 17	10 12	10 12	11 13
Ten years	3 4	2 3	3 4	1 1	1 1	2 3	2 3	4 6	4 6	3 4	3 4	3 4
Women	*(202)*	*(175)*	*(191)*	*(171)*	*(132)*	*(1,072)*	*(216)*	*(226)*	*(234)*	*(218)*	*(169)*	*(1,120)*
One year	62 64	47 49	50 52	54 56	41 43	52 54	49 51	50 52	50 53	48 50	47 49	49 51
Five years	16 17	10 11	9 10	11 13	10 12	12 13	12 14	10 12	13 15	12 14	11 13	11 13
Ten years	4 5	2 2	3 3	1 2	1 1	2 2	4 6	3 4	2 3	3 4	2 3	3 4
Adults	*(392)*	*(355)*	*(410)*	*(370)*	*(267)*	*(2,210)*	*(442)*	*(502)*	*(495)*	*(462)*	*(377)*	*(2,390)*
One year	57 60	49 51	48 51	51 53	44 47	50 53	52 54	51 54	48 51	48 51	44 47	49 52
Five years	13 15	9 11	10 11	10 12	10 12	11 12	12 14	10 12	14 16	11 13	11 13	11 13
Ten years	4 5	2 3	3 4	1 2	1 1	2 3	3 4	3 4	3 4	3 4	2 3	3 4

	Men	*Women*	*Adults*				*Men*	*Women*	*Adults*
Northern & Yorkshire	*(160)*	*(124)*	*(284)*				*(164)*	*(157)*	*(321)*
One year	46 49	55 57	50 53				42 45	44 46	43 46
Five years	12 14	14 17	13 15				9 12	10 12	10 12
Ten years	2 3	2 2	2 3				4 5	2 3	3 4
Trent	*(98)*	*(105)*	*(203)*				*(136)*	*(101)*	*(237)*
One year	54 56	55 57	55 57				54 57	49 51	52 54
Five years	13 15	16 17	14 16				13 15	11 12	12 14
Ten years	2 3	2 2	2 2				1 1	1 2	1 2
Anglia & Oxford	*(123)*	*(96)*	*(219)*				*(154)*	*(108)*	*(262)*
One year	60 64	62 65	61 64				60 64	67 71	63 67
Five years	10 11	12 13	11 12				10 13	17 21	13 16
Ten years	2 2	0 0	1 1				4 6	4 6	4 6
North Thames	*(126)*	*(120)*	*(246)*				*(122)*	*(124)*	*(246)*
One year	54 57	58 62	56 59				58 61	52 55	55 58
Five years	11 14	15 17	13 15				16 19	6 7	11 13
Ten years	3 4	3 4	3 4				5 8	2 2	4 5
South Thames	*(142)*	*(151)*	*(293)*				*(134)*	*(133)*	*(267)*
One year	52 55	45 47	48 51				49 52	43 46	46 49
Five years	7 9	10 11	8 10				15 20	10 11	13 15
Ten years	1 2	1 2	1 2				3 5	3 4	3 4
South & West	*(157)*	*(122)*	*(279)*				*(195)*	*(160)*	*(355)*
One year	48 50	56 58	51 54				41 44	49 52	45 47
Five years	10 12	13 14	11 13				9 11	16 19	12 14
Ten years	3 4	4 5	3 4				3 4	5 7	4 5
West Midlands	*(104)*	*(100)*	*(204)*				*(118)*	*(84)*	*(202)*
One year	44 46	53 55	48 51				44 47	52 55	47 50
Five years	6 8	7 9	7 8				6 8	13 15	9 11
Ten years	*1 2*	1 2	1 2				1 1	2 2	1 1
North & West	*(156)*	*(188)*	*(344)*				*(173)*	*(181)*	*(354)*
One year	47 49	43 45	45 47				47 51	45 47	46 49
Five years	9 11	9 11	9 11				10 12	9 10	9 11
Ten years	1 1	2 3	2 2				3 3	3 3	3 3
WALES	*(72)*	*(66)*	*(138)*				*(74)*	*(72)*	*(146)*
One year	32 33	*48 49*	40 41				44 46	43 44	43 45
Five years	11 13	10 11	11 12				12 14	10 12	11 13
Ten years	1 1	*3 4*	2 2				*5 9*	*2 4*	4 6

[1] Including those for whom no deprivation category was assigned. C - crude survival, R - relative survival (see Chapter 3).
Number of patients contributing to each analysis in parentheses.

ENGLAND & WALES

	1981-85						1986-90						
	Affluent	*2*	*3*	*4*	*Deprived*	*All[1]*	*Affluent*	*2*	*3*	*4*	*Deprived*	*All[1]*	
	C R	C R	C R	C R	C R	C R	C R	C R	C R	C R	C R	C R	
	(273)	*(255)*	*(258)*	*(322)*	*(235)*	*(1,375)*	*(337)*	*(360)*	*(336)*	*(294)*	*(228)*	*(1,567)*	**Men**
One year	58 62	54 57	60 64	55 59	50 54	55 59	60 64	57 61	58 62	57 61	52 55	57 61	
Five years	14 17	15 18	16 20	13 16	12 16	14 17	18 23	15 20	19 24	17 22	14 18	17 22	
	3 4	4 7	5 7	5 7	5 8	4 6							
	(216)	*(243)*	*(237)*	*(218)*	*(199)*	*(1,142)*	*(261)*	*(291)*	*(296)*	*(275)*	*(185)*	*(1,329)*	**Women**
One year	51 53	59 62	58 63	51 54	50 53	54 57	62 65	55 58	57 60	56 59	53 57	57 60	
Five years	14 17	16 19	15 18	14 17	11 13	14 17	20 25	18 23	17 22	22 27	14 18	19 23	
	3 5	7 9	5 6	3 5	3 5	5 6							
	(489)	*(498)*	*(495)*	*(540)*	*(434)*	*(2,517)*	*(598)*	*(651)*	*(632)*	*(569)*	*(413)*	*(2,896)*	**Adults**
One year	55 58	56 60	59 63	53 57	50 53	55 58	61 64	56 60	58 61	57 60	52 56	57 61	
Five years	14 17	16 19	16 19	13 16	12 15	14 17	19 24	17 22	18 23	19 24	14 18	18 22	
	3 4	6 8	5 7	4 6	4 7	4 6							

	Men	*Women*	*Adults*	*Men*	*Women*	*Adults*	
	(205)	*(174)*	*(379)*	*(210)*	*(178)*	*(388)*	**Northern & Yorkshire**
One year	54 58	59 63	56 60	58 62	57 60	58 61	
Five years	16 20	13 16	14 18	17 22	17 22	17 22	
	5 8	3 5	4 6				
	(156)	*(108)*	*(264)*	*(154)*	*(144)*	*(298)*	**Trent**
One year	60 63	55 58	58 61	53 57	51 55	52 56	
Five years	15 19	16 18	15 19	11 15	16 21	14 18	
	5 7	7 10	6 8				
	(120)	*(97)*	*(217)*	*(187)*	*(127)*	*(314)*	**Anglia & Oxford**
One year	45 48	52 55	48 51	52 55	48 50	50 53	
Five years	11 13	12 14	11 14	14 18	15 19	15 19	
	2 4	1 1	2 2				
	(100)	*(89)*	*(189)*	*(137)*	*(111)*	*(248)*	**North Thames**
One year	60 63	59 63	59 63	67 73	66 71	67 72	
Five years	20 24	13 16	17 20	15 20	28 34	21 27	
	7 9	5 7	6 8				
	(140)	*(129)*	*(269)*	*(151)*	*(159)*	*(310)*	**South Thames**
One year	62 66	58 63	60 65	57 60	63 67	60 64	
Five years	16 21	18 23	17 22	16 21	20 26	18 24	
	6 10	5 8	6 9				
	(209)	*(166)*	*(375)*	*(313)*	*(259)*	*(572)*	**South & West**
One year	60 65	53 57	57 61	62 67	62 66	62 66	
Five years	14 18	19 22	16 20	18 23	21 27	19 25	
	4 5	9 13	6 9				
	(164)	*(164)*	*(328)*	*(141)*	*(113)*	*(254)*	**West Midlands**
One year	50 54	55 58	53 56	58 62	60 63	59 62	
Five years	11 14	18 21	14 18	18 22	15 18	17 20	
	4 6	4 6	4 6				
	(225)	*(159)*	*(384)*	*(177)*	*(161)*	*(338)*	**North & West**
One year	54 57	41 43	48 51	48 51	44 47	46 49	
Five years	12 15	6 8	9 12	17 23	13 17	15 20	
	4 6	2 2	3 4				
	(56)	*(56)*	*(112)*	*(97)*	*(77)*	*(174)*	**WALES**
One year	50 53	59 63	54 58	60 63	59 61	59 62	
Five years	8 10	10 11	9 11	26 34	25 32	26 33	
	3 4	3 3	3 4				

[1] Including those for whom no deprivation category was assigned. C - crude survival, R - relative survival (see Chapter 3). Number of patients contributing to each analysis in parentheses.

Fig. 47.1   Relative survival up to ten years: trends by calendar period of diagnosis (1971-90) and NHS Region

ADULTS

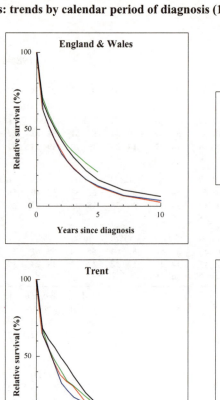

England & Wales

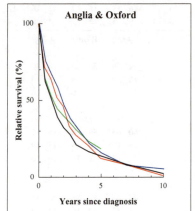

Period of diagnosis
- 1986-90
- 1981-85
- 1976-80
- 1971-75

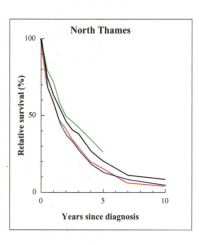

Northern & Yorkshire

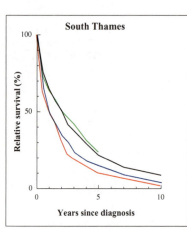

Trent

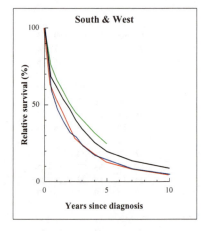

Anglia & Oxford

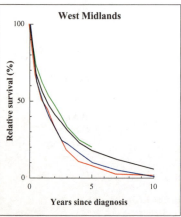

North Thames

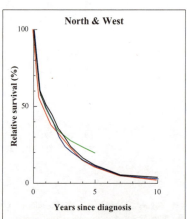

South Thames

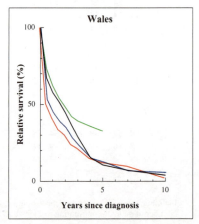

South & West

West Midlands

North & West

Wales

Fig. 47.2 Relative survival up to five years, by deprivation category: England and Wales, patients diagnosed 1986-90

ADULTS

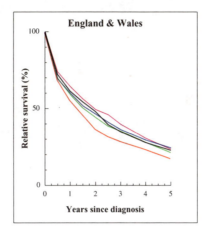

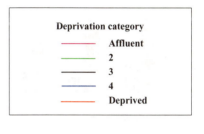

Fig. 47.3 Relative survival at five years, by deprivation category and period of diagnosis (1981-90): England and Wales

ADULTS
(Note vertical scale)

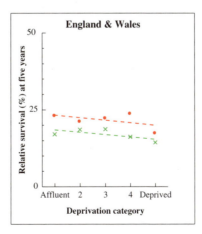

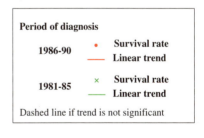

Fig. 47.4 Relative survival (%) by NHS Region and sex:
one-year and five-year survival for patients diagnosed 1986-90, with 95% confidence intervals

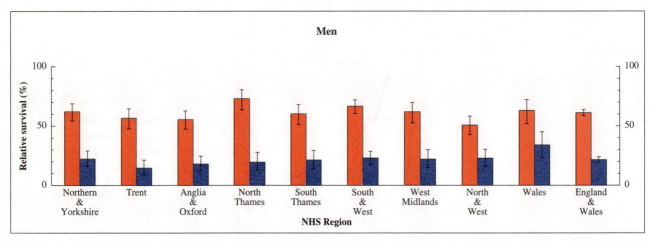

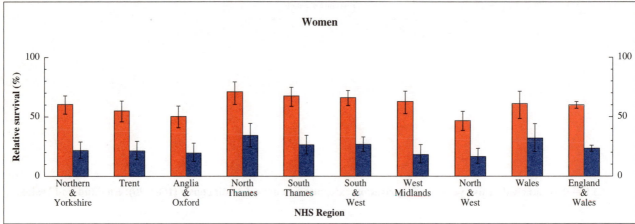

Fig. 47.5 Relative survival (%) at five years, with 95% confidence interval, by age at diagnosis and sex:
England and Wales, patients diagnosed 1986-90

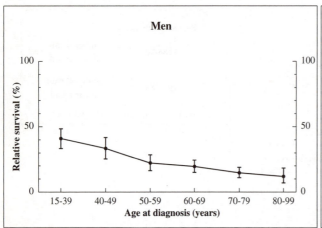

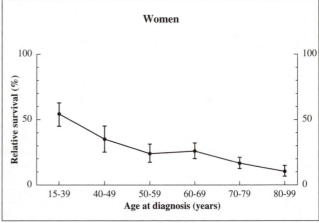

Chapter 48

MONOCYTIC LEUKAEMIA

Monocytic leukaemia is very rare, accounting for less than 3% of all leukaemias in adults, and less than 0.05% of all malignancies. In England and Wales in 1990, there were just 40 cases in men and 34 in women. Three-quarters of all cases occurred at age 60 or over. Age-specific rates were generally higher in men than in women. The recorded incidence of monocytic leukaemia in England and Wales has fallen slightly since the 1970s, and age-standardised (Europe) incidence rates in the late 1980s were about 0.2 per 100,000 in men and less than 0.1 in women. Mortality showed a similar pattern, but the decline in the 1980s was slightly steeper than for incidence. In 1995, there were 20 deaths in men and 16 in women.

In the eighth (1968) and ninth (1979) revisions of the International Classification of Diseases, under which cancer registration data have been coded, monocytic leukaemia was classified as a distinct rubric, but in 1976 the French-American-British committee[1] classified it on the basis of cell lineage and maturation as a poorly differentiated subtype of acute myeloid leukaemia (M5). Survival from all leukamias combined is discussed in Chapter 49.

The survival analyses reported here are based on almost 1,700 adults who were diagnosed with monocytic leukaemia in England and Wales during the period 1971-90 and followed up to the end of 1995. After exclusion of some 90 cases in children under 15 at diagnosis (4%), these adults represented about 84% of all eligible patients (Table 48.1). Almost 4% of adults were excluded because their vital status was unknown at 6 July 1997. A further 12% were excluded because their survival was zero or unknown, and the proportion excluded for this reason reached 16% during the 1980s.

Almost two-thirds (65%) of the monocytic leukaemias were classified as acute, and one in eight (13%) as chronic; for most of the remainder, the type was unspecified (Table 48A). All cases had been assigned to one of the morphology categories for monocytic leukaemia.

Survival trends

Relative survival from monocytic leukaemia for adults diagnosed in England and Wales during 1986-90 was just under 20% one year after diagnosis, and about 10% at five years. The corresponding survival rates for patients diagnosed during the early 1970s were 10% at one year and only 2% at five years (Table 48.2).

The improvement in short-term and medium-term survival for England and Wales as a whole is shown in Figure 48.1. Regional survival patterns also show this general trend, but there is considerable random fluctuation between successive time periods: the regional survival curves for each five-year period are typically based on fewer than 50 patients (Table 48.2). It is nevertheless clear that five-year survival was less than 5% in all regions during the early 1970s, but closer to 10% for patients diagnosed during the late 1980s. Differences in survival between the regions for patients diagnosed during 1986-90 cannot readily be interpreted (Figure 48.4).

There is only limited information about survival rates at each age. A missing value was inserted in the graph of age-specific rates for men (Figure 48.5), because all 56 men diagnosed in their seventies during the period 1986-90 had died before the fourth anniversary of diagnosis: this was set to zero in the table of age-specific rates (see Table 4.7). All the age-specific survival estimates for adults diagnosed during 1986-90 are imprecise because of small numbers of patients. It does appear safe to say that in patients under the age of 60 at diagnosis, five-year relative survival is above 10%, while for patients aged 60 or more, five-year survival is close to zero.

The sparsity of data on age-specific survival in turn complicates the quantitative estimation of survival trends. National survival trends since the early 1970s were obtained by standardising the survival rate for each sex in each five-year period to the age distribution of men and women diagnosed during 1986-90. The average change between successive five-year periods was then estimated by linear regression of the age-standardised survival rates for each period, weighted for their variance (see Chapter 3). For such an uncommon malignancy, this approach is limited, however, because the age distribution used to provide the standard weights is itself subject to considerable random fluctuation (see Table 3.4). The age-standardised rates may differ markedly from the unstandardised rates for this reason: the five-year relative survival for men diagnosed during 1971-75 was 1% (Table 48.2), but the corresponding age-standardised rate was 4% (see Table 4.4).

With this caveat, relative survival from monocytic leukaemia one year after diagnosis improved by about 8%, from about 10% in 1971-75 to 18% for patients diagnosed during 1986-90 (see Table 4.4). The average gain between successive quinquennia was 2-3%, although most of the improvement occurred between the early 1970s and the late 1970s. Between 1971-75 and 1986-90, age-

Table 48A Sub-type distribution of monocytic leukaemia

ICD-9 code	Description	1971-75 No.	1971-75 %	1976-80 No.	1976-80 %	1981-85 No.	1981-85 %	1986-90 No.	1986-90 %	Total No.	Total %
206.0	Acute	315	67.9	290	59.1	277	66.7	214	69.3	1,096	65.3
206.1	Chronic	46	9.9	51	10.4	68	16.4	50	16.2	215	12.8
206.2	Subacute	–	–	2	0.4	4	1.0	2	0.6	8	0.5
206.8	Other	–	–	0	0.0	0	0.0	0	0.0	0	0.0
206.9	Unspecified	103	22.2	148	30.1	66	15.9	43	13.9	360	21.4
	Total	**464**		**491**		**415**		**309**		**1,679**	

standardised five-year survival rose by 7% in men, and by 3% in women.

There was no systematic difference in survival between men and women.

Deprivation

Survival curves for patients diagnosed during 1986-90 in England and Wales in each of the five deprivation groups are shown in Figure 48.2. They largely overlap, and the low curve for the most deprived group is based on only 25 patients (Table 48.2).

For patients diagnosed during 1981-85, there was a significant deprivation gradient in survival. One-year survival was 11% lower for patients living in the most deprived areas than for the most affluent group, and five-year survival was 8% lower (see Table 4.9). For patients diagnosed during 1986-90, the estimated differences in

relative survival between deprivation groups were smaller, and not statistically significant. During the 1980s, five-year survival improved slightly more among patients in deprived groups than for those in the most affluent group. The change in the slope and significance of the deprivation gradient can be seen in Figure 48.3 (note expanded vertical scale).

International comparison

Five-year survival from monocytic leukaemia in England and Wales (9%) is the same as in the territories covered by the SEER programme in the USA (see Table 4.12). Monocytic leukaemia was not included in the analyses for the EUROCARE study, and no comparable data are available for Europe.

1.   Bennett JM, Catovsky D, Daniel M-T, Flandrin G, Galton DAG, Gralnick HR, Sultan C. Proposals for the classification of the acute leukaemias. French-American-British (FAB) Co-operative Group. Br J Haem 1976; 33: 451-458

Table 48.1 Ineligible and excluded records, and patients included in analyses, by calendar period of diagnosis

	Calendar period of diagnosis														
	1971-75			1976-80			1981-85			1986-90			Overall		
Total registered	569			592			552			423			2,136		
Ineligible	*Records*	*Patients*	*%*	*Records*	*Patients*	*%*	*Records*	*Patients*	*%*	*Records*	*Patients*	*%*	*Records*	*Patients*	*%*
Incomplete data[1]	3	3		2	2		5	5		2	2		12	**12**	
Resident outside England and Wales	6	5		6	6		3	3		2	1		17	**15**	
In situ neoplasm[2]	0	0		0	0		0	0		0	0		0	**0**	
Benign or uncertain behaviour[2]	0	0		0	0		0	0		0	0		0	**0**	
Metastatic[3]	-	-		-	-		-	-		-	-		-	**-**	
Otherwise ineligible[3]	-	-		-	-		-	-		-	-		-	**-**	
Lymphoma[4]	-	-		-	-		-	-		-	-		-	**-**	
Leukaemia or myeloma[4]	-	-		-	-		-	-		-	-		-	**-**	
		8			**8**			**8**			**3**			**27**	
Total eligible		**561**	**100**		**584**	**100**		**544**	**100**		**420**	**100**		**2,109**	**100**
Aged under 15 or 100 years or over at diagnosis	26	26	4.6	19	19	3.3	27	26	4.8	21	21	5.0	93	**92**	4.4
Vital status unknown[5]	19	17	3.0	25	20	3.4	19	17	3.1	26	22	5.2	89	**76**	3.6
Duplicate registration[6]	0	0	0.0	0	0	0.0	0	0	0.0	0	0	0.0	0	**0**	0.0
Multiple primary[7]	0	0	0.0	0	0	0.0	0	0	0.0	0	0	0.0	0	**0**	0.0
Synchronous tumours[8]	0	0	0.0	3	2	0.3	0	0	0.0	1	0	0.0	4	**2**	<0.1
Sex not known	0	0	0.0	0	0	0.0	0	0	0.0	0	0	0.0	0	**0**	0.0
Sex incompatible with site code[9]	-	-	-	-	-	-	-	-	-	-	-	-	-	**-**	-
Invalid dates[10]	0	0	0.0	0	0	0.0	0	0	0.0	1	1	0.2	1	**1**	<0.1
Zero survival[11]	60	54	9.6	58	52	8.9	95	86	15.8	75	67	16.0	288	**259**	12.3
Total exclusions		**97**	**17.3**		**93**	**15.9**		**129**	**23.7**		**111**	**26.4**		**430**	**20.4**
Patients included in analysis		**464**	**82.7**		**491**	**84.1**		**415**	**76.3**		**309**	**73.6**		**1,679**	**79.6**

---

[1] Main data item(s) invalid or incompatible with one another: name, sex, date of birth, date of diagnosis and (if present) date of death, postcode, site, morphology and behaviour

[2] Tumour either *in situ* (behaviour code 2), benign (0), uncertain if benign or malignant (1) or metastatic (6 or 9)

[3] Not applicable for this malignancy

[4] Not applicable for this malignancy

[5] Tracing of vital status at National Health Service Central Register incomplete at 6 July 1997

[6] Same site code, sex, cancer registry and cancer registry number as an earlier registration

[7] Same site code and person number as an earlier registration(s): mostly confirmed multiple primary tumours at this site, some unresolved duplicate registrations

[8] Same site code, sex, date of birth and date of diagnosis as another registration(s): mostly synchronous or (in paired organs) bilateral tumours in same anatomic site in one individual, not previously linked: also some duplicate registrations

[9] Not applicable for this malignancy

[10] Impossible sequence of dates (birth, diagnosis and, if present, emigration or death); or date of death after 6 July 1997

[11] Date of diagnosis same as date of death: some are patients who did die on the day of diagnosis, but most were registered solely from a death certificate (death certificate only, or DCO), and their survival time is thus unknown

Table 48.2   Crude and relative survival (%) at one, five and ten years since diagnosis: NHS Region, deprivation category and calendar period of diagnosis

		1971-75						1976-80					
		Affluent	*2*	*3*	*4*	*Deprived*	*All*[1]	*Affluent*	*2*	*3*	*4*	*Deprived*	*All*[1]
		C R	C R	C R	C R	C R	C R	C R	C R	C R	C R	C R	C R
ENGLAND & WALES													
Men		*(50)*	*(34)*	*(41)*	*(41)*	*(28)*	*(243)*	*(58)*	*(64)*	*(51)*	*(50)*	*(29)*	*(260)*
	One year	12 13	7 7	7 7	8 9	8 8	9 10	21 23	24 26	9 10	13 14	7 8	16 18
	Five years	0 0	1 2	0 0	2 2	0 0	1 1	3 4	10 13	1 1	2 3	0 0	4 5
	Ten years	0 0	1 2	0 0	2 2	0 0	1 1	3 4	7 9	1 1	0 0	0 0	2 4
Women		*(27)*	*(35)*	*(42)*	*(41)*	*(31)*	*(221)*	*(52)*	*(39)*	*(51)*	*(40)*	*(34)*	*(231)*
	One year	16 17	14 15	17 18	7 8	3 3	10 10	17 18	14 15	8 9	24 25	6 6	14 14
	Five years	4 4	6 6	2 3	2 2	0 0	2 3	7 9	5 5	4 4	9 13	5 5	6 8
	Ten years	0 0	6 6	2 3	2 2	0 0	2 2	6 9	5 5	1 1	7 11	3 4	4 6
Adults		*(77)*	*(69)*	*(83)*	*(82)*	*(59)*	*(464)*	*(110)*	*(103)*	*(102)*	*(90)*	*(63)*	*(491)*
	One year	14 14	11 11	12 12	8 8	5 5	10 10	19 20	20 21	9 9	18 19	6 6	15 16
	Five years	1 1	4 4	1 1	2 2	0 0	2 2	5 6	9 11	2 2	5 7	3 3	5 6
	Ten years	0 0	4 4	1 1	2 2	0 0	1 2	4 6	6 9	1 1	3 5	2 2	3 5

		Men	*Women*	*Adults*		*Men*	*Women*	*Adults*
Northern & Yorkshire		*(27)*	*(24)*	*(51)*		*(26)*	*(28)*	*(54)*
	One year	17 18	4 4	10 11		7 8	6 6	7 7
	Five years	8 9	0 0	3 4		0 0	2 2	1 1
	Ten years	5 5	0 0	2 2		0 0	0 0	0 0
Trent		*(27)*	*(34)*	*(61)*		*(28)*	*(17)*	*(45)*
	One year	11 11	9 9	11 11		16 17	13 13	15 15
	Five years	0 0	1 1	1 1		0 0	0 0	0 0
	Ten years	0 0	1 1	1 1		0 0	0 0	0 0
Anglia & Oxford		*(17)*	*(17)*	*(34)*		*(30)*	*(19)*	*(49)*
	One year	18 19	9 10	13 14		40 43	9 9	27 28
	Five years	0 0	4 4	2 2		19 25	3 4	13 17
	Ten years	0 0	4 4	2 2		16 22	3 4	11 15
North Thames		*(21)*	*(20)*	*(41)*		*(29)*	*(26)*	*(55)*
	One year	7 8	19 19	12 13		25 28	23 24	24 26
	Five years	0 0	3 4	1 2		6 8	13 18	10 13
	Ten years	0 0	0 0	0 0		3 4	9 13	6 8
South Thames		*(54)*	*(38)*	*(92)*		*(38)*	*(42)*	*(80)*
	One year	7 7	15 16	10 10		3 3	13 13	8 8
	Five years	0 0	4 5	2 2		1 1	4 4	2 2
	Ten years	0 0	4 5	2 2		0 0	4 4	1 2
South & West		*(36)*	*(24)*	*(60)*		*(38)*	*(32)*	*(70)*
	One year	17 17	10 11	14 15		16 18	26 28	20 22
	Five years	0 0	3 4	1 1		0 0	15 17	7 8
	Ten years	0 0	3 4	1 1		0 0	13 17	5 8
West Midlands		*(17)*	*(20)*	*(37)*		*(31)*	*(16)*	*(47)*
	One year	7 7	0 1	3 3		29 30	40 40	32 34
	Five years	0 0	0 0	0 0		5 5	29 33	13 16
	Ten years	0 0	0 0	0 0		3 3	17 26	7 12
North & West		*(33)*	*(30)*	*(63)*		*(28)*	*(39)*	*(67)*
	One year	3 3	11 12	6 7		4 4	2 2	3 3
	Five years	2 2	4 6	2 4		3 3	0 0	1 1
	Ten years	2 2	1 2	1 2		1 1	0 0	1 1
WALES		*(11)*	*(14)*	*(25)*		*(12)*	*(12)*	*(24)*
	One year	1 1	9 9	5 5		23 25	20 23	22 23
	Five years	0 0	5 5	3 3		5 5	0 0	2 3
	Ten years	0 0	0 0	0 0		5 5	0 0	2 3

[1] Including those for whom no deprivation category was assigned. C - crude survival, R - relative survival (see Chapter 3). Number of patients contributing to each analysis in parentheses.

ENGLAND & WALES

	1981-85						1986-90						
	Affluent	2	3	4	Deprived	All[1]	Affluent	2	3	4	Deprived	All[1]	
	C R	C R	C R	C R	C R	C R	C R	C R	C R	C R	C R	C R	
	(43)	*(54)*	*(59)*	*(50)*	*(26)*	*(237)*	*(33)*	*(43)*	*(33)*	*(29)*	*(17)*	*(155)*	**Men**
One year	22 24	10 11	17 18	12 12	4 4	13 14	14 14	27 29	15 16	16 18	3 3	17 18	
Five years	6 10	4 4	3 4	5 6	0 0	4 5	4 4	15 18	5 6	5 8	0 0	6 9	
	4 9	2 3	1 2	1 2	0 0	2 3							
	(27)	*(37)*	*(38)*	*(40)*	*(31)*	*(178)*	*(30)*	*(37)*	*(38)*	*(36)*	*(8)*	*(154)*	**Women**
One year	18 19	10 11	18 19	5 5	11 11	13 13	26 27	8 8	21 22	25 26	25 26	19 19	
Five years	8 8	4 4	4 6	1 1	0 0	3 4	9 12	3 3	11 12	5 5	25 26	7 9	
	8 8	0 0	2 4	1 1	0 0	2 3							
	(70)	*(91)*	*(97)*	*(90)*	*(57)*	*(415)*	*(63)*	*(80)*	*(71)*	*(65)*	*(25)*	*(309)*	**Adults**
One year	21 23	10 11	17 18	8 9	7 7	13 14	19 20	18 19	18 19	21 22	10 10	18 19	
Five years	7 11	4 4	3 4	3 3	0 0	3 5	6 8	9 10	8 10	5 6	5 6	7 9	
	6 11	1 2	1 3	1 2	0 0	2 3							

	1981-85				1986-90			
	Men	Women	Adults		Men	Women	Adults	
	(30)	*(23)*	*(53)*		*(21)*	*(17)*	*(38)*	**Northern & Yorkshire**
One year	19 20	10 10	15 15		5 5	3 3	4 4	One year
Five years	8 11	2 2	6 8		3 4	0 0	2 2	Five years
	5 8	2 2	4 6					
	(42)	*(29)*	*(71)*		*(20)*	*(21)*	*(41)*	**Trent**
One year	21 22	9 9	16 17		13 14	16 17	15 15	One year
Five years	8 10	0 0	4 5		4 4	6 7	5 6	Five years
	4 7	0 0	2 4					
	(23)	*(14)*	*(37)*		*(22)*	*(23)*	*(45)*	**Anglia & Oxford**
One year	9 9	13 13	10 11		17 18	20 21	18 20	One year
Five years	0 0	6 6	2 2		7 7	1 1	4 4	Five years
	0 0	0 0	0 0					
	(17)	*(6)*	*(23)*		*(3)*	*(8)*	*(11)*	**North Thames**
One year	11 12	- -	17 18		- -	22 22	15 15	One year
Five years	4 5	- -	10 12		- -	12 12	8 8	Five years
	4 5	- -	10 12					
	(23)	*(15)*	*(38)*		*(12)*	*(9)*	*(21)*	**South Thames**
One year	7 7	16 17	10 11		48 49	14 16	33 36	One year
Five years	1 1	0 0	1 1		8 9	8 9	8 10	Five years
	0 0	0 0	0 0					
	(41)	*(32)*	*(73)*		*(38)*	*(36)*	*(74)*	**South & West**
One year	13 14	18 18	15 16		20 21	29 30	25 26	One year
Five years	0 0	7 9	3 4		13 16	11 12	12 15	Five years
	0 0	2 4	1 2					
	(25)	*(16)*	*(41)*		*(17)*	*(14)*	*(31)*	**West Midlands**
One year	17 18	8 8	13 14		23 24	13 14	18 20	One year
Five years	8 9	0 0	5 6		0 0	3 4	2 2	Five years
	3 4	0 0	2 3					
	(29)	*(33)*	*(62)*		*(16)*	*(20)*	*(36)*	**North & West**
One year	8 8	6 6	7 7		6 7	15 15	11 12	One year
Five years	0 0	0 0	0 0		3 4	11 13	7 9	Five years
	0 0	0 0	0 0					
	(7)	*(10)*	*(17)*		*(6)*	*(6)*	*(12)*	**WALES**
One year	- -	27 30	17 19		- -	- -	32 34	One year
Five years	- -	8 10	8 10		- -	- -	20 26	Five years
	- -	8 10	4 5					

[1]   Including those for whom no deprivation category was assigned. C - crude survival, R - relative survival (see Chapter 3). Number of patients contributing to each analysis in parentheses.

Fig. 48.1   Relative survival up to ten years: trends by calendar period of diagnosis (1971-90) and NHS Region

ADULTS

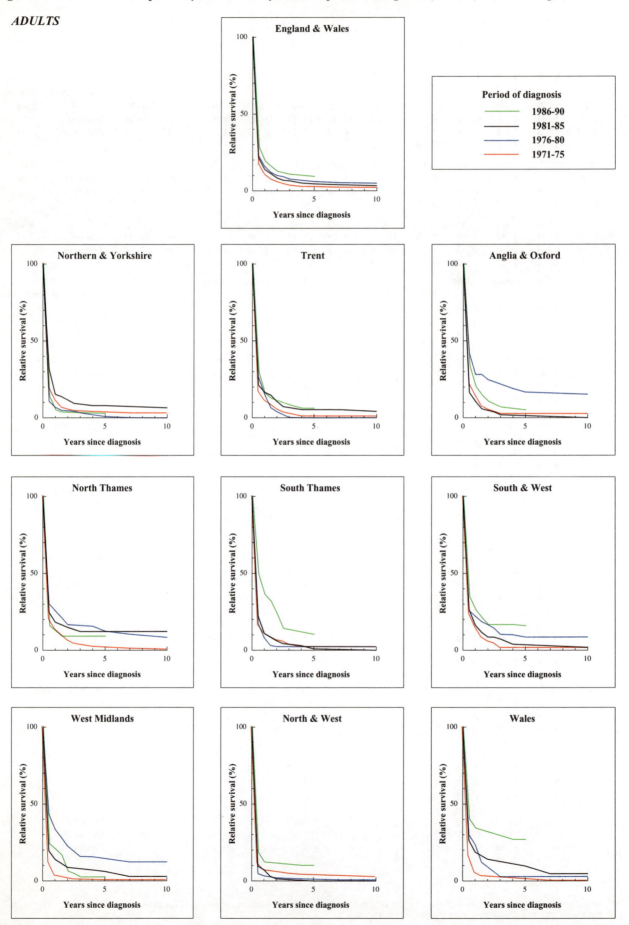

Fig. 48.2 Relative survival up to five years, by deprivation category: England and Wales, patients diagnosed 1986-90

ADULTS

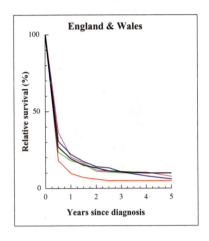

Fig. 48.3 Relative survival at five years, by deprivation category and period of diagnosis (1981-90): England and Wales

ADULTS
(Note vertical scale)

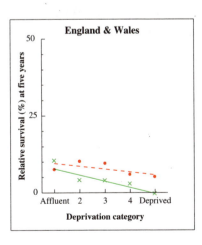

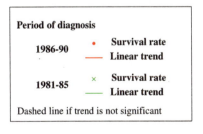

Fig. 48.4 Relative survival (%) by NHS Region and sex:
one-year and five-year survival for patients diagnosed 1986-90, with 95% confidence intervals

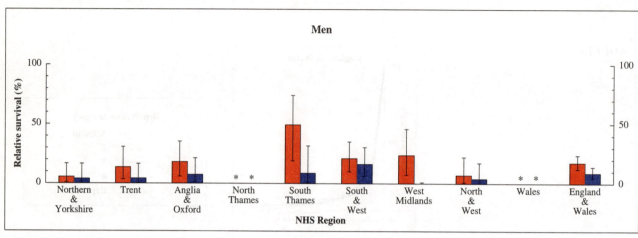

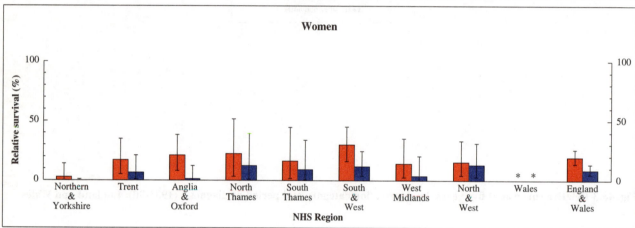

Fig. 48.5 Relative survival (%) at five years, with 95% confidence interval, by age at diagnosis and sex:
England and Wales, patients diagnosed 1986-90

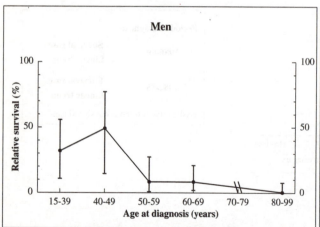

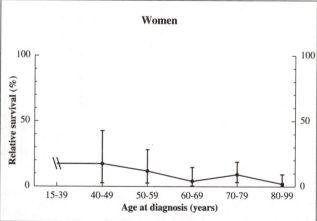

Chapter 49

ALL LEUKAEMIAS

The five types of leukaemia presented in Chapters 44-48 comprise a biologically and clinically diverse group of malignancies derived from the blood precursor cells. Collectively, lymphoid and myeloid leukaemias that are specified as either acute or chronic, together with monocytic leukaemia, have accounted for about 90% of all leukaemias diagnosed in adults in England and Wales during the period 1971-90. Some are more responsive than others to the chemotherapy regimes that have been introduced since the 1970s. Current patterns of survival from these leukaemias (or survival trends since the early 1970s) therefore differ widely. Information on survival for all adult leukaemias combined remains of some interest, however, even though the results are a weighted average of current survival rates (or a combination of the survival trends) for the different types of leukaemia.

For most tumours of solid organs, surgery has been the mainstay of treatment for more than 100 years, with the addition of external radiotherapy and radioactive agents since the first half of the twentieth century. By contrast, no effective treatments for leukaemia were available until the advent of chemotherapy regimes in the 1960s and early 1970s, and they have become the mainstay of treatment for all the leukaemias since then. The subsequent increase in survival has outstripped the improvement for most of the 36 common solid tumours examined here: only for cancers of colon, bone, testis, bladder and thyroid, and for melanoma, has survival improved as much (see Table 4.4). Combined analyses facilitate the comparison. In addition, trends in incidence and mortality have often been presented for all leukaemias combined, because the different types of leukaemia were only reported separately after the introduction of the eighth revision of the International Classification of Diseases in 1968. It is thus of some interest to have corresponding data on long-term trends in survival.

As a group, the leukaemias rank ninth among the most common malignancies in men in England and Wales, and twelfth in women. Some 4,600 new cases are diagnosed in adults each year, just over 2% of all cancers. Leukaemia in England and Wales is slightly less common than in much of western Europe, but has been increasing similarly at 1-2% a year since the 1970s. Age-standardised (Europe) incidence rates in 1990 were about 11 per 100,000 for men and 7 for women, around 20% higher than in 1971. In contrast, the decline in leukaemia mortality from the peak around 1970 has been less marked in England and Wales than in other western European countries, affecting mainly young adults aged 15-44. Mortality in 1995 was 7 per 100,000 in men and 4 in women.

The age-specific incidence of all leukaemias combined has two peaks, the first in early childhood, followed by a trough in young adulthood (age 15-29) and a steady rise above age 30. Chronic lymphoid leukaemia is a disease of the elderly; myeloid leukaemia increases with age; and acute lymphoid leukaemia occurs predominantly in young adults, but is less common than the other types. Ionising radiation, many chemotherapeutic drugs and occupational exposure to benzene can all cause leukaemia, and some leukaemias may be induced by viruses, but these causes are likely to account for only a small proportion of cases.

The survival analyses reported here are based on 65,600 adults diagnosed with one of the leukaemias in England and Wales during the period 1971-90 and followed up to the end of 1995. After exclusion of 9% of the cases, arising in children under 15 (see Chapter 50), these patients represented about 82% of all eligible adults (75% of all patients; Table 49.1). About 4% of adults were excluded because their vital status was not known on 6 July 1997, and a further 12% because their survival was zero or unknown. Half the leukaemias (51%) were myeloid (including monocytic) leukaemias, and two in five (42%) were lymphoid leukaemias: these proportions have been fairly stable since the early 1970s (Table 49A).

Survival trends

Relative survival from leukaemia among adults diagnosed in England and Wales during 1986-90 was 51% at one year and 27% at five years, substantially higher than the corresponding survival rates in the early 1970s: 37% at one year and 15% at five years (Table 49.2). For adults diagnosed during the period 1981-85 and followed up to the end of 1995, ten-year relative survival was 15%. Survival has increased both nationally and in most regions, with the notable exception of Anglia & Oxford (Figure 49.1).

After adjustment for differences in the age distribution of patients over time and between men and women, one-year survival for adults diagnosed during 1986-90 was about 50%, some 17% higher than for those diagnosed in the early 1970s. This represents a national average improvement of 6% between successive quinquennia, (see Table 4.4). Survival at five years doubled from 12-13% to about 25% over the same period, an average gain of 4-5% between successive five-year periods. One-year survival continued to improve every five years, but most of the gains in longer-term survival occurred around 1980, and five-year survival improved more slowly in the 1980s (Table 49.2, Figure 49.1).

Table 49A Distribution of leukaemias

ICD-9 code	Description	1971-75 No.	%	1976-80 No.	%	1981-85 No.	%	1986-90 No.	%	Total No.	%
204	Lymphoid leukaemia[1]	5,911	40.3	6,455	40.6	7,496	43.3	7,608	42.8	**27,470**	**41.8**
205	Myeloid leukaemia[1]	7,012	47.8	7,568	47.6	8,049	46.5	8,868	49.9	**31,497**	**48.0**
206	Monocytic leukaemia	479	3.3	503	3.2	415	2.4	309	1.7	**1,706**	**2.6**
207	Other specified leukaemia	1,260	8.6	935	5.9	232	1.3	140	0.8	**2,567**	**3.9**
208	Leukaemia of unspecified cell type[2]	-	-	447	2.8	1,129	6.5	832	4.7	**2,408**	**3.7**
	Total	**14,662**		**15,908**		**17,321**		**17,757**		**65,648**	

[1] Includes other and unspecified types: numbers exceed by 2-3% the sum of acute and chronic lymphoid or myeloid leukaemias (Chapters 44-47).
[2] Rubric 208 was assigned to polycythaemia vera in ICD-8 (1971-78): this neoplasm is not included.

In Anglia & Oxford Region, one-year survival for patients diagnosed during the 1970s was 44-47%, about 5% higher than the national average, but survival in this region has barely changed, and these levels were 5% below the national average for patients diagnosed during the 1980s. This reflects the patterns for several types of leukaemia in one or both sexes, and it is difficult to explain. Artefact seems unlikely: the data for this region derive from two independent cancer registries of good quality; leukaemia incidence is close to the national average; survival estimates for each period are based on 1,400-1,700 patients, the proportion of leukaemias registered solely from a death certificate is low and has not changed, and in East Anglia, active follow-up complements the national passive follow-up system (see Chapter 2). Increasing survival trends for many other cancers in Anglia & Oxford are a notable contrast. Higher survival from the leukaemias in Anglia & Oxford in the 1970s may well have been due to earlier or more effective use of chemotherapy than elsewhere, but the anomalous survival trends in this region should be explored in more detail.

In South Thames and South & West, five-year survival increased by 7-10% during the late 1980s, but there was little change in other regions (Figure 49.1). One-year survival for patients diagnosed in 1986-90 ranged from 44% in North & West to 60% in South Thames, while five-year survival ranged from 20% in Anglia & Oxford to 33% in South Thames and South & West (Table 49.2, Figure 49.4).

For leukaemias as a group, survival does not vary greatly with age (Figure 49.5). In adults diagnosed under the age of 70 in England and Wales during 1986-90, relative survival at five years was generally in the range 30-40%, falling to about 15% for those aged 80-99 at diagnosis (see Table 4.7). Women aged 15-39 at diagnosis have an 8% survival advantage at five years: a similar difference is seen for each of the four main types of leukaemia.

Deprivation

Since the early 1970s, one-year survival has generally been 5-7% lower among patients living in more deprived areas than among more affluent groups (Table 49.2). Survival curves for patients in the various deprivation groups who were diagnosed during 1986-90 are shown in Figure 49.2. In North Thames and South & West, the survival curves overlap, and five-year survival does not differ systematically between the groups; but the deprivation gap is 15% in Anglia & Oxford.

During both the early and late 1980s, the deprivation gap in survival between the most affluent and the most deprived groups in England and Wales was 7-8% at one year and 4-5% at five years, based on 17,000 patients of known deprivation status in each period (see Table 4.9). The deprivation gradient in England and Wales was clear-cut and statistically significant, both nationally and in several regions (Figure 49.3; note expanded vertical scale).

International comparison

Five-year relative survival for adults diagnosed with leukaemia in England and Wales during 1986-90 (27%) was broadly similar to that in Scotland but 7% lower than the average for Europe (see Table 4.12). The deficit compared to survival rates in the USA was 11-14%.

Table 49.1 Ineligible and excluded records, and patients included in analyses, by calendar period of diagnosis

	Calendar period of diagnosis														
	1971-75			**1976-80**			**1981-85**			**1986-90**			**Overall**		
Total registered	**18,886**			**20,808**			**24,189**			**25,435**			**89,318**		
Ineligible	*Records*	*Patients*	*%*	*Records*	*Patients*	*%*	*Records*	*Patients*	*%*	*Records*	*Patients*	*%*	*Records*	*Patients*	*%*
Incomplete data[1]	22	22		125	125		520	520		415	415		1,082	**1,082**	
Resident outside England and Wales	196	194		211	208		206	177		99	57		712	**636**	
In situ neoplasm[2]	0	0		0	0		0	0		0	0		0	**0**	
Benign or uncertain behaviour[2]	0	0		0	0		0	0		0	0		0	**0**	
Metastatic[3]	-	-		-	-		-	-		-	-		-	**-**	
Otherwise ineligible[3]	-	-		-	-		-	-		-	-		-	**-**	
Lymphoma[4]	-	-		-	-		-	-		-	-		-	**-**	
Leukaemia or myeloma[4]	-	-		-	-		-	-		-	-		-	**-**	
	216			**333**			**697**			**472**			**1,718**		
Total eligible	**18,670**	**100**		**20,475**	**100**		**23,492**	**100**		**24,963**	**100**		**87,600**	**100**	
Aged under 15 or 100 years or over at diagnosis	2,129	2,073	11.1	1,957	1,898	9.3	1,945	1,873	8.0	1,951	1,914	7.7	7,982	**7,758**	8.9
Vital status unknown[5]	755	558	3.0	1,000	760	3.7	1,139	848	3.6	1,512	1,213	4.9	4,406	**3,379**	3.9
Duplicate registration[6]	1	1	<0.1	2	2	<0.1	1	0	0.0	1	1	<0.1	5	**4**	<0.1
Multiple primary[7]	0	0	0.0	4	2	<0.1	53	40	0.2	92	77	0.3	149	**119**	0.1
Synchronous tumours[8]	38	32	0.2	63	44	0.2	62	43	0.2	93	58	0.2	256	**177**	0.2
Sex not known	0	0	0.0	0	0	0.0	0	0	0.0	0	0	0.0	0	**0**	0.0
Sex incompatible with site code[9]	-	-		-	-	-	-	-		-	-		-	**-**	-
Invalid dates[10]	2	2	<0.1	2	2	<0.1	12	9	<0.1	19	11	<0.1	35	**24**	<0.1
Zero survival[11]	1,490	1,342	7.2	2,066	1,859	9.1	3,725	3,358	14.3	4,230	3,932	15.8	11,511	**10,491**	12.0
Total exclusions	**4,008**	21.5		**4,567**	22.3		**6,171**	26.3		**7,206**	28.9		**21,952**	25.1	
Patients included in analysis	**14,662**	78.5		**15,908**	77.7		**17,321**	73.7		**17,757**	71.1		**65,648**	74.9	

[1] Main data item(s) invalid or incompatible with one another: name, sex, date of birth, date of diagnosis and (if present) date of death, postcode, site, morphology and behaviour

[2] Tumour either *in situ* (behaviour code 2), benign (0), uncertain if benign or malignant (1) or metastatic (6 or 9)

[3] Not applicable for this malignancy

[4] Not applicable for this malignancy

[5] Tracing of vital status at National Health Service Central Register incomplete at 6 July 1997

[6] Same site code, sex, cancer registry and cancer registry number as an earlier registration

[7] Same site code and person number as an earlier registration(s): mostly confirmed multiple primary tumours at this site, some unresolved duplicate registrations

[8] Same site code, sex, date of birth and date of diagnosis as another registration(s): mostly synchronous or (in paired organs) bilateral tumours in same anatomic site in one individual, not previously linked: also some duplicate registrations

[9] Not applicable for this malignancy

[10] Impossible sequence of dates (birth, diagnosis and, if present, emigration or death); or date of death after 6 July 1997

[11] Date of diagnosis same as date of death: some are patients who did die on the day of diagnosis, but most were registered solely from a death certificate (death certificate only, or DCO), and their survival time is thus unknown

Table 49.2 Crude and relative survival (%) at one, five and ten years since diagnosis: NHS Region, deprivation category and calendar period of diagnosis

MEN	1971-75 Affluent C R	2 C R	3 C R	4 C R	Deprived C R	All[1] C R	1976-80 Affluent C R	2 C R	3 C R	4 C R	Deprived C R	All[1] C R
ENGLAND & WALES	*(1,317)*	*(1,490)*	*(1,330)*	*(1,047)*	—	*(8,060)*	*(1,681)*	*(1,812)*	*(1,794)*	*(1,711)*	*(1,341)*	*(8,710)*
One year	38 40	39 41	36 38	35 37	35 37	**37** 38	42 44	42 44	38 40	38 41	38 40	**40** 42
Five years	12 14	13 17	11 14	12 15	12 15	**12** 15	15 19	15 19	15 19	14 18	13 17	**15** 19
Ten years	5 7	6 10	5 8	5 8	5 7	**5** 8	7 11	7 12	7 11	7 11	5 9	**7** 11
ENGLAND	*(1,268)*	*(1,311)*	*(1,372)*	*(1,218)*	*(964)*	*(7,552)*	*(1,622)*	*(1,737)*	*(1,666)*	*(1,581)*	*(1,272)*	*(8,170)*
One year	39 40	39 41	37 39	36 38	35 37	**37** 39	43 45	42 44	38 40	38 40	38 40	**40** 42
Five years	12 14	13 17	12 15	13 16	12 15	**12** 15	15 19	15 19	15 19	14 18	13 17	**14** 18
Ten years	5 7	6 10	5 9	5 9	5 8	**5** 8	7 11	7 12	7 11	7 11	5 8	**7** 11
Northern & Yorkshire	*(96)*	*(124)*	*(143)*	*(154)*	*(183)*	*(1,012)*	*(125)*	*(184)*	*(222)*	*(284)*	*(319)*	*(1,166)*
One year	39 41	43 46	40 43	35 38	27 28	**36** 38	45 47	38 41	37 39	31 33	37 40	**39** 42
Five years	12 15	15 20	12 16	17 21	8 11	**12** 16	16 21	13 17	13 17	13 17	12 17	**14** 19
Ten years	6 9	8 14	4 8	3 5	2 5	**4** 7	6 10	7 13	4 8	6 11	5 10	**6** 11
Trent	*(95)*	*(124)*	*(166)*	*(205)*	*(136)*	*(804)*	*(133)*	*(173)*	*(189)*	*(239)*	*(166)*	*(923)*
One year	45 46	37 39	30 31	38 40	32 34	**38** 40	41 43	41 43	39 41	30 33	37 39	**39** 41
Five years	16 19	15 18	14 17	11 13	11 15	**14** 17	15 18	16 19	17 22	11 15	13 16	**15** 19
Ten years	5 7	5 8	8 11	5 8	5 9	**6** 9	5 8	7 11	6 10	4 7	5 8	**5** 9
Anglia & Oxford	*(181)*	*(158)*	*(153)*	*(108)*	*(39)*	*(765)*	*(253)*	*(240)*	*(224)*	*(135)*	*(61)*	*(947)*
One year	34 36	45 47	39 41	32 33	*41 44*	**42** 45	45 47	49 52	34 36	45 47	*44 48*	**45** 48
Five years	11 13	13 17	14 17	16 21	*14 17*	**15** 18	14 17	16 19	12 15	15 18	*21 26*	**15** 19
Ten years	4 5	7 11	7 13	8 14	*1 2*	**6** 9	6 9	8 14	7 10	4 6	*7 12*	**7** 11
North Thames	*(227)*	*(186)*	*(175)*	*(179)*	*(132)*	*(1,005)*	*(180)*	*(163)*	*(155)*	*(156)*	*(103)*	*(788)*
One year	43 45	42 44	33 34	39 41	31 33	**40** 42	45 46	53 55	51 53	49 51	42 44	**49** 52
Five years	14 17	17 22	11 14	14 17	13 16	**14** 18	19 23	21 26	20 26	20 26	15 19	**20** 25
Ten years	5 7	6 11	7 11	8 12	6 9	**7** 11	14 20	10 16	10 17	12 20	6 12	**11** 18
South Thames	*(295)*	*(242)*	*(226)*	*(178)*	*(84)*	*(1,133)*	*(326)*	*(259)*	*(219)*	*(192)*	*(80)*	*(1,137)*
One year	38 40	28 30	35 37	35 37	39 42	**37** 39	42 44	33 35	33 35	35 37	30 33	**37** 40
Five years	10 13	11 14	9 11	10 14	16 20	**12** 15	16 19	12 15	13 18	13 16	10 14	**14** 18
Ten years	4 6	5 8	3 5	6 10	*9 14*	**6** 9	8 11	5 8	5 9	7 11	*5 8*	**6** 10
South & West	*(211)*	*(282)*	*(270)*	*(151)*	*(44)*	*(1,057)*	*(299)*	*(406)*	*(333)*	*(167)*	*(53)*	*(1,314)*
One year	33 34	40 42	35 37	28 29	*38 40*	**38** 39	39 41	38 40	36 38	35 37	22 23	**39** 41
Five years	10 13	12 15	14 17	8 11	*16 20*	**12** 16	15 19	15 20	16 20	13 16	*11 13*	**16** 20
Ten years	5 8	6 9	5 9	3 6	*9 14*	**5** 9	7 12	7 12	9 14	6 9	*1 2*	**7** 12
West Midlands	*(38)*	*(53)*	*(73)*	*(58)*	*(90)*	*(697)*	*(130)*	*(120)*	*(121)*	*(169)*	*(179)*	*(721)*
One year	25 26	27 28	39 41	35 37	37 39	**35** 36	41 43	32 34	30 32	33 35	30 32	**36** 37
Five years	*10 11*	8 11	6 7	*12 16*	9 12	**9** 11	13 16	11 14	14 18	10 14	10 13	**12** 16
Ten years	2 2	5 7	0 0	4 7	3 5	**3** 5	4 6	5 8	7 13	5 10	4 7	**5** 9
North & West	*(125)*	*(142)*	*(166)*	*(185)*	*(256)*	*(1,079)*	*(176)*	*(192)*	*(203)*	*(239)*	*(311)*	*(1,174)*
One year	25 27	34 37	31 33	30 32	33 35	**34** 36	31 32	41 43	30 32	37 39	37 39	**37** 39
Five years	5 6	8 10	7 9	8 11	11 13	**9** 11	9 11	13 17	8 11	12 15	11 14	**11** 14
Ten years	2 4	3 4	2 4	4 6	4 6	**3** 6	2 3	6 11	4 7	6 9	3 6	**4** 7
WALES	*(49)*	*(52)*	*(118)*	*(112)*	*(83)*	*(508)*	*(59)*	*(75)*	*(128)*	*(130)*	*(69)*	*(540)*
One year	*32 33*	27 29	23 24	23 24	30 31	**29** 30	*33 35*	32 34	31 32	40 43	42 44	**39** 41
Five years	*17 22*	8 11	4 5	5 6	7 9	**7** 10	*15 20*	11 15	11 13	17 24	*15 22*	**16** 21
Ten years	*4 6*	5 7	1 1	2 4	1 1	**3** 4	*8 11*	4 8	7 10	7 13	*8 13*	**8** 13

[1] Including those for whom no deprivation category was assigned. C - crude survival, R - relative survival (see Chapter 3).
Number of patients contributing to each analysis in parentheses.

		1981-85												1986-90												MEN
		Affluent		*2*		*3*		*4*		*Deprived*		*All¹*		*Affluent*		*2*		*3*		*4*		*Deprived*		*All¹*		
		C	R	C	R	C	R	C	R	C	R	C	R	C	R	C	R	C	R	C	R	C	R	C	R	
		(1,989)		*(1,983)*		*(1,925)*		*(1,916)*		*(1,463)*		*(9,508)*		*(2,043)*		*(2,227)*		*(2,191)*		*(1,796)*		*(1,434)*		*(9,779)*		**ENGLAND & WALES**
	One year	47	50	46	49	46	49	46	49	44	47	46	49	53	56	53	56	49	52	47	50	48	50	50	53	*One year*
	Five years	20	25	19	24	18	23	18	23	17	23	19	24	23	28	22	28	20	26	19	25	20	25	21	27	*Five years*
		10	15	9	15	8	14	8	14	8	13	9	14													
		(1,937)		*(1,908)*		*(1,813)*		*(1,793)*		*(1,382)*		*(9,016)*		*(1,948)*		*(2,137)*		*(2,038)*		*(1,636)*		*(1,353)*		*(9,184)*		**ENGLAND**
	One year	47	50	46	49	46	49	46	49	44	46	46	49	53	56	53	56	50	53	47	50	48	51	51	53	*One year*
	Five years	20	25	19	24	18	24	18	23	17	23	19	24	23	29	22	28	20	26	19	25	20	26	21	27	*Five years*
		10	15	9	15	8	14	8	14	7	13	9	14													
		(180)		*(222)*		*(236)*		*(297)*		*(332)*		*(1,285)*		*(189)*		*(261)*		*(274)*		*(290)*		*(314)*		*(1,332)*		**Northern & Yorkshire**
	One year	46	49	44	47	47	50	47	51	40	43	46	49	48	51	53	57	44	48	43	46	47	51	48	51	*One year*
	Five years	21	27	18	22	18	24	19	25	16	20	18	24	23	32	27	36	19	25	15	21	19	26	21	27	*Five years*
		10	16	8	14	6	11	8	15	5	10	8	13													
		(206)		*(232)*		*(245)*		*(304)*		*(181)*		*(1,177)*		*(156)*		*(207)*		*(240)*		*(216)*		*(213)*		*(1,037)*		**Trent**
	One year	43	45	49	53	46	49	45	48	40	43	46	49	48	51	53	56	45	48	44	47	45	49	48	51	*One year*
	Five years	18	22	23	30	21	26	18	24	14	19	19	25	15	19	23	31	16	22	20	27	19	25	19	25	*Five years*
		9	16	7	13	8	14	7	13	5	10	8	13													
		(254)		*(229)*		*(207)*		*(116)*		*(39)*		*(856)*		*(271)*		*(285)*		*(237)*		*(128)*		*(47)*		*(978)*		**Anglia & Oxford**
	One year	38	40	36	38	36	39	33	35	*46*	*50*	39	41	47	50	43	46	45	48	43	47	*33*	*35*	45	48	*One year*
	Five years	15	19	16	20	13	17	9	12	*22*	*30*	15	19	18	23	14	19	15	21	15	21	*10*	*12*	16	20	*Five years*
		8	14	6	11	4	9	5	8	*9*	*20*	7	12													
		(178)		*(155)*		*(128)*		*(154)*		*(90)*		*(714)*		*(184)*		*(206)*		*(217)*		*(176)*		*(156)*		*(955)*		**North Thames**
	One year	54	57	53	56	44	47	49	51	55	59	52	55	57	61	61	65	57	61	59	63	53	57	58	61	*One year*
	Five years	27	34	22	29	20	25	21	26	27	38	24	30	23	31	22	28	20	26	27	37	23	29	23	29	*Five years*
		14	21	11	19	13	20	11	20	*14*	*25*	13	20													
		(354)		*(293)*		*(239)*		*(197)*		*(60)*		*(1,196)*		*(347)*		*(281)*		*(208)*		*(155)*		*(69)*		*(1,077)*		**South Thames**
	One year	46	49	44	47	42	45	48	52	*40*	*42*	45	48	61	65	57	61	50	54	59	63	*63*	*67*	58	62	*One year*
	Five years	20	25	17	22	18	24	22	28	*23*	*30*	20	25	30	38	25	32	23	32	23	31	*20*	*26*	26	33	*Five years*
		12	20	7	12	8	15	14	24	*12*	*24*	10	17													
		(331)		*(414)*		*(336)*		*(186)*		*(67)*		*(1,386)*		*(385)*		*(483)*		*(441)*		*(232)*		*(70)*		*(1,620)*		**South & West**
	One year	52	55	42	46	47	51	44	47	*43*	*45*	48	51	54	58	56	60	51	55	49	54	*46*	*50*	54	57	*One year*
	Five years	20	26	18	24	18	24	16	21	*18*	*25*	19	24	25	32	23	32	24	32	23	30	*25*	*35*	24	31	*Five years*
		8	14	11	20	11	18	6	11	*7*	*12*	9	16													
		(189)		*(163)*		*(206)*		*(241)*		*(207)*		*(1,021)*		*(193)*		*(177)*		*(207)*		*(179)*		*(183)*		*(943)*		**West Midlands**
	One year	49	52	48	50	51	54	44	47	39	41	47	50	53	57	48	51	48	52	44	47	53	56	51	54	*One year*
	Five years	22	30	21	26	18	24	18	24	14	19	19	24	21	26	17	24	21	30	14	19	25	33	20	26	*Five years*
		7	13	8	13	7	14	6	11	9	15	8	13													
		(245)		*(200)*		*(216)*		*(298)*		*(406)*		*(1,381)*		*(223)*		*(237)*		*(214)*		*(260)*		*(301)*		*(1,242)*		**North & West**
	One year	41	44	46	49	44	47	40	44	42	45	44	47	42	45	45	47	46	50	33	35	39	42	42	45	*One year*
	Five years	16	20	17	22	16	21	13	19	16	22	16	21	19	25	17	23	16	22	10	17	15	22	16	21	*Five years*
		8	13	10	18	6	12	6	12	6	11	7	13													
		(52)		*(75)*		*(112)*		*(123)*		*(81)*		*(492)*		*(95)*		*(90)*		*(153)*		*(160)*		*(81)*		*(595)*		**WALES**
	One year	*47*	*50*	45	48	43	46	41	44	46	49	46	49	47	50	43	46	43	46	47	50	38	41	45	48	*One year*
	Five years	*20*	*25*	22	30	16	20	19	27	17	23	19	25	19	27	19	27	25	33	18	24	14	20	20	27	*Five years*
		13	*23*	*11*	*22*	8	17	10	20	*11*	*20*	11	20													

---

¹ Including those for whom no deprivation category was assigned. C - crude survival, R - relative survival (see Chapter 3).
  Number of patients contributing to each analysis in parentheses.

Table 49.2   **Crude and relative survival (%) at one, five and ten years since diagnosis: NHS Region, deprivation category and calendar period of diagnosis**

	1971-75						1976-80					
WOMEN	*Affluent* **C R**	*2* **C R**	*3* **C R**	*4* **C R**	*Deprived* **C R**	*All[1]* **C R**	*Affluent* **C R**	*2* **C R**	*3* **C R**	*4* **C R**	*Deprived* **C R**	*All[1]* **C R**
ENGLAND & WALES	*(1,083)*	*(1,187)*	*(1,186)*	*(1,087)*	*(857)*	*(6,602)*	*(1,366)*	*(1,486)*	*(1,526)*	*(1,366)*	*(1,134)*	*(7,198)*
One year	39 40	35 36	34 35	33 34	30 31	34 36	41 43	39 41	39 41	38 39	33 35	38 40
Five years	15 17	13 15	12 14	11 13	10 12	12 14	17 20	14 18	16 19	15 19	13 16	15 18
Ten years	7 10	6 8	5 8	5 8	4 7	5 8	8 12	8 12	7 11	8 12	7 10	8 11
ENGLAND	*(1,051)*	*(1,123)*	*(1,097)*	*(985)*	*(814)*	*(6,184)*	*(1,308)*	*(1,411)*	*(1,422)*	*(1,261)*	*(1,062)*	*(6,730)*
One year	39 40	35 36	33 35	34 35	30 31	**34 36**	41 43	39 41	38 40	38 40	34 35	**38 40**
Five years	14 17	13 15	12 14	11 13	10 13	**12 14**	17 20	14 18	15 18	15 19	13 17	**15 18**
Ten years	7 10	6 8	5 8	5 8	4 7	**5 8**	9 12	8 12	7 10	8 12	7 10	**8 11**
Northern & Yorkshire	*(76)*	*(109)*	*(123)*	*(129)*	*(151)*	*(807)*	*(122)*	*(158)*	*(199)*	*(190)*	*(252)*	*(953)*
One year	43 45	35 37	35 37	33 34	32 33	**35 37**	34 35	37 39	38 40	31 32	31 32	**36 38**
Five years	18 21	15 18	15 17	10 12	9 11	**12 15**	17 19	12 14	12 15	10 13	11 14	**13 16**
Ten years	7 10	6 9	5 9	2 4	5 7	**5 7**	8 12	5 9	4 6	4 6	6 10	**6 9**
Trent	*(85)*	*(108)*	*(118)*	*(127)*	*(124)*	*(615)*	*(90)*	*(124)*	*(159)*	*(180)*	*(140)*	*(710)*
One year	34 35	40 41	26 26	33 35	31 32	**35 37**	39 41	34 35	36 37	35 36	33 34	**38 39**
Five years	16 18	13 15	11 13	9 11	13 16	**13 16**	15 19	14 17	17 20	16 19	13 18	**16 20**
Ten years	8 10	5 6	4 7	7 9	5 9	**6 9**	7 9	6 9	7 10	11 16	6 11	**8 12**
Anglia & Oxford	*(136)*	*(156)*	*(125)*	*(81)*	*(35)*	*(626)*	*(176)*	*(168)*	*(176)*	*(123)*	*(30)*	*(711)*
One year	39 40	42 44	39 41	31 32	*31 32*	**42 43**	42 44	42 45	43 45	49 51	*36 37*	**45 47**
Five years	15 17	17 20	14 17	13 16	*13 15*	**16 19**	14 17	14 19	17 20	18 24	*15 16*	**16 20**
Ten years	9 12	8 11	5 8	6 10	*7 15*	**7 11**	8 12	8 13	10 14	8 13	*4 5*	**8 13**
North Thames	*(154)*	*(139)*	*(154)*	*(145)*	*(103)*	*(781)*	*(158)*	*(157)*	*(149)*	*(123)*	*(109)*	*(722)*
One year	39 40	34 35	43 45	37 38	35 37	**40 42**	48 50	39 41	33 34	34 36	38 39	**41 43**
Five years	14 16	14 16	13 15	14 16	13 16	**15 17**	19 22	13 16	10 12	10 12	21 26	**15 19**
Ten years	7 10	6 8	6 8	7 10	6 9	**6 9**	10 15	9 13	4 6	5 8	11 17	**9 13**
South Thames	*(284)*	*(212)*	*(201)*	*(175)*	*(80)*	*(1,047)*	*(295)*	*(240)*	*(225)*	*(179)*	*(73)*	*(1,069)*
One year	33 34	26 27	25 26	29 31	22 23	**30 32**	40 42	36 38	32 34	33 35	36 38	**39 41**
Five years	13 15	11 13	11 13	12 14	10 12	**13 15**	18 21	12 16	14 17	17 21	13 16	**16 20**
Ten years	6 9	6 9	7 11	6 10	5 8	**7 11**	8 11	8 12	9 14	9 15	6 8	**9 13**
South & West	*(163)*	*(237)*	*(186)*	*(108)*	*(41)*	*(810)*	*(239)*	*(288)*	*(262)*	*(142)*	*(53)*	*(1,031)*
One year	41 42	31 33	34 35	35 37	*16 17*	**36 38**	40 41	38 39	41 43	51 54	*37 37*	**42 44**
Five years	16 18	10 13	12 15	11 14	*11 13*	**13 15**	15 18	16 19	19 24	20 25	*17 21*	**18 22**
Ten years	8 11	5 9	6 8	7 11	*0 1*	**6 10**	9 12	11 16	6 9	10 16	*10 14*	**10 15**
West Midlands	*(31)*	*(38)*	*(44)*	*(59)*	*(68)*	*(554)*	*(85)*	*(98)*	*(89)*	*(127)*	*(117)*	*(520)*
One year	*33 34*	*32 34*	*31 32*	*38 40*	23 24	**35 36**	28 29	31 32	34 36	35 36	30 31	**34 36**
Five years	*8 9*	*5 6*	*9 11*	*6 7*	1 1	**8 10**	11 13	10 12	19 21	11 14	9 10	**13 15**
Ten years	*6 7*	*3 3*	*1 2*	*3 4*	1 1	**3 4**	7 10	4 5	6 8	6 10	6 9	**6 9**
North & West	*(122)*	*(124)*	*(146)*	*(161)*	*(212)*	*(944)*	*(143)*	*(178)*	*(163)*	*(197)*	*(288)*	*(1,014)*
One year	33 34	22 23	16 16	22 23	22 23	**25 27**	36 37	33 35	28 29	27 28	27 28	**32 33**
Five years	8 9	6 8	3 4	7 8	7 9	**7 9**	13 15	14 17	8 9	12 14	11 13	**12 15**
Ten years	6 7	1 2	1 1	1 2	1 2	**2 3**	5 7	9 13	4 5	5 9	5 7	**6 9**
WALES	*(32)*	*(64)*	*(89)*	*(102)*	*(43)*	*(418)*	*(58)*	*(75)*	*(104)*	*(105)*	*(72)*	*(468)*
One year	*50 51*	39 39	33 34	25 26	27 27	**37 38**	*37 40*	42 44	46 48	29 31	23 25	**38 40**
Five years	*22 25*	*13 16*	6 8	8 9	2 2	**10 13**	*17 21*	16 18	23 27	15 17	6 7	**16 20**
Ten years	*12 16*	*5 13*	- -	6 8	2 2	**6 10**	*8 13*	4 6	14 20	7 11	*3 5*	**8 12**

[1] Including those for whom no deprivation category was assigned. C - crude survival, R - relative survival (see Chapter 3).
Number of patients contributing to each analysis in parentheses.

	1981-85							1986-90						
	Affluent	*2*	*3*	*4*	*Deprived*	*All[1]*		*Affluent*	*2*	*3*	*4*	*Deprived*	*All[1]*	**WOMEN**
	C R	C R	C R	C R	C R	C R		C R	C R	C R	C R	C R	C R	
	(1,548)	*(1,641)*	*(1,691)*	*(1,530)*	*(1,228)*	*(7,813)*		*(1,605)*	*(1,751)*	*(1,790)*	*(1,556)*	*(1,180)*	*(7,978)*	**ENGLAND & WALES**
	47 50	47 49	43 46	41 43	38 40	43 46		50 52	50 52	46 48	46 48	42 44	47 49	*One year*
	22 26	21 26	19 24	17 21	17 21	19 24		24 29	23 28	21 26	21 26	19 23	21 27	*Five years*
	12 18	11 17	10 16	10 15	9 13	10 16								
	(1,505)	*(1,572)*	*(1,580)*	*(1,418)*	*(1,174)*	*(7,376)*		*(1,550)*	*(1,673)*	*(1,695)*	*(1,440)*	*(1,116)*	*(7,545)*	**ENGLAND**
	48 50	47 49	43 46	41 43	38 40	**43 46**		50 53	50 52	47 49	46 49	42 44	**47 49**	*One year*
	22 27	21 26	19 23	17 21	17 21	**19 24**		24 29	23 28	21 26	21 26	18 23	**21 27**	*Five years*
	12 18	11 17	10 16	9 14	9 13	**10 16**								
	(127)	*(182)*	*(231)*	*(255)*	*(279)*	*(1,087)*		*(173)*	*(189)*	*(235)*	*(222)*	*(260)*	*(1,083)*	**Northern & Yorkshire**
	44 47	51 53	40 42	37 39	40 42	**44 46**		42 44	49 51	39 41	48 51	43 45	**45 48**	*One year*
	19 25	18 22	19 24	15 20	19 23	**19 24**		25 33	20 26	17 22	22 28	18 23	**21 27**	*Five years*
	10 17	7 11	6 11	9 15	10 15	**9 14**								
	(136)	*(149)*	*(208)*	*(204)*	*(154)*	*(857)*		*(104)*	*(172)*	*(202)*	*(210)*	*(172)*	*(864)*	**Trent**
	50 52	41 43	42 44	44 46	36 38	**44 46**		42 43	45 47	40 42	40 42	30 31	**41 43**	*One year*
	20 23	26 32	19 24	19 24	15 20	**20 25**		21 26	25 30	19 25	17 21	11 14	**19 24**	*Five years*
	15 20	14 21	12 19	9 13	7 11	**11 18**								
	(199)	*(166)*	*(167)*	*(102)*	*(37)*	*(686)*		*(228)*	*(215)*	*(174)*	*(124)*	*(37)*	*(783)*	**Anglia & Oxford**
	43 46	34 36	34 36	30 32	*34 37*	**38 40**		42 44	42 44	37 38	34 36	28 29	**40 42**	*One year*
	20 25	15 19	11 13	10 12	*9 14*	**15 19**		18 23	17 20	12 16	14 17	*7 8*	**16 20**	*Five years*
	12 17	6 10	5 8	8 12	*2 4*	**8 13**								
	(152)	*(139)*	*(119)*	*(94)*	*(65)*	*(576)*		*(157)*	*(165)*	*(152)*	*(157)*	*(115)*	*(769)*	**North Thames**
	51 53	54 57	47 49	43 45	*48 50*	**51 53**		55 57	49 50	56 58	46 48	54 55	**53 55**	*One year*
	28 32	30 34	18 22	17 20	22 25	**24 29**		25 30	22 27	29 35	27 32	29 33	**27 32**	*Five years*
	17 24	19 25	11 17	11 15	*13 18*	**15 21**								
	(294)	*(235)*	*(204)*	*(147)*	*(72)*	*(987)*		*(250)*	*(219)*	*(209)*	*(157)*	*(71)*	*(926)*	**South Thames**
	49 51	48 50	47 49	47 51	34 36	**48 50**		61 64	57 60	52 54	51 53	*45 47*	**55 58**	*One year*
	22 26	23 28	20 25	23 29	21 25	**22 28**		33 41	26 31	23 30	23 29	26 30	**27 33**	*Five years*
	10 15	14 22	10 17	12 17	15 21	**12 19**								
	(284)	*(341)*	*(315)*	*(161)*	*(56)*	*(1,187)*		*(317)*	*(396)*	*(378)*	*(198)*	*(72)*	*(1,368)*	**South & West**
	47 49	49 51	41 42	40 42	*48 50*	**46 48**		54 57	53 56	55 57	51 53	*53 55*	**55 58**	*One year*
	22 27	22 26	22 26	17 21	*22 28*	**22 27**		24 30	28 35	26 33	27 33	26 33	**27 34**	*Five years*
	12 18	11 17	12 18	12 18	*13 21*	**12 19**								
	(121)	*(157)*	*(178)*	*(207)*	*(179)*	*(846)*		*(136)*	*(122)*	*(143)*	*(156)*	*(113)*	*(673)*	**West Midlands**
	48 50	42 44	44 47	49 51	36 37	**45 47**		49 52	47 49	38 40	45 47	39 41	**45 47**	*One year*
	21 26	15 19	17 19	21 25	15 18	**18 22**		19 23	15 18	15 17	18 21	15 19	**17 20**	*Five years*
	11 18	9 13	9 12	8 13	5 8	**8 13**								
	(192)	*(203)*	*(158)*	*(248)*	*(332)*	*(1,150)*		*(185)*	*(195)*	*(202)*	*(216)*	*(276)*	*(1,079)*	**North & West**
	38 40	40 42	39 41	29 31	28 30	**36 38**		40 41	40 42	35 36	41 43	36 38	**40 42**	*One year*
	19 24	15 19	17 21	10 13	12 15	**15 19**		15 18	15 19	12 16	17 22	15 19	**16 20**	*Five years*
	8 13	7 12	11 16	6 10	6 10	**8 13**								
	(43)	*(69)*	*(111)*	*(112)*	*(54)*	*(437)*		*(55)*	*(78)*	*(95)*	*(116)*	*(64)*	*(433)*	**WALES**
	32 33	*51 53*	*43 46*	*39 41*	*34 35*	**44 46**		*47 48*	*52 55*	*38 39*	*42 45*	*43 45*	**46 48**	*One year*
	13 16	*25 29*	*20 28*	*17 21*	*20 23*	**21 26**		*21 27*	*26 35*	*16 19*	*17 21*	*24 27*	**22 29**	*Five years*
	8 13	*17 24*	*10 18*	*13 19*	*8 12*	**12 20**								

[1] Including those for whom no deprivation category was assigned. C - crude survival, R - relative survival (see Chapter 3).
Number of patients contributing to each analysis in parentheses.

Table 49.2   Crude and relative survival (%) at one, five and ten years since diagnosis: NHS Region, deprivation category and calendar period of diagnosis

ADULTS	1971-75 Affluent C R	2 C R	3 C R	4 C R	Deprived C R	All[1] C R	1976-80 Affluent C R	2 C R	3 C R	4 C R	Deprived C R	All[1] C R
ENGLAND & WALES	*(2,400)*	*(2,550)*	*(2,676)*	*(2,417)*	*(1,904)*	*(14,662)*	*(3,047)*	*(3,298)*	*(3,320)*	*(3,077)*	*(2,475)*	*(15,908)*
One year	39 40	37 39	35 37	34 36	33 34	**36 37**	42 44	41 43	38 40	38 40	36 38	**39 41**
Five years	13 16	13 16	11 14	11 14	11 14	**12 15**	16 19	15 18	15 19	15 19	13 17	**15 18**
Ten years	6 8	6 9	5 8	5 8	4 7	**5 8**	8 11	8 12	7 11	7 12	6 9	**7 11**
ENGLAND	*(2,319)*	*(2,434)*	*(2,469)*	*(2,203)*	*(1,778)*	*(13,736)*	*(2,930)*	*(3,148)*	*(3,088)*	*(2,842)*	*(2,334)*	*(14,900)*
One year	39 40	37 39	35 37	35 37	33 34	**36 38**	42 44	41 43	38 40	38 40	36 38	**39 41**
Five years	13 15	13 16	12 15	12 15	11 14	**12 15**	16 19	15 19	15 19	14 18	13 17	**15 18**
Ten years	6 8	6 9	5 8	5 9	5 7	**5 8**	8 11	8 12	7 11	7 11	6 9	**7 11**
Northern & Yorkshire	*(172)*	*(233)*	*(266)*	*(283)*	*(334)*	*(1,819)*	*(247)*	*(342)*	*(421)*	*(474)*	*(571)*	*(2,119)*
One year	41 43	39 41	38 40	34 36	29 31	**36 37**	40 41	38 40	38 40	31 33	35 37	**38 40**
Five years	15 18	15 19	13 17	14 17	8 11	**12 16**	17 20	13 16	12 16	12 15	12 16	**13 17**
Ten years	7 9	7 11	5 8	3 4	3 6	**4 7**	7 12	6 11	4 7	6 9	6 10	**6 10**
Trent	*(180)*	*(232)*	*(284)*	*(332)*	*(260)*	*(1,419)*	*(223)*	*(297)*	*(348)*	*(419)*	*(306)*	*(1,633)*
One year	40 41	38 40	28 29	36 38	32 33	**37 38**	40 42	38 39	38 39	32 34	35 37	**38 40**
Five years	16 18	14 16	13 15	10 13	12 15	**14 16**	15 18	15 18	17 21	13 17	13 17	**15 19**
Ten years	6 9	5 7	6 9	6 9	5 9	**6 9**	6 8	7 10	6 10	7 11	5 9	**7 10**
Anglia & Oxford	*(317)*	*(314)*	*(278)*	*(189)*	*(74)*	*(1,391)*	*(429)*	*(408)*	*(400)*	*(258)*	*(91)*	*(1,658)*
One year	36 38	43 45	39 41	31 33	37 38	**42 44**	44 46	46 49	38 40	47 49	41 44	**45 47**
Five years	13 15	15 18	14 17	15 19	13 17	**15 18**	14 17	15 19	14 17	16 21	19 23	**15 19**
Ten years	6 8	7 11	6 10	7 12	4 6	**7 10**	7 10	8 13	8 12	6 9	6 9	**8 12**
North Thames	*(381)*	*(325)*	*(329)*	*(324)*	*(235)*	*(1,786)*	*(338)*	*(320)*	*(304)*	*(279)*	*(212)*	*(1,510)*
One year	41 43	39 40	38 40	38 40	33 34	**40 42**	46 48	46 48	42 44	42 44	40 41	**45 47**
Five years	14 17	15 19	12 14	14 17	13 16	**14 17**	19 22	17 21	15 19	16 20	18 22	**18 22**
Ten years	6 8	6 10	7 10	7 11	6 9	**7 10**	12 18	10 14	7 11	9 14	9 14	**10 15**
South Thames	*(579)*	*(454)*	*(427)*	*(353)*	*(164)*	*(2,180)*	*(621)*	*(499)*	*(444)*	*(371)*	*(153)*	*(2,206)*
One year	35 37	27 28	30 32	32 34	30 32	**34 35**	41 43	35 37	33 35	34 36	33 35	**38 40**
Five years	11 14	11 14	10 13	11 14	13 16	**12 15**	17 20	12 15	14 18	15 19	12 15	**15 19**
Ten years	5 7	6 8	5 8	6 10	7 11	**6 10**	8 11	6 10	7 11	8 13	5 9	**7 12**
South & West	*(374)*	*(519)*	*(456)*	*(259)*	*(85)*	*(1,867)*	*(538)*	*(694)*	*(595)*	*(309)*	*(106)*	*(2,345)*
One year	36 37	36 38	34 36	31 32	27 28	**37 39**	39 41	38 40	38 40	42 45	29 30	**40 42**
Five years	13 15	11 14	13 16	10 12	13 18	**12 16**	15 18	15 20	17 22	16 20	14 17	**17 21**
Ten years	6 10	6 9	6 9	5 8	4 8	**6 9**	8 12	9 14	8 12	8 12	5 9	**8 13**
West Midlands	*(69)*	*(91)*	*(117)*	*(117)*	*(158)*	*(1,251)*	*(215)*	*(218)*	*(210)*	*(296)*	*(296)*	*(1,241)*
One year	29 30	29 30	36 38	37 39	31 32	**35 36**	36 37	32 33	32 34	34 35	30 32	**35 37**
Five years	9 10	7 10	7 9	9 11	5 7	**9 11**	12 15	11 13	16 19	10 14	9 12	**12 16**
Ten years	4 5	4 7	0 1	4 5	2 4	**3 4**	5 8	4 6	7 10	6 9	5 8	**6 9**
North & West	*(247)*	*(266)*	*(312)*	*(346)*	*(468)*	*(2,023)*	*(319)*	*(370)*	*(366)*	*(436)*	*(599)*	*(2,188)*
One year	29 30	29 30	24 25	26 28	28 30	**30 31**	33 34	37 39	29 30	32 34	32 34	**35 36**
Five years	7 8	7 9	5 6	8 10	9 11	**8 10**	11 13	14 17	8 10	12 15	11 14	**12 15**
Ten years	4 5	2 3	1 2	3 4	3 4	**3 4**	3 5	7 12	4 6	6 9	4 7	**5 8**
WALES	*(81)*	*(116)*	*(207)*	*(214)*	*(126)*	*(926)*	*(117)*	*(150)*	*(232)*	*(235)*	*(141)*	*(1,008)*
One year	39 41	33 35	28 29	24 25	29 30	**32 34**	36 37	37 39	37 39	35 38	32 34	**39 41**
Five years	19 23	11 14	5 6	6 7	5 7	**9 11**	16 21	13 17	16 19	16 21	10 14	**16 20**
Ten years	7 10	5 10	2 4	4 6	1 2	**4 6**	8 12	4 7	10 15	7 12	5 9	**8 12**

[1]   Including those for whom no deprivation category was assigned. C - crude survival, R - relative survival (see Chapter 3).
     Number of patients contributing to each analysis in parentheses.

	1981-85						1986-90						ADULTS
	Affluent	2	3	4	Deprived	All[1]	Affluent	2	3	4	Deprived	All[1]	
	C R	C R	C R	C R	C R	C R	C R	C R	C R	C R	C R	C R	
	(3,537)	*(3,624)*	*(3,616)*	*(3,446)*	*(2,691)*	*(17,321)*	*(3,648)*	*(3,978)*	*(3,981)*	*(3,352)*	*(2,614)*	*(17,757)*	**ENGLAND & WALES**
	47 50	46 49	45 47	44 46	41 43	45 47	52 54	52 54	48 50	47 49	45 47	49 51	*One year*
	21 26	20 25	19 24	18 22	17 22	19 24	23 29	22 28	20 26	20 25	19 24	21 27	*Five years*
	11 16	10 16	9 15	9 14	8 13	9 15							
	(3,442)	*(3,480)*	*(3,393)*	*(3,211)*	*(2,556)*	*(16,392)*	*(3,498)*	*(3,810)*	*(3,733)*	*(3,076)*	*(2,469)*	*(16,729)*	**ENGLAND**
	48 50	46 49	45 48	44 46	41 43	45 47	52 54	52 54	48 51	47 49	45 48	49 52	*One year*
	21 26	20 25	19 24	17 22	17 22	19 24	23 29	22 28	20 26	20 26	19 24	21 27	*Five years*
	11 16	10 16	9 15	9 14	8 13	9 15							
	(307)	*(404)*	*(467)*	*(552)*	*(611)*	*(2,372)*	*(362)*	*(450)*	*(509)*	*(512)*	*(574)*	*(2,415)*	**Northern & Yorkshire**
	45 48	47 50	44 46	42 45	40 42	45 48	45 47	51 54	42 45	45 48	45 48	47 50	*One year*
	20 26	18 22	18 24	17 23	17 22	18 24	24 32	24 32	18 24	18 24	19 24	21 27	*Five years*
	10 17	8 12	6 11	8 15	8 13	8 13							
	(342)	*(381)*	*(453)*	*(508)*	*(335)*	*(2,034)*	*(260)*	*(379)*	*(442)*	*(426)*	*(385)*	*(1,901)*	**Trent**
	46 48	46 49	44 47	44 47	38 41	45 48	45 48	49 52	43 45	42 45	38 40	45 47	*One year*
	19 23	24 31	20 25	18 24	14 19	20 25	17 22	24 31	18 23	18 24	15 20	19 25	*Five years*
	12 17	10 16	10 16	8 13	6 11	9 15							
	(453)	*(395)*	*(374)*	*(218)*	*(76)*	*(1,542)*	*(499)*	*(500)*	*(411)*	*(252)*	*(84)*	*(1,761)*	**Anglia & Oxford**
	40 42	35 37	35 38	32 33	40 44	38 41	44 47	43 45	41 44	39 41	31 32	43 45	*One year*
	17 21	15 20	12 15	9 12	16 22	15 19	18 23	15 20	14 19	15 19	8 10	16 20	*Five years*
	10 15	6 11	5 8	6 10	*6 11*	7 12							
	(330)	*(294)*	*(247)*	*(248)*	*(155)*	*(1,290)*	*(341)*	*(371)*	*(369)*	*(333)*	*(271)*	*(1,724)*	**North Thames**
	53 55	53 56	45 48	47 49	52 55	52 54	56 59	56 58	57 60	53 56	53 56	56 58	*One year*
	28 33	26 31	19 24	20 23	25 32	24 29	24 31	22 28	24 30	27 35	25 31	25 30	*Five years*
	16 22	15 22	12 19	11 18	14 22	14 20							
	(648)	*(528)*	*(443)*	*(344)*	*(132)*	*(2,183)*	*(597)*	*(500)*	*(417)*	*(312)*	*(140)*	*(2,003)*	**South Thames**
	47 50	46 48	44 47	48 51	36 39	46 49	61 64	57 61	51 54	55 58	54 57	57 60	*One year*
	21 25	20 25	19 25	23 28	22 27	21 26	31 39	25 32	23 31	23 30	23 30	26 33	*Five years*
	11 18	10 17	9 16	13 20	14 23	11 18							
	(615)	*(755)*	*(651)*	*(347)*	*(123)*	*(2,573)*	*(702)*	*(879)*	*(819)*	*(430)*	*(142)*	*(2,988)*	**South & West**
	50 52	45 48	44 47	42 44	45 47	47 50	54 57	55 58	53 56	50 53	49 53	54 57	*One year*
	21 27	20 25	20 25	16 21	20 26	20 25	25 31	25 33	25 32	25 32	26 35	26 33	*Five years*
	10 16	11 19	11 18	9 15	10 16	11 17							
	(310)	*(320)*	*(384)*	*(448)*	*(386)*	*(1,867)*	*(329)*	*(299)*	*(350)*	*(335)*	*(296)*	*(1,616)*	**West Midlands**
	48 51	45 47	48 51	46 49	37 39	46 49	51 55	47 50	44 47	45 47	48 50	48 51	*One year*
	22 28	18 23	18 22	19 24	14 18	19 23	20 24	16 21	19 24	16 20	21 27	19 24	*Five years*
	9 15	8 13	8 13	7 12	7 12	8 13							
	(437)	*(403)*	*(374)*	*(546)*	*(738)*	*(2,531)*	*(408)*	*(432)*	*(416)*	*(476)*	*(577)*	*(2,321)*	**North & West**
	40 42	43 45	42 45	35 38	36 39	40 43	41 43	43 45	41 43	37 39	37 40	41 44	*One year*
	17 22	16 21	16 21	12 16	14 19	15 20	17 22	16 21	14 19	13 19	15 20	16 21	*Five years*
	8 13	8 15	8 14	6 11	6 11	7 13							
	(95)	*(144)*	*(223)*	*(235)*	*(135)*	*(929)*	*(150)*	*(168)*	*(248)*	*(276)*	*(145)*	*(1,028)*	**WALES**
	40 42	48 50	43 46	40 42	41 43	45 48	47 50	47 50	41 44	45 48	40 43	46 48	*One year*
	16 21	23 29	18 24	18 23	18 24	20 26	20 27	22 31	21 28	17 23	18 24	21 28	*Five years*
	11 18	14 23	9 17	11 19	10 17	12 20							

[1] Including those for whom no deprivation category was assigned. C - crude survival, R - relative survival (see Chapter 3).
Number of patients contributing to each analysis in parentheses.

Fig. 49.1 Relative survival up to ten years: trends by calendar period of diagnosis (1971-90) and NHS Region

ADULTS

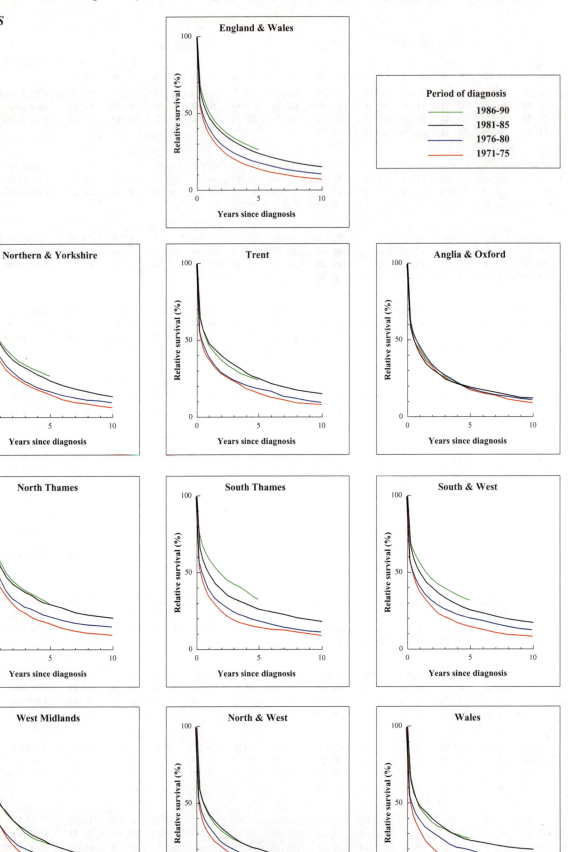

Fig. 49.2 Relative survival up to five years, by deprivation category and NHS Region: patients diagnosed 1986-90

ADULTS

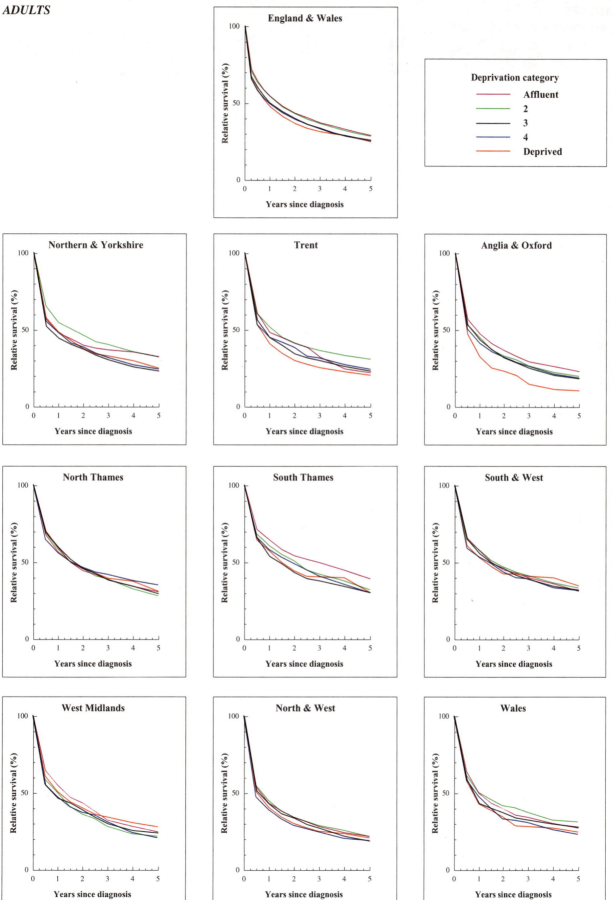

Fig. 49.3   Relative survival at five years by deprivation category, period of diagnosis (1981-90) and NHS Region

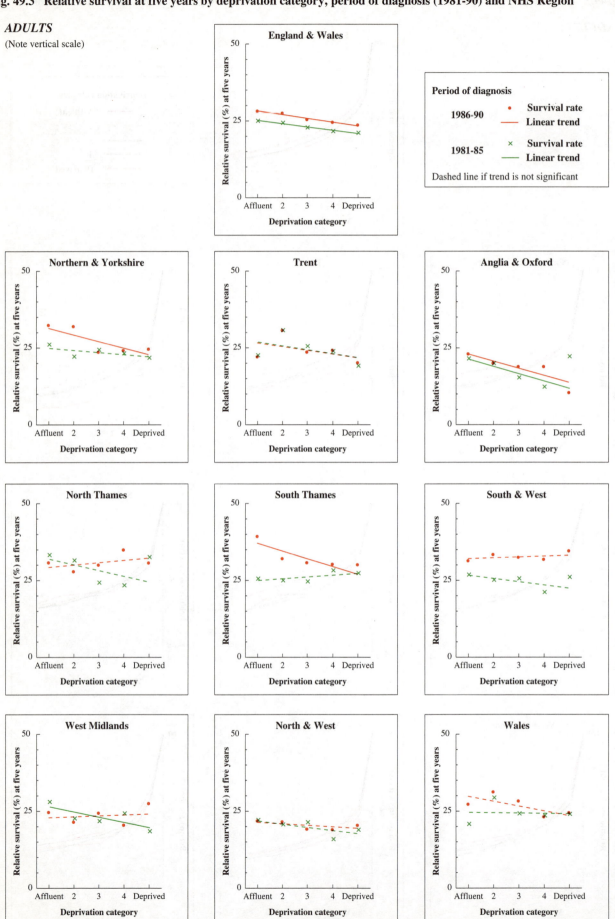

Fig. 49.4 Relative survival (%) by NHS Region and sex:
   one-year and five-year survival for patients diagnosed 1986-90, with 95% confidence intervals

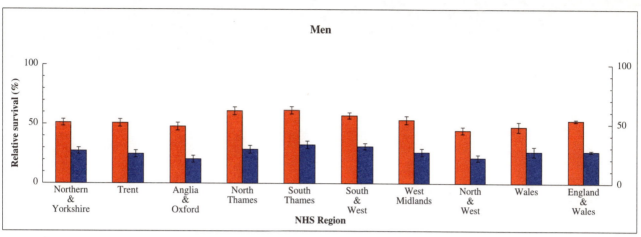

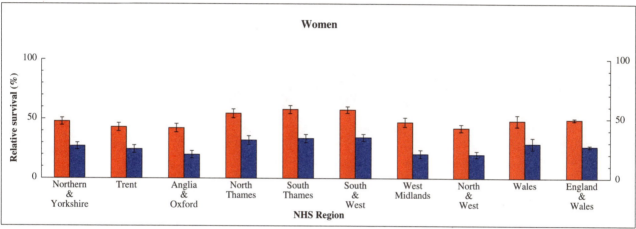

Fig. 49.5 Relative survival (%) at five years, with 95% confidence interval, by age at diagnosis and sex:
   England and Wales, patients diagnosed 1986-90

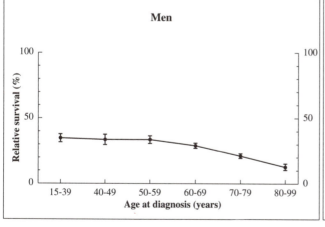

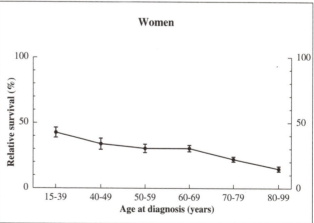

Chapter 50

ALL LEUKAEMIAS *Children*

Leukaemias account for one third of all childhood malignancies. Four out of five are acute lymphoblastic leukaemia (see Chapter 51).

Leukaemias are slightly more common in boys than girls in almost all populations: the ratio of cumulative risks for the age range 0-14 years was 1.1 to 1 in the UK in 1985. The incidence of childhood leukaemia in the UK is lower than elsewhere in western Europe, but has been increasing in both sexes. In boys, incidence fell from an overall peak in 1975, then rose again by 12% between 1979 and 1990. The peak in boys in the early 1970s was not seen in all regions of England[1,2]. Incidence in girls increased by 57% between 1979 and 1990, in parallel with the increase in the Nordic countries. In contrast, incidence in girls in other western European countries fell by more than a third between the mid-1970s and the mid-1980s[3]. In England and Wales in 1990, the incidence rate was 4.9 per 100,000 in boys under 15, and 4.4 in girls: the age group of highest risk is 1 to 4 years.

In common with most western European countries, there has been an impressive decline in mortality from childhood leukaemia in England and Wales. Mortality rates have fallen by more than half since the late 1960s despite the overall increase in incidence; the average rate of decline has been about 12% every five years. The widespread decline in mortality is clearly attributable to improvements in treatment.

The survival analyses reported here are based on almost 7,000 children diagnosed with leukaemia during the period 1971-90 and followed up to the end of 1995, about 90% of those eligible (Table 50.1). Almost 500 children (6%) were excluded because their vital status was unknown at 6 July 1997, and a further 200 (3%) because their survival was zero or unknown. Almost 80% of the leukaemias were classified as acute lymphocytic and 13% as acute myeloid in ICD-9 (Table 50A). These sub-types were reflected in the morphological codes assigned from the International Classification of Diseases for Oncology at the time of registration of the leukaemia: 77% had been coded to acute lymphocytic and 12% to acute myeloid.

Because of the particular interest in survival trends for leukaemias in children, full results are presented by deprivation category, NHS Region, sex and calendar period of diagnosis, even though the overall precision of the survival estimates is lower than for more common malignancies. This is an exception to the rule ("West Midlands rule": see Chapter 3) adopted for all other cancers, and many survival rates are imprecise: these are

italicised in Table 50.2, and should be interpreted accordingly. No analysis was done if fewer patients (usually less than 10) were available for analysis in a given NHS Region, deprivation group, sex and calendar period than the number of time intervals since diagnosis for which survival would be analysed (see Chapter 3). In each case, the number of patients is shown in Table 50.2.

Survival trends

Relative survival[a] for children diagnosed with leukaemia in England and Wales during 1986-90 was 86% at one year and 69% five years after diagnosis. The improvement in these national average survival rates since the early 1970s has been little short of dramatic: a 20% gain in one-year survival and a 36% improvement in survival at five years (Table 50.2).

Five-year survival jumped 13% (from 33% to 46%) between 1971-75 and 1976-80, and rose a further 16% for children diagnosed during 1981-85. Survival at ten years also increased sharply, from 26% for children diagnosed during the early 1970s up to 40% (1976-80), and then to 56% for those diagnosed during 1986-90. The magnitude of these improvements in survival is clear from Figure 50.1, which also shows that all regions have seen huge improvements, although not all at the same time.

Survival is generally slightly higher for girls than for boys, but the increase has been slightly more rapid for boys. The average improvement in one-year survival between successive quinquennia from 1971-75 to 1986-90 has been 5% for girls and 7% for boys (see Table 4.4). The corresponding rate of improvement in five-year survival was 12-13% every five years.

For children diagnosed during 1986-90, one-year survival rates range from 82% to 90% in the various regions, and five-year survival rates range from 62% to 75%, although none of the regional survival rates is significantly different from the national average (Figure 50.4).

Survival at five and ten years increased more in Anglia & Oxford than in other regions during the 1970s, and subsequent gains have been smaller: this is in marked contrast with the pattern in this region for survival from leukaemia in adults (Figure 50.1, cf. Figure 49.1). Large

---

[a] Children have very low mortality from other causes of death, and adjustment for this mortality therefore makes very little difference to the survival estimates. Crude and relative survival rates up to ten years after diagnosis are usually the same or within one per cent of each other, and can be considered as interchangeable.

Table 50A Distribution of leukaemias in children

ICD-9 code	Description	1971-75 No.	%	1976-80 No.	%	1981-85 No.	%	1986-90 No.	%	Total No.	%
204	Lymphoid leukaemia[1]	1,413	73.2	1,395	80.0	1,323	80.2	1,306	79.6	**5,437**	**78.1**
205	Myeloid leukaemia[1]	333	17.2	257	14.7	240	14.6	255	15.5	**1,085**	**15.6**
206	Monocytic leukaemia	25	1.3	17	1.0	24	1.5	17	1.0	**83**	**1.2**
207	Other specified leukaemia	160	8.3	54	3.1	15	0.9	17	1.0	**246**	**3.5**
208	Leukaemia of unspecified cell type[2]	-	-	21	1.2	47	2.9	45	2.7	**113**	**1.6**
	Total	**1,931**		**1,744**		**1,649**		**1,640**		**6,964**	

[1] Includes acute, chronic, other and unspecified. For lymphoid leukaemia, all except 148 (2.7%) were acute (see Chapter 51).
[2] Rubric 208 was assigned to polycythaemia vera in ICD-8 (1971-78): this neoplasm is not included.

increases in survival occurred in several regions around 1980. Children diagnosed in West Midlands during 1986-90 had five-year relative survival of 75%, an increase of 19% compared with children diagnosed in the early 1980s. The corresponding gain in five-year survival was only 1% in North Thames and South & West (Table 50.2, Figure 50.1). Regional survival estimates are each based on 150-250 children, and are precise.

Deprivation

Survival is generally highest in the most affluent groups, although the survival curves for children in the various deprivation categories who were diagnosed during 1986-90 are close (Figure 50.2). Differences in survival between deprivation categories are generally small (Table 50.2). Among 1,600 children of known deprivation status diagnosed in England and Wales during 1981-85, the estimated difference in survival between children in the most affluent and the most deprived groups was 4% at one year and 2% at five years, but these differences are not significant (see Table 4.9). The deprivation gap for 1,600 children diagnosed during 1986-90 was slightly larger: 5%

at one year and 4% at five years. The difference in one-year survival is significant, but the difference in five-year survival is not: this is reflected in Figure 50.3, which shows the national and regional gradients in five-year survival for patients diagnosed during the 1980s. National survival estimates for each deprivation category are based on about 300 children.

Regional estimates of survival for each deprivation category are almost all imprecise. The discrepant curves in South Thames and Wales should be discounted: they are based on 11 or 12 patients each (Table 50.2). Regional patterns of survival between deprivation groups are very variable (Figure 50.3).

1. Bell CMJ. Temporal trends in the incidence of childhood leukaemia and in radiation exposure of parents. J Radiol Prot 1992; 12: 3-8
2. Craft AW, Amineddine JES, Scott JES, Wagget J. The Northern Region Children's Malignant Disease Registry 1968-82: incidence and survival. Br J Cancer 1987; 56: 853-858
3. Coleman MP, Estève J, Damiecki P, Arslan A, Renard H. *Trends in cancer incidence and mortality (IARC Scientific Publications No. 121).* Lyon: International Agency for Research on Cancer, 1993

Table 50.1 Ineligible and excluded records, and patients included in analyses, by calendar period of diagnosis

	Calendar period of diagnosis														
	1971-75			**1976-80**			**1981-85**			**1986-90**			**Overall**		
Total registered	**2,121**			**1,951**			**1,939**			**1,937**			**7,948**		
Ineligible	*Records*	*Patients*	*%*	*Records*	*Patients*	*%*	*Records*	*Patients*	*%*	*Records*	*Patients*	*%*	*Records*	*Patients*	*%*
Incomplete data[1]	10	10		15	15		36	36		30	30		91	**91**	
Resident outside England and Wales	46	46		42	42		37	36		7	5		132	**129**	
In situ neoplasm[2]	0	0		0	0		0	0		0	0		0	**0**	
Benign or uncertain behaviour[2]	0	0		0	0		0	0		0	0		0	**0**	
Metastatic[3]	-	-		-	-		-	-		-	-		-	**-**	
Otherwise ineligible[3]	-	-		-	-		-	-		-	-		-	**-**	
Lymphoma[4]	-	-		-	-		-	-		-	-		-	**-**	
Leukaemia or myeloma[4]	-	-		-	-		-	-		-	-		-	**-**	
		56			**57**			**72**			**35**			**220**	
Total eligible		**2,065**	**100**		**1,894**	**100**		**1,867**	**100**		**1,902**	**100**		**7,728**	**100**
Aged under 15 or 100 years or over at diagnosis[4]	-	-	-	-	-	-	-	-	-	-	-	-	-	**-**	-
Vital status unknown[5]	112	77	3.7	128	89	4.7	160	134	7.2	203	196	10.3	603	**496**	6.4
Duplicate registration[6]	0	0	0.0	0	0	0.0	1	1	<0.1	0	0	0.0	1	**1**	<0.1
Multiple primary[7]	0	0	0.0	2	2	0.1	6	5	0.3	5	4	0.2	13	**11**	0.1
Synchronous tumours[8]	1	1	<0.1	9	7	0.4	9	8	0.4	13	7	0.4	32	**23**	0.3
Sex not known	0	0	0.0	0	0	0.0	0	0	0.0	1	0	0.0	1	**0**	0.0
Sex incompatible with site code[9]	-	-	-	-	-	-	-	-	-	-	-	-	-	**-**	-
Invalid dates[10]	0	0	0.0	0	0	0.0	0	0	0.0	1	1	<0.1	0	**1**	<0.1
Zero survival[11]	60	56	2.7	68	52	2.7	92	70	3.7	65	54	2.8	285	**232**	3.0
Total exclusions		**134**	**6.5**		**150**	**7.9**		**218**	**11.7**		**262**	**13.8**		**764**	**9.9**
Patients included in analysis		**1,931**	**93.5**		**1,744**	**92.1**		**1,649**	**88.3**		**1,640**	**86.2**		**6,964**	**90.1**

[1] Main data item(s) invalid or incompatible with one another: name, sex, date of birth, date of diagnosis and (if present) date of death, postcode, site, morphology and behaviour

[2] Tumour either *in situ* (behaviour code 2), benign (0), uncertain if benign or malignant (1) or metastatic (6 or 9)

[3] Not applicable for this malignancy

[4] Not applicable for this malignancy

[5] Tracing of vital status at National Health Service Central Register incomplete at 6 July 1997

[6] Same site code, sex, cancer registry and cancer registry number as an earlier registration

[7] Same site code and person number as an earlier registration(s): mostly confirmed multiple primary tumours at this site, some unresolved duplicate registrations

[8] Same site code, sex, date of birth and date of diagnosis as another registration(s): mostly synchronous or (in paired organs) bilateral tumours in same anatomic site in one individual, not previously linked: also some duplicate registrations

[9] Not applicable for this malignancy

[10] Impossible sequence of dates (birth, diagnosis and, if present, emigration or death); or date of death after 6 July 1997

[11] Date of diagnosis same as date of death: some are patients who did die on the day of diagnosis, but most were registered solely from a death certificate (death certificate only, or DCO), and their survival time is thus unknown

Table 50.2   Crude and relative survival (%) at one, five and ten years since diagnosis: NHS Region, deprivation category and calendar period of diagnosis

BOYS	1971-75						1976-80					
	Affluent C R	2 C R	3 C R	4 C R	Deprived C R	All[1] C R	Affluent C R	2 C R	3 C R	4 C R	Deprived C R	All[1] C R
ENGLAND & WALES	(210)	(190)	(166)	(190)	(157)	(1,134)	(224)	(187)	(185)	(164)	(193)	(985)
One year	68 68	60 60	63 63	69 69	70 70	65 65	73 73	74 74	75 75	76 76	69 69	74 74
Five years	27 27	33 33	32 32	34 34	34 34	30 30	42 42	50 50	44 44	50 50	38 38	45 45
Ten years	21 21	24 24	21 21	26 26	27 27	22 22	38 38	42 42	36 36	44 44	31 32	38 38
ENGLAND	(198)	(184)	(156)	(168)	(148)	(1,064)	(212)	(175)	(177)	(156)	(183)	(929)
One year	67 67	60 60	61 61	72 72	69 69	65 65	73 73	75 75	75 75	77 77	68 68	73 74
Five years	28 28	32 32	31 31	36 36	33 33	30 30	41 41	51 51	44 44	51 51	36 36	44 44
Ten years	22 22	24 24	22 22	28 28	25 26	22 22	37 37	42 42	36 36	44 45	29 29	38 38
Northern & Yorkshire	(12)	(14)	(17)	(16)	(31)	(137)	(22)	(15)	(15)	(20)	(39)	(114)
One year	74 74	49 49	52 52	81 81	73 73	65 65	75 75	80 80	80 80	71 71	66 67	72 72
Five years	33 33	34 34	11 11	56 56	41 41	29 29	44 44	59 59	39 39	47 47	36 36	43 43
Ten years	25 25	18 18	11 11	43 43	34 35	22 22	40 40	31 31	33 33	41 41	28 28	34 34
Trent	(24)	(21)	(23)	(33)	(22)	(139)	(17)	(16)	(22)	(20)	(27)	(103)
One year	66 67	47 47	60 60	66 66	73 73	61 61	69 69	65 65	68 68	75 75	66 66	69 69
Five years	32 32	24 24	42 42	42 42	23 23	32 32	38 38	40 40	36 36	45 45	40 40	39 39
Ten years	24 24	14 14	21 21	33 33	19 19	23 23	32 32	33 33	36 36	35 35	37 37	34 35
Anglia & Oxford	(30)	(20)	(26)	(15)	(6)	(119)	(25)	(31)	(29)	(13)	(10)	(111)
One year	64 64	60 60	66 66	63 64	- -	63 63	79 80	80 80	72 72	76 76	69 69	77 77
Five years	29 30	39 39	39 39	14 14	- -	29 29	47 47	58 58	45 45	54 54	50 50	51 51
Ten years	26 26	34 34	27 28	7 7	- -	23 23	47 47	52 52	41 41	47 47	50 50	47 48
North Thames	(31)	(27)	(29)	(21)	(18)	(138)	(26)	(21)	(17)	(22)	(12)	(101)
One year	81 81	59 59	68 68	62 62	78 78	67 67	73 73	86 86	82 82	72 72	74 75	77 77
Five years	39 39	30 30	37 37	28 28	43 43	35 35	53 54	66 66	47 47	36 36	25 25	48 48
Ten years	36 36	22 22	30 30	28 28	26 26	29 29	53 54	62 62	47 47	32 32	9 9	44 44
South Thames	(40)	(25)	(23)	(33)	(8)	(136)	(43)	(26)	(12)	(20)	(9)	(115)
One year	53 54	68 68	55 55	68 68	- -	63 63	76 76	80 81	43 43	69 69	- -	70 70
Five years	17 17	32 32	12 12	32 32	- -	25 25	37 37	45 45	16 16	49 49	- -	37 37
Ten years	10 10	28 28	12 12	22 23	- -	17 17	30 30	45 45	8 8	38 38	- -	31 32
South & West	(36)	(33)	(15)	(20)	(10)	(132)	(36)	(33)	(40)	(18)	(15)	(147)
One year	68 68	56 57	50 50	85 85	70 70	64 64	80 80	70 70	88 88	67 67	44 44	75 75
Five years	28 28	30 30	31 31	36 36	31 31	32 32	55 55	48 48	54 55	37 37	18 18	48 49
Ten years	22 22	21 21	25 25	36 36	31 31	28 28	47 47	36 37	45 45	37 37	18 18	41 41
West Midlands	(9)	(15)	(10)	(7)	(10)	(107)	(21)	(16)	(20)	(16)	(31)	(105)
One year	- -	72 72	59 59	- -	37 37	64 64	62 62	75 75	63 63	88 88	61 61	68 68
Five years	- -	46 46	28 28	- -	18 18	29 29	20 20	57 57	33 33	68 68	35 35	40 40
Ten years	- -	26 26	12 12	- -	18 18	17 17	15 15	50 50	23 24	68 68	32 32	35 35
North & West	(16)	(29)	(13)	(23)	(43)	(156)	(22)	(17)	(22)	(27)	(40)	(133)
One year	82 82	65 65	69 69	78 78	61 61	69 69	60 60	60 60	86 87	93 93	80 80	78 78
Five years	24 24	32 32	38 38	39 39	31 31	30 30	28 28	29 29	59 59	70 70	40 40	46 46
Ten years	11 11	28 28	23 23	26 26	23 24	20 20	28 28	17 17	35 36	58 58	27 27	33 33
WALES	(12)	(6)	(10)	(22)	(9)	(70)	(12)	(12)	(8)	(8)	(10)	(56)
One year	74 74	- -	100 100	42 42	- -	67 67	75 75	68 68	- -	- -	100 100	75 75
Five years	16 16	- -	41 41	20 20	- -	32 33	59 59	43 43	- -	- -	71 71	50 50
Ten years	16 16	- -	9 9	13 13	- -	23 23	51 51	43 43	- -	- -	71 71	45 45

[1] Including those for whom no deprivation category was assigned. C - crude survival, R - relative survival (see Chapter 3). Number of patients contributing to each analysis in parentheses.

BOYS — Survival (%), C - crude survival, R - relative survival. Number of patients in parentheses.

Region / Period	Affluent C	Affluent R	2 C	2 R	3 C	3 R	4 C	4 R	Deprived C	Deprived R	All[1] C	All[1] R
1981-85												
ENGLAND & WALES *(n)*	(229)	(183)	(171)	(143)	(171)		(923)					
One year	83	83	82	82	85	85	69	69	81	81	81	81
Five years	61	61	61	61	63	63	57	57	59	60	61	61
	52	52	52	52	56	56	53	53	54	54	53	54
ENGLAND *(n)*	(221)	(180)	(162)	(132)	(164)		(881)					
One year	83	83	82	82	84	85	69	69	81	81	81	81
Five years	60	61	61	62	64	64	56	56	60	60	61	61
	51	51	53	53	57	57	51	51	53	54	53	53
Northern & Yorkshire *(n)*	(17)	(22)	(20)	(19)	(33)		(115)					
One year	70	70	91	91	71	71	57	57	76	76	74	74
Five years	45	45	68	68	61	61	41	41	54	54	54	55
	40	40	68	68	55	56	36	36	51	51	50	50
Trent *(n)*	(14)	(23)	(18)	(20)	(14)		(90)					
One year	93	93	91	91	89	89	65	65	93	93	85	86
Five years	65	65	61	61	72	72	60	60	57	57	63	63
	57	58	40	40	61	61	50	50	44	44	50	50
Anglia & Oxford *(n)*	(37)	(22)	(14)	(11)	(8)		(98)					
One year	72	72	73	73	72	72	72	72	-	-	75	75
Five years	58	58	55	55	58	58	63	63	-	-	60	60
	50	50	50	50	51	51	63	63	-	-	54	54
North Thames *(n)*	(26)	(21)	(18)	(20)	(24)		(112)					
One year	92	92	86	86	83	83	69	69	87	87	85	85
Five years	69	69	76	76	67	68	54	54	62	62	67	67
	62	62	66	66	50	51	54	54	54	54	59	59
South Thames *(n)*	(39)	(25)	(20)	(14)	(12)		(113)					
One year	79	79	92	92	90	90	86	86	92	92	87	87
Five years	53	53	60	60	59	60	71	71	75	76	60	60
	38	38	52	52	49	50	56	56	75	76	49	49
South & West *(n)*	(43)	(24)	(22)	(10)	(10)		(112)					
One year	91	91	75	75	81	81	79	79	90	91	84	84
Five years	72	72	58	59	58	58	69	69	80	81	67	67
	67	68	50	50	54	54	69	69	70	71	61	61
West Midlands *(n)*	(21)	(22)	(21)	(20)	(28)		(112)					
One year	81	81	68	68	95	95	70	70	78	78	78	78
Five years	62	62	59	59	76	76	45	45	46	46	57	57
	47	48	50	50	67	67	40	40	46	46	50	50
North & West *(n)*	(24)	(21)	(29)	(18)	(35)		(129)					
One year	84	84	81	81	90	90	62	62	71	71	78	78
Five years	50	50	56	57	62	62	56	56	65	65	59	59
	38	38	47	48	62	62	56	56	53	54	52	52
WALES *(n)*	(8)	(3)	(9)	(11)	(7)		(42)					
One year	-	-	-	-	-	-	72	72	-	-	86	86
Five years	-	-	-	-	-	-	72	72	-	-	62	62
	-	-	-	-	-	-	72	72	-	-	62	62

Region / Period	Affluent C	Affluent R	2 C	2 R	3 C	3 R	4 C	4 R	Deprived C	Deprived R	All[1] C	All[1] R
1986-90												
ENGLAND & WALES *(n)*	(190)	(204)	(163)	(158)	(168)		(894)					
One year	89	90	88	88	85	85	90	91	85	85	88	88
Five years	70	70	68	68	65	65	63	63	65	65	66	66
ENGLAND *(n)*	(185)	(192)	(156)	(150)	(162)		(854)					
One year	90	90	89	89	87	87	90	90	84	85	88	88
Five years	70	70	69	69	67	67	63	63	65	65	67	67
Northern & Yorkshire *(n)*	(24)	(11)	(26)	(23)	(33)		(117)					
One year	96	96	81	81	88	89	91	92	72	73	85	85
Five years	87	88	40	40	73	73	61	61	52	52	65	65
Trent *(n)*	(11)	(22)	(12)	(19)	(28)		(92)					
One year	91	91	95	95	92	92	74	74	89	89	88	88
Five years	64	64	64	64	66	66	58	58	79	79	67	68
Anglia & Oxford *(n)*	(25)	(21)	(14)	(16)	(8)		(84)					
One year	84	84	91	91	71	71	87	87	75	75	83	83
Five years	60	60	76	77	64	64	56	56	62	62	64	64
North Thames *(n)*	(25)	(28)	(27)	(28)	(34)		(146)					
One year	96	96	89	89	89	89	93	93	91	91	91	91
Five years	72	72	75	75	55	55	68	68	70	70	68	69
South Thames *(n)*	(39)	(29)	(18)	(21)	(3)		(113)					
One year	95	95	86	86	100	100	100	100	-	-	92	92
Five years	72	72	79	80	67	67	76	76	-	-	70	70
South & West *(n)*	(22)	(42)	(22)	(12)	(7)		(106)					
One year	86	86	88	88	86	86	92	92	-	-	89	89
Five years	72	72	59	59	68	68	76	76	-	-	66	66
West Midlands *(n)*	(19)	(12)	(20)	(13)	(16)		(81)					
One year	89	89	100	100	90	90	92	92	100	100	94	94
Five years	69	69	92	92	70	70	60	60	94	94	76	77
North & West *(n)*	(20)	(27)	(17)	(18)	(33)		(115)					
One year	75	75	85	85	77	77	89	89	81	82	82	82
Five years	54	54	67	67	71	71	45	45	54	54	58	58
WALES *(n)*	(5)	(12)	(7)	(8)	(6)		(40)					
One year	-	-	72	72	-	-	100	100	-	-	80	80
Five years	-	-	48	48	-	-	74	74	-	-	57	57

[1] Including those for whom no deprivation category was assigned. C - crude survival, R - relative survival (see Chapter 3).
Number of patients contributing to each analysis in parentheses.

Table 50.2   Crude and relative survival (%) at one, five and ten years since diagnosis: NHS Region, deprivation category and calendar period of diagnosis

GIRLS	1971-75												1976-80											
	Affluent		2		3		4		Deprived		All[1]		Affluent		2		3		4		Deprived		All[1]	
	C	R	C	R	C	R	C	R	C	R	C	R	C	R	C	R	C	R	C	R	C	R	C	R
ENGLAND & WALES	(153)		(123)		(120)		(107)		(128)		(797)		(195)		(151)		(136)		(136)		(121)		(759)	
One year	72	72	74	74	61	61	68	68	65	65	67	67	77	77	82	82	70	70	80	80	70	70	76	76
Five years	38	38	39	39	36	36	41	41	39	39	36	36	50	50	52	52	42	42	51	51	42	42	48	48
Ten years	34	34	31	31	32	32	36	36	34	34	31	31	44	45	48	48	38	38	43	44	38	38	43	43
ENGLAND	(149)		(118)		(108)		(99)		(122)		(755)		(185)		(145)		(125)		(131)		(117)		(717)	
One year	73	73	75	75	62	62	66	66	65	65	68	68	78	78	81	81	71	71	79	79	71	71	76	76
Five years	37	37	39	40	37	37	37	37	38	38	36	36	50	50	51	51	43	43	50	50	43	43	48	48
Ten years	34	34	32	32	32	32	32	32	33	33	31	31	45	45	47	48	38	38	42	42	39	39	43	43
Northern & Yorkshire	(6)		(9)		(14)		(10)		(23)		(89)		(15)		(12)		(10)		(19)		(30)		(88)	
One year	-	-	-	-	62	63	44	44	50	50	56	56	73	73	92	92	50	50	73	73	73	73	74	74
Five years	-	-	-	-	55	55	16	16	33	33	28	28	40	40	40	40	39	39	53	53	56	56	49	49
Ten years	-	-	-	-	48	48	16	16	25	25	25	25	33	33	40	40	28	28	47	47	49	50	43	43
Trent	(14)		(15)		(17)		(17)		(19)		(91)		(20)		(22)		(19)		(26)		(13)		(101)	
One year	100	100	93	93	49	49	69	69	73	73	74	74	65	65	82	82	53	54	89	89	77	77	74	74
Five years	48	48	47	47	21	21	46	46	36	36	38	38	50	50	50	50	32	32	50	50	38	38	45	45
Ten years	48	48	40	40	21	21	46	46	30	30	36	36	45	45	46	46	20	20	43	43	38	38	39	40
Anglia & Oxford	(28)		(18)		(14)		(7)		(3)		(83)		(29)		(24)		(19)		(10)		(4)		(88)	
One year	79	79	78	78	78	78	-	-	-	-	77	77	76	76	83	83	90	90	70	70	-	-	79	79
Five years	43	43	56	56	42	42	-	-	-	-	41	41	49	49	51	51	53	53	60	60	-	-	49	49
Ten years	32	32	44	44	28	29	-	-	-	-	30	30	45	45	51	51	53	53	39	39	-	-	46	46
North Thames	(23)		(19)		(16)		(20)		(8)		(100)		(21)		(10)		(13)		(12)		(10)		(69)	
One year	69	69	84	84	55	55	66	66	-	-	71	71	80	80	68	68	85	85	84	84	70	70	78	78
Five years	39	39	24	24	36	37	46	46	-	-	39	39	52	52	39	39	54	54	51	51	29	29	48	48
Ten years	39	39	20	20	36	37	41	41	-	-	36	36	47	47	39	39	47	47	51	51	18	18	43	44
South Thames	(24)		(18)		(10)		(17)		(21)		(99)		(37)		(24)		(15)		(17)		(10)		(104)	
One year	70	71	60	60	90	90	52	52	67	67	65	65	78	78	87	87	61	61	76	76	79	80	77	77
Five years	37	37	20	20	51	51	29	29	24	24	30	30	49	49	54	54	38	38	47	47	31	31	46	46
Ten years	33	33	15	15	40	40	23	23	24	24	25	25	43	43	50	50	38	38	41	41	19	19	41	41
South & West	(27)		(20)		(15)		(7)		(10)		(88)		(28)		(27)		(22)		(11)		(4)		(95)	
One year	70	70	75	75	57	57	-	-	47	47	67	67	71	71	88	88	67	67	82	82	-	-	76	77
Five years	33	33	65	65	31	31	-	-	38	38	39	40	39	39	63	63	31	31	36	36	-	-	44	44
Ten years	30	30	45	45	25	25	-	-	38	38	32	32	39	39	55	55	26	26	36	36	-	-	41	41
West Midlands	(8)		(4)		(5)		(2)		(4)		(70)		(16)		(9)		(13)		(12)		(18)		(69)	
One year	-	-	-	-	-	-	-	-	-	-	62	62	87	87	-	-	76	76	68	68	59	59	70	70
Five years	-	-	-	-	-	-	-	-	-	-	31	31	75	75	-	-	68	69	51	51	48	48	60	60
Ten years	-	-	-	-	-	-	-	-	-	-	27	27	62	62	-	-	68	69	35	35	43	43	53	53
North & West	(19)		(15)		(17)		(19)		(34)		(135)		(19)		(17)		(14)		(24)		(28)		(103)	
One year	74	74	73	73	52	52	79	79	71	71	70	70	100	100	70	70	85	85	80	80	68	68	80	80
Five years	37	37	39	39	23	23	47	47	41	42	39	39	59	59	41	41	37	37	50	50	38	38	45	45
Ten years	37	37	39	39	23	23	36	36	35	36	35	35	48	48	35	35	29	29	42	42	38	38	39	39
WALES	(4)		(5)		(12)		(8)		(6)		(42)		(10)		(6)		(11)		(5)		(4)		(42)	
One year	-	-	-	-	48	48	-	-	-	-	54	54	58	58	-	-	62	62	-	-	-	-	76	76
Five years	-	-	-	-	32	32	-	-	-	-	40	40	39	39	-	-	34	34	-	-	-	-	47	47
Ten years	-	-	-	-	32	32	-	-	-	-	40	40	39	39	-	-	34	34	-	-	-	-	47	47

[1] Including those for whom no deprivation category was assigned. C - crude survival, R - relative survival (see Chapter 3). Number of patients contributing to each analysis in parentheses.

	1981-85 Affluent C R	2 C R	3 C R	4 C R	Deprived C R	All[1] C R	1986-90 Affluent C R	2 C R	3 C R	4 C R	Deprived C R	All[1] C R	GIRLS
(N)	(181)	(150)	(144)	(136)	(106)	(726)	(151)	(175)	(121)	(147)	(142)	(746)	**ENGLAND & WALES**
One year	78 78	81 81	78 78	83 83	74 74	**79 79**	91 91	83 83	83 83	84 84	82 82	**85 85**	*One year*
Five years	62 62	64 64	61 61	65 65	59 59	**62 62**	73 74	72 72	71 71	71 71	72 72	**72 72**	*Five years*
	57 57	63 63	55 55	63 63	57 57	59 59							
(N)	(177)	(142)	(137)	(127)	(101)	(690)	(144)	(166)	(117)	(134)	(132)	(702)	**ENGLAND**
One year	78 78	81 81	78 79	82 82	74 74	**79 79**	91 91	82 82	84 84	84 84	82 82	**85 85**	*One year*
Five years	61 61	65 65	61 61	66 66	59 59	**62 62**	74 74	70 70	72 72	69 69	71 71	**71 71**	*Five years*
	56 56	65 65	55 55	63 64	57 57	59 59							
(N)	(20)	(23)	(19)	(20)	(21)	(103)	(10)	(20)	(19)	(17)	(24)	(91)	**Northern & Yorkshire**
One year	85 85	78 78	84 84	84 84	59 59	**78 78**	100 100	84 84	95 95	82 83	75 75	**86 86**	*One year*
Five years	75 75	69 69	73 73	70 70	46 46	**66 67**	90 90	80 80	84 84	70 70	67 67	**77 77**	*Five years*
	70 70	69 69	73 73	70 70	46 46	65 66							
(N)	(14)	(12)	(19)	(21)	(6)	(72)	(10)	(17)	(10)	(16)	(17)	(70)	**Trent**
One year	100 100	100 100	73 73	76 76	- -	**83 83**	100 100	82 82	70 70	87 88	77 77	**83 83**	*One year*
Five years	93 93	75 75	52 52	47 47	- -	**64 64**	90 90	71 71	49 50	63 63	77 77	**70 70**	*Five years*
	93 93	75 75	47 47	47 47	- -	62 62							
(N)	(22)	(18)	(20)	(8)	(4)	(72)	(27)	(18)	(7)	(17)	(6)	(77)	**Anglia & Oxford**
One year	63 63	66 66	79 79	- -	- -	**72 72**	88 88	59 59	- -	83 83	- -	**80 80**	*One year*
Five years	54 54	33 33	65 65	- -	- -	**54 54**	74 74	43 43	- -	77 77	- -	**69 69**	*Five years*
	50 50	33 33	60 60	- -	- -	51 51							
(N)	(20)	(22)	(19)	(14)	(11)	(86)	(21)	(20)	(20)	(23)	(22)	(108)	**North Thames**
One year	85 85	82 82	79 79	93 93	72 72	**82 83**	86 86	80 80	68 68	83 83	91 91	**81 81**	*One year*
Five years	60 60	82 82	58 58	93 93	63 63	**71 71**	62 62	80 80	59 59	74 74	82 82	**71 71**	*Five years*
	55 55	82 82	47 47	93 93	63 63	67 68							
(N)	(34)	(14)	(21)	(16)	(6)	(93)	(21)	(18)	(15)	(16)	(9)	(82)	**South Thames**
One year	73 73	93 93	76 76	87 87	- -	**78 78**	91 91	95 95	94 94	81 81	67 67	**88 88**	*One year*
Five years	53 53	86 86	71 71	68 68	- -	**63 63**	81 81	89 89	73 73	62 62	56 56	**74 75**	*Five years*
	47 47	86 86	71 71	68 68	- -	61 61							
(N)	(32)	(27)	(21)	(19)	(6)	(106)	(20)	(32)	(25)	(15)	(3)	(95)	**South & West**
One year	94 94	82 82	76 76	78 78	- -	**83 83**	95 95	91 91	92 92	72 72	- -	**89 90**	*One year*
Five years	72 72	67 67	57 57	73 73	- -	**68 68**	75 75	69 69	84 84	46 46	- -	**70 71**	*Five years*
	65 66	67 67	52 52	62 63	- -	63 63							
(N)	(18)	(13)	(8)	(15)	(21)	(76)	(10)	(18)	(10)	(14)	(24)	(77)	**West Midlands**
One year	61 61	77 77	- -	72 72	86 86	**74 74**	78 78	72 72	100 100	93 93	83 83	**84 84**	*One year*
Five years	39 39	62 62	- -	52 52	61 61	**53 54**	59 59	66 66	79 79	93 93	71 71	**74 74**	*Five years*
	39 39	62 62	- -	52 52	61 61	52 52							
(N)	(17)	(13)	(10)	(14)	(26)	(82)	(25)	(23)	(11)	(16)	(27)	(102)	**North & West**
One year	59 59	76 76	90 90	86 86	80 80	**78 78**	92 92	82 82	70 70	94 94	80 81	**85 85**	*One year*
Five years	47 47	38 38	50 50	64 64	61 61	**54 54**	72 72	65 65	52 52	69 69	63 63	**65 65**	*Five years*
	40 41	38 38	28 28	57 57	53 53	46 46							
(N)	(4)	(8)	(7)	(9)	(5)	(36)	(7)	(9)	(4)	(13)	(10)	(44)	**WALES**
One year	- -	- -	- -	- -	- -	**83 83**	- -	100 100	- -	84 84	80 80	**84 84**	*One year*
Five years	- -	- -	- -	- -	- -	**63 63**	- -	100 100	- -	84 84	80 80	**79 79**	*Five years*
	- -	- -	- -	- -	- -	55 55							

[1] Including those for whom no deprivation category was assigned. C - crude survival, R - relative survival (see Chapter 3). Number of patients contributing to each analysis in parentheses.

Table 50.2   Crude and relative survival (%) at one, five and ten years since diagnosis: NHS Region, deprivation category and calendar period of diagnosis

CHILDREN	1971-75 Affluent C R	2 C R	3 C R	4 C R	Deprived C R	All[1] C R	1976-80 Affluent C R	2 C R	3 C R	4 C R	Deprived C R	All[1] C R
ENGLAND & WALES	(363)	(313)	(286)	(297)	(285)	(1,931)	(419)	(338)	(321)	(300)	(314)	(1,744)
One year	70 70	66 66	62 62	69 69	68 68	66 66	75 75	78 78	73 73	77 77	69 70	75 75
Five years	32 32	35 35	34 34	36 36	36 36	33 33	46 46	51 51	43 43	51 51	39 39	46 46
Ten years	27 27	27 27	26 26	30 30	30 30	26 26	41 41	45 45	37 37	44 44	34 34	40 40
ENGLAND	(347)	(302)	(264)	(267)	(270)	(1,819)	(397)	(320)	(302)	(287)	(300)	(1,646)
One year	70 70	66 66	61 61	70 70	67 67	66 66	75 75	78 78	74 74	78 78	69 69	75 75
Five years	32 32	35 35	33 34	36 36	35 35	32 33	46 46	51 51	44 44	50 51	39 39	46 46
Ten years	27 27	27 27	26 26	29 29	29 29	26 26	41 41	45 45	37 37	43 44	33 33	40 40
Northern & Yorkshire	(18)	(23)	(31)	(26)	(54)	(226)	(37)	(27)	(25)	(39)	(69)	(202)
One year	71 71	52 52	57 57	67 67	63 63	61 61	74 74	85 85	68 68	72 72	69 69	72 73
Five years	32 32	30 30	32 32	41 41	38 38	28 29	43 43	51 51	39 39	49 49	45 45	45 45
Ten years	27 27	20 20	28 28	33 33	30 31	23 23	37 37	36 36	31 31	44 44	37 38	38 38
Trent	(38)	(36)	(40)	(50)	(41)	(230)	(37)	(38)	(41)	(46)	(40)	(204)
One year	79 79	66 66	56 56	67 67	73 73	66 66	67 67	74 74	61 61	83 83	70 70	71 71
Five years	38 39	33 33	34 34	43 44	29 29	34 34	45 45	45 45	34 34	48 48	40 40	42 42
Ten years	33 33	25 25	21 21	37. 38	24 24	28 28	39 39	40 40	29 29	39 39	37 37	37 37
Anglia & Oxford	(58)	(38)	(40)	(22)	(9)	(202)	(54)	(55)	(48)	(23)	(14)	(199)
One year	71 71	68 68	70 70	70 70	- -	69 69	77 78	81 81	79 79	73 74	62 63	78 78
Five years	36 36	47 47	40 40	16 16	- -	34 34	48 48	55 55	48 48	56 57	35 35	50 50
Ten years	29 29	39 39	28 28	9 9	- -	26 26	46 46	51 51	46 46	43 43	35 35	47 47
North Thames	(54)	(46)	(45)	(41)	(26)	(238)	(47)	(31)	(30)	(34)	(22)	(170)
One year	76 76	69 69	63 63	64 64	77 77	69 69	76 76	80 80	83 83	77 77	72 73	77 78
Five years	39 39	28 28	37 37	37 37	49 49	36 36	53 53	58 58	50 50	42 42	27 27	48 48
Ten years	37 37	21 21	32 33	34 35	37 37	32 32	50 51	55 55	46 47	39 39	13 13	44 44
South Thames	(64)	(43)	(33)	(50)	(29)	(235)	(80)	(50)	(27)	(37)	(19)	(219)
One year	60 60	65 65	65 65	62 62	76 76	64 64	77 77	84 84	53 53	72 72	74 74	73 73
Five years	24 25	27 27	24 24	31 31	28 28	27 27	43 43	50 50	28 28	48 48	32 32	41 41
Ten years	19 19	23 23	21 21	23 23	24 25	21 21	36 36	48 48	24 24	39 39	26 26	36 36
South & West	(63)	(53)	(30)	(27)	(20)	(220)	(64)	(60)	(62)	(29)	(19)	(242)
One year	69 69	64 64	53 53	78 78	59 59	65 65	76 76	78 78	80 80	73 73	56 56	76 76
Five years	30 30	43 43	31 31	31 31	35 35	35 35	48 48	55 55	46 46	37 37	31 31	47 47
Ten years	25 25	30 30	25 25	31 31	35 35	29 29	43 44	45 45	38 39	37 37	31 31	41 41
West Midlands	(17)	(19)	(15)	(9)	(14)	(177)	(37)	(25)	(33)	(28)	(49)	(174)
One year	47 47	73 73	66 66	- -	39 39	63 63	73 73	68 68	68 69	79 79	61 61	69 69
Five years	17 17	36 36	39 39	- -	25 26	30 30	44 44	56 56	47 47	61 61	40 40	48 48
Ten years	11 11	20 20	27 27	- -	19 19	21 21	35 35	52 52	41 41	54 54	36 36	42 42
North & West	(35)	(44)	(30)	(42)	(77)	(291)	(41)	(34)	(36)	(51)	(68)	(236)
One year	78 78	68 68	60 60	78 79	65 65	70 70	78 78	65 65	86 86	86 86	75 75	79 79
Five years	31 31	34 34	30 30	43 43	35 35	34 34	42 42	35 35	50 50	60 61	39 40	46 46
Ten years	26 26	32 32	23 23	31 31	29 29	27 27	37 37	26 26	33 33	50 51	32 32	36 36
WALES	(16)	(11)	(22)	(30)	(15)	(112)	(22)	(18)	(19)	(13)	(14)	(98)
One year	69 69	62 62	72 72	58 58	73 73	62 63	68 68	78 78	67 67	70 70	86 86	75 75
Five years	25 25	36 36	36 36	38 38	54 54	35 35	50 50	51 51	36 36	54 54	50 50	49 49
Ten years	25 25	26 26	22 23	32 32	54 54	29 29	45 45	51 51	31 31	54 54	50 50	46 46

[1] Including those for whom no deprivation category was assigned. C - crude survival, R - relative survival (see Chapter 3). Number of patients contributing to each analysis in parentheses.

Columns grouped under two periods — **1981-85** and **1986-90**. Each period has deprivation categories: *Affluent*, 2, 3, 4, *Deprived*, *All[1]*. C = crude survival, R = relative survival. Numbers in parentheses = number of patients contributing to each analysis.

Region	Follow-up	Affl C	Affl R	2 C	2 R	3 C	3 R	4 C	4 R	Depr C	Depr R	All C	All R	Affl C	Affl R	2 C	2 R	3 C	3 R	4 C	4 R	Depr C	Depr R	All C	All R
ENGLAND & WALES	(n)	(410)		(333)		(315)		(279)		(277)		(1,649)		(341)		(379)		(284)		(305)		(310)		(1,640)	
	One year	81	81	81	81	82	82	76	76	78	79	80	80	90	90	86	86	84	84	87	88	83	84	86	86
	Five years	61	61	62	62	62	62	61	61	59	59	61	62	72	72	70	70	67	68	67	67	68	68	69	69
		54	54	57	57	55	56	58	58	55	55	56	56												
ENGLAND	(n)	(398)		(322)		(299)		(259)		(265)		(1,571)		(329)		(358)		(273)		(284)		(294)		(1,556)	
	One year	81	81	82	82	82	82	75	75	78	78	80	80	90	90	86	86	86	86	87	87	83	83	86	87
	Five years	61	61	63	63	63	63	61	61	59	59	61	62	72	72	70	70	69	69	66	66	68	68	69	69
		53	54	58	58	56	56	57	57	55	55	56	56												
Northern & Yorkshire	(n)	(37)		(45)		(39)		(39)		(54)		(218)		(34)		(31)		(45)		(40)		(57)		(208)	
	One year	78	78	84	84	77	77	71	71	69	70	76	76	97	97	84	84	91	91	87	88	73	73	85	86
	Five years	62	62	68	69	67	67	56	56	51	51	60	60	88	88	67	67	78	78	65	65	58	58	70	70
		57	57	68	69	64	64	53	54	49	49	57	58												
Trent	(n)	(28)		(35)		(37)		(41)		(20)		(162)		(21)		(39)		(22)		(35)		(45)		(162)	
	One year	96	96	94	94	81	81	71	71	85	85	84	85	95	95	90	90	82	82	80	80	84	85	86	86
	Five years	79	79	66	66	62	62	53	54	60	60	63	64	76	76	67	67	59	59	60	60	78	78	68	69
		75	75	52	52	54	54	49	49	50	50	55	56												
Anglia & Oxford	(n)	(59)		(40)		(34)		(19)		(12)		(170)		(52)		(39)		(21)		(33)		(14)		(161)	
	One year	69	69	70	70	76	77	79	79	84	84	74	74	86	86	77	77	71	71	85	85	86	86	82	82
	Five years	57	57	45	45	62	62	63	63	59	59	57	57	67	67	61	61	67	67	67	67	78	78	66	66
		50	50	42	43	56	56	63	63	59	59	53	53												
North Thames	(n)	(46)		(43)		(37)		(34)		(35)		(198)		(46)		(48)		(47)		(51)		(56)		(254)	
	One year	89	89	84	84	81	81	79	79	83	83	84	84	91	91	85	85	80	80	88	88	91	91	87	87
	Five years	65	65	79	79	62	62	70	70	62	63	68	69	68	68	77	77	57	57	71	71	75	75	70	70
		59	59	74	74	48	49	70	70	57	57	62	63												
South Thames	(n)	(73)		(39)		(41)		(30)		(18)		(206)		(60)		(47)		(33)		(37)		(12)		(195)	
	One year	77	77	92	92	83	83	87	87	84	84	83	83	93	93	89	90	97	97	92	92	59	59	90	90
	Five years	53	53	69	69	66	66	70	70	67	67	61	62	75	75	83	83	70	70	70	70	42	42	72	72
		42	42	64	64	61	61	63	63	67	67	55	55												
South & West	(n)	(75)		(51)		(43)		(29)		(16)		(218)		(42)		(74)		(47)		(27)		(10)		(201)	
	One year	92	92	78	79	78	78	79	79	82	82	83	83	90	90	89	89	89	89	81	81	100	100	89	89
	Five years	72	72	63	63	58	58	72	72	75	76	67	67	74	74	63	64	77	77	59	59	57	57	68	68
		67	67	59	59	53	53	65	65	69	69	62	62												
West Midlands	(n)	(39)		(35)		(29)		(35)		(49)		(188)		(29)		(30)		(30)		(27)		(40)		(158)	
	One year	72	72	71	71	90	90	71	71	81	81	77	77	86	86	83	83	93	93	92	92	90	90	89	89
	Five years	51	51	60	60	69	69	48	48	52	53	56	56	65	65	76	77	73	73	77	77	80	80	75	75
		43	43	54	55	58	58	45	45	52	53	51	51												
North & West	(n)	(41)		(34)		(39)		(32)		(61)		(211)		(45)		(50)		(28)		(34)		(60)		(217)	
	One year	73	74	79	79	90	90	72	72	75	75	78	78	84	84	84	84	74	74	91	91	81	81	83	83
	Five years	49	49	49	50	59	59	59	59	64	64	57	57	64	64	66	66	64	64	56	56	58	58	62	62
		39	39	44	44	53	54	56	56	53	54	49	50												
WALES	(n)	(12)		(11)		(16)		(20)		(12)		(78)		(12)		(21)		(11)		(21)		(16)		(84)	
	One year	92	92	73	73	87	87	80	80	83	83	85	85	83	83	85	85	45	45	90	90	87	87	82	82
	Five years	84	84	41	41	51	51	65	65	58	59	63	63	67	67	71	71	36	36	80	80	75	75	69	69
		76	76	31	31	44	44	65	65	58	59	59	59												

[1] Including those for whom no deprivation category was assigned. C - crude survival, R - relative survival (see Chapter 3).
Number of patients contributing to each analysis in parentheses.

Fig. 50.1   Relative survival up to ten years: trends by calendar period of diagnosis (1971-90) and NHS Region

CHILDREN

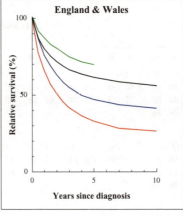

England & Wales

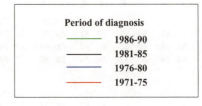

Period of diagnosis
1986-90
1981-85
1976-80
1971-75

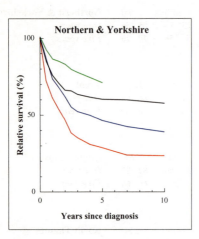

Northern & Yorkshire

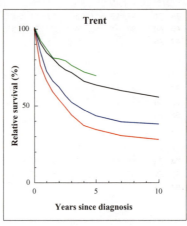

Trent

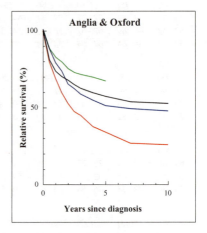

Anglia & Oxford

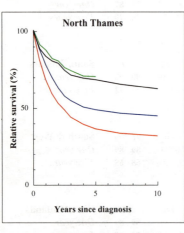

North Thames

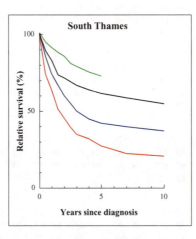

South Thames

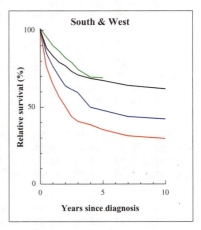

South & West

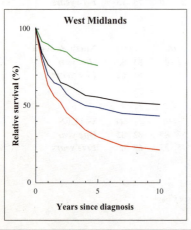

West Midlands

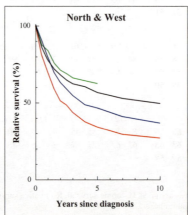

North & West

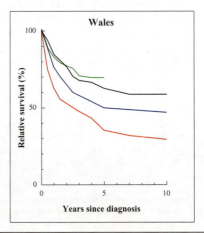

Wales

Fig. 50.2   Relative survival up to five years, by deprivation category and NHS Region: patients diagnosed 1986-90

CHILDREN

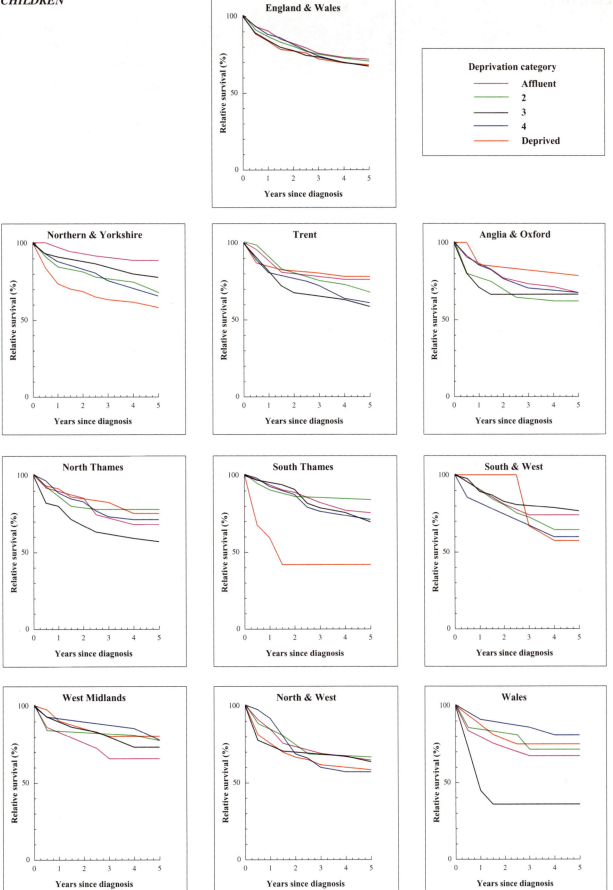

Fig. 50.3 Relative survival at five years by deprivation category, period of diagnosis (1981-90) and NHS Region

CHILDREN

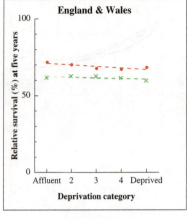

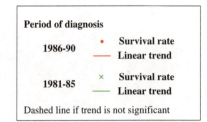

Period of diagnosis

1986-90	● Survival rate
	—— Linear trend
1981-85	✕ Survival rate
	—— Linear trend

Dashed line if trend is not significant

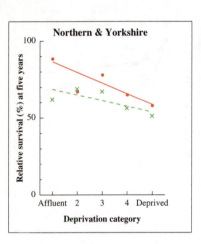

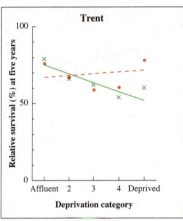

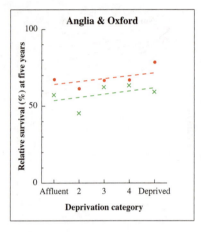

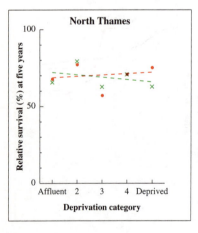

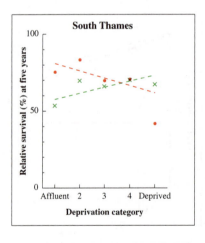

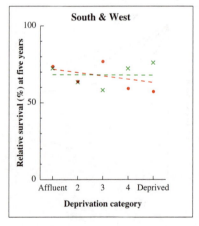

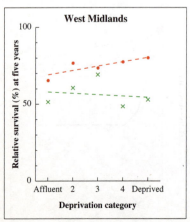

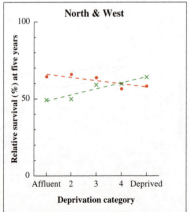

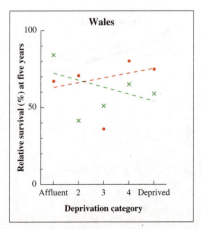

Fig. 50.4 **Relative survival (%) by NHS Region and sex:**
       one-year and five-year survival for children diagnosed 1986-90, with 95% confidence intervals

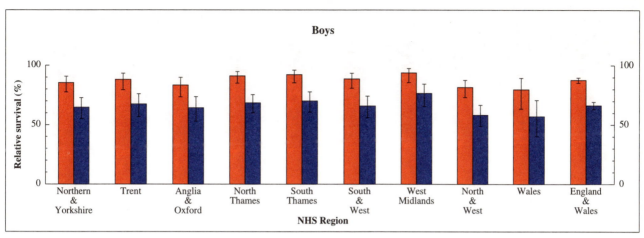

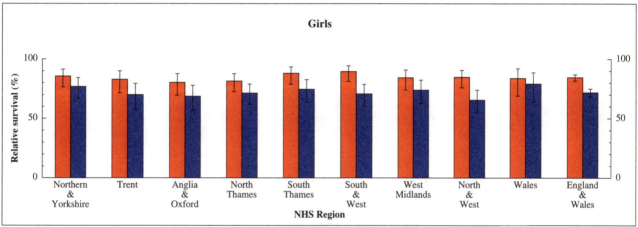

Chapter 51

ACUTE LYMPHOID LEUKAEMIA *Children*

Acute lymphoid leukaemia is the most common malignancy in children in England and Wales, accounting for 30% of all cancers and 80% of all leukaemias. About 300 new cases are diagnosed each year. Acute lymphoid leukaemia is slightly more common in boys than girls, but incidence rose more steeply between 1971 and 1990 in girls (by 83%, compared with an overall increase of 32% in boys), and the sex ratio fell from 1.4 to 1 during the 1970s to about 1.2 to 1 during the 1980s. In England and Wales in 1990, incidence rates were 3.7 per 100,000 in boys and 3.3 in girls, broadly similar to the rates in other western European countries. Peak incidence occurs in boys aged 4 and girls aged 2; almost two-thirds of the cases occur in children aged 2 to 6 years.

In direct contrast to the increase in incidence since the 1970s, mortality from childhood leukaemia has fallen by more than half since the late 1960s, a decline that is clearly attributable to improvements in treatment.

The survival analyses reported here are based on 5,300 children who were diagnosed with acute lymphoid leukaemia in England and Wales between 1971 and 1990 and followed up to the end of 1995, about 91% of eligible cases (Table 51.1). Overall, about 6% of eligible children were excluded because their vital status was unknown on 6 July 1997: the proportion reached 9% for the period 1986-90. Another 2% of the children were excluded because their survival was recorded as zero, or was unknown. The two categories cannot be distinguished in these data (see Chapter 2).

Because of the particular interest in survival trends for leukaemias in children, full results are presented by deprivation category, NHS Region, sex and calendar period of diagnosis, even though the overall precision of the survival estimates is lower than for more common malignancies. This is an exception to the rule on the global precision of survival estimates ("West Midlands rule": see Chapter 3) adopted for all other cancers, and many of the survival rates presented are therefore imprecise: these are italicised in Table 51.2, and should be interpreted accordingly. No analysis was done if fewer patients (usually less than 10) were available for analysis in a given deprivation group, NHS Region, sex and calendar period than the number of time intervals since diagnosis for which survival would be analysed (see Chapter 3). In each case, the number of patients is shown in Table 51.2.

Survival trends

Relative survival[a] for children diagnosed with acute lymphoid leukaemia in England and Wales during 1986-90 was 93% at one year and 76% at five years. The improvement in national average survival rates compared with those for patients diagnosed during 1971-75 has been quite dramatic: a 17% improvement at one year and a 36% increase in five-year survival (Table 51.2). Five-year survival increased by 12% in the 1970s, and it jumped by 18% (from 52% to 70%) between 1976-80 and 1981-85 alone. Survival at ten years after diagnosis has also risen rapidly, by 14% during the 1970s and by a further 18% (from 46% to 64%) between 1976-80 and 1981-85. The national improvement was greatest in the early 1980s, but in four regions (Northern & Yorkshire, Anglia & Oxford, West Midlands and Wales) there were gains of 10% to 20% in short-term and long-term survival for children diagnosed during the 1970s.

The improvements in survival are shown in Figure 51.1. The patterns are of course closely similar to those for all leukaemias combined (see Figure 50.1), but survival rates for acute lymphocytic leukaemia are higher than for all leukaemias combined (80% of which are acute lymphocytic), because other types of leukaemia in children have had a substantially poorer prognosis[1,2]. Very large improvements in survival occurred in all regions. The average improvement in one-year survival between successive five-year periods up to 1986-90 was 6% for boys and 4% for girls (see Table 4.4). The pace of improvement in national average five-year survival has been still more striking: 12-13% every five years.

These remarkable increases in national average survival at both five and ten years confirm that the large gains in one-year survival were not achieved at the cost of long-term fatal adverse effects, and that the proportion of children effectively cured of their disease - with little further excess mortality relative to children their own age - rose substantially during this period.

Improvements have generally been most marked in regions where survival was lowest in the 1970s, and five-year survival rates varied less for children diagnosed during 1986-90 (absolute range 10% for boys, 15% for girls) than for those diagnosed during the early 1970s (20% range; Table 51.2). One-year survival rose by an

---

[a] See footnote in Chapter 50.

average 9-11% every five years in West Midlands, but by only 2-3% every five years in North and West (see Table-Figure 4.5T). Five-year survival improved more in South Thames (42% in boys and 49% in girls) than in North Thames (31%, 21%) over the same timespan, and differences between these regions are now smaller than they were. None of the regional five-year survival rates for patients diagnosed during 1986-90 is significantly different from the national average (Figure 51.4, see also Table-Figure 4.5T).

Girls have consistently had an advantage over boys in survival at five years, but not for one-year survival.

The results presented here are consistent with those published for Great Britain by the National Registry of Childhood Tumours at the Childhood Cancer Research Group in Oxford, in which five-year actuarial survival from acute lymphoid leukaemia was 37% for children diagnosed in 1971-73 and 73% for those diagnosed during 1986-88[2,3]. The substantial improvement in survival has been ascribed to the higher proportion of children entered into trials or managed at specialist centres treating larger numbers of new patients[4], combined with more effective chemotherapy regimens and supportive care[5].

Deprivation

Differences between deprivation categories in survival from acute lymphoid leukaemia among children in England and Wales are small (Table 51.2, Figure 51.2). Among 1,300 children of known deprivation status who were diagnosed during 1981-85, the estimated deficits in one-year and five-year survival for the most deprived group were 3-4%. Among the 1,300 children of known status who were diagnosed during 1986-90, the survival

rates had risen by 5% and the difference between deprivation groups was smaller (2-3%). These slight gradients in five-year survival in the 1980s are shown in Figure 51.3: they are not statistically significant. More detailed analyses of these data confirmed the absence of a socioeconomic survival gradient[5].

Survival by deprivation group shows no consistent patterns between the regions (Figure 51.2, Figure 51.3). Only in Northern & Yorkshire is there a difference in five-year survival of more than 10% between the most affluent and most deprived groups in both the early and late 1980s, but these survival rates are based on about 30 children in each group, and they are statistically imprecise.

1.   Stiller CA, Eatock EM. Survival from acute non-lymphocytic leukaemia, 1971-88: a population based study. Arch Dis Child 1994; 70: 219-223

2.   Stiller CA. Population-based survival rates for childhood cancer in Britain, 1980-91. Br Med J 1994; 309: 1612-1616

3.   Stiller CA, Bunch KJ. Trends in survival for childhood cancer in Britain diagnosed 1971-85. Br J Cancer 1990; 62: 806-815

4.   Stiller CA, Draper GJ. Treatment centre size, entry to trials, and survival in acute lymphoblastic leukaemia. Arch Dis Child 1989; 64: 657-661

5.   Schillinger JA, Grosclaude PC, Honjo S, Quinn MJ, Sloggett A, Coleman MP. Survival from acute lymphocytic leukaemia: socioeconomic status and geographic region. Arch Dis Child 1999; (in press)

Table 51.1 Ineligible and excluded records, and patients included in analyses, by calendar period of diagnosis

	Calendar period of diagnosis														
	1971-75			**1976-80**			**1981-85**			**1986-90**			**Overall**		
Total registered	**1,436**			**1,505**			**1,484**			**1,466**			**5,891**		
Ineligible	*Records*	*Patients*	*%*	*Records*	*Patients*	*%*	*Records*	*Patients*	*%*	*Records*	*Patients*	*%*	*Records*	*Patients*	*%*
Incomplete data[1]	7	7		4	4		2	2		3	3		16	**16**	
Resident outside England and Wales	30	30		32	32		22	21		7	5		91	**88**	
In situ neoplasm[2]	0	0		0	0		0	0		0	0		0	**0**	
Benign or uncertain behaviour[2]	0	0		0	0		0	0		0	0		0	**0**	
Metastatic[3]	-	-		-	-		-	-		-	-		-	**-**	
Otherwise ineligible[3]	-	-		-	-		-	-		-	-		-	**-**	
Lymphoma[4]	-	-		-	-		-	-		-	-		-	**-**	
Leukaemia or myeloma[4]	-	-		-	-		-	-		-	-		-	**-**	
		37			36			23			8			104	
Total eligible		**1,399**	**100**		**1,469**	**100**		**1,461**	**100**		**1,458**	**100**		**5,787**	**100**
Aged under 15 or 100 years or over at diagnosis[4]	-	-	-	-	-	-	-	-	-	-	-	-	-	-	-
Vital status unknown[5]	83	55	3.9	102	73	5.0	119	105	7.2	134	130	8.9	438	**363**	6.3
Duplicate registration[6]	0	0	0.0	0	0	0.0	1	1	<0.1	0	0	0.0	1	**1**	<0.1
Multiple primary[7]	0	0	0.0	1	1	<0.1	6	5	0.3	4	3	0.2	11	**9**	0.2
Synchronous tumours[8]	1	1	<0.1	5	4	0.3	3	2	0.1	10	4	0.3	19	**11**	0.2
Sex not known	0	0	0.0	0	0	0.0	0	0	0.0	0	0	0.0	0	**0**	0.0
Sex incompatible with site code[9]	-	-	-	-	-	-	-	-	-	-	-	-	-	**-**	-
Invalid dates[10]	0	0	0.0	0	0	0.0	0	0	0.0	1	1	<0.1	1	**1**	<0.1
Zero survival[11]	29	27	1.9	31	24	1.6	45	36	2.5	30	26	1.8	135	**113**	2.0
Total exclusions		**83**	**5.9**		**102**	**6.9**		**149**	**10.2**		**164**	**11.2**		**498**	**8.6**
Patients included in analysis		**1,316**	**94.1**		**1,367**	**93.1**		**1,312**	**89.8**		**1,294**	**88.8**		**5,289**	**91.4**

[1]   Main data item(s) invalid or incompatible with one another: name, sex, date of birth, date of diagnosis and (if present) date of death, postcode, site, morphology and behaviour

[2]   Tumour either *in situ* (behaviour code 2), benign (0), uncertain if benign or malignant (1) or metastatic (6 or 9)

[3]   Not applicable for this malignancy

[4]   Not applicable for this malignancy

[5]   Tracing of vital status at National Health Service Central Register incomplete at 6 July 1997

[6]   Same site code, sex, cancer registry and cancer registry number as an earlier registration

[7]   Same site code and person number as an earlier registration(s): mostly confirmed multiple primary tumours at this site, some unresolved duplicate registrations

[8]   Same site code, sex, date of birth and date of diagnosis as another registration(s): mostly synchronous or (in paired organs) bilateral tumours in same anatomic site in one individual, not previously linked: also some duplicate registrations

[9]   Not applicable for this malignancy

[10]   Impossible sequence of dates (birth, diagnosis and, if present, emigration or death); or date of death after 6 July 1997

[11]   Date of diagnosis same as date of death: some are patients who did die on the day of diagnosis, but most were registered solely from a death certificate (death certificate only, or DCO), and their survival time is thus unknown

Table 51.2 **Crude and relative survival (%) at one, five and ten years since diagnosis: NHS Region, deprivation category and calendar period of diagnosis**

BOYS	1971-75 Affluent C	R	2 C	R	3 C	R	4 C	R	Deprived C	R	All[1] C	R	1976-80 Affluent C	R	2 C	R	3 C	R	4 C	R	Deprived C	R	All[1] C	R
ENGLAND & WALES	(128)		(136)		(119)		(133)		(108)		(785)		(171)		(152)		(139)		(134)		(153)		(780)	
One year	80	80	70	70	69	69	82	82	79	79	74	75	82	82	84	84	81	81	79	79	78	78	81	81
Five years	34	34	38	38	38	39	42	42	39	39	36	36	48	48	55	55	51	51	57	57	44	44	51	51
Ten years	28	28	26	26	26	26	33	33	29	30	27	27	43	44	46	46	40	40	50	50	37	37	43	43
ENGLAND	(123)		(131)		(111)		(122)		(103)		(745)		(165)		(146)		(136)		(130)		(147)		(749)	
One year	80	80	69	69	67	67	83	83	79	79	75	75	81	81	84	84	81	81	79	79	77	78	80	81
Five years	33	33	38	38	38	38	44	44	39	39	36	36	47	47	55	55	50	50	58	58	43	43	50	50
Ten years	28	28	27	27	27	27	34	34	29	29	27	27	43	43	46	46	40	40	50	50	35	35	42	43
Northern & Yorkshire	(8)		(13)		(12)		(14)		(23)		(102)		(16)		(14)		(10)		(18)		(33)		(94)	
One year	-	-	52	52	50	50	86	86	82	83	72	72	94	94	85	85	80	80	68	68	76	76	78	78
Five years	-	-	37	37	16	16	58	58	47	47	35	35	56	56	63	63	50	50	46	46	42	42	49	49
Ten years	-	-	20	20	16	16	42	42	38	39	26	26	50	50	33	33	40	40	40	40	33	33	38	38
Trent	(19)		(13)		(15)		(21)		(12)		(89)		(16)		(14)		(20)		(14)		(21)		(86)	
One year	69	69	70	70	65	65	76	76	75	75	69	69	73	73	67	67	74	75	71	72	75	76	73	73
Five years	36	36	31	31	45	45	48	48	26	26	37	37	40	40	38	38	40	40	50	50	52	52	44	44
Ten years	31	31	16	16	25	25	43	43	26	26	29	29	34	34	30	30	40	40	35	35	47	47	38	38
Anglia & Oxford	(16)		(17)		(18)		(14)		(5)		(85)		(21)		(27)		(21)		(11)		(9)		(92)	
One year	87	87	65	65	68	68	61	61	-	-	69	69	85	86	89	89	80	81	91	91	-	-	85	85
Five years	42	42	46	46	45	45	15	15	-	-	35	35	57	57	63	63	52	53	64	64	-	-	58	58
Ten years	36	36	40	40	29	29	7	7	-	-	26	26	57	57	56	56	47	48	56	56	-	-	53	53
North Thames	(20)		(18)		(20)		(13)		(15)		(92)		(18)		(17)		(14)		(19)		(9)		(79)	
One year	90	90	67	67	68	68	85	85	87	87	78	78	89	89	94	94	85	85	74	74	-	-	86	86
Five years	41	41	33	33	48	48	45	45	53	53	44	44	72	72	70	71	56	56	42	42	-	-	58	58
Ten years	41	41	21	22	38	39	45	45	31	31	36	37	72	72	65	65	56	56	37	37	-	-	53	53
South Thames	(22)		(18)		(17)		(24)		(8)		(94)		(33)		(22)		(8)		(18)		(8)		(94)	
One year	67	67	83	83	70	70	83	83	-	-	78	78	82	82	95	95	-	-	70	70	-	-	79	79
Five years	19	19	39	39	17	17	41	41	-	-	32	32	40	40	54	54	-	-	48	48	-	-	41	41
Ten years	14	14	33	34	17	17	28	28	-	-	22	22	33	34	54	54	-	-	36	36	-	-	35	36
South & West	(23)		(21)		(10)		(13)		(6)		(86)		(30)		(25)		(34)		(10)		(11)		(115)	
One year	82	82	62	62	80	80	100	100	-	-	80	80	83	83	76	76	91	91	70	70	64	64	82	82
Five years	35	35	29	29	50	50	47	47	-	-	41	41	53	53	52	52	58	58	60	60	26	26	53	54
Ten years	26	26	16	16	40	40	47	47	-	-	34	34	43	43	40	40	47	47	60	60	26	26	45	45
West Midlands	(5)		(12)		(7)		(5)		(5)		(82)		(16)		(13)		(12)		(15)		(24)		(81)	
One year	-	-	74	74	-	-	-	-	-	-	69	69	63	64	85	85	64	64	93	94	75	75	77	77
Five years	-	-	51	51	-	-	-	-	-	-	32	32	20	20	62	62	30	30	73	73	46	46	46	46
Ten years	-	-	25	25	-	-	-	-	-	-	19	19	20	20	54	54	15	15	73	73	41	41	41	41
North & West	(10)		(19)		(12)		(18)		(29)		(115)		(15)		(14)		(12)		(25)		(32)		(108)	
One year	90	90	79	79	75	75	89	90	69	69	78	78	75	75	72	72	82	83	92	92	88	88	84	85
Five years	18	18	43	43	41	41	50	51	35	35	35	35	34	34	35	35	65	65	76	76	44	44	53	53
Ten years	18	18	38	38	25	25	34	34	24	24	24	24	34	34	21	21	34	34	63	63	31	31	37	38
WALES	(5)		(5)				(11)		(5)		(40)		(6)				(6)				(6)		(31)	
One year	-	-	-	-	-	-	60	61	-	-	74	74	-	-	-	-	-	-	-	-	-	-	93	93
Five years	-	-	-	-	-	-	24	24	-	-	35	35	-	-	-	-	-	-	-	-	-	-	65	65
Ten years	-	-	-	-	-	-	17	17	-	-	23	23	-	-	-	-	-	-	-	-	-	-	55	55

[1] Including those for whom no deprivation category was assigned. C - crude survival, R - relative survival (see Chapter 3).
Number of patients contributing to each analysis in parentheses.

Table 51.2 Crude and relative survival (%) at one, five and ten years since diagnosis: NHS Region, deprivation category and calendar period of diagnosis

BOYS

	1981-85 Affluent		2		3		4		Deprived		All[1]		1986-90 Affluent		2		3		4		Deprived		All[1]		
	C	R	C	R	C	R	C	R	C	R	C	R	C	R	C	R	C	R	C	R	C	R	C	R	
	(187)		*(151)*		*(136)*		*(107)*		*(138)*		*(740)*		*(152)*		*(173)*		*(123)*		*(122)*		*(134)*		*(711)*		**ENGLAND & WALES**
One year	90	90	86	86	90	90	80	80	89	89	88	88	95	95	94	94	93	93	96	96	92	92	94	94	
Five years	71	71	67	67	71	71	68	68	66	66	69	69	76	76	74	75	73	73	68	68	71	71	73	73	
Ten years	60	60	58	58	62	62	63	64	58	59	60	61													
	(182)		*(151)*		*(128)*		*(99)*		*(131)*		*(710)*		*(148)*		*(165)*		*(120)*		*(115)*		*(130)*		*(685)*		**ENGLAND**
One year	91	91	86	86	89	89	79	79	89	89	88	88	95	95	94	94	93	93	96	96	91	92	94	94	
Five years	71	71	67	67	72	72	67	67	66	66	69	69	76	76	76	76	73	73	68	68	71	72	73	73	
Ten years	60	60	58	58	62	63	61	62	59	59	60	60													
	(13)		*(20)*		*(15)*		*(11)*		*(24)*		*(85)*		*(20)*		*(9)*		*(17)*		*(19)*		*(20)*		*(85)*		**Northern & Yorkshire**
One year	85	85	90	91	80	80	82	82	79	79	83	84	100	100	100	100	100	100	100	100	95	95	99	99	
Five years	61	61	70	70	73	73	63	63	57	57	64	65	90	90	49	50	82	82	63	63	80	80	76	76	
Ten years	54	54	70	70	66	67	54	54	53	53	60	60													
	(12)		*(18)*		*(15)*		*(15)*		*(12)*		*(73)*		*(8)*		*(21)*		*(11)*		*(16)*		*(26)*		*(82)*		**Trent**
One year	100	100	100	100	86	87	74	74	100	100	92	92	100	100	95	95	100	100	75	76	92	92	91	92	
Five years	75	75	72	72	73	73	67	67	67	67	71	71	62	62	66	66	72	72	63	63	81	81	71	71	
Ten years	67	67	45	45	60	60	61	61	51	51	56	57													
	(30)		*(19)*		*(12)*		*(8)*		*(7)*		*(82)*		*(19)*		*(21)*		*(11)*		*(11)*		*(7)*		*(69)*		**Anglia & Oxford**
One year	86	86	74	74	76	76	-	-	-	-	84	84	100	100	91	91	80	80	91	91	-	-	91	91	
Five years	73	73	63	63	67	67	-	-	-	-	71	71	74	74	76	77	80	80	54	54	-	-	72	73	
Ten years	63	63	58	58	59	59	-	-	-	-	63	64													
	(26)		*(18)*		*(14)*		*(14)*		*(21)*		*(96)*		*(23)*		*(24)*		*(19)*		*(21)*		*(25)*		*(116)*		**North Thames**
One year	92	92	89	89	100	100	86	86	95	95	93	93	96	96	92	92	89	90	100	100	96	97	94	94	
Five years	69	69	83	83	86	86	71	71	71	71	76	76	74	74	83	83	58	58	81	81	75	76	75	75	
Ten years	62	62	77	77	64	64	71	71	62	62	67	68													
	(29)		*(21)*		*(16)*		*(10)*		*(11)*		*(90)*		*(28)*		*(23)*		*(14)*		*(15)*		*(2)*		*(83)*		**South Thames**
One year	90	90	100	100	94	94	90	90	91	91	93	93	93	93	96	96	100	100	100	100	-	-	95	95	
Five years	65	65	66	67	62	62	80	80	73	73	67	67	72	72	91	91	71	71	67	67	-	-	74	74	
Ten years	48	48	57	57	50	50	59	59	73	73	54	55													
	(34)		*(20)*		*(14)*		*(10)*		*(9)*		*(90)*		*(19)*		*(35)*		*(19)*		*(11)*		*(6)*		*(91)*		**South & West**
One year	94	94	85	85	93	93	79	79	-	-	90	90	100	100	91	91	89	89	100	100	-	-	95	95	
Five years	82	82	70	70	72	72	69	69	-	-	77	77	84	84	65	65	69	69	82	82	-	-	71	71	
Ten years	76	77	60	60	65	65	69	69	-	-	69	69													
	(20)		*(17)*		*(17)*		*(17)*		*(21)*		*(92)*		*(14)*		*(9)*		*(16)*		*(10)*		*(14)*		*(64)*		**West Midlands**
One year	80	80	64	65	94	94	64	64	86	86	78	78	100	100	100	100	94	94	100	100	100	100	98	98	
Five years	65	65	59	59	77	77	46	47	47	47	59	59	79	79	89	89	75	75	58	58	93	93	80	80	
Ten years	50	50	53	53	65	65	41	41	47	47	51	51													
	(18)		*(18)*		*(25)*		*(14)*		*(26)*		*(102)*		*(17)*		*(23)*		*(13)*		*(12)*		*(30)*		*(95)*		**North & West**
One year	100	100	83	83	88	88	79	79	84	84	87	87	77	77	96	96	92	92	100	100	83	83	88	88	
Five years	66	67	55	55	68	68	71	71	76	77	68	68	65	65	79	79	85	85	67	68	53	53	67	67	
Ten years	50	50	44	44	68	68	71	71	60	61	59	60													
	(5)		*(0)*		*(8)*		*(8)*		*(7)*		*(30)*		*(4)*		*(8)*		*(3)*		*(7)*		*(4)*		*(26)*		**WALES**
One year	-	-	-	-	-	-	-	-	-	-	90	90	-	-	87	87	-	-	-	-	-	-	92	92	
Five years	-	-	-	-	-	-	-	-	-	-	64	64	-	-	49	49	-	-	-	-	-	-	65	65	
Ten years	-	-	-	-	-	-	-	-	-	-	64	64													

[1] Including those for whom no deprivation category was assigned. C - crude survival, R - relative survival (see Chapter 3). Number of patients contributing to each analysis in parentheses.

Table 51.2   Crude and relative survival (%) at one, five and ten years since diagnosis: NHS Region, deprivation category and calendar period of diagnosis

	1971-75												1976-80											
GIRLS	Affluent		2		3		4		Deprived		All[1]		Affluent		2		3		4		Deprived		All[1]	
	C	R	C	R	C	R	C	R	C	R	C	R	C	R	C	R	C	R	C	R	C	R	C	R
ENGLAND & WALES	(101)		(90)		(72)		(74)		(80)		(531)		(149)		(124)		(107)		(103)		(87)		(587)	
One year	79	79	86	86	77	77	81	81	76	76	78	78	84	84	89	89	79	79	85	85	80	80	84	84
Five years	45	45	48	49	51	51	48	48	51	51	46	46	56	56	57	57	50	50	57	57	50	50	54	54
Ten years	41	42	41	41	44	44	43	43	45	45	40	40	52	52	54	54	45	45	50	51	44	45	50	50
ENGLAND	(98)		(88)		(65)		(67)		(76)		(505)		(147)		(121)		(100)		(99)		(85)		(565)	
One year	81	81	85	85	81	81	79	79	76	76	78	79	84	84	88	88	80	80	85	85	81	81	84	84
Five years	46	46	48	49	53	54	44	44	51	51	46	46	56	56	57	57	51	51	56	56	51	52	54	54
Ten years	42	42	41	41	46	46	38	39	45	45	40	40	52	52	54	54	46	46	48	49	46	46	49	50
Northern & Yorkshire	(3)		(7)		(11)		(8)		(12)		(59)		(13)		(10)		(4)		(14)		(22)		(65)	
One year	-	-	-	-	82	82	-	-	66	66	65	65	76	76	100	100	-	-	78	78	95	95	89	89
Five years	-	-	-	-	72	72	-	-	58	58	38	38	38	39	49	49	-	-	57	57	72	72	60	60
Ten years	-	-	-	-	63	63	-	-	50	50	35	35	38	39	49	49	-	-	50	50	63	64	54	54
Trent	(9)		(13)		(9)		(13)		(10)		(58)		(15)		(19)		(16)		(17)		(13)		(81)	
One year	-	-	100	100	-	-	76	76	60	60	80	80	74	74	84	84	57	57	100	100	77	77	79	79
Five years	-	-	54	54	-	-	54	54	40	40	44	44	53	53	53	53	38	38	59	59	38	38	49	49
Ten years	-	-	46	46	-	-	54	54	29	29	41	41	53	53	48	48	24	24	53	53	38	38	44	44
Anglia & Oxford	(22)		(15)		(9)		(6)		(2)		(62)		(24)		(22)		(16)		(8)		(3)		(75)	
One year	82	82	86	86	-	-	-	-	-	-	82	82	83	83	86	86	94	94	-	-	-	-	87	87
Five years	50	50	60	60	-	-	-	-	-	-	50	50	55	55	55	55	57	57	-	-	-	-	55	55
Ten years	36	36	53	53	-	-	-	-	-	-	37	37	51	51	55	55	57	57	-	-	-	-	51	51
North Thames	(12)		(12)		(7)		(15)		(5)		(58)		(18)		(8)		(11)		(11)		(6)		(57)	
One year	75	75	92	92	-	-	74	74	-	-	86	86	89	89	-	-	82	82	82	82	-	-	81	81
Five years	49	49	29	29	-	-	48	48	-	-	55	55	61	61	-	-	55	55	54	54	-	-	56	56
Ten years	49	49	22	22	-	-	41	41	-	-	50	50	55	56	-	-	55	55	54	54	-	-	53	53
South Thames	(14)		(14)		(7)		(9)		(12)		(64)		(26)		(22)		(12)		(13)		(5)		(79)	
One year	71	71	72	72	-	-	-	-	76	76	75	75	84	85	91	91	76	76	76	76	-	-	82	82
Five years	42	43	26	26	-	-	-	-	24	25	36	36	58	58	55	55	47	47	38	38	-	-	51	51
Ten years	36	36	20	20	-	-	-	-	24	25	28	28	54	54	50	50	47	47	38	38	-	-	47	48
South & West	(18)		(13)		(7)		(4)		(6)		(54)		(22)		(21)		(18)		(6)		(3)		(72)	
One year	84	84	85	85	-	-	-	-	-	-	84	84	77	77	95	95	71	71	-	-	-	-	83	83
Five years	45	45	85	85	-	-	-	-	-	-	57	58	46	46	72	72	33	33	-	-	-	-	50	50
Ten years	45	45	61	62	-	-	-	-	-	-	48	48	46	46	62	62	27	27	-	-	-	-	46	46
West Midlands	(6)		(3)		(4)		(2)		(2)		(51)		(14)		(7)		(12)		(12)		(12)		(58)	
One year	-	-	-	-	-	-	-	-	-	-	70	71	85	85	-	-	83	83	68	68	73	73	79	79
Five years	-	-	-	-	-	-	-	-	-	-	35	35	78	78	-	-	75	75	51	51	57	57	67	67
Ten years	-	-	-	-	-	-	-	-	-	-	31	31	63	64	-	-	75	75	35	35	48	49	58	58
North & West	(14)		(11)		(11)		(10)		(27)		(99)		(15)		(12)		(11)		(18)		(21)		(78)	
One year	86	86	91	91	65	65	100	100	82	82	82	82	100	100	91	92	100	100	89	89	72	72	89	89
Five years	50	50	53	53	37	37	70	70	45	45	50	50	61	61	50	50	47	47	61	61	37	37	50	50
Ten years	50	50	53	53	37	37	60	60	37	37	44	45	54	54	50	50	37	37	56	56	37	37	46	46
WALES	(3)		(2)		(7)		(7)		(4)		(26)		(2)		(3)		(7)		(4)		(2)		(22)	
One year	-	-	-	-	-	-	-	-	-	-	66	66	-	-	-	-	-	-	-	-	-	-	86	86
Five years	-	-	-	-	-	-	-	-	-	-	46	46	-	-	-	-	-	-	-	-	-	-	54	54
Ten years	-	-	-	-	-	-	-	-	-	-	46	46	-	-	-	-	-	-	-	-	-	-	54	54

[1]   Including those for whom no deprivation category was assigned. C - crude survival, R - relative survival (see Chapter 3). Number of patients contributing to each analysis in parentheses.

GIRLS — C = crude survival, R = relative survival. The first six data columns (Affluent, 2, 3, 4, Deprived, All[1]) are for **1981-85**; the last six are for **1986-90**. Each cell shows "C R". Figures in parentheses are numbers of patients.

	Affluent	2	3	4	Deprived	All[1]	Affluent	2	3	4	Deprived	All[1]
ENGLAND & WALES												
(n)	(138)	(118)	(110)	(114)	(84)	(572)	(125)	(134)	(85)	(114)	(116)	(583)
One year	88 88	89 89	88 88	88 89	82 82	87 87	93 93	91 91	95 95	90 90	90 90	92 92
Five years	73 73	73 73	72 72	69 69	68 68	71 71	78 78	79 79	82 82	77 77	79 79	79 79
	68 68	72 72	64 64	66 67	67 67	68 68						
ENGLAND												
(n)	(136)	(113)	(103)	(106)	(80)	(544)	(118)	(127)	(84)	(106)	(107)	(550)
One year	88 88	89 89	89 89	88 88	81 81	87 87	93 93	90 90	95 95	91 91	91 91	92 92
Five years	73 73	74 74	73 73	70 70	67 67	71 72	79 79	78 78	82 82	76 76	79 80	79 79
	68 68	74 74	66 66	67 67	66 66	68 68						
Northern & Yorkshire												
(n)	(18)	(16)	(17)	(20)	(16)	(87)	(9)	(17)	(16)	(15)	(22)	(80)
One year	89 89	94 94	88 88	84 84	81 81	87 87	100 100	88 88	100 100	87 87	82 82	90 90
Five years	78 78	87 87	76 76	70 70	62 62	75 75	100 100	82 82	87 87	73 73	72 73	81 81
	73 73	87 87	76 76	70 70	62 62	73 74						
Trent												
(n)	(11)	(9)	(17)	(18)	(6)	(61)	(9)	(15)	(7)	(14)	(13)	(58)
One year	100 100	- -	76 76	83 83	- -	85 85	100 100	87 87	- -	93 93	92 92	93 93
Five years	100 100	- -	58 58	49 50	- -	69 69	100 100	74 74	- -	72 72	92 92	81 81
	100 100	- -	52 52	49 50	- -	67 67						
Anglia & Oxford												
(n)	(16)	(13)	(14)	(6)	(4)	(53)	(24)	(14)	(5)	(13)	(6)	(63)
One year	81 81	85 85	93 93	- -	- -	87 87	87 87	77 77	- -	100 100	- -	90 90
Five years	68 68	46 46	86 86	- -	- -	68 68	79 79	56 56	- -	93 93	- -	81 81
	62 62	46 46	78 78	- -	- -	64 64						
North Thames												
(n)	(17)	(18)	(15)	(12)	(8)	(70)	(19)	(17)	(14)	(16)	(18)	(86)
One year	94 94	89 89	86 87	92 92	- -	91 92	85 85	88 89	93 93	81 81	94 94	87 87
Five years	65 65	89 89	66 66	92 92	- -	79 79	58 58	88 89	79 79	69 69	89 89	76 76
	59 59	89 89	52 53	92 92	- -	74 75						
South Thames												
(n)	(24)	(12)	(16)	(10)	(4)	(68)	(14)	(15)	(8)	(12)	(6)	(58)
One year	83 84	100 100	94 94	100 100	- -	91 92	93 93	100 100	100 100	92 92	- -	97 97
Five years	67 67	92 92	87 87	79 79	- -	76 77	86 86	93 93	88 88	75 75	- -	84 85
	62 63	92 92	87 87	79 79	- -	75 75						
South & West												
(n)	(27)	(22)	(12)	(17)	(4)	(83)	(19)	(21)	(21)	(9)	(2)	(72)
One year	96 96	86 86	91 91	82 82	- -	89 89	100 100	100 100	91 91	76 76	- -	94 94
Five years	78 78	77 77	75 75	76 76	- -	77 77	79 79	77 77	81 81	53 53	- -	75 75
	70 70	77 77	75 75	64 64	- -	72 72						
West Midlands												
(n)	(12)	(11)	(5)	(12)	(17)	(58)	(5)	(10)	(6)	(13)	(16)	(51)
One year	76 76	82 82	- -	82 82	82 82	82 83	- -	90 90	- -	100 100	94 94	96 96
Five years	59 59	64 64	- -	57 58	64 64	62 62	- -	79 79	- -	100 100	81 81	86 86
	59 59	64 64	- -	57 58	64 64	60 60						
North & West												
(n)	(11)	(12)	(7)	(11)	(21)	(64)	(19)	(18)	(7)	(14)	(24)	(82)
One year	73 73	82 82	- -	91 91	76 76	83 83	95 95	88 88	- -	93 93	87 87	90 90
Five years	73 73	41 41	- -	73 73	62 62	61 61	79 79	72 72	- -	72 72	67 67	71 71
	63 64	41 41	- -	63 63	57 57	53 53						
WALES												
(n)	(2)	(5)	(7)	(8)	(4)	(28)	(7)	(7)	(1)	(8)	(9)	(33)
One year	- -	- -	- -	- -	- -	89 89	- -	- -	- -	86 86	77 77	88 88
Five years	- -	- -	- -	- -	- -	63 63	- -	- -	- -	86 86	77 77	81 82
	- -	- -	- -	- -	- -	56 56						

[1] Including those for whom no deprivation category was assigned. C - crude survival, R - relative survival (see Chapter 3). Number of patients contributing to each analysis in parentheses.

Table 51.2   Crude and relative survival (%) at one, five and ten years since diagnosis: NHS Region, deprivation category and calendar period of diagnosis

CHILDREN	1971-75												1976-80											
	Affluent		2		3		4		Deprived		All[1]		Affluent		2		3		4		Deprived		All[1]	
	C	R	C	R	C	R	C	R	C	R	C	R	C	R	C	R	C	R	C	R	C	R	C	R
ENGLAND & WALES	(229)		(226)		(191)		(207)		(188)		(1,316)		(320)		(276)		(246)		(237)		(240)		(1,367)	
One year	79	80	76	76	72	72	81	81	78	78	76	76	83	83	86	86	80	80	82	82	79	79	82	82
Five years	39	39	42	42	43	43	44	44	44	44	40	40	52	52	56	56	50	51	57	57	46	47	52	52
Ten years	34	34	32	32	33	33	36	37	36	36	32	32	47	47	50	50	42	42	50	50	40	40	46	46
ENGLAND	(221)		(219)		(176)		(189)		(179)		(1,250)		(312)		(267)		(236)		(229)		(232)		(1,314)	
One year	80	80	76	76	72	72	82	82	78	78	76	76	82	82	86	86	80	80	82	82	79	79	82	82
Five years	39	39	42	42	43	43	44	44	44	44	40	40	51	51	56	56	51	51	57	57	46	46	52	52
Ten years	34	34	32	32	34	34	36	36	36	36	32	32	47	47	49	50	42	43	49	49	39	39	45	46
Northern & Yorkshire	(11)		(20)		(23)		(22)		(35)		(161)		(29)		(24)		(14)		(32)		(55)		(159)	
One year	91	91	59	59	65	65	77	77	77	77	70	70	86	86	92	92	86	86	72	72	84	84	82	82
Five years	54	54	34	34	43	43	45	45	51	51	36	36	48	48	58	58	64	64	50	50	54	54	53	53
Ten years	46	46	23	23	39	39	36	36	42	43	29	29	45	45	40	40	50	50	44	44	45	45	44	45
Trent	(28)		(26)		(24)		(34)		(22)		(147)		(31)		(33)		(36)		(31)		(34)		(167)	
One year	79	79	85	85	64	64	76	76	68	68	74	74	74	74	76	76	67	67	87	87	76	76	76	76
Five years	38	38	42	42	35	35	50	50	32	32	40	40	47	47	46	46	39	39	55	55	47	47	46	46
Ten years	34	34	30	30	23	23	47	47	27	27	34	34	44	44	40	40	33	33	45	45	44	44	41	41
Anglia & Oxford	(38)		(32)		(27)		(20)		(7)		(147)		(45)		(49)		(37)		(19)		(12)		(167)	
One year	84	84	75	75	71	71	67	67	-	-	74	75	84	84	88	88	86	87	89	89	65	65	85	86
Five years	47	47	53	53	52	52	13	13	-	-	41	41	56	56	59	59	54	54	69	69	32	33	56	56
Ten years	36	36	46	46	34	34	5	5	-	-	30	31	53	54	55	55	51	52	53	53	32	33	52	52
North Thames	(32)		(30)		(27)		(28)		(20)		(150)		(36)		(25)		(25)		(30)		(15)		(136)	
One year	85	85	76	77	77	77	79	79	90	90	81	81	89	89	83	83	83	83	77	77	87	87	84	84
Five years	44	44	32	32	54	54	47	47	65	65	48	48	67	67	64	64	56	56	47	47	40	40	57	57
Ten years	44	44	23	23	47	47	43	43	49	49	42	42	64	64	60	60	56	56	44	44	19	19	53	53
South Thames	(36)		(32)		(24)		(33)		(20)		(158)		(59)		(44)		(20)		(31)		(13)		(173)	
One year	69	69	78	78	79	79	81	81	85	85	77	77	83	83	93	93	71	71	73	73	78	78	80	80
Five years	28	28	34	34	30	30	42	42	30	30	34	34	47	47	54	54	38	38	44	44	46	46	46	46
Ten years	23	23	28	28	25	25	29	29	25	25	25	25	42	42	52	52	32	32	37	37	38	38	41	41
South & West	(41)		(34)		(17)		(17)		(12)		(140)		(52)		(46)		(52)		(16)		(14)		(187)	
One year	83	83	71	71	88	88	94	94	83	83	81	81	81	81	85	85	84	84	81	81	71	71	82	82
Five years	39	39	50	50	53	53	42	42	51	51	47	47	50	50	61	61	50	50	56	57	36	36	52	52
Ten years	34	34	33	33	40	41	42	42	51	51	39	39	44	44	50	50	40	40	56	57	36	36	45	46
West Midlands	(11)		(15)		(11)		(7)		(7)		(133)		(30)		(20)		(24)		(27)		(36)		(139)	
One year	55	55	73	73	63	63	-	-	-	-	70	70	73	73	85	85	74	74	82	82	74	75	78	78
Five years	18	18	40	40	35	35	-	-	-	-	33	33	47	47	65	65	53	53	63	63	50	50	55	55
Ten years	9	9	19	19	26	27	-	-	-	-	24	24	40	40	60	60	45	45	56	56	44	44	48	48
North & West	(24)		(30)		(23)		(28)		(56)		(214)		(30)		(26)		(28)		(43)		(53)		(186)	
One year	88	88	84	84	70	70	93	94	75	75	80	80	87	87	81	81	89	90	91	91	81	81	86	86
Five years	37	37	47	47	39	39	57	58	40	40	42	42	47	47	42	42	57	58	70	70	41	41	52	52
Ten years	37	37	44	44	31	31	43	43	30	31	33	33	44	44	34	34	35	35	60	60	34	34	41	41
WALES	(8)		(7)		(15)		(18)		(9)		(66)		(8)		(9)		(10)		(8)		(8)		(53)	
One year	-	-	-	-	72	72	76	77	-	-	71	71	-	-	-	-	79	79	-	-	-	-	91	91
Five years	-	-	-	-	40	40	48	49	-	-	39	39	-	-	-	-	49	49	-	-	-	-	60	60
Ten years	-	-	-	-	19	19	43	44	-	-	32	32	-	-	-	-	39	39	-	-	-	-	55	55

[1] Including those for whom no deprivation category was assigned. C - crude survival, R - relative survival (see Chapter 3). Number of patients contributing to each analysis in parentheses.

	1981-85						1986-90						
	Affluent	*2*	*3*	*4*	*Deprived*	*All*[1]	*Affluent*	*2*	*3*	*4*	*Deprived*	*All*[1]	***CHILDREN***
	C R	C R	C R	C R	C R	C R	C R	C R	C R	C R	C R	C R	
	(325)	*(269)*	*(246)*	*(221)*	*(222)*	*(1,312)*	*(277)*	*(307)*	*(208)*	*(236)*	*(250)*	*(1,294)*	**ENGLAND & WALES**
One year	89 89	87 87	89 89	84 84	86 86	**87 88**	94 94	92 93	94 94	93 93	91 91	**93 93**	*One year*
Five years	72 72	70 70	71 71	69 69	67 67	**70 70**	77 77	76 77	77 77	72 72	75 75	**76 76**	*Five years*
	63 64	64 64	63 63	65 65	62 62	**63 64**							
	(318)	*(264)*	*(231)*	*(205)*	*(211)*	*(1,254)*	*(266)*	*(292)*	*(204)*	*(221)*	*(237)*	*(1,235)*	**ENGLAND**
	89 89	87 87	89 89	83 83	86 86	**87 87**	94 94	92 93	94 94	93 93	91 91	**93 93**	*One year*
	72 72	70 70	72 72	68 68	67 67	**70 70**	77 77	77 77	77 77	72 72	75 75	**76 76**	*Five years*
	63 63	65 65	64 64	64 64	61 62	**64 64**							
	(31)	*(36)*	*(32)*	*(31)*	*(40)*	*(172)*	*(29)*	*(26)*	*(33)*	*(34)*	*(42)*	*(165)*	**Northern & Yorkshire**
	87 87	92 92	84 84	83 84	80 80	**85 85**	100 100	92 92	100 100	94 94	88 88	**95 95**	*One year*
	71 71	78 78	75 75	67 68	59 60	**70 70**	93 93	72 72	85 85	67 68	76 76	**79 79**	*Five years*
	65 65	78 78	72 72	64 64	57 57	**67 67**							
	(23)	*(27)*	*(32)*	*(33)*	*(18)*	*(134)*	*(17)*	*(36)*	*(18)*	*(30)*	*(39)*	*(140)*	**Trent**
	100 100	100 100	81 81	79 79	89 89	**89 89**	100 100	92 92	100 100	83 84	92 92	**92 92**	*One year*
	87 87	78 78	65 65	58 58	67 67	**70 70**	82 82	69 70	72 72	67 67	85 85	**75 75**	*Five years*
	83 83	60 60	56 56	55 55	56 56	**61 61**							
	(46)	*(32)*	*(26)*	*(14)*	*(11)*	*(135)*	*(43)*	*(35)*	*(16)*	*(24)*	*(13)*	*(132)*	**Anglia & Oxford**
	84 84	78 78	85 85	93 93	91 91	**85 85**	93 93	86 86	87 87	96 96	92 92	**91 91**	*One year*
	71 71	56 56	77 77	72 72	64 64	**70 70**	77 77	68 69	87 87	75 75	85 85	**76 77**	*Five years*
	62 63	53 53	69 69	72 72	64 64	**64 64**							
	(43)	*(36)*	*(29)*	*(26)*	*(29)*	*(166)*	*(42)*	*(41)*	*(33)*	*(37)*	*(43)*	*(202)*	**North Thames**
	93 93	89 89	93 93	89 89	97 97	**92 92**	91 91	90 90	91 91	92 92	95 96	**91 91**	*One year*
	68 68	86 86	76 76	81 81	76 76	**77 77**	67 67	85 85	67 67	76 76	81 82	**75 75**	*Five years*
	60 61	83 83	58 58	81 81	69 69	**70 71**							
	(53)	*(33)*	*(32)*	*(20)*	*(15)*	*(158)*	*(42)*	*(38)*	*(22)*	*(27)*	*(8)*	*(141)*	**South Thames**
	87 87	100 100	94 94	95 95	94 94	**92 93**	93 93	97 97	100 100	96 96	88 88	**96 96**	*One year*
	66 66	76 76	75 75	80 80	73 74	**71 71**	76 76	92 92	77 77	70 71	62 62	**78 78**	*Five years*
	55 55	70 70	69 69	70 70	73 74	**63 64**							
	(61)	*(42)*	*(26)*	*(27)*	*(13)*	*(173)*	*(38)*	*(56)*	*(40)*	*(20)*	*(8)*	*(163)*	**South & West**
	95 95	86 86	92 92	81 81	93 93	**90 90**	100 100	95 95	90 90	90 90	100 100	**94 95**	*One year*
	80 80	74 74	73 73	74 74	85 85	**77 77**	82 82	70 70	75 75	70 70	46 46	**73 73**	*Five years*
	74 74	69 69	69 69	66 66	77 77	**70 71**							
	(32)	*(28)*	*(22)*	*(29)*	*(38)*	*(150)*	*(19)*	*(19)*	*(22)*	*(23)*	*(30)*	*(115)*	**West Midlands**
	78 78	71 71	95 95	72 72	84 84	**80 80**	100 100	95 95	95 95	100 100	97 97	**97 97**	*One year*
	63 63	61 61	73 73	51 51	55 55	**60 60**	74 74	84 84	82 82	82 82	87 87	**83 83**	*Five years*
	53 53	57 57	59 59	48 48	55 55	**54 55**							
	(29)	*(30)*	*(32)*	*(25)*	*(47)*	*(166)*	*(36)*	*(41)*	*(20)*	*(26)*	*(54)*	*(177)*	**North & West**
	90 90	83 83	91 91	84 84	80 80	**85 86**	86 86	93 93	90 90	96 96	85 85	**89 89**	*One year*
	69 69	49 49	65 65	72 72	70 70	**65 66**	72 72	76 76	75 75	70 70	59 59	**69 69**	*Five years*
	55 55	43 43	59 59	67 67	59 59	**57 57**							
	(7)	*(5)*	*(15)*	*(16)*	*(11)*	*(58)*	*(11)*	*(15)*	*(4)*	*(15)*	*(13)*	*(59)*	**WALES**
	- -	- -	87 87	94 94	91 91	**90 90**	91 91	93 93	- -	93 93	84 84	**90 90**	*One year*
	- -	- -	54 54	75 75	64 64	***64 64***	73 73	73 73	- -	78 78	69 69	**74 74**	*Five years*
	- -	- -	47 48	75 75	64 64	**60 60**							

[1] Including those for whom no deprivation category was assigned. C - crude survival, R - relative survival (see Chapter 3).
Number of patients contributing to each analysis in parentheses.

Fig. 51.1   Relative survival up to ten years: trends by calendar period of diagnosis (1971-90) and NHS Region

CHILDREN

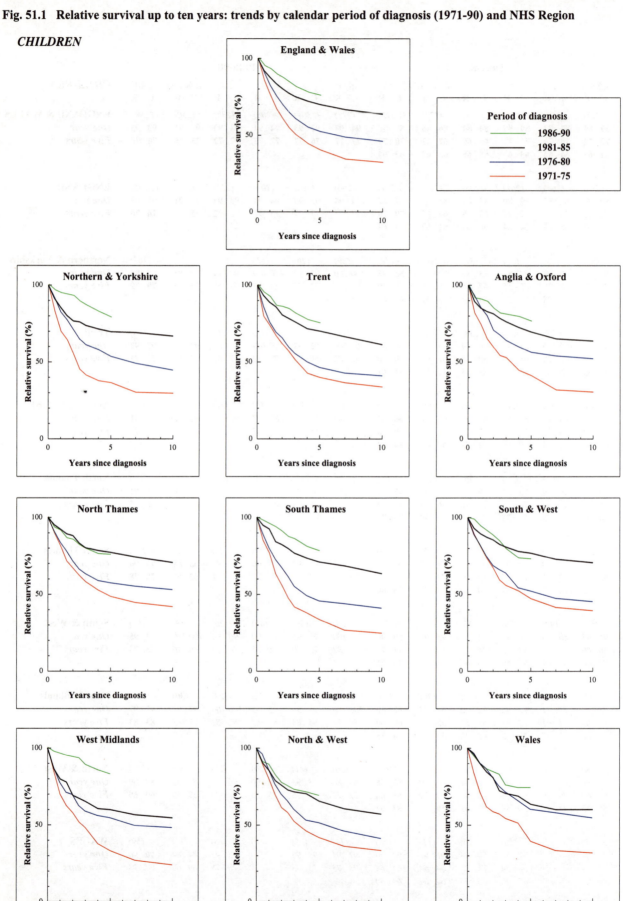

Fig. 51.2 Relative survival up to five years, by deprivation category and NHS Region: patients diagnosed 1986-90

CHILDREN

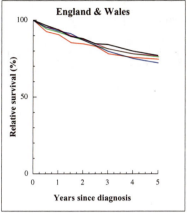

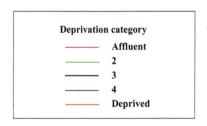

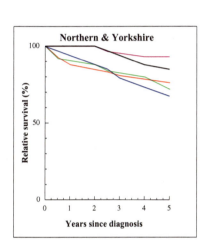

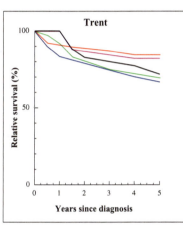

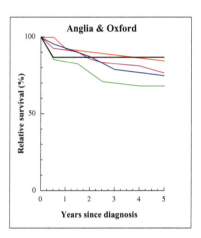

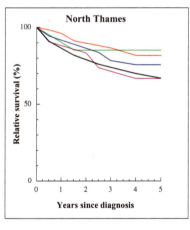

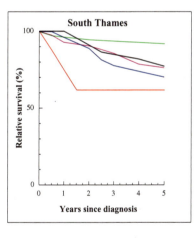

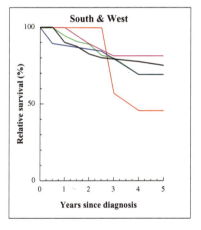

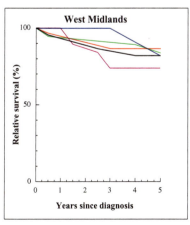

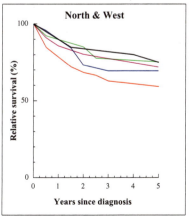

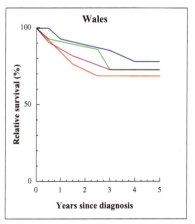

Fig. 51.3 Relative survival at five years by deprivation category, period of diagnosis (1981-90) and NHS Region

CHILDREN

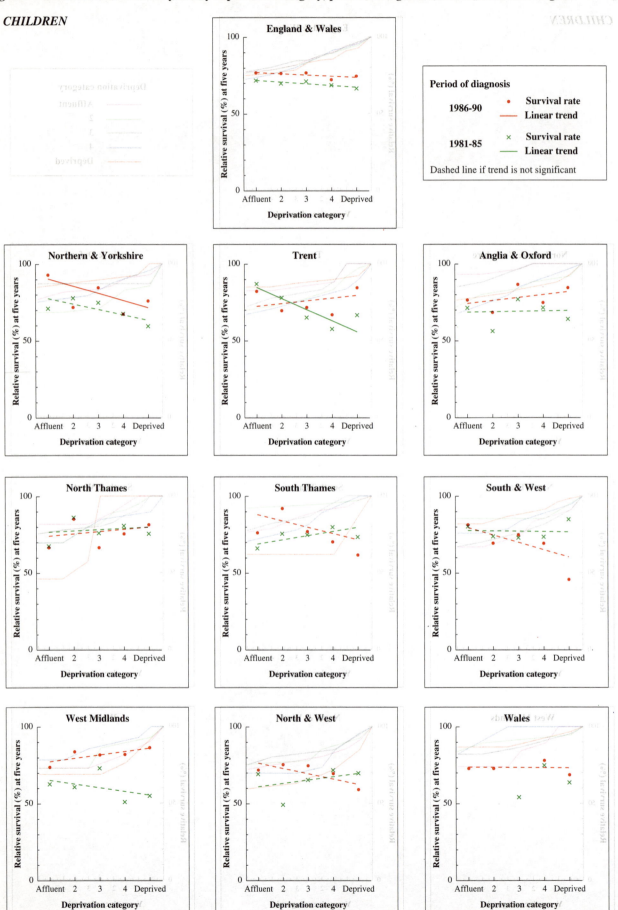

Fig. 51.4 Relative survival (%) by NHS Region and sex:

        one-year and five-year survival for children diagnosed 1986-90, with 95% confidence intervals

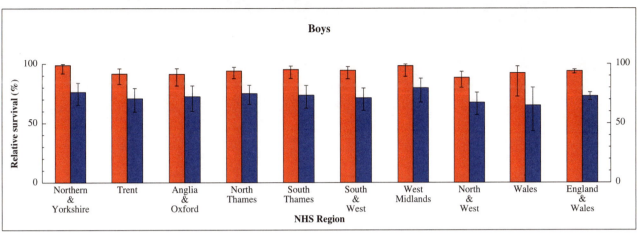

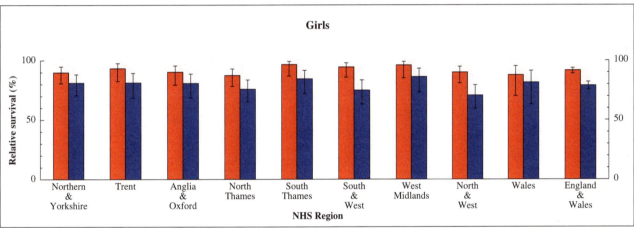

Chapter 52

HODGKIN'S DISEASE *Children*

Hodgkin's disease accounts for one-third of lymphomas in children in England and Wales. It is rare before five years of age, but accounts for 5-10% of cancers in older children. Hodgkin's disease is twice as common in boys. Incidence is lower in England and Wales (0.5 per 100,000 in boys) than in many other countries. Those with the highest incidence include Costa Rica (1.7), India (0.9), Australia (0.8) and the USA (whites, 0.7), while countries with the lowest incidence include China (0.2) and Japan (0.1).

The survival analyses reported here are based on 900 children diagnosed with Hodgkin's disease in England and Wales between 1971 and 1990 and followed up to the end of 1995, about 91% of those eligible. Almost 9% of eligible children were excluded because their vital status was unknown on 6 July 1997.

The different types of Hodgkin's disease were not separately coded until 1979. Among those diagnosed during 1986-90, 38% showed nodular sclerosis, 13% had mixed cellularity and 12% lymphocytic-histiocytic predominance, but for more than a third (36%) of cases the type of Hodgkin's disease was not specified (Table 52A).

The morphology was coded as Hodgkin's disease with nodular sclerosis in 23% of cases (data not shown); 11% were of mixed cellularity and 9% showed lymphocytic predominance, broadly reflecting the ICD-9 type classification, but for more than half (55%) the cases, the morphology was coded simply as Hodgkin's disease, not otherwise specified. Only 1% of cases were coded under the Parker-Jackson classification.

Survival trends

Relative survival[a] from Hodgkin's disease in children diagnosed in England and Wales during 1986-90 was 99% at one year and 92% at five years, compared to 93% and 79%, respectively, for children diagnosed during 1971-75 (Table 52.2). Most of the improvement in five-year survival occurred during the 1970s, while further improvement in one-year survival occurred during the 1980s (Figure 52.1).

The national pattern of trends in survival is broadly reflected in all regions, but regional survival curves for each calendar period are all based on fewer than 50 cases, and there is considerable fluctuation.

One-year survival in England and Wales increased by an average of about 2% between successive five-year periods for both boys and girls. Little further improvement in survival is possible, since very few children now die within the first year after diagnosis.

Five-year survival increased by an average of 4% every five years (to 96%) in boys and by about 3% every five years (to 85%) in girls (see Table 4.4). Survival is generally higher in boys, but the differences are small: national survival rates for Hodgkin's disease in girls are based on fewer than 80 cases every five years, and they are imprecise (see Chapter 3).

These results are similar to those from the National Registry of Childhood Tumours for Great Britain, which has published actuarial survival estimates of 76% for

[a] See footnote in Chapter 50.

Table 52A Sub-site distribution of Hodgkin's disease in children[1]

ICD-9 code	Site description	1971-75 No.	%	1976-80 No.	%	1981-85 No.	%	1986-90 No.	%	Total No.	%
201.0	Hodgkin's paragranuloma	-	-	0	0.0	1	0.4	0	0.0	1	0.1
201.1	Hodgkin's granuloma	-	-	0	0.0	0	0.0	0	0.0	0	0.0
201.2	Hodgkin's sarcoma	-	-	0	0.0	0	0.0	0	0.0	0	0.0
201.4	Lymphocytic-histiocytic	-	-	8	3.1	22	9.5	23	12.0	53	5.8
201.5	Nodular sclerosis	-	-	25	9.6	67	28.9	73	38.2	165	18.1
201.6	Mixed cellularity	-	-	17	6.5	33	14.2	25	13.1	75	8.2
201.7	Lymphocytic depletion	-	-	3	1.2	3	1.3	1	0.5	7	0.8
201.9	Unspecified	229	100.0	207	79.6	106	45.7	69	36.1	611	67.0
	Total	**229**		**260**		**232**		**191**		**912**	

[1] Hodgkin's disease was first classified separately from other lymphomas only in 1968 (ICD-8), but was not subdivided until 1979, when ICD-9 codes allowed the alternative Parker-Jackson (201.0-201.2) and Rye (Lukes-Butler) classifications (201.4-201.7).

children diagnosed in 1971-73[1], 90% for 1980-82 and 93% for those diagnosed in 1986-88[2].

Relative survival from Hodgkin's disease in children is 10% higher than for adults at one year and 10-20% higher at five years, even after adjustment for other causes of death, which affect mainly adults (see Chapter 42). The overall rate of improvement in five-year survival since the 1970s has been similar in adults and children (see Table 4.4).

Deprivation

Differences in survival between deprivation groups are not systematic (Table 52.2, Figure 52.2). The estimated differences in one-year and five-year survival during the 1980s are small and not statistically significant (see Table 4.9). This pattern is reflected in Figure 52.3, which shows no significant gradient in five-year survival from Hodgkin's disease in childhood during the 1980s.

International comparison

Five-year survival from Hodgkin's disease for children diagnosed in England and Wales during 1986-90 (92%) was similar to that for children diagnosed in the territory covered by the SEER programme in the USA (88%)[3]. For children aged 5-14 who were diagnosed in Scotland during 1983-87 (23 cases), the five-year relative survival rate was 84%[4].

1. Stiller CA, Bunch KJ. Trends in survival for childhood cancer in Britain diagnosed 1971-85. Br J Cancer 1990; 62: 806-815
2. Stiller CA. Population-based survival rates for childhood cancer in Britain, 1980-91. Br Med J 1994; 309: 1612-1616
3. Miller BA, Ries LAG, Hankey BF, Kosary CL, Edwards BK. *Cancer statistics review: 1973-1989. NIH Publication No. 92-2789.* Bethesda: National Cancer Institute, 1992
4. Black RJ, Sharp L, Kendrick SW. *Trends in cancer survival in Scotland, 1968-1990.* Edinburgh: Information and Statistics Division, 1993

Table 52A Sub-site distribution of Hodgkin's disease in children.

ICD-9 code	Site description	1971-75 No.	%	1976-80 No.	%	1981-85 No.	%	1986-90 No.	%	Total No.	%
201.0	Hodgkin's paragranuloma	-		0	0.0	1	0.4	0	0.0	1	0.1
201.1	Hodgkin's granuloma	-		0	0.0	0	0.0	0	0.0	0	0.0
201.2	Hodgkin's sarcoma	-		0	0.0	0	0.0	0	0.0	0	0.0
201.4	Lymphocytic-histiocytic	-		8	3.1	22	9.5	23	12.0	53	5.8
201.5	Nodular sclerosis	-		25	9.6	67	28.9	73	38.2	165	18.1
201.6	Mixed cellularity	-		17	6.5	33	14.2	25	13.1	75	8.2
201.7	Lymphocytic depletion	-		3	1.2	3	1.3	1	0.5	7	0.8
201.9	Unspecified	229	100.0	207	79.6	106	45.7	69	36.1	611	67.0
Total		**229**		**260**		**232**		**191**		**912**	

Hodgkin's disease was first classified separately from other lymphomas only in 1968 (ICD-8), but was not subdivided until 1979, when ICD-9 codes allowed the alternative Parker-Jackson (201.0-201.2) and Rye (Lukes-Butler) classifications (201.4-201.7).

Table 52.1 Ineligible and excluded records, and patients included in analyses, by calendar period of diagnosis

	Calendar period of diagnosis															
	1971-75			1976-80			1981-85			1986-90			Overall		
Total registered	**265**			**302**			**282**			**227**			**1,076**		
Ineligible	*Records*	*Patients*	*%*	*Records*	*Patients*	*%*	*Records*	*Patients*	*%*	*Records*	*Patients*	*%*	*Records*	*Patients*	*%*
Incomplete data[1]	9	9		6	6		15	15		14	14		44	44	
Resident outside England and Wales	10	10		10	10		8	7		0	0		28	27	
In situ neoplasm[2]	0	0		0	0		0	0		0	0		0	0	
Benign or uncertain behaviour[2]	0	0		0	0		0	0		0	0		0	0	
Metastatic[3]	-	-		-	-		-	-		-	-		-	-	
Otherwise ineligible[3]	-	-		-	-		-	-		-	-		-	-	
Lymphoma[4]	-	-		-	-		-	-		-	-		-	-	
Leukaemia or myeloma[4]	-	-		-	-		-	-		-	-		-	-	
		19			**16**			**22**			**14**			**71**	
Total eligible	**246**	**100**		**286**	**100**		**260**	**100**		**213**	**100**		**1,005**	**100**	
Aged under 15 or 100 years or over at diagnosis[4]	-	-	-	-	-	-	-	-	-	-	-	-	-	-	-
Vital status unknown[5]	30	17	6.9	34	25	8.7	35	26	10.0	24	21	9.9	123	89	8.9
Duplicate registration[6]	0	0	0.0	0	0	0.0	0	0	0.0	0	0	0.0	0	0	0.0
Multiple primary[7]	0	0	0.0	0	0	0.0	1	0	0.0	0	0	0.0	1	0	0.0
Synchronous tumours[8]	0	0	0.0	0	0	0.0	1	0	0.0	1	0	0.0	2	0	0.0
Sex not known	0	0	0.0	0	0	0.0	0	0	0.0	0	0	0.0	0	0	0.0
Sex incompatible with site code[9]	-	-	-	-	-	-	-	-	-	-	-	-	-	-	-
Invalid dates[10]	0	0	0.0	0	0	0.0	0	0	0.0	0	0	0.0	0	0	0.0
Zero survival[11]	0	0	0.0	1	1	0.3	2	2	0.8	1	1	0.5	4	4	0.4
Total exclusions		**17**	**6.9**		**26**	**9.1**		**28**	**10.8**		**22**	**10.3**		**93**	**9.3**
Patients included in analysis		**229**	**93.1**		**260**	**90.9**		**232**	**89.2**		**191**	**89.7**		**912**	**90.7**

[1] Main data item(s) invalid or incompatible with one another: name, sex, date of birth, date of diagnosis and (if present) date of death, postcode, site, morphology and behaviour

[2] Tumour either *in situ* (behaviour code 2), benign (0), uncertain if benign or malignant (1) or metastatic (6 or 9)

[3] Not applicable for this malignancy

[4] Not applicable for this malignancy

[5] Tracing of vital status at National Health Service Central Register incomplete at 6 July 1997

[6] Same site code, sex, cancer registry and cancer registry number as an earlier registration

[7] Same site code and person number as an earlier registration(s): mostly confirmed multiple primary tumours at this site, some unresolved duplicate registrations

[8] Same site code, sex, date of birth and date of diagnosis as another registration(s): mostly synchronous or (in paired organs) bilateral tumours in same anatomic site in one individual, not previously linked: also some duplicate registrations

[9] Not applicable for this malignancy

[10] Impossible sequence of dates (birth, diagnosis and, if present, emigration or death); or date of death after 6 July 1997

[11] Date of diagnosis same as date of death: some are patients who did die on the day of diagnosis, but most were registered solely from a death certificate (death certificate only, or DCO), and their survival time is thus unknown

Table 52.2   Crude and relative survival (%) at one, five and ten years since diagnosis: NHS Region, deprivation category and calendar period of diagnosis

		1971-75												1976-80										
		Affluent		*2*		*3*		*4*		*Deprived*		*All[1]*		*Affluent*		*2*		*3*		*4*		*Deprived*		*All[1]*	
		C	R	C	R	C	R	C	R	C	R	C	R	C	R	C	R	C	R	C	R	C	R	C	R
ENGLAND & WALES																									
Boys		*(25)*		*(18)*		*(20)*		*(27)*		*(35)*		*(162)*		*(42)*		*(27)*		*(24)*		*(40)*		*(43)*		*(187)*	
	One year	96	96	100	100	90	90	96	96	86	86	94	94	93	93	96	96	100	100	98	98	100	100	96	96
	Five years	80	80	83	83	85	85	89	89	71	71	78	78	91	91	96	96	96	96	83	83	93	93	90	91
	Ten years	80	80	77	78	80	80	81	82	52	52	71	71	83	84	89	89	96	96	80	81	91	91	87	87
Girls		*(9)*		*(14)*		*(7)*		*(12)*		*(9)*		*(67)*		*(20)*		*(16)*		*(10)*		*(10)*		*(16)*		*(73)*	
	One year	-	-	93	93	-	-	100	100	-	-	93	93	80	80	88	88	100	100	90	90	94	94	89	89
	Five years	-	-	86	86	-	-	66	66	-	-	81	81	55	55	75	75	90	90	80	80	88	88	75	75
	Ten years	-	-	86	86	-	-	58	58	-	-	76	76	55	55	62	62	90	90	80	80	81	81	71	71
Children		*(34)*		*(32)*		*(27)*		*(39)*		*(44)*		*(229)*		*(62)*		*(43)*		*(34)*		*(50)*		*(59)*		*(260)*	
	One year	94	94	97	97	93	93	97	97	86	86	93	93	89	89	93	93	100	100	96	96	98	98	94	94
	Five years	79	80	84	84	81	82	82	82	75	75	79	79	79	79	88	88	94	94	82	82	92	92	86	86
	Ten years	79	80	81	81	78	78	74	74	57	57	72	73	74	75	79	79	94	94	80	81	88	88	82	83

		Boys		*Girls*		*Children*								*Boys*		*Girls*		*Children*		
Northern & Yorkshire		*(20)*		*(8)*		*(28)*								*(31)*		*(9)*		*(40)*		
	One year	95	95	-	-	93	93							94	94	-	-	90	90	
	Five years	75	75	-	-	75	75							90	90	-	-	88	88	
	Ten years	75	75	-	-	71	72							90	90	-	-	88	88	
Trent		*(11)*		*(7)*		*(18)*								*(13)*		*(11)*		*(24)*		
	One year	82	82	-	-	84	84							92	92	91	91	92	92	
	Five years	73	74	-	-	72	73							77	77	73	73	75	75	
	Ten years	64	64	-	-	67	67							77	77	63	63	71	71	
Anglia & Oxford		*(18)*		*(10)*		*(28)*								*(27)*		*(10)*		*(37)*		
	One year	100	100	100	100	100	100							100	100	100	100	100	100	
	Five years	94	94	90	90	93	93							96	96	80	80	92	92	
	Ten years	94	94	80	80	89	89							89	89	80	80	86	87	
North Thames		*(26)*		*(4)*		*(30)*								*(26)*		*(10)*		*(36)*		
	One year	92	92	-	-	90	90							100	100	90	90	97	97	
	Five years	85	85	-	-	83	83							92	92	69	69	86	86	
	Ten years	77	77	-	-	77	77							88	89	69	69	83	84	
South Thames		*(12)*		*(7)*		*(19)*								*(18)*		*(5)*		*(23)*		
	One year	92	92	-	-	95	95							89	89	-	-	87	87	
	Five years	84	84	-	-	90	90							84	84	-	-	78	79	
	Ten years	76	76	-	-	84	84							78	78	-	-	70	70	
South & West		*(20)*		*(5)*		*(25)*								*(20)*		*(10)*		*(30)*		
	One year	95	95	-	-	92	92							100	100	90	90	97	97	
	Five years	80	80	-	-	80	80							100	100	70	70	90	90	
	Ten years	70	70	-	-	72	72							100	100	70	70	90	90	
West Midlands		*(17)*		*(13)*		*(30)*								*(18)*		*(4)*		*(22)*		
	One year	100	100	92	92	97	97							94	94	-	-	95	95	
	Five years	58	58	84	84	70	70							89	89	-	-	91	91	
	Ten years	53	53	77	77	63	63							78	78	-	-	82	82	
North & West		*(29)*		*(6)*		*(35)*								*(23)*		*(9)*		*(32)*		
	One year	97	97	-	-	97	97							100	100	-	-	97	97	
	Five years	79	79	-	-	77	77							87	87	-	-	85	85	
	Ten years	69	69	-	-	69	69							83	83	-	-	78	79	
WALES		*(9)*		*(7)*		*(16)*								*(11)*		*(5)*		*(16)*		
	One year	-	-	-	-	87	87							91	91	-	-	88	88	
	Five years	-	-	-	-	62	62							91	91	-	-	88	88	
	Ten years	-	-	-	-	56	56							91	91	-	-	88	88	

[1] Including those for whom no deprivation category was assigned. C - crude survival, R - relative survival (see Chapter 3).
Number of patients contributing to each analysis in parentheses.

ENGLAND & WALES

	1981-85 Affluent C R	2 C R	3 C R	4 C R	Deprived C R	All[1] C R	1986-90 Affluent C R	2 C R	3 C R	4 C R	Deprived C R	All[1] C R	
	(34)	*(35)*	*(36)*	*(25)*	*(31)*	*(164)*	*(30)*	*(31)*	*(18)*	*(28)*	*(24)*	*(132)*	**Boys**
One year	100 100	91 91	100 100	100 100	100 100	98 98	100 100	100 100	100 100	100 100	100 100	100 100	*One year*
Five years	94 94	86 86	83 83	88 88	93 94	88 89	100 100	97 97	95 95	93 93	92 92	95 96	*Five years*
	91 91	77 77	83 83	84 84	90 91	85 85							
	(20)	*(18)*	*(9)*	*(10)*	*(10)*	*(68)*	*(13)*	*(12)*	*(13)*	*(12)*	*(8)*	*(59)*	**Girls**
One year	95 95	95 95	- -	100 100	100 100	97 97	100 100	92 92	100 100	91 91	100 100	97 97	*One year*
Five years	90 90	89 89	- -	90 90	90 90	91 91	85 85	92 92	70 70	83 83	100 100	85 85	*Five years*
	90 90	83 83	- -	90 90	79 79	88 88							
	(54)	*(53)*	*(45)*	*(35)*	*(41)*	*(232)*	*(43)*	*(43)*	*(31)*	*(40)*	*(32)*	*(191)*	**Children**
One year	98 98	92 92	100 100	100 100	100 100	98 98	100 100	98 98	100 100	97 97	100 100	99 99	*One year*
Five years	93 93	87 87	87 87	89 89	93 93	89 89	95 95	95 95	84 84	90 90	94 94	92 92	*Five years*
	91 91	79 79	87 87	86 86	88 88	86 86							

	1981-85 Boys C R	Girls C R	Children C R	1986-90 Boys C R	Girls C R	Children C R	
	(29)	*(5)*	*(34)*	*(11)*	*(3)*	*(14)*	**Northern & Yorkshire**
One year	97 97	- -	94 94	100 100	- -	100 100	*One year*
Five years	90 90	- -	88 88	100 100	- -	100 100	*Five years*
	86 86	- -	85 85				
	(13)	*(11)*	*(24)*	*(19)*	*(6)*	*(25)*	**Trent**
One year	100 100	100 100	100 100	100 100	- -	100 100	*One year*
Five years	100 100	91 91	96 96	84 84	- -	88 88	*Five years*
	100 100	91 91	96 96				
	(12)	*(10)*	*(22)*	*(8)*	*(4)*	*(12)*	**Anglia & Oxford**
One year	100 100	90 90	96 96	100 100	- -	100 100	*One year*
Five years	100 100	90 90	96 96	88 88	- -	83 83	*Five years*
	100 100	90 90	96 96				
	(21)	*(12)*	*(33)*	*(21)*	*(7)*	*(28)*	**North Thames**
One year	100 100	100 100	100 100	100 100	- -	96 96	*One year*
Five years	81 81	91 91	85 85	90 90	- -	86 86	*Five years*
	71 72	83 83	76 76				
	(25)	*(9)*	*(34)*	*(14)*	*(12)*	*(26)*	**South Thames**
One year	100 100	- -	100 100	100 100	100 100	100 100	*One year*
Five years	84 84	- -	88 88	100 100	92 92	96 96	*Five years*
	80 80	- -	85 86				
	(13)	*(5)*	*(18)*	*(14)*	*(8)*	*(22)*	**South & West**
One year	92 92	- -	94 94	100 100	100 100	100 100	*One year*
Five years	69 69	- -	67 67	100 100	100 100	100 100	*Five years*
	69 69	- -	67 67				
	(18)	*(10)*	*(28)*	*(15)*	*(8)*	*(23)*	**West Midlands**
One year	100 100	100 100	100 100	100 100	100 100	100 100	*One year*
Five years	100 100	100 100	100 100	100 100	88 88	96 96	*Five years*
	95 95	90 90	93 93				
	(21)	*(4)*	*(25)*	*(24)*	*(8)*	*(32)*	**North & West**
One year	95 95	- -	96 96	100 100	87 87	97 97	*One year*
Five years	86 86	- -	88 88	100 100	51 51	88 88	*Five years*
	86 86	- -	88 88				
	(12)	*(2)*	*(14)*	*(6)*	*(3)*	*(9)*	**WALES**
One year	100 100	- -	100 100	- -	- -	100 100	*One year*
Five years	91 91	- -	93 93	- -	- -	100 100	*Five years*
	83 83	- -	85 86				

[1] Including those for whom no deprivation category was assigned. C - crude survival, R - relative survival (see Chapter 3).
Number of patients contributing to each analysis in parentheses.

Fig. 52.1 Relative survival up to ten years: trends by calendar period of diagnosis (1971-90) and NHS Region

CHILDREN

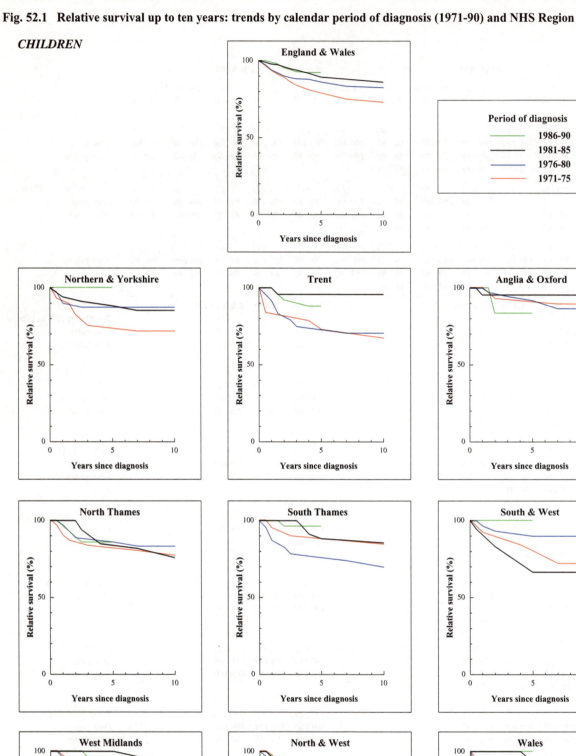

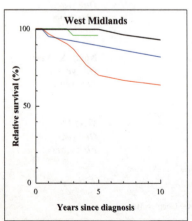

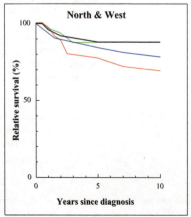

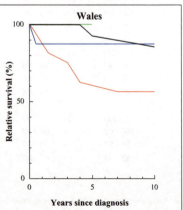

Fig. 52.2 Relative survival up to five years, by deprivation category: England and Wales, patients diagnosed 1986-90

CHILDREN

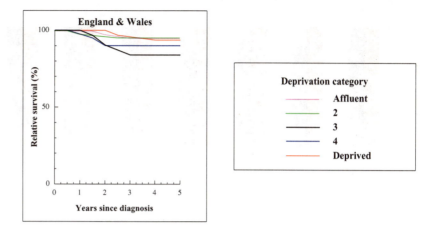

Fig. 52.3 Relative survival at five years, by deprivation category and period of diagnosis (1981-90): England and Wales

CHILDREN

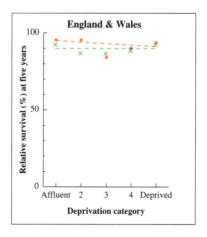
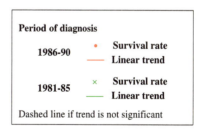

**Fig. 52.4 Relative survival (%) by NHS Region and sex:
one-year and five-year survival for children diagnosed 1986-90, with 95% confidence intervals**

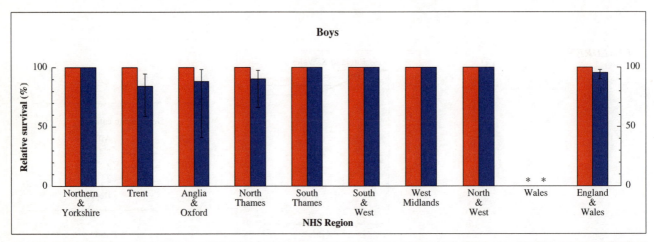

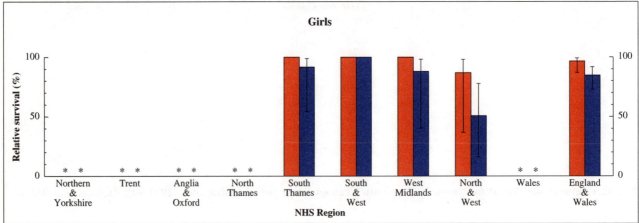

Chapter 53

NON-HODGKIN LYMPHOMA *Children*

The lymphomas are the third most common group of malignancies in children, after the leukaemias and tumours of the central nervous system. Hodgkin's disease accounts for a third of the lymphomas, but about 100 new cases of non-Hodgkin lymphoma are diagnosed in children in England and Wales each year. They are a diverse group of malignancies whose clinical behaviour, management and prognosis vary widely according to the histological type, stage and bulk of the disease. As for lymphomas in adults, the international classifications of type and morphology under which cancer registries classified the lymphomas between 1971 and 1990 do not reflect current clinical prognostic groups for lymphoma in children, which are based on immunophenotype. Lymphoma incidence and mortality are usually reported for the same broad groups used here for survival.

Lymphomas are more common in boys than girls, with a sex ratio of 2.7 to 1 in England and Wales. Incidence is higher in children aged 5-14, in contrast with the leukaemias, which are more common under the age of 5. The incidence of lymphoma in boys doubled between 1971 and 1990 from 7 to 14 per million, but remained fairly level at 5.5 in girls. Mortality fell in both sexes between 1971 and 1995, from 4.5 to 3.1 per million in boys and from 3.5 to 1.4 in girls. The substantial fall in mortality, despite stable or rising incidence, is consistent with large improvements in survival since the late 1960s and early 1970s, after the introduction of combination chemotherapy regimes.

The survival analyses reported here are based on 1,200 children diagnosed with a non-Hodgkin lymphoma in England and Wales between 1971 and 1990 and followed up to the end of 1995, some 89% of eligible children (Table 53.1). About 8% of cases were excluded because their vital status was unknown at 6 July 1997, and 3% because their recorded survival was zero or was unknown: the two groups cannot be distinguished in the available data.

Information on the different types of lymphoma included in the analyses is given in Table 53A. As with non-Hodgkin lymphoma in adults, there has been a major shift in classification and coding of the lymphomas, with a huge decline in neoplasms coded as reticulosarcoma and lymphosarcoma (ICD-9 200) and a corresponding increase in other types of lymphoma (202). Separate analysis of trends in survival from lymphomas assigned to these two rubrics would be misleading.

All the extra-nodal lymphomas diagnosed in the 1980s

had been reassigned as lymphomas in the original data. Only 23 extra-nodal lymphomas, diagnosed in the 1970s, were reassigned from solid tumour sites (see Chapter 2). These involved the pharynx, soft tissue, brain and other organs.

For almost 90% of the cases, the morphology was coded as diffuse or unspecified lymphoma. This proportion was higher than for adults (80%), but also increased steadily, from 83% in 1971-75 to 94% in 1986-90. Reticulosarcoma accounted for 9% overall (15% in 1971-75 but only 2% in 1986-90). Only 2% were coded as either nodular or follicular lymphoma (compared to 9% in adults). None of the lymphomas was classified as malignant histiocytosis.

Survival trends

Relative survival[a] from non-Hodgkin lymphoma in children diagnosed in England and Wales during 1986-90 was 85% at one year and 76% at five years, and was similar in boys and girls (Table 53.2). Survival has improved by no less than 40-50% since the early 1970s. For children diagnosed in 1971-75, one-year relative survival was 45% and five-year survival just 26%. These dramatic increases in survival are clearly shown in Figure 53.1. Even in the early 1970s, survival fell very little after the fifth anniversary of diagnosis, and ten-year survival has remained very similar to five-year survival, implying that most children who survive five years are effectively cured, and can be expected to have similar mortality to that of other children. For children diagnosed during 1981-85, ten-year survival up to the end of 1995 was 65%, only 3% lower than the survival rate among the same children at five years.

The average improvement in survival between successive five-year periods from 1971-75 to 1986-90 in England and Wales has been impressive: one-year survival rose by 12-13% every five years, and the corresponding rate of increase in five-year survival has been 18% every five years (see Table 4.4). These are the largest increases in survival among all the childhood malignancies examined here.

The large and progressive increases in national survival rates between successive five-year periods are consonant with stable or rising incidence and the decline in mortality, and they are so striking that the only plausible explanation is improved treatment. By the same token, and with due

---

[a] See footnote in Chapter 50.

Table 53A Sub-type distribution of non-Hodgkin lymphoma in children

ICD-9 code	Site description	1971-75		1976-80		1981-85		1986-90		Total	
		No.	%	No.	%	No.	%	No.	%	No.	%
200.0	Reticulosarcoma	42	13.3	38	11.3	9	3.4	6	2.2	95	7.9
200.1	Lymphosarcoma	200	63.5	149	44.2	56	21.1	26	9.4	431	36.1
200.2	Burkitt's tumour	–	–	0	0.0	0	0.0	0	0.0	0	0.0
200.8	Other named variants	–	–	1	0.3	4	1.5	2	0.7	7	0.6
200	**Lymphosarcoma and reticulosarcoma**	**242**	*76.8*	**188**	*55.8*	**69**	*25.9*	**34**	*12.3*	**533**	*44.6*
202.0	Nodular lymphoma	8	2.5	14	4.2	4	1.5	5	1.8	31	2.6
202.1	Mycosis fungoides	1	0.3	0	0.0	3	1.1	5	1.8	9	0.8
202.8	Other lymphomas	–	–	68	20.2	190	71.4	232	83.8	490	41.0
202.9	Other and unspecified	48	15.2	60	17.8	0	0.0	1	0.4	21	1.8
202	**Other malignant neoplasm of lymphoid and histiocytic tissue**	**57**	*18.1*	**142**	*42.1*	**197**	*74.1*	**243**	*87.7*	**639**	*53.5*
140-199	Extranodal	16	5.1	7	2.1	0	0.0	0	0.0	23	1.9
	Total	**315**		**337**		**266**		**277**		**1,195**	

[1] The ICD-8 code 202.2 (other primary malignant forms of lymphoma) is not compatible with the ICD-9 code 202.2: the 88 neoplasms so coded during 1971-80 were reassigned to 202.9 for this table. No lymphoma in a child was coded in the range 202.3-202.6 (see Table 41A).

allowance for the imprecision of regional survival rates for each calendar period, which are based on an average of some 30 children (Figure 53.4), the absence of any improvement in survival in some regions in some five-year periods in the past may raise questions about the speed with which improved treatments became available to all children with lymphoma in England and Wales.

These estimates of five-year relative survival in England and Wales as a whole are consistent with the actuarial five-year survival rates reported by the National Registry of Childhood Tumours for Great Britain: for example 28% for children diagnosed in 1974-76[1] and 73% for those diagnosed in 1986-88[2].

Survival trends for non-Hodgkin lymphoma in adults and children have been very different. In the early 1970s, survival up to five years after diagnosis was 7-8% lower for children than for adults, although the difference is largely due to lower survival in boys. The most common lymphomas in children are very chemosensitive, and survival for children has risen much more quickly than for adults. For children diagnosed during 1986-90, relative survival at one year was 20% higher than for adults, and five-year survival was 30% higher (Tables 53.2 and 41.2, see also Table 4.4).

Deprivation

Survival from non-Hodgkin lymphoma in children did not differ systematically between deprivation groups (Table 53.2, Figure 53.2).

Among 260 children of known deprivation status who were diagnosed during 1981-85, survival at one and five years was an estimated 7% higher among children in the most deprived group than among those in the most affluent group, but the differences were in the opposite direction for a similar number of children who were diagnosed during 1986-90, and none is statistically significant (Figure 53.3; see Table 4.9).

International comparison

Relative survival for children diagnosed with non-Hodgkin lymphoma in England and Wales during 1986-90 was 76%, compared to 68% for children living in the parts of the USA covered by the SEER programme. Children diagnosed aged 5-14 in Scotland during 1983-87 (34 cases) also had a five-year relative survival rate of 68%[3].

1.   Stiller CA, Bunch KJ. Trends in survival for childhood cancer in Britain diagnosed 1971-85. Br J Cancer 1990; 62: 806-815

2.   Stiller CA. Population-based survival rates for childhood cancer in Britain, 1980-91. Br Med J 1994; 309: 1612-1616

3.   Black RJ, Sharp L, Kendrick SW. *Trends in cancer survival in Scotland, 1968-1990.* Edinburgh: Information and Statistics Division, 1993

Table 53.1 Ineligible and excluded records, and patients included in analyses, by calendar period of diagnosis

							Calendar period of diagnosis								
	1971-75			**1976-80**			**1981-85**			**1986-90**			**Overall**		
Total registered	**368**			**415**			**311**			**348**			**1,442**		
Ineligible	*Records*	*Patients*	*%*	*Records*	*Patients*	*%*	*Records*	*Patients*	*%*	*Records*	*Patients*	*%*	*Records*	*Patients*	*%*
Incomplete data[1]	7	7		5	5		8	8		18	18		38	**38**	
Resident outside England and Wales	15	15		29	28		9	9		3	1		56	**53**	
In situ neoplasm[2]	0	0		0	0		0	0		0	0		0	**0**	
Benign or uncertain behaviour[2]	5	0		2	2		0	0		0	0		7	**2**	
Metastatic[2]	0	0		0	0		0	0		0	0		0	**0**	
Otherwise ineligible[3]	-	-		-	-		-	-		-	-		-	**-**	
Lymphoma[4]	-	-		-	-		-	-		-	-		-	**-**	
Leukaemia or myeloma[4]	-	-		-	-		-	-		-	-		-	**-**	
		22			**35**			**17**			**19**			**93**	
Total eligible		**346**	**100**		**380**	**100**		**294**	**100**		**329**	**100**		**1,349**	**100**
Aged under 15 or 100 years or over at diagnosis[4]	-	-	-	-	-	-	-	-	-	-	-	-	-	**-**	-
Vital status unknown[5]	36	21	6.1	58	31	8.2	25	15	5.1	44	38	11.6	163	**105**	7.8
Duplicate registration[6]	0	0	0.0	0	0	0.0	0	0	0.0	0	0	0.0	0	**0**	0.0
Multiple primary[7]	0	0	0.0	2	2	0.5	5	4	1.4	0	0	0.0	7	**6**	0.4
Synchronous tumours[8]	1	1	0.3	3	2	0.5	0	0	0.0	2	1	0.3	6	**4**	0.3
Sex not known	0	0	0.0	0	0	0.0	0	0	0.0	0	0	0.0	0	**0**	0.0
Sex incompatible with site code[9]	-	-	-	-	-	-	-	-	-	-	-	-	-	**-**	-
Invalid dates[10]	0	0	0.0	0	0	0.0	0	0	0.0	0	0	0.0	0	**0**	0.0
Zero survival[11]	11	9	2.6	21	8	2.1	13	9	3.1	15	13	4.0	60	**39**	2.9
Total exclusions		**31**	**9.0**		**43**	**11.3**		**28**	**9.5**		**52**	**15.8**		**154**	**11.4**
Patients included in analysis		**315**	**91.0**		**337**	**88.7**		**266**	**90.5**		**277**	**84.2**		**1,195**	**88.6**

[1] Main data item(s) invalid or incompatible with one another: name, sex, date of birth, date of diagnosis and (if present) date of death, postcode, site, morphology and behaviour

[2] Tumour either *in situ* (behaviour code 2), benign (0), uncertain if benign or malignant (1) or metastatic (6 or 9)

[3] Not applicable for this malignancy

[4] Not applicable for this malignancy

[5] Tracing of vital status at National Health Service Central Register incomplete at 6 July 1997

[6] Same site code, sex, cancer registry and cancer registry number as an earlier registration

[7] Same site code and person number as an earlier registration(s): mostly confirmed multiple primary tumours at this site, some unresolved duplicate registrations

[8] Same site code, sex, date of birth and date of diagnosis as another registration(s): mostly synchronous or (in paired organs) bilateral tumours in same anatomic site in one individual, not previously linked: also some duplicate registrations

[9] Not applicable for this malignancy

[10] Impossible sequence of dates (birth, diagnosis and, if present, emigration or death); or date of death after 6 July 1997

[11] Date of diagnosis same as date of death: some are patients who did die on the day of diagnosis, but most were registered solely from a death certificate (death certificate only, or DCO), and their survival time is thus unknown

Table 53.2   Crude and relative survival (%) at one, five and ten years since diagnosis: NHS Region, deprivation category and calendar period of diagnosis

		1971-75 Affluent C R	2 C R	3 C R	4 C R	Deprived C R	All[1] C R	1976-80 Affluent C R	2 C R	3 C R	4 C R	Deprived C R	All[1] C R
ENGLAND & WALES													
Boys		*(32)*	*(40)*	*(31)*	*(34)*	*(34)*	*(208)*	*(72)*	*(45)*	*(47)*	*(37)*	*(37)*	*(243)*
	One year	39 39	63 63	37 37	36 36	43 43	41 41	63 63	68 68	75 75	67 67	68 68	67 67
	Five years	26 26	35 35	20 20	21 21	31 31	25 25	39 39	50 51	54 54	37 37	49 49	45 45
	Ten years	26 26	33 33	16 16	18 18	31 31	23 23	37 38	46 46	47 47	35 35	46 46	42 42
Girls		*(17)*	*(13)*	*(21)*	*(21)*	*(11)*	*(107)*	*(27)*	*(16)*	*(17)*	*(16)*	*(15)*	*(94)*
	One year	58 58	43 44	53 53	47 47	47 47	53 53	67 67	43 43	62 62	50 50	66 66	60 60
	Five years	34 34	36 36	15 15	34 34	37 37	28 28	40 40	25 25	33 33	38 38	40 40	37 37
	Ten years	34 34	29 29	15 15	34 34	37 37	27 27	40 40	13 13	33 33	38 38	40 40	35 35
Children		*(49)*	*(53)*	*(52)*	*(55)*	*(45)*	*(315)*	*(99)*	*(61)*	*(64)*	*(53)*	*(52)*	*(337)*
	One year	45 45	58 58	43 43	40 40	44 44	45 45	64 64	62 62	72 72	62 62	67 67	65 65
	Five years	29 29	36 36	18 18	26 26	32 32	26 26	39 39	44 44	48 48	37 37	46 46	43 43
	Ten years	29 29	32 32	16 16	24 24	32 32	25 25	38 38	37 37	44 44	35 36	44 44	40 40

		Boys	Girls	Children				Boys	Girls	Children
Northern & Yorkshire		*(26)*	*(19)*	*(45)*				*(29)*	*(14)*	*(43)*
	One year	53 53	73 73	61 61				51 51	71 72	58 58
	Five years	31 31	43 43	36 36				34 34	35 35	35 35
	Ten years	27 27	43 43	33 34				23 23	35 35	27 27
Trent		*(19)*	*(17)*	*(36)*				*(28)*	*(15)*	*(43)*
	One year	59 59	60 60	59 60				69 69	39 39	58 58
	Five years	49 49	19 19	34 34				43 43	20 20	35 35
	Ten years	43 43	19 19	31 32				40 40	20 20	33 33
Anglia & Oxford		*(26)*	*(15)*	*(41)*				*(18)*	*(5)*	*(23)*
	One year	39 40	46 46	41 41				62 62	- -	62 62
	Five years	20 20	19 19	20 20				40 40	- -	36 36
	Ten years	20 20	19 19	20 20				40 40	- -	36 36
North Thames		*(29)*	*(17)*	*(46)*				*(30)*	*(7)*	*(37)*
	One year	40 40	47 47	42 42				63 63	- -	64 65
	Five years	23 23	41 41	29 29				37 37	- -	40 40
	Ten years	23 23	35 35	27 27				33 33	- -	38 38
South Thames		*(29)*	*(10)*	*(39)*				*(30)*	*(12)*	*(42)*
	One year	36 36	41 41	37 37				84 84	56 56	76 76
	Five years	23 23	19 19	22 22				64 64	48 48	60 60
	Ten years	23 23	19 19	22 22				64 64	41 41	57 57
South & West		*(16)*	*(9)*	*(25)*				*(36)*	*(18)*	*(54)*
	One year	40 40	- -	44 44				77 77	55 55	70 70
	Five years	25 25	- -	24 24				52 52	38 38	48 48
	Ten years	18 18	- -	20 20				52 52	33 33	46 46
West Midlands		*(20)*	*(8)*	*(28)*				*(32)*	*(2)*	*(34)*
	One year	11 11	- -	18 18				58 58	- -	55 55
	Five years	11 11	- -	15 15				40 40	- -	38 38
	Ten years	11 11	- -	15 15				40 40	- -	38 38
North & West		*(29)*	*(9)*	*(38)*				*(31)*	*(15)*	*(46)*
	One year	45 45	- -	45 45				64 64	66 66	65 65
	Five years	21 21	- -	21 21				38 38	32 32	36 36
	Ten years	21 21	- -	21 21				32 32	32 32	32 32
WALES		*(14)*	*(3)*	*(17)*				*(9)*	*(6)*	*(15)*
	One year	52 52	- -	55 55				- -	- -	86 86
	Five years	29 29	- -	30 30				- -	- -	73 73
	Ten years	29 29	- -	30 30				- -	- -	67 67

[1] Including those for whom no deprivation category was assigned. C - crude survival, R - relative survival (see Chapter 3). Number of patients contributing to each analysis in parentheses.

1981-85   1986-90

	Affluent	*2*	*3*	*4*	*Deprived*	*All*[1]	*Affluent*	*2*	*3*	*4*	*Deprived*	*All*[1]	
	C R	C R	C R	C R	C R	C R	C R	C R	C R	C R	C R	C R	**ENGLAND & WALES**
	(49)	*(32)*	*(36)*	*(34)*	*(37)*	*(191)*	*(32)*	*(52)*	*(31)*	*(42)*	*(35)*	*(196)*	*Boys*
One year	79 79	68 68	74 74	82 82	81 81	78 78	91 91	82 82	90 90	81 81	80 80	84 84	
Five years	71 71	62 62	63 63	65 65	70 70	67 67	84 84	73 73	84 84	69 69	74 74	76 76	
	65 65	59 59	61 61	65 65	67 67	64 64							
	(19)	*(13)*	*(20)*	*(10)*	*(12)*	*(75)*	*(18)*	*(15)*	*(18)*	*(14)*	*(16)*	*(81)*	*Girls*
One year	68 68	83 83	95 95	90 90	75 75	82 82	89 89	87 87	89 89	93 93	74 74	86 86	
Five years	57 57	54 54	80 80	90 90	75 75	70 71	78 78	80 80	78 78	79 79	68 68	77 77	
	57 57	46 46	75 75	90 90	75 75	68 68							
	(68)	*(45)*	*(56)*	*(44)*	*(49)*	*(266)*	*(50)*	*(67)*	*(49)*	*(56)*	*(51)*	*(277)*	*Children*
One year	76 76	73 73	82 82	84 84	79 79	79 79	90 90	83 83	90 90	84 84	78 78	85 85	
Five years	67 67	59 59	69 69	71 71	71 71	68 68	82 82	74 74	82 82	71 71	72 72	76 76	
	63 63	55 55	66 66	71 71	69 69	65 65							

	Boys	*Girls*	*Children*		*Boys*	*Girls*	*Children*	
	(28)	*(7)*	*(35)*		*(26)*	*(10)*	*(36)*	**Northern & Yorkshire**
One year	78 78	- -	80 80		81 81	80 80	80 80	
Five years	54 54	- -	60 60		77 77	80 80	78 78	
	50 50	- -	57 57					
	(18)	*(6)*	*(24)*		*(20)*	*(6)*	*(26)*	**Trent**
One year	77 77	- -	74 74		90 90	- -	88 88	
Five years	60 60	- -	62 62		85 85	- -	80 80	
	55 55	- -	58 58					
	(23)	*(9)*	*(32)*		*(17)*	*(12)*	*(29)*	**Anglia & Oxford**
One year	74 74	- -	72 72		88 88	83 83	86 86	
Five years	65 65	- -	59 59		82 82	58 58	72 72	
	65 65	- -	59 59					
	(23)	*(7)*	*(30)*		*(29)*	*(12)*	*(41)*	**North Thames**
One year	74 74	- -	77 77		83 83	76 76	81 81	
Five years	69 69	- -	73 73		72 72	76 76	73 73	
	69 69	- -	73 73					
	(20)	*(8)*	*(28)*		*(26)*	*(10)*	*(36)*	**South Thames**
One year	80 80	- -	78 78		85 85	80 80	83 83	
Five years	70 70	- -	68 68		85 85	80 80	83 83	
	60 61	- -	61 61					
	(26)	*(11)*	*(37)*		*(34)*	*(14)*	*(48)*	**South & West**
One year	80 80	91 91	84 84		82 82	93 93	85 85	
Five years	80 80	73 73	78 78		70 70	72 72	71 71	
	80 80	73 73	78 78					
	(15)	*(4)*	*(19)*		*(12)*	*(10)*	*(22)*	**West Midlands**
One year	86 86	- -	89 89		83 83	100 100	91 91	
Five years	80 80	- -	79 79		75 75	100 100	86 86	
	80 80	- -	73 73					
	(20)	*(12)*	*(32)*		*(24)*	*(4)*	*(28)*	**North & West**
One year	69 70	92 92	78 78		79 79	- -	82 82	
Five years	55 55	83 83	65 65		58 58	- -	61 61	
	50 50	83 83	62 62					
	(18)	*(11)*	*(29)*		*(8)*	*(3)*	*(11)*	**WALES**
One year	83 83	80 80	82 82		100 100	- -	100 100	
Five years	77 77	63 63	72 72		100 100	- -	100 100	
	72 72	54 54	65 65					

[1] Including those for whom no deprivation category was assigned. C - crude survival, R - relative survival (see Chapter 3).
Number of patients contributing to each analysis in parentheses.

Fig. 53.1   Relative survival up to ten years: trends by calendar period of diagnosis (1971-90) and NHS Region

CHILDREN

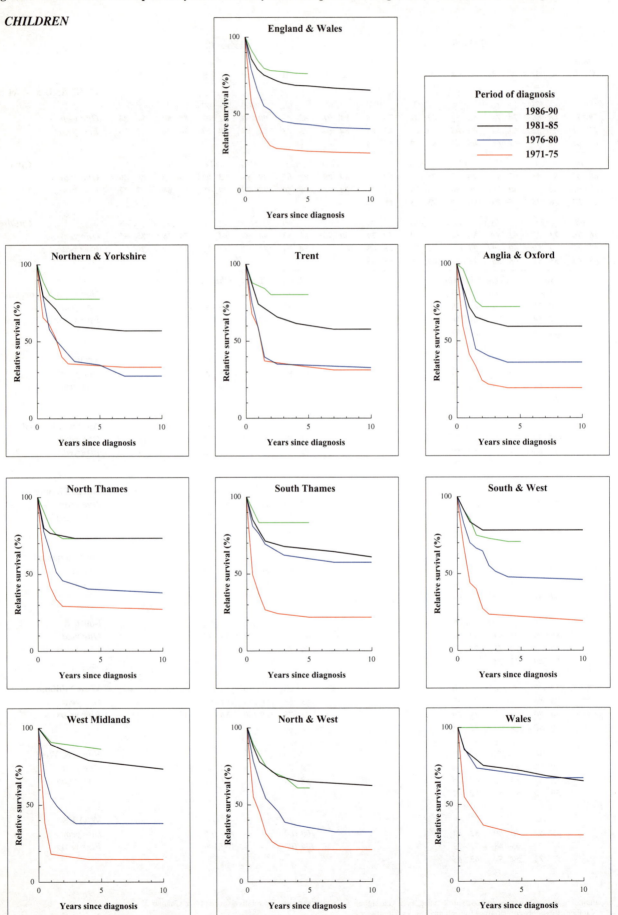

Fig. 53.2 Relative survival up to five years, by deprivation category: England and Wales, patients diagnosed 1986-90

CHILDREN

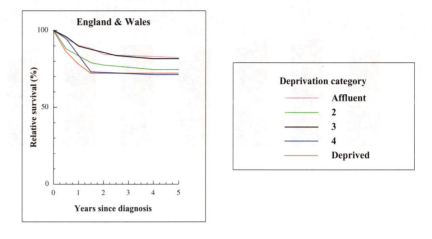

Fig. 53.3 Relative survival at five years, by deprivation category and period of diagnosis (1981-90): England and Wales

CHILDREN

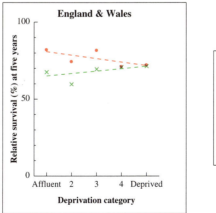
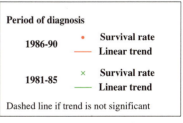

Fig. 53.4 Relative survival (%) by NHS Region and sex:
one-year and five-year survival for children diagnosed 1986-90, with 95% confidence intervals

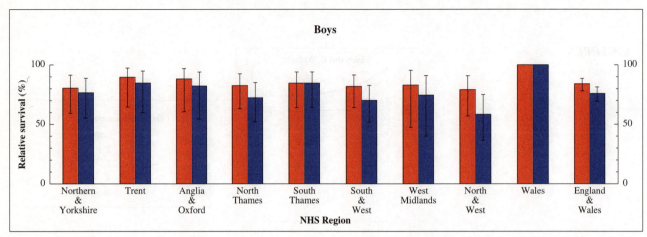

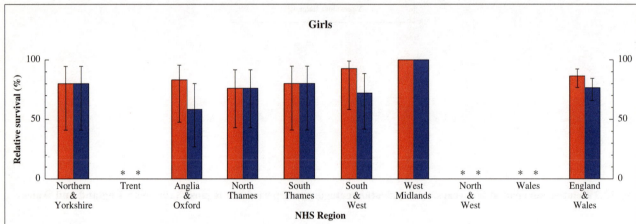

Chapter 54

BRAIN AND SPINAL CORD *Children*

Tumours of the brain and spinal cord are the commonest solid tumours in children, second in overall incidence only to the leukaemias. They account for 20% of all malignant neoplasms in children in England and Wales. Gliomas are the most common, including astrocytoma (30-50% of all brain tumours) and ependymoma. Central nervous system tumours occur more frequently in boys than girls, with an overall sex ratio of 1.2 to 1, although for medulloblastoma the sex ratio is 1.7 to 1. The tumours for which survival is analysed here comprise group III of the Birch and Marsden classification for childhood tumours[1] (see Chapter 2 and Table 2.5). This group includes malignant tumours of the brain and central nervous system, but also those of benign and uncertain behaviour, as well as tumours of the pituitary and pineal.

About 250 new cancers of the brain and spinal cord are diagnosed in children in England and Wales each year. Between 1979 and 1990 incidence rose by 27% to 2.8 per 100,000 in boys, and by 60% to 2.4 in girls. These rates are somewhat lower than in the Nordic countries, the Netherlands and the USA, where incidence is in the range 3-6 for boys and 3-5 for girls, and in Scotland, where the rate for girls (3.3) slightly exceeds that for boys (2.9).

The survival analyses reported here are based on 4,400 children who were diagnosed with a malignant tumour of the central nervous system in England and Wales between 1971 and 1990 and followed up to the end of 1995, about 87% of the eligible cases (Table 54.1). Almost 7% of children were excluded because their vital status was unknown at 6 July 1997, and a further 5% because their survival was either zero or unknown (see Chapter 2). Most were malignant tumours of the brain (77%) and the cranial nerves, spinal cord and meninges (8%), but there were also some 300 tumours of the pituitary and pineal (7%; Table 54A).

Gliomas accounted for 58% of the tumours, including astrocytoma or astroblastoma (30%), ependymoma (5%) and oligodendroglioma (2%), and glioma not otherwise specified (16%). Some 19% were medulloblastoma, 5% craniopharyngioma, and about 1% each were classified as pinealoma, germ cell tumours or malignant meningeal tumours; non-specific morphologies accounted for 12%.

Survival trends

Relative survival[a] for children diagnosed with a tumour of the brain or spinal cord in England and Wales during 1986-90 was 76% at one year and 59% at five years (Table 54.2). These survival rates are some 16% higher than those for children diagnosed during the early 1970s. The largest improvement occurred around 1980. On average, one-year survival increased by about 6% between successive five-year periods (see Table 4.4). Five-year survival also improved by 16%, from 43% to 59%, also an average gain of about 6% between successive quinquennia. The general improvement in survival is shown graphically in Figure 54.1.

Survival was similar for boys and girls. There were similar substantial improvements in all regions. Regional variation in survival for children diagnosed during 1986-90 was not marked (Figure 54.4).

Actuarial survival rates have been published by the National Registry of Childhood Tumours for Great Britain for several tumours of the central nervous system[2,3]. Five-year survival for astrocytoma improved from 61% to 71% for children diagnosed in 1971-73 and 1986-88, respectively, while the corresponding increase in survival from ependymoma was from 38% to 50%, and for medulloblastoma from 24% to 42%.

[a] See footnote in Chapter 50.

Table 54A Tumours of the brain and central nervous system in children

ICD-9 code	Site description (behaviour)	1971-75 No.	1971-75 %	1976-80 No.	1976-80 %	1981-85 No.	1981-85 %	1986-90 No.	1986-90 %	Total No.	Total %
191	Brain (malignant)	869	72.7	825	72.3	825	82.1	863	81.2	**3,382**	**76.8**
192	Other parts of CNS (malignant)	123	10.3	105	9.2	56	5.6	75	7.1	**359**	**8.2**
	Pituitary, pineal (any)[1]	69	5.8	96	8.4	72	7.2	76	7.1	**313**	**7.1**
	CNS (benign/uncertain)[2]	128	10.7	111	9.7	52	5.2	47	4.4	**338**	**7.7**
	Other sites	6	0.5	4	0.4	0	0.0	2	0.2	**12**	**0.3**
	Total	**1,195**		**1,141**		**1,005**		**1,063**		**4,404**	

[1] Includes pituitary tumours (22 malignant, 75 benign, 124 uncertain behaviour) and pineal tumours (45, 25, 22, respectively).
[2] Includes 119 benign tumours of the brain and spinal cord (ICD-9 225x) and 219 of uncertain behaviour (237.5-237.6).

Deprivation

Survival rates for children diagnosed in England and Wales during a given five-year period are based on some 200 children in each deprivation category. Survival from tumours of the central nervous system in children does not differ systematically between deprivation groups (Figure 54.2). Estimated differences in one-year and five-year survival between children in the most affluent and the most deprived groups were 3-6%, but none of the deprivation gradients is significant (Figure 54.3, see Table 4.9).

International comparison

Survival in England and Wales is broadly comparable with that in Scotland. For 62 children under the age of 5 who were diagnosed with a brain tumour in Scotland during 1983-87, relative survival at five years was 50%; for the 86 children aged 5-14 at diagnosis, five-year survival was 59%[4].

1. Birch JM, Marsden HB. A classification scheme for childhood cancer. Int J Cancer 1987; 40: 620-624

2. Stiller CA, Bunch KJ. Trends in survival for childhood cancer in Britain diagnosed 1971-85. Br J Cancer 1990; 62: 806-815

3. Stiller CA. Population-based survival rates for childhood cancer in Britain, 1980-91. Br Med J 1994; 309: 1612-1616

4. Black RJ, Sharp L, Kendrick SW. *Trends in cancer survival in Scotland, 1968-1990.* Edinburgh: Information and Statistics Division, 1993

Table 54.1 Ineligible and excluded records, and patients included in analyses, by calendar period of diagnosis

	Calendar period of diagnosis														
	1971-75			**1976-80**			**1981-85**			**1986-90**			**Overall**		
Total registered	**1,432**			**1,353**			**1,200**			**1,280**			**5,265**		
Ineligible	*Records*	*Patients*	*%*	*Records*	*Patients*	*%*	*Records*	*Patients*	*%*	*Records*	*Patients*	*%*	*Records*	*Patients*	*%*
Incomplete data[1]	32	32		14	14		17	17		9	9		72	**72**	
Resident outside England and Wales	59	56		43	42		31	31		23	22		156	**151**	
In situ neoplasm[2]	1	0		0	0		0	0		0	0		1	**0**	
Benign or uncertain behaviour[2]	25	2		7	1		0	0		0	0		32	**3**	
Metastatic[2]	0	0		0	0		0	0		0	0		0	**0**	
Otherwise ineligible[3]	-	-		-	-		-	-		-	-		-	**-**	
Lymphoma[4]	-	-		-	-		-	-		-	-		-	**-**	
Leukaemia or myeloma[4]	-	-		-	-		-	-		-	-		-	**-**	
		90			**57**			**48**			**31**			**226**	
Total eligible	**1,342**	**100**		**1,296**	**100**		**1,152**	**100**		**1,249**	**100**		**5,039**	**100**	
Aged under 15 or 100 years or over at diagnosis[4]	-	-		-	-		-	-		-	-		-	**-**	-
Vital status unknown[5]	144	74	5.5	120	74	5.7	98	73	6.3	144	122	9.8	506	**343**	6.8
Duplicate registration[6]	0	0	0.0	0	0	0.0	0	0	0.0	0	0	0.0	0	**0**	0.0
Multiple primary[7]	9	9	0.7	9	9	0.7	4	4	0.3	5	5	0.4	27	**27**	0.5
Synchronous tumours[8]	7	6	0.4	6	3	0.2	6	5	0.4	13	6	0.5	32	**20**	0.4
Sex not known	0	0	0.0	0	0	0.0	0	0	0.0	0	0	0.0	0	**0**	0.0
Sex incompatible with site code[9]	-	-	-	-	-	-	-	-	-	-	-	-	-	**-**	-
Invalid dates[10]	0	0	0.0	0	0	0.0	1	0	0.0	0	0	0.0	1	**0**	0.0
Zero survival[11]	71	58	4.3	76	69	5.3	78	65	5.6	56	53	4.2	281	**245**	4.9
Total exclusions		**147**	**11.0**		**155**	**12.0**		**147**	**12.8**		**186**	**14.9**		**635**	**12.6**
Patients included in analysis		**1,195**	**89.0**		**1,141**	**88.0**		**1,005**	**87.2**		**1,063**	**85.1**		**4,404**	**87.4**

[1] Main data item(s) invalid or incompatible with one another: name, sex, date of birth, date of diagnosis and (if present) date of death, postcode, site, morphology and behaviour

[2] Tumour either *in situ* (behaviour code 2), benign (0), uncertain if benign or malignant (1) or metastatic (6 or 9)

[3] Not applicable for this malignancy

[4] Not applicable for this malignancy

[5] Tracing of vital status at National Health Service Central Register incomplete at 6 July 1997

[6] Same site code, sex, cancer registry and cancer registry number as an earlier registration

[7] Same site code and person number as an earlier registration(s): mostly confirmed multiple primary tumours at this site, some unresolved duplicate registrations

[8] Same site code, sex, date of birth and date of diagnosis as another registration(s): mostly synchronous or (in paired organs) bilateral tumours in same anatomic site in one individual, not previously linked: also some duplicate registrations

[9] Not applicable for this malignancy

[10] Impossible sequence of dates (birth, diagnosis and, if present, emigration or death); or date of death after 6 July 1997

[11] Date of diagnosis same as date of death: some are patients who did die on the day of diagnosis, but most were registered solely from a death certificate (death certificate only, or DCO), and their survival time is thus unknown

Table 54.2 Crude and relative survival (%) at one, five and ten years since diagnosis: NHS Region, deprivation category and calendar period of diagnosis

	1971-75						1976-80					
	Affluent	2	3	4	Deprived	All[1]	Affluent	2	3	4	Deprived	All[1]
	C R	C R	C R	C R	C R	C R	C R	C R	C R	C R	C R	C R
ENGLAND & WALES												
Boys	(121)	(113)	(103)	(101)	(86)	(662)	(135)	(118)	(104)	(122)	(120)	(629)
One year	66 66	62 62	64 64	58 58	55 55	61 61	65 65	65 66	65 65	63 63	60 60	64 64
Five years	45 45	48 48	42 42	44 44	34 34	43 43	45 45	47 47	43 43	42 42	46 46	45 45
Ten years	42 42	42 42	40 40	43 44	26 26	39 39	43 43	39 40	*41 41*	40 40	*45 46*	42 42
Girls	(97)	(69)	(86)	(86)	(80)	(533)	(120)	(113)	(77)	(70)	(106)	(512)
One year	59 59	62 62	69 69	66 66	46 46	59 59	70 70	67 67	57 57	66 66	58 58	64 64
Five years	43 43	47 48	43 43	51 51	32 32	42 43	55 55	51 51	46 46	50 50	51 51	51 51
Ten years	41 41	*46 46*	42 42	41 41	30 30	39 39	52 52	48 48	41 41	44 44	45 45	46 47
Children	(218)	(182)	(189)	(187)	(166)	(1,195)	(255)	(231)	(181)	(192)	(226)	(1,141)
One year	63 63	62 62	66 66	61 62	51 51	60 60	67 67	66 66	62 62	64 64	59 59	64 64
Five years	44 44	48 48	43 43	47 47	33 33	43 43	50 50	49 49	45 45	45 45	48 49	48 48
Ten years	42 42	44 44	41 41	42 43	28 28	39 39	47 47	43 44	41 42	41 41	45 46	44 44

	Boys	Girls	Children				Boys	Girls	Children
Northern & Yorkshire	(82)	(63)	(145)				(95)	(72)	(167)
One year	59 59	57 57	58 58				62 62	60 60	61 61
Five years	37 36	36 36	37 36				46 46	43 43	45 45
Ten years	*36 35*	*33 33*	*35 34*				*44 44*	*36 37*	*41 41*
Trent	(73)	(57)	(130)				(67)	(66)	(133)
One year	61 61	55 55	58 58				65 65	59 59	62 62
Five years	*44 44*	*43 43*	43 43				44 44	*51 51*	48 48
Ten years	*42 42*	*39 40*	41 41				41 41	47 47	44 44
Anglia & Oxford	(67)	(68)	(135)				(64)	(51)	(115)
One year	57 57	60 60	59 59				67 67	71 71	69 69
Five years	43 43	50 50	47 47				48 48	51 51	50 50
Ten years	42 42	44 44	43 43				44 44	45 45	44 44
North Thames	(85)	(73)	(158)				(60)	(40)	(100)
One year	66 66	65 65	66 66				70 70	72 73	71 71
Five years	41 41	46 46	43 43				47 48	68 68	56 56
Ten years	36 36	45 45	40 40				42 43	60 60	50 50
South Thames	(59)	(53)	(112)				(75)	(62)	(137)
One year	69 70	*64 64*	67 67				68 68	62 62	65 65
Five years	*54 54*	*48 49*	51 52				51 51	*50 51*	51 51
Ten years	*46 46*	*47 47*	46 46				47 47	*46 46*	*46 47*
South & West	(83)	(58)	(141)				(78)	(62)	(140)
One year	62 62	74 74	67 67				57 57	70 70	63 63
Five years	52 52	57 57	54 54				39 39	56 56	46 46
Ten years	47 47	50 50	48 48				*38 38*	*56 56*	*46 46*
West Midlands	(83)	(54)	(137)				(56)	(54)	(110)
One year	53 53	43 43	49 49				47 47	60 60	53 53
Five years	37 37	*34 34*	36 36				30 30	*44 44*	37 37
Ten years	37 37	*34 34*	36 36				28 28	*42 42*	35 35
North & West	(93)	(78)	(171)				(89)	(77)	(166)
One year	62 62	50 50	56 56				66 66	64 64	65 65
Five years	40 40	29 29	35 35				49 50	50 50	50 50
Ten years	31 32	27 27	29 29				44 44	45 45	44 45
WALES	(37)	(29)	(66)				(45)	(28)	(73)
One year	52 52	69 69	60 60				*71 71*	*64 64*	*68 68*
Five years	*41 41*	*42 42*	42 42				*51 51*	*50 50*	*50 50*
Ten years	*33 33*	*38 39*	36 36				*48 49*	*46 46*	*47 48*

[1] Including those for whom no deprivation category was assigned. C - crude survival, R - relative survival (see Chapter 3). Number of patients contributing to each analysis in parentheses.

ENGLAND & WALES

	1981-85						1986-90						
	Affluent	2	3	4	Deprived	All[1]	Affluent	2	3	4	Deprived	All[1]	
	C R	C R	C R	C R	C R	C R	C R	C R	C R	C R	C R	C R	**Boys**
	(126)	(101)	(102)	(100)	(120)	(560)	(133)	(123)	(106)	(118)	(93)	(579)	
One year	73 74	69 69	82 82	75 75	73 73	74 75	79 79	80 80	74 74	77 77	75 75	77 77	
Five years	60 59	59 59	63 64	57 57	58 58	59 59	59 59	61 61	55 55	54 54	57 57	57 57	
	54 54	53 53	53 53	53 53	52 52	53 53							
	(96)	(70)	(71)	(104)	(94)	(445)	(116)	(109)	(69)	(98)	(85)	(484)	**Girls**
One year	83 83	71 71	66 66	70 70	73 73	74 74	71 71	78 78	84 84	66 66	74 74	74 74	
Five years	64 64	56 56	49 49	53 53	61 61	57 57	61 61	70 70	64 64	50 50	62 62	62 62	
	55 55	46 46	46 46	52 52	58 58	52 52							
	(222)	(171)	(173)	(204)	(214)	(1,005)	(249)	(232)	(175)	(216)	(178)	(1,063)	**Children**
One year	78 78	70 70	75 75	72 72	73 73	74 74	75 75	79 79	78 78	72 72	74 74	76 76	
Five years	62 62	58 58	58 58	55 55	59 60	58 58	60 60	65 65	58 58	52 52	59 59	59 59	
	55 54	50 50	50 50	52 52	55 55	52 53							

	1981-85			1986-90			
	Boys	Girls	Children	Boys	Girls	Children	
	(88)	(52)	(140)	(67)	(63)	(130)	**Northern & Yorkshire**
One year	67 68	66 66	67 67	83 84	79 79	81 81	
Five years	55 55	47 47	52 52	61 61	65 65	63 63	
	51 51	42 42	47 47				
	(50)	(53)	(103)	(70)	(48)	(118)	**Trent**
One year	77 77	77 77	77 77	67 67	68 68	68 68	
Five years	57 57	62 62	60 60	48 49	62 62	54 54	
	53 53	58 58	56 56				
	(55)	(47)	(102)	(62)	(54)	(116)	**Anglia & Oxford**
One year	76 76	68 68	72 72	77 77	68 68	73 73	
Five years	61 60	53 53	57 56	58 58	55 55	57 57	
	52 50	51 51	51 51				
	(41)	(26)	(67)	(66)	(46)	(112)	**North Thames**
One year	78 78	92 92	83 84	72 72	69 70	71 71	
Five years	66 66	69 69	67 67	47 47	47 47	47 47	
	54 54	65 65	58 58				
	(54)	(55)	(109)	(55)	(50)	(105)	**South Thames**
One year	75 75	79 79	77 77	84 84	78 78	81 81	
Five years	64 64	63 63	64 64	62 62	70 70	66 66	
	59 59	53 54	56 56				
	(69)	(68)	(137)	(79)	(66)	(145)	**South & West**
One year	78 78	72 72	75 75	77 77	77 77	77 77	
Five years	61 61	59 59	60 60	55 55	64 64	59 59	
	55 55	54 54	55 55				
	(77)	(53)	(130)	(65)	(63)	(128)	**West Midlands**
One year	70 70	76 76	72 72	83 83	69 69	76 76	
Five years	54 55	51 51	53 53	63 63	57 57	60 60	
	45 45	42 42	44 44				
	(97)	(69)	(166)	(95)	(65)	(160)	**North & West**
One year	77 77	68 68	73 73	77 78	80 80	78 79	
Five years	59 59	58 58	58 58	63 63	71 71	66 66	
	55 55	55 55	55 55				
	(29)	(22)	(51)	(20)	(29)	(49)	**WALES**
One year	76 76	76 77	76 76	64 64	80 80	73 73	
Five years	62 62	59 59	60 61	64 64	62 62	63 63	
	55 55	59 59	57 57				

[1] Including those for whom no deprivation category was assigned. C - crude survival, R - relative survival (see Chapter 3).
Number of patients contributing to each analysis in parentheses.

Fig. 54.1   Relative survival up to ten years: trends by calendar period of diagnosis (1971-90) and NHS Region

CHILDREN

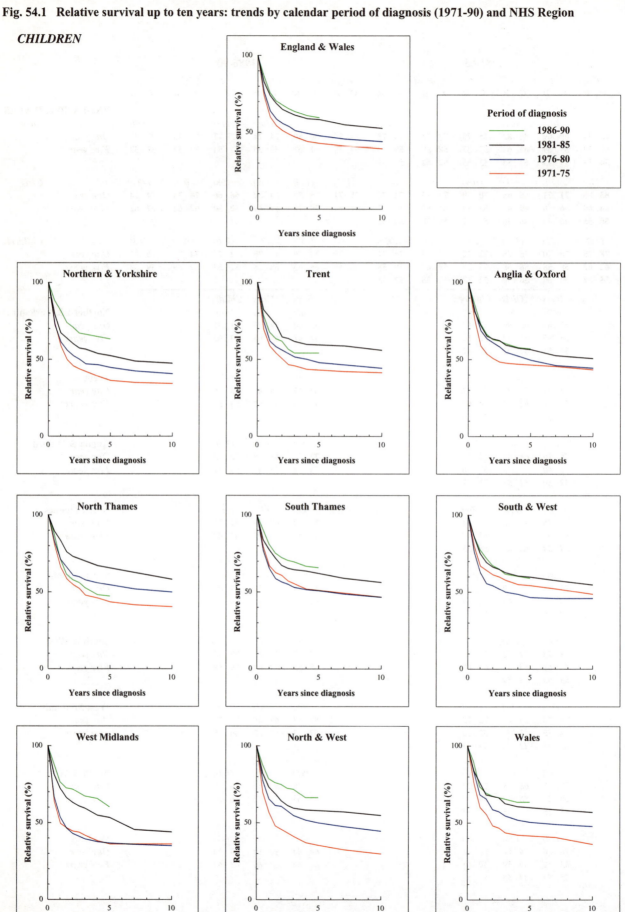

Fig. 54.2 Relative survival up to five years, by deprivation category: England and Wales, patients diagnosed 1986-90

CHILDREN

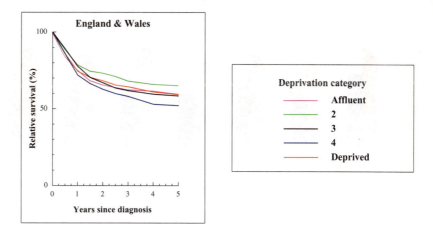

Fig. 54.3 Relative survival at five years, by deprivation category and period of diagnosis (1981-90): England and Wales

CHILDREN

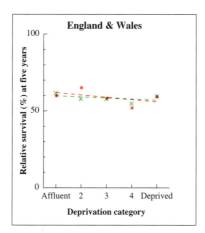
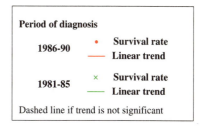

Fig. 54.4 Relative survival (%) by NHS Region and sex:

　　　　one-year and five-year survival for children diagnosed 1986-90, with 95% confidence intervals

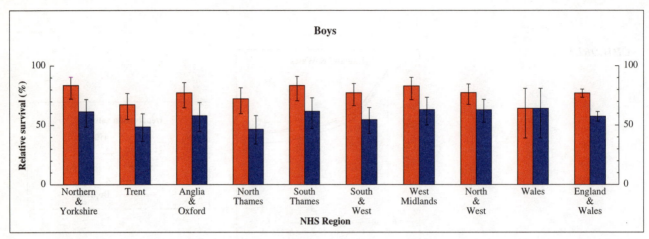

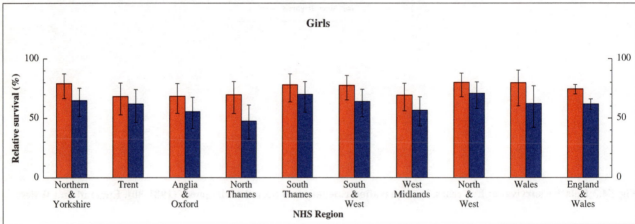

Chapter 55

RETINOBLASTOMA *Children*

Retinoblastoma is a malignant tumour of the eye arising from retinal precursor cells. It is one of the primitive neuroectodermal tumours of childhood, with medulloblastoma and neuroblastoma[1]. Retinoblastoma is extremely rare, and only 20-30 new cases are diagnosed each year in England and Wales. It accounts for almost all ocular tumours in children, and about 2% of all childhood cancers, as in many Caucasian populations[2]. Retinoblastomas mostly occur in children under 5, and they are equally common in boys and girls.

The symptoms of retinoblastoma may include reduced vision, white pupils, squint, and pain. Small tumours may be destroyed by laser coagulation or cryotherapy. Larger tumours may require enucleation of the eye and radiotherapy to the orbit. The probability of cure is very high, especially if the tumour has not spread beyond the orbit or to the optic nerve, and sight can often be preserved.

The survival analyses reported here are based on just over 500 children who were diagnosed with retinoblastoma in England and Wales between 1971 and 1990 and followed up to the end of 1995, about 87% of eligible cases. Sixty eligible children (10%) were excluded from the analyses because their vital status was not known at 6 July 1997, but only four because their survival was zero or unknown (Table 55.1). Laterality was not included in the national cancer registry for cases diagnosed before 1993 (see Table 2.1), and only 12 children were explicitly recorded as having synchronous (multiple or bilateral) tumours in the registry data. A child with two registrations was retained only once in the analysis, from the date of diagnosis of the first tumour.

Since 1979, when the parts of the eye were first classified separately in ICD-9, all retinoblastomas have been assigned to the retina (Table 55A). The morphology of all the tumours was coded as retinoblastoma.

Survival trends

Relative survival[a] among 106 children diagnosed with retinoblastoma in England and Wales during 1986-90 was 98% at one year and 93% at five years (Table 55.2). For 171 children diagnosed during 1971-75, survival was already high: one-year survival of 94% and 88% at five years. The progressive improvement in the survival curves for England and Wales is shown in Figure 55.1.

The regional survival estimates mostly depend on 10-30 children diagnosed with retinoblastoma in a given five-

year period, and they are imprecise. The survival curves in Figure 55.1 reflect this, and in most regions no estimate at all was made for one or more quinquennia because there were too few cases – fewer than 10 (1971-85) or fewer than 8 during 1986-90 (see Chapter 3). The survival curve for children diagnosed in Trent during 1986-90 is superposed on the curve for 1981-85, at 100%. When regional survival rates were examined separately for boys and girls diagnosed during 1986-90, only four estimates could be made (Figure 54.4). It is abundantly clear that five-year survival must be high in all regions, however, since the national average survival rate is 93%, and very few deaths were recorded within five years of diagnosis in this group of children.

Ten-year survival for children diagnosed in 1981-85 was 88%, only slightly lower than five-year survival. The plateau in the relative survival curves after the fifth anniversary of diagnosis reflects very low excess mortality up to ten years after diagnosis (Table 55.2, Figure 55.1).

Survival was slightly higher for boys than for girls, and the increases in survival were also slightly larger in boys, but the differences are generally small (see Table 4.4).

These results are consistent with actuarial survival estimates published by the National Registry of Childhood Tumours for Great Britain. Five-year survival was 87% for children diagnosed in 1971-73[3], rising to 89% in 1983-85 and to 94% in 1986-88[4]. Even for children diagnosed during 1971-73, survival up to 15 years (84%) was no lower than 10-year survival. Forty (91%) of the 44 children registered with retinoblastoma in the Northern Regional Health Authority area during 1968-82 survived five years[5].

Deprivation

Five-year survival from retinoblastoma was lowest in the most deprived group for children diagnosed in the 1980s, but even in England and Wales as a whole there are too few cases to draw conclusions (Table 55.2, Figure 55.2). Regression analysis of the deprivation gradient is not reported.

International comparison

Survival from retinoblastoma in England and Wales is comparable to that in the USA: relative survival at five years in the areas covered by the SEER programme

---

[a] See footnote in Chapter 50.

Table 55A Distribution of retinoblastoma in children

ICD-9 code	Site description	1971-75 No.	%	1976-80 No.	%	1981-85 No.	%	1986-90 No.	%	Total No.	%
190.5	Retina	-	-	42	32.8	102	100.0	106	100.0	**250**	**49.3**
190.9	Eye, part unspecified[1]	170	99.4	86	67.2	0	0.0	0	0.0	**256**	**50.5**
199.0	Site unspecified	1	0.6	0	0.0	0	0.0	0	0.0	**1**	**0.2**
	Total	**171**		**128**		**102**		**106**		**507**	

[1] Special NCR codes used up to 1978 (190x; see Chapter 2) are not compatible with ICD-9: tumours were assigned to 190.9 for this table.

increased from 88% for children diagnosed during 1978-82 to 96% for those diagnosed during 1983-87[6].

1. Kingston JE, Hungerford J. Retinoblastoma. In: Peckham M, Pinedo HM, Veronesi U, (eds.) *Oxford textbook of oncology*. Oxford: Oxford Medical Publications, 1995

2. Parkin DM, Stiller CA, Draper GJ, Bieber CA, Terracini B, Young JL. *International Incidence of Childhood Cancer (IARC Scientific Publications No. 87)*. Lyon: International Agency for Research on Cancer, 1988

3. Stiller CA, Bunch KJ. Trends in survival for childhood cancer in Britain diagnosed 1971-85. Br J Cancer 1990; 62: 806-815

4. Stiller CA. Population-based survival rates for childhood cancer in Britain, 1980-91. Br Med J 1994; 309: 1612-1616

5. Craft AW, Amineddine JES, Scott JES, Wagget J. The Northern Region Children's Malignant Disease Registry 1968-82: incidence and survival. Br J Cancer 1987; 56: 853-858

6. Polednak AP, Flannery JT. Brain, other central nervous system, and eye cancer. Cancer 1995; 75: 330-337

Table 55.1 Ineligible and excluded records, and patients included in analyses, by calendar period of diagnosis

	Calendar period of diagnosis														
	1971-75			**1976-80**			**1981-85**			**1986-90**			**Overall**		
Total registered	**223**			**171**			**164**			**158**			**716**		
Ineligible	*Records*	*Patients*	*%*	*Records*	*Patients*	*%*	*Records*	*Patients*	*%*	*Records*	*Patients*	*%*	*Records*	*Patients*	*%*
Incomplete data[1]	2	2		5	5		36	36		33	33		76	**76**	
Resident outside England and Wales	26	26		21	21		12	10		0	0		59	**57**	
In situ neoplasm[2]	0	0		0	0		0	0		0	0		0	**0**	
Benign or uncertain behaviour[2]	2	0		2	0		0	0		0	0		4	**0**	
Metastatic[2]	0	0		0	0		0	0		0	0		0	**0**	
Otherwise ineligible[3]	-	-		-	-		-	-		-	-		-	**-**	
Lymphoma[4]	-	-		-	-		-	-		-	-		-	**-**	
Leukaemia or myeloma[4]	-	-		-	-		-	-		-	-		-	**-**	
	28			**26**			**46**			**33**			**133**		
Total eligible	**195**	**100**		**145**	**100**		**118**	**100**		**125**	**100**		**583**	**100**	
Aged under 15 or 100 years or over at diagnosis[4]	-	-	-	-	-	-	-	-	-	-	-	-	-	**-**	-
Vital status unknown[5]	43	20	10.3	32	13	9.0	30	11	9.3	21	16	12.8	126	**60**	10.3
Duplicate registration[6]	0	0	0.0	0	0	0.0	0	0	0.0	0	0	0.0	0	**0**	0.0
Multiple primary[7]	0	0	0.0	0	0	0.0	0	0	0.0	0	0	0.0	0	**0**	0.0
Synchronous tumours[8]	2	2	1.0	4	3	2.1	5	5	4.2	2	2	1.6	13	**12**	2.1
Sex not known	0	0	0.0	0	0	0.0	0	0	0.0	0	0	0.0	0	**0**	0.0
Sex incompatible with site code[9]	-	-	-	-	-	-	-	-	-	-	-	-	-	**-**	-
Invalid dates[10]	0	0	0.0	0	0	0.0	0	0	0.0	1	0	0.0	1	**0**	0.0
Zero survival[11]	2	2	1.0	3	1	0.7	0	0	0.0	1	1	0.8	6	**4**	0.7
Total exclusions		**24**	12.3		**17**	11.7		**16**	13.6		**19**	15.2		**76**	13.0
Patients included in analysis		**171**	**87.7**		**128**	**88.3**		**102**	**86.4**		**106**	**84.8**		**507**	**87.0**

[1] Main data item(s) invalid or incompatible with one another: name, sex, date of birth, date of diagnosis and (if present) date of death, postcode, site, morphology and behaviour

[2] Tumour either *in situ* (behaviour code 2), benign (0), uncertain if benign or malignant (1) or metastatic (6 or 9)

[3] Not applicable for this malignancy

[4] Not applicable for this malignancy

[5] Tracing of vital status at National Health Service Central Register incomplete at 6 July 1997

[6] Same site code, sex, cancer registry and cancer registry number as an earlier registration

[7] Same site code and person number as an earlier registration(s): mostly confirmed multiple primary tumours at this site, some unresolved duplicate registrations

[8] Same site code, sex, date of birth and date of diagnosis as another registration(s): mostly synchronous or (in paired organs) bilateral tumours in same anatomic site in one individual, not previously linked: also some duplicate registrations

[9] Not applicable for this malignancy

[10] Impossible sequence of dates (birth, diagnosis and, if present, emigration or death); or date of death after 6 July 1997

[11] Date of diagnosis same as date of death: some are patients who did die on the day of diagnosis, but most were registered solely from a death certificate (death certificate only, or DCO), and their survival time is thus unknown

Table 55.2 Crude and relative survival (%) at one, five and ten years since diagnosis: NHS Region, deprivation category and calendar period of diagnosis

		1971-75						1976-80					
		Affluent	*2*	*3*	*4*	*Deprived*	*All*[1]	*Affluent*	*2*	*3*	*4*	*Deprived*	*All*[1]
		C R	C R	C R	C R	C R	C R	C R	C R	C R	C R	C R	C R
ENGLAND & WALES													
Boys		(15)	(13)	(16)	(17)	(17)	(90)	(13)	(6)	(13)	(19)	(9)	(62)
	One year	100 100	100 100	100 100	88 89	88 88	96 96	100 100	- -	100 100	95 95	- -	97 97
	Five years	87 87	92 92	94 94	83 83	82 82	89 89	100 100	- -	92 92	95 95	- -	94 94
	Ten years	87 87	92 92	94 94	83 83	82 82	88 88	100 100	- -	92 92	95 95	- -	94 94
Girls		(15)	(12)	(15)	(6)	(10)	(81)	(11)	(14)	(11)	(15)	(14)	(66)
	One year	87 87	100 100	87 87	- -	100 100	93 93	100 100	100 100	91 91	100 100	79 79	94 94
	Five years	73 73	100 100	87 87	- -	79 79	86 86	100 100	82 82	100 100	82 82	58 58	85 85
	Ten years	73 73	90 90	74 74	- -	79 79	83 83	82 82	100 100	82 82	86 86	58 58	82 82
Children		(30)	(25)	(31)	(23)	(27)	(171)	(24)	(20)	(24)	(34)	(23)	(128)
	One year	93 93	100 100	94 94	91 91	92 93	94 94	100 100	100 100	96 96	97 97	83 83	95 96
	Five years	80 80	96 96	90 90	87 87	81 81	88 88	92 92	100 100	88 88	97 97	70 70	89 89
	Ten years	80 80	91 91	84 84	87 87	81 81	85 86	92 92	100 100	88 88	91 91	70 70	87 88

		1971-75			1976-80		
		Boys	*Girls*	*Children*	*Boys*	*Girls*	*Children*
Northern & Yorkshire		(12)	(18)	(30)	(8)	(10)	(18)
	One year	100 100	100 100	100 100	- -	90 90	89 89
	Five years	91 91	94 94	93 93	- -	80 80	83 84
	Ten years	91 91	94 94	93 93	- -	80 80	83 84
Trent		(10)	(10)	(20)	(9)	(6)	(15)
	One year	100 100	90 90	95 95	- -	- -	94 94
	Five years	90 90	79 79	85 85	- -	- -	80 80
	Ten years	90 90	79 79	85 85	- -	- -	73 73
Anglia & Oxford		(11)	(7)	(18)	(6)	(11)	(17)
	One year	100 100	- -	100 100	- -	100 100	100 100
	Five years	91 91	- -	94 94	- -	82 82	82 83
	Ten years	91 91	- -	94 94	- -	82 82	82 83
North Thames		(8)	(5)	(13)	(5)	(8)	(13)
	One year	- -	- -	93 93	- -	- -	93 93
	Five years	- -	- -	85 85	- -	- -	93 93
	Ten years	- -	- -	85 85	- -	- -	93 93
South Thames		(10)	(7)	(17)	(8)	(8)	(16)
	One year	90 90	- -	82 82	- -	- -	94 94
	Five years	80 81	- -	76 77	- -	- -	94 94
	Ten years	80 81	- -	76 77	- -	- -	94 94
South & West		(13)	(13)	(26)	(12)	(11)	(23)
	One year	100 100	92 92	96 96	100 100	100 100	100 100
	Five years	93 93	92 92	92 92	100 100	91 91	96 96
	Ten years	93 93	84 84	88 88	100 100	91 91	96 96
West Midlands		(6)	(9)	(15)	(5)	(3)	(8)
	One year	- -	- -	87 87	- -	- -	- -
	Five years	- -	- -	80 80	- -	- -	- -
	Ten years	- -	- -	60 60	- -	- -	- -
North & West		(11)	(6)	(17)	(1)	(5)	(6)
	One year	100 100	- -	100 100	- -	- -	- -
	Five years	100 100	- -	94 94	- -	- -	- -
	Ten years	100 100	- -	94 94	- -	- -	- -
WALES		(9)	(6)	(15)	(8)	(4)	(12)
	One year	- -	- -	87 87	- -	- -	100 100
	Five years	- -	- -	80 80	- -	- -	100 100
	Ten years	- -	- -	80 80	- -	- -	100 100

[1] Including those for whom no deprivation category was assigned. C - crude survival, R - relative survival (see Chapter 3). Number of patients contributing to each analysis in parentheses.

ENGLAND & WALES

Values shown as C R (C – crude survival, R – relative survival). Number of patients contributing in parentheses.

	1981-85						1986-90					
	Affluent	*2*	*3*	*4*	*Deprived*	*All¹*	*Affluent*	*2*	*3*	*4*	*Deprived*	*All¹*
Boys												
(n)	(8)	(12)	(8)	(10)	(14)	(53)	(8)	(12)	(12)	(9)	(14)	(55)
One year	- -	100 100	- -	100 100	100 100	100 100	100 100	100 100	100 100	100 100	100 100	100 100
Five years	- -	100 100	- -	80 80	86 86	91 91	100 100	100 100	91 91	100 100	100 100	98 98
	- -	100 100	- -	70 70	86 86	87 87						
Girls												
(n)	(11)	(7)	(8)	(4)	(18)	(49)	(9)	(12)	(7)	(9)	(12)	(51)
One year	91 91	- -	- -	- -	94 94	96 96	100 100	91 91	- -	100 100	91 91	96 96
Five years	91 91	- -	- -	- -	78 78	90 90	100 100	83 83	- -	89 89	75 75	88 88
	91 91	- -	- -	- -	78 78	90 90						
Children												
(n)	(19)	(19)	(16)	(14)	(32)	(102)	(17)	(24)	(19)	(18)	(26)	(106)
One year	95 95	100 100	100 100	100 100	97 97	98 98	100 100	96 96	100 100	100 100	96 96	98 98
Five years	89 90	100 100	100 100	86 86	81 81	90 90	100 100	92 92	95 95	94 94	88 88	93 93
	84 84	100 100	100 100	78 79	81 81	88 88						

	1981-85			1986-90			
	Boys	*Girls*	*Children*	*Boys*	*Girls*	*Children*	
(n)	(6)	(2)	(8)	(10)	(6)	(16)	**Northern & Yorkshire**
One year	- -	- -	- -	100 100	- -	100 100	
Five years	- -	- -	- -	100 100	- -	100 100	
	- -	- -	- -				
(n)	(8)	(3)	(11)	(3)	(5)	(8)	**Trent**
One year	- -	- -	100 100	- -	- -	100 100	
Five years	- -	- -	100 100	- -	- -	100 100	
	- -	- -	91 91				
(n)	(8)	(6)	(14)	(3)	(4)	(7)	**Anglia & Oxford**
One year	- -	- -	100 100	- -	- -	- -	
Five years	- -	- -	86 86	- -	- -	- -	
	- -	- -	86 86				
(n)	(1)	(1)	(2)	(6)	(3)	(9)	**North Thames**
One year	- -	- -	- -	- -	- -	100 100	
Five years	- -	- -	- -	- -	- -	89 89	
	- -	- -	- -				
(n)	(6)	(2)	(8)	(7)	(5)	(12)	**South Thames**
One year	- -	- -	- -	- -	- -	100 100	
Five years	- -	- -	- -	- -	- -	100 100	
	- -	- -	- -				
(n)	(9)	(13)	(22)	(6)	(7)	(13)	**South & West**
One year	- -	92 92	96 96	- -	- -	100 100	
Five years	- -	92 92	91 91	- -	- -	100 100	
	- -	92 92	86 86				
(n)	(5)	(12)	(17)	(7)	(9)	(16)	**West Midlands**
One year	- -	100 100	100 100	- -	100 100	100 100	
Five years	- -	83 83	76 76	- -	78 78	88 88	
	- -	83 83	76 76				
(n)	(5)	(4)	(9)	(10)	(10)	(20)	**North & West**
One year	- -	- -	- -	100 100	89 90	95 95	
Five years	- -	- -	- -	100 100	89 90	95 95	
	- -	- -	- -				
(n)	(5)	(6)	(11)	(3)	(2)	(5)	**WALES**
One year	- -	- -	100 100	- -	- -	- -	
Five years	- -	- -	100 100	- -	- -	- -	
	- -	- -	100 100				

¹ Including those for whom no deprivation category was assigned. C - crude survival, R - relative survival (see Chapter 3). Number of patients contributing to each analysis in parentheses.

Fig. 55.1 Relative survival up to ten years: trends by calendar period of diagnosis (1971-90) and NHS Region

CHILDREN

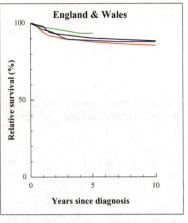

England & Wales

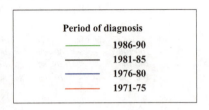

Period of diagnosis
1986-90
1981-85
1976-80
1971-75

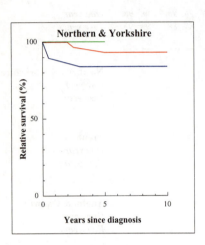

Northern & Yorkshire

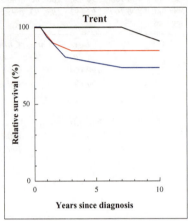

Trent

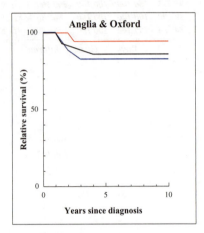

Anglia & Oxford

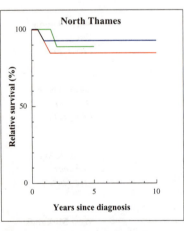

North Thames

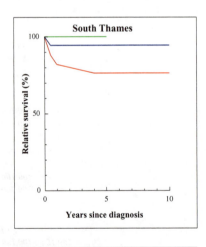

South Thames

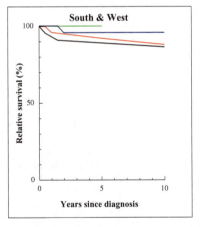

South & West

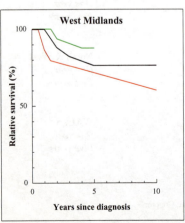

West Midlands

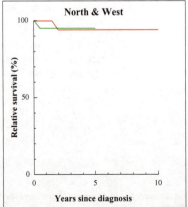

North & West

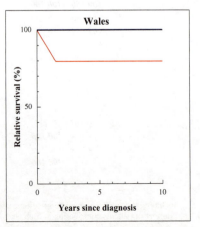

Wales

Fig. 55.2 Relative survival up to five years, by deprivation category: England and Wales, patients diagnosed 1986-90

CHILDREN

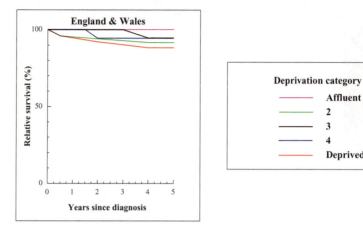

Fig. 55.4 Relative survival (%) by NHS Region and sex:
one-year and five-year survival for children diagnosed 1986-90, with 95% confidence intervals

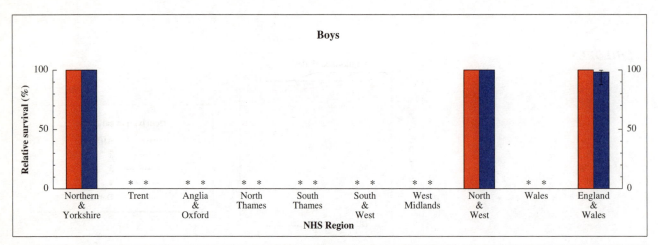

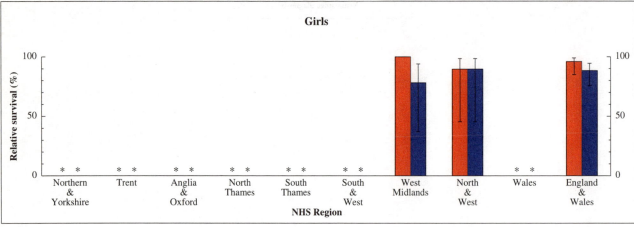

Chapter 56

WILMS' TUMOUR *Children*

Wilms' tumour (nephroblastoma) is a rare malignancy of the kidney arising from embryonic renal cells. In 1990, there were 35 cases of Wilms' tumour registered in England and Wales in boys and 32 in girls. Age-standardised incidence fluctuated in the range 0.4 to 0.9 per 100,000 per year during the period 1971-90, and the rate was 0.7 in both sexes in 1990. About half the cases occur between the ages of one and three. Most malignant renal tumours in children are nephroblastoma; it represents some 7-8% of cancers in children. Very rarely, nephroblastoma arises in sites other than the kidney. Mortality from cancer of the kidney in children under 15 can be regarded as a good indicator of mortality from Wilms' tumour: between 1971 and 1995, age-standardised mortality fell to about a quarter of the level in 1971. Only 4 boys and 5 girls died from kidney cancer in 1995.

Children usually present with an abdominal mass. Wilms' tumour metastasises to lung and liver, usually via the bloodstream. Treatment involves surgical resection, followed by post-operative radiotherapy and chemotherapy, based on operative evaluation of the extent of spread and on the histological type, using the staging system of the National Wilms' Tumor Study Group in the USA[1]. Wilms' tumour is sensitive to radiotherapy and chemotherapy, and the outlook for tumours limited to the kidney and with favourable histology is good.

The survival analyses reported here include about 1,200 children who were diagnosed with Wilms' tumour between 1971 and 1990 and followed up to the end of 1995, about 91% of those eligible (Table 56.1). Six per cent were excluded because their vital status was unknown at 6 July 1997 and 2% because their survival was zero or unknown. All cases of Wilms' tumour were coded to the kidney, except two reported during the 1970s (cervix, adrenal). By definition, the morphology of all cases was nephroblastoma (see Chapter 2 and Table 2.5).

Survival trends

Relative survival[a] from Wilms' tumour for children diagnosed in England and Wales during 1986-90 was 94% at one year and 84% at five years, increases of some 20% from the survival rates observed for children diagnosed during the early 1970s (76% at one year and 61% at five years; Table 56.2). Ten-year survival is very similar to survival at five years, and the plateau in the relative survival curves reflects very low excess mortality among children who survive three years or more (Figure 56.1). In effect, most children who survive more than three years will be considered as cured[2].

The largest gains in one-year and five-year survival rates in England and Wales occurred between 1971-75 and 1976-80. Regional estimates of survival in each five-year period are based on 10-40 cases, and they are imprecise, but there was little change in survival in South Thames and South & West until the early 1980s (Figure 56.1). Regional differences in survival for children diagnosed during 1986-90 are not marked, however (Figure 56.4). For West Midlands and Wales (boys), there were too few cases during 1986-90 for reliable analysis (fewer than eight; see Chapter 3), and this is reflected in Table 56.2[b] and Figure 56.4.

Differences in survival between boys and girls diagnosed during 1986-90 were small: survival was slightly higher for boys during the early 1970s, but it improved rather more for girls (see Table 4.5). The average increase every five years for one-year survival was about 4% for boys and 6% for girls; five-year survival rose by about 7% between successive quinquennia for all children.

Trends in the actuarial survival rates, reported by the National Registry of Childhood Tumours for Great Britain, are consistent with these results. Five-year survival (period of diagnosis) improved from 58% (1971-73)[3] to 76% (1980-82) and 84% (1986-88)[4].

Deprivation

Survival in England and Wales did not differ systematically between children in the various deprivation groups (Table 56.2). Survival curves for children diagnosed during 1986-90 overlap (Figure 56.2). The estimated deprivation gradients are variable and not significant (see Table 4.9), and this is reflected in the graphic for the early and late 1980s (Figure 56.3).

International comparison

Survival from Wilms' tumour in England and Wales is broadly comparable with that in other countries. Survival rates have often been reported for cancers of the kidney, but the great majority of kidney tumours in children are nephroblastoma. For children under 5 with malignant renal tumours in Scotland, five-year relative survival (period of diagnosis) increased from 39% (1968-72) to 84% (1983-87). These survival rates are each based on about 30 children[5].

---

[a] See footnote in Chapter 50.

[b] Non-convergence occurred in the analysis of ten-year survival for girls, England and Wales, 1976-80 (see Chapter 3).

Relative survival from tumours of the kidney in Europe has been reported for 633 children who were diagnosed in one of 11 European countries during 1978-85 and included in the EUROCARE study[6]. The mean age at diagnosis was 3.3 years, and 474 (75%) of the children were younger than five. Average survival was 87% at one year, 75% at five years and 70% at ten years.

In the USA, relative survival from Wilms' tumour at five years increased from 65% (children diagnosed during 1967-73) to 76% (1973-81)[7].

1.   Mitchell CD. Wilms' tumour. In: Peckham M, Pinedo HM, Veronesi U, (eds.) *Oxford textbook of oncology*. Oxford: Oxford Medical Publications, 1995

2.   Hawkins MM. Long term survival and cure after childhood cancer. Arch Dis Child 1989; 64: 798-807

3.   Stiller CA, Bunch KJ. Trends in survival for childhood cancer in Britain diagnosed 1971-85. Br J Cancer 1990; 62: 806-815

4.   Stiller CA. Population-based survival rates for childhood cancer in Britain, 1980-91. Br Med J 1994; 309: 1612-1616

5.   Black RJ, Sharp L, Kendrick SW. *Trends in cancer survival in Scotland, 1968-1990*. Edinburgh: Information and Statistics Division, 1993

6.   Berrino F, Sant M, Verdecchia A, Capocaccia R, Hakulinen T, Estève J, (eds.) *Survival of cancer patients in Europe: the EUROCARE study. (IARC Scientific Publications No. 132)*. Lyon: International Agency for Research on Cancer, 1995

7.   Young JL, Gloeckler Ries LA, Silverberg E, Horm JW, Miller RW. Cancer incidence, survival and mortality for children younger than age 15 years. Cancer 1986; 58: 598-602

Table 56.1 Ineligible and excluded records, and patients included in analyses, by calendar period of diagnosis

	Calendar period of diagnosis														
	1971-75			**1976-80**			**1981-85**			**1986-90**			**Overall**		
Total registered	**355**			**348**			**317**			**314**			**1,334**		
Ineligible	*Records*	*Patients*	*%*	*Records*	*Patients*	*%*	*Records*	*Patients*	*%*	*Records*	*Patients*	*%*	*Records*	*Patients*	*%*
Incomplete data[1]	1	1		8	8		5	5		8	8		22	**22**	
Resident outside England and Wales	6	6		7	7		4	4		2	2		19	**19**	
In situ neoplasm[2]	0	0		0	0		0	0		0	0		0	**0**	
Benign or uncertain behaviour[2]	1	0		3	0		0	0		0	0		4	**0**	
Metastatic[2]	0	0		0	0		0	0		0	0		0	**0**	
Otherwise ineligible[3]	-	-		-	-		-	-		-	-		-	**-**	
Lymphoma[4]	-	-		-	-		-	-		-	-		-	**-**	
Leukaemia or myeloma[4]	-	-		-	-		-	-		-	-		-	**-**	
		7			**15**			**9**			**10**			**41**	
Total eligible	**348**	**100**		**333**	**100**		**308**	**100**		**304**	**100**		**1,293**	**100**	
Aged under 15 or 100 years or over at diagnosis[4]	-	-	-	-	-	-	-	-	-	-	-	-	-	-	-
Vital status unknown[5]	20	17	4.9	24	14	4.2	13	10	3.2	41	40	13.2	98	**81**	6.3
Duplicate registration[6]	0	0	0.0	0	0	0.0	0	0	0.0	0	0	0.0	0	**0**	0.0
Multiple primary[7]	0	0	0.0	0	0	0.0	2	2	0.6	0	0	0.0	2	**2**	0.2
Synchronous tumours[8]	0	0	0.0	3	2	0.6	2	2	0.6	0	0	0.0	5	**4**	0.3
Sex not known	0	0	0.0	0	0	0.0	0	0	0.0	0	0	0.0	0	**0**	0.0
Sex incompatible with site code[9]	-	-	-	-	-	-	-	-	-	-	-	-	-	**-**	-
Invalid dates[10]	0	0	0.0	0	0	0.0	0	0	0.0	0	0	0.0	0	**0**	0.0
Zero survival[11]	9	9	2.6	4	3	0.9	9	8	2.6	5	5	1.6	27	**25**	1.9
Total exclusions		**26**	7.5		**19**	5.7		**22**	7.1		**45**	14.8		**112**	8.7
Patients included in analysis		**322**	92.5		**314**	94.3		**286**	92.9		**259**	85.2		**1,181**	91.3

[1] Main data item(s) invalid or incompatible with one another: name, sex, date of birth, date of diagnosis and (if present) date of death, postcode, site, morphology and behaviour

[2] Tumour either *in situ* (behaviour code 2), benign (0), uncertain if benign or malignant (1) or metastatic (6 or 9)

[3] Not applicable for this malignancy

[4] Not applicable for this malignancy

[5] Tracing of vital status at National Health Service Central Register incomplete at 6 July 1997

[6] Same site code, sex, cancer registry and cancer registry number as an earlier registration

[7] Same site code and person number as an earlier registration(s): mostly confirmed multiple primary tumours at this site, some unresolved duplicate registrations

[8] Same site code, sex, date of birth and date of diagnosis as another registration(s): mostly synchronous or (in paired organs) bilateral tumours in same anatomic site in one individual, not previously linked: also some duplicate registrations

[9] Not applicable for this malignancy

[10] Impossible sequence of dates (birth, diagnosis and, if present, emigration or death); or date of death after 6 July 1997

[11] Date of diagnosis same as date of death: some are patients who did die on the day of diagnosis, but most were registered solely from a death certificate (death certificate only, or DCO), and their survival time is thus unknown

WILMS' TUMOUR *Children*

Table 56.2 Crude and relative survival (%) at one, five and ten years since diagnosis: NHS Region, deprivation category and calendar period of diagnosis

	1971-75 Affluent		2		3		4		Deprived		All¹		1976-80 Affluent		2		3		4		Deprived		All¹	
	C	R	C	R	C	R	C	R	C	R	C	R	C	R	C	R	C	R	C	R	C	R	C	R
ENGLAND & WALES																								
Boys	(41)		(19)		(35)		(23)		(27)		(171)		(32)		(25)		(31)		(30)		(29)		(155)	
One year	81	81	73	73	64	65	87	87	81	81	79	79	84	85	84	84	90	90	84	84	86	86	84	85
Five years	64	64	52	53	50	50	78	78	55	55	62	63	72	72	72	72	84	84	74	74	79	79	75	76
Ten years	61	61	52	53	47	48	78	78	55	55	61	63	72	72	72	72	74	74	74	74	79	79	74	74
Girls	(22)		(28)		(20)		(27)		(22)		(151)		(33)		(27)		(28)		(37)		(27)		(159)	
One year	81	81	72	72	68	69	67	68	77	77	73	73	82	82	85	85	89	89	87	87	89	89	87	87
Five years	68	68	46	46	58	58	64	64	59	59	59	59	76	76	70	70	68	68	76	76	81	81	74	74
Ten years	68	68	46	46	53	54	56	56	59	59	57	57	73	73	66	66	68	68	76	76	81	81	-	-
Children	(63)		(47)		(55)		(50)		(49)		(322)		(65)		(52)		(59)		(67)		(56)		(314)	
One year	81	81	73	73	66	66	76	77	79	79	76	76	83	83	84	84	90	90	85	85	87	88	86	86
Five years	65	65	49	49	53	54	70	71	57	57	61	61	74	74	71	71	76	76	75	75	80	81	75	75
Ten years	63	64	49	49	50	50	66	67	57	57	59	59	72	73	69	69	71	71	75	75	80	81	73	73

	Boys		Girls		Children		Boys		Girls		Children	
	C	R	C	R	C	R	C	R	C	R	C	R
Northern & Yorkshire	(13)		(20)		(33)		(21)		(16)		(37)	
One year	70	70	75	75	73	73	85	86	88	88	87	87
Five years	38	38	60	60	51	52	81	81	57	57	70	71
Ten years	38	38	60	60	51	52	76	76	57	57	68	68
Trent	(17)		(13)		(30)		(22)		(25)		(47)	
One year	82	82	55	55	70	70	77	78	88	88	83	83
Five years	64	64	55	55	60	60	59	59	80	80	70	70
Ten years	64	64	48	48	57	57	59	59	80	80	70	70
Anglia & Oxford	(18)		(17)		(35)		(15)		(13)		(28)	
One year	71	71	77	77	74	74	100	100	85	85	93	93
Five years	60	60	71	71	65	66	81	81	61	61	72	72
Ten years	60	60	71	71	65	66	81	81	53	53	68	68
North Thames	(18)		(15)		(33)		(20)		(20)		(40)	
One year	94	94	86	87	91	91	85	85	95	95	90	90
Five years	66	66	73	73	69	70	80	80	90	90	85	85
Ten years	66	66	67	67	66	67	80	80	85	85	82	83
South Thames	(17)		(21)		(38)		(12)		(14)		(26)	
One year	88	88	76	76	81	82	83	83	65	65	73	73
Five years	64	64	62	62	63	63	75	75	57	57	65	65
Ten years	58	58	62	62	60	60	67	67	57	57	61	61
South & West	(24)		(16)		(40)		(25)		(30)		(55)	
One year	68	68	61	61	65	65	68	68	83	83	76	76
Five years	60	60	55	55	58	58	60	60	66	66	63	63
Ten years	60	60	55	55	58	58	60	60	66	66	63	63
West Midlands	(18)		(18)		(36)		(16)		(10)		(26)	
One year	100	100	83	83	92	92	88	88	80	80	85	85
Five years	83	83	55	55	69	70	82	82	80	80	81	81
Ten years	83	83	55	55	69	70	82	82	80	80	81	81
North & West	(30)		(20)		(50)		(20)		(24)		(44)	
One year	76	76	74	74	75	75	95	95	96	96	95	95
Five years	66	66	54	54	61	61	90	90	88	88	89	89
Ten years	66	66	54	54	61	61	85	85	88	88	86	86
WALES	(16)		(11)		(27)		(4)		(7)		(11)	
One year	62	62	64	64	63	63	-	-	-	-	100	100
Five years	50	50	35	35	44	44	-	-	-	-	91	91
Ten years	43	44	26	26	37	37	-	-	-	-	91	91

¹ Including those for whom no deprivation category was assigned. C - crude survival, R - relative survival (see Chapter 3).
Number of patients contributing to each analysis in parentheses.

1981-85 & 1986-90 — ENGLAND & WALES

	1981-85						1986-90						
	Affluent C R	2 C R	3 C R	4 C R	*Deprived* C R	*All*[1] C R	*Affluent* C R	2 C R	3 C R	4 C R	*Deprived* C R	*All*[1] C R	
Boys	(28)	(28)	(30)	(27)	(24)	(141)	(28)	(29)	(14)	(31)	(15)	(118)	
One year	89 89	89 89	93 93	85 85	92 92	89 89	93 93	97 97	93 93	94 94	74 74	91 92	
Five years	79 79	86 86	83 83	78 78	83 84	81 81	82 82	90 90	86 86	84 84	74 74	84 84	
	75 75	86 86	80 80	78 78	79 80	79 79							
Girls	(32)	(29)	(29)	(26)	(25)	(145)	(33)	(33)	(27)	(32)	(15)	(141)	
One year	100 100	86 86	79 79	84 84	92 92	88 88	97 97	94 94	93 93	100 100	93 93	96 96	
Five years	91 91	58 58	72 73	69 69	76 76	74 74	70 70	88 88	85 85	91 91	87 87	84 84	
	88 88	58 58	72 73	69 69	72 72	72 73							
Children	(60)	(57)	(59)	(53)	(49)	(286)	(61)	(62)	(41)	(63)	(30)	(259)	
One year	95 95	88 88	86 87	85 85	92 92	89 89	95 95	95 95	93 93	97 97	84 84	94 94	
Five years	85 85	72 72	78 78	74 74	80 80	77 77	75 75	89 89	85 86	87 87	80 80	84 84	
	82 82	72 72	76 76	74 74	75 76	76 76							

Regions

	1981-85 Boys	Girls	Children	1986-90 Boys	Girls	Children	
	(12)	(21)	(33)	(17)	(28)	(45)	**Northern & Yorkshire**
One year	74 75	91 91	85 85	94 94	100 100	98 98	*One year*
Five years	74 75	76 76	76 76	94 94	89 89	91 91	*Five years*
	74 75	71 71	73 73				
	(12)	(9)	(21)	(8)	(8)	(16)	**Trent**
One year	92 92	- -	95 95	87 87	100 100	94 94	*One year*
Five years	92 92	- -	95 95	87 87	100 100	94 94	*Five years*
	83 83	- -	90 91				
	(22)	(15)	(37)	(14)	(13)	(27)	**Anglia & Oxford**
One year	87 87	100 100	92 92	78 78	100 100	89 89	*One year*
Five years	73 73	73 73	73 73	55 55	77 77	66 66	*Five years*
	73 73	73 73	73 73				
	(20)	(27)	(47)	(15)	(25)	(40)	**North Thames**
One year	90 90	78 78	83 83	100 100	100 100	100 100	*One year*
Five years	75 75	74 74	75 75	93 93	92 92	93 93	*Five years*
	70 70	74 74	72 73				
	(21)	(16)	(37)	(16)	(19)	(35)	**South Thames**
One year	95 95	88 88	92 92	87 88	85 85	86 86	*One year*
Five years	86 86	74 75	81 81	87 88	74 74	80 80	*Five years*
	86 86	74 75	81 81				
	(18)	(19)	(37)	(17)	(14)	(31)	**South & West**
One year	89 89	95 95	92 92	94 94	93 93	93 94	*One year*
Five years	83 83	64 64	73 73	88 88	64 64	77 77	*Five years*
	83 83	64 64	73 73				
	(13)	(15)	(28)	(6)	(6)	(12)	**West Midlands**
One year	92 92	79 79	85 86	- -	- -	92 92	*One year*
Five years	85 85	66 66	75 75	- -	- -	66 66	*Five years*
	77 77	66 66	71 71				
	(18)	(17)	(35)	(19)	(18)	(37)	**North & West**
One year	95 95	94 94	94 94	90 90	95 95	92 92	*One year*
Five years	89 89	82 82	86 86	84 84	89 89	87 87	*Five years*
	89 89	76 76	83 83				
	(5)	(6)	(11)	(6)	(10)	(16)	**WALES**
One year	- -	- -	71 71	- -	100 100	100 100	*One year*
Five years	- -	- -	53 53	- -	90 90	88 88	*Five years*
	- -	- -	53 53				

[1] Including those for whom no deprivation category was assigned. C - crude survival, R - relative survival (see Chapter 3).
Number of patients contributing to each analysis in parentheses.

Fig. 56.1   Relative survival up to ten years: trends by calendar period of diagnosis (1971-90) and NHS Region

CHILDREN

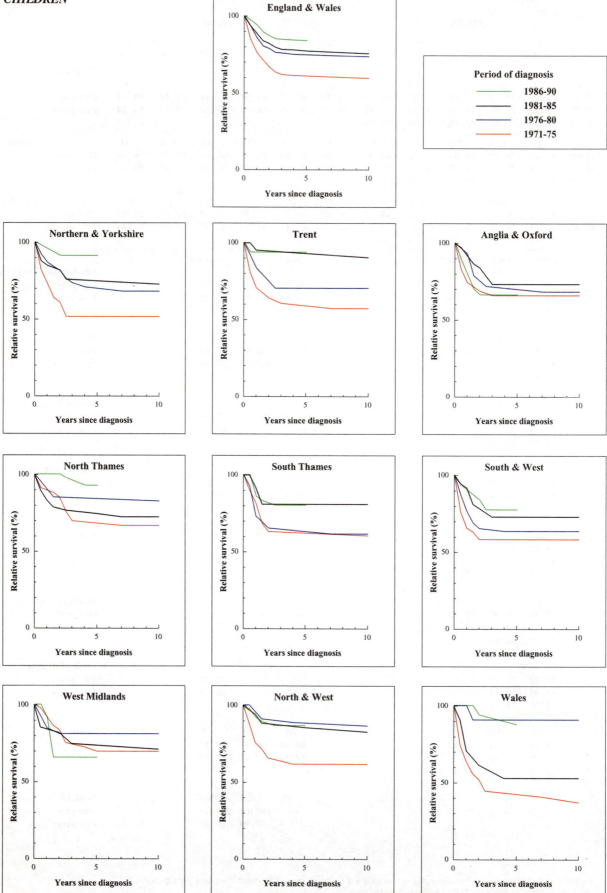

Fig. 56.2 Relative survival up to five years, by deprivation category: England and Wales, patients diagnosed 1986-90

CHILDREN

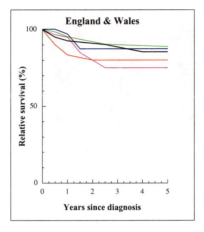

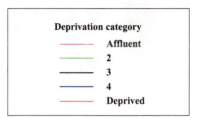

Fig. 56.3 Relative survival at five years, by deprivation category and period of diagnosis (1981-90): England and Wales

CHILDREN

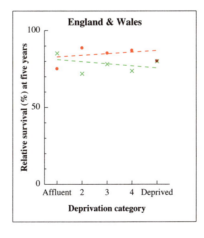

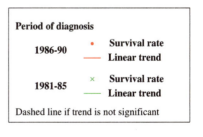

Fig. 56.4 Relative survival (%) by NHS Region and sex:
  one-year and five-year survival for children diagnosed 1986-90, with 95% confidence intervals

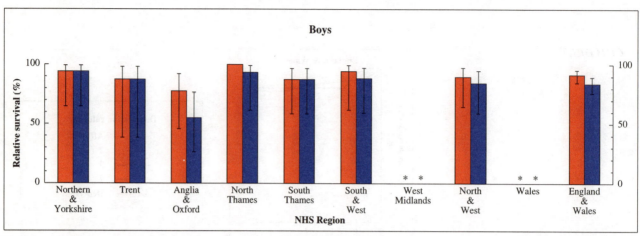

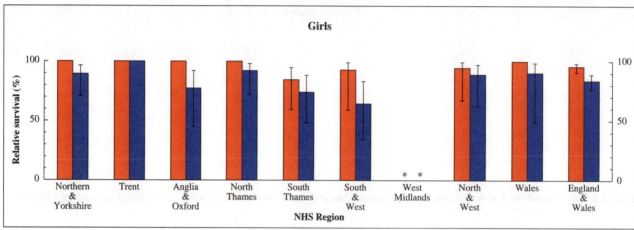

Chapter 57

HEPATOBLASTOMA *Children*

Hepatoblastoma is a malignant tumour of the liver arising from embryonal or foetal liver cells. It occurs mostly in children under the age of one year, and is very rare, accounting for 0.5-2% of all cancers in children[1]: fewer than 10 cases are registered each year in England and Wales. Incidence is similar in boys and girls. For the period 1986-90, overall incidence in children under 15 was less than 1 per million per year, but 4 per million in children less than one year old.

Children with hepatoblastoma may present with an abdominal mass. Surgery is usually preceded by combined chemotherapy with cisplatinum and doxorubicin to reduce tumour bulk, and up to 60% can now be cured, including some children with metastatic disease at diagnosis[2].

The survival analyses reported here are based on 116 children who were diagnosed with hepatoblastoma in England and Wales between 1971 and 1990 and followed up to the end of 1995, about 88% of those eligible (Table 57.1). Thirteen children (10%) were excluded because their survival was zero or unknown, and two because their vital status was not known at 6 July 1997. The cases were defined for analysis by the morphology code for hepatoblastoma (see Chapter 2); each case had the anatomic site code for a primary malignant tumour of the liver.

Survival trends

Hepatoblastoma is the rarest tumour for which survival has been examined here. Even national survival estimates are based on an average of 25-30 children every five years, and they are all imprecise. Only overall trends in national survival for boys and girls can be examined. Very few survival rates were calculated for deprivation groups in England and Wales, and only one regional survival rate.

For children diagnosed with hepatoblastoma during 1986-90, relative survival[a] was 59% at one year and 43% at five years. Survival has doubled since the early 1970s, when relative survival rates were 27% at one year and 15% at five years (Table 57.2).

Five-year survival for children with hepatoblastoma increased by an average of 11% between successive five-year periods from 1971-75 to 1986-90 (see Table 4.4). Almost all the overall increase in survival occurred between 1976-80 and 1981-85 (Table 57.2, Figure 57.1).

Ten-year survival for children diagnosed during 1981-85 was 42%, the same as five-year survival. The plateau in the relative survival curves suggests that for children who survive beyond three to five years after diagnosis, excess mortality relative to other children of the same age is extremely low. This resembles the pattern for other embryonal tumours, such as (sporadic) retinoblastoma (Chapter 55), Wilms' tumour (Chapter 56), neuroblastoma and several other childhood tumours[3].

These results are comparable with survival rates reported by the National Registry of Childhood Tumours for Great Britain. Actuarial five-year survival rates for hepatoblastoma were 8% for children diagnosed in 1971-73, about 20% in the late 1970s and 40% for children diagnosed during 1983-85[4].

Deprivation

Survival patterns by deprivation category (Figure 57.2) cannot be interpreted because the analyses are based on so few cases.

---

[a] See footnote in Chapter 50.

Non-convergence occurred in the analysis of ten-year survival for girls, England and Wales, 1976-80 (see Chapter 3).

1. Voûte PA, Pinkerton R. Liver tumours. In: Peckham M, Pinedo HM, Veronesi U, (eds.) *Oxford textbook of oncology*. Oxford: Oxford Medical Publications, 1995

2. Pinkerton R. Paediatric solid tumours. In: Horwich A (ed.) *Oncology: a multi-disciplinary textbook*. London: Chapman and Hall, 1995, pp417-440

3. Hawkins MM. Long term survival and cure after childhood cancer. Arch Dis Child 1989; 64: 798-807

4. Stiller CA, Bunch KJ. Trends in survival for childhood cancer in Britain diagnosed 1971-85. Br J Cancer 1990; 62: 806-815

Table 57.1 Ineligible and excluded records, and patients included in analyses, by calendar period of diagnosis

	Calendar period of diagnosis														
	1971-75			1976-80			1981-85			1986-90			Overall		
Total registered	**39**			**35**			**32**			**36**			**142**		
Ineligible	*Records*	*Patients*	*%*	*Records*	*Patients*	*%*	*Records*	*Patients*	*%*	*Records*	*Patients*	*%*	*Records*	*Patients*	*%*
Incomplete data[1]	1	1		2	2		3	3		0	0		6	**6**	
Resident outside England and Wales	1	1		1	1		2	2		0	0		4	**4**	
In situ neoplasm[2]	0	0		0	0		0	0		0	0		0	**0**	
Benign or uncertain behaviour[2]	0	0		0	0		0	0		0	0		0	**0**	
Metastatic[2]	0	0		0	0		0	0		0	0		0	**0**	
Otherwise ineligible[3]	-	-		-	-		-	-		-	-		-	**-**	
Lymphoma[4]	-	-		-	-		-	-		-	-		-	**-**	
Leukaemia or myeloma[4]	-	-		-	-		-	-		-	-		-	**-**	
		2			**3**			**5**			**0**			**10**	
Total eligible		**37**	**100**		**32**	**100**		**27**	**100**		**36**	**100**		**132**	**100**
Aged under 15 or 100 years or over at diagnosis[4]	-	-	-	-	-	-	-	-	-	-	-	-	-	**-**	-
Vital status unknown[5]	1	1	2.7	1	1	3.1	2	0	0.0	0	0	0.0	4	**2**	1.5
Duplicate registration[6]	0	0	0.0	0	0	0.0	0	0	0.0	0	0	0.0	0	**0**	0.0
Multiple primary[7]	0	0	0.0	0	0	0.0	0	0	0.0	0	0	0.0	0	**0**	0.0
Synchronous tumours[8]	0	0	0.0	1	1	3.1	0	0	0.0	0	0	0.0	1	**1**	0.8
Sex not known	0	0	0.0	0	0	0.0	0	0	0.0	0	0	0.0	0	**0**	0.0
Sex incompatible with site code[9]	-	-	-	-	-	-	-	-	-	-	-	-	-	**-**	-
Invalid dates[10]	0	0	0.0	0	0	0.0	0	0	0.0	0	0	0.0	0	**0**	0.0
Zero survival[11]	4	4	10.8	4	3	9.4	3	2	7.4	4	4	11.1	15	**13**	9.8
Total exclusions		**5**	**13.5**		**5**	**15.6**		**2**	**7.4**		**4**	**11.1**		**16**	**12.1**
Patients included in analysis		**32**	**86.5**		**27**	**84.4**		**25**	**92.6**		**32**	**88.9**		**116**	**87.9**

[1] Main data item(s) invalid or incompatible with one another: name, sex, date of birth, date of diagnosis and (if present) date of death, postcode, site, morphology and behaviour

[2] Tumour either *in situ* (behaviour code 2), benign (0), uncertain if benign or malignant (1) or metastatic (6 or 9)

[3] Not applicable for this malignancy

[4] Not applicable for this malignancy

[5] Tracing of vital status at National Health Service Central Register incomplete at 6 July 1997

[6] Same site code, sex, cancer registry and cancer registry number as an earlier registration

[7] Same site code and person number as an earlier registration(s): mostly confirmed multiple primary tumours at this site, some unresolved duplicate registrations

[8] Same site code, sex, date of birth and date of diagnosis as another registration(s): mostly synchronous or (in paired organs) bilateral tumours in same anatomic site in one individual, not previously linked: also some duplicate registrations

[9] Not applicable for this malignancy

[10] Impossible sequence of dates (birth, diagnosis and, if present, emigration or death); or date of death after 6 July 1997

[11] Date of diagnosis same as date of death: some are patients who did die on the day of diagnosis, but most were registered solely from a death certificate (death certificate only, or DCO), and their survival time is thus unknown

HEPATOBLASTOMA *Children*

Table 57.2 Crude and relative survival (%) at one, five and ten years since diagnosis: NHS Region, deprivation category and calendar period of diagnosis

	1971-75																			1976-80																		
	Affluent		2		3		4		Deprived		All¹									Affluent		2		3		4		Deprived		All¹	
	C	R	C	R	C	R	C	R	C	R	C	R								C	R	C	R	C	R	C	R	C	R	C	R
ENGLAND & WALES																															
Boys	*(1)*		*(2)*		*(3)*		*(1)*		*(6)*		*(15)*									*(3)*		*(0)*		*(5)*		*(1)*		*(1)*		*(10)*	
One year	-	-	-	-	-	-	-	-	-	-	42	42								-	-	-	-	-	-	-	-	-	-	14	14
Five years	-	-	-	-	-	-	-	-	-	-	13	13								-	-	-	-	-	-	-	-	-	-	7	7
Ten years	-	-	-	-	-	-	-	-	-	-	13	13								-	-	-	-	-	-	-	-	-	-	7	7
Girls	*(2)*		*(0)*		*(5)*		*(2)*		*(6)*		*(17)*									*(5)*		*(4)*		*(1)*		*(2)*		*(4)*		*(17)*	
One year	-	-	-	-	-	-	-	-	-	-	15	15								-	-	-	-	-	-	-	-	-	-	24	24
Five years	-	-	-	-	-	-	-	-	-	-	15	15								-	-	-	-	-	-	-	-	-	-	19	19
Ten years	-	-	-	-	-	-	-	-	-	-	15	15								-	-	-	-	-	-	-	-	-	-	19	19
Children	*(3)*		*(2)*		*(8)*		*(3)*		*(12)*		*(32)*									*(8)*		*(4)*		*(6)*		*(3)*		*(5)*		*(27)*	
One year	-	-	-	-	-	-	-	-	32	32	27	27								-	-	-	-	-	-	-	-	-	-	20	20
Five years	-	-	-	-	-	-	-	-	6	6	15	15								-	-	-	-	-	-	-	-	-	-	14	14
Ten years	-	-	-	-	-	-	-	-	6	6	15	15								-	-	-	-	-	-	-	-	- · ·		14	14

	Boys	Girls	Children				Boys	Girls	Children
Northern & Yorkshire	*(1)*	*(3)*	*(4)*				*(0)*	*(3)*	*(3)*
One year	-	-	-				-	-	-
Five years	-	-	-				-	-	-
Ten years	-	-	-				-	-	-
Trent	*(1)*	*(1)*	*(2)*				*(0)*	*(2)*	*(2)*
One year	-	-	-				-	-	-
Five years	-	-	-				-	-	-
Ten years	-	-	-				-	-	-
Anglia & Oxford	*(0)*	*(2)*	*(2)*				*(2)*	*(0)*	*(2)*
One year	-	-	-				-	-	-
Five years	-	-	-				-	-	-
Ten years	-	-	-				-	-	-
North Thames	*(2)*	*(1)*	*(3)*				*(1)*	*(3)*	*(4)*
One year	-	-	-				-	-	-
Five years	-	-	-				-	-	-
Ten years	-	-	-				-	-	-
South Thames	*(2)*	*(1)*	*(3)*				*(0)*	*(2)*	*(2)*
One year	-	-	-				-	-	-
Five years	-	-	-				-	-	-
Ten years	-	-	-				-	-	-
South & West	*(0)*	*(3)*	*(3)*				*(4)*	*(5)*	*(9)*
One year	-	-	-				-	-	-
Five years	-	-	-				-	-	-
Ten years	-	-	-				-	-	-
West Midlands	*(3)*	*(3)*	*(6)*				*(1)*	*(2)*	*(3)*
One year	-	-	-				-	-	-
Five years	-	-	-				-	-	-
Ten years	-	-	-				-	-	-
North & West	*(5)*	*(3)*	*(8)*				*(2)*	*(0)*	*(2)*
One year	-	-	-				-	-	-
Five years	-	-	-				-	-	-
Ten years	-	-	-				-	-	-
WALES	*(1)*	*(0)*	*(1)*				*(0)*	*(0)*	*(0)*
One year	-	-	-				-	-	-
Five years	-	-	-				-	-	-
Ten years	-	-	-				-	-	-

[1] Including those for whom no deprivation category was assigned. C - crude survival, R - relative survival (see Chapter 3).
Number of patients contributing to each analysis in parentheses.

	1981-85						1986-90						
	Affluent	*2*	*3*	*4*	*Deprived*	*All*[1]	*Affluent*	*2*	*3*	*4*	*Deprived*	*All*[1]	**ENGLAND & WALES**
	C R	C R	C R	C R	C R	C R	C R	C R	C R	C R	C R	C R	
	(3)	*(0)*	*(1)*	*(4)*	*(4)*	*(12)*	*(5)*	*(3)*	*(3)*	*(4)*	*(5)*	*(20)*	**Boys**
One year	- -	- -	- -	- -	- -	48 48	- -	- -	- -	- -	- -	55 55	*One year*
Five years	- -	- -	- -	- -	- -	48 48	- -	- -	- -	- -	- -	39 39	*Five years*
	- -	- -	- -	- -	- -	48 48							
	(2)	*(2)*	*(3)*	*(3)*	*(3)*	*(13)*	*(3)*	*(2)*	*(2)*	*(1)*	*(4)*	*(12)*	**Girls**
	- -	- -	- -	- -	- -	44 44	- -	- -	- -	- -	- -	65 66	*One year*
	- -	- -	- -	- -	- -	37 37	- -	- -	- -	- -	- -	49 49	*Five years*
	- -	- -	- -	- -	- -	37 37							
	(5)	*(2)*	*(4)*	*(7)*	*(7)*	*(25)*	*(8)*	*(5)*	*(5)*	*(5)*	*(9)*	*(32)*	**Children**
	- -	- -	- -	- -	- -	46 46	87 87	- -	- -	- -	66 66	59 59	*One year*
	- -	- -	- -	- -	- -	42 42	61 61	- -	- -	- -	44 44	43 43	*Five years*
	- -	- -	- -	- -	- -	42 42							

	Boys	Girls	Children				Boys	Girls	Children				
	(3)	*(1)*	*(4)*				*(1)*	*(3)*	*(4)*				**Northern & Yorkshire**
One year	- -	- -	- -				- -	- -	- -				*One year*
Five years	- -	- -	- -				- -	- -	- -				*Five years*
	- -	- -	- -										
	(1)	*(4)*	*(5)*				*(1)*	*(1)*	*(2)*				**Trent**
	- -	- -	- -				- -	- -	- -				*One year*
	- -	- -	- -				- -	- -	- -				*Five years*
	- -	- -	- -										
	(1)	*(1)*	*(2)*				*(2)*	*(1)*	*(3)*				**Anglia & Oxford**
	- -	- -	- -				- -	- -	- -				*One year*
	- -	- -	- -				- -	- -	- -				*Five years*
	- -	- -	- -										
	(2)	*(0)*	*(2)*				*(3)*	*(2)*	*(5)*				**North Thames**
	- -	- -	- -				- -	- -	- -				*One year*
	- -	- -	- -				- -	- -	- -				*Five years*
	- -	- -	- -										
	(1)	*(1)*	*(2)*				*(4)*	*(1)*	*(5)*				**South Thames**
	- -	- -	- -				- -	- -	- -				*One year*
	- -	- -	- -				- -	- -	- -				*Five years*
	- -	- -	- -										
	(0)	*(2)*	*(2)*				*(2)*	*(1)*	*(3)*				**South & West**
	- -	- -	- -				- -	- -	- -				*One year*
	- -	- -	- -				- -	- -	- -				*Five years*
	- -	- -	- -										
	(2)	*(1)*	*(3)*				*(1)*	*(1)*	*(2)*				**West Midlands**
	- -	- -	- -				- -	- -	- -				*One year*
	- -	- -	- -				- -	- -	- -				*Five years*
	- -	- -	- -										
	(1)	*(3)*	*(4)*				*(6)*	*(2)*	*(8)*				**North & West**
	- -	- -	- -				- -	- -	47 47				*One year*
	- -	- -	- -				- -	- -	24 24				*Five years*
	- -	- -	- -										
	(1)	*(0)*	*(1)*				*(0)*	*(0)*	*(0)*				**WALES**
	- -	- -	- -				- -	- -	- -				*One year*
	- -	- -	- -				- -	- -	- -				*Five years*
	- -	- -	- -										

[1] Including those for whom no deprivation category was assigned. C - crude survival, R - relative survival (see Chapter 3).
Number of patients contributing to each analysis in parentheses.

Fig. 57.1   Relative survival up to ten years: trends by calendar period of diagnosis (1971-90) and NHS Region

CHILDREN

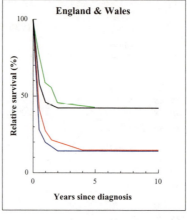

England & Wales

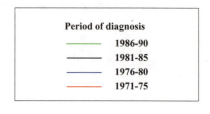

Period of diagnosis
- 1986-90
- 1981-85
- 1976-80
- 1971-75

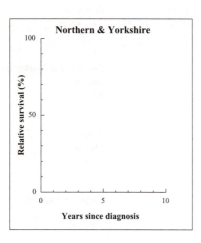

Northern & Yorkshire

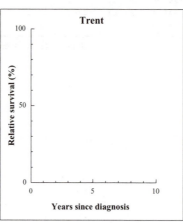

Trent

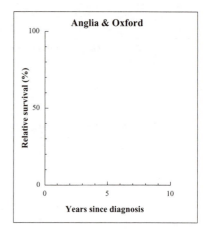

Anglia & Oxford

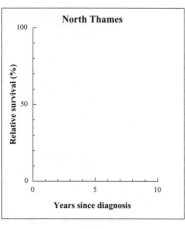

North Thames

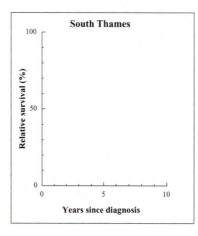

South Thames

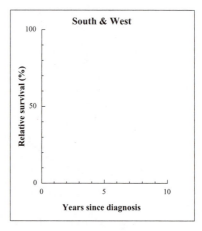

South & West

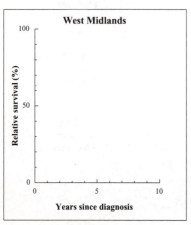

West Midlands

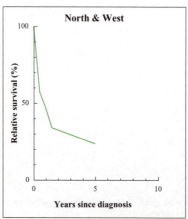

North & West

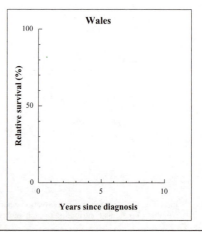

Wales

Fig. 57.2 Relative survival up to five years, by deprivation category: England and Wales, patients diagnosed 1986-90

CHILDREN

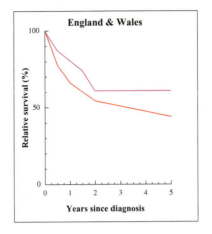

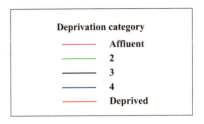

Fig. 57.4 Relative survival (%) by NHS Region and sex:
one-year and five-year survival for children diagnosed 1986-90, with 95% confidence intervals

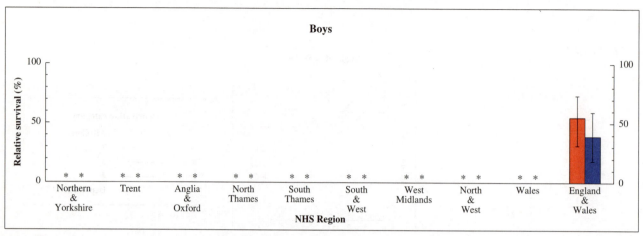

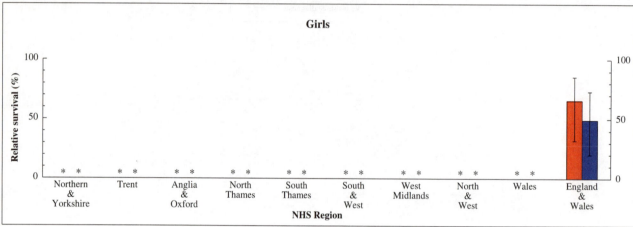

Chapter 58

OSTEOSARCOMA *Children*

Malignancies of bone are uncommon, accounting for only 6% of cancers in children. Osteosarcoma is the most common of these, and about 25 new cases are diagnosed each year in England and Wales. The highest incidence in children is at 10-14 years of age, and osteosarcoma is rare under the age of five; incidence is similar in boys and girls. Most cases occur in the long limb bones.

Osteosarcoma in children is associated with several hereditary conditions, including the Li-Fraumeni syndrome. Ionising radiation for a first malignancy, including genetic retinoblastoma, can cause osteosarcoma in children, as can various alkylating agents, but these causes only account for a small proportion of cases. Most osteosarcomas have no known predisposing cause[1].

The survival analyses reported here are based on 559 children diagnosed with osteosarcoma in England and Wales between 1971 and 1990 and followed up to the end of 1995, about 94% of those eligible (Table 58.1). Five per cent of eligible children were excluded because their vital status was unknown at 6 July 1997, and 1% because their survival was zero or unknown. Almost nine out of ten osteosarcomas arose in one of the long bones of the leg (76%) or arm (13%) (Table 58A): most of these would have been tumours of the lower end of the femur and the upper end of the tibia or humerus[1]. In almost all cases (95%), the morphology was simply coded as osteosarcoma, not otherwise specified; juxtacortical (2%), chondroblastic (1%) and fibroblastic (1%) osteosarcomas accounted for the remainder.

Survival trends

Relative survival[a] from osteosarcoma in England and Wales has improved substantially. For children diagnosed during 1971-75, one-year survival was 55% and five-year survival 17%. Those diagnosed during 1986-90 had one-year survival of 88% and five-year survival of 51%, an absolute increase of more than 30% (Table 58.2). The largest improvement, an increase of 20% in five-year survival from 28% to 48%, occurred between 1976-80 and 1981-85. Five-year survival increased by only a further 3% between children diagnosed in 1981-85 and those diagnosed during 1986-90.

For England and Wales as a whole, the absolute increase in relative survival between successive five-year periods from 1971-75 to 1986-90 averaged 11% for one-year survival and 12-13% for five-year survival (see Table 4.4). There were no consistent differences in survival between boys and girls.

Ten-year survival for children diagnosed during 1981-85 and followed up to the end of 1995 was 45%, similar to the survival rate five years after diagnosis. This is shown in the survival curves (Figure 58.1), and it reflects the relatively low excess mortality, compared to children of the same age in the general population, among children who have survived more than five years following the diagnosis of an osteosarcoma[2].

Regional estimates of survival are based on no more than 30 children for a given five-year period, and they are imprecise (Figure 58.4). In some regions there were too few cases for analysis (see Chapter 3). Survival nevertheless increased substantially in every NHS Region of England (Table 58.2, Figure 58.1).

The results presented here are effectively identical with those reported by the National Registry of Childhood Tumours for Great Britain. Five-year actuarial survival (period of diagnosis) increased from 17% (1971-73)[3] to 35% (1980-82) and then 51% (1986-88). One-year survival improved significantly in the early 1980s, but not thereafter[4].

Deprivation

There was no consistent pattern in one-year or five-year survival between the various deprivation groups (Table 58.2). The survival curves for children diagnosed in England and Wales during 1986-90 overlap irregularly (Figure 58.2). Estimates of the difference in one-year and five-year survival between the most affluent and the most deprived groups are based on only 120-130 children in each five-year period: the differences are irregular and not statistically significant (see Table 4.9). The pattern is shown graphically in Figure 58.3.

International comparison

Survival from osteosarcoma in the areas covered by the SEER programme cancer registries in the USA has been reported[5]. For children diagnosed during 1967-73, five-year relative survival was 26%, similar to the 28% five-year survival rate for children diagnosed in England and Wales during 1976-80.

Among 131 children diagnosed in the USA during 1973-81, one-year survival was 81% and five-year survival 43%: these survival rates are broadly comparable with those for

---

[a] See footnote in Chapter 50.

Table 58A Sub-site distribution of osteosarcoma in children

ICD-9 code	Site description	1971-75 No.	%	1976-80 No.	%	1981-85 No.	%	1986-90 No.	%	Total No.	%
170.0	Bones of skull and face	3	1.8	2	1.4	1	0.7	5	4.3	11	2.0
170.1	Lower jaw	0	0.0	1	0.7	2	1.4	0	0.0	3	0.5
170.2	Vertebral column	0	0.0	0	0.0	3	2.2	3	2.6	6	1.1
170.3	Ribs, sternum and clavicle	1	0.6	1	0.7	3	2.2	0	0.0	5	0.9
170.4	Upper limb (long bones), scapula	22	13.5	18	12.8	15	10.9	15	12.8	70	12.5
170.5	Upper limb, short bones	2	1.2	0	0.0	0	0.0	0	0.0	2	0.4
170.6	Pelvic bones, sacrum, coccyx	2	1.2	3	2.1	4	2.9	1	0.9	10	1.8
170.7	Lower limb, long bones	129	79.1	105	74.5	103	74.6	85	72.6	422	75.5
170.8	Lower limb, short bones	1	0.6	4	2.8	1	0.7	2	1.7	8	1.4
170.9	Site unspecified	3	1.8	7	5.0	6	4.3	6	5.1	22	3.9
	Total	**163**		**141**		**138**		**117**		**559**	

1981-85 in England and Wales (82% and 48%, respectively).

1. Westbury G, Pinkerton R. Bone tumours. In: Horwich A (ed.) *Oncology: a multi-disciplinary textbook.* London: Chapman and Hall, 1995, pp365-378

2. Hawkins MM. Long term survival and cure after childhood cancer. Arch Dis Child 1989; 64: 798-807

3. Stiller CA, Bunch KJ. Trends in survival for childhood cancer in Britain diagnosed 1971-85. Br J Cancer 1990; 62: 806-815

4. Stiller CA. Population-based survival rates for childhood cancer in Britain, 1980-91. Br Med J 1994; 309: 1612-1616

5. Young JL, Gloeckler Ries LA, Silverberg E, Horm JW, Miller RW. Cancer incidence, survival and mortality for children younger than age 15 years. Cancer 1986; 58: 598-602

Table 58.1 Ineligible and excluded records, and patients included in analyses, by calendar period of diagnosis

	Calendar period of diagnosis														
	1971-75			**1976-80**			**1981-85**			**1986-90**			**Overall**		
Total registered	**181**			**156**			**155**			**127**			**619**		
Ineligible	*Records*	*Patients*	*%*	*Records*	*Patients*	*%*	*Records*	*Patients*	*%*	*Records*	*Patients*	*%*	*Records*	*Patients*	*%*
Incomplete data[1]	0	0		0	0		0	0		0	0		0	**0**	
Resident outside England and Wales	8	8		3	3		9	9		0	0		20	**20**	
In situ neoplasm[2]	0	0		0	0		0	0		0	0		0	**0**	
Benign or uncertain behaviour[2]	2	2		0	0		0	0		0	0		2	**2**	
Metastatic[2]	0	0		0	0		0	0		0	0		0	**0**	
Otherwise ineligible[3]	-	-		-	-		-	-		-	-		-	**-**	
Lymphoma[4]	-	-		-	-		-	-		-	-		-	**-**	
Leukaemia or myeloma[4]	-	-		-	-		-	-		-	-		-	**-**	
	10			**3**			**9**			**0**			**22**		
Total eligible	**171**	**100**		**153**	**100**		**146**	**100**		**127**	**100**		**597**	**100**	
Aged under 15 or 100 years or over at diagnosis[4]	-	-	-	-	-	-	-	-	-	-	-	-	-	-	-
Vital status unknown[5]	15	8	4.7	13	10	6.5	13	6	4.1	7	7	5.5	48	**31**	5.2
Duplicate registration[6]	0	0	0.0	0	0	0.0	1	0	0.0	0	0	0.0	1	**0**	0.0
Multiple primary[7]	0	0	0.0	0	0	0.0	1	1	0.7	0	0	0.0	1	**1**	0.2
Synchronous tumours[8]	0	0	0.0	0	0	0.0	1	1	0.7	0	0	0.0	1	**1**	0.2
Sex not known	0	0	0.0	0	0	0.0	0	0	0.0	0	0	0.0	0	**0**	0.0
Sex incompatible with site code[9]	-	-	-	-	-	-	-	-	-	-	-	-	-	**-**	-
Invalid dates[10]	0	0	0.0	0	0	0.0	0	0	0.0	0	0	0.0	0	**0**	0.0
Zero survival[11]	0	0	0.0	3	2	1.3	0	0	0.0	3	3	2.4	6	**5**	0.8
Total exclusions		**8**	**4.7**		**12**	**7.8**		**8**	**5.5**		**10**	**7.9**		**38**	**6.4**
Patients included in analysis		**163**	**95.3**		**141**	**92.2**		**138**	**94.5**		**117**	**92.1**		**559**	**93.6**

[1] Main data item(s) invalid or incompatible with one another: name, sex, date of birth, date of diagnosis and (if present) date of death, postcode, site, morphology and behaviour

[2] Tumour either *in situ* (behaviour code 2), benign (0), uncertain if benign or malignant (1) or metastatic (6 or 9)

[3] Not applicable for this malignancy

[4] Not applicable for this malignancy

[5] Tracing of vital status at National Health Service Central Register incomplete at 6 July 1997

[6] Same site code, sex, cancer registry and cancer registry number as an earlier registration

[7] Same site code and person number as an earlier registration(s): mostly confirmed multiple primary tumours at this site, some unresolved duplicate registrations

[8] Same site code, sex, date of birth and date of diagnosis as another registration(s): mostly synchronous or (in paired organs) bilateral tumours in same anatomic site in one individual, not previously linked: also some duplicate registrations

[9] Not applicable for this malignancy

[10] Impossible sequence of dates (birth, diagnosis and, if present, emigration or death); or date of death after 6 July 1997

[11] Date of diagnosis same as date of death: some are patients who did die on the day of diagnosis, but most were registered solely from a death certificate (death certificate only, or DCO), and their survival time is thus unknown

OSTEOSARCOMA *Children*

Table 58.2 Crude and relative survival (%) at one, five and ten years since diagnosis: NHS Region, deprivation category and calendar period of diagnosis

ENGLAND & WALES

		1971-75 Affluent C R	2 C R	3 C R	4 C R	Deprived C R	All[1] C R	1976-80 Affluent C R	2 C R	3 C R	4 C R	Deprived C R	All[1] C R
Boys		(15)	(13)	(16)	(13)	(21)	(104)	(20)	(10)	(14)	(8)	(7)	(63)
	One year	80 81	47 47	75 75	70 70	30 30	54 54	70 70	71 71	70 71	- -	- -	67 67
	Five years	47 47	7 7	25 25	13 13	4 4	18 18	40 40	11 11	22 22	- -	- -	22 22
	Ten years	47 47	7 7	25 25	13 13	4 4	16 16	40 40	11 11	22 22	- -	- -	22 22
Girls		(9)	(6)	(12)	(9)	(13)	(59)	(15)	(17)	(15)	(21)	(10)	(78)
	One year	- -	- -	65 65	- -	52 52	57 57	79 80	76 76	54 54	65 65	71 71	69 69
	Five years	- -	- -	22 22	- -	8 8	17 17	40 41	41 41	40 40	18 18	29 29	33 33
	Ten years	- -	- -	22 22	- -	8 8	13 13	40 41	24 24	40 40	18 18	29 29	29 29
Children		(24)	(19)	(28)	(22)	(34)	(163)	(35)	(27)	(29)	(29)	(17)	(141)
	One year	71 71	54 54	71 71	73 73	38 38	55 55	74 74	74 74	62 62	62 62	66 66	68 68
	Five years	38 38	15 15	24 24	12 12	5 6	17 17	40 40	30 30	31 31	17 17	23 23	28 28
	Ten years	34 34	10 10	24 24	12 12	5 6	15 15	40 40	19 19	31 31	17 17	23 23	26 26

		1971-75 Boys C R	Girls C R	Children C R	1976-80 Boys C R	Girls C R	Children C R
Northern & Yorkshire		(17)	(10)	(27)	(10)	(12)	(22)
	One year	52 52	61 61	56 56	60 60	57 57	59 59
	Five years	17 17	15 15	17 17	19 19	30 30	26 26
	Ten years	11 12	15 15	14 14	19 19	30 30	26 26
Trent		(15)	(5)	(20)	(8)	(6)	(14)
	One year	67 67	- -	59 59	- -	- -	85 85
	Five years	27 27	- -	19 19	- -	- -	28 28
	Ten years	27 27	- -	19 19	- -	- -	28 28
Anglia & Oxford		(7)	(7)	(14)	(12)	(6)	(18)
	One year	- -	- -	63 63	68 68	- -	78 78
	Five years	- -	- -	16 16	17 17	- -	28 28
	Ten years	- -	- -	9 9	17 17	- -	16 16
North Thames		(16)	(6)	(22)	(8)	(12)	(20)
	One year	44 44	- -	54 55	- -	49 49	54 54
	Five years	13 13	- -	23 23	- -	11 11	15 15
	Ten years	13 13	- -	23 23	- -	11 11	15 15
South Thames		(11)	(8)	(19)	(3)	(9)	(12)
	One year	83 83	- -	63 63	- -	- -	60 60
	Five years	35 35	- -	19 19	- -	- -	26 26
	Ten years	35 35	- -	19 19	- -	- -	26 26
South & West		(8)	(8)	(16)	(7)	(13)	(20)
	One year	- -	- -	70 70	- -	55 55	66 66
	Five years	- -	- -	19 19	- -	55 55	51 51
	Ten years	- -	- -	13 13	- -	55 55	51 51
West Midlands		(10)	(6)	(16)	(6)	(5)	(11)
	One year	31 31	- -	40 40	- -	- -	46 46
	Five years	10 10	- -	14 14	- -	- -	9 9
	Ten years	0 0	- -	8 8	- -	- -	9 9
North & West		(12)	(7)	(19)	(8)	(11)	(19)
	One year	35 36	- -	31 31	- -	91 91	84 84
	Five years	16 16	- -	16 16	- -	36 36	36 36
	Ten years	16 16	- -	16 16	- -	36 36	36 36
WALES		(8)	(2)	(10)	(1)	(4)	(5)
	One year	- -	- -	71 71	- -	- -	- -
	Five years	- -	- -	9 9	- -	- -	- -
	Ten years	- -	- -	9 9	- -	- -	- -

[1] Including those for whom no deprivation category was assigned. C - crude survival, R - relative survival (see Chapter 3). Number of patients contributing to each analysis in parentheses.

1981-85

ENGLAND & WALES — Boys

	Affluent C	R	2 C	R	3 C	R	4 C	R	Deprived C	R	All[1] C	R
(N)	*(16)*		*(10)*		*(16)*		*(12)*		*(9)*		*(64)*	
One year	62	62	80	80	100	100	75	75	-	-	81	81
Five years	38	38	51	51	81	81	32	32	-	-	54	55
	38	38	40	40	75	75	32	32	-	-	51	52

Girls

	Affluent C	R	2 C	R	3 C	R	4 C	R	Deprived C	R	All[1] C	R
(N)	*(17)*		*(12)*		*(14)*		*(12)*		*(16)*		*(74)*	
One year	77	77	84	84	79	79	83	83	87	87	82	82
Five years	41	41	33	33	35	35	32	32	56	56	42	42
	41	41	25	25	35	35	23	23	56	56	39	39

Children

	Affluent C	R	2 C	R	3 C	R	4 C	R	Deprived C	R	All[1] C	R
(N)	*(33)*		*(22)*		*(30)*		*(24)*		*(25)*		*(138)*	
One year	70	70	82	82	90	90	79	79	92	92	82	82
Five years	39	39	41	41	60	60	32	32	64	64	48	48
	39	39	32	32	57	57	28	28	64	64	45	45

1986-90

ENGLAND & WALES — Boys

	Affluent C	R	2 C	R	3 C	R	4 C	R	Deprived C	R	All[1] C	R
(N)	*(16)*		*(12)*		*(13)*		*(14)*		*(7)*		*(62)*	
One year	81	81	100	100	77	77	93	93	-	-	85	85
Five years	51	51	57	58	54	54	43	43	-	-	47	47

Girls

	Affluent C	R	2 C	R	3 C	R	4 C	R	Deprived C	R	All[1] C	R
(N)	*(19)*		*(9)*		*(10)*		*(7)*		*(10)*		*(55)*	
One year	95	95	100	100	100	100	-	-	70	70	91	91
Five years	47	47	67	67	68	68	-	-	40	40	56	56

Children

	Affluent C	R	2 C	R	3 C	R	4 C	R	Deprived C	R	All[1] C	R
(N)	*(35)*		*(21)*		*(23)*		*(21)*		*(17)*		*(117)*	
One year	89	89	100	100	87	87	91	91	70	70	88	88
Five years	49	49	62	62	60	60	52	53	28	28	51	51

Regions, 1981-85

Region		Boys C	R	Girls C	R	Children C	R
Northern & Yorkshire	*(N)*	*(6)*		*(13)*		*(19)*	
	One year	-	-	100	100	89	89
	Five years	-	-	70	70	64	64
		-	-	54	54	53	53
Trent	*(N)*	*(5)*		*(4)*		*(9)*	
	One year	-	-	-	-	-	-
	Five years	-	-	-	-	-	-
		-	-	-	-	-	-
Anglia & Oxford	*(N)*	*(4)*		*(7)*		*(11)*	
	One year	-	-	-	-	81	81
	Five years	-	-	-	-	36	36
		-	-	-	-	36	36
North Thames	*(N)*	*(10)*		*(12)*		*(22)*	
	One year	60	60	75	75	69	69
	Five years	49	49	23	23	35	35
		40	40	23	23	31	31
South Thames	*(N)*	*(10)*		*(12)*		*(22)*	
	One year	90	90	75	75	82	82
	Five years	40	40	18	18	28	28
		40	40	18	18	28	28
South & West	*(N)*	*(5)*		*(6)*		*(11)*	
	One year	-	-	-	-	91	91
	Five years	-	-	-	-	65	65
		-	-	-	-	65	65
West Midlands	*(N)*	*(10)*		*(12)*		*(22)*	
	One year	90	90	83	83	86	86
	Five years	80	80	40	40	59	59
		70	70	40	40	54	54
North & West	*(N)*	*(11)*		*(7)*		*(18)*	
	One year	100	100	-	-	89	89
	Five years	82	82	-	-	67	67
		82	82	-	-	67	67
WALES	*(N)*	*(3)*		*(1)*		*(4)*	
	One year	-	-	-	-	-	-
	Five years	-	-	-	-	-	-
		-	-	-	-	-	-

Regions, 1986-90

Region		Boys C	R	Girls C	R	Children C	R
Northern & Yorkshire	*(N)*	*(7)*		*(9)*		*(16)*	
	One year	-	-	78	78	75	75
	Five years	-	-	44	44	50	50
Trent	*(N)*	*(7)*		*(10)*		*(17)*	
	One year	-	-	90	90	83	83
	Five years	-	-	60	60	52	52
Anglia & Oxford	*(N)*	*(10)*		*(5)*		*(15)*	
	One year	70	70	-	-	80	80
	Five years	51	51	-	-	54	54
North Thames	*(N)*	*(12)*		*(9)*		*(21)*	
	One year	92	92	100	100	95	95
	Five years	48	48	31	31	42	42
South Thames	*(N)*	*(8)*		*(10)*		*(18)*	
	One year	87	87	90	90	89	89
	Five years	47	47	70	70	60	60
South & West	*(N)*	*(8)*		*(2)*		*(10)*	
	One year	100	100	-	-	100	100
	Five years	36	36	-	-	49	49
West Midlands	*(N)*	*(2)*		*(2)*		*(4)*	
	One year	-	-	-	-	-	-
	Five years	-	-	-	-	-	-
North & West	*(N)*	*(6)*		*(7)*		*(13)*	
	One year	-	-	-	-	100	100
	Five years	-	-	-	-	55	55
WALES	*(N)*	*(2)*		*(1)*		*(3)*	
	One year	-	-	-	-	-	-
	Five years	-	-	-	-	-	-

[1] Including those for whom no deprivation category was assigned. C - crude survival, R - relative survival (see Chapter 3).
Number of patients contributing to each analysis in parentheses.

Fig. 58.1   Relative survival up to ten years: trends by calendar period of diagnosis (1971-90) and NHS Region

CHILDREN

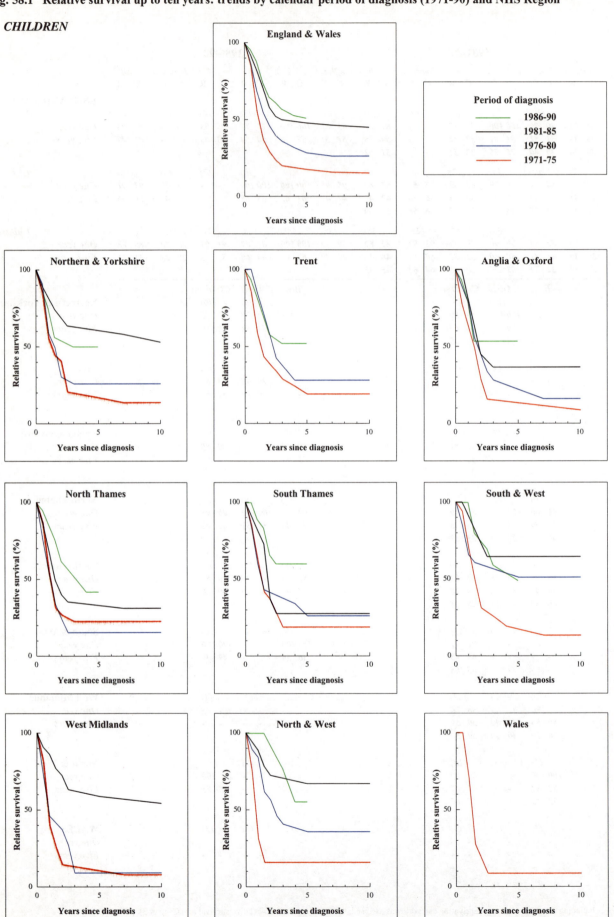

Fig. 58.2 Relative survival up to five years, by deprivation category: England and Wales, patients diagnosed 1986-90

CHILDREN

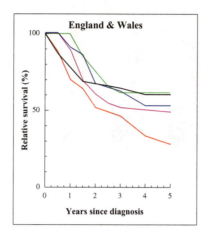

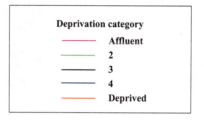

Fig. 58.3 Relative survival at five years, by deprivation category and period of diagnosis (1981-90): England and Wales

CHILDREN

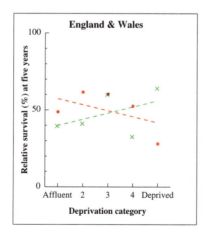

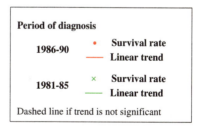

Fig. 58.4 Relative survival (%) by NHS Region and sex:
one-year and five-year survival for children diagnosed 1986-90, with 95% confidence intervals

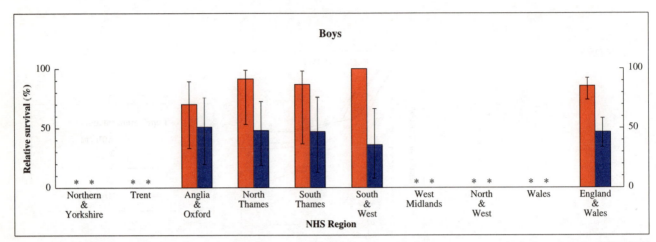

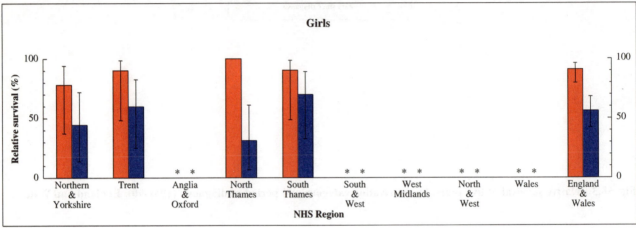

Chapter 59

EWING'S SARCOMA *Children*

Ewing's sarcoma is the second most common malignancy of bone in children, after osteosarcoma. It is rare, and only about 20 children are diagnosed each year in England and Wales. As with osteosarcoma, the peak age in children is 10-14 years, but it is more common than osteosarcoma in the first decade of life[1]. Incidence is similar in boys and girls, with rates of 2-3 per million in 1990. These rates are similar to those among whites in North America and in other parts of Europe, but Ewing's sarcoma is rare in blacks and in East Asians[2].

Ewing's tumour presents with local swelling and pain. Any bone may be affected, but the femur and the pelvis are the most common sites. Unlike osteosarcoma, Ewing's sarcoma in long bones tends to arise in the central part of the shaft. Metastases occur in the lung and in other bones. Very rarely, tumours identical with Ewing's may arise in soft tissue. Ewing's sarcoma is sensitive to radiotherapy and to more recent chemotherapy regimens, which have greatly improved survival[1].

The survival analyses reported here are based on 413 children who were diagnosed with Ewing's sarcoma in England and Wales between 1971 and 1990 and followed up to the end of 1995, about 89% of those eligible (Table 59.1). About 8% of eligible children were excluded from the analyses because their vital status was unknown at 6 July 1997, and another 2% because their survival was zero or unknown. The cases were defined for analysis by the morphology code for Ewing's sarcoma (M9260/3) and by the site code for bone (or ill-defined sites; see Table 2.5).

All the tumours had been coded to bone. Ewing's tumour in soft tissue was registered in 14 children: these cases are included with other soft tissue tumours (see Table 2.5 and Chapter 60). The anatomical distribution of Ewing's sarcoma differed from that for osteosarcoma (Table 59A; cf. Table 58A). The long bones of the leg were most often involved (37%; cf. 76% for osteosarcoma), followed by the pelvic bones, sacrum and coccyx (18%, cf. 2%), the long bones of the arm (16%), and the ribs, sternum and clavicle (10%, cf. 1% for osteosarcoma).

Survival trends

Relative survival[a] for children diagnosed with Ewing's sarcoma in England and Wales during 1986-90 was 89% at one year and 61% at five years. This represents an improvement of 11% from the one-year survival rate for children diagnosed in 1971-75 (78%), and a 28% gain over the five-year survival rate for that period (33%; Table 59.2).

Much the largest component of the overall increase in five-year survival (23% of the total 28% gain) occurred between children diagnosed during 1981-85 and those diagnosed during 1986-90 (Figure 59.1).

For England and Wales as a whole, this represents an average increase of 5% in one-year survival between successive five-year periods from 1971-75 to 1986-90 (see Table 4.4). For five-year survival, the corresponding average pace of improvement between successive five-year periods was 11% for boys and 5% for girls, but these estimates are imprecise; survival has clearly improved substantially in both sexes. There were no consistent differences in survival between boys and girls.

Ten-year survival for children diagnosed during 1981-85 and followed up to the end of 1995 was 35%, not far below the five-year survival rate of 38% (Table 59.2).

Regional estimates of survival are based on fewer than 20 children in any five-year period, and they are imprecise (Figure 59.4). In some regions there were too few cases for analysis in certain time periods (see Chapter 3). Survival nevertheless increased substantially in most regions of the country (Table 59.2, Figure 59.1).

These trends are similar to the trends in actuarial survival reported by the National Registry of Childhood Tumours for Great Britain. Five-year survival (period of diagnosis) for Ewing's sarcoma in children increased slightly from 39% (1971-73)[3] to 43% (1983-85), and then jumped to 59% (1986-88); survival improved significantly in the late 1980s[4].

Deprivation

The survival rates for children in each category of deprivation in England and Wales are all based on fewer than 40 cases in a given five-year period, and they are imprecise. There was no consistent pattern in one-year or five-year survival between deprivation groups (Table 59.2). The survival curves for children diagnosed during 1986-90 overlap irregularly, although the curve for the most deprived group is the lowest (Figure 58.2).

Estimates of the difference in one-year and five-year survival between the most affluent and the most deprived groups are based on only about 100 children of known deprivation status in each five-year period: the differences are irregular and not statistically significant (see Table

---

[a] See footnote in Chapter 50.

Table 59A Sub-site distribution of Ewing's sarcoma in children

ICD-9 code	Site description	1971-75 No.	1971-75 %	1976-80 No.	1976-80 %	1981-85 No.	1981-85 %	1986-90 No.	1986-90 %	Total No.	Total %
170.0	Bones of skull and face	0	0.0	2	1.8	2	1.6	3	3.1	**7**	**1.7**
170.1	Lower jaw	0	0.0	1	0.9	0	0.0	2	2.1	**3**	**0.7**
170.2	Vertebral column	2	2.6	2	1.8	11	8.7	5	5.2	**20**	**4.8**
170.3	Ribs, sternum and clavicle	6	7.7	13	11.7	10	7.9	8	8.2	**37**	**9.0**
170.4	Upper limb (long bones), scapula	18	23.1	20	18.0	17	13.4	11	11.3	**66**	**16.0**
170.5	Upper limb, short bones	0	0.0	2	1.8	0	0.0	1	1.0	**3**	**0.7**
170.6	Pelvic bones, sacrum, coccyx	11	14.1	23	20.7	21	16.5	18	18.6	**73**	**17.7**
170.7	Lower limb, long bones	32	41.0	40	36.0	47	37.0	33	34.0	**152**	**36.8**
170.8	Lower limb, short bones	3	3.8	3	2.7	7	5.5	3	3.1	**16**	**3.9**
170.9	Site unspecified	6	7.7	5	4.5	12	9.4	13	13.4	**36**	**8.7**
	Total	**78**		**111**		**127**		**97**		**413**	

4.9). The pattern is shown graphically in Figure 58.3, from which it is clear that the large national increase in five-year survival between children diagnosed in 1981-85 and 1986-90 was reflected in four of the five deprivation groups. The 9-10% deprivation gap in one-year and five-year survival for children diagnosed during 1986-90 is not statistically significant.

International comparison

Survival from Ewing's sarcoma in the areas covered by the SEER programme cancer registries in the USA has been reported[5]. Five-year survival for children diagnosed during 1973-81 was 23%. This was lower than five-year survival for children diagnosed in England and Wales during 1971-75 (33%).

Among 106 children diagnosed in the USA during 1973-81, one-year relative survival was 85% and five-year survival 48%. This may be compared with one-year and five-year survival rates of 88% and 38%, respectively, for children diagnosed in England and Wales during 1981-85.

1. Westbury G, Pinkerton R. Bone tumours. In: Horwich A (ed.) *Oncology: a multi-disciplinary textbook.* London: Chapman and Hall, 1995, pp365-378

2. Miller RW, Boice JD, Curtis RE. Bone cancer. In: Schottenfeld D, Fraumeni JF, (eds.) *Cancer epidemiology and prevention.* 2nd edn. Oxford: Oxford University Press, 1996, pp971-983

3. Stiller CA, Bunch KJ. Trends in survival for childhood cancer in Britain diagnosed 1971-85. Br J Cancer 1990; 62: 806-815

4. Stiller CA. Population-based survival rates for childhood cancer in Britain, 1980-91. Br Med J 1994; 309: 1612-1616

5. Young JL, Gloeckler Ries LA, Silverberg E, Horm JW, Miller RW. Cancer incidence, survival and mortality for children younger than age 15 years. Cancer 1986; 58: 598-602

Table 59.1 Ineligible and excluded records, and patients included in analyses, by calendar period of diagnosis

	1971-75			1976-80			1981-85			1986-90			Overall		
Total registered		84			122			152			120			478	
Ineligible	*Records*	*Patients*	*%*	*Records*	*Patients*	*%*	*Records*	*Patients*	*%*	*Records*	*Patients*	*%*	*Records*	*Patients*	*%*
Incomplete data[1]	0	0		1	1		0	0		0	0		1	1	
Resident outside England and Wales	4	4		1	1		7	7		3	3		15	15	
In situ neoplasm[2]	0	0		0	0		0	0		0	0		0	0	
Benign or uncertain behaviour[2]	0	0		1	0		0	0		0	0		1	0	
Metastatic[2]	0	0		0	0		0	0		0	0		0	0	
Otherwise ineligible[3]	-	-		-	-		-	-		-	-		-	-	
Lymphoma[4]	-	-		-	-		-	-		-	-		-	-	
Leukaemia or myeloma[4]	-	-		-	-		-	-		-	-		-	-	
		4			2			7			3			16	
Total eligible		80	100		120	100		145	100		117	100		462	100
Aged under 15 or 100 years or over at diagnosis[4]	-	-	-	-	-	-	-	-	-	-	-	-	-	-	-
Vital status unknown[5]	6	2	2.5	9	8	6.7	19	12	8.3	18	17	14.5	52	39	8.4
Duplicate registration[6]	0	0	0.0	0	0	0.0	0	0	0.0	0	0	0.0	0	0	0.0
Multiple primary[7]	0	0	0.0	0	0	0.0	1	1	0.7	1	1	0.9	2	2	0.4
Synchronous tumours[8]	0	0	0.0	0	0	0.0	1	1	0.7	0	0	0.0	1	1	0.2
Sex not known	0	0	0.0	0	0	0.0	0	0	0.0	0	0	0.0	0	0	0.0
Sex incompatible with site code[9]	-	-	-	-	-	-	-	-	-	-	-	-	-	-	-
Invalid dates[10]	0	0	0.0	0	0	0.0	0	0	0.0	0	0	0.0	0	0	0.0
Zero survival[11]	0	0	0.0	2	1	0.8	5	4	2.8	3	2	1.7	10	7	1.5
Total exclusions		2	2.5		9	7.5		18	12.4		20	17.1		49	10.6
Patients included in analysis		78	97.5		111	92.5		127	87.6		97	82.9		413	89.4

[1] Main data item(s) invalid or incompatible with one another: name, sex, date of birth, date of diagnosis and (if present) date of death, postcode, site, morphology and behaviour

[2] Tumour either *in situ* (behaviour code 2), benign (0), uncertain if benign or malignant (1) or metastatic (6 or 9)

[3] Not applicable for this malignancy

[4] Not applicable for this malignancy

[5] Tracing of vital status at National Health Service Central Register incomplete at 6 July 1997

[6] Same site code, sex, cancer registry and cancer registry number as an earlier registration

[7] Same site code and person number as an earlier registration(s): mostly confirmed multiple primary tumours at this site, some unresolved duplicate registrations

[8] Same site code, sex, date of birth and date of diagnosis as another registration(s): mostly synchronous or (in paired organs) bilateral tumours in same anatomic site in one individual, not previously linked: also some duplicate registrations

[9] Not applicable for this malignancy

[10] Impossible sequence of dates (birth, diagnosis and, if present, emigration or death); or date of death after 6 July 1997

[11] Date of diagnosis same as date of death: some are patients who did die on the day of diagnosis, but most were registered solely from a death certificate (death certificate only, or DCO), and their survival time is thus unknown

Table 59.2 Crude and relative survival (%) at one, five and ten years since diagnosis: NHS Region, deprivation category and calendar period of diagnosis

	1971-75												1976-80											
	Affluent		2		3		4		Deprived		All[1]		Affluent		2		3		4		Deprived		All[1]	
	C	R	C	R	C	R	C	R	C	R	C	R	C	R	C	R	C	R	C	R	C	R	C	R
ENGLAND & WALES																								
Boys	(13)		(8)		(3)		(5)		(5)		(41)		(16)		(17)		(13)		(9)		(7)		(64)	
One year	76	77	-	-	-	-	-	-	-	-	83	83	64	64	100	100	84	84	-	-	-	-	77	77
Five years	38	38	-	-	-	-	-	-	-	-	36	37	19	19	35	35	38	38	-	-	-	-	31	32
Ten years	30	30	-	-	-	-	-	-	-	-	34	34	19	19	29	29	23	23	-	-	-	-	25	25
Girls	(7)		(5)		(7)		(9)		(5)		(37)		(11)		(10)		(4)		(4)		(14)		(47)	
One year	-	-	-	-	-	-	-	-	-	-	74	74	100	100	62	62	-	-	-	-	100	100	87	87
Five years	-	-	-	-	-	-	-	-	-	-	30	30	45	45	33	33	-	-	-	-	52	52	43	43
Ten years	-	-	-	-	-	-	-	-	-	-	22	22	45	45	20	20	-	-	-	-	52	52	39	39
Children	(20)		(13)		(10)		(14)		(10)		(78)		(27)		(27)		(17)		(13)		(21)		(111)	
One year	75	75	76	76	81	81	71	71	81	81	78	78	78	78	85	85	82	82	78	78	76	76	81	81
Five years	31	31	15	15	39	39	50	50	40	40	33	33	30	30	34	34	35	35	48	48	39	39	36	36
Ten years	26	26	15	15	39	39	36	36	30	30	28	28	30	30	26	26	24	24	39	39	39	39	31	31

	Boys		Girls		Children								Boys		Girls		Children			
Northern & Yorkshire	(4)		(2)		(6)								(6)		(5)		(11)			
One year	-	-	-	-	-	-							-	-	-	-	80	80		
Five years	-	-	-	-	-	-							-	-	-	-	26	26		
Ten years	-	-	-	-	-	-							-	-	-	-	26	26		
Trent			(6)		(2)		(8)								(9)		(4)		(13)	
One year			-	-	-	-	-	-							-	-	-	-	85	85
Five years			-	-	-	-	-	-							-	-	-	-	47	47
Ten years			-	-	-	-	-	-							-	-	-	-	47	47
Anglia & Oxford			(3)		(4)		(7)								(10)		(6)		(16)	
One year			-	-	-	-	-	-							100	100	-	-	94	94
Five years			-	-	-	-	-	-							41	41	-	-	46	46
Ten years			-	-	-	-	-	-							41	41	-	-	46	46
North Thames			(2)		(3)		(5)								(10)		(9)		(19)	
One year			-	-	-	-	-	-							90	90	-	-	95	95
Five years			-	-	-	-	-	-							40	40	-	-	37	37
Ten years			-	-	-	-	-	-							12	12	-	-	22	22
South Thames			(5)		(5)		(10)								(7)		(6)		(13)	
One year			-	-	-	-	80	80							-	-	-	-	59	59
Five years			-	-	-	-	41	41							-	-	-	-	2	2
Ten years			-	-	-	-	19	20							-	-	-	-	0	0
South & West			(7)		(7)		(14)								(10)		(3)		(13)	
One year			-	-	-	-	72	72							79	80	-	-	77	77
Five years			-	-	-	-	36	36							40	40	-	-	46	47
Ten years			-	-	-	-	36	36							29	29	-	-	31	31
West Midlands			(3)		(3)		(6)								(4)		(4)		(8)	
One year			-	-	-	-	-	-							-	-	-	-	-	-
Five years			-	-	-	-	-	-							-	-	-	-	-	-
Ten years			-	-	-	-	-	-							-	-	-	-	-	-
North & West			(10)		(10)		(20)								(4)		(6)		(10)	
One year			80	80	62	62	71	71							-	-	-	-	81	81
Five years			12	12	31	31	21	21							-	-	-	-	41	41
Ten years			12	12	20	20	16	16							-	-	-	-	30	30
WALES			(1)		(1)		(2)								(4)		(4)		(8)	
One year			-	-	-	-	-	-							-	-	-	-	-	-
Five years			-	-	-	-	-	-							-	-	-	-	-	-
Ten years			-	-	-	-	-	-							-	-	-	-	-	-

[1] Including those for whom no deprivation category was assigned. C - crude survival, R - relative survival (see Chapter 3).
Number of patients contributing to each analysis in parentheses.

ENGLAND & WALES

1981-85						1986-90						
Affluent	2	3	4	Deprived	All¹	Affluent	2	3	4	Deprived	All¹	
C R	C R	C R	C R	C R	C R	C R	C R	C R	C R	C R	C R	
(16)	(17)	(9)	(4)	(12)	(59)	(16)	(17)	(12)	(8)	(4)	(57)	**Boys**
94 94	77 77	- -	- -	83 83	85 85	94 94	88 88	100 100	100 100	- -	93 93	*One year*
38 38	29 29	- -	- -	35 35	35 35	56 56	71 71	92 92	75 75	- -	69 69	*Five years*
32 32	29 29	- -	- -	26 26	32 32							
(15)	(19)	(14)	(7)	(12)	(68)	(4)	(11)	(5)	(7)	(13)	(40)	**Girls**
94 94	100 100	93 93	- -	75 75	91 91	- -	82 82	- -	- -	77 77	83 83	*One year*
40 40	42 42	29 29	- -	49 49	40 40	- -	36 36	- -	- -	46 46	50 50	*Five years*
40 40	36 36	29 29	- -	49 49	38 38							
(31)	(36)	(23)	(11)	(24)	(127)	(20)	(28)	(17)	(15)	(17)	(97)	**Children**
94 94	89 89	87 87	91 91	79 79	88 88	95 95	86 86	94 94	93 93	77 77	89 89	*One year*
39 39	36 36	35 35	42 42	42 42	37 38	59 60	57 57	83 83	66 66	42 42	61 61	*Five years*
36 36	33 33	35 35	42 42	37 38	35 35							

1981-85			1986-90			
Boys	Girls	Children	Boys	Girls	Children	
(7)	(9)	(16)	(9)	(7)	(16)	**Northern & Yorkshire**
- -	- -	82 82	89 89	- -	76 76	*One year*
- -	- -	44 44	67 67	- -	51 51	*Five years*
- -	- -	38 38				
(9)	(2)	(11)	(6)	(6)	(12)	**Trent**
- -	- -	100 100	- -	- -	100 100	*One year*
- -	- -	29 29	- -	- -	58 58	*Five years*
- -	- -	19 19				
(9)	(9)	(18)	(4)	(6)	(10)	**Anglia & Oxford**
- -	- -	100 100	- -	- -	100 100	*One year*
- -	- -	42 42	- -	- -	70 70	*Five years*
- -	- -	36 36				
(7)	(9)	(16)	(11)	(7)	(18)	**North Thames**
- -	- -	69 69	91 91	- -	95 95	*One year*
- -	- -	18 18	73 73	- -	83 83	*Five years*
- -	- -	18 18				
(6)	(13)	(19)	(9)	(3)	(12)	**South Thames**
- -	100 100	95 95	78 78	- -	76 76	*One year*
- -	38 38	32 32	66 66	- -	58 58	*Five years*
- -	38 38	32 32				
(6)	(8)	(14)	(7)	(2)	(9)	**South & West**
- -	- -	93 93	- -	- -	100 100	*One year*
- -	- -	64 64	- -	- -	67 67	*Five years*
- -	- -	64 64				
(5)	(6)	(11)	(5)	(1)	(6)	**West Midlands**
- -	- -	91 91	- -	- -	- -	*One year*
- -	- -	36 36	- -	- -	- -	*Five years*
- -	- -	36 36				
(4)	(9)	(13)	(4)	(6)	(10)	**North & West**
- -	- -	85 85	- -	- -	80 80	*One year*
- -	- -	23 23	- -	- -	21 21	*Five years*
- -	- -	23 23				
(6)	(3)	(9)	(2)	(2)	(4)	**WALES**
- -	- -	- -	- -	- -	- -	*One year*
- -	- -	- -	- -	- -	- -	*Five years*
- -	- -	- -				

¹ Including those for whom no deprivation category was assigned. C - crude survival, R - relative survival (see Chapter 3).
Number of patients contributing to each analysis in parentheses.

Fig. 59.1   Relative survival up to ten years: trends by calendar period of diagnosis (1971-90) and NHS Region

CHILDREN

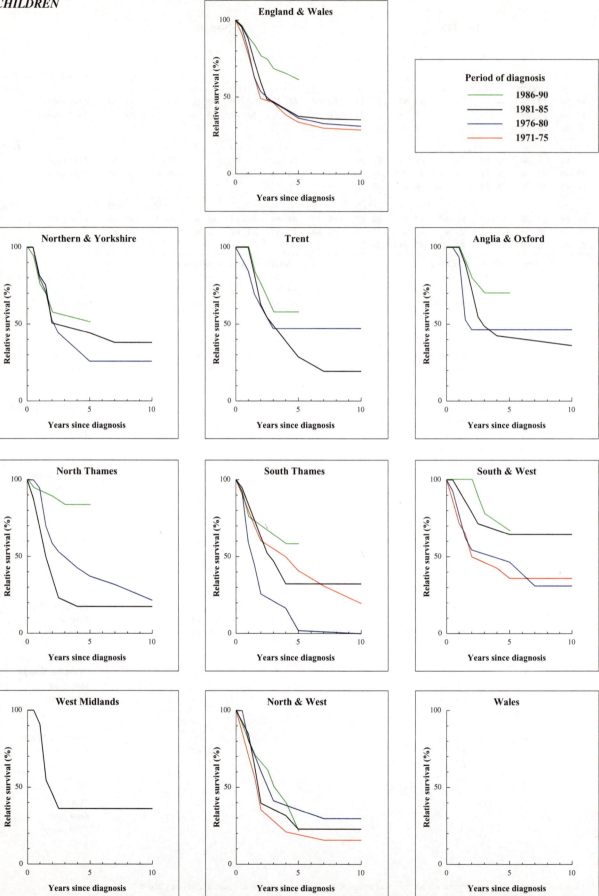

Fig. 59.2 Relative survival up to five years, by deprivation category: England and Wales, patients diagnosed 1986-90

CHILDREN

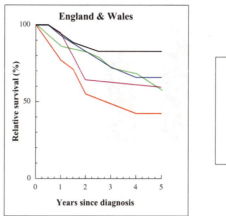

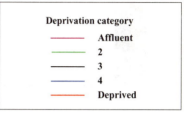

Fig. 59.3 Relative survival at five years, by deprivation category and period of diagnosis (1981-90): England and Wales

CHILDREN

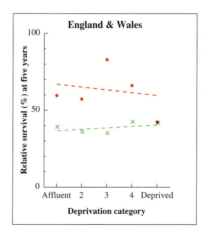

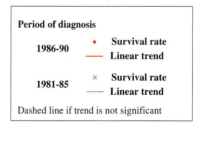

Fig. 59.4 Relative survival (%) by NHS Region and sex:
  one-year and five-year survival for children diagnosed 1986-90, with 95% confidence intervals

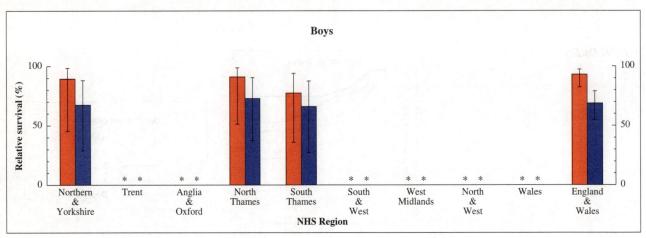

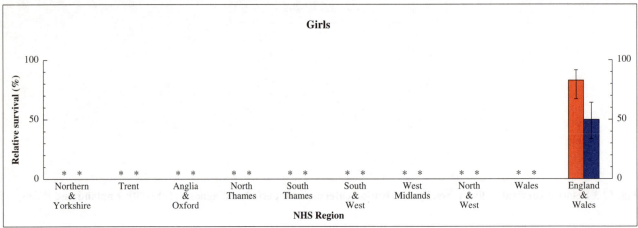

Chapter 60

SOFT TISSUE SARCOMA *Children*

Soft tissue sarcomas are a group of malignancies arising from cells in muscle, fat, fibrous tissue, blood vessels and nerve sheath. They account for some 6% of all malignant neoplasms in children, with about 80 new cases each year in England and Wales. The most common type is rhabdomyosarcoma, a tumour of embryonal striated muscle cells, followed by fibrosarcoma. Soft tissue sarcomas may arise anywhere but the head and neck region and the genitourinary tract are the most frequently affected. They occur most often in the first five years of life, accounting for about half of all cancers in children at this age. They are slightly more common in boys, with a sex ratio of 1.8 to 1. Incidence rates in England and Wales in 1990 were 10 per million in boys and 6 in girls. Mortality rates are difficult to interpret because histology is rarely reported at death certification and is not coded.

Treatment patterns have changed with the introduction of combination chemotherapy regimes. Surgery and early radiotherapy have given way to primary chemotherapy for tumour shrinkage, delayed resection and conservative radiotherapy if remission is not achieved[1].

The survival analyses reported here are based on 1,259 children who were diagnosed with a soft tissue sarcoma between 1971 and 1990 and followed up to the end of 1995, about 88% of those eligible (Table 60.1). Eight per cent of the children were excluded from analyses because their vital status was unknown at 6 July 1997, and a further 2% because their survival was zero or unknown. More than half (55%) the sarcomas had been assigned to soft tissues (Table 60A), principally those of the trunk (18% of all cases; detail not shown); leg (12%); head, face and neck (12%) and arm (6%). Most of the remainder had been assigned to specific organs in the genitourinary tract, the mouth and pharynx, and other sites.

Sixty per cent of the tumours were rhabdomyosarcoma (one-third of these were specified as embryonal) and 13% fibrosarcoma. Sarcomas of synovial tissue, nerve sheath and blood vessels each accounted for 3%: only one tumour, diagnosed in 1982, was classified as Kaposi sarcoma. About 11% were coded only as soft tissue sarcoma, not otherwise specified.

Survival trends

For children diagnosed with soft tissue sarcoma in England and Wales during 1986-90, relative survival[a] was 86% at one year and 66% at five years. These figures represent an increase of some 25% over the corresponding survival rates for children diagnosed in the early 1970s,

60% at one year and 41% at five years (Table 60.2). Survival improved progressively every five years (Figure 60.1). On average, survival rose by an average of 8-9% between successive five-year periods (see Table 4.4). The pace of improvement was similar for boys and girls, and survival rates were very similar.

Ten-year survival was 57% for children diagnosed during 1981-85, about 3% lower than five-year survival. The survival curves reflect this, and even for children diagnosed in the early 1970s there was only a small further decline in relative survival beyond the fifth anniversary of diagnosis (Figure 60.1).

Regional survival rates are based on 20-50 children in any five-year period, and are imprecise (see Chapter 3); regional differences for children diagnosed in 1986-90 are not significant (Figure 60.4). It is nevertheless clear that survival has increased substantially in most regions.

The results reported here may be compared with those from the National Registry of Childhood Tumours for Great Britain, which has reported actuarial survival rates for rhabdomyosarcoma, embryonal sarcoma and soft-tissue Ewing's tumour: together, these comprise 61% of the tumours analysed here. Five-year survival (period of diagnosis) for this group of tumours rose from 26% (1971-73)[2] to 56% for children diagnosed during 1983-85[3].

Deprivation

The survival rates for children in each category of deprivation in England and Wales are all based on fewer than 80 cases in a given five-year period, and they are imprecise. There was no consistent pattern in one-year or five-year survival between the various deprivation groups (Table 60.2). The survival curves for children diagnosed in England and Wales during 1986-90 overlap irregularly (Figure 60.2).

Estimates of the difference in one-year and five-year survival between the most affluent and the most deprived groups are based on about 300 children of known deprivation status in each five-year period. The fitted estimates for one-year and five-year survival are up to 8% higher among children in the most deprived group than for those in the most affluent group, but none is statistically significant (see Table 4.9). The pattern is shown graphically in Figure 60.3.

---

[a] See footnote in Chapter 50.

Table 60A Anatomic distribution of soft tissue sarcoma in children

ICD-9 code	Site description	1971-75 No.	1971-75 %	1976-80 No.	1976-80 %	1981-85 No.	1981-85 %	1986-90 No.	1986-90 %	Total No.	Total %
140-149	Lip, oral cavity, pharynx	29	9.7	24	7.9	32	9.7	18	5.6	**103**	**8.2**
150-159	Digestive organs, peritoneum	19	6.3	14	4.6	18	5.4	19	5.9	**70**	**5.6**
160-164	Respiratory tract, thorax	15	5.0	9	3.0	16	4.8	14	4.3	**54**	**4.3**
170	Bone, articular cartilage	3	1.0	8	2.6	8	2.4	8	2.5	**27**	**2.1**
171	Connective and soft tissue	164	54.7	174	57.2	176	53.2	176	54.3	**690**	**54.8**
173	Skin	4	1.3	4	1.3	10	3.0	15	4.6	**33**	**2.6**
174-189	Breast, genitourinary organs	33	11.0	43	14.1	34	10.3	41	12.7	**151**	**12.0**
190-199	Other and unspecified sites	33	11.0	28	9.2	37	11.2	33	10.2	**131**	**10.4**
	Total	**300**		**304**		**331**		**324**		**1,259**	

International comparison

Survival in the areas covered by the SEER programme cancer registries in the USA has been reported[4]. For children diagnosed during 1967-73, five-year relative survival from soft tissue sarcoma was 44% (34% for rhabdomyosarcoma alone), similar to the 41% five-year survival rate for children with soft tissue sarcoma diagnosed in England and Wales during 1971-75.

Among 364 children diagnosed with soft tissue sarcoma in the USA during 1973-81, one-year relative survival was 79% and five-year survival 58%: these are comparable with the survival rates for children diagnosed in England and Wales during 1981-85 (81% and 60%, respectively).

1.   Pinkerton R. Paediatric solid tumours. In: Horwich A (ed.) *Oncology: a multi-disciplinary textbook.* London: Chapman and Hall, 1995, pp417-440

2.   Stiller CA, Bunch KJ. Trends in survival for childhood cancer in Britain diagnosed 1971-85. Br J Cancer 1990; 62: 806-815

3.   Stiller CA. Population-based survival rates for childhood cancer in Britain, 1980-91. Br Med J 1994; 309: 1612-1616

4.   Young JL, Gloeckler Ries LA, Silverberg E, Horm JW, Miller RW. Cancer incidence, survival and mortality for children younger than age 15 years. Cancer 1986; 58: 598-602

Table 60.1 Ineligible and excluded records, and patients included in analyses, by calendar period of diagnosis

	Calendar period of diagnosis														
	1971-75			1976-80			1981-85			1986-90			Overall		
Total registered	375			368			401			395			1,539		
Ineligible	*Records*	*Patients*	*%*	*Records*	*Patients*	*%*	*Records*	*Patients*	*%*	*Records*	*Patients*	*%*	*Records*	*Patients*	*%*
Incomplete data[1]	3	3		7	7		13	13		6	6		29	29	
Resident outside England and Wales	18	17		10	10		14	13		2	1		44	41	
In situ neoplasm[2]	0	0		0	0		0	0		0	0		0	0	
Benign or uncertain behaviour[2]	35	32		12	10		0	0		0	0		47	42	
Metastatic[2]	0	0		0	0		0	0		0	0		0	0	
Otherwise ineligible[3]	-	-		-	-		-	-		-	-		-	-	
Lymphoma[4]	-	-		-	-		-	-		-	-		-	-	
Leukaemia or myeloma[4]	-	-		-	-		-	-		-	-		-	-	
		52			27			26			7			112	
Total eligible		323	100		341	100		375	100		388	100		1,427	100
Aged under 15 or 100 years or over at diagnosis[4]	-	-	-	-	-	-	-	-	-	-	-	-	-	-	-
Vital status unknown[5]	30	14	4.3	37	27	7.9	41	26	6.9	50	47	12.1	158	114	8.0
Duplicate registration[6]	0	0	0.0	0	0	0.0	0	0	0.0	0	0	0.0	0	0	0.0
Multiple primary[7]	1	1	0.3	1	1	0.3	9	9	2.4	5	3	0.8	16	14	1.0
Synchronous tumours[8]	4	1	0.3	2	2	0.6	4	3	0.8	7	3	0.8	17	9	0.6
Sex not known	0	0	0.0	0	0	0.0	0	0	0.0	0	0	0.0	0	0	0.0
Sex incompatible with site code[9]	-	-	-	-	-	-	-	-	-	-	-	-	-	-	-
Invalid dates[10]	1	0	0.0	1	0	0.0	0	0	0.0	0	0	0.0	2	0	0.0
Zero survival[11]	10	7	2.2	9	7	2.1	12	6	1.6	13	11	2.8	44	31	2.2
Total exclusions		23	7.1		37	10.9		44	11.7		64	16.5		168	11.8
Patients included in analysis		300	92.9		304	89.1		331	88.3		324	83.5		1,259	88.2

[1] Main data item(s) invalid or incompatible with one another: name, sex, date of birth, date of diagnosis and (if present) date of death, postcode, site, morphology and behaviour

[2] Tumour either *in situ* (behaviour code 2), benign (0), uncertain if benign or malignant (1) or metastatic (6 or 9)

[3] Not applicable for this malignancy

[4] Not applicable for this malignancy

[5] Tracing of vital status at National Health Service Central Register incomplete at 6 July 1997

[6] Same site code, sex, cancer registry and cancer registry number as an earlier registration

[7] Same site code and person number as an earlier registration(s): mostly confirmed multiple primary tumours at this site, some unresolved duplicate registrations

[8] Same site code, sex, date of birth and date of diagnosis as another registration(s): mostly synchronous or (in paired organs) bilateral tumours in same anatomic site in one individual, not previously linked: also some duplicate registrations

[9] Not applicable for this malignancy

[10] Impossible sequence of dates (birth, diagnosis and, if present, emigration or death); or date of death after 6 July 1997

[11] Date of diagnosis same as date of death: some are patients who did die on the day of diagnosis, but most were registered solely from a death certificate (death certificate only, or DCO), and their survival time is thus unknown

Table 60.2 Crude and relative survival (%) at one, five and ten years since diagnosis: NHS Region, deprivation category and calendar period of diagnosis

	1971-75						1976-80					
	Affluent C R	2 C R	3 C R	4 C R	Deprived C R	All[1] C R	Affluent C R	2 C R	3 C R	4 C R	Deprived C R	All[1] C R
ENGLAND & WALES												
Boys	*(28)*	*(28)*	*(22)*	*(39)*	*(23)*	*(176)*	*(45)*	*(28)*	*(33)*	*(30)*	*(39)*	*(185)*
One year	60 60	61 61	49 49	66 67	70 70	61 61	78 78	78 78	85 85	62 63	67 67	75 75
Five years	28 28	46 46	35 36	48 49	34 34	41 41	53 53	53 53	67 67	29 29	46 46	50 50
Ten years	28 28	43 43	35 36	46 46	30 30	39 39	53 53	53 53	61 61	29 29	46 46	48 49
Girls	*(20)*	*(19)*	*(21)*	*(24)*	*(14)*	*(124)*	*(27)*	*(20)*	*(20)*	*(24)*	*(22)*	*(119)*
One year	46 46	53 53	66 66	62 62	51 51	58 58	71 71	75 75	80 80	71 71	67 67	72 72
Five years	21 21	43 43	57 57	37 37	36 36	40 40	60 60	51 51	50 50	50 50	54 54	52 52
Ten years	21 21	43 43	52 52	33 33	36 36	37 37	56 56	45 46	45 45	46 46	54 54	48 48
Children	*(48)*	*(47)*	*(43)*	*(63)*	*(37)*	*(300)*	*(72)*	*(48)*	*(53)*	*(54)*	*(61)*	*(304)*
One year	54 54	58 58	57 57	65 65	62 63	60 60	75 75	77 77	83 83	66 67	67 67	74 74
Five years	25 25	45 45	46 46	44 44	35 35	41 41	56 56	52 52	61 61	38 38	49 49	51 51
Ten years	25 25	43 43	43 44	41 41	32 32	38 38	54 54	50 50	55 55	36 37	49 49	48 48

	Boys	Girls	Children				Boys	Girls	Children			
Northern & Yorkshire	*(18)*	*(12)*	*(30)*				*(23)*	*(15)*	*(38)*			
One year	50 50	52 52	51 51				73 73	80 80	76 76			
Five years	38 39	29 29	34 34				38 38	60 60	47 47			
Ten years	38 39	17 17	31 31				38 38	52 52	44 44			
Trent	*(14)*	*(22)*	*(36)*				*(19)*	*(13)*	*(32)*			
One year	72 72	68 68	70 70				68 69	63 63	66 66			
Five years	50 50	50 50	50 50				42 43	56 56	48 48			
Ten years	43 43	45 45	44 45				42 43	48 48	44 45			
Anglia & Oxford	*(17)*	*(10)*	*(27)*				*(22)*	*(18)*	*(40)*			
One year	47 47	43 43	45 45				68 68	83 83	75 75			
Five years	40 40	21 21	34 34				37 37	55 55	45 45			
Ten years	29 29	21 21	27 27				37 37	50 50	43 43			
North Thames	*(36)*	*(17)*	*(53)*				*(19)*	*(16)*	*(35)*			
One year	55 55	58 58	56 56				79 79	52 52	66 66			
Five years	33 33	35 35	34 34				58 58	32 32	46 46			
Ten years	33 33	35 35	34 34				52 52	25 25	40 40			
South Thames	*(30)*	*(16)*	*(46)*				*(16)*	*(11)*	*(27)*			
One year	52 52	68 68	57 57				87 87	72 72	81 81			
Five years	32 32	42 42	36 36				75 75	54 54	66 66			
Ten years	29 29	42 42	34 34				69 69	54 54	63 63			
South & West	*(14)*	*(17)*	*(31)*				*(31)*	*(15)*	*(46)*			
One year	57 58	52 52	55 55				70 71	74 74	71 72			
Five years	29 29	52 52	42 42				54 54	48 48	52 52			
Ten years	29 29	52 52	42 42				51 51	48 48	50 50			
West Midlands	*(15)*	*(19)*	*(34)*				*(26)*	*(8)*	*(34)*			
One year	80 80	75 75	77 77				73 73	- -	77 77			
Five years	60 60	43 43	50 50				42 42	- -	50 50			
Ten years	60 60	38 38	47 47				42 42	- -	50 50			
North & West	*(27)*	*(3)*	*(30)*				*(26)*	*(15)*	*(41)*			
One year	74 74	- -	70 70				85 85	80 80	83 83			
Five years	44 44	- -	43 43				61 61	53 53	58 58			
Ten years	44 44	- -	43 43				61 61	53 53	58 58			
WALES	*(5)*	*(8)*	*(13)*				*(3)*	*(8)*	*(11)*			
One year	- -	- -	54 54				- -	- -	55 55			
Five years	- -	- -	54 54				- -	- -	37 37			
Ten years	- -	- -	54 54				- -	- -	37 37			

[1] Including those for whom no deprivation category was assigned. C - crude survival, R - relative survival (see Chapter 3). Number of patients contributing to each analysis in parentheses.

	1981-85							1986-90						
	Affluent C R	2 C R	3 C R	4 C R	*Deprived* C R	*All*[1] C R	*Affluent* C R	2 C R	3 C R	4 C R	*Deprived* C R	*All*[1] C R	**ENGLAND & WALES**
	(38)	(40)	(40)	(34)	(27)	(182)	(33)	(37)	(35)	(44)	(26)	(179)	**Boys**
One year	92 92	70 70	85 85	73 74	89 90	81 81	85 85	81 81	92 92	93 93	77 77	87 87	*One year*
Five years	66 66	50 50	67 67	47 48	81 82	60 61	70 70	62 62	68 68	73 73	65 65	68 68	*Five years*
	60 60	50 50	65 65	47 48	81 82	59 59							
	(31)	(27)	(31)	(29)	(28)	(149)	(35)	(27)	(42)	(17)	(23)	(145)	**Girls**
	72 72	78 78	90 90	76 76	89 89	81 81	86 86	78 78	76 76	94 94	96 96	84 84	*One year*
	62 62	59 59	58 58	58 58	53 53	58 58	60 60	56 56	64 64	65 65	74 74	63 64	*Five years*
	59 59	56 56	51 51	54 54	53 53	55 55							
	(69)	(67)	(71)	(63)	(55)	(331)	(68)	(64)	(77)	(61)	(49)	(324)	**Children**
	83 83	73 73	87 87	74 75	89 90	81 81	85 85	80 80	83 83	94 94	86 86	85 86	*One year*
	64 64	54 54	63 63	52 52	67 68	59 60	65 65	59 60	66 66	71 71	69 69	66 66	*Five years*
	59 60	52 52	59 59	51 51	67 68	57 57							

	Boys	Girls	Children		Boys	Girls	Children	
	(24)	(11)	(35)		(31)	(17)	(48)	**Northern & Yorkshire**
	75 75	72 72	74 74		84 84	76 76	81 81	*One year*
	63 63	64 64	63 63		71 71	59 59	67 67	*Five years*
	63 63	64 64	63 63					
	(18)	(19)	(37)		(26)	(15)	(41)	**Trent**
	94 94	79 79	86 87		74 74	80 80	76 76	*One year*
	61 61	58 58	59 59		62 62	67 67	64 64	*Five years*
	55 55	41 42	48 48					
	(31)	(15)	(46)		(21)	(13)	(34)	**Anglia & Oxford**
	80 80	70 70	77 77		76 76	69 69	74 74	*One year*
	48 48	41 41	46 46		52 52	54 54	53 53	*Five years*
	45 45	41 41	44 44					
	(20)	(11)	(31)		(25)	(22)	(47)	**North Thames**
	95 95	91 91	94 94		92 92	95 95	94 94	*One year*
	65 65	55 55	61 61		68 68	67 68	68 68	*Five years*
	65 65	55 55	61 61					
	(23)	(28)	(51)		(24)	(18)	(42)	**South Thames**
	74 74	89 89	82 82		91 91	77 78	86 86	*One year*
	57 57	68 68	63 63		70 70	50 50	62 62	*Five years*
	52 52	68 68	61 61					
	(22)	(20)	(42)		(18)	(22)	(40)	**South & West**
	64 65	80 80	72 72		100 100	91 91	95 95	*One year*
	55 55	65 65	60 60		72 72	77 77	75 75	*Five years*
	55 55	65 65	60 60					
	(13)	(16)	(29)		(11)	(9)	(20)	**West Midlands**
	92 92	94 94	93 93		91 91	89 89	90 90	*One year*
	77 77	74 74	76 76		91 91	68 68	80 80	*Five years*
	77 77	68 68	72 72					
	(18)	(18)	(36)		(16)	(17)	(33)	**North & West**
	83 83	72 72	78 78		87 88	88 88	88 88	*One year*
	72 72	55 55	64 64		69 69	65 65	67 67	*Five years*
	72 72	50 50	61 61					
	(13)	(11)	(24)		(7)	(12)	(19)	**WALES**
	77 77	69 69	74 74		- -	84 84	90 90	*One year*
	61 61	25 25	45 45		- -	58 58	63 63	*Five years*
	61 61	25 25	45 45					

[1] Including those for whom no deprivation category was assigned. C - crude survival, R - relative survival (see Chapter 3). Number of patients contributing to each analysis in parentheses.

Fig. 60.1   Relative survival up to ten years: trends by calendar period of diagnosis (1971-90) and NHS Region

CHILDREN

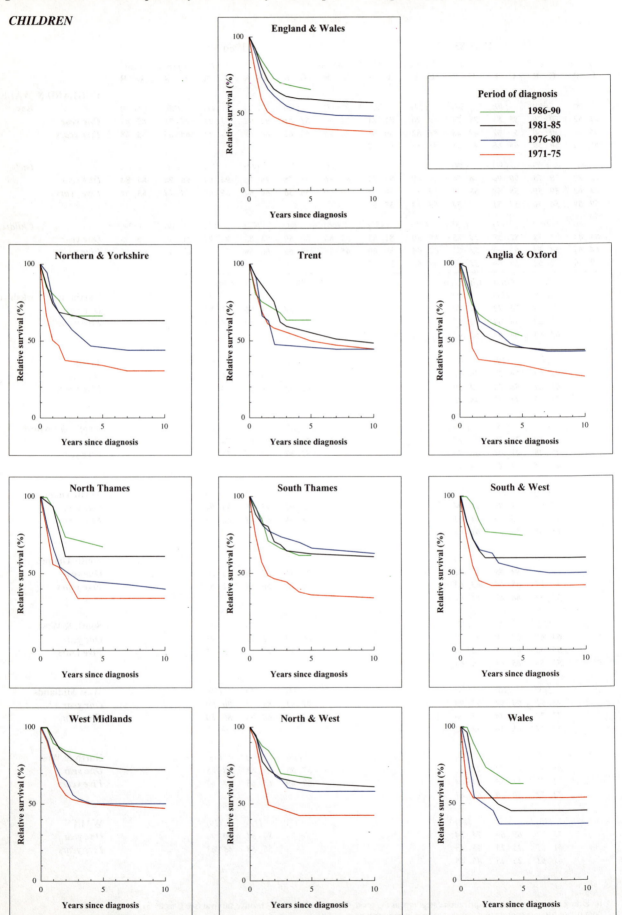

Fig. 60.2 Relative survival up to five years, by deprivation category: England and Wales, patients diagnosed 1986-90

CHILDREN

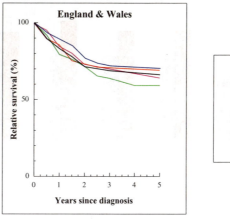

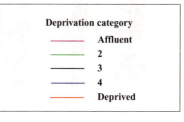

Fig. 60.3 Relative survival at five years, by deprivation category and period of diagnosis (1981-90): England and Wales

CHILDREN

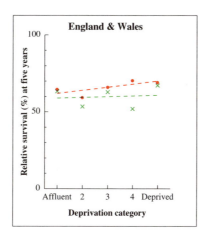

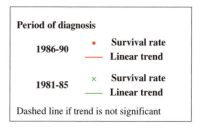

Fig. 60.4 Relative survival (%) by NHS Region and sex:
one-year and five-year survival for children diagnosed 1986-90, with 95% confidence intervals

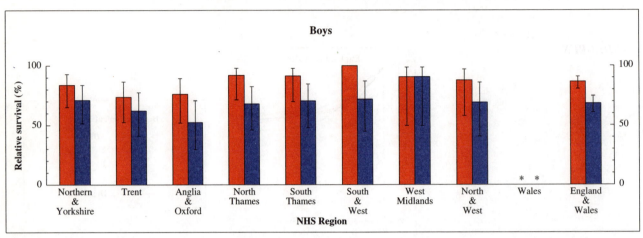

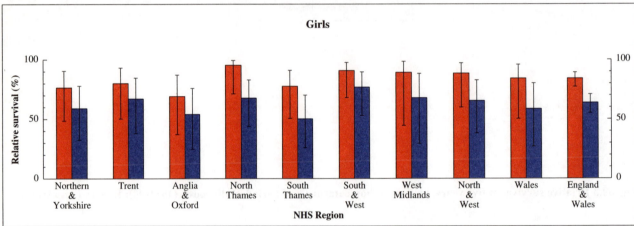

Chapter 61

GERM CELL AND GONADAL NEOPLASMS *Children*

Germ cell neoplasms in children include embryonic tumours of the ovary and testis and germ cell tumours arising in the brain, pelvis and other sites. They are rare, and only about 30 children are diagnosed in England and Wales each year. Overall incidence is similar in boys and girls at 2.5-3.0 cases per million, but the age of onset differs between the sexes. In boys, the highest incidence of malignant germ cell tumours of the testis is in the first year of life, and most occur before the age of five. Incidence is increasing: the rate was 14.2 per million in 1988-91, compared with 4.4 in 1971-73. In girls under 15, the highest incidence of malignant ovarian tumours is in the age group 10-14. Incidence is also increasing: the rate was 5.0 per million in 1988-91, compared with 2.1 in 1971-73. Benign germ cell tumours are relatively common, but this chapter is restricted to malignant germ cell and gonadal tumours.

Testicular germ cell tumours are more common in boys with testicular maldescent. Germ cell tumours are associated with some congenital malformations and intersex states, and some are familial. Increasing incidence suggests an environmental influence on risk[1].

The survival analyses reported here are based on 469 children who were diagnosed with a germ cell tumour in England and Wales between 1971 and 1990 and followed up to the end of 1995, about 92% of those eligible (Table 61.1). Almost 5% of eligible children were excluded because their vital status was not known at 6 July 1997, and a further 3% because their survival was zero or unknown. More than 70% of the tumours were gonadal: 39% arose in the testis (representing 76% of the 241 tumours in boys) and 32% arose in the ovary (67% of the 228 tumours in girls) (Table 61A). A further 5% each arose in the brain, in the pineal and in bone or connective tissue, and 2% were retroperitoneal. For the great majority of tumours (89%), the morphology had been coded as germ cell neoplasm: malignant teratoma (32%), endodermal sinus tumours (26%) and dysgerminoma (14%) were the most common types.

Survival trends

Relative survival[a] for children diagnosed with a gonadal or germ cell neoplasm in England and Wales during 1986-90 was 95% at one year and 83% at five years. Survival has improved by around 35%: children diagnosed during 1971-75 had one-year survival of 60% and five-year survival of 47% (Table 61.2). The largest improvements occurred during the 1970s (Figure 61.1).

Survival for girls was lower during the 1970s, but it has improved more than for boys, and survival was broadly similar for boys and girls diagnosed during the 1980s. The average gain in five-year survival was 10-13% every five years (see Table 4.4).

Ten-year survival for children diagnosed during 1981-85 and followed up to the end of 1995 was 74% for boys and 79% for girls, in each case no lower than the five-year survival rate. The plateaux in the relative survival curves for children diagnosed in the 1980s reflect this in all regions: after three to five years, there is almost no further excess mortality (Figure 61.1).

Regional survival estimates are based on fewer than 25 children in any five-year period, and they are all imprecise (see Chapter 3). In some regions, there were too few cases for analysis in certain time periods, and no analyses were done for Wales (Table 61.2, Figure 61.4). Five-year survival in South & West was an estimated 76% for children diagnosed during 1971-75, more than double the national average, but has changed very little. Survival has nevertheless risen substantially in most regions (Figure 61.1).

These results bear some comparison with the trends in actuarial survival reported by the National Registry of Childhood Tumours for Great Britain, although survival rates were presented separately for gonadal tumours and for germ cell tumours in other sites[2,3]. Five-year survival from testicular tumours in boys rose from 59% for those diagnosed during 1971-73 to 97% for those diagnosed during 1986-88. The corresponding increase in survival from ovarian tumours in girls was from 43% to 85%. Survival for germ cell tumours arising in brain and other organs was much lower, and did not improve during the 1980s.

Deprivation

The survival rates for children in each category of deprivation in England and Wales are based on fewer than 30 cases in a given five-year period, and they are imprecise. There was no consistent pattern in one-year or five-year survival between deprivation groups (Table 61.2). The survival curves for children diagnosed during 1986-90 overlap (Figure 61.2). Estimates of the difference in one-year and five-year survival between the most affluent and the most deprived groups are based on about 120 children of known deprivation status in each five-year

---

[a] See footnote in Chapter 50.

Table 61A Anatomic distribution of germ cell and gonadal neoplasms in children

ICD-9 code	Site description	1971-75 No.	%	1976-80 No.	%	1981-85 No.	%	1986-90 No.	%	Total No.	%
150-159	Digestive organs, retroperitoneum	1	0.9	5	4.2	1	0.8	4	3.2	11	2.3
160-164	Respiratory tract, mediastinum	7	6.5	1	0.8	6	5.1	4	3.2	18	3.8
170-171	Bone, connective and soft tissue	10	9.3	5	4.2	1	0.8	5	4.0	21	4.5
183	Ovary	35	32.4	39	32.8	42	35.6	35	28.2	151	32.2
186	Testis	43	39.8	47	39.5	49	41.5	44	35.5	183	39.0
191	Brain	3	2.8	10	8.4	5	4.2	7	5.6	25	5.3
194.3	Pineal	2	1.9	6	5.0	5	4.2	9	7.3	22	4.7
	Other and unspecified sites	7	6.5	6	5.0	9	7.6	16	12.9	38	8.1
	Total	**108**		**119**		**118**		**124**		**469**	

period: the differences are not significant (see Table 4.9). The pattern is shown graphically in Figure 61.3.

International comparison

Survival from gonadal and germ cell tumours has been reported for the areas covered by the SEER programme cancer registries in the USA[4]. Among 101 children diagnosed during 1973-81, one-year relative survival was 95% and five-year survival 83%, the same as for children diagnosed about ten years later in England and Wales (1986-90).

1.   Mann JR. Germ cell tumours of childhood. In: Peckham M, Pinedo HM, Veronesi U, (eds.) *Oxford textbook of oncology*. Oxford: Oxford Medical Publications, 1995

2.   Stiller CA, Bunch KJ. Trends in survival for childhood cancer in Britain diagnosed 1971-85. Br J Cancer 1990; 62: 806-815

3.   Stiller CA. Population-based survival rates for childhood cancer in Britain, 1980-91. Br Med J 1994; 309: 1612-1616

4.   Young JL, Gloeckler Ries LA, Silverberg E, Horm JW, Miller RW. Cancer incidence, survival and mortality for children younger than age 15 years. Cancer 1986; 58: 598-602

Table 61.1 Ineligible and excluded records, and patients included in analyses, by calendar period of diagnosis

	Calendar period of diagnosis														
	1971-75			**1976-80**			**1981-85**			**1986-90**			**Overall**		
Total registered	**150**			**150**			**151**			**160**			**611**		
Ineligible	*Records*	*Patients*	*%*	*Records*	*Patients*	*%*	*Records*	*Patients*	*%*	*Records*	*Patients*	*%*	*Records*	*Patients*	*%*
Incomplete data[1]	4	4		5	5		26	26		13	13		48	**48**	
Resident outside England and Wales	4	4		4	4		2	1		1	1		11	**10**	
In situ neoplasm[2]	0	0		0	0		0	0		0	0		0	**0**	
Benign or uncertain behaviour[2]	28	27		14	14		0	0		1	1		43	**42**	
Metastatic[2]	0	0		0	0		0	0		0	0		0	**0**	
Otherwise ineligible[3]	-	-		-	-		-	-		-	-		-	**-**	
Lymphoma[4]	-	-		-	-		-	-		-	-		-	**-**	
Leukaemia or myeloma[4]	-	-		-	-		-	-		-	-		-	**-**	
		35			**23**			**27**			**15**			**100**	
Total eligible		**115**	**100**		**127**	**100**		**124**	**100**		**145**	**100**		**511**	**100**
Aged under 15 or 100 years or over at diagnosis[4]	-	-	-	-	-	-	-	-	-	-	-	-	-	**-**	**-**
Vital status unknown[5]	8	4	3.5	10	5	3.9	10	4	3.2	13	12	8.3	41	**25**	4.9
Duplicate registration[6]	0	0	0.0	0	0	0.0	0	0	0.0	0	0	0.0	0	**0**	0.0
Multiple primary[7]	0	0	0.0	0	0	0.0	0	0	0.0	0	0	0.0	0	**0**	0.0
Synchronous tumours[8]	1	0	0.0	1	1	0.8	1	1	0.8	2	1	0.7	5	**3**	0.6
Sex not known	0	0	0.0	0	0	0.0	0	0	0.0	0	0	0.0	0	**0**	0.0
Sex incompatible with site code[9]	-	-	-	-	-	-	-	-	-	-	-	-	-	**-**	**-**
Invalid dates[10]	0	0	0.0	0	0	0.0	0	0	0.0	0	0	0.0	0	**0**	0.0
Zero survival[11]	5	3	2.6	4	2	1.6	3	1	0.8	12	8	5.5	24	**14**	2.7
Total exclusions		**7**	**6.1**		**8**	**6.3**		**6**	**4.8**		**21**	**14.5**		**42**	**8.2**
Patients included in analysis		**108**	**93.9**		**119**	**93.7**		**118**	**95.2**		**124**	**85.5**		**469**	**91.8**

[1] Main data item(s) invalid or incompatible with one another: name, sex, date of birth, date of diagnosis and (if present) date of death, postcode, site, morphology and behaviour

[2] Tumour either *in situ* (behaviour code 2), benign (0), uncertain if benign or malignant (1) or metastatic (6 or 9)

[3] Not applicable for this malignancy

[4] Not applicable for this malignancy

[5] Tracing of vital status at National Health Service Central Register incomplete at 6 July 1997

[6] Same site code, sex, cancer registry and cancer registry number as an earlier registration

[7] Same site code and person number as an earlier registration(s): mostly confirmed multiple primary tumours at this site, some unresolved duplicate registrations

[8] Same site code, sex, date of birth and date of diagnosis as another registration(s): mostly synchronous or (in paired organs) bilateral tumours in same anatomic site in one individual, not previously linked: also some duplicate registrations

[9] Not applicable for this malignancy

[10] Impossible sequence of dates (birth, diagnosis and, if present, emigration or death); or date of death after 6 July 1997

[11] Date of diagnosis same as date of death: some are patients who did die on the day of diagnosis, but most were registered solely from a death certificate (death certificate only, or DCO), and their survival time is thus unknown

Table 61.2 Crude and relative survival (%) at one, five and ten years since diagnosis: NHS Region, deprivation category and calendar period of diagnosis

	1971-75												1976-80											
	Affluent		2		3		4		Deprived		All¹		Affluent		2		3		4		Deprived		All¹	
	C	R	C	R	C	R	C	R	C	R	C	R	C	R	C	R	C	R	C	R	C	R	C	R
ENGLAND & WALES																								
Boys	(9)		(4)		(11)		(12)		(8)		(56)		(10)		(12)		(18)		(7)		(7)		(55)	
One year	-	-	-	-	91	91	40	40	-	-	68	68	100	100	76	76	83	83	-	-	-	-	84	84
Five years	-	-	-	-	63	63	31	31	-	-	53	53	90	90	68	68	83	83	-	-	-	-	73	73
Ten years	-	-	-	-	63	63	31	31	-	-	53	53	90	90	68	68	83	83	-	-	-	-	73	73
Girls	(5)		(8)		(11)		(11)		(8)		(52)		(16)		(12)		(8)		(11)		(14)		(64)	
One year	-	-	-	-	55	55	42	42	-	-	51	51	63	63	64	64	-	-	61	61	71	71	68	68
Five years	-	-	-	-	46	46	25	25	-	-	40	40	50	50	33	33	-	-	61	61	56	56	56	56
Ten years	-	-	-	-	46	46	25	25	-	-	40	40	50	50	33	33	-	-	61	61	56	56	54	54
Children	(14)		(12)		(22)		(23)		(16)		(108)		(26)		(24)		(26)		(18)		(21)		(119)	
One year	64	64	83	83	73	73	41	41	51	51	60	60	77	77	71	71	84	85	71	71	71	71	75	76
Five years	56	56	58	58	55	55	28	28	44	45	47	47	65	66	50	50	84	85	66	66	46	47	64	64
Ten years	56	56	58	58	55	55	28	28	44	45	47	47	65	66	50	50	81	81	66	66	46	47	63	63

	1971-75						1976-80					
	Boys		Girls		Children		Boys		Girls		Children	
	C	R	C	R	C	R	C	R	C	R	C	R
Northern & Yorkshire	(5)		(6)		(11)		(12)		(7)		(19)	
One year	-	-	-	-	74	74	100	100	-	-	89	89
Five years	-	-	-	-	55	55	75	75	-	-	68	68
Ten years	-	-	-	-	55	55	75	75	-	-	63	63
Trent	(3)		(6)		(9)		(2)		(5)		(7)	
One year	-	-	-	-	-	-	-	-	-	-	-	-
Five years	-	-	-	-	-	-	-	-	-	-	-	-
Ten years	-	-	-	-	-	-	-	-	-	-	-	-
Anglia & Oxford	(2)		(7)		(9)		(7)		(6)		(13)	
One year	-	-	-	-	-	-	-	-	-	-	61	61
Five years	-	-	-	-	-	-	-	-	-	-	61	61
Ten years	-	-	-	-	-	-	-	-	-	-	61	61
North Thames	(6)		(2)		(8)		(1)		(5)		(6)	
One year	-	-	-	-	-	-	-	-	-	-	-	-
Five years	-	-	-	-	-	-	-	-	-	-	-	-
Ten years	-	-	-	-	-	-	-	-	-	-	-	-
South Thames	(10)		(8)		(18)		(5)		(11)		(16)	
One year	49	49	-	-	49	49	-	-	54	54	62	62
Five years	39	39	-	-	33	33	-	-	45	45	56	56
Ten years	39	39	-	-	33	33	-	-	45	45	56	56
South & West	(9)		(8)		(17)		(7)		(4)		(11)	
One year	-	-	-	-	82	82	-	-	-	-	82	82
Five years	-	-	-	-	76	76	-	-	-	-	82	82
Ten years	-	-	-	-	76	76	-	-	-	-	82	82
West Midlands	(8)		(2)		(10)		(9)		(9)		(18)	
One year	-	-	-	-	52	52	-	-	-	-	83	83
Five years	-	-	-	-	52	52	-	-	-	-	49	49
Ten years	-	-	-	-	52	52	-	-	-	-	49	49
North & West	(11)		(9)		(20)		(10)		(15)		(25)	
One year	43	43	-	-	47	47	79	79	81	81	80	80
Five years	35	35	-	-	38	38	69	69	74	74	72	72
Ten years	35	35	-	-	38	38	69	69	74	74	72	72
WALES	(2)		(4)		(6)		(2)		(2)		(4)	
One year	-	-	-	-	-	-	-	-	-	-	-	-
Five years	-	-	-	-	-	-	-	-	-	-	-	-
Ten years	-	-	-	-	-	-	-	-	-	-	-	-

¹ Including those for whom no deprivation category was assigned. C - crude survival, R - relative survival (see Chapter 3). Number of patients contributing to each analysis in parentheses.

ENGLAND & WALES

	1981-85						1986-90					
	Affluent	2	3	4	Deprived	All[1]	Affluent	2	3	4	Deprived	All[1]
	C R	C R	C R	C R	C R	C R	C R	C R	C R	C R	C R	C R
Boys	(25)	(14)	(12)	(9)	(6)	(66)	(13)	(15)	(11)	(7)	(18)	(64)
One year	84 84	79 80	100 100	- -	- -	85 85	93 93	100 100	91 91	- -	100 100	94 94
Five years	72 72	72 72	100 100	- -	- -	74 74	77 77	100 100	82 82	- -	100 100	88 88
Five years	72 72	72 72	100 100	- -	- -	74 74						
Girls	(11)	(4)	(10)	(10)	(17)	(52)	(10)	(14)	(10)	(11)	(12)	(60)
One year	73 73	- -	100 100	80 80	76 77	81 81	90 90	100 100	90 90	91 91	100 100	95 95
Five years	73 73	- -	100 100	70 70	76 77	79 79	79 79	72 72	70 70	81 82	84 84	78 78
Five years	73 73	- -	100 100	70 70	76 77	79 79						
Children	(36)	(18)	(22)	(19)	(23)	(118)	(23)	(29)	(21)	(18)	(30)	(124)
One year	80 80	78 78	100 100	79 79	78 78	83 83	91 91	100 100	91 91	83 83	100 100	94 95
Five years	72 72	72 73	100 100	63 63	74 74	76 76	78 78	86 86	76 76	72 72	93 93	83 83
Five years	72 72	72 73	100 100	63 63	74 74	76 76						

	1981-85			1986-90			
	Boys	Girls	Children	Boys	Girls	Children	
	(5)	(8)	(13)	(8)	(13)	(21)	**Northern & Yorkshire**
	- -	- -	85 85	100 100	92 92	95 95	One year
	- -	- -	77 77	88 88	77 77	81 81	Five years
	- -	- -	77 77				
	(9)	(3)	(12)	(8)	(5)	(13)	**Trent**
	- -	- -	83 83	100 100	- -	100 100	One year
	- -	- -	75 75	88 88	- -	92 92	Five years
	- -	- -	75 75				
	(5)	(7)	(12)	(8)	(8)	(16)	**Anglia & Oxford**
	- -	- -	65 65	87 87	100 100	94 94	One year
	- -	- -	65 65	87 87	100 100	94 94	Five years
	- -	- -	65 65				
	(7)	(4)	(11)	(6)	(10)	(16)	**North Thames**
	- -	- -	82 83	- -	90 90	94 94	One year
	- -	- -	82 83	- -	81 81	88 88	Five years
	- -	- -	82 83				
	(8)	(7)	(15)	(9)	(3)	(12)	**South Thames**
	- -	- -	93 93	100 100	- -	100 100	One year
	- -	- -	87 87	89 89	- -	83 83	Five years
	- -	- -	87 87				
	(12)	(4)	(16)	(6)	(9)	(15)	**South & West**
	92 92	- -	94 94	- -	100 100	93 93	One year
	75 75	- -	81 81	- -	78 78	80 80	Five years
	75 75	- -	81 81				
	(8)	(11)	(19)	(6)	(2)	(8)	**West Midlands**
	- -	71 71	78 78	- -	- -	88 88	One year
	- -	71 71	67 67	- -	- -	75 75	Five years
	- -	71 71	67 67				
	(11)	(6)	(17)	(13)	(7)	(20)	**North & West**
	74 75	- -	77 77	92 92	- -	90 90	One year
	64 65	- -	71 71	92 92	- -	80 80	Five years
	64 65	- -	71 71				
	(1)	(2)	(3)	(0)	(3)	(3)	**WALES**
	- -	- -	- -	- -	- -	- -	One year
	- -	- -	- -	- -	- -	- -	Five years
	- -	- -	- -				

[1] Including those for whom no deprivation category was assigned. C - crude survival, R - relative survival (see Chapter 3).
Number of patients contributing to each analysis in parentheses.

Fig. 61.1   Relative survival up to ten years: trends by calendar period of diagnosis (1971-90) and NHS Region

CHILDREN

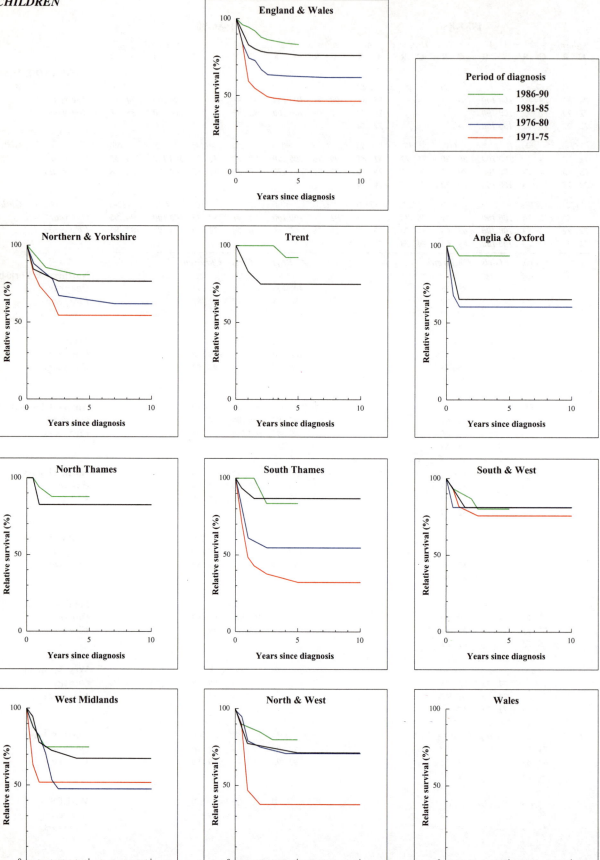

Fig. 61.2 Relative survival up to five years, by deprivation category: England and Wales, patients diagnosed 1986-90

CHILDREN

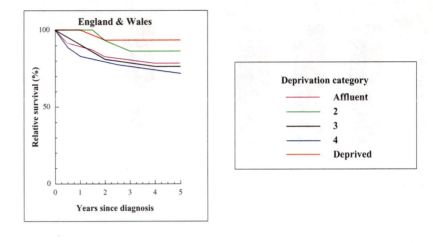

Fig. 61.3 Relative survival at five years, by deprivation category and period of diagnosis (1981-90): England and Wales

CHILDREN

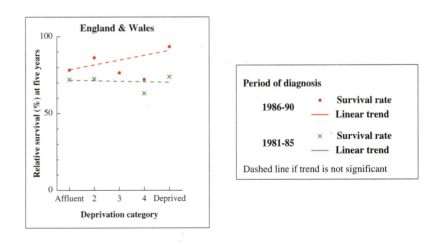

Fig. 61.4 Relative survival (%) by NHS Region and sex:
one-year and five-year survival for children diagnosed 1986-90, with 95% confidence intervals

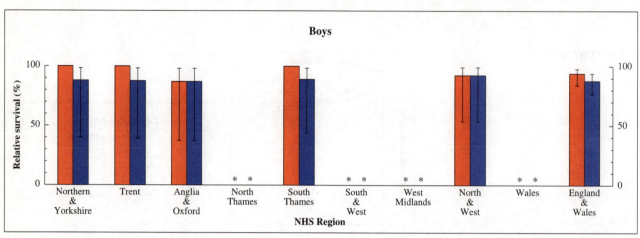

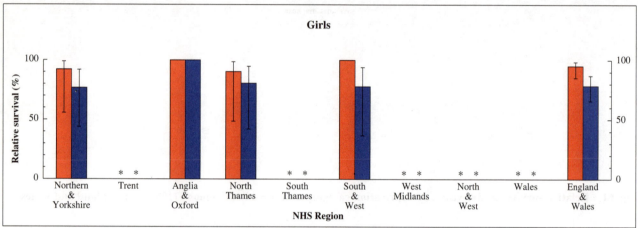

Chapter 62

SPINAL CORD, ADRENAL, PITUITARY

In this chapter, survival results for tumours of the spinal cord, adrenal and pituitary gland will be briefly outlined. Survival rates are presented for adults in England and Wales only, by calendar period of diagnosis, age, sex and deprivation category. Survival rates are not presented by NHS Region.

Primary tumours of the spinal cord are rare: metastatic deposits from tumours elsewhere, either to the vertebrae or to the cord itself, are more common causes of malignant spinal compression.

Malignant tumours of the adrenal gland are also rare, and it is sometimes the site for metastases from cancers in other organs. Adrenal cortical carcinoma, neuroblastoma and malignant phaeochromocytoma are the most common types.

Benign adenomas of the pituitary are more common than malignant tumours. About 500 are registered in England and Wales each year, as opposed to some 200 malignant tumours. By virtue of their location in the skull, pituitary adenomas present a more serious clinical problem than benign tumours in most other organs, but they are not included here.

Spinal cord

The analyses relate to 715 patients diagnosed with a primary malignancy of the spinal cord in England and Wales during 1971-90 and followed up to the end of 1995. After the exclusion of children (15% of all patients), these patients represent 86% of eligible adults (Table 62.1). Two-thirds (67%) were specified as gliomas, including ependymoma (24%) and astrocytoma (22%); 7% were neurofibrosarcoma or malignant neurilemmoma, and 2% chordoma. Most of the remainder were poorly specified.

Relative survival for men diagnosed during 1986-90 was 85% at one year and 68% at five years; for women, the corresponding survival rates were 74% and 58%. Survival increased by about 10% in both sexes between the early and late 1970s (Table 62.2, Figure 62.1a). Five-year survival increased by an average of 6% every five years in England and Wales, after adjustment for differences in the age distribution of patients over time and between men and women (see Table 4.4), and similar gains occurred in most regions (data not shown).

For patients aged less than 50 at diagnosis during 1986-90, relative survival at five years was in the range 70-80%; survival was lower for patients aged 50-69 (Figure 62.5a).

There were too few patients to estimate age-specific survival for older patients (see Table 4.7).

Survival rates varied between deprivation groups, but there was no systematic pattern (Table 62.2). The survival curves for adults diagnosed during 1986-90 in each deprivation group overlap irregularly (Figure 62.2a). The survival gradients were estimated among about 200 adults of known deprivation status who were diagnosed during each five-year period 1981-85 and 1986-90: the gradient was positive in the early 1980s and negative for 1986-90, but neither gradient was statistically significant (see Table 4.9). This pattern is reflected in Figure 62.3a.

Adrenal

The analyses include over 1,000 patients diagnosed with a primary malignant tumour of the adrenal in England and Wales during the period 1971-90 and followed up to the end of 1995. After the exclusion of children, who accounted for more than a quarter of all adrenal malignancies, these patients represent 70% of eligible adults (Table 62.1). Some 4% of adults were excluded because their vital status was unknown at 6 July 1997, but 18% because their survival was zero or unknown: this proportion is unusually high, particularly for the early 1970s (20%). A fifth (22%) of the tumours were specified as phaeochromocytoma, 15% adrenocortical carcinoma and only 1% neuroblastoma, but the morphology was poorly specified for half (51%) of the tumours.

Relative survival for adults diagnosed during 1986-90 was 53% at one year and 35% at five years. Survival increased sharply during the 1980s: one-year survival was more than 10% higher for patients diagnosed during 1986-90 than for those diagnosed during 1981-85, an average of only five years earlier (Table 62.2, Figure 62.1b). One-year survival rose by 20%, an average gain of 6-9% every five years, after adjustment for age (see Table 4.4). Five-year survival also rose by 15-20%: similar gains occurred in most regions (data not shown). Women had a 10% survival advantage during the early 1970s, but survival has risen more in men, and among adults diagnosed during 1986-90, survival rates were similar in both sexes.

Five-year survival for adults diagnosed during 1986-90 declined with age, from over 50% for those aged 15-39 at diagnosis to 20% or less for those aged 70 or over (Figure 62.5b; see Table 4.7).

Survival varied widely between deprivation groups, but national survival rates in each group are based on fewer than 30 patients in any five-year period, and they are imprecise: there was no systematic pattern (Table 62.2). The survival curves for adults diagnosed during 1986-90 in each deprivation group overlap irregularly (Figure 62.2b). The survival gradients were estimated among about 300 adults of known deprivation status who were diagnosed during each five-year period 1981-85 and 1986-90. One-year survival was 26% lower in the most deprived group (20% vs. 46%) for the period 1981-85, a large and significant deficit, and five-year survival was 11% lower (see Table 4.9). The differences were smaller for patients diagnosed during 1986-90, and most of the increase in survival in England and Wales between these two periods occurred among the less affluent groups. The deprivation gradients in five-year survival are not statistically significant (Figure 62.3b).

Pituitary

The analyses reported here are based on 800 adults diagnosed with a malignant tumour of the pituitary gland in England and Wales during the period 1971-90 and followed up to the end of 1995, about 77% of those eligible. Pituitary tumours classified as benign or of uncertain behaviour had largely been removed from the data for the 1980s, but 250 such tumours registered during the 1970s were also removed as ineligible (Table 62.1). Some 10% of adults were excluded because their vital status was unknown at 6 July 1997, and a further 11% because their survival was zero or unknown. For two-thirds of the tumours, the morphology was either poorly specified (46%) or absent (20%). Almost all the tumours for which there was no information on the morphology were diagnosed during the 1970s, when radiotherapy would commonly have been used without surgery or biopsy. All of them had been assigned a behaviour code for primary malignant tumour, but some may have been benign. A specific morphology was available for only 34% of the tumours, including chromophobe carcinoma (10%), acidophil carcinoma (6%) and adenocarcinoma without further specification (14%).

Relative survival for adults diagnosed with a malignant pituitary tumour in England and Wales during 1986-90 was 85% at one year and 76% at five years. Survival was already 79% at one year and 72% at five years for patients diagnosed during the period 1971-75 (Table 62.2). Five-year survival improved to 81% for patients diagnosed during 1981-85, but fell slightly to 76% for those diagnosed during 1986-90 (Figure 62.1c). This is deceptive: after standardisation for differences in age between patients diagnosed in each five-year period and between men and women, five-year survival rates were essentially unchanged for patients diagnosed during the 1980s (see Table 4.4). Most of the overall increase in survival occurred during the 1970s. Survival was about 10% higher for men diagnosed during the early 1970s, but it increased by 14% for women and 2-6% for men. For adults diagnosed during 1986-90, there was no difference in survival between men and women.

Excess mortality over that of the general population continues beyond the first year after diagnosis, but relative survival declines fairly slowly with time since diagnosis. Ten-year survival for patients diagnosed during 1981-85 and followed up to the end of 1995 was 77%, only marginally below the 81% survival rate at five years (Table 62.2, Figure 62.1c).

Survival falls sharply with age at diagnosis (Figure 62.5c). For adults diagnosed during 1986-90, relative survival at five years was 85-95% for adults under 50 at diagnosis, 65-75% for those aged 50-69, and 40% or less for older patients (see Table 4.7).

National survival rates for any five-year period are generally based on fewer than 25 patients in each deprivation category, and they are imprecise. Variation in survival between deprivation groups was wide (Table 62.2). The survival curves for adults diagnosed during 1986-90 in each deprivation group overlap irregularly (Figure 62.2c). The survival gradients were estimated among 150-200 adults of known deprivation status for the five-year periods 1981-85 and 1986-90. Five-year survival was an estimated 13-17% lower for the most deprived group than for the most affluent group in both periods, although these differences are not statistically significant (see Table 4.9). The deprivation gradients in five-year survival are shown in Figure 62.3c.

Table 62.1 Ineligible and excluded records, and patients included in analyses, 1971-90

	Calendar period of diagnosis								
	Spinal cord			Adrenal			Pituitary		
Total registered	1,126			2,102			1,385		
Ineligible	Records	Patients	%		Patients	%	Records	Patients	%
Incomplete data[1]	37	37			22	22		24	24	
Resident outside England and Wales	23	22			22	21		45	44	
In situ neoplasm[2]	0	0			0	0		0	0	
Benign or uncertain behaviour[2]	81	79			36	33		277	250	
Metastatic[2]	2	2			15	15		0	0	
Otherwise ineligible[3]	-	-			-	-		-	-	
Lymphoma[4]	11	9			1	0		1	0	
Leukaemia or myeloma[4]	0	0			1	1		0	0	
		149			92			318		
Total eligible		977	100		2,010	100		1,067	100
Aged under 15 or 100 years or over at diagnosis	160	146	14.9		564	535	26.6	40	17	1.6
Vital status unknown[5]	106	67	6.9		136	81	4.0	161	107	10.0
Duplicate registration[6]	0	0	0.0		0	0	0.0	0	0	0.0
Multiple primary[7]	0	0	0.0		2	1	<0.1	0	0	0.0
Synchronous tumours[8]	4	3	0.3		9	4	0.2	9	0	0.0
Sex not known	0	0	0.0		0	0	0.0	0	0	0.0
Sex incompatible with site code[9]	-	-	-		-	-	-	-	-	-
Invalid dates[10]	0	0	0.0		1	0	0.0	1	1	<0.1
Zero survival[11]	65	46	4.7		413	351	17.5	131	118	11.1
Total exclusions		262	26.8		972	48.4		243	22.8
Patients included in analysis		715	73.2		1,038	51.6		824	77.2
Percentage of eligible adults			86.0			70.4			78.5

[1] Main data item(s) invalid or incompatible with one another: name, sex, date of birth, date of diagnosis and (if present) date of death, postcode, site, morphology and behaviour

[2] Tumour either *in situ* (behaviour code 2), benign (0), uncertain if benign or malignant (1) or metastatic (6 or 9)

[3] Not applicable for these malignancies

[4] Morphology that of lymphoma (included with non-Hodgkin lymphoma, Ch. 41) or myeloma (included with multiple myeloma, Ch. 43), or leukaemia (excluded)

[5] Tracing of vital status at National Health Service Central Register incomplete at 6 July 1997

[6] Same site code, sex, cancer registry and cancer registry number as an earlier registration

[7] Same site code and person number as an earlier registration(s): mostly confirmed multiple primary tumours at this site, some unresolved duplicate registrations

[8] Same site code, sex, date of birth and date of diagnosis as another registration(s): mostly synchronous or (in paired organs) bilateral tumours in same anatomic site in one individual, not previously linked: also some duplicate registrations

[9] Not applicable for these malignancies

[10] Impossible sequence of dates (birth, diagnosis and, if present, emigration or death); or date of death after 6 July 1997

[11] Date of diagnosis same as date of death: some are patients who did die on the day of diagnosis, but most were registered solely from a death certificate (death certificate only, or DCO), and their survival time is thus unknown

Table 62.2 Crude and relative survival (%) at one, five and ten years since diagnosis: Deprivation category and calendar period of diagnosis

SPINAL CORD (ICD-9 192.2)

		1971-75						1976-80					
		Affluent	2	3	4	Deprived	All[1]	Affluent	2	3	4	Deprived	All[1]
		C R	C R	C R	C R	C R	C R	C R	C R	C R	C R	C R	C R
ENGLAND & WALES													
Men		*(13)*	*(15)*	*(13)*	*(12)*	*(11)*	*(82)*	*(18)*	*(25)*	*(18)*	*(17)*	*(10)*	*(91)*
	One year	66 67	57 58	32 32	44 45	71 71	61 62	64 65	71 73	83 85	51 52	70 71	68 70
	Five years	36 37	45 46	12 12	36 37	34 36	41 44	43 45	63 68	56 60	45 48	50 52	52 56
	Ten years	36 37	38 40	12 12	29 31	34 36	34 39	43 45	47 52	45 50	28 30	50 52	42 48
Women		*(8)*	*(17)*	*(11)*	*(12)*	*(5)*	*(62)*	*(17)*	*(20)*	*(11)*	*(13)*	*(10)*	*(78)*
	One year	- -	56 60	53 54	74 74	- -	61 63	82 83	85 88	64 65	51 52	69 69	74 75
	Five years	- -	39 43	35 37	42 43	- -	38 41	53 54	70 73	36 38	43 44	41 41	54 56
	Ten years	- -	33 42	35 37	35 37	- -	33 39	42 49	65 70	28 29	23 24	41 41	45 50

ADRENAL (ICD-9 194.0)

		1971-75						1976-80					
		Affluent	2	3	4	Deprived	All[1]	Affluent	2	3	4	Deprived	All[1]
		C R	C R	C R	C R	C R	C R	C R	C R	C R	C R	C R	C R
ENGLAND & WALES													
Men		*(20)*	*(15)*	*(11)*	*(14)*	*(17)*	*(92)*	*(20)*	*(26)*	*(23)*	*(15)*	*(26)*	*(111)*
	One year	30 31	15 15	9 10	34 35	4 4	20 20	52 53	33 33	6 6	28 29	22 23	26 27
	Five years	13 14	6 6	9 10	19 20	0 0	8 9	34 35	22 24	2 2	17 17	16 18	17 19
	Ten years	9 10	6 6	5 6	13 15	0 0	6 7	20 21	19 22	0 0	6 7	7 8	10 12
Women		*(16)*	*(18)*	*(16)*	*(29)*	*(19)*	*(114)*	*(29)*	*(22)*	*(25)*	*(16)*	*(20)*	*(118)*
	One year	32 32	30 30	43 43	42 42	29 30	33 34	38 42	35 36	41 41	25 25	42 43	37 37
	Five years	16 17	20 20	22 22	32 33	8 8	18 18	23 26	24 24	26 27	20 20	24 24	22 24
	Ten years	16 17	15 15	16 17	26 29	4 4	14 16	14 17	24 24	22 25	15 16	24 24	19 22

PITUITARY (ICD-9 194.3)

		1971-75						1976-80					
		Affluent	2	3	4	Deprived	All[1]	Affluent	2	3	4	Deprived	All[1]
		C R	C R	C R	C R	C R	C R	C R	C R	C R	C R	C R	C R
ENGLAND & WALES													
Men		*(21)*	*(20)*	*(12)*	*(16)*	*(8)*	*(98)*	*(26)*	*(25)*	*(23)*	*(15)*	*(16)*	*(107)*
	One year	80 81	90 90	75 77	88 88	- -	81 82	88 90	84 85	82 83	79 80	81 83	83 86
	Five years	71 73	59 67	67 71	63 78	- -	67 77	73 79	80 81	69 82	53 60	56 66	69 77
	Ten years	66 69	59 67	- -	63 78	- -	61 75	58 71	63 76	56 74	53 60	44 65	57 73
Women		*(19)*	*(20)*	*(24)*	*(21)*	*(13)*	*(123)*	*(28)*	*(26)*	*(18)*	*(22)*	*(12)*	*(108)*
	One year	95 96	79 79	78 78	42 42	75 77	75 76	66 70	76 77	76 77	86 87	63 64	74 76
	Five years	79 83	69 70	58 60	33 34	75 77	63 68	56 59	65 72	71 75	72 80	39 42	62 71
	Ten years	57 68	64 65	45 60	33 34	67 71	53 63	46 49	61 70	65 75	63 79	31 36	54 68

Table 1

	1981-85						1986-90						
	Affluent	2	3	4	Deprived	All[1]	Affluent	2	3	4	Deprived	All[1]	
	C R	C R	C R	C R	C R	C R	C R	C R	C R	C R	C R	C R	**ENGLAND & WALES**
													Men
	(33)	(23)	(21)	(22)	(17)	(118)	(19)	(23)	(27)	(18)	(26)	(114)	
One year	63 65	70 71	71 72	76 77	76 77	70 71	89 89	82 83	93 93	71 72	80 81	84 85	
Five years	54 56	66 67	38 40	58 62	53 55	54 57	84 85	56 58	67 71	66 67	57 58	65 68	
	45 49	52 58	29 31	45 51	53 55	45 51							
	(20)	(16)	(14)	(22)	(18)	(91)	(21)	(18)	(12)	(19)	(9)	(79)	*Women*
One year	90 90	75 76	78 81	53 55	83 86	75 76	77 77	77 79	83 83	54 54	88 93	73 74	
Five years	65 67	50 53	57 59	40 42	83 86	58 62	62 67	44 46	58 63	43 45	77 82	54 58	
	65 67	37 46	57 59	36 39	61 71	50 58							

Table 2

	1981-85						1986-90						
	Affluent	2	3	4	Deprived	All[1]	Affluent	2	3	4	Deprived	All[1]	
	C R	C R	C R	C R	C R	C R	C R	C R	C R	C R	C R	C R	**ENGLAND & WALES**
													Men
	(30)	(24)	(33)	(22)	(21)	(137)	(43)	(30)	(33)	(32)	(23)	(164)	
One year	61 63	20 21	39 40	28 29	21 22	37 38	52 54	37 39	66 67	51 51	35 36	50 52	
Five years	42 43	7 7	25 28	20 21	11 13	24 26	26 28	21 23	48 51	27 29	27 30	30 34	
	25 27	3 3	22 27	20 21	4 5	18 21							
	(31)	(30)	(24)	(31)	(20)	(139)	(30)	(36)	(36)	(37)	(23)	(163)	*Women*
One year	50 51	29 29	44 45	23 23	21 22	35 36	69 71	42 43	54 56	47 47	59 61	54 55	
Five years	25 26	14 15	29 34	17 18	9 10	21 23	46 48	25 27	30 32	34 35	38 39	34 36	
	22 23	14 15	16 22	15 17	5 6	17 20							

Table 3

	1981-85						1986-90						
	Affluent	2	3	4	Deprived	All[1]	Affluent	2	3	4	Deprived	All[1]	
	C R	C R	C R	C R	C R	C R	C R	C R	C R	C R	C R	C R	**ENGLAND & WALES**
													Men
	(24)	(7)	(14)	(19)	(13)	(81)	(33)	(27)	(23)	(26)	(8)	(118)	
One year	88 92	- -	93 98	78 87	75 76	84 88	94 97	74 76	87 90	69 73	75 81	81 85	
Five years	75 82	- -	78 93	38 49	52 59	63 76	82 91	66 71	65 74	54 61	62 68	68 77	
	64 77	- -	78 93	32 49	44 55	53 74							
	(21)	(17)	(13)	(10)	(14)	(80)	(31)	(18)	(27)	(14)	(15)	(109)	*Women*
One year	85 85	65 67	69 71	100 100	93 97	81 83	90 91	100 100	85 87	46 47	93 94	84 86	
Five years	80 83	65 67	61 69	100 100	65 74	72 82	71 75	83 87	59 66	46 47	80 81	67 74	
	76 83	65 67	40 56	90 91	40 48	61 76							

SPINAL CORD

Fig 62.1a

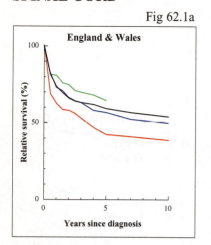

Fig 62.2a

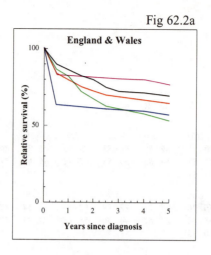

Fig 62.3a

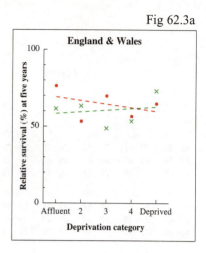

ADRENAL

Fig 62.1b

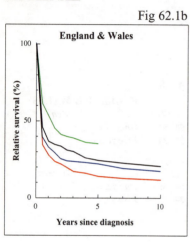

Fig 62.2b

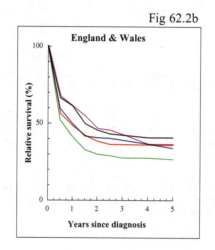

Fig 62.3b

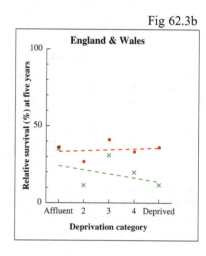

PITUITARY

Fig 62.1c

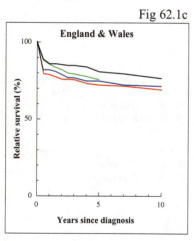

Fig 62.2c

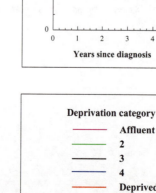

Fig 62.3c

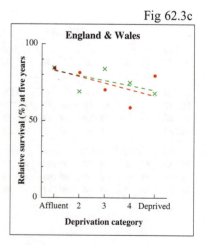

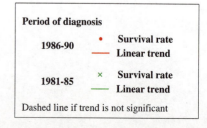

Period of diagnosis

1986-90	
1981-85	
1976-80	
1971-75	

Deprivation category

Affluent
2
3
4
Deprived

Period of diagnosis

1986-90 — Survival rate, Linear trend

1981-85 — Survival rate, Linear trend

Dashed line if trend is not significant

Fig 62.5 Relative survival (%) at five years, with 95% confidence interval, by age at diagnosis and sex; England and Wales, patients diagnosed 1986-90

Fig 62.5a

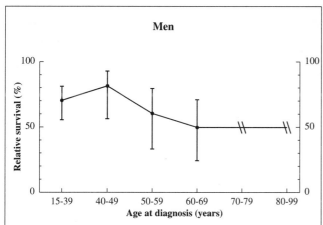

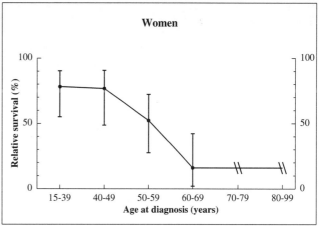

Fig 62.5b

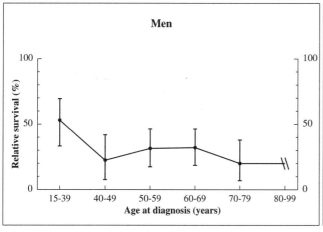

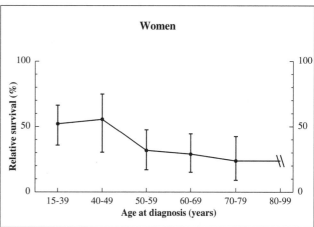

Fig 62.5c

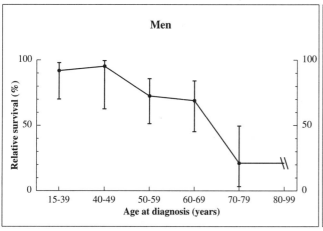

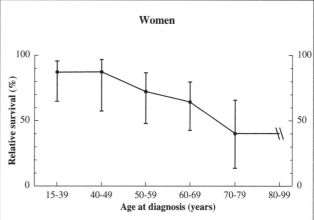